SUMMARY OF CONTENTS

ENVIRONMENTAL INORGANIC CHEMISTRY

Properties, Processes, and Estimation Methods

Pergamon Titles of Related Interest

Bergman, Kimmerle & Maki ENVIRONMENTAL HAZARD ASSESSMENT OF EFFLUENTS

Kickson, Maki and Brungs FATE & EFFECTS OF SEDIMENT-BOUND CHEMICALS IN AQUATIC SYSTEMS

Cairns MULTISPECIES TOXICITY TESTING

Vandermeulen and Hrudey OIL IN FRESHWATER: Chemistry, Biology, Countermeasure Technology

Huck and Toft TREATMENT OF DRINKING WATER FOR ORGANIC CONTAMINANTS

Grisham HEALTH ASPECTS OF THE DISPOSAL OF WASTE CHEMICALS

Related Journals

(Sample copies available upon request)

ENVIRONMENTAL TOXICOLOGY AND CHEMISTRY

ENVIRONMENTAL INORGANIC CHEMISTRY

Properties, Processes, and Estimation Methods

Edited by
Itamar Bodek, Ph.D., Arthur D. Little, Inc.
Warren J. Lyman, Ph.D., Arthur D. Little, Inc.
William F. Reehl, Arthur D. Little, Inc.
David H. Rosenblatt, Ph.D., U.S. Army Biomedical Research and Development Laboratory

SETAC Special Publications Series
Series Editors
Barbara T. Walton,
Oak Ridge National Laboratory
Richard Ashley Conway,
Union Carbide Corporation

PERGAMON PRESS
New York • Oxford • Beijing • Frankfurt
São Paulo • Sydney • Tokyo • Toronto

Pergamon Press Offices:

U.S.A.	Pergamon Press, Maxwell House, Fairview Park, Elmsford, New York 10523, U.S.A.
U.K.	Pergamon Press, Headington Hill Hall, Oxford OX3 0BW, England
PEOPLE'S REPUBLIC OF CHINA	Pergamon Press, Qianmen Hotel, Beijing, People's Republic of China
FEDERAL REPUBLIC OF GERMANY	Pergamon Press, Hammerweg 6, D-6242 Kronberg, Federal Republic of Germany
BRAZIL	Pergamon Editora, Rua Eça de Queiros, 346, CEP 04011, São Paulo, Brazil
AUSTRALIA	Pergamon Press (Aust.) Pty., P.O. Box 544, Potts Point, NSW 2011, Australia
JAPAN	Pergamon Press, 8th Floor, Matsuoka Central Building, 1-7-1 Nishishinjuku, Shinjuku-ku, Tokyo 160, Japan
CANADA	Pergamon Press Canada, Suite 271, 253 College Street, Toronto, Ontario M5T 1R5, Canada

Library of Congress Cataloging in Publication Data

Environmental inorganic chemistry.

(SETAC special publications series)
Bibliography: p.
Includes index.
1. Environmental chemistry. 2. Chemistry, Inorganic.
I. Bodek, Itamar. II. Series.
TD193.E565 1988 574.5′222 88-25281
ISBN 0-08-036833-6

Printed in the United States of America

SUMMARY TABLE OF CONTENTS

(A detailed Table of Contents, followed by a List of Tables and List of Figures, is given at the beginning of each Chapter and each Appendix.)

3. KINETICS OF SELECTED PROCESSES

4. UPTAKE BY BIOTA

5. MATHEMATICAL ENVIRONMENTAL FATE MODELING

6. MATRIX ELEMENTS AND LIGANDS

7. TRACE METALS

8. POLLUTANT/TRACE LIGANDS

9. RADIONUCLIDES

10. CHEMICAL CLASSES

Introduction to the SETAC Special Publications Series

The SETAC Special Publications series was established by the Society of Environmental Toxicology and Chemistry to provide in-depth reviews and critical appraisals on scientific subjects relevant to understanding the impacts of chemicals and technology on the environment. The series consists of single and multiple authored/edited books on topics selected by the Board of Directors for their importance, their timeliness, and their contribution to multidisciplinary approaches to solving environmental problems. The diversity and breadth of subjects covered in this series will reflect the wide range of disciplines encompassed by environmental toxicology, environmental chemistry, and hazard assessment. Despite this diversity, the goals of these volumes are similar; they are to present the reader with authoritative coverage of the literature, as well as paradigms, methodologies, controversies, research needs, and new developments specific to the featured topics. All books in the series are peer reviewed for SETAC by acknowledged experts.

SETAC Special Publications will be useful to scientists and engineers in environmental research, research management, chemical manufacturing, product design, regulation, and education, as well as to students preparing for careers in these areas, for keeping abreast of recent developments in familiar areas and for rapid introduction to principles and approaches in new subject areas.

Environmental Inorganic Chemistry: Processes, Properties, and Estimation Methods, the fourth book published in this series, compiles in one place data and other information on environmentally important physicochemical properties of inorganic chemicals for use by environmental scientists, engineers, and managers. Previous books in the series dealt with effluent hazard assessment, multispecies toxicity testing, and sediment-bound chemicals.

We wish to thank Dr. Theodore Mill for his contributions to the review and production of this volume by SETAC.

Series Editors
Barbara T. Walton,
Oak Ridge National Laboratory
Richard Ashley Conway,
Union Carbide Corporation

Preface

The primary purpose of this project was to compile, in one book, information on environmentally important physicochemical properties of inorganic chemicals for use by environmental scientists and managers. This book, in a sense, is a companion document to the report we previously prepared for environmentally important properties of organic chemicals, and which was subsequently published under the title: Handbook of Chemical Property Estimation Methods. Like the "organics" book, this "inorganics" book describes available estimation methods for various properties. Such information supports one of our primary objectives, which is to assist environmental scientists and managers (many of whom may not have a detailed knowledge or understanding of inorganic environmental chemistry) in overcoming the common problem of property data gaps and in developing timely responses to environmental problems.

In marked contrast to the "organics" book, however, this "inorganics" book is not limited to a presentation of selected estimation methods for estimable properties. Because so few properties are easily estimable, it was decided to cover environmentally important properties, without regard to estimability, and to present not only a generic discussion of these properties, but also a summary of environmentally important data for the most common elements and pollutants.

This book is composed of two parts. Part I provides a generic description of sixteen properties and processes with special focus on the aqueous phase. To the extent possible, it gives specific instructions to assist in estimating (or selecting) property values and/or assessing the importance of a particular property or process for a specific site or pollutant. Where even this kind of assistance is difficult or impossible, tabulated data, illustrations, and examples are presented to demonstrate the range of possible values or behavior. Part I also contains major sections on uptake by biota and environmental fate models.

Part II of the book presents environmentally important property data for several groups of elements or coumpounds: (a) major elements and anions in the earth's crust and surface waters; (b) metals; (c) ligands; (d) radionuclides; and (e) selected chemical classes, including organometallics. In each case, data are provided for specific elements, ligands or compounds along with text describing, where pertinent, how these chemicals generally behave in the environment.

Itamar Bodek
Warren J. Lyman
William F. Reehl
David H. Rosenblatt

Acknowledgments

This book is the result of a project undertaken by Arthur D. Little, Inc., under contract to the U.S. Army Medical Research and Development Command. The Army's willingness to support this work indicates a real concern for environmental protection.

The U.S. Army Medical Research and Development Command funded both Phase I of this program, a preliminary problem definition study, and a major portion of Phase II, the writing of the book. Mr. Jesse Barkley (U.S. Army Biomedical Research and Development Laboratory) was very supportive and encouraging, and was especially instrumental in obtaining funding for this project. Financial support for Phase II was also provided by the U.S. Environmental Protection Agency's Office of Toxic Substances (Washington, DC).

The contributors to this handbook are listed below; the first page of each chapter lists the specific authors of that chapter. Dr. Brian Barnett contributed significantly in the early planning phase of this project and reviewed several major sections in the writing phase.

All of the chapters in this book were reviewed by one or more extramural reviewers selected from various universities, research and consulting firms, and government offices and laboratories. A list of reviewers is provided on the following page.

Itamar Bodek
Warren J. Lyman
William F. Reehl
David H. Rosenblatt

CONTRIBUTORS

James D. Birkett, Ph.D.
Itamar Bodek, Ph.D.
Marc Bonazountas, Ph.D.
Aviva Brecher, Ph.D.
Sara E. Bysshe
Deborah L. Cerundolo
Susan F. Coons
Daniel J. Ehntholt, Ph.D.
Elizabeth D. Gibson
Anita E. Glazer
Clark F. Grain
Janice W. Green
Denise Hayes
Abraham Lerman, Ph.D.
Northwestern University
Evanston, IL
Donald B. Lindsay
Independent consultant
Lexington, MA
Christopher P. Loreti
Warren J. Lyman, Ph.D.
Madeline F. McComish
Joo Hooi Ong
Jan Shim
Robert G. Vranka, Ph.D.

INITIAL REVIEWERS[a]

Dr. Marc Anderson, University of Wisconsin, Madison, WI
Dr. Robert Boethling, U.S. EPA, Washington, DC
Dr. Fred C. Brinckman, National Bureau of Standards, Gaithersburg, MD
Dr. Rufus Chaney, U.S. Department of Agriculture, Beltsville, MD
Dr. Jim Dragun, E.C. Jordan Co., Smithfield, MI
Dr. Stephen C. Hern, U.S. EPA, Las Vegas, NV
Dr. Donald Langmuir, Colorado School of Mines, Golden, CO
Dr. Henry Lee, U.S. EPA, Newport, OR
Dr. Asa Leifer, U.S. EPA, Washington, DC
Dr. Abraham Lerman, Northwestern University, Evanston, IL
Dr. Donald Mackay, University of Toronto, Toronto, Ontario
Dr. Constantine Maletskos, consultant, Gloucester, MA
Dr. Theodore Mill, SRI International, Menlo Park, CA
Dr. Darrell Nelson, University of Nebraska, Lincoln, NE
Dr. Albert Page, University of California, Riverside, CA
Dr. David Ryan, New England Aquarium, Boston, MA
Dr. Gary Thom, U.S. EPA, Washington, DC
Dr. Marvin Wesely, Argonne National Laboratory, Argonne, IL
Mr. William Wood, U.S. EPA, Washington, DC

(a) The book later was peer-reviewed anonymously by SETAC

1. INTRODUCTION

Warren J. Lyman

LIST OF TABLES

LIST OF FIGURES

1.1 OVERVIEW

This report provides basic information on important processes and properties relating to the environmental fate, transformation and mobility of inorganic and organometallic chemicals. The general style and organization of the report is that of a handbook, in order to provide easy access to specific information, data and instructions. The report is divided into two major parts. Part I provides a generic description and discussion of several environmentally important processes, while Part II provides data and information for a variety of inorganic species. The contents are summarized in Table 1-1.

1.1.1 Content of Part I

Table 1-2 lists the processes covered in Part I. The rationale for the selection of these processes, as well as the overall organization and content of the report, evolved from a preliminary problem definition study conducted for the U.S. Army.[1] The sections that describe these processes, which are primarily in Chapters 2 and 3, give specific instructions to assist in estimating (or selecting) property values and/or assessing the importance of a particular process for a specific pollutant or set of environmental conditions. For some processes, it is impractical or impossible to provide such assistance here; in those cases, tabulated data, illustrations, and examples are presented to demonstrate the range of possible values or behaviors.

A major focus of Chapter 2, which covers 15 properties or processes, is on equilibrium conditions. For example, there is significant coverage of acid/base, complexation, redox, and solubility equilibria and of equilibrium sorption on soils. The distribution coefficients or equilibrium constants corresponding to these equilibria can, in several instances, be estimated using the information provided in this report. In many cases the estimates may appear to have a large uncertainty (e.g., an order of magnitude), but this may be quite acceptable in preliminary assessments of pollution problems.

One section of Chapter 2 (§ 2.13) describes how the consideration of various aquatic speciation processes, such as complexation, redox, acid/base and precipitation reactions, can be integrated to evaluate their simultaneous effect on the distribution of species present in the system. While it is convenient in many environmental assessments to assume equilibrium conditions, it should be realized that many reactions may approach equilibrium only very slowly relative to the time scale of interest (e.g., minutes to a few years) and that erroneous conclusions could result from such an assumption.

Chapter 3 focuses on process kinetics, i.e., the rates at which reactions approach equilibrium. Due in part to project limitations, as well as to a lack of available

---

1. W.J. Lyman and B.M. Barnett, "Methods for Estimating Physicochemical Properties of Inorganic Chemicals of Environmental Concern," Final Report, Phase I, Contract No. DAMD17-83-C-3274 U.S. Army Medical Research and Development Command, Fort Detrick, Frederick, Md. (1984).

TABLE 1-1

Summary of Contents

PART I	GENERIC DESCRIPTION OF PROPERTIES AND PROCESSES
1	Introduction
2	Description of Individual Processes
3	Kinetics of Selected Processes
4	Uptake by Biota
5	Mathematical Environmental Fate Modeling
PART II	INFORMATION ON SPECIFIC CHEMICALS
6	Matrix Elements and Ligands
7	Trace Metals
8	Pollutant/Trace Ligands
9	Radionuclides
10	Commercial Compounds
APPENDICES	
A	Notations and Symbols
B	Chemical Composition of the Environment
C	Classification and Properties of Soils
D	Pollutant Criteria and Standards
SUBJECT INDEX	

information on the kinetics of some reactions, only three processes are treated in detail: (1) ligand exchange, (2) oxidation and reduction, and (3) dissolution, precipitation and crystallization. (Some information on process kinetics is also provided in Chapter 2, principally in the sections on Atmospheric Deposition, Photolysis in Air, Photolysis in Water, Diffusion Coefficient, and Radioactive Processes).

The mathematical treatment of the kinetics of these processes focuses on the rate constant and methods for estimating its approximate value. Even when the rates can only be roughly estimated, they are helpful in assessing the importance of reaction kinetics for a specific environmental scenario. For example, if a reaction is very fast relative to the time scale of a scenario (e.g., discharge and dilution of a wastewater stream in a river), a detailed evaluation of the kinetics would not be warranted, and only the equilibrium reaction products need be considered.

TABLE 1-2

Properties and Processes Covered[a]

Vapor Pressure	Attenuation on Soils
Photolysis in Air	Aqueous Photolysis
Gas Adsorption on Solids	Integration of Aqueous Processes
Atmospheric Deposition	Microbial Transformations
Activity Coefficient	Diffusion of Ions
Polymerization	Radioactive Processes and Properties
Complexation[b]	Uptake by Biota[c]
Electron Transfer Reactions[b]	
Solubility and Precipitation[b]	

a. Generic discussions of these processes and properties, including available estimation methods, are provided in Chapter 2.

b. Chapter 3 of the report describes the kinetics of these processes. (Complexation is discussed in terms of ligand exchange.)

c. Covered in Chapter 4.

Chapter 4 of Part I describes the uptake of inorganic chemicals by biota, particularly biotic groups of importance in the human food chain (terrestrial plants and freshwater and saltwater fish), and gives special attention to the uptake of selected heavy metals by these groups. Although detailed coverage of other biotic groups and chemical classes was not possible within the scope of this project, uptake and loss processes are discussed in sufficient detail to permit appropriate questions to be asked and some answers to be inferred.

There are no generally applicable mathematical techniques for estimating the extent of bioaccumulation of inorganics; their development awaits further research on chemical speciation in the (aquatic) environment and mechanisms and rates of uptake and loss for specific chemical species. In the absence of these estimation methods, Chapter 4 describes a "decision process" that may help the reader to choose an appropriate focus for more detailed review and use of the scientific literature, or to structure an appropriately inclusive laboratory or field experiment to investigate a potential uptake problem.

Chapter 5 provides a non-mathematical description of available environmental fate models that can predict the equilibrium speciation of many elements in aquatic environments and the transport and (generalized) fate of individual pollutants in air, water and soil media. Its primary objectives are to help readers in (1) understanding the complexities of modeling, (2) selecting specific modeling packages for their own use, and (3) evaluating other people's modeling results when, for example, that information is part of a permit application for a plant site and is used in the design of waste treatment, storage or disposal facilities.

A background section in Chapter 5 (§ 5.2) includes a general discussion of environmental pathways (for both transport and fate) to guide environmental managers in the preliminary identification and assessment of those that are potentially important in a given scenario. This is followed by descriptions of modeling transport and fate in the air, soil/groundwater, and surface water compartments (subsections 5.3, 5.4, and 5.5, respectively). A final section (§ 5.6) is devoted to the highly specialized subject of modeling the transport and fate of radionuclides in the soil/groundwater system.

1.1.2 Content of Part II

Part II presents environmentally important data and information for several groups of elements or compounds. These have been separated into chapters that focus on: (1) major elements and anions in the earth's crust and surface waters, (2) pollutant metals and metalloids, (3) pollutant ligands, (4) radionuclides, and (5) selected classes of commercial chemicals, including organometallics.

A detailed list of the chemicals and species discussed in Part II is given in Table 1-3. The specific chemicals covered in this report were selected because of their importance as environmental pollutants, workplace hazards, hazardous chemicals used in industry, and/or chemicals that might affect (e.g., through complexation) the aquatic speciation and, thus, the behavior of trace pollutants. Further details of the selection process are provided in the introductions to the various sections and in the preliminary problem definition report footnoted earlier.

Chapters 6 and 7 typically contain information under the following headings:

- Occurrence, Production, and Uses
- Speciation Reactions in Water
  - Acid/Base Dissociation
  - Complexation
  - Redox Reactions
  - Solubility and Precipitation
- Sorption on Soils and Sediments
- Biotransformation (where appropriate)
- Volatilization (where appropriate)

Chapter 8, which addresses compounds containing N, S, P and F, provides information similar to that in Chapters 6 and 7 plus a discussion of the environmental cycles for these elements and the species of environmental importance.

The coverage of radionuclides in Chapter 9 (like that of radioactive processes and properties in § 2.17 and radionuclide migration modeling in § 5.6) is limited. Understanding of the environmental fate and transport of radionuclides has advanced significantly in recent years, not only as a result of early studies of

TABLE 1-3

Chemicals and Species Covered in Part II

Chapter	Coverage	
6	Na, K, Ca, Mg, Si, Al, Fe H^+/OH^-, Cl^-, SO_4^{-2}, HCO_3^-/CO_3^{-2}	
7	Sb, As, Ba, Be, Cd, Cr, Cu, Pb, Mn, Hg, Ni, Se, Ag, Tl, Zn	
8	Compounds of: N, S, P, F	
9	^{3}H, ^{14}C, ^{32}P, ^{35}S, ^{40}K, ^{50}V, ^{51}Cr, ^{87}Rb, $^{99m/99}Tc$, ^{115}In, $^{125/129/131}I$, ^{133}Xe, ^{137}Cs, ^{147}Pm, ^{154}Eu, ^{176}Lu, ^{187}Re, ^{201}Tl, $^{210/214}Pb$, $^{210/214}Bi$, $^{210/214/218}Po$, $^{220/222}Rn$, ^{226}Ra, $^{232/234}Th$, ^{234}Pa, $^{235/238}U$, ^{239}Pu, ^{241}Am	
10	Organometallics	Sulfur-containing Compounds
	Acids	Phosphorus-containing Compounds
	Bases (Hydroxides)	Boron-containing Compounds
	Salts	Silicon-containing Compounds
	Oxides	Carbonyls
	Gases	Carbides
	Hydrides	Cyanides
	Peroxides	Hypochlorites and Chlorates

radioactive fallout, but also due to extensive analyses of potential problems associated with the disposal of radioactive wastes from commercial products, industrial wastes, and nuclear power plants. Due to the availability of detailed books on this subject, Chapter 9 is primarily designed to provide an introduction to the problem and present some data on radiation emission and environmental behavior.

The discussion of chemical classes in Chapter 10 focuses on specific chemicals of commercial or environmental importance. Their major commercial uses and the physical and chemical properties that govern their fate in the environment are described. Information is also provided on the processes and transformations that chemicals in each class are likely to undergo under environmental conditions, such as (1) volatilization, (2) reaction with air (i.e., oxidation), (3) dissolution in and/or reaction with water (solubility, acid-base dissociation, hydrolysis), (4) aqueous reactions (complexation, oxidation/reduction, polymerization), and (5) mobility and degradation in soils.

Organometallics (§ 10.1) are covered more extensively than any of the other classes in Chapter 10. Although their inclusion is somewhat out of keeping with the primary focus on inorganics in this report, it reflects an awareness of their growing commercial and environmental importance.

1.1.3 Content of Appendices

This report concludes with five appendices and a subject index. Appendix A is a list of commonly used notations and symbols. Background concentrations of the elements and a few other chemical species in air, water, soil and sediments are tabulated in Appendix B. Because the amount of this data is limited, the values should be interpreted with caution; however readers should be able to estimate the approximate extent of pollution by comparing data from a polluted site with the listed values.

Appendix C is a primer on the classification and properties of soils, focusing on those properties that affect the movement of water and pollutants through the soil/groundwater system. Appendix D summarizes federal pollutant criteria and standards for inorganic chemicals, such as the EPA's ambient water quality standards (for the protection of aquatic life) and drinking water standards (for protection of human health).

1.2 BACKGROUND

The last 15 years have brought an ever-growing awareness in this country of the need for environmental protection, especially where the use of chemicals is involved. The decade of the 1970s was especially important in the United States for the passage of key government acts or actions that mandate a minimum level of protection. Included were the following:

- National Environmental Policy Act (1969)
- Creation of the U.S. Environmental Protection Agency (1970)
- Occupational Safety and Health Act (1970)
- Clean Water Act Amendments (1972)
- Marine Protection Research and Sanctuaries Act (1972)
- Safe Drinking Water Act (1973)
- Toxic Substances Control Act (1976)
- Resource Conservation and Recovery Act (1976)
- EPA/NRDC Settlement Agreement on Toxics ("Priority Pollutants") (1976)
- Clean Water Act Amendments ("Toxics" added) (1977)
- Comprehensive Emergency Response, Compensation and Liability Act (Superfund) (1980)

More recently, much of the regulatory activity at the federal level has been geared to amending these acts; at the state and local levels, it has been geared to implementing state acts, regulations and standards as required or suggested by the federal legislation.

The heightened concern and regulatory activity have, of necessity, created a multitude of environmental scientists and managers that may number in the several tens of thousands. This army of specialists is responsible for the development and implementation of regulations, for research into control technologies, for the operation of pollution control facilities and other related activities; it is these people to whom this report is directed.

1.3 THE NEED FOR GUIDANCE

There is a real need today for a wide variety of guidance manuals directed toward environmental scientists and managers in government, industry and universities. These individuals come to their positions with varied training (e.g., a B.S. in Biology, an M.S. in Civil Engineering, or a 20-year job as a chemical process engineer), little of which helps them to make timely decisions related to environmental protection where toxic chemicals are involved.

Manuals are now available for designing wastewater treatment plants, cleaning up oil spills, monitoring stack gas emissions, and preparing river management plans, for example; however, there is virtually no guidance on how to assess the very complex problems that relate more directly to the fate and transport of chemicals in the environment. What are needed are: (a) guidance in assessing what physicochemical properties and processes are environmentally important, (b) advice on how to use the properties to maximum advantage, and (c) some actual values of these properties. In 1982 we prepared a Handbook[2] designed to fill part of this need by providing a compilation of carefully described estimation methods for environmentally important properties of organic chemicals. The need for estimation methods was described in the introduction to this "organics" Handbook as follows:

> "Over the past decade, the chemical contamination of our environment has justifiably aroused growing concern. A proper assessment of the risk — to man and the environment — created by exposure to these chemicals generally includes attempts to measure or predict the concentrations in various environmental compartments in conjunction with toxicological data. Frequently, however, neither the concentration data nor the toxicological data are adequate for any realistic assessment. In addition, basic physical and chemical data are often unavailable, especially for new organic chemicals being considered for bulk manufacture. If, however, the most important physical and chemical properties of these chemicals could be estimated, their transport and fate in the environment could be better understood — even modeled in some cases — and the environmental concentrations might be estimated."

2. W.J. Lyman, W.F. Reehl and D.H. Rosenblatt (eds.), *Handbook of Chemical Property Estimation Methods: Environmental Behavior of Organic Chemicals,* McGraw-Hill Book Co., New York (1982).

It was clear at the time the above Handbook was prepared that a separate program, and perhaps quite a different handbook, would also be required for the coverage of inorganic chemicals and special classes such as organometallics. The current program was instituted to address this need.

As noted above, the preparation of this report was preceded by a problem definition study in which the needs of environmental scientists and managers were considered in conjunction with the available information on inorganic environmental chemistry (Lyman and Barnett, *op. cit.*). Many individuals contributed to the preliminary study, and the outline and format of the present report reflect their collective wisdom.

1.4 OBJECTIVES

The primary purpose of this project was to compile, in one report, information on environmentally important physicochemical properties of inorganic chemicals for use by environmental scientists and managers. In a sense, the report complements the "organics" Handbook; like that document, it describes available estimation methods for various properties. This information supports one of our primary objectives, which is to assist environmental scientists and managers (many of whom may not have a detailed knowledge or understanding of inorganic environmental chemistry) in overcoming the common problem of property data gaps and developing timely responses to environmental problems.

In marked contrast to the "organics" report, however, this "inorganics" report is not limited to the description of estimation methods. Because so few properties of inorganic species are easily estimable, the authors decided to cover environmentally important properties without regard to estimability, and to present not only a generic discussion of these properties but also a summary of environmentally important data for the most common elements and pollutants.

With this report as guidance, environmental scientists and managers should be able to undertake preliminary environmental assessments of the fate and transport of inorganic and organometallic compounds for such situations or scenarios as the following:

- NPDES, FIFRA, RCRA (etc.) permit applications
- R&D for wastewater treatment (e.g., metal removal)
- Modeling or assessment of chemical spills
- Modeling or assessment of soil/groundwater contamination
- Modeling the speciation and fate of metals in water
- Assessment of the consequences of agricultural use of various chemicals
- Assessment of land treatment of hazardous wastes or land application of municipal wastes or sewage sludge

- Assessment of environmental releases of pollutants to all media
- Remedial action at old chemical dump sites
- Premanufacture Notice (PMN) review
- Assessment of potential food contamination incidents
- Utilization of wastes
- Migration of food additives
- Ocean disposal
- Coal combustion
- Assessing impacts of acid rain.

Each case may present a variety of naturally occurring and pollutant chemical species to contend with, several fate and transport pathways to consider, and a host of environmental variables (e.g., temperature, pH, Eh, solid types) with which to contend. Where can the scientist or manager turn to for help today? Little reliable guidance is easily and quickly available, and research or field studies to obtain specific answers are both costly and time-consuming. This report was prepared to help meet that need.

1.5 USING THE HANDBOOK

As discussed earlier, this report is divided into (1) a generic discussion of properties and processes and (2) information on specific chemicals and chemical classes. Thus, for any given environmental scenario, an initial choice of process- (or property-) specific versus chemical-specific sections is available:

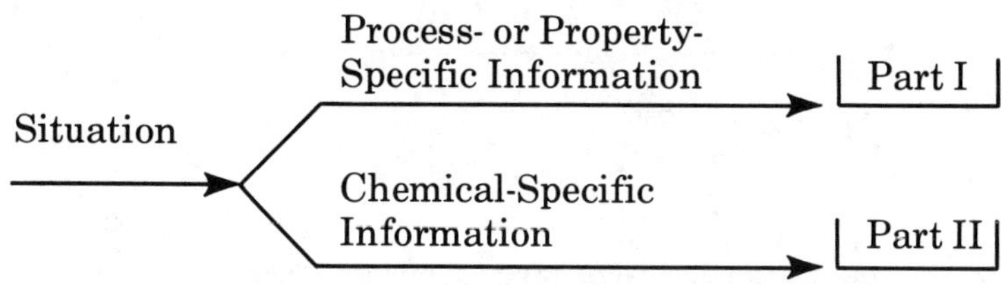

For example, to evaluate a metal-contaminated sediment, the reader can go to Part I to obtain a basic understanding of the processes that may be important (speciation in water, sediment sorption, diffusion of ions in water-filled pores of sediment, etc.). Then he can go to Part II to obtain some of the speciation and sorption information (measured constants) for the specific chemicals of concern and to determine the applicability of the data to the situation at hand. If data are not available in this report or in other readily available sources, the reader can return to Part I to estimate missing constants (or their reasonable limits) and/or modify literature values to take into account situation-specific factors such as temperature, species concentration, pH and ionic strength. A significant amount of chemical-specific information is also available in Part I; this may be found by referring to the Table of Contents or Subject Index, or by a perusal of the section of interest.

SITUATION

↓

Evaluation of Applicable Processes (Chapter 2)

Dry Environment

- Phase of chemical (solid, liquid, gas)
- Volatilization
- In atmosphere, consider:
  - photolysis
  - oxidation
  - gas adsorption
  - wet and dry deposition
- In soil, consider:
  - entrainment of particles by wind
  - gas sorption
  - loss by volatilization
  - redox reactions
  - dissolution potential

Wet/Aqueous Environment

- Dissolution in, or reaction with, water; precipitation.
- Equilibrium speciation in water (complexation, redox, polymerization)
- Use of speciation model
- Sorption on soils, sediments
- Volatilization to air
- Microbial transformations
- Aqueous photolysis
- Diffusion of ions through water-filled pores of soil or sediment
- Evaluate need to consider kinetics of any process (Chapter 3)

↓

- Evaluate impact of speciation and particle association on mobility (in air, water)

↓

Evaluation of Environmental Transport (Chapter 5)

- Estimate concentrations of chemical in various compartments (as a function of time, if necessary). May use coupled transport-speciation model.

↓

Evaluation of Uptake by Biota (Chapter 4)

- Evaluate impact of speciation, particle association, compartment concentration and other factors on uptake by biota.

FIGURE 1-1 Approach for Evaluating Situations Involving Inorganic Pollutants

Deciding or determining what processes are important requires some work by the reader, as each pollutant and each scenario may entail consideration of a different set of processes. Figure 1-1 outlines an approach that may be helpful. It is essential to distinguish between dry and wet environments; only in the latter is it necessary to consider aqueous speciation reactions and sorption processes, which take up a major portion of Chapter 2. The introduction to Chapter 3 (§ 3.1) will help evaluate the need to consider reaction kinetics for selected aqueous processes.

Section 2.13 describes (with examples) ways to consider the net effect of several processes (e.g., complexation, redox reaction and precipitation) on a specified environmental system. The approach is practical only for simple systems (i.e., a small number of species and processes); computerized speciation models (described in sections 5.4 and 5.5) must be used for complex systems.

The individual sections of Chapters 2 and 3 typically describe the influence of key environmental variables such as temperature, pH, chemical concentration, and ionic strength on a given process. In general, however, extreme environments (e.g., very high temperatures or pressures) are not considered. The reader should be aware of the controlling variables and, where possible, specifically account for them by modifying the equilibrium and rate constants used.

The general goal of all such assessments is to evaluate the persistence, mobility and bioavailability of a chemical. The answers clearly require knowledge of the speciation of the chemical and of the biological species present.

Before evaluating uptake potential, one may have to evaluate the transport of the chemical over the relevant intervals of time and space. These can range from minutes to decades and meters to hundreds of kilometers, respectively. Transport models provide approximate concentrations of the pollutant as a function of location — usually, the distance from the source — and (possibly) time. The selection of a model, or the evaluation of model results presented by others, is aided by the material in Chapter 5.

However, sophisticated models can be misleading. They can provide the illusion of power over complexity and, for all but the most expert, can easily mask unwarranted or inappropriate assumptions, poor input data, and errors in calculation. This puts an added burden on the user (or reviewer) to understand the model's weaknesses and, if necessary, to support the results by independent assessments and hand calculations (perhaps only order-of-magnitude calculations) of the most fundamental processes or rate-limiting factors.

This report is severely limited in some aspects, largely as a result of the attempt to cover so many subjects within a given budget, scope and report size. It will clearly not be adequate for detailed investigations in which time and money allow more specific and thorough literature searches, and possibly laboratory and field experiments. In many cases, the information was obtained principally from secondary sources; this limitation on literature review and citation may disappoint those who hope to find references to the latest journal articles on a particular process, property or pollutant. We hope that such omissions will be corrected in subsequent revisions.

2. DESCRIPTION OF INDIVIDUAL PROCESSES

James D. Birkett, Itamar Bodek, Anita E. Glazer, Clark F. Grain, Denise Hayes, Abraham Lerman, Donald B. Lindsay, Christopher P. Loreti, Joo Hooi Ong

LIST OF TABLES

VAPOR PRESSURE

PHOTOLYSIS IN AIR

GAS ADSORPTION ON SOLIDS

ATMOSPHERIC DEPOSITION

ACTIVITY COEFFICIENT

ACID/BASE EQUILIBRIA

POLYMERIZATION

COMPLEXATION

ELECTRON TRANSFER REACTIONS

SOLUBILITY AND PRECIPITATION EQUILIBRIA

ATTENUATION ON SOILS

INTEGRATION AND ANALYSIS OF AQUEOUS PROCESSES

PHOTOLYSIS IN WATER

MICROBIAL TRANSFORMATIONS

DIFFUSION COEFFICIENTS

RADIOACTIVE PROCESSES AND PROPERTIES

LIST OF FIGURES

PHOTOLYSIS IN AIR

GAS ADSORPTION ON SOLIDS

ATMOSPHERIC DEPOSITION

ACTIVITY COEFFICIENT

ACID/BASE EQUILIBRIA

POLYMERIZATION

ELECTRON TRANSFER REACTIONS

SOLUBILITY AND PRECIPITATION EQUILIBRIA

ATTENUATION ON SOILS

PHOTOLYSIS IN WATER

MICROBIAL TRANSFORMATIONS

RADIOACTIVE PROCESSES AND PROPERTIES

2.1 INTRODUCTION

Chapter 2 provides detailed information on 15 physicochemical properties or processes that are the chief determinants of the speciation, fate and transport of inorganic chemicals in the environment. (The properties covered were listed in Table 1-2.) The chapter is, in a sense, the keystone of the whole report, as it deals with the major subject — the description of fundamental (environmentally important) properties and available estimation methods.

With information and tools such as those provided here, we expect that preliminary environmental assessments can be made in a timely fashion without the delays and expense of detailed literature searches, or of laboratory and field studies that may not represent the exact (or changing) environmental conditions at the problem site. The assessments described in this report require the user to identify the fundamental processes that may be acting on the pollutant(s) of interest (e.g., dissolution in water, complexation, photolytic degradation), and then use the guidance provided on those processes to obtain (usually) quantitative evaluations of their importance in the form os species abundance, extent of sorption, rates of photolysis, etc.

It should be emphasized that the numbers generated, especially estimates, can have high uncertainties associated with them. However, they should at least be adequate to bracket the probable environmental behavior of the pollutant so that reasonable upper or lower limits on mobility, concentrations, and bioavailability can be set and used in preliminary exposure and risk assessments.

A major focus of Chapter 2 is on equilibrium conditions. For example, there is significant coverage of acid/base, complexation, redox and solubility equilibria and of equilibrium sorption on soils. The distribution coefficients or equilibrium constants corresponding to these equilibria can, in several instances, be estimated with the information provided in this report. Section 2.13 describes how the simultaneous action of several such speciation equilibria can be evaluated to determine the ultimate distribution of species present in the system. For complex situations, the computerized speciation models described in Chapter 5 would be required. The kinetics of several of the most important speciation reactions (complexation, oxidation/reduction, and dissolution/precipitation) are covered in Chapter 3.

In addition to the above, Chapter 2 also covers other physicochemical processes that cannot be treated as equilibrium processes. Included, for example, are photolytic degradation (in both water and air), atmospheric deposition, diffusion of ions in water, and radioactive decay along with the (sometimes important) recoil of alpha-emitting radionuclides. Biologically mediated processes, especially the microbial methylation of heavy metals, are also covered in § 2.15. (A detailed discussion of uptake of inorganic pollutants by biota is provided in Chapter 4.)

Each section of Chapter 2 is somewhat different in content; however, most provide: (1) a description of the property, including its environmental importance, range of values, units, and sensitivity to environmental variables such as temperature and ionic strength; (2) a description of the mathematical and/or graphical representation of the property; (3) tabulated data (e.g., complexation constants); (4) a description of selected available estimation methods; (5) worked-out examples of various calculations; and (6) a bibliography of literature cited.

2.2 VAPOR PRESSURE

2.2.1 Introduction

The evaporative loss of a chemical, volatilization, depends on its vapor pressure, heat of vaporization, and on environmental conditions that influence diffusion from a surface. Volatilization is an important source of material for airborne transport and may lead to its distribution over areas far removed from the point of release. Vapor pressure values provide indications of the tendency of pure materials to vaporize in an unperturbed situation. For example, knowledge of vapor pressure and solubility (or, more precisely, activity coefficient) permits calculations of evaporation rates of dissolved species from water using Henry's Law constants, as discussed by Thomas [16]. Similarly, estimation of the transport of species from soils into the air can be made if data on soil adsorption coefficients is available [17].

The vapor pressure of a chemical determines the maximum concentration of the chemical that can be present in the air phase. In evaporation from water, an important parameter is the ratio of the concentration in the atmosphere to that in water, Henry's Law constant, H, which may be defined as the product of vapor pressure P (atm) and activity coefficient γ (mole fraction). For low solubilities, H is approximated as the ratio of P to concentration C (mol/m^3). Compounds of high H tend to partition predominantly into the atmosphere. For low-H materials, partitioning is predominantly into the water, and the evaporation rate tends to be controlled by the concentration in the air-water boundary layer.

Whereas volatilization is an extremely important transport pathway for many organic chemicals, the majority of inorganic materials have relatively high solubilities (salts) or extremely low vapor pressures, so that volatilization is unlikely to play a significant role in their transport. Organometallics, however, exhibit more of the properties of organic materials (e.g., low water solubility and non-ionic character) and probably would be transported like them through the environment. These materials are of increasing commercial importance, particularly in the semiconductor industry, where they are used as precursors in the preparation of semiconductive thin films. They are also used as catalysts in a variety of manufacturing processes. The bioalkylation of heavy metals by bacteria under aerobic conditions has been recognized as a significant environmental source of organometallics such as $Hg(CH_3)_2$.

2.2.2 Experimental Determination

Vapor pressure can be expressed in a variety of units. The most common are mm Hg, atmospheres (1 atm = 760 mm Hg), and pascals, defined as newtons per square meter (N/m^2). One mm Hg is equal to 133.3 pascals.

The experimental method used to measure vapor pressure depends on the range involved. For values exceeding 1 mm Hg, the isoteniscope can be used; the procedure is described as ANSI/ASTM Method D-2879-75. [15]

For low-vapor-pressure materials (10^{-9} to 10^{-1} mm Hg) the gas saturation method is widely used. Details of the method are described by Ambrose [2] and in the *Federal Register* [18]. While the method is straightforward, meticulous experimental technique is required; the flow rates of carrier gas through the sample of interest must be measured with great care. Moreover, the equipment is expensive, and the collection of enough material for analysis can take days if the material has a very low vapor pressure.

Many inorganic solids have high boiling points and melting points; hence, their vapor pressures at environmental temperatures can range from 10^{-9} mm to 10^{-17} mm. If such a solid is a salt, its solubility in water would be high relative to that of organic materials, and volatilization would generally not be a significant transport mechanism. However, a substantial number of inorganic solids and liquids, such as MoF_6, $AsBr_3$, $Fe(CO)_5$ and $TiCl_4$, have appreciable vapor pressures at environmental temperatures.

2.2.3 Estimation Methods

All of the equations used to represent and estimate vapor pressure begin with the Clausius-Clapeyron equation:

$$\frac{d \ln P}{dT} = \frac{\Delta H_v}{RT^2} \tag{1}$$

where
- P = vapor pressure, mm Hg
- ΔH_v = heat of vaporization, cal/mol
- T = temperature, K
- R = gas constant = 1.987 cal/mol K

To obtain an equation relating vapor pressure to temperature, one can assume an analytical form for ΔH_v and integrate equation 1. Table 2.2-1 lists a few analytical expressions (eqs. 2-5) for ΔH_v and the resulting integrated forms of equation 1.

For any of these equations to be useful, the empirical constants must be related to some readily accessible molecular property. Quite often, this property is ΔH_{vb}, the heat of vaporization at the normal boiling point (T_b). Equation 2 is of limited use for liquids and solids, being valid only below 50°C; for gases, however, it may be used if a reasonable estimate of ΔH_v can be made. Mackay [10] has successfully used equation 3 to predict the vapor pressures of organic hydrocarbons and halogenated hydrocarbons, both liquid and solid, but the correlation is not applicable to other molecules (alcohols, ketones, etc.). Equation 4 is useful only in fitting data, as the number of constants is too great for predictive purposes. Equation 5 is the recommended predictive equation, as all of the parameters can be related to structural properties of materials.

TABLE 2.2-1

Analytical Expressions for ΔH_v and Integrated Forms of Equation 1

Analytical Form for ΔH_v	Integrated Form of Eq. 1	Eq.	Reference
$\Delta H_v =$ constant A	$\ln P = \frac{1}{R}\left[-\frac{A}{T} + K\right]$	(2)	[1]
$\Delta H_v = H_{vb}\left[1 + B(1 - T_\rho)\right]$	$\ln P = \frac{\Delta H_{vb}}{R}\left[1 + B\left(\frac{1}{T_\rho} - 1\right) + B \ln T_\rho\right]$	(3)	[10]
$\Delta H_v = C + DT + ET^2$	$\ln P = \frac{1}{R}\left[-\frac{C}{T} + D \ln T + ET + K\right]$	(4)	[12]
$\Delta H_v = \Delta H_{vb}(3\text{–}2\,T_\rho)^m$	$\ln P \approx \frac{\Delta H_{vb}}{RT_b \Delta Z_b}\left[1 - \frac{(3\text{–}2\,T_\rho)^m}{T_\rho} - 2m\,(3\text{–}2\,T_\rho)^{m-1} \ln T_\rho\right]$	(5)	[6]

A, B, C, D, E = empirical constants

K = constant of integration, empirically determined

ΔH_{vb} = heat of vaporization at the normal boiling point

$T_\rho = T/T_b$

T_b = normal boiling point

$m = f(T_\rho)$ (See text below.)

ΔZ_b = compressibility factor

2.2.4 Recommended Methods

LIQUIDS AND SOLIDS ($T < T_b$)

Grain [7] has related the heat of vaporization at constant values of P and T to structural parameters for organic chemicals as follows:

$$\frac{\Delta H_v}{RT} = \ln RT - a\,[\exp(bP)]\ln P \tag{6}$$

where P is in atmospheres and 'exp' implies the exponential, base e. This relationship is also applicable to inorganic and organometallic materials. The values of a and b are related to the structure and polarity of the material, as shown in Table 2.2-2. When $T = T_b$ (the boiling point), ln P becomes zero, and all of the structural information contained in the values of a and b is lost; hence, a reference temperature and pressure must be chosen. A reference temperature $T_1 = 0.99\ T_b$ was arbitrarily chosen. At this temperature the vapor pressure can vary from about 700 mm to 650 mm, depending upon the structure of the material; an intermediate value of 680 mm was used as a fixed reference pressure (P_1). Combining equations 5 and 6 and substituting $T_1 = 0.99\ T_b$, one obtains **for liquids**:

$$\ln P_\ell = \frac{(\ln RT_1 - a\exp(bP_1)\ln P_1)}{\Delta Z_b}\left[\frac{1-(3-2T_{\rho1})^m}{T_{\rho1}} - 2m\,(3-2T_{\rho1})^{m-1}\ln T_{\rho1}\right] + \ln P_1 \tag{7}$$

where

P_ℓ	=	liquid vapor pressure (atm)
ΔZ_b	=	compressibility factor = 0.97
T_1	=	reference temperature (K) = 0.99 T_b
P_1	=	reference pressure = 0.895 atm
$T_{\rho1}$	=	T/T_1
T	=	system temperature (K)
m	=	$0.4133 - 0.2575\ T_{\rho1}$
R	=	gas constant = 82.05 cc/atm/K
a,b	=	adjustable parameters (see Table 2.2-2)

Equation 7 works equally well for any reduced boiling point datum (T_2, P_2) using the same values of a and b. Thus, if a normal boiling point is given, then $T_1 = 0.99\ T_b$ and $P_1 = 680$ mm $= 0.895$ atm. If a reduced boiling point is given, T_2 and P_2 may be substituted directly for T_1 and P_1.

For **solids**, another term is added to equation 7 such that

$$\ln P_s = \ln P_\ell + \ln \Delta P_s \tag{8}$$

TABLE 2.2-2

Values of a and b for Equation 6

	a	b
Liquids ($T < T_b$)		
Metallic elements	0.80	0
Nonmetallic elements, Inorganic Compounds	1.00	0
Organometallics	1.67	3.50
Gases ($T > T_b$)		
Elements	1.00	0
Other Gases	1.00	2.45[a]

a. Default value. If the molar refraction, R_D, is known or can be estimated, then b is expressed as follows:

$$b = 4.9 - 0.37\,R_D \quad (R_D \leq 13)$$

$$b = 0 \quad (R_D > 13)$$

The additional term, which results in a decrease of vapor pressure as compared to the liquid, is defined as

$$\ln \Delta P_s = \ln(RT_m)\,(1 - 1/T_{\rho m}) \tag{9}$$

where
- P_s = solid vapor pressure (atm)
- ΔP_s = decrease in solid vapor pressure vs that of supercooled liquid (atm)
- R = gas constant = 1.987 cal/mol K
- T_m = melting point (K)
- $T_{\rho m}$ = T/T_m

For solids, two cases must be considered if a reduced vapor pressure datum (i.e., reduced boiling point) is to be used as input. If $T_1 > T_m$, $\ln \Delta P_s$ is given by equation 9 and

$$\ln P_s = \ln P_\ell\ \textit{[eq. 7]} + \ln \Delta P_s\ \textit{[eq. 9]} \tag{10}$$

If $T_1 < T_m$, $\ln \Delta P_s$ is given by

$$\ln \Delta P_s = \ln\,(RT_m)\,(T_m/T_1 - T_m/T) \tag{11}$$

and

$$\ln P_s = \ln P_\ell\ \textit{[eq. 7]} + \ln \Delta P_s\ \textit{[eq. 11]} \tag{12}$$

GASES AND VAPORS ($T > T_b$)

When the chemical is a gas or vapor (i.e., when the temperature of interest is greater than the normal boiling point), equation 2 coupled with the procedure for estimating ΔH_v (eq. 6) is adequate for estimating vapor pressures above 1 atmosphere. The estimation equation becomes

$$\ln P_g = \frac{(\ln RT_1 - a\exp(bP_1)\ln P_1)}{\Delta Z_b}\left(1 - \frac{1}{T_{\rho 1}}\right) + \ln P_1 \qquad (13)$$

where the symbols have the same meaning as in equation 7. Values of a and b for gases were listed in Table 2.2-2.

If neither a reduced datum nor normal boiling point is available, a hypothetical boiling point must be estimated. (This is a frequent problem with organometallics, because they generally decompose or sublime before boiling.) Banks [3] gives a simple correlation between the normal boiling point and the molecular weight, M:

$$\log T_b = 3 - 4/\sqrt{M} \qquad (14)$$

Equation 14 is reasonably good for materials with molecular weights greater than 150 and should be especially suited for high-molecular-weight organometallics. For example, it yields an estimated boiling point of 600 K for aluminum acetyl acetonate, which is only 2% above the measured value of 588 K [14].

2.2.5 Method Error

Table 2.2-3 compares estimated vapor pressures with literature values. Measured vapor pressures are listed when available. However, when data had to be extrapolated, care was taken to ascertain that the extrapolation was valid. (For example, extrapolated data on solids were used only if the extrapolation was from higher temperature measurements in the solid region, not the liquid region.) In some instances normal boiling points were estimated using equation 14; these are noted in the table. Table 2.2-4 gives average method errors as a function of vapor pressure range.

The errors listed in Table 2.2-4 were correlated with ln P in an attempt to determine the errors to be expected for a given pressure range. This was done separately for gases, liquids, and solids. The correlations were not particularly good, indicating essentially random scattering; however, they were useful in at least determining the average error to be expected.

TABLE 2.2-3

Experimental vs Calculated Vapor Pressures
(temperatures are deg. K and pressures are mm Hg)

Compound	T_b	T_m	T	T_2	@	P_2	P_v(exper)	P_v(calc)	% Error	Ref
Inorganic Liquids[a]										
$POCl_3$	378	———	300	———		———	40.0	38.8	− 3.0	[19]
MoF_6	309	———	290	———		———	400.0	377.2	− 5.7	[19]
$PSCl_3$	397	———	289	———		———	10.0	9.3	− 7.0	[19]
$SiCl_4$	330	———	278	———		———	100.0	95.5	− 4.5	[19]
SiH_2I_2	423	———	291	———		———	10.0	3.2	− 67.9	[19]
$SiHBr_3$	385	———	303	———		———	40.0	33.9	− 15.2	[19]
$SnCl_4$	386	———	283	———		———	10.0	10.6	+ 6.0	[19]
$TiCl_4$	409	———	294	———		———	10.0	7.3	− 26.8	[19]
BBr_3	365	———	287	———		———	40.0	33.9	− 15.2	[19]
$VOCl_3$	400	———	285	———		———	10.0	6.3	− 37.0	[19]
$SOCl_2$	348	———	294	———		———	100.0	98.3	− 1.7	[19]
S_2Cl_2	411	———	301	———		———	10.0	10.1	+ 1.0	[19]
CrO_2Cl_2	390	———	287	———		———	10.0	11.3	+ 13.0	[19]
PCl_3	347	———	275	———		———	40.0	38.9	− 2.8	[19]
$SbCl_5$	———	———	296	387		100	1.0	1.4	+ 40.0	[19]
Br_2	331	———	282	———		———	100.0	112.0	+ 12.0	[19]
CSSe	358	———	301	———		———	100.0	90.6	− 9.4	[19]
Hg	———	———	293	399		1.0	1.2×10^{-3}	1.6×10^{-3}	+ 29.6	[19]
CCl_2NO_2	385	———	281	———		———	10.0	9.7	− 3.0	[19]
$Fe(CO)_5$	382	———	278	———		———	10.0	9.2	− 8.0	[19]
$SOBr_2$	413	———	304	———		———	10.0	11.3	+ 13.0	[19]

(Continued)

TABLE 2.2-3 (Continued)

Compound	T_b	T_m	T	T_2	@	P_2	P_v(exper)	P_v(calc)	% Error	Ref
Inorganic Solids										
$AsBr_3$	529	371	354	---		---	1.0	1.2	+ 20.0	[19]
NH_4N_3	---	433	302	407[b]		760	1.0	1.0	0	[19]
$NH_4CO_2NH_2$	---	333[b]	300	331		760	100.0	125.8	+ 28.8	[19]
$SbBr_3$	---	370	323	403		4.2	1.9×10^{-2}	2.2×10^{-2}	+ 15.8	[8]
$TaCl_5$	---	494	298	473		193	3.8×10^{-4}	6.9×10^{-4}	+ 80.6	[5]
$F_3B_2N_3H_3$	---	385[b]	293	385[b]		760	1.6	2.5	+ 54.2	[9]
$Cl_3B_3N_3H_3$	458	357	302	---		---	2.9×10^{-1}	4.2×10^{-1}	+ 44.0	[9]
Organometallics[c]										
$Ge(CH_3)_4$	320	---	301	---		---	389.1	380	− 2.3	[13]
$Ge(C_2H_5)_4$	439	---	337	---		---	21.4	21.4	0	[13]
	439	---		---		---	2.0	2.3	+ 17.0	[13]
$Ge(C_3H_7)_4$	---	---	390	498		676	19.5	23.5	+ 20.0	[13]
Fe(Ac–Ac)$_3$	613 (est)[d]	452	374	---		---	2.0×10^{-3}	4.4×10^{-3}	+119.0	[14]
Al(Ac–Ac)$_3$	588	462	389	---		---	4.1×10^{-2}	3.1×10^{-2}	− 23.8	[14]
Zn(Ac–Ac)$_2$	567 (est)	411	338	---		---	4.0×10^{-3}	2.2×10^{-3}	− 45.9	[14]
U(thd)$_4$	720 (est)	498	372	---		---	6.8×10^{-6}	5.3×10^{-6}	− 22.8	[4]
Th(Ac–Ac)$_4$	679 (est)	445	373	---		---	3.2×10^{-4}	6.9×10^{-5}	− 78.6	[11]

a. At environmental temperatures and pressures

b. Sublimes

c. Ac-Ac = acetylacetonate; thd = tetrakis (2,2,6,6-tetramethylheptane-3,5-dionato)

d. est = estimated via equation 14

TABLE 2.2-4

Average Method Errors as a Function of Pressure

Pressure Range	No. Cpds	Average Error (%)
1 atm-40 atm	2	5
>10 mm - 760 mm	13	8
>1 mm - 10 mm	12	20
>10^{-3} mm - 1 mm	6	40
10^{-6} mm - 10^{-3} mm	3	60

2.2.6 Sample Calculations

Example 1 Estimate the vapor pressure of $POCl_3$ at 27°C. The normal boiling point is 105°C.

$$T_b = 105 + 273 = 378 \text{ K}$$
$$T = 27 + 273 = 300 \text{ K}$$

From Table 2.2-2, a = 1.0, b = 0

$$T_1 = 0.99 \times T_b = 374.2 \text{ K}$$
$$P_1 = 680 \text{ mm} = 0.895 \text{ atm}$$

Substituting in eq. 6 gives:

$$\frac{\Delta H_v}{RT_1} = \ln RT_1 - ae^{bP_1} \ln P_1$$

$$= \ln(82.05 \times 374.2) - 1 \times e^{0 \times 0.895} \times \ln(0.895)$$
$$= 10.33 + 0.11 = 10.44$$
$$T_{\rho 1} = T/T_1 = 0.802$$
$$m = 0.4133 - 0.2575\, T_{\rho 1} = 0.207$$
$$(3 - 2\,T_{\rho 1}) = 1.396$$

From eq. 7:

$$\ln P = \frac{10.44}{0.97}\left[1 - \frac{(1.396)^{0.207}}{0.802} - 2\,(0.207) \times (1.396)^{-0.793} \ln(0.802)\right] + \ln(0.895)$$

$$= -2.973$$

P = antiln (−2.973) = 0.0512 atm (38.9 mm)

Since the literature value is 40 mm, the error is −2.8%.

Example 2 Estimate the vapor pressure of solid $SbBr_3$ at 323 K, given that P_2 = 4.2 mm at 403 K and that T_m = 370 K.

$$P_2 = 4.2/760 = 5.53 \times 10^{-3} \text{ atm}$$

From Table 2.2-2: a = 1.0, b = 0

$$\begin{aligned}
\frac{\Delta H_{v2}}{RT_2} &= \ln RT_2 - a\, e^{bP_2} \ln P_2 \\
&= \ln(82.05 \times 403) - \ln(5.53 \times 10^{-3}) \\
&= 10.41 + 5.20 = 15.61 \\
T_{\rho 2} &= 323/403 = 0.801 \\
m &= 0.4133 - 0.2575\,(0.801) = 0.207 \\
(3 - 2T_{\rho 2}) &= 1.397
\end{aligned}$$

From eq. 7:

$$\ln P_\ell = \frac{15.61}{0.97}\left[1 - \frac{(1.397)^{0.207}}{0.801} - 2\,(0.207) \times (1.397)^{-0.793} \ln(0.801)\right] + \ln(5.53 \times 10^{-3})$$

$$= -4.29 - 5.20 = -9.49$$

As $SbBr_3$ is a solid at the temperature of interest and $T_2 > T_m$, we must use eq. 9 and add the result to $\ln P_\ell$:

$$T_{\rho m} = 323/370 = 0.873$$

From eq. 9: $\ln \Delta P_s = \ln(1.987 \text{ x } 370)\,(1 - 1/0.873)$
$= 6.60\,(-0.146) = -0.960$

From eq. 8: $\ln P_s = -9.49 - 0.960 = -10.450$

P_s = antiln (-10.454)
$= 2.89 \times 10^{-5}$ atm
$= 2.20 \times 10^{-2}$ mm

Since the literature value is 1.9×10^{-2} mm, the error is +16%

Example 3 Estimate the vapor pressure of NH_3 at 352 K (79°C). The normal boiling point is 239 K.

$$\begin{aligned}
T_1 &= 0.99\,T_b = 237\text{ K} \\
P_1 &= 0.895\text{ atm} \\
T_{\rho 1} &= 352/237 = 1.49
\end{aligned}$$

The molar refraction of NH_3 is 5.54. From Table 2.2-2, a = 1 and b = 4.9 − 0.37(5.54) = 2.84.

Substituting in eq. 13:

$$\ln P_g = \frac{\ln(82.05 \times 237) - 1 \exp(2.84 \times 0.895) \ln(0.895)}{0.97} \times (1 - 1/1.49) + \ln(0.895)$$

$$= 3.694$$

$$P_g = \text{antiln}\, 3.94 = 40.2 \text{ atm}$$

The literature value is 40 atm, indicating an error of 0.5%.

2.2.7 Literature Cited

1. Adam, N.K., *Physical Chemistry*, Oxford Univ. Press, London, 234 (1958) *or any other standard text on physical chemistry.*
2. Ambrose, D., "Vapor Pressures," *Chemical Thermodynamics*, **1**, 218 (1973).
3. Banks, W.H., "Considerations of a Vapour Pressure Temperature Equation, and Their Relation to Burnop's Boiling Point Function," *J. Chem. Soc.*, **1939**, 292 (1939).
4. Belford, R. and E. Huss, "The Vapor Pressure of Tetrakis (2,2,6,6-tetramethyl-heptane-3,5, dianato) Uranium," *J. Chem. Eng. Data*, **22**, 239 (1977).
5. Brink, T.M. and F. Stevenson, "Vapor Pressure of Tantalum Pentachloride," *J. Chem. Eng. Data*, **17**, 143 (1972).
6. Grain, C.F., "Vapor Pressure" in *Handbook of Chemical Property Estimation Methods*, W.J. Lyman, W.F. Reehl and D.H. Rosenblatt (eds.), McGraw-Hill Book Co., New York (1982).
7. Grain, C.F. and W.J. Lyman, "Enhancements to CHEMEST Program: Vapor Pressure," Interim Report, EPA Contract No. 68-01-6271 (October 7, 1983).
8. Gray, G.A. and R. Sime, "Vapor Pressure of Antimony (III) Bromide," *J. Chem. Eng. Data*, **17**, 176 (1972).
9. Laubengayer, A.W. and C. Scaife, "The Vapor Pressures and Some Thermodynamic Functions of B-Trifluoro, B-Trichloro, and B-Tribromoborazine," *J. Chem. Eng. Data*, **11**, 172 (1966).
10. Mackay, D., A. Bobra, D. Chen, and W. Shiu, "Vapor Pressure Correlations for Low Volatility Environmental Chemicals," *Environ. Sci. Technol.*, **16**, 645 (1982).
11. MacKenzie Chemical Works, "Metal Acetylacetonates," Technical Data Bulletin, Central Islip, NY (undated).
12. Miller, D.G., "Estimating Vapor Pressures — A Comparison of Equations," *Ind. Eng. Chem.*, **56**, 46 (March 1964).
13. Mogul, P.H., M. Hochberg, R. Michiel, G. Nestei, B. Warmsley and S. Coren, "Physical Properties of Tetra-n-alkylgermanes (C_1 - C_6)," *J. Chem. Eng. Data*, **19**, 4 (1974).

14. Sachindis, J. and J.O. Hill, "A Re-evaluation of the Enthalpy of Sublimation of Some Metal Acetylacetonate Complexes," *Thermochim. Acta*, **35**, 59 (1980).

15. "Standard Test Method for Vapor Pressure-Temperature Relationship and Initial Decomposition Temperature of Liquids by Isoteniscope," *Annual Book of ASTM Standards*, Sec. 5, V 5.02, 747 (1983).

16. Thomas, R.G., "Volatilization from Soil" in *Handbook of Chemical Property Estimation Methods*, W.J. Lyman, W.F. Reehl and D.H. Rosenblatt (eds.), McGraw-Hill Book Co., New York (1982).

17. Thomas, R.G., "Volatilization from Water," *ibid.*

18. "Vapor Pressure," *Fed. Reg.*, **145**, 77345 (Nov. 21, 1980).

19. Weast, R.C. (ed.), *CRC Handbook of Chemistry and Physics*, 58th ed. , CRC Press, Inc., Boca Raton, FL, D-182 (1977-1978).

2.3 PHOTOLYSIS IN AIR

2.3.1 Introduction

This section provides an overall view of some of the complexities associated with the formation or transformation of atmospheric pollutants by photolysis. The evaluation of inorganic gaseous species with respect to their potential photolysis rate is explained, and reaction mechanisms leading to the generation of additional pollutants are qualitatively described.

It has long been recognized that certain air pollution smogs are caused by photochemical reactions among pollutants from automobile exhaust [9]. The direct photolysis of NO_2 is a major source of other oxidants in the atmosphere [6, 7]. "Direct photolysis" refers to the process in which a molecule absorbs solar radiation and is either dissociated or is promoted to an excited electronic state amenable to further reactions with other species present in the atmosphere. "Sensitized photolysis," in which energy is transferred from some other species in the atmosphere, is not known as an important source of pollutants.

While the direct photolysis of NO_2 is one of the most important photolytic sources of pollutants, other inorganic species subject to direct photolysis include SO_2, O_3, H_2O_2 and HONO. We shall attempt to assess the relative importance of these additional species as sources of pollutant molecules.

Photochemical reactions begin with the absorption of light radiation. The immediate product is an excited state of the absorbing molecule, with energy in excess of the normal state equal to the energy of the absorbed radiation. Light that produces only rotational or vibrational changes is chemically inactive in the atmosphere; only when the light absorbed produces electronic transitions leading to dissociation will a photochemical reaction occur. As most vibrational and rotational spectra lie in the infrared and most electronic spectra in the visible and ultraviolet, it is the latter regions which are of photochemical importance.

A photochemical primary process may yield stable molecules directly, or it may yield unstable products that undergo secondary reactions. In many instances the primary process consists of dissociation of the absorbing molecule into free radicals or atoms, which then take part in secondary reactions.

For environmental assessment purposes, the following information is required with respect to the photochemical process: (1) an understanding of the intensity and spectral distribution of the incident radiation, (2) the absorption characteristics of the molecular species, (3) the kinetics of the primary and secondary processes, and (4) transport mechanisms of products of the primary and secondary processes.

We have restricted our discussion to photolysis in the lower atmosphere. This is not meant to diminish the importance of upper-atmosphere photolysis, particularly as it relates to the preservation of the ozone protective layer, but simply to reflect the general concern with respect to air quality in the lower atmosphere.

2.3.2 Units for Rate Constants and Concentrations

Current practice expresses incident solar radiation in terms of photons · cm^{-2} · s^{-1} · $(10\ nm)^{-1}$. The absorption coefficients of gaseous species are generally expressed in terms of liters · mole $^{-1}$ · cm^{-1}. In studies of polluted air, it has become customary to express concentrations in terms of ppm (parts per million) or sometimes pphm (parts per hundred million). Table 2.3-1 lists factors for converting air concentrations (moles · l^{-1}) and rate constants.

TABLE 2.3-1

Conversion Factors
(Relative to air at 1 atm, 25°C)

CONCENTRATIONS:

$$\text{mole} \cdot l^{-1} \times 2.445 \times 10^{7} = \text{ppm}$$
$$\text{mole} \cdot l^{-1} \times 2.445 \times 10^{9} = \text{pphm}$$

RATE CONSTANTS

Bimolecular

$$l \cdot \text{mole}^{-1} \cdot s^{-1} \times 2.45 \times 10^{-6} = \text{ppm}^{-1} \cdot \text{min}^{-1}$$
$$l \cdot \text{mole}^{-1} \cdot s^{-1} \times 1.47 \times 10^{-6} = \text{pphm}^{-1} \cdot \text{hr}^{-1}$$

Termolecular

$$l^{2} \cdot \text{mole}^{-2} \cdot s^{-1} \times 1.005 \times 10^{-13} = \text{ppm}^{-2} \cdot \text{min}^{-1}$$
$$l^{2} \cdot \text{mole}^{-2} \cdot s^{-1} \times 6.03 \times 10^{-16} = \text{pphm}^{-2} \cdot \text{hr}^{-1}$$

2.3.3 Solar Irradiance and Absorption Rates

The lower atmosphere receives solar radiation not only directly from the sun but also from the sky via scattering, and in the ultraviolet the latter may be the more important of the two. Estimating the intensity of incident solar radiation involves consideration of several factors, including the spectral irradiance outside the atmosphere, the solar zenith angle, the nature and amount of scattering, diffusion and absorption of radiation by the atmosphere, and the albedo of the earth's surface. Leighton [9] has discussed these factors at length; Table 2.3-2 (from Leighton's book) lists estimated values of solar irradiance at midday, $W(\lambda)$, in the lower atmosphere after all of the factors listed above have been accounted for. An updated and presumably more accurate listing is given by Dermerjian *et al.* [3]. (See § 2.14.3 for additional discussion of solar irradiance.)

TABLE 2.3-2

Estimated Solar Irradiance, W(λ), in the Lower Atmosphere[a]
($\times 10^{15}$ photon · cm^{-2} · s^{-1} · $(10\ nm)^{-1}$)

Wavelength λ(nm)	Solar Zenith Angle, Z 0°	20°	40°	60°	80°
290	0.0014	0.0009	0.0002	—	—
300	0.12	0.10	0.05	0.01	—
310	0.65	0.60	0.43	0.13	0.01
320	1.17	1.10	0.91	0.53	0.10
330	1.83	1.75	1.48	0.96	0.26
340	1.88	1.82	1.55	1.05	0.32
350	2.06	1.98	1.72	1.18	0.36
360	2.10	2.02	1.77	1.24	0.38
370	2.48	2.40	2.11	1.50	0.46
380	2.36	2.28	2.02	1.45	0.44
390	2.20	2.13	1.90	1.38	0.42
400	3.10	3.01	2.70	2.00	0.62
410	4.01	3.90	3.51	2.63	0.82
420	4.06	3.95	3.57	2.71	0.86
430	3.86	3.76	3.42	2.62	0.85
440	4.50	4.39	4.02	3.11	1.03
450	5.00	4.88	4.48	3.51	1.19
460	5.00	4.89	4.50	3.56	1.22
470	5.13	5.02	4.64	3.68	1.30
480	5.20	5.11	4.72	3.79	1.38
490	4.88	4.78	4.44	3.59	1.33
500	4.95	4.85	4.51	3.68	1.40
525	5.14	5.05	4.73	3.91	1.56
550	5.30	5.21	4.90	4.09	1.69
575	5.32	5.23	4.92	4.44	1.75
600	5.32	5.24	4.95	4.21	1.86
625	5.27	5.19	4.94	4.25	2.01
650	5.22	5.16	4.92	4.30	2.16
675	5.16	5.10	4.89	4.32	2.29
700	5.05	5.00	4.82	4.29	2.38
750	4.80	4.75	4.59	4.10	2.47
800	4.55	4.51	4.37	3.98	2.47

Source: Leighton [9]. *(Copyright 1961, Academic Press. Reprinted with permission.)*

Any discussion of a photochemical primary process must begin with a determination of the species that exhibit absorption in the region of interest. Of the inorganic species present in the lower atmosphere, just oxygen, ozone, nitrogen dioxide, sulfur dioxide, nitric and nitrous acids and hydrogen peroxide absorb to any measurable degree. To date, these species are the only ones that have been of any interest in the determination of photolysis mechanisms for inorganics. The organic gases or vapors that are of some importance include aldehydes, ketones, peroxides and various organic nitrogen compounds.

To assess the photodegradation potential of a species, we must have some means of determining its average absorption rate. The usual starting point is the Lambert-Beer law: if a surface layer of height h contains an absorbing substance at concentration c, with an absorption coefficient ϵ, then the average rate of absorption per unit value for direct and sky radiation is

$$I_a(\lambda) = \frac{I_d(\lambda)\ (1 - 10^{-\epsilon(\lambda)ch \sec Z}) + I_s(\lambda)\ (1 - 10^{-\epsilon(\lambda)\, cih})}{h} \qquad (1)$$

where

$I_a(\lambda)$	=	average absorption rate (photons $\cdot$ cm^{-2} $\cdot$ s^{-1})
$I_d(\lambda)$	=	direct irradiance (photons $\cdot$ cm^{-2} $\cdot$ s^{-1})
$I_s(\lambda)$	=	sky radiance (photons $\cdot$ cm^{-2} $\cdot$ s^{-1})
$\epsilon(\lambda)$	=	molar absorptivity (l $\cdot$ mol^{-1} $\cdot$ cm^{-1})
c	=	concentration of absorbing species (mol $\cdot$ l^{-1})
h	=	height of absorbing layer (cm)
Z	=	solar zenith angle
i	=	constant = 2

As c is low (typically 10^{-14} mol $\cdot$ l^{-1}), equation 1 is reduced to

$$I_a(\lambda) = 2.303c\epsilon(\lambda)\ [I_d(\lambda) \sec Z + I_s(\lambda)\ i] \qquad (2)$$

The specific rate constant, $k_a(\lambda)$, is then given by

$$k_a(\lambda) = \frac{I_a(\lambda)}{jc} \qquad (3)$$

where j is a conversion factor for concentration in l $\cdot$ mol^{-1}, numerically equal to Avogadro's number $\times\ 10^{-3} = 6.023 \times 10^{20}$.

The average absorption rate $I_a(\lambda)$ is related to $W(\lambda)$, the total solar irradiance in the lower atmosphere (Table 2.3-2), through the relationship

$$I_a(\lambda) = 2.303\ \epsilon(\lambda)c\ W(\lambda) \qquad (4)$$

Hence,

$$k_a(\lambda) = \frac{2.303\ \epsilon(\lambda)\ W(\lambda)}{j} \qquad (5)$$

Thus, by multiplying the extinction coefficient by the estimated solar irradiance at each wavelength, we obtain the rate constant $k_a(\lambda)$ of a species as a function of λ. Figure 2.3-1 shows absorption rate constants calculated from equation 5 for three of the absorbers mentioned earlier. Clearly, NO_2 is the most significant absorber of the three and a major inorganic species to consider with respect to primary photochemical processes.

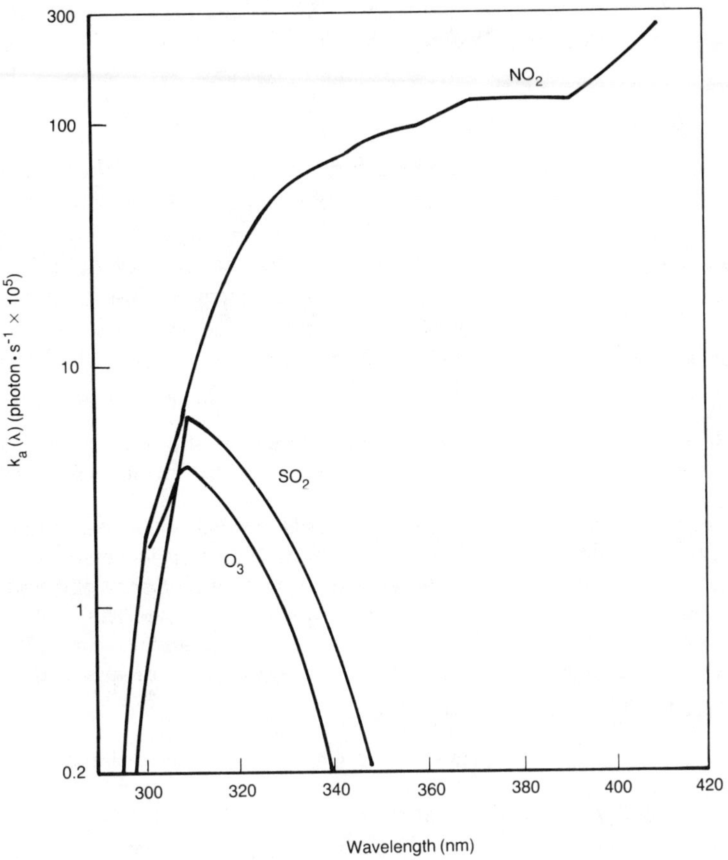

Source: Leighton [9]. *(Copyright 1961, Academic Press. Reprinted with permission.)*

FIGURE 2.3-1 Absorption Rates for Three Inorganic Pollutants

Formation and reaction of hydroxy radicals are also important in oxidant formation processes. The species that are photolyzed to yield HO radicals are O_3 and H_2O_2, through the following reactions:

$$O_3 + H_2O \xrightarrow{h\nu} 2\ HO + O_2 \tag{6}$$

$$H_2O_2 \xrightarrow{h\nu} 2\ HO \tag{7}$$

The absorption of solar radiation in the lower atmosphere by O_3 and H_2O_2 is low, but rates of HO formation can be significant if the initial concentrations are high enough. We shall now examine how these rates can be estimated.

The efficiency of any primary photochemical process is expressed in terms of the quantum yield, ϕ, which may be defined as the ratio of the number of molecules reacting through a primary process to the number of quanta absorbed. The maximum possible value for ϕ is unity, as the number of molecules reacting cannot exceed the number of quanta absorbed. The rate of product formation is then given by:

$$\text{Rate} = \phi\, c \sum_{\lambda} k_a(\lambda) \qquad (8)$$

Before considering possible mechanisms for a photochemical reaction, one can assess the probability that a species will undergo photolysis and the relative importance of this process as a source of pollution. Examples 1 and 2 below show how this is done. The minimum information required is (1) the free energy of bond formation or dissociation, (2) the light absorption characteristics of the material, and (3) its approximate concentration. A useful relationship between the wavelength λ (nm) and the energy E (kcal mol^{-1}) of a photon that may cause bond dissociation is $E \cdot \lambda = 2.9 \times 10^4$.

Example 1 What is the likelihood that photolysis of sulfur dioxide would generate products that could enhance the parent compound's pollution potential?

The bond dissociation energy for the reaction $SO_2 \rightarrow SO + O$ is -135 kcal $\cdot$ mol^{-1} [9]. The wavelength equivalent of this energy is 218 nm. As this is considerably shorter than ultraviolet wavelengths, we can immediately conclude that sunlight will not cause bond dissociation of SO_2. However, it might generate excited states that could undergo secondary reactions. The rate of absorption of solar radiation can be estimated from equation 5 using data from Tables 2.3-2 and -3; Table 2.3-4 shows the results of this calculation for various wavelengths.

The rate of product formation is equal to $\phi k_a(\lambda)c$, where ϕ is the quantum yield for the primary photochemical process. For SO_2, ϕ has been estimated to range from 1.0×10^{-2} to 0.3 [9]. Based on the average value of $k_a(\lambda)$ from Table 2.3-4 (1.7×10^{-4} photon $\cdot$ s^{-1}) and an assumed concentration (c) of 1 ppm, the approximate rate of product formation is found to be 6.0×10^{-3} to 0.2 ppm $\cdot$ hr^{-1}. Coupled with the knowledge that photolysis cannot produce bond dissociation of SO_2 but only excited states that could be quenched, the calculated rate indicates that photolysis of SO_2 plays a minor role in the generation of other pollutants.

Example 2 Is the photolysis of NO_2 a significant source of oxidizing pollutants?

The dissociation energy of NO_2 is 72 kcal $\cdot$ mol^{-1} [9]. This is equivalent to light with a wavelength of 400 nm, which is well within the portion of solar spectrum that penetrates the atmosphere. Estimates of the rate of absorption for NO_2 in the range of 290-400 nm are listed in Table 2.3-5. The average $k_a(\lambda)$ is 5.0×10^{-3} photon $\cdot$ s^{-1}, which is 29 times that of SO_2; this, plus the fact that the photolysis of NO_2 causes its dissociation into active products, suggests that the process could be a significant source of pollutants. The

TABLE 2.3-3

Absorption Coefficients, ϵ (λ), at Various Wavelengths
($l \cdot mol^{-1} \cdot cm^{-1}$)

	ϵ (λ)			
λ(nm)	**SO_2**	**NO_2**	**O_3**	**H_2O_2**
290	156.0	20.0	384.0	3.9
300	121.0	37.0	102.0	2.6
310	46.0	57.0	28.0	1.8
320	13.0	78.0	7.4	1.3
330	3.7	98.0	0.5	1.0
340	1.1	119.0		0.8
350		136.0		0.5
360		149.0		0.3
370		158.0		0.2
380		163.0		
390		167.0		
400		171.0		
Bond Dissociation Energy (kcal/mol)	135	72	25[a]	51[b]

a. Third body required. For comparison, value for O_2 is 119 kcal/mol [14].
b. For 0-0 bond in H_2O_2 [14].
Source: Leighton [9]

TABLE 2.3-4

Calculated Rates of Absorption of Solar Radiation by SO_2
(at Z = 40°)

λ(nm)	**$k_a(\lambda)$ (photon · s^{-1})**
290	1.2×10^{-7}
300	2.3×10^{-5}
310	7.6×10^{-5}
320	4.6×10^{-5}
330	2.1×10^{-5}
340	6.4×10^{-6}

$k_a = \Sigma k_a(\lambda) = 1.7 \times 10^{-4}$ photon · s^{-1}

TABLE 2.3-5

Calculated Rates of Absorption of Solar Radiation by NO_2
(at Z = 40°)

λ(nm)	**$k_a(\lambda)$ (photon · s^{-1})**
290	2.0×10^{-8}
320	2.7×10^{-4}
340	7.1×10^{-4}
360	8.9×10^{-4}
380	1.3×10^{-3}
400	1.8×10^{-3}

$k_a = \Sigma k_a(\lambda) = 5.0 \times 10^{-3}$ photon · s^{-1}

quantum yield has been estimated at $\phi = 0.9$ [9]; using the same value of c as in Example 1,

$$\begin{aligned}\text{Rate of product formation} &= \phi k_a(\lambda)c \\ &= (0.9)(5.0 \times 10^{-3})(1) \\ &= 4.5 \times 10^{-3}\ \text{ppm} \cdot \text{s}^{-1} \\ &= 16.2\ \text{ppm} \cdot \text{hr}^{-1}\end{aligned}$$

Similar reasoning, assuming $\phi = 0.9$ and the same conditions as in the preceding examples, leads to estimating product yields of 0.7 ppm • hr^{-1} for O_3 and 0.2 ppm • hr^{-1} for H_2O_2.

We thus see that some reasonable conclusions can be drawn as to whether a given material should be considered a possible source of pollutants through a primary photochemical process. These conclusions were based on some elementary considerations of the energy requirements for dissociation by light and the absorption characteristics of the species.

2.3.4 Photochemical Mechanisms

The oxidant found in the largest quantity as a result of photolysis is ozone. Although oxygen is photolyzed in the upper atmosphere, the short wavelengths required do not reach the lower atmosphere. Therefore, photochemical generation of the observed amounts of ozone [7] must take place by some process other than direct photolysis of oxygen.

Comparison of the absorption characteristics of the major inorganic pollutants (Figure 2.3-1) indicates that NO_2 is the most efficient absorber of the ultraviolet portion of sunlight that reaches the lower atmosphere. The absorption of UV light (300-400 nm) initiates the following reactions:

$$NO_2 + h\nu \xrightarrow{k_1} NO + O \tag{9}$$

$$O + O_2 + M \xrightarrow{k_2} O_3 + M \tag{10}$$

$$O_3 + NO \xrightarrow{k_3} NO_2 + O_2 \tag{11}$$

In equation 10, M is a third body (air). k_1, k_2 and k_3 are reaction rate constants, with $k_1 = \phi k_a$

where ϕ = quantum yield
k_a = average rate of absorption = $\Sigma k_a(\lambda)$

The rate equations resulting from these reactions are:

$$\frac{d(NO_2)}{dt} = k_3[O_3][NO] - \phi k_a[NO_2] \tag{12}$$

$$\frac{d(NO)}{dt} = \phi k_a[NO_2] - k_3[O_3][NO] \tag{13}$$

$$\frac{d(O_3)}{dt} = k_2[O][O_2][G] - k_3[O_3][NO] \tag{14}$$

If we assume the steady state for these rates, the ozone and oxygen atom concentrations are given by

$$[O_3] = \frac{\phi k_a}{k_3} \frac{[NO_2]}{[NO]} \tag{15}$$

$$[O] = \frac{\phi k_a}{k_2} \frac{[NO_2]}{[O_2]\ [G]} \tag{16}$$

The NO_2 photolytic cycle explains the initial formation of ozone in polluted atmospheres; however, it cannot account for the relatively high concentrations of O_3 that have been measured. At steady state, O_3 and NO would be formed and consumed in equal quantities, and there would be no O_3 accumulation. Laboratory data and atmospheric measurements indicate that hydrocarbons are involved [6, 7]. Certain hydrocarbons, notably olefins and aromatics that are introduced into the atmosphere from automobile exhausts, enter into the NO_2 cycle. Apparently, oxygen atoms attack the hydrocarbons, and the resulting free radicals react with NO to form more NO_2; thus, O_3 and NO_2 levels increase while NO is depleted.

The mechanisms that have been suggested to account quantitatively for the myriad of observed products are quite complex. For example, the inclusion of just one hydrocarbon, propylene, into the nitrogen cycle has led to the postulation of 150 elementary reactions, of which 60 are considered important [11]. The effects can be studied only through the use of a computer, as each of the non-steady-state rate equations must be numerically integrated to yield concentration vs. time data. Another difficulty in assessing complex mechanisms is the lack of reliable data on rate constants; a recent publication by Atkinson and Lloyd [1] has helped to alleviate this problem.

A highly simplified mechanism, first introduced by Friedlander and Seinfeld [5], provides a qualitative picture of the NO_2-hydrocarbon interaction. In this mechanism, all hydrocarbons are considered as a single species (RH). No detailed description of products is given, except that free-radical species (RO) are postulated that cause a chain reaction with initiation, propagation and terminating steps. Overall, only seven reactions are considered — the three reactions of the NO_2 cycle (equations 9-11) plus the following four:

$$RH + O \xrightarrow{k_4} RO + Products \tag{17}$$

$$RH + O_3 \xrightarrow{k_5} RO + Products \tag{18}$$

$$NO + RO \xrightarrow[O_2]{k_6} NO_2 + RO \tag{19}$$

$$NO_2 + RO \xrightarrow{k_7} Products \tag{20}$$

With certain assumptions and simplifications, the final rate equations are:

$$\frac{d[NO_2]}{dt} = [NO_2][RH]\left(\frac{k_1}{k_2} k_6 k_4'[NO] - \frac{k_1}{k_3} k_4'[NO_2]\right) \qquad (21)$$

$$\frac{d[NO]}{dt} = -\frac{k_1}{k_2} k_6 k_4'[NO_2][RH][NO] \qquad (22)$$

$$\frac{d[RH]}{dt} = -[NO_2][RH]\left(\frac{k_1}{k_2} k_4 + \frac{k_1 k_5}{k_3[NO]}\right) \qquad (23)$$

where $$k_4' = \frac{k_4}{k_6[NO] + k_7[NO_2]}$$

Using the initial concentrations and rate constants listed in Table 2.3-6, one can integrate these equations numerically, yielding the results shown in Figure 2.3-2.

If quantitative information on NO_2-hydrocarbon interaction is desired, the detailed mechanisms must be considered. These mechanisms have been recently updated by Atkinson *et al.* [2].

Attempts have been made to simplify and consolidate the reaction mechanisms without losing too much of their quantitative aspects. Notable among these are the carbon bond approach of Whitten *et al.* [15] and the counter species concept of Leone and Seinfeld [10].

TABLE 2.3-6

Initial Concentrations and Rate Constants

$[NO_2]_0$	= 0.2 ppm
$[NO]_0$	= 1.0 ppm
$[RH]_0$	= 2.0 ppm
$\frac{k_1}{k_2} k_6 k_4'$	= 0.1 ppm^{-2} min^{-1}
$\frac{k_1}{k_3} k_4'$	= 0.02 ppm^{-2} min^{-1}
$\frac{k_1}{k_2} k_4$	= 1.83×10^{-3} ppm^{-1} min^{-1}
$\frac{k_1 k_5}{k_3}$	= 2.45×10^{-4} min^{-1}

Source: Friedlander and Seinfeld [5]

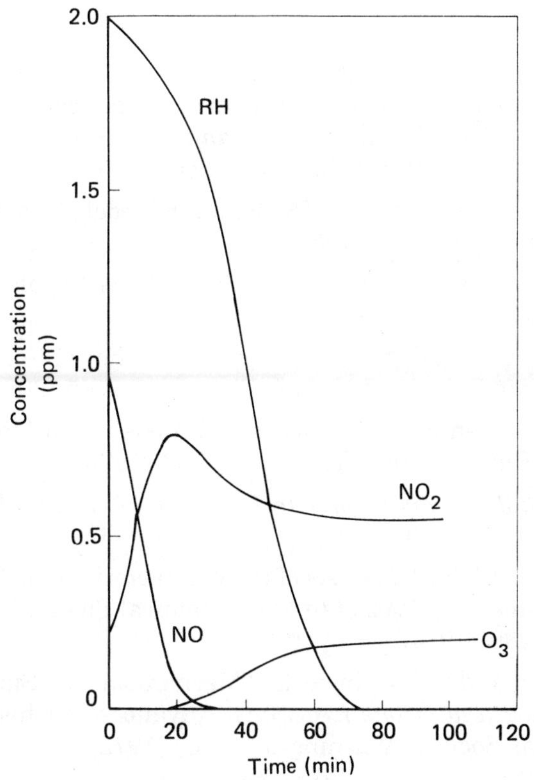

FIGURE 2.3-2 Concentration of Reacting Species vs Time

The analysis of even the most complex reaction mechanisms is only the initial step in modeling the production and transport of atmospheric pollutants resulting from photolysis. The rate equations for each species must be incorporated in a general model that describes the transport of these species over large distances.

Photochemical transport models have been reviewed by Eschenroeder and Martinez [4], Lamb and Seinfeld [8] and Seinfeld *et al.* [13]. A more recent evaluation of these models has been presented by Ruff *et al.* [12].

2.3.5 Literature Cited

1. Atkinson, R., and A. Lloyd, "Evaluation of Kinetic and Mechanistic Data for Modeling of Photochemical Smog," *J. Phys. Chem. Ref. Data*, **13**, 315 (1984).
2. Atkinson, R., A. Lloyd and L. Winges, "An Updated Chemical Mechanism for Hydrocarbon/NO_x/SO_2 Photoxidations Suitable for Inclusion in Atmospheric Simulation Models," *Atmos. Environ.*, **16**, 1341 (1982).
3. Dermerjian, K.L., K. Schere, and J. Peterson, "Theoretical Estimates of Actinic (Spherically Integrated) Flux and Photolytic Rate Constants of Atmospheric Species in the Lower Troposphere," in *Advances in Environmental Science and Technology*, **10**, pp. 369-459, Pitts, J.N., R. Metcalf and D. Grosjean (eds.), Wiley-Interscience, John Wiley & Sons, New York (1980).

4. Eschenroeder, A. and J. Martinez, "Concepts and Applications of Photochemical Smog Models," in *Photochemical Smog and Ozone Reactions*, Advances in Chemistry Series 113, American Chemical Society, Washington, D.C. (1972).
5. Friedlander, S.K. and J.H. Seinfeld, "A Dynamic Model of Photochemical Smog," *Environ. Sci. Technol.*, **3**, 1175 (1969).
6. Haagen-Smit, A.J., "Chemistry and Physiology of Los Angeles Smog," *Ind. Eng. Chem.*, **44**, 13422 (1952).
7. Haagen-Smit, A.J., C. Bradley and M. Fox, "Ozone Formation in Photochemical Oxidation of Organic Substances," *Ind. Eng. Chem.*, **45**, 2086 (1953).
8. Lamb, R.G. and J.H. Seinfeld, "Mathematical Modeling of Urban Air Pollution: General Theory," *Environ. Sci. Technol.*, **7**, 253 (1973).
9. Leighton, P.A., *Photochemistry of Air Pollution,* Academic Press, New York (1961).
10. Leone, J.A. and J. Seinfeld, "Analysis of the Characteristics of Complex Chemical Reaction Mechanisms: Application to Photochemical Smog Chemistry," *Environ. Sci. Technol.*, **18**, 280 (1984).
11. Niki, H., E. Daby and B. Weinstock, "Mechanisms of Smog Reactions," in *Photochemical Smog and Ozone Reactions,* Advances in Chemistry Series 113, American Chemical Society, Washington, D.C. (1972).
12. Ruff, R.E., K.C. Nitz, F.L. Ludwig, C.M. Bhumralkar, J.D. Shannon, C.M. Sheih, I.Y. Lee, R. Kumar and D.J. McNaughton, "Evaluation of Three Regional Air Quality Models," *Atmos. Environ.*, **19**, 1103 (1985).
13. Seinfeld, J.H., S.D. Reynolds and P.M. Roth, "Simulation of Air Pollution," in *Photochemical Smog and Ozone Reactions*, Advances in Chemistry Series 113, American Chemical Society, Washington, D.C. (1972).
14. Weast, R.C. (ed.), *CRC Handbook of Chemistry and Physics,* 54th ed., CRC Press, Cleveland (1973).
15. Whitten, C.Z., H. Hogo and J. Kills, "The Carbon Bond Mechanism: A Condensed Kinetic Mechanism for Photochemical Smog," *Environ. Sci. Technol.*, **14**, 691 (1980).

2.4 GAS ADSORPTION ON SOLIDS

2.4.1 Introduction

The extent to which a gas is adsorbed on a solid surface affects the rate at which it is released from the surface and transported to the environment. Even if it has a vapor pressure greater than 1 atmosphere under environmental conditions, a gas that is strongly adsorbed on a solid surface may remain there for some time. As the volatilization rate is related to the vapor pressure of the gas species, the "effective" vapor pressure at the soil-air interface is determined by, among other properties, the adsorption behavior of the gas. Therefore, to determine the persistence of a gas and the rate at which it is released into the atmosphere, we must be able to estimate how strongly it is adsorbed on a surface.

The association between a gas and a solid surface can be either physical or chemical in character. Physical or Van der Waals adsorption is characterized by relatively low heats of adsorption ($\leq$5 kcal/mol), which are of about the same order as heats of vaporization. The equilibrium between solid and gas is reversible and is rapidly attained when the temperature or pressure changes. This form of adsorption is usually favored at low temperature.

Van der Waals adsorption occurs to some extent whenever a gas is in contact with a solid. However, when the gas and the solid can react chemically, the other type of adsorption (chemisorption) is sometimes more significant, especially at elevated temperatures. The heats of adsorption are in the range of 20-100 kcal/mol, indicating that chemical reactions are involved. Chemisorption occurs chiefly on metal surfaces. It can also occur on other surfaces, but in such cases the temperature must be elevated, because the process then requires a substantial energy input. For these reasons, a first approximation of the extent of adsorption need not consider chemisorption.

Furthermore, one can usually ignore the physical adsorption of gases that boil below 200K, since only small amounts would be adsorbed at normal temperatures and they would not persist on solid surfaces.

The environmental implications of gas and vapor adsorption are important, because high-technology industries are making increasing use of organometallic gases and low-boiling liquids, and the likelihood of their release into the environment is therefore growing.

Some of the gases and vapors now used by industry, such as PH_3, AsH_3, Ti(i-propoxide)$_4$ and SiH_4, rapidly decompose when exposed to air, forming oxides or hydroxides. While these may cause environmental problems, gas adsorption is not involved. However, other materials such as $Si(CH_3)_4$, $Sn(CH_3)_4$ and $Si(OCH_3)_4$ may persist for finite times in the atmosphere, so an estimate of their potential adsorption might be useful in assessing their persistence.

2.4.2 Data Values

Experimentally, the amount of gas adsorbed on a given substance is determined in apparatus containing a chamber whose volume is variable and can be measured accurately. The pressure and temperature are also adjustable. A weighed sample of adsorbent is placed in the chamber, and a measured quantity of gas is introduced. The subsequent loss in volume indicates the amount of gas adsorbed. So long as there is no chemical reaction between the two components, one solid will adsorb the same amount of gas as another solid of equal surface area.

The adsorption at a given temperature (adsorption isotherm) is usually plotted as the amount adsorbed, X, vs. the relative pressure, P/P_o, where P is the system pressure and P_o is the saturated vapor pressure of the gas or vapor (atm), at the given temperature. In this discussion, P is always assumed to be 1 atmosphere. The quantity X may be expressed in units of volume (ml adsorbate/g adsorbent), weight (g adsorbate/g adsorbent), or mol adsorbate/g adsorbent. If the surface area per gram of adsorbent is known, X can be expressed in terms of weight or volume of gas per unit area, thereby facilitating estimates of adsorption on soil or some other substance. To do this, however, one must also know the unit surface area of the other substance; this is, in fact, the purpose of most adsorption measurements.

Figure 2.4-1, which shows isotherms for SO_2 on silica gel, illustrates the significant influence of temperature on the extent of adsorption. The temperature dependence of the amount adsorbed can be generally described by a function of the type $X \propto e^{-A/T}$, where A is proportional to the difference between the heat of adsorption and the latent heat of vaporization, and T is the temperature in degrees K.

Table 2.4-1 shows typical adsorption data for a variety of gases at two different temperatures. Some anomalies will be noted, in the form of an apparent decrease in adsorption at the lower temperature for some gases (N_2O, CO_2, CO and N_2). This could be simply due to the fact that the two sets of data were obtained with nonidentical kinds of charcoal, which may have had different surface areas or surface treatments.

2.4.3 Estimating Methods

Methods of calculating the extent of physical adsorption to be expected for a given gas have not been developed; however, a way may be found to obtain at least rough estimates, such as by correlating the quantity adsorbed with other physical properties. For several organic and inorganic gases, for example, the logarithm of the volume adsorbed per gram of adsorbent varies generally in a linear fashion as a function of the normal boiling point of the gas. Figure 2.4-2 illustrates this relationship using adsorption data from Table 2.4-1; the correlation coefficient of the linear plot is 0.969.

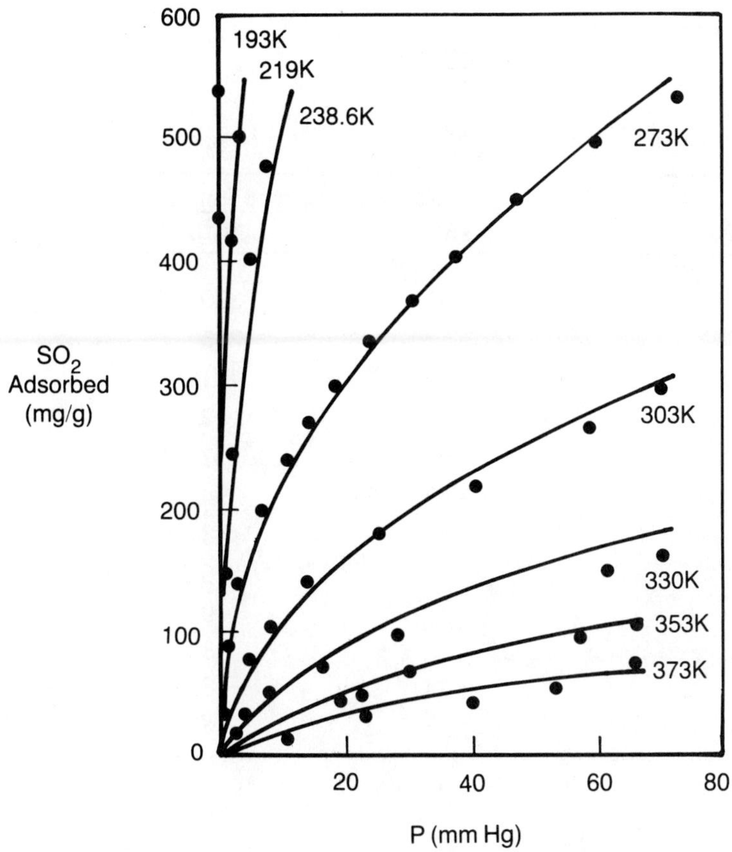

Source: Adamson [2]. *(Copyright 1960, Wiley Interscience. Reprinted with permission.)*

FIGURE 2.4-1 Adsorption Isotherms for SO_2 on Silica Gel

Theoretical treatments of physical adsorption were made by Langmuir [8] and Brunauer, Emmett and Teller [5]. Langmuir derived the following relationship between the volume of gas adsorbed and the gas pressure:

$$\frac{X}{X_m} = \frac{K(P/P_o)}{1 + K(P/P_o)} \tag{1}$$

where X is the volume of gas adsorbed per weight of solid adsorbent, X_m is the volume of gas adsorbed with monolayer coverage, K is a temperature-dependent constant, and P/P_o is the relative pressure. The equation is based on the assumptions that equilibrium has been established between the gas and the solid and that only a monomolecular layer of gas molecules has been adsorbed. These assumptions are probably valid for gases at environmental temperatures and pressures; multilayer adsorption occurs only at very low temperatures and at pressures approaching the saturated vapor pressure of the gas.

TABLE 2.4-1

Adsorption of Gases on Charcoal

Gas	Volume Adsorbed (ml/g) 25°C [2]	Volume Adsorbed (ml/g) 15°C [6]
SO_2	98	380
Cl_2	—	235
NH_3	136	181
H_2S	—	99
HCl	—	72
N_2O	67	54
CO_2	60	48
CH_4	—	16.2
CO	14	9.3
O_2	—	8.2
N_2	11	8.0
H_2	2	4.7

Brunauer, Emmett and Teller extended Langmuir's treatment to include multilayer adsorption. Their final equation takes the form

$$\frac{X}{X_m} = \frac{c(P/P_o)}{(1-P/P_o)[1+(c-1)(P/P_o)]} \tag{2}$$

where c is a temperature-dependent constant and the other symbols have the same meaning as in equation 1. Equations 1 and 2 are special cases of a more generalized equation for n layers; equation 1 applies when $n=1$, and equation 2 applies when $n=\infty$.

Equation 1 is generally applicable when environmental temperatures are above the normal boiling point of gases and vapors. Thus, we would expect only monolayer coverage for gases, particularly since the relative pressure P/P_o rarely exceeds 0.3 for these materials. For the vapors of liquids whose normal boiling points, T_b, are near or well above environmental temperatures, we can no longer expect monolayer coverage, so equation 2 is more applicable.

As they stand, equations 1 and 2 are not very useful for estimating directly the extent to which a gas or vapor will be adsorbed on a given solid. Furthermore, in discussing estimation methods we must clearly distinguish between (a) gases and vapors with

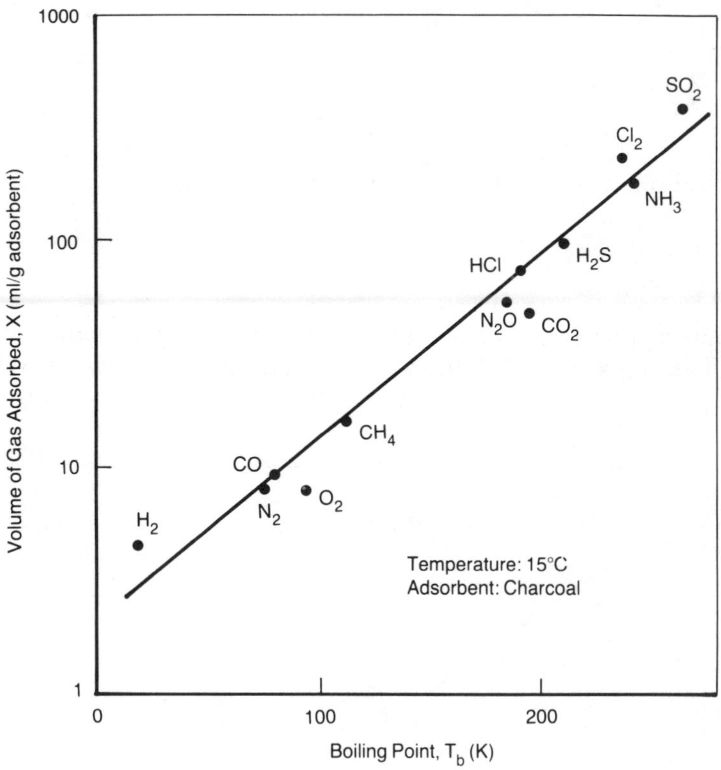

FIGURE 2.4-2 Adsorption of Various Gases as a Function of Their Normal Boiling Points

boiling points below environmental temperatures and (b) vapors whose normal boiling points are above environmental temperatures. Each of these categories is discussed below.

GASES AND VAPORS ($T>T_b$)

The correlation between X and T_b (Figure 2.4-2) implies a relationship between the extent of physical adsorption and saturated vapor pressure, P_o. Such a relationship would not be surprising, as it is natural to suppose that as a gas becomes more condensable its adsorption potential should increase. If we replot the 15°C data in Table 2.4-1 as a function of the saturated vapor pressure of each gas, we should observe a regular decrease in adsorption potential as P_o increases. Such a plot is shown in Figure 2.4-3 as ln X vs ln P_o/P. The correlation is quite good (r^2 = 0.988). The relationship is given by

$$\ln X = a - b \ln(P_o/P) \tag{3}$$

where X = volume adsorbed per gram of adsorbent
a = 6.62 at 288 K
b = 0.69 at 288 K
P = 1 atm

The saturated vapor pressure values were obtained via extrapolation of log P_o vs 1/T calculated from handbook data [12]. In some instances the extrapolated values are in

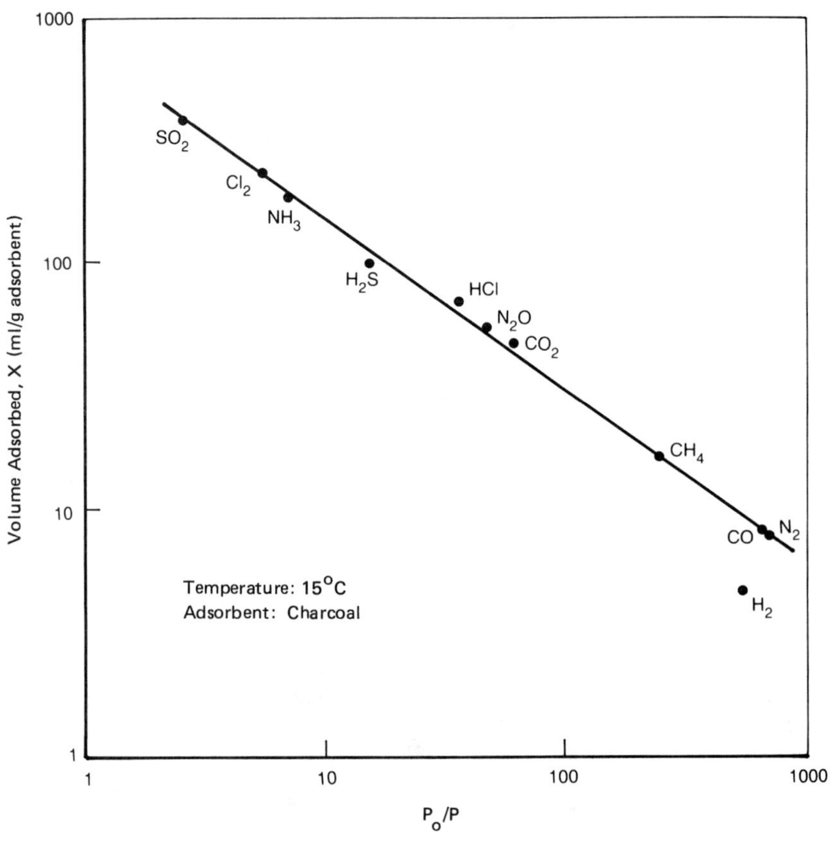

FIGURE 2.4-3 Amount Adsorbed vs Saturated Vapor Pressures of Various Gases

the supercritical region; while these are hypothetical, they are useful for estimation purposes, because they should be proportional to the tendency of the gas to condense and, hence, its adsorption potential.

Equation 3 has the same form as the Freundlich isotherm, an expression that has been found to fit isotherm data for monolayer adsorption [3]:

$$X = K(P/P_o)^{1/n} \tag{4}$$

or

$$\ln X = \ln K - (1/n) \ln (P_o/P) \tag{5}$$

In normal use, K and n are adjusted to fit the isotherm data. In the use proposed here, the values of K, n and P are fixed, so any variation in X is related to the value of P_o. Some of the temperature dependence is contained in P_o, as it is a strong function of temperature; however, we have correlated b (eq. 3) with temperature as follows:

$$b = 0.24 + (1.57 \times 10^{-3}T) \tag{6}$$

for which

$$r^2 = 0.9989$$

The task remains to obtain or estimate values of P_o as a function of temperature for the gas in question. Where possible, extrapolated experimental data should be used; if these are not available, the estimation method for gases and vapors ($T>T_b$) outlined in section 2.2 can be used.

In Table 2.4-2, experimental and estimated values of X are compared for the gases listed in Table 2.4-1. The estimates were obtained with saturated vapor pressures calculated by the method shown in section 2.2. In general, the agreement is good; the average method error is ± 20%. Similar comparisons are shown in Figure 2.4-4 for a single gas (CO_2) at three different temperatures and several values of P below atmospheric pressure.

The above correlations apply only to adsorption on clean charcoal. While the estimates have no quantitative value for other surfaces, they are useful for obtaining some indication of the relative adsorption of a gas if its boiling point is known.

Direct comparisons of the adsorption estimates obtained here with data obtained on natural surfaces (soils, clay, etc.) are complicated by the fact that interaction with these surfaces is affected by the presence of chemicals, microbes and water. For example, the uptake of SO_2 on soil surfaces has been highly correlated with soil pH and the presence of carbonate and other basic species [3]. Other studies have attributed the uptake of CO [7] and NO_2 [1] primarily to microbial action, as considerably less adsorption occurred with sterilized soil. The uptake of SO_2 and NO_2 is also enhanced by the presence of moisture [4], as these gases are quite soluble and react with water.

If we were to assume that a soil had the same unit surface area as charcoal, we would estimate its capacity for adsorbing SO_2 at 25°C to be about 800 mg/g of soil. Data

TABLE 2.4-2

Experimental vs. Estimated Adsorption on Charcoal
($t = 15°C$, $P = 760$ mm)

Gas	T_b(K)	Volume Adsorbed, X (ml/g) Experimental [6]	Estimated	Error (%)
SO_2	262	380	393	+ 3
NH_3	240	181	198	+ 9
Cl_2	238	235	222	− 7
H_2S	211	99	113	+14
CO_2	195	48	68	+42
HCl	189	72	60	−17
N_2O	183	54	54	0
CH_4	109	16	13	−19
O_2	90	8	10	+25
CO	82	9	5	−80
N_2	63	8	8	0
H_2	14	5	6	+20

reported by Bremner [4] yield values ranging from 0.042 mg/g to 60.1 mg/g, depending upon pH, water content, and surface area. We may be able to account for some of the large discrepancy in the data by examining the effects of surface area and water content. The typical specific surface areas of charcoals used in laboratory studies are of the order of 1000 m^2/g [10]. For soils, the specific surface area can range from about 20 m^2/g to 200 m^2/g [9]. Thus if we assume that the actual amount of SO_2 adsorbed on soils will be proportional to the ratio of specific surface areas of soil and charcoal, we can estimate adsorption values within the range of 16 mg/g to 160 mg/g. For soils with low pH values we would expect that these estimates would be lower [3, 4]. The solubility of SO_2 in water is approximately 200 mg/g of water. If, for example, typical soils contain 5% water, we would add to the above numbers an additional 10 mg/g of soil, yielding estimated adsorption values within the range 26 mg/g to 170 mg/g. These data illustrate the difficulty of making direct comparisons between estimated values and experimental measurements.

VAPORS ($T<T_b$)

The apparent success of the correlation for gases (eq. 3) is related to the ratio P/P_o. Since this ratio is usually less than 0.3 at environmental temperatures and pressures, only monolayer coverage of the adsorbent has to be considered in most cases. Hence, isotherms can generally be characterized by equations similar to equation 3. For vapors of liquids whose boiling points are equal to or greater than environmental

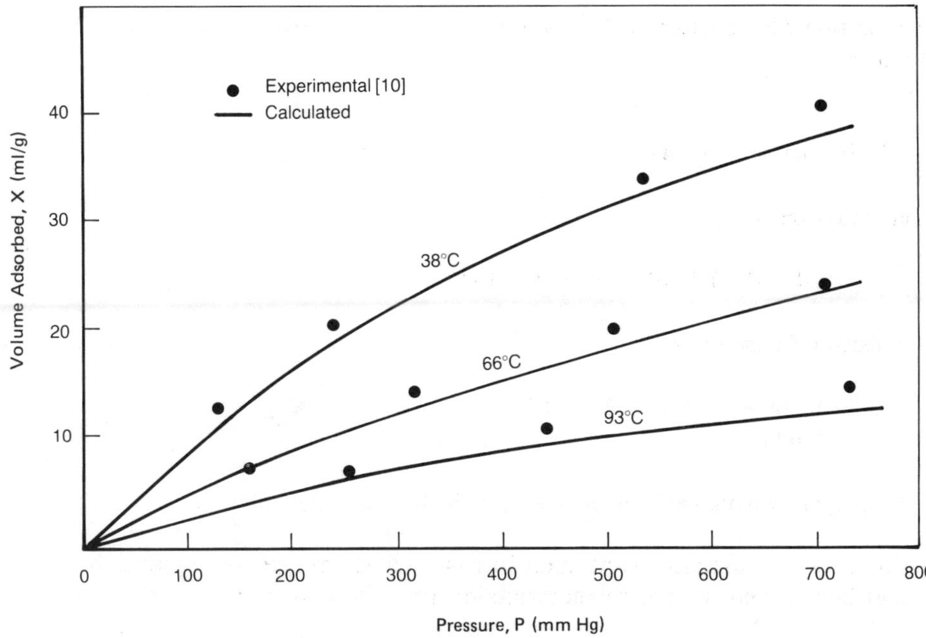

FIGURE 2.4-4 Experimental vs Calculated Adsorption Isotherms for CO_2 on Charcoal

temperatures, however, P/P_o is 1.0 or more. This means that multilayer adsorption has occurred and the surface is saturated — i.e., liquid has condensed not only on the surface but also in any pores that may exist.

The extent to which a vapor is adsorbed depends, in fact, almost exclusively on the pore volume of the solid, because the mechanism of adsorption at the saturation pressure causes the vapor to condense within the pores of the adsorbent and fill them. Subsequent evaporation of the condensed liquid is also affected by the pores: those with narrow constrictions, for example, tend to retain the liquid longer.

Except for its vapor pressure and solubility in water, the nature of the liquid in this case is irrelevant [11].

2.4.4 Method Error

The method error is listed in Table 2.4-2. The average error for the few gases shown is ±20%. Again, this refers only to adsorption on charcoal.

2.4.5 Sample Calculations

Example 1 Estimate the adsorption of CO_2 on charcoal at a temperature of 25°C and a pressure of 760 mm Hg. The normal boiling point of CO_2 is 195 K.

$T_b = 195\ K$
$T = 25 + 273 = 298\ K$

From section 2.2 (equation 12), the estimated saturated vapor pressure of CO_2 at 298 K is 41.5 atm.

$$P_o/P = 41.5/1 = 41.5$$

From equation 6,

$$b = 0.24 + (1.57 \times 10^{-3} \times 298) = 0.71$$

Substituting in equation 3,

$$\ln X = 6.62 - 0.71 \ln 41.5 = 3.97$$
$$X = 53 \text{ ml/g}$$

The experimental value is 60 ml/g [2], indicating an error of −12%.

The next example illustrates that even in cases of adsorption on surfaces other than charcoal, estimates to within factors considerably less than 2 can be made.

Example 2 Estimate the adsorption of SO_2 on silica gel at 273 K and a pressure of 60 mm Hg.

$$T_b = 262 \text{ K}$$
$$T = 273 \text{ K}$$

The saturated vapor pressure of SO_2 at 273 K is estimated to be 1.37 atm, using the method for gases described in section 2.2 (equation 12).

$$b = 0.24 + (1.57 \times 10^{-3} \times 273) = 0.67$$
$$a = 6.62$$
$$P = 60 \text{ mm}/760 \text{ mm} = 0.08 \text{ atm}$$
$$P_o/P = 1.37/0.08 = 17.13$$

From equation 3,

$$\ln X = 6.62 - 0.67 \ln 17.13 = 4.72$$
$$X = 112 \text{ ml/g}$$

The experimental value is 175 ml/g, indicating an error of −36%.

2.4.6 Literature Cited

1. Abeles, F.B., L. Cruker, L. Forence and G. Leather, "Fate of Air Pollutants: Removal of Ethylene, Sulfur Dioxide, and Nitrogen by Soil," *Science*, **173**, 914 (1972).
2. Adamson, A.N., *Physical Chemistry of Surfaces*, Interscience (Wiley), New York (1960).
3. Bohn, H.L., "Soil Adsorption of Air Pollutants," *J. Environ. Qual.*, **1**, 372 (1972).
4. Bremner, J.M., "Sorption of Pollutant Gases by Soils," Final Report to U.S. Dept. of Energy, Contract DE-AC02-79EV02530 (1981).
5. Brunauer, S., P. Emmett and E. Teller, "Adsorption of Gases in Multimolecular Layers," *J. Am. Chem. Soc.*, **60**, 309 (1938).
6. Glasstone, S., *Textbook of Physical Chemistry*, 2nd ed., D. Van Nostrand, New York (1946).
7. Inman, R.E., R. Ingersoll and E.A. Lery, "Soil: A Natural Sink for Carbon Monoxide," *Science*, **172**, 1229 (1971).
8. Langmuir, I., "The Constitution and Fundamental Properties of Solids and Liquids, Part I: Solids," *J. Am. Chem. Soc.*, **38**, 2269 (1916).
9. Puri, A.N., *Soils, Their Physics and Chemistry*, Chapter IX (p. 54), Reinhold Publishing Corp., New York (1949).
10. Ray, G. and E. Box, "Adsorption of Gases on Activated Charcoal," *Ind. Eng. Chem.*, **42**, 1315 (1950).
11. Thomas, R.G., "Volatilization from Water," in *Handbook of Chemical Property Estimation Methods*, W.J. Lyman, W.F. Reehl and D.H. Rosenblatt (eds.), McGraw-Hill Book Co., New York (1982).
12. Weast, R.C. (ed.), *CRC Handbook of Chemistry and Physics*, 58th ed., CRC Press, Cleveland (1977).

2.5 ATMOSPHERIC DEPOSITION

2.5.1 Introduction

The deposition of gaseous and particulate air pollutants, like their ambient concentration, is a determinant of their environmental effects. The ambient concentration is of primary interest for such problems as the health effects of ozone on humans; often, however, as in the effects of air pollutants on terrestrial or aquatic ecosystems, deposition may be more important.

The mechanisms by which air pollutants are deposited from the atmosphere are described below. For convenience, they have been divided into wet (i.e., occurring in precipitation) and dry deposition of gases and particulate matter. While these divisions may seem straightforward, their distinctions are often blurred. The deposition of fog droplets is considered wet deposition by some researchers but dry deposition by others. Chemical transformations also complicate the process, as in the oxidation of sulfur dioxide gas to sulfuric acid, an aerosol (particle).

2.5.2 Dry Deposition

Dry deposition may be defined as the turbulent transfer of gaseous and particulate species from the atmosphere to the earth's surface. This includes gravitational settling but excludes deposition of gases and particles absorbed or entrained in precipitation. The deposition of pollutants with fog droplets is considered here as wet deposition.

As shown in Figure 2.5-1, many of the factors influencing the rate of dry deposition differ for gases and particles. Because of the differences between the mechanisms that most strongly affect the deposition of gases and large particles, they are described separately below. This figure emphasizes the physical factors that affect the rate of deposition rather than the chemical properties and reactions of the depositing material.

DEPOSITION OF PARTICLES

As the rate of dry deposition of particles (and gases) is extremely difficult to measure, a relationship between dry deposition and ambient concentration, which can be measured relatively easily, is often employed. The factor relating the ambient concentration to the deposition rate is known as the *deposition velocity*, since its dimensions are length/time. It is determined for a given gas or aerosol, based on a particular reference height; when multiplied by the concentration (mass/volume) at that height, the product is the deposition rate [mass/(area-time)]. Because of its importance in characterizing the removal of pollutants from the atmosphere and its use in predictive models, the deposition velocity serves as the focus of the discussion that follows. First, the processes embodied in the deposition velocity are described.

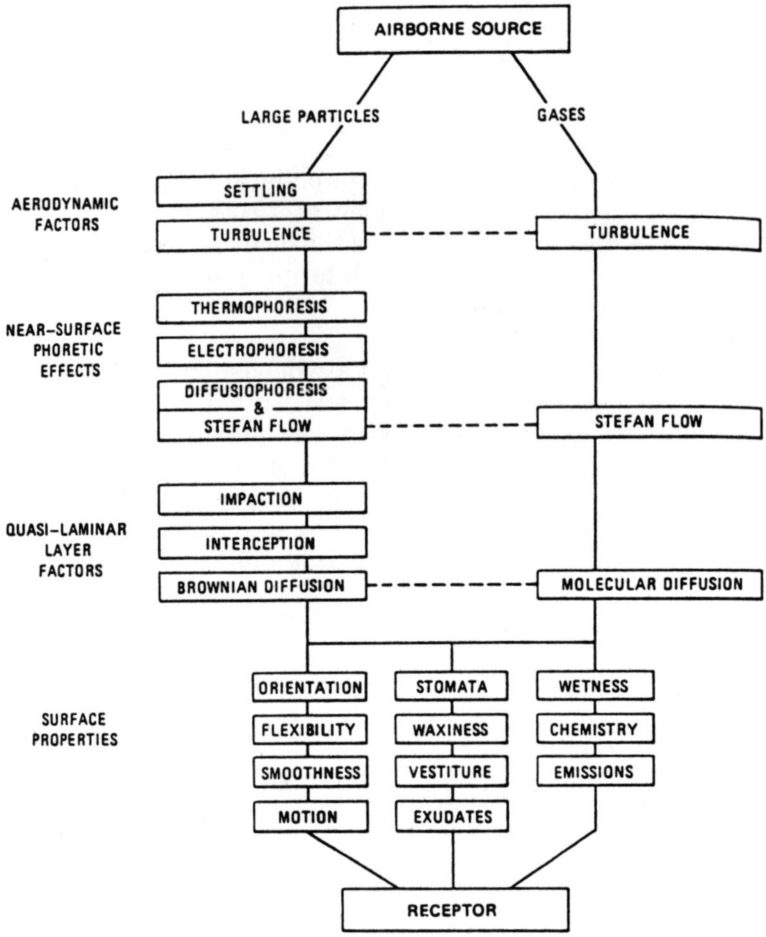

Source: National Research Council [16]

FIGURE 2.5-1 Schematic Representation of Processes Likely to Influence the Rate of Dry Deposition of Gases and Particles

Particles are transported vertically above the earth's surface by eddy diffusion (turbulence) and gravitational settling (the "aerodynamic factors" in Figure 2.5-1). Like other falling objects, particles accelerate until frictional drag forces equal the force of gravity, at which point they have reached their terminal velocity. The terminal settling velocity of falling particles (based on Stoke's law) can be expressed as [32]:

$$V_t = \frac{2a^2\rho g}{9\mu} \tag{1}$$

where:

V_t = particle's terminal settling velocity (m/s)
ρ = particle density (kg/m^3)
g = acceleration due to gravity (9.8 m/s^2)
μ = viscosity of air (1.78×10^{-5} kg/(m • s) at 18°C, 1 atm)
a = radius of the particle (m)

Equation 1 is applicable to particles having radii up to about 50 μm, after which the turbulence in the wake of the falling particle must be considered [32]. It is assumed that the falling particles are solid spheres, although in reality they are often not spherical and may contain void spaces [20].

As equation 1 shows, V_t is more strongly a function of the particle's radius than its density, since the R term is squared. Furthermore, the densities of aerosol particles cover a much narrower range than their radii, varying by a factor of only 2-3 and averaging 2-2.5 g/cm^3 [32].

The other aerodynamic factor listed in Figure 2.5-1 is turbulence. Conceptually, turbulent transfer is quite simple: particles are carried towards the earth's surface with the turbulent eddies of the air. The effect a turbulent eddy has on a pollutant plume is shown in Figure 2.5-2. The mathematical description of eddy diffusion is much more complicated, however, and turbulent eddy diffusivities for mass transport may range over three orders of magnitude, depending on the height above the ground and the local hydrostatic stability of the air [13].

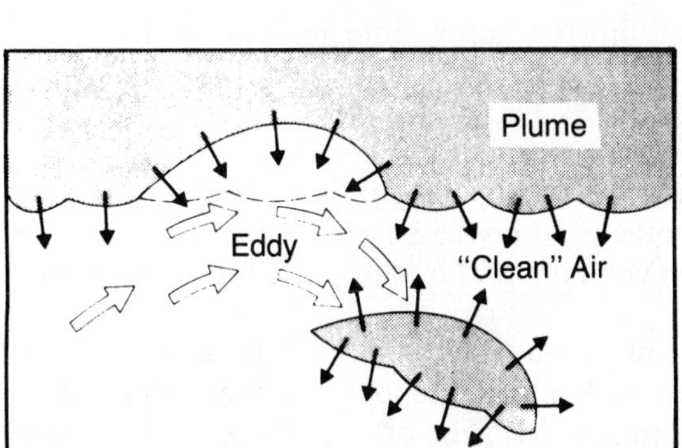

Source: Williamson [32]. *(Copyright 1973, Addison-Wesley Publishing Co., Inc. Reprinted with permission.)*

FIGURE 2.5-2 Grossly Simplified View of the Breakup of a Pollutant Plume by a Turbulent Eddy, Causing the Mixing of Pollutants with "Clean" Air

The stability of the atmosphere is indicated by the change in temperature with height. The atmosphere is said to be "neutrally stable" when the temperature decreases with height at the same rate as the adiabatic lapse rate. The adiabatic lapse rate is the decrease in temperature that a parcel of air would undergo if it were raised in the atmosphere with no heat flow to or from its surroundings (or the temperature increase it would undergo if it were lowered through the atmosphere). For dry air, the lapse rate is approximately -9.8°C/km; but for warm air saturated with water vapor it may be as little as -3.5°C/km; the average rate in the troposphere is about -6.5°C/km [32]. When the temperature decrease with altitude is greater than the adiabatic lapse rate, as typically occurs in the middle of a sunny day, the atmosphere is said to be "unstable," and vertical mixing is promoted. A parcel of warm air cooling at the adiabatic lapse rate will become more buoyant as it rises because the temperature difference between itself and the air surrounding it increases. (The parcel cools at the adiabatic lapse rate, but the surrounding air cools an even greater amount for a given increase in height.) Similarly, a cool, heavy parcel of air will tend to accelerate as it falls since its temperature increases at a lesser rate than the air around it.

"Stable" conditions (typical of a calm night) occur when the rate of temperature decrease with height is less than the adiabatic lapse rate. In this case, the fall of a cool parcel of air, or the rise of a warm one is inhibited, for just the opposite reasons that their motions are promoted under unstable conditions. Poor vertical mixing of the air results with the consequent decrease in the removal of air pollutants.

Near the receiving surface, various phoretic effects (i.e., transmission due to a gradient) may influence the rate of particle deposition. *Thermophoresis* refers to the tendency of a particle to be driven away from a hot surface; it can be thought of as the consequence of "hotter," more energetic air molecules hitting the side of the particle facing the hot surface [16] and therefore applies primarily to small particles (diameter $<0.03\ \mu m$). For larger particles, forces due to thermal radiation can be important if a temperature gradient develops across particles that are poor heat conductors. As a rule of thumb, for particles of diameter less than 0.03 μm, the thermophoretic velocity is likely to be about 0.03 cm/s, while for larger particles ($>1\ \mu m$ diameter) the velocity will be only one fourth this amount [16].

Electrophoresis, as the name implies, occurs in electric fields, and serves to promote the retention as well as the deposition of particles due to the attraction of opposite electrical charges [16]. The effect of electrophoresis is greatly dependent on the particle size. Very small particles (diameter of about 0.001 μm) attain a velocity of 2 cm/s per volt/cm of field strength, compared with a velocity of only 0.0003 cm/s per volt/cm for a particle of 0.1 μm diameter [6]. Since typical values of field strengths under fair-weather conditions are less than 10 volts/cm [16], electrophoresis is expected to be significant only for very small particles, which usually account for a very small fraction of the total aerosol mass.

Particles in intermixing gases undergo *diffusiophoresis*, which is the tendency to move toward the lighter of the two gases. In the atmosphere, the intermixing gases are generally water vapor and air; particles move toward the lighter of the two (water vapor), since they collide with fewer of the heavier (air) molecules on that side. A related phenomenon is *Stefan flow*, which results from the introduction into a gaseous medium of new gas molecules at an evaporating or subliming surface [16]. Since a mole of any gas occupies 22.41 liters at standard temperature and pressure, a Stefan flow velocity of 22.41 mm/s results as 18 g of water (one mole) evaporates from one square meter in one second.

While it is important to recognize the existence of phoretic effects and their significant influence on deposition rates in certain cases, often they need not be considered. Both diffusiophoresis and thermophoresis are usually small, and their effects on dry deposition can be ignored [16].

In the quasi-laminar layer (the layer less than 1 mm thick next to a surface), the mechanisms of impaction, interception, and Brownian diffusion become important. *Impaction* occurs when particles heavier than air are carried to a surface by wind or eddies. Those particles that are unable to flow around the surface, due to their inertia, will collide with it. This process is known as impaction when the particles are considered as point masses; it is important primarily for particles with diameters of between 2 and 20 μm, as larger particles tend to bounce off the surfaces and smaller ones are able to follow the air flow around them [16]. When the size of the particle is taken into account (i.e., the particle is considered larger than a point) and its trajectory relative to the surface is calculated, the process is referred to as *interception* [4].

Brownian diffusion occurs as particles are randomly struck by air molecules. The motion of smaller particles is greater than that of larger particles [32], thus increasing the possibility that the particle will be trapped by impaction. Equation 2 can be used (as described below) to estimate the Brownian diffusion coefficient [22].

$$D = \frac{kT}{6\pi\mu a}\{1 + \frac{10^{-4}}{pa}[6.32 + 2.01 \exp(-2190\ pa)]\} \tag{2}$$

where:

D = Brownian diffusion coefficient, cm^2/s
k = Boltzmann's constant, 1.38×10^{-16} erg/K
T = temperature, K (291 K used)
π = 3.14159
μ = air viscosity, g/(cm•s) [1.78×10^{-4} g/(cm•s) used]
a = particle radius, cm
p = pressure, cm Hg (76.0 cm used)

(Note that the constants within the braces have been derived such that the combined term is rendered dimensionless when p and a are employed as specified above.)

For particles with diameters of less than about 0.1 μm, the gravitational settling velocity is quite low compared with the deposition velocity [21], since turbulent transfer and Brownian motion in the quasi-laminar layer are more important. For particles with diameters greater than 20 μm, gravitational settling is the controlling factor in deposition; impaction is not significant with particles of this size [16]. Atmospheric sulfates, nitrates, and ammonium aerosols tend to occur in a size range of 0.1 to 2 μm diameter, and thus both gravitational settling and impaction play a role in their deposition [16]. The deposition velocity of a particle is at least as great as its terminal settling velocity, regardless of size [21], since impaction and Brownian diffusion act to enhance deposition.

Figure 2.5-1 lists a number of surface properties that influence the deposition of gases and particles. Chamberlain *et al.* [3], have shown that surface texture can have a pronounced effect on the deposition velocity for particles in the sub-micron to several micron range; the deposition velocity to filter paper was observed to be 2-3 orders of magnitude larger than that to a smooth surface. The effect of surface wetness on deposition rate was investigated by Jenkin [15], who used hygroscopic particles of 2.5-8.4 μm dry diameter (sodium chloride aerosol), and found a doubling of the deposition velocity when filter paper was moistened but only a 14% enhancement when artificial grass was moistened. Evaluating two models for predicting deposition to the sea, Arimoto and Duce [1] found that surface waves had little effect on the deposition rate of sodium and aluminum aerosols (mass mean diameter of approximately 2 and 10 μm, respectively) while the deposition rate of lead-210 (mass mean diameter of 0.4 μm) was as much as five times as great in the presence of waves, thus suggesting that the deposition of small particles may be greatly enhanced by broken water surfaces. Plant vestiture (surface covering) also affects the deposition rate; the presence of leaf hairs has been found by several researchers to enhance the deposition of small particles (< 5 μm) [16].

Ideally, all of the factors affecting the deposition rate would be easily described in mathematical terms that could be combined to give an overall equation for the deposition velocity. However, no generally accepted equation for predicting the deposition velocity of particles exists, and the actual deposition rates of particles, especially those of sub-micron diameter, remain controversial. The work of Sehmel [20, 22], Sehmel and Hodgson [23], and Slinn [26] shows relatively small deposition velocities for particles in this size range, while field studies of others, e.g., Wesely *et al.* [30] and Cadle *et al.* [2] indicate much greater values.

As an example of one method for predicting the deposition velocity of particles, the correlations of Sehmel [21, 22] and Sehmel and Hodgson [23] are described below. Sehmel and Hodgson's original correlation was published in 1978 [23] and was presented in graphical form in a 1980 review by Sehmel [21]. This work, which is based on wind tunnel studies, has the advantage of being in an easily usable engineering form, but as noted above, field measurements have indicated significantly higher deposition velocities for particles in the 0.1 μm to 1.0 μm diameter range. Thus the results of the correlation should not be taken as absolute, but rather as indicative of the mechanisms affecting deposition rates. An alternative method for predicting the deposition velocities of sulfate particles (which are usually unimodally dispersed with the mode in the 0.1-1 μm range) and nitrate particles (bimodally dispersed with one mode in the same range and one in the 2-10 μm range) is described by Voldner *et al.* [28]; this method is similar to that described below for gases based on a resistance analogue, which includes a correction function for the effects of atmospheric stability.

Sehmel and Hodgson's work [23] indicates that atmospheric stability has little effect on the deposition velocity of particles, and the plots based on their correlation [21] are for stable conditions. However, Sehmel [22] notes that atmospheric stability must be considered in applying predictive equations to field situations, and others (e.g., Wesely *et al.* [30]) have observed decreased deposition velocities in the field for particles under neutral and stable conditions.

For particles of known density and diameter, two meteorological parameters known as the friction velocity (u_*) and the aerodynamic surface roughness (or roughness height, z_0) are needed to use Sehmel and Hodgson's correlation. The roughness height and friction velocity are used in describing the wind profile (wind speed as a function of height). As shown in Figure 2.5-3, under conditions of neutral atmospheric stability, wind speed is a log-linear function of height. Mathematically, this relationship can be expressed as [17]:

$$u = \ln (z/z_0)\, u_*/k \tag{3}$$

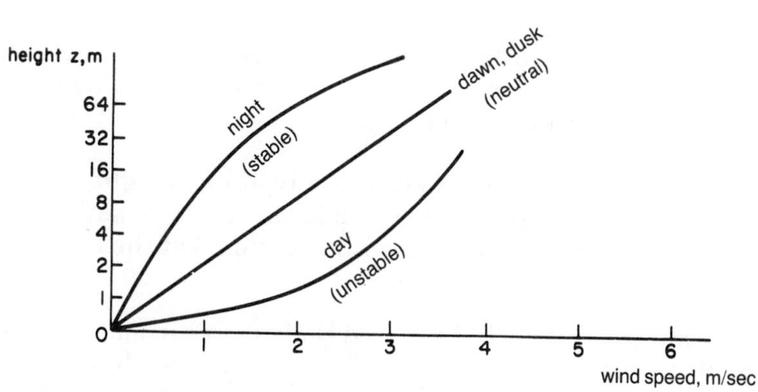

Source: Panofsky and Dutton [17]. *(Copyright 1984, John Wiley & Sons. Reprinted with permission.)*

FIGURE 2.5-3 Diurnal Variation of Wind Profiles (Schematic) for the Same Day

Here, u is the wind speed at height z, and k is the von Karman constant (dimensionless), typically taken to be 0.4. If the wind profile of interest occurs over roughness elements such as trees, houses, or crops, rather than bare soil or calm water, a displacement length, d, is employed, and the term z - d is substituted for z in equation 3.

When d is small, it can be estimated from the wind profile. If the profile is measured for neutral conditions over a surface where the displacement length must be considered, the wind profile falls off sharply below the average height of the roughness elements (i.e., the plot for neutral stability in Figure 2.5-3 would drop off on the left side of the figure, below this height). To determine d, the linear portion of the profile above the height of the roughness elements is extrapolated to the ordinate (wind speed = 0); the intercept is z_0 + d. Over tall roughness elements such as trees or houses, a displacement length equal to 80% of the average height of the roughness elements should be assumed, rather than using the wind profile to make the estimate, since small errors in the profile can result in large errors in d [17].

Methods for determining z_0 and u_* are described below for cases where d is insignificant. (In cases where it is significant, z - d is substituted for z in the equations that follow.) If the wind velocity is known at a number of heights, z_0 and u_* can be determined by plotting the wind profile (as ln z versus u). The slope of the resulting straight line is k/u_* and the intercept is ln z_0.

More likely, the wind speed is known only at one or two heights. If u is known at two heights, then:

$$u_* = \frac{k(u_2 - u_1)}{\ln (z_2/z_1)} \qquad (4)$$

where the subscripts one and two denote the two points at which data are available. By substituting u_* calculated from equation 4 into equation 3, z_0 can be solved for with u and z data for either of the two points.

In the case where the wind velocity is known only at one height, a value of z_0 is assumed. Figure 2.5-4 lists roughness lengths for a variety of surfaces. Once the appropriate z_0 is selected, u_* is solved for with equation 3 at the known u and z.

Once u_* and z_0 are known, plots of Sehmel and Hodgson's correlation (in which the deposition velocity at a given friction velocity is plotted as a function of particle diameter for a series of roughness heights and particle densities) can be used. An example of one of these plots is shown in Figure 2.5-5 where the deposition velocity, $v_{1\text{-}m}$, is based on a reference height of 1 m. A friction velocity of 30 cm/s is used in this figure; plots for other values can be found in Sehmel's review [21].

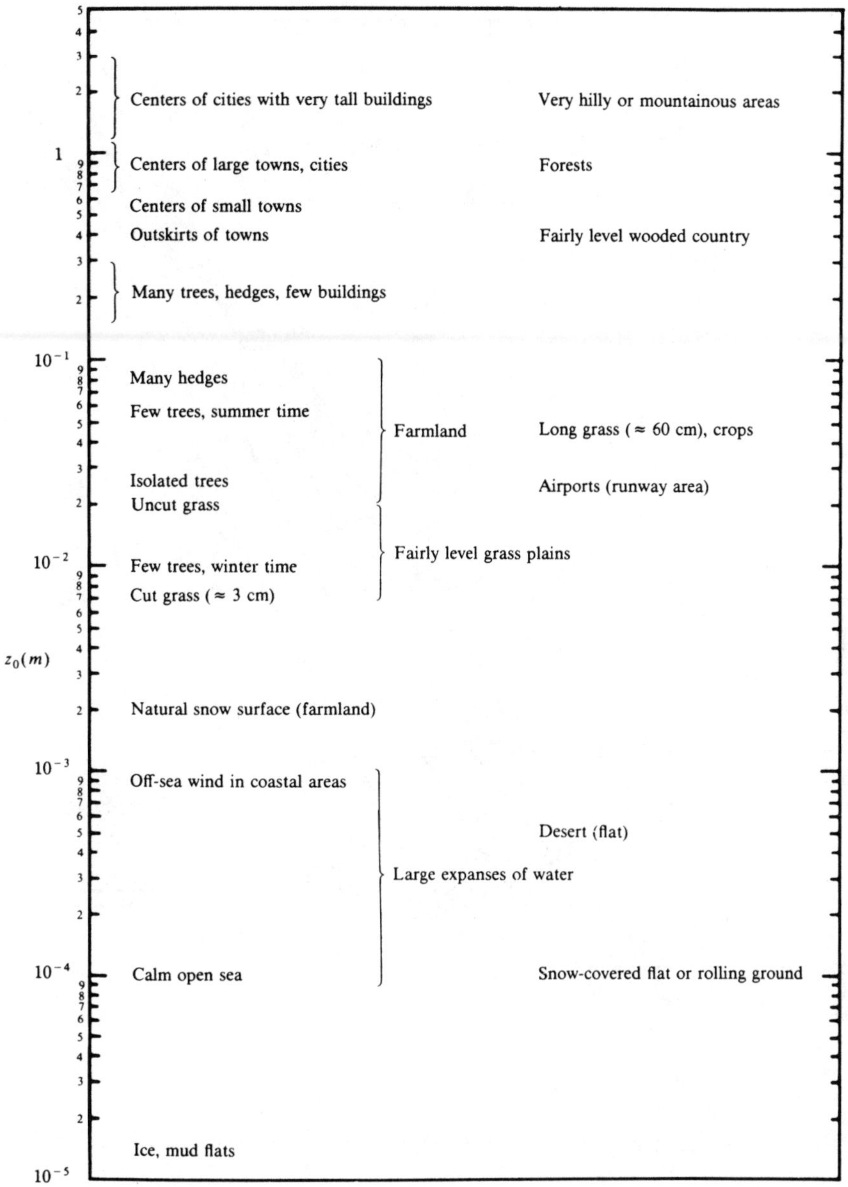

Source: Panofsky and Dutton [17]

FIGURE 2.5-4 Roughness Lengths, z_0, for Various Terrain Types

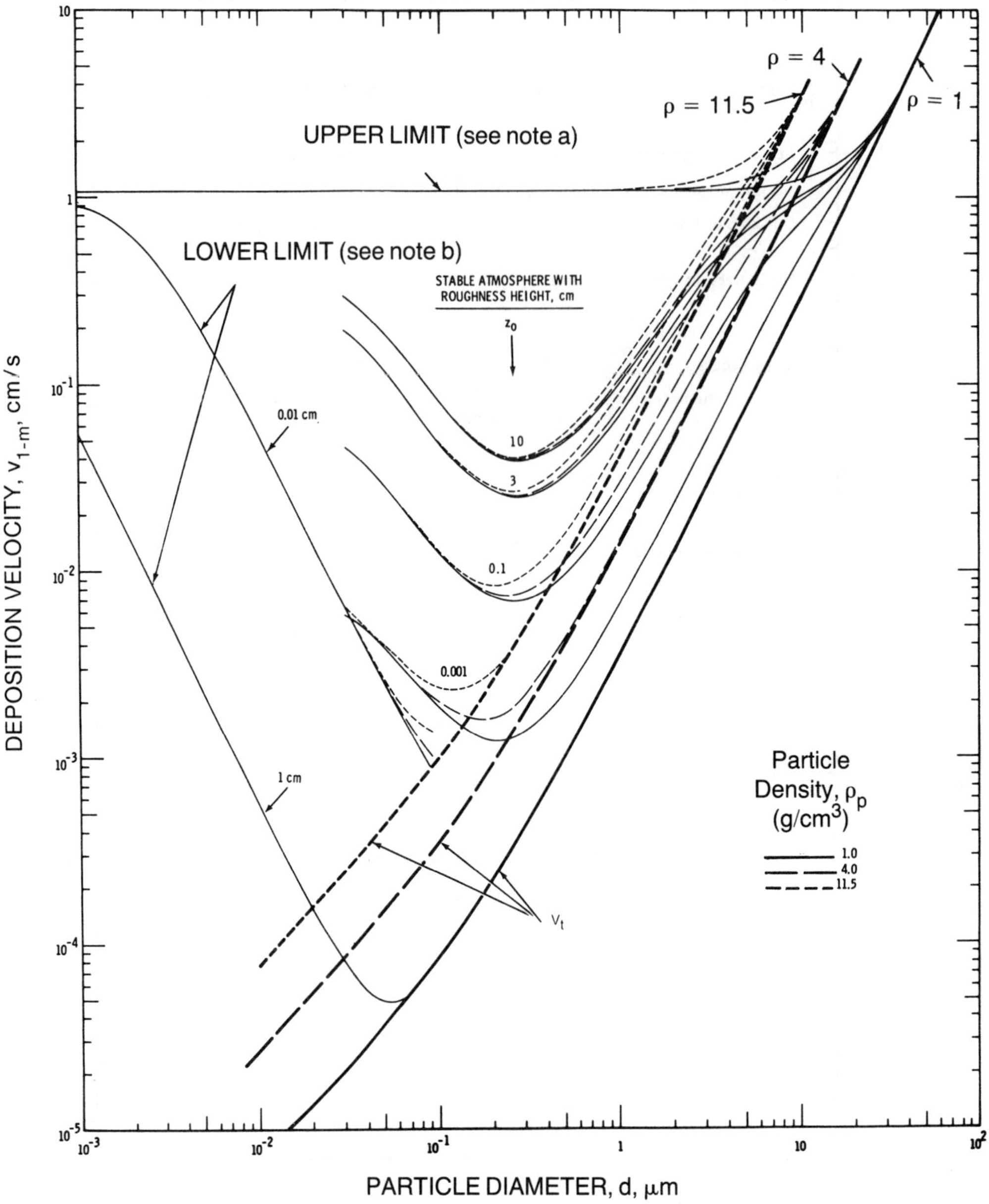

a. Based on gravity settling and atmospheric diffusion between 1 cm and 1 m from the surface; no resistance to deposition assumed within 1 cm of the surface.

b. Between the indicated height and the receiving surface, mass transfer is assumed to be caused by Browian diffusion only; above indicated height, both turbulent and Brownian diffusion occur.

Source: Sehmel [21]. *(Copyright 1980, Pergamon Press, Inc. Reprinted with permission.)*

FIGURE 2.5-5 Predicted Particle Deposition Velocities at 1 meter, $v_{1\text{-}m}$, for a Friction Velocity, u_*, of 30 cm/s and Various Particle Densities

To use the plots, z_0 and u_* are determined as described above. After selecting the plot that best corresponds to this friction velocity, the appropriate curve for the particle density and the calculated z_0 is located. With the aid of this curve, the deposition velocity at one meter is read off the ordinate.

The shapes of the curves reflect the importance of gravitational settling, turbulent diffusion, and particle inertia in enhancing the deposition of larger particles ($>1\ \mu$m) and the effect of Brownian diffusion in promoting the deposition of smaller particles [21].

The lower limits of the deposition velocity are indicated by the two curves on the left side of Figure 2.5-5. Below the height indicated, only Brownian diffusion is assumed to cause mass transfer, while above the specified height, both turbulent and Brownian diffusion occur. The upper limits of deposition velocity are also shown in the figure; these assume no resistance to deposition in the 1-cm layer next to the surface and are based on gravity settling and atmospheric diffusion from 1 cm to 1 m. The three heavy curves show the deposition velocities when they are assumed to equal the terminal settling velocities, V_t.

An improved correlation based on the data used to derive the correlation illustrated in Figure 2.5-5 was published by Sehmel in 1984 [22]. This correlation is given in equation 5 based on a 1-centimeter reference height:

$$\begin{aligned} v_{1\text{-cm}} = \exp\{&7.1108 + [\ln(Sc)\ln(d/z_0)][0.2529 - 0.00273\ln(d/z_0)] \\ &+ [\ln(\tau^+)][0.03239\ln(\tau^+) - 0.09177\ln(d/z_0)] \\ &- 0.14919[\ln(d)]^2 - 4.180\ln(d/z_0)\}\ \text{in cm/s} \end{aligned} \tag{5}$$

where:

$v_{1\text{-cm}}$ = deposition velocity at 1 cm, cm/s
d = particle diameter, cm
Sc = Schmidt number, ν/D
u_* = friction velocity, cm/s
z_0 = aerodynamic surface roughness, cm
ν = kinematic viscosity, μ/ρ
ρ = air density, g/cm^3 (1.2×10^{-3} g/cm^3 used)
ρ_p = particle density, g/cm^3
τ^+ = dimensionless relaxation time, $(\rho_p d^2 u_*^2)/(18\,\mu\nu)$

The term μ is defined after equation 2, and D, the Brownian diffusion coefficient, is calculated using that equation. The values of u_* and z_0 are determined as described above.

This new correlation gives estimates of the deposition velocity based on a 1-centimeter reference height whereas the earlier one was combined with resistances to mass transfer between one centimeter and one meter before being plotted in Figure 2.5-5, and thus uses a 1-meter height. (The earlier correlation for a 1-centimeter reference height can be found in Sehmel and Hodgson's original publication [23].) Aerodynamic factors (settling and turbulence) become increasingly important in the transport of large particles from one meter to one centimeter, and the deposition velocities of large particles based on a 1-meter reference height are similar to their terminal settling velocities [22]. Because aerodynamic factors between one centimeter and one meter are not considered in equation 5, the estimated deposition velocity may be much greater than the terminal settling velocity for particles as large as 50 μm in diameter [22].

Although Sehmel considers the new correlation improved, except for very small particles and roughness heights greater than three centimeters, the results differ little from the earlier correlation, given the uncertainties involved; on average, the estimates of the earlier correlation are 13% higher and range from a maximum of 3.9 times as large to 0.44 as large [22]. The use of equation 5 is illustrated in example 1.

Example 1 Calculate the deposition velocity of sulfate aerosol to a field of grass 45 cm tall, where the average wind speed has been measured as 4 m/s at a height of 6 m, and the temperature is 18°C. Assume that the particle has a diameter of 0.4 μm and a density of 1.5 g/cm^3, and that the atmosphere has neutral stability.

To predict the deposition velocity using equation 5, the friction velocity and roughness height must be determined. Since the wind speed is known at only one height, z_0 is assumed. Figure 2.5-4 gives a value for z_0 of approximately 3.5×10^{-2} m for the surface described. A displacement length must be employed to calculate the friction velocity because the surface (tall grass) is not smooth. We will denote this displacement length as d_g. As described above, the displacement length is taken as 80% of the height of the roughness elements:

$$0.8 \times 0.45 \text{ m} = 0.36 \text{ m}$$

Equation 3 (including the term d_g) is rearranged to solve for u_*

$$\begin{aligned} u_* &= k\,u/\ln\,[(z - d_g)/z_0] \\ &= (0.4 \times 4 \text{ m/s})/\ln\,[(6 \text{ m} - 0.36 \text{ m})/0.035 \text{ m}] \\ &= 0.315 \text{ m/s} \end{aligned}$$

The Brownian diffusion coefficient, D, is also required to evaluate equation 5. It is determined by use of equation 2:

$$D = \frac{1.38 \times 10^{-16} \text{ erg/K} \cdot 291 \text{ K}}{6\,\pi \cdot 1.78 \times 10^{-4} \text{ g/(cm s)} \cdot 2 \times 10^{-5} \text{ cm}} \left\{1 + \frac{10^{-4}}{p \cdot a}[6.32 + 2.01 \exp(-2190\,p \cdot a)]\right\}$$

where:

$$p \cdot a = 76.0 \text{ cm Hg} \cdot 2 \times 10^{-5} \text{ cm}$$
$$D = 8.50 \times 10^{-7} \text{ cm}^2/\text{s}$$

The particle deposition velocity is expressed in terms of ln(Sc), $\ln(d/z_0)$, $\ln(\tau^+)$, and $[\ln(d)]^2$ in equation 5. Evaluating each term using the values of u_*, z_0, and D determined above:

$$Sc = \nu/D = (\mu/\rho)/D$$
$$= \frac{(1.78 \times 10^{-4} g/(cm \cdot s)/(1.2 \times 10^{-3} g/cm^3)}{(8.50 \times 10^{-7}\ cm^2/s)}$$
$$\ln Sc = 12.07$$

$$d/z_0 = 4 \times 10^{-5}\ cm/3.5\ cm$$
$$\ln(d/z_0) = -11.38$$

τ^+ is calculated with the value of u_* determined above (31.5 cm/s)

$$\tau^+ = \rho_p\ d^2\ u_*^2/[18 \cdot (\mu\nu)]$$

$$\tau^+ = \frac{1.5 g/cm^3\ (4 \times 10^{-5}\ cm)^2\ (31.5\ cm/s)^2}{18\ (1.78 \times 10^{-4}\ g/cm \cdot s \times 1.78 \times 10^{-4}\ g/cm \cdot s/1.2 \times 10^{-3}\ g/cm^3)}$$

$$= 5.01 \times 10^{-3}$$
$$\ln(\tau^+) = -5.30$$

$$d = 4 \times 10^{-5}\ cm$$
$$[\ln(d)]^2 = 102.55$$

Substituting these terms into equation 5 yields:

$$v_{1\text{-}cm} = 0.014\ cm/s$$

The conditions chosen for the example above were taken from a field study of particulate sulfur deposition by Wesely *et al.* [30] to illustrate the difference between the results of field studies and those obtained based on wind tunnel studies. In that study, a long term mean deposition velocity of 0.22 ± 0.06 cm/s with a day-to-day variation of over 50% was measured. This average is based on the concentration measured at 6 meters, and thus is not directly comparable to the deposition velocity obtained in this example, which is based on a 1-cm reference height. However, the value of 0.014 cm/s obtained in this example is considerably less than that found by Wesely *et al.* [30], and would be even smaller (though perhaps only slightly so) if the aerodynamic resistance to transport above 1 cm were included as it is for the 6-meter reference height.

It should be clear from the discussion above that the size of particles greatly affects the magnitude of the various processes contributing to their deposition. Particles are generally classified according to their radius (a) as giant ($a > 1.0\ \mu m$), large ($0.1 < a < 1.0\ \mu m$, the size range most effective for light scattering), or Aitken ($a < 0.1\ \mu m$) [4,32]. Giant particles are also referred to as "coarse" and large and Aitken particles as "fine" [16]. In most aerosols over continental land masses, the concentration (on a number basis) of Aitken particles is much greater than that of large and giant particles, though they normally make up only 10-20% of the aerosol mass [32].

The size distribution of particles in the atmosphere is not static. Smaller particles may collide by any of the deposition mechanisms described above and then adhere together, a process known as coagulation. Particulates are also formed by chemical reactions such as the oxidation of gaseous SO_2 or NO_x to SO_4^{-2} or NO_3^- (either as aqueous acids or their salts), and they may grow as gases sorb to them. Large, mechanically-generated particles (e.g., dust entrained by wind or anthropogenic sources) tend to be short-lived in the troposphere, since they are removed effectively by gravity [16].

DRY DEPOSITION OF GASES

In general, the dry deposition behavior of gases is similar to that of very small particles, except that the gravitational settling of gases is insignificant. As Figure 2.5-1 shows, gases are not susceptible to thermophoretic or electrophoretic effects, because they are not distinct points and are not readily charged. Gas molecules tend to follow the movements of the air with which they are mixed; in the quasi-laminar layer, however, Brownian (molecular) diffusion becomes important, because gas molecules, like small particles, are easily bounced around by molecular collisions.

The effects of the receiving surface on gases are quite different from those on particles, primarily because of the solubility of gases in water and their aqueous-phase reactions [21]. Solubility in water is important in assessing deposition not only on bodies of water but also on surfaces wetted by rain or dew. Furthermore, the interaction of gases with vegetation can involve such processes as the dissolution of gases in leaf fluids after passing through the stomata (the tiny pores covering the surfaces of leaves through which gaseous exchange with the atmosphere occurs), or the reaction with leaf exudates.

Gases also interact with particles in the atmosphere. When sorbed onto particles, they affect both their own deposition velocities and those of the particles. (See section 2.4.) A gas sorbed onto a very small particle (diameter $< 0.1\ \mu m$) has a decreased deposition velocity because the Brownian diffusion coefficient for the particle is much less than that for the gas [21].

The dry deposition of gases (and particles) is often analyzed by using an analogue to Ohm's law in which the overall resistance to deposition is equal to the inverse of the deposition velocity, v_d, and the concentration at a reference height is analogous to the electrical potential. The deposition rate (current) is equal to the concentration (potential) divided by the resistance.

The basic form of the resistance analogue is given by equation 6, where the total resistance, $1/v_d$, is equal to the sum of three resistances in series [31]:

$$1/v_d = r_a + r_s + r_c \qquad (6)$$

The three resistances on the right hand side of the equation are in decreasing order of distance from the point of deposition. The aerodynamic resistance, r_a, describes the resistance to transfer between the reference height at which the concentration is measured and the quasi-laminar layer next to the receiving surface [14]. The surface resistance, r_s, is associated with the transfer across the quasi-laminar layer next to the surface [29]. The canopy resistance, r_c, is related to the adsorptivity of the surface for the gas and the opening and closing of stomata [14].

The aerodynamic resistance can be evaluated as [31]:

$$r_a = \frac{\ln(z/z_0) - \psi_c}{k\,u_*} \tag{7}$$

where ψ_c is a correction function for atmospheric stability, which serves to increase r_a for stable conditions and decrease it for unstable conditions. This function is employed in the study of the transfer of heat and water vapor (ψ_H and ψ_W), and ψ_c for the transfer of trace gases can be approximated by use of the same equations as for ψ_H and ψ_W [31]. The correction function ψ_M is used in the study of momentum transfer; ψ_M is subtracted from the term $\ln(z/z_0)$ in equation 3 to obtain the wind profile when the atmosphere is not neutrally stable [17]. The equations for the correction functions are [31]:

$$\psi_c = \psi_M = \psi_H = \psi_W = -5\,z/L \tag{8}$$

for $0 < z/L < 1$, (stable conditions) and

$$\psi_c = \psi_H = \psi_W = \exp\{0.0598 + 0.39\ln(-z/L) - 0.090(\ln(-z/L))^2\} \tag{9}$$

$$\psi_M = \exp\{0.032 + 0.448\ln(-z/L) - 0.132(\ln(-z/L))^2\} \tag{9a}$$

for $0 > z/L > -1$ (unstable conditions)

The correction functions are expressed in terms of a stability parameter z/L, in which z is the height and L is the Obukhov scale length. The scale length can be evaluated directly by use of the equation given in Shieh *et al.* [24] and Wesely [29], for example. However, its evaluation requires such information as the latent heat flux, which usually is not available. Instead, L can be approximated by use of stability classes such as those of Pasquill [19] or Turner [27], which are based on readily available meteorological variables (airport weather observations in the case of Turner classes). Pasquill classes range from A to G and Turner classes from 1 to 7 (extremely unstable to extremely stable) with D and 4 representing neutral conditions in the respective classes. Golder [10,11] explains how to estimate L with either stability class.

Although Turner classes are less commonly used than Pasquill classes in air pollution work, they provide a better estimate of L [17]. The determination of the Turner class and the estimation of L are described below. If Pasquill classes are already known, they can be converted to Turner classes as follows (Pasquill→Turner): A→1, B→2, C→3, D→4, E→6, F→7, G→7 [11].

TABLE 2.5-1

Turner Classes

DAYTIME[a]					NIGHTTIME[b,c]		
Wind Speed (knots)[d]	**Solar Altitude**				**Wind Speed (knots)**	**Cloud Cover[e]**	
	>60°	**60°–35°**	**35°–15°**	**≤15°**		**>4/10**	**≤4/10**
0-1	1	1	2	3	0-3	6	7
2-3	1	2	2	3	4-5	5	5
4-5	1	2	3	4	6	5	6
6-7	2	2	3	4	7-10	4	5
8-9	2	3	3	4	≥11	4	4
10-11	3	3	4	4			
≥12	3	4	4	4			

a. The use of the daytime table is modified as follows when the cloud cover is > 5/10 *and*:

1) Cloud ceiling is < 7000 ft: read 2 columns to the right of the indicated column (or the rightmost column). For the special case of 10/10 cloud cover with a ceiling of < 7000 ft, use Turner class 4 regardless of wind speed.

2) Cloud ceiling is from 7000 – 16,000 ft (or > 16,000 ft with 10/10 cloud cover): read 1 column to the right of the indicated column (or the rightmost column).

3) Otherwise, read the table as it appears.

b. The use of the nighttime table is modified in the following cases:

1) Urban areas classified as 6 or 7 should be made class 5 because they do not become as stable in the lower layers of the atmosphere as rural areas.

2) For complete cloud cover (10/10), with a ceiling of < 7000 ft, use Turner class 4 regardless of wind speed.

c. Sunset to sunrise

d. 1 knot = 1.151 miles/hr = 0.514 m/s

e. Fraction of the sky covered by clouds

Source: Adapted from Turner [27].

Turner classes as a function of meteorological conditions are given in Table 2.5-1. The surface wind speed, cloud cover (fraction of the sky covered by clouds), cloud ceiling height, and solar altitude (angle of the sun above the horizon) are needed to use this table. The solar altitude depends on the time of day, day of the year, and the line of latitude of the observer. It can be expressed mathematically as [15a]:

$$\alpha = \arcsin(\sin\phi \sin\delta + \cos\phi \cos\delta \cos h) \tag{10}$$

α = solar altitude

ϕ = latitude of the observer

δ = declination of the sun (see Figure 2.5-6)

h = hour angle of the sun

(all in degrees)

The declination of the sun reflects the seasonal change in the height of the sun in the sky, and ranges from -23.45° at the start of winter (Dec. 22) to +23.45° at the start of summer (June 22). It is plotted as a function of the day of the year in Figure 2.5-6.

The hour angle of the sun is a means of expressing the time of day. For the central meridian of a time zone, it can be calculated as:

$$h = \frac{\text{hours before noon (-) or hours after noon (+)}}{\text{24 hrs}} \cdot 360° \tag{11}$$

h = hour angle of the sun, degrees

The hours before or after noon are based on local, standard time. For example, 8:15 am Eastern standard time equals -3.75 hours, but 2:30 pm Eastern daylight time equals +1.5 hours to reflect that standard time has been set ahead an hour for daylight savings. Since equation 11 is for the central meridian of the time zone, for each degree of longitude the observer lies to the east of this meridian, one degree should be added to h, and for each degree west of the meridian, one degree should be subtracted. In the United States, the central meridians for the Eastern, Central, Mountain, and Pacific time zones are 75°, 90°, 105°, and 120° W longitude, respectively.

An additional correction to h has been omitted here for the sake of simplicity. Because of this, its true value may deviate from the calculated value by as much as approximately 4°, depending on the time of year. However, it should be sufficiently accurate to find the appropriate range of α for use of Table 2.5-1, since α will never deviate by more than about 4° from its true value either. The omitted correction for the mean solar time can be found in List [15a].

With ϕ, δ, and h determined, α is evaluated with equation 10 and the Turner class read from Table 2.5-1. Once the Turner class is known, 1/L can be approximated for a given roughness length by use of Figure 2.5-7. Once L is determined, ψ_c is calculated from either equation 8 or 9, as appropriate, and equation 7 is then evaluated. The calculation of r_a is illustrated in example 2 below, which follows the discussion of r_s and r_c.

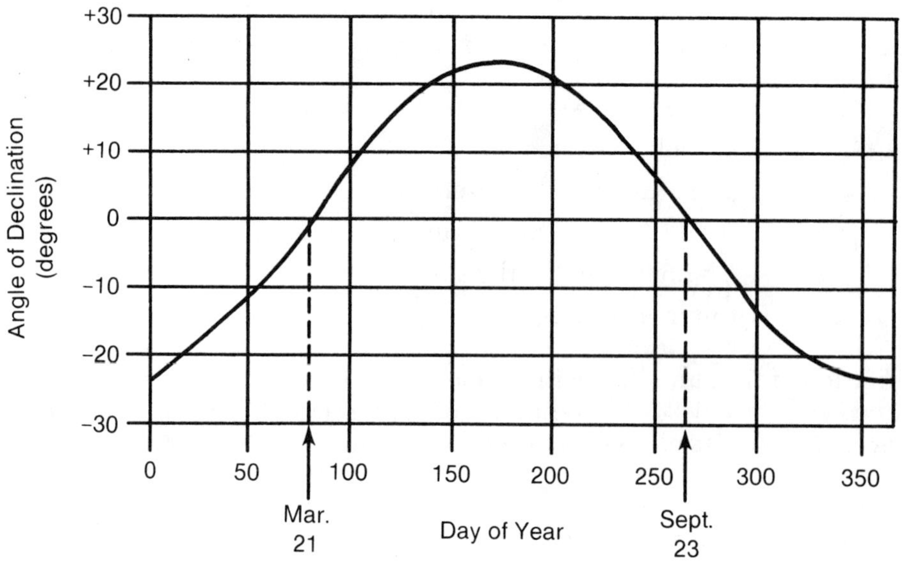

Source: After List [15a].

FIGURE 2.5-6 Declination of the Sun Over the Year

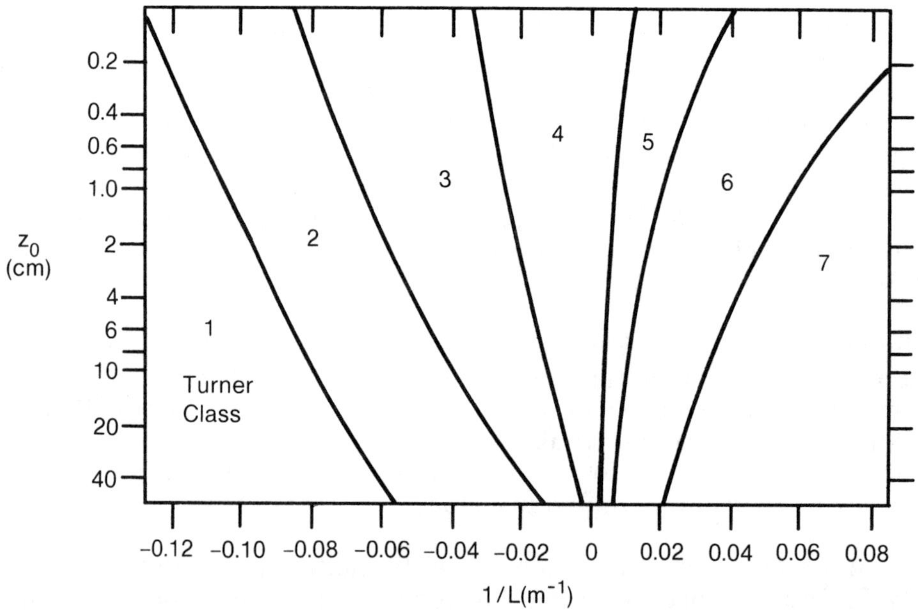

Source: Adapted from Golder [11]. *(Copyright 1972, D. Reidel Publishing Co. Reprinted with permission.)*

FIGURE 2.5-7 1/L as a Function of Turner Class and z_0

To evaluate the surface resistance, r_s, for deposition to vegetation, Wesely and Hicks [31], like others, use a surface transfer function, B^{-1}, where:

$$k B^{-1} = 2 (\kappa/D_c)^{2/3} \tag{12}$$

Again, k is the von Karman constant (typically 0.4); κ is the thermal diffusivity of air (e.g., 0.186 cm^2/s @ 0°C, 0.213 cm^2/s @ 20°C [28a]) and D_c is the molecular diffusivity for the gas of interest in air (cm^2/s). (Expressions for B^{-1} used by other researchers in various situations are described in Garland [9] and Wesely and Hicks [31].) For SO_2 and ammonia (NH_3), D_c has been measured as 0.122 and 0.198 cm^2/s, respectively, at 0°C and 1 atm [28a]; nitric oxide (NO), nitrogen dioxide (NO_2), and ozone (O_3) are estimated to have values of 0.175, 0.115, and 0.124 cm^2/s at the same temperature and pressure. Because D_c is roughly proportional to the absolute temperature raised to the 3/2 power, its value at some temperature T_2 can be estimated by multiplying its value at T_1 by $(T_2/T_1)^{3/2}$ (T_1, T_2 both absolute temperatures) [28a].

The surface resistance is then expressed as:

$$r_s = B^{-1}/u_* \tag{13}$$

For transfer to smooth surfaces (small z_0, e.g., water) Wesely [29] evaluates r_s as:

$$r_s = (u_* k)^{-1} \ln (k z_0 u_*/D_c) \tag{14}$$

where all the variables are as previously defined.

The canopy resistance, r_c, is taken as the residual resistance in field studies of deposition rates (whether a canopy of vegetation is present or not). Figure 2.5-8 shows the resistance scheme of Fowler and Unsworth [7] for SO_2 deposition to wheat. Here, the canopy resistance comprises, in parallel, the resistances to SO_2 uptake by stomata, vegetation surfaces, moisture, and soil. For the purposes of estimating the overall resistance, the total canopy resistance is more important than its individual components. Nevertheless, Figure 2.5-8 is useful for illustrating the factors affecting r_c, and their temporal variations. For example, stomata close during darkness and periods of water stress (to prevent the plant from desiccation); plants may be covered with dew at night and during mornings whereby $r_{c,w}$ becomes important (e.g., of negligible value in the case of SO_2 deposition).

Wesely and Hicks [31] have suggested a minimum r_c to for SO_2 transfer to field crops at roughly 0.7 s/cm and Sheih *et al.* [22] give rough approximations of r_c for SO_2 for various land use categories and Pasquill stability classes, which serve as an indicator of the surface conditions. Wesely [29] provides equivalent estimates for ozone. Voldner *et al.* [28] give estimates of r_c for SO_2 and NO_x for various conditions, and Huebert and Robert [14] provide values for HNO_3 based on field measurements.

The dependence of r_c on the depositing gas is shown by comparing the resistances of SO_2, O_3 and NO_x (NO and NO_2 combined) transfer to water bodies. Sulfur dioxide is estimated to have no resistance to lakes or oceans [24] whereas ozone is estimated to have a large resistance—100 s/cm to fresh water lakes and 20 s/cm to the ocean [29], while NO_x is estimated to have a resistance of 70 s/cm to fresh, open water [28].

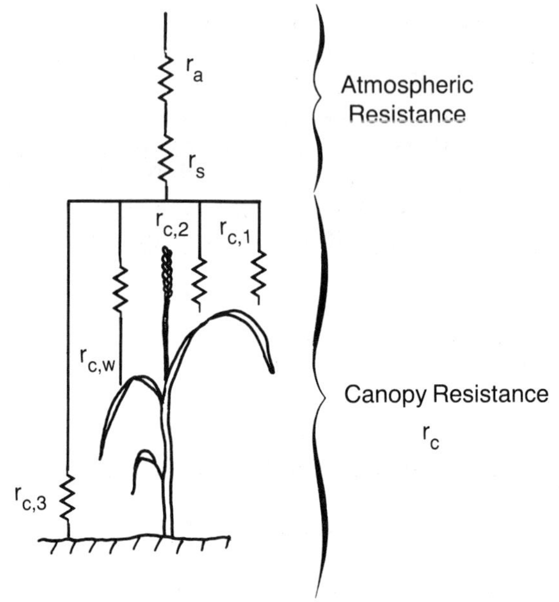

$r_{c,1}$, $r_{c,2}$, $r_{c,3}$, and $r_{c,w}$ are the canopy resistances to stomatal uptake, deposition on the plant surface, deposition on the soil, and uptake by surface moisture, respectively; r_a is the aerodynamic resistance and r_s the surface resistance.

Source: Adapted from Fowler and Unsworth [7]. *(Copyright 1979, Royal Meteorological Society. Reprinted with permission.)*

FIGURE 2.5-8 Resistance Analogue of Dry Deposition of SO_2 to a Wheat Canopy

The following example shows how the three resistances to gaseous deposition can be estimated, and then combined to give a deposition velocity.

Example 2 Estimate the deposition velocity of sulfur dioxide, SO_2, to a large corn field, 2 meters high, near Burlington, Vermont during mid-afternoon in the second week of August if the wind speed at 10 meters is 8 miles per hour and the day is mostly sunny (cloud cover of 2/10).

The deposition velocity is solved for by use of equation 6, but first, each of the resistances r_a, r_s, and r_c must be estimated. r_a is evaluated by use of equation 7, which requires knowing the value of z_0, the roughness length, u_*, the friction velocity, and ψ_c, the correction function. Because z_0 is not specified, a value of 5×10^{-2} m, or 5 cm, is read off of Figure 2.5-4. As in example 1, z is adjusted by a displacement length equal to 80% of the crop height. Thus z is taken as 8.4 m (10 m - 0.8•2 m) for the rest of this example.

The correction function for atmospheric stability, ψ_c, is determined next. The procedure used here is first to calculate the solar altitude, which is needed to find the Turner class, and then to read 1/L from Figure 2.5-7. ψ_c can then be calculated with equation 8 or 9 as appropriate.

The variables ϕ, δ, and h are required to find the solar altitude, α, by use of equation 10. From an atlas or almanac, the position of Burlington, Vermont is found to be 44° 28.5′ N, 73° 12.8′ W. Thus $\phi = 44° + 28.5'/[60('/°)] = 44.48°$. The angle of declination, δ, is read off of Figure 2.5-6. Assuming "the second week in August" corresponds to August 12, the 224th day of the year, δ equals +15°. Taking mid-afternoon to be 3:30 pm local time (daylight savings time) then h is evaluated by use of equation 11 as:

$$h = (+2.5 \text{ hrs}/24 \text{ hrs}) \cdot 360° = 37.5°$$

Since our location lies at 73° 12.8′ W, or 1° 47.2′ east of the central meridian for the Eastern standard time zone, the adjustment to h is small, but is included here for illustrative purposes. Thus, 1° 47.2′ or 1.79° is added to h since we are east of the central meridian.

$$h = 37.5° + 1.79° = 39.29°$$

The solar altitude can now be calculated with equation 10:

$$\begin{aligned}\alpha &= \arcsin(\sin 44.48° \sin 15° + \cos 44.48° \cos 15° \cos 39.29°)\\ &= 45.6°\end{aligned}$$

The Turner class is given as a function of wind speed and solar altitude in Table 2.5-1. A wind speed of 8 mph is equivalent to 6.95 knots (1 mph = 0.869 knots). Reading down the second column (α falls between 35° and 60°), Turner class 2 is found to correspond to this wind speed. Now, by use of Figure 2.5-7 for Turner class 2 and z_0 of 5×10^{-2} m, 1/L is estimated to be -0.07 m^{-1}.

Thus, $z/L = 8.4 \text{ m} \times -0.07 \text{ m}^{-1} = -0.59$, indicating that equation 9 should be used to find ψ_c ($-1 < z/L < 0$).

$$\begin{aligned}\psi_c &= \exp(0.0598 + 0.39 \ln(0.59) - 0.090 [\ln(0.59)]^2)\\ &= 0.843\end{aligned}$$

Equation 3 is used to evaluate u_*, but first ψ_M must be subtracted from $\ln(z/z_0)$ because the atmosphere is unstable (Turner class 2); from equation 9a:

$$\begin{aligned}\psi_M &= \exp\{0.032 + 0.448 \ln(0.59) - 0.132 (\ln(0.59))^2\}\\ &= 0.786\end{aligned}$$

$$\begin{aligned}u_* &= k\, u/[\ln(z/z_0) - \psi_M]\\ &= \frac{0.4 \times 8 \text{ mph} \times 0.4470 \text{ (m/s)/mph}}{\ln[(8.4 \text{ m})/(5 \times 10^{-2} \text{ m})] - 0.786}\\ &= 0.330 \text{ m/s (or 33.0 cm/s)}\end{aligned}$$

Now, r_a can be evaluated with equation 7:

$$r_a = \frac{\ln[8.4 \text{ m}/(5 \times 10^{-2} \text{ m})] - 0.843}{0.4 \times 0.330 \text{ m/s}}$$
$$= 32.4 \text{ s/m (or } 0.324 \text{ s/cm)}$$

To obtain r_s, equations 12 and 13 are combined. Taking D_c for SO_2 and κ at 20°C (from the text above), $\kappa = 0.213 \text{ cm}^2/\text{s}$ and $D_c = (293 \text{ K}/273 \text{ K})^{3/2} \cdot 0.122 \text{ cm}^2/\text{s} = 0.136 \text{ cm}^2/\text{s}$.

$$r_s = \frac{2\ [(0.213 \text{ cm}^2/\text{s})/(0.136 \text{ cm}^2)]^{2/3}}{0.4 \times 0.330 \text{ m/s}}$$
$$= 20.4 \text{ s/m (or } 0.204 \text{ s/cm)}$$

As noted in the text, 0.7 s/cm has been observed as a minimum value of r_c for SO_2 to field crops [31], and we will use that value here. Now that r_a, r_s, and r_c are known, v_d can be evaluated with equation 6.

$$1/v_d = 0.324 \text{ s/cm} + 0.204 \text{ s/cm} + 0.7 \text{ s/cm} = 1.228 \text{ s/cm}$$
$$v_d = 0.81 \text{ cm/s}$$

In this example it can be seen that the crop canopy, r_c, offers the greatest resistance to SO_2 transfer. Had the field been wet (from a rain shower or spray irrigation, for example) this resistance would have been taken to be zero. The total resistance would have been more than halved resulting in over a doubling of the deposition velocity.

2.5.3 Wet Deposition

Wet deposition is the removal of gaseous and particulate pollutants through such processes as rain, snow, and fog. It can account for a significant fraction of the total atmospheric deposition. For toxic metals, the mean fraction of total atmospheric deposition that occurs as wet deposition has been estimated to be between 0.4 and 0.7 [8]. At locations distant from SO_2 emitters, most of the deposited sulfur from these sources may be deposited in wet form as sulfate [16]. The effects of wet deposition can be important temporally, since precipitation is generally episodic and short lived; they can also be important spatially, because wet deposition may exceed dry deposition in rainy areas.

The wet removal of pollutants from the atmosphere by precipitation is sometimes classified into rainout and washout. *Rainout* is the removal of pollutants within a cloud by such processes as nucleation and interception, while *washout* is the removal of pollutants that occurs below the cloud base. This removal can occur as falling hydrometeors such as rain or snow, combine with gases or particles through any of the mechanisms described above for dry deposition on a surface.

Compared to dry deposition, wet deposition is relatively easy to measure experimentally, except in convective storms. For a given precipitation event, only representative precipitation samples are needed for both volume and concentration measurements. The total deposition for the precipitation event is simply the product of the precipitation volume and the concentration of the constituent in that volume. Because the measurement of wet deposition is straightforward (and because the processes leading up to the deposition are complicated and incompletely understood), the discussion that follows is brief.

Figure 2.5-9, which illustrates the scavenging sequence for atmospheric pollutants, shows the reversibility of many of the processes. "Type 1 mixing" refers to relative movement of the unmixed pollutant and condensed water (e.g., rain falling through a plume), while "Type 2 mixing" occurs as water vapor condenses in the immediate vicinity of pollutants [16]. A more complete description of the processes shown in Figure 2.5-9 is given in *Acid Deposition: Atmospheric Processes in Eastern North America* [16].

PARTICLE SCAVENGING

The removal of particles by precipitation (without considering the specific mechanisms involved) is known as *scavenging*. (This term is sometimes used to mean merely the attachment of particles to falling precipitation, whether or not the particles are actually removed from the atmosphere.) A scavenging coefficient, Λ, which represents the fractional removal of particulates (e.g., on a mass or number basis) per unit time under given precipitation conditions, is used to quantify the process. Figure 2.5-10 illustrates the effect of particle size and size distribution on the scavenging coefficient. In this figure the scavenging coefficient Λ, on a mass basis (hr^{-1}) divided by the rainfall rate, J (mm/hr), is plotted against the geometric mean radius of the particle, R_g (μm). Each curve represents a different distribution of particle sizes, as described by the geometric standard deviation of the particle radii, σ_g. For $\sigma_g = 1$, all the particles in the aerosol have the same radius. The figure is derived for a typical frontal rain storm with droplets having a geometric mean radius of 0.02 cm and a geometric standard deviation of 1.86. The use of Figure 2.5-10 is illustrated by the following example:

Example 3 Calculate the fraction of the aerosol mass concentration, m, remaining after a three-hour frontal rainstorm if the rainfall rate, J, is 2 mm/hr and the particle geometric mean radius, a_g, is 0.20 μm with a geometric standard deviation, σ_g, of 1.75.

As shown in Figure 2.5-10, an aerosol with the characteristics described above would have a normalized washout coefficient, Λ/J, of about 0.02 mm^{-1}.

Since J = 2 mm/hr, the fractional rate of aerosol removal is 0.04/hr, i.e.:

$$\frac{dm/m}{dt} = (-\Lambda/J)\bullet J = -0.02\ mm^{-1}\bullet 2\ mm/hr = -0.04\ hr^{-1}$$

or

$$dm/m = -0.04\ dt$$

Integrating this expression,

$$m/m_o = e^{-0.04t}$$

where m_o is the initial aerosol mass concentration and t is the time since the start of the storm, in hours. Evaluating the equation above for t = 3, the fraction of aerosol remaining, m/m_o, equals 0.89 of the initial value.

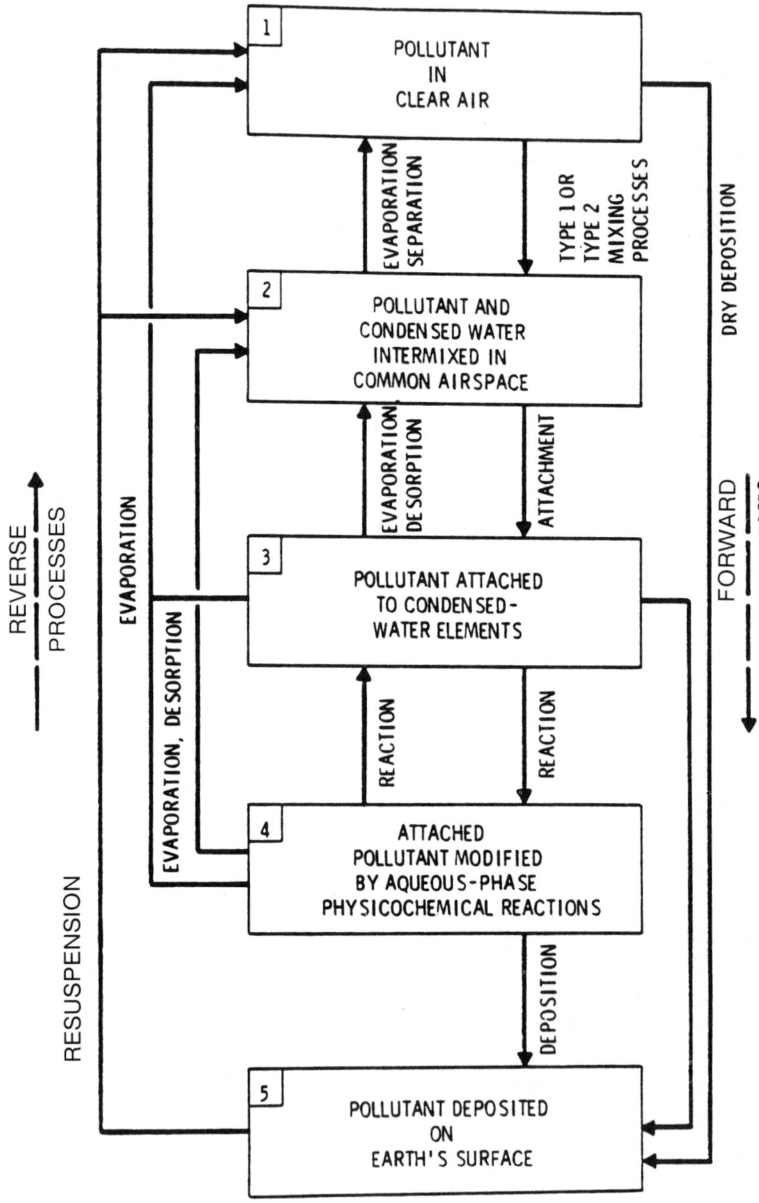

Source: National Research Council [16]

FIGURE 2.5-9 Wet Deposition Scavenging Sequence

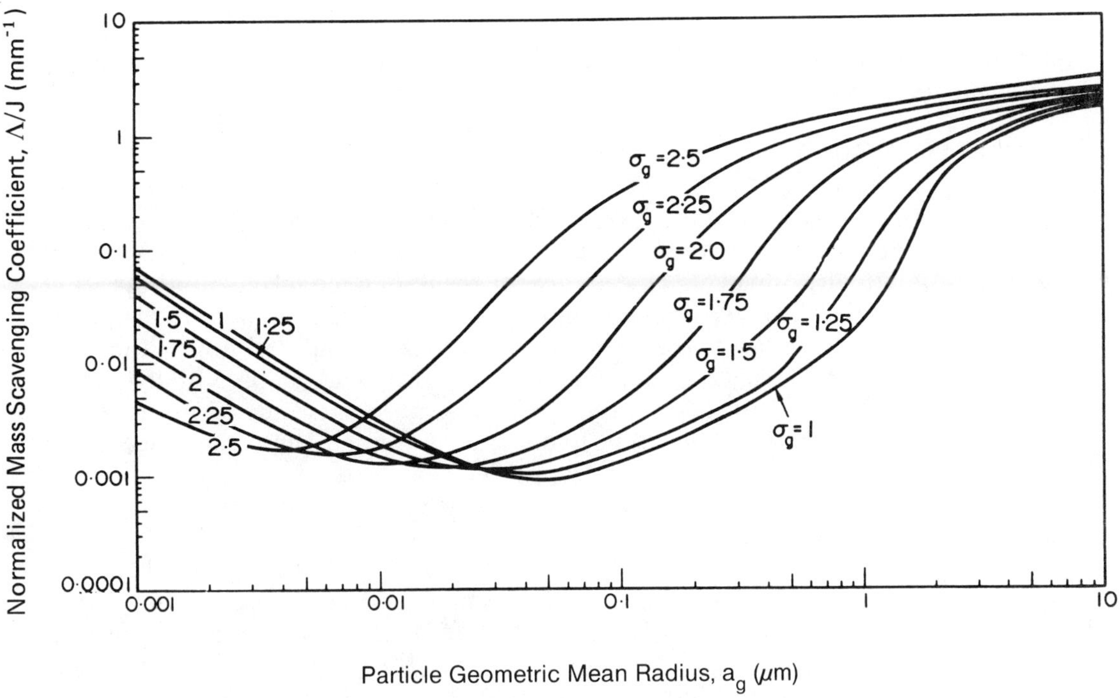

Particle density = 1 g/cm^3; σ_g = geometric std. deviation of particle radii

Source: Dana and Hales[5]. *(Copyright 1976, Pergamon Press, Inc. Reprinted with permission.)*

FIGURE 2.5-10 Theoretical Mass Scavenging Coefficients Normalized to Rainfall Rate, Based on a Typical Frontal Raindrop Size Spectrum

For small particles (the left-hand side of Figure 2.5-10), Brownian diffusion is effective in moving them to the surface of suspended and falling droplets [16]. As discussed above, the Brownian movement of particles decreases with increasing size; this is why Λ is lower toward the center of the plots.

For large particles, inertial effects become important in scavenging, so Λ increases toward the right. Larger particles are unable to escape a falling hydrometeor by moving with the air flow around it, and thus become attached by collision. Similarly, particles moved through the air by turbulent eddies may collide with droplets in the same way they are dry-deposited on a surface.

The most notable feature of Figure 2.5-10 is the minimum in the scavenging coefficient for particles between 0.01 and 0.1 μm radius (for $\sigma_g = 1$). This region is known as the Greenfield gap and corresponds to the range of particle sizes where neither Brownian diffusion nor inertial impaction is significant [19]. Phoretic effects tend to fill the gap somewhat, but it is still pronounced.

Particles that fall within the Greenfield gap are removed from the atmosphere by other mechanisms. Particles in this size range most readily serve as cloud condensation and ice nuclei and are therefore removed from the atmosphere if cloud formation is followed by precipitation [19].

GAS SCAVENGING

For modeling purposes, it would be desirable to know the gas scavenging ratio ξ, defined as the concentration of a substance in a precipitation sample divided by its concentration in air. However, ξ is usually the result of a model calculation rather than an input [16]. While particle scavenging may be approximated by the procedure described above, gas scavenging is more complex and will not be included here. Hales [12] developed a flow chart for scavenging calculations, which has also been published by the NRC [16]. Slinn [25] has described a method for approximating the scavenging of gases by rain. Readers may consult these references for a description of gas scavenging calculations.

2.5.4 Deposition Velocity Data

Due to the difficulty of measuring and estimating deposition velocities, a value of 1 cm/s is often assumed for predicting either gaseous or particulate deposition rates [21]. Galloway *et al.* [8] have compiled data on the annual deposition quantities of trace metals and on their concentrations in precipitation. Sehmel [21] has compiled deposition velocity data for particles and gases and a variety of surfaces, and Voldner *et al.* [28] list data and experimental conditions from observations of the deposition of sulfur and nitrogen oxides. Some of these data (and those of other researchers) are given in Tables 2.5-2 and 2.5-3; since they are specific to the experimental conditions under which they were determined, caution should be used in applying them to other conditions.

TABLE 2.5-2

Dry Deposition Velocities for Particles

Depositing Material	Deposition Velocity[a] (cm/s)
Particles 0.03 – 30 μm diameter, z_0 from 0.001 – 10 cm	10^{-3} – 40
Pollen 20 μm diameter, to microscope slides in grassland	4.5
Sulfate	
to vegetation	0.1 (typically used)[b]
to water	0.4 ± 0.2[c]

(Continued)

TABLE 2.5-2 (Continued)

Dry Deposition Velocities for Particles

Depositing Material	Deposition Velocity[a] (cm/s)
to snow, unstable conditions, $u_* = 0.17$ m/s, 0.1 m reference height	0.10[c]
to beech forest, summer	1.1[c]
to loblolly pine forest, summer	0.5[c]
to grass and crops, for both unstable and stable conditions, $u_* = 0.025$, 0.30 m/s, resp.	0.18[c]
ZnS	0 – 31
5 μm mass mean diameter, to snow	
Pb auto exhaust (aggregated) to whole leaves	0.008 – 0.12
Lead, bromine, zinc	
Urban St. Louis 1973	2.2 ± 1.4
1975	1.8 ± 0.7
Natural aerosol	0.8
1 – 10 μm diameter, to grass sward	
Ag	<0.3 – 1.1
Al	0.9 – 2.7
As	<0.1 – <0.6; 0.70
Br	0.1 – 2.6; 2.3
Ca	0.4 – 1.4; 2.6
Cd	<0.4 – >8
Ce	0.5 – 1.9
Cl	0.2 – 6.3; 6.4
Co	0.3 - 1.9
Cr	0.6 – 6.8; 0.60
Cs	0.2 – 0.6
Cu	<0.6 – 1.1; 3.0
Eu	0.3 – >2
Fe	1.0 – 2.5; 0.37
I	<0.3 – <2
In	0.3 – >6
K	0.6 – 13; 2.6

(Continued)

TABLE 2.5-2 (Continued)

Depositing Material	Deposition Velocity[a] (cm/s)
La	>0.6 – 3.5
Mg	0.6 – > 3
Mn	0.4 – 0.9; 0.59
Na	0.2 – 4.3
Ni	0.7 – <2; 4.0
Pb	0.38
Rb	<0.9 – 3.0
Sb	0.06 – <0.4
Sc	0.6 – 2.5
Se	0.1 – 0.6; 0.24
Sm	0.6 – 4.1
Th	0.7 – 2.2
Ti	0.7 – 2.2; 0.67
V	0.2 – <0.7; 0.24
Zn	0.4 – 4.5; 1.8
Trace elements to open bucket	0.2 – 6.4

Source: Sehmel [21], except as indicated. *(Copyright 1980, Pergamon Press, Inc. Reprinted with permission.)*

a. Values following a semicolon represent those of a different researcher.

b. From NRC [16]

c. From Voldner, *et al.* [28]

TABLE 2.5-3

Dry Deposition Velocities for Gases

Depositing Material	Deposition Surface	Deposition Velocity (cm/s)	Source
Cl_2	Alfalfa	1.8, 2.1	[21]
CO_2	Alfalfa	0.3	[21]
Fluorides	Forage	1.9 ± 0.5	[21]
HF	Alfalfa	3.5	[21]
H_2S	Sandy loam soil	0.015	[21]
Iodine	Grass	7.2 ± 0.9 (upper limit)	[21]
NO_x	Pine needles (summer day)	0.4-0.8 avg.	[28]
NO_x	Crops and grassland (soybeans, summer day, unstable conditions)	0.6	[28]
HNO_3	Crops and grassland (summer day, 1 m reference height)	2.5 0.9	[28]
O_3	Deciduous forest (leafless, during winter day) (during winter night)	0.19-0.37 0.05	[29]
O_3	Soybean canopy (midday, appr. 25% of soil exposed) (windy conditions)	0.85 1.2	[29]
SO_2	Water (neutral stability, 0.2 m reference height)	0.5 ± 0.1	[28]
SO_2	Snow (unstable conditions)	0.17	[28]
SO_2	Forest (dry) (wet)	0.8 1.5	[28]
SO_2	Grass and crops (unstable conditions, $u_* = 0.11 - 0.26$ m/s, 2 m reference height)	0.33 ± 0.30	[28]

2.5.5 Literature Cited

1. Arimoto, R. and R.A. Duce, "Dry Deposition Models and the Air/Sea Exchange of Trace Elements," *J. Geophys. Res.*, **91**, 2787-2792 (1986).
2. Cadle, S.H., J. Muhlbaier Dasch, and P.A. Mulawa, "Atmospheric Concentrations, and the Deposition Velocity to Snow of Nitric Acid, Sulfur Dioxide and Various Particulate Species," *Atmos. Environ.*, **19**, 1819-1827 (1985).
3. Chamberlain, A.C., J.A. Garland, and A.C. Wells, "Transport of Gases and Particles to Surfaces with Widely Spaced Roughness Elements," *Boundary-Layer Meteorol.*, **29**, 343-360 (1984).
4. Corn, M., "Aerosols and the Primary Air Pollutants — Nonviable Particles: Their Occurrence, Properties and Effects," in *Air Pollutants, Their Transformations and Transport*, Vol. 1 of *Air Pollution*, 3rd ed., A.C. Stern (ed.), Academic Press, New York (1976).
5. Dana, M.T. and J.M. Hales, "Statistical Aspects of the Washout of Polydisperse Aerosols," *Atmos. Environ.*, **10**, 45-50 (1976).
6. Davies, C.N., "Aerosol Properties Related to Surface Contamination," in *Surface Contamination*, B.R. Fish (ed.), Pergamon Press, New York (1967), as cited in National Research Council [16].
7. Fowler, D. and M.H. Unsworth, "Turbulent Transfer of Sulfur Dioxide to a Wheat Crop," *J. R. Meteorol. Soc.* **105**, 767-783 (1979).
8. Galloway, J.N., J.D. Thornton, S.A. Norton, H.L. Volchok, and R.A.N. McClean, "Trace Metals in Atmospheric Deposition: A Review and Assessment," *Atmos. Environ.*, **16**, 1677-1700 (1982).
9. Garland, J.A., "The Dry Deposition of Sulfur Dioxide to Land and Water Surfaces," *Proc. R. Soc. London, A.*, **354**, 245-268 (1977).
10. Golder, D.G., "A Comparison of Stability Parameters," M.S. Thesis, Department of Meteorology, The Pennsylvania State U., University Park, Pennsylvania (1970).
11. Golder, D.G., "Relations Among Stability Parameters in the Surface Layer," *Boundary-Layer Meteorol.*, **3**, 47-58 (1972).
12. Hales, J.M., "Precipitation Chemistry: Its Behavior and Calculations," in *Air Pollutants and Their Effects on the Terrestrial Ecosystem*, S.V. Krupper and A.H. Legge (eds.), John Wiley & Sons, New York (1983).
13. Hidy, G.M., "Removal Processes of Gaseous and Particulate Pollutants," in *Chemistry of the Lower Atmosphere*, S.I. Rasool (ed.), Plenum Press, New York (1973).
14. Huebert, B.J. and C.H. Robert, "The Dry Deposition of Nitric Acid to Grass," *J. Geophys. Res.*, **90**, 2085-2090 (1985).
15. Jenkin, M.E., "An Investigation into the Enhancement of Deposition of Hygroscopic Aerosols to Wet Surfaces in a Wind Tunnel," *Atmos. Environ.*, **18**, 1017-1024 (1984).

15a. List, R.J., *Smithsonian Meteorological Tables*, 6th revised ed., Smithsonian Institution, Washington, D.C. (1951).

16. National Research Council, *Acid Deposition: Atmospheric Processes in Eastern North America*, National Academy Press, Washington, D.C. (1983).

17. Panofsky, H.A. and J.A. Dutton, *Atmospheric Turbulence: Models and Methods for Engineering Applications*, John Wiley & Sons, New York (1984).

18. Pasquill, F., "The Estimation of the Dispersion of Windborne Material," *Meterol. Mag.*, **3**, 47-58 (1972).

19. Pruppacher, H.R. and J.D. Klett, *Microphysics of Clouds and Precipitation*, D. Reidel Publishing Co., Dordrecht, Holland (1978).

20. Sehmel, G.A., "Particle Diffusivities and Deposition Velocities over a Horizontal Smooth Surface," *J. Colloid and Interface Sci.*, **37**, 891-906 (1971).

21. Sehmel, G.A., "Particle and Gas Dry Deposition: A Review," *Atmos. Environ.*, **14**, 983-1011 (1980).

22. Sehmel, G.A., "Dry Deposition Velocities," Paper Prepared for Presentation at the California Air Resources Board Workshop on Dry Acid Deposition, March 26, 1984, San Francisco, Calif. (March, 1984), NTIS PNL-SA-12156.

23. Sehmel, G.A. and W.H. Hodgson, "A Model for Predicting Dry Deposition of Particles and Gases to Environmental Surfaces," Battelle Pacific Northwest Laboratory, Richland, Washington, (January, 1978) PNL-SA-6721 (and in revised form October, 1979, PNL-SA-6721-REV-1).

24. Shieh, C.M., M.L. Wesely, and B.B. Hicks, "Estimated Dry Deposition Velocities of Sulfur over the Eastern United States and Surrounding Regions," *Atmos. Environ.*, **13**, 1361-1368 (1979).

25. Slinn, W.G.N., "Some Approximations for the Wet and Dry Removal of Particles and Gases from the Atmosphere," *Water Air Soil Pollut.*, **7**, 513-43 (1977).

26. Slinn, W.G.N., "Predictions for Particle Deposition to Vegetative Canopies," *Atmos. Environ.*, **16**, 1785-1794 (1982).

27. Turner, D.B., "A Diffusion Model for an Urban Area," *J. Appl. Meteorol.*, **3**, 83-91 (1964).

28. Voldner, E.C., L.A. Barrie and A. Sirois, "A Literature Review of Dry Deposition of Oxides of Sulfur and Nitrogen with Emphasis on Long-Range Transport Modelling in North America," *Atmos. Environ.*, **20**, 2101-2123 (1986).

28a. Welty, J.R., C.E. Wicks, and R.E. Wilson, *Fundamentals of Momentum, Heat, and Mass Transfer*, 2nd ed., John Wiley & Sons, New York (1976).

29. Wesely, M.L., "Turbulent Transport of Ozone to Surfaces Common in the Eastern Half of the United States," in *Trace Atmospheric Constituents: Properties, Transformations, and Fates*, S.E. Schwartz (ed.), John Wiley & Sons, New York (1983).

30. Wesely, M.L., D.R. Cook, R.L. Hart, and R.E. Speer, "Measurement and Parameterization of Particulate Sulfur Dry Deposition Over Grass," *J. Geophys. Res.*, **90**, 2131-2143 (1985).

31. Wesely, M.L., and B.B. Hicks, "Some Factors that Affect the Deposition Rates of Sulfur Dioxide and Similar Gases on Vegetation," *J. Air Pollut. Control Assoc.*, **27**, 1110-1116 (1977).

32. Williamson, S.J., *Fundamentals of Air Pollution*, Addison-Wesley Publishing Co., Reading, Mass. (1973).

2.6 ACTIVITY COEFFICIENT

2.6.1 Introduction

As applied to dissolved species in solution, an activity coefficient, γ, is a factor that corrects for nonideal behavior relating to chemical reactivity.[1] In an ideal solution, the components coexist with no volume effects and no interaction energies. In such ideal systems, chemical interactions are described by relationships of species concentrations. As solutions approach infinite dilution, the behavior of their components approaches this ideal behavior. However, in nonideal systems, the actual chemical "activity" of an ion is smaller than the measured concentration by a factor represented by the activity coefficient. For an ion A, we define:

$$(C_A) = \gamma_A [C_A] \quad (1)$$

where

(C_A) = activity of ion A in solution
γ_A = activity coefficient of ion A in solution ($0 < \gamma_A \leq 1$ in most cases)
$[C_A]$ = concentration of ion A in solution

Systems found in the environment (fresh waters, seawater, concentrated saline brines, leachates) are not infinitely dilute; therefore, the dissolved species that they contain cannot be assumed to behave ideally. Activity coefficients may be needed for the calculation of equilibrium reaction constants (e.g., solubility constants) and equilibrium concentrations for such systems. The use of activity coefficients is especially important and necessary in systems that deviate significantly from ideal behavior, i.e., concentrated or high-ionic-strength solutions such as saline brines. Seawater is a unique solution of practically constant ionic composition; as a result, activity coefficients for species in seawater may be specially described in relation to this constancy.

This section covers the following topics:

- Description of the activity coefficient and its relationship to activities, concentrations and equilibrium constants;
- The importance and use of the activity coefficient in calculations for environmental systems;
- Key variables (ionic strength and temperature) that affect the value of the activity coefficient; and
- Available methods for estimating activity coefficients, their limitations and examples of their use.

1. See section 2.6.2, which discusses the alternative "activity coefficient" relating to nonideal behavior of organics in solution (i.e., nonideal with regard to phase partitioning).

2.6.2 Description of Activity Coefficient

Chemical species in solution interact in a variety of ways. Some of these interactions are strong and are considered to be "chemical reactions"; others are defined as "nonideal" effects that cause a system to depart from ideal behavior and affect the chemical reactions occurring in the solution. In dilute electrolyte mixtures (with ionic strength $I < 10^{-2}$ *M*), deviations from ideal behavior are primarily due to long-range electrostatic interactions among ions, i.e., attractions of ions of opposite charge and repulsion of ions of like charge. In more concentrated solutions ($I > 10^{-2}$ *M*), the deviations are due to the general electric fields of the ions, solute-water interactions and specific ionic interactions (e.g., ion-pair formation, complex formation)[19]. All these interactions on an ion serve to stabilize it and therefore decrease its chemical reactivity in the solution. The activity coefficient may be perceived as a measure of this decreased reactivity.

DEFINITION

The activity coefficient, γ_A, of ion A in solution is defined by equation 1 above or by:

$$\gamma_A = \frac{(C_A)}{[C_A]} \tag{2}$$

where

(C_A) indicates activity of species A
$[C_A]$ indicates concentration of species A

Since the activity and concentration are expressed in the same units (e.g., molar or molal), the activity coefficient is a dimensionless factor.
For a reaction:

$$aA + bB = cC + dD \tag{3}$$

the activity equilibrium constant is defined as:

$$K_a = \frac{(C)^c\,(D)^d}{(A)^a\,(B)^b} \tag{4}$$

The parentheses signify the activity of each species at equilibrium. In terms of concentrations:

$$K_a = \frac{\gamma_C^c\,[C]^c\,\gamma_D^d\,[D]^d}{\gamma_A^a\,[A]^a\,\gamma_B^b\,[B]^b} = \frac{\gamma_C^c\,\gamma_D^d}{\gamma_A^a\,\gamma_B^b}\,\frac{[C]^c\,[D]^d}{[A]^a\,[B]^b} = \frac{\gamma_C^c\,\gamma_D^d}{\gamma_A^a\,\gamma_B^b}\,K_c \tag{5}$$

where K_c is the equilibrium constant expressed in terms of equilibrium concentrations:

$$K_c = \frac{[C]^c [D]^d}{[A]^a [B]^b} \tag{6}$$

It is important to know if values of equilibrium constants reported in the literature and used in calculations are based on activities or concentrations. Values are sometimes reported as a mixture of both activities and concentrations.

$$\left.\begin{array}{ll} \text{In the ideal system,} & \gamma_i = 1 \\ \text{and} & K_a = K_c \end{array}\right\} \tag{7}$$

CONVENTIONS

There are basically two conventions by which the reference state (or ideal system at which $\gamma_i = 1$ and activity equals concentration) is defined. These are the Infinite Dilution Scale and the Ionic Medium Scale. The choice of one convention over the other is primarily based on the solution under consideration.

The Infinite Dilution Scale

In this convention, the activity coefficient of species A, γ_A (equation 2), is defined such that γ_A approaches 1 as the concentration of all solutes in solution approaches 0.

$$\gamma_A \rightarrow 1 \text{ as } [C_A] + \sum_i [C_i] \rightarrow 0 \tag{8a}$$

where $[C_i]$ = concentration of all solutes in solution.

For the solvent,

$$\left.\begin{array}{l} (\text{solvent}) = \gamma_{\text{solvent}} X_{\text{solvent}} \\ \gamma_{\text{solvent}} \rightarrow 1 \text{ as } X_{\text{solvent}} \rightarrow 1 \end{array}\right\} \tag{8b}$$

where X_{solvent} = mole fraction of the solvent
(solvent) = activity of the solvent

In dilute solutions ($I < 10^{-2}$ *M*), e.g., in many fresh waters, the infinite dilution activity convention is used. The relationship of the equilibrium constant based on activities (K_a) to the equilibrium constant based on concentrations (K_c) has been shown above (equation 5). The activity coefficients should be calculated for the conditions of ionic strength and temperature of the solution under consideration.

The Ionic Medium Scale

In this convention, the reference state is a specified ionic medium, instead of the infinitely dilute solution above. It is useful in solutions (e.g., seawater) where there is a background concentration of inert electrolyte that provides the constant ionic medium. In this case, the activity coefficient, γ', approaches 1 as the concentrations of all ions *other than the medium ions* approach 0; i.e.,

$$\gamma_A' \rightarrow 1 \text{ as } [C_A] \rightarrow 0 \tag{8c}$$

where $[C_A]$ = concentration of all ions except the major medium ions.

If the concentrations of the medium ions are more than approximately 10 times the concentration of the solute species under consideration, the activity coefficients of the species may be approximated as 1 [19].

Seawater is a solution of practically constant ionic composition. The ionic medium activity scale is thus particularly useful for equilibrium calculations in seawater systems and more concentrated solutions [19].

The value of the activity coefficient is usually greater than 0 and less than or equal to 1. In certain cases, γ may go above 1 (see below in § 2.6.3 on neutral components in solution and on very high ionic strength solutions).

The equilibrium constant for a reaction using the constant ionic medium convention is called the *apparent equilibrium constant*, K_{app}. For example, the second dissociation constant for carbonate in solution,

$$HCO_3^- \rightleftharpoons H^+ + CO_3^{-2} \tag{9}$$

is written as:

$$K_{app} = \frac{(H^+)\,[CO_{3T}^{-2}]}{[HCO_{3T}^-]} \tag{10}$$

where the subscript T refers to the total concentration of ions, both free and associated. If the concentration of the constant medium is very high compared with that of the ions under consideration, the activity coefficients of the ions may be approximated as 1. However, the apparent equilibrium constant must be determined for the medium under consideration.

The conventions above may be contrasted with the activity coefficient for organic solute solutions (including organic solutes in water), which is defined as:

$$\gamma_A = P_A / X_A P_A^0 \tag{11}$$

where

γ_A = activity coefficient of component A

P_A = partial pressure of component A

x_A = mole fraction of component A in solution

P_A^0 = vapor pressure of pure liquid of component A

The activity coefficients for organic solute solutions range in value from 0.4 up to 10^7 [2].

For all dilute solutions, using a standard state of pure water, the activity of water is defined as 1. In seawater, the activity of water is defined as the ratio of vapor pressure of the seawater, P_s, to that of pure water, P_{H_2O}. It is close to 1 and does not usually fall below 0.98 [19].

When a solid is in equilibrium with a solution, the pure solid phase is selected as the reference state, and its activity is set to 1. For gases in equilibrium with a solution, the activity coefficient, γ_g (or fugacity coefficient) is defined as:

$$\gamma_g = \frac{(f_g)}{P_g} \tag{12}$$

where

P_g = partial pressure of gas in atmosphere

(f_g) = fugacity of the gas (the equivalent of activity)

As the total pressure decreases, $\gamma_g \rightarrow 1$. Gases deviate from ideal behavior only at very high pressures. Reactions at atmospheric pressure are considered to occur at sufficiently low pressures for the assumption of ideal behavior, and the activity of a gas is estimated by its partial pressure.

SINGLE ION ACTIVITY AND MEAN ION ACTIVITY COEFFICIENTS

The single ion activity coefficient, also known as the individual ion activity coefficient, is the activity coefficient of a cation or anion by itself, without its companion anion or cation respectively. Because anions cannot be added to solutions without an equivalent number of cations (and vice versa), the single ion activity (γ_+ or γ_-) cannot be determined experimentally. Instead, the product $\gamma_+ \gamma_-$ is experimentally measured. The mean ion activity coefficient, $\gamma_\pm$, for a univalent symmetrical compound (e.g., KCl) is defined as:

$$\gamma_\pm = (\gamma_+\gamma_-)^{1/2} \tag{13}$$

The single ion activities of the ions of a univalent symmetrical compound are approximately equal to each other and to the mean ion activity coefficient:

$$\gamma_\pm \approx \gamma_+ \approx \gamma_- \tag{14}$$

Stumm and Morgan [19] have provided relationships between single and mean ion activity coefficients for other types of compounds.

FREE ACTIVITY AND TOTAL ACTIVITY COEFFICIENTS

Many ions, particularly major anions such as SO_4^{-2}, HCO_3^- and CO_3^{-2}, often exist in solution both as free ions and associated with other ions in complexes or ion pairs. It is difficult to establish unambiguously the types of interactions that occur with these ions; therefore, a total activity coefficient, γ_A^T, has been defined:

$$\gamma_A^T = \frac{(C_A)}{[C_A]_T} \tag{15}$$

where

(C_A) = activity of A in solution

$[C_A]_T$ = total (free and associated) ion concentration of A in solution

The free activity coefficient, γ_A^F, is defined as:

$$\gamma_A^F = \frac{(C_A)}{[C_{A\,free}]} \tag{16}$$

where "free" represents aquo-associated ions of SO_4^{-2}, CO_3^{-2}, etc.

Total activity coefficients are dependent on the ionic composition of the system. The use of total activity coefficients is particularly appropriate for seawater because of the constancy in ionic composition and the elimination of ion pair formation constants from the calculations. Also, the total activity coefficient is more amenable to experimental measurement than individual ion pair or complex formation constants.

IMPORTANCE IN ENVIRONMENTAL CALCULATIONS

The activity coefficient is important in equilibrium calculations for environmental systems, because natural systems are rarely ideal. Corrections from concentrations to the lower activity values may sometimes be smaller than the errors associated with use of equilibrium constants from the literature, where it is not unusual to find cited values that span orders of magnitude.

Example 1 Determine the error in the calculated solubility of CaF_2 in a solution of $I = 0.1M$, if non-ideality is ignored. At this ionic strength, $\gamma_{Ca^{+2}} = 0.38$ and $\gamma_{F^-} = 0.79$ (using the Davies method described in § 2.6.3).

The activity equilibrium solubility constant, $K = 10^{-10.41}$ (mole/liter)^3 [7]

$$K_a = (Ca^{+2})(F^-)^2 = \gamma_{Ca^{+2}}\,\gamma_{F^-}^2\,[Ca^{+2}][F^-]^2 \tag{17}$$

If we assume ideal solutions and all $\gamma = 1$,

$$K_c = [Ca^{+2}]\,[F^-]^2 = 10^{-10.41} \tag{18}$$

which gives solution concentrations in equilibrium with CaF_2 (s) as follows:

$$[Ca^{+2}] = 2 \times 10^{-4} M$$
$$[F^-] = 4 \times 10^{-4} M$$

If the activity coefficients were considered,

$$K = \gamma_{Ca^{+2}}\,[Ca^{+2}]\,\gamma^2_{F^-}\,[F^-]^2 \tag{19}$$
$$10^{-10.41} = (0.38)\,[Ca^{+2}]\,(0.79)^2\,[F^-]^2$$
$$[Ca^{+2}]\,[F^-]^2 = 1.64 \times 10^{-10}$$

which gives these concentrations in equilibrium with CaF_2 (s):

$$[Ca^{+2}] = 3.4 \times 10^{-4} M$$
$$[F^-] = 6.9 \times 10^{-4} M$$

Thus, when activity coefficients are not included, the calculated saturation concentration is lower by around 40%.

VARIABLES AFFECTING THE VALUE OF ACTIVITY COEFFICIENTS

Other than those values intrinsic to the ion (e.g., charge and ion size), the key variables that affect the magnitude of the activity coefficient of an ion are solution ionic composition and concentrations (which determine the ionic strength) and temperature. Pressure affects the activity coefficient only at extremely high pressures, which are not normally encountered under environmental conditions; these effects will be neglected here. A brief discussion of pressure effects is given in Stumm and Morgan [19].

The effect of temperature, T, on the infinite dilution scale activity coefficient is described by:

$$\frac{d \ln \gamma_i}{dT} = \frac{\overline{H}_i^0 - \overline{H}^i}{RT^2} \tag{20}$$

where:

$\overline{H}_i$ = partial molal enthalpy in actual solution
$\overline{H}_i^0$ = standard-state value of partial molal enthalpy
R = gas constant

An activity coefficient is defined for a particular temperature. The effect of temperature on the activity coefficients of electrolytes becomes appreciable at concentrations greater than 0.1 molal, when $\overline{H}_i^0 - \overline{H}_i$ becomes significant [19]. In seawater, the activity coefficients of ions are significantly influenced by temperature. The magnitude and direction of the effect is dependent on the magnitude and sign of the heat of reaction as a function of temperature [19]. However, few measurements of partial molal enthalpies exist.

The ionic composition of the species in solution determines the ionic strength of the solution. The ionic strength, I, is defined by:

$$I = \frac{1}{2} \sum_i C_i Z_i^2 \tag{21}$$

where:

C_i = concentration of ionic species, i, in solution
Z_i = charge of species, i, in solution

Values of ionic strength in environmental systems range from:

Fresh water	$< 0.01M$ to $< 0.1M$
Seawater	$\sim 0.7M$
Saline brines	$\geq 5M$

Example 2 Calculate the ionic strength of a solution containing:

$[Na^+] = 10^{-4}M$ $[Cl^-] = 2 \times 10^{-4}M$
$[H^+] = 10^{-4}M$ $[HCO_3^-] = 2 \times 10^{-5}M$
$[Mg^{+2}] = 10^{-5}M$

From equation 21,

$$I = \frac{1}{2}\left\{[Na^+](1)^2 + [Cl^-](-1)^2 + [Mg^{+2}](2)^2 + [H^+](1)^2 + [HCO_3^-](-1)^2\right\}$$

$$= \frac{1}{2}\left[10^{-4} + 2 \times 10^{-4} + (4 \times 10^{-5}) + 10^{-4} + 2 \times 10^{-5}\right]$$

$$= 2.3 \times 10^{-4}M$$

In many natural waters, the ionic composition may be unknown and the ionic strength cannot be calculated by equation 21. In such cases, empirical relationships can be used to approximate the ionic strength from measured parameters. The two following relationships for natural waters, for example, are based on total dissolved solids and specific conductance [18]:

$$I = 2.5 \times 10^{-5} \times \text{TDS} \tag{22}$$

and

$$I = 1.6 \times 10^{-5} \times \text{specific conductance} \tag{23}$$

where:

I = ionic strength (*M*)
TDS = total dissolved solids (mg/l)
Specific conductance is in μmho/cm

Horne [4] has suggested an empirical relationship to estimate the ionic strength of seawater:

$$I = 0.0054 + 0.01840\,(S‰) + 1.78 \times 10^{-5}\,(S‰)^2 - 3.0 \times 10^{-4}\,(25 - T) + 7.6 \times 10^{-6}\,(P) \quad (24)$$

where:

I = ionic strength in molal units (mol/kg)
S = salinity (34-36‰ in open ocean water, where ‰ = parts per thousand)
T = temperature (°C)
P = atmospheric pressure (atm)

2.6.3 Estimation Methods

Methods for estimating activity coefficients are of two basic types: those for low-ionic-strength solutions (<0.5*M*) and those for high-ionic-strength solutions (>1*M*).

Methods in the first category are based on the Debye-Hückel assumption that deviations from ideal behavior in dilute solutions are primarily caused by long-range electrostatic interactions. All these methods have the general form:

$$\log \gamma_i = -\frac{AZ_i^2\sqrt{I}}{1 + Ba_i\sqrt{I}} - b_i I \quad (25)$$

where:

γ_i = single ion activity coefficient
A, B = Debye-Hückel constants
I = ionic strength of the solution (*M*)
Z_i = charge of ion i
a_i = empirical parameter representing the distance of closest approach between centers of adjacent ions
b_i = adjustable parameter

Four methods based on the Debye-Hückel assumption are frequently used: Debye-Hückel, extended Debye-Hückel, Guntelberg, and Davies. These are described in many chemistry texts [11, 12, 19].

At higher ionic strengths, two methods are used. One of these is based on the Brønsted-Guggenheim specific ion interaction model, which leads to a virial equation of the general form:

$$\log \gamma_i = \log \gamma_{iDH} + \sum_j B_{ij}[j] + \sum_j \sum_k C_{ijk}[j][k] + \ldots \quad (26)$$

where:

γ_i = single ion activity coefficient of ion i

γ_{iDH} = Debye-Hückel activity coefficient

[j], [k] = molar concentrations of species j and k

B_{ij} = second virial coefficients for specific interactions among pairs of ions

C_{ijk} = third virial coefficients for specific interactions among three ions

Pitzer and co-workers [10] have provided expressions to calculate the virial coefficients for equation 26 for a variety of ions.

The other method applicable to high-ionic-strength solutions was developed by Kusik and Meissner [6]. It is based on experimental observations of the behavior of activity coefficients with ionic strength, which were then developed into empirical relationships.

Table 2.6-1 summarizes all of the above estimation methods and the range of their applicability. Each method is described in greater detail below.

DEBYE-HÜCKEL METHOD

The Debye-Hückel approximation is given by:

$$\log \gamma_i = -AZ_i^2\sqrt{I} \qquad (27)$$

where:

A = ion-size parameter for water (in aqueous solutions)

Z_i = ionic charge of ion i

I = ionic strength of solution (*M*)

The parameter A is related to the solvent dielectric constant, D, and temperature by:

$$A = 1.823 \times 10^6/(DT^{3/2}) \qquad (28)$$

where:

D = dielectric constant of solvent

T = temperature (K)

For water at 25°C, A is approximately 0.5. Values of A at other temperatures are given by Pagenkopf [12].

The Debye-Hückel estimation method is applicable to dilute solutions with ionic strength less than $5 \times 10^{-3}M$.

TABLE 2.6-1

Summary of Methods for Estimating Activity Coefficients

Method	Equation[a]	Range of I Over Which Method is Valid (*M*)
For low-ionic-strength solutions:		
Debye-Hückel	$\log \gamma_i = -AZ_i^2 \sqrt{I}$	$0 - 10^{-2.3}$
Extended Debye-Hückel	$\log \gamma_i = -AZ_i^2 \dfrac{\sqrt{I}}{1 + Ba_i \sqrt{I}}$	$0 - 10^{-1}$
Güntelberg[b]	$\log \gamma_i = -AZ_i^2 \dfrac{\sqrt{I}}{1 + \sqrt{I}}$	$0 - 10^{-1}$
Davies	$\log \gamma_i = -AZ^2 \left(\dfrac{\sqrt{I}}{1 + \sqrt{I}} - 0.3\,I \right)$	0 – 0.5
For high-ionic-strength solutions:		
Pitzer *et al.*	Binary Solution:	
	$\log \gamma_{\pm} (MX) = Z_M Z_X f + m (\beta^0 + f^1 \beta^1) + 1.5\, m^2 c^{\phi}$	0 – 6
	Mixed Electrolyte Solution:	
	$\log \gamma_M = Z_M^2 f + 2 \sum m_a (B_{Ma} + EC_{Ma}) + Z_M^2 \sum_c \sum_a m_c m_a B^1_{ca} + Z_M \sum_c \sum_a m_c m_a C_{ca}$	
	$\log \gamma_X = Z_X^2 f + 2 \sum m_c (B_{cX} + EC_{cX}) + Z_X^2 \sum_c \sum_a m_c m_a B^1_{ca} + Z_X \sum_c \sum_a m_c m_a C_{ca}$	
Meissner *et al.*	$\gamma_r = \gamma_{\pm}^{1/Z_1 Z_2}$	
	$\gamma_r = \{1 + B(1 + 0.1\,I)^q - B\} \gamma^*$	0 – 100
	$B = 0.75 - 0.065\,q$	
	$\log \gamma^* = \dfrac{-0.5107 \sqrt{I}}{1 + C \sqrt{I}}$	
	$C = 1 + 0.055\,q \exp(-0.023\, I^3)$	
	Mixed Electrolyte Solution:	
	$\log \gamma_{ij} = \dfrac{Z_i}{Z_i + Z_j} (V_{i2} I_2 \log \gamma^0_{i2} + V_{i4} I_4 \log \gamma^0_{i4} + \ldots)/I + \dfrac{Z_j}{Z_i + Z_j} (V_{j1} I_1 \log \gamma^0_{ji} + V_{j3} I_3 \log \gamma^0_{j3} + \ldots)/I$	

a. See § 2.6.3 for definitions of parameters.

b. Useful in solutions of several electrolytes.

EXTENDED DEBYE-HÜCKEL METHOD

The extended Debye-Hückel approximation is given by:

$$\log \gamma_i = -AZ_i^2 \frac{\sqrt{I}}{I + Ba_i\sqrt{I}} \tag{29}$$

In addition to the parameters used in the Debye-Hückel approximation, two others are used in the extended Debye-Hückel:

B = parameter dependent on solvent dielectric constant and absolute temperature
a_i = empirical parameter representing distance of closest approach between centers of adjacent ions (Å)

The parameter B is given by:

$$B = 50.3\,(DT)^{-1/2} \tag{30}$$

where:

D = dielectric constant of solvent
T = temperature (K)

For water at 25°C, B is approximately 0.33.

The parameter a_i corresponds roughly to the radius of the hydrated ion, i, in angstroms and has a value between 3 and 9. Table 2.6-2 lists values of "a" for various ions.

The extended Debye-Hückel estimation method is applicable to dilute solutions with ionic strength less than 0.1 *M*.

Example 3 What is the activity coefficient of CO_3^{-2} in lake water with an ionic strength of 0.02 *M*?

Using the extended Debye-Hückel method:

$$\log \gamma_i = -AZ_i^2 \frac{\sqrt{I}}{1 + Ba_i\sqrt{I}}$$

For CO_3^{-2}, a = 5 (see Table 2.6-2)
For water, A = 0.5, B = 0.33

TABLE 2.6-2
Values of the Parameter "a" for Some Ions

Empirical Parameter, a (Å)	Cations	Anions
3	K^+, Ag^+, NH_4^+, Rb^+, Cs^+	OH^-, F^-, Cl^-, I^-, Sr^-, CN^-, NO_2^-, NO_3^-, ClO_3^-, ClO_4^-, MnO_4^-
4	Na^+, Hg_2^{+2}, $CdCl^+$	HCO_3^-, $H_2PO_4^-$, HSO_3^-, SO_4^{-2}, HPO_4^{-2}, SeO_4^{-2}, CrO_4^{-2}, PO_4^{-3}, $H_2AsO_4^-$, $S_2O_3^{-2}$
5	Sr^{+2}, Ba^{+2}, Ra^{+2}, Cd^{+2}, Hg^{+2}, Pb^{+2}	CO_3^{-2}, S^{-2}, $S_2O_4^{-2}$, SO_3^{-2}, $Fe(CN)_6^{-4}$, MoO_4^{-2}
6	Ca^{+2}, Zn^{+2}, Cu^{+2}, Sn^{+2}, Mn^{+2}, Fe^{+2}, Ni^{+2}, Co^{+2}, Li^{+2}	
8	Mg^{+2}, Be^{+2}	
9	H^+, Al^{+3}, Fe^{+3}, Cr^{+3}, La^{+3}, Ce^{+3}, Y^{+3}	

Source: Adapted from Stumm and Morgan [19], Pagenkopf [12] and Morel [11].

$$\log \gamma_{CO_3^{-2}} = -0.5\,(-2)^2 \frac{\sqrt{0.02}}{1 + (0.33)\,(5)\,\sqrt{0.02}}$$

$$= -0.229$$

$$\gamma_{CO_3^{-2}} = 0.59$$

GÜNTELBERG METHOD

For mixtures of several electrolytes, Güntelberg proposed that a value of 3 Å be used for the empirical parameter "a" in the extended Debye-Huckel. For water at normal temperatures, Ba ≈ 1; this simplifies equation 29 to:

$$\log \gamma_i = -AZ_i^2 \frac{\sqrt{I}}{1 + \sqrt{I}} \tag{31}$$

The Güntelberg estimation method is applicable to dilute solutions with ionic strength less than 0.1 *M*.

Example 4 What is the activity coefficient of Ca^{+2} in river water with an ionic strength of 0.01 *M*?

Using the Güntelberg method:

$$\log \gamma_{Ca^{+2}} = -AZ_i^2 \frac{\sqrt{I}}{1 + \sqrt{I}}$$

$$= -0.5\,(2)^2 \frac{\sqrt{0.01}}{1 + \sqrt{0.01}}$$

$$= -0.182$$

$$\gamma_{Ca^{+2}} = 0.66$$

DAVIES METHOD

The Davies approximation is given by:

$$\log \gamma_i = -AZ^2 \left(\frac{\sqrt{I}}{1 + \sqrt{I}} - 0.3\,I \right) \qquad (32)$$

This empirically derived method differs from that of Güntelberg by its additional term of $0.3AZ^2I$. (Davies earlier used a coefficient of 0.2 instead of 0.3.) The method is applicable to solutions of ionic strength less than 0.5 *M*.

METHOD OF PITZER *et al.*

The Pitzer method is based on the specific ion interaction model (equation 26). For a *binary* solution of a 1-1 electrolyte, MX, the reduced activity coefficient given by the Pitzer equation is [10]:

$$\log \gamma_{\pm}\,(MX) = Z_M\,Z_X\,f + m\,(\beta^0 + f^1\,\beta^1) + 1.5\,m^2\,c^{\phi} \qquad (33)$$

where:

m = concentration of the cation (or anion)
f = $-0.392\,[\sqrt{I}/(1 + 1.2\,\sqrt{I}) + (2/1.2)\log(1 + 1.2\,\sqrt{I})]$
f^1 = $(\frac{1}{2I})\,[1 - \exp(-2\,\sqrt{I})\,(1 + 2\sqrt{I} - 2I)]$
I = ionic strength (*M*)

Some values of the adjustable parameters β^0, β^1 and c^{ϕ} are given in Table 2.6-3. More comprehensive listings are given by Pitzer and Mayorga [15, 16] and Pitzer [13].

For a *mixed electrolyte solution*, the single ion activity coefficients of a particular cation, *M*, and particular anion, X, are affected by the presence of all the other cations (c) and anions (a) in solution. The relationships are given by equations 34 and 35.

$$\log \gamma_M = Z_M^2 f + 2 \sum_a m_a (B_{Ma} + EC_{Ma}) + Z_M^2 \sum_c \sum_a m_c m_a B_{ca}^1 + Z_M \sum_c \sum_a m_c m_a C_{ca} \quad (34)$$

$$\log \gamma_X = Z_X^2 f + 2 \sum_c m_c (B_{cX} + EC_{cX}) + Z_X^2 \sum_c \sum_a m_c m_a B_{ca}^1 + Z_X \sum_c \sum_a m_c m_a C_{ca} \quad (35)$$

where:

m_i = concentration of cations (indicated by subscript c) or anions (subscript a), in molal units

E = the equivalent molality, defined as $E = 1/2 \sum_i m_i |Z_i|$

The second and third virial coefficients in equations 34 and 35 are:

$$\left.\begin{aligned} B_{MX} &= \beta_{MX}^0 + (\beta_{MX}^1/2I)\,[1 - (1 + 2\sqrt{I})\exp(-2\sqrt{I})] \\ B_{MX}^1 &= (\beta_{MX}^1/2I^2)\,[-1 + (1 + 2\sqrt{I} + 2I)\exp(-2\sqrt{I})] \\ C_{MX} &= c_{MX}^{\phi}/(2\,|Z_M Z_X|^{1/2}) \end{aligned}\right\} \quad (36)$$

Additional terms may be added to account for interactions between like-charged ions and among triple ions [3, 14].

The Pitzer estimation method is applicable to solutions of high ionic strength (up to 6*M*). The deficiency of the Pitzer method lies in the general lack of parameter values for virial coefficients to calculate the activity coefficients.

METHOD OF MEISSNER *et al.*

In this method for an electrolyte that dissociates into cations of charge Z_1 and anions of charge Z_2, Kusik and Meissner [6] used the definition:

$$\gamma_r = \gamma_{\pm}^{1/Z_1 Z_2} \quad (37)$$

where:

γ_r = reduced activity coefficient

$\gamma_{\pm}$ = mean ion activity coefficient

TABLE 2.6-3

Parameters β^0, β^1, and c^Φ for Some Inorganic Compounds

1:1 Compounds	β^0	β^1	c^Φ	2:1 Compounds[a]	$4/3\ \beta^0$	$4/3\ \beta^1$	$(2^{5/2}/3)\ c^\Phi$
HCl	0.1775	0.2945	0.00080	$MgCl_2$	0.4698	2.242	0.00979
NaCl	0.0765	0.2664	0.00127	$CaBr_2$	0.5088	2.151	−0.00485
NaOH	0.0864	0.253	0.0044	$Ba(ClO_4)_2$	0.4819	2.101	−0.05894
KF	0.08089	0.2021	0.00093	$NiCl_2$	0.4639	2.108	−0.00702
KNO_3	−0.0816	0.0494	0.00660	$Cu(NO_3)_2$	0.4224	1.907	−0.04136
NH_4Br	0.0624	0.1947	−0.00436	$Zn(NO_3)_2$	0.4641	2.255	−0.02955
NH_4NO_3	−0.0154	0.1120	−0.00003	K_2HAsO_4	0.1728	2.198	−0.0336
3:1 Compounds[a]	$3/2\ \beta^0$	$3/2\ \beta^1$	$(3^{3/2}/2)\ c^\Phi$	**4:1 Compounds[a]**	$8/5\ \beta^0$	$8/5\ \beta^1$	$16/5\ c^\Phi$
$AlCl_3$	1.0490	8.767	0.0071	$ThCl_4$	1.622	21.33	−0.3309
$Cr(NO_3)_3$	1.0560	7.777	−0.1533	$K_4P_2O_7$	0.977	17.88	−0.2418
Na_3AsO_4	0.3582	5.895	−0.1240	$K_4Fe(CN)_6$	1.021	16.23	−0.5579
K_3PO_4	0.5594	5.958	−0.2255	$K_4Mo(CN)_8$	0.854	18.53	−0.3499

a. The values provided for 2:1, 3:1 and 4:1 compounds are for single electrolyte solutions, and in such cases the numeric values shown are used in Equation 33 without modification. The multiplicative factors (e.g., 3/2) must be removed before the parameter values are used for solutions of mixed electrolytes.

Source: Adapted from Pitzer and Mayorga [15]

These investigators found from experimental observations that the reduced activity coefficient of a strong electrolyte is related to the ionic strength by an empirical relationship, as shown graphically in Figure 2.6-1. With a known value of γ_r at a given ionic strength, the value at another ionic strength may be extrapolated.

Each of the curves in Figure 2.6-1 is an isotherm that can be represented by the empirical equation:

$$\gamma_r = [1 + B\,(1 + 0.1\,I)^q - B]\,\gamma^* \tag{38}$$

where:

$$B = 0.75 - 0.065\,q \tag{39}$$

$$\log \gamma^* = \frac{-0.5107\sqrt{I}}{1 + C\sqrt{I}} \tag{40}$$

$$C = 1 + 0.055\,q\,\exp\,(-0.023\,I^3) \tag{41}$$

q = a parameter whose value is dependent on the electrolyte and temperature

exp = the exponent to the base e

From the equations above, if γ_r for an electrolyte is known for a given ionic strength, the parameter q may be calculated. Using this calculated value of the parameter q, the γ_r at any other ionic strength may be calculated. Note that this method is based on molal units for all concentrations and ionic strengths.

Kusik and Meissner have developed values of q for various electrolytes (Table 2.6-4). To calculate γ_r for a given electrolyte, the value of q is substituted into equations 39, 41, and then 38 with the appropriate ionic strength.

Where no experimental values are available, γ_r may be estimated from data on vapor pressure lowering, as described in Kusik and Meissner [5]. Other methods by Meissner and his co-workers may also be used if there are no data on vapor pressure lowering [9].

Kusik and Meissner developed relationships to calculate γ_r at various temperatures; the reader is referred to their paper [6] for details.

An equation was also developed by Kusik and Meissner [6] to estimate the activity coefficient in a multi-component system from activity coefficients (as calculated above) for single-electrolyte aqueous solutions. This relationship is:

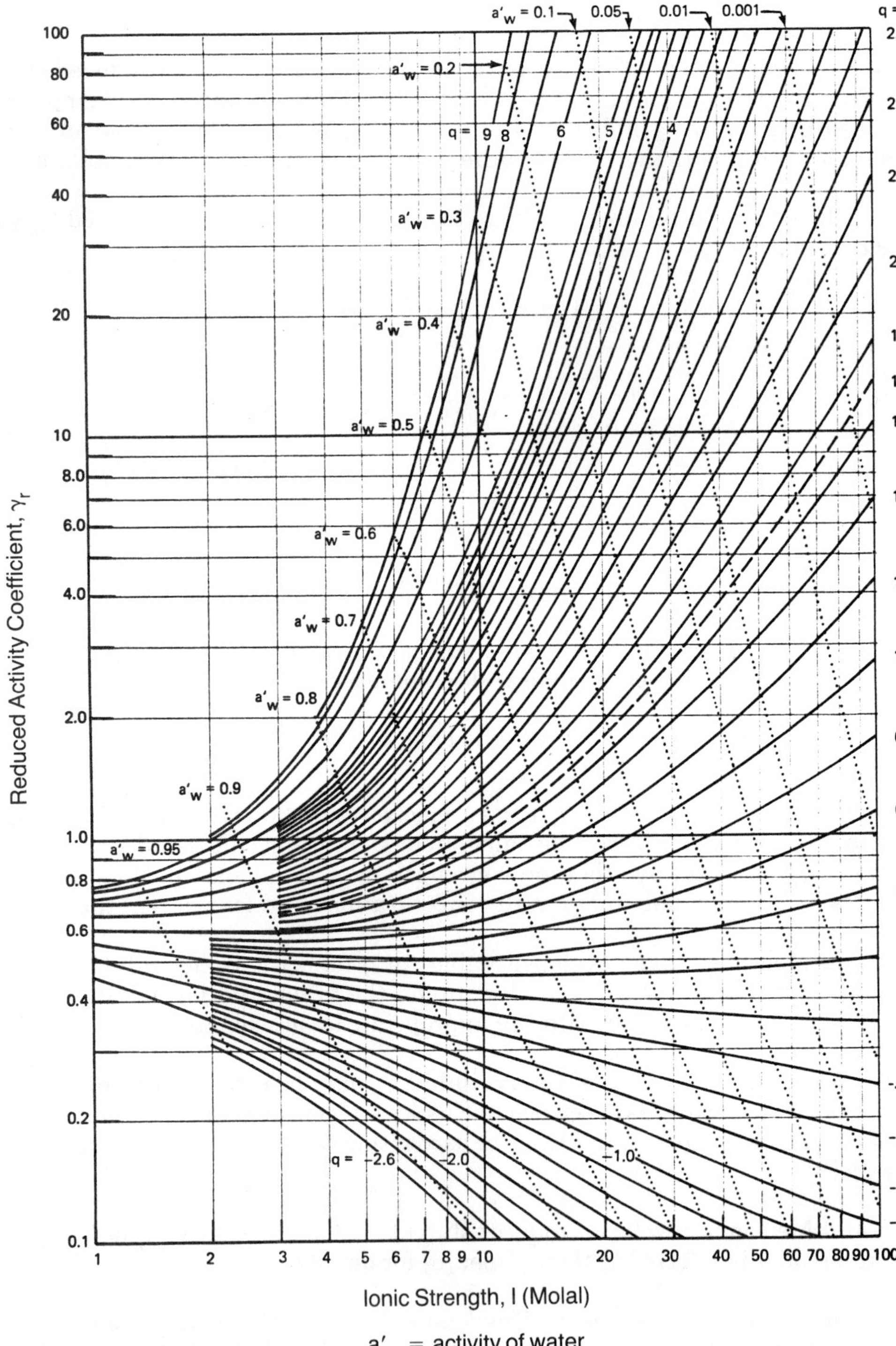

Source: Kusik and Meissner [6]. *(Copyright 1978, American Institute of Chemical Emgineers. Reprinted with permission.)*

FIGURE 2.6-1 Relationship of Reduced Activity Coefficient to Ionic Strength

TABLE 2.6-4
Average Values of the Parameter "q" for Selected Electrolytes

1:1 Electrolyte	q	I	1:1 Electrolyte	q	I	Higher Electrolyte	q	I	Higher Electrolyte	q	I
$AgNO_3$	−2.55	*b*	LiAC	2.81	*a*	$AlCl_3$	1.92	*d*	$Mg(ClO_4)_2$	4.15	*b*
CsAC	5.59	*a*	LiBr	7.27	*b*	$Al_2(SO_4)_3$	0.36	*d*	MgI_2	4.04	*b*
CsBr	−0.06	*b*	LiCl	5.62	*b*	$BaBr_2$	1.92	*b*	$Mg(NO_3)_2$	2.32	*d*
CsCl	0.16	*b*	LiOH	−0.08	*a*	$BaCl_2$	1.48	*b*	$MgSO_4$	0.15	*c*
CsI	−0.41	*a*	$LiNO_3$	3.80	*b*	$Ba(ClO_4)_2$	1.90	*c*	$MnSO_4$	0.14	*b*
CsOH	7.34	*u*	LiTOL	0.84	*b*	BaI_2	2.84	*b*	$MnCl_2$	1.60	*d*
$CsNO_3$	−2.62	*u*	NaAC	4.20	*a*	$Ba(NO_3)_2$	−0.52	*c*	Na_2CrO_4	0.41	*c*
HCl	6.69	*b*	NaBr	2.98	*a*	$CaCl_2$	2.40	*d*	Na_2FUM	0.88	*b*
HNO_3	3.66	*a*	$NaBrO_3$	−0.68	*u*	CaI_2	3.27	*b*	Na_2MAL	0.12	*b*
KAC	5.05	*a*	NaCl	2.23	*a*	$Ca(NO_3)_2$	0.93	*d*	$Na_2S_2O_3$	0.18	*c*
KBr	1.15	*b*	$NaClO_3$	0.41	*a*	$Cd(NO_3)_2$	1.53	*b*	Na_2SO_4	−0.19	*c*
$KBrO_3$	−2.00	*u*	$NaClO_4$	1.30	*b*	$CdSO_4$	0.016	*b*	$NdCl_3$	1.42	*c*
KCl	0.92	*b*	NaCNS	2.94	*a*	$CeCl_3$	1.41	*c*	$(NH_4)_2SO_4$	−0.25	*c*
$KClO_3$	−1.70	*u*	NaF	0.37	*u*	$CoBr_2$	3.08	*d*	$NiCl_2$	2.33	*d*
KCNS	0.61	*b*	NaFORM	1.83	*a*	$CoCl_2$	2.25	*b*	$NiSO_4$	0.025	*c*
KF	2.13	*a*	NaHMAL	0.01	*b*	CoI_2	3.87	*d*	$Pb(ClO_4)_2$	2.25	*d*
KHMAL	−0.72	*b*	NaHSUC	0.60	*b*	$Co(NO_3)_2$	2.08	*d*	$Pb(NO_3)_2$	−0.97	*b*
KH_2PO_4	−2.54	*a*	NaH_2PO_4	−1.59	*a*	$CrCl_3$	1.72	*b*	$PrCl_3$	1.40	*c*
KHSUC	0.02	*b*	NaI	4.06	*a*	$Cr(NO_3)_3$	1.51	*c*	Rb_2SO_4	0.007	*b*
KI	1.62	*b*	$NaNO_3$	−0.39	*b*	$Cr_2(SO_4)_3$	0.43	*d*	$ScCl_3$	1.68	*c*
KNO_3	−2.33	*a*	NaOH	3.00	*a*	Cs_2SO_4	0.16	*b*	$SmCl_3$	1.47	*c*
KOH	4.77	*b*	NaPROP	5.54	*a*	$CuCl_2$	1.40	*b*	$SrBr_2$	2.34	*b*
KTOL	−1.75	*a*	NaTOL	−0.80	*a*	$Cu(NO_3)_2$	1.83	*d*	$SrCl_2$	1.95	*c*
			NH_4Cl	0.82	*b*	$CuSO_4$	0.00	*c*	$Sr(ClO_4)_2$	2.84	*d*
			NH_4NO_3	−1.15	*b*	$EuCl_3$	1.49	*c*	SrI_2	3.03	*b*
			RbAC	5.39	*a*	$FeCl_2$	2.16	*b*	$Sr(NO_3)_2$	0.30	*c*
			RbBr	0.46	*b*	K_2CrO_4	0.16	*c*	UO_2Cl_2	2.40	*b*
			RbCl	0.62	*b*	K_2SO_4	−0.25	*u*	$UO_2(ClO_4)_2$	5.64	*d*
			RbI	0.45	*b*	$LaCl_3$	1.41	*c*	$UO_2(NO_3)_2$	2.90	*b*
			$RbNO_3$	−2.49	*b*	Li_2SO_4	0.57	*c*	UO_2SO_4	0.066	*b*
			TlAC	−0.73	*b*	$MgAC_2$	0.83	*c*	YCl_3	1.55	*c*
						$MgBr_2$	3.50	*d*	$Zn(ClO_4)_2$	4.30	*a*
						$MgCl_2$	2.90	*d*	$Zn(NO_3)_2$	2.28	*d*
									$ZnSO_4$	0.05	*c*

Legend:

AC = Acetate
FORM = Formate
FUM = Fumarate
MAL = Maleate

PROP = Propionate
SUC = Succinate
TOL = p-toluene sulphonate

a = 3-4 Molal
b = 4.5-6 Molal
c = 9 Molal
d = 15 Molal
u = limited data

Source: Kusik and Meissner [6]

$$\log \gamma_{ij} = \frac{Z_i}{Z_i + Z_j} (V_{i2}\, I_2 \log \gamma^0_{i2} + V_{i4}\, I_4 \log \gamma^0_{i4} + ...)/I$$

$$+ \frac{Z_j}{Z_i + Z_j} (V_{j1}\, I_1 \log \gamma^0_{j1} + V_{j3}\, I_3 \log \gamma^0_{j3} + ...)/I \qquad (42)$$

where:

ij denotes the electrolyte consisting of cation i and anion j

Z_i = absolute charge on cation i

Z_j = absolute charge on anion j

γ_{ij} = reduced activity coefficient of ij in multi-component solution

γ^0 = reduced activity coefficient in single-electrolyte solution

Even subscripts denote anions in the multi-component system

Odd subscripts denote cations in the multi-component system

V_{xy} = $0.5\,(Z_x + Z_y)^2/(Z_x Z_y)$

I_x = ionic strength of individual ion x

= 0.5 (conc. of ion x) $(Z_x)^2$, molal

I = total ionic strength in solution, molal

Example 5 Calculate the mean ion activity coefficient, $\gamma_{\pm}$, of $FeCl_2$ in an aqueous solution of ionic strength 5 molal.

From Table 2.6-4, the value of q for $FeCl_2$ is 2.16. Substituting this value in equation 39,

$$B = 0.75 - 0.065\,q$$
$$= 0.75 - 0.065\,(2.16) = 0.6096$$

From equation 41,

$$C = 1 + 0.055\,q\,\exp(-0.023\,I^3)$$
$$= 1 + 0.055\,(2.16)\,\exp(-0.023 \times 5^3)$$
$$= 1.007$$

The values of C and I are next substituted in equation 40:

$$\log \gamma^* = -\frac{0.5107\sqrt{I}}{1 + C\sqrt{I}}$$

$$= -\frac{0.5107\sqrt{5}}{1 + 1.007\,(\sqrt{5})}$$

$$= -0.351$$

$$\gamma^* = 0.445$$

The reduced activity coefficient is now calculated by equation 38:

$$\gamma_r = [1 + B\,(1 + 0.1\,I)^q - B]\,\gamma^*$$
$$= [1 + 0.6096\,(1 + 0.1 \times 5)^{2.16} - 0.6096]\,0.445$$
$$= 0.825$$

According to equation 37,

$$\gamma_r = \gamma_\pm^{1/Z_1Z_2} = \gamma_\pm^{1/2}$$

Therefore,

$$\gamma_\pm = 0.681$$

The Meissner *et al.* method is applicable to solutions of high ionic strength from above 1 *M* to very high ionic strengths (up to 100 molal).

Compared with that of Pitzer *et al.*, the method of Meissner *et al.* is superior for estimating activity coefficients in solutions of very high ionic strength (Table 2.6-5); it is also easier to use, and the necessary data are more readily available.

TABLE 2.6-5

Comparison of Pitzer *et al.* and Meissner *et al.* Methods for High-Ionic-Strength Solutions

	Pitzer et al.	Meissner et al.
Applicability		
Ionic strength 1-6	Yes	Yes
Ionic strength 6-20	Limited	Verified for a large number of systems
Ionic strength 20-100	No	Yes
Temperature effect	No	Yes (see ref. 6)
Database for coefficients	Limited; requires extensive effort to enlarge	Limited, but easily expanded to required components
Complexity of calculation and number of parameters required (particularly with increasing number of ions in multi-component system)	Relatively high (complexity and number of parameters increase exponentially)	Relatively low (complexity and number of parameters increase linearly)

OTHER METHODS

Helgeson [8] proposed a method found to be suitable for calculating activity coefficients for ions present in low concentrations in multi-component electrolyte solutions that contain sodium chloride as the dominant component:

$$\log \gamma_i = \frac{AZ_i^2 \sqrt{I}}{1 + Ba_i \sqrt{I}} + B^0 I \qquad (43)$$

The value of B^0 is 0.041 at 25°C. The Helgeson equation can be used for trace ions in sodium chloride solutions of ionic strength up to 3 *M*. It is appropriate for use with saline soil solutions and brines.

Additional methods for estimating activity coefficients, which are not discussed in detail here, include the following:

- Sun, Harriss and Magnuson [20] developed a modified version of the Debye-Hückel method in the form of equation 25.
- Van Luik and Jurinak [21] describe other methods by Reilly, Wood and Robinson [17] and by Glueckauf [1].

NEUTRAL SPECIES

For neutral species in solution, such as gaseous oxygen, activity coefficients can be estimated by the following empirical equation [18]:

$$\log \gamma_i = k_s I \qquad (44)$$

where:

k_s = empirically determined coefficient ranging from 0.01 to 0.15
I = ionic strength (*M*)

The activity coefficient of non-ionic species is approximately 1 for *M*<1. The coefficient k_s is normally around 0.1. Equation 44 holds for solutions of ionic strength up to 5 *M* and solute concentrations in the range of 0.1 *M* to 0.5 *M* [12]. Methods for estimating activity coefficients for neutral organic species are discussed by Grain [2].

2.6.4 Literature Cited

1. Glueckauf, E., "Electrostatic Interactions in Electrolyte Solutions," *Proc. R. Soc. London, Ser. A,* **310**, 449-62 (1969), as cited by Van Luik and Jurinak [21].
2. Grain, C.F., "Activity Coefficient," Chap. 11 in *Handbook of Chemical Property Estimation Methods,* W.J. Lyman, W.F. Reehl and D.H. Rosenblatt (eds.), McGraw-Hill Book Co., New York (1982).

3. Harvie, C.E. and J.H. Weare, "The Prediction of Mineral Solubilities in Natural Waters: The Na-K-Mg-Cl-SO_4-H_2O System from Zero to High Concentrations at 25°C," *Geochim. Cosmochim. Acta,* **44**, 981-97 (1980).

4. Horne, R.A., *The Chemistry of Our Environment,* John Wiley & Sons, New York (1978).

5. Kusik, C.L. and H.P. Meissner, "Vapor Pressures of Water over Aqueous Solutions of Strong Electrolytes," *Ind. Eng. Chem. Process Des. Dev.,* **12**, 112-15 (1973).

6. Kusik, C.L. and H.P. Meissner, "Electrolyte Activity Coefficients in Inorganic Processing," in *Fundamental Aspects of Hydrometallurgical Processes,* T.W. Chapman, L.L. Tavlarides, G.L. Hubred, and R.M. Wellek (eds.), Symposium Series 173, **74**, American Institute of Chemical Engineers, New York, 14-20 (1978).

7. Martell, A.E. and R.M. Smith, *Inorganic Complexes,* Vol. 4 of *Critical Stability Constants,* Plenum Press, New York (1976).

8. Mattigod, S.V. and G. Sposito, "Chemical Modeling of Trace Metal Equilibria in Contaminated Soil Solutions Using the Computer Program GEOCHEM," in *Chemical Modeling in Aqueous Systems,* E.A. Jenne (ed.), ACS Symposium Series 93, American Chemical Society, Washington, D.C., 837-56 (1979).

9. Meissner, H.P. and J.W. Tester, "Activity Coefficients of Strong Electrolytes in Aqueous Solutions," *Ind. Eng. Chem. Process Des. Dev.,* **11**, 128-33 (1972).

10. Millero, F.J., "The Activity of Metal Ions at High Ionic Strengths," in *Complexation of Trace Metals in Natural Waters,* C.J.M. Kramer and J.C. Duinker (eds.), Proceedings of the International Symposium, May 2-6, 1983, Texel, The Netherlands, Martinus Nijhoff/Dr. W. Junk Publishers, The Hague, 187-200 (1984).

11. Morel, F.M.M., *Principles of Aquatic Chemistry,* John Wiley & Sons, New York (1983).

12. Pagenkopf, G.K., *Introduction to Natural Water Chemistry,* Marcel Dekker, New York (1978).

13. Pitzer, K.S., "Ion Interaction Approach," in *Activity Coefficients in Electrolyte Solutions,* R.M. Pytkowicz (ed.), CRC Press, Boca Raton, Fla., 157-208 (1979).

14. Pitzer, K.S. and J.J. Kim, "Thermodynamics of Electrolytes. IV: Activity and Osmotic Coefficients for Mixed Electrolytes," *J. Am. Chem. Soc.,* **96**, 5701-707 (1974).

15. Pitzer, K.S. and G. Mayorga, "Thermodynamics of Electrolytes. II: Activity and Osmotic Coefficients for Strong Electrolytes with One or Both Ions Univalent," *J. Phys. Chem.,* **77**, 2300-308 (1973).

16. Pitzer, K.S. and G. Mayorga, "Thermodynamics of Electrolytes. III: Activity and Osmotic Coefficients for 2:2 Electrolytes," *J. Solution Chem.,* **3**, 539-46 (1974).

17. Reilly, P.J., R.H. Wood and R.A. Robinson, "The Prediction of Osmotic and Activity Coefficients in Mixed-Electrolyte Solutions," *J. Phys. Chem.,* **75**, 1305-315 (1971).

18. Snoeyink, V.L. and D. Jenkins, *Water Chemistry,* John Wiley & Sons, New York (1980).

19. Stumm, W. and J.J. Morgan, *Aquatic Chemistry*, 2nd ed., John Wiley & Sons, New York (1981).

20. Sun, M.S., D.K. Harriss and V.R. Magnuson, "Activity Corrections for Ionic Equilibria in Aqueous Solutions," *Can. J. Chem.*, **58**, 1253-57 (1980).

21. Van Luik, A.E. and J.J. Jurinak, "Equilibrium Chemistry of Heavy Metals in Concentrated Electrolyte Solution," in *Chemical Modeling in Aqueous Systems,* E.A. Jenne (ed.), ACS Symposium Series 93, American Chemical Society, Washington, D.C., 683-710 (1979).

2.7 ACID/BASE EQUILIBRIA

2.7.1 Introduction

This section addresses acid/base equilibria of inorganic substances in aqueous systems, focusing on the acid dissociation constant as the functional parameter that describes the extent of these types of equilibria. The following discussion defines the acid dissociation constant and explains how it is used, where its value can be obtained, how much it varies under different conditions, and the practicality of using an estimate. The use of the constant to calculate relative concentrations of various hydrolyzed species is also described and illustrated.

2.7.2 Description of Property

According to the definition proposed by Brønsted [3], an acid is a species that can donate a proton. Therefore, to behave as an acid, a species must be associated with a hydrogen atom that it can release to form hydronium ions (H^+_{aq}) in the aqueous solvent. Two common representations of inorganic acids are H-X, where the hydrogen atom is bonded to a single atom (e.g., HCl, H_2S), and M-X-H, where it is bonded to an atom that is, in turn, bonded to others — e.g., $[(H_2O)_5Fe\text{-}O\text{-}H]^{+2}$ (with H bonded to the O) and >As-O-H.

Species like >As-O-H are generally referred to as oxy acids.

In aqueous environmental systems, acid dissociation occurs when H_2O (acting as a base) reacts with the hydrogen ion (proton) in an acid. This results in a species that is often referred to as H^+ but in reality is a hydrated ion containing several water molecules: H_3O^+, $H_5O_2^+$, $H_9O_4^+$, etc., sometimes referred to as the hydronium ion. For this reason, acid dissociation reactions are commonly written as in equation 1, where K_a is the acid dissociation equilibrium constant.

$$\text{M-X-H} + H_2O \rightleftharpoons \text{M-X}^- + H_3O^+ \qquad K_a \tag{1}$$

In inorganic reaction mechanism chemistry, the water reactant is commonly omitted and protonated water molecules are denoted simply as H^+. The acid dissociation of hydrogen iodide, for example, is written as:

$$HI_{aq} \rightleftharpoons H^+_{aq} + I^-_{aq} \tag{2}$$

The subscript "aq" refers to the aqueous hydration sphere surrounding each ion in the water solvent; several H_2O molecules may be loosely associated with the central ion in this sphere as for example with I^- in the case of equation 2. It is often convenient to drop even the "aq" notation, which is understood to be present when reactions in aqueous systems are discussed. We have adopted this convention in the text that follows.

The extent of acid dissociation can be represented mathematically in terms of the activities of products and reactants of the dissociation reaction. The acid dissociation constant, K_a, for a reaction of the kind represented by equation 1 is expressed as

$$K_a = \frac{(a_{MX^-})(a_{H^+})}{(a_{MXH})} \tag{3}$$

where "a" is the activity of the component named in the subscript. Since the activity of water is essentially constant in aqueous systems, it is incorporated in the value of K_a. Activities are related to the molar concentrations of the components as follows:

$$a_i = \gamma_i[C_i] \tag{4}$$

where γ_i is the activity coefficient of component i, and $[C_i]$ is its molar concentration. (See Section 2.6 for methods of calculating activity coefficients.) Thus, equation 3 can be rewritten as

$$K_a = \frac{(\gamma_{MX^-})(\gamma_{H^+})}{(\gamma_{MXH})} \frac{[MX^-][H^+]}{[MXH]} \tag{5}$$

Dissociation constants are often quoted in the literature at an ionic strength of zero, where the γ values are 1. These data are obtained either from measurements at several ionic strengths and subsequent extrapolation to zero ionic strength or by calculation of the γ values and correction of the value of K_a measured at ionic strengths greater than zero. When K_a values are given, they are often "conditional" constants defined as:

$$K_a = \frac{a_{H^+}[MX^-]}{[MXH]} \tag{6}$$

This definition arises from the experimental procedures used to determine these constants. The activity of H^+ can be directly measured (e.g., by a pH electrode), but the activities of the other components may not be known, although their concentrations can be determined. It is highly desirable that one know how a particular constant was derived before using it in calculations.

The behavior in aqueous systems of many inorganic species of environmental concern (e.g., trace metals) under a variety of pH conditions is affected by the hydroxo ($-OH$) or aquo ($-OH_2$) ligands coordinated to the central metal. In general, two or more of these ligands are bonded to the central metal; this leads to successive acid dissociation processes, each with a different dissociation constant. For example, phosphoric acid (H_3PO_4), which has three hydroxy (OH) and one oxy (O) bonded to the phosphorus (V) center, and structurally is better represented as $PO(OH)_3$ undergoes the following acid dissociation reactions:

$$PO(OH)_3 \rightleftharpoons PO_2(OH)_2^- + H^+ \qquad K_{a1} \tag{7}$$

$$PO_2(OH)_2^- \rightleftharpoons PO_3(OH)^{-2} + H^+ \qquad K_{a2} \tag{8}$$

$$PO_3(OH)^{-2} \rightleftharpoons PO_4^{-3} + H^+ \qquad K_{a3} \tag{9}$$

For a trace metal ion in an aqueous environment, successive acid dissociation reactions of this kind can give rise to aquo-hydroxo species. For example, for iron (III). which is hexacoordinated with water molecules, the first three (stepwise) acid dissociation reactions are:

$$Fe(OH_2)_6^{+3} \rightleftharpoons Fe(OH_2)_5(OH)^{+2} + H^+ \qquad K_1 \qquad (10)$$

$$Fe(OH_2)_5OH^{+2} \rightleftharpoons Fe(OH_2)_4(OH)_2^+ + H^+ \qquad K_2 \qquad (11)$$

$$Fe(OH_2)_4(OH)_2^+ \rightleftharpoons Fe(OH_2)_3(OH)_3^0 + H^+ \qquad K_3 \qquad (12)$$

Acid dissociation data in the literature are commonly in the form of either K_a values or of overall (cumulative) reaction equilibrium constants (β) that are products of individual K_a values [2,17]. For example, for the aqueous iron (III) ion, the relationship of β values to the acid dissociation values is as follows:

$$Fe(OH_2)_6^{+3} \rightleftharpoons Fe(OH_2)_5OH^{+2} + H^+ \qquad \beta_1 = K_1 \qquad (13)$$

$$Fe(OH_2)_6^{+3} \rightleftharpoons Fe(OH_2)_4(OH)_2^+ + 2H^+ \qquad \beta_2 = K_1K_2 \qquad (14)$$

$$Fe(OH_2)_6^{+3} \rightleftharpoons Fe(OH_2)_3(OH)_3 + 3H^+ \qquad \beta_3 = K_1K_2K_3 \qquad (15)$$

Data are also provided in the literature in the form of base association constants, K_b:

$$A^- + H_2O \rightleftharpoons HA + OH^- \qquad K_b \qquad (16)$$

where A^- is the conjugate base of HA, the acid. The relationship between K_a and K_b is:

$$K_a = K_w/K_b \qquad (17)$$

where K_w is the dissociation constant of water ($10^{-14.0}$ at 25°C). Because K_a values tend to be very small numbers, they are most conveniently expressed in logarithmic form, $pK_a = -\log K_a$.

2.7.3 Environmental Importance

From an environmental standpoint, the acidic and basic behavior of inorganic species in solution is important for two reasons. First, the acidic (or basic) nature of major inorganic species affects the properties and pH of the solution. Second, the solution pH can play a major role in the speciation of the dissolved trace inorganic species. Modification of the solution pH (which, in turn, affects speciation of other components) is the more important of these effects when the substance is strongly acidic or basic or is relatively concentrated; the matrix impact on speciation is of greater importance when the substance is weakly acidic or basic and at low concentrations, as is the case with many trace metals. In comparison with other aqueous environmental inorganic reactions, acid dissociation reactions are usually considered to be instantaneous.

Since acid dissociation involves loss of one or more protons, the overall charge of the metal ion complex is correspndingly reduced. The dissociated ion (conjugate base) differs from the undissociated acid with regard to environmentally important properties such as solubility, soil attenuation, biological uptake, complexation, and kinetics of redox reactions. The formation of hydroxy compounds through acid dissociation is the first step in precipitation of many species from solution. The precipitates formed are often either hydroxide, oxyhydroxide, or polymeric hydroxo and oxo species.

Thus, assessment of environmental impact and mobility requires a knowledge of the speciation of the elements of interest. As noted, acid dissociation often plays a major role in this process, determining whether the prevailing ion is AsO_4^{-3} or $HAsO_4^{-2}$, for example. It affects the speciation of an element even in a system that contains no species (e.g., potential ligands) other than water that can react with the element. Accordingly, environmentalists must consider the extent of acid dissociation when dealing with most dissolved species that act as acids; these include dissolved gases (e.g., CO_2, H_2S), trace metals (e.g., including transition metals) and other pollutants of environmental concern (e.g., compounds of As or Se).

The behavior of undissociated versus dissociated metal ions (i.e., M^{+x} vs. $M(OH)_y^{x-y}$) with regard to the extent of sorption, solubility, redox reactivity, competitive complexation, and uptake reactions is described in other sections of this report.

It should be noted that in environmental systems pH controls can be dominated by the presence of solid phases (i.e., in heterogeneous systems) or by the formulation of solid precipitates. For example, calcite ($CaCO_3$) can act as a base (i.e., react with H^+) to consume acidity while jarosite ($KFe_3(SO_4)_2(OH)_6$) or alunite ($K_3Al3(SO_4)_2(OH)_6$) can precipitate from acidic soil pore waters when the levels of the respective ions (e.g., K^+, Fe^{+2}, SO_4^{-2} and Al^{+3}) rise as H_2SO_4 is released during oxidation of sulfide in the soil. The presence of those solids can thus provide a long term pH control in heterogeneous environmental systems [10].

2.7.4 Mathematical and Graphical Representation

In a situation involving several interrelated equilibria, such as where several hydroxy complexes of an ion are formed by successive acid dissociation reactions, a graphical representation is useful. A graph can show the relative abundance of the various species as a function of some variable (in this case, pH) that affects which species are formed. The relative abundance of a particular hydroxy species or unprotonated form of the acid can be calculated from the acid dissociation constants and a mathematical representation of all the equilibria present in the aqueous system. In the case where dimers or polymers can be formed total concentration of the analyte is also needed.

In homogeneous reaction solution media free of ligands that might participate in further complexation equilibria or redox reactions, without gas/liquid equilibria, without precipitation/dissolution equilibria and without liquid/solid equilibria, the acid dissociation reactions (summarized in Table 2.7-1) govern the speciation of a dissociable species. While the other equilibria are obviously important and can determine concentrations and speciation in many cases, they can sometimes be ignored; for example, if the volatility of dissolved components in the system is very low, gas/liquid equilibria may be neglected. Similarly, if concentration levels in solution are much lower than the solubility limits, it may be unnecessary to consider liquid/solid equilibria, and the presence of very weakly complexing ions (relative to that of OH^- ligand) may allow representation of the system by the equations given in the table.

TABLE 2.7-1

Equations Used to Calculate f_i, the Ratio of Concentrations of a Particular Species to the Total Present in Acid/Base Equilibria[a]

Acid System	Equations
$HL \rightleftharpoons H^+ + L^-$, K_1	$f_L = \left[\frac{(H)}{K_1} + 1\right]^{-1}$
	$f_{HL} = \left[1 + \frac{K_1}{(H)}\right]^{-1}$
$H_2L \rightleftharpoons HL^- + H^+$, K_1	$f_L = \left[\frac{(H)^2}{K_1 K_2} + \frac{(H)}{K_2} + 1\right]^{-1}$
$HL^- \rightleftharpoons L^{-2} + H^+$, K_2	$f_{HL} = \left[\frac{(H)}{K_1} + 1 + \frac{K_2}{(H)}\right]^{-1}$
	$f_{H_2L} = \left[1 + \frac{K_1 K_2}{(H)^2} + \frac{K_1}{(H)}\right]^{-1}$
$H_3L \rightleftharpoons H_2L^- + H^+$, K_1	$f_L = \left[\frac{(H)^3}{K_1K_2K_3} + \frac{(H)^2}{K_2K_3} + \frac{(H)}{K_3}\right]^{-1}$
$H_2L^- \rightleftharpoons HL^{-2} + H^+$, K_2	$f_{HL} = \left[\frac{(H)^2}{K_1K_2} + \frac{(H)}{K^2} + 1 + \frac{K_3}{(H)}\right]^{-1}$
$HL^{-2} \rightleftharpoons L^{-3} + H^+$, K_3	$f_{H2L} = \left[\frac{H}{K_1} + 1 + \frac{K_2}{(H)} + \frac{K_2K_3}{(H)^2}\right]^{-1}$
	$f_{H3L} = \left[1 + \frac{K_1}{(H)} + \frac{K_1K_2}{(H)^2} + \frac{K_1K_2K_3}{(H)^3}\right]^{-1}$

(Continued)

TABLE 2.7-1 (Continued)

Acid System	Equations
$H_4L \rightleftharpoons H_3L^- + H^+,\ K_1$	$f_L = \left[\dfrac{(H)^4}{K_1K_2K_3K_4} + \dfrac{(H)^3}{K_2K_3K_4} + \dfrac{(H)^2}{K_3K_4} + \dfrac{(H)}{K_4} + 1\right]^{-1}$
$H_3L^- \rightleftharpoons H_2L^{-2} + H^+,\ K_2$	$f_{HL} = \left[\dfrac{(H)^3}{K_1K_2K_3} + \dfrac{(H)^2}{K_2K_3} + \dfrac{(H)}{K_3} + 1 + \dfrac{K_4}{(H)}\right]^{-1}$
$H_2L^{-2} \rightleftharpoons HL^{-3} + H^+,\ K_3$	$f_{H2L} = \left[\dfrac{(H)^2}{K_1K_2} + \dfrac{(H)}{K_2} + 1 + \dfrac{K_3}{(H)} + \dfrac{K_3K_4}{(H)^2}\right]^{-1}$
$HL^{-3} \rightleftharpoons L^{-4} + H^+,\ K_4$	$f_{H3L} = \left[\dfrac{(H)}{K_1} + 1 + \dfrac{K_2}{(H)} + \dfrac{K_2K_3}{(H)^2} + \dfrac{K_2K_3K_4}{(H)^3}\right]^{-1}$
	$f_{H4L} = \left[1 + \dfrac{K_1}{(H)} + \dfrac{K_1K_2}{(H)^2} + \dfrac{K_1K_2K_3}{(H)^3} + \dfrac{K_1K_2K_3K_4}{(H)^4}\right]^{-1}$

a. f_i = Fraction of total concentration of acid present in the *ith* form. (Total concentration = sum of concentrations of all species containing central atom(s) of concern.)

(H) = Activity of hydrogen ion

K_i = Acid dissociation constant

Activity coefficient for all species assumed to be 1.

It is often difficult to determine which of the various equilibria predominate in determining the speciation in a complex chemical system. The effect of equilibria other than acid/base equilibria on speciation is discussed in the sections of this report that deal with those processes. Computer models help in understanding which of the various equilibria are predominant.

These simple systems are generally represented by plots of fraction (f) or percent ($100 \times f$) of the ion present in the various forms (e.g., H_2L, HL^-, L^{-2}) as a function of pH. An example of this plot is given in Figure 2.7-1 for Zn^{+2} in solution, under the assumption that this ion is a tetraprotic acid [4]. Such plots can be generated from the equations in Table 2.7-1, which were derived from the definitions of each of the K_i values (e.g., $K_2 = [L^{-2}][H^+]/[HL]$ for the diprotic acid system) and an assumption that the total concentration of the various species is a constant. The fraction is then defined; for H_3L, for example, it is $f_{H_3L} = [H_3L]/[\text{Acid Total}]$. Examination of the table shows that the f values are a function of (H^+) and the respective acid dissociation constants. If we can ignore the influence of other equilibria, these plots are valid at the particular temperature and ionic strength at which the (conditional)

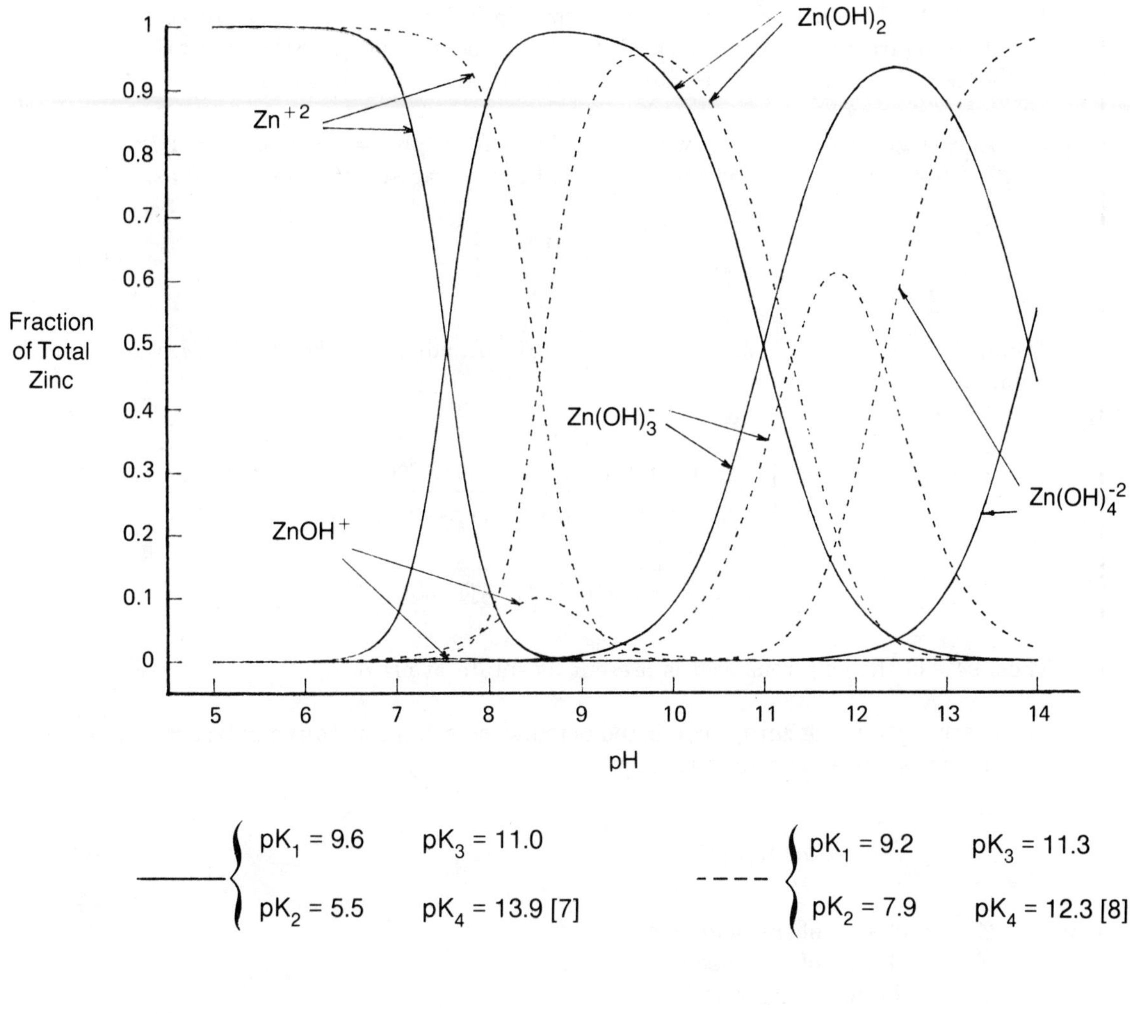

—— { pK_1 = 9.6 pK_3 = 11.0; pK_2 = 5.5 pK_4 = 13.9 [7] }

- - - - { pK_1 = 9.2 pK_3 = 11.3; pK_2 = 7.9 pK_4 = 12.3 [8] }

Source: Calculated by Arthur D. Little, Inc.

FIGURE 2.7-1 Effect of pK Values on Calculated Distribution of Zn(II) Species with pH

dissociation constants were obtained. It is somewhat unusual for a higher order pK_a (e.g., pK_1) to be less than a lower order pK_a (e.g., pK_2); yet, such is true for Zn^{+2}. This is why the concentrations of $Zn(OH)^+$ are so low. In addition, the values of the pK_a's have a marked impact on the relative abundance of the individual hydroxy species calculated (see Figure 2.7-1).

Example 1 To illustrate how the equations in Table 2.7-1 are used, we shall calculate the fraction of sulfurous acid (H_2SO_3) present as sulfite (SO_3^{-2}) at pH 8 and ionic strength of zero. It is assumed that no other equilibria apply.

Sulfurous acid is a diprotic acid with $K_1 = 1.58 \times 10^{-2}$ ($pK_1 = 1.8$) and $K_2 = 1.58 \times 10^{-7}$ ($pK_2 = 6.8$) [3]. The fraction of acid present as sulfite is expressed as:

$$f_{SO_3^{-2}} = \left[\frac{(H^+)^2}{K_1 K_2} + \frac{(H^+)}{K_2} + 1 \right]^{-1} \tag{18}$$

Using $(H^+) = 1 \times 10^{-8}\ M$, $K_1 = 1.58 \times 10^{-2} M$ and $K_2 = 1.58 \times 10^{-7} M$, we obtain

$$f_{SO_3^{-2}} = \left[\frac{(1\times10^{-8})\,(1\times10^{-8})}{(1.58\times10^{-2})\,(1.58\times10^{-7})} + \frac{(1\times10^{-8})}{(1.58\times10^{-7})} + 1 \right]^{-1}$$

$$= [4.00\times10^{-8} + 6.33\times10^{-2} + 1]^{-1} = [1.063]^{-1} = 0.94$$

Thus, 94% of the sulfurous acid is present as sulfite at pH 8.

If the ionic strength is not zero, a corrected acid dissociation constant can be estimated by an equation such as the following:

$$pK' = -\log K' = pK + \frac{0.5\,(Z_A^2 - Z_B^2)\,\sqrt{I}}{1 + \sqrt{I}} \tag{19}$$

where
Z_A = charge of the acid form
Z_B = charge of the base form
I = ionic strength *(M)*
K = acidity constant at ionic strength of zero
K' = corrected acidity constant at ionic strength I.

This equation utilizes the Güntelberg approximation for calculation of ion activity coefficients [19]. Other equations can be generated using the variety of approaches available for calculation of activity coefficients as discussed in section 2.6.

Equation 19 is useful for calculating ionic strength corrections up to 0.1 *M*. Other correction factors may be used if the activity is calculated by a different method, such as Debye-Hückel, Davies or extended Debye-Hückel. (See Section 2.6.)

The magnitude of the correction term in equation 19 ranges from 0.005 at an ionic strength of 10^{-4} *M* for a 1,0 electrolyte (e.g., $H_2S \rightleftharpoons HS^- + H^+$) to 0.60 at an ionic strength of 0.1 *M* for a 2,3 electrolyte (e.g., $HPO_4^{-2} \rightleftharpoons PO_4^{-3} + H^+$).[1] Use of the corrected pK′ value permits calculations using concentrations for all species except H^+, which is incorporated as activity, (H^+) [19].

Example 2 The use of equation 19 will be illustrated by calculating the effect of a 0.1 *M* ionic strength on the pK values of arsenic acid (H_3AsO_4) obtained at an ionic strength of 10^{-3} *M*.

The acid dissociation reactions and constants for H_3AsO_4 at an ionic strength of zero [17] are:

$$H_3AsO_4 \rightleftharpoons H_2AsO_4^- + H^+ \qquad pK_1 = 2.19 \tag{20}$$

$$H_2AsO_4^- \rightleftharpoons HAsO_4^{-2} + H^+ \qquad pK_2 = 6.89 \tag{21}$$

$$HAsO_4^{-2} \rightleftharpoons AsO_4^{-3} + H^+ \qquad pK_3 = 11.49 \tag{22}$$

The values of Z_A and Z_B for these reactions are:

Eq.	Z_A	Z_B
20	0	1
21	1	2
22	2	3

It can be shown by equation 19 that the pK values at $I \approx 10^{-3}$ are approximately the same as at I = 0. The corrected values for an ionic strength of 0.1 *M* are calculated as follows:

$$pK'_1 = pK_1 - 0.12 = 2.19 - 0.12 = 2.07$$
$$pK'_2 = pK_2 - 0.36 = 6.89 - 0.36 = 6.53$$
$$pK'_3 = pK_3 - 0.6 = 11.49 - 0.6 = 10.89$$

One should keep in mind, however, that the uncertainty in the magnitude of the pK values may be greater than the correction: the values reported by various sources can differ by 0.5 unit or more. The need to make the correction should be assessed on the basis of the uncertainty in the pK values and the accuracy required from the calculation. Variations in the reported pK values for Zn^{+2} and their effect on the calculated speciation are shown in Figure 2.7-1. The use of critical sources [11,12,18] of stability constants may alleviate this problem somewhat.

2.7.5 Effect of Temperature on Acidity Constants

Acid/base equilibria are affected by temperature in various ways, depending on the magnitude and sign of the enthalpy of reaction (ΔH). As Table 2.7-2 shows, the magnitude of ΔH for acid-base reaction can be either positive or negative but is seldom large, typically ranging from −15 to +15 kcal/mole. Given a reaction such as

$$H^+ + A^- \rightleftharpoons HA \tag{23}$$

1. "1,0" refers to $Z_A = 0$, $Z_B = \pm 1$ or vice versa; "2,3" refers to $Z_A = \pm 2$, $Z_B = \pm 3$ or vice versa.

TABLE 2.7-2

Temperature Effects on Acid/Base Reaction Equilibria[a]

Reaction	log K (25°C)[b]	log K (t°C)	ΔH (25°C)
$H^+ + CN^- \rightleftharpoons HCN$	9.21	9.29 (15°C)	-10.1
$H^+ + OH^- \rightleftharpoons H_2O$	14.00	14.34 (15°C)	-13.48
$H^+ + CO_3^{-2} \rightleftharpoons HCO_3^-$	10.25	10.59 (0°C)	—
$H^+ + HCO_3^- \rightleftharpoons H_2CO_3$	6.35	6.29 (50°C)	—
$H^+ + NO_2^- \rightleftharpoons HNO_2$	3.14	3.22 (15°C)	- 4.48
$H^+ + AsO_2^- \rightleftharpoons HAsO_2$	—	9.614 – 0.0184t[c]	—
$H^+ + HAsO_2^{-2} \rightleftharpoons H_2AsO_4^-$	—	$6.971 + 5 \times 10^{-5}(t-39.4)^2$	- 0.69
$H^+ + H_2AsO_4^- \rightleftharpoons H_3AsO_4$	—	$2.014 + 5 \times 10^{-5}(t+40)^2$	2.65
$H^+ + S^{-2} \rightleftharpoons HS^-$	d	12.9 – 0.031 (t–25)	- 7.6
$H^+ + HS^- \rightleftharpoons H_2S$	—	7.05 – 0.0125 (t–25)	- 4.42
$H^+ + SO_4^{-2} \rightleftharpoons HSO_4^-$	1.99	1.80 (15°C)	5.4
$H^+ + SeO_4^{-2} \rightleftharpoons HSeO_4^-$	1.88	1.84 (15°C)	—
$H^+ + {}_{3}^{-2} \rightleftharpoons HSO_3^-$	7.30	7.17 (10°C)	2.9
$H^+ + HSO_3^- \rightleftharpoons H_2SO_3$	2.00	1.92 (10°C)	3.9
$H^+ + F^- \rightleftharpoons HF$	3.17	3.40 (50°C)[e]	3.20

a. All data are experimentally derived and corrected to an ionic strength of zero unless noted. K refers to reaction in column 1.

b. The relationship between K_a (as defined by equation 1) and K as written for these reactions is given by equation 25.

c. Data at ionic strength 0.1 *M;* 14< t <45°C.

d. Value of K in dispute.

e. Data at ionic strength 0.5 *M*.

Sources: Sillén and Martell [17], Martell and Smith [11,12] and Smith and Martell [18].

with an equilibrium constant, K, and an enthalpy change of ΔH that is independent of temperature, we can calculate the effect on the equilibrium constant of a change in temperature from T_1 to T_2 by the following equation:

$$\log K_{(T_1)} - \log K_{(T_2)} = \frac{\Delta H}{2.3R}\left(\frac{1}{T_1} - \frac{1}{T_2}\right) \tag{24}$$

where K = equilibrium constant at specified temperature
ΔH = enthalpy of reaction (cal/mol)
R = gas constant = 1.987 cal/mol K
T = temperature (K)

The relationship between the pK_a of HA and K, the association equilibrium constants in Table 2.7-2 is

$$pK_a = \log K \tag{25}$$

At the highest ΔH values (about 15 kcal, assumed constant over the temperature range), equation 24 would predict a variation of 0.58 pK units over a 15-degree temperature range (283 K to 298 K).

Example 3 To illustrate the use of equation 24, we shall calculate the second acid association constant (K) of sulfurous acid (H_2SO_3) at 10°C if the value is 7.30 at 25°C.

The acid/base reaction of interest is

$$H^+ + SO_3^{-2} \rightleftharpoons HSO_3^- \qquad K,\ \Delta H \tag{26}$$

From Table 2.7-2, ΔH is +2.9 kcal/mole at 25°C. Assuming that the enthalpy is independent of temperature in the 10°C to 25°C range, equation 24 shows that

$$\log K_{(25°C)} - \log K_{(10°C)} = \frac{2900}{(2.3)(1.987)}\left(\frac{1}{298} - \frac{1}{283}\right)$$

$$7.30 - \log K_{(10°C)} = 634\ (-0.000178)$$

$$\log K_{(10°C)} = +0.11 + 7.30 = 7.41$$

The calculated second association constant is thus $10^{7.41}$ at 10°C. Using equation 25, the value of pK_a at 10°C is 7.41.

More rigorous approaches (e.g., when ΔH is not constant but the change in heat capacity is, or when neither is constant) are available and discussed in references 5 and 19.

2.7.6 Values and Estimation Methods for Acid Dissociation Constants

Much experimental data is available on the acid dissociation constants of aqueous inorganic species. Useful compendia of these and related thermodynamic parameters are given by Sillén and Martell [17], Martell and Smith [11,12] and subsequent supplements [18] and by Perrin [14]. Table 2.7-3 lists pK_a values for a variety of inorganic substances.[2] The following general observations on acidity (which are not without exceptions) are suggested by the data in this table:

- For cationic species, the higher the central ion oxidation state, the higher the acidity of the first water ligand (lower pK_a value).

2. Because these data come from different sources than those in Table 2.7-2, some variations will be noted for corresponding species.

TABLE 2.7-3

Values of Acid Dissociation Constants for a Variety of Inorganic Species

Species[a]	pK_1	pK_2	pK_3	pK_4	Ref.[b]
Ag^{+}	11.9	11.9	—	—	[14]
Al^{+3}	5.4	4.6	5.7	7.9	[14]
H_3AsO_4	2.1	6.7	11.2	—	[16]
H_3AsO_3	9.1	12.1	13.4	—	[16]
Ba^{+2}	13.2	—	—	—	[14]
Be^{+2}	5.7	8.2	9.9	14.0	[14]
H_3BO_3	9.1	12.7	13.8	—	[6]
HCN	9.0	—	—	—	[12]
H_2CO_3	6.2	10.0	—	—	[12]
H_2CrO_4	0.74	6.5	—	—	[6]
Cr^{+3}	4.2	6.2	8.3	9.1	[14]
Cd^{+2}	10.3	10.3	13.2	13.1	[14]
Co^{+2}	9.9	8.9	12.7	—	[14]
Cu^{+2}	8.2	9.3	10.3	11.3	[14]
HF	2.9	—	—	—	[12]
Fe^{+2}	9.7	11.1	10.2	15.0	[14]
Fe^{+3}	2.6	3.6	3.8	11.9	[14]
Hg^{+2}	3.8	2.4	14.9	—	[14]
Mn^{+2}	10.8	11.6	12.4	13.1	[14]
Ni^{+2}	10.2	9.0	10.8	14.0	[14]
H_3PO_4	2.0	6.7	11.7	—	[12]
Pb^{+2}	7.9	9.4	10.7	—	[14]
H_2S	7.0	12.0	—	—	[6]
H_2SO_3	1.8	6.8	—	—	[6]
Sn^{+2}	3.6	3.7	9.3	—	[8]
Tl^{+3}	0.94	1.2	1.7	—	[8]
V^{+3}	2.7	3.8	7.0	—	[8]
Zn^{+2}	9.2	7.9	11.3	12.3	[8]

(Continued)

TABLE 2.7-3 (Continued)

a. The terminology assumes an aq subscript on each species, which refers to the fact that the ion is in aqueous solution. The acid dissociation constants thus refer to loss of H^+ from successive water molecules, e.g.,

$$Fe(H_2O)_6^{+3} \rightleftharpoons Fe(H_2O)_5\,OH^{+2} + H^+ \qquad K_1$$

$$Fe(H_2O)_5\,OH^{+2} \rightleftharpoons Fe(H_2O)_4\,(OH)_2 + H^+ \qquad K_2$$

Values are given as $pK_i = -\log K_i$. Temperature is 20-25°C and ionic strength is 0.1 M.

b. Data obtained from Kragten [8] was used to calculate values as follows: $K_1 = {*\beta_1}$, $K_2 = {*\beta_2}/{*\beta_1}$, $K_3 = {*\beta_3}/({*\beta_1}\,K_2)$, and $K_4 = {*\beta_4}/({*\beta_1}\,K_2\,K_3)$.

- The magnitude of successive acid dissociation constants for a single species generally decreases (i.e., pK's generally increase).
- Small ions tend to be stronger acids than larger ions with the same charge.

Because of the diversity of inorganic species and the wealth of information available on inorganic reactions, the best sources for these constants are compilations and individual articles in the literature. In addition, five methods of estimation are described below, but all are of limited utility:

- The **Ricci formula** is useful only for oxy acids of a particular structure.
- The **thermodynamic cycle method** is complex and requires data on other reactions that may not be available. However, if a cycle can be written that includes other reactions for which data are available, the desired values can be estimated.
- Calculation from **hydroxide complexation constants** is possible when a procedure is available for estimating the latter constants.
- The **Bayless equations** apply to nonmetal hydrides and other acids.

These methods are discussed in detail below.

ESTIMATION BY THE RICCI FORMULA

The Ricci formula [15] is an empirical relationship for calculating the pK_a of an acid such as $O_nM(OH)_x$:

$$pK_a = 8.0 - m(9.0) + n(4.0) \qquad (27)$$

where m is the formal charge on the central atom, M, and n is the number of non-hydrogenated oxygen atoms. The value of m is calculated as follows:

$$m = P.G.N. - E_u - \frac{1}{2}E_s \tag{28}$$

where P.G.N. = periodic group number of the atom
E_u = number of unshared electrons around atom
E_s = number of shared electrons

Example 4 To derive the first pK of arsenious acid (H_3AsO_3), for example, we first calculate m. The structure HO-Äs-OH (with a third OH on As) indicates that $E_s = 6$ and $E_u = 2$. Since arsenic is in Group V,

$$m = 5 - 2 - 6/2 = 0$$

Therefore, from equation 27,

$$pK_a = 8.0 - 0(9.0) + 0(4.0)$$
$$= 8.0$$

The experimental value is 9.1.

ESTIMATION BY THE THERMODYNAMIC CYCLE METHOD

Although this procedure as described below is likely to be of little use for other chemicals of environmental concern, it provides a means of estimating the acidity of diatomic hydrogen halide acids and is therefore included here. However, where data on kinetics and on thermodynamic equilibria are available for trace elements, similar cycles could be prepared and might be useful in estimating acid dissociation and other data for trace metal complexes in aqueous systems.

The acid dissociation reaction of an acid HA is:

$$HA \rightleftharpoons H^+ + A^- \qquad K_a \tag{29}$$

The equilibrium constant, K_a, is related to the free energy change of this reaction by the Gibbs Equation:

$$\Delta G_a = -RT \ln K_a \tag{30}$$

where R is the universal gas constant and T is the temperature (K). The free energy change in equation 29 at constant temperature can be calculated from the changes in enthalpy (ΔH) and entropy (ΔS) for this reaction as follows:

$$\Delta G_a = \Delta H_a - T\Delta S_a \tag{31}$$

As illustrated below, thermodynamic cycles that express the conservation of energy can be used to estimate free energy or enthalpy changes for an unknown reaction if the values for other reactions in the thermodynamic cycle are known or can be estimated.

Example 5 Figure 2.7-2 shows a thermodynamic cycle that includes the dissociation of HA [13].

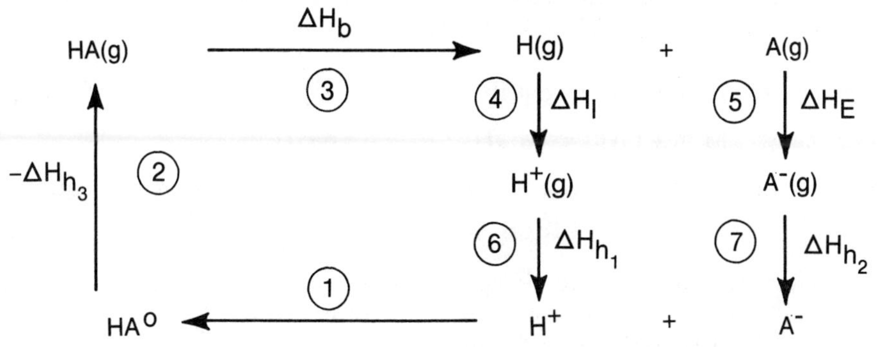

FIGURE 2.7-2 Thermodynamic Cycle which Includes Aqueous Acid Dissociation

In this cycle, (1) is the reverse of equation 23, (2) is the reverse of hydration of the neutral gas molecule, (3) is the dissociation of the gaseous molecule into atoms in the gas phase, (4) is the ionization potential of hydrogen, (5) is the enthalpy associated with the addition of an electron of the counter element (− electron affinity), (6) is the hydration of the gaseous hydrogen ion, and (7) is the hydration of the counter gaseous ion.

ΔH_a, the enthalpy change in (1), is thus:

$$\Delta H_a = \Delta H_D + \Delta H_I + H_E + \Delta H_{h_1} + \Delta H_{h_2} - \Delta H_{h_3} \tag{32}$$

ΔS_a, the entropy change in (1), is calculated as follows:

$$\Delta S_a = (S^0{}_{H^+\,aq} + S_{A^-\,aq}) - S^0{}_{HAaq} \tag{33}$$

This calculation requires some thermodynamic parameters that are unknown and must therefore be estimated. In the case of hydrogen halide acids, the enthalpy of hydration of undissociated HA gas (ΔH_{h3}) can be approximated using the values for substances that are similar in size but do not ionize in solution (e.g., Ar and CH_3Cl for HCl, Kr and CH_3Br for HBr, and Xe and CH_3I for HI). $\Delta S^0{}_{HAaq}$, the standard entropy of the undissociated acid, can be calculated from the standard entropy of the gas at 1 atmosphere, if the change in entropy from gas at one atmosphere to solute at unit molality is known. Values of the other parameters for the hydrogen halide acids are known [20].

Since much of the needed thermodynamic data may be unavailable for other acids (e.g., polyatomic acids), the general utility of this method is limited.

CALCULATION FROM ESTIMATED OR EXPERIMENTALLY DERIVED HYDROXIDE COMPLEXATION CONSTANTS

K_a can be calculated from K_{OH}, the equilibrium constant for reaction with hydroxide:

$$M^{n+} + OH^- \rightleftharpoons MOH^{(n-1)+} \qquad K_{OH} \tag{34}$$

The acid dissociation constant, K_a, is related to K_{OH} by the following equation:

$$K_{a_i} = K_w K_{OH_i} \tag{35}$$

where K_{ai} refers to the *ith* acid dissociation constant (e.g., see equations 7-9), $K_w = [H^+][OH^-] = 10^{-14.0}$ (at 25°C), and K_{OHi} refers to the *ith* hydroxide equilibrium constant, which is defined as follows:

$$M(OH)^{X}_{i-1} + OH^- \rightleftharpoons M(OH)^{y}_{i} \qquad K_{OH_i} \tag{36}$$

Example 6 To illustrate the use of this method, we will calculate the first acid dissociation constant of $Tl^{+3}{}_{aq}$ ion (at 25°C), given a hydroxide equilibrium constant of $10^{15.4}$[17]. From equation 35, the value of K_a is

$$K_A = 10^{-14.0} \times 10^{15.4} = 10^{1.4}$$

Thus, $pK_a = 1.4$.

Note that this value is higher than that given in Table 2.7-3 (0.94). The difference of 0.46 pK units reflects the differnces due to choice in K_{OH} used. This is another example of the variation in values of acid/base equilibrium constants (and other equilibrium constants) published in the literature.

ESTIMATION BY THE BAYLESS EQUATIONS

Bayless [1] has derived an empirical relationship for calculating the aqueous pK_a of an acid in the form $[YH_{(8-g)}]^q$:

$$pK_a = 73.7 - 11.71g - 0.29gh - \frac{378.6\,(q-1)}{gp^2} \tag{37}$$

where

g = periodic table group number of element Y
h = number of hydrogen atoms on Y
p = period number of element Y
q = charge on the acid

Normally, hydride acids are uncharged (e.g., NH_3, HF) so q = 0, but the equation is also useful for charged species such as NH_4^+ and SeH^-. Reasonable agreement (average deviation of 0.4 pK_a units) was observed for a set of 17 acids [1].

Example 7 To derive the pK_a of H_2Se, we use equation 37 with the following parameter values:

$$g = 6$$
$$h = 2$$
$$p = 4$$
$$q = 0$$

Thus,

$$pK_a = 73.7 - 11.71\,(6) - 0.29\,(6)\,(2) - \frac{(378.6)\,(-1)}{(6)\,(4)^2}$$

$$= 3.90$$

The experimental value is 4.0 [1].

Bayless [1] has also extended the Ricci equation to bases of the form $(HA)_nXB_m$, where H represents hydrogen atoms, and A and B are atoms bonded to central atom X:

$$pK_b = -59.7 + 12g + 0.29gh + \frac{55.2q_a + 98.4tq_a + 225q_x}{gp^2} \tag{38}$$

where
- g = periodic group number of atom A
- h = number of H atoms on all A atoms capable of being protonated (acting as bases)
- p = period number of atom A
- q_a = formal charge on atom A
- q_x = formal charge on atom X
- t = number of A and B atoms capable of being protonated (usually, but not always, equal to n + m).

The formal charge, q, is calculated as follows:

$$q = g - (\text{no. of unshared electrons}) - \tfrac{1}{2}\,(\text{no. of shared electrons}) \tag{39}$$

When using equation 38, it is helpful to draw resonance structures of the base that puts the most negative formal charge on the A and B atoms.

The equation has been tested for only a limited number of cases where A and B are nitrogen atoms, such as

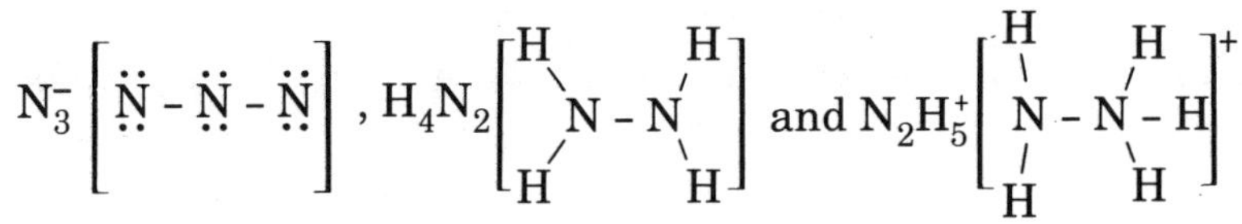

Example 8 To derive the pK_b of $HN_2O_2^-$, we use equation 38 with the following parameter values derived from the resonance structure

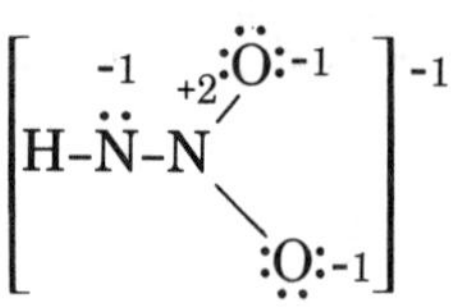

$g \quad = 5$
$p \quad = 2$
$h \quad = 1$
$q_a \quad = 5 - 4 - \frac{1}{2}(4) = -1$ (for N) and $6 - 6 - \frac{1}{2}(2) = -1$ (for O)
$t \quad = 3$ (the two oxygen and nitrogen atoms are considered equivalent, each having the same formal charge of -1)
$q_x \quad = 5 - 0 - \frac{1}{2}(6) = 2$

Thus,

$$pK_b = -59.7 + 12(5) + 0.29(5)(1) + \frac{55.2(-1) + 98.4(3)(-1) + 225(2)}{(5)(2)^2}$$

$$= 6.73$$

The reported value is 7.42 [1].

2.7.7 Literature Cited

1. Bayless, P.L., "Empirical pK_b and pK_a for Nonmetal Hydrides from Periodic Table Position," *J. Chem. Ed.*, **60**, 546 (1983).
2. Beck, M.T., *Chemistry of Complex Equilibrium,* Von Nostrand Reinhold Company, New York (1970).
3. Brønsted, J., *Rec. Trav. Chim.*, **42**, 718 (1923).
4. Butler, J.N., *Ionic Equilibrium, a Mathematical Approach,* Addison-Wesley Publishing Company, Reading, Mass. (1964).
5. Garrels, R.M. and C.L. Christ, *Solutions, Minerals and Equilibria,* Freeman, Cooper & Company, San Francisco, Calif. (1965).
6. Gordon, A.J. and R.A. Ford, *The Chemist's Companion*, John Wiley & Sons, New York (1972).
7. Hahne, H.C.H. and W. Kroontje, "Significance of pH and Chloride Concentration on Behavior of Heavy Metal Pollutants: Mercury (II), Cadmium (II), Zinc (II) and Lead (II)," *J. Environ. Qual.*, **2**, 444 (1973).
8. Kragten, J., *Atlas of Metal-Ligand Equilibria in Aqueous Solutions*, Halsted Press (Wiley), New York (1978).

9. Langmuir, D., *Techniques for Estimating Thermodynamic Properties for Some Aqueous Complexes of Geochemical Interest,* ACS Symposium Series No. 93, E.A. Terine (ed.), American Chemical Society (1976).

10. Lindsay, W.L., *Chemical Equilibria in Soils,* John Wiley & Sons, New York (1979).

11. Martell, A.E. and R.M. Smith, *Critical Stability Constants, Vol. 3: Other Organic Ligands,* Plenum Press, New York (1977).

12. Martell, A.E. and R.M. Smith, *Critical Stability Constants, Vol. 5: First Supplement,* Plenum Press, New York (1982).

13. McCoubrey, J.C., "The Acid Strength of the Hydrogen Halides," *Trans. Faraday Soc.*, **51**, 743 (1955).

14. Perrin, D.D., *Ionization Constants of Inorganic Acids and Bases in Aqueous Solutions*, IUPAC Chemical Data Series, No. 29, Pergamon Press, New York (1981).

15. Ricci, J.E., "The Aqueous Ionization Constants of Inorganic Oxygen Acids," *J. Am. Chem. Soc.*, **70**, 109 (1948).

16. Ringbom, A., *Complexation in Analytical Chemistry*, Interscience (Wiley), New York (1963).

17. Sillén, L.G. and A.E. Martell, *Stability Constants of Metal-Ion Complexes*, Special Publication No. 17, The Chemical Society, Burlington House, London (1964).

18. Smith, R.M. and A.E. Martell, *Critical Stability Constants, Vol.4: Inorganic Complexes,* Plenum Press, New York (1982).

19. Stumm, W. and J.J. Morgan, *Aquatic Chemistry*, 2nd ed., John Wiley & Sons, New York (1981).

20. Wagman, D.D., W.H. Evans, V.B. Parker, R.H. Schumm, I. Halow, S.M. Bailey, K.L. Churney and R.L. Nuttall, "The NBS Tables of Chemical Thermodynamic Properties," *J. Phys. Chem. Ref. Data*, Vol. II, Supplement No. 2 (1982).

2.8 POLYMERIZATION

2.8.1 Introduction

Polymerization is the process whereby several small molecules (monomers) react chemically with one another to form larger molecules (polymers). These polymers may be linear, linear with cross-linkages, or nonlinear.

Three types of polymers are discussed in this section: polymeric metal hydroxo complexes, polysulfides, and polyphosphates. Both metal hydroxo polymers and polysulfides can form spontaneously in aqueous systems. Polyphosphates, on the other hand, are artificially manufactured and are of environmental interest because of their degradation rather than their formation. Two additional polymerization processes, which are discussed elsewhere in this report, are the formation of polyborates (section 10.9, Boron-Containing Compounds) and the polymerization of hydrogen cyanide (section 10.2, Cyanides).

2.8.2 Hydroxo- and Oxo-Bridged Metal Polymers

When the hydroxide ion concentration of an aqueous solution containing dissolved metals is increased, the metal cations may be hydrolyzed, forming complexes with hydroxide ions. Some of the complexes formed are mononuclear — i.e., they include only one metal ion — while others, referred to as polynuclear hydroxo complexes, involve several metal ions. The formation of such polymeric species should be taken into account when one estimates the activity of various metal species in solution.

A solvated metal ion reacts with a molecule of water as follows:

$$[M(H_2O)_a]^{+z} + H_2O \rightleftharpoons [M(H_2O)_{a-1}\,OH]^{+(z-1)} + H_3O^+ \qquad (1)$$

Two of the resulting partially hydrolyzed species may then react, with the elimination of water, to form a dinuclear hydroxo complex:

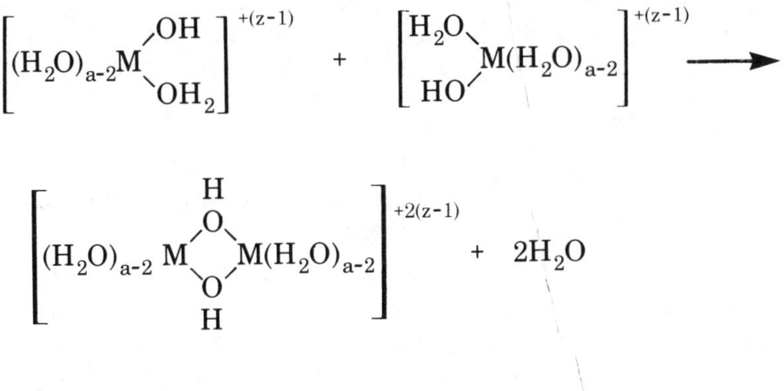

(2)

If these hydrolysis and condensation processes are repeated, larger polynuclear hydroxo complexes may be formed:

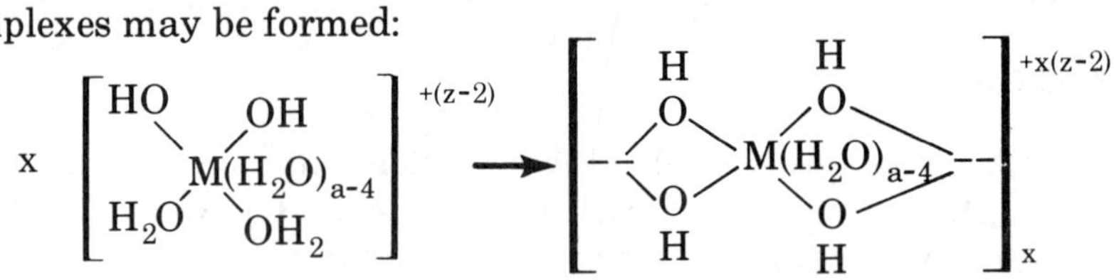

(3)

This formation of -OH- bridges is sometimes called "olation." Under conditions of high temperature, prolonged aging and/or high pH, it may be followed by "oxolation," a process in which the -OH- bridges are converted to -O- bridges, leading eventually to the precipitation of hydrated metal oxides [6].

The linear chain structures described above are referred to as "core-plus-links" structures. They have the general formula $M[(OH)_tM]_{x-1}^{+(xz-xt+t)}$ representing a chain of x M^{+z} ions in which each pair of adjacent M^{+z} ions is linked by t OH-groups. The "core-plus-links" structure is in good agreement with experimental data for many metal cations [14]. However, a number of cyclic and polyhedral structures (also having hydroxide bridges connecting adjacent metal ions) have been proposed for certain polynuclear hydroxo species, and several have been confirmed by diffraction studies [1].

Formation of polynuclear hydroxo species has been reported for most metals, although in some cases the predominant species in solution is a monomer. Some metals form only dimers or trimers, while a few form much larger polynuclear species. (See Table 2.8-1.) Equilibria for the formation of simple hydrolysis species are usually established quickly, but polynuclear species are often formed more slowly. Many of these polynuclear complexes are not present under equilibrium conditions; they are often formed only under conditions of oversaturation with respect to the metal hydroxide or oxide and are kinetic intermediates in the transition from free metal ions to a solid precipitate. However, they can persist as metastable species for years and thus can be significant in natural water systems [15].

Increasing the pH of a metal ion solution, by shifting the position of the hydrolysis equilibrium (reaction 1), results in an increased concentration of hydrolyzed species M-OH, which in turn causes increased formation of polymeric species (as in reaction 2). Diluting a solution has two opposing effects on the formation of polymeric species:

(1) Because dilution of acidic solutions causes a decrease in H^+ concentration (i.e., an increase in pH), it causes a shift in the hydrolysis equilibrium toward formation of hydrolyzed species.

(2) On the other hand, dilution decreases the ratio of polynuclear to mononuclear complexes in solution. For metals that form both mononuclear and polynuclear complexes, this means that the mononuclear species predominate beyond a certain level of dilution [15].

Equilibrium constants for the formation of some polynuclear hydrolysis species are given in Table 2.8-1. An example of equilibrium calculations for the hydrolysis of iron (III), taking into account the formation of $Fe_2(OH)_2^{+4}$, is given in Stumm and Morgan [15].

2.8.3 Polysulfides

Polysulfide ions, S_n^{-2}, may be formed in natural waters by the interaction of aqueous sulfide with elemental sulfur. Elemental sulfur may be formed initially either by bacterial oxidation of sulfide or by the partial oxidation of sulfide by dissolved oxygen [3]:

$$2\,HS^- + O_2^0 \longrightarrow 2S(s) + 2OH^- \tag{4}$$

The reaction of hydrosulfide with sulfur then proceeds as follows [3]:

$$HS^- + (n\text{-}1)S(s) \rightleftharpoons S_n^{-2} + H^+ \tag{5}$$

Once initiated, the formation of polysulfides from sulfur is autocatalytic, and equilibrium is attained rapidly [2].

TABLE 2.8-1

Formation Quotients for Polynuclear Metal Hydrolysis Products at 25°C, Corrected to Zero Ionic Strength[a]

$$\log K_{xy} = \log \frac{[M_x(OH)_y^{+(xz-y)}]\,[H^+]^y}{[M^{+z}]^x}$$

M	M_2OH^{+3}	$M_2(OH)_2^{+2}$	$M_3(OH)_3^{+3}$	$M_3(OH)_4^{+2}$	$M_3(OH)_5^{+}$	$M_4(OH)_4^{+4}$	$M_6(OH)_8^{+4}$	$M_x(OH)_y^{+(2x-y)}$[x,y]
Divalent Cations								
Be^{+2}	-3.97		-8.92				-27.2	
Mg^{+2}						-39.71		
Mn^{+2}	-10.56							-23.90 [2,3]
Co^{+2}	-11.2					-30.53		
Ni^{+2}	-10.7					-27.74		
Cu^{+2}		-10.36						
Zn^{+2}	-9.0							-57.8 [2,6]
Cd^{+2}	-9.390					-32.85		
Hg^{+2}	-3.33		-6.42					
Sn^{+2}		-4.77		-6.88				
Pb^{+2}	-6.36			-23.88		-20.88	-43.61	
UO_2^{+2}		-5.62		(-11.75)[b]	-15.63			
NpO_2^{+2}		-6.39			-17.49			
PuO_2^{+2}		-8.36			-21.65			
VO^{+2}		-6.67						

(Continued)

TABLE 2.8-1 (Continued)

M	$M_2(OH)_2^{+4}$	$M_3(OH)_4^{+5}$	$M_3(OH)_5^{+4}$	$M_x(OH)_y^{+(3x-y)}$[x,y]
Trivalent Cations				
Al^{+3}	−7.7	−13.94		−98.73 [13,32][c]
Sc^{+3}	−6.0		−16.34	
Y^{+3}	−14.23		−31.6	
La^{+3}	$<$−17.5		$<$−38.3	−71.2 [5,9]
Ce^{+3}	$<$−15.5		−33.5	
Nd^{+3}	−13.86		$<$−28.5	
Er^{+3}	−13.65		$<$−29.3	
Ti^{+3}	-3.6			
V^{+3}	-3.8			
Cr^{+3}	-5.06	-8.15		
Fe^{+3}	-2.95	-6.3		
Ga^{+3}				−139.1 [26,65][d]
In^{+3}		−5.82[e]		
Bi^{+3}				f

M	$M_2(OH)_2^{+6}$	$M_3(OH)_4^{+8}$	$M_3(OH)_5^{+7}$	$M_4(OH)_8^{+8}$	$M_6(OH)_{15}^{+9}$
Tetravalent Cations					
Zr^{+4}		(-0.6)[g]	(3.70)[g]	6.0	
Th^{+4}	-6.14			-21.1	-36.76
U^{+4}					-17.2[d]

M	$Sb_{12}(OH)_{64}^{-4}$	$Sb_{12}(OH)_{65}^{-5}$	$Sb_{12}(OH)_{66}^{-6}$	$Sb_{12}(OH)_{67}^{-7}$
Pentavalent Cation				
Sb^{+5}	20.34	16.72	11.89	6.07

(Continued)

TABLE 2.8-1 (Continued)

a. Expressions representing the variation of these formation constants with ionic strength are given in Baes and Mesmer [1].
b. Apparent constant which reproduces data at high chloride concentrations.
c. The species formed is actually $Al_{13}O_4(OH)_{24}^{+7}$.
d. Approximate formula of one or more species formed.
e. This species is consistent with the evidence, but its composition is not well established.
f. There is evidence for the existence of $Bi_6(OH)_{12}^{+6}$, $Bi_9(OH)_{22}^{+5}$, $Bi_9(OH)_{21}^{+6}$, and $Bi_9(OH)_{20}^{+7}$, but insufficient data are available for the estimation of formation quotients.
g. One or more of these trimeric species are believed to form. The constants are calculated from the data assuming that only one is formed.

Source: Adapted from Baes and Mesmer [1]

The equilibrium constant for reaction 5 can be written as

$$K_n = \frac{[H^+][S_n^{-2}]}{[HS^-]} \tag{6}$$

since the S(s) activity is taken as constant (unity) in a saturated solution [5]. Equilibrium constants K_n for the formation of the species S_2^{-2} through S_6^{-2} are given in Table 2.8-2. However, there is some disagreement in the literature as to which polysulfides are actually formed. Some authors consider S_5^{-2} the largest polysulfide formed [5,8], while others include S_6^{-2} as an important species [2,3,10]. Nevertheless, there is general agreement that the dominant S_n^{-2} species are those with $n \geq 4$.

TABLE 2.8-2

Equilibrium Constants for the Formation of Polysulfide Ions

Reaction[a]	pK Median[b]	pK Range
$S(s) + HS^- \rightleftharpoons S_2^{-2} + H^+$	12.68	12.27-14.43
$2\,S(s) + HS^- \rightleftharpoons S_3^{-2} + H^+$	11.75	10.96-13.19
$3\,S(s) + HS^- \rightleftharpoons S_4^{-2} + H^+$	9.74	9.35-10.07
$4\,S(s) + HS^- \rightleftharpoons S_5^{-2} + H^+$	9.41	9.29-9.52
$5\,S(s) + HS^- \rightleftharpoons S_6^{-2} + H^+$	9.62	9.43-9.79
$4\,S_4^{-2} + H_2O \rightleftharpoons OH^- + HS^- + 3S_5^{-2}$	3.54	1.95-5.16
$HS_4^- \rightleftharpoons S_4^{-2} + H^+$	7.0[c]	—
$HS_5^- \rightleftharpoons S_5^{-2} + H^+$	6.1[c]	—

(Continued)

TABLE 2.8-2 (Continued)

a. For the reactions involving elemental sulfur (S), the sulfur activity is taken as unity and the equilibrium constant is given as $K = [S_n^{-2}]\,[H^+]/[HS^-]$.

b. Except where noted, values are medians of 3 to 5 values given in cited source.

c. Single measurements.

Source: Adapted from Boulègue & Michard [2]

This result can be demonstrated by calculating the equilibrium concentrations of the individual polysulfide species. First, the above expression for K_n can be rearranged to give the following:

$$\log [S_n^{-2}] = \log K_n + pH + \log [HS^-] \tag{7}$$

Next, using the expression for the first acid dissociation constant (K_{a1}) for H_2S and neglecting the second acid dissociation, we have

$$[HS^-] = \frac{K_{a1}[H_2S]}{[H^+]} = \frac{K_{a1}(\Sigma\, S^{-2} - [HS^-])}{[H^+]} \tag{8}$$

where $\Sigma\, S^{-2}$ is the sum of the concentrations of H_2S and HS^-. Solving this equation for $[HS^-]$ and substituting into equation 7 yields an expression for the concentration of a given polysulfide species as a function of pH [5]:

$$\log [S_n^{-2}] = \log K_n + pH + \log(\Sigma S^{-2}) - \log \left(1 + \frac{[H^+]}{K_{a1}}\right) \tag{9}$$

Using the values for K_n given in Table 2.8-2, equation 9 predicts the following concentrations of polysulfide ions in a solution at pH 7 containing a total S(-II) concentration of 10^{-5} *M*:

$$[S_2^{-2}] = 1.05 \times 10^{-11}\ M$$
$$[S_3^{-2}] = 8.91 \times 10^{-11}\ M$$
$$[S_4^{-2}] = 9.12 \times 10^{-9}\ M$$
$$[S_5^{-2}] = 1.95 \times 10^{-8}\ M$$
$$[S_6^{-2}] = 1.20 \times 10^{-8}\ M$$

By means of similar calculations (but using slightly different K_n values), Boulègue and Michard [2] plotted the distribution of sulfur species between sulfide and various polysulfides at equilibrium with colloidal elemental sulfur, as shown in Figure 2.8-1.

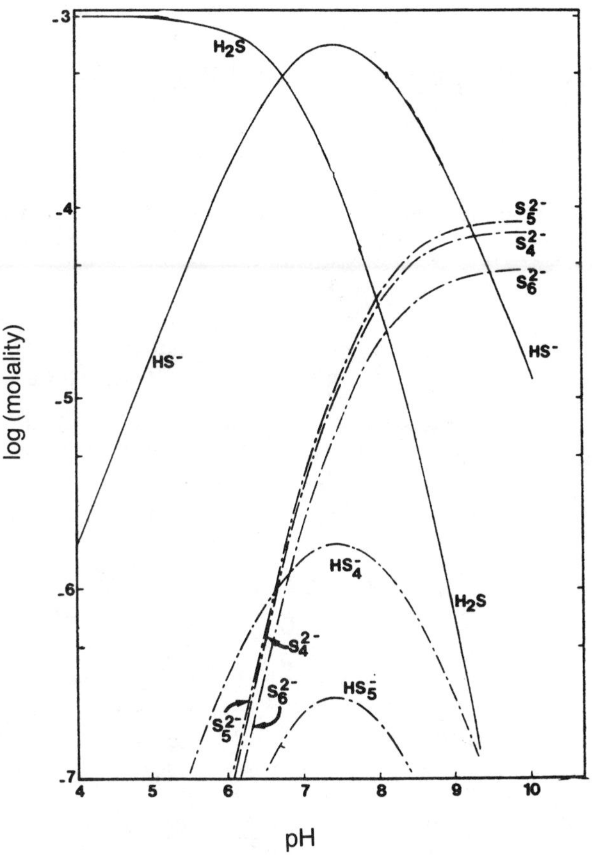

Source: Boulègue & Michard [3]. *(Copyright 1979, American Chemical Society. Reprinted with permission.)*

FIGURE 2.8-1 Distribution of Dissolved Sulfur Species in the System H_2S-S_8 (colloid) – H_2O as a Function of pH for $\Sigma[S] = 10^{-3}$ mol/kg with 0.7 *M* NaCl at 298 K

The above calculation is based on the assumption of a saturation concentration of elemental sulfur. Under actual reaction conditions, when sulfur must be formed by the oxidation of sulfide, polysulfide concentrations will be slightly less than the values given above. At higher pH, the formation of sulfur is not favored, so polysulfide concentrations will be much lower [5].

Table 2.8-3 gives thermodynamic data useful in calculations of equilibria involving polysulfides.

Example Is a solution containing 0.5×10^{-3} *M* HS^- and 0.5×10^{-3} *M* S_5^{-2} at pH 9.41 at equilibrium with respect to oxidation by atmospheric oxygen?

We first write the reaction of interest as

$$5HS^- + 3H^+ + 2O_2 \rightleftharpoons S_5^{-2} + 4H_2O$$

The equilibrium constant for this reaction can be calculated from the free energy change as follows:

$$\begin{aligned}\Delta G_r^0 &= \Delta G_f^0(S_5^{-2}) + 4\Delta G_f^0(H_2O) - 5\Delta G_f^0(HS^-) - 3\Delta G_f^0(H^+) - 2\Delta G_f^0(O_2)\\ &= 15.7 + 4(-56.69) - 5(2.88) - 0 - 0\\ &= -225.46 \text{ kcal/mol}\end{aligned}$$

where the ΔG_f^0 values are free energies of formation of the various species from Table 2.8-3.

TABLE 2.8-3

Free Energy of Formation Values for Species Involved in Aqueous Polysulfide Equilibria

Species	ΔG_f^0 (kcal/mol)
S_2^{-2}	19.0
S_3^{-2}	17.6
S_4^{-2}	16.5
S_5^{-2}	15.7
HS^-	2.88
H_2S	-6.66
$S_2O_3^{-2}$	-122.8
$H_2O(l)$	-56.687

Source: Nriagu & Hem [11] *(Copyright 1978, John Wiley & Sons. Reprinted with permission.)*

According to equation 6 of section 2.10.4 we can then write the following expression for the redox equilibrium constant:

$$\log K_{eq} = \frac{\Delta G_r^0}{-2.3RT} = -0.733\Delta G_r^0 = 165.26$$

The partial pressure of oxygen above this solution at equilibrium can be computed from the expression for K_{eq} together with the numerical values of the pH and concentrations of the dissolved species:

$$K_{eq} = \frac{[S_5^{-2}]}{[HS^-]^5[H^+]^3(pO_2)^2}$$

with $K_{eq} = 10^{165.26}\ M^{-7}\cdot\text{atm}^{-2}$

$[S_5^{-2}] = 0.5 \times 10^{-3}\ M$

$[HS^-] = 0.5 \times 10^{-3}\ M$

and $[H^+] = 10^{-9.41}\ M$

Note that since 9.41 is the pK value for the reaction of HS^- with sulfur to form S_5^{-2}, at pH 9.41 the concentrations of these two species should be approximately equal, as given above. Rearranging to solve for pO_2, we have:

$$pO_2 = \left(\frac{[S_5^{-2}]}{[HS^-]^5[H^+]^3K_{eq}}\right)^{1/2} = \left(\frac{10^{-3.3}}{(10^{-3.3})^5(10^{-9.41})^3 10^{165.3}}\right)^{1/2}$$

$$= 10^{-62} \text{ atm}$$

The result shows an equilibrium partial pressure of oxygen of 10^{-62} atm, a partial pressure that is vanishingly small for all practical purposes. Oxygen pressures higher than the equilibrium value should drive the reaction to the right, producing sulfur in higher oxidation states.

A similar calculation shows that about the same pressure of oxygen is an equilibrium value for reaction 4, in which HS^- is oxidized to elemental sulfur. Thus, we see that sulfide and polysulfide are not equilibrium species in water open to the atmosphere.

As demonstrated in the preceding example, polysulfides are not stable species under aerobic conditions. However, they are sometimes found in the natural environment, especially in hydrothermal or heavily polluted water [4,11]. They are formed as intermediates in the chemical oxidation of sulfide by oxygen, as shown in equations 4 and 5. Polysulfide ions may subsequently be oxidized to thiosulfate as follows [3]:

$$2S_n^{-2} + O_2^0 + H_2O \rightarrow S_2O_3^{-2} + S_{n-1}^{-2} + 2H^+ \tag{10}$$

However, the oxygenation of sulfide occurs very slowly in the absence of catalysts; thus, sulfides may coexist with dissolved oxygen for long periods of time [4]. Anaerobic or reducing conditions favor the persistence of sulfides and polysulfides.

The rate of polysulfide formation, and therefore the overall rate of oxygenation of sulfide, shows a complex dependence on pH. Figure 2.8-2 shows the time dependence of polysulfide formation in dilute aqueous solution at various pH values (as determined in laboratory experiments in which sodium sulfide was added to water containing dissolved oxygen). As shown, maximum polysulfide formation occurs near pH 7. The fact that polysulfides are not formed at low pH, which is predicted by calculations of the equilibrium distribution of sulfur species (Figure 2.8-1), may be attributed to the protonation of HS^- to form H_2S. Polysulfide formation also tapers off above pH 9, presumably because elemental sulfur is not formed by the chemical oxidation of sulfide at high pH [5].

2.8.4 Polyphosphates

Polyphosphoric acids and polyphosphates consist of multiple PO_4 units. They are formed by the condensation of two or more molecules of orthophosphoric acid (H_3PO_4, also known as simply phosphoric acid), with the loss of water. Inorganic polyphosphates do not occur naturally, but they are manufactured in large quantities. They are used primarily as fertilizers, components of synthetic detergents, and water-softening agents in water treatment. They are also used in a variety of industrial processes, in the petroleum industry, in pigment manufacture, and in food processing. They are thus released to the environment both directly (e.g., as fertilizers) and in waste streams [7].

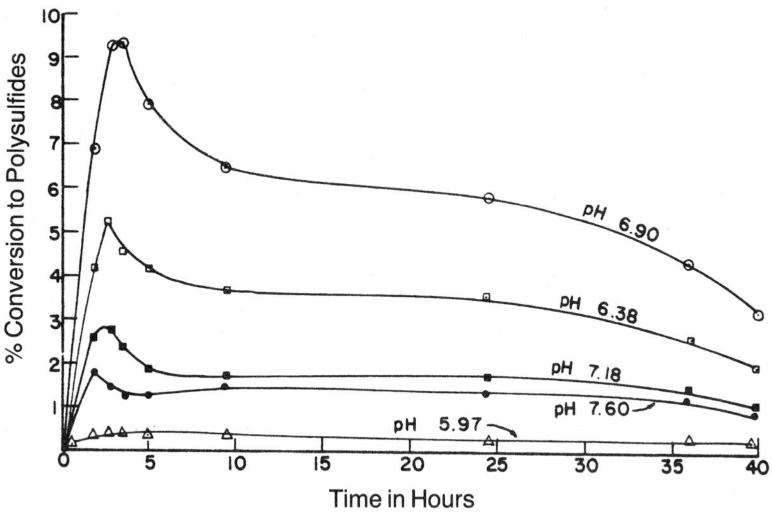

Source: Chen [4]. *(Copyright 1974, Ann Arbor Science Publishers. Reprinted with permission from Butterworth Publishers, current holders of the copyright.)*

FIGURE 2.8-2 Time Dependence of Polysulfide Formation in Aqueous Solution With an Initial Total S(-II) Concentration of 3.0 × 10^{-3}*M*, and p(O_2) = 1 atm

Structurally, each PO_4 unit of a polyphosphate is a phosphorus atom surrounded by a tetrahedron of four oxygen atoms; in polyphosphates, these tetrahedra are linked at their corners by shared oxygen atoms. Linear polyphosphates, with straight chains of P-O-P linkages, range from the simple diphosphate (referred to as pyrophosphate) to long-chain structures with molecular weights in the millions. The linear polyphosphate anion may be represented by the general formula

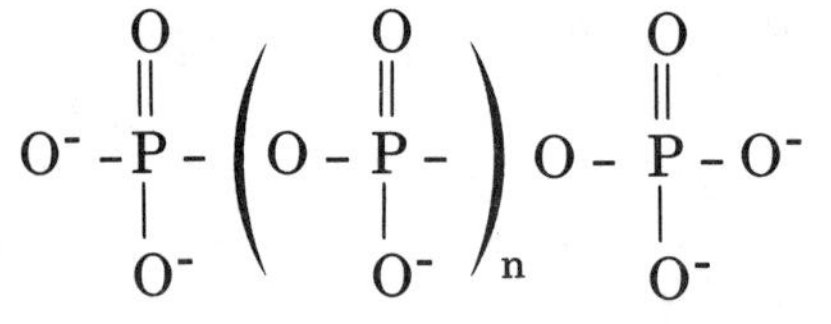

The linear polyphosphates include most commercially important phosphate salts. However, cyclic structures (known as metaphosphates), branched-chain structures (ultraphosphates), and compounds with incompletely defined structures are also known [7].

Pyrophosphoric acid ($H_4P_2O_7$) and tripolyphosphoric acid ($H_5P_3O_{10}$) are major components of household synthetic detergents: they have the ability to complex the hardness ions Ca^{+2} and Mg^{+2}, thus preventing these ions from combining with detergent molecules. Largely as a result of their use in detergents, condensed phosphates comprise 50-60% of the phosphates in domestic wastewater [13].

Once released to the environment, however, polyphosphates are gradually hydrolyzed. Their hydrolysis ultimately yields orthophosphate, as in the following equations for the hydrolysis of tripolyphosphate and pyrophosphoric acid [13]:

$$HP_3O_{10}^{-4} + 2\,H_2O \longrightarrow 3\,HPO_4^{-2} + 2H^+ \qquad (11)$$

$$H_4P_2O_7^0 + H_2O \longrightarrow 2H_3PO_4^0 \qquad (12)$$

The mechanisms and initial products of hydrolysis differ for short-chain, long-chain, and cyclic polyphosphates, and the rates of hydrolysis are influenced by several factors, including pH, temperature, metal-ion catalysis, and enzymatic catalysis.

Short-chain polyphosphates (up to four phosphate units) are hydrolyzed primarily by the splitting of phosphorus-oxygen bonds at one end of the polyphosphate chain. Thus, for example, tetrapolyphosphate is converted first to tripolyphosphate, then to pyrophosphate, and finally to orthophosphate. (There is some evidence, however, that the degradation proceeds by other routes as well.) The hydrolysis reaction follows first-order kinetics in more concentrated solutions; apparent first-order degradation rate constants for sodium polyphosphates are shown in Figure 2.8-3. As shown, the hydrolysis rates for the short-chain polyphosphates increase with decreasing pH. The rates also increase with increasing temperature [12].

Long-chain polyphosphates (more than seven or eight phosphates) may be hydrolyzed by several routes: cleavage of chain-end bonds, cleavage of bonds in the interior of the chain, and splitting off of trimetaphosphate rings from either the end or the middle of the chain. Three rate constants can be calculated from experimental data: k_t, overall disappearance of a polyphosphate species; k_e, clipping of the end group to form orthophosphate; and k_m, formation of trimetaphosphate. All three rate constants follow first-order kinetics, so the degradation rate can be represented as follows:

$$-dC/dt = k_tC = (k_e + k_m)C \qquad (13)$$

where C is the concentration of the polyphosphate species being studied. Variations in pH have different effects on the various hydrolysis mechanisms. The formation of orthophosphate by end-group clipping increases at low pH, while the formation of trimetaphosphate is inhibited at low pH. As a result, the ratio of orthophosphate to trimetaphosphate produced increases with decreasing pH, and the overall degradation rate reaches a maximum at pH 7. Increasing temperature results in a decrease in the ratio of orthophosphate to trimetaphosphate [12].

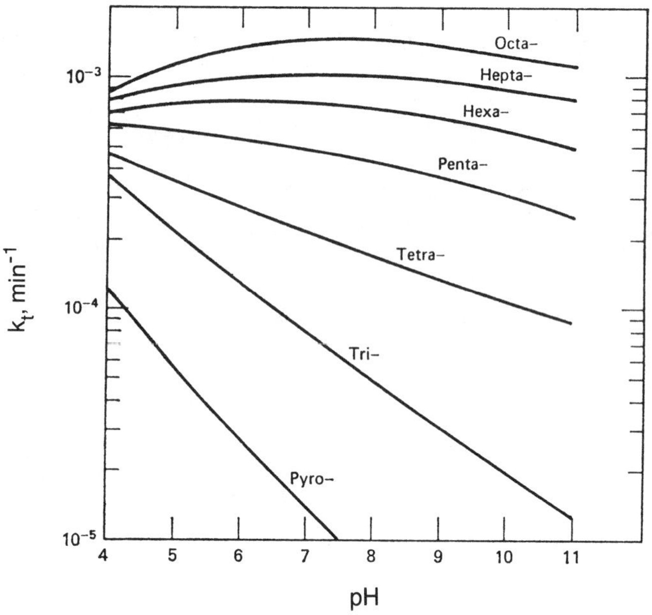

Source: Shen and Morgan [12]. *(Copyright 1973, John Wiley & Sons. Reprinted with permission.)*

FIGURE 2.8-3 Apparent First-order Rate Constants for the Hydrolytic Degradation of Sodium Polyphosphates at 60°C.

Of the cyclic metaphosphates, the most common are tri- and tetrametaphosphates. These compounds are initially hydrolyzed to the corresponding linear polyphosphates, which are then degraded to lower phosphates. The hydrolysis to the linear polyphosphates is catalyzed by both acid and base, proceeding most slowly between pH 6 and 10 [12].

Hydrolysis of both linear polyphosphates and cyclic metaphosphates is enhanced by the presence of metal cations. In the case of long chain polyphosphates, the formation of trimetaphosphate is accelerated more than the clipping of chain ends. The degradation rate for linear polyphosphates increases with increasing chain length, approaching a limiting value for chain lengths greater than about ten [12].

Polyphosphate hydrolysis is very slow in sterile solutions. At 10°C, for example, the time required for 5% hydrolysis of a pyrophosphate solution would be about a year at pH 4, many years at pH 7, and over a century at pH 10. Hydrolysis rates are substantially increased at higher temperatures: 5% hydrolysis of a pyrophosphate solution at pH 7 would take under a month at 40°C and only a few days at 60°C [13]. However, much faster hydrolysis can result from catalysis by a number of enzymes, known as phosphatases, which are present in most microorganisms [13]. Their catalytic action is strongly affected by pH and the presence of metal ions. Such enzymes increase rates of hydrolysis by a factor of up to 10^6 [12]. Because this rapid hydrolysis may be catalyzed by microorganisms in raw sewage, much of the condensed phosphate in wastewater is converted to orthophosphate before it even reaches a waste treatment plant. Secondary biological treatment usually ensures the hydrolysis of all condensed phosphate in wastewater [13]. Environmental concerns related to phosphate discharges are discussed in § 8.3.2.

The direct application of polyphosphate fertilizers to the soil is another source of polyphosphates in the environment. Various cations in soil, such as Ca^{+2} and Fe^{+2}, react with liquid polyphosphate fertilizers and precipitate as complex polyphosphates [9]. However, the adsorption of polyphosphate on clay has been observed to increase its degradation rate [12]. Calculations of the equilibrium concentrations of various polyphosphate species show that pyrophosphate is unstable in soils, eventually reverting to orthophosphate. Longer-chain polyphosphates are thought to be even less stable [9].

2.8.5 Literature Cited

1. Baes, C.F., Jr. and R.E. Mesmer, *The Hydrolysis of Cations*, John Wiley & Sons, New York (1976).
2. Boulègue, J. and G. Michard, "Constantes de Formation des Ions Polysulfurés S_6^{-2}, S_5^{-5}, et S_4^{-2} en Phase Aqueuse," *J. Fr. Hydrol.*, **9**, 27-34 (1978).
3. Boulègue, J. and G. Michard, "Sulfur Speciations and Redox Processes in Reducing Environments," in *Chemical Modeling in Aqueous Systems*, E.A. Jenne (ed.), ACS Symposium Series 93, American Chemical Society, Washington, D.C. (1979).
4. Chen, K.Y., "Chemistry of Sulfur Species and Their Removal from Water Supply," in *Chemistry of Water Supply, Treatment and Distribution*, A.J. Rubin (ed.), Ann Arbor Science Publishers, Ann Arbor, Michigan (1974).
5. Chen, K.Y. and J.C. Morris, "Kinetics of Oxidation of Aqueous Sulfide by O_2," *Environ. Sci. Technol.*, **6**, 529-37 (1972).
6. Gimblett, F.G.R., *Inorganic Polymer Chemistry*, Butterworth and Co., London (1963).
7. Hudson, R.B. and M.J. Dolan, "Phosphoric Acid and Phosphates," in *Kirk-Othmer Encyclopedia of Chemical Technology*, 3rd ed., Vol. 17, John Wiley & Sons, New York (1982).
8. Licht, S., G. Hodes and J. Manassen, "Numerical Analysis of Aqueous Polysulfide Solutions and Its Application to Cadmium Chalcogenide/Polysulfide Photoelectrochemical Solar Cells," *Inorg. Chem.*, **25**, 2486-89 (1986).
9. Lindsay, W.L., *Chemical Equilibria in Soils*, John Wiley & Sons, New York (1979).
10. Morel, F.M.M., *Principles of Aquatic Chemistry*, John Wiley & Sons, New York (1983).

11. Nriagu, J.O. and J.D. Hem, "Chemistry of Pollutant Sulfur in Natural Waters," in *Sulfur in the Environment*, J.O. Nriagu (ed.), John Wiley & Sons, New York (1978).

12. Shen, C.Y. and F.W. Morgan, "Hydrolysis of Phosphorus Compounds," in *Environmental Phosphorus Handbook*, E.J. Griffith, A. Beeton, J.M. Spencer and D.T. Mitchell (eds.), John Wiley & Sons, New York (1973).

13. Snoeyink, V.R. and D. Jenkins, *Water Chemistry*, John Wiley & Sons, New York (1980).

14. Stumm, W., "Metal Ions in Aqueous Solutions," in *Principles and Applications of Water Chemistry; Proceedings*, S.D. Faust and J.V. Hunter (eds.), John Wiley & Sons, New York (1967).

15. Stumm, W. and J.J. Morgan, *Aquatic Chemistry: An Introduction Emphasizing Chemical Equilibria in Natural Waters*, 2nd ed., John Wiley & Sons, New York (1981).

2.9 COMPLEXATION

This section addresses the formation of aqueous complexes between metal ions and complexing agents (ligands). The various types of aqueous complexes, the importance of complexation in environmental systems and a mathematical representation (the complex formation or equilibrium constant) of the process under equilibrium conditions are presented. Values, trends, and methods for estimating the complex formation constant are provided.

2.9.1 Environmental Importance

Inorganic substances adapt to the particular aqueous chemical environments to which they are exposed, subject to kinetic limitations on the processes. Chemical environments include many potential ligands that can complex strongly or form loose ion pairs. The behavior of an aquo complexed metal ion varies considerably from that of other complexes of the same central ion in many chemical reactions. Chemical properties such as solubility, attenuation behavior on soils, bioconcentration factors and toxicity are modified through complexation.

2.9.2 Description of Property

Metal ions exist in aqueous systems in the form of complex ions, which are formed by reaction of the metal ions (e.g., Fe^{+2}) with ligands (usually neutral or anions such as H_2O or Cl^-). Ligands surround metal ions in what are termed "coordination spheres." A coordination sphere can be divided into two parts:

- The *primary* or *inner coordination sphere* consists of the ligands that are directly bonded to the metal ion.
- The *secondary* or *outer coordination sphere* consists of ligands associated (polarized or loosely coordinated) with the metal ion.

The *coordination number* of the metal ion (typically 6 or 4) refers to the number of ligand positions in the primary coordination sphere. Since the boundary between the secondary coordination sphere and the bulk solvent is vague, the total number of associated ligands is difficult to define and can greatly exceed the coordination number.

In aqueous systems, water molecules generally occupy most of the locations in both coordination spheres. In an *aquated metal ion* (designated M^{+x}, with the subscript aq assumed), all the ligands are water molecules. Complexation of these "aquo ions" by other ligands refers to either of the following processes:

$$M^{+x} + L^{-y} \rightleftharpoons (M,L)^{+x-y} \qquad K_{os} \tag{1}$$

$$(M,L)^{+x-y} \rightleftharpoons (M\text{-}L)^{+x-y} \qquad K_{is} \tag{2}$$

Equation 1 refers to substitution of one or more of the outer-sphere water molecules by a ligand (L^{-y}) to form an outer-sphere complex ion, $(M, L)^{+x-y}$. Equation 2 refers to substitution of one or more of the inner-sphere water ligands to form an inner-sphere complex, $(M\text{-}L)^{+x-y}$. The *outer-sphere* or *ion pair association constant* (at low ionic strength, where activity coefficients are 1), is defined as

$$K_{os} = \frac{[(M,L)^{+x-y}]}{[M^{+x}]\,[L^{-y}]} \tag{3}$$

where $[(M,L)^{+x-y}]$, $[M^{+x}]$, and $[L^{-y}]$ are the concentrations of the respective species. Similarly, the *inner-sphere complexation constant* is defined as:

$$K_{is} = \frac{[(M\text{-}L)^{+x-y}]}{[(M,L)^{+x-y}]} \tag{4}$$

The *complexation equilibrium constant* or *stability constant* as experimentally measured for many metal ions refers to the reaction:

$$M^{+x} + L^{-y} \rightleftharpoons (M\text{-}L)^{+x-y} \qquad K_1 \tag{5}$$

where

$$K_1 = \frac{[(M\text{-}L)^{+x-y}]}{[M^{+x}]\,[L^{-y}]} = K_{is}\,K_{os}$$

For many ions, the exchange between inner and outer coordination ligands is so rapid that it is not possible to distinguish between outer- and inner-sphere complexes; in this case, the measured equilibrium constant (K_1) actually refers to mixed inner and outer sphere complex formation. For other ions, the exchange between inner and outer spheres is so slow that only outer-sphere equilibria are observed.

The ligand water molecules in the inner coordination sphere of the metal ion can be successively replaced by other potential ligands. The equilibria are designated by individual equilibrium constants (called *stepwise stability constants*):

$$M + L \rightleftharpoons ML \qquad K_1 \tag{6}$$

$$ML + L \rightleftharpoons ML_2 \qquad K_2 \tag{7}$$

$$ML_2 + L \rightleftharpoons ML_3 \qquad K_3 \tag{8}$$

The *overall stability constant*, β, refers to the following reaction:

$$M + nL \rightleftharpoons ML_n \qquad \beta_n \tag{9}$$

where

$$\beta_n = K_1\,K_2\,K_3 \ldots K_n \tag{10}$$

Ligands are often classified as to the number of donor atoms that can form a bond to the metal ion. A *monodentate* ligand (e.g., $-OH_2$, $-NH_3$) has only one atom that can coordinate to the metal; *bidentate* and *tridentate* indicate that two and three atoms of the ligand, respectively, can coordinate to the metal at two or three ligand positions (e.g., ethylenediamine and diethylenetriamine). The formation of a complex between a polydentate ligand (more than one coordinating atom) and a metal ion is referred to as a *chelate complex ion.* As chelation markedly affects chemical behavior of the central metal ion, many polydentate ligands are used commercially to "sequester," i.e., prevent precipitation.

2.9.3 Values of Complexation Equilibrium Constants

Several compilations of complexation equilibrium constants are available [11,15,22,23]. Representative values for metals with common monodentate ligands are given in Table 2.9-1. Data for polydentate ligands used in commercial applications are summarized in Table 2.9-2. It is not unusual for reported values of complexation constants to differ by 0.25 to 0.5 log units, depending on the methods by which they were calculated or experimentally derived. The coordination numbers of metal ions (Table 2.9-3) are useful in determining how many ligand equilibria should be considered when assessing their complex speciation. For many kinetically labile ions, e.g., Ag^{+}, Cd^{+2}, Hg^{+2}, the coordination numbers in solution are not known [1]; for others (e.g., Ni^{+2}, V^{+3}, Fe^{+3}, Mn^{+2}, Fe^{+2}, Cu^{+2}, Ti^{+3}), spectral interpretations are consistent with a value of 6. There is some evidence [1] that rare earths have coordination numbers of 9 in solution. Values of some ion pair association constants are given in Table 2.9-4.

Naturally occurring organic compounds such as humic and fulvic acids, carbohydrates and proteinaceous material play an important role in controlling the solubilities and activities of metals in natural waters [14]. Many trace organics derived from degradation of plant and animal tissues have functional groups that can form chelates with metals. Humic and fulvic acids form quite stable complexes, especially when chelation rings such as the following can occur:

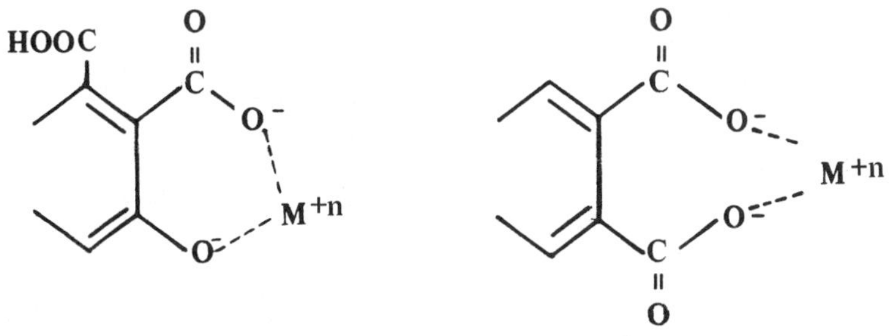

Humic acids are macromolecular materials of variable composition consisting of core units (having carboxyl, hydroxyl, carbonyl, and methoxy groups) plus hydrolyzed protein and carbohydrate fractions. Detailed complexation research is complicated by the lack of information on the structure of the macroligands and the fact that their composition varies with the geographic location and types of systems from which they are obtained. Thus, complexation equilibria constants in the literature are useful only for the particular systems tested.

A somewhat similar situation exists for fulvic acids, which are also macromolecular-weight acids and represent the alkali-soluble degradation products of humic acids. They contain carboxyl, phenolic, and aliphatic hydroxyl groups, but their exact structures are unknown. Data for their complexation constants have been reported [6,13,19,24], but, as with humic acids, their nature varies with the sources and conditions of formation. The pH of a solution often markedly affects the extent of

TABLE 2.9-1

Logarithms of Complexation Equilibrium Constants (K) and Overall Stability Constants (β_2) for Metals with Commonly Occurring Ligands
(at 25°C and zero ionic strength, unless noted)

Metal	M-Cl[a]	$M\text{-}Cl_2$[b]	$M\text{-}SO_4$[a]	M-OH[a]	$M(OH)_2$[b]
Na^+	−1.85	—	1.06	−0.2	—
K^+	−0.5	—	0.96	−0.5	—
Ca^{+2}	—	—	2.31	1.15	—
Mg^{+2}	—	—	2.36	2.56	—
Ba^{+2}	−0.13[d]	—	2.7	0.5	—
Cr^{+3}	0.6	—	3.0	10.0	18.3
Al^{+3}	—	—	—	9.0	18.7
Fe^{+3}	1.5	2.1	4.0	11.8	22.3
Mn^{+2}	0.6	—	2.3	3.4	5.8
Fe^{+2}	0.36[c]	0.40[c]	2.2	4.5	7.4
Co^{+2}	0.5	—	2.2	4.3	5.1
Ni^{+2}	0.6	—	2.3	4.1	9.0
Cu^{+2}	0.4	−0.4	2.4	6.5	11.8
Zn^{+2}	0.4	0.0	2.1	5.0	11.1
Pb^{+2}	1.6	1.8	2.8	6.3	10.9
Hg^{+2}	7.2	14.0	2.5	10.6	21.8
Cd^{+2}	2.0	2.6	2.3	3.9	7.6
Ag^+	3.3	5.3	1.3	2.0	4.0

Sources: Kotrlý and Šůcha [11]; Morel [15]; Sillén and Martell [22]

a. Refers to K values for the reaction $M + L \rightleftharpoons ML$ where $L = Cl^-, OH^-$ or SO_4^{-2}
b. Refers to β_2 values for the reaction $M + 2L \rightleftharpoons ML_2$ where $L = Cl^-$ (or OH^-)
c. 2 *M* ionic strength
d. 18°C

complexation of an organic species with a trace metal since H^+ competes for the basic coordination site. In addition, complexation constants based on purified samples are believed to be only approximations of those for natural samples [14]. Data for humic and fulvic acids complexes of trace metals are given in Chapter 7.

TABLE 2.9-2

Logarithms of Complexation Equilibrium Constants (K) for Metals with Sequestering Agents of Commercial Interest
(1:1 complex formation at 25°C and 0.10 *M* ionic strength, unless noted)

Metal	Sequestering Agent[a]					
	NTA	HIDA	EDTA	HEDTA	TPP	CIT[b]
Mg^{+2}	5.47	3.46	8.83	7.0	5.8	3.40
Ca^{+2}	6.39	4.77	10.61	8.2	5.2	3.55
Mn^{+2}	7.46	5.56	13.81	10.8	7.0	3.7
Fe^{+2}	8.82	6.77	14.27	12.2	—	4.4
Fe^{+3}	15.9	11.6	25.0	19.8	—	11.40
Co^{+2}	10.38	8.02	16.26	14.5	6.9	5.00
Ni^{+2}	11.50	9.33	18.52	17.1	6.7	5.40
Cu^{+2}	12.94	11.72	18.70	17.5	8.1	5.90
Zn^{+2}	10.66	8.45	16.44	14.6	7.3	4.98
Cd^{+2}	9.78	7.24	16.36	13.1	6.5	3.75
Hg^{+2}	14.6	5.4	21.5	20.0	—	—
Pb^{+2}	11.34	9.5	17.88	15.5	—	4.08[c]

a. NTA = nitrilotriacetic acid, $N(CH_2COOH)_3$
HIDA = Ethanolaminediacetic acid, $(HOCH_2CH_2)N(CH_2COOH)_2$
EDTA = ethylenediaminetetraacetic acid, $(HOOCCH_2)_2NCH_2CH_2N(CH_2COOH)_2$
HEDTA = N-(2-hydroxyethyl)-ethylene-dinitrilotriacetic acid, $(HOCH_2CH_2)(HOOCCH_2)NCH_2CH_2N(CH_2COOH)_2$
TPP = tripolyphosphoric acid, $(HO)_2OPOPO_2(OH)PO(OH)_2$
CIT = citric acid, $(HOOCCH_2(OH)(COOH)C(CH_2COOH)$

b. 20°C

c. 2.0 *M* ionic strength

Source: Martell [14]

TABLE 2.9-3

Experimental Coordination Numbers of Various Ions in Aqueous Media

Ion	Coordination Number[a]	Ion	Coordination Number[a]
Al^{+3}	6	Ni^{+2}	4 or 6
Be^{+2}	4	Sc^{+3}	5
Co^{+2}	6	Zn^{+2}	6
Ga^{+3}	6	Rh^{+3}	6
Mg^{+2}	6	Co^{+3}	6
Cr^{+3}	6	Li^{+}	4

a. Rounded to nearest integer

Sources: Basolo and Pearson [1]; Lincoln [12]

TABLE 2.9-4

Experimental Values of Ion Pair Association Constants at 25°C

Reaction	Ionic Strength (*M*)	K_{os} (M^{-1})
$Co(NH_3)_5H_2O^{+3} + SCN^- \rightleftharpoons [Co(NH_3)_5H_2O, SCN]^{+2}$	0.5	0.65
$Co(NH_3)_5H_2O^{+3} + N_3^- \rightleftharpoons [Co(NH)_3)_5H_2O, N_3]^{+2}$	2.0	0.26
$Ni(H_2O)_6{}^{+2} + MePO_4{}^{-2} \rightleftharpoons [Ni(H_2O)_6, MePO_4]$	0.1	41
$Ni(H_2O)_6{}^{+2} + NH_3{}^{+2} \rightleftharpoons [Ni(H_2O)_6, NH_3]^{+2}$	0.1	0.15
$Cr(H_2O)_6{}^{+3} + SCN^- \rightleftharpoons [Cr(H_2O)_6, SCN]^{+2}$	0	7
$Cr(H_2O)_6{}^{+3} + Cl^- \rightleftharpoons [Cr(H_2O)_6, Cl]^{+2}$	0	13
$Co(NH_3)_6{}^{+3} + SO_4{}^{-2} \rightleftharpoons [Co(NH_3)_6, SO_4]^{+}$	0	2.2×10^3

Sources: Basolo and Pearson [1]; Sillén and Martell [23]; Wilkins [27]

2.9.4 Estimation of Complexation Equilibrium Constants

Owing to the diversity of inorganic metals and complexing ions, and to the volume of experimental work that has been performed, the best sources of complexation data are individual studies and compilations. However, a few relationships have been derived to estimate these constants for a limited group of complexes. The following methods will be described:

- *Eigen-Fuoss equation* for estimation of outer-sphere (ion-pair) association constants;
- *Hancock and Marsicano approach* for estimating inner-sphere complexation constants for monodentate ligands, polydentate ligands containing nitrogen donor atoms, and aminocarboxylate ligands; and
- *Tanaka equations* for estimation of inner-sphere complexation constants for formation of ML_2 complexes and mixed complexes (two different ligands, excluding H_2O) of nickel and copper.

OUTER-SPHERE ASSOCIATION CONSTANTS

Outer-sphere (ion pair) association constants can be estimated with the following equation, which was derived independently by Eigen [3,4] and Fuoss [5]:

$$K_{os} = \frac{4\pi Na^3}{3000} \exp[-U(a)/kT] \tag{11}$$

where K_{os} = ion pair association constant for spherically symmetrical ions, i.e., all ligands identical (M^{-1})

N = Avogadro's number (6.023×10^{23}/mole)
a = distance of closest approach between ions (cm)
e = electron charge (4.80×10^{-10} esu)
k = Boltzmann constant (1.38×10^{-16} erg/K)
T = temperature (K)
U(a) = Debye-Hückel interionic potential = $(Z_1Z_2e^2/aD) - [Z_1Z_2e^2\theta/D(1+a\theta)]$
θ = $(8\pi Ne^2I/1000\ DkT)^{1/2}$
Z_1 = charge on Ion 1
Z_2 = charge on Ion 2
D = dielectric constant of the solvent (e.g., 78.5 for H_2O at 25°C)
I = ionic strength (mol/l)

As noted by Hammes and Steinfeld [7], this method for calculating K_{os} involves many approximations, including the following:

- The distance of closest approach (a) can only be estimated.
- Only electrostatic binding forces are considered.

- A macroscopic dielectric constant (D) is used at the molecular level.
- The term for calculating the Debye-Hückel interionic potential (U(a)) is known to become inaccurate at higher ionic strengths.

Values of K_{os} calculated by the above method are generally regarded in the literature as accurate to within a factor of 4 [7,16].

Establishing the value of "a" is one of the chief problems in using equation 11. An approximation can be obtained by the following equation:

$$a = \frac{|Z_1 Z_2| e^2}{3DkT} \tag{12}$$

Table 2.9-5 lists values of "a" estimated by equation 12 for various values of Z_1 and Z_2. (Note that the tabulated values must be converted to centimeters for use in equation 11). Some investigators have used crystallographic radii or empirically derived values that fit experimentally observed results; these values of "a" typically range from 3 to 6 Å.

TABLE 2.9-5

Initial Estimates of Parameter "a"
(using equation 12)

Z_1	Z_2	a (Å)
+1	−1	2.4
+1	−2	4.8
+2	−2	9.5
+1	−3	7.1
+2	−3	14.3
+3	−3	21.4

Another limitation of equation 11 is that ion pairing may not always occur — i.e., in some cases where ion pairs are predicted, none have been detected. This is true, for example, of a large class of 2:1 electrolytes such as the alkaline earth halides [17].

The ion pair association constant for an unsymmetrically coordinated metal ion (one having other ligands besides H_2O) with a total coordination number of 6 can be estimated as follows [7]:

$$K'_{os} = SK_{os} \tag{13}$$

where
- S = steric correction factor = n/6
- n = number of sites occupied by water molecules
- K_{os} = ion pair association constant for a spherically symmetrical ion

Calculated values of K_{os} are given in Table 2.9-6 for a variety of ion charges, ionic strengths and values of "a". The strong dependence of K_{os} on "a" is evident: the calculated value can change by nearly three orders of magnitude with a 2 Å variation in "a". The magnitude of the error depends on the charges of the ions. Changes in ionic strength have no effect on K_{os} if one of the ions in the pair has a zero charge, but deviations of four orders of magnitude can occur with highly charged species (+3 and −3) when the ionic strength increases from zero to 0.5 *M*.

Example 1 Calculate the outer-sphere association constant for the reaction between $Cr(H_2O)_6{}^{+3}$ and Cl^- at 25°C and an ionic strength of zero. The following parameter values are used with equation 11:

$$
\begin{aligned}
I &= 0\,M \\
D &= 78.5 \text{ (dielectric constant of water at 25°C)} \\
Z_1 &= +3 \text{ (charge on } Cr(H_2O)_6{}^{+3}\text{, Ion 1)} \\
Z_2 &= -1 \text{ (charge on } Cl^-\text{, Ion 2)} \\
a &= 7.1 \times 10^{-8} \text{ cm (using equation 12)} \\
T &= 298 \text{ K}
\end{aligned}
$$

Thus,

$$\theta = \sqrt{\frac{8\,(3.14)(6.023\times10^{23})(4.80\times10^{-10})^2\,(0)}{1000\,(78.5)(138\times10^{-16})(298)}}$$

$$= 0$$

and

$$U(a) = \frac{(+3)(-1)(4.80\times10^{-10})^2}{(7.1\times10^{-8})(78.5)} - \frac{(+3)(-1)(4.80\times10^{-10})^2(0)}{(78.5)[1 + (7.1\times10^{-8})(0)]}$$

$$= -1.24\times10^{-13}$$

Therefore,

$$K_{os} = \frac{4\,(3.14)(6.023\times10^{23})(7.1\times10^{-8})^3}{3000}\ \exp\left[\frac{-(-1.24\times10^{-13})}{(1.38\times10^{-16})(298)}\right]$$

$$= 0.902\exp(3.02) = 18.5\,M^{-1}$$

The experimental value is 13 [1].

INNER-SPHERE COMPLEXATION CONSTANTS (HANCOCK AND MARSICANO APPROACH)

Monodentate Ligands

The approach to estimation of inner-sphere complexation equilibrium constants (formation constants) is based upon the presence of linear free energy relationships between properties of the metal and the ligand. Such relationships are observed, for example, in logarithmic plots of the formation constant of a metal with a particular ligand (K_1) *versus* equilibrium constant of the reaction of the ligand with another metal ion or with H^+ (K_2), according to the equation

$$\log K_1 = A \log K_2 \tag{14}$$

TABLE 2.9-6

Outer-sphere Formation Constants Calculated by Eigen-Fuoss Equation
(in aqueous solution at 298 K; see eq. 11 for definitions of symbols)

Ion Pair	Z_1	Z_2	I(*M*)	a(Å)	$K_{os}(M^{-1})$
$(M^{+x})(L^0)$	+x	0	0	3	0.068
	+x	0	0	4	0.161
	+x	0	0	5	0.315
	+x	0	0	6	0.545
	+x	0	0.5	5	0.315
$(M^{+1})(L^{-1})$	+1	−1	0	2.4	0.68
	+1	−1	0	5	1.30
	+1	−1	0.5	4	0.407
$(M^{+2})(L^{-1})$	+2	−1	0	4.8	5.45
	+2	−1	0.5	4.8	1.14
$(M^{+3})(L^{-1})$	+3	−1	0	5	22.7
	+3	−1	0	7.1	18.40
	+3	−1	0.5	5	2.58
$(M^{+2})(L^{-2})$	+2	−2	0	5	94.8
	+2	−2	0	9.5	94.8
	+2	−2	0.5	4	6.51
$(M^{+3})(L^{-2})$	+3	−2	0	5	1642
	+3	−2	0	14.3	147
	+3	−2	0.5	5	16.5
$(M^{+3})(L^{-3})$	+3	−3	0	3	1.33×10^8
	+3	−3	0.5	3	2.04×10^4
	+3	−3	0	4	1.50×10^6
	+3	−3	0.5	4	661
	+3	−3	0	5	1.19×10^5
	+3	−3	0	21.4	496

X = charge does not matter

Equation 14 is not applicable to all metals and ligands, since this would imply common behavior of all ligands with all metals (e.g., only ionic or covalent bonding and no steric or desolvation effects). Equation 14 is likely to apply to one metal with a variety of ligands of similar structure, such as Cu(II) with meta- and para-substituted pyridines.

Edwards [2] modified equation 14 to introduce the relative tendency to form ionic or covalent bonds:

$$\log K_1 = E_A E_B + C_A C_B \quad (15)$$

where A and B refer to the Lewis acid (metal) and Lewis base (ligand), and E_i and C_i (where i = A or B) are empirically derived parameters (some initially based on physical measurements, such as the pK_a of the ligand), adjusted to give the best overall fit to stability constant data. E and C were chosen to represent the extent of ionic or covalent bonding.

A recent application of equation 15 with all parameters adjustable [9] showed that its power of prediction was unsatisfactory, unless confined to a particular class of ligands having first-row donor atoms (e.g., F^-, amines, alkyl-oxygen type ligands). This shortcoming was attributed to the neglect of specific solvation effects, such as the disruption of coordinated water molecules around the Lewis acid (metal) by the close approach of Lewis base (ligand). These solvation effects were expected to be more severe with smaller Lewis acids and larger Lewis bases, the conditions where equation 15 appeared to be less useful. Accordingly, a correction term (parameters D_A and D_B) was introduced by Hancock and Marsicano [8,9]:

$$\log K_1 = E_A E_B + C_A C_B - D_A D_B \quad (16)$$

The parameters E, C, and D are derived empirically from relationships between formation constants and by setting initial values for some parameters; the subscripts A and B refer to the Lewis acid (metal) and Lewis base (ligand), respectively. Values of E, C and D are given in Table 2.9-7 for metals and Table 2.9-8 for ligands. (See the former table for an explanation of how the values are derived). Good agreement has been observed between calculated and experimentally observed values for a variety of complexes.

Example 2 Calculate the formation constant of the monochloro complex of Hg^{+2}. The reaction of interest is:

$$Hg^{+2} + Cl^- \rightleftharpoons HgCl^+ \qquad K_1$$

where Hg^{+2} is the "acid" and Cl^- is the "base." The following parameter values are obtained from Tables 2.3.4-7 and -8:

$E_A = 1.346$
$C_A = 0.826$
$D_A = 0.0$

TABLE 2.9-7

Values of Empirical Parameters E_A, C_A, and D_A for Cations
(for calculation of aqueous formation constants by equation 16)

Cation	E_A	C_A	D_A
Ag^{+}	−1.516	0.143	0.0
Al^{+3}	6.899	0.657	2.0
Bi^{+3}	5.917	0.926	0.0
Ca^{+2}	0.985	0.081	0.0
Cd^{+2}	0.993	0.300	0.6
Co^{+2}	1.198	0.276	3.0
Co^{+3}	3.299	0.875	7.0
Cr^{+3}	5.148	0.721	1.5
Cu^{+}	−0.559	0.43	2.5
Cu^{+2}	1.249	0.466	6.0
Fe^{+2}	1.521	0.256	2.0
Fe^{+3}	6.072	0.841	1.5
Ga^{+3}	6.060	0.788	1.5
H^{+}	3.067	1.099	20
Hg^{+2}	1.346	0.826	0.0
CH_3Hg^{+}	1.600	0.64	0.0
In^{+3}	4.498	0.714	0.5
La^{+3}	3.904	0.379	0.0
Lu^{+3}	4.572	0.454	0.0
Mg^{+2}	1.862	0.178	1.5
Mn^{+2}	1.581	0.223	1.0
Ni^{+2}	1.011	0.300	4.5
Pb^{+2}	2.763	0.413	0.0
Sc^{+3}	7.039	0.671	0.0
Sn^{+2}	5.649	0.700	0.0
Tl^{+3}	2.554	0.96	0.0
U^{+4}	7.550	0.968	3.0
UO_2^{+2}	4.948	0.589	1.0
VO_2^{+2}	3.964	0.664	3.5
Y^{+3}	2.235	0.447	0.0
Zn^{+2}	1.329	0.312	4.0

Source: Calculated from data derived by Hancock and Marsicano [9], who give values for C_A, D_A, and H_A (where $H_A = E_A/C_A$). Hancock and Marsicano used the following procedure to derive these parameters: *(see next page)*

TABLE 2.9-7 (Continued)

(1) Set E_B for F^- = 1, C_B for F^- = 0, E_B for OH^- = 0 and C_B for OH^- = 14.

(2) From Equation 15, $\log K_1 (F^-) = E_A$ and $\log K_1 (OH^-) = 14C_A$, which allows calculation of E_A and C_A for the metal (acid) ions. The experimentally obtained complexation constants (K) are corrected for outer-sphere contribution by the equation $\log K_{is} = \log K - \log K_{os}$, where K_{os} is calculated by equation 11.

(3) Rewriting equation 15 as $\frac{\log K_1}{C_A} = \frac{E_A}{C_A}(E_B) + C_B$, plot $\frac{\log K_1(NH_3)}{C_A}$ versus $\frac{E_A}{C_A}$ to derive the slope of E_B and intercept of C_B for the ligands (bases).

(4) E_A and C_A can now be recalculated to give a better overall fit of log K(OH), log K(F) and log $K(NH_3)$ data.

(5) Using the revised values of E_A and C_A and rewriting equation 16 as $\frac{\log K\text{ (ligand)}}{C_A} = \frac{E_A}{C_A}(E_B) + C_B - \frac{D_A D_B}{C_A}$, plot values of $\frac{\log K\text{ (ligand)}}{C_A}$ versus E_A/C_A. E_B is then obtained from the slope and $(C_B - \frac{D_A D_B}{C_A})$ from the intercept.

(6) Set D_B = 1 for Br^-.

(7) Calculate D_A, D_B and C_B values for other metals and ligands from the results of Step 5.

TABLE 2.9-8

Values for Empirical Parameters E_B, C_B, and D_B for Ligands
(for calculation of aqueous formation constants by equation 16)

Ligand	E_B	C_B	D_B
$As(C_6H_4SO_3)_3^{-3}$	−1.93	14.3	—
Br^-	−1.54	14.2	1.0[a]
CH_3COO^-	0.0	4.76	0.0
C_5H_5N	−0.74	7.0	0.0
Cl^-	−1.04	10.4	0.6
CN^-	−4.43	30.0	0.38
F^-	1.00[a]	0.0[a]	0.0
I^-	−2.43	20.0	1.7
$N{=}N{=}N^-$	−0.67	10.4	0.2
$N{\equiv}C{-}S^-$	−1.83	14.3	1.0
$(NH_2)_2C{=}S$	−2.46	18.2	0.6
$(HOCH_2CH_2)_2S$	−1.36	10.1	0.6
NH_3	−1.08	12.34	0.0
SCN^-	−0.76	9.3	0.2
SO_3^{-2}	−1.94	18.2	0.4
$S_2O_3^{-2}$	−3.15	26.5	1.1
OH^-	0.0[a]	14.00[a]	0.0

a. Arbitrarily set

Source: Hancock and Marsicano [9]

$E_B = -1.04$
$C_B = 10.4$
$D_B = 0.6$

Substituting these values in equation 16,

$$\log K_1 = (1.346)(-1.04) + (0.826)(10.4) - (0.0)(0.6) = 7.19$$

The experimental value is 7.37 [9].

Polydentate Ligands Containing Nitrogen Donor Atoms

Equation 17 relates the complexation constants of polyamine ligands to those of ammonia, where the polyamines are of the type $H(NHCH_2CH_2)_{n-1}NH_2$.

$$\log K_1 \text{ (polyamine)} = 1.152n \log K_1 \text{ (ammonia)} - \left(\sum_{i=1}^{n-1} i\right)\lambda_N + (n-1)\log 55.5 \qquad (17)$$

where $\log K_1$ (polyamine) = complexation constant for mono-polyamine complex
$\log K_1$ (ammonia) = complexation constant for mono-ammonia complex
n = defined by the structure of the ligand[1] ($n \geq 2$)
λ_N = 0.5 (for all complexes of polyamines forming five-membered chelate rings)

As indicated by the definition of λ_N, this equation applies only to complexes with five-membered chelate rings. Complexation constants for ligands with larger chelate rings are generally less than predicted via this equation. The equation is also not valid for situations where the polydentate ligand is unable to assume the same coordination geometry as the monodentate analog. The equation thus is not well suited for linearly coordinated Hg^{+2}, tetrahedral Cd^{+2} and Zn^{+2}, and tetragonally distorted ions such as VO^{+2}, UO_2^{+2} and Cu^{+2} [8].

Example 3 Estimate the complex formation constant of Pb^{+2} with NH_3 if the complex formation constant is 10.4 for TRIEN and 10.9 for TETREN. The following parameters are used:

$\log K_1$ (TRIEN) = 10.4

$n = 4$

$\log K_1$ (ammonia) = unknown

$$\sum_{i=0}^{n-1} i = 1 + 2 + 3 = 6$$

$\lambda_N = 0.5$

Substituting in equation 17,

1 Ethylenediamine (EN) = 2; diethylenetriamine (DIEN) = 3; triethylenetetramine (TRIEN) = 4; tetraethylenepentamine (TETREN) = 5.

$$10.4 = 1.152\,(4)\log K_1\,(\text{ammonia}) - 6\,(0.5) + (3)\log 55.5$$
$$\log K_1\,(\text{ammonia}) = (10.4 + 3.0 - 5.23)/4.6$$
$$= 1.77$$

If we had used the complex formation constant for TETREN (10.9), the corresponding calculation would have been

$$10.9 = 1.152\,(5)\log K_1\,(\text{ammonia}) - 10\,(0.5) + (4)\log 55.5$$
$$\log K_1\,(\text{ammonia}) = (10.9 + 5.0 - 6.98)/5.76$$
$$= 1.55$$

The average value calculated for the $Pb^{+2}(NH_3)$ complexation constant is thus 1.66. The value has not been determined experimentally.

Aminocarboxylate Chelating Ligands

Hancock [8] developed an equation that relates the logarithm of the complexation constant for aminocarboxylate ligands to the logarithms of the complexation constants of acetate and ammonia with various metals. The relationship is:

$$\log K_1\,(\text{aminocarboxylate}) = 1.152n\log K_1\,(\text{ammonia}) + m\log K_1\,(\text{acetate}) - \left(\sum_{i=0}^{n-1} i\right)\lambda_N - \left(\sum_{i=1}^{m} i\right)\lambda_O + (n + m - 1)\log 55.5 \qquad (18)$$

where K_1(aminocarboxylate) = complexation constant for mono-aminocarboxylate complex formation

K_1(ammonia) = complexation constant for mono-ammonia complex formation

n = number of nitrogen donor groups on the aminocarboxylate ligand[2]

m = number of carboxylate groups on the aminocarboxylate ligand[2]

K_1(acetate) = complexation constant for monoacetate complex formation

λ_O = empirical parameter set equal to 0.26 log K_1(acetate) − 0.19

λ_N = 0.5 (for all complexes containing five-membered chelate rings)

2.

Ligand	n	m
NTA	1	3
EDDA	2	2
DTMA	3	1
HEDTRA	2	3
EDTA	2	4

EDDA = Ethylenediamine-N,N′-diacetic acid. Other abbreviations defined in footnote 1.

Example 4 Estimate the formation constant of Pb^{+2} with nitrilotriacetate (NTA).

The following parameters apply:

n = number of nitrogen groups in NTA = 1
m = number of carboxylate groups in NTA = 3
log K_1(ammonia) = 1.66 (from Example 3)
log K_1(acetate) = 2.68 [21]
λ_O = 0.26 (2.68) − 0.19 = 0.51

$$\sum_{i=0}^{n-1} i = 0$$

$$\sum_{i=1}^{m} i = 1 + 2 + 3 = 6$$

λ_N = 0.5

Substituting in equation 18,

$$\log K_1 \text{ (NTA)} = 1.152(1)(1.66) + 3(2.68) - 0(0.5) - 6(0.51) + 3(1.744)$$
$$= 12.14$$

Observed values are 11.34 (Table 2.9-2) and 11.8 [21].

TANAKA EQUATIONS FOR OVERALL FORMATION CONSTANTS OF ML_2 AND MAL COMPLEXES

Tanaka [25,26] derived the following equation for predicting NiL_2 complexation constants from NiL data:

$$\log K_{ML_2} = \log K_{ML} + \Delta + \log\left(\frac{\text{No. of } H_2O \text{ in ML}}{\text{No. of } H_2O \text{ in M}}\right) + \delta_{NN} N^2$$
$$+ \delta_{NO} N\underline{O} + \delta_{ON} \underline{O}N + \delta_{OO} \underline{O}^2 - \log 2 \quad (19)$$

where K_{ML_2} = complexation constant of ML with L
K_{ML} = complexation constant of M with L
Δ = $\log K_{os(ML,L)} - \log K_{os(M.L)}$ = difference between logarithms of outer-sphere complexation constants
δ_{NN} = parameter for interaction of amine N in L with amine nitrogen in second ligand L
δ_{ON} = as above for carboxylate in L with amine nitrogen in second L
δ_{OO} = as above for carboxylate in L with carboxylate in second L
δ_{NO} = as above for amine N in L with a carboxylate in second ligand L
N = number of amine nitrogens coordinated in L
$\underline{O}$ = number of carboxylate groups coordinated in L

If L is uncharged, the value of Δ is about zero; if L is mono-anionic, $\Delta = 0.4$ at $\mu = 0.1\ M$ (using a = 5Å). For Ni complexes:[3]

$$\delta_{NN} = -0.25, \qquad \delta_{NO} = 0.11$$
$$\delta_{ON} = 0.11, \qquad \delta_{OO} = -0.16$$

This equation and the related parameters were derived for ligands with no steric hindrance in their formation. Also note that all ligands contained only aliphatic nitrogen donor atoms.

Example 5 Calculate the second stability constant of Ni^{+2} complexation with NH_3. The following parameters apply:

$\log K_{Ni(NH_3)} = 2.80$ [21]
No. of H_2O in $Ni(NH_3)^{+2} = 5$
No. of H_2O in $Ni^{+2} = 6$
$\Delta = \log K_{os(Ni(NH_3),NH_3)} - \log K_{os(Ni,NH_3)} \approx 0$
$N = 1$
$\underline{O} = 0$

Using equation 19,

$$\log K_{Ni(NH_3)_2} = \log K_{Ni(NH_3)} + \Delta + \log\left[\frac{\text{No. of } H_2O \text{ in } Ni(NH_3)}{\text{No. of } H_2O \text{ in } Ni}\right]$$

$$- 0.25(1)^2 + 0.11(1)(0) + 0.11(0)(1) - 0.16(0)^2 - \log 2$$

$$= 2.80 + 0 + \log(5/6) - 0.25 + 0 + 0 - 0 - 0.30 = 2.17.$$

The observed value is 2.05 [22].

For predicting CuL_n and CuAL (i.e. where A is a different ligand than L) complexation constants from data available for CuL, Tanaka [24] derived the following equations:

$$\log K_{CuAL} = \log K_{CuL} + \Delta_1 + \log\left[\frac{\text{No. of } H_2O \text{ in } CuA}{\text{No. of } H_2O \text{ in } Cu}\right] + \sum_i \sum_j \delta_{ij} X_i(A)\ Y_j(L) \quad (20)$$

with $\Delta_1 = [\log K_{os(CuA,L)} - \log K_{os(Cu,L)}]$ and

$$\log K_{CuL_n} = \log K_{CuL} + \Delta_2 + \log\left(\frac{\text{No. of } H_2O \text{ in } CuL_{n-1}}{\text{No. of } H_2O \text{ in } Cu}\right)$$

$$+ \sum_i \sum_j \delta_{ij} X_i(L)\ Y_j(L) - \log n \quad (21)$$

with $\Delta_2 = [\log K_{os(CuL_{n-1},L)} - \log K_{os(Cu,L)}]$

3. Ligands used to derive δ_{NN}, δ_{ON}, and δ_{NO} were NH_3, NTA, EDDA, HEDTA, EN, TETREN, GLA, and OX. Those used to derive δ_{OO} were glycine, alanine, β-alanine, sarcosine, serine, asparagine, N, N-dimethylglycine, α-amino-α-methylpropionic acid and N-ethylglycine.

where K_{CuL} = complexation constant for Cu with L

K_{os} = outer-sphere complexation constant of pair in ( )

δ_{ij} = empirical parameter denoting effect of donor atoms i in one ligand on donor atom j in a second ligand

X_i = number of donor atoms i in one ligand

Y_j = number of donor atoms j in other ligand

A = first ligand

L = second ligand

n = number of L ligands on metal

Equation 20 becomes equation 22 for a copper complex containing ligands having aromatic nitrogens [N(AR)], aliphatic nitrogen [N(AL)] and oxygen donor atoms [$\underline{O}$] as follows (see also example 6):

$$\log K_{CuAL} = \log K_{CuL} + \Delta_1 + \log\left(\frac{\text{No. of } H_2O \text{ in CuA}}{\text{No. of } H_2O \text{ in Cu}}\right)$$

$$+ \delta_{N(AL)N(AL)} \times N_{N(AL)}(A) \times N_{N(AL)}(L) + \delta_{N(AR)N(AR)} \times N_{N(AR)}(A) \times N_{N(AR)}(L)$$
$$+ \delta_{N(AL)N(AR)} \times N_{N(AL)}(A) \times N_{N(AR)}(L) + \delta_{N(AR)\underline{O}} \times N_{N(AR)}(A) \times \underline{O}_o(L)$$
$$+ \delta_{N(AL)O} \times N_{N(AL)}(A) \times \underline{O}_o(L) + \delta_{\underline{O}O(\pi)}\, \underline{O}_{o(\pi)}(A) \times \underline{O}_{o(\pi)}(L)$$
$$+ \delta_{\underline{O}O(\sigma)} \underline{O}_{o(\sigma)}(A) \times \underline{O}_{o(\sigma)}(L) \qquad (22)$$

where

$\delta_{N(AL)N(AL)}$ = −0.35 (aliphatic N interactions)

$\delta_{N(AR)N(AR)}$ = −0.39 (aromatic N interactions)

$\delta_{N(AL)N(AR)}$ = −0.25 (aromatic and aliphatic N interaction)

$\delta_{N(AR)\underline{O}}$ = +0.09 (aromatic N and O interaction)

$\delta_{N(AL)\underline{O}}$ = −0.26 (aliphatic N and O interaction)

$\delta_{\underline{O}O(\sigma)}$ = −0.29 (for O in a σ bonding system)

$\delta_{\underline{O}O(\pi)}$ = 0.10 (for O in a π bonding system)

N,$\underline{O}$ = number of N and O donor atoms respectively, on the ligand (A or L, as noted in parentheses); subscript notation is as defined above for the various parameters

Equation 21 similarly becomes equation 23 for Cu complexes having aromatic or aliphatic nitrogens and oxygen donor atoms in the various ligands as follows:

$$\log K_{CuL_n} = \log K_{CuL} + \Delta_2 + \log\left(\frac{\text{No. of } H_2O \text{ in } CuL_{n-1}}{\text{No. of } H_2O \text{ in Cu}}\right)$$

$$+ \delta_{N(AL)N(AL)} \times N_{N(AL)}(L) \times N_{N(AL)}(L) + \delta_{N(AR)N(AR)} \times N_{N(AR)}(L) \times N_{N(AR)}(L)$$

$$+ \delta_{N(AL)N(AR)} \times N_{N(AL)}(L) \times N_{N(AR)}(L) + \delta_{N(AR)\underline{O}} \times N_{N(AR)}(L) \times \underline{O}_o(L)$$

$$+ \delta_{N(AL)\underline{O}} \times N_{N(AL)}(L) \times \underline{O}_o(L) + \delta_{\underline{OO}(\pi)} \times \underline{O}_{o(\pi)}(L) \times \underline{O}_{o(\pi)}(L)$$

$$+ \delta_{\underline{OO}(\sigma)} \times \underline{O}_{o(\sigma)}(L) \times \underline{O}_{o(\sigma)}(L) \quad (23)$$

The definitions and parameter values derived for Cu(II) complexes are the same as given for equation 22. Based upon these relationships, the values of the ligand interaction terms (δ_{ij}) given above provide reasonably close estimates of formation constants of mixed and higher Cu(II) complexes. These relationships have been used for calculating complexes where at least one non-charged amine is a ligand; this reduces the term Δ to about zero, which eliminates the problem of calculating outer-sphere formation constants and the need to assume a valid interatomic distance (parameter "a") for the ion pair.

The manner in which the empirical parameters (δ_{ij}) were derived is summarized below, to indicate the kinds of ligands for which equations 22 and 23 are useful.

- $\delta_{N(AL)N(AL)}$ was derived from data for complexes containing only aliphatic nitrogen donor atoms (e.g., 1,3-diaminopropane, NH_3, and N-n-propyl-1,2-diaminoethane).
- $\delta_{N(AR)N(AR)}$ was derived from data for complexes containing aromatic nitrogen donor atoms only (e.g., pyridine, imidazole, and 2,6 lutidine).
- $\delta_{N(AR)N(AL)}$ was estimated from data for complexes containing mixed aliphatic nitrogen and aromatic nitrogen ligands (e.g., triethylenetriamine and pyridine; 1,2-diaminoethane and bipyridine).
- $\delta_{N(AR)\underline{O}}$ and $\delta_{N(AL)\underline{O}}$ were estimated from complexation data for aromatic nitrogen and oxygen donor (e.g., formate, acetate, benzoate) mixed complexes and aliphatic nitrogen and oxygen donor ligands, respectively.
- $\delta_{\underline{OO}}$ was derived from oxygen donor atom ligands, both for those having no pi system (e.g., formate, methoxyacetate) and those having pi system(s) directly connected to the oxygen donor (e.g., acetylacetonate, salicylaldehyde).

Example 6 Estimate the complexation constant for the (oxalato) (1,2-diaminoethane) copper(II) complex if the logarithm of the formation constant for the oxalato copper(II) complex is 5.7.

The structure of the oxalato ligand (L) is 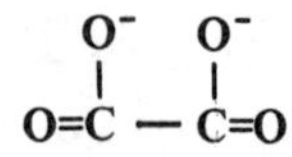, and the structure of the 1,2-diaminoethane ligand (A) is $NH_2-CH_2-CH_2-NH_2$.

Thus, the Cu-oxalato (Cu-L) complex is

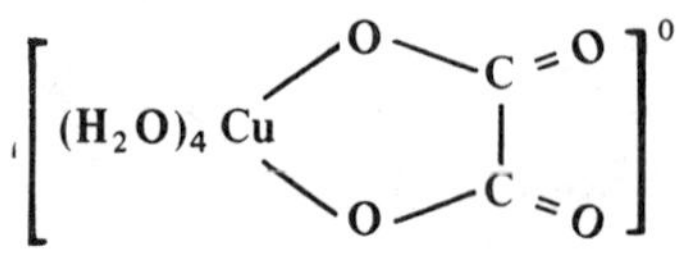

The Cu-1,2 diaminoethane complex is

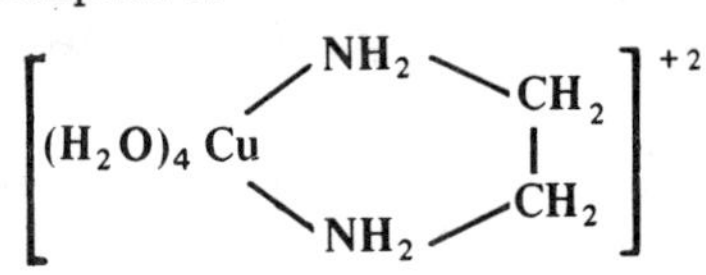

and the mixed Cu (oxalato) (1,2-diaminoethane) complex is

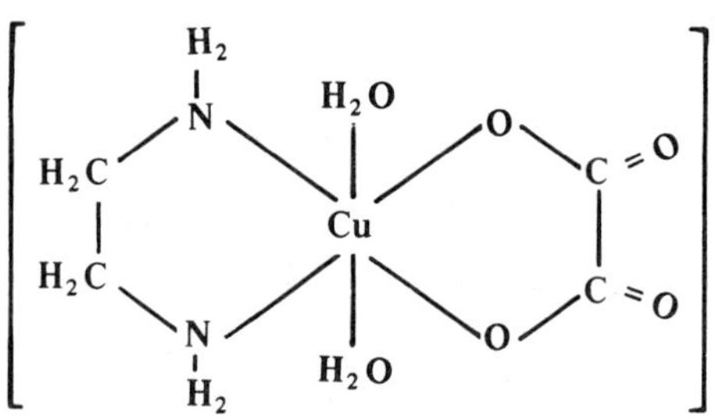

We now use equation 22 with the following parameter values (which are specific to copper):

(A = 1,2 diaminoethane, L = oxalate)

$\log K_{CuL} = 5.7$

$\Delta_1 = 0$, since the charge product ($Z_1 Z_2$) is both $+4$ for CuA,L and Cu,L (A has no charge); $K_{os(CuA,L)} \approx K_{os(Cu,L)}$

No. of H_2O in CuA = 4

No. of H_2O in Cu = 6 (Cu exists as $Cu(H_2O)_6{}^{+2}$)

$\delta_{N(AL)N(AL)}$ = -0.35 (interaction between aliphatic nitrogens)

$N_{N(AL)}(A)$ = 2 (no. of aliphatic nitrogen electrons in ligand A)

$N_{N(AL)}(L)$ = 0 (no. of aliphatic nitrogen atoms in ligand L)

$\delta_{N(AR)N(AR)}$ = -0.39 (interaction between aromatic nitrogens)

$N_{N(AR)}(A)$ = 0 (no. of aromatic nitrogens in ligand A)

$N_{N(AR)}(L)$ = 0 (no. of aromatic nitrogen atoms in ligand L)

$\delta_{N(AL)N(AR)}$ = −0.25 (interaction between aliphatic and aromatic nitrogens)

$\delta_{N(AR)\underline{O}}$ = 0.09 (interaction between aromatic nitrogen and oxygen)

$\underline{O}_o(L)$ = 2 (no. of carboxylate groups in ligand L)

$\delta_{N(AL)\underline{O}}$ = −0.26 (interaction between aliphatic nitrogen and oxygen)

$\delta_{\underline{O}O(\pi)}$ = −0.10 (interaction between oxygens in a π system)

$\underline{O}_{o(\pi)}(A)$ = 0 (no. of oxygen atoms in a π system in ligand A)

$\underline{O}_{o(\pi)}(L)$ = 0 (no. of oxygen atoms in a π system in ligand L)

$\delta_{\underline{O}O(\sigma)}$ = 0.10 (interaction between oxygens in a σ system)

$\underline{O}_{o(\sigma)}(A)$ = 0 (no. of oxygen atoms in a σ system in ligand A)

$\underline{O}_{o(\sigma)}(L)$ = 2 (no. of oxygen atoms in ligand L in a σ system)

Therefore,

$$\log K_{CuAL} = 5.7 + 0 + \log(4/6) + 0 + 0 + 0 + 0 + (-0.26)(2)(2) + 0 + 0 = 4.48.$$

The observed value is 4.6 [23].

2.9.5 Qualitative Trends for Complexation of Ions

Studies of the complexing ability of metals by Schwarzenbach [20] and Petrucci [17] have revealed the following trends, which allow qualitative estimation of the relative magnitude of complexation constants within particular series of metals and ligand donor atoms:

(1) For metal ions with noble gas configurations such as the alkali and alkali earth metal ions and Al^{+3}, Sc^{+3}, La^{+3}, Ti^{+4}, Zr^{+4}, Hf^{+4}, Th^{+4}, Nb^{+5}, and Ta^{+5}, purely electrostatic bonding predominates. The association constants generally are found to increase with decreasing cationic size and increasing charge for a particular ligand according to

$$K^+ < Na^+ < Li^+ \text{ and } Ba^{+2} < Sr^{+2} < Ca^{+2} < Mg^{+2}$$

For oxyacid anions (e.g., NO_3^-, SO_4^{-2}, $S_2O_3^{-2}$), the sequence for magnitude of the stability constants is reversed. For example, complexation comstants ($\log \beta_1$ at zero ionic strength) for acetate follow the sequence Ca^{+2} (1.18) > Sr^{+2} (1.14) > Ba^{+2} (1.07) [1]. The sequence for complexation by nitrate is Ba^{+2} (0.9) > Sr^{+2} (0.8) > Ca^{+2} (0.7) [1]. Ligands with F and O donor groups (e.g., OH^-) are preferred by these metal ions over N-donor groups (e.g., NH_3, CN^-).

(2) For metal ions with completely filled d subshells (18 electrons) such as Cu^+, Ag^+, Au^+, Tl^+, Zn^{+2}, Cd^{+2}, and Hg^{+2}, the bonding with ligands is mostly covalent. The properties that determine complexation are the electronegativity of the metal and of the ligand donor atom. An increase in the association constant

is observed with increasing electronegativity of the metal for a particular ligand. For a particular metal, an increasing association constant is observed with an increase in electron donating (decrease of electronegativity) of the ligand. The association constants thus generally decrease from S~I>Br>Cl>N>O>F. NH_3 is preferred over H_2O and CN^- over OH^-.

(3) The association constants for metal ions with incomplete subshells (such as Mn^{+2}, Fe^{+2}, Co^{+2}, Ni^{+2}, and Cu^{+2}) depend on both ionization potential and ionic radii. These ions follow the Irving-Williams series [10] for a particular ligand. The order of complexing is:

$$Mn^{+2} < Fe^{+2} < Co^{+2} < Ni^{+2} < Cu^{+2} > Zn^{+2}$$

with nitrogen, carbon, and sulfur ligand donor atoms. For example, the sequence for complexation constants (log β_1 at zero ionic strength) for NH_3 is Mn^{+2} (1.00) < Fe^{+2} (1.40) < Co^{+2} (1.99) < Ni^{+2} (2.72) < Cu^{+2} (4.04) > Zn^{+2} (2.21) [1]. Additional discussion of this series is given in Reference 18.

2.9.6 Literature Cited

1. Basolo, F. and R.G. Pearson, *Mechanisms of Inorganic Reactions*, 2nd. ed., John Wiley & Sons, New York, 37 (1967).
2. Edwards, J.O., "Correlation of Relative Rates and Equilibria with Double Basicity Scale," *J. Am. Chem. Soc.*, **76,** 1540 (1954).
3. Eigen, M., "Kinetics of High Speed Ion Reactions in Aqueous Solution," *Z. Phys. Chem.*, **1,** 176 (1954).
4. Eigen, M., "Metal Complexes and Protolytic Reactions in Aqueous Solution," *Z. Electrochem.*, **64,** 115 (1960).
5. Fuoss, R.M., "Ionic Association: III. The Equilibrium Between Ion Pairs and Free Ions," *J. Am. Chem. Soc.*, **80,** 5059 (1958).
6. Gamble, D., M. Schnitzer and I. Hoffman, "Cu(II)-Fulvic Acid Chelation Equilibrium in 0.1 *M* Potassium Chloride at 25°C," *Can. J. Chem.*, **48,** 3197 (1970).
7. Hammes, G.G. and J.I. Steinfeld, "Relaxation Spectra of Some Nickel(II) and Cobalt(II) Complexes," *J. Am. Chem. Soc.*, **84,** 4639 (1962).
8. Hancock, R.D. and F. Marsicano, "Parametric Correlation of Formation Constants in Aqueous Solution — Ligands with Small Donor Atoms," *Inorg. Chem.*, **17,** 560 (1978).
9. Hancock, R.D. and F. Marsicano, "Patterns in Lewis Acid-Base Interactions in Aqueous Solution," *S. Afr. J. Chem.*, **33,** 77 (1980).
10. Irving, H. and R.J.P. Williams, "The Stability of Transition Metal Complexes," *J. Chem. Soc.*, 3192 (1953).
11. Kotrlý, S. and L. Šůcha, *Handbook of Chemical Equilibria in Analytical Chemistry,* John Wiley & Sons, New York (1985).

12. Lincoln, S.F., "Solvent Coordination Numbers of Metal Ions in Solution," *Coord. Chem. Rev.,* **6,** 309 (1971).
13. Manning, P.G. and S. Ramamoorthy, "Equilibrium Studies of Metal-Ion Complexes of Interest to Natural Waters — VII," *J. Inorg. Nucl. Chem.,* **35,** 1577 (1973).
14. Martell, A.E., "The Influence of Natural and Synthetic Ligands on the Transport and Function of Metal Ions in the Environment," *Pure Appl. Chem.,* **44,** 81 (1975).
15. Morel, F.M.M., *Principles of Aquatic Chemistry,* John Wiley & Sons, New York, 242 (1983).
16. Pearson, R.G. and P. Allgen, "Rates and Mechanism of Formation of Some Nickel(II) Complexes with Methanol," *Inorg. Chem.,* **6,** 1379 (1967).
17. Petrucci, S., *Ionic Interactions,* Academic Press, New York, 147 (1971).
18. Prue, J.E., *Ionic Equilibrium,* Pergamon Press, Oxford (1966).
19. Schnitzer, M. and E.H. Hansen, "Organometallic Interactions in Soils: 8. Evaluation of Methods for the Determination of Stability Constants of Metal Fulvic Acid Complexes," *Soil Sci.,* **109,** 333 (1970).
20. Schwarzenbach, G., *Experientia Suppl.,* **5,** 162 (1956).
21. Schwarzenbach, G., G. Anderegg, W. Schneider and H. Senn, "On the Coordination Tendency of N-Substituted Iminodiacetic Acids," *Helv. Chim. Acta,* **38,** 1147 (1955).
22. Sillén, L.G. and A.E. Martell, *Stability Constants of Metal-Ion Complexes,* Special Publication No. 17, The Chemical Society, London (1964).
23. Sillén, L.G. and A.E. Martell, *Stability Constants of Metal-Ion Complexes — Supplement No. 1,* Special Publication No. 25, The Chemical Society, London, 64 (1971).
24. Stevenson, F.J., *Humus Chemistry,* Wiley-Interscience, New York, 1982.
25. Tanaka, M., "A Mechanistic Consideration of the Formation Constant of Metal Complexes with Special Reference to the Mixed Ligand Complexes of Nickel Involving Amines and Aminocarboxylates," *J. Inorg. Nucl. Chem.,* **35,** 965 (1973).
26. Tanaka, M., "The Mechanistic Consideration on the Formation Constants of Copper(II) Complexes," *J. Inorg. Nucl. Chem.,* **36,** 151 (1974).
27. Wilkins, R.G., *The Study of Kinetics and Mechanisms of Reactions of Transition Metal Complexes,* Allyn and Bacon, Boston, 183 (1974).

2.10 ELECTRON TRANSFER REACTIONS

2.10.1 Introduction

This section deals with electron transfer (redox) reactions, primarily homogeneous reactions in aqueous systems. It explains some of the parameters and graphical techniques used to represent redox equilibria and summarizes important environmental redox couples. Typical values of the redox status of various environmental waters are given in terms of parameters such as Eh and pe. Limitations on the use of experimentally determined Eh values in relation to the redox status of the solution are discussed. Finally, methods are described for calculating redox equilibria under various reaction conditions and for estimating the redox half-potential for a complexed species.

2.10.2 Environmental Importance

Many inorganic species of environmental concern can undergo electron transfer or redox reactions with species that are present naturally or as pollutants in environmental systems. The product of an electron transfer reaction is typically the reacting species in a modified oxidation state, but it can also be a product of the reaction of water (often very rapid) with the modified oxidation state species. This product species in the modified oxidation state differs greatly from the original reactant species in terms of chemical behavior. Thus, homogeneous (e.g., acid/base dissociation, complexation) and heterogeneous (e.g., solubility, attenuation) reactivity properties are modified; the result of a redox reaction is a species that differs in environmental mobility and impact from the original species.

Like other reactions, such as acidic or basic dissociation and complexation, redox reactions are fundamental to determining speciation in solution. Figure 2.10-1 summarizes the chemical identity and valence states of elements that are capable of undergoing redox reactions in the environment.

2.10.3 Mechanisms of Redox Reactions

An oxidation-reduction reaction involves a transfer of electrons between two species — the oxidizing agent (oxidant) and the reducing agent (reductant). In this transfer, the oxidant gains electrons and the reductant loses them. For example, in the reaction

$$4Fe^{+2} + O_2 + 4H^+ \longrightarrow 4Fe^{+3} + 2H_2O \qquad (1)$$

the Fe(II) is oxidized to Fe(III) by loss of electrons, and the O_2 is reduced to the -2 oxidation state by gaining electrons to form H_2O. (In aqueous systems, hydronium or hydroxide ions or water may be used to balance redox reactions.) Electrons are not shown in equation 1, since free electrons usually do not exist in aqueous solution; electron transfer occurs by direct interaction of the reactants (see below) or other intermediate chemical species.

The initial reaction products are the oxidized form of the reducing agent and the reduced form of the oxidizing agent. In aqueous solution, these initial products can

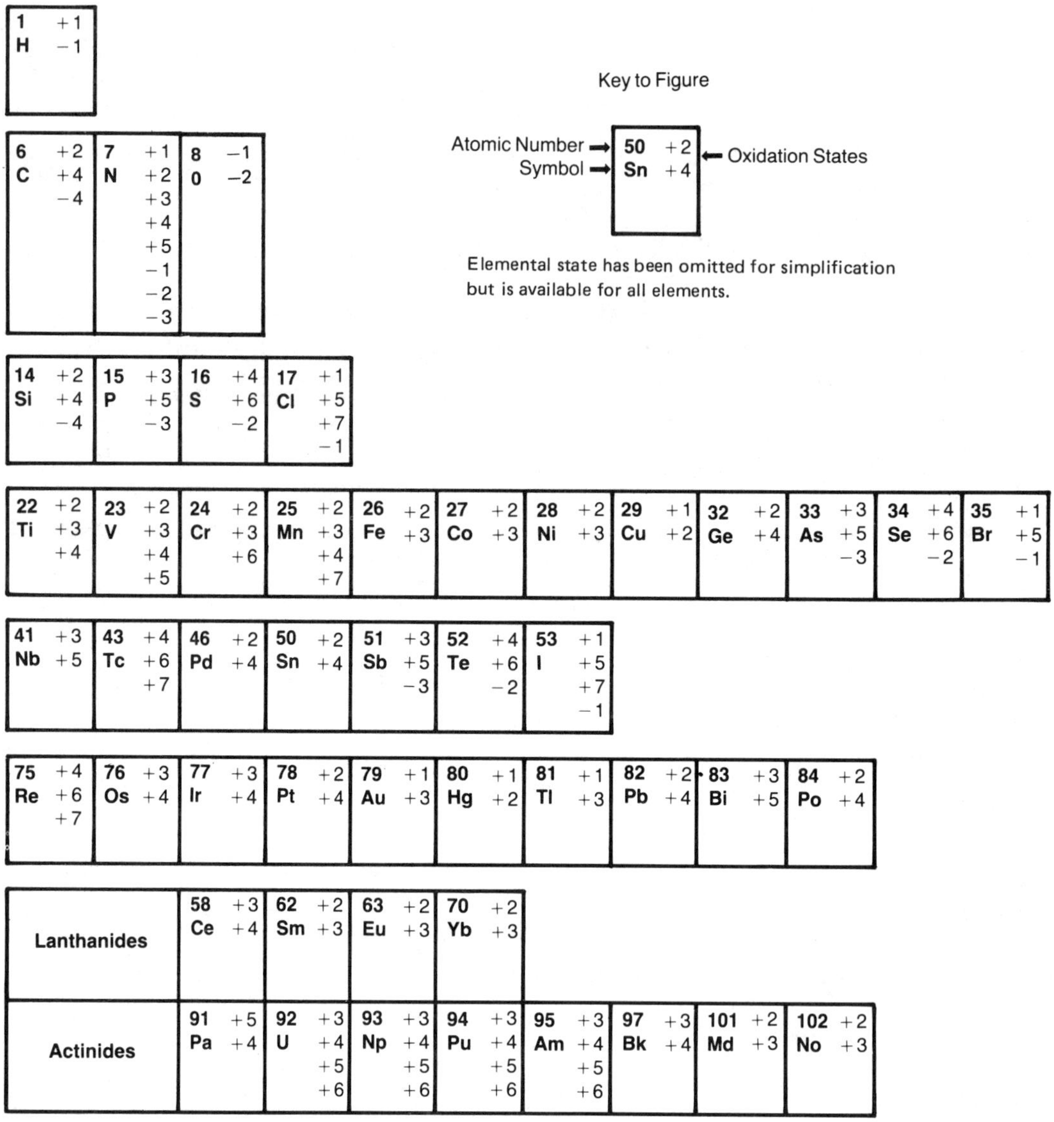

FIGURE 2.10-1 Summary of Elements That Can Undergo Redox Reactions And Their Valence States

then undergo further reactions. The latter include acid-base equilibria and dissociation of complexes with subsequent substitution of coordinated ligands by water.

The mechanism by which a redox reaction occurs is important, since it can determine the nature of the products as well as the rate of the reaction. There are two kinds of homogeneous electron transfer reactions: outer sphere and inner sphere. The reaction between Fe(III) and V(II) shown in Figure 2.10-2 is an example of an outer-sphere redox reaction [23]. In such reactions, the bonds in the primary or inner coordination sphere[1] remain intact during the transition through species I and II. (The species separated by commas are ion pairs.) The initial products of this reaction have oxidation states different from those of the reactants but have the same primary coordination spheres as the reactants. These coordination spheres may, however, be modified by subsequent rapid reactions with the aqueous environment, such as acid/base equilibria or ligand exchange.

In an inner-sphere reaction, the two reacting species are coordinated to a common "bridging" ligand in the transition state (species III and IV in Figure 2.10-3). The illustrated reaction between Fe(II) and $Co(NH_3)_5NTA$ is an example of an inner-sphere redox reaction with NTA as the bridging ligand [5]. Electron transfer occurs in the bridged intermediate (species III). This kind of redox reaction may involve transfer of the bridging ligand from one metal center to the other, yielding a product that differs not only in oxidation state but also in composition of the primary coordination sphere. Figure 2.10-4, for example, shows an inner-sphere redox reaction between $[Co(NH_3)_5Cl]^{+2}$ and Cr^{+2}, in which the Cl ligand is transferred from the Co(III) metal center to the product Cr(III) center because of the kinetically inert nature of the Cr(III) and the kinetically labile nature of the Co(II) products.

$$Fe(H_2O)_6^{+3} + V(H_2O)_6^{+2} \rightleftharpoons [Fe(H_2O)_6^{+3}, V(H_2O)_6^{+2}]$$

$$\text{(I)}$$

$$\rightleftharpoons$$

$$[Fe(H_2O)_6^{+2}, V(H_2O)_6^{+3}]$$

$$\text{(II)}$$

$$\rightleftharpoons$$

$$Fe(H_2O)_6^{+2} + V(H_2O)_6^{+3}$$

FIGURE 2.10-2 Sample Redox Mechanism: Outer-Sphere Electron Transfer Reaction Between Fe^{+3} and V^{+2}

When both products of electron transfer are kinetically labile (see section 3.2), the initial products of the electron transfer reaction will rapidly equilibrate with the aqueous environment (via ligand exchange) and form predominantly aquo complexes as well as hydroxo and other complexes resulting from acid-base equilibria. As shown in Figure 2.10-3, the product Fe(III) and Co(II) centers generated in the inner-sphere intermediate (species IV) are labile and thus exchange the ammonia and NTA ligands

1. See §2.9.2 for definition.

$$Fe(H_2O)_6^{+2} + Co(NH_3)_5\,NTA \rightleftharpoons$$

$$[(H_2O)_5\,Fe(II) - NTA - Co(III)\,(NH_3)_5]^{+2}$$
$$\text{(III)}$$
$$\rightleftharpoons$$
$$[(H_2O)_5\,Fe(III) - NTA - Co(II)\,(NH_3)_5]^{+2}$$
$$\text{(IV)}$$
$$\rightleftharpoons$$
$$Fe(H_2O)_6^{+3} + Co(H_2O)_6^{+2} + 5NH_3 + NTA$$

$$NTA \equiv \text{Nitrilotriacetate ion } [N(CH_2CO_2)_3^{-3}]$$

FIGURE 2.10-3 Sample Redox Mechanism: Inner-Sphere Electron Transfer Reaction Between Fe^{+2} and $Co(NH_3)_5$ NTA

(Gain or loss of free water is not shown.)

$$Co(NH_3)_5\,Cl^{+2} + Cr(H_2O)_6^{+2} \rightleftharpoons$$

$$[(NH_3)_5\,Co(III) - Cl - Cr(II)\,(H_2O)_5]^{+4}$$
$$\rightleftharpoons$$
$$[(NH_3)_5\,Co(II) - Cl - Cr(III)\,(H_2O)_5]^{+4}$$
$$\rightleftharpoons$$
$$Co(H_2O)_6^{+2} + Cr(H_2O)_5\,Cl^{+2} + 5NH_3$$

FIGURE 2.10-4 Sample Redox Mechanism: Inner-Sphere Electron Transfer Reaction Between Cr^{+2} and $Co(NH_3)_5\,Cl^{+2}$

(Gain or loss of free water is not shown.)

rapidly to form aquo complexes. In Figure 2.10-4, the product Cr(III) center is inert and retains the Cl ligand, while the Co(II) center is labile and exchanges the ammonia and chloride ligands rapidly to form aquo complexes.

It is therefore necessary to distinguish between inner- and outer-sphere redox processes if one wishes to determine the rate of a reaction (see section 3.3). In addition, one must also have information about the inert or labile character of the reactant and product metal or other element centers involved (see section 3.3) and the time frame of interest to determine the final products of the reaction.

In a predominantly aqueous environment, and when the redox reaction of concern is between entirely aquated species (i.e., aquo ions), an inner-sphere redox reaction with a water-bridged intermediate that produces one inert and one labile product would cause the products to have essentially the same primary coordination spheres as those of the reactants. It would not matter if the reactants or products were kinetically labile or inert. If a water molecule were exchanged between the reactants, it would be noticed only if it were labeled (e.g., with isotopically labeled oxygen). However, an inner-sphere redox reaction leading to reaction products with inert centers may produce a bridged species similar to IV in Figure 2.10-3.

In environmental systems, electron transfer occurs only between species that are capable of donating and accepting one or more electrons. In contrast to H^+, which can exist in finite concentrations in aqueous systems (as expressed by the pH value), the aqueous electron is not detectable under typical environmental conditions. Thus, an oxidizing agent that is added to a solution reacts directly with a reducing agent rather than with an aqueous electron. The concept of pe, which is equal to $-\log(e^-)$, where (e^-) is the electron activity, is very useful for calculating equilibrium conditions and simplifies treatment of redox reactions in a manner analogous to that of pH [21]; nevertheless, this concept should not be interpreted as representing the mechanism of electron transfer, except in the presence of electrodes.

2.10.4 Mathematical Representation of Redox Equilibria

EQUILIBRIUM CONSTANT, FREE ENERGIES, ELECTRODE POTENTAL AND ELECTRON ACTIVITY

The overall electron transfer reaction typified by equation 1 can be thought of as two coupled half-reactions, representing reduction and oxidation, respectively. The half-reactions for equation 1 would be as follows:

$$Fe^{+2} \longrightarrow Fe^{+3} + e^- \tag{2}$$

$$O_2 + 4H^+ + 4e^- \longrightarrow 2H_2O \tag{3}$$

Either of these equations can be considered a "redox buffer" (analogous to a pH buffer): the activity of the electron (e^-) affects the distribution between the oxidized and reduced state, just as the activity of H^+ affects the distribution between the dissociated and undissociated forms of an acid.

Hypothetical equilibrium constants can be assigned to each half-reaction. For an oxidation reaction, the equilibrium expression would be

$$K_{red,\,oxid} = [oxid]\,[e^-]/[red] \tag{4}$$

Thus, the equilibrium constant for equation 2 would be

$$K_{Fe^{+2},\,Fe^{+3}} = [Fe^{+3}]\,[e^-]/[Fe^{+2}] \tag{5a}$$

Similarly, for the reduction half-reaction (equation 3),

$$K_{O_2,H_20} = 1/[pO_2]\,[H^+]^4\,[e^-]^4 \tag{5b}$$

These equilibrium constants are related to the free energy change of the reaction according to

$$\log K = -\Delta G^0/2.303\ R\ T \tag{6}$$

where R = gas constant (1.986 cal/deg-mole)
T = absolute temperature (K)
ΔG^0 = free energy change (kcal/mole)

At 25°C, this relationship is

$$\Delta G^0 = -1.364 \log K \tag{7}$$

Each half-reaction can be described by a half-cell potential, E^0, referenced to the hydrogen electrode. The half-cell potential is related to the free energy change according to

$$\Delta G^0 = -n\ F\ E^0 = -n\,(23.06)\ E^0 \tag{8}$$

where n = number of electrons transferred
F = Faraday constant (23.06 kcal/V-mole-equiv)
ΔG^0 = free energy change (kcal/mole)
E^0 = half-cell potential (V)

(Note that equation 8 is written based on the American sign convention for E^0 values which gives a positive sign to any complete reaction which goes spontaneously as written [10].)

Equations 7 and 8 allow calculation of K, ΔG^0, or E^0 if one of these is known. Since ΔG^0 values are a function of temperature, the value of E^0 varies with temperature. This is discussed in greater detail in Section 2.11. If pe is defined as $-\log\,[e^-]$, equation 4 can be written in the form

$$\log K_{red,\ oxid} = \log\,[oxid]/[red] + \log\,[e^-] \tag{9}$$

$$= \log\,[oxid]/[red] - pe$$

This relationship allows determination of the equilibrium concentrations (or ratio) of the oxidized and reduced forms of the redox couple at a given pe value (similar to pH governing the ratio of protonated and unprotonated forms in a buffer).

Example 1 What is the relationship between pe and the ratio of oxidized and reduced forms of dissolved iron(II) and iron(III)?

For equation 2 (using equation 9),

$$\log K_{Fe^{+2}, Fe^{+3}} = \log [Fe^{+3}]/[Fe^{+2}] - pe \tag{10}$$

From equations 7 and 8, with n=1

$$\log K_{Fe^{+2}, Fe^{+3}} = 16.91\ E_{Fe^{+3}/Fe^{+2}} \tag{11}$$

Using a value of −0.771V for the half-cell potential, E^0 (see §2.10.7), we obtain

$$\log K_{Fe^{+2}, Fe^{+3}} = 16.95\,(-0.771) = -13.07 \tag{12}$$

Whence from equation 10,

$$pe = 13.07 + \log [Fe^{+3}]/[Fe^{+2}] \tag{13}$$

A plot of equation 13 is shown in Figure 2.10-5. Figure 2.10-6 is a plot of similar equations for arsenic(III)/arsenic(V) ratio at three pH values.

When a redox half-reaction contains H^+ or OH^- as reactants or products, the pe depends on pH. The mathematical dependence of pe on pH is a function of the reaction and of the form of the equilibrium constant. Similarly, if either a reactant or product is a gas, the pe is a function of the partial pressure of the gas. These variables are illustrated in Example 2 below. It is important to use the equilibrium constant that corresponds to the equation as written (i.e., with the same coefficients of the reactants) in the original source of the measurement. This applies to reactions where pure phases (e.g., H_2O) or gases are products or reactants. For such reactions, the values of the equilibrium constants depend on the coefficients.

Example 2 What is the dependence of pe on the pH and partial pressure of oxygen for the reduction of oxygen to water according to the following reaction?

$$1/4\ O_2 + H^+ + e^- \longrightarrow 1/2\ H_2O \tag{14}$$

Using the convention that $[H_2O]$ is a constant, we can write the equilibrium expression for this half-reaction as

$$K_{O_2, H_2O} = \frac{1}{[pO_2]^{1/4}\,[H^+]\,[e^-]} \tag{15}$$

In terms of logarithms,

$$\log K_{O_2, H_20} = \log 1 - 1/4 \log [pO_2] - \log [H^+] - \log [e^-] \tag{16}$$

$$= -1/4 \log [pO_2] + pH + pe \tag{17}$$

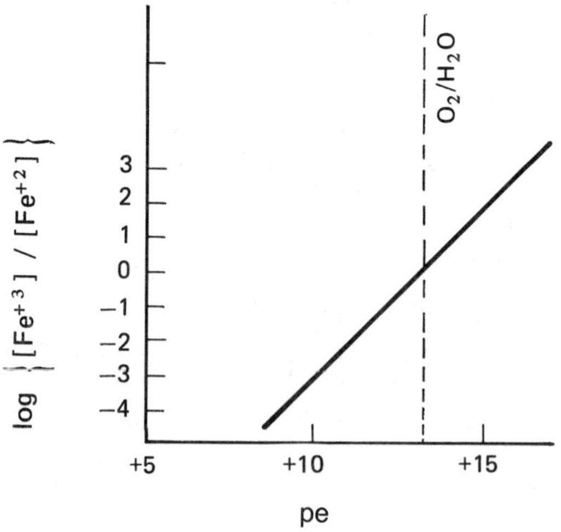

Dashed line indicates position of oxygen/water couple at pH 7.5.

Source: Equation 13

FIGURE 2.10-5 Theoretical Distribution of Iron(II) and Iron(III) at Various pe Values

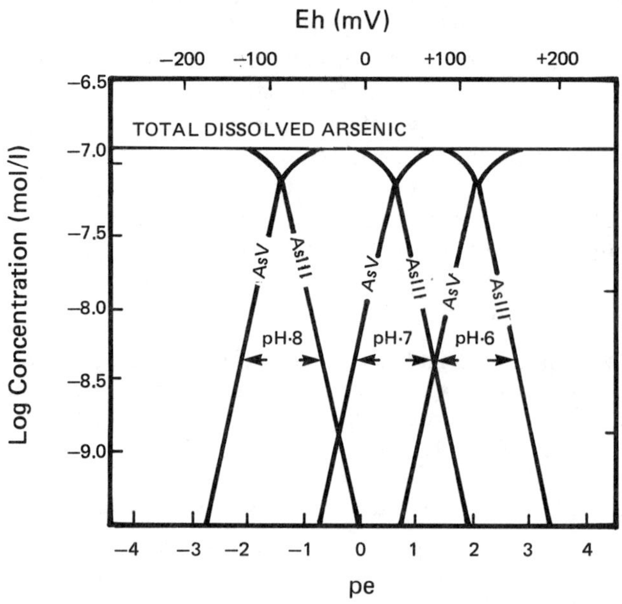

Source: Adapted from Cherry [7]. *(Copyright 1979, New Zeland Hydrological Society. Reproduced with permission.)*

FIGURE 2.10-6 Variation in Concentrations of Arsenic(III) and Arsenic(V) with pe at pH values of 6, 7 and 8. (Derived from pH + ($^2/_3$)pe = 6.6 for the $AsO_4^{-3}/HAsO_3^{-2}$ Half Reaction)

The E^0 for the water/oxygen couple is 1.229 V (see §2.10.7). Since equation 14 involves one electron, n = 1. Using these values, equations 7 and 8 yield

$$\log K_{O_2, H_2O} = 16.95(1.229) = 20.83 \tag{18}$$

Hence, equation 17 becomes

$$pe + pH = 20.83 + 1/4 \log[pO_2] \tag{19}$$

A plot of pe vs pH at various partial pressures of oxygen is shown in Figure 2.10-7.

Note that if equation 14 had been written

$$O_2 + 4H^+ + 4e^- \longrightarrow 2H_2O$$

a different value of the equilibrium constant (four times as large) would have been used. The result would have been equivalent to equation 19 but multiplied by 4 throughout.

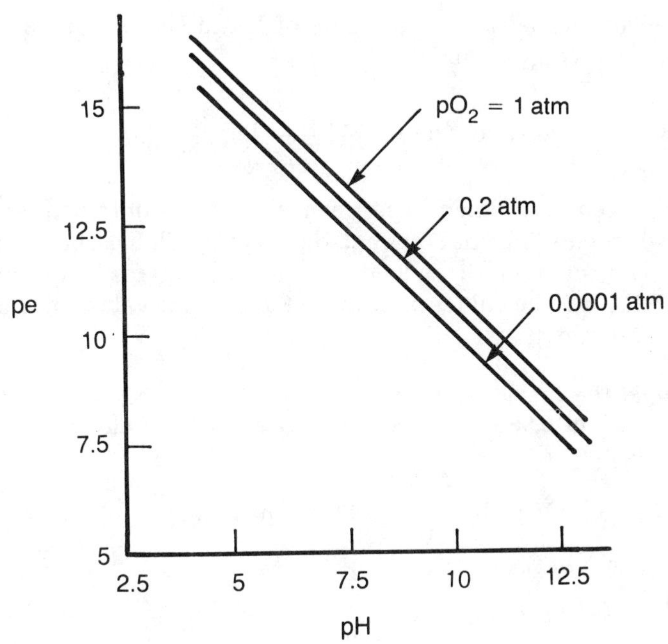

Source: Equation 19

FIGURE 2.10-7 pe Versus pH Diagram for the Oxygen/Water Half Reaction at Various Partial Pressures of Oxygen

The relationships described above can be used to calculate the concentration ratios between oxidized and reduced forms that would be expected under equilibrium conditions, given particular values of pe and the other parameters. Values of pe can be obtained experimentally, although they are subject to uncertainties and limitations (see § 2.10.5). In some cases, these values are fairly representative of redox conditions. One experimental method of determining pe is to measure potentials with a platinum sensing electrode, using the following relationship:

$$E(Pt) = E^0_{Pt} - B \log[e^-] = E^0_{Pt} + B\ pe \tag{20}$$

where

$E(Pt)$ = potential measured using a platinum electrode (equal to Eh)

E^0_{Pt} = zero by convention, so that $Eh = B\ pe$

$B = 2.3RT/nF$ (0.0592 V if n= 1 at 25°C)

The value of E(Pt) is also referred to as the value of Eh in the literature. Alternatively, we can use the following equation:

$$Eh = E^0 \text{ (redox couple)} + \frac{2.3\ RT}{nF} \log \frac{[oxid]}{[red]} \tag{21}$$

Values of Eh are more often used when experimental data are obtained whereas pe is convenient to use when equilibrium constant data are available and calculations are made from them. The range of Eh values in aqueous systems is approximately from −0.8 to +1.2V (§ 2.10.5). The values of pe are related to the values of equilibrium constants and can vary greatly.

Example 3 What is the predicted distribution between reduced and oxidized dissolved iron at an E^0 of 0.900 V (measured with a platinum electrode and a calomel reference electrode)?

Substituting E(Pt) = 0.900V and B = 0.0592 in equation 20,

$$0.900 = 0.0592\ pe \qquad pe = 15.21 \tag{22}$$

From equation 13 (or Figure 2.10-5),

$$15.21 = 13.07 + \log [Fe^{+3}]/[Fe^{+2}] \tag{23}$$

Therefore,

$$\log [Fe^{+3}]/Fe^{+2}] = 2.14$$

or

$$[Fe^{+3}]/[Fe^{+2}] = 138$$

Example 4 What is the pe of water in equilibrium with the atmosphere at pH 7?

The partial pressure of oxygen in the atmosphere is 0.21 atm. From equation 19,

$$\begin{aligned} pe &= 1/4 \log [pO_2] - pH + 20.83 \\ &= 1/4 (\log 0.21) - 7 + 20.83 \\ &= -0.17 - 7 + 20.83 \\ &= 13.66 \end{aligned} \tag{24}$$

Example 5 Calculate the expected equilibrium pe of water at pH 7.4 containing Cr(III) at 0.5 nmol/l and Cr(VI) at 0.3 nmol/l, using the ratio of the chromium species concentrations.

We first identify the species of Cr(III) and Cr(VI) that are expected to be present at pH 7.4. Using the acid dissociation constants for the aquo ions (see §7.6.2), we calculate that the dominant species are expected to be $Cr(OH)_2^+$ and CrO_4^{-2}. The redox reaction of concern is therefore,

$$CrO_4^{-2} + 6H^+ + 3e^- \longrightarrow Cr(OH)_2^+ + 2H_2O \tag{25}$$

Equation 25 yields the following expressions for the equilibrium constant as a function of pe and pH:

$$K = \frac{[Cr(OH)_2{}^+]}{[CrO_4^{-2}]\,[H^+]^6\,[e^-]^3} \tag{26}$$

or

$$\log K = \log\{[Cr(OH)_2^+]/[CrO_4^{-2}]\} + 6\ pH + 3\ pe \tag{27}$$

The following values are now substituted in equation 27:

$\log K = 66.1$ [11]
$pH = 7.4$
$[Cr(OH)_2^+] = 0.5$ nmol/l
$[CrO_4^{-2}] = 0.3$ nmol/l

$66.1 = \log (0.5/0.3) + 6(7.4) + 3\ pe$
$pe = 7.16$

The parameters K, ΔG^0, E^0 and their relationships are applicable to the overall redox reaction

$$A(oxid) + B(red) \longrightarrow A(red) + B(oxid)$$

but pe is a parameter useful only in the half reactions and not for the overall reaction. Further discussions of these parameters are given in references 10 and 12.

STABILITY FIELD DIAGRAMS

Stability field diagrams are another useful method for presenting results of calculations of redox reactions under equilibrium conditions. These diagrams, which include plots of Eh (or pe) against pH for the most thermodynamically stable species, show equilibrium conditions and important chemical reactions that may dominate the behavior of an element under environmental conditions. Diagrams and procedures for their preparation have been described by several authors [8,10,11,17]. When using these diagrams to decide whether a particular phase or species is present, one should remember that they reflect the following assumptions:

- Thermodynamic equilibrium applies (i.e., kinetic limitations do not exist),
- The system is completely described (i.e., all applicable reactions are included),
- All solids are at unit activity and the activity coefficients of dissolved species are each one,
- The concentration of every dissolved species is assumed to be $10^{-6}M$ in all calculations.

For fast reactions and the existence of equilibrium, little error is introduced by using the diagrams, provided all the possible reactions have been considered. Where conditions of ionic strength are such that unit activities are not expected, the possible impact of this variation should be taken into account.

To familiarize the reader with stability field diagrams, the preparation of an Eh-pH diagram for the iron-water system is illustrated below.

Example 6 Construct the $Fe-H_2O$ Eh–pH diagram.

The following procedure and data are adapted from Cloke[8]:

(1) Consider all known valence states, solids and dissolved states as well as pH-dependent precipitation reactions. Scan tables that give free energies to locate compounds that can exist. Recognize that compounds that can form in a water system are those with water, OH^- and O^{-2}.

The list for iron is as follows:

Valence of Iron	Solid Substance	Dissolved Species
0	Fe	
2	FeO	Fe^{+2} and
	$Fe(OH)_2$	Fe(II) complexes
2 and 3	Fe_3O_4	
3	Fe_2O_3	Fe^{+3} and
	$Fe(OH)_3$	Fe(III) complexes
	FeOOH	

(2) Write balanced reactions for oxidation of the most reduced form of the substance to each of the other oxidation states given above, using only water as reactant with the most reduced form.

For the six oxidized forms of iron, one may consider the following reactions, starting with the most reduced form (Fe):

$$Fe + H_2O = FeO + 2H^+ + 2e^- \quad E^0 = -0.037V \tag{28}$$
$$Fe + 2H_2O = Fe(OH)_2 + 2H^+ + 2e^- \quad E^0 = -0.047V \tag{29}$$
$$3Fe + 4H_2O = Fe_3O_4 + 8H^+ + 8e^- \quad E^0 = -0.084V \tag{30}$$
$$2Fe + 3H_2O = Fe_2O_3 + 6H^+ + 6e^- \quad E^0 = -0.051V \tag{31}$$
$$Fe + 3H_2O = Fe(OH)_3 + 3H^+ + 3e^- \quad E^0 = +0.059V \tag{32}$$
$$Fe + 2H_2O = FeOOH + 3H^+ + 3e^- \quad E^0 = -0.052V \tag{33}$$

(E^0 values are derived or obtained using the procedure described in the next step)

(3) Calculate E^0 values for each of the reactions above. (Depending on the form of the data available, equations 7 and 8 may be useful.) Write the values of E^0 next to the reactions, as shown above in step 2.

(4) We can express the general reaction by means of the following equation:

$$aA + bB + ... \longrightarrow cC + dD + ... \tag{34}$$

for which [8]

$$Eh = E^0 + (0.059/n)\log[C]^c [D]^d ... / [A]^a[B]^b .... \tag{35}$$

Write equation 35 for all of the reactions listed in step 2. In the case of reaction 28, for example, a=1, b=1, c=2, d=1, A=[Fe], B=$[H_2O]$, C=$[H^+]$, D=[FeO], n=2, and $E^0 = -0.037$. Thus,

$$Eh = -0.037 + (0.059/2) \log [H^+]^2 [FeO]/[Fe] [H_2O] \tag{36}$$

This simplifies to the following equation:

$$Eh = -0.037 - 0.059 \text{ pH} \tag{37}$$

since the activities of the solid phases as well as of the water are defined as unity.

Similarly, for equations 29 to 33,

$$Eh = -0.047 - 0.059 \text{ pH} \tag{38}$$
$$Eh = -0.084 - 0.059 \text{ pH} \tag{39}$$
$$Eh = -0.051 - 0.059 \text{ pH} \tag{40}$$
$$Eh = +0.059 - 0.059 \text{ pH} \tag{41}$$
$$Eh = -0.052 - 0.059 \text{ pH} \tag{42}$$

(5) Plot equations 37 to 42 using Eh as the y axis and pH as the x axis as shown in Figure 2.10-8.

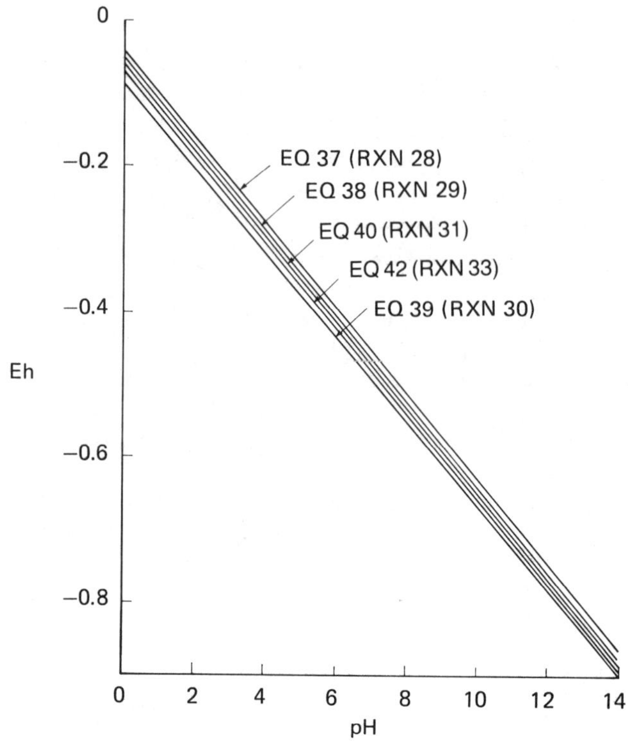

FIGURE 2.10-8 Eh Versus pH for the Fe-H_2O System: Diagram for Step 5

(6) Use the line that is lowest (reaction 30) and disregard the others, because this line represents the reaction that uses up all the Fe and prevents any of the other reactions involving Fe from occurring.

If the lowest lines intersect, one of the lines would be used above the intersection and the other line below the intersection. The general rule is to use the line or line segment corresponding to the more negative value of Eh for a given pH. An additional line would also be drawn through the point of intersection according to an equation derived from the reaction of the two oxidized species, using only these two species and H_2O and H^+ (not O_2, H_2 or OH^-). Further details of this procedure are given in step 13.

(7) Now consider oxidation to higher valences from the oxidized product of the reaction retained in step 6. The species that can be produced were identified in step 2.

For the iron system, we consider what species can result from the oxidation of Fe_3O_4:

$$2Fe_3O_4 + H_2O = 3Fe_2O_3 + 2H^+ + 2e^- \qquad E^0 = 0.221V \qquad (43)$$

$$Fe_3O_4 + 5H_2O = 3Fe(OH)_3 + H^+ + e^- \qquad E^0 = 1.21V \qquad (44)$$

$$Fe_3O_4 + 2H_2O = 3FeOOH + H^+ + e^- \qquad E^0 = 0.208V \qquad (45)$$

These lead to the following three equations for the relationship between Eh and pH:

$$Eh = 0.221 - 0.059\ pH \tag{46}$$

$$Eh = 1.21 - 0.059\ pH \tag{47}$$

$$Eh = 0.208 - 0.059\ pH \tag{48}$$

When these equations are plotted, we find that the line corresponding to equation 48 is the lowest of the three, indicating that FeOOH is the product of oxidation of Fe_3O_4 at equilibrium. (See Figure 2.10-9.)

(8) Repeat step 7 until the highest oxidation state is reached.

(9) Draw two lines on the diagram defining the Eh-pH region where water is stable, as follows:

- For the water/oxygen half reaction,

$$Eh = 1.23 + 0.015 \log [pO_2] - 0.059\ pH \tag{49}$$

which at one atmosphere of oxygen corresponds to

$$Eh = 1.23 - 0.059\ pH \tag{50}$$

- For the hydrogen/proton couple,

$$Eh = -0.059\ pH - 0.0295 \log [pH_2] \tag{51}$$

which at one atmosphere of hydrogen gas is

$$Eh = -0.059\ pH \tag{52}$$

Plots of equations 50 and 52 are included on Figure 2.10-9.

(10) Discard all lines outside the area of water stability defined above — i.e., above the line for equation 50 and below the line for equation 52. Thus the plot for equation 39 is discarded, but the region between equations 52 and 48 remains a valid stability range for Fe_3O_4.

(11) Add lines for equilibria between solids and dissolved states of the species shown in the diagram. In this example, the following reactions are applicable:

$$Fe^{+3} + 2H_2O \rightleftharpoons FeOOH + 3H^+ \qquad K = 6.3 \tag{53}$$

$$Fe^{+2} + 2H_2O \longrightarrow FeOOH + 3H^+ + e^- \qquad E^0 = 0.723V \tag{54}$$

$$3Fe^{+2} + 4H_2O \longrightarrow Fe_3O_4 + 8H^+ + 2e^- \qquad E^0 = 0.962V \tag{55}$$

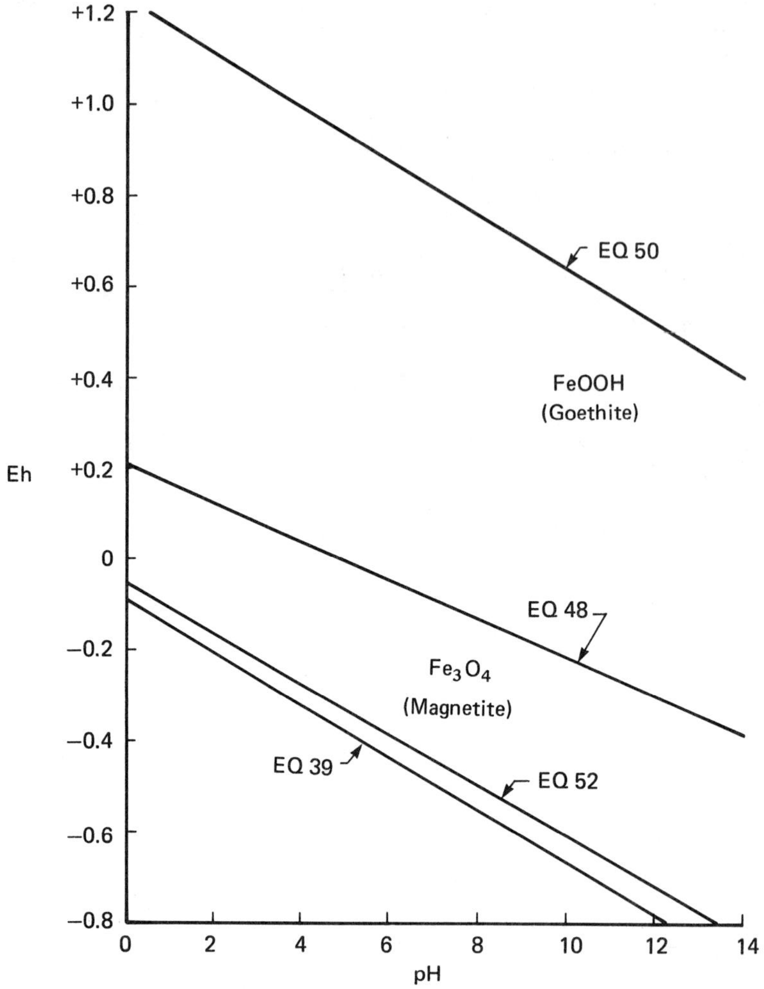

FIGURE 2.10-9 Eh Versus pH for the Fe-H_2O System: Diagram for Steps 7 and 9

For reaction 53,

$$K = [H^+]^3 / [Fe^{+3}] = 6.3 \quad (56)$$

or

$$pH = -(\log[Fe^{+3}]) / 3 - 0.266 \quad (57)$$

Since the stability field diagrams are drawn assuming a $10^{-6}M$ concentration of dissolved ions, $[Fe^{+3}] = 10^{-6}M$, and equation 57 reduces to pH = 1.734.

For equation 54, $E^0 = 0.723V$ and $n = 1$. Therefore, we can rewrite equation 35 as

$$Eh = 0.723 + (0.059/1)(\log[H^+]^3) - (0.059/1)(\log[Fe^{+2}]) \quad (58)$$

or

$$Eh = 0.723 - 0.177\ pH - 0.059 \log [Fe^{+2}] \quad (59)$$

For an $[Fe^{+2}] = 10^{-6}M$,

$$Eh = 1.077 - 0.177\ pH \quad (60)$$

Similarly, for equation 55,

$$Eh = 0.962 - 0.236\ pH - 0.0885 \log [Fe^{+2}] \quad (61)$$

and at $[Fe^{+2}] = 10^{-6}M$,

$$Eh = 1.493 - 0.236\ pH \quad (62)$$

Figure 2.10-10 shows the addition of lines corresponding to pH = 1.734 (equation 57) and equations 60 and 62 to Figure 2.10-9.

(12) Add any redox reactions between the dissolved species. For this example, the reaction is

$$Fe^{+2} \longrightarrow Fe^{+3} + e^- \qquad E^0 = 0.771V \quad (63)$$

The redox potential for this reaction is

$$Eh = 0.771 + 0.059 \log [Fe^{+3}]/[Fe^{+2}] \quad (64)$$

At $[Fe^{+3}] = [Fe^{+2}] = 10^{-6}M$, equation 64 reduces to $Eh = 0.771V$ (referenced to hydrogen electrode). This is also plotted in Figure 2.10-10.

(13) Erase all lines according to the following procedure. Where two lines intersect, delete the portion of each line that corresponds to the higher (more positive) Eh values at a given pH. Erase lines that no longer identify proper equilibrium reactions.

For this example, the lines to be erased are dashed in Figure 2.10-10. The portion of line for equation 57 within the Fe^{+2} area is erased because it no longer represents the proper species on either side.

The final diagram is shown in Figure 2.10-11.

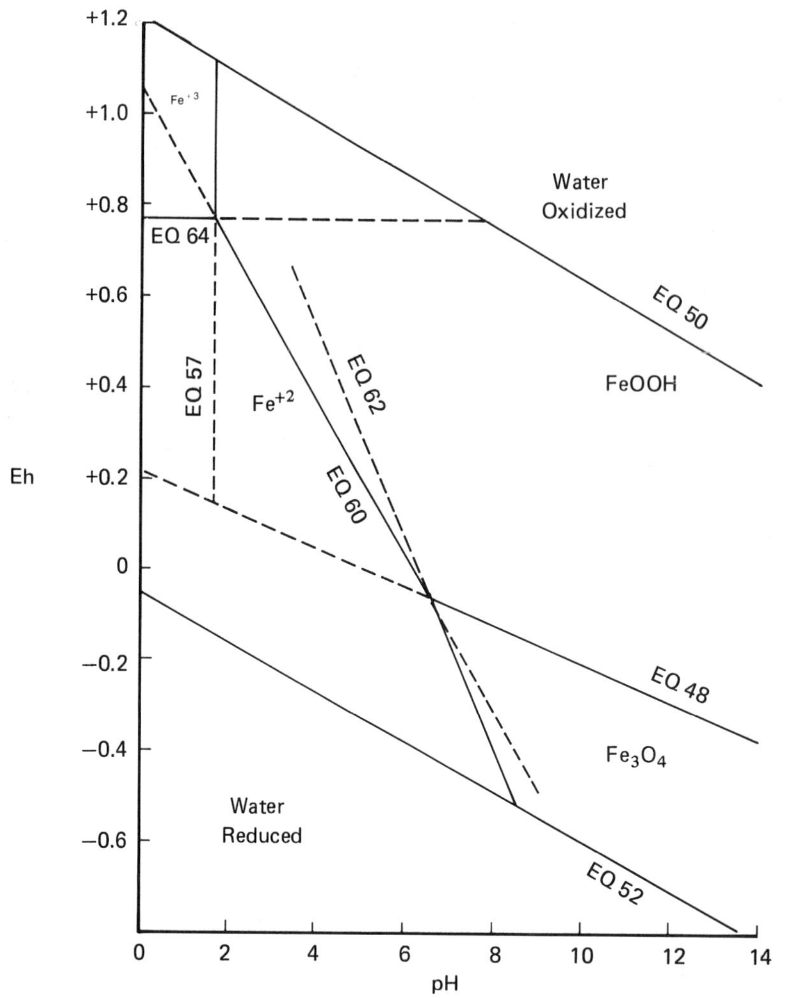

FIGURE 2.10-10 Eh Versus pH for the Fe-H_2O System: Diagram for Steps 11 to 13

The line shown on the Eh versus pH diagram for the Iron-Water System (Figure 2.10-11) separating the Fe^{+2} and Fe^{+3} (at Eh=0.77 and pH=1) represents the conditions where the concentrations (or activities) of these two dissolved species are equal. Within the region labelled Fe^{+2} this species predominates, and similarly for the Fe^{+3} region. The line separating regions of solid phases (e.g. Fe_3O_4 and FeOOH) provides the boundary where relative stability of one solid becomes greater than that of the other. A line that separates a solid phase and a dissolved species (e.g. Fe^{+3} and FeOOH at Eh=1 and pH=2) indicates the values of Eh and pH where the dissolved species is 10^{-6} *M* (by convention for the way this diagram was drawn). Similar diagrams could be drawn with these lines representing other dissolved ion concentrations (e.g. 10^{-4} or 10^{-2} *M*). The lines that would represent higher dissolved Fe^{+3} concentrations would occur parallel to the line shown on Fig. 2.10-11 at similar Eh values but at lower pH values.

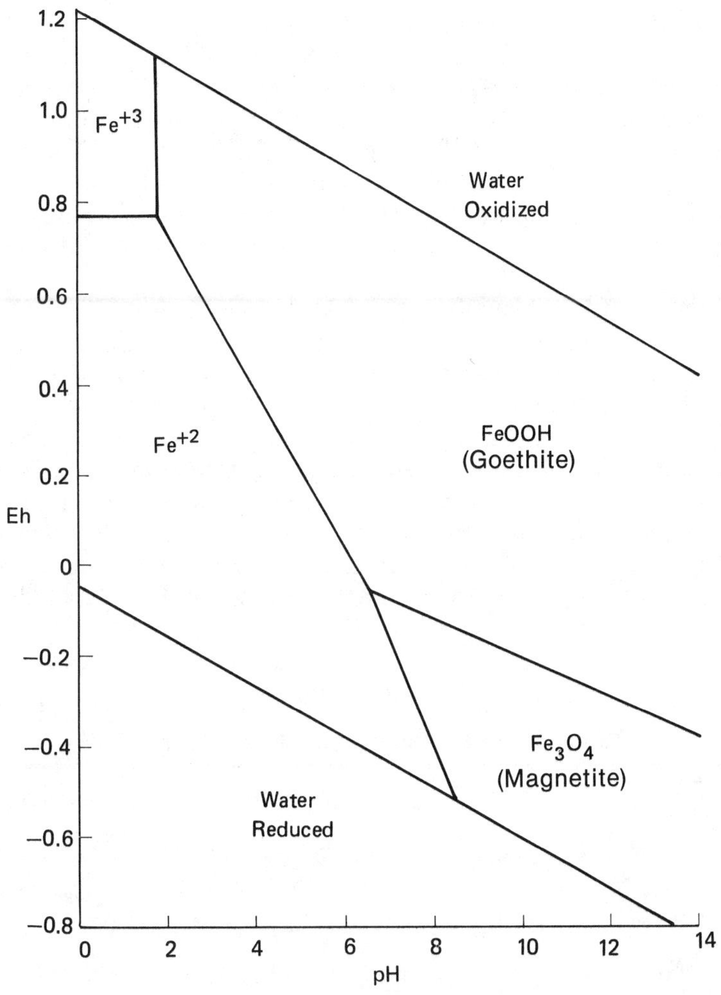

FIGURE 2.10-11 Eh Versus pH for the Fe-H_2O System: Final Diagram

2.10.5 Redox in Environmental Systems

IMPORTANT REDOX PROCESSES

Redox reactions in environmental systems can occur between species of elements that occur naturally, between pollutant species and naturally occurring species, and between pollutant species. Of the naturally occurring species, only a few commonly participate in redox processes; the typical oxidizing agents in waters (in order of decreasing oxidizing strength) are oxygen, nitrate, nitrite, ferric hydroxide, ferric phosphate, sulfate, sulfur, carbon dioxide and bicarbonate. The most important naturally occurring reducing agents in waters (in order of decreasing reducing strength) are organic substances, hydrogen sulfide, sulfur, iron sulfide, ammonium and nitrite. (This excludes kinetic considerations.)[20]

Due to kinetic limitations, local inhomogeneity, and local catalytic effects of organisms, redox equilibrium conditions may not be attained in many environmental systems, and several of the above listed species may coexist. For example, in water in equilibrium with air at pH 7 (pe = 13.6), all carbon should theoretically be present as carbon dioxide or carbonate-type species, all nitrogen as nitrate, all sulfur as sulfate, all iron as iron(III) species and all manganese as Mn(IV) species. In fact, nitrogen, reduced forms of carbon (e.g., organic matter and CH_4), sulfides, iron(II) and Mn(II) all exist under various conditions, attesting to a lack of equilibrium.

Nonequilibrium conditions usually occur when environments are isolated from the oxidizing species (e.g., oxygen) and appropriate catalyzing bacteria are present. Important redox half-reactions that are considered in the chemical equilibrium model WATEQF[16] are listed in Table 2.10-1.

All of the various forms of the naturally occurring elements (e.g. as shown in Table 2.10-1) can participate in reactions with specific species, leading to partial equilibrium (with respect to particular reactions) because of the limitations imposed on reactions with other species present. Thus, if the redox reaction and kinetics are favorable, pollutants introduced into that environment can react with these and other naturally present species.

TABLE 2.10-1

Redox Half-Reactions Considered in WATEQF

$Fe^{+2} = Fe^{+3} + e^-$	$Fe_3O_4 + 8H^+ = 3Fe^{+3} + 4H_2O + e^-$
$Fe^{+2} + SO_4^{-2} = FeSO_4^+ + e^-$	$FeS_2 + 2H^+ + 2e^- = Fe^{+2} + 2HS^-$
$Fe^{+2} + Cl^- = FeCl^{+2} + e^-$	$Fe_3S_4 + 4H^+ + 2e^- = 3Fe^{+2} + 4HS^-$
$Fe^{+2} + 2Cl^- = FeCl_2^+ + e^-$	$Fe^{+2} + HPO_4^{-2} = FeHPO_4^+ + e^-$
$Fe^{+2} + 3Cl^- = FeCl_3^0 + e^-$	$Fe^{+2} + H_2PO_4^- = FeH_2PO_4^{+2} + e^-$
$SO_4^{-2} + 10H^+ + 8e^- = H_2S + 4H_2O$	$Mn^{+2} + 4H_2O = MnO_4^- + 8H^+ + 5e^-$
$2H_2O = O_2 + 4H^+ + 4e^-$	$Mn^{+2} + 4H_2O = MnO_4^{-2} + 8H^+ + 4e^-$
$HCO_3^- + 8e^- + 9H^+ = CH_4 + 3H_2O$	δ-MnO_2(birnessite) $+ 4H^+ + e^- = Mn^{+3} + 2H_2O$
$Fe^{+2} + 2H_2O = Fe(OH)_2^+ + 2H^+ + e^-$	δ-MnO_2 (insutite) $+ 4H^+ + e^- = Mn^{+3} + 2H_2O$
$Fe^{+2} + 3H_2O = Fe(OH)_3^0 + 3H^+ + e^-$	$Mn_3O_4 + 8H^+ + 2e^- = 3Mn^{+2} + 4H_2O$
$Fe^{+2} + 4H_2O = Fe(OH)_4^- + 4H^+ + e^-$	

Source: Plummer [16]

In addition to the major elements, many trace components present in the environment can undergo redox reactions. This is important, because pollutants are generally present at very low levels, and the kinetic and redox properties of the trace substances that occur naturally may favor their reactions with the pollutants rather than with the major species such as those in Table 2.10-1. Trace elements can also catalyze redox reactions.

Some other redox species that may need consideration are shown in Figure 2.10-1. Many of these species, if present at elevated levels, would be considered pollutants.

The other category of redox interactions in environmental systems that may have to be addressed is the reaction between two or more pollutants introduced into the system. In some situations, the likelihood of a redox reaction in a mixture of two or more pollutants increases after it enters an environmental system such as groundwater or seawater. The environmental system can make the conditions for reaction more favorable or enhance the kinetics by modifying the speciation of the pollutants (e.g., by modifying the pH), or it can introduce chemical or biological species that can enhance the rate of the redox reaction. Examples of redox reactions that can be biologically mediated are given in Table 2.10-2.

APPLICABILITY OF Eh (OR pe) MEASUREMENTS

The measurement and interpretation of Eh or pe in environmental systems can be difficult [15,22]. Potentiometric measurements of Eh in environmental systems usually do not agree with those predicted by theory, due to the low concentration of redox-active species and their low exchange currents.[2] The most important problem in applying the Eh concept to natural waters is the lack of redox equilibrium between all the redox-active species. Reactions between the major environmentally important redox species (listed in Figure 2.10-12) often appear to be very slow; thus, for a system in disequilibrium, an Eh value that is meaningful with regard to rapid exchange between a redox couple is not expected.

Lindberg and Runnells [15] have examined data from about 30 groundwater analyses and compared the measured Eh or pe values to those predicted from calculations through the ratio of concentrations of many of the redox-active couples (iron(II)/iron(III), oxygen/water, bisulfide/sulfate, etc.) in these solutions. The expected Eh values and those calculated by use of these redox couples do not agree and vary over a wide range, as shown in Figure 2.10-12. These data show that none of the representative waters chosen for analysis exhibited equilibrium and that the computed Eh values span a wide range (as much as 1000 mV). They also suggest that indicator species (e.g., iodide/iodate or arsenic(III)/arsenic(V)) ratios may not be more representative of the redox state than the measurement using an electrode. It may be reasonably concluded that a master value of Eh or pe cannot be used to predict redox equilibria in groundwater environments and that computer models utilizing pe or Eh to calculate species ratios would most likely make erroneous predictions.

Lindberg and Runnells [15] have suggested an alternative to using measured Eh values, namely, to measure concentrations of certain redox-active species (e.g., dissolved oxygen, total sulfide species and methane) as a qualitative guide to the redox status of waters (see next section for discussion of this approach); the concentrations of the redox components of immediate interest in the water should be measured if there is a need to consider their possible redox effect on pollutants introduced into that environment.

2. The species do not readily give up or take up electrons at platinum or gold electrode surfaces.

TABLE 2.10-2

Microbially Mediated Oxidation and Reduction Reactions

	Reaction	pe^0_{pH7} (note a)
Oxidation		
(1)[b]	$CH_2O + H_2O = CO_2(g) + 4H^+ + 4e^-$	−8.20
(2)	$HS^- + 4H_2O = SO_4^{-2} + 9H^+ + 8e^-$	−3.75
(3)	$NH_4^+ + 3H_2O = NO_3^- + 10H^+ + 8e^-$	+6.16
(4)[c]	$FeCO_3(s) + 2H_2O = FeOOH(s) + HCO_3^-(10^{-3}M) + 2H^+ + e^-$	−1.67
(5)[c]	$MnCO_3(s) + 2H_2O = MnO_2(s) + HCO_3^-(10^{-3}M) + 3H^+ + 2e^-$	+8.5
(1a)[b]	$HCOO^- = CO_2(g) + H^+ + 2e^-$	−8.73
(1b)	$CH_2O + H_2O = HCOO^- + 3H^+ + 2e^-$	−7.68
(1c)	$CH_3OH = CH_2O + 2H^+ + 2e^-$	−3.01
(1d)	$CH_4(g) + H_2O = CH_3OH + 2H^+ + 2e^-$	+2.88
Reduction		
(A)	$O_2(g) + 4H^+ + 4e^- = 2H_2O$	+13.75
(B)[d]	$2NO_3^- + 12H^+ + 10e^- = N_2(g) + 6H_2O$	+12.65
(C)	$NO_3^- + 10H^+ + 8e^- = NH_4^+ + 3H_2O$	+6.16
(D)	$SO_4^{-2} + 9H^+ + 8e^- = HS^- + 4H_2O$	−3.75
(E)	$CO_2(g) + 8H^+ + 8e^- = CH_4(g) + 4H_2O$	−4.13
(F)[b]	$CH_2O + 2H^+ + 2e^- = CH_3OH$	−3.01

(Continued)

Despite the limitations of Eh and pe, redox equilibrium models (including Eh/pH diagrams) can be valuable for interpreting observed data and identifying possible regulating mechanisms for pe buffering of a few aqueous systems [20]. The discrepancies between equilibrium calculations and experimental data can give insight as to whether the reactions are well understood and whether the systems are at equilibrium. They may also provide clues concerning which species will predominate as redox conditions change and what solid phase(s) may be expected.

Eh AND pe VALUES ENCOUNTERED IN VARIOUS ENVIRONMENTS

For waters in equilibrium with atmospheric oxygen, there is no correlation of the measured Eh and dissolved oxygen levels [6]; the water/oxygen half-reaction (equation 3) is too slow to affect the Eh value measured. The measurement of Eh is only considered valid and useful when dissolved oxygen (DO) is below the minimum detection level of 0.01 ppm. For waters where DO is above this level, measurement of dissolved oxygen is a better indication of redox conditions.

TABLE 2.10-2 (Continued)

Thermodynamic Sequence of Biologically Mediated Redox Reactions

Model 1: Excess organic material. (Water contains incipiently O_2, NO_3^-, SO_4^{-2}, HCO_3^-.) Examples: Hypolimnetic layers of eutrophic lake, sediments; digester.

	Combination	(1/n)log K_{pH7}[e]
Aerobic respiration	(1) + (A)	21.95
Dentrification[f]	(1) + (B)	20.85
Nitrate reduction	(1) + (C)	14.36
Fermentation[b,g]	(1b) + (F)	4.67
Sulfate reduction	(1) + (D)	4.45
Methane fermentation	(1) + (E)	4.07

Model 2: Excess O_2. (Water contains incipiently organic material, SH^-, NH_4^+ and possibly Fe(II) and Mn(II).) Examples: Aerobic waste treatment, self-purification in streams; epilimnetic waters.

	Combination	(1/n)log K_{pH7}[e]
Aerobic respiration	(A) + (1)	21.95
Sulfide oxidation	(A) + (2)	17.50
Nitrification[f]	(A) + (3)	7.59
Ferrous oxidation[c]	(A) + (4)	15.42
Mn(II) oxidation	(A) + (5)	5.25

a. pe^0_{pH7} is the pe if all reaction participants with exception of (H^+) are present at unit activity (for gases, partial pressure = 1). (H^+) has been set equal to 10^{-7}. $pe^0 = (1/n)\log K_{reduction}$ (n = no. of electrons). All constants given are for 25°C.

b. CH_2O is used here as a general symbol for organic material. Examples for organic carbon with formal oxidation states of +II, 0, −II, and −IV are given in reactions 1a to 1d, respectively. $H_2(g)$ is an almost equally strong reductant as organic material and in some mediations it can be substituted for organic material: $H_2(g) = 2H^+ + 2e^-$; $pe^0_{pH7} = -7.00$.

c. The autotrophic nature of iron and manganese bacteria is in dispute.

d. $N_2O(g)$ may appear as an intermediate; some facultative anaerobes catalyze the reduction to NO_2.

e. (1/n)log K is the equilibrium constant for the combined redox reaction per electron transferred. With exception of (H^+), all reaction participants are present at unit activity. [(H^+) = 10^{-7}]. (1/n)log K is proportional to the free energy decrease of the redox reaction per electron transferred: (1/n)log K = $-\Delta G^\circ_{pH7}/(nRT \ln 10)$.

f. It has been assumed, rather arbitrarily, that the direct oxidation of NH_4^+ into N_2 is hindered. Conversion of NH_4^+ to N_2 is thus achieved by nitrification/denitrification.

g. Fermentation is interpreted as an organic redox reaction where one organic substance is reduced by oxidizing another organic substance (e.g., alcohol fermentation). The participants in such a reaction are themselves thermodynamically unstable.

Source: Stumm [20] (Copyright 1967, Water Pollution Control Federation. Reprinted with permission.)

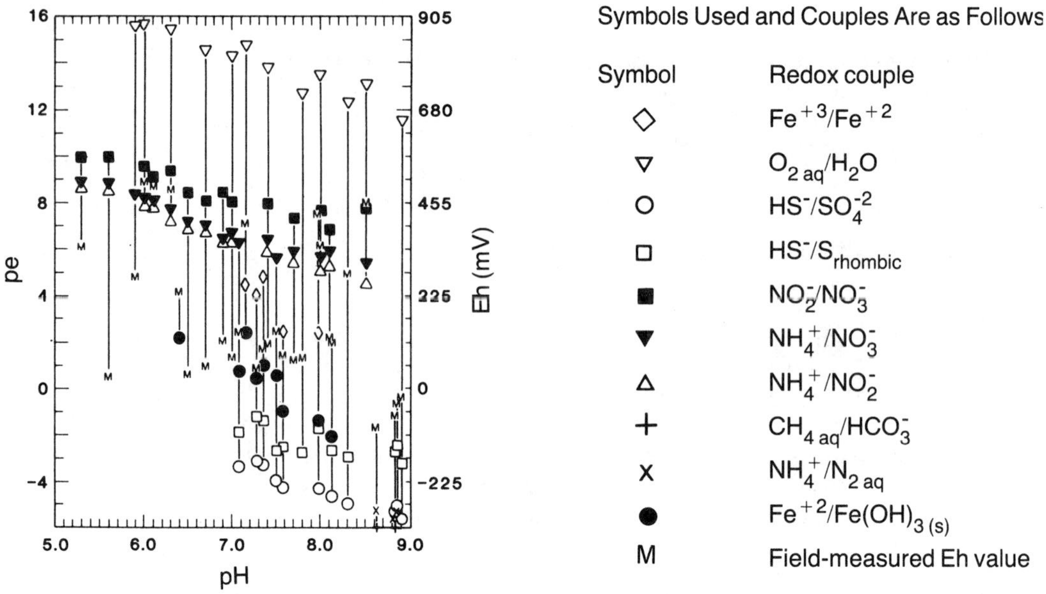

Source: Lindberg and Runnells [13]. *(Copyright 1984, American Association for the Advancement of Science. Reprinted with permission.)*

FIGURE 2.10-12 Comparison of Measured and Computed Eh Versus pH Values in 30 Groundwater Systems

The relative positions of some natural environments as functions of Eh and pH are shown in Figure 2.10-13. The validity of Eh measurement for various conditions encountered in the environment is summarized in Table 2.10-3. Reproducible and meaningful Eh values are generally obtained in well-poised systems (i.e., those that contain high concentrations of both the reduced and oxidized forms of the redox couple). This occurs in some acid mine waters that contain high levels of dissolved iron. For the other waters listed in Table 2.10-3, Eh measurements are not considered reliable because of their lack of reproducibility and stability [6].

Although, as discussed above, dissolved oxygen levels do not readily provide the redox status of oxygenated waters, concentrations of sulfide in many sediment water environments do provide an indication. This is because the Eh values of many sediments containing H_2S have been found to be controlled by the following reversible half-reactions [1]:

$$HS^- \rightleftharpoons S\ (\text{rhombic}) + H^+ + 2e^- \qquad (65)$$

or

$$S^{-2} \longrightarrow S\ (\text{rhombic}) + 2e^- \qquad (66)$$

according to the equation (at 25°C), where

$$Eh = -0.475 + 0.0295\ p[S^{-2}] \qquad (67)$$

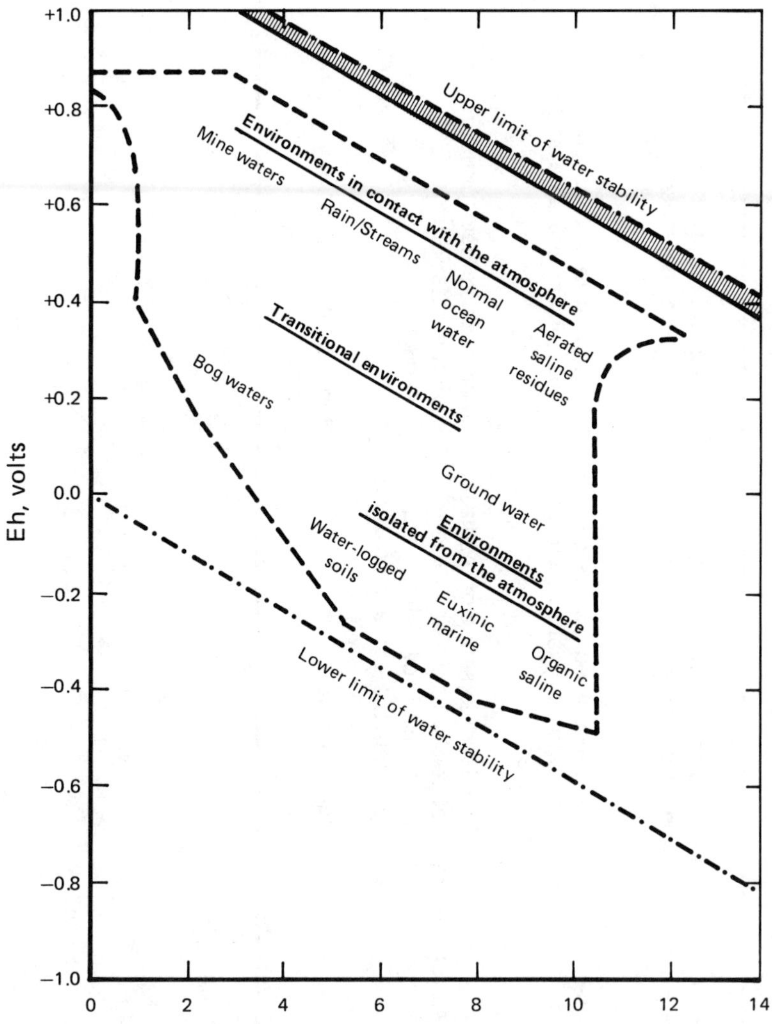

Dashed lines represent limits of measurements in natural systems. Crosshatched area represents the Eh and pH of waters containing more than .01 ppm of oxygen, based on the oxygen/water half-reaction.

Source: Langmuir[13] *(Copyright 1971, Wiley Interscience Publishers. Reprinted with permission.)*

FIGURE 2.10-13 Approximate Relative Position of Eh and pH for Natural Environments

TABLE 2.10-3

Redox Characteristics and Eh Measurements for Various Aqueous Environments

Environment	Redox Characteristics	Eh Response/Comments
Oxygenated Waters (e.g., rain, streams, some groundwater, shallow lakes)	Poorly poised, low or no concentration of electroactive species	Eh measurement drifts; D.O. levels may be more useful
Acid mine water, groundwater rich in Fe at pH below 4	May be well poised; reversible redox reaction present at significant levels	Uncommon in natural environments; Eh measurable and useful
Organic rich systems, soil waters, stagnant polluted water with low oxygen levels	Poorly poised; absence of oxidized form of redox species; mixed potentials and irreversible reactions present; biological mediation	Lowest Eh values found in these environments; drift in Eh observed
Heterogeneous environments	Variable condition	Reproducible value may not be possible

Source: Adapted from Langmuir [13]

Empirical data for values of Eh and S^{-2} correlate well with this half-reaction, as shown in Figure 2.10-14. When electrodes are inserted into these sediment environments, such measurements are rapidly attained and are stable at low Eh values, indicating a well-poised system.

OTHER INDICATORS OF REDOX CONDITIONS

A classification of sediment redox environments proposed by Berner [2] may be more practical than the Eh approach, because it is directly related to measured concentrations. In this scheme, Eh is not considered to be a measurable quantity, owing to the difficulties described previously. The primary basis for this classification is that dissolved oxygen (DO) and total dissolved sulfide species (H_2S, HS^-, S^{-2}) are measurable and are intimately involved in redox reactions. The scheme is summarized in Figure 2.10-15, and the concentration basis for each classification is given in Table 2.10-4. The classification depends on the ability to identify naturally occurring (authigenic) Mn and Fe phases in the solids and is thus useful only for undisturbed sediment environments.

The best (but most tedious) approach to understanding redox properties of the aqueous environment and their possible impact on added pollutants is to consider the reaction of the added pollutant with each of the species present in the environment. This characterization would include determining identity, speciation, relative concentration and assessment of the rate and equilibrium position of all the possible redox reactions, and would be performed either by calculation or by experiment. The initial candidates for consideration would be the most commonly found redox active species in solution (dissolved oxygen, sulfide species, iron(II), iron(III), manganese(II) and possibly nitrite), followed by species commonly found in the solid phase (if present) — iron(II), iron (III), manganese(II) and manganese(IV).

As discussed in Section 3.3, "Kinetics of Redox Reactions," catalysis by trace species as well as biological organisms is possible; thus, after the major components are considered, possible redox reaction with trace components should be addressed. The speciation of the reactants involved will depend on the environment and the effects of the added pollutants.

An example of the concentration profile observed for a variety of species in a pore water system that progresses from being oxic to anoxic (various depths in a sediment) is shown in Figure 2.10-16. Concentrations of many metals in the dissolved state above the interface are low (mostly existing in the solid state). Release from the solid into solution after reduction is observed below the O_2/H_2S interface.

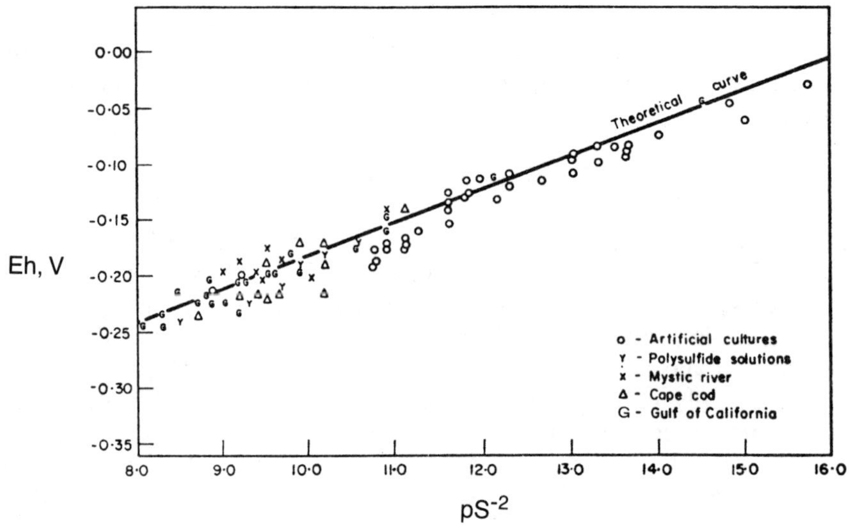

"Theoretical Curve" represents the half-reaction sulfide ⇌ sulfur (rhomb) + 2 e^-

Source: Berner [1]. *(Copyright 1963, Pergamon Press, Inc. Reprinted with permission.)*

FIGURE 2.10-14 Eh Versus p(Sulfide) in Natural Sediment Environments and Polysulfide Solutions

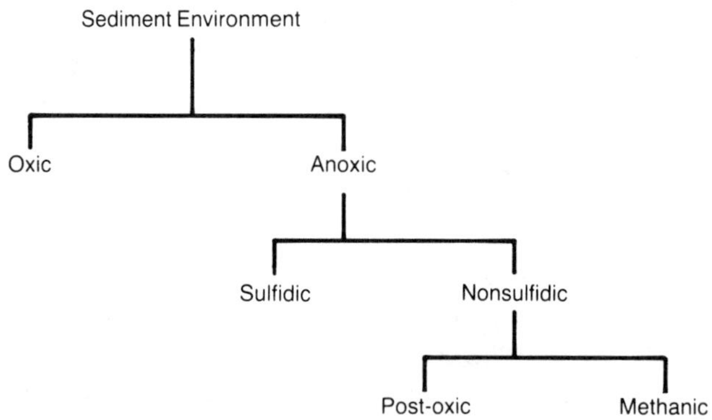

See Table 2.10-4 for clarification of terms

Source: Berner [2]

FIGURE 2.10-15 Classification of Sediment Environments

TABLE 2.10-4

Redox Classification Scheme Proposed by Berner

Environment Class of Water/ Sediment	$[O_2]_{aq}$ (*M*)	$[\sum H_2S]_{aq}$ (*M*)	P_{CO_2} (atm)	Characteristic Typical of Solid Phase in Equilibrium[a]	Other Characteristics	Progression of Biologically Mediated Organic Matter Decomposition
OXIC	$\geq 10^{-6}$	$\ll 10^{-6}$	≤ 1	FeOOH, MnO_2, Fe_2,Fe_2O_3, Relative absence of decomposable organic matter (MnO_2 is a better indicator than Fe(III) solids; latter persist under anoxic conditions, since they are not reduced by any organic species)	Lack of bacteria capable of producing H_2S	Oxygen consumption
ANOXIC	$< 10^{-6}$					
Sulfidic	$< 10^{-6}$	$\geq 10^{-6}$	$> 10^{-4}$	Fe(II)S_2, $MnCO_3$, Mn(II)S	Lack of aerobic organisms and oxidized minerals	Sulfate reduction
Nonsulfidic	$< 10^{-6}$	$< 10^{-6}$				
• Post-Oxic	$< 10^{-6}$	$< 10^{-6}$		Fe^{+2}/Fe^{+3} - containing alumino silicates, $FeCO_3$, $Fe_3(PO_4)_2$, $MnCO_3$. No sulfide minerals, minor amount of decomposable organic matter		Nitrate reduction
•Methanic	$< 10^{-6}$	$< 10^{-6}$		$FeCO_3$, $Fe_3(PO_4)_2$, $MnCO_3$. Earlier formed sulfide minerals, decomposable organic matter present		Methane formation

a. Fe(III)OOH = goethite; Fe(III)$_2O_3$ = hematite; FeS_2 = pyrite or marcasite; Fe(II)CO_3 = siderite; $Fe_3(PO_4)_2$ = ninanite; $MnCO_3$ = rhodochrosite

Source: Adapted from Berner [2]

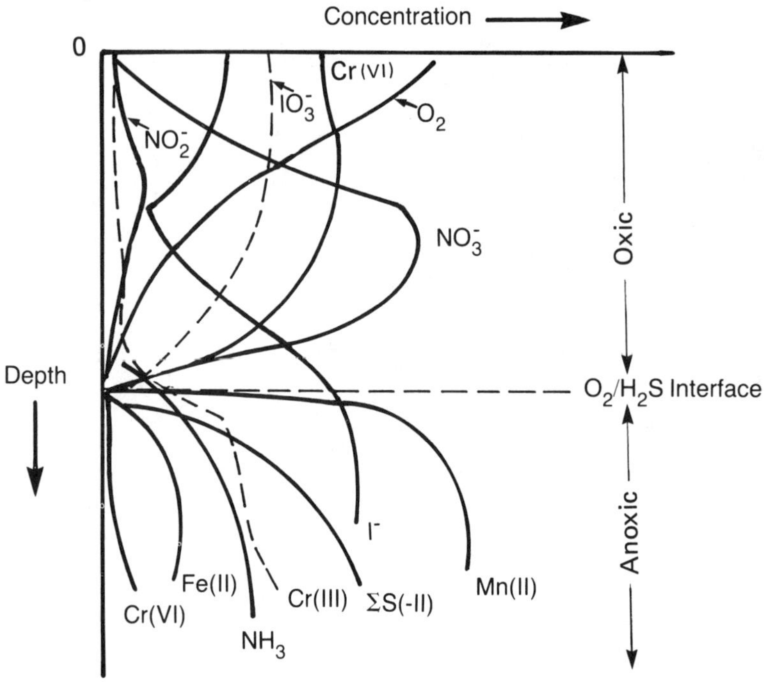

Source: Adapted from Emerson [9]

FIGURE 2.10-16 Variation in Concentration with Depth for Various Redox Active Species in a Fjord

2.10.6 Estimation of Equilibrium Parameters for Redox Reactions

The preceding sections of this chapter have discussed methods for estimating:

- Relative concentrations of oxidized versus reduced forms of a species from pe (or Eh) values;
- Relationship of pe to Eh values and field measurements;
- Equilibrium constants from free energy data or redox half-reaction potentials; and
- pe values from concentrations of other species, including dissolved oxygen and sulfide levels.

This section addresses the estimation of the equilibrium constant and the electrode potential for a half-reaction under limited conditions. The following are discussed:

- Estimation of K for a redox reaction from E^0 values when no reactions after the redox reaction occur;
- Estimation of K for a redox reaction from E^0 and stability constants when a complex dissociation reaction occurs after redox; and
- Estimation of E^0 for a half-reaction of a complexed ion from that of an aquo ion and the complex stability constants.

ESTIMATION OF K FOR REDOX WHEN NO FURTHER REACTION TAKES PLACE

If the half-reaction potentials (or equilibrium constants or free energies) that together represent an overall redox reaction are known, the overall equilibrium constant for the reaction can be calculated as follows:

$$\text{A(oxid)} + \text{n e}^- \longrightarrow \text{A(red)} \qquad \Delta\text{G(A)} \tag{68}$$

$$-\text{n/m}\{\text{B(oxid)} + \text{m e}^- \longrightarrow \text{B(red)}\} \qquad -\text{n/m}\,\{\Delta\text{G(B)}\} \tag{69}$$

Taken together, equations 68 and 69 represent the overall reaction

$$\text{A(oxid)} + \text{n/m B(red)} \longrightarrow \text{A(red)} + \text{n/m B(oxid)} \qquad \Delta\text{G (overall)} \tag{70}$$

where

$$\Delta\text{G(overall)} = \Delta\text{G(A)} - \text{n/m}\,\{\Delta\text{G(B)}\} \tag{71}$$

This ΔG(overall) is valid as long as A(red) and B(oxid) are the final products of the redox reaction (i.e., if no post-redox equilibria are involved).

If the data for equations 68 and 69 are given in terms of half-reaction potentials, the following expressions apply:

$$\Delta\text{G(A)} = -\text{n F E(A)} \tag{72}$$

$$\Delta\text{G(B)} = -\text{m F E(B)} \tag{73}$$

Thus,

$$\Delta\text{G(overall)} = -\text{n F [E(A)} - \text{E(B)]} = -\text{n F E(overall)} \tag{74}$$

where E(A) and E(B) are both expressed as reduction potentials.

The equilibrium constant at 25°C for reaction 70 is therefore related to the above parameters as follows:

$$\Delta G(\text{overall}) = -1.36 \log K\ (\text{overall}) \tag{75}$$

and

$$\Delta G(\text{overall}) = -n\ (23.06)\ E(\text{overall}) \tag{76}$$

where ΔG values are in kcal/mole and E values are in volts. The equilibrium constant at 25°C for reaction 70 is given by

$$\begin{aligned} \log K(\text{overall}) &= -\Delta G(\text{overall})/1.36 \\ &= +n\ (23.06/1.36)\ E(\text{overall}) \\ &= +n\ (16.96)\ E(\text{overall}) \end{aligned} \tag{77}$$

Example 7 Calculate the equilibrium constant for the reaction of copper(II) with bromide at 25°C according to the reaction,

$$Cu^{+2} + Br^{-} \rightleftharpoons Cu^{+} + 1/2\ Br_2 \tag{78}$$

The half-reactions that can represent the above reaction are

$$Cu^{+2} + e^{-} \longrightarrow Cu^{+} \qquad E(A) = 0.167\ V \tag{79}$$

$$1/2\ Br_2 + e^{-} \longrightarrow Br^{-} \qquad E(B) = 1.066\ V \tag{80}$$

The overall E value is:

$$E(\text{overall}) = E(A) - E(B) = 0.167 - 1.066 = -0.899\ V \tag{81}$$

Using equation 77,

$$\begin{aligned} \log K(\text{overall}) &= n\ (16.96)\ E(\text{overall}) \\ &= (1)\ (16.96)\ (-0.899) \\ &= -15.25 \end{aligned}$$

Since the equilibrium constant is very small (6×10^{-16}), the reaction is highly unfavorable.

Example 8 Calculate the equilibrium constant for the reaction of tetraaquodiamminecobalt(III) ion with bromide, assuming no dissociation of the product cobalt(II) complex. (This is unrealistic, but it represents a useful hypothetical case for comparison with the more realistic Example 9.)

The reaction of interest is

$$Co(NH_3)_2(H_2O)_4^{+3} + Br^- \longrightarrow Co(NH_3)_2(H_2O)_4^{+2} + 1/2\ Br_2 \qquad (82)$$

The half reactions are

$$Co(NH_3)_2(H_2O)_4^{+3} + e^- \longrightarrow Co(NH_3)_2(H_2O)_4^{+2} \qquad E(A) = 1.22\ V \qquad (83)$$

$$1/2\ Br_2 + e^- \longrightarrow Br^- \qquad E(B) = 1.066\ V \qquad (84)$$

The overall E is

$$E(\text{overall}) = 1.22 - 1.066 = 0.154$$

Thus, according to equation 77, the equilibrium constant should be

$$\begin{aligned} \log K(\text{overall}) &= +n\ (16.96)\ E(\text{overall}) \\ &= (1)\ (16.96)\ (0.154) \\ &= 2.61 \end{aligned}$$

The reaction is thus favorable, with an equilibrium constant of 4.07×10^2.

ESTIMATION OF K FOR A REDOX REACTION WITH SUBSEQUENT REACTIONS

Many redox reactions, especially those between complexed metal ions, lead to products that can undergo subsequent dissociation or other reactions after the electron transfer. These subsequent processes can provide additional driving force for the overall reaction to occur (including the redox process).

To illustrate, we shall consider a case where the post-redox reaction is a dissociation of a labile metal complex:

$$\text{Complex-A(oxid)} + ne^- \longrightarrow \text{Complex}-\text{A(red)} \qquad \Delta G(A) \qquad (85)$$

$$(-n/m)\{\text{B(oxid)}\} + me^- \longrightarrow \text{B(red)} \qquad \Delta G(B) \qquad (86)$$

$$\text{Complex}-\text{A(red)} \longrightarrow \text{Aquo}-\text{A(reduced)} + \text{free ligands} \qquad \Delta G(C) \qquad (87)$$

Taken together, these equations represent the overall reaction

$$\text{Complex} - \text{A(oxid)} + \text{n/m}\ \{\text{B(red)}\} \longrightarrow \text{Aquo} - \text{A(reduced)} + \text{n/m}\ \{\text{B(oxid)}\} + \text{free ligands} \quad \Delta\text{G(overall)} \qquad (88)$$

where

$$\Delta\text{G(overall)} = \Delta\text{G(A)} - \text{n/m}\ \{\Delta\text{G(B)}\} + \Delta\text{G(C)} \qquad (89)$$

An important aspect of this follow-on reaction is that it can provide sufficient driving force for the overall reaction (equation 88) to proceed, even though the relative free energy changes for reactions 85 and 86 (or relative E^0 values) indicate that redox will not proceed substantially. The effect is illustrated in Example 9. When the stability constants for the oxidized metal are very large compared with those for the reduced metal, the effect can be much larger than in this example.

Example 9 Calculate the equilibrium constant for the redox reaction in Example 8 between the cobalt diammine aquo complex and bromide, including consideration of the dissociation of the cobalt(II) product into aquo cobalt(II) and NH_3. (Assume that NH_3 does not undergo any further equilibria.)

The overall reaction is

$$Co(NH_3)_2\,(H_2O)_4^{+3} + e^- \longrightarrow Co(NH_3)_2\,(H_2O)_4^{+2} \qquad \Delta G(A) \qquad (90)$$

$$-(1/2\ Br_2 + e^- \longrightarrow Br^-) \qquad \Delta G(B) \qquad (91)$$

$$Co(NH_3)_2\,(H_2O)_4^{+2} \longrightarrow Co(H_2O)_6^{+2} + 2NH_3 \qquad \Delta G(C) \qquad (92)$$

$$Co(NH_3)_2\,(H_2O)_4^{+3} + Br^- \longrightarrow Co(H_2O)_6^{+2} + 2\ NH_3 + 1/2\ Br_2 \qquad \Delta G(\text{overall}) \qquad (93)$$

where

$$\Delta G(\text{overall}) = \Delta G(A) - \Delta G(B) + \Delta G(C) \qquad (94)$$

The value of $\Delta G(A)$ is calculated as follows, using equation 8:

$$\begin{aligned}\Delta G(A) &= -\ n\ F\ E\ (A)\\ &= -\ 1\ (23.06)\ (1.22)\\ &= -\ 28.13\ \text{kcal/mole}\end{aligned}$$

Similarly, the value of $\Delta G(B)$ is:

$$\begin{aligned} \Delta G(B) &= -\ n\ F\ E(B) \\ &= -\ 1\ (23.06)\ (1.066) \\ &= -\ 24.58\ \text{kcal/mole} \end{aligned}$$

The value of $\Delta G(C)$ is calculated as follows, using data obtained from Bjerrum[3] for β_2, the stability constant for the diammine complex of cobalt(II):

$$\begin{aligned} \Delta G(C) &= -1.36 \log K(C) \\ &= -1.36\ (1/\beta_2) \\ &= -1.36\ (1/3.62) \\ &= -0.376\ \text{kcal/mole} \end{aligned}$$

Thus, with equation 94,

$$\begin{aligned} \Delta G(\text{overall}) &= -28.13 + 24.58 + (-0.376) \\ &= -3.93 \end{aligned}$$

Therefore, with equation 77,

$$\begin{aligned} \log K(\text{overall}) &= -\Delta G(\text{overall})/1.36 \\ &= -(-3.93/1.36) \\ &= 2.89 \end{aligned}$$

and

$$K(\text{overall}) = 7.76 \times 10^2$$

ESTIMATION OF E FOR THE HALF-REACTION OF A COMPLEXED ION

Values of the potential for a half reaction as shown for the following equation

$$M\,L_n^{+y} + e^- \longrightarrow M\,L_n^{+y-1} \qquad E_c = E(\text{complexed ion}) \tag{95}$$

can be calculated from the half-reaction potential for the aquo ions

$$M(H_2O)_n^{+y} + e^- \longrightarrow M(H_2O)_n^{+y-1} \qquad E_a = E(\text{aquo ion}) \tag{96}$$

according to the following equation [3]:

$$E_c = E(\text{complexed ion}) = -0.059[\beta_n(y) - \beta_n(y-1)] + E_a \tag{97}$$

where

E_c = half-reaction potential for the complex ion

E_a = half-reaction potential for the aquo ion

$\beta_n(y)$ = stability constant for the complex ion with the overall charge on the complex of $+y$

$\beta_n(y-1)$ = stability constant for the complex ion with the overall charge on the complex of $y-1$.

The estimate of the half-reaction potential is obviously related to the accuracy of the stability constants used in the calculation.

Example 10 Estimate the reduction potential of the following half reaction:

$$Co(NH_3)_2(H_2O)_4{}^{+3} + e^- \longrightarrow Co(NH_3)_2(H_2O)_4{}^{+2} \qquad E_c(\text{cobalt}) \qquad (98)$$

The reduction potential of the aquo ion of cobalt(III) is, according to Bjerrum[3]:

$$Co(H_2O)_6^{+3} + e^- \longrightarrow Co(H_2O)_6^{+2} \qquad E_a(\text{cobalt}) = 1.83V \qquad (99)$$

Next we obtain data for the equilibrium constants for the following reactions:

$$Co(H_2O)_6^{+3} + NH_3 \longrightarrow Co(H_2O)_5NH_3^{+3} \qquad \log K_1 = 7.3 \qquad (100)$$

$$Co(H_2O)_5NH_3 + NH_3 \longrightarrow Co(H_2O)_4(NH_3)_2{}^{+3} \qquad \log K_2 = 6.7 \qquad (101)$$

$$Co(H_2O)_6^{+3} + 2NH_3 \longrightarrow Co(H_2O)_4(NH_3)_2^{+3} \qquad \beta_2(3) = 14.0 \qquad (102)$$

$$Co(H_2O)_6^{+2} + NH_3 \longrightarrow Co(H_2O)_5NH_3^{+2} \qquad \log K'_1 = 2.11 \qquad (103)$$

$$Co(H_2O)_5\,NH_3^{+2} + NH_3 \longrightarrow Co\,(H_2O)_4\,(NH_3)_2^{+2} \qquad \log K'_2 = 1.51 \qquad (104)$$

$$Co(H_2O)_6^{+2} + 2\,NH_3 \longrightarrow Co(H_2O)_4\,(NH_3)_2^{+2} \qquad \beta_2(2) = 3.62 \qquad (105)$$

The following data are now used in equation 97:

$\beta_n(y) = \beta_2\,(3) = 14.0$

$\beta_n(y-1) = \beta_2(2) = 3.62$

$E_a = 1.83$ volts

yielding

$$E_c(\text{cobalt}) = -\,0.059\,(14.0-3.62) + 1.83 = 1.22 \text{ volts}$$

The value has not been experimentally determined [4].

2.10.7 Sources of Redox Data

Values of electrode potentials and equilibrium constants for redox reactions can be found in a variety of articles and compendia such as those given in References 12, 14, 18, and 19. Table 2.10-5 lists values of electrode potentials for a variety of species.

TABLE 2.10-5

Electrode Potentials of Redox Half Reactions[a]

Oxidation — Reduction Reaction	E^0(V)
$Ag^{+2} + e^- \longrightarrow Ag^+$	+2.00
$Ag^{+2} + e^- \longrightarrow Ag(s)$	+0.799
$Al^{+3} + 3e^- \longrightarrow Al(s)$	−1.66
$Am^{+4} + e^- \longrightarrow Am^{+3}$	+2.40
$H_3AsO_4 + 2H^+ + 2e^- \longrightarrow HAsO_2 + H_2O$	+0.56
$As(s) + 3H^+ + 3e^- \longrightarrow AsH_3(g)$	−0.61
$BrO_3^- + 3H_2O + 6e^- \longrightarrow Br^- + 6OH^-$	+0.61
$\frac{1}{2} Br_2(aq) + e^- \longrightarrow Br^-$	+1.08
$Ce^{+4} + e^- \longrightarrow Ce^{+3}$	+1.74
$ClO_3^- + 2H^+ + e^- \longrightarrow ClO_2(g) + H_2O$	+1.15
$ClO_2 + e^- \longrightarrow ClO_2^-$	+0.93
$\frac{1}{2} Cl_2(aq) + e^- \longrightarrow Cl^-$	+1.39
$Co^{+3} + e^- \longrightarrow Co^{+2}$	+1.95
$HCrO_4^- + 7H^+ + 3e^- \longrightarrow Cr^{+3} + 4H_2O$	+1.20
$Cr^{+3} + e^- \longrightarrow Cr^{+2}$	−0.41
$Cu^{+3} + e^- \longrightarrow Cu^{+2}$	+2.3
$Cu^{+2} + e^- \longrightarrow Cu^+$	+0.17
$Cu^+ + e^- \longrightarrow Cu(s)$	−0.52
$\frac{1}{2} F_2(g) + e^- \longrightarrow F^-$	+2.87
$Fe^{+3} + e^- \longrightarrow Fe^{+2}$	+0.771
$2\,Hg^{+2} + 2e^- \longrightarrow Hg_2^{+2}$	+0.907
$Hg_2^{+2} + 2e^- \longrightarrow 2Hg(l)$	+0.792
$\frac{1}{2} I_2(aq) + e^- \longrightarrow I^-$	+0.621
$I_3^- + 2e^- \longrightarrow 3I^-$	+0.536
$MnO_4^- + e^- \longrightarrow MnO_4^{-2}$	+0.57

(Continued)

TABLE 2.10-5

Electrode Potentials of Redox Half Reactions

Oxidation — Reduction Reaction	E^0(V)
$MnO_4^- + 4H^+ + 3e^- \longrightarrow MnO_2(s) + 2H_2O$	+1.68
$MnO_2(s) + 4H^+ + 2e^- \longrightarrow Mn^{+2} + 2H_2O$	+1.23
$Mn^{+3} + e^- \longrightarrow Mn^{+2}$	+1.488
$Mn^{+2} + 2e^- \longrightarrow Mn(s)$	−1.17
$Mo(VI) + e^- \longrightarrow Mo(V)$	+0.53
$NO_3^- + 3H^+ 2e^- \longrightarrow HNO_2 + H_2O$	+0.94
$Ni^{+2} + 2e^- \longrightarrow Ni(s)$	−0.25
$O_2(g) + 4H^+ + 4e^- \longrightarrow 2H_2O$	+1.229
$O_2(g) + 2H^+ + 2e^- \longrightarrow H_2O_2$	+0.69
$H_3PO_3 + 2H^+ + 2e^- \longrightarrow H_3PO_2 + H_2O$	−0.50
$Pb(IV) + 2e^- \longrightarrow Pb^{+2}$	+1.655
$Pb^{+2} + 2e^- \longrightarrow Pb(s)$	−0.126
$Pu^{+4} + e^- \longrightarrow Pu^{+3}$	+0.967
$Ru^{+3} + e^- \longrightarrow Ru^{+2}$	+0.249
$SO_4^{-2} + 4H^+ + 2e^- \longrightarrow H_2SO_3 + H_2O$	+0.17
$S(s) + 2H^+ + 2e^- \longrightarrow H_2S$	+0.141
$S(s, Rhombic) + 2e^- \longrightarrow S^{-2}$	−0.48
$2S(s) + 2e^- \longrightarrow S_2^{-2}$	−0.49
$3S(s) + 2e^- \longrightarrow S_3^{-2}$	−0.45
$4S(s) + 2e^- \longrightarrow S_4^{-2}$	−0.36
$5S(s) + 2e^- \longrightarrow S_5^{-2}$	−0.34
$6S(s) + 2e^- \longrightarrow S_6^{-2}$	−0.36
$Sb(v) + 2e^- \longrightarrow Sb(III)$	+0.75
$SeO_4^{-2} + 4H^+ + 2e^- \longrightarrow H_2SeO_3 + H_2O$	+1.15
$H_2SeO_3 + 4H^+ + 4e^- \longrightarrow Se(s) + 3H_2O$	+0.74
$Se(s) + 2H^+ + 2e^- \longrightarrow H_2Se(g)$	−0.37
$Sn(IV) + 2e^- \longrightarrow Sn(II)$	+0.144
$Ti(IV) + e^- \longrightarrow Ti(III)$	+0.130
$Tl^{+3} + 2e^- \longrightarrow Tl^+$	+1.26
$U^{+4} + e^- \longrightarrow U^{+3}$	−0.609
$VO_2^+ + 2H^+ + e^- \longrightarrow V^{+3} + H_2O$	+0.34

a. Gaseous concentrations are in atmospheres.

Source: Kotrlý and Ŝůcha [12]

2.10.8 Literature Cited

1. Berner, R.A., "Electrode Studies of Hydrogen Sulfide in Marine Sediments," *Geochim. Cosmochim. Acta*, **27**, 563-75 (1963).
2. Berner, R.A., "A New Geochemical Classification of Sedimentary Environments," *J. Sediment. Petrol.*, **51**, 359 (1981).
3. Bjerrum, J., *Metal-Ammine Formation in Aqueous Solution*, P. Haase and Son, Copenhagen, 241 (1941).
4. Bodek, I., "Reactions of cis-Tetraaquodiamminecobalt(III) Ion in Aqueous Perchlorate Media," Ph.D. Thesis, Northeastern University (1976).
5. Cannon, R.D. and J. Gardiner, "A Binuclear Intermediate Preceding the Cobalt(III)-Iron(II) Electron Transfer Process," *J. Am. Chem. Soc.*, **92**, 3800 (1970).
6. Carver, R.E., *Procedures in Sedimentary Petrology*, Wiley-Interscience, New York (1971).
7. Cherry, J.A., A.U. Shaikh, D.E. Tallman and R.V. Nicholson, "Arsenic Species as an Indicator of Redox Conditions in Groundwater," *J. Hydrol. (Amsterdam)*, **43**, 373 (1979).
8. Cloke, P.L., "The Geochemical Application of Eh-pH Diagrams," *J. Geol. Educ.*, **14**, 140-48 (1966).
9. Emerson, S., R.E. Cranston and P.S. Liss, "Redox Species in a Reducing Fjord: Equilibrium and Kinetic Considerations," *Deep-Sea Res.*, **26A**, 859-78 (1979).
10. Garrels, R.M. and C.L. Christ, *Solutions, Minerals and Equilibria*, Freeman, Cooper & Co., San Francisco (1965)
11. Hem, J.D., "Reactions of Metal Ions at Surfaces of Hydrous Ion Oxides," *Geochim. Cosmochim. Acta*, **41**, 527-38 (1977), as cited by Emerson [9].
12. Kotrlý, S. and L. Ŝůcha, *Handbook of Chemical Equilibria in Analytical Chemistry,* John Wiley & Sons, New York (1985).
13. Langmuir, D., "Eh-pH Determinations," Chapter 26 in Carver [6].
14. Latimer, W.M., *Oxidation Potentials*, 2nd ed., Prentice-Hall, New York (1952).
15. Lindberg, R.D. and D.D. Runnells, "Groundwater Redox Reactions: An Analysis of Equilibrium State Applied to Eh Measurements and Geochemical Modeling," *Science*, **225**, 925 (1984).
16. Plummer, L.N., B.F. Jones and A.H. Truesdell, "WATEQF — A Fortran IV Version of WATEQ — A Computer Program for Calculating Chemical Equilibrium of Natural Waters," *Water-Resour. Invest. (U.S. Geol. Surv.)*, **76-13**, 60 (rev. and reprinted January 1984).
17. Pourbaix, M., *Atlas of Electrochemical Equilibria in Aqueous Solutions*, Pergamon Press, Oxford, England (1966).
18. Sillén, L.G. and A.E. Martell, *Stability Constants of Metal-Ion Complexes,* Special Publication No. 17, The Chemical Society, London (1964).
19. Sillén, L.G. and A.E. Martell, *Stability Constants of Metal-Ion Complexes — Supplement No. 1,* Special Publication No. 25, The Chemical Society, London, 64 (1971).

20. Stumm, W., "Redox Potentials as an Environmental Parameter: Conceptual Significance and Operational Limitations," *Adv. Water Pollut. Res., Proc. Int. Conf. 3rd, 1966 (Munich),* **1**, 283 (1967).

21. Truesdell, A.H., "The Advantage of Using pE Rather than Eh in Redox Calculations," *J. Geol. Educ.*, **16**, 17-20 (1968).

22. Whitfield, M., "Thermodynamic Limitations on the Use of the Platinum Electrode in Eh Measurements," *Limnol. Oceanogr.*, *19*, 857 (1974).

23. Wilkins, R.G., *The Study of Kinetics and Mechanism of Reactions of the Transition Metal Complexes*, Allyn and Bacon, Boston (1974).

2.11 SOLUBILITY AND PRECIPITATION EQUILIBRIA

2.11.1 Introduction

In studies of environmental water quality, the question often arises as to whether a given solid is likely to dissolve and, if so, how much. Likewise, investigators often want to know if a solid phase can precipitate from a water of given composition and what the chemical consequences would be. Such questions can often be resolved through consideration of the chemical equilibria between solids and aqueous solutions.

On a macroscopic scale of observation, a solution saturated with respect to a substance neither dissolves more of the substance nor precipitates it. Put in different terms, the solution contains dissolved species in a state of thermodynamic equilibrium with the solid phase — or, at least, with some of the components of that phase. Departures from equilibrium may occur either in a direction of supersaturation or of undersaturation. Such departures may be caused by changes in the environmental characteristics of the system, as well as by a variety of internal chemical processes. When precipitation occurs in a supersaturated solution, the solution is brought towards saturation or equilibrium "from above." In contrast, dissolution may occur in an undersaturated solution that comes in contact with a particular solid phase, bringing the solute concentration towards saturation "from below." Thus, the state of a chemical thermodynamic equilibrium between a solid and an aqueous solution indicates two environmentally important quantities:

(1) The current state of the water, with respect to its ability either to dissolve or precipitate certain solids, and

(2) The expected limits of change in concentrations of dissolved substances, if certain solid phases were to dissolve in, or precipitate from, the water.

This chapter addresses chemical equilibria between inorganic solids and aqueous solutions. Although it is practically impossible to find an inorganic solid whose solubility in water is truly zero, not every class of inorganic solids coming in contact with waters in the environment can be considered in this chapter. Excluded from the discussion of the solid-solution equilibria are pure metals, alloys, glasses, and other industrial solids of complex composition whose chemical behavior in the environmental context usually falls outside the field of the solid-solution equilibria. The focus is on substances that can be broadly classified as ionic or partially ionic solids, such as metal oxides, hydroxides, halides, sulfides, and other compounds with anions in oxidized states (sulfates, carbonates, arsenates, selenites, selenates, and a number of silicate and aluminosilicate minerals).

The environmental importance of the classes of compounds listed above is twofold: (1) for some substances, dissolution in natural waters may never reach saturation, although they will constitute a lasting environmental load on the water systems; and (2) for others, their metallic and anionic components may enter solution from different sources to create a condition of supersaturation with respect to solids that may eventually precipitate. The examples in this chapter deal primarily with solid-solution equilibria involving heavy-metal ions,[1] alkali metal ions, and alkaline earth ions among the cationic components of the solids.

This chapter also addresses the following issues:

- Definitions of the solubility product, K_{sp}, and the equilibrium constant, K;
- Establishment of either undersaturated, saturated, or supersaturated conditions by the criteria of ion-concentration product and ion-activity product;
- Effects of temperature and pressure on solubility;
- Effects of ionic strength, pH, and complexation on solubility;
- Effects of particle size on solubility; and
- Estimation of the solubility values (K_{sp} or K) from the free energy of formation, and from empirical correlations between solubilities and such crystal properties as electronegativities of the ions in the lattice, lattice energy, ionization potentials, and interfacial energies of solids.

The process of adsorption at solid-liquid interfaces is discussed in Section 2.12 (Attenuation on Soils). Other important solid precipitation processes (e.g., co-precipitation) and dissolution of impure phases (e.g., trace element leaching) are not easily described mathematically and thus their discussion has been left to books and articles which focus on them.

2.11.2 Description of Property

SOLUBILITY PRODUCT (K_{sp})

A solid made of a metal ion (M) and an anionic ligand (A) dissolves in solution and dissociates according to an overall reaction that can be written as

$$M_pA_q\ (s) = pM^{z+}(aq) + qA^{z-}(aq) \tag{1}$$

where (s) denotes a solid phase, (aq) denotes an aqueous species, p and q are stoichiometric coefficients, and z^+ and z^- are the positive and negative valence electrical charges. The principle of electrical neutrality of the ions in solution requires a charge balance between the positively and negatively charged species:

$$pz^+ + qz^- = 0$$

---

1. Heavy metals are those with atomic numbers between 21 (scandium) and 92 (uranium). Within this series, the transition elements, such as As and Se, and the heavy halogens (atomic numbers 33-36, 52-54, 85-86) are not heavy metals. Aluminum (atomic number 13), although not a heavy metal in the physical sense, is included in many environmental chemical studies of the heavy metals in water, probably because its low abundance and potential toxicity are similar to those of the trace heavy-metals.

A solution is said to be saturated with respect to the solid M_pA_q when the solubility product (K_{sp}), defined as follows, is maintained in solution:

$$K_{sp} = [M]^p[A]^q \quad (mol^{p+q}/liter^{p+q}) \tag{2}$$

where the brackets [ ] denote concentrations of the species in solution. In equation 2, concentrations of M and A are given in moles per liter (abbreviated *M*). Another unit of concentration in common use is moles per kg H_2O, known as molality. The differences between mol/liter and mol/kg H_2O can often be neglected for dilute solutions (i.e., concentrations of fractions of a mole per liter or per kg H_2O). At higher concentrations, such as in natural saline brines, the differences between molar and molal concentrations must generally be considered for accurate results.

EQUILIBRIUM CONSTANT (K)

A more accurate definition of a state of saturation or equilibrium between a chemical species in a solid phase and in an aqueous solution is based on the concept of the chemical potential and the concept of the thermodynamic activities of the reactant and product species. (The activities of aqueous and solid-phase species are defined in section 2.6.) The basic relationship between the activity of a chemical species in solution and its concentration is defined as

$$a_i = \gamma_i \, m_i \tag{3}$$

where subscript i denotes a chemical species, γ is its activity coefficient, and m is its molal concentration. The value of γ depends on the nature of the ion and its charge, the ionic strength of the solution, the nature of the solvent, and the temperature. Methods for computing ionic activity coefficients are outlined in section 2.6. Activities are dimensionless quantities, referred to $a = 1$ in a standard state.

For reaction 1 at equilibrium, the following relationship is obeyed:

$$K = \frac{a_M^p \, a_A^q}{a_{M_pA_q}} \tag{4}$$

where the equilibrium constant K is defined as a quotient of the activities of the products, raised to the powers of their stoichiometric coefficients, and the activity of the reactant. The reactant, solid phase of composition M_pA_q, is a pure solid. By definition, the activity of a pure solid in a standard state (25°C, 1 atm pressure) is unity, and equation 4 can therefore be written as

$$K = a_M^p \, a_A^q \tag{5}$$

Thus, for a dissolution equilibrium between a pure solid and its aqueous solution, we consider only the product of the activities of the aqueous species.

The concept of a pure solid, whose activity is defined as unity in a standard state, can be briefly demonstrated by reference to a common substance, such as NaCl. In a solid phase, NaCl is a pure solid: the proportions of the metal and the anion cannot vary. However, if some of the Cl atoms in the lattice were replaced by Br atoms, a solid of composition $NaCl_xBr_{1-x}$ would result. If the proportions of the Cl and Br atoms can vary in the crystal lattice within some limits, then the solid is not a pure solid but a *solid solution* of two components — NaCl and NaBr. In this case, the activities of the solid components are not necessarily unity. Discussion of the solubility of solid solutions is not within the scope of this chapter; for the specialized methods of evaluating the activities of the components in solid phases, texts dealing with this subject should be consulted [10, 21, 30].

Equation 5 for an equilibrium constant can be written in terms of concentrations and the activity coefficients of the species in solution, using equation 3:

$$K = [M]^p\,[A]^q\,\gamma_M^p\;\gamma_A^q \tag{6}$$

Finally, the equilibrium constant K for reaction 1 and the solubility product K_{sp} are interrelated as follows:

$$K = K_{sp}\,\gamma_M^p\;\gamma_A^q \tag{7}$$

The values of the activity coefficients (γ) for aqueous species in dilute solutions are usually less than unity. Therefore, K is numerically smaller than K_{sp} —sometimes much smaller, because of the exponents p and q raising the fractional values of γ to a power. Computation of the activity coefficients of aqueous species is described in section 2.6.

CONDITIONS OF EQUILIBRIUM

The conditions of a state of saturation (or equilibrium) in a solution can be defined with the aid of the relationships for K_{sp} and K.

The product of the concentrations of dissolved species on the right-hand side of equation 2 is called the *ion-concentration product* (ICP):

$$ICP = [M]^p[A]^q \tag{8}$$

In a given solution, the ICP may be less than, equal to, or greater than the value of K_{sp} for the stoichiometrically equivalent solid phase, such as shown in reaction 1. If the ICP is less than K_{sp}, the solution is undersaturated with respect to the solid, and dissolution may take place. If the two quantities are equal, the solution is saturated. If the ICP is greater than K_{sp}, the solution is supersaturated with respect to the solid phase. This can be summarized as follows:

$$\left.\begin{array}{rl} \text{Undersaturation:} & ICP < K_{sp} \\ \text{Saturation:} & ICP = K_{sp} \\ \text{Supersaturation:} & ICP > K_{sp} \end{array}\right\} \tag{9}$$

From the definition of K based on activities rather than concentrations (equation 5), the product of the activities of the species in solution is called the *ion-activity product* (IAP):

$$IAP = a_M^p \, a_A^q \tag{10}$$

Thus, the three states — undersaturation, saturation, and supersaturation — can also be defined in terms of the IAP and the equilibrium constant K as follows:

$$\left.\begin{aligned} \text{Undersaturation:} &\quad IAP < K \\ \text{Saturation:} &\quad IAP = K \\ \text{Supersaturation:} &\quad IAP > K \end{aligned}\right\} \tag{11}$$

In very dilute solutions where the activity coefficients (γ_i) are close to unity, the differences between K_{sp} and K may be very small. Thus, the state of a solution with respect to saturation may be defined by either equation 10 or equation 11 without significant loss of accuracy. In more concentrated solutions, particularly in the presence of complexing agents, reliance on the IAP and K criteria should be encouraged; the values of the activity coefficients in such cases (section 2.6) assures consistency and constancy in the value of K for a reaction, whereas the solubility product K_{sp} may vary from one solution to another.

CONCENTRATIONS OF DISSOLVED SPECIES FROM K_{sp}

Solubilities of natural and man-made substances in water vary widely. For some poorly soluble substances, K_{sp} is expressed in high negative powers of 10, while highly soluble solids can have a K_{sp} higher than 10. On the lower end of the solubility scale, values smaller than 10^{-12} mol/liter are encountered; on the upper end, some strongly soluble electrolytes may be characterized by values of about 20 mol/liter.

If the value of K_{sp} for a given solid and its aqueous solution is known, concentrations of the aqueous species may be estimated. In the notation of equilibrium reaction 1, a solid of stoichiometric composition M_pA_q produces in solution a total of p moles of M and q moles of A. Concentrations [M] and [A], both coming from the same solid, are interrelated as follows:

$$[M] = \frac{p}{q}\,[A] \tag{12}$$

$$[A] = \frac{q}{p}\,[M] \tag{13}$$

In terms of the anion concentration [A],

$$K_{sp} = \left(\frac{p}{q}\right)^p [A]^{p+q} \tag{14}$$

If the value of K_{sp} is known, the concentration of either the cationic or anionic species in solution at equilibrium with the solid can now be calculated from equation 14:

$$[M] = \left((p/q)^q \ K_{sp}\right)^{1/(p+q)} \tag{15}$$

$$[A] = \left((q/p)^p \ K_{sp}\right)^{1/(p+q)} \tag{16}$$

These equations are valid under conditions where the solutions are of low ionic strength and do not contain any of the ions of interest (i.e., there is no "common ion" effect). In real systems one needs to consider the activity as well as the presence of common ions and thus equation (5) may be more appropriate.

The preceding relationships between K_{sp}, K, and concentrations of the dissolved species are demonstrated below.

Example 1 What are the relationships between the solubility product, equilibrium constant, concentrations, and activity coefficients for fluorite (CaF_2) in pure water containing no Ca^{+2} or F^- at room temperature?

The equilibrium dissolution reaction is:

$$CaF_2(s) = Ca^{+2}(aq) + 2F^-(aq)$$

The solubility product for this reaction at 25°C, in a dilute solution, is [24]:

$$K_{sp} = 10^{-10.31}$$

$$= 5.01 \times 10^{-11} \text{ mol}^3/\text{liter}^3$$

Concentrations of Ca^{+2} and F^- in solution at equilibrium with solid CaF_2 can be estimated from equations 15 and 16, with p = 1 and q = 2:

$$[Ca^{+2}] = 2.32 \times 10^{-4} \text{ mol/liter}$$
$$[F^-] = 4.64 \times 10^{-4} \text{ mol/liter}$$

For the equilibrium dissolution reaction given above, the value of K at 25°C can be computed from the tabulated data of the free energies of formation of the solid and aqueous species. Such computations are explained in section 2.11.3; here, only the value of K will be given:

$$K = 10^{-10.5}$$
$$= 3.22 \times 10^{-11}$$

This value is slightly smaller than that of K_{sp}, as may be expected from equation 7. Ideally, the values of K_{sp} reported in the literature are the solubility values that were extrapolated to the value at zero ionic strength (I = 0), and they are, theoretically, identical to K. In practice, small differences between the reported values of K_{sp} and K often reflect differences between the measurement techniques of the investigators.

Disregarding the differences between K_{sp} and K for the fluorite-water solution equilibrium, concentrations of the calcium and fluoride ions in solution at equilibrium with CaF_2 are (using $K = 3.22 \times 10^{-11}$ in equations 15 and 16):

$$[Ca^{+2}] = 2.00 \times 10^{-4} \text{ mol/liter}$$
$$[F^-] = 4.00 \times 10^{-4} \text{ mol/liter}$$

Note that although K_{sp} is about 50% greater than K, the computed concentrations of the calcium and fluoride ions differ only by about 15% when estimated from either the K_{sp} or K values. It should be noted that in natural waters which contain both free and complexed ions, and other ions such that the ionic strength is significantly greater than zero, the calculation would be more complex (see other examples in this section).

Example 2 How would the presence of 0.1 *M* NaCl in a solution at equilibrium with CaF_2 affect the concentrations of Ca^{+2} and F^- calculated above?

To a good approximation, the ionic strength of the solution will be determined by the NaCl concentration: 0.1 *M* NaCl is much higher than the values of 10^{-4} to 10^{-3} *M* derived for Ca^{+2} and F^- in a pure solution. Thus, the ionic strength of the solution is, approximately, $I = 0.1$ *M*. The activity coefficients of the Ca^{+2} and F^- ions at this ionic strength are (see section 2.6 on estimation of the ionic strength):

$$\gamma_{Ca} = 0.40 \text{ and } \gamma_F = 0.76$$

From equation 7, the solubility product is:

$$K_{sp} = \frac{3.22 \times 10^{-11}}{0.40 \times 0.76^2} = 1.39 \times 10^{-10}$$

From equations 15 and 16, the ionic concentrations of Ca^{+2} and F^- in solution are:

$$[Ca] = 3.27 \times 10^{-4} \text{ mol/liter}$$
$$[F] = 6.53 \times 10^{-4} \text{ mol/liter}$$

The results indicate that the solubility of calcium fluoride in the presence of 0.1 *M* NaCl will be about 40% greater than its solubility in pure water.

DEGREE OF SATURATION

The preceding two sections (equations 2, 8 and 9) showed how the conditions of either an equilibrium or a non-equilibrium were defined by means of the relationships between the solubility product K_{sp} and the ion-concentration product ICP. It is useful to introduce a parameter (Ω) defining the degree of saturation of a solution as the following quotient:

$$\Omega = \frac{ICP}{K_{sp}} = \frac{[M]^p[A]^q}{[M]^p_{eq}\ [A]^q_{eq}} \qquad (17)$$

where subscript eq denotes concentrations of the aqueous species in a solution at equilibrium with the solid of composition M_pA_q.

Instead of the two aqueous species in the above equation, only one can be used, if the concentration value of the other is expressed as in equation 12 or 13. Thus, the degree of saturation, written in terms of the metal-ion concentration [M], becomes

$$\Omega = \left(\frac{[M]}{[M]_{eq}}\right)^{p+q} \tag{18}$$

In an undersaturated solution, $\Omega < 1$, and at equilibrium, $\Omega = 1$; this is analogous to the definitions of undersaturation, saturation, and supersaturation in relationships 9 (for concentrations) and 11 (for activities).

EFFECT OF TEMPERATURE ON SOLUBILITY

The effect of temperature on solubility of a solid can be expressed by a relationship between the equilibrium constant, temperature, and the standard enthalpy change of the reaction. Over relatively small ranges of temperature, when the standard enthalpy of a reaction at equilibrium remains approximately constant, the dependence of K on temperature is given by the following:

$$\log \frac{K_2}{K_1} = \frac{\Delta H_r{}^0}{2.3R} \left(\frac{1}{T_1} - \frac{1}{T_2}\right) \tag{19}$$

where ΔH_r^0 is the standard enthalpy of the reaction, R is the gas constant, and K_1 and K_2 are the equilibrium constants at temperatures T_1 (K) and T_2 (K), respectively.

The algebraic sign of ΔH_r^0 determines whether solubility, as expressed by the value of K, will increase or decrease if $T_2 > T_1$. To an order of magnitude, a 40°C rise in temperature increases or decreases the value of the equilibrium constant K by a factor of about 2. This approximation is based on the following considerations. For many inorganic solids, the change in the standard enthalpy of dissolution at equilibrium is of the order of a few kilocalories per mole; ±5 kcal/mol would be a reasonable estimate. The sources of data on ΔH^0 are the same as those for the standard free energies of formation, as detailed in § 2.11.5. At ambient temperatures, the value of $1/T_1 - 1/T_2$ for a change of 40°C is approximately 4×10^{-4} deg^{-1}. Substitution of the above estimate of ΔH_r^0, the reciprocal temperature difference term, and the value of the gas constant ($R \approx 2 \times 10^{-3}$ kcal mol^{-1} deg^{-1}) in equation 19 gives

$$\log K_2/K_1 \approx \pm 0.4$$

Therefore, K_2 may range from $0.4K_1$ to $2.5K_1$.

An illustration of a use of equation 19 is given in the following example.

Example 3 How are temperature changes likely to affect the equilibrium solubilities of the following three solids in water: calcium carbonate ($CaCO_3$, calcite), cadmium carbonate ($CdCO_3$), and barium selenate ($BaSeO_4$)?

For each of the three solids, an equilibrium dissolution reaction should be written, values of the standard enthalpies of formation (ΔH_f^0) obtained from the literature (see section 2.11.5), and the standard enthalpy change for the reaction computed. This procedure is illustrated below for $CaCO_3$.

$$CaCO_3 = Ca^{+2} + CO_3^{-2}$$

Calcite $CaCO_3$: $\Delta H_f^0 = -288.57$ kcal/mol
Ca^{+2}(aq): $\Delta H_f^0 = -129.74$ kcal/mol
CO_3^{-2}(aq): $\Delta H_f^0 = -161.84$ kcal/mol
(Source: Robie *et al.* [22])

Reaction: $\Delta H_r^0 = -129.74 - 161.84 + 288.57$
$= -3.0$ kcal/mol

Similar computations of ΔH_r^0 for the reactions at equilibrium with the two other solids, using the data from reference 22, give the following results:

$$CdCO_3: \quad \Delta H_r^0 = -0.6 \text{ kcal/mol}$$
$$BaSeO_4: \quad \Delta H_r^0 = +6.0 \text{ kcal/mol}$$

The values of the enthalpy change for the three solids should be considered in the context of the algebraic sign of the right-hand side of equation 19. The positive value of ΔH_r^0 for $BaSeO_4$ indicates that the solubility product will be higher at a higher temperature. For $CdCO_3$, the change in ΔH_r^0 is small, so there will be only a small change in K with a changing temperature. For calcite, the negative value of ΔH_r^0 indicates that the solubility of $CaCO_3$ will decrease with increasing temperature.

The above statement regarding the solubility of calcite in water is confirmed by experimental data. For $BaSeO_4$, experimental data (Figure 2.11-1) also affirm that its solubility increases by an amount corresponding to a value of $\Delta H_r^0 \approx +6$ kcal/mol, but only up to 35°C. From 35° to 95°C the solubility decreases, possibly indicating a change in the reaction mechanism. The solubility of another selenate phase, $PbSeO_4$, is also plotted in Figure 2.11-1; these values increase by a factor of about 2.5 with a temperature increase of 50°C.

In general, if the solubility values determined at a series of temperature points form a straight line in a graph of log K against 1/T, it indicates that ΔH_r^0 for the reaction remains constant within the experimental accuracy of the data.

EFFECT OF PRESSURE ON SOLUBILITY

Under most environmental conditions, the pressure is around 1 atmosphere; much higher values are found in nature only far below a water surface or beneath an overburden of crustal rocks. An increase in the pressure by 1 atm is equivalent to the pressure under a column of water 10 m high or a column of rock about 3.5 m thick. Thus, the pressure at the bottom of one of the bigger lakes of the North American continent, Lake Superior, is only about 40 atm, as the maximum depth of the lake is about 400 m.

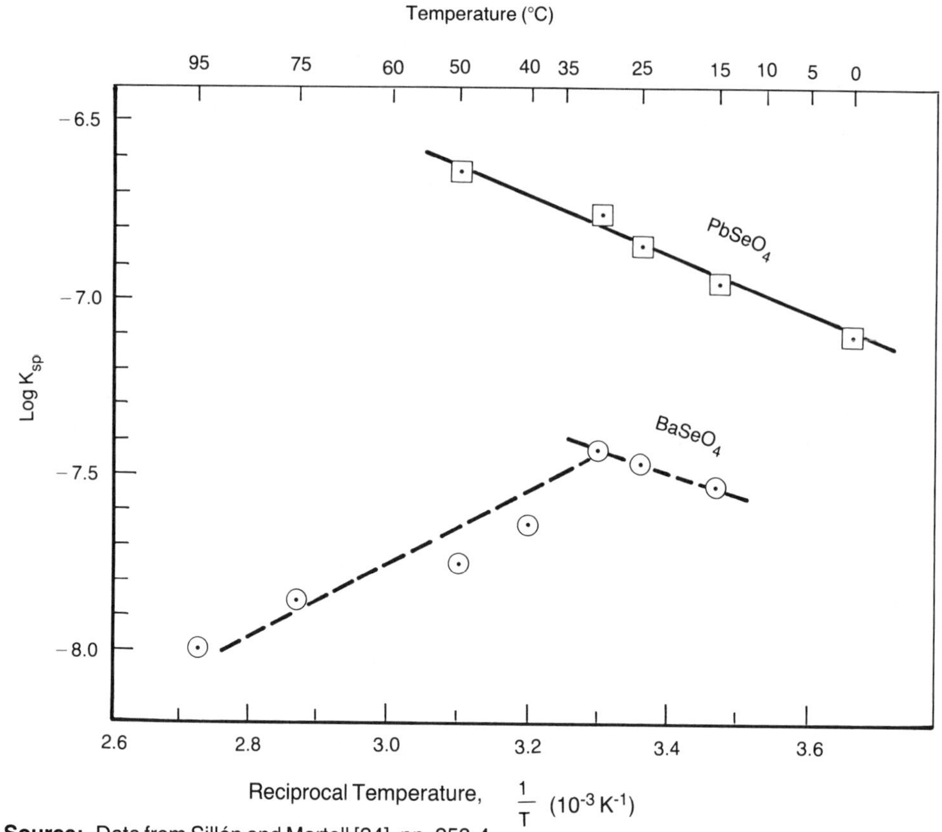

Source: Data from Sillén and Martell [24], pp. 253-4

FIGURE 2.11-1 Solubilities of $BaSeO_4$ and $PbSeO_4$ as a Function of Temperature

In general, the effect of pressure on the solubility of minerals and inorganic solids in water is small, and it is relatively much smaller than the effect of temperature within the limits of its variation in the environment. The dependence of solubility, as represented by the equilibrium constant K of a reaction, such as equation 1, is given by the following relationship:

$$\log \frac{K_2}{K_1} = - \frac{\Delta \bar{V}_r^0}{2.3RT}(P - 1) \tag{20}$$

where K_2 and K_1 are the values of the equilibrium constant at pressures P and 1 atm respectively, R is the gas constant, and T is absolute temperature. $\Delta \bar{V}_r^0$ is the standard volume change for the reaction; sources of data on the molar volumes of solids and aqueous species are given in section 2.11.5.

The relatively small effect of pressure on solubility under environmental conditions can be illustrated by the following consideration. The standard volume change in dissolution reaction at equilibrium is usually a negative quantity, typically about -40 cm^3/mol [19].

To estimate the increase in the equilibrium constant caused by an increase in pressure from 1 to 100 atm, the following values are substituted in equation 20:

$$\Delta\bar{V}_r^0 \approx -40\ \text{cm}^3\ \text{mol}^{-1}$$
$$R \approx 82\ \text{atm cm}^3\ \text{mol}^{-1}\ \text{deg}^{-1}$$
$$T \approx 300\ \text{K}$$
$$P \approx 100\ \text{atm}$$

The result is:

$$\log \frac{K_2}{K_1} \approx 0.07$$

or, alternatively,

$$K_2 \approx 1.17\ K_1$$

This indicates an increase in the equilibrium constant K of 17%. For the individual species in solution at equilibrium with a solid, the increase in their concentrations will be even smaller, as can be estimated from equations 15 and 16.

Pressures of several tens or hundreds of atmospheres, which are very high by environmental standards, are not likely to produce any significant changes in the molar volumes of many minerals and inorganic solids. A measure of the volume change of a solid, due to pressure, is its compressibility. Near room temperature, compressibilities of the more common minerals and inorganic solids are so small that a 1% change in volume requires pressures of an order of thousands to tens of thousands of atmospheres.

Solids that formed at higher pressures are, in general, more soluble than the low-pressure phases of identical composition at the same temperature [16]. Phase transformation in the solid state at environmental temperatures may be caused by strong mechanical stresses, either natural or man-made.

EFFECT OF PARTICLE SIZE ON SOLUBILITY

Very small particles have large surface areas in relation to their volumes. The energy needed to form that surface in the process of nucleation of the particle can amount to a substantial contribution to the total free energy of formation of the small particle. Thus, in principle, very small particles are more soluble than larger ones, because the free energy of formation of the small particles is higher [29, 30].

The increase in solubility of a particle of small size, or of small diameter, relative to one whose surface has "an infinitely large" radius of curvature, is given by the following relationship:

$$\log \frac{K_r}{K} = \frac{4\sigma V}{2.3RTd} \tag{21}$$

where K = equilibrium constant for solubility of the bulk solid
K_r = equilibrium constant for the small particle
d = linear dimension of the small particle (diameter of spherical particle or edge length of a cube)
σ = interfacial energy[2] of small particle surface in solution, in units of energy per unit area
V = molar volume of the solid
R = gas constant
T = temperature (K)
(Units of all parameters are given in Example 4.)

Equation 21 is valid for spheres or cubes. For other shapes, the factor 4 in the numerator will change, but the present form of the equation is adequate for many of the cases of interest, where the particle sizes are of primary concern.

Equation 21 shows that the solubility of small particles increases exponentially as the particle size and density of the solid decrease (i.e., smaller d and larger V). In general, as shown in Example 4 below, the effect of solid solubility becomes pronounced for particles smaller than about 100 Å in diameter (10^{-6} cm or 0.01 micron; for comparison, the diameters of atoms in crystal lattices are roughly 2 – 4 Å).

Our ability to estimate the solubility of very small particles of a given solid hinges on the availability of data on the interfacial energy (σ). Figure 2.11-2 gives this information for many minerals and inorganic solids in aqueous solutions. Note that solids of higher solubility are characterized by lower values of σ and that the limiting value of zero surface energy ($\sigma = 0$) corresponds to a solubility of 40 mol/liter.

A change in the solution composition owing to the presence of other dissolved species may affect the value of the interfacial energy reported for a solid phase in its saturated aqueous solution. For some oxides and hydroxides, errors in the measured values of the interfacial energy were reported as ±25-50% [23]. (Additional values of interfacial energies in aqueous solutions at 25° and 70°C are summarized in Table 2.11-1.) For the solids shown in Figure 2.11-2, the interfacial energies in solution cluster mostly within the range from 50 to 150 erg/cm^2 (equivalent to about 0.5 to 2 kcal/mol for particles of 100 Å diameter).

2. This quantity is sometimes referred to in the literature as "interfacial tension" or "surface energy."

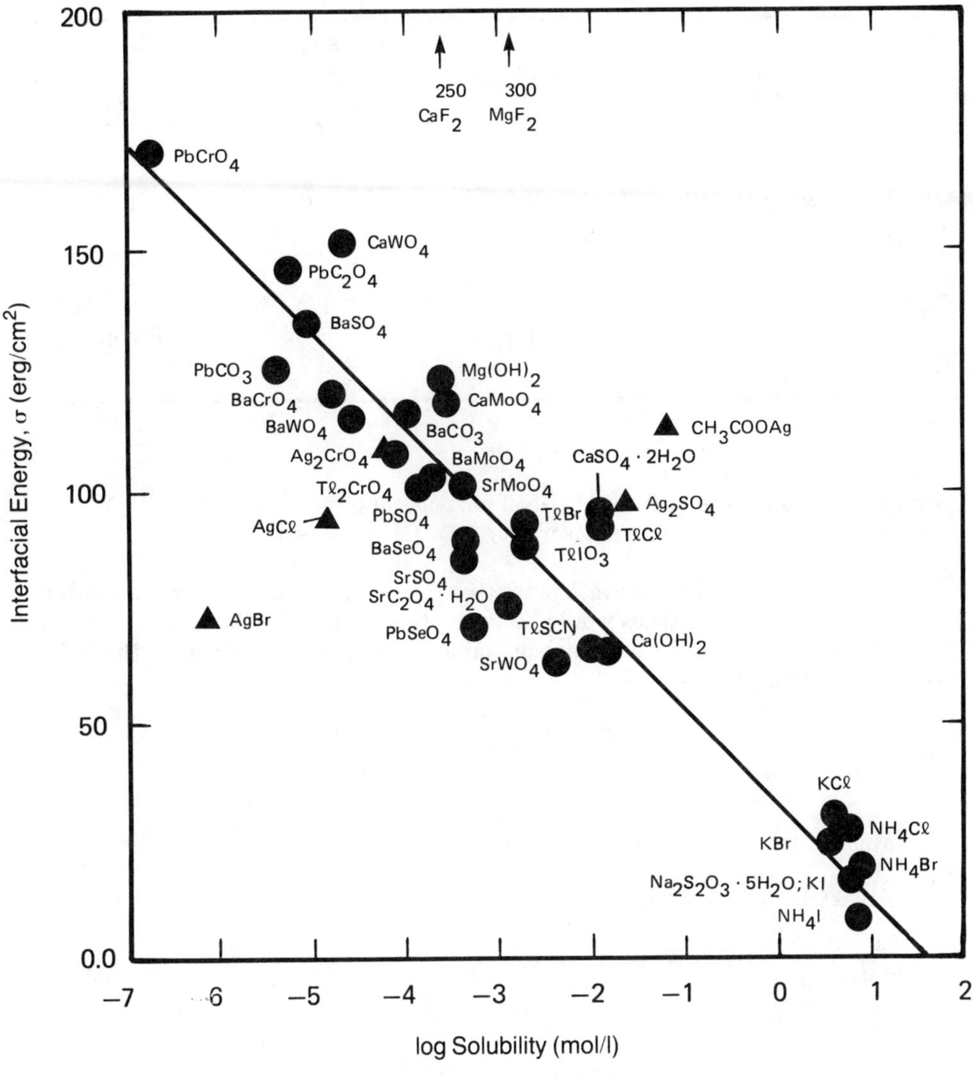

Source: Nielsen [20]. *(Copyright 1983, John Wiley & Sons. Reprinted with permission.)*

FIGURE 2.11-2 Correlation of Interfacial Tension of Solids in Solution with Their Solubility

TABLE 2.11-1

Interfacial Energies of Oxides and Hydroxides in Aqueous Solutions

Solid	Solution Composition	Temperature (°C)	Interfacial Energy, σ (erg/cm^2)	Source
ZnO	$0.2M$ $NaClO_4$	25	770 ± 330	[23]
CuO	$0.2M$ $NaClO_4$	25	690 ± 150	[23]
$Cu(OH)_2$	$0.2M$ $NaClO_4$	25	410 ± 130	[23]
α-Fe_2O_3	?	70	770	Ferrier (1966), cited in [15]
α-FeOOH	?	70	1250	Ferrier (1966), cited in [15]

Example 4 What is the maximum size of particles of barium molybdate ($BaMoO_4$) that may produce in solution a supersaturation of a factor of two?

"Supersaturation of a factor of two" means that the solubility product (or the equilibrium constant) of the small particles would be twice that of the bulk phase. Thus, we seek to determine from equation 21 the particle diameter (d) that corresponds to the following increase in solubility:

$$\frac{K_r}{K} = 2 \quad \text{or} \quad \log\frac{K_r}{K} = 0.3$$

The values of the other parameters needed are:

$$V = \text{mol.wt./density}$$
$$= \frac{297.28 \text{ g/mol}}{4.65 \text{ g/cm}^3} \text{ (from Weast [33])}$$
$$= 63.9 \text{ cm}^3\text{/mol}$$

σ = 100 erg/cm^2 (from Figure 2.11-2)
R = 8.31×10^7 erg mol^{-1} deg^{-1}
T = 300 K (assumed for this example)

Substitution of the above values in equation 21 gives:

$$d = \frac{4 \times 100 \times 63.9}{0.3 \times 2.3 \times 8.3 \times 10^7 \times 300}$$

$$= 1.49 \times 10^{-6} \text{ cm} \quad \text{or} \quad 149 \text{ Å}$$

As an approximation, concentrations of the Ba^{+2} and MoO_4^{-2} ions at equilibrium with the small particles can be estimated by equations 15 and 16, with p = 1 and q = 1. Writing $[Ba]_r$ for the barium concentration at equilibrium with the small particles and [Ba] for its concentration at equilibrium with the bulk solid $BaMoO_4$, we have:

$$\frac{[Ba]_r}{[Ba]} = \left(\frac{K_r}{K}\right)^{1/2}$$

$$= 2^{1/2} = 1.41$$

Thus, the concentrations of barium and molybdate in solution at equilibrium with the small particles would be about 40% higher than the concentrations at equilibrium with the bulk solid phase.

EFFECTS OF SOME SOLUTION PROPERTIES ON SOLUBILITY

Two characteristics of natural waters may have pronounced effects on the solubilities of solids: (1) the presence of complexing agents in solution, and (2) the acidity (pH) of the solution. Both of these characteristics are briefly discussed with the aid of examples in this section.

Solubility in the Presence of Complexing Ions

Inorganic and organic ions in solution tend to form a great variety of charged and neutral complexes among themselves. Manual calculation of the effect of a given complexing (or chelating) agent in solution on the solubility of a given solid can be difficult, because of the many association or dissociation equilibria that must be considered. Anyone extensively involved with the problems of chemical speciation in solution and its effects on the solubilities of different solids should acquire a familiarity with the computer methods available for solving these problems. The field is extensively covered by a number of textbooks and reference works, some of which are listed in section 2.11.5 of this chapter and others in section 2.13.

Example 5 At an equilibrium between CaF_2 and its aqueous solution, does calcium form significant concentrations of complexes with any of the negatively charged ions?

In water free of dissolved CO_2 and closed to the atmosphere, dissolution of CaF_2 is likely to produce at least the following species in solution: Ca^{+2}, H^+, F^-, OH^-, and two ionic complexes — CaF^+ and $CaOH^+$. (If the system contained dissolved CO_2, one would include such complexes as $CaHCO_3^+$ and $CaCO_3^0$.)

The solubility product of CaF_2 and the association constants for the complexes CaF^+ and $CaOH^+$ in solution are as follows:

$$[Ca^{+2}][F^-]^2 = K_{sp}; \quad K_{sp} = 5\times10^{-11} \quad [24] \tag{22}$$

$$\frac{[CaF^+]}{[Ca^{+2}][F^-]} = K_1; \quad K_1 = 10 \quad [3] \tag{23}$$

$$\frac{[CaOH^+]}{[Ca^{+2}][OH^-]} = K_2; \quad K_2 = 25 \quad [3] \tag{24}$$

At equilibrium between CaF_2 and its aqueous solution, a condition of electroneutrality or charge balance must be maintained for the ions in solution:

$$2[Ca^{+2}] + [CaF^+] + [CaOH^+] + [H^+] = [F^-] + [OH^-] \tag{25}$$

Having set the equation for the charge balance in solution, we have six unknowns — the concentrations of the six aqueous species above — but only four equations. A further simplification can be made: in nearly neutral solutions, the terms $[H^+]$ and $[OH^-]$ are nearly equal, so they cancel out of equation 25:

$$2[Ca^{+2}] + [CaF^+] + [CaOH^+] = [F^-] \tag{26}$$

There remain five unknowns in four equations (22, 23, 24 and 26). Consecutive substitution of equations 22-24 in equation 26 gives a final equation in terms of the fluoride-ion concentration, hydroxyl-ion concentration (or activity), and the stability constants:

$$[F^-]^3 - K_1K_{sp}[F^-] - K_2K_{sp}[OH^-] - 2K_{sp} = 0 \tag{27}$$

Substituting the numerical values of the solubility product and the stability constants in this equation gives

$$[F^-]^3 - 5\times10^{-10}[F^-] - 12.5\times10^{-10}[OH^-] - 1\times10^{-10} = 0 \tag{28}$$

The value of $[OH^-]$, usually taken as $pOH = 14 - pH$, can be measured more easily than the fluoride or calcium-ion concentration. For any reasonable value of pOH, except for the conditions of extreme alkalinity ($pOH < 2$ or $pH > 10$), the OH^--containing term in equation 28 is negligible in comparison with the last term, 1×10^{-10}. Thus, the equation to be solved for the unknown fluoride-ion concentration becomes:

$$[F^-]^3 - 5\times10^{-10}[F^-] - 1\times10^{-10} = 0 \tag{29}$$

A solution that reasonably satisfies this equation can be obtained by trial and error, substituting numerical values for $[F^-]$:

$$[F^-] = 4.64\times10^{-4} \text{ mol/liter}$$

This result is practically identical to the fluoride-ion concentration at equilibrium with CaF_2 obtained in Example 1. The fraction of the fluoride-ion in solution that is tied up in complexes must therefore be very small and, consequently, the concentration of the ion-complex CaF^+ must be very low.

As the small effects due to the ionic strength on concentrations at equilibrium in a dilute solution are disregarded in this example, concentrations of the two ionic complexes with calcium will be computed. From equation 22, the calcium-ion concentration is

$$[Ca^{+2}] = (5\times10^{-11})/(4.64\times10^{-4})^2$$

$$= 2.32 \times 10^{-4} \text{ mol/liter}$$

The concentration of the CaF^+ complex is, from equation 23 and the preceding results,

$$[CaF^+] = 10 \times 2.32 \times 10^{-4} \times 4.64 \times 10^{-4}$$

$$= 1.08 \times 10^{-6} \text{ mol/liter}$$

For the complex $CaOH^+$, a value of the hydroxyl-ion concentration in solution must be known. At pH = 7, we have

$$[H^+] = [OH^-] = 10^{-7}$$

and the ion-complex concentration from equation 24 and the above value of $[Ca^{+2}]$ is

$$[CaOH^+] = 25 \times 2.32 \times 10^{-4} \times 1 \times 10^{-7}$$

$$= 5.8 \times 10^{-10} \text{ mol/liter.}$$

From the preceding results, it follows that the concentration of the ion-complex $CaOH^+$ at a solution pH of 7 is negligibly small in comparison to the concentrations of the other species. The concentration of the other ionic complex, CaF^+, is somewhat higher, but it also amounts to only a very small fraction of either the calcium-ion or fluoride-ion concentration: $[CaF^+]$ represents about 0.5% of the calcium ion in solution and about 0.25% of the fluoride ion.

The answer to the question posed by this example is that the aqueous complexes of the calcium and fluoride species at equilibrium with CaF_2 in a closed system are present at concentrations that are negligible for most practical purposes.

The pH and Solubility of Oxides

Hydrogen ion activity in solution is one of the more easily measurable characteristics. Therefore, chemical reactions are often studied by observing changes in rate with solution pH. Experimentally, the pH of a solution can usually be controlled by addition of acids, alkalis or buffers. Thus, in the laboratory, pH is regarded as one of the main or master environmental variables.

In natural systems, however, the pH of water solutions is often determined by numerous and complex reactions involving minerals and organic materials, as well as some of the atmospheric gases. Controlling the pH of a solution in a large tank, of a sludge in a large basin, or of water in a lake or stream is not as simple as in a glass vessel indoors. Moreover, the pH in an environmental system consisting of several solids and water may not be easily and simply estimated from one or more equilibrium reactions. The extensive buffering of natural water systems by chemical reactions that are not always fully understood makes one look at the pH of such systems as one of their characteristic parameters, rather than as a master variable that changes in a predictable manner, such as the temperature of the environment.

Despite these difficulties, pH is an important and useful characteristic of aqueous solutions, even if only due to our accumulated knowledge of the role of hydrogen-ion concentration in equilibrium solubility reactions.

A metal oxide of general composition Me_pO_q, where p and q are the stoichiometric coefficients, can react with water to produce a solid metal-hydroxide:

$$Me_pO_q + qH_2O = pMe(OH)_{2q/p} \quad (30)$$

Considering a Fe^{III}-oxide of composition Fe_2O_3, the preceding general formula of a hydration reaction becomes

$$Fe_2O_3 + 3\ H_2O = 2Fe(OH)_3 \quad (31)$$

Dissolution of a metal hydroxide in water and establishment of equilibrium between the solid and solution involve some or all of the following types of reactions:

$$Fe(OH)_3 + 3H^+ = Fe^{+3} + 3H_2O \quad (32)$$

$$Fe(OH)_3 + 2H^+ = FeOH^{+2} + 2H_2O \quad (33)$$

$$Fe(OH)_3 + H^+ = Fe(OH)_2^+ + H_2O \quad (34)$$

$$Fe(OH)_3 + OH^- = Fe(OH)_4^- \quad (35)$$

The preceding four equilibria show a dissolved species of iron as a product and the hydrogen or hydroxyl-ion as one of the reactants. From the known values of the equilibrium constants of each of the four reactions, the concentrations of ferric ion and of the other iron complexes can be estimated for various values of the pH. The same procedure applies to other metal hydroxides at equilibrium with their aqueous solutions, in the absence of any chelating agents that could form additional complexes with the dissolved iron species.

Figure 2.11-3 shows concentrations of metal ions as a function of the solution pH for reactions such as equation 32. The steeper the slope of the lines, the greater is the response of concentration to change in the pH of the solution. The response is greater for cations of higher valence.

A different representation of certain oxide and hydroxide solubilities is shown in Figure 2.11-4. The concentration of the dissolved metal as a function of the pH is high in acidic solutions, goes through a minimum in slightly acidic to mildly alkaline solutions, and increases again in strongly alkaline media.

Different sections of the solubility curves represent regions of dominance of one of the dissolved species. In acidic solutions, free ferric ion Fe^{+3} is the dominant species, but in alkaline solutions, the negatively charged $Fe(OH)_4^-$ predominates.

The pH regions where the minima of the solubility curves occur should be noted, as they help explain the differences in the solubility and precipitation behavior of metal hydroxides in general.

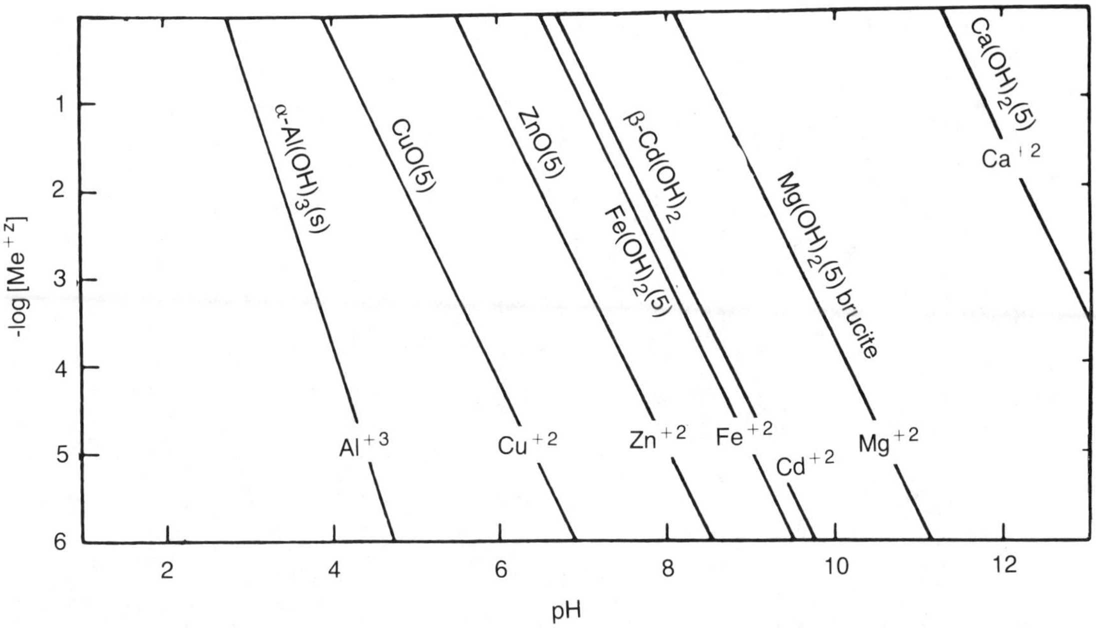

Source: Stumm and Morgan [29]. *(Copyright 1981, Wiley-Interscience. Reprinted with permission.)*

FIGURE 2.11-3 Free Metal-ion Concentration in Equilibrium with Solid Oxides or Hydroxides. (The occurrence of hydroxo metal complexes must be considered for evaluation of complete solubility)

Other more complex and detailed correlations between the solubilities of metal oxides and hydroxides, and the stability constants of their aqueous complexes, are given by Stumm and Morgan [29] and by Baes and Mesmer [2].

2.11.3 Estimation of Equilibrium Constant K and Solubility

INTRODUCTION

The preceding sections of this chapter showed how the values of a solubility product K_{sp} and an equilibrium constant K can be used to estimate concentrations of the species in solution at equilibrium with a solid. For determining the solubilities of solids, K is an important parameter whose generality and usefulness are far broader than those of K_{sp}; the latter depends on the chemical composition and the ionic strength of the solution, as shown by the relationship between K and K_{sp} in equation 7.

To reiterate briefly, the essentials of the method of estimating the solubilities of solids in water, based on the equilibrium constants of their dissolution reactions, are as follows:

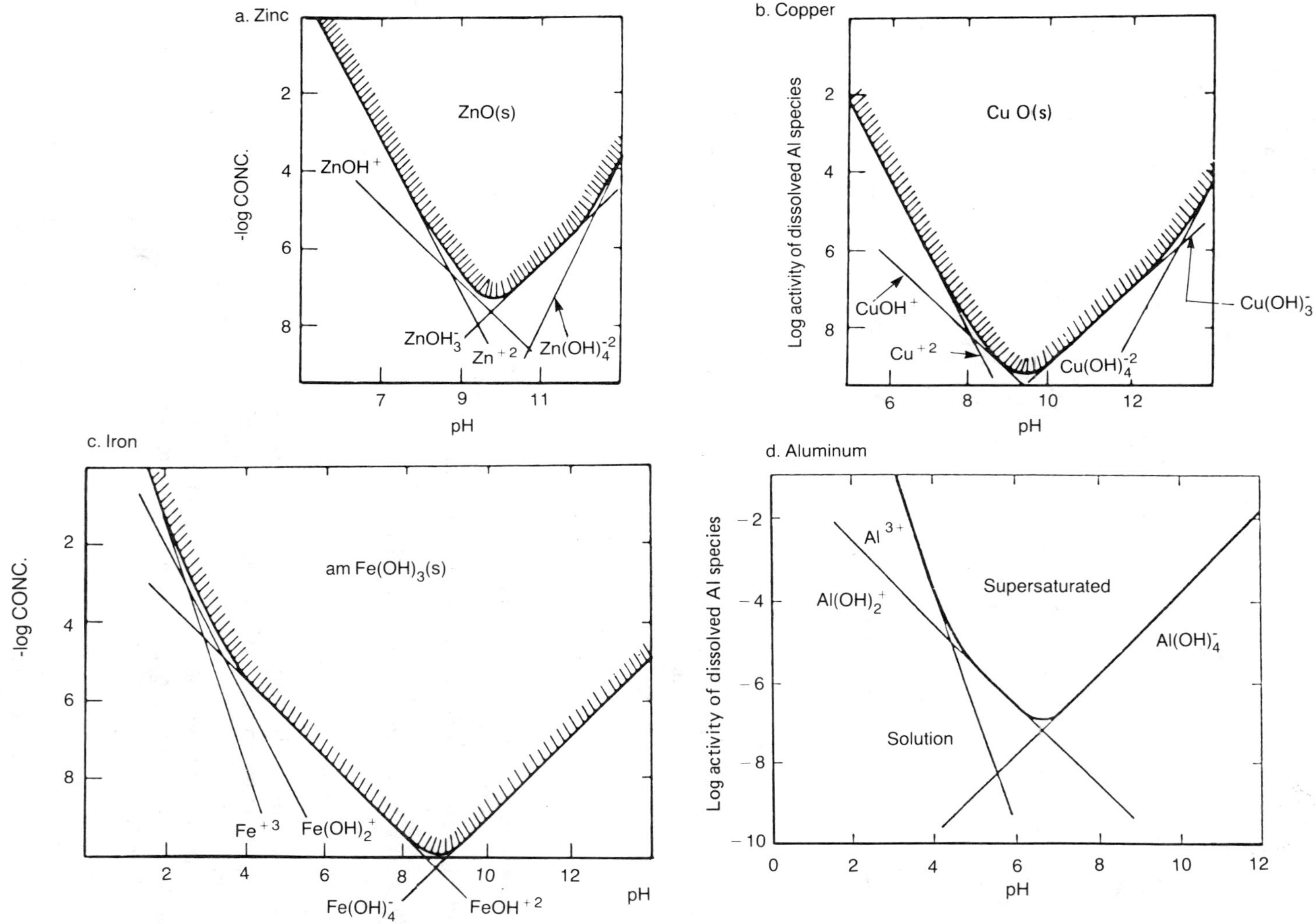

Source: a, b and c from Stumm and Morgan [29]. *(Copyright 1981, Wiley-Interscience. Reprinted with permission.)* d from Drever [7]. *(Copyright 1982, Prentice Hall. Reprinted with permission.)*

FIGURE 2.11-4 Solubilities of Metal Oxides and Hydroxides as a Function of the Solution pH

(1) Determine the composition of the solid and write a reaction for an equilibrium between this solid and its aqueous solution;

(2) Determine the equilibrium constant (K) for the reaction;

(3) Determine the bulk chemical composition of the solution and the ionic complexes that are likely to be present;

(4) Calculate the ionic strength of the solution and the activity coefficients of the dissolved species;

(5) Calculate the concentrations of the individual species in solution, which should indicate the solubility of the solid phase.

Steps (3) through (5) were covered briefly in section 2.11.2. For solutions containing several dissolved species, these steps may require fairly extensive and iterative computations, whose complexity may depend on the chosen model of chemical speciation. More detailed treatment of chemical speciation in solutions containing several dissolved species, as well as of the solubilities of solids in them, is given in section 2.13 and in a number of textbooks [3, 7, 10, 29].

The remainder of this section describes a few simple steps for estimating the equilibrium constant K and the solubility of some classes of solids.

The estimation methods described are:

- Estimation using Gibbs free energy data;
- Correlations with crystal properties (lattice energy and crystal ionic radii); and
- Correlation with cation electronegatives.

Of these, the most widely useful method is that which utilizes the Gibbs free energy data. In instances where these data are not available for the species of interest, the other more empirical relationships may be useful.

ESTIMATION OF K FROM THE GIBBS FREE ENERGIES OF FORMATION

The classical textbook method of computing the equilibrium constant is based on a relationship between K and the change in the Gibbs free energy for a reaction at equilibrium, when the reactants and the products are considered in their standard states. (For definitions and discussions of the standard states of solids and aqueous solutions, see the references cited two paragraphs earlier.)

A reaction at equilibrium between a solid of composition M_pA_q and its aqueous solution containing the aqueous species M(aq) and A(aq) can be written as

$$\underset{\text{solid state}}{M_pA_q} = \underset{\text{aqueous state}}{pM_{aq} + qA_{aq}} \tag{36}$$

The standard *free energy change* for this reaction (that is, at the standard state of 25°C and 1 atm of total pressure), denoted ΔG_r^0, is the sum of the *standard free energies of formation* of the products less the sum of the *standard free energies of formation* of the reactants:

$$\Delta G_r^0 = p\,\Delta G_f^0(M_{aq}) + q\,\Delta G_f^0(A_{aq}) - \Delta G_f^0(M_pA_{q(s)}) \tag{37}$$

The equilibrium constant K for reaction 36 is related to ΔG_r^0 as follows:

$$\log K = -\frac{\Delta G_r^0}{2.3RT} \tag{38}$$

Example 6 illustrates the use of the preceding equation.

Example 6 Compute an equilibrium constant for dissolution or precipitation of fluorite (CaF_2) in water at 25°C and 1 atm total pressure. The reaction at equilibrium is:

$$CaF_2(s) = Ca^{+2}(aq) + 2F^-(aq)$$

The standard free energies of formation for the solid and aqueous species are:[3]

CaF_2: $\Delta G_f^0 = -281.29$ kcal/mol [22]

Ca^{+2}: $\Delta G_f^0 = -132.30$ kcal/mol [7]

F^-: $\Delta G_f^0 = -67.34$ kcal/mol [7]

According to equation 37, the standard free energy change for the reaction is:

$$\Delta G_r^0 = -132.30 + 2(-67.34) - (-281.29)$$
$$= 14.31 \text{ kcal/mol}$$

The equilibrium constant can now be calculated from equation 38, using the computed value of ΔG_r^0, $R = 1.987 \times 10^{-3}$ kcal mol^{-1} deg^{-1}, and $T = 298.15$ K:

$$\log K = -\frac{14.31}{1.364} = -10.49$$
$$K = 3.22 \times 10^{-11}$$

This is the value used in Example 1.

Relatively small differences of about 1 kcal/mol in the values of ΔG_f^0 given by different sources may result in greater differences between the values of K computed from them. From the data listed in Table 2.11-2, an equilibrium constant for CaF_2 and its aqueous solution can be calculated as shown above:

$$\log K = -\frac{13.42}{1.364} = -9.84$$
$$K = 14.5 \times 10^{-11}$$

The latter value of K is about four times greater than the preceding.

3. Values were obtained from references noted below and differ slightly from those in Table 2.11-2.

Values of the Gibbs free energy of formation for a number of representative solids and aqueous ions are given in Table 2.11-2. Also listed in the table are enthalpies of formation at the standard state (ΔH_f^0), which can be used as in Example 3 to compute K at higher temperatures.

TABLE 2.11-2

Gibbs Free Energy of Formation (ΔG_f^0) and Enthalpy of Formation (ΔH_f^0) for Selected Solids, Liquids, and Aqueous Ions
(at 25°C and 1 atm total pressure)

Formula	ΔG_f^0 (kcal/mol)	ΔH_f^0 (kcal/mol)
Crystalline Solids		
$BaSeO_4$	−249.7	−274.0
CaF_2	−279.0, −281.29[a]	−291.5
$Ca_3(AsO_4)_2$	−732.1	−788.4
NaF	−129.902	−137.105
$PbSeO_4$	−120.7	−145.6
Liquids		
D_2O	−58.195	−70.411
HDO	−57.817	−69.285
H_2O	−56.687	−68.315
H_2O_2	−28.78	−44.88
Aqueous Ions		
Ba^{+2}	−134.02	−128.50
Ca^{+2}	−132.30	−129.74
$CaOH^+$	−171.7	—
H^+	0	0
Na^+	−62.593	−57.39
Pb^{+2}	−5.83	−0.4
AsO_4^{-3}	−155.00	−212.27
F^-	−66.64	−79.50
OH^-	−37.594	−54.970
SeO_3^{-2}	−88.38[b]	−121.70[a]
SeO_4^{-2}	−105.47[b]	−143.19[a]

a. From Robie *et al.* [22]

b. From Wagman *et al.* [32]

Source: National Bureau of Standards data reprinted in [33] unless otherwise noted.

If the values of the standard free energy (ΔG_f^0), enthalpy (ΔH_f^0), and entropy of formation (ΔS_f^0) cannot be found in the literature, they can be estimated for certain types of compounds; methods for various classes of solid phases are described in a number of publications [27, 30, 31].

ESTIMATION OF K AND SOLUBILITY FROM CORRELATIONS WITH CRYSTAL PROPERTIES

The relationship between an equilibrium constant K and the Gibbs free energy of a reaction in a standard state, as given in equation 38, is based on a well-established theory of chemical thermodynamics. Beyond the domain of this accurate relationship, it is possible to construct a number of *empirical* correlations between the values of K for solid-solution reactions and the values of certain properties characteristic of the solid crystalline phase. Such correlations are also possible between the solubility values of certain classes of solids and some of their crystalline properties. In this section, empirical correlations and their possible use are demonstrated for the solubility or K values and for the values of such characteristics as the lattice energy of the solid, the electronegativity, ionization potential, and crystal radius of the ions that comprise the crystalline solid.

To reiterate, the correlations between the solubilities of solids and their crystal properties are strictly empirical. As will be shown, the resulting estimates of solubilities or equilibrium constants are often no better than order-of-magnitude, but they can serve an orientational purpose in the absence of more reliable measured or computed data.

Correlation between Solubilities and Lattice Energies

Energy liberated in the process of formation of an ionic solid from its gaseous ions is the *lattice energy* of the solid, ΔU_o [6, 8, 12]. For typical ionic crystals, such as alkali halides[4], good correlations exist between such properties as the metal-to-anion bond length and various physical properties of the solids, such as their melting and boiling points. For the fluorides and chlorides of the alkali metals, the lattice energies (Table 2.11-3) show a systematic increase from the Li-halogenide to Cs-halogenide within each group. This trend is similar to the increase in the molar volume of the solid, the latter being the result of the increase in the crystal ionic radius of the metal as one progresses from Li to Cs.

The solubilities of the alkali-metal fluorides and chlorides, also listed in Table 2.11-3, show somewhat uneven trends. A plot of solubility versus lattice energy (Figure 2.11-5) gives no useful correlation for the chlorides; a trend of increasing solubility with increasing lattice energy is clearer with the fluorides, although the correlation is anything but simple.

Another set of solubility and lattice energy values is given in Table 2.11-4. These data refer to three classes of compounds: alkali-metal selenites ($M_2^+SeO_3$), alkali-metal selenates ($M_2^+SeO_4$), and alkaline-earth selenates ($M^{+2}\,SeO_4$). In Figure 2.11-6, the data are plotted as the square root of solubility versus the lattice energy. Example 7 illustrates a possible application of these three sets of data.

4. Compounds (1:1 type) of the alkali metals Li, Na, K, Rb, and Cs with the halogens F, Cl, Br, and I.

TABLE 2.11-3

Solubilities in Water and Crystal Properties of Alkali Fluorides and Chlorides

Solid Phase	Solubility[a] (mol/kg H_2O)	Molar Volume[b] V (cm^3/mol)	Lattice Energy[c] (kcal/mol)	Crystal Ionic Radius (Å) of Cation for Coordination Number:[d]	
				6	8
LiF	0.075 ± 0.025	9.83	−244	0.82	
NaF	0.99	15.7 ± 0.7	−215	1.10	1.24
KF	17.5	23.2 ± 0.2	−193	1.46	1.59
RbF	20.6 ± 8.2 (18°C)	29.4	−183	1.57	1.68
CsF	22.7 ± 1.5 (18°C)	34.9 ± 2.0	−176	1.78	1.82
LiCl	20.0	20.5	−200	(same values as above)	
NaCl	6.15	27.0	−184		
KCl	4.8	37.6	−168		
RbCl	7.8	42.9	−162		
CsCl	11.4	42.2	−153		

Sources:

a. Stephen and Stephen [28]; Weast [33]. Values are at 25°C unless otherwise indicated.

b. Computed from the molecular weights and densities of the solids [33].

c. Evans [8], p. 48.

d. Frye [9], p. 26.

Example 7 Figure 2.11-6 shows data points for the selenites of Li, Na and K only. Can the solubilities of two heavier members of the series, Rb_2SeO_3 and Cs_2SeO_3, be estimated from the graph?

A rough approximation of these solubilities can be obtained by extrapolating the line for the selenites beyond the point for K_2SiO_3 and finding its intersection with ordinates representing the lattice energies of the Rb and Cs selenites listed in Table 2.11-4. The solubilities of Rb_2SiO_3 and Cs_2SeO_3 are found to be about 12.5 and 15.5 mol/kg H_2O respectively. Given the empirical nature of the correlation and the data set of only three points, these estimates are not very credible. The true solubilities might deviate from the linear relationship as in the fluoride curve of Figure 2.11-5, in which case the extrapolated values would be much too high.

A similarly rough estimate of the solubility of lithium selenate (Li_2SeO_4) can be made. From the plot of the four points representing the alkali-metal selenates in Figure 2.11-6, the line is extrapolated below the point for Na_2SeO_4 to the lattice energy value $\Delta U_o = -491$ kcal/mol that is characteristic of Li_2SeO_4 (Table 2.11-4). From the graph, the solubility value that corresponds to this point is about 0.6 mol/kg H_2O.

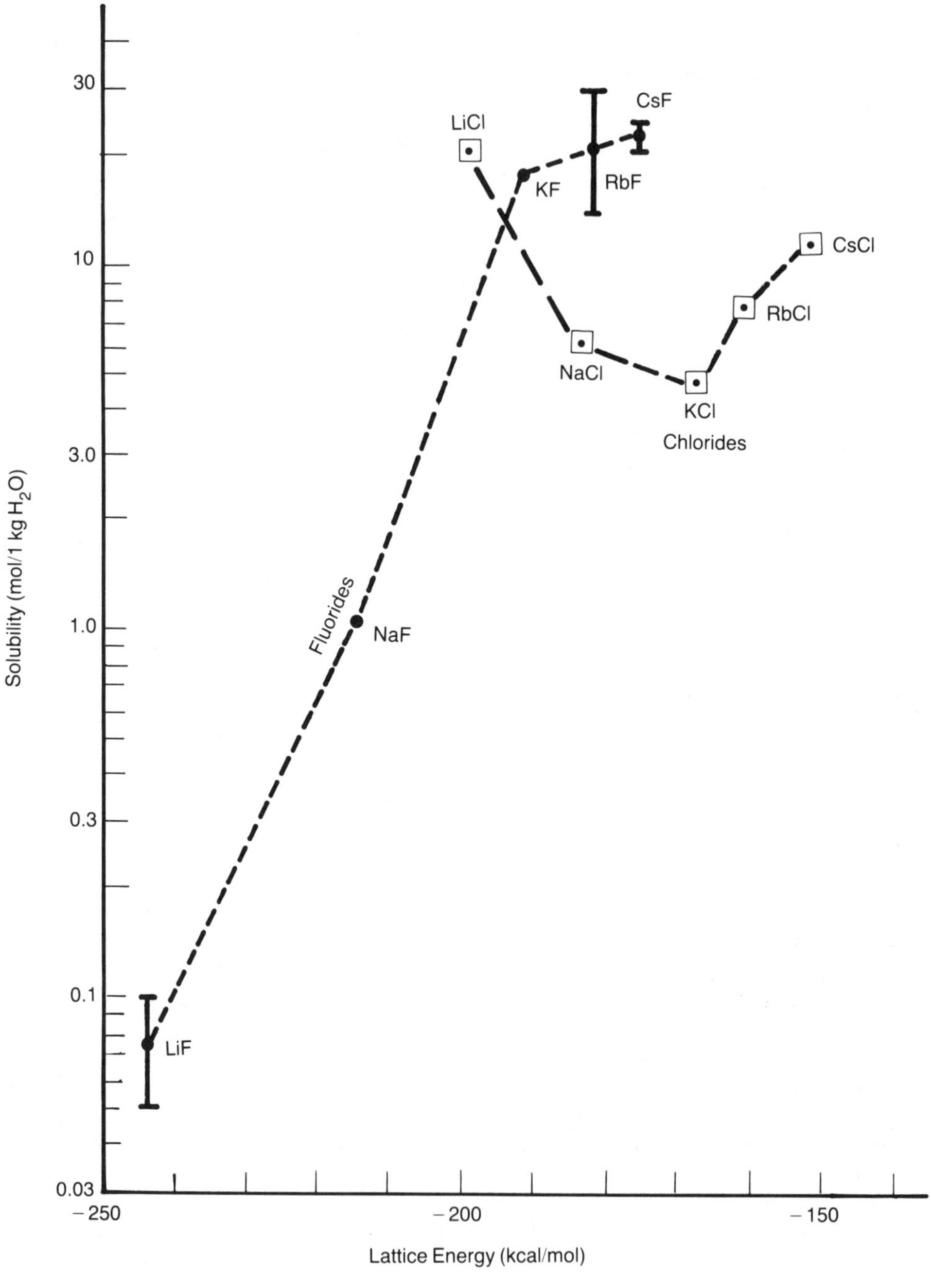

Source: Table 2.11-3

FIGURE 2.11-5 Solubilities as a Function of the Lattice Energy of Alkali Fluorides and Chlorides

TABLE 2.11-4

Solubility and Lattice Energy Values for Some Solid-Phase Selenites and Selenates

Solid	Solubility[a] (mol/kg H_2O)	Temperature[a] for Solubility Value (°C)	Lattice Energy[b] ΔU_0 (kcal/mol)
Li_2SeO_3	1.43	24-25	−519
Na_2SeO_3	5.19	24-25	−466
K_2SeO_3	10.60	24-25	−424
Rb_2SeO_3	—		−410
Cs_2SeO_3	—		−392
Li_2SeO_4	—		−491
Na_2SeO_4	1.76	15	−449
K_2SeO_4	4.94	12	−414
Rb_2SeO_4	5.06	12	−403
Cs_2SeO_4	5.99	12	−386
$BeSeO_4$	3.7[c]	25	−824
$MgSeO_4$	2.27	25	−692
$CaSeO_4$	0.404	25	−629
$SrSeO_4$	—		−595
$BaSeO_4$	2.9×10^{-4}	25	−570

a. **Source**: Stephen and Stephen [28]

b. **Source**: Jenkins [12]

c. Hydrated phase

Finally, in the series of alkaline-earth selenates, there are solubility and lattice energy data for three solids: Mg-, Ca-, and Ba-selenate. (The given value for Be-selenate appears to be anomalous.) We may try to estimate the solubilities of the Be- and Sr-selenates by extrapolation. For Be-selenate, whose lattice energy is −824 kcal/mol, we obtain a solubility of about 9 mol/kg H_2O, in marked contrast to that of the hydrated phase ($BeSeO_4 \cdot 4H_2O$), which is about 3.7 mol/kg H_2O. For Sr-selenate, the value of the lattice energy ($\Delta U_o = -595$ kcal/mol) corresponds to a solubility of about 0.08 mol/kg H_2O. Weast [33] describes Sr-selenate as "insoluble in water", but also lists the solubility of Ba-selenate as about double the value given in Table 2.11-4. Clearly, not only the empirical correlations in Figures 2.11-5 and -6 but also some of the data behind them should not be accepted unquestioningly.

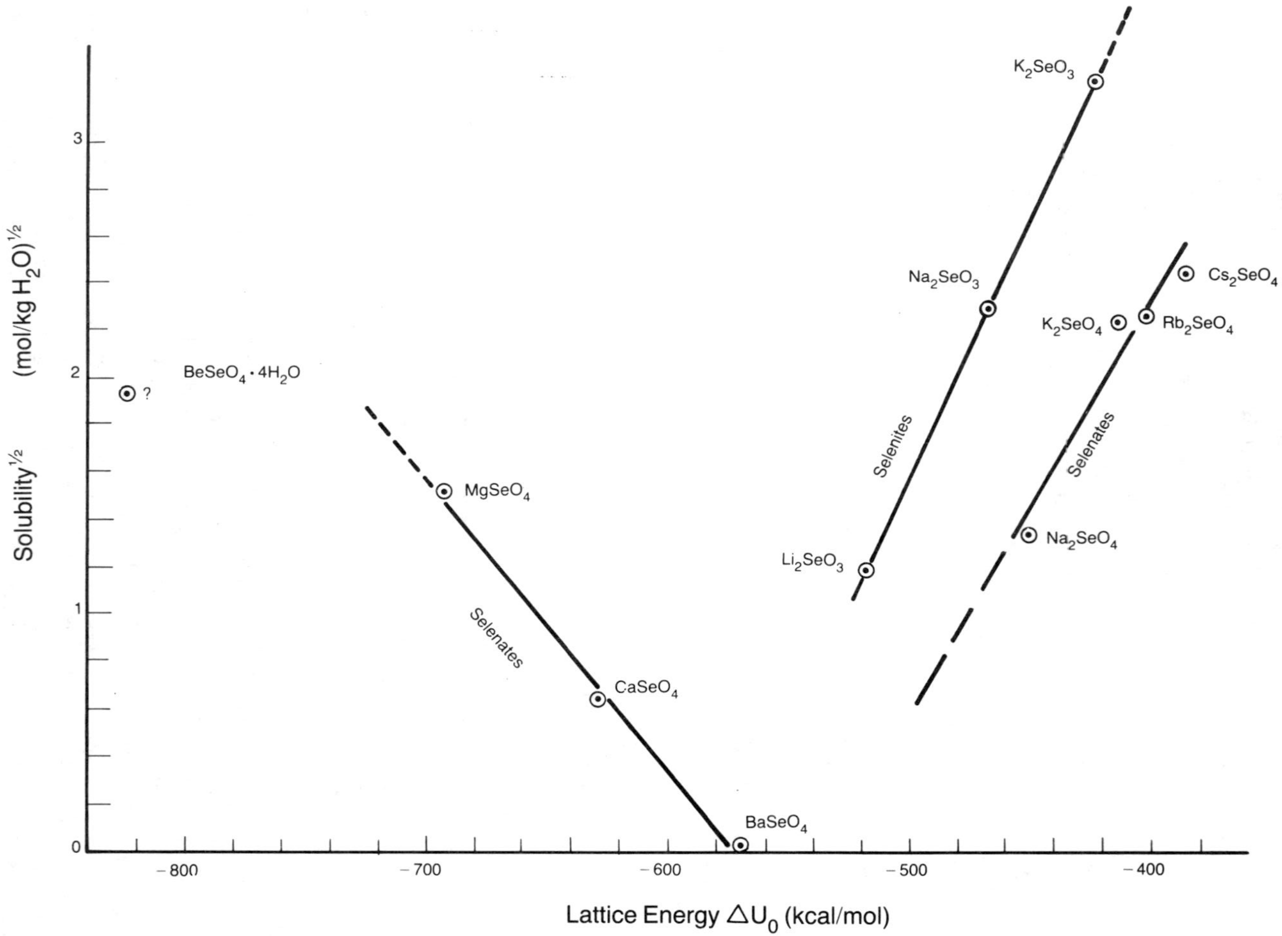

Source: Solubilities from Stephen and Stephen [28]; lattice energies from Jenkins [12]

FIGURE 2.11-6 Correlation Between the Square Root of Solubility and the Lattice Energy for Selected Selenites and Selenates

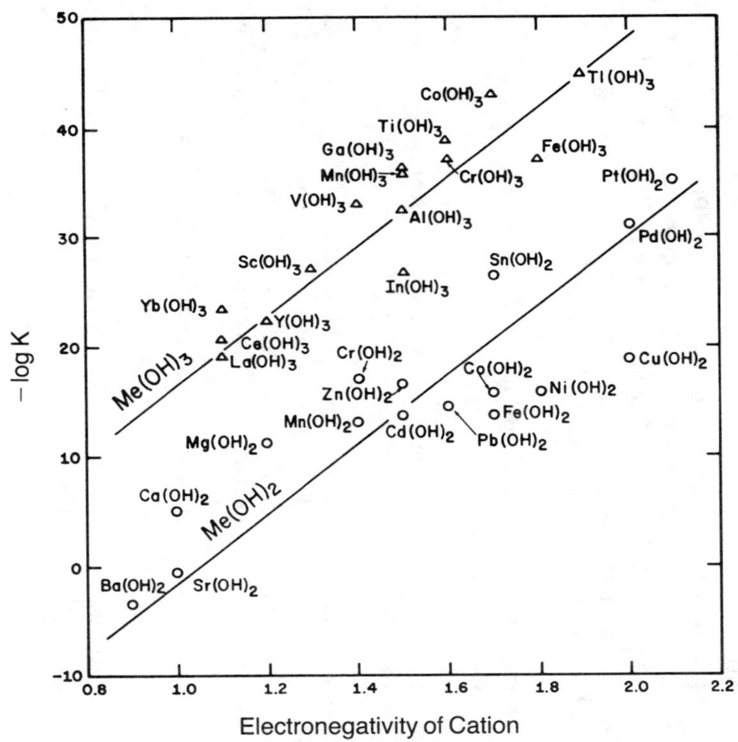

Source: Garrels and Christ [10]. *(Copyright 1965, Harper and Row. Reprinted with permission.)*

FIGURE 2.11-7 Solubilities of Metal Hydroxides Plotted Against the Cation Electronegativities at 25°C

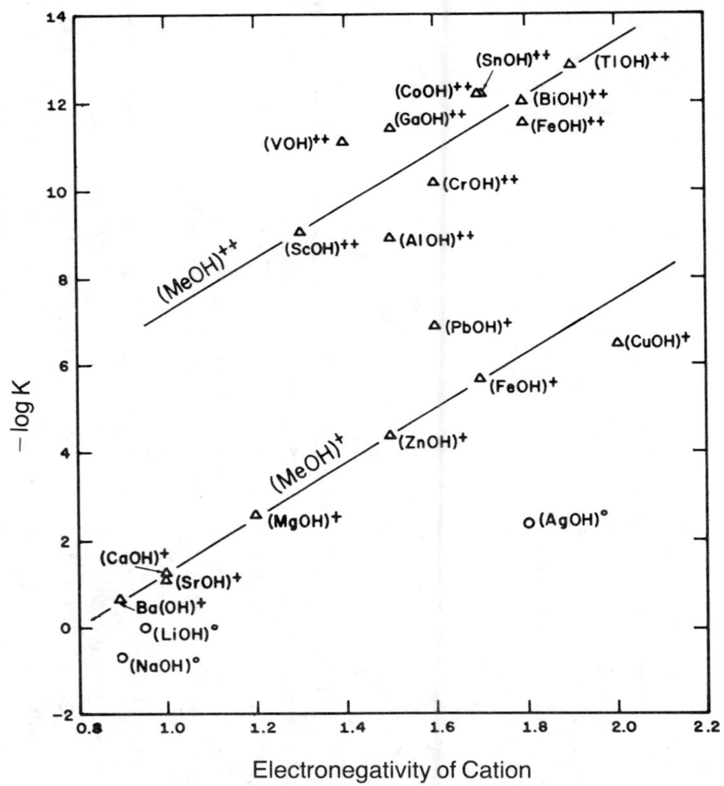

Source: Garrels and Christ [10]. *(Copyright 1965, Harper and Row. Reprinted with permission.)*

FIGURE 2.11-8 Dissociation Constants of Singly and Doubly Charged Hydroxide Complexes in Aqueous Solutions at 25°C Plotted Against Cation Electronegativities

Correlation between Solubilities and Cation Electronegativities

Clifford [4, 5] demonstrated and discussed a correlation between the electronegativity values of cations and the solubility values of poorly soluble solids of those cations. One such set of correlations is shown in Figure 2.11-7, a plot of the solubilities of metal hydroxides (expressed as the negative logarithms of their equilibrium constants, $-\log K$) versus the metal-cation electronegativities. Figure 2.11-8 is a similar plot for ion complexes in aqueous solution and includes some of the same metals that are considered in the hydroxide form in the preceding figure.

Figure 2.11-9 illustrates a correlation between the solubilities of divalent-metal arsenates, $(M^{+2})_3(AsO_4)_2$, and the cation electronegativity values.

For the three types of solids shown in Figures 2.11-7 and -9 — namely, divalent-metal hydroxides, trivalent-metal hydroxides, and divalent metal arsenates — the solubility generally decreases with increasing cation electronegativity. (Note the difference in the algebraic signs on the vertical axes of the two figures; the trend of the log K values as a function of the cation electronegativity is the same in both figures.)

The values on the horizontal axes of Figures 2.11-7 and -9 are the Pauling electronegativities, and their numerical values have been subject to corrections and revisions over a period of years. The plotted values stem from the 1950s. In Figure 2.11-9, two values of the cation electronegativity are shown for each metal arsenate: one from Clifford [4] and another from Evans [8]. An additional set of revised electronegativities for the elements in different valence states was produced by Allred [1].

All the values of K_{sp} plotted in Figure 2.11-9 are from the tables in Sillén and Martell [24], except for that of Ba-arsenate. The solubility product for this compound was estimated from the values given in Stephen and Stephen [28] and Weast [33]: $K_{sp} \approx 10^{-13.5}$. The value given by Sillén and Martell is many orders of magnitude lower ($K_{sp} = 10^{-50.1}$), which is far removed from the trend of the other data points. The estimate of $10^{-13.5}$ fits the trend, as shown in the figure.

Some uses of the correlation between solubility and cation electronegativity are discussed in the following examples.

Example 8 What is the approximate value of the association constant of the ion-complex $CdOH^+$ in an aqueous solution at 25°C?

Among the single-charged ionic complexes shown in Figure 2.11-8, $CdOH^+$ is not given. The electronegativity value of the Cd^{+2} ion can be read off the graph in Figure 2.11-7: it is 1.5. The value of $-\log K$ for dissociation of a complex $MeOH^+$, where the cation Me^{+2} has an electronegativity of 1.5, can next be read off the graph in Figure 2.11-8: the value of $-\log K$ is similar to that of the $ZnOH^+$ complex, which is $K \approx 10^{-4}$. The association constant sought is the reciprocal of the dissociation constant. Thus, the association constant for the $CdOH^+$ complex in aqueous solution is $K \approx 10^4$.

The latter value compares well with the range of values for $CdOH^+$ association constant ($10^{3.4} - 10^{5.0}$) reported in the literature [24].

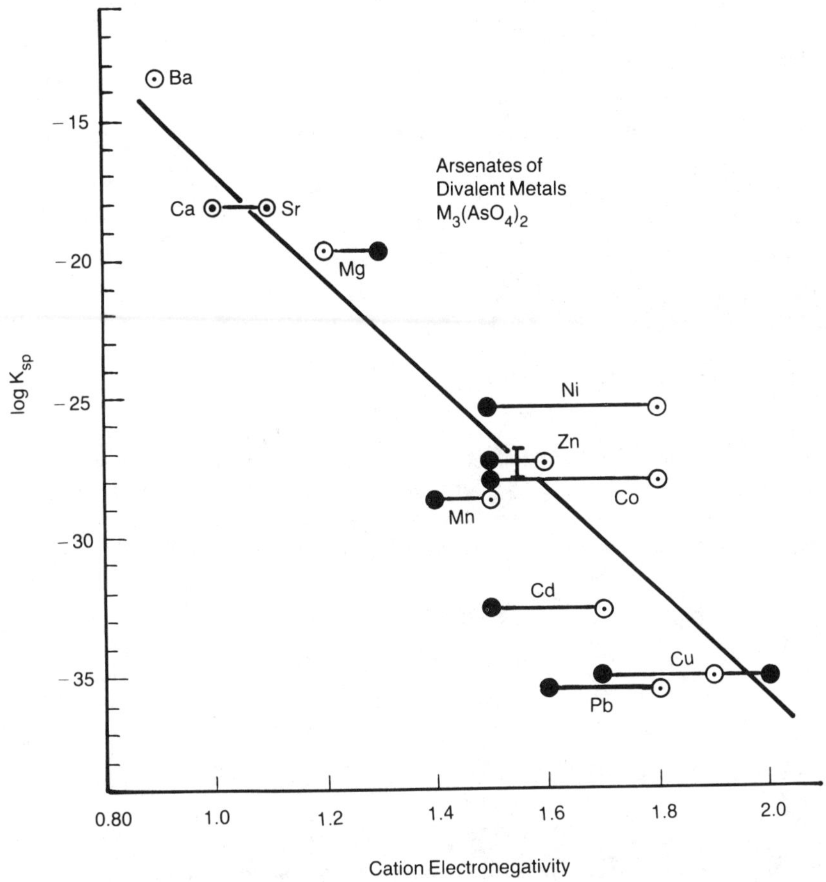

Explanatory notes: for comment on solubility of Ba-arsenate see text; electronegativities of Ca and Sr, each varies from 1.0 to 1.1; for Cu, three electronegativity values are shown.

Source: Solubilities — Sillén and Martell [24]
Electronegativity values — Evans [8]
Clifford [4]

FIGURE 2.11-9 Solubilities of Divalent-Metal Arsenates, $M_3(AsO_4)_2$, in Water at Room Temperatures

Example 9 What is the solubility product of Cd-arsenate from the empirical correlation line drawn in Figure 2.11-9, and what error is this estimate likely to introduce in the estimated concentration of arsenic in solution?

First, using the Cd^{+2}-ion electronegativity values of 1.5 and 1.7, the solubility product of Cd-arsenate (read off the straight line in Figure 2.11-9) is between $K_{sp} = 10^{-26}$ and $K_{sp} = 10^{-30}$. The difference between the K_{sp} value shown in the figure (10^{-33}) and the mean K_{sp} (10^{-28}) that corresponds to the electronegativity of 1.6 is about five orders of magnitude. We now determine the error in the arsenic concentration in solution that is introduced by this uncertainty of 10^5 in the estimated K_{sp}. Equations 15 and 16 are used, with p = 3 and q = 2, insofar as the compound $Cd_3(AsO_4)_2$ is a solid of the type M_3A_2. In equation 16, [A] stands for concentration of the arsenate-ion AsO_4^{-3} in solution; the two estimates of [A] for the two values of K_{sp} are:

$$\text{Lower estimate of } [AsO_4^{-3}] = \left(\frac{8}{27} \times 10^{-33}\right)^{1/5} = 1.97 \times 10^{-7} \text{ mol/liter}$$

$$\text{Upper estimate of } [AsO_4^{-3}] = \left(\frac{8}{27} \times 10^{-28}\right)^{1/5} = 1.96 \times 10^{-6} \text{ mol/liter}$$

The quotient of the two estimates gives:

$$\frac{\text{Lower estimate}}{\text{Upper estimate}} = (10^{-5})^{1/5} = 0.1$$

Thus, despite a difference of a factor of 10^5 between the two estimates of the K_{sp} of the solid, the resulting uncertainty in the computed concentration of the arsenate ion in solution is a factor of 10. (Note, however, that in this example no complexes of cadmium and arsenate were considered and neither cadmium nor arsenate were initially present in solution.)

Correlation between Solubilities and Crystal Ionic Radii

Intuitively, it might be supposed that the size of the cation in the crystalline lattice of a solid would affect its solubility. In fact, this is the case, as shown in Figure 2.11-10 for divalent-metal sulfates. The sulfates characterized by similar values of the crystal ionic radii of the cations (Co^{+2}, . . . , Fe^{+2}) vary greatly in solubility; however, for the sulfates of the alkaline earths (Ca, . . . , Ra) and of lead, a very good correlation is found between solubility and cation crystal radius. For the aqueous complexes of some of these metal ions, either negatively charged or electrically neutral, there is no correlation between the cation radius and the dissociation constant of the complex.

A fairly good correlation between solubility and cation radius is shown in Figure 2.11-11 for a series of solid uranyl arsenates. The variable constituent of the series is the monovalent cation: H^+, Li^+, Na^+, K^+, and NH_4^+. An increase in the cation radius from hydrogen to ammonium correlates with a decreasing solubility of the solid at room temperature.

Example 10 Evaluate the solubility product of Na-uranyl arsenate.

The crystal ionic radius of the Na^+ ion, for a sixfold coordination, is given in Table 2.11-3: $r_{Na} = 1.10$Å. In Figure 2.11-11, the value of log 1.10 = 0.04 on the horizontal axis corresponds to a solubility product value (read off the empirical correlation line) of $K_{sp} \approx 10^{-21}$. The value for the solubility of Na-uranyl arsenate shown in this figure is $K_{sp} \approx 10^{-22}$, one order of magnitude lower.

Example 11 What would be the solubilities of Rb- and Cs-uranyl arsenates (if they exist)?

As in Examples 8 and 9 in the preceding section, this can only be a crude estimate. To illustrate the technique, the crystal ionic radii of Rb^+ and Cs^+ (Table 2.11-3) are: r_{Rb} = 1.57 Å and r_{Cs} = 1.78 Å. Entering Figure 2.11-11 with the logarithms of these values ($\log r_{Rb}$ = 0.20 and $\log r_{Cs}$ = 0.25), we find the corresponding intersections with the straight line at $K_{sp} \approx 10^{-24}$ and $K_{sp} \approx 10^{-25}$. This method of estimating the solubility product assumes that uranyl arsenates exist with cations bigger than NH_4^+, and that the trend of decrease in solubility with increasing cation radius is unchanged.

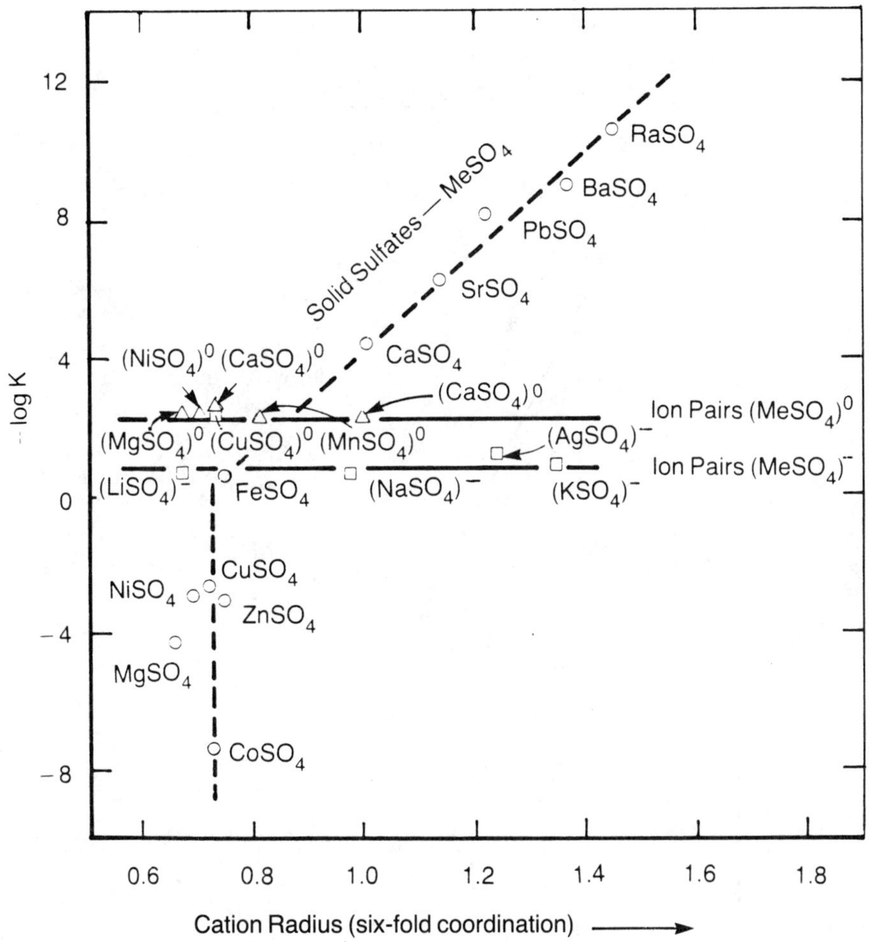

Source: Garrels and Christ [10]. *(Copyright 1965, Harper and Row. Reprinted with permission.)*

FIGURE 2.11-10 Equilibrium Solubility Constants (25°C) for Divalent Metal Sulfates Plotted Against the Crystal Radii of the Cations for Sixfold Coordination

Correlation of Solubilities with Ionization Potentials

The ionization potential is a measure of the energy needed to remove an electron from an atom, making it a positively charged ion. Ionization potentials, I, are commonly given in the literature in units of electron-volts per atom. One electron-volt (eV) is a very small quantity of energy, equivalent to 1.6×10^{-12} erg or 3.83×10^{-20} calories. An ionization potential of about 20 eV may represent the energy required for removal of two electrons from a metal atom; this quantity of energy corresponds to about 460 kcal/mol, which is of the same order of magnitude as the Gibbs free energy of formation and the lattice energy of various solids.

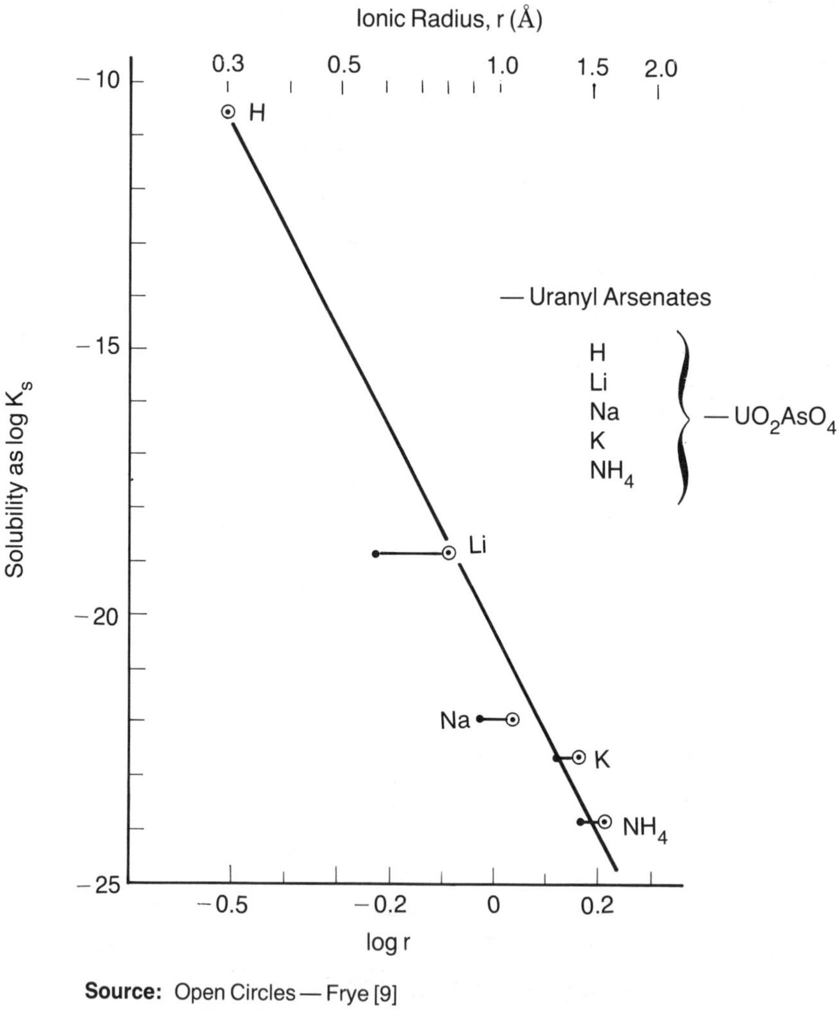

Source: Open Circles — Frye [9]
Closed Circles — Evans [8]

FIGURE 2.11-11 Solubility Products of Uranyl Arsenates as a Function of the Ionic Radius of the Other Cation

The quotient of the ionization potential and the crystal ionic radius, I/r, has the dimensions of force (such as eV/Å, erg/cm or dyne). For two series of divalent-metal fluorides — fluorides of the alkaline earths and of other heavy metals — the relationships between the solubilities of the solid phases and the parameter I/r for the metal ion are plotted in Figure 2.11-12. The data on which the plots are based, and additional information on the molar volumes of the solids, are listed in Table 2.11-5. The plot of the alkaline-earth series in Figure 2.11-12 shows a variable relationship between solubility and the I/r value of the metal. For the heavy-metal series, however, there is a fairly consistent correlation. A use of this rough correlation is demonstrated below.

Example 12 Estimate the solubilities at room temperature of ferrous fluoride (FeF_2) and mercuric fluoride (HgF_2). Weast [33] indicates that FeF_2 is "slightly soluble" in water and HgF_2 "decomposes" in water; no solubility value for either compound is given in this source.

From Table 2.11-5,

$$\text{For } Fe^{+2}: \quad I/r = 28.0 \text{ eV/Å}$$

$$\text{For } Hg^{+2}: \quad I/r = 26.4 \text{ eV/Å}$$

In Figure 2.11-12, ordinates drawn at these values of I/r intersect the "heavy-metals" line at points corresponding to

$$FeF_2 \text{ solubility} \approx 6 \times 10^{-2} \text{ mol/liter}$$

$$HgF_2 \text{ solubility} \approx 4 \times 10^{-2} \text{ mol/liter}$$

The solubility value for FeF_2 is of the same order of magnitude as that of MnF_2 (2×10^{-2} mol/liter). Although the results may be purely coincidental, Fe^{+2} and Mn^{+2} behave quite similarly in aqueous solutions.

2.11.4 Summary

Equilibria between solid phases and their aqueous solutions are defined in terms of one of the two following parameters:

- The solubility product, K_{sp}, defined as the product of the concentrations of the dissolved species at equilibrium with the solid;
- The equilibrium constant, K, defined as the quotient of the products of the thermodynamic activities of the species in solution and in the solid.

The solubility product may vary with the composition of the solution, whereas the equilibrium constant does not. Both K_{sp} and K depend on temperature and, to a much smaller degree for solids and aqueous solutions, on pressure.

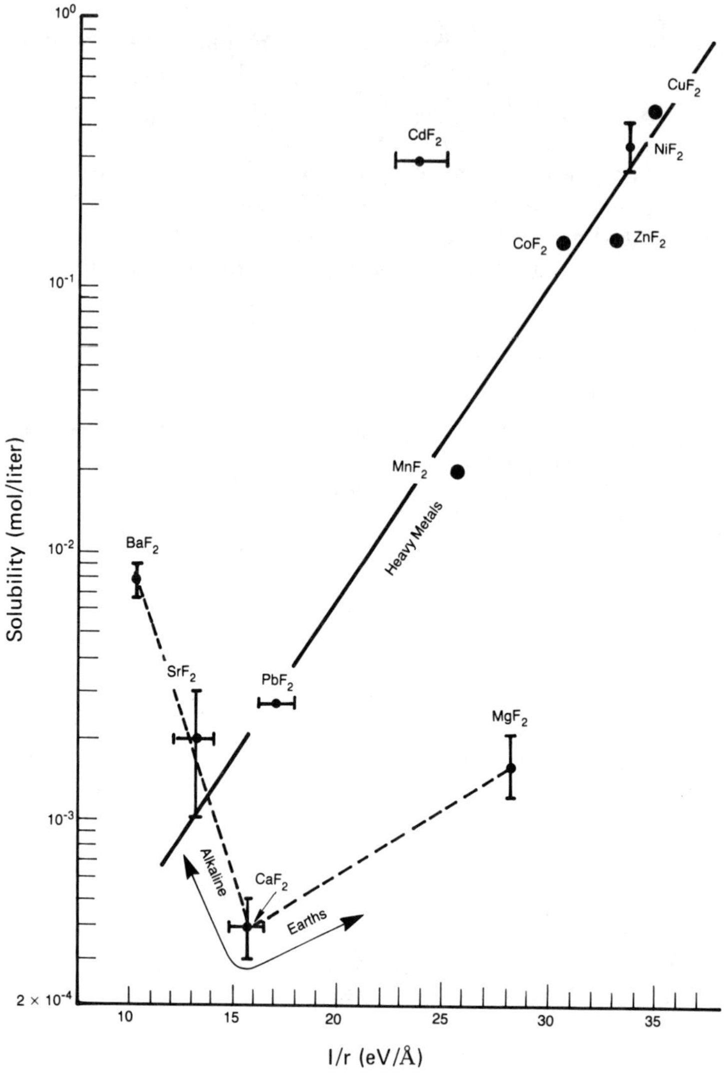

Source of data: Table 2.11-5

FIGURE 2.11-12 Solubilities of Divalent-Metal Fluorides Plotted Against the Quotient of Ionization Potential and Cation Radius

Solubilities of the solids and concentrations of dissolved species at equilibrium with the solids can be estimated from either K_{sp} or K for a given set of environmental conditions, such as the temperature, composition of the aqueous solution, the pH of the solution, and the sizes of the solid particles reacting with the solution.

TABLE 2.11-5

Solubilities and Ionization Potentials Per Unit Radius of Some Alkaline-Earth Fluorides and Heavy-Metal Fluorides

Atomic Number of Metal	Composition of Solid	Molar Volume V (cm^3/mol)	Metal^{II} Crystal Radius for Coordination No. 6 r (Å)	Ionization Potential for M^{+2} I (eV)	Ionization Potential per Unit Radius I/r (eV/Å)	Solubility (mol/liter)
Alkaline Earths						
12	MgF_2	19.6	0.80	22.67	28.3	$1.65 \pm 0.45 \times 10^{-3}$
20	CaF_2	24.6	1.08	17.98	16.7	$0.4 \pm 0.1 \times 10^{-3}$
38	SrF_2	29.6	1.21	16.67	13.8	$2.0 \pm 1.0 \times 10^{-3}$
56	BaF_2	35.9	1.44	15.16	10.5	$8.0 \pm 1.0 \times 10^{-3}$
Heavy Metals						
25	MnF_2	23.3	0.91	23.13	25.4	0.02
26	FeF_2	22.9	0.86	24.06	28.0	—
27	CoF_2	21.7	0.83	25.16	30.3	0.145
28	NiF_2	20.9	0.77	28.83	33.6	0.34 ± 0.07
29	CuF_2	24.0	0.81	28.06	34.6	0.46
30	ZnF_2	20.9	0.83	27.28	32.9	0.15
48	CdF_2	22.7	1.03	25.83	25.1	0.29
80	HgF_2	50.3	1.10	29.06	26.4	—
82	PbF_2	29.8	1.26	22.38	17.8	0.0027

Sources:

V calculated from mol. wts. and densities listed by Weast [33].

I, r: Frye [9]; values of I are for reaction $M \rightarrow M^{+2} + 2e^-$

Solubilities: Stephen and Stephen [28]

Tabulations of K_{sp} and K are available in the literature for a great variety of naturally occurring and man-produced inorganic solids in aqueous solutions. It is also possible to estimate the value of K for solid-solution reactions at equilibrium, at a specified temperature and pressure. The most reliable method is to compute the values of K from the Gibbs free energies of formation of the reactants and products in a standard state (such as at 25°C and 1 atm total pressure). Additional methods for estimating the equilibrium constants for solid-solution reactions are based on empirical correlations between the solubilities of different solids and some of their solid crystal properties. Such empirical correlations between the solubility and crystal properties have been demonstrated for a number of classes of inorganic solids and for the following properties: lattice energy, cation electronegativity, crystal cation radius, and the ionization potential of the metal.

2.11.5 Sources of Information

Extensive treatment of dissolution reactions and chemical speciation in solution is given by Butler [3], Garrels and Christ [10], and Stumm and Morgan [29].

Compilations of data on the solubility of solids at various temperatures are available in *International Critical Tables* [11], Weast [33], Landolt-Börnstein [14], Linke [17, 18], and Stephen and Stephen [28].

For computations of the equilibrium constant, K, the main source of data for the Gibbs free energies of formation of the solids and aqueous species (ΔG_f^0) is Wagman *et al.* [32], who give the data for the standard state of 25°C and 1 atm total pressure. Compilations of the Gibbs free energy of formation and of other useful thermodynamic data for solids at temperatures from 25°C and up are also given by Robie *et al.* [22]. The Gibbs free energy data of Wagman and others are also compiled by Weast [33]. Some of these data, with additions that are particularly relevant to natural waters and mineral-water or rock-water systems, are also tabulated in publications by Garrels and Christ [10], Stumm and Morgan [29], and Drever [7].

The books and publications listed in the preceding paragraph also list values of the standard enthalpies of formation (ΔH_f^0) needed for estimating the temperature effect on the equilibrium constant K.

Estimation of the values of K at elevated pressures requires data on the molar volumes of the solids and aqueous species. For the solids, conveniently arranged tabulations of the molar volumes are given by Robie *et al.* [22] and Weast [33]. (The latter lists the molecular weight and the density of the solid; the molar volume is the ratio of the two.) For aqueous ions, an extensive compilation of molar volumes at temperatures of environmental interest is given by Millero [19], which also includes references to earlier works on methods of estimation of the partial molar volumes in solution.

Solubility products for solid-solution equilibria and association constants for ion-complexes in aqueous solutions are extensively tabulated by Sillén and Martell [24, 25], Smith and Martell [26], and Kotrlý and Šůcha [13]. In addition to the data on inorganic cation-anion complexes, references 13, 24 and 25 list data for metal complexes with organic ligands.

2.11.6 Literature Cited

1. Allred, A.L., "Electronegativity Values from Thermochemical Data," *J. Inorg. Nucl. Chem.*, **17**, 215 (1961).
2. Baes, C.F. and R.E. Mesmer, *The Hydrolysis of Cations*, Wiley-Interscience, New York (1976).
3. Butler, J.N., *Ionic Equilibrium*, Addison-Wesley, Reading, Mass. (1964).
4. Clifford, A.F., "The Electronegativity of Groups," *J. Phys. Chem.*, **63**, 1227 (1959).
5. Clifford, A.F., "The Prediction of Solubility Product Constant," *J. Am. Chem. Soc.*, **79**, 5404 (1957).
6. Dasent, W.E., *Inorganic Energetics*, 2nd ed., Cambridge Univ. Press (1982).
7. Drever, J.I., *The Geochemistry of Natural Waters*, Prentice-Hall, Englewood Cliffs, N.J. (1982).
8. Evans, R.C., *An Introduction to Crystal Chemistry*, 2nd ed., Cambridge Univ. Press (1964).
9. Frye, K., *Modern Mineralogy*, Prentice-Hall, Englewood Cliffs, N.J. (1974).
10. Garrels, R.M. and C.L. Christ, *Solutions, Minerals, and Equilibria*, Harper and Row, New York (1965).
11. *International Critical Tables*, 7 vols., McGraw-Hill Book Co., New York (1926-1930).
12. Jenkins, H.D.B., "Lattice Energies," in Weast [33], D-101.
13. Kotrlý, S. and L. Šůcha, *Handbook of Chemical Equilibria in Analytical Chemistry*, John Wiley & Sons, New York (1985).
14. Landolt-Börnstein, *Zahlenwerte und Funktionen. Lösungsgleichgewichte I*, Vol. 2, Part 2, Sec. B, Springer-Verlag, Heidelberg (1962).
15. Langmuir, D. and D.O. Whittemore, "Variations in the Stability of Precipitated Ferric Oxyhydroxides," *Adv. Chem. Ser.*, **106**, 209 (1971).
16. Lerman, A., *Geochemical Processes — Water and Sediment Environments*, Wiley-Interscience, New York (1979).
17. Linke, W.F., *Solubilities — Inorganic and Metal-Organic Compounds, A-Ir*, 4th ed., Vol. I, Van Nostrand, New York (1958).
18. Linke, W.F., *Solubilities — Inorganic and Metal-Organic Compounds, K-Z*, 4th ed., Vol. II, American Chemical Society, Washington, D.C. (1965)
19. Millero, F.J., "The Molal Volumes of Electrolytes," *Chem. Rev.*, **71**, 147 (1971).
20. Nielsen, A.E., "Precipitates: Formation, Coprecipitation, and Aging," in *Treatise on Analytical Chemistry*, 2nd ed., I.M. Kolthoff and P.J. Elving (eds.), Part I, Vol. 3, John Wiley & Sons, New York (1983).
21. Prigogine, I. and R. Defay, *Chemical Thermodynamics*, John Wiley & Sons, New York (1962).

22. Robie, R.A., B.S. Hemingway and J.R. Fisher, "Thermodynamic Properties of Minerals and Related Substances at 298.15 K and 1 Bar (10^5 Pascals) Pressure and at Higher Temperatures," *U.S. Geol. Surv. Bull.*, No. 1452, 1 (1978).
23. Schindler, P.W., "Heterogeneous Equilibria Involving Oxides, Hydroxides, Carbonates, and Hydroxide Carbonates," *Adv. Chem. Ser.*, **67**, 196 (1967).
24. Sillén, L.G. and A.E. Martell, *Stability Constants of Metal-Ion Complexes*, Spec. Pub. No. 17, The Chemical Society, London (1964).
25. Sillén, L.G. and A.E. Martell, *Stability Constants of Metal-Ion Complexes — Supplement No. 1*, Spec. Pub. No. 25, The Chemical Society, London (1971).
26. Smith, R.M. and A.E. Martell, *Inorganic Ligands*, Vol. 4 of *Critical Stability Constants*, Plenum Press, New York (1976).
27. Sposito, G., *The Thermodynamics of Soil Solutions*, Oxford Univ. Press (1981).
28. Stephen, H. and T. Stephen, *Solubilities of Inorganic and Organic Compounds*, Vol. 1, Part 1, Pergamon Press, Oxford (1963).
29. Stumm, W. and J.J. Morgan, *Aquatic Chemistry*, 2nd ed., Wiley-Interscience, New York (1981).
30. Swalin, R.A., *Thermodynamics of Solids*, 2nd ed., Wiley-Interscience, New York (1972).
31. Tardy, Y., "Relationships among Gibbs Energies of Formation of Compounds," *Am. J. Sci.*, **279**, 217 (1979).
32. Wagman, D.D., W.H. Evans, V.B. Parker, R.H. Schumm, I. Halow, S.M. Bailey, K.L. Churney and R.L. Nuttall, *The NBS Tables of Chemical Thermodynamic Properties*, Journal of Physical and Chemical Reference Data, **11**, Supplement 2 (1982).
33. Weast, R.C. (ed.), *Handbook of Chemistry and Physics*, 65th ed., CRC Press, Boca Raton, Fla. (1984).

2.12 ATTENUATION ON SOILS

2.12.1 Introduction

Attenuation of aqueous species via removal onto, and in the form of, solid phases is an important phenomenon in environmental systems. Attenuation reduces mobility by decreasing the concentration of species in the aqueous phase. The bioavailabilities of dissolved versus sorbed and insoluble species also differ: dissolved species are much more available to both plants and animals. Therefore, an understanding of sorption and attenuation is necessary in assessing the fate of inorganic compounds in the environment.

Although numerous experiments have been conducted using various sorbents to study the sorption of inorganics, the environmentally important sorbents include sediments, soils, and their components. This section therefore focuses on these sorbents and discusses the extent of sorption that occurs under equilibrium conditions. Although not discussed in detail, the importance of kinetics is pointed out where relevant. The emphasis of the discussion is on cations, particularly trace heavy metals, because they are the major pollutants in the environment.

The attenuation or sorption of an inorganic species is highly influenced by the environmental conditions of the system under consideration. Thus, unlike the case for most organic chemicals, *no species-unique attenuation or sorption constants can be applied to a broad range of soils and sediments.* Depending on such factors as soil texture, soil chemistry, pH, redox potential, solute and ligand concentrations in the water, a particular inorganic species may be highly mobile in one soil environment and essentially immobile in another. The effective sorption or attenuation constants for an inorganic species may differ by as much as two or three orders of magnitude. For roughly similar environments, however, the differences are usually less than one order of magnitude.

A further result of this variability (and complexity) is that *the extent of attenuation or sorption of an inorganic pollutant on a particular soil cannot be estimated, ab initio,* with the degree of accuracy and reliability usually desired for quantitative transport modeling or even for semiquantitative assessments. Computerized models of aqueous speciation do exist, and some of them include routines that represent sorption; nevertheless, there is as yet no proven methodology for using these speciation calculations, along with information on such properties as molecular structure and ionic size and charge, to predict sorption for a wide range of natural soils, sediments and minerals. The investigator is left with three general approaches for estimating the extent of attenuation of sorption in a real soil or sediment environment:[1]

(1) Rough estimates based on published data (including, for example, sorption constants, attenuation factors, and empirical correlations) for the pollutant in similar soil environments.

1. These approaches are discussed in more detail in § 2.12.6.

(2) Rough estimates of boundaries (i.e., upper or lower limits for sorption) using calculations based on the assumption that one or more mechanisms predominate (e.g., precipitation, ion exchange, physical absorption).

(3) Laboratory experiments.

Accordingly, this section discusses:

- attenuation mechanisms for inorganics in soils and sediments (§ 2.12.2);
- the effect of various factors (pH, Eh, ionic composition, etc.) on the attenuation of inorganics in natural environments (§ 2.12.3);
- the general procedures used to measure attenuation of pollutants in soils and sediments (§ 2.12.4);
- attempts made to model the attenuation of inorganics in natural environments (§ 2.12.5);
- empirical representations (distribution coefficient, Langmuir and Freundlich) that have been used with some success to describe the attenuation of inorganics in natural environments (§ 2.12.6);
- examples of data for parameters used in the empirical representations (§ 2.12.6);
- sample calculations of attenuation using measured data and these empirical equations (§ 2.12.6);
- soil column studies by Fuller and co-workers (§ 2.12.7);
- approaches to estimating the probable attenuation of an inorganic pollutant in a natural system (§ 2.12.8); and
- brief examples of two approaches (§ 2.12.8).

2.12.2 Attenuation Mechanisms

As used here, *sorption* refers generally to the removal of solute components from the aqueous phase of an environmental system by *the solid phase(s) at the surface of the solid phase(s)*. A number of mechanisms and processes are involved. The term is often used interchangeably with *attenuation*, which is used in this section to refer to the overall increase in concentration in undissolved phases, including precipitates (and a corresponding decrease in concentration in the dissolved phase), regardless of the mechanisms.

Attenuation processes can be classified into two categories: physical and chemical. The distinction between the two is defined by the attractive forces responsible for ionic interactions between the surface and the solute. Physical adsorption is due to van der Waal's forces and is characterized by low heats of adsorption (5-10 kcal/mole) [32]. Chemical attenuation refers to chemical interactions and/or reactions with higher heats of reaction. Because of the low energy of adsorption, physical adsorption is easily reversible.

Within these two broad categories of attenuation, there are various contributing processes in natural systems. These include physical adsorption and processes involving chemical interactions: ion exchange, organic complexation, precipitation and co-precipitation, solid-state diffusion, and isomorphic substitution.

Physical adsorption, as described above, is due to long-range van der Waal's forces. It does not involve direct interaction between the sorbent surface and the solute; the forces responsible for physical adsorption are intermolecular, in contrast to the valency forces in chemical interaction. Physical adsorption occurs because the surface charges on particles attract dissolved species with opposite charges. As adsorption proceeds, the charges become neutralized, physical adsorption slows down and equilibrium is eventually reached. Also, as the surface becomes filled, fewer sites are available for adsorption. Therefore, the extent of physical adsorption is closely related to surface charge and the surface area available on the sorbent. Surface charges on minerals occur due to isomorphic substitution (see below for further discussion). Clay minerals tend to have a negative surface charge and will attract positively-charged ions.

An important chemical process responsible for the sorption of ionic species is ion exchange. In this process, the sorption of one or more ionic species is accompanied by the desorption of an amount of previously sorbed ionic species equivalent in total ionic charge [16]. When anions on the surface are replaced by dissolved anions, it is called *anion exchange*. Similarly, *cation exchange* occurs when cationic species are exchanged between the sorbent surface and the solute at the sorbent-solution interface. The cation-exchange capacity (CEC) can be calculated from a test designed to measure the amount of cations that may be exchanged by a solid phase from a solution. Similarly, an anion-exchange capacity (AEC) may be calculated. The CEC of a solid phase is specified for a particular temperature, ionic strength, pH of solution and ionic species involved; it is most frequently expressed in meq/100g or μeq/g of solid phase [16].

The formation of complexes between dissolved and surface materials is another important sorption mechanism. Complexation in the aqueous phase has been described in section 2.9 of this report; in an analogous fashion, carboxyl, phenolic, sulfhydryl and other functional groups on the surface of particles can coordinate to metal ions. The surface complexation reaction involves the loss of one or more water molecules from the inner coordination sphere of the metal ion for one or more ligand coordinating atoms on the surface of the particle. An excess of the metal over the ligand on the surface would result in metal-like coordination properties on the surface, while a ligand excess would cause ligand-like behavior of the surface toward solutes at the solid-solute interface [27]. Conceptually, on the surface of a solid, not all the coordination possibilities of the metal or the ligand atoms are satisfied [27]. The difference between this type of adsorption and aqueous complex formation is that the former occurs at heterogeneous surfaces, and the sorbed solutes are therefore no longer mobile. The process of complexation is used to model surface-solute interactions for both organic and inorganic surfaces. Mathematical equations have been used to explain and model sorption at solid-liquid interfaces [27,28,42]; these were based on the assumption that certain surface species are formed in the complexation process.

Precipitation and co-precipitation are also important processes for removing inorganics, particularly heavy metals, from solution. Precipitation is the separation of the solute as a solid from the aqueous phase under conditions of saturation; section 2.11 of this report describes these processes in greater detail. Co-precipitation, on the other hand, refers to the removal of a solute from solution via sorption onto a solid phase as the solid phase is precipitating out of solution. In natural systems, it is difficult to distinguish between precipitation and co-precipitation. The various conditions, mechanisms, and reactions involved are not easily delineated.

During the formation of clay minerals, cations may be co-precipitated in the lattice in a process called *isomorphic substitution.* Kinniburgh and Jackson [23] described nonstoichiometric isomorphic substitution in soil silicate minerals. The most common is the substitution of Al^{+3} for Si^{+4} and Mg^{+2} for Al^{+3}. Other possible substitutions include Fe^{+3} for Si^{+4} [23]. Ions that are similar to each other will have a greater tendency to substitute for each other. Arsenate, for example, may replace phosphate on surfaces of clays and amorphous hydroxides and may substitute for phosphate during precipitation of minerals [39].

One of the processes that may cause the “precipitation” of metals in natural systems is coagulation or flocculation. Most suspended particulates have a natural tendency to coagulate when placed close enough together (due to van der Waals forces). Charged particulates may repel similarly charged particles such that the relatively small distances required for van der Waals forces to become important may never be reached. These suspensions are called stable. If adsorption of oppositely charged ions occurs, interparticle repulsion may decrease and coagulation may occur. Metals associated with fine particulates (e.g., dust) suspended in water may be settled out of solution when attracted to particles with opposite surface charges. The surface charge on suspended particles may change as the pH changes. A pH value may be defined when the net surface charge of the particle equals zero, i.e., the point of zero charge (PZC). The stability of the particle-pairs thus decreases as the pH approaches its PZC value. Changes in pH could destabilize particles and cause them to coagulate. The physical adsorption of metal ions on the surface of suspended particles may also destabilize these particles and cause coagulation [16]. This process is used extensively in wastewater treatment by adding agents like alum, iron salts or polyelectrolytes to flocculate fine suspended particles.

The process of coagulation is similar to the occlusion of metals in precipitates. Freshly precipitated hydrous oxides exhibit a high degree of disorder, being frequently gelatinous and having large surface areas and high surface reactivity [23]. Metal ions are sorbed onto these surfaces and essentially become occluded or co-precipitated with the hydrous oxides. Jenne [21] suggested that the occlusion of heavy metals in growing solid phases, such as Fe and Mn oxides, is an important process for immobilizing metals in soils. At sufficiently high concentrations of a metal ion, precipitation could also occur. Kinniburgh and Jackson [23] suggested that metal hydroxides can precipitate at substoichiometric hydroxide-to-metal ratios as basic salts or as solids with a range of ill-defined stoichiometry.

A relatively minor process is that of *lattice penetration* or *solid-state diffusion*. "Holes" may occur on the surfaces of natural zeolites and other minerals, forming the "open" surface structures of these minerals [16]. Exchangeable divalent cations such as Mg, Ni, Cu, Co, and Zn may become embedded in these "holes". Upon dehydration, they become "stuck" in these holes and then diffuse into empty octahedral positions in layer silicates [9,16]. Several authors believe that this is probably not a significant process for immobilizing heavy metals [21,26]. There is little penetration (only a few atomic planes), and it occurs only at the edges and along fractures [9].

Although sorption may be measured using natural soils and sediments and natural water samples, it is difficult to determine from the data the mechanisms that are responsible. The complexity of natural systems and the diverse mechanisms of sorption and attenuation make it hard to attribute an effect to a specific mechanism. However, numerous studies of simpler, more controlled experimental systems have illuminated the individual mechanisms contributing to attenuation in natural systems. This attenuation is often found to be the result of more than one process.

The processes described above could occur as single-step reactions, parts of multi-step reactions or simultaneously with each other. As a hypothetical example, cations could be sorbed by a single-step ion exchange process on clays, and they could also be attenuated by a two-step process of physical adsorption followed by co-precipitation or precipitation. Several authors [16,19,24,45] have observed that physical adsorption probably precedes bulk precipitation, followed by dehydration and rearrangement of the precipitated phase. Similarly, nonstoichiometric isomorphic substitution usually results in a net negative surface charge, which could further result in ion exchange or other forms of sorption. Because of the heterogeneous structure of natural soils and sediments, several processes could occur simultaneously. For many heavy metals, cation exchange is important on clay minerals, co-precipitation on hydrous Fe and Mn oxides, and organic complexation on organic particles. All of these processes occur simultaneously. The attenuation of individual metals is described in Chapter 7 (Part II).

2.12.3 Factors Affecting Attenuation

The attenuation behavior of an inorganic species in soil is a complex function of the chemical form and properties of the solute, the chemical composition of the soil solution and the physical/chemical properties of the soil. Soil properties such as bulk density, compaction characteristics, and permeability influence attenuation through their effect on the rate of movement of water through the soil and the surface area of the soil exposed to the aqueous phase. Appendix C, which describes soil types and properties, includes a list (Table C-2) of the site and soil characteristics that affect the migration of inorganics. The important soil, soil solution, and solute physical/chemical properties (not including engineering properties) affecting attenuation are also listed in Table 2.12-1.

Soil composition is significant because each component differs in its ability to sorb solutes from the soil solution. Soils and sediments are heterogeneous mixtures of

TABLE 2.12-1

Physical and Chemical Properties of Soil, Soil Solution, and Solute that Affect Attenuation[a]

Soil Solids	Soil Solution	Solute
Composition:	pH	Chemical identity
Hydrous oxides (Fe, Mn, Al)	Eh	Chemical behavior:
Silicates — clay content, type of clay	Temperature	Charge
Organic material	Ionic strength	Size
Carbonate minerals	Ionic composition:	Complexation chemistry
Specific area	Competing ions	Solubility
Cation exchange capacity (CEC)	Complexing ions	Precipitation chemistry
pH (content of basic species)		Redox chemistry (oxidation states)
Eh (content of redox active species, e.g., Fe(II), Mn(IV)		
Aeration status (saturated, unsaturated)		
Microbial type and population		
Temperature		

a. Does not include engineering properties; see Appendix C.

many different solid and dissolved phases that provide a variety of mineral and organic surfaces. The primary sorbents in natural soils and sediments are amorphous aluminosilicates, organic materials and hydrous oxides, oxyhydroxides and hydroxides of aluminum, iron and manganese [34]. Metal oxides are also important, because they often occur as coatings on other soil component surfaces [23]. Organisms, dead remains of organisms and condensed humic material provide organic surfaces in the environment; organics also exist as coatings on inorganic oxide particles [27]. Table C-11 (Appendix C) shows the ranges of organic matter content in soils. Clay minerals and other aluminosilicates provide surfaces on which other sorbents can form and are also significant sorbents in their own right. As described in Appendix C, soils from different sources contain different amounts of the various components.

Similarly, sediments contain a range of solid components that sorb solutes from aquatic systems. In one study of sediment sorption data by Pavlou and Weston [33], a statistically significant relationship was found between the sorption coefficient (K_d) of Cu, Cd and Pb (but not Zn, As or Hg) and the sediment organic carbon content. Their data are shown in Table 2.12-2. The solution pH for these equations was not specified. Other solid components may also have contributed to the sorption of these trace metals, explaining the low correlation to organic carbon.

The *specific area*, usually expressed as m^2/g of soil, is a measure of the surface area available for sorption. This property is related to the soil size distribution, pore

TABLE 2.12-2

Relationship of Sediment-Water Partition Coefficient of Trace Metals to the Organic Carbon in Sediments

Metal	Regression Equation[a]	n	r
Cu	$\log K_d = 0.33\,(TOC) + 3.28$	22	0.74[b]
Zn	$\log K_d = 0.074\,(TOC) + 3.29$	21	0.19
As	$\log K_d = -0.05\,(TOC) + 2.46$	21	−0.19
Cd	$\log K_d = 0.21\,(TOC) + 2.34$	21	0.55[b]
Pb	$\log K_d = 0.20\,(TOC) + 3.10$	22	0.47[b]
Hg	$\log K_d = 0.05\,(TOC) + 1.87$	18	0.21

a. K_d = sediment/water partition coefficient
TOC = percent organic carbon in sediment
n = number of data points used to derive regression
r = correlation coefficient

b. r values indicate significant correlation.

Source: Pavlou and Weston [33]

structure and pore size distribution. Normally, it is inversely proportional to the size of the soil particles. Colloidal-sized particles make up a major part of most clays (clay particles defined as < 0.002 mm in size) and present an extremely large surface area for sorption. Specific areas of three major types of clay are shown in Table C-10, which also lists their cation exchange capacities. As defined earlier, the CEC is directly related to the maximum amount of cations that may be exchanged onto the soil surface at a specified pH. The listed values range from 3-15 meq/100g for kaolinite to 80-100 meq/100g for montmorillonite. The values for some surface soils (Table C-12) range from 2.0 meq/100g dry soil for a New Jersey sand to 57.5 meq/100g for a California clay.

The pH of the soil surface is strongly related to sorption. It affects the net surface charge on amphoteric sorbents like hydrous oxides and amorphous silicates. It also affects the surface speciation and therefore the reactions on the soil surface, e.g., ion exchange, hydrolysis, complexation, and particularly the exchange of H^+. The pH ranges in soils (as measured by mixing specified amounts of soil and water) are shown in Figure C-4. Surface pH values are normally 2 units lower than the pH of soil-water mixtures.

Solution pH is one of the primary parameters controlling solute speciation, chemical nature of protolyzable surface species and surface charge. The pH dependence of attenuation via ionic interaction is different for anions and cations. In the case of an anion, sorption is increased with increasing pH as dissociation increases. However, with increasing pH, the sorbent surface charge of an oxide becomes increasingly negative. At pH values higher than the pK_a of the anion, sorption decreases [18]. A good correlation has been found between the pK_a values for conjugate acids and points of inflection in sorption curves reflecting maxima in sorption with varying pH; this is illustrated for FeO(OH) and $Al(OH)_3$ in Figure 2.12-1.

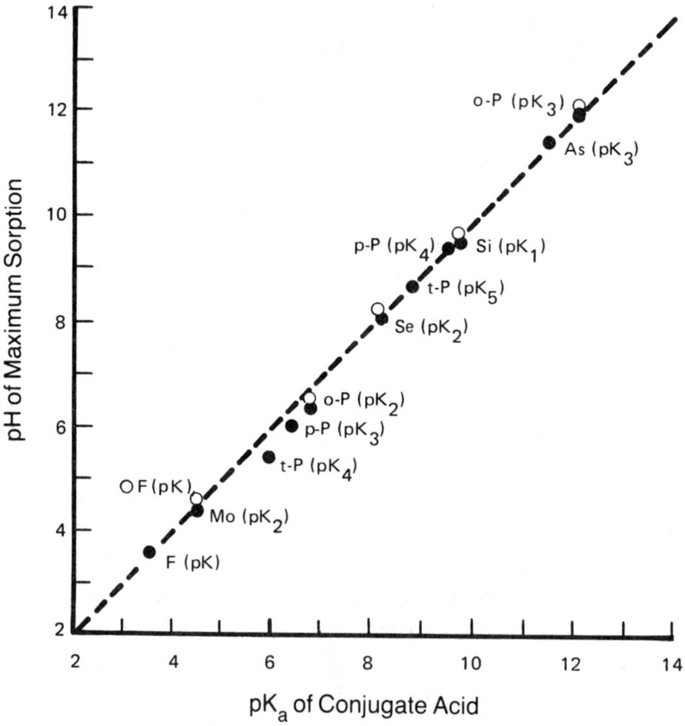

Sorbents: ● goethite (FeO(OH)), ○ gibbsite ($Al(OH)_3$).
Sorbates: F, fluoride; Mo, molybdate; t-P, tripolyphosphate; p-P, pyrophosphate; o-P, orthophosphate; Se, selenite; Si, silicate; As, arsenate.

Source: Hingston [18]. *(Copyright 1981, Ann Arbor Science Publishers. Reprinted with permission from M. Anderson and A. Rubin, current holders of the copyright.)*

FIGURE 2.12-1 Relationship Between pK_a and pH at the Point of Maximum Sorption

The attenuation of most cations increases with increasing pH. At the point of zero charge, there is little surface ionic interaction with the cations, but as the pH increases above the PZC, the surface becomes more negatively charged and the sorption of cations increases [23,34]. Precipitation of many divalent metals also increases with pH.

Another reason for the pH-dependence of cation attenuation may be the preference of hydrolyzed species over the free ions as sorbed species [3,23]. The increase in attenuation appears to be very rapid, increasing from 0 at low pH to almost 100% with an increase of 1-2 pH units [38]. Attenuation usually reaches a plateau, and desorption occurs if the pH is increased much further; Figure 2.12-2 illustrates this behavior

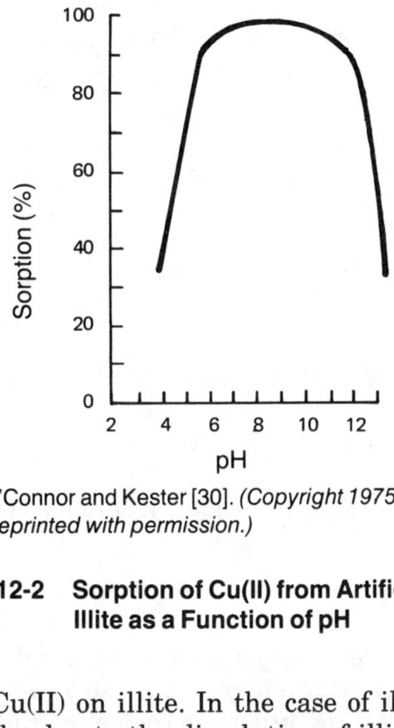

Source: O'Connor and Kester [30]. *(Copyright 1975. Pergamon Press, Inc. Reprinted with permission.)*

FIGURE 2.12-2 Sorption of Cu(II) from Artificial River Water on Illite as a Function of pH

for the sorption of Cu(II) on illite. In the case of illite, the decreased sorption with increasing pH may be due to the dissolution of illite at high pH ($SiO_2(s) \rightarrow H_4SiO_4^0$ above pH 9). An additional example of the effect of pH (and ionic strength) on metal sorption by soils is given in Figure 2.12-3: the sorption coefficient for both cadmium and lead increases significantly as the pH rises from 4 to 8.

The redox conditions in the soil environment (e.g., as measured by the Eh) affect the speciation of the chemical constituents in the soil, which in turn affects the attenuation of the solutes present. Redox conditions also influence precipitation: under reducing conditions, sulfide precipitates of mercury, cadmium and some other heavy metals are significant species. Attenuation of these metals could be increased under reducing conditions in the presence of sulfur. Whether the soil is saturated or non-saturated with water may also affect attenuation, because changes in Eh are common in soil that is subject to repeated flooding and draining.

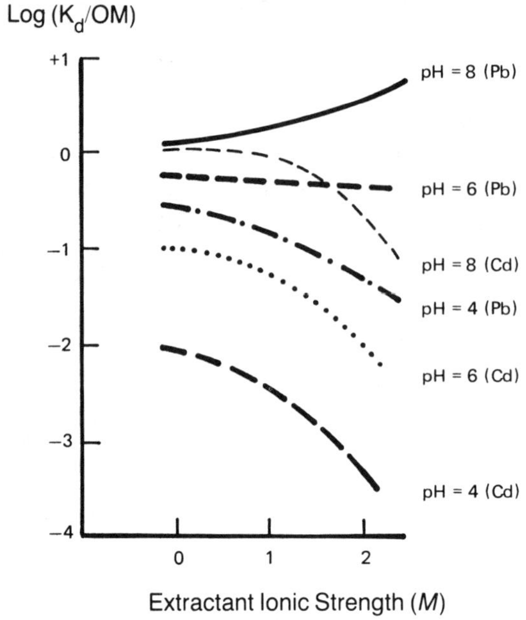

Plots of logarithm of the distribution coefficient expressed per unit of soil organic matter (log K_d/OM) versus ionic strength. K_d (l/g) is the ratio between the concentration of the metal in the soil and in the solution, OM is in % of dry matter. Curves were derived from averaged data from studies on 20 Dutch soils.

Source: Gerritse *et al.* [14]. *(Copyright 1983. Reprinted with permission.)*

FIGURE 2.12-3 Effect of pH and Ionic Strength on Sorption of Lead and Cadmium

The attenuation of inorganics in soil may be affected by sorption on microbial cell walls. Microorganisms may also transform inorganics into other species that differ in attenuation behavior under the soil conditions or that alter the attenuation of still other species. The transformation of SO_4^{-2} to S^{-2} by biological reduction under anaerobic conditions may cause the precipitation of metals from a soil system.

The presence of complexing species also affects the attenuation of an inorganic solute. The process may be described as a competition between surface and dissolved ligands for coordination sites on the cation [38]. The effect of complex formation on the sorption of a metal ion depends on the relative stabilities of the cation-ligand in solution and the cation-surface and on whether the surface has an affinity for the complexed form [23]. An example is the effect of Cl^- on Hg sorption: in studies with sediments, a decrease in the sorption of Hg has been found with increasing Cl^- concentration [7,36]. This is likely due to the formation of very stable Hg-Cl complexes, which are poorly sorbed [22].

Sposito [43] classified the ligand effects of metal-complexing ligands in a soil solution into five general categories:

(1) The ligand has a high affinity for the metal, and the soluble complex formed has a high affinity for the sorbent, resulting in increased sorption;

(2) The ligand has a high affinity for the sorbent, and the sorbed ligand has a high affinity for the metal, resulting in increased sorption;

(3) The ligand has a high affinity for the metal, but the soluble complex formed has a low affinity for the sorbent, resulting in decreased sorption;

(4) The ligand has a high affinity for the sorbent, but the sorbed ligand has a low affinity for the metal, resulting in decreased sorption; and

(5) The ligand has low affinity for both metal and sorbent, resulting in little effect on sorption.

Complexing ions influence the precipitation of many metals. The complexing of metal ions effectively increases the overall solubility of the metal, as further discussed in section 2.9 of this report. In the presence of complexing ions, precipitation is less likely to occur because of the lowered concentration of the uncomplexed ion.

The relative affinity of a sorbent for different species is called its *selectivity*. This is a measure of the competitiveness among various species for surface sites on the sorbent. Kinniburgh and Jackson [23] have reported various selectivity sequences for cation sorption. Generally, divalent transition and heavy-metal cations are more strongly sorbed than alkaline-earth cations on hydrous oxides. On clays, however, there is not much difference in their selectivity. These authors state that the selectivity and competition among ions for surface sites is extremely dependent on the pH and concentration of ions present. Noting the wide separation of sorption curves between alkaline-earth cations and transition and heavy-metal ions, they suggest that Ca^+ would not be expected to interfere significantly with the sorption of these ions, even if the concentration of Ca^+ were several orders of magnitude greater.

Changes in temperature appear to affect sorption, although the relationship is not clear [23]. Sorption can result from several processes, each of which is affected by temperature changes in a different way; thus, the overall temperature dependence of the sorption process is difficult to predict for natural systems.

DiToro *et al.* [8] have investigated the effect of concentration of suspended sorbent on heavy metal sorption. In experiments with nickel and cobalt sorbed onto montmorillonite and quartz, they found that the reversible component partition coefficient was a function of particle concentration. Sorption of heavy metals does not appear to be completely reversible and is affected by the particle concentration.

2.12.4 Measurement of Attenuation

Three laboratory methods are available for measuring the extent of attenuation by soils: (1) thin-layer chromatography with soils, (2) batch tests that generate soil adsorption isotherms and (3) column tests that measure capacity and breakthrough. The EPA's Office of Toxic Substances has established some procedures for testing attenuation under the Toxic Substances Control Act [47]. These procedures are soil thin-layer chromatography and the sediment/soil adsorption isotherm (Guidelines CG-1700 and CG-1710 respectively). The Organization for Economic Cooperation and Development (OECD) has also established procedures for the testing of chemicals [30]. The procedure for determining attenuation in soils and sediments entails batch equilibrium tests and extraction, generating adsorption isotherms (Guideline No. 106). While the procedures given by both the EPA and the OECD appear to be focused on the testing of organics, they may be appropriate for inorganics with modifications. Examples of various extraction procedures to measure attenuation in soils have been described by Conway and Malloy [6]. The American Society for Testing and Materials (ASTM) has developed a new method for batch experiments to determine sorption by soils and sediments [1].

Soil thin-layer chromatography, as described by the EPA, is a qualitative screening tool for estimating the leaching potential of a chemical. This test is analogous to conventional thin-layer chromatography (TLC) but uses soil instead of adsorbents like silica gel and alumina. A slurry containing fine particles of the soil is used to produce the soil TLC plates. The plates are then spotted with the chemical and certain reference chemicals, and the movement along the TLC plate is subsequently used to estimate the relative mobility of the chemical concerned. The absolute movement on these soil TLC plates is not translatable into actual field leaching or mobility of the chemical, because natural soil systems are much more complex; however, a qualitative measure of relative leachability is provided by this test.

Data from batch equilibration tests of aqueous solutions containing environmentally realistic concentrations can be plotted to produce an "adsorption isotherm", a curve relating the concentration left in solution to the amount sorbed. EPA Guideline CG-1710 [47] uses the Freundlich isotherm (see "Empirical Models" below). A linear plot of the log of the equilibrium concentration in the solid phase (sorbed) versus the log of the equilibrium concentration in the solution phase will produce an intercept of log K_F and a slope of 1/N. (This is true if sorption of the inorganic follows the Freundlich isotherm under the conditions of the test; other equations may be used if this isotherm is found to be unsuitable.)

Another way to measure attenuation is by the use of column studies. The approach to predicting pollutant movement suggested by Amoozegar-Fard *et al.* [2] used column studies to predict an equivalent of the relative velocity of metals in leachate through columns of soil. Samples representing a range of soil properties were packed into columns, 5.2 cm in diameter and 10 cm long. Leachates containing known initial concentrations of Cd, Ni, and Zn were perfused through the soil columns at a constant rate, and the effluent was partitioned on a fraction collector. Each fraction of the effluent was analyzed for the metal(s) in the original leachate. The relative concentrations (concentration in effluent fraction/initial leachate concentration) were then plotted against time or pore volumes to generate breakthrough curves. Details of this procedure are given under "Empirical Models" below.

2.12.5 Mathematical Representations

Numerous models have been proposed to describe the attenuation of solutes in soils. As may be deduced from the foregoing discussion, no one mathematical representation can describe the attenuation of all solutes under different environmental conditions. The models may be divided into two types: theoretical and empirical. The theoretical models attempt to account for the specific physical and/or chemical interactions occurring between the solutes and the solid surface. The empirical models, on the other hand, are based on the overall attenuation behavior of a solute and usually assume equilibrium conditions. In this section, theoretical models will be briefly described; these are more complicated than empirical representations and are not easily applied to natural soil or sediment environments, which are very complex.

Kinetic models, which may be either theoretical or empirical, do not assume equilibrium conditions and are time-dependent. They are not discussed in detail in this section; further information on the kinetics of processes is given in Chapter 3.

THEORETICAL MODELS

Theoretical models are related to the specific processes leading to sorption in a system. They attempt to model interactions due to the occurrence of surface charges (which attract opposite charges) and also the interactions with surface species (which result in formation of surface complexes with the solutes in solution).

The Gouy-Chapman electric double-layer model is an example of a mechanistic model for the interactions of surface charges. Ravina and Gur [35] applied this model to the sorption of alkali metals onto charged surfaces from chloride solutions. It is very complicated mathematically, even for the simplest systems. Application of this theory to natural systems containing many components would be extremely difficult. Westall and Hohl [48] have shown that electrostatic models of the current generation are not necessarily more accurate than the Stern model, which makes some very simple assumptions regarding surface electrostatic interactions.

Sposito [42-44] has given excellent compilations of models that attempt to account for surface complexation; he described three surface complexation models: the Constant Capacitance Model, the Triple-Layer Model and the Objective Model. These differ generally in the hypotheses used to derive the equations in the models. For example, one of the differences between the Constant Capacitance and Triple-Layer models is that the former assumes that only inner-sphere surface complexes are formed, whereas the Triple-Layer model assumes that the proton (and/or hydroxide ions) form outer-sphere surface complexes with the sorbed species. In a surface complexation model, the equations for complexation of ions in solution may be adapted to describe sorption on surfaces. Morel [27] illustrated the use of a surface complexation model to describe sorption in simple heterogeneous systems, such as the sorption of lead on alumina.

A primary aspect of surface complexation models is that they hypothesize the existence of one or more complexed surface species. These species are often constructs rather than real species. In simple systems (e.g., artificial laboratory systems) the surface complexation model may be applied, because equations may be derived to describe the limited number of surface complexes in such a system. In natural systems, however, numerous inorganic and organic species occur on the sorbents and the soil solution. The identities of the surface complexation species that can occur are often unknown. Also, because of the number of surface species, the set of mathematical equations is large, and their solution is complicated. Surface complexation models are therefore difficult to apply to real systems.

KINETIC MODELS

Kinniburgh and Jackson [23] reported that the sorption of cations by hydrous metal oxides is often extremely rapid; most of the sorption occurs within minutes. However, depending on the time frame under which other reactions are occurring, kinetics may become important. Cases have also been found in which more than one step is involved in the process of sorption: Jardine *et al.* [20] stated that the sorption of aluminum on kaolinite appeared to occur on two different kinds of surface sites: (1) instantaneous equilibrium sites and (2) sites following first-order reversible kinetics. It should therefore not be assumed that sorption equilibrium exists in any particular situation. The rate of desorption is very important in determining whether equilibrium can be assumed; however, there is little information on desorption.

The consideration of non-equilibrium sorption has led to the formulation of kinetic models. One of these is the Elovich equation [18,46], which is shown below in its simplest form.

$$dq/dt = Ae^{-Bq} \tag{1}$$

where q = fraction of sites occupied by solute at time t
A = constant relating to initial velocity of reaction
B = constant relating to activation energy for sorption

Travis and Etnier [46] reviewed a number of first-order kinetic models. Their use in environmental systems obviously requires that the values of the parameters in the equations be known. A review of the literature shows that few parameter values for sorption-desorption kinetic models are available for inorganics in soils, but several of the models have been used to describe phosphorus sorption in soils [46].

EMPIRICAL MODELS

The complexity of natural systems militates against the application of theoretical models. Individual mechanisms for attenuation are not readily or easily distinguishable, and, even if they were, modeling each mechanism could lead to complex mathematical equations containing many unknown parameters. For example, the use of any of the surface complexation models requires complexation constants for the surface complexation species for each sorbent phase. The electrostatic interaction energy between the surface and the solute must also be described.

Empirical models that ignore the mechanisms of attenuation but describe the overall behavior using simple equations are presently more appropriate and applicable to real-life conditions. The next section discusses commonly used adsorption isotherms for describing soil-solute interactions. These are the linear, Langmuir, and Freundlich isotherms. Also described is an empirical, non-equilibrium method demonstrated by Fuller and co-workers utilizing soil column studies [2,9,10].

Table 2.12-3 summarizes the empirical representations and their limitations. Tables 2.12-4, -5 and -6 show some data for the parameters; these tables by no means compile all the existing data for attenuation of inorganics, but the EPRI report on which they are based [34] includes a comprehensive review of available data for some elements. Readers should also refer to the sections on individual elements in Part II of this report for additional data. Because attenuation depends on environmental conditions, the tables include some examples of the variation in parameter values with changing environmental conditions (e.g., pH and ionic strength).

2.12.6 Adsorption Isotherms

Numerous adsorption isotherms were originally developed to describe the physical adsorption of gases onto solid surfaces. Investigators found that the experimental data fitted certain curves that were based on assumptions about the behavior of the gas molecules and the solid surface — e.g., the existence of a maximum number of uniform sites (Langmuir), or heterogeneous sites with heats of adsorption decreasing exponentially with adsorption density (Freundlich). These isotherms have been adapted to describing the attenuation of solutes in soils and sediments. Although physical adsorption is not the only mechanism that contributes to attenuation of solutes in natural environments, the isotherms are useful for describing the overall attenuation process, particularly at dilute concentrations of solutes. We will therefore use the term "adsorption isotherm" to refer to the mathematical equation describing solute sorption and attenuation in soils.

An adsorption isotherm is defined as the relationship between the amount of a substance sorbed and its concentration in solution under conditions of equilibrium and at constant temperature [23]. The shape of the isotherm curve is related to the energy of sorption and the number of sorption sites. Essentially, an adsorption isotherm is obtained by batch equilibrium experiments to obtain data points (amounts sorbed versus equilibrium concentration in solution) at increasing concentrations. An equation is then fitted to the data.

TABLE 2.12-3

Summary of Empirical Models of Sorption

Model	Equation	Definitions	Typical Units
Langmuir	$S = K_L A_m C/[1 + K_L C]$	S = mass sorbed at equilibrium per mass sorbent	mol/g
	Linear: $C/S = 1/[K_L A_m] + C/A_m$	A_m = maximum sorption capacity of sorbent	mol/g
		K_L = Langmuir sorption constant related to binding energy	liter/mol
		C = sorbate concentration in solution at equilibrium	mol/liter
Freundlich	$S = K_F C^{1/N}$	S = mass sorbed at equilibrium per mass sorbent	mol/g
	Linear: $\log S = \log K_F + (1/N) \log C$	K_F = Freundlich isotherm constant	liter/g
		N = Freundlich isotherm constant, $N \geqslant 1$	—
		C = sorbate concentration in solution at equilibrium	mol/liter
Distribution Coefficient	$S = K_d C$	S = mass sorbed at equilibrium per mass sorbent	mol/g
		K_d = distribution coefficient	liter/g
		C = sorbate concentration in solution at equilibrium	mol/liter
Fuller and Co-workers	For given $[Me]_{ss}/[Me]_i$,	$[Me]_i$ = initial concentration of metal in leachate	mg/liter
	Cadmium:	$[Me]_{ss}$ = final steady state concentration of metal in leachate	mg/liter
	$V = (v_p/25)\{a[\%\ clay]^{-1} + b[\%\ sand] +$	V = velocity for a given $[Me]_{ss}/[Me]_i$ through soil	cm/day
	$c[\%\ sand]^2 + d[\%\ FeO]^2 + e[\%\ FeO]^{-1} +$	v_p = pore water velocity = v_d/θ_v	cm/day
	$f[\%\ TSS] + g[\%\ TSS]^2 + h[\%\ TOC] + i\}$	v_d = average water flow (Darcian velocity through soil)	cm/day
	Nickel:	θ_v = porosity (water or leachate volumetric content)	m/m
	$V = (v_p/25)\{a[\%\ clay]^{-1} + b[\%\ sand]^{-1} +$	% clay = soil clay content; 1% = 10g/kg	—
	$c[\%\ FeO] + d[\%\ FeO]^{-1} + e[\%\ TSS] +$	% sand = soil sand content; 1% = 10g/kg	—
	$f[\%\ TSS]^2 + g[\%\ TOC] + h[\%\ TOC]^2 + i\}$	% silt = soil silt content; 1% = 10g/kg	—
	Zinc:	% FeO = iron oxide content; 1% = 10g/kg	—
	$V = (v_p/25)\{a[\%\ sand] + b[\%\ silt]^2 +$	% TOC = total organic carbon in leachate; 1% = 10g/liter	—
	$c[\%FeO]^{-1} + d[\%\ FeO]^2 + e[\%\ TOC] +$	% TSS = total soluble salts in leachate; 1% = 10g/liter	—
	$f[\%TOC]^2 + g[\%\ TSS] + h[\%\ TSS]^2 + i\}$	a,........,i = empirical parameters, different values for each metal	—

(Continued)

TABLE 2.12-3 (Continued)

Model	Variations	Applicability and Limitations	Experimental Method
Langmuir	*Multiple-Site:* $S = \sum_i K_{Li}A_{mi}C/[1 + K_{Li}C]$	Empirical; does not relate to mechanism of sorption.	Batch equilibration
	Subscript i denotes site type	Equilibrium assumed. Applicable to dilute systems.	
	Competitive: $S_j = [A_{mj}K_{Lj}C_j]/[1 + K_{Lj}C_j + \sum_i n_iK_{Li}C_i]$ Subscript i denotes species competing with species j. $n_i = A_{mj}/A_{mi}$	Predicts saturation of monolayer; therefore, maximum implied. Sorption sites assumed uniform; no lateral interactions among sorbed species.	
	Competitive: $[C_j/C_i]/S = A_{mi}/[K_{Lj}A_{mj}] + C_j/A_{mi}C_i$		
	Sorbable and non-sorbable species of element: $S_a = K_{La}A_{ma}[C_m - C_n]/[1 + K_{La}(C_m - C_n)]$ Subscript a denotes sorbed species, m denotes total and n denotes non-sorbed species.		
Freundlich	*Phosphate Model:* Above C_c: $S = aC + X_o$ linear Below C_c: $S = aC + KC^n$ Freundlich where: $X_o = KC_c^{\ n}$ $C_c = [X_o/K]^{1/n}$ X_o and a are intercepts and slope of linear equation. C_c is the equilibrium concentration at which isotherm becomes linear.	Empirical; does not relate to mechanism of sorption. Equilibrium assumed. Applicable to dilute systems. Accounts for heterogeneity of surfaces; heat of sorption decreases exponentially with coverage. Extrapolation beyond data points not advisable; no maximum predicted, and sorption not linear at dilute concentrations.	Batch equilibration
Distribution Coefficient	K_d often given for a single data point, which should not be extrapolated to generate a line.	Empirical; does not relate to mechanism of sorption. Equilibrium assumed. Applicable to dilute systems. No maximum predicted.	Batch equilibration
Fuller and Co-workers	—	Empirical; does not relate to mechanism of sorption. Based on physically measurable soil properties. Data available for cadmium, nickel, and zinc only.	Column tests

TABLE 2.12-4

Distribution Coefficients for Some Metals

Metal	Sorbent	Sorbate Conc. (*M*)	Electrolyte Identity	Electrolyte Conc. (*M*)	pH	K_d (ml/g)
Ba	River sediment	$10^{-5.9}$	Seawater	≈0.7	8	530
	River sediment	$10^{-6.4}$	River water	—	—	2,800
Cd	Montmorillonite (Na-form)	$10^{-6}-10^{-7}$	$NaNO_3$ + NaOAc	1.0,0.01	5.0	8
				1.0,0.01	6.5	100
				0.01,0.01	5.0	210
				0.01,0.01	6.5	900
Cu	Bentonite (Ca-form)	$0\text{-}10^{-5.3}$	Seawater	≈0.7	8	43
	Kaolinite	Trace	Humic acid added	0 μg/l	6.4	43
				0.5 μg/l	6.4	3.2
				1.0 μg/l	6.4	2.5
				71.5 μg/l	6.4	2.2
	$Fe_2O_3 \cdot H_2O$(am)	$10^{-6.5}$	Seawater	≈0.7	8	7,000
	MnO_2(hydrous)	$0\text{-}10^{-5.5}$	Seawater	≈0.7	8	7,300
Mn	Fe oxide(hydrous)	$10^{-7.8}-10^{-5.8}$	Seawater	≈0.7	8	20,000
Hg	$Fe_2O_3 \cdot H_2O$(am)	2.5×10^{-5}	$NaNO_3$	1.0	4.5	59,200
					5.95	2,550,000
	Bentonite	10^{-6}	$Ca(NO_3)_2$	0.01	6.7	408,000
					7.9	179,000
					8.9	119,000
Ni	Montmorillonite	$10^{-6.5}$	Seawater	≈0.7	8	200
	Fe oxide(hydrous)	$10^{-6.5}$	Seawater	≈0.7	8	100,000
Zn	δ-MnO_2	$10^{-8}\text{-}10^{-6}$	Seawater	≈0.7	8	800,000

Definitions:

$S = K_dC$

where: S = mass sorbed at equilibrium per mass of sorbent (μmol/g)

K_d = distribution coefficient (ml/g)

C = sorbate concentration in solution at equilibrium (μmol/ml)

Source: Data (from various sources) adapted from Rai *et al*. [34]

TABLE 2.12-5

Langmuir Constants for Some Metals

Metal	Sorbent	Sorbate Conc. (*M*)	Electrolyte Identity	Electrolyte Conc. (*M*)	pH	Constants A_m (μmol/g)	Constants K_L (log M^{-1})
As	Hydroxide, $Al(OH)_3$ (am)	As(V)	—	—	5	1,600	5.08
					6	1,487	5.17
					7	1,179	5.24
					8	538	5.12
					8.5	681	4.85
					9	501	4.82
	Oxyhydroxide, $Fe_2O_3 \cdot H_2O$(am)	As(III) as AsO_2^- $10^{-7} - 10^{-5}$	$NaNO_3$	1.0	4.0	457	5.98
					5.7	490	6.26
					7.0	513	6.36
					8.8	417	6.18
	Montmorillonite	$10^{-4} - 10^{-3}$ As(V) + $10^{-4} - 10^{-3}$ As(III)	Leachate	—	5	9.9[a]	3.57[a]
Ba	MnO_2 (hydrous)	$10^{-5} - 10^{-3}$	$NaClO_4$	0.01	5	2,050	4.6
Cd	Montmorillonite (Na-form)	$10^{-6.9} - 10^{-6}$	$NaClO_4$	0.01	6.5-7	0.5	7.7
				0.03	6.5-7	0.4	7.2
				0.05	6.5-7	0.4	6.8
				0.17	6.5-7	0.2	6.9
				1.0	6.5-7	0.3	6.7
	River sediment	$10^{-6.7} - 10^{-3.8}$	—	—	~7.5	10-173	4.4-6.0
Cr	Kaolinite	$10^{-4} - 10^{-2.2}$ Cr(VI)	Leachate	—	3	3.64	—
					4	2.5	—
					7	0.98	—
		$10^{-3.3} - 10^{-1.8}$ Cr(III)	Leachate	—	3	96.3	—
					4	283	—

(Continued)

TABLE 2.12-5 (Continued)

Metal	Sorbent	Sorbate Conc. (*M*)	Electrolyte Identity	Electrolyte Conc. (*M*)	pH	Constants A_m (μmol/g)	Constants K_L (log M^{-1})
Cu	Kaolinite	$10^{-5.8} - 10^{-3.8}$	$CaCl_2$	0.05	5.5	1.9	4.5
	Peat	$0 - 10^{-3.8}$	$CaCl_2$	0.05	5.5	184	5.0
	$Fe_2O_3 \cdot H_2O$(am)	$0 - 10^{-4.5}$	Seawater	~0.7	8	1,120	5.3
Pb	Montmorillonite	$10^{-4.3} - 10^{-2.7}$	Landfill Leachate	~0.1	5.0	8.8 (A_{m1}) 5.38 (A_{m2})	4.3 (K_{L1}) 2.1 (K_{L2})
	Goethite	$10^{-3} - 10^{-2.2}$	KNO_3	0.1	5.0	85	2.9
	Mn oxide(hydrous)	$10^{-3.5} - 10^{-2.8}$	—	—	6	7,000	3.7
	Various soils	$10^{-4.6} - 10^{-3.3}$	KCl	0.1	7	7.0-23	3.7-4.9
Mn	Montmorillonite	0.005-0.05	—	—	5-6	370,610	2.7,2.8
	Cryptomelane (α-MnO_2 as $K_2Mn_8O_{16}$)	$10^{-3} - 10^{-2.2}$	KNO_3	0.1	5	1,130	3.5
	Clay loam	$0 - 10^{-2.9}$	Na_2SO_4	0.01	5.4	14.7,19.3	3.4
Hg	Sediments	$10^{-6} - 10^{-4}$	—	—	—	11.9-237	5.5-6.5
Ni	2	$10^{-3} - 10^{-2.2}$	KNO_3	0.1	5	690	3.6
Se	Kaolinite	$10^{-3} - 10^{-4}$ Se(IV)	Landfill Leachate	—	3	4.20	3.23
					5	2.93	3.37
					7	2.40	3.26

(Continued)

TABLE 2.12-5 (Continued)

Metal	Sorbent	Sorbate Conc. (*M*)	Electrolyte Identity	Electrolyte Conc. (*M*)	pH	Constants A_m (μmol/g)	Constants K_L (log M^{-1})
Zn	Montmorillonite,	$10^{-3.6} - 10^{-2.6}$	Landfill	~0.1	5	29	3.0
	Ca-saturated		Leachate		6	35	3.1
	$Fe(OH)_3$(am)	$10^{-6.7} - 10^{-5}$	—	—	6	25	5.4
					7	170	5.9
					8	440	6.7
	Calcite	$10^{-7} - 10^{-5.3}$	—	—	8.4	59	5.2
	Clay loam	$0 - 10^{-5.5}$	$CaCl_2$	0.016	7.7^b	9.4	5.5

a. Values apply to As(V)

b. Soil pH

Definitions:

$S = K_L A_m C/[1 + K_L C]$

where: S = mass sorbed at equilibrium per mass of sorbent (μmol/g)

A_m = maximum sorption capacity of sorbent (μmol/g)

K_L = Langmuir constant related to binding energy of sorbate (liter/μmol), given above as log [liter/mol]

C = sorbate concentration in solution at equilibrium (μmol/liter)

A_{m1}, K_{L1}, A_{m2}, K_{L2} refer to two sorption sites

Source: Data (from various sources) adapted from Rai *et al.* [34]

TABLE 2.12-6

Freundlich Constants for Some Metals

Metal	Sorbent	Sorbate Conc. (*M*)	Electrolyte Identity	Electrolyte Conc. (*M*)	pH	Constants K_F (l/g)	Constants 1/N
Cd	Montmorillonite (Na-form)	$10^{-6.9} - 10^{-6}$	Na_2SO_4	0.03	6.5-7	3.9	1.03
				0.05	6.5-7	0.74	0.78
				0.075	6.5-7	0.63	0.82
	$Fe_2O_3 \cdot H_2O$(am)	$0 - 10^{-7.7}$	$NaNO_3$	0.1	6.8	1.1	1.0
		$10^{-7.7} - 10^{-4}$	$NaNO_3$	0.1	6.6	0.065	0.67
		$10^{-7.7} - 10^{-4}$	$NaNO_3$	0.1	6.9	0.15	0.67
		$10^{-7.7} - 10^{-4}$	$NaNO_3$	0.1	7.2	0.32	0.67
	Four soils	$10^{-6.9} - 10^{-5.4}$	$NaNO_3$	0.1	5.2-8.4[a]	3.6-23.2	0.63-0.95
Cu	Forest soil	$10^{-5.1} - 10^{-3.5}$	Simulated sludge leachate	—	5.4-6.4	2.5-6.1	1.0
Pb	Various soils	$10^{-6} - 10^{-4.4}$	$CaCl_2$	0.05	7.7	12.6,18.8	0.33,0.22
					8.2	4.3	0.22
					8.6	8.9	0.18
Mn	Bentonite	$0 - 10^{-3}$	—	—	4.9	0.011	0.9
					6.2	7.8	0.26
					7.6	23	0.18
	Illite	$0 - 10^{-3}$	—	—	8.8	4.3	0.48
	Kaolinite	$0 - 10^{-2.9}$	—	—	4.9	0.015	1.13
					6.2	0.011	1.20
					7.6	0.018	1.17
					8.8	0.043	1.07

(Continued)

TABLE 2.12-6 (Continued)

Metal	Sorbent	Sorbate Conc. (*M*)	Electrolyte Identity	Electrolyte Conc. (*M*)	pH	Constants K_F (l/g)	Constants 1/N
Hg	$Fe(OH)_3$(am)	$10^{-7} - 10^{-6}$	—	—	5-8	90.8	0.76
Ni	Silt loam	$10^{-6.8} - 10^{-3.8}$	$CaCl_2$	0.01	6.1[a]	0.30	0.95
			$CaCl_2$	0.1	6.1[a]	0.057	0.91
			$Ca(ClO_4)_2$	0.1	6.1[a]	0.083	0.94
	Sandy loam	$10^{-3.8} - 10^{-1.8}$	$CaCl_2$	0.01	6.1[a]	0.27	0.57
Zn	$Fe_2O_3 \cdot H_2O$(am)	$10^{-7.5} - 10^{-4.0}$	$NaNO_3$	0.01	6.4	8.9	1.0
	Clay loam	$0 - 10^{-5.5}$	$CaCl_2$	0.016	7.7[a]	1.8	0.94
	Clay	$0 - 10^{-5.8}$	$CaCl_2$	0.016	8.5[a]	13	0.85

a. Soil pH

Definitions:

$$S = K_F C^{1/N}$$

where: S = mass sorbed at equilibrium per mass of sorbent (μmol/g)

K_F = Freundlich isotherm constant (liter/g)

N = Freundlich isotherm constant; $N \geq 1$

C = sorbate concentration in solution at equilibrium (μmol/liter)

Source: Data (from various sources) adapted from Rai *et al.* [34]

Oakley *et al.* [29] obtained adsorption isotherms for Cu and Cd in seawater using several sorbents. Their measurements with Cu and $Fe(OH)_3$ in seawater are discussed below to illustrate how an adsorption isotherm is derived. Similar procedures could be used to derive adsorption isotherms for other solutes and sorbents (including soils and sediments) under different conditions. The experiments were carried out in 50-ml polyethylene centrifuge tubes containing a known amount of solid (~10 mg) and 20 ml of seawater. The required amounts of stock solution of $Cu(NO_3)_2$ were added to provide a range of initial Cu concentrations, and the pH was adjusted to 8.0 ± 0.1 with NaOH or HNO_3. The tubes were agitated for 24 hours. The solid/liquid mixtures were centrifuged to separate the dissolved and solid phases. The soluble Cu concentration was measured by atomic absorption spectrophotometry. The quantity of Cu sorbed on the $Fe(OH)_3$ was calculated by difference from the amount of metal added and the final soluble concentration. The adsorption isotherm (Figure 2.12-4) was obtained by plotting the concentration of dissolved Cu at equilibrium on the x-axis and the amount of Cu sorbed per gram of $Fe(OH)_3$ on the y-axis.

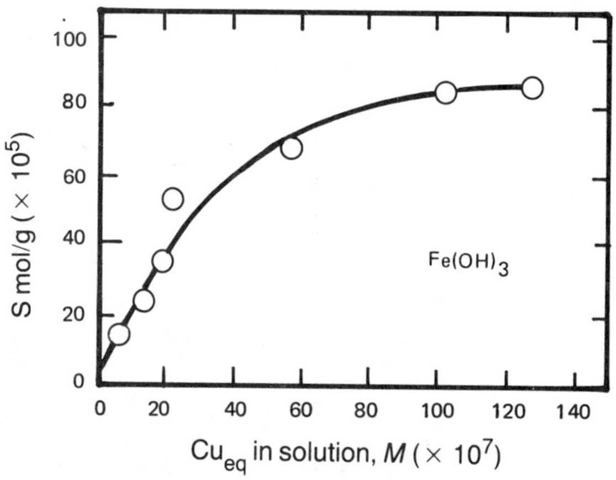

Salinity = 32 ‰, pH = 8.0, Temperature = 20°C.

Source: Oakley, Nelson and Williamson [29]. *(Copyright 1981, American Chemical Society. Reprinted with permission.)*

FIGURE 2.12-4 Adsorption Isotherm of Cu on $Fe(OH)_3$ in Seawater

Different authors have found that either the Langmuir or Freundlich isotherm would fit their experimental data. Cavallaro and McBride [5] reported that the Langmuir model described the sorption of Cu and Cd in samples of acidic and neutral silt loam soils, while Sidle and Kardos [41] found that data for Cu fitted the Freundlich isotherm better than the Langmuir in experiments with forest soil. In some cases, both models have been used to fit the same experimental data.

As Travis and Etnier have pointed out, the fit of sorption data to a particular isotherm has no bearing on the mechanism of attenuation [46]. Also, data used to generate isotherms are obtained from batch experiments in which the sorbent and solute are in prolonged contact (usually 24 hours); this may not be representative of environmental conditions.

DISTRIBUTION COEFFICIENT

The simplest mathematical representation is the linear isotherm with a distribution coefficient:

$$S = K_d C \tag{2}$$

where S = mass sorbed at equilibrium per mass of sorbent (mol/g)
K_d = distribution coefficient (l/g)
C = sorbate concentration in solution at equilibrium (mol/l)

The linear isotherm does not imply a maximum. The distribution coefficient is often given as a single point measured at a particular equilibrium concentration and defined as the mass sorbed per gram at equilibrium divided by the equilibrium concentration in mol/liter. Extrapolating this point to a different concentration could result in large errors and should not be attempted.

Example 1 How much Cd would be sorbed onto one gram of soil with a main component of montmorillonite at equilibrium concentration of 10^{-6} *M* in the liquid phase? What would the final conditions in the system be if the initial solution had a concentration of 10^{-5} *M* Cd?

Table 2.12-4 lists a K_d value of 210 ml/g for montmorillonite (Na-form) for a Cd concentration of $10^{-6} - 10^{-7}$ *M* using 0.01M $NaNO_3$ and 0.01 *M* NaOAc as electrolyte at pH of 5.0. At an equilibrium concentration of 10^{-6} *M*, C = 1.124×10^{-4} g/l. Therefore,

$$\begin{aligned} S &= K_d C \\ &= 0.210 \text{ l/g} \times 10^{-6} \text{ mol/l} \\ &= 0.210\ \mu\text{mol/g} \end{aligned}$$

To calculate the final conditions in the system, let Y *M* be the Cd concentration attenuated. (It is assumed that no Cd was adsorbed initially.) The Cd concentration in the final solution is then C = $(10^{-5} - Y)$ mol/l.

At equilibrium, the mass of Cd sorbed onto 1 gram of soil from 1 liter of solution is

$$S = Y \text{ mol/l} \times \text{l/g} = Y \text{ mol/g}$$

Using equation 2,

$$\begin{aligned} S &= K_dC \\ Y \text{ mol/g} &= (210 \text{ ml/g})(10^{-5} - Y) \text{ mol/l } (\text{l}/1000 \text{ ml}) \\ Y &= 0.21 \times 10^{-5} - 0.21\, Y \text{ mol/g} \\ 1.21\, Y &= 0.21 \times 10^{-5} \text{ mol/g} \\ Y &= 1.74 \times 10^{-6} \text{ mol/g} \\ &= 1.74\ \mu\text{mol Cd/g soil} \end{aligned}$$

The final solution concentration of Cd is thus

$$\begin{aligned} C &= 10^{-5} - Y \\ &= 8.3 \times 10^{-6}\, M \end{aligned}$$

LANGMUIR ISOTHERM

The Langmuir isotherm was first developed to describe the physical adsorption of gases by solids. It is based on the assumption that there are a finite number of sorption sites, resulting in a maximum sorption that is possible on the sorbent surface. As given by Rai *et al.* [34], the Langmuir isotherm is expressed as:

$$S = K_LA_mC/(1 + K_LC) \qquad (3)$$

where:
- S = moles sorbed at equilibrium per mass of sorbent (mol/g)
- A_m = maximum sorption capacity of the sorbent (mol/g)
- K_L = Langmuir sorption constant, related to binding energy of the sorbate (l/mol)
- C = sorbate concentration in solution at equilibrium (mol/l)

A_m is a constant, representing the maximum number of sites on the sorbent surface. K_L is a measure of the bond strength between the sorbate and the surface site. In the Langmuir model, this bond strength is assumed to be constant. The heat of sorption is assumed to be independent of the number of occupied sites or the different types of surface sites. Essentially, there are no interactions between the sorbed molecules (or ions), and the sites on the surface are assumed to be uniform. Related to these two assumptions is that sorption is independent of the number of sites already occupied (which implies that sorption occurs on only a single layer on the sorbent surface) and that the sorption of molecules (ions) on the surface does not generate new sites on the sorbate layer.

The Langmuir isotherm may be expressed linearly in the following form [34]:

$$C/S = 1/[K_LA_m] + C/A_m \qquad (4)$$

Using this equation, sorption data (C/S vs. C) are plotted to produce a line with a slope of $1/A_m$ and an intercept of $1/[K_L A_m]$.

The assumptions used in deriving the Langmuir isotherm are often not true for real systems. In soils, the sites are unlikely to be uniform because of the different types of sorbents present. As a result, K_L may not be constant. A range of different sites may be filled simultaneously. Over a large concentration range, interactions between sorbed molecules (ions) and multilayer sorption are more likely. Also, as the sorption density (fraction of occupied sites) increases, the heat of sorption changes and it becomes more difficult for subsequent molecules (ions) to become sorbed.

There are a number of ways to compensate for the imperfections of the Langmuir isotherm. Besides using another model (e.g., the Freundlich, which provides for heterogeneity of the sorbent surface), multiple-site Langmuir isotherms and the competitive Langmuir model have also been proposed and used.

The multiple-site variation of the Langmuir isotherm is based on the postulate that several energetically distinct sites are present or that different mechanisms are occurring on the same sites [34]. The total sorption arising from either of these alternatives may be represented by:

$$S = \sum_i K_{Li} A_{mi} C/[1 + K_{Li} C] \quad (5)$$

The total sorption is summed over all the possible types of sites; i.e., A_{mi} and K_{Li} are the Langmuir constants as they apply to site type i. The subscript i may also be used to denote different mechanisms, with the assumption that each of the mechanisms may be modeled using the Langmuir isotherm.

If there are two types of sites, the multiple-site variation of the Langmuir isotherm is written as:

$$S = K_{L1} A_{m1} C/[1 + K_{L1} C] + K_{L2} A_{m2} C/[1 + K_{L2} C] \quad (6)$$

In an environmental system with more than one sorbate, competition for surface sites may explain non-linearity in the Langmuir isotherm [15,17]. Rai *et al.* [34] introduced another term into the Langmuir equation to account for this factor. The sorption of the sorbate, j, in a competitive system is represented by:

$$S_j = [A_{mj} K_{Lj} C_j]/[1 + K_{Lj} C_j + \sum_i n_i K_{Li} C_i] \quad (7)$$

where i denotes the species competing with species j and $n_i = A_{mj}/A_{mi}$.

n_i compensates for the size differences of competing species. The larger the species j is, the more space it occupies compared with i. Therefore, n_i represents the ratio of the surface areas occupied by molecules (ions) j and i.

For competing species of similar size, n = 1. The final term in the denominator of equation 7 effectively reduces the sorption of species j in keeping with the bond strengths of sorption and equilibrium concentrations of competing species. For two species of similar size, the equation reduces to:

$$S_j = [A_{mj}K_{Lj}C_j]/[1 + K_{Lj}C_j + K_{Li}C_i] \qquad (8)$$

If C_i is assumed to be constant, equation 8 may be expressed linearly as:

$$C_j/S_j = [1 + K_{Li}C_i]/[A_{mj}K_{Lj}] + C_j/A_{mj} \qquad (9)$$

If species i is present at sufficiently high concentrations compared with species j (e.g., if it is a major cation like Ca^{+2} or a major anion like Cl^-, which is present in natural environments at concentrations several orders of magnitude larger than trace metals or ligands), equation 9 will be applicable. K_{Li} will have to be determined separately under experimental conditions (pH, ionic strength, etc.) similar to those of the experiments used for the competitive system.

Boyd *et al.* [4] described the simultaneous competitive sorption of two equally charged cations by:

$$[C_j/C_i]/S = A_{mi}/[K_{Lj}A_{mj}] + C_j/A_{mi}C_i \qquad (10)$$

A plot of $[C_j/C_i]/S$ against C_j/C_i generates a straight line with a slope of $1/A_{mi}$ and an intercept of $A_{mi}/[K_{Lj}A_{mj}]$.

The Langmuir isotherm can be adapted to include cases in which a solute is present in more than one form and only one form is sorbed by the sorbent [16]. For example, if the measured concentration (C_m) includes both the species that can be sorbed (C_a) and a non-sorbable species (C_n), the Langmuir model can include:

$$C_a = C_m - C_n \qquad (11)$$

Therefore,

$$S_a = K_{La}A_{ma}[C_m - C_n]/[1 + K_{La}(C_m - C_n)] \qquad (12)$$

Example 2 Using the Langmuir model, calculate the amount of Cd sorbed onto soil at an equilibrium concentration of 10^{-6} *M* in the liquid phase.

Sidle and Kardos [41] reported the following Langmuir constants for sorption of Cd in forest soil at a depth of 0-7.5 cm:

K_L = 47.19 l/mg
A_m = 0.0024 mg/g soil

These values were obtained for initial Cd concentrations of 0.005-0.2 mg/l (or $10^{-7.4}$ – $10^{-5.8}$ mol/l).

To find S in terms of mol/g, the above constants must be converted to the proper units by applying the conversion factor of 112,400 mg/mol Cd:

$$K_L = 47.19 \text{ l/mg} \times 112{,}400 \text{ mg/mol}$$
$$= 5.30 \times 10^6 \text{ l/mol or } 5.30 \text{ l/}\mu\text{mol}$$
$$A_m = 0.0024 \text{ mg/g} \times \text{mol/}112{,}400 \text{ mg}$$
$$= 2.14 \times 10^{-8} \text{ mol/g or } 0.021\ \mu\text{mol/g}$$
$$C = 10^{-6}\ M = 1\ \mu\text{mol/l}$$

Substituting these values into equation 3, we obtain

$$S = K_L A_m C / [1 + K_L C]$$
$$= \frac{5.3 \text{ l/}\mu\text{mol} \times 0.021\ \mu\text{mol/g} \times 1\ \mu\text{mol/l}}{1 + (5.3 \text{ l/}\mu\text{mol} \times 1\ \mu\text{mol/l})}$$
$$= 0.018\ \mu\text{mol/g sorbed}$$

Example 3 Determine the final conditions in the above system if the initial solution had a concentration of 10^{-5} *M* Cd. (Assume that no Cd was sorbed initially.)

Let the Cd concentration attenuated = Y *M*
Final concentration, C = initial conc. − attenuated conc.
$= (10^{-5} - Y)\ M$

At equilibrium, the mass of Cd sorbed onto 1 g of soil from 1 liter of solution is

$$S = Y \text{ mol/l} \times \text{l/g soil} = Y \text{ mol/g}$$

Then, from equation 3,

$$S = K_L A_m C / [1 + K_L C]$$
$$= \frac{[5.3 \times 10^6 \text{ l/mol}]\,[2.14 \times 10^{-8} \text{ mol/g}]\,[C \text{ mol/l}]}{1 + (5.3 \times 10^6 \text{ l/mol})(C \text{ mol/l})]}$$
$$= 0.113\ C / [1 + 5.3 \times 10^6\ C] \text{ mol/g}$$

Substituting $C = (10^{-5} - Y)\ M$ and $S = Y$ mol/g,

$$Y = \frac{0.113\,(10^{-5} - Y)}{1 + (5.3 \times 10^6)(10^{-5} - Y)}$$
$$(1.13 \times 10^{-6}) - 0.113\ Y = [54 - (5.3 \times 10^6\ Y)]\ Y$$
$$54\ Y - (5.3 \times 10^6)\ Y^2 = 1.13 \times 10^{-6} - 0.113\ Y$$
$$(5.3 \times 10^6)\ Y^2 - 54.11\ Y + 1.13 \times 10^{-6} = 0$$

The two solutions to this quadratic equation are found to be:

$$Y = 1.02 \times 10^{-5} \text{ mol/g or } 2.08 \times 10^{-8} \text{ mol/g}$$

Since Y is always $\leqslant A_m$, the second solution is the appropriate value.

Accordingly, $C = (10^{-5} - Y)\ M = 10^{-5} - 2.08 \times 10^{-8}\ M$
$= 9.98 \times 10^{-6}\ M$

The final conditions are:

Solution concentration = $9.98 \times 10^{-6}\ M$
Amount attenuated = $0.02\ \mu$mol/g

FREUNDLICH ISOTHERM

This isotherm attempts to account for the heterogeneity of surfaces by assuming that the heat of sorption decreases exponentially with coverage. In essence, it is assumed that the initial sites are the more favorable ones and that it becomes increasingly difficult to form a bond between the sorbate and sorbent surface as more of the surface sites are occupied. The Freundlich isotherm form is [34]:

$$S = K_F C^{1/N} \tag{13}$$

where
S = moles sorbed at equilibrium per mass of sorbent (mol/g)
K_F = Freundlich isotherm constant (l/g)
N = Freundlich isotherm constant; $N \geq 1$
C = sorbate concentration in solution at equilibrium (mol/l)

The Freundlich isotherm is an empirical relationship. From equation 13, it can be seen that the Freundlich does not imply a maximum sorption. At low concentrations, it generally does not become linear. Because of these two conditions, it is generally unwise to extrapolate this isotherm beyond the data points of measurement [23]. The Freundlich isotherm having $N = 1$ becomes a linear equation with K_F equal to K_d, the distribution coefficient.

The linear form of the Freundlich isotherm is:

$$\log S = \log K_F + (1/N) \log C \tag{14}$$

A plot of log S versus log C provides a straight line with a slope of 1/N and an intercept of log K_F.

In their study of the behavior of phosphate, Shayan and Davey [40] proposed a modified Freundlich isotherm that included a linear portion at concentrations above a critical concentration (C_c) and a modified Freundlich isotherm below this concentration. The two equations are:

$$\text{Above } C_c\text{: } S = aC + X_o \text{ (linear)} \tag{15}$$

$$\text{Below } C_c\text{: } S = aC + KC^n \text{ (Freundlich)} \tag{16}$$

At C_c, these equations are equal; therefore, $X_o = KC_c^{\,n}$ and $C_c = [X_o/K]^{1/n}$

The Freundlich isotherm successfully describes sorption at low concentrations. The Langmuir and Freundlich isotherms are sometimes indistinguishable at low concentrations and low surface coverage [16].

Example 4 Calculate the amount of Cd sorbed onto soil at an equilibrium concentration of 10^{-6} M in the liquid phase.

Sidle and Kardos [41] provided the following Freundlich constants for Cd in forest soil: K_F' (equal to S) = 32.4 μg/g at an equilibrium concentration of 1 μg/ml and 1/N = 0.82.

Substituting these parameters in equation 13,

$$32.4\ \mu g/g = K_F(1)^{0.82}\ \mu g/ml$$
$$K_F' = 32.4\ \mu g/g \times ml/\mu g = 32.4\ ml/g$$

At an equilibrium concentration of 10^{-6} M Cd, C = 1.124×10^{-4} g/l. Again using equation 13,

$$\begin{aligned} S &= 32.4\ ml/g\ (1.124 \times 10^{-4})^{0.82}\ g/l\ (10^{-3}\ l/ml) \\ &= 0.0187 \times 10^{-3}\ g/g \\ &= 18.7\ \mu g/g \text{ or } 0.17\ \mu mol/g \text{ sorbed} \end{aligned}$$

Example 5 Determine the final conditions in the above system if the initial solution had a concentration of 10^{-5} M Cd. (Assume that no Cd was sorbed initially.)

Let the Cd concentration attenuated = Y M
Final concentration, C = initial conc. − attenuated conc.
= $(10^{-5} - Y)$ M

At equilibrium, the mass of Cd sorbed onto 1 g of soil from 1 liter of solution is

$$S = Y\ mol/l \times 1\ l/g\ soil = Y\ mol/g$$

Then, from equation 13,

$$\begin{aligned} S &= K_F C^{1/N} \\ &= 0.0324\ l/g\ (C\ mol/l)^{0.82} \\ &= 0.0324\ C^{0.82} \end{aligned}$$

Substituting C = $(10^{-5} - Y)$ M and S = Y mol/g,

$$Y = 0.0324\ (10^{-5} - Y)^{0.82}$$
$$f(Y) = Y - 0.0324\ (10^{-5} - Y)^{0.82} = 0$$

To solve this equation, we can use Newton's Method, which is an iterative method:

$$x_{n+1} = x_n - [f(x_n)/f'(x_n)]$$

$f'(Y)$ is the first derivative of $f(Y)$

$$\begin{aligned} f'(Y) &= 1 - 0.027\ (10^{-5} - Y)^{-0.18}\ (-1) \\ &= 1 + 0.027\ (10^{-5} - Y)^{-0.18} \end{aligned}$$

Picking $Y_1 = 2 \times 10^{-7}$ as a first approximation and substituting it into $f(Y_1)$ and $f'(Y_1)$, we obtain:

$$\begin{aligned} Y_2 &= 2 \times 10^{-7} - [-2.33 \times 10^{-6}]/1.215 \\ &= 2.12 \times 10^{-6} \\ Y_3 &= 2.12 \times 10^{-6} - [3.1 \times 10^{-9}]/1.22 \\ &= 2.12 \times 10^{-6} \end{aligned}$$

The solution has converged at $Y = 2.12 \times 10^{-6}$ mol/g.

$$C = (10^{-5} - Y) = 7.88 \times 10^{-6}\ M$$

Thus,

Final solution concentration $= 7.88 \times 10^{-6}\ M$
Amount attenuated $= 2.12\ \mu$mol/g

2.12.7 Soil Column Studies by Fuller and Co-workers

Simple, field-oriented regression equations have been developed to predict the movement of Cd, Ni and Zn through soils of known composition [2]. For example, the Cd equation can be used to predict the length of time it will take a given concentration of Cd to reach a particular soil depth when the initial leachate level is known. The equations can also be used to compute a concentration profile of Cd, Ni or Zn with depth of soil for periods of time after the application of a leachate containing a known quantity of the metal.

The equations were derived from data obtained in laboratory experiments with columns packed with disturbed soils that encompassed a broad spectrum of physical and chemical characteristics. The soils chosen contained little organic matter; most of the soluble organic compounds were removed by perfusion with deionized water before the leachate was applied. Leachates of known total organic carbon (TOC) and total soluble salts (TSS) content were enriched with 100 mg/l or 0.001 mol/l of Cd, Ni or Zn and were perfused through the soil columns at a constant rate. Effluent fractions were analyzed for the added metal. The data collected from these experiments were used in connection with the Lapidus and Amundson model [25] with a non-equilibrium adsorption term. This model, which provides an analytical solution to the solute flow for a step input, was used to generate the regression equations for predicting movement of the three metals through soils.

The three equations are valid only for (a) leachates that contain a single metal and have a pH of ~5.5-6.0 and (b) soils that are low in organic matter. The flux rate has little effect on the movement of Cd, Ni and Zn.

Use of these equations to calculate the propagation velocity, V(cm/day), for given concentrations of these metals requires determination of the following variables:

Average water flow (Darcian velocity through soil) $= v_d$(cm/day)
Porosity (water or leachate volumetric content) $= \theta_v(m^3/m^3)$

Pore-water velocity = $v_p = v_d/\theta_v$(cm/day)
Soil clay content = % clay[2] (mass/mass)
Soil sand content = % sand[2] (mass/mass)
Soil silt content = % silt[2] (mass/mass)
Iron oxide content (FeO) = % FeO[2] (mass/mass)
Total organic carbon in leachate = % TOC[2] (mass/volume)
Total soluble salts in leachate = % TSS[2] (mass/volume)
Initial concentration of metal in leachate = $[Me]_i$ (mg/l)
Final steady-state concentration of metal in leachate = $[Me]_{ss}$ (mg/l)

For a given $[Cd]_{ss}/[Cd]_i$, the empirical equation for Cd is:

$$V = (v_p/25)\,[a_{Cd}(\%\ clay)^{-1} + b_{Cd}(\%\ sand) + c_{Cd}(\%\ sand)^2 + d_{Cd}(\%\ FeO)^2 + e_{Cd}(\%\ FeO)^{-1} + f_{Cd}(\%\ TSS) + g_{Cd}(\%\ TSS)^2 + h_{Cd}(\%\ TOC) + i_{Cd}] \quad (17)$$

For a given $[Ni]_{ss}/[Ni]_i$, the empirical equation for Ni is:

$$V = (v_p/25)\,[a_{Ni}(\%\ clay)^{-1} + b_{Ni}(\%\ sand)^{-1} + C_{Ni}(\%\ FeO) + d_{Ni}(\%\ FeO)^{-1} + e_{Ni}(\%\ TSS) + f_{Ni}(\%\ TSS)^2 + g_{Ni}(\%\ TOC) + h_{Ni}(\%\ TOC)^2 + i_{Ni}] \quad (18)$$

For a given $[Zn]_{ss}/[Zn]_i$, the empirical equation for Zn is:

$$V = (v_p/25)\,[a_{Zn}(\%\ sand) + b_{Zn}(\%\ silt)^2 + c_{Zn}(\%\ FeO)^{-1} + d_{Zn}(\%\ FeO)^2 + e_{Zn}(\%\ TOC) + f_{Zn}(\%\ TOC)^2 + g_{Zn}(\%\ TSS) + h_{Zn}(\%\ TSS)^2 + i_{Zn}] \quad (19)$$

where a_{Me} to i_{Me} are empirical parameters derived from a regression fit (analysis) of data from a variety of soils and metal concentrations of 100 mg/l or 0.001 mole/l. Values of a to i for various values of $[Me]_{ss}/[Me]_i$ are given in Table 2.12-7 (cadmium), 2.12-8 (nickel) and 2.12-9 (zinc).

For more detailed data and examples, readers should refer to several papers that describe the method used by Fuller and co-workers [9-12]. These investigators have prepared diagrams of the kind shown here as Figures 2.12-5 and -6, which portray the relative mobilities of cations and anions in various soils.

The data reported by Fuller *et al.* were obtained from column experiments, not measurements in the field. The advantage of this approach is that the data are based on physically measurable soil properties; however, detailed data are available only for Zn, Ni and Cd.

2. For clay, silt, sand, and FeO, 1% = 10 g/kg; for TSS and TOC, 1% = 10 g/liter.

TABLE 2.12-7

Empirical Parameters for Calculating the Propagation Velocity of Cadmium by Equation 17

Parameter	Ratio of $[Cd]_{ss}/[Cd]_l$								
	0.1	**0.2**	**0.3**	**0.4**	**0.5**	**0.6**	**0.7**	**0.8**	**0.9**
a	32.02	31.07	30.65	29.54	29.91	29.52	29.18	28.65	28.10
b	−0.233	−0.227	−0.223	−0.217	−0.217	−0.214	−0.212	−0.209	−0.204
c	0.00117	0.00114	0.00112	0.00109	0.00108	0.00107	0.00106	0.00105	0.00102
d	0.0116	0.0111	0.0106	0.00916	0.0101	0.00996	0.00970	0.00970	0.00939
e	9.846	9.385	9.057	8.817	8.532	8.293	8.048	7.761	7.349
f	90.83	89.19	87.32	92.20	84.13	83.06	81.63	79.10	77.81
g	−218.6	−216.4	−212.3	−234.1	−205.1	−203.1	−199.8	−194.9	−193.9
h	0.322	0.370	0.339	−0.148	0.442	0.494	0.511	0.559	0.426
i	−2.090	−2.015	−1.923	−2.069	−1.760	−1.702	−1.585	−1.388	−1.307
r^2	0.844	0.843	0.844	0.845	0.844	0.842	0.840	0.838	0.838

Source: Amoozegar-Fard, Fuller and Warrick [2]. *(Copyright 1984, Reprinted with permission.)*

TABLE 2.12-8

Empirical Parameters for Calculating the Propagation Velocity of Nickel by Equation 18

	$[Ni]_{ss}/[Ni]_i$								
Parameter	**0.1**	**0.2**	**0.3**	**0.4**	**0.5**	**0.6**	**0.7**	**0.8**	**0.9**
a	17.18	15.71	14.72	14.04	13.46	13.09	12.70	12.62	12.85
b	29.29	27.37	26.28	24.78	23.85	23.06	21.49	20.89	19.00
c	0.350	0.333	0.323	0.312	0.304	0.297	0.290	0.280	0.263
d	7.541	7.026	6.775	6.413	6.200	5.972	5.734	5.448	5.013
e	82.19	76.34	73.00	71.06	69.32	67.59	65.56	66.12	65.15
f	−266.9	−246.5	−234.4	−228.6	−222.9	−217.3	-210.0	−212.7	−210.5
g	−40.13	−37.44	−36.28	−35.63	−34.71	−34.20	−32.33	−32.19	−32.57
h	132.2	122.9	120.5	117.5	114.8	113.6	106.4	112.7	116.6
i	−6.791	−6.298	−6.052	−5.711	−5.534	−5.350	−5.195	−5.130	−4.927
r^2	0.908	0.905	0.907	0.892	0.885	0.874	0.861	0.839	0.795

Source: Amoozegar-Fard, Fuller and Warrick [2]. *(Copyright 1984, Reprinted with permission.)*

TABLE 2.12-9

Empirical Parameters for Calculating the Propagation Velocity of Zinc by Equation 19

	$[Zn]_{ss}/[Zn]_i$								
Parameter	**0.1**	**0.2**	**0.3**	**0.4**	**0.5**	**0.6**	**0.7**	**0.8**	**0.9**
a	0.0210	0.0173	0.0161	0.0149	0.0136	0.0111	0.00954	0.00803	0.00487
b	0.000657	0.000591	0.000570	0.000550	0.000526	0.000482	0.000466	0.000443	0.000398
c	2.784	2.747	2.697	2.660	2.637	2.638	2.660	2.648	2.698
d	0.00373	0.00336	0.00324	0.00310	0.00288	0.00255	0.00254	0.00236	0.00211
e	−15.38	−14.92	−14.86	−14.72	−14.75	−14.57	−14.76	−14.96	−15.82
f	82.37	80.25	79.18	78.42	77.79	77.12	76.71	76.57	78.93
g	67.11	60.90	58.93	56.97	54.66	51.12	51.03	49.37	46.16
h	−188.2	−167.5	−162.1	−156.3	−148.2	−135.9	−137.1	−132.3	−122.0
i	−6.44	−5.88	−5.68	−5.48	−5.27	−4.92	−4.82	−4.61	−4.22
r^2	0.841	0.844	0.846	0.848	0.847	0.848	0.848	0.848	0.846

Source: Amoozegar-Fard, Fuller and Warrick [2]. *(Copyright 1984, Reprinted with permission.)*

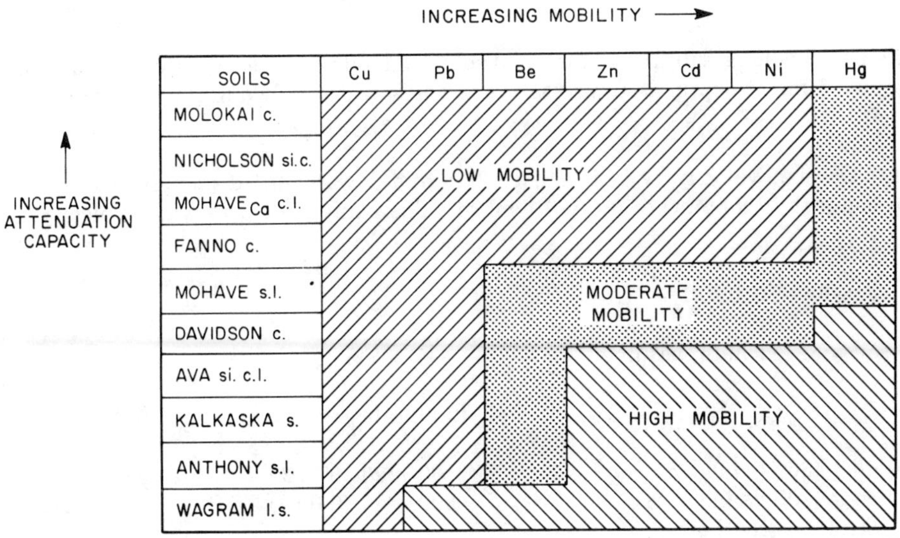

Source: Fuller [10]

FIGURE 2.12-5 Relative Mobility of Cation-Forming Elements in Soils

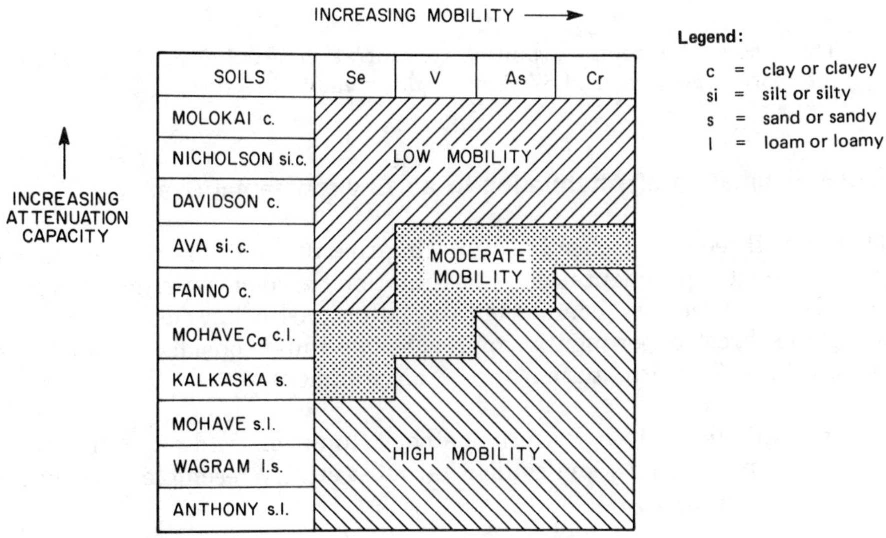

Source: Fuller [10]

FIGURE 2.12-6 Relative Mobility of Anion-Forming Elements in Soils

Example 6 How long will it take for a soil solution having a concentration of 3 mg/l of Cd to reach a 5-meter depth, assuming that the concentration of Cd in the leachate is 15 mg/l?[3]

Assume a Darcian velocity (v_d) of 0.15 cm/day and a water-filled porosity (water or leachate volumetric content) of $\theta_v = 0.1\ m^3/m^3$. The equivalent pore-water velocity is then:

$$v_p = v_d/\theta_v = 0.15/0.1 = 1.5 \text{ cm/day}.$$

For the soil and leachate, assume:

Soil	Leachate
300 g/kg (30%) clay	1100 mg/l (0.11%) TSS
200 g/kg (20%) sand	900 mg/l (0.09%) TOC
40 g/kg (4%) free FeO	

The relative concentration or attenuation factor of interest is $[Cd]_{ss}/[Cd]_i = 3/15 = 0.2$. Substituting the corresponding values for the empirical parameters from Table 2.12-7 into equation 17:

$$\begin{aligned} V_{0.2} &= (1.5/25)\ [(31.07/30) - (0.227)(20) + (0.00114)(20)^2 + \\ &\quad (0.0111)(4)^2 + (9.385/4) + (89.19)(0.11) - \\ &\quad (216.4)(0.11)^2 + (0.37)(0.09) - 2.015] \\ &= 0.281 \text{ cm/day} \end{aligned}$$

The time for the relative solution concentration of 0.2 to reach a depth of 5 meters is therefore 1780 days or 4.87 years under steady-state flow.

2.12.8 Estimation of Attenuation in an Environmental System

There are three approaches to estimating the attenuation of inorganic pollutants in soil: (1) direct experimentation, (2) estimation based on experiments with other soils, and (3) calculation of "boundaries" based on particular mechanisms. Except when using the direct experimental approach, one must consider each of the important factors that affect attenuation in the environment.

- **Approach 1** — Perform batch, column or field tests of attenuation, using solutions and solids that most closely resemble those of concern. (See § 2.12.4.)

- **Approach 2**

  (a) Determine soil and soil-solution properties (physical and chemical) plus other liquor (leachate) properties that affect attenuation. (See Table 2.12-1 and Appendix C of this report.)

3. This example is taken from Amoozegar-Fard, Fuller and Warrick [2].

(b) Determine species of interest in solution. The speciation of an element is affected by the solution equilibria; see section on each element in Part II of this report and relevant chapters in Part I.

(c) Review literature for similar conditions, referring to section on each element in Part II of this report as a start.

(d) Pick some values for similar conditions. Calculate attenuation that may be expected.

This approach is illustrated in more specific terms in Example 7 below.

- **Approach 3** — Determine the boundaries of expected attenuation on the basis of calculations for the following possible contributing processes:

  (a) Precipitation equilibria (determine least soluble salt and its soluble level for the conditions of the soil environment). Refer to Part I, section 2.11.

  (b) Ion exchange capacity (for soil CEC or components of soil).

  (c) Linear, Langmuir, Freundlich, or other empirical equation.

  It is also possible to assume that a single component of the soil, such as manganite, is responsible for most of the attenuation. The properties of this material as a sorbent may then be used to calculate the attenuation expected as a boundary.

  Example 8 illustrates the use of Approach 3 in more specific terms.

Example 7 Estimate the attenuation of cadmium using Approach 2.

(a) Measure soil and soil-solution properties. Refer to Table 2.12-1.

(b) The properties and environmental behavior of Cd are discussed in section 7.5 (Part II) of this report. Refer to relevant portions of Part I for more information, particularly on complexation and solubility/precipitation, which appear to be important processes affecting attenuation of Cd.

(c) Refer to section 7.5.4 on the attenuation of Cd in soils and sediments and the literature cited there. (Additional sources of information are available.)

(d) If conditions are similar to those of the Pennsylvania forest soil found by Sidle and Kardos [41] as described in Examples 2-5 of section 2.12.6, Langmuir and Freundlich parameters are available. As shown in these examples, S = 0.02 μmol/g sorbed (Langmuir) and 0.17 μmol/g sorbed (Freundlich). While the Freundlich value is 8.5 times that of the Langmuir, this range is not too great to prevent good estimates of the probable attenuation.

Example 8 Estimate the attenuation of cadmium using Approach 3.

Precipitation — Section 2.11 presents examples and calculations of precipitation equilibria.

Ion Exchange — Suppose the soil has a CEC of ~10 meq/100g of soil. For a crude estimate, assume that all the CEC is taken by the Cd ions: 0.1 meq of Cd^{+2} would then be sorbed by 1 g of soil, which is equal to 50 μmol/g.

Distribution Coefficient — Ranges of distribution coefficients for soils were not found for Cd in the sources reviewed. However, for montmorillonite (Na-form) in the concentration range under consideration, values of K_d ranging from 8 to 900 ml/g have been obtained (see Table 2.12-4). Using these values for an equilibrium concentration of 10^{-6} *M* (or 1.124×10^{-4} g/l),

$$\text{Lower range: } S = K_dC = 8 \text{ ml/g} \times 1.124 \times 10^{-4} \text{ g/l} = 8.992 \times 10^{-7} \text{ g/g or } 0.9\ \mu\text{g/g or } 0.008\ \mu\text{mol/g}$$

$$\text{Upper range: } S = 900 \text{ ml/g} \times 1.124 \times 10^{-4} \text{ g/l} = 1011.6 \times 10^{-7} \text{ g/g or } 101.2\ \mu\text{g/g or } 0.9\ \mu\text{mol/g}$$

Langmuir — Rai *et al.* [34] reported many data values for soils using the Langmuir isotherm. Some of the data were for sorbate concentrations much higher than 10^{-6} *M*, and these may be eliminated to provide a reasonable range for our purpose. The range of Langmuir constants for Cd in soils for sorbate concentrations around 10^{-6} *M* are: A_m = 0.01 – 89 μmol/g and $K_L = 10^3 - 10^{7.1}$ l/mol or 0.001 – 12.6 l/μmol. To calculate sorption for an equilibrium soil solution concentration of 10^{-6} *M* (1 μmol/l), we use equation 3:

$$S = K_LA_mC / [1 + K_LC]$$

$$\text{Lower range: } S = [0.001 \times 0.01 \times 1]/[1 + (0.001 \times 1)] = 10^{-5}\ \mu\text{mol/g}$$

$$\text{Upper range: } S = [12.6 \times 89 \times 1]/[1 + (12.6 \times 1)] = 82\ \mu\text{mol/g}$$

Freundlich — Garcia-Miragaya [13] reported Freundlich constants for four soils containing 0.72-16.3% organic matter and a CEC of 23.8-60 meq/100g (see Table 2.12-6). These values are K_F = 3.6 – 23.2 l/g and 1/N = 0.63 – 0.95. We now use equation 13 to calculate sorption for an equilibrium soil solution concentration of 10^{-6} *M* (1 μmol/l):

$$S = K_FC^{1/N}$$

$$\text{Lower range: } S = 3.6 \times (1)^{0.63} = 3.6\ \mu\text{mol/g}$$

$$\text{Upper range: } S = 23.2 \times (1)^{0.95} = 23.2\ \mu\text{mol/g}$$

Fuller and Co-workers [2] — For cadmium, equation 17 may also be used (see Example 6).

Summarizing the above results,

	Ion Exchange	Distribution Coeff.	Langmuir	Freundlich
S (μmol/g)	50	0.008-0.9	10^{-5}-82	3.6-23.2

The range obtained from the above calculations is very large ($82:10^{-5}$ = 8,200,000:1), because wide ranges of soils, equilibrium concentrations, pHs, etc., were included in the data values selected. If only one main component of the soil is responsible for sorption, the boundaries of attenuation may be calculated using a narrower range of values for sorption constants.

2.12.9 Summary

Attenuation is a complex phenomenon consisting of many different processes and reactions within the environmental system. Estimation methods based on mechanistic considerations are extremely limited by their mathematical complexity and applicability to "real" systems. Empirical methods are more applicable to "real" systems, but there are no methods for estimating the associated parameters. The "estimation" methods consist of direct measurement, obtaining data for soil environments similar to the one under consideration, or obtaining ranges by postulating mechanisms for attenuation for the system under consideration.

2.12.10 Literature Cited

1. American Society for Testing and Materials, *Emergency Standard Method for 24-Hour Batch-Type Distribution Ratio (R_d) for Contaminant Sorption by Soils and Sediments*, ASTM ES10-85, Philadelphia, Pa. (1985).
2. Amoozegar-Fard, A., W.H. Fuller and A.W. Warrick, "An Approach to Predicting the Movement of Selected Polluting Metals in Soils," *J. Environ. Qual.*, **13**, 290-96 (1984).
3. Baes, C.F., Jr. and R.E. Mesmer, *The Hydrolysis of Cations*, John Wiley & Sons, New York (1976).
4. Boyd, G.E., J. Schubert and A.W. Adamson, "The Exchange Adsorption of Ions from Aqueous Solutions by Organic Zeolites: I: Ion Exchange Equilibria," *J. Am. Chem. Soc.*, **69**, 2818-29 (1947), as cited by Travis and Etnier [46].
5. Cavallaro, N. and M.B. McBride, "Copper and Cadmium Adsorption Characteristics of Selected Acid and Calcareous Soils," *Soil Sci. Soc. Am. J.*, **42**, 550-56 (1978).
6. Conway, R.A. and B.C. Malloy (eds.), *Hazardous Solid Waste Testing: First Conference*, A Symposium Sponsored by ASTM Committee D-34 on Waste Disposal, American Society for Testing and Materials, Ft. Lauderdale, Fla., Jan. 14-15, 1981, ASTM Special Technical Publication 760, ASTM Publication Code No. 04-760000-16, Philadelphia, Pa. (1981).

7. DeGroot, A.J., J.J.M. DeGoeij and C. Zegers, "Contents and Behavior of Mercury as Compared with Other Heavy Metals in Sediments from the Rivers Rhine and Ems," *Geol. Mijnbouw*, **50**, 393 (1971), as cited by Reimers *et al.* [37].
8. DiToro, D.M., J.D. Mahony, P.R. Kirchgraber, A.L. O'Byrne, L.R. Pasquale and D.C. Piccirilli, "Effects of Nonreversibility, Particle Concentration, and Ionic Strength on Heavy Metal Sorption," *Environ. Sci. Technol.*, **20**, 55-61 (1986).
9. Fuller, W.H., *Movement of Selected Metals, Asbestos, and Cyanide in Soil: Applications to Waste Disposal Problems*, EPA-600/2-77-020, NTIS PB 266 905, Municipal Environmental Research Laboratory, Office of Research and Development, U.S. EPA (1977).
10. Fuller, W.H., *Investigation of Landfill Leachate Pollutant Attenuation by Soils*, EPA-600/2-78-158, Municipal Environmental Research Laboratory, Office of Research and Development, U.S. EPA (1978).
11. Fuller, W.H., A. Amoozegar-Fard, E.E. Niebla and M. Boyle, "Behavior of Cd, Ni, and Zn in Single and Mixed Combinations in Landfill Leachates," in *Land Disposal: Hazardous Waste*, D. Shultz (ed.), Proc. Seventh Annual Research Symposium, EPA-600/9-81-002b, Municipal Environmental Research Laboratory, Office of Research and Development, U.S. EPA, 18-28 (1981).
12. Fuller, W.H., A. Amoozegar-Fard, E.E. Niebla and M. Boyle, "Influence of Leachate Quality on Soil Attenuation of Metals," in *Disposal of Hazardous Waste*, D. Shultz (ed.), Proc. Sixth Annual Research Symposium, EPA-600/9-80-010, Municipal Environmental Research Laboratory, Office of Research and Development, U.S. EPA, 108-17 (1980).
13. Garcia-Miragaya, J., "Specific Sorption of Trace Amounts of Cadmium by Soils," *Commun. Soil Sci. Plant Anal.*, **2**, 1157-66 (1980), as cited by Rai *et al.* [34].
14. Gerritse, R.G., W. Van Driel, K.W. Smilde and B. Van Luit, "Uptake of Heavy Metals by Crops in Relation to Their Concentration by Soil Solution," *Plant Soil*, **75**, 393-404 (1983).
15. Griffin, R.A. and A.K. Au, "Lead Adsorption by Montmorillonite Using a Competitive Langmuir Equation," *Soil Sci. Soc. Am. J.*, **41**, 880-82 (1977), as cited by Travis and Etnier [46].
16. Harmsen, K., *Behaviour of Heavy Metals in Soils*, Agricultural Research Reports 866, Center for Agricultural Publishing and Documentation, Wageningen, The Netherlands (1977).
17. Harter, R.D. and D.E. Baker, "Applications and Misapplications of the Langmuir Equation to Soil Adsorption Phenomena," *Soil Sci. Soc. Am. J.*, **41**, 1077-80 (1977), as cited by Travis and Etnier [46].
18. Hingston, F.J., "A Review of Anion Adsorption," in *Adsorption of Inorganics at Solid-Liquid Interfaces*, M.A. Anderson and A.J. Rubin (eds.), Ann Arbor Science Publishers, Ann Arbor, Mich., 51-90 (1981).
19. James, R.O. and T.W. Healy, "Adsorption of Hydrolyzable Metal Ions at the Oxide-Water Interface. I: Co(II) Adsorption on SiO_2 and TiO_2 as Model Systems," *J. Colloid Interface Sci.*, **40**, 42-52 (1972), as cited by Kinniburgh and Jackson [23].

20. Jardine, P.M., J.C. Parker and L.W. Zelazny, "Kinetics and Mechanisms of Aluminum Adsorption on Kaolinite Using a Two-Site Nonequilibrium Transport Model," *Soil Sci. Soc. Am. J.*, **49**, 867-72 (1985).
21. Jenne, E.A., "Controls on Mn, Fe, Co, Ni, Cu, and Zn Concentrations in Soils and Water: The Significant Role of Hydrous Mn and Fe Oxides," *Adv. Chem. Ser.*, **73**, 337-87 (1968), as cited by Harmsen [16].
22. Kinniburgh, D.G. and M.L. Jackson, "Adsorption of Mercury (II) by Iron Hydrous Oxide Gel," *Soil Sci. Soc. Am. J.*, **42**, 45-47 (1978), as cited by Rai *et al.* [34].
23. Kinniburgh, D.G. and M.L. Jackson, "Cation Adsorption by Hydrous Metal Oxides and Clays," in *Adsorption of Inorganics at Solid-Liquid Interfaces*, M.A. Anderson and A.J. Rubin (eds.), Ann Arbor Science Publishers, Ann Arbor, Mich., 91-160 (1981).
24. Kinniburgh, D.G., M.L. Jackson and J.K. Syers, "Adsorption of Alkaline Earth, Transition, and Heavy Metal Cations by Hydrous Oxide Gels of Iron and Aluminum," *Soil Sci. Soc. Am. J.*, **40**, 796-99 (1976), as cited by Kinniburgh and Jackson [23].
25. Lapidus, L. and N.R. Amundson, "Mathematics of Absorption in Beds. VI: The Effect of Longitudinal Diffusion in Ion Exchange and Chromatographic Columns," *J. Phys. Chem.*, **56**, 984-88 (1952), as cited by Amoozegar-Fard, Fuller and Warrick [2].
26. McBride, M.B. and M.M. Mortland, "Copper(II) Interactions with Montmorillonite: Evidence from Physical Methods," *Soil Sci. Soc. Am. Proc.*, **38**, 408-15 (1974), as cited by Harmsen [16].
27. Morel, F.M.M., *Principles of Aquatic Chemistry*, John Wiley & Sons, New York (1983).
28. Morel, F.M.M., J.C. Westall and J.G. Yeasted, "Adsorption Models: A Mathematical Analysis in the Framework of General Equilibrium Calculations," in *Adsorption of Inorganics at Solid-Liquid Interfaces,* M.A. Anderson and A.J. Rubin (eds.), Ann Arbor Science Publishers, Ann Arbor, Mich., 263-94 (1981).
29. Oakley, S.M., P.O. Nelson and K.J. Williamson, "Model of Trace-Metal Partitioning in Marine Sediments," *Environ. Sci. Technol.*, **15**, 474-80 (1981).
30. O'Connor, T.P. and D.R. Kester, "Adsorption of Copper and Cobalt from Fresh and Marine Systems," *Geochim. Cosmochim. Acta*, **39**, 1531-43 (1975).
31. Organization for Economic Co-Operation and Development, *OECD Guidelines for Testing of Chemicals*, OECD, Paris (1981).
32. Pagenkopf, G.K., *Introduction to Natural Water Chemistry*, Marcel Dekker, New York (1978).
33. Pavlou, S.P. and D.P. Weston, "Initial Evaluation of Alternatives for Development of Sediment Related Criteria for Toxic Contaminants in Marine Waters (Puget Sound). Phase II: Development and Testing of the Sediment-Water Equilibrium Partitioning Approach," Report No. EPA 910/0-83-117, U.S. EPA, Washington, D.C. (1984).

34. Rai, D., J. Zachara, A. Schwab, R. Schmidt, D. Girvin and J. Rogers, *Chemical Attenuation Rates, Coefficients, and Constants in Leachate Migration. Vol. I: A Critical Review*, Report EA-3356 to EPRI by Pacific Northwest Laboratories (Battelle Institute), Richland, Wash. (1984).

35. Ravina, L. and Y. Gur, "Application of the Electrical Double Layer Theory to Predict Ion Adsorption in Mixed Ionic Systems," *Soil Sci.*, **125**, 204-9 (1978).

36. Reimers, R.S. and P.A. Krenkel, "Kinetics of Mercury Adsorption and Desorption in Sediments," *J. Water Pollut. Control Fed.*, **46**, 352-65 (1974), as cited by Reimers *et al.* [37].

37. Reimers, R.S., P.A. Krenkel, M. Eagle and G. Tragitt, "Sorption Phenomenon in the Organics of Bottom Sediments," in *Heavy Metals in the Aquatic Environment*, P.A. Krenkel (ed.), Pergamon Press, Oxford, England, 117-29 (1975).

38. Schindler, P.W., "Surface Complexes at Oxide-Water Interfaces," in *Adsorption of Inorganics at Solid-Liquid Interfaces*, M.A. Anderson and A.J. Rubin (eds.), Ann Arbor Science Publishers, Ann Arbor, Mich., 1-49 (1981).

39. Scow, K., M. Byrne, M. Goyer, L. Nelken, J. Perwak, M. Wood and S. Young, *An Exposure and Risk Assessment for Arsenic*, Final Draft Report to U.S. Environmental Protection Agency, EPA Contract 68-01-6160, Monitoring and Data Support Division, Office of Water Regulations and Standards, Washington, D.C. (1981).

40. Shayan, A. and B.G. Davey, "A Universal Dimensionless Phosphate Adsorption Isotherm for Soil," *Soil Sci. Soc. Am. J.*, **42**, 878-82 (1978), as cited by Travis and Etnier [46].

41. Sidle, R.C. and L.T. Kardos, "Adsorption of Copper, Zinc, and Cadmium by a Forest Soil," *J. Environ. Qual.*, **6**, 313-17 (1977).

42. Sposito, G., "On the Surface Complexation Model of the Oxide-Aqueous Solution Interface," *J. Colloid Interface Sci.*, **91**, 329-40 (1983).

43. Sposito, G., *The Surface Chemistry of Soils*, Oxford University Press, New York (1984).

44. Sposito, G., "Chemical Models of Inorganic Pollutants in Soils," *CRC Crit. Rev. Environ. Control*, **15**, 1-24 (1985).

45. Tewari, P.H., A.B. Campbell and W. Lee, "Adsorption of Co^{2+} by Oxides from Aqueous Solutions," *Can. J. Chem.*, **50**, 1642-48 (1972), as cited by Kinniburgh and Jackson [23].

46. Travis, C.C. and E.L. Etnier, "A Survey of Sorption Relationships for Reactive Solutes in Soil," *J. Environ. Qual.*, **10**, 8-17 (1981).

47. U.S. Environmental Protection Agency, *Chemical Fate Test Guidelines*, EPA-560/6-82-003, Office of Pesticides and Toxic Substances, U.S. EPA (1982).

48. Westall, J. and H. Hohl, "A Comparison of Electrostatic Models for the Oxide/Solution Interface," *Advances in Colloid and Interface Science*, **12**, 265-94 (1980).

2.13 INTEGRATION AND ANALYSIS OF AQUEOUS PROCESSES

2.13.1 Introduction

A variety of chemical processes in aqueous solutions have been described individually in other sections of this chapter. This section demonstrates how some of the methods of estimation and analysis of the results for individual chemical processes can be integrated. In the case study considered here, the individual methods are integrated to permit a partial analysis of the state of an aqueous system—specifically, water from a fly-ash disposal site.

As many environmental processes may affect the behavior of chemicals in an aqueous system, the first steps are to establish the nature and the kinds of the chemical species present, i.e., the chemical speciation in the system. Because of the importance of these steps in any analysis of aqueous systems, the fundamentals of chemical speciation are summarized in section 2.13.2. The theory is simplified for straightforward application, and examples are presented later.

In a typical aqueous system, several processes occur simultaneously. All of them must be considered if one is to understand the chemical state of the system under specified environmental conditions and predict how changes in that environment are likely to affect the system. This is the chief reason for attempting to integrate the methods that apply to the individual chemical processes. The results can be used to examine such aspects of aqueous species as their states of oxidation, rates of oxidation reactions, chemical speciation of the major and trace components of the solution, and the degree of saturation with respect to a number of commonly occurring solids.

2.13.2 Chemical Speciation in Solution

Chemical speciation in solution—i.e., the chemical species that exist in a solution of a given chemical composition under specified environmental conditions—is usually reported as concentrations of elements and some of the more obvious ions. For example, concentrations of sulfate and bicarbonate ions are commonly reported in analyses of water instead of, or in addition to, concentrations of the total sulfur and total carbon in solution.

The cations and anions of the components of the solution can combine to form ionic complexes (section 2.9). A cation M_i and an anion L_j can form an ionic complex M_iL_j according to a reaction of the following type:

$$M_i + L_j \rightleftharpoons M_iL_j \tag{1}$$

In terms of concentrations, the equilibrium of this reaction can be expressed as:

$$K_{ij} = \frac{[M_iL_j]}{[M_i]\,[L_j]} \tag{2}$$

where K_{ij} is the *concentration equilibrium constant* (also called the *stability constant* or *association constant*). If expressed in terms of the activities rather than the concentrations of the dissolved species, K_{ij} is called the *activity equilibrium constant* (section 2.11).

The total concentration of M_i in solution, $[M_{iT}]$, is the sum of the concentrations of the "free" species, $[M_i]$, and of all the complexes involving M_i.[1] For the case of reaction 1, representing one cation, one anion, and one 1:1 ionic complex (one metal to one ligand), the total concentrations of M_i and L_j are as follows:

$$[M_{iT}] = [M_i] + [M_iL_j] \quad (3)$$

$$[L_{jT}] = [L_j] + [M_iL_j] \quad (4)$$

The complex M_iL_j can be either a positively, negatively, or uncharged species, depending on the valence charges of M_i and L_j.

The subscripts i and j are introduced for convenience, to denote each of the many cationic and anionic species in solution. In the presence of one cationic species (M_{1T}) and two different anionic ligands (L_{1T} and L_{2T}), the following equilibria hold for the two different complexes of a 1:1 type:

$$K_{11} = \frac{[M_1L_1]}{[M_1]\,[L_1]} \quad \text{and} \quad K_{12} = \frac{[M_1L_2]}{[M_1]\,[L_2]} \quad (5)$$

The case of calcium fluoride (CaF_2) in its aqueous solution, considered in section 2.11, corresponds closely to a case of one cation (Ca^{+2}) and two ligands (F^- and OH^-), with the complexes formed being CaF^+ and $CaOH^+$.

A 1:1 ionic complex is the simplest variety, but complexes containing more than one stoichiometric unit of the metal (M_i) and/or the ligand (L_j) are also common (e.g., AgI_4^{-3}). In solutions of many different dissolved species, aqueous complexes may contain more than one kind of ligand in the formula (e.g., $Pt(OH)_2Cl_2^{-2}$). A multiligand complex containing more than one L_j in its formula can be represented by the following reactions:

	REACTION	EQUILIBRIUM CONSTANT	
	$M_i + L_j \rightleftharpoons M_iL_j$	$K_{ij} = \beta_{ij.1}$	(6)
	$M_iL_j + L_j \rightleftharpoons M_i(L_j)_2$	$\beta_{ij.2} / \beta_{ij.1}$	
	•		
	•		
	•		
	$M_i(L_j)_{p-1} + L_j \rightleftharpoons M_i(L_j)_p$	$\beta_{ij.p} / \beta_{ij.p-1}$	
Overall reaction:	$M_i + pL_j = M_i(L_j)_p$	$\beta_{ij.p}$	(7)

1. The term "free" species, which actually refers to the aquocomplexed monomeric ion in solution, is used throughout this section for simplicity.

From the preceding notation of the overall stability constants $\beta_{ij.p}$, it is clear that subscript p is the stoichiometric number of the ligands L_j in the formula, and $\beta_{ij.p}$ is defined from equation 7 as

$$\beta_{ij.p} = \frac{[M_i(L_j)_p]}{[M_i]\,[L_j]^p} \tag{8}$$

The number of possible ionic complexes in an aqueous system containing only a few cationic and anionic components can be very large. In a solution containing m metal ions M_i (i=1, ..., m) and n anion-ligands L_j (j=1, ..., n), there are m×n possible ionic complexes of 1:1 type, M_iL_j. There are also m×n theoretically possible complexes of 1:2 type, $M_i(L_j)_2$. The total number of the 1:1, 1:2, and up to 1:p types of complexes rapidly increases to p×m×n. Computations of the concentrations of the individual species in such a system can be very time-consuming, even for a computer (see Chapter 5). Fortunately, some simple approximations make computations of chemical speciation models possible by hand, as will be demonstrated in § 2.13.3. Here, we shall simply summarize the equations for chemical speciation for three cases:

- Free ions and 1:1 complexes;
- Free ions and multiligand 1:p complexes; and
- The somewhat special case in which either the ligand or the cation concentration is much greater than those of the other species. (This case is particularly useful for computing concentrations of the major ligands with trace metals, or of the trace ligands with major cationic components of solutions.)

FREE IONS AND 1:1 COMPLEXES

Equations 3 and 4 described chemical speciation in a solution containing one metal, one anionic ligand, and one complex. A general case for 1:1 complexes can be represented by mathematically similar formulations for m cations M_i, n ligands L_j, and m × n ionic complexes of a 1:1 type, M_iL_j [4]. The total concentration of each of the metals (or cations) in solution, $[M_{iT}]$, is:

$$[M_{iT}] = [M_i]\left(1 + \sum_{j=1}^{n} K_{ij}[L_j]\right) \tag{9}$$

where the notation of the parameters is as defined in equations 3-5. Similarly, the total concentration of each of the ligands in solution, $[L_{jT}]$, is:

$$[L_{jT}] = [L_j]\left(1 + \sum_{i=1}^{m} K_{ij}[M_i]\right) \tag{10}$$

If some of the metals do not form any 1:1 complexes with some of the ligands, the corresponding terms would be zeroes in the summation series in the preceding equations.

In an aqueous system containing m metals and n ligands, there would be m + n simultaneous equations like 9 and 10 above, in m + n unknowns: m unknown concentrations $[M_i]$ and n unknown concentrations $[L_j]$. In general, equations 9 and 10 are nonlinear algebraic equations; methods for their solution are described later in this section under "Mathematical Solutions."

FREE IONS AND MULTILIGAND COMPLEXES

In an aqueous solution of silver iodide, the following ionic complexes and ion pairs may have to be considered: HI^0, AgI^0, AgI_2^-, AgI_3^{-2}, and AgI_4^{-3}. In the absence of any additional information on the total concentrations of the silver and iodide, we should consider all five. The total concentrations of silver $[Ag_T]$ and iodide $[I_T]$ are then given by the following mass-balance equations:

$$[Ag_T] = [Ag^+] + [AgI^0] + [AgI_2^-] + [AgI_3^{-2}] + [AgI_4^{-3}]$$

$$[I_T] = [I^-] + [HI^0] + [AgI^0] + 2[AgI_2^-] + 3[AgI_3^{-2}] + 4[AgI_4^{-3}]$$

In this case, there are two cations (Ag^+ and H^+) and one ligand (I^-) in solution. For any number of metals (m) and ligands (n), and for any type of multiligand complex (1:p), the relationships for the total cation (metal) concentrations $[M_{iT}]$ and total ligand concentrations $[L_{jT}]$ become:

$$[M_{iT}] = [M_i] \left(1 + \sum_{k=1}^{p} \sum_{j=1}^{n} K_{ij.k} [L_j]^k \right) \tag{11}$$

$$[L_{jT}] = [L_j] \left(1 + \sum_{k=1}^{p} k \sum_{i=1}^{m} K_{ij.k} [M_i] [L_j]^{k-1} \right) \tag{12}$$

where, as in the preceding sections, i designates a metal-ion (i = 1, ..., m), j designates an anionic ligand (j = 1, ..., n), and k designates the number of the L_j-ligands in the stoichiometric formula of a complex $M_i (L_j)_k$, with k = 1, ..., p. If some of the complexes $M_i (L_j)_k$ do not exist, the number of terms in the summation series in equations 11 and 12 is less than the total of p × n and p × m, respectively.

LARGE EXCESS OF ONE COMPONENT

In the special case of aqueous systems containing trace metals as well as major metals and anions, concentrations of the trace constituents are often one or more orders of magnitude lower than those of the anionic ligands. The systems of the simultaneous equations 9-10 or 11-12 can be considerably simplified in this case, from nonlinear to linear equations, by using the following approximation: When computing the concentrations of the trace metals with a large excess of a free ligand, one may assume the concentration of the free ligand to be constant and not affected by the formation of complexes with the trace metals; i.e.,

$$[L_j] = \text{constant } (j = 1, \ldots, n) \tag{13}$$

This assumption is justified insofar as only a very small fraction of the free ligand in solution, $[L_j]$, will be taken up by the total concentrations of trace metals, $[M_{iT}]$.

However, the degree of complexation of the trace metals may be large, and the differences between the total trace-metal concentrations $[M_{iT}]$ and their free-ion concentrations $[M_i]$ may be significant. With reference to a system containing free ions and complexes of only the 1:1 type between trace metals and anionic ligands at much higher concentrations, a condition additional to equation 13 is that:

$$K_{ij}[M_i] < < 1, \text{ for all } M_i \text{ forming complexes with all } L_j$$

In this case, chemical speciation of the trace-metal complexes with a ligand L_j can be determined using a set of equations 9—one for each of the m trace metals M_i (i = 1, ..., m) that form 1:1 complexes with the ligands L_j at a constant concentration:

$$[M_{iT}] = [M_i] \left(+ \sum_{j=1}^{n} K_{ij} \, [L_j] \right) \tag{14}$$

When the ligand concentrations, $[L_j]$, are practically constant and not affected by complexation of trace metals occurring at much lower concentrations, there are up to m equations of this kind. Equation 14 can be immediately solved for $[M_i]$ when the values of the total trace-metal concentration $[M_{iT}]$, ligand concentrations $[L_j]$, and stability constants of the 1:1 complexes K_{ij} are known. In this special case the equations are linear, and the unknowns are the concentrations of the free metal-ions in solution.

The preceding case of chemical speciation, summarized in equation 14, refers to an anionic ligand that forms complexes with trace metals. Similarly, the case of a metal ion forming 1:1 complexes with trace ligands L_j can be represented by the following equation:

$$[L_{jT}] = [L_j] \left(1 + \sum_{i=1}^{m} K_{ij} \, [M_i] \right) \tag{15}$$

where the condition $[M_i]$ = constant (i = 1,..., m) holds for the aqueous system.

A variation of this special case is when either the metal or the ligand concentration is constant, but not necessarily much greater than that of the other species. A constant low concentration of an L_j or an M_i may be maintained by a solubility equilibrium with a solid phase. In such a case, equation 14 or 15 can also be used for the aqueous system, as will be shown in the next section.

MATHEMATICAL SOLUTIONS OF THE GENERAL CASES

In the sets of the simultaneous equations 9-10 or 11-12, the unknown parameters are concentrations of the free metal-ions $[M_i]$ and free anionic ligands $[L_j]$. The values of the other parameters — the total concentrations of the metals $[M_{iT}]$ and the ligands $[L_{jT}]$, and the stability constants K_{ij} for the metal-ligand complexes — should be known. Each of the equations 9-10 or 11-12 can be written as a difference between the terms to the right of the equal sign and those to the left. This difference is some algebraic function of the known and the unknown parameters, which can be denoted $F_c(M_i,L_j)$. Thus, the solutions sought are the values of $[M_i]$ and $[L_j]$ from a set of equations

$$F_c(M_i,L_j) = 0 \tag{16}$$

where subscript c indicates the number of the equation, from 1 up to i + j. For example, in an aqueous system containing m = 5 metals (i = 1, ..., 5) and n = 3 ligands (j = 1, 2, 3), computation of the free metal-ion and ligand concentrations requires solution of a system of c=5+3=8 simultaneous equations.

For a dilute aqueous system, when the activity coefficients of the chemical species in solution may be taken as $\gamma \approx 1$, simultaneous nonlinear algebraic equations, such as sets of c equations 15, can be solved with the aid of standardized routines available for many types of minicomputers and mainframe computers. A computer-program library IMSL,[2] available for use with mainframe computers of many manufacturers, contains several routines for solutions of simultaneous nonlinear algebraic equations. One such routine, designated by the code ZSCNT in the IMSL list of programs, was used to verify some of the solutions given in §2.13.3. (The results were practically identical with those obtained by simplified methods, as demonstrated in §2.13.3.)

For aqueous systems at higher concentrations, particularly those involving equilibria with solid phases, there exist in the public domain highly sophisticated programs that can be used to compute concentrations of the individual chemical species in multicomponent aqueous systems, determine degrees of saturation with respect to many solid phases, and estimate changes in the chemical composition of a solution where either precipitation or dissolution of solids takes place. Some of the existing programs have been reviewed and compared by Drever [2] and in Chapter 5 of this report.

METHODS AND STEPS IN INTEGRATION OF CHEMICAL PROCESSES IN SOLUTION

Determination of the chemical speciation of an aqueous solution entails many steps, the first and most basic of which is to ascertain the bulk chemical composition of the solution. The number of steps and the amount of detail in each depend on the desired accuracy of the calculated concentrations of the individual aqueous species. Table 2.13-1 lists the steps for a relatively simple case, with additional references to background information in other sections of this report and in the case study described in section 2.13.3.

---

2. IMSL Library Reference Manuals, published by IMSL Inc.,7500 Bellaire Blvd., Houston, TX 77036.

TABLE 2.13-1

Methods and Steps in Determination of Chemical Speciation of an Aqueous System

Steps to Consider	Comments	Report Sections with Additional Information
(1) Determine bulk chemical composition of solution: a. Dissolved inorganic components	Results of chemical analysis commonly reported as concentrations of elements and some ionic species	
•Major components •Minor or trace components	Concentrations of major components usually 10^2 or more times higher than concentrations of minor components	Chapters 6, 7, 8, Appendix B
b. Dissolved gases	Reactive gases, such as oxygen, carbon dioxide, hydrogen sulfide, and methylated trace-metals.	6.12, 2.15, 10.6, Chapter 8
c. Dissolved organics	Results usually reported as concentrations of organic carbon, nitrogen, and, less commonly, S and P	
d. Hydrogen-ion activity	pH values should characterize the system under environmental conditions	2.7, 6.9
(2) First look at chemical speciation: a. Different oxidation states	For some elements, their different oxidation states may be reported in analysis, such as SO_4^{-2} and S^{-2}, Fe^{+2} and Fe^{+3}, Cr^{+3} and Cr^{+6}	2.10
b. Anionic species	For elements forming anions of polyprotic acids (H_nL), such as P or As, the main anion(s) can be determined from the acid pK_n values and solution pH.	2.7, 6.12, Chapters 7, 8, 10
(3) Cation-anion complexes	List the possible complexes, using information from (1) and (2)	Chapters 6, 7 and 8, 2.7, 2.9, 3.2, 2.13
(4) Determine the most abundant complexes	The most abundant complexes among those in (3) may be identified through use of stability constants K_{ij} and the solution pH, from data in (1) and (2)	2.7, 2.9, 3.2, 2.13

(Continued)

TABLE 2.13-1 (Continued)

Steps to Consider	Comments	Report Sections with Additional Information
(5) Oxidation state of ionic species: a. From Eh or pe measurements	Such measurements often give an indication of either oxidizing or reducing conditions in solution.	2.10
b. From oxidation-reduction equilibria	Computed concentrations or concentration ratios of oxidized and reduced forms of different components of solution.	2.10, 2.13.3
c. From kinetics of oxidation reactions	Results give concentrations of oxidized species, forming from reduced species, computed from reaction-rate equations.	3.3, 3.4, 2.13
(6) Determine ionic strength of solution	Nominal ionic strength can be computed from concentrations of major ionic components, (1). The result indicates how dilute the solution is.	2.6
(7) Calculate speciation of major components	Results are concentrations of aquo-ions and cation-anion complexes—(1), (3) and (4).	2.9, 2.13.3
	For dilute solutions (ionic strength not over 0.02 mol/l), concentrations may be computed by approximation methods.	2.13.2, 2.13.3
	For more concentrated solutions, use detailed computer programs.	2.13.2, Chapter 5
(8) Calculate speciation of minor components	Results are concentrations of aquo-ions and cation-anion complexes involving minor components.	2.9, 7.1-7.15
	For dilute solutions (ionic strength less than about 0.02 mol/l), approximation methods can be used.	2.13.2, 2.13.3
	For solutions of higher ionic strength, use detailed computer programs.	5.4

(Continued)

TABLE 2.13-1 (Continued)

Steps to Consider	Comments	Report Sections with Additional Information
(9) Determine degrees of saturation of solution with respect to solid phases	Results of such computations indicate whether solution is, or is not, supersaturated with respect to solids being considered.	2.11, 3.4, 2.13.3
(10) Estimate response of aqueous system to environmental changes	Changes in the state of the aqueous system (for example, temperature, composition of the gaseous phase, introduction of new components in solution) affect its chemical speciation. Effects of such changes may be estimated by methods listed under (7), (8), and (9).	2.13.3

2.13.3 Case Study of an Aqueous System

This section demonstrates the application of methods developed in other sections of the report to an aqueous system containing a number of major and minor dissolved components. It considers the chemical speciation of some of the main components of the solution, chemical speciation of some trace metals and nutrient elements, the degree of saturation of the aqueous system with respect to some common solids and, finally, possible effects on the system due to addition of other complexing ligands, such as humic acids.

DESCRIPTION OF THE AQUEOUS SYSTEM

The chemical composition of a sample of water taken from a coal fly-ash disposal site is summarized in Table 2.13-2. Certain characteristics of this solution should be particularly noted.

The water is of a calcium-sulfate type, as calcium and sulfate are the main dissolved components of the solution. The solution is slightly alkaline (pH = 8.0) and contains very little dissolved oxygen (0.7 ppm). No concentrations of dissolved carbonate species (HCO_3^- and CO_3^{-2}) are reported. Chloride and nitrate, which are common anionic constituents of natural waters, are present at very low concentrations. Among the numerous trace constituents, 14 elements are listed (alphabetically); another 10 elements are below their detection limits.

The low concentration of dissolved oxygen and the absence of reported data for the bicarbonate ion may indicate that the water was not in contact with the atmosphere.

A dilute solution such as this, at equilibrium with the atmosphere, should contain much more dissolved oxygen (between 8 and 10 ppm), and it should also contain about 10^{-3} mol HCO_3^-/liter and about 10^{-5} mol CO_3^{-2}/liter.

TABLE 2.13-2
Chemical Analysis of Water from a Screened Piezometer at a Coal Fly-Ash Disposal Site

pH = 8.0
Dissolved oxygen = 0.7 ppm O_2 (2.19×10^{-5} mol/l)[a] at 25° C and 1 atm total pressure

Cation	ppm	mol/l[a]
Ca^{+2}	126	3.14×10^{-3}
K^+	11.0	2.81×10^{-4}
Mg^{+2}	10.5	4.32×10^{-4}
Na^+	5.0	2.15×10^{-4}

Anion	ppm	mol/l[a]
SO_4^{-2}	320	3.33×10^{-3}
Cl^-	6.8	1.92×10^{-4}
NO_3^-	1.0	1.61×10^{-5}

Trace Component	ppm	mol/l[a]
As	1.5	2.00×10^{-5}
B	3.7	3.40×10^{-4}
Ba	0.37	2.69×10^{-6}
Cr	0.01	1.92×10^{-7}
F	0.85	4.47×10^{-5}
Fe	0.02	3.58×10^{-7}
Mn	0.06	1.09×10^{-6}
Mo	0.90	9.38×10^{-6}
Ni	0.02	3.41×10^{-7}
P	0.60	1.94×10^{-5}
Se	0.003	3.80×10^{-8}
Si	1.65	5.87×10^{-5}
Sr	3.7	4.22×10^{-5}
V	0.035	6.87×10^{-7}

Trace components at concentrations below detection limits as shown (in ppm)	
Ag	<0.005
Al	<0.3
Be	<0.005
Br	<5.0
Cd	<0.007
Co	<0.05
Cu	<0.008
Pb	<0.05
Zn	<0.02
Zr	<0.05

a. In converting ppm to mol/l, no allowance was made for the difference between 1 liter and 1 kg of solution, owing to its dilute nature.

Source: Arthur D. Little, Inc. [1]

Taking the chemical analysis at face value, the nominal ionic strength of the solution (see section 2.6), computed from the listed concentrations of the cations and anions, is low: $I \approx 0.014$ mol/liter. It will be shown below that about 30% of the calcium, magnesium, and sulfate are tied up as zero-charged ion-pair complexes in solution. Thus, only the remaining 70% are free ions that contribute to the value of the ionic strength, making it somewhat lower than the preceding estimate: $I \approx 0.01$ mol/liter.

OXIDATION STATE OF IRON IN SOLUTION

The total concentration of dissolved iron is listed as 0.02 ppm. At such low concentrations, it is common to report only the total concentration, without identifying the proportions of Fe^{+2} and Fe^{+3}. The proportions of ferrous and ferric iron will be computed in this section.

Because of the tendency of Fe^{+2} and Fe^{+3} ions to hydrolyze in solution, we must first establish the most abundant species of each form of iron in solution at the pH of 8.0. The species to consider are the following (*see also* section 2.11.3):[3]

For Fe(II): Fe^{+2}, $FeOH^{+}$, $Fe(OH)_2^0$

For Fe(III): Fe^{+3}, $FeOH^{+2}$, $Fe(OH)_2^+$, $Fe(OH)_4^-$

From consideration of complexation equilibria between either species of Fe and the hydroxyl ion, using the complex-stability constants listed in Tables 2.13-3 and 6.8-3, the most abundant species in the fly ash water would be Fe^{+2} for Fe(II), and $Fe(OH)_2^+$ for Fe(III).

To determine the main oxidation state of iron in solution, we must compute the concentration ratio of the two main species of dissolved iron:

$$\frac{[Fe(OH)_2^+]}{[Fe^{+2}]} \approx \frac{[Fe\ (III)]}{[Fe\ (II)]}$$

The following oxidation reaction at equilibrium can be used:

$$2Fe^{+2} + 2OH^- + 2H_2O + \tfrac{1}{2}O_2 \rightleftharpoons 2Fe(OH)_2^+ \tag{17}$$

All the species in this reaction are aqueous species, including dissolved oxygen. From the values of the standard free energies of formation of the aqueous species [2, 6], the equilibrium constant of the reaction is found (using the methods of section 2.11.3) to be $\log K = 28.5$.

3. The zero-charged hydroxo complex, $Fe(OH)_3^0$, was excluded from the calculation because of inconsistencies in the open literature on its presence and in the relevant thermodynamic data. Such inconsistencies were noted also for other metal-hydroxo species (e.g., $Al(OH)_3^0$). As such, many published sources do not include them in the speciation calculation although others do depending on their assessment of the validity of the available thermodynamic data. See, for example, Figure 2.11-4, page 2.11-20.

Thus, in a system at pH = 8.0 and containing dissolved oxygen at a concentration of 2.19×10^{-5} mol O_2/liter, the value of the concentration ratio [Fe (III)] / [Fe (II)] is:

$$\frac{[Fe(OH)_2^+]}{[Fe^{+2}]} = \left(K\,[OH^-]^2\,[O_2]^{1/2} \right)^{1/2}$$

$$= \left(10^{28.5} \times (10^{-6.0})^2 \times (10^{-4.66})^{1/2}\right)^{1/2}$$

$$= 10^{7.1}$$

The answer clearly shows that the oxidized form, $Fe(OH)_2^+$, is the expected equilibrium oxidation state of iron in solution.

The preceding computation assumes that an equilibrium exists between Fe (II) and Fe (III) in the fly-ash solution. If this were not the case, and if the main form of dissolved iron were the reduced ion Fe^{+2}, how long would it take the Fe^{+2} to oxidize?

An equation for the rate of oxidation of Fe^{+2} in solution was given in section 3.4 (equation 76), in the following form:

$$-\frac{d[Fe^{+2}]}{dt} = k\,pO_2\,[OH^-]^2\,[Fe^{+2}] \qquad (18)$$

This reaction-rate equation shows that higher partial pressures of oxygen and alkaline conditions increase the rate of oxidation of ferrous iron. The values of the parameters are:[4]

$k = 2 \times 10^{13}\ l^2 \cdot mol^{-2} \cdot atm^{-1} \cdot min^{-1}$ (as given in section 3.4)

$pO_2 = 0.016$ atm

$[OH^-] = 10^{-6.0}$ (from pH = 8.0, given in Table 2.13-2)

Accordingly, equation 18 reduces to

$$-\frac{d\,[Fe^{+2}]}{dt} = k'\,[Fe^{+2}] = 0.32\,[Fe^{+2}]\ \ (mol \cdot liter^{-1} \cdot min^{-1}) \qquad (19)$$

where $k' = 2 \times 10^{13} \times 0.016 \times (10^{-5.7})^2$

4. pO_2 is the partial pressure of oxygen at equilibrium with an aqueous solution containing 2.19×10^{-5} mol O_2/liter, at 25°C and 1 atm total pressure. Based on the equilibrium constant for a reaction between gaseous and aqueous O_2, $pO_2 = 10^{2.86}\,[O_2]$, where pO_2 is in atm and $[O_2]$ in mol/kg H_2O.

The half-life of the reaction can be estimated from equation 5 of section 3.2.3:

$$t_{1/2} = \frac{0.693}{k'} = \frac{0.693}{0.32\,(\text{min}^{-1})} = 2.2 \text{ min} \tag{20}$$

This result shows that the rate of oxidation of Fe^{+2} in the fly-ash leachate is relatively fast (in the absence of complexation); this strengthens the conclusion, based on the oxidation-reduction equilibrium of equation 17, that Fe (III) is the dominant form of dissolved iron in solution.

CHEMICAL SPECIATION OF MAJOR COMPONENTS

The analytical data in Table 2.13-2 indicate that sulfate is the main anionic species in solution and calcium is the main cationic species. Concentrations of Mg^{+2} and K^{+} are about equal, but Mg^{+2} has a much stronger tendency than K^{+} to form ionic complexes with SO_4^{-2}. Therefore, only calcium, magnesium, and sulfate were considered in computing the concentrations of the free ions and metal-sulfate complexes in the fly-ash leachate.

The stability constants of the ionic complexes of the metals and anions dealt with in this section are summarized in Table 2.13-3. In an aqueous solution containing calcium, magnesium and sulfate as the major components, the ionic complexes to be considered are: HSO_4^-, $MgSO_4^0$, $CaSO_4^0$, $MgOH^+$, and $CaOH^+$.

From the values of the stability constants listed in Table 2.13-3, it follows that at the pH value of the solution (8.0), which is much higher than the pK_a value of HSO_4^-, the concentration of HSO_4^- can be neglected. The complexes $MgOH^+$ and $CaOH^+$ may also be neglected, for the following reason: at pH = 8.0, the concentration of the hydroxide ion is 10^{-6} mol/liter, which is much lower than the sulfate concentration of about 3×10^{-3} mol/liter (Table 2.13-2); fractions of the total Ca and Mg in solution that may be complexed by OH^- are therefore much smaller than the fractions complexed by SO_4^{-2}.

Thus, the problem at hand reduces to computing concentrations of three free ions (Ca^{+2}, Mg^{+2}, and SO_4^{-2}) and two ionic complexes ($CaSO_4^0$ and $MgSO_4^0$).

Using equations 9 and 10 and the notation of Table 2.13-3, we can express simultaneous equations for the concentration values of the three free ions:

$$[Ca_T] = [Ca^{+2}]\left(1 + K_{33}[SO_4^{-2}]\right) \tag{21}$$

$$[Mg_T] = [Mg^{+2}]\left(1 + K_{23}[SO_4^{-2}]\right) \tag{22}$$

$$[SO_{4T}] = [SO_4^{-2}]\left(1 + K_{23}[Mg^{+2}] + K_{33}[Ca^{+2}]\right) \tag{23}$$

TABLE 2.13-3

Stability Constants of Aqueous Ion-Complexes

(Tabulated values are log K_{ij}, where i designates metal ion M_i and j designates ligand L_j. Most values are for ionic strength I=0 and 25°C. Stability constants are for equilibrium reactions of the type $M_i + L_j = M_iL_j$, where a 1:1 complex is formed.)

i	M_i	j= 1 L_j= OH^-	2 HPO_4^{-2}	3 SO_4^{-2}
1	H^+	14.0	7.2	2.0
2	Mg^{+2}	2.58	2.91	2.23
3	Ca^{+2}	1.30	2.74	2.31
4	Ba^{+2}	0.60	—	2.70
5	Ni^{+2}	4.10	2.08	2.32
6	Fe^{+2}	4.50	3.60	2.20
7	Fe^{+3}	11.81[a]	8.30	4.04

a. For the reaction $Fe^{+3} + 2OH^- \rightleftharpoons Fe(OH)_2^+$, log $\beta_{71.2} = 22.3$ (for $\beta_{71.2}$ see equation 8). See also footnote 3.

Source: Kotrlý and Šůcha [3]

After inserting the numerical values of the total concentrations (from Table 2.13-2) and stability constants (Table 2.13-3) in these equations, we solve equation 21 for $[Ca^{+2}]$ and equation 22 for $[Mg^{+2}]$. When the resulting expressions are substituted in equation 23, the following third-degree equation is obtained:

$$[SO_4^{-2}]^3 + 1.103\times10^{-2}\,[SO_4^{-2}]^2 + 1.355\times10^{-5}\,[SO_4^{-2}] - 9.604\times10^{-8} = 0 \qquad (24)$$

Equation 24 can be solved by trial and error[5], and the resulting value of $[SO_4^{-2}]$ is then substituted in equations 21 and 22 to find the concentrations of the other two species. The results of this solution are:

$[SO_4^{-2}] = 2.23 \times 10^{-3}$ mol/l; 67% free ion

$[Ca^{+2}] = 2.16 \times 10^{-3}$ mol/l; 69% free ion

$[Mg^{+2}] = 3.13 \times 10^{-4}$ mol/l; 72% free ion

The remainder of the total calcium and magnesium is thus present as sulfate ion pairs for this solution; their respective fractions of the total concentration are 100% minus the free ion concentration:

$[CaSO_4^0] = 31\%$

5. That is, guessing a value of $[SO_4^{-2}]$ and subsequently adjusting it until the left-hand terms total zero.

$$[MgSO_4^0] = 28\%$$

These values lead to the following concentrations of the ion pairs:

$$[CaSO_4^0] = 0.31\ (3.14 \times 10^{-3}) = 9.7 \times 10^{-4}$$

$$[MgSO_4^0] = 0.28\ (4.32 \times 10^{-4}) = 1.2 \times 10^{-4}$$

MOST ABUNDANT SPECIES OF IRON AND NICKEL

The procedure used to determine the most abundant chemical species of ferric iron and nickel in the fly-ash solution is similar to that described above. The mass-balance equation for the total dissolved iron is:

$$[Fe_T] = [Fe(OH)_2^+] + FeHPO_4^+ + FeSO_4^+$$

$$= \beta_{71.2}[Fe^{+3}]\,[OH^-]^2 + K_{72}[Fe^{+3}]\,[HPO_4^{-2}] + K_{73}[Fe^{+3}]\,[SO_4^{-2}]$$

Using the values of $[Fe_T]$, $[OH^-]$ and $[SO_4^{-2}]$ from the preceding subsections, a value of $10^{-5.1}$ for $[HPO_4^{-2}]$ (calculated as 7.95×10^{-6} in the following subsection), and stability constants K_{ij} from Table 2.13-3, the solution of the mass-balance equation in terms of $[Fe^{+3}]$ is

$$[Fe^{+3}] = \frac{3.58 \times 10^{-7}}{10^{22.3}\,(10^{-6.0})^2 + 10^{8.3} \times 10^{-5.1} + 10^{4.04} \times 10^{-2.65}}$$

$$= 1.79 \times 10^{-17}\ \text{mol/liter}$$

Concentrations of the three iron-ligand complexes are, from the above result:

$$[Fe(OH)_2^+] = 3.58 \times 10^{-7}\ \text{mol/liter}$$

$$[FeHPO_4^+] \approx 0$$

$$[FeSO_4^+] \approx 0$$

Thus, practically all of the dissolved iron occurs in the form of the hydroxide complex, $Fe(OH)_2^+$.

For nickel in the fly-ash solution, the mass-balance equation can be written to include the following aqueous species (see also section 7.11):

$$[Ni_T] = [Ni^{+2}] + [NiOH^+] + [NiHPO_4^0] + NiSO_4^0]$$

$$= [Ni^{+2}]\left(1 + K_{51}[OH^-] + K_{52}[HPO_4^{-2}] + K_{53}[SO_4^{-2}]\right) \qquad (25)$$

This equation is now solved in terms of $[Ni^{+2}]$ as an unknown, substituting the values of the other parameters from Table 2.13-2, Table 2.13-3, and the preceding parts of this section:

$$[Ni^{+2}] = \frac{3.41 \times 10^{-7}}{1 + 10^{4.1} \times 10^{-6.0} + 10^{2.08} \times 10^{-5.1} + 10^{2.32} \times 10^{-2.65}} \tag{26}$$

$$= \frac{3.41 \times 10^{-7}}{1.48} = 2.30 \times 10^{-7} \text{ mol/liter}; \qquad 67\% \text{ of total Ni is free ion}$$

The concentrations of the three Ni-ligand complexes that appear in the mass-balance equation (25) can now be calculated:

$$[NiOH^{+}] = 5.74 \times 10^{-9} \text{ mol/liter}; \qquad 2\% \text{ of total Ni in solution}$$

$$[NiHPO_4^0] = 2.18 \times 10^{-10} \text{ mol/liter}; \qquad < 0.1\% \text{ of total Ni in solution}$$

$$[NiSO_4^0] = 1.07 \times 10^{-7} \text{ mol/liter}; \qquad 31\% \text{ of total Ni in solution}$$

Thus, the free Ni^{+2} ion is the most abundant form in the fly-ash solution, followed by the sulfate complex, with the hydroxide a distant third.

COMPLEXATION OF PHOSPHORUS

The total concentration of dissolved phosphorus in the fly-ash solution is 0.6 ppm, or 1.94×10^{-5} mol P/liter. In the absence of any further information, P will be assumed to be present as orthophosphate. At the pH value of the fly-ash solution (8.0), the main phosphate species is HPO_4^{-2} (see Figure 8.3-2, page 8.3-4). Examination of the stability constants of the metal-phosphate complexes (Table 2.13-3) shows that strong complexes exist with all the metal-ion species (except barium, for which no data were available).

A mass balance for total phosphorus in solution can be written as a sum of concentrations of the free ion $[HPO_4^{-2}]$ and of the phosphate complexes of the metals listed in Table 2.13-3:

$$[P_T] = [HPO_4^{-2}] + [CaHPO_4^0] + [MgHPO_4^0] + [NiHPO_4^0] + [FeHPO_4^{+}] \tag{27}$$

As shown previously, Fe^{+3} rather than Fe^{+2} is the stable form of iron in solution; therefore, the ferrous iron-phosphate complex, $FeHPO_4^0$, is not included in equation 27.

Next, examining the concentration values of the metals in solution (Table 2.13-2), it should be noted that the total phosphate concentration, 10^{-5} mol/liter, is much higher than the concentration of the free Ni^{+2} ion (10^{-7} mol/liter) and free Fe^{+3} ion (10^{-18} mol/liter) as determined in the preceding subsection. As a result, the fraction of the dissolved phosphate that may be complexed by Ni and Fe in solution would be very small in comparison to its total concentration. Thus, as an approximation, we can ignore the terms $[NiHPO_4^0]$ and $[FeHPO_4^+]$ in equation 27. Using the notation in equations 9 and 10 and Table 2.13-3, the mass balance becomes:

$$[P_T] = [HPO_4^{-2}]\left(1 + K_{32}[Ca^{+2}] + K_{22}[Mg^{+2}]\right) \tag{28}$$

The free ionic concentrations of Ca^{+2} and Mg^{+2} calculated earlier are of the order of 10^{-3} and 10^{-4} mol/liter, both much higher than the dissolved phosphate concentration, 10^{-5} mol/liter. This corresponds to a special case of a solution containing one of the components in excess of another, discussed in §2.13.2. Therefore, equation 28 may be used alone, without two additional equations for $[Ca^{+2}]$ and $[Mg^{+2}]$. Substitution of the calculated values of $[Ca^{+2}]$ and $[Mg^{+2}]$, the K_{ij} values from Table 2.13-3 and $[P_T] = 1.94 \times 10^{-5}$ mol/liter in equation 28 gives the free phosphate-ion concentration sought:

$$[HPO_4^{-2}] = \frac{1.94 \times 10^{-5}}{1 + 10^{2.74} \times 2.16 \times 10^{-3} + 10^{2.91} \times 3.13 \times 10^{-4}}$$

$$= 7.95 \times 10^{-6} \text{ mol/l; 41\% of total P is free } HPO_4^{-2} \text{ ion}$$

The remaining 59% of total dissolved phosphorus is complexed by calcium and magnesium in the fly-ash solution. This relatively high fraction reflects the tendency of phosphate to form strong complexes with calcium and magnesium, as indicated by the values of their stability constants (Table 2.13-3).

DEGREE OF SATURATION OF THE AQUEOUS SYSTEM

The presence in the fly-ash solution of such components as barium, sulfate, ferric iron, phosphate, and hydroxide ions poses a question of whether the aqueous system is, or is not, supersaturated with respect to such poorly soluble solids as $BaSO_4$, $Fe(OH)_3$, and ferric phosphate.

To determine the degree of saturation of the solution with respect to each of the three solids, the criteria of the ion-activity product (IAP) and of the ion-concentration product (ICP) will be used, as explained in section 2.11.

Barium Sulfate: Barite, $BaSO_4(s)$

Concentrations of the free Ba^{+2} ion and of the complex $BaSO_4^0$ in the fly-ash solution can be computed from the following equation (using data in Tables 2.13-2 and -3):

$$[Ba_T] = [Ba^{+2}]\left(1 + K_{43}[SO_4^{-2}]\right)$$

$$[Ba^{+2}] = \frac{2.69 \times 10^{-6}}{1 + 10^{2.7} \times 2.23 \times 10^{-3}}$$

$$= 1.27 \times 10^{-6} \text{ mol/liter; 47\% of total Ba is free ion}$$

$$[BaSO_4^0] = 1.42 \times 10^{-6} \text{ mol/liter; } 53\% \text{ of total Ba is complexed}$$

The ion-activity product (IAP) of the barium and sulfate ions in solution is:

$$IAP = [Ba^{+2}]\,[SO_4^{-2}]\,\gamma_{Ba}\,\gamma_{SO_4} \tag{29}$$

The activity coefficients (γ) of the Ba^{+2} and SO_4^{-2} ions must be determined at the ionic strength of the fly-ash solution ($I \approx 0.01$ mol/liter). This is done by using the extended Debye-Hückel equation (section 2.6), yielding $\gamma_{Ba} = 0.67$ and $\gamma_{SO_4} = 0.66$. Thus, the IAP for barium sulfate in the fly-ash solution is,

$$IAP = 1.27 \times 10^{-6} \times 2.23 \times 10^{-3} \times 0.67 \times 0.66$$

$$= 1.25 \times 10^{-9}$$

From Kotrlý and Šůcha [3], the solubility product of $BaSO_4(s)$ at 25° C and I = 0 is $K = 1.1 \times 10^{-10}$. Therefore, the degree of saturation of the solution with respect to the solid phase is

$$\frac{IAP}{K} = \frac{1.25 \times 10^{-9}}{0.11 \times 10^{-9}} = 11$$

A degree of saturation greater than unity indicates that the solution is supersaturated with respect to the solid phase. Therefore, one can conclude that barium sulfate may precipitate from the solution. Table 2.13-2 does not indicate the presence of any other anionic ligands that may form strong ionic complexes with Ba^{+2}, a process that could lower the free Ba^{+2} ion concentration and, consequently, make the fly-ash solution undersaturated with respect to solid $BaSO_4$. (A condition of undersaturation is $IAP/K < 1$.)

Ferric Hydroxide: Amorphous Solid, $Fe(OH)_3(s)$

To determine the degree of saturation of the solution with respect to a ferric hydroxide precipitate, a reaction between the most abundant aqueous species of iron, $Fe(OH)_2^+$, and the hydroxyl ion will be considered:

$$Fe(OH)_3(s) = Fe(OH)_2^+ + OH^- \tag{30}$$

The equilibrium constant of this reaction at 25°C, computed from the standard free energy data [6], is $K = 10^{-17.7}$ (*see also* section 2.11.3).

The ion-concentration product (ICP) of the ions in solution, as given in equation 30, is:

$$ICP = [Fe(OH)_2^+][OH^-] = 10^{-6.45} \times 10^{-6.0}$$

$$= 10^{-12.45}$$

The degree of saturation of the solution with respect to $Fe(OH)_3(s)$ is equal to the ratio ICP/K (section 2.11):

$$\frac{\text{ICP}}{\text{K}} = \frac{10^{-12.45}}{10^{-17.7}} = 1.8 \times 10^5$$

The fact that ICP/K is much greater than 1 indicates that the fly-ash solution is strongly supersaturated with respect to the solid ferric hydroxide phase.[6]

Ferric Phosphate: Strengite, $FePO_4{\cdot}2H_2O$ (s)

Dissociation of ferric phosphate (mineral strengite) in an aqueous solution can be represented by the following reaction at equilibrium:

$$FePO_4{\cdot}2H_2O\ (s) = Fe(OH)_2^+ + HPO_4^{-2} + H^+ \tag{31}$$

From the standard free energy of formation data [6], the equilibrium constant of this reaction is $K = 10^{-23.76}$. The ion-concentration product of the species in solution, as given in the right-hand side of equation 31, is

$$\frac{\text{ICP}}{\text{K}} = \frac{10^{-19.85}}{10^{-23.76}} = 8 \times 10^3$$

As ICP/K is much greater than 1, the solution is strongly supersaturated with respect to the hydrous ferric phosphate, as well as with respect to ferric hydroxide, dealt with in the preceding computation.

EFFECT OF HUMIC ACIDS ON COMPLEXATION OF METALS

Humic acids are organic substances of complex composition and high molecular weight that occur in soils, sediments, and in ground and surface waters. Humic acids can form strong complexes with heavy metals and with polyvalent cations in aqueous solutions. From the extensive characteristics of humic substances detailed by Stevenson [5], a few that are relevant to the complexation of metals are summarized below.

Molecular weight:	20,000 to 100,000
Carbon content (% dry basis):	50-60
Oxygen content (in COOH, CO, and OH groups; % dry basis):	30-35
Stability constants of 1:1 complexes (with Cu and Ni):	$K_{11} \approx 10^8$ near pH=8; $K_{11} \approx 10^4$ near pH=5

6. Although this result is somewhat surprising, extreme supersaturation of $Fe(OH)_3$ has beeen observed in laboratory systems [7]. It is also possible that the original Fe_T analysis included dispersed colloidal and dissolved iron.

Acid dissociation constants
(K_a), dilute solutions: $pK_a \approx 5.0 \pm 0.5$

Although the chemical analysis of the fly ash leachate (Table 2.13-1) shows no dissolved organic carbon in the water sample, a possible effect on complexation of the trace metals in this water will be demonstrated by considering a hypothetical case of small amounts of dissolved carbon in the system. Organic carbon, in the form of humic substances, may appear in the fly-ash leachate from its contact with a soil, or from its mixing with other surface or ground waters. The large reservoir of organic matter in the soil may provide a low constant concentration of dissolved organic matter in the soil waters.

For the humic acids in the fly-ash leachate, it will be assumed that

(a) the solution contains 1.2 mg/liter dissolved organic carbon (1×10^{-4} mol C/liter),

(b) all the dissolved organic carbon is contained in a humic acid of molecular weight 30,000, and

(c) the humic acid contains, by weight, 55% total C and 35% O, with all of the oxygen contained in the acidic COOH groups.

From the preceding assumptions, the concentration of dissolved organic C attributable to the complex-forming groups (COOH) of the humic acid is found as follows:

$$\text{Moles of O in humic acid} = \frac{(\text{fraction O})(\text{m.w. of humic acid})}{\text{m.w. of O}} = \frac{(0.35)(30{,}000)}{16} = 656$$

Moles of O in humic acid $= 2 \times$ moles of $-$COOH in humic acid; thus,

$$\text{moles of COOH} = \frac{\text{moles of O}}{2} = \frac{656}{2} = 328$$

$$\text{Moles of total C in acid} = \frac{(\text{fraction C})(\text{m.w. humic acid})}{\text{m.w. C}} = \frac{(0.55)(30{,}000)}{12} = 1375$$

$$\text{Fraction of total C present as COOH} = \frac{\text{moles of COOH}}{\text{moles of total C}} = \frac{328}{1375} = 0.24$$

Concentration of COOH units in solution that can complex

$$= \text{fraction total C as COOH} \times \text{conc. total C in solution}$$

$$= 0.24 \times 1 \times 10^{-4}$$

$$= 2.4 \times 10^{-5}$$

This is the concentration of carbon as a complexing ligand, L_j, in the fly-ash leachate;

i.e., $[L_j] = 2.4 \times 10^{-5}$ mol/liter.

The effect of the humic acid ligand on the chemical speciation of a heavy metal will be computed for dissolved nickel. In equation 26 the concentration of free Ni^{+2} ion in the fly-ash leachate was computed from the total nickel concentration, $[Ni_T] = 3.41 \times 10^{-7}$ mol/liter, and concentrations of three Ni-anion complexes. To add an additional complex $[NiL_j]$, and to compute the new concentration of the free Ni^{+2} ion, only one more $K_{ij}[L_j]$ term must be added to the denominator of equation 26:

$$K_{ij}[L_j] = 10^8 \times 2.4 \times 10^{-5} = 2400, \text{ near pH} = 8$$

or

$$K_{ij}[L_j] = 10^4 \times 2.4 \times 10^{-5} = 0.24, \text{ near pH} = 5$$

(The values of the stability constants at pH = 8 and pH = 5 were given at the beginning of this section.)

Substitution of each result in equation 26 gives:

$$\text{at pH} = 8\text{: } [Ni^{+2}] = \frac{3.41 \times 10^{-7}}{1.494 + 2400} = 1.4 \times 10^{-10} \text{ mol/liter}$$

$$\text{at pH} = 5\text{: } [Ni^{+2}] = \frac{3.41 \times 10^{-7}}{1.468 + 0.24} = 2.0 \times 10^{-7} \text{ mol/liter}$$

Near the solution pH value of 8, practically all of the dissolved nickel is taken up in the Ni-humate complex. Near the value of pH = 5, only 42% of the total nickel concentration is complexed, mainly in $[NiSO_4^0]$ and in the nickel-humate complex (the complexes $NiOH^+$ and $NiHPO_4^0$ can be neglected at pH = 5). Concentrations of the Ni-sulfate and Ni-humate complexes under the more acidic conditions at pH = 5 are:

$$[NiSO_4^0] = 9.4 \times 10^{-8} \text{ mol/liter; 28\% of total Ni in solution}$$

$$[\text{Ni-humate}] = 4.8 \times 10^{-8} \text{ mol/liter; 14\% of total Ni in solution}$$

The large differences between the proportions of nickel in the nickel-humate complexes at the two pH values reflect a greater tendency of humic acids to form complexes with heavy metals under mildly alkaline conditions. Under slightly acidic conditions in the fly-ash leachate, much smaller fractions of dissolved Ni may be expected to occur as Ni-humate complexes.

2.13.4 Summary

A model analysis of an aqueous system, based on the concepts and techniques described for various processes taking place in aqueous solutions, was done on leachate from a fly-ash site. The principal conclusions were as follows:

(1) The leachate is a calcium-sulfate type water, of low ionic strength ($I \approx 0.01$ mol/liter), mildly alkaline (pH = 8.0), and low in dissolved oxygen (0.7 ppm O_2).

(2) Iron in solution is present in an oxidized state, Fe (III), and the main aqueous species of iron is the hydroxide complex $Fe(OH)_2^+$. This conclusion is based on an assumption of an oxidation-reduction equilibrium between Fe (II) and Fe(III) in solution. However, aqueous Fe(II), if present, may be rapidly oxidized to Fe(III) at the conditions of the fly-ash site (alkaline pH and dissolved oxygen in water).

(3) Among the three major constituents of the fly-ash leachate — Ca^{+2}, Mg^{+2}, and SO_4^{-2} — about 70% of the total concentration of each constituent occurs as free ions, and the remaining 30% occurs as metal-sulfate complexes.

(4) About 60% of the total dissolved phosphorus is combined in calcium-phosphate and magnesium-phosphate aqueous complexes.

(5) The leachate is supersaturated with respect to solid $BaSO_4$, $Fe(OH)_3$, and $FePO_4 \cdot 2H_2O$.

(6) Nickel, occurring as a trace constituent of solution (10^{-7} mol/liter), exists primarily as a free Ni^{+2} ion (about 67% of total Ni in solution). An aqueous complex of importance to the chemical speciation of Ni in the leachate is the ion-pair $NiSO_4^0$ (about 31% of total Ni).

(7) Estimates were made of the consequences of adding small amounts of humic acids to the leachate (1×10^{-4} mol C/liter or 1.2 ppm). Under mildly acidic conditions (pH = 5), only about 15% of the total Ni in solution would be complexed in a Ni-humate complex. However, at an alkaline pH of 8, as in the fly-ash leachate, practically all of the dissolved Ni would be combined in the Ni-humate complex.

2.13.5 Literature Cited

1. Arthur D. Little, Inc., *Full-Scale Field Evaluation of Waste Disposal from Coal-Fired Electric Generating Plants. Appendix F (Part 4)*, EPA Contract No. 68-02-3167, Final Report to U.S. EPA, Office of Research and Development, Washington, D.C. (1984).

2. Drever, J.I., *The Geochemistry of Natural Waters,* Prentice-Hall, Englewood Cliffs, N.J. (1982).

3. Kotrlý, S. and L. Šůcha, *Handbook of Chemical Equilibria in Analytical Chemistry*, John Wiley & Sons, New York (1985).

4. Lerman, A. and C.W. Childs, "Metal-organic Complexes in Natural Waters: Control of Distribution by Thermodynamic, Kinetic and Physical Factors," in *Trace Metals and Metal-Organic Interactions in Natural Waters*, Ann Arbor Science Publishers, Ann Arbor, Mich. (1973).

5. Stevenson, F.J., *Humus Chemistry*, Wiley-Interscience, New York (1982).

6. Wagman, D.D., W.H. Evans, V.B. Parker, R.H. Schumm, I. Halow, S.M. Bailey, K.L. Churney, and R.I. Nutall, "The NBS Tables of Chemical Thermodynamic Properties," *J. Phys. Chem. Ref. Data,* **11**, Supplement 2 (1982).

7. White, A.F. and A. Yee, "Aqueous Oxidation-Reduction Kinetics Associated with Coupled Electron Cation Transfer from Iron-Containing Silicates at 25°C," *Geochim. Cosmochim. Acta,* **49**, 1263 (1985).

2.14 PHOTOLYSIS IN WATER

2.14.1 Introduction

This section discusses the processes involved when inorganic or organometallic species are exposed to sunlight in natural waters. The resulting photochemical degradation, which competes with other processes such as biotransformation and hydrolysis, is of two kinds:

- *Direct photolysis*, in which the species absorbs solar radiation, is promoted to an excited state, and either decomposes or undergoes secondary reactions; and
- *Indirect photolysis*, in which energy or electrons are transferred from some other photolytically sensitized species. The sensitized species may be dissolved [30] or in the form of a solid suspension [9,13,19,20].

Models for predicting direct photochemical reactivity generally include methods of calculating the net rate at which the chemical absorbs light. The approach, which is discussed by Zepp and Cline [29], Harris [12] and Leighton [16], is to evaluate the degree of overlap between the ultraviolet/visible absorption spectrum of each molecule or ionic species and the solar spectrum to which it is exposed. Section 2.3 of this volume presents Leighton's method; although applied to gaseous species, it is also applicable to absorbers in solution.

2.14.2 Units

Concentrations are generally given in terms of $mol \cdot l^{-1}$ or $mol \cdot cm^{-3}$. Molar absorptivities are given in terms of $l \cdot mol^{-1} \cdot cm^{-1}$, while the decadic attenuation coefficient of water is in cm^{-1}. Linear measures are expressed in cm. The intensity of solar radiation is usually written in units of $photon \cdot cm^{-2} \cdot s^{-1}$ or in $millieinsteins \cdot cm^{-2} \cdot s^{-1}$ (1 einstein = 6.02×10^{23} photons). As described below, the symbols Z_λ, W_λ and L_λ are used to represent solar radiation intensity; each has a different definition.

2.14.3 Solar Irradiance and Absorption Rates

The lower atmosphere receives solar radiation directly from the sun and through sky reflectance. In the ultraviolet region the latter may be more important. Estimating the intensity of incident solar irradiance involves consideration of solar zenith angle, the path length of solar radiation, the amount of scattering, the absorption and diffusion of radiation by the atmosphere, and the albedo of the earth's surface. For water, one must consider additional factors, such as reflectance at the air-water interface, the pathlength of the radiation in water, and the attenuation coefficient of water (Figure 2.14-1). These topics are discussed at length in the literature [12,16,29]; hence, only the results will be given here.

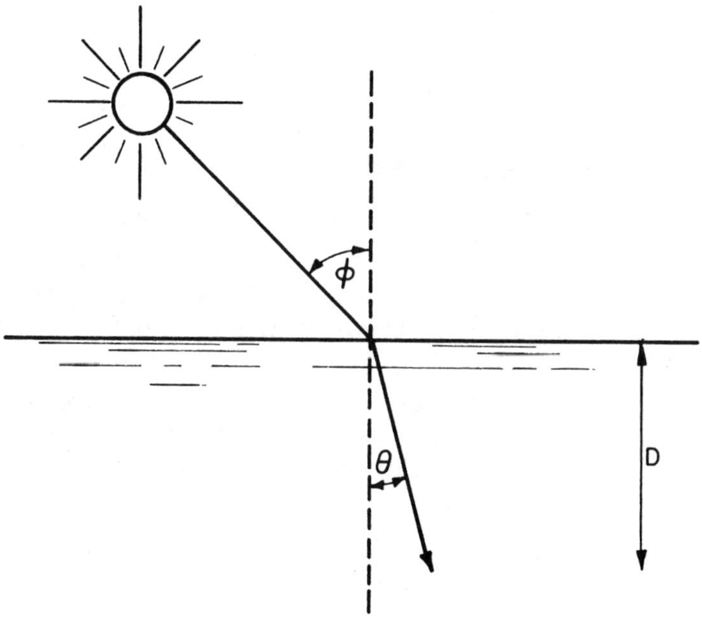

FIGURE 2.14-1 Diagram of Passage of a Beam of Sunlight through Atmosphere into Water Body

For direct photolysis, Zepp and Cline [29] derived equations relating the rate of absorption to the intensity of sunlight in the water. They considered the following two cases:

- *Case 1 — Essentially all sunlight is absorbed by the system (i.e., water plus chemical).*

$$k_{a\lambda} = \frac{W_\lambda \varepsilon_\lambda}{jD\alpha_\lambda} \tag{1}$$

where

$k_{a\lambda}$ = rate of absorption for light of wavelength λ (s^{-1})

W_λ = $I_{d\lambda} + I_{s\lambda}$

ε_λ = molar absorption coefficient of absorbing species ($M^{-1} \cdot cm^{-1}$)

$I_{s\lambda}$ = intensity of solar sky radiation at midday ($photon \cdot cm^{-2} \cdot s^{-1}$)

$I_{d\lambda}$ = intensity of direct solar radiation at midday (photon $\cdot$ $cm^{-2} \cdot s^{-1}$)

j = conversion factor for concentration in
$l \cdot mol^{-1}$ = Avogadro's number $\times 10^{-3}$
= 6.023×10^{20} photons $\cdot$ mol^{-1}

D = depth of sunlight penetration in water body (cm)

α_λ = decadic attenuation coefficient of water (cm^{-1})
(Note that α_λ increases with decreasing water clarity. Values range from about 6×10^{-3} cm^{-1} to 0.2 cm^{-1} at a wavelength of 400 nm [29].)

Equation 1 is general, hence it is applicable to any depth where strong absorption occurs.

- *Case 2 — Less than 5% of the sunlight is absorbed by the system.*

$$k_{a\lambda} = \frac{2.303\ \varepsilon_\lambda\ Z_\lambda}{j} \qquad (2)$$

or

$$k_{a\lambda} = \varepsilon_\lambda\ L_\lambda \qquad (d^{-1}) \qquad (3)$$

where

Z_λ = $I_{d\lambda}$ sec θ + 1.2 $I_{s\lambda}$ (photon $\cdot$ $cm^{-2} \cdot s^{-1}$)

θ = angle of refraction (degrees)

L_λ = solar irradiance in water (10^{-3} einsteins $\cdot$ $cm^{-2} \cdot day^{-1}$) under clear sky conditions, averaged to account for diurnal fluctuations. (Note that $k_{a\lambda}$ in equation 3 is obtained with units of $days^{-1}$ when L_λ is in units of 10^{-3} einsteins $\cdot$ $cm^{-2} \cdot day^{-1}$ and ε_λ is in units of $M^{-1} \cdot cm^{-1}$.)

Equations 2 and 3 are applicable to shallow depths in any natural water and depths of up to 0.5 m in distilled water.

Zepp and Cline [29] have published extensive tables of W_λ and Z_λ over the wavelength range of 300 nm to 800 nm. Essentially the same tables can be found in the chapter by Harris [12]. More recent and more accurate values are given by Demerjian *et al.* [4]. Figure 2.14-2 shows the general variation of both W_λ and Z_λ with wavelength for the winter and summer seasons at 40° N latitude.

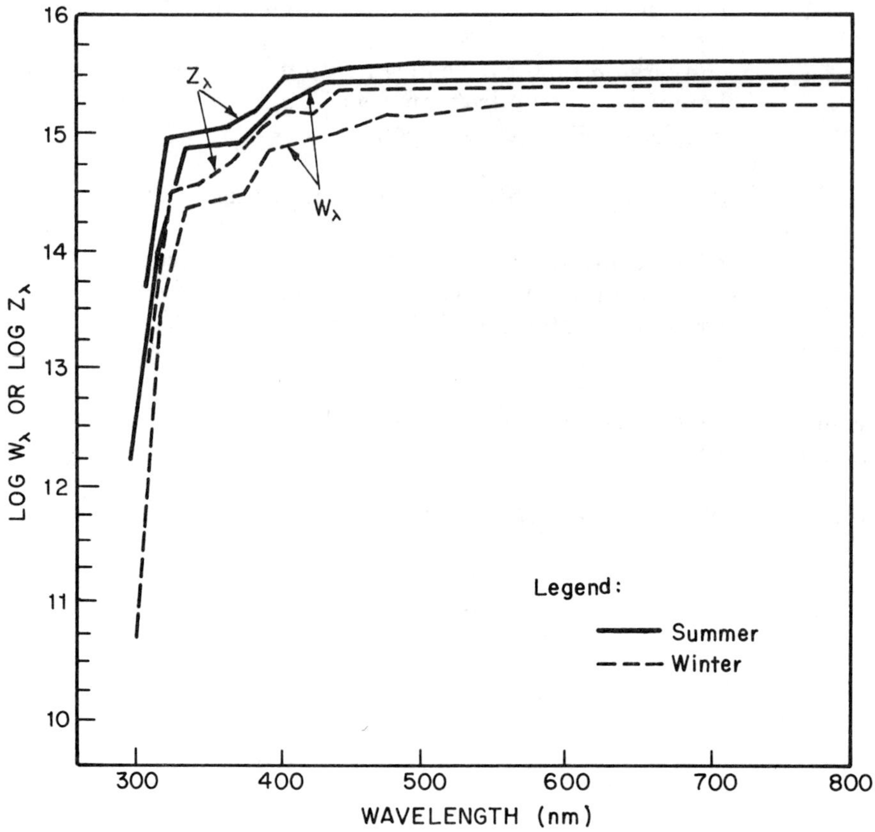

Source: Harris [12]

FIGURE 2.14-2 Variations in Log Z_λ and Log W_λ with Wavelength for 2 Seasons

Note that the value of the solar irradiance W_λ corresponds to the condition in which essentially all the sunlight is absorbed (by the chemical and the water) while Z_λ corresponds to very weak absorption of sunlight. Both terms correspond to the solar irradiance at midday (i.e., at noon), are a function of latitude and season of the year, and are in the units of photons $\cdot$ cm^{-2} $\cdot$ sec^{-1}. While one can use Z_λ or W_λ for calculating rate constants for direct photolysis, this is not the best approach, particularly when a compound photolyzes over several days or weeks. For calculating rate constants and half-lives for direct photolysis, it is preferable to use L_λ solar irradiance data obtained from the GCSOLAR computer program of Zepp [28]. The term L_λ corresponds to the day-averaged solar irradiance in millieinsteins $\cdot$ cm^{-2} $\cdot$ day^{-1} for shallow depths, weak absorbance, and is a function of latitude and season of the year. Tables of L_λ from the GCSOLAR program of Zepp have been published by the EPA for North lattitudes 20°, 30°, 40° and 50° [8]. Those for 40° N are given here in Table 2.14-1.

TABLE 2.14-1

L_λ Values for Latitude 40° N[a]

		L_λ 10^{-3} einsteins cm^{-2} day^{-1}			
λcenter (nm)	**$\Delta\lambda$(nm)[b]**	**Spring**	**Summer**	**Fall**	**Winter**
297.5	2.5	1.85(−5)	6.17(−5)	7.83(−6)	5.49(−7)
300.0	2.5	1.06(−4)	2.70(−4)	4.76(−5)	5.13(−6)
302.5	2.5	3.99(−4)	8.30(−4)	1.89(−4)	3.02(−5)
305.0	2.5	1.09(−3)	1.95(−3)	5.40(−4)	1.19(−4)
307.5	2.5	2.34(−3)	3.74(−3)	1.19(−3)	3.38(−4)
310.0	2.5	4.17(−3)	6.17(−3)	2.19(−3)	7.53(−4)
312.5	2.5	6.51(−3)	9.07(−3)	3.47(−3)	1.39(−3)
315.0	2.5	9.18(−2)	1.22(−2)	4.97(−3)	2.22(−3)
317.5	2.5	1.20(−2)	1.55(−2)	6.57(−3)	3.19(−3)
320.0	2.5	1.48(−2)	1.87(−2)	8.18(−3)	4.23(−3)
323.1	3.8	2.71(−2)	3.35(−2)	1.51(−2)	8.25(−3)
330.0	10.0	9.59(−2)	1.16(−1)	5.44(−2)	3.16(−2)
340.0	10.0	1.23(−1)	1.46(−1)	7.09(−2)	4.31(−2)
350.0	10.0	1.37(−1)	1.62(−1)	8.04(−2)	4.98(−2)
360.0	10.0	1.52(−1)	1.79(−1)	9.02(−2)	5.68(−2)
370.0	10.0	1.63(−1)	1.91(−1)	9.77(−2)	6.22(−2)
380.0	10.0	1.74(−1)	2.04(−1)	1.05(−1)	6.78(−2)
390.0	10.0	1.64(−1)	1.93(−1)	9.86(−2)	6.33(−2)
400.0	10.0	2.36(−1)	2.76(−1)	1.42(−1)	9.11(−2)
410.0	10.0	3.10(−1)	3.64(−1)	1.87(−1)	1.20(−1)
420.0	10.0	3.19(−1)	3.74(−1)	1.93(−1)	1.24(−1)
430.0	10.0	3.08(−1)	3.61(−1)	1.87(−1)	1.20(−1)
440.0	10.0	3.65(−1)	4.26(−1)	2.22(−1)	1.43(−1)
450.0	10.0	4.11(−1)	4.80(−1)	2.51(−1)	1.61(−1)
460.0	10.0	4.16(−1)	4.85(−1)	2.54(−1)	1.64(−1)
470.0	10.0	4.30(−1)	5.02(−1)	2.63(−1)	1.69(−1)
480.0	10.0	4.40(−1)	5.14(−1)	2.70(−1)	1.74(−1)
490.0	10.0	4.16(−1)	4.86(−1)	2.56(−1)	1.65(−1)
500.0	10.0	4.25(−1)	4.96(−1)	2.62(−1)	1.68(−1)
525.0	25	1.12	1.31	6.93(−1)	4.45(−1)
550.0	25	1.16	1.36	7.21(−1)	4.61(−1)
575.0	25	1.17	1.37	7.22(−1)	4.61(−1)
600.0	25	1.18	1.38	7.39(−1)	4.69(−1)
625.0	25	1.20	1.40	7.50(−1)	4.82(−1)
650.0	25	1.21	1.41	7.62(−1)	4.95(−1)
675.0	25	1.22	1.41	7.68(−1)	5.03(−1)
700.0	25	1.21	1.40	7.66(−1)	5.05(−1)
750.0	50	2.33	2.69	1.48	9.84(−1)
800.0	50	2.25	2.59	1.43	9.56(−1)

a. The second number in the columns in parentheses is the power of ten by which the first number is multiplied

b. Wavelength interval

Source: Environmental Protection Agency [8]

2.14.4 Direct Photolysis

For direct photolysis, the average rate is proportional to the quantum yield, ϕ_λ, for the reaction and the absorption rate $k_{a\lambda}$. The kinetic expression for direct photolysis is:

$$-\left(\frac{d[C]}{dt}\right)_\lambda = \phi_\lambda k_{a\lambda} [C] \quad (4)$$

where [C] = species concentration (mol • l^{-1})

Since ϕ is approximately independent of wavelength, equation 4 becomes:

$$-\frac{d[C]}{dt} = \phi k_a [C] \quad (5)$$

where $k_a = \sum_\lambda k_{a\lambda}$

For Case 1 above, the half-life of this reaction is given by:

$$t_{1/2} = \frac{j\,D[C_o]}{2\,\phi \sum_\lambda W_\lambda} \quad (6)$$

where $[C_o]$ is the initial concentration and the other symbols are as defined earlier. For Case 2, the half-life is

$$t_{1/2} = \frac{0.693}{\phi k_a} \quad (7)$$

When L_λ solar irradiance is used, the direct photolysis rate constant in the environment, k_{dE}, is defined by the equation

$$k_{dE} = \phi \sum_\lambda \varepsilon_\lambda L_\lambda, \quad (8)$$

and the half-life is given by the equation

$$t_{1/2} = 0.693/k_{dE} \quad (9)$$

These equations give the rate constant and half-life for direct photolysis in the units days^{-1} and days, respectively.

Studies of the direct photolysis of inorganic species are rare. Furthermore, the simple inorganic components of natural waters are generally transparent to sunlight. However, nitrite and nitrate have been extensively studied; nitrogen oxides and OH • radicals have been identified as the photoproducts [24, 27]. Nitrate has been found to be unreactive, while nitrite shows a net loss of about 10% per day. A few additional weak chromophores with known photochemistry include iodate, uranyl ion (UO_2^+), hydrogen peroxide (H_2O_2), and ferrous ions [1]; their environmental photochemistry has not been well studied.

Simple metal ligands such as OH^-, CO_3^{-2}, Cl^- and SO_4^{-2} generally absorb at wavelengths < 290 nm. Inorganic metal complexes that do react photochemically follow one of two pathways [26]:

- Charge transfer (ligand to metal or metal to ligand)
- Ligand exchange (photodissociation).

Table 2.14-2 lists typical inorganic and organometallic complexes that undergo direct photolysis. The following example shows how one can estimate the half-life of this process.

TABLE 2.14-2

Examples of Direct Photolysis Involving Inorganic or Organometallic Complexes

Species	Products	Probable Mechanism	Ref.
NO_2^-	NO + OH	Decomposition	[24]
MnO_2	Mn(II)	Electron transfer to metal	[22]
Cu(II)	Cu(I)	Reduction; dissociation	[15]
Cu(II)-Organic	Cu(I)	Reduction; dissociation	[15]
Fe(III)-Organic	Fe(II) + CO_2	Oxidation of organics; reduction of O_2; dissociation	[17]
Fe(III)-Organic	Fe(II), CO_2, amine	Electron transfers to metal; decomposition	[25]
Organic mercurials	Elemental Hg, Hg salts	Decomposition	[10, 32]
$Fe(CN)_6^{-4}$	$Fe(CN)_5^{-3}$ + CN^-	Reduction; decomposition	[2]
$Pb(CH_2CH_3)_4$	Ethane and, eventually, inorganic Pb salts	Decomposition	[3]
BrO^-	Br + O^-; Br^- + O	Decomposition	[21a]
ClO_2	ClO + O; or Cl + O_2	Decomposition; cage recombination of photolysis products an intermediate	[16a]
ClO^-	Cl^- + O; Cl + O^-	Decomposition	[2a]
ClO_2^-	ClO^- + O; ClO + O^-	Decomposition	[2a]

Example Estimate the half-life for the direct photolysis of Chemical B (hypothetical) at shallow depths. Case 2 will be assumed. Table 2.14-3 lists the information needed for this calculation. (Values of ε_λ were taken from an example in reference 8.)

TABLE 2.14-3

Data Needed to Evaluate the Rate of Direct Photolysis in Example 1

λ Center (nm)	ε_λ $(l \cdot mol^{-1} \cdot cm^{-1})$[a]	L_λ $(10^{-3}$ einsteins $\cdot cm^{-2} \cdot d^{-1})$[b]	$\varepsilon_\lambda L_\lambda$
297.5	1684	6.17(-5)	0.10
300.0	1434	2.70(-4)	0.39
302.5	1221	8.30(-4)	1.01
305.0	919	1.95(-3)	1.79
307.5	742	3.74(-3)	2.78
310.0	208	6.17(-3)	1.28
312.5	138	9.07(-3)	1.25
315.0	94	1.22(-2)	1.15
317.5	57	1.55(-2)	0.88
320.0	9	1.87(-2)	0.17
323.1	2	3.35(-2)	0.07
330.0	0	1.16(-1)	0.00
		$\sum_\lambda \varepsilon_\lambda L_\lambda =$	10.87

a. From reference 8.
b. From Table 2.14-1.

From Table 2.14-3, $k_\alpha = \Sigma\ \varepsilon_\lambda L_\lambda = 10.87\ d^{-1}$

Assuming a value of 0.1 for ϕ, we can then use equation 8 to calculate k_{dE}:

$$k_{dE} = (0.1)(10.87) = 1.08\ d^{-1}$$

and from equation 9

$$t_{1/2} = 0.693/1.08 = 0.64\ d$$

2.14.5 Indirect Photolysis — Homogeneous

In indirect photolysis, a reaction is initiated when light is absorbed by a chromophore (sensitizer) other than the reacting material. If the chromophore is regenerated, it plays a photocatalytic role; if it changes irreversibly, it has undergone direct photolysis itself, even while causing indirect photolysis of other materials. For dissolved chromophores the photolysis is homogeneous. The exact nature of the sensitizers is unknown, but they are believed to consist of dissolved organic materials of unidentified and variable structure [18,30]. "Humic" and "fulvic" acids apparently account for a large fraction of the sensitizers.

The known sensitized photoreactions involve initial excitation of chromophores, followed either by energy transfer or the transfer of electrons or hydrogen atoms to or from other components in the system. The energy transfer reaction was first recognized for inorganics by Joussot-Dubien and Kadiri [14], who studied the sensitized oxidation of ammonia to nitrite and nitrate ions. The mechanism of the reaction was believed to involve the photo formation of long-lived triplet states. The quantum yield of the original photolysis was approximately 0.01, representing a large flux of excited species [26]. In laboratory studies, dyes such as eosin were used as sensitizers, but the authors speculated that several aromatic species in seawater and wastewater could act as photosensitizers [14]. Among the prime candidates was riboflavin, despite its low concentration levels. The importance of photochemically generated singlet oxygen as an oxidant in natural waters was clearly demonstrated in subsequent work by Zepp *et al.* [31].

The known homogeneous sensitizers appear to be primarily organic in nature, although there is some evidence that nitrate can be an important source of OH • radicals in water [10a], and that metallo-organic complexes act as sensitizers. For example, Miles and Brezonik [17] hypothesized that free Fe[II] and a Fe[II]-humic complex are first oxidized by oxygen to free Fe[III] and a Fe[III]-humic complex respectively. The Fe[III]-humic complex was slowly reduced to Fe[II] in the dark. The reduction was significantly accelerated in the light through a ligand-to-metal charge transfer, producing CO_2 and other oxidation products. The Fe[II]-humic complex could dissociate and replace the free Fe[II] originally oxidized. The overall result was a catalytic redox cycle whereby iron and light caused the production of a half-molecule of CO_2 for every turn of the reaction cycle. The light-accelerated oxidation of other organic complexing agents such as citric acid and histidine by Fe[III] was also demonstrated.

2.14.6 Indirect Photolysis — Heterogeneous

Heterogenous solutions contain both a solution and a solid phase. The solid phase — which may consist of living or nonliving matter, mineral particles and other colloidal materials — may alter the rate of photolysis of sorbed species. Much of the evidence for the general importance of heterogeneous photolysis comes from studies showing spectral shifts, excited state quenching, variation in luminescence yields, and enhancement of electron transfer rates for sorbed species [26]. Although important, such reactions are difficult to study, and relatively little information is available on the mechanisms.

Several investigators [9,13,19,20] have reported, for example, that insoluble semiconductor materials suspended in water can catalyze the complete photolysis of halogenated hydrocarbons to CO_2 and HX and the oxidation of CN^- and SO_2. Notable among these catalysts are TiO_2 [9,19], ZnO [9] and, to a lesser extent, Fe_2O_3 and CdS [9]. These findings are significant, because they may lead to an important means of improving the quality of water containing small amounts of oxidizable organic and inorganic pollutants.

Those who have analyzed the data have generally approached it from the standpoint of first-order kinetics. This treatment is adequate for determining relative reactivities, but it does not clarify the mechanism of the i.e., the role of the solid catalyst and its relationship to the properties of the decomposed species. In the hope of encouraging further research into this matter, we briefly discuss the possible relationship between catalyst and reactants below.

The catalytic materials mentioned above all have semiconductive properties, as manifested by phenomena such as photoconductivity. The oxide materials (TiO_2, ZnO, Fe_2O_3) in particular chemisorb molecular oxygen and convert it to O_2^- and O^-. The kinetics of the chemisorption/desorption phenomena has been discussed at length by Halsey [11]. The hypothesis put forth leads to the conclusion that both chemisorption and desorption obey the following rate law:

$$\pm \frac{dQ}{dt} \propto \frac{1}{t} \tag{10}$$

where

$$Q \propto \pm \ln t \tag{11}$$

Q = quantity chemisorbed (+) or desorbed (-)
t = time

The form of the rate law is a consequence of the fact that as material is chemisorbed (desorbed), the probability of further chemisorption (desorption) decreases. Equation 11 has been confirmed many times [5,6,7,23].

In the dark, if an organic molecule is adsorbed on the surface nothing will happen. Upon exposure of the sorbent to light of suitable wavelength, electron-hole pairs are generated, the holes moving toward the surface and combining with the electron trapped by oxygen molecules. With the loss of the trapped electron, oxygen is released which then reacts with the adsorbed organic species nearby.

Photoconductive measurements indicate that equation 11 is obeyed for both ZnO and TiO_2 [23], hence we would expect the oxidation kinetics of organic species near these surfaces to be linear in ln t. Figure 2.14-3 shows that with the available data equation 11 is obeyed in every case. All of the data are fitted by an equation of the form

$$\frac{[C]}{[C_o]} = 1 + k \ln (t/t_o) \tag{12}$$

where

[C] = concentration of unoxidized species in solution at time t (ppm)
$[C_o]$ = initial concentration of unoxidized species in solution (ppm)

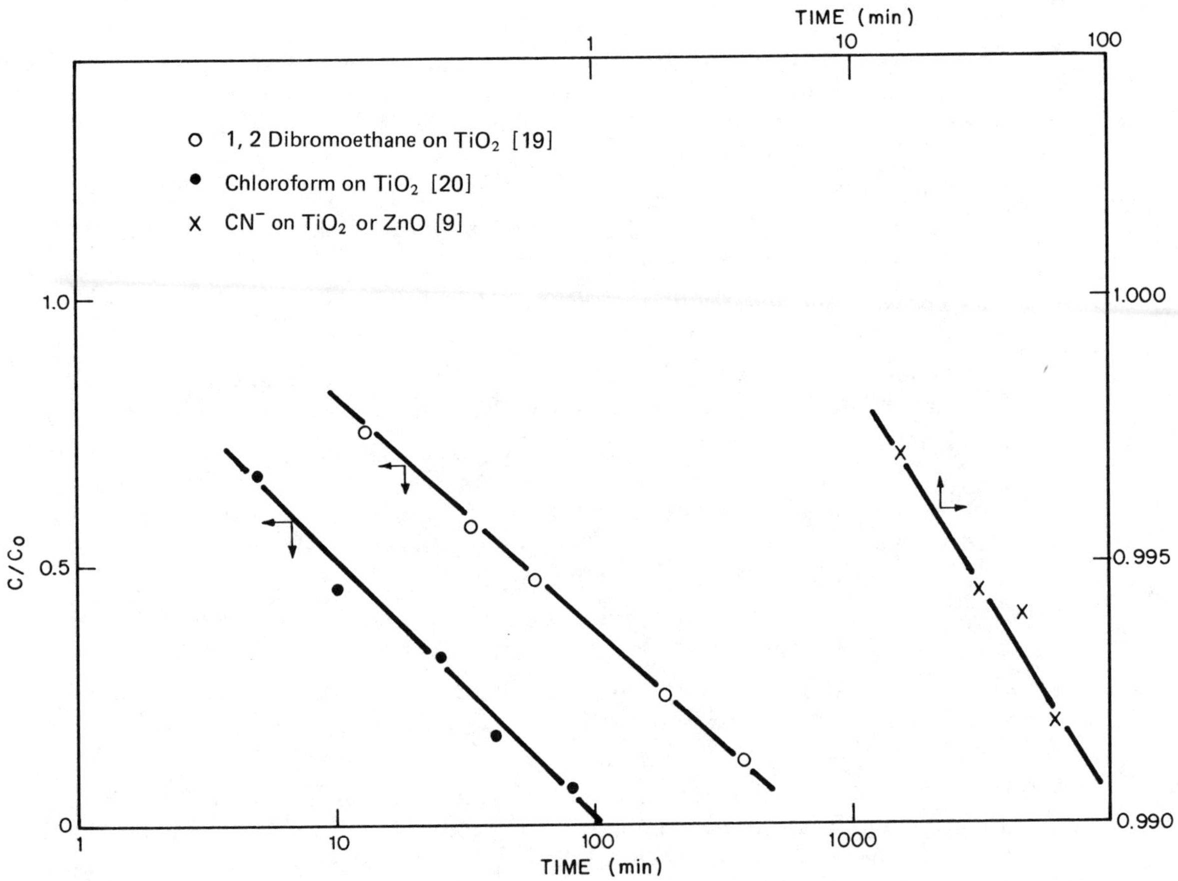

FIGURE 2.14-3 Kinetic Data for Catalytically Oxidized Species

(See text, equation 11)

t_o = characteristic time related to the nature of the reactive species (min)
k = rate constant (dimensionless) (k is a function of light intensity)

According to Stone [21], t_o is proportional to the time interval in which no chemisorption (desorption) should occur. In gas-phase reactions, this time interval is so short as to be neglected; in a water body, however, we would expect t_o to be significant. Furthermore, we would expect t_o to be related to the ease with which the chemical species can be oxidized.

2.14.7 Literature Cited

1. Brateman, P.S., A. Cairns-Smith, R. Sloper, T. Trusealt and M. Craw, "Photoxidation of Iron(II) in Water between pH 7.5 and 4.0," *J. Chem. Soc., Dalton Trans.*, **1984**, 1441 (1984).

2. Broderios, S.J. and L.L. Smith, Jr., "Direct Photolysis of Hexacyanferrate Complexes: Proposed Applications to Aquatic Environment," Report No. EPA 600/3-80-003, U.S. Environmental Protection Agency, Washington, D.C. (1980).

2a. Buxton, G.V. and M.S. Subhani, "Radiation Chemistry and Photochemistry of Oxychlorine Ions," (Parts 2 and 3), *Trans. Faraday Soc. I*, **68**, 958-77 (1972).

3. Charlou, J.L., M. Caprais, G. Blanchard and G. Martin, "Degradation of TEL (Tetraethyllead) in Seawater," *Environ. Technol. Lett.*, **3**, 415-24 (1982).

4. Demerjian, K.L., K.L. Schere and J.T. Peterson, "Theoretical Estimates of Actinic (Spherically Integrated) Flux and Photolytic Rate Constants of Atmospheric Species in the Lower Troposphere," in *Advances in Environmental Science and Technology*, Volume 10, J.N. Pitts, Jr. and R.L. Metcalf (eds.), John Wiley & Sons, New York (1980).

5. Eley, D., "The Kinetics of Unimolecular Processes in Condensed Phases, Including Adsorption," *Trans. Faraday Soc.*, **49**, 643 (1953).

6. Elovich, S. and G. Zhabrova, "Mechanism of the Catalytic Hydrogenation of Ethylene on Nickel. I: Kinetics of the Process," *J. Phys. Chem. (U.S.S.R.)*, **13**, 1761-64 (1939).

7. Elovich, S. and G. Zhabrova, "Role of the Activated Adsorption of Ethylene and Hydrogen in the Hydrogenation Process: General Scheme of the Process," *J. Phys. Chem. (U.S.S.R.)*, **13**, 1775 (1939).

8. Environmental Protection Agency, "40 CFR Parts 796, 797, and 798: Toxic Substances Control Act Test Guidelines, Final Rules," *Fed. Regist.*, **50**, 39252 (September 27, 1985). (See especially § 796.3700.)

9. Frank, S. and A. Bard, "Heterogeneous Photocatalytic Oxidation of Cyanide and Sulfite in Aqueous Solutions at *[sic]* Semiconductor Powders," *J. Phys. Chem.*, **81**, 1484 (1977).

10. Gowenlock, B.G. and J. Trofman, "UV Absorption Spectrum of Diphenylmercury in Ethanol," *Chem. Ind. (London)*, 309 (March 13, 1954).

10a. Haag, W.R. and J. Hoigné, "Photo-sensitized Oxidation in Natural Water Via OH Radicals," *Chemosphere*, **14**, 1659-71 (1985).

11. Halsey, G., "The Rate of Absorption on a Nonuniform Surface," *J. Phys. Chem.*, **55**, 21 (1951).

12. Harris, J., "Rate of Aqueous Photolysis" in *Handbook of Chemical Property Estimation Methods*, W.J. Lyman, W.F. Reehl and D.H. Rosenblatt (eds.), McGraw-Hill Book Co., New York (1982).

13. Hsiao, C.Y., C. LiLee and D. Ollis, "Heterogeneous Photocatalysis: Degradation of Dilute Solutions of CH_2Cl_2, $CHCl_3$ and CCl_4 with Illuminated TiO_2 Photocatalyst," *J. Catal.*, **82**, 418 (1983).

14. Joussot-Dubien, J. and A. Kadiri, "Photosensitized Oxidation of Ammonia by Singlet Oxygen in Aqueous Solution and in Seawater," *Nature*, **227**, 700 (1970).

15. Langford, C., M. Wingham and V. Sastri, "Ligand Photooxidation in Copper(II) Complexes of Nitrilotriacetic Acid," *Environ. Sci. Technol.*, **17**, 821 (1983).

16. Leighton, P.A., *Photochemistry of Air Pollution*, Academic Press, New York (1961).

16a. Mialocq, J.C., F. Barat, L. Gilles, B. Hickel, and B. Lesigne, "Flash Photolysis of Chlorine Dioxide in Aqueous Solution," *J. Phys. Chem.*, **77**, 742-9 (1973).

17. Miles, C. and P. Brezonik, "Oxygen Consumption in Humic-Colored Waters by a Photochemical Ferrous-Ferric Catalytic Cycle," *Environ. Sci. Technol.*, **15**, 1089 (1981).

18. Norwood, D., J. Johnson, R. Christman, J. Hass and M. Bobenrieth, "Reactions of Chlorine with Selected Aromatic Models of Aquatic Humic Material," *Environ. Sci. Technol.*, **14**, 187 (1980).

19. Ollis, D., "Contaminant Degradation in Water," *Environ. Sci. Technol.*, **19**, 480 (1985).

20. Proden, A. and D. Ollis, "Degradation of Chloroform by Photoassisted Heterogeneous Catalysis in Dilute Aqueous Suspensions of Titanium Dioxide," *Environ. Sci. Technol.*, **17**, 628 (1983).

21. Stone, F.S., *Chemistry of the Solid State*, Ch. 15, W.E. Garner (ed.), Academic Press, New York (1955).

21a. Subhani, M.S. and S.F. Lodhi, "A Mechanism for the Photolytic Decomposition of Hypobromite in Aqueous Solution at 365 nm, 313 nm and 253.7 nm," *Rev. Roum. Chim.*, **25**, 1567-78 (1980).

22. Sunda, W., S. Huntsman and G. Harvey, "Photoreduction of Manganese Oxides in Seawater and Its Geochemical and Biological Implications," *Nature*, **301**, 234 (1983).

23. Taylor, H. and N. Thon, "Kinetics of Chemisorption," *J. Am. Chem. Soc.*, **74**, 4169 (1952).

24. Treinin, A. and E. Hayon, "Absorption Spectra and Reaction Kinetics of NO_2, N_2, N_2O_3 and N_2O_5 in Aqueous Solution," *J. Am. Chem. Soc.*, **92**, 5821 (1970).

25. Trott, T., R. Henwood and C. Langford, "Sunlight Photochemistry of Ferric (NTA) Complexes," *Environ. Sci. Technol.*, **6**, 367 (1972).

26. Zafiriou, O., J. Joussot-Dubien, R. Zepp and R. Zika, "Photochemistry of Natural Waters," *Environ. Sci. Technol.*, **18**, 358 (1984).

27. Zafiriou, O.C., M. McFarland and R. Bromond, "Nitric Oxide in Seawater," *Science*, **207**, 637 (1980).

28. Zepp, R.G., Environmental Research Laboratory, U.S. Environmental Protection Agency, College Station Road, Athens, GA.

29. Zepp, R.G. and D. Cline, "Rates of Direct Photolysis in Aquatic Environment," *Environ. Sci. Technol.*, **II(4)**, 359 (1977).

30. Zepp, R.G., P. Schlotzhaver and R. Sink, "Photosensitized Transformations Involving Electronic Energy Transfer in Natural Waters: Role of Humic Substances," *Environ. Sci. Technol.*, **19**, 74 (1985).

31. Zepp, R.G., N.L. Wolfe, G.L. Baughman and R.C. Hollis, "Singlet Oxygen in Natural Waters," *Nature*, **267**, 421-3 (1977).

32. Zepp, R.G., N.L. Wolfe and J. Gordon, "Photodecomposition of Phenylmercury Compounds in Sunlight," *Chemosphere*, **3**, 93 (1973).

2.15 MICROBIAL TRANSFORMATIONS

2.15.1 General Considerations

Microorganisms such as bacteria, yeast and fungi catalyze the modification of inorganics in the environment. These transformations can affect the mobility and toxicity of these substances. The high metabolic activity and versatility of diverse populations of microorganisms, combined with their rapid reproduction and mutation rates, enable them to adapt and develop enzyme systems that detoxify a variety of substances.

Microbes can be classified by their oxygen requirements: aerobes, which grow in the presence of air or oxygen, and anaerobes, which grow in the absence of air or oxygen. Facultative organisms have the ability to adapt and grow under either oxygen-containing or excluding atmospheres. Organisms in these groups may also hydrolyze, hydrate or dehydrate substances in the environment.

The rates of transformation and the transformation products are dependent on the environmental conditions and the population of microorganisms available. These transformations produce dynamic systems in the environment.

This section describes the microbial transformation of metals and metalloids and gives specific examples of reactions that can be mediated by microbes. The mechanisms involved in the transformation of one form of an element to another are important to consider, because they affect the concentration and distribution of elements in the environment. Table 2.15.1 summarizes the processes discussed in this section and the elements known to be affected by them.

TABLE 2.15.1

Elements Affected by Microbial Transformation Processes

Process	Elements Known to Be Affected
Oxidation/Reduction (Sect. 2.15.2)	Hg, As, Fe, Mn, Sb, Te, Se, S
Sulfide Precipitation (Sect. 2.15.2)	Cu, Pb, Zn, Ag, others
Methylation (Sect. 2.15.3)	Hg, As, Se, Pb, Cd, Sn, Te, Pu, Tl(?), Pt(?) Sb(?)
Dealkylation (Sect. 2.15.4)	Hg, Sn, As

2.15.2 Oxidation and Reduction

Metals and metalloids can exist in the environment in several valence forms, each of which is available for chemical and biochemical reactions. Each form may differ in its toxicity and mobility in the environment. By means of oxidative and reductive reactions, microorganisms can control the relative amount of each of the forms in the environment. Changes in temperature, pH and the composition of the microbial population may shift the equilibrium among the different forms.

For example, mercury exists in three forms shown in the following equilibrium [31]:

$$Hg_2^{+2} \rightleftharpoons Hg^{+2} + Hg^0 \qquad (1)$$

It has been found that bacteria and yeasts can convert $HgCl_2$ to elemental mercury. From a stream, Brunker and Bott [5] isolated a yeast of the genus *Cryptococcus* that was capable of reducing $HgCl_2$ to Hg^0; *Escherichia coli* and strains of *Pseudomonas* sp. were also able to catalyze this reaction [1]. Nicotinamide adenine dinucleotide, reduced form (NADH), acts as a coenzyme in this reaction [31]:

$$Hg^{+2} + NADH + H^+ \rightleftharpoons Hg^0 + 2H^+ + NAD^+ \qquad (2)$$

Reduced nicotinamide adenine dinucleotide phosphate (NADPH) has also been indicated as a coenzyme for similar transformations [1]. The elemental mercury formed escapes from the system due to its high volatility and shifts the equilibrium in equation 1 to the right.

A *Pseudomonas* strain was able to produce elemental mercury following incubation with phenylmercuric acetate, ethylmercuric phosphate and methylmercuric chloride [1]. The reductive dealkylation of organic mercury compounds is described in section 2.15.4.

The reverse reaction, the oxidation of Hg^0 to Hg^{+2}, has also been reported to occur in soils and sediments. Although microbes appear to mediate this reaction, the extent to which they are involved is not clear [22]. The process has important toxicologic significance, because Hg^{+2} can be methylated to form methylmercury, which is more toxic to higher organisms than elemental mercury or inorganic divalent mercury. The methylation of Hg^{+2} is described in section 2.15.3.

Microorganisms can also catalyze the oxidation and reduction of arsenic compounds. The oxidation of trivalent arsenite, As(III), to pentavalent arsenate, As(V), has been reported to occur in cultures of pseudomonads and in soil [1]. Quastel and Scholefield [21] suggest the following reaction to describe the oxidation of As(III) to As(V) in soil:

$$AsO_2^- + H_2O + \tfrac{1}{2}O_2(g) \longrightarrow H_2AsO_4^- \qquad (3)$$

Pure cultures of several organisms are also able to reduce As(V) to As(III), including *Chlorella, Micrococcus* sp., the yeast *Pichia guillermondii* and marine bacteria [1]. McBride and Wolfe [15] described the reductive methylation of arsenate to dimethyl arsine by *Methanobacterium* in which the first step in the overall reaction was the reduction of As(V) to As(III). Reductive methylation of arsenic compounds is discussed in Section 2.15.3.

Environmental conditions are important in the interconversion of inorganic forms of arsenic. Myers *et al.* [16] showed that microbial populations in activated sludge oxidized arsenite to arsenate under aerobic conditions; under anaerobic conditions, arsenate was reduced to arsenite and to an even lower oxidation state when anaerobic incubation was continued. The form of arsenic produced is of environmental and toxicological importance, because arsenite, As(III), is generally more toxic to higher organisms than arsenate, As(V) [1]. For example, the apparent no-effect doses in rat carcinogenicity studies were 12.5 mg/kg/day and 23 mg/kg/day for sodium arsenite and sodium arsenate, respectively. In general the toxicity of the various forms of arsenic are related to their rate of clearance from the body and thus to their degree of accumulation in the tissues [23].

The oxidation of Fe(II) to Fe(III) is catalyzed by bacteria such as *Ferrobacillus* and *Gallionella*. The energy released in a reaction such as the following is used by the bacteria in their metabolism [9]:

$$4\ FeCO_3(s) + O_2(g) + 6H_2O \longrightarrow 4\ Fe(OH)_3(s) + 4\ CO_2(g) \qquad (4)$$

Manganese is oxidized by microbes in soil, and bacterial enzymes have been reported to function in the oxidation and reduction of manganese in marine sediments. Microorganisms also appear capable of oxidizing antimony and reducing tellurite and selenite to the metallic form [27].

Sulfide minerals of metals such as Cu, Pb, Zn and Ag are produced by the action of *Desulphovibrio desulphuricans* and other sulfate-reducing bacteria, which are found in mud, swamps and drained pools. The reactions catalyzed by these microorganisms are inhibited under aerobic and acidic conditions [1].

The first step in the formation of insoluble metal sulfides in the environment is the reduction of sulfate catalyzed by the microorganisms, followed by the precipitation of the sulfide [9]:

$$SO_4^{-2} + \underset{\text{carbohydrate}}{2(CH_2O)} + 2H^+ \longrightarrow H_2S^0 + 2\ CO_2(g) + 2\ H_2O \qquad (5)$$

$$Pb^{+2} + H_2S^0 \longrightarrow PbS(s) + 2H^+ \qquad (6)$$

Hydrogen sulfide, H_2S, can also be formed by the cleavage of organic molecules such as proteins, polypeptides and amino acids by bacteria. An example is the transformation of cysteine to pyruvic acid:

$$HSCH_2CH(NH_2)COOH + H_2O \longrightarrow CH_3C(O)COOH + H_2S^0 + NH_3(g) \qquad (7)$$

2.15.3 Methylation

Microorganisms can catalyze the methylation of inorganic metals and metalloids. This may be a detoxification mechanism for the organisms, but the methylated products formed can be more toxic to higher organisms. Investigators have reported that mercury, arsenic, selenium, lead, tin, tellurium, plutonium and cadmium can be methylated by microorganisms, and it has been suggested that thallium, antimony and platinum should undergo a similar reaction [12,22,28,31].

Mercury can be methylated by both anaerobic and aerobic microorganisms. Jensen and Jernelöv [13] reported the formation of monomethyl and dimethylmercury in the sediment of freshwater aquaria, for which they postulated the following two reactions:

$$Hg^{+2} + 2R-CH_3 \longrightarrow CH_3HgCH_3^0 \longrightarrow CH_3Hg^+ \quad (8)$$

$$Hg^{+2} + R-CH_3 \longrightarrow CH_3Hg^+ \xrightarrow{RCH_3} CH_3HgCH_3^0 \quad (9)$$

Methylation of mercury has also been detected in river and marine sediments and soil [2,3,11]. D'Itri [8] states that the mechanism involves the reaction of mercuric ion with methylcobalamin, a methylated vitamin B-12 derivative, through a combination of the reactions described in equations 8 and 9. For methylation to occur, mercury must be in the form of the divalent inorganic species. (However, HgS is not available for methylation due to its low solubility and the normally anaerobic conditions [14].) Most forms of mercury released into the environment, including elemental mercury, phenyl mercuries, alkyl mercuries and alkoxy-alkyl mercuries, can be directly or indirectly transformed into inorganic mercuric ions (Figure 2.15-1).

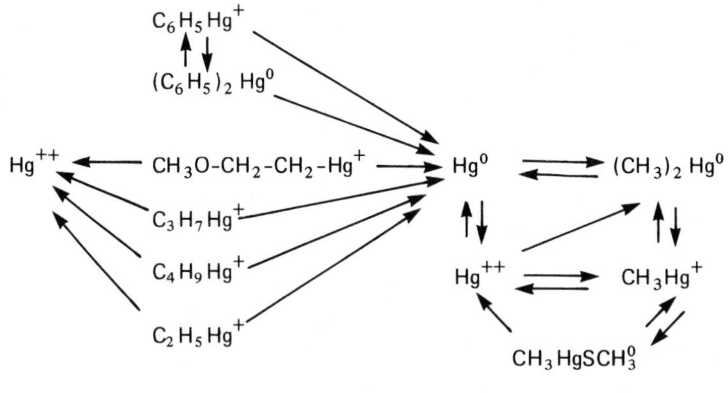

Source: D'Itri [8]. *(Copyright 1972, CRC Press. Reprinted with permission.)*

FIGURE 2.15-1 Hypothetical Model for the Conversion of Mercury in the Aquatic Environment

The relative amounts of mono- and dimethylmercury formed are dependent on the initial mercury concentration and the temperature and pH of the system. The formation of dimethylmercury is favored in neutral to alkaline waters and when the initial mercury concentrations are low [30]. Dimethylmercury is volatile; when released to the atmosphere, it undergoes photolysis to elemental mercury plus methyl free radicals, as shown below [8]; the methyl radicals are then reduced to methane or combine to form ethane.

$$(CH_3)_2Hg(g) \xrightarrow{\text{light}} Hg^0 + 2CH_3\cdot \qquad (10)$$

Acidic pH and higher initial mercury concentrations favor the formation of monomethylmercury. Its rate of synthesis is about 6000 times faster than that of dimethylmercury [31].

Aerobic and facultative anaerobic bacterial species capable of methylating mercury include *Klebsiella pneumoniae*, *Escherichia coli* and *Clostridium cochlearum* [20].

The biological cycle for mercury suggested by Wood [32] is shown in Figure 2.15-2. Sulfide is important to volatilization and precipitation of mercury through disproportionation chemistry in the aqueous environment.

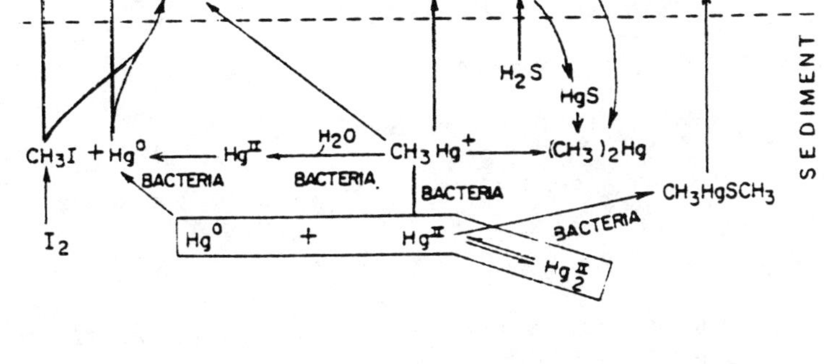

Source: Wood [32]. *(Copyright 1984, Gordon and Breach Science Publishers, Inc. Reprinted with permission.)*

FIGURE 2.15-2 The Biological Cycle for Mercury

Arsenic compounds are methylated by bacteria and fungi to yield dimethyl and trimethylarsines by a mechanism that involves the replacement of substituent oxygen atoms by methyl groups [6,23]. Methylation is thought to be a detoxification mechanism for the microorganisms [4] and is important in the transfer of arsenic from the sediment to the water and atmosphere.

McBride and Wolfe [15] reported that arsenate could be reductively methylated to dimethylarsine by *Methanobacterium* under anaerobic conditions. The reaction, shown below as given by Saxena and Howard [22], involves the initial reduction of arsenate to arsenite. Methylarsonic acid, formed by the methylation of arsenite, is reductively methylated to dimethylarsinic acid, which is reduced to dimethylarsine. Methylcobalamin is the methyl-donor in the reaction:

(11)

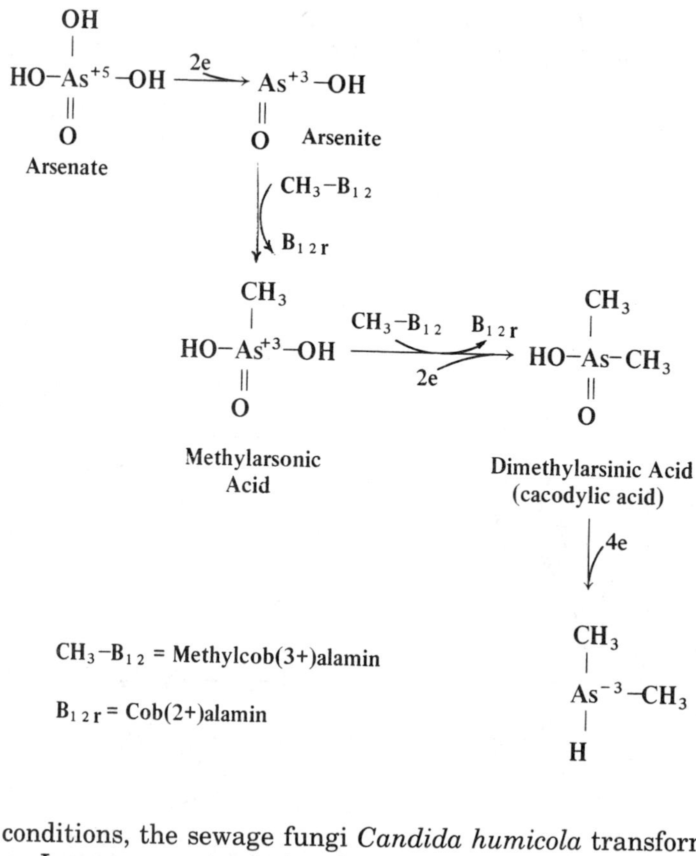

CH_3-B_{12} = Methylcob(3+)alamin

B_{12r} = Cob(2+)alamin

Under acidic conditions, the sewage fungi *Candida humicola* transform arsenate to trimethylarsine. Lesser amounts of trimethylarsine are also formed by the incubation of the fungi with arsenite, methylarsonate, and dimethylarsinate. These may be intermediates in the reduction and alkylation of arsenate to trimethylarsine [7]. Fungi capable of generating trimethylarsine from the pesticides monomethylarsonate and dimethylarsine include *Candida humicola*, *Gliocladium roseum*, and a strain of *Penicillium* [1].The rate of formation of trimethylarsine by *Candida humicola* and other molds decreases as the mold reaches its resting stage [28].

The toxic alkylarsines are volatile, with a distinct garlic-like odor, and are rapidly oxidized to less toxic products in the atmosphere. One such product, dimethylarsinic acid (cacodylic acid), has been shown to be an intermediate in the synthesis of dimethylarsine from arsenic salts [31]. The biological cycle for arsenic is shown in Figure 2.15-3.

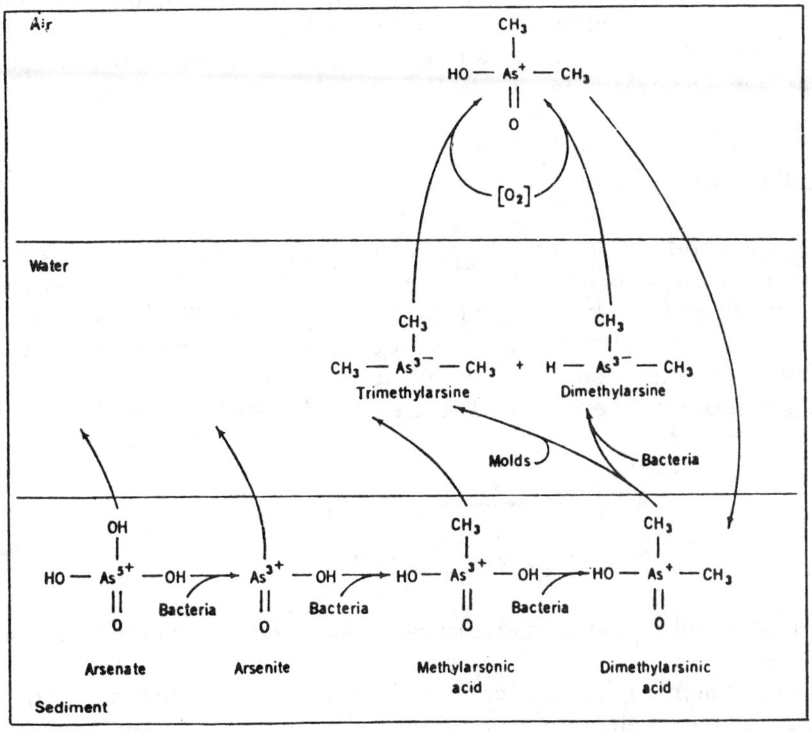

Source: Wood [31]. *(Copyright 1974, American Association for the Advancement of Science. Reproduced with permission.)*

FIGURE 2.15-3 The Biological Cycle for Arsenic

Selenium can also be acted on by microorganisms to produce methylated compounds. Inorganic selenium compounds have been transformed to dimethylselenide by a strain of *Penicillium* isolated from raw sewage and by the fungus *Candida humicola* [7,10]. Methylation of selenium has also been shown to occur in soil and lake sediment [1,22]. Dimethylselenide, $(CH_3)_2$ Se, dimethyldiselnide, $(CH_3)_2Se_2$, and dimethylselenone, $(CH_3)_2SeO_2$, were formed following incubation of selenium with sludge microorganisms [28].

Wong *et al.* [29] have reported the transformation of inorganic and organic lead compounds into volatile tetramethyl lead by microorganisms in lake sediments under anaerobic conditions. Tetramethyl lead has been formed by the incubation of sediment with inorganic lead nitrate, lead chloride, or organic trimethyllead acetate. The pathways for biological transformation of lead are not well understood. Species of *Pseudomonas*, *Alcaligenes*, *Acinetobacter* and *Aeromonas* isolated from lake sediment were able to convert $(CH_3)_3Pb^+$ to $(CH_3)_4Pb$ but were ineffective in methylating inorganic lead in chemically defined media; this conversion probably requires specific physical, chemical and biological conditions to proceed.

Evidence of the methylation of inorganic cadmium and tin to a volatile organic species by *Pseudomonas* sp. has also been reported [12,22]. In the presence of Sn(IV), volatile methylated forms of tin were produced. These biomethylated tin species were subsequently demonstrated to abiotically methylate Hg(II) to form methylmercury [15].

2.15.4 Dealkylation

Microbial dealkylation of alkylmercury compounds has been studied. Aerobic, anaerobic and facultative anaerobic species isolated from soil, sewage and lake sediments are capable of dealkylating methylmercury [20,24,26]. It appears that these methylmercury-resistant strains have developed enzyme systems capable of hydrolyzing the metal-carbon bond, resulting in the formation of methane and mercuric ion; the latter is then reduced to elemental mercury as described in section 2.15.2.

$$CH_3Hg^+ \xrightarrow{\text{bacteria}} CH_4(g) + Hg^{+2} \qquad (12)$$

$$Hg^{+2} \longrightarrow Hg^0 \qquad (13)$$

The high volatility of elemental mercury permits it to escape from the system [20,26].

While the rate of demethylation of mercury is reported to be much slower than that of methylation, it nevertheless helps to offset the formation of methylmercury [17,20]. Recent studies have also shown that bacteria from soil, sediment, and water were capable of dealkylating monosodium methanearsonate [25].

Dealkylation of the fungicide-slimicide phenylmercuric acetate by bacteria isolated from water and sediment samples has also been reported. Nelson *et al.* [18] isolated mercury-resistant bacteria from the Chesapeake Bay that were capable of degrading phenylmercuric acetate to benzene and metallic mercury.

Dealkylation of tributylin oxide by several fungi has also been reported, with a dibutyltin compound the primary product in the studies [19]. Cleavage of aryltin bonds by bacteria acting on triphenyltin acetate has also been reported [12].

2.15.5 Literature Cited

1. Alexander, M., "Microbial Formation of Environmental Pollutants," *Adv. Appl. Microbiol.,* **18**, 1-73 (1974).
2. Beckert, W.F., A.A. Moghissi, F.H.F. Au, E.W. Bretthauer and J.C. McFarlane, "Formation of Methylmercury in a Terrestrial Environment," *Nature*, **249**, 674-75 (1974).
3. Berdicevsky, I., H. Shoyerman and S. Yannai, "Formation of Methylmercury in the Marine Sediment," *Environ. Res.*, **20**, 325-34 (1979).
4. Braman, R.S. and C.C. Forback, "Methylated Forms of Arsenic in the Environment," *Science*, **183**, 1247-49 (1973).
5. Brunker, R.L. and T.L. Bott, "Reduction of Mercury to the Elemental State by a Yeast," *Appl. Microbiol.*, **27**, 870-73 (1974).
6. Challenger, F., "Biological Methylation," *Sci. Progress,* **35**, 396-416 (1947), as cited by Scow *et al.* [23].
7. Cox, D.P. and M. Alexander, "Effect of Phosphate and Other Anions on Trimethylarsine Formation by *Candida humicola*," *Appl. Microbiol.*, **25**, 408-13 (1973).
8. D'Itri, F.M., "Biological Methylation of Mercury in the Environment," in *The Environmental Mercury Problem*, CRC Press, Cleveland (1972).
9. Fergusson, J.E., *Inorganic Chemistry and the Earth*, Pergamon Press, Elmsford, N.Y. (1982).
10. Fleming, R.W. and M. Alexander, "Dimethylselenide and Demethyltelluride Formation by a Strain of *Penicillium*," *Appl. Microbiol.*, **24**, 424-29 (1972).
11. Hamdy, M.K and O.R. Noyes, "Formation of Methyl Mercury by Bacteria," *Appl. Microbiol.*, **30**, 424-432 (1975).
12. Iverson, W.P. and F.E. Brinckman, "Microbial Metabolism of Heavy Metals," in *Water Pollution Microbiology*, Vol. 2, Mitchell, R. (ed.), John Wiley & Sons, New York (1978).
13. Jensen, S. and A. Jernelöv, "Biological Methylation of Mercury in Aquatic Organisms," *Nature*, **223**, 753-54 (1969).
14. Lexmond, T.M., F.A.M. deHaan and M.J. Frissel, "On the Methylation of Inorganic Mercury and the Decomposition of Organo-mercury Compounds: A Review," *Neth. J. Agric. Sci.*, **24**, 79-97 (1976).
15. McBride, B.C. and R.S. Wolfe, "Biosynthesis of Dimethylarsine by Methanobacterium," *Biochemistry*, **10**, 4312-17 (1971).
16. Myers, D.J., M.E. Heimbrook, J. Osteryoung and S.M. Morrison, "Arsenic Oxidation State in the Presence of Microorganisms. Examination by Differential Pulse Polarography," *Environ. Lett.*, **5**, 53-61 (1973).
17. National Research Council, *An Assessment of Mercury in the Environment*, National Academy of Sciences, Washington, D.C. (1977).

18. Nelson, J.D., W. Blair, F.E. Brinckman, R.R. Colwell and W.P. Iverson, "Biodegradation of Phenylmercuric Acetate by Mercury-Resistant Bacteria," *Appl. Microbiol.*, **26**, 321-26 (1973).
19. Orsler, R.J. and G.E. Holland, "Degradation of Tributyltin Oxide by Fungal Culture Filtrates," *Int. Biodeterior. Bull.*, **18**, 95-98 (1982).
20. Perwak, J., M. Goyer, L. Nelken, K. Scow, M. Wald, D. Wallace and G. Kew, "An Exposure and Risk Assessment for Mercury," Final Draft Report, U.S. Environmental Protection Agency, Office of Water Regulations and Standards (WH-553), Washington, D.C. (1980).
21. Quastel, J. H. and P.G. Scholefield, "Arsenite Oxidation in Soil," *Soil Sci.*, **75**, 279-85 (1953).
22. Saxena, J. and P.H. Howard, "Environmental Transformation of Alkylated and Inorganic Forms of Certain Metals," *Adv. Appl. Microbiol.*, **21**, 185-226 (1977).
23. Scow, K., M. Byrne, M. Goyer, L. Nelken, J. Perwak, M. Wood, S. Young, P. Cruse and S. Kroner, "An Exposure and Risk Assessment for Arsenic," Final Draft Report, U.S. Environmental Protection Agency, Office of Water Regulations and Standards (WH-553), Washington, D.C. (1981).
24. Shariat, M., A.C. Anderson and J.W. Mason, "Screening of Common Bacteria Capable of Demethylation of Methylmercuric Chloride," *Bull. Environ. Contam. Toxicol.*, **21**, 155-261 (1979).
25. Shariatpanahi, M., A.C. Anderson and A.A. Abdelghani, "Microbial Demethylation of Monosodium Methanearsonate," *Trace Subst. Environ. Health*, **15**, 383-87 (1981).
26. Spangler, W.J., J.L. Spigarelli, J.M. Rose, R.S. Flippin and H.H. Miller, "Degradation of Methylmercury by Bacteria Isolated from Environmental Samples," *Appl. Microbiol.*, **25**, 488-93 (1973).
27. Summers, A.O. and S. Silver, "Microbial Transformation of Metals," *Annu. Rev. Microbiol.*, **32**, 637-72 (1978).
28. Thayer, J.S. and F.E. Brinckman, "The Biological Methylation of Metals and Metalloids," *Adv. Organomet. Chem.*, **20**, 313-356 (1982).
29. Wong, P.T.S., Y.K. Chau and P.L. Luxon, "Methylation of Lead in the Environment," *Nature*, **253**, 263-64 (1975).
30. Wood, J.M., F.S. Kennedy and C.G. Rosen, "Synthesis of Methylmercury Compounds by Extracts of a Methanogenic Bacterium," *Nature*, **220**, 173-74 (1968).
31. Wood, J.M., "Biological Cycles for Toxic Elements in the Environment," *Science*, **183**, 1049-52 (1974).
32. Wood, J.M., "Alkylation of Metals and the Activity of Metal Alkyls," *Toxicol. Environ. Chem.*, **7**, 229-40 (1984).

2.15.6 Other Sources of Information

Compeau, G. and R. Bartha, "Methylation and Demethylation of Mercury Under Controlled Redox, pH, and Salinity Conditions," *Appl. Environ. Microbiol.*, **48**, 1203-1207 (1984).

Craig, P.J., "Metal Cycles and Biological Methylation," in *The Handbook of Environmental Chemistry*, Vol. I, Part A, The Natural Environment and the Biogeochemical Cycles, O. Huntzinger (ed.), Springer-Verlag (1980).

Huey, C., F.E. Brinckman, S. Grim and W.P. Iverson, "Role of Tin in Bacterial Methylation of Mercury," *Proc. Int. Conf. Transp. Persistent Chem. Aquat. Ecosyst.*, 73-78 (1974).

Jernelov, A. and A. Martin, "Ecological Implications of Metal Metabolism by Microorganisms," *Annu. Rev. Microbiol.*, **29**, 61-77 (1975).

Rapsomanikis, S. and J.H. Weber, "Environmental Implications of Methylation of Tin(II) and Methyltin(IV) Ions in the Presence of Manganese Dioxide," *Environ. Sci. Technol.*, **19**, 352-356 (1985).

Robinson, J.B. and O.H. Tuovinen, "Mechanisms of Microbial Resistance and Detoxification of Mercury and Organomercury Compounds: Physiological, Biochemical, and Genetic Analyses," *Microbiol. Rev.*, **48**, 95-124 (1984).

2.16 DIFFUSION COEFFICIENTS

2.16.1 Introduction

The methods by which diffusion coefficients are estimated, measured and used vary considerably with both the nature of the diffusing material and the environment in which it diffuses. The diffusion of neutral and nonreactive organic species in both air and water has been comprehensively discussed in recent literature [5,7], and the approximation methods offered may be applied to neutral and nonreactive inorganic materials as well. This section focuses primarily on the diffusion in water of electrolytes, which probably include most inorganics of interest to the reader.

2.16.2 Description of Property

To describe diffusion or diffusion coefficients, one must begin by considering the motion of dissolved species in fluids. Under virtually all conditions encountered in the environment (gases at ambient pressure or liquids), a species acted upon by a force accelerates almost instantly to a limiting velocity determined by drag or viscous forces. Thus, the Newtonian approach, $f = ma$ (force equals mass times acceleration), is for practical purposes supplemented by $f = v/u$, where v is the limiting velocity and u is defined as mobility. Quite obviously, u is a rather complex function of the mobile species and, especially, of the fluid through which it moves.

From the above concept of mobility, various related but differing theories of diffusion, electrophoresis and conduction have been developed. In diffusion, the mobile species may be neutral or charged, and the acting force is a concentration gradient alone. In electrophoresis, the species is charged and the acting force is an electric field. In conduction, the species is charged, but the acting force is the concentration gradient resulting from the passage of an electric current.

Since velocity is neither easy to measure nor particularly useful, diffusion is usually defined in terms of Fick's law [6]:

$$F = -\mathcal{D} \frac{dC}{dX} \tag{1}$$

In this equation, which is compatible with the mobility concept, F is the flux or rate of transfer of the species per unit of cross-section area, dC/dX is the concentration gradient, and $\mathcal{D}$ is defined as the diffusion coefficient of the species in that medium. Dimensionally, F has the units of mass (or moles) per unit time, C is mass (or moles) per unit volume, and X is unit length. Thus, $\mathcal{D}$ must be expressed as area per unit time, usually cm^2/s. (The negative sign in equation 1 reflects the fact that diffusion takes place in the direction of decreasing concentration.[1])

1. Fick's law is analogous to Ohm's law, $I = E/R$, if we consider the current (I) to represent flux, voltage (E) to represent the driving force, and the reciprocal of resistance (1/R) to represent the diffusion coefficient. The mobile electron also moves with a limiting velocity, in this case the speed of light.

Up to this point, the description of diffusion appears straightforward. Regrettably, it infers several conditions that are rarely true in the real world, such as:

- there is only one diffusing species;
- the species is present only in very dilute solutions;
- no chemical interactions occur in the solution or at interfaces;
- the solution is isothermal;
- concentration and activity are considered the same; and
- the diffusion takes place in unrestricted space.

If there were only one diffusing species in very dilute solution, a simple diffusion coefficient could be used to describe its behavior. However, to be rigorous, we must use a second coefficient to describe the diffusion of the solvent into the space vacated by the solute. (Otherwise there would be net movement of the system.) If more than one solute species is present, there must be additional coefficients describing diffusion of pure components, and cross-term coefficients reflecting interaction between the components. Fortunately, however, a single value of $\mathcal{D}$ will usually suffice for all practical applications in dilute solutions.

When the mobile species are electrolytes, the situation is even more interesting. A highly dissociated 1:1 electrolyte, such as KCl or NaCl, can in fact be described by a single diffusion coefficient; although two separately definable ionic species are present, diffusion of one ion must be accompanied by diffusion of an ion of opposite charge in the same direction, to avoid the establishment of a charge separation in the fluid.[2] As might be expected, the assignment, measurement, and calculation of diffusion coefficients becomes increasingly complex with electrolytes of mixed charge, multicomponent electrolytes, and partially dissociated electrolytes, and a rigorous description becomes impossible.

From the above comments, one might assume that diffusion coefficients are only of academic interest and that their measurement or estimation is impractical. However, arguments that diffusion phenomena cannot be treated rigorously should be weighed against two other points:

- Overall, diffusion coefficients do not vary greatly. (See Tables 2.16-1 and -2.) The coefficients of many simple salts generally do not change by more than a factor of 2 over a range of concentrations and temperatures, even though their conductivities vary greatly. (This is related to the fact that each diffusing ion is accompanied by a diffusing counter ion, which tends to compensate for possibly large differences in specific ion mobilities.)

2. In this respect, diffusion differs from conductance, wherein ions of opposite charge move in opposite directions and the concept of transference numbers is added to that of mobility.

TABLE 2.16-1

Measured Diffusion Coefficients of Dilute Aqueous Electrolyte Solutions
($\mathcal{D}$ expressed in $10^{-5} cm^2/s$, c expressed in mol/l)

		Diffusion Coefficient, $\mathcal{D}$, at Indicated Concentration[a]						
Salt	**Temp. (°C)**	**c = 0[b]**	**0.001**	**0.002**	**0.003**	**0.005**	**0.007**	**0.010**
LiCl	25	1.366	1.345	1.337	1.331	1.323	1.318	1.312
NaCl	25	1.610	1.585	1.576	1.570	1.560	1.555	1.545
KCl	20	1.763	1.739	1.729	1.722	1.708	—	1.692
KCl	25	1.993	1.964	1.954	1.945	1.934	1.925	1.917
KCl	30	2.230	—	—	2.174	2.161	2.152	2.144
RbCl	25	2.051	—	2.011	2.007	1.995	1.984	1.973
CsCl	25	2.044	2.013	2.000	1.992	1.978	1.969	1.958
$LiNO_3$	25	1.336	—	—	1.296	1.289	1.283	1.276
$NaNO_3$	25	1.568	—	1.535	—	1.516	1.513	1.503
$KClO_4$	25	1.871	1.845	1.841	1.835	1.829	1.821	1.790
KNO_3	25	1.928	1.899	1.884	1.879	1.866	1.857	1.846
$AgNO_3$	25	1.765	—	—	1.719	1.708	1.698	—
$MgCl_2$	25	1.249	1.187	1.169	1.158	—	—	—
$CaCl_2$	25	1.335	1.249	1.225	1.201	1.179	—	—
		1.335	1.263	1.243	1.230	1.213	1.201	1.188
$SrCl_2$	25	1.334	1.269	1.248	1.236	1.219	1.209	—
$BaCl_2$	25	1.385	1.320	1.298	1.283	1.265	—	—
Li_2SO_4	25	1.041	0.990	0.974	0.965	0.950	—	—
Na_2SO_4	25	1.230	1.175	1.160	1.147	1.123	—	—
Cs_2SO_4	25	1.569	1.489	1.454	1.437	1.420	—	—
$MgSO_4$	25	0.849	0.768	0.740	0.727	0.710	0.704[c]	—
$ZnSO_4$	25	0.846	0.748	0.733	0.724	0.705	—	—
$LaCl_3$	25	1.293	1.175	1.145	1.126	1.105	1.084	1.021[d]
$K_4Fe(CN)_6$	25	1.468	—	—	1.213	1.183	—	—

a. $\mathcal{D}$ values for round concentrations have been interpolated graphically from the original data.

b. Values in this column are limiting values derived from the limiting mobilities of the ions.

c. At 0.006 mol/l

d. At 0.026 mol/l

Source: Robinson and Stokes [6]. *(Copyright 1959, Academic Press. Reprinted with permission.)*

TABLE 2.16-2

Measured Diffusion Coefficients of Concentrated Aqueous Electrolyte Solutions at 25°C

($\mathcal{D}$ expressed in $10^{-5} cm^2/s$, c expressed in mol/l)

c	HCl	HBr	LiCl	LiBr	NaCl	CsCl	NaBr	NaI	KCl	KBr	KI	NH_4Cl	NH_4NO_3	$LiNO_3$	$CaCl_2$	$(NH_4)_2SO_4$	$BaCl_2$
0[a]	3.336	3.400	1.366	1.377	1.610	2.044	1.625	1.614	1.993	2.016	1.999	1.994	1.929	1.336	1.335	1.530	1.385
0.05	3.07	3.15	1.28	1.30	1.507	—	1.53	1.52	1.864	1.89	1.89	—	1.788	—	1.121	0.802	1.179
0.1	3.05	3.14	1.26	1.27	1.483	1.871	1.51	1.52	1.844	1.87	1.86	1.838	1.769	1.240	1.110	0.825	1.159
0.2	3.06	3.19	1.26	1.28	1.475	1.857	1.50	1.53	1.838	1.87	1.85	1.836	1.749	1.243	1.107	0.867	1.150
0.3	3.09	3.24	1.26	1.29	1.475	1.855	1.51	1.54	1.838	1.87	1.88	1.841	1.739	1.248	1.116	0.897	1.151
0.5	3.18	3.38	1.27	1.32	1.474	1.860	1.54	1.58	1.850	1.88	1.95	1.861	1.724	1.260	1.140	0.938	1.160
0.7	3.28	3.55	1.28	1.36	1.475	1.871	1.56	1.61	1.866	1.91	2.00	1.883	1.709	1.274	1.168	0.972	1.168
1.0	3.43	3.87	1.30	1.40	1.484	1.902	1.59	1.66	1.892	1.97	2.06	1.921	1.690	1.293	1.203	1.011	1.179
1.5	3.74	—	1.33	1.47	1.495	—	1.62	1.75	1.943	2.06	2.16	1.986	1.661	1.317	1.263	1.047	1.180
2.0	4.04	—	1.36	1.54	1.516	2.029	1.66	1.84	1.999	2.13	2.25	2.051	1.633	1.332	1.307	1.069	—
2.5	4.33	—	1.39	1.59	—	—	1.70	1.92	2.057	2.19	2.34	2.113	1.605	1.336	1.306	1.088	—
3.0	4.65	—	1.43	1.65	1.565	2.175	—	1.99	2.112	2.28	2.44	2.164	1.578	1.332	1.265	1.106	—
3.5	4.92	—	1.46	1.69	—	—	—	—	2.160	2.35	2.53	2.203	—	—	1.195	1.122	—
4.0	5.17	—	—	—	1.594	2.291	—	—	2.196[b]	2.43	—	2.235	1.524	1.292	—	1.135	—
4.5	—	—	—	—	—	—	—	—	—	—	—	2.257	—	—	—	—	—
5.0	—	—	—	—	1.590	2.364	—	—	—	—	—	2.264	1.472	1.238	—	—	—
6.0	—	—	—	—	—	2.335	—	—	—	—	—	—	1.421	1.157	—	—	—
7.0	—	—	—	—	—	—	—	—	—	—	—	—	1.370	—	—	—	—

a. Limiting values
b. At 3.9 mol/l

Source: Robinson and Stokes [6]. *(Copyright 1959, Academic Press. Reprinted with permission.)*

- In most practical applications involving use of the diffusion coefficient, there are much larger uncertainties and inaccuracies in the equations employed (such as those that describe the environment) than those introduced by even approximated values of the diffusion coefficients.

2.16.3 Environmental Importance

Diffusion in bulk fluids is rarely a primary factor in dispersion of materials in the environment; hydraulic flow, convection, and mechanical mixing are generally orders of magnitude more important. However, there are some special circumstances in which diffusion does assume a significant role. These include diffusion in a fluid constrained in a porous medium, diffusion very near a gas/liquid interface that limits the rate of evaporation/condensation, diffusion at a liquid/solid interface, which may limit transfer of nutrients to the roots of a plant, or the rate of dissolution of a mineral, for example. In each of these cases, either bulk flow is limited (as in the case of porous media) or a stagnant layer only a few molecules thick exists at an interface.

However, even in the case of saturated porous media, such as groundwater in an aquifer, the overall dispersion of a component has been calculated as 100 times that of the true diffusion contribution if the pore velocities exceed 0.002 cm/s (5.7 ft/day) [7]. (Slower pore velocities than this may exist in deep artesian ground waters [3].) For the case of a single non-volatile solute in a saturated porous medium, and where the ions are not adsorbed on the walls [4],

$$F = -\mathcal{D}\theta\lambda \frac{dC}{dX} \tag{2}$$

where θ is the void fraction of the medium (an effective cross-section for diffusion) and λ is a "tortuosity" or impedance factor. One can readily appreciate that in a given soil, for example, both θ and λ may vary considerably over the area involved, and that cracks, firmness, or other contributions to non-uniform diffusion or hydraulic flow can have significant impact on the observed flux. If there is interaction such as ion exchange or adsorption on the medium, an additional complex flux term (F_E) is added to account for that factor. Also, if the porous medium is not fully saturated (i.e., it has some vapor volume) and the solute is volatile, further terms must be employed to account for diffusion in the vapor space.

Thus, while a working estimate of $\mathcal{D}$ is important for flux calculations, many uncertainties appear elsewhere in the equation; therefore, close estimates are usually not required. In bulk solution, $\mathcal{D}$ will be found to vary with concentration, solution viscosity, temperature, and ionic strength, but in porous media these variations can often be ignored.

2.16.4 Estimation Methods

The diffusion coefficient of a neutral molecule is often described in terms of the Stokes-Einstein equation [6]:

$$\mathcal{D} = \frac{RT}{6\pi\eta r} \qquad (3)$$

in which R is the universal gas constant, T is temperature (K), η is solution viscosity in centipoise (cP), and r is the effective radius of the solute. This expression is not generally useful for direct estimation of $\mathcal{D}$, as r is not readily known for a hydrated ion, η is rarely available, and the equation itself does not always give a good fit with experimental data. (For example, it seems empirically that η should be raised by a power of -0.45 to -0.66 to achieve the best fit, but this exponent is not predicted by theory.)

If conductance data are available, better estimates can be obtained from an equation such as the following [1]:

$$\mathcal{D}^0 = \frac{2RT}{\mathcal{F}^2} \tau_+^0 \tau_-^0 \Lambda^0 \qquad (4)$$

Equation 4 relates $\mathcal{D}^0$ (the diffusion coefficient at infinite dilution) to R, T, $\mathcal{F}$ (the Faraday), τ_+^0 and τ_-^0 (the transference numbers of the positive and negative ions at infinite dilution), and Λ^0 (the limiting specific conductance).[3] While Λ^0 may be available or measured without great difficulty, τ_+ and τ_- may be more difficult to find in the literature. Therefore, the Nernst-Haskel equation [5] may be preferred:

$$\mathcal{D}^0 = \frac{RT}{\mathcal{F}^2} \frac{1/n^+ + 1/n^-}{1/\Lambda_+^0 + 1/\Lambda_-^0} \qquad (5)$$

where n^+ and n^- are the charges on the positive and negative ion, and Λ_+^0 and Λ_-^0 are the limiting specific conductances of these ions. The data in Table 2.16-3 are useful for estimating some values of Λ^0. As can be seen by inspection of equations 4 and 5,

$$1/\Lambda^0 = (1/\Lambda_+^0 + 1/\Lambda_-^0)(\tau_-\tau_+)(n^+n^-)/(n^+ + n^-) \qquad (6)$$

---

3. The equation as written applies to 1:1 or monovalent electrolytes. If a 2:1 or 1:2 salt such as $CaCl_2$ were under study, the equation would become:

$$\mathcal{D}^0 = \left(\frac{3}{2}\right) \frac{RT}{\mathcal{F}^2} \tau_+^0 \tau_-^0 \Lambda^0$$

in which 3/2 is a stoichiometric factor. The generalized form of the stoichiometric factor is $(n^+ + n^-)/n^+n^-$.

TABLE 2.16-3

Limiting Ionic Conductance[a] in Water at 25°C

[Units = A/(cm^2)(V/cm)(g-equiv/cm^3) = cm^2/(ohm)(g-equiv)]

Cation	Λ_+^0	Anion	Λ_-^0
H^+	349.8	OH^-	197.8
Li^+	38.7	Cl^-	76.4
Na^+	50.1	Br^-	78.2
K^+	73.5	I^-	76.9
Cs^+	77.3	NO_3^-	71.4
NH_4	73.4	ClO_4^-	67.3
Ag^+	61.9	ClO_3^-	64.6
Tl^+	74.7	BrO_3^-	55.8
½Mg^{+2}	53.1	HCO_3^-	44.5
½Ca^{+2}	59.5	HCO_2^-	54.6
½Sr^{+2}	50.5	$CH_3CO_2^-$	40.9
½Ba^{+2}	63.6	$ClCH_2CO_2^-$	39.8
½Cu^{+2}	54	$CNCH_2CO_2^-$	41.8
½Zn^{+2}	53	$CH_3CH_2CO_2^-$	35.8
⅓La^{+3}	69.5	$C_6H_5CO_2^-$	32.3
⅓Ce^{+3}	69.9	$HC_2O_4^-$	40.2
		½$C_2O_4^{-2}$	74.2
		½SO_4^{-2}	80

a. When describing limiting ionic conductances, it is customary to use the convention 1/n (ion)^n in naming a polyvalent ion, e.g., ½Zn^{+2}. This reflects the fact that the values given are calculated on a gram-equivalent, not a molar concentration basis, and so the ionic entity is described in this manner. The values of Λ^0 given here are to be used as is in equations 4, 5 and 6.

Source: Harned and Owen [2].

It is often necessary to modify the calculated value of $\mathcal{D}^0$ for a different concentration or temperature. Diffusion coefficients at temperatures other than 25°C are usually estimated by assuming that

$$\mathcal{D}^0_t = \mathcal{D}^0_{25} \frac{\eta_{25}}{\eta_t} \qquad (7)$$

in which $\mathcal{D}^0_t$ and $\mathcal{D}^0_{25}$ are the diffusion coefficients at temperatures t and 25°C, and η_t and η_{25} are the viscosities of the solvent (water) at these temperatures. Equation 7 is based on the Walden Rule [2], which states that the product of viscosity and limiting mobility (or conductance) is relatively constant. Table 2.16-4 lists water viscosity data that may be used for estimation by this method.

TABLE 2.16-4

Water Viscosity as a Function of Temperature

°C	η(cP)	°C	η(cP)	°C	η(cP)
0	1.787	35	0.7194	70	0.4042
5	1.519	40	0.6529	75	0.3781
10	1.307	45	0.5960	80	0.3547
15	1.139	50	0.5468	85	0.3337
20	1.002	55	0.5040	90	0.3147
25	0.8904	60	0.4665	95	0.2975
30	0.7975	65	0.4335	100	0.2818

Source: Weast [8]

Different methods are used to estimate diffusion coefficients at concentrations other than high dilution. The most general of these was developed by Falkenhagen [2]:

$$\eta_c/\eta_0 = 1 + A\sqrt{C} + BC \qquad (8)$$

in which η_c and η_0 are viscosities at concentration C and for pure water respectively, and A and B are empirical constants. This viscosity ratio is then used to correct $\mathcal{D}^0$ in a manner analogous to equation 7 above.

$$\mathcal{D} = \mathcal{D}^0 (\eta_c/\eta_o) \qquad (9)$$

Values of some A and B coefficients are shown in Table 2.16-5. (The fact that some B coefficients are positive and others negative is due to the different "structure-making and structure-breaking" characteristics of the salts in question.)

TABLE 2.16-5

Viscosity A and B Coefficients for Some Common Electrolytes

(for use with equation 8)

Electrolyte	Coefficients A	Coefficients B	Concentration Limit (mol/l)
NH_4Cl	0.0057	−0.0144	0.2
NaCl	0.0067	0.0244	0.2
KCl	0.0052	−0.0140	0.2
KBr	0.00474	−0.0480	0.1
KNO_3	0.0050	−0.053	0.1
$KClO_3$	0.0050	−0.031	0.1
$KBrO_3$	0.0058	−0.001	0.1
$KMnO_4$	0.0058	−0.066	0.1
CsI	0.0039	−0.118	0.2
$CsNO_3$	0.0043	−0.092	0.02
$AgNO_3$	0.0063	0.045	0.1
K_2SO_4	0.01406	0.194	0.1
K_2CrO_4	0.0133	0.152	0.1
$BaCl_2$	0.0201	0.207	0.1
$LaCl_3$	0.0304	0.567	0.1
$K_4Fe(CN)_6$	0.0370	0.366	0.1
$MgSO_4$	0.0225	a	b
$MnSO_4$	0.0231	a	b
$CuSO_4$	0.0229	a	b
$CdSO_4$	0.0232	a	b
$Cr_2(SO_4)_3$	0.0495	a	b
$Ca_3[Fe(CN)_6]_2$	0.0467	a	b
$Ca_2Fe(CN)_6$	0.0495	a	b

a. B coefficients not available for these salts. n_c/n_o must be estimated on the basis of the A coefficient alone.
b. No concentration limits listed in source.

Source: Harned and Owen [2].

Example 1 Estimate the diffusion coefficient of cesium chloride at 25°C by means of the Nernst-Haskel equation.

The following parameters apply:

$$
\begin{aligned}
R &= 8.31 \text{ joules/mol-deg} \\
T &= 298 \text{ K} \\
\mathcal{F} &= 96{,}500 \text{ coulombs/g-equiv} \\
\left.\begin{matrix} n^+ \\ n^- \end{matrix}\right\} &= 1 \text{ for 1:1 salts} \\
\Lambda_+^0 &= 77.3 \text{ cm}^2/(\text{ohm})(\text{g-equiv}) \\
\Lambda_-^0 &= 76.3 \text{ cm}^2/(\text{ohm})(\text{g-equiv})
\end{aligned}
\quad \text{from Table 2.16-3}
$$

Use of the above units will yield $\mathcal{D}^0$ in cm^2/s.

Substituting in equation 5,

$$\mathcal{D}^0 = \left(\frac{8.31 \times 298}{(96{,}500)^2}\right)\left(\frac{1/1 + 1/1}{1/77.3 + 1/76.3}\right)$$

$$= 2.04 \times 10^{-5} \text{ cm}^2/\text{s}$$

This matches the measured value listed in Table 2.16-1.

Example 2 The measured diffusion coefficient for KCl at 25°C is 1.994×10^{-5} cm^2/s [2]. What is the estimated value at 4°C?

In order to use equation 7, we need the viscosities (η) of water at the two temperatures. From Table 2.16-4, η_{25} = 0.8904 cP and η_4 (interpolated) = 1.573 cP. Substituting in equation 7,

$$\mathcal{D}_4^0 = 1.994 \times 10^{-5}\left(\frac{0.8904}{1.573}\right)$$

$$= 1.129 \times 10^{-5} \text{ cm}^2/\text{s}$$

The measured value [2] is 1.134×10^{-5} cm^2/s.

2.16.5 Literature Cited

1. Cleary, J.G., G.E. Von Neida and W.T. Lindsay, Jr., "Diffusion and Hideout in Crevices," Report No. EPRI-NP-2979 to Electric Power Research Institute, Palo Alto, Calif. (1983).
2. Harned, H.S. and B.B. Owen, *The Physical Chemistry of Electrolyte Solutions,* 3rd ed., Reinhold Publishing Corp., New York (1958).
3. Lerman, A., *Geochemical Processes*, John Wiley & Sons, New York (1979).
4. Nye, P.H., "Diffusion of Ions and Uncharged Solutes in Soils and Soil Clays," *Adv. Agron.*, **31**, 225-69 (1979).

5. Reid, R.C., J.M. Prausnitz and T.K. Sherwood, *The Properties of Gases and Liquids,* 3rd ed., McGraw-Hill Book Co., New York (1977).
6. Robinson, R.A. and R.H. Stokes, *Electrolyte Solutions,* 2nd ed., Academic Press, New York (1959).
7. Tucker, W.A. and L.H. Nelken, "Diffusion Coefficients in Air and Water" in *Handbook of Chemical Property Estimation Methods,* W.J. Lyman, W.F. Reehl and D.H. Rosenblatt (eds.), McGraw-Hill Book Co., New York (1982).
8. Weast, R.C., M.J. Astle and W.H. Beyer (eds.), *CRC Handbook of Chemistry and Physics,* 67th ed., CRC Press, Inc., Boca Raton, FL (1986).

2.17 RADIOACTIVE PROCESSES AND PROPERTIES

Radioactive decay can affect both the transport and transformation of materials in the environment. The decay occurs spontaneously and is independent of all external physical and chemical influence. In general, the environmental effects produced are so small or so rarely encountered as to be essentially of academic interest.

In some special situations, however, radioactive decay is of practical significance in the context of this handbook. This section discusses three aspects of radioactivity that could be of importance to the reader:

- The spontaneous creation of new chemical species through radioactive decay can have troublesome consequences when the newly generated species are themselves radioactive (section 2.17.1).
- The phenomenon of nuclear recoil can transport individual atoms over unusually large distances through solid and liquid materials (section 2.17.2).
- Materials occurring at the extremely low concentrations characteristic of many radioactive materials exhibit changes in chemical properties that often result in anomalous chemical behavior (section 2.17.3).

2.17.1 Creation of New Chemical Species

Almost all types of radioactive decay lead to the formation of atoms of another element. Except in the case of spontaneous fission, in which process the large parent nucleus divides into two smaller fragments of about equal mass, the atomic number of the resulting nucleus is only one or two units different from that of the parent. Nevertheless, the chemical properties of the newly formed atoms are likely to be substantially different from those of the parent and can present special problems to users of radionuclides.

SINGLE TRANSMUTATIONS

Radioactive decay that is limited to a single transmutation results in conversion of the initial species to the product species at an exponential rate, reflecting the probabilistic nature of the process. The probability, p, that an individual atom of a particular radionuclide will decay in a time interval Δt is independent of everything except the length of that time interval and the decay constant, λ, that characterizes the disintegration probability for that nuclide. Thus, for sufficiently short time intervals, $p = \lambda\Delta t$. The probability that the atom will *not* disintegrate during the interval Δt is $1-p$, or $1-\lambda\Delta t$. Similarly, for any subsequent number, n, of such intervals, the probability of survival is $(1-\lambda\Delta t)^n$. For a total time $t = n\Delta t$, $1-p = (1-\lambda t/n)^n$ as Δt becomes infinitely small and n approaches infinity. Because the general mathematical expression for such a process is $e^x = \lim_{n\to\infty}(1 + x/n)^n$, the substitution of λt for x gives $e^{-\lambda t}$ for the limiting value as $n\to\infty$:

$$e^{-\lambda t} = \lim_{n\to\infty}\left(1 + \frac{\lambda t}{n}\right)^n \qquad (1)$$

For a large initial population, N_0, of atoms of a given nuclide, the number of atoms remaining after time t is $N=N_0e^{-\lambda t}$. This, of course, is similar to the rate law for any monomolecular reaction. A convenient way of expressing the decay constant λ for a given species is to show it as the time required for any given (large) number of atoms to decay to half its initial value. Thus, for $t=T_{1/2}$, $N=N_0/2$ and $\lambda = \ln 2/T_{1/2} \approx 0.693/T_{1/2}$.

Example 1 32-Phosphorus, a widely useful radiotracer, has a half-life of 14.3 days and decays by beta emission to form 32-sulfur. One micromole of ^{32}P has been made in a nuclear reactor. How much ^{32}S will be created when it decays?

The number of ^{32}P atoms in 1 micromole is

$$\frac{6.02 \times 10^{23}}{10^6} = 6.02 \times 10^{17}$$

After 14.3 days, the number of ^{32}P atoms remaining is

$$\frac{6.02 \times 10^{17}}{2} = 3.01 \times 10^{17}$$

and an equal number of ^{32}S atoms will have formed.

After 10 half-lives (143 days), the number of ^{32}P atoms remaining will have been reduced to

$$\frac{N_0}{2^{10}} = \frac{6.02 \times 10^{17}}{1024} = 5.88 \times 10^{14}$$

and the number of ^{32}P atoms will be

$$(6.02 \times 10^{17}) - (5.88 \times 10^{14}) = 6.01 \times 10^{17}$$

In the above example, as in most instances of radioactive decay, the product atom is not radioactive and is therefore stable. Furthermore, the amounts created by the process are infinitesimal in comparison with those already in the environment.

MULTIPLE SEQUENTIAL DECAY

In some instances radioactive decay produces a daughter atom that is also radioactive, with its own characteristic half-life and type of emission (alpha, beta, etc.). Occasionally, the process continues through several unstable nuclides before reaching a stable end-product. Each of the nuclides formed in this sequential decay process has a different elemental identity from its immediate parent and therefore demonstrates different chemical and physical properties. Furthermore, their radioactivity makes them potentially hazardous to human health, even in the small quantities that may be generated. For these reasons, they constitute a special problem for environmental managers.

For example, consider a composite solid waste containing an isotope such as 99-molybdenum; this isotope decays with a 67-hour half-life to form 99m-technetium,[1] a metastable isotope with a 5.9-hour half-life, which in turn decays (through isomeric transition and gamma emission) to long-lived 99-technetium:

$$^{99}\text{Mo} \xrightarrow[T_{1/2} = 67\text{ hr}]{\beta^-} {}^{99m}\text{Tc} \xrightarrow[T_{1/2} = 5.9\text{ hr}]{\gamma} {}^{99}\text{Tc}$$

The composition of the resulting mixture varies as the parent species is depleted and its progeny accumulate and successively decay. The composition is also susceptible to change by selective chemical leaching, which can favor either technetium or molybdenum, depending on the chemical form of each element and the chemical environment to which it is exposed [7,10-12].

Although the third nuclide in this multiple-decay sequence is also radioactive, its half-life (2.12×10^5 years) is so long that it can be considered stable for purposes of illustrating two special types of equilibrium that can exist between the concentrations of the nuclides in a successive decay process.[2] These are called "transient equilibrium" and "secular equilibrium" and are illustrated schematically in Figure 2.17-1. (Note that values in these figures are arbitrary ones, used only for purpose of illustration.)

Figure 2.17-1a represents schematically the relationship between a longer-lived parent nuclide and a shorter-lived daughter. The ratio of numbers of parent atoms (N_1) to numbers of daughter atoms (N_2) is given by:

$$\frac{N_1}{N_2} = \frac{\lambda_2 - \lambda_1}{\lambda_1} \qquad (2)$$

(As discussed earlier, for $t = T_{1/2}$, $\lambda = 0.693/T_{1/2}$.)

During this "transient equilibrium," the decay rate of the daughter appears to be identical with that of the parent, because a daughter atom cannot decay until it is formed. Of course, if the supply of new daughter atoms is cut off by physical removal (e.g., chemical separation) of the parent, the true decay rate of the daughter is immediately seen.

An example of transient equilibrium that has practical significance, in that the product nuclide is widely used in medical research and chemical analysis, is ^{99m}Tc. In commercial practice, ^{99m}Tc is generated in cartridges of ion-exchange resins onto which quantities of ^{99}Mo have been loaded. The relatively short-lived metastable technetium nuclide is allowed to grow into radioactive equilibrium with its parent nuclide, a process that requires about five times the half-life of ^{99m}Tc to exceed 95% of

1. The letter "m" following the atomic number indicates a metastable species that does not change its isotopic or elemental identity when it decays; it undergoes only a loss of energy through photon radiation. This process is known as "isomeric transition."
2. Note that this "equilibrium" is unlike chemical equilibrium, in that it refers to the relative abundances of radionuclides formed by a series of completely irreversible transformations. It is analogous to "steady state" concentrations in sequential chemical reactions.

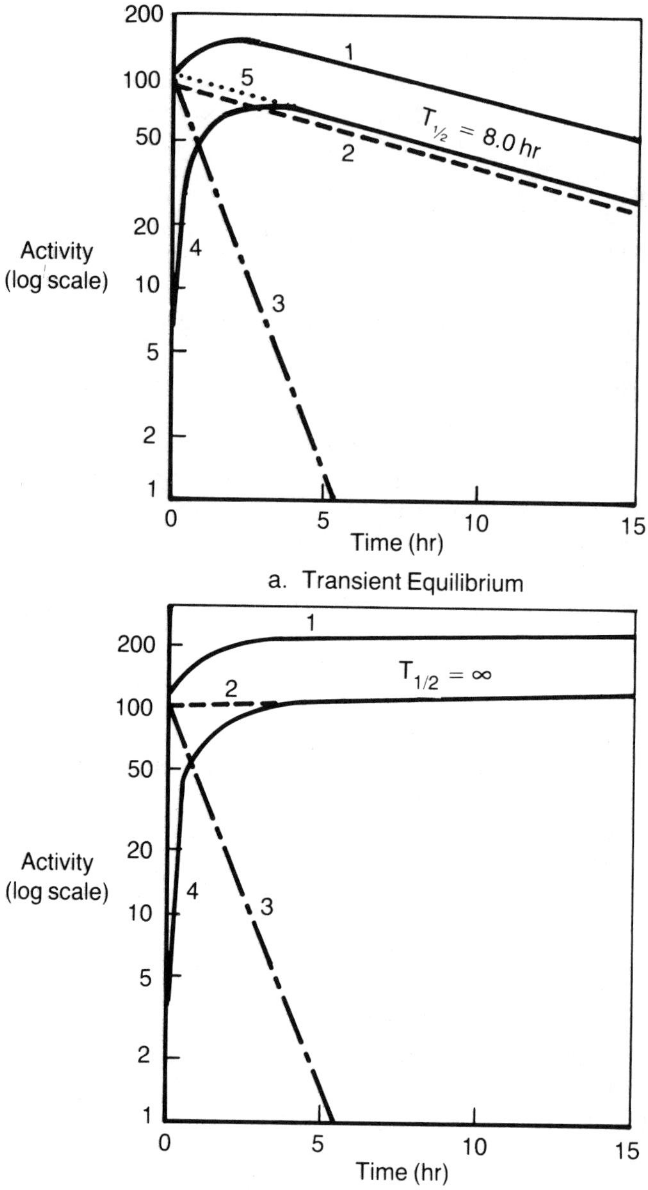

b. Secular Equilibrium

(1) Total activity of an initially pure parent fraction.

(2) Activity due to parent. (In secular equilibrium this is also the total daughter activity in parent-plus-daughter fractions.)

(3) Decay of freshly isolated daughter fraction ($T_{1/2}$ = 0.8 hr).

(4) Daughter activity growing in freshly purified parent fraction.

(5) Total daughter activity in parent-plus-daughter fractions. (Transient equilibrium only.)

Source: Adapted from Friedlander *et al.* [5]. *(Copyright 1981, John Wiley & Sons. Reprinted with permission.)*

FIGURE 2.17-1 Equilibria in Multiple Sequential Decay Processes

its maximum potential concentration. The desired technetium tracer may then be extracted, free of the parent molybdenum, by the ion-exchange process. The re-supply of daughter atoms proceeds constantly, governed only by the remaining supply of parent atoms, which ultimately drops to a level that is not practically useful. Meanwhile, the daughter-atoms can be repeatedly "milked" from the mixture.[3]

The other form of radioactive equilibrium, known as "secular equilibrium" (Figure 2.17-1b), is actually a limiting case of transient equilibrium in which the parent is so long-lived as to exhibit no measurable decrease in its activity over a time that is long compared to that of the daughter's half-life. A case in point is that of naturally occurring 238-uranium, whose half-life is 4.51×10^9 years; this element gives rise to a sequence of 17 radioactive daughter-nuclides, all shorter-lived than ^{238}U itself, leading ultimately to stable ^{206}Pb. (See Table 2.17-2.) The figure depicts a simple two-step case of secular equilibrium, representing the relationship between ^{238}U and any of its shorter-lived daughter products. When a radioactive daughter product has a much shorter half-life than that of the parent and remains physically comingled with the parent, it appears to decay at the same rate as the parent once equilibrium is attained. Only if the mixture is disturbed (such as by the separation of ^{226}Ra from the parent ^{238}U by a chemical extraction process, or by the escape of volatile ^{222}Rn to the atmosphere) is the natural balance of nuclide abundances upset.

Because gamma radiation is easily detected by remote sensors, prospectors for uranium have traditionally depended on the detection of such late decay products as ^{214}Bi and ^{214}Pb. A crucial assumption that must be made in the use of this technique is that the secular equilibrium is undisturbed by any natural or artificial process, an assumption that often proves to be invalid. Geologists have discovered that radium and thorium are sometimes displaced from each other and/or from parent uranium by natural geochemical processes [10,12,13]. The spontaneous escape of ^{222}Rn from rocks, minerals and soil removes the telltale ^{214}Bi and ^{214}Pb nuclides from their original sites and can produce environmental health hazards that would not otherwise exist.

Evans [4] has examined the problems of calculating the relative concentrations of the physiologically hazardous, short-lived daughters of ^{222}Rn. Figure 2.17-2 illustrates one way of portraying the complex radioactive equilibria created by the "grow-in" of short-lived daughters in an initially pure quantity of longer-lived ^{222}Rn ($T_{1/2} = 3.8$ days). With respect to the parent activity, the equilibria of the first four daughter nuclides is secular, because each of their half-lives is quite short compared with that of ^{222}Rn. The equilibrium between the second and third daughters, ^{214}Pb and ^{214}Bi, whose half-lives are 26.8 and 19.7 minutes respectively, is transient.

3. Technetium has no stable isotopes. Its presence in the sun has been detected, but all of the technetium on earth is man-made; some 18 isotopes of the element have been created and recognized [7].

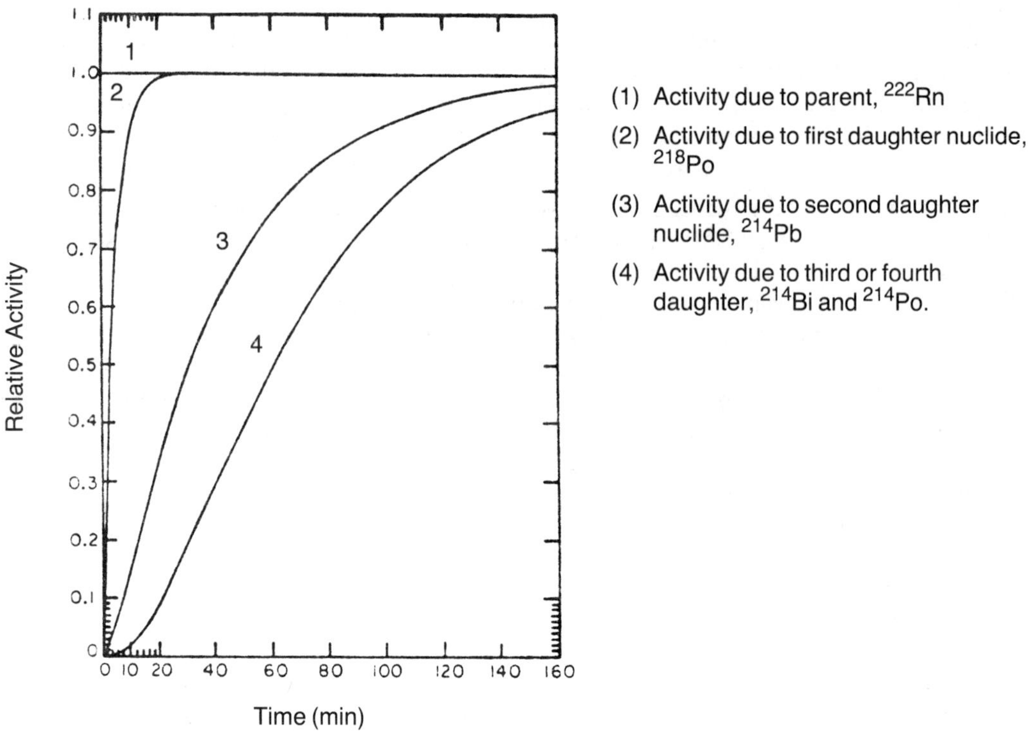

Source: Adapted from Evans [4]. *(Copyright 1969, Pergamon Press. Reprinted with permission.)*

FIGURE 2.17-2 Growth of Activity of Individual Short-lived Decay Products in Constant Source of Radon Having Unit Activity

Sequential radioactive decay need not result in either of the forms of equilibrium described above. Returning to the example of the relatively short-lived parent-daughter pair of ^{99}Mo$\rightarrow$^{99m}Tc producing the much longer lived nuclide ^{99}Tc, it is clear that there is never a constant proportionality between the decay rates of ^{99}Mo and ^{99}Tc (ignoring, for the sake of illustration, the metastable intermediate). Figure 2.17-3 illustrates this lack of equilibrium; although, as in Figure 2.17-1, the values are only suggestive, curve 2 might represent the combined activity of the ^{99}Mo $\rightarrow$ ^{99m}Tc pair and curve 4 that of the ^{99}Tc daughter. The latter eventually becomes the only activity remaining after the short-lived parent has decayed to an insignificant level. Similarly, the fifth successive decay product of ^{222}Rn is ^{210}Po, whose half-life is 138 days. Thus, a condition of "no equilibrium" exists between ^{222}Rn, together with its shorter-lived daughters, and the longer-lived ^{210}Po.

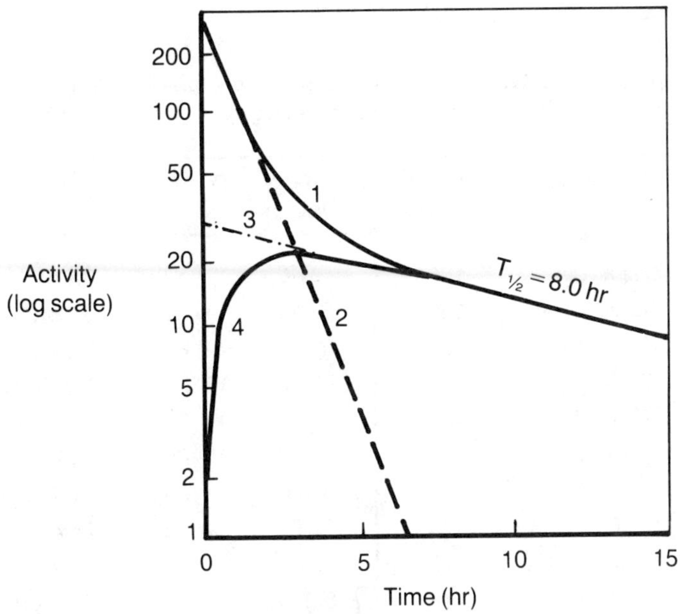

(1) Total activity
(2) Activity due to parent ($T_{1/2} = 0.8$ hr)
(3) Extrapolation of final decay curve to time zero
(4) Daughter activity in initially pure parent

Source: Adapted from Friedlander *et al.* [5]. *(Copyright 1981, John Wiley & Sons. Reprinted with permission.)*

FIGURE 2.17-3 Radioactive Decay Without Equilibrium

2.17.2 Recoil Transport

Radioactive decay processes involve spontaneous losses of energy from atomic nuclei, which can occur in several ways. In most such processes there is little or no loss of mass in the parent nuclei, since the emissions are usually in the form of beta and/or gamma radiation and therefore cause a change of only one electron mass unit. (Beta decay involves a one-electron loss; gamma and x-radiation are photons and involve no mass change. A few nuclei decay by positron emission or by electron-capture, processes that result, effectively, in one-electron gains rather than losses.) Some elements, however —principally those higher than lead in the periodic table — emit an alpha particle, which consists of two neutrons and two protons; still more rarely, heavy elements in the transuranic series undergo spontaneous fission, yielding two quite massive fragments of roughly equal mass. Energies of some of the alpha-emitting radionuclides are listed in Table 2.17-1. Note that, in general, the half-life of a nuclide decreases as the energy of its alpha emission increases.

TABLE 2.17-1

Principal Naturally Occurring and Transuranic Alpha Emitters[a] (in order of increasing energy)

Energy (MeV)	Source	Half-life	Yield[b] (%)	Energy (MeV)	Source	Half-life	Yield[b] (%)
1.83	^{144}Nd	2.1×10^{15} y	100	4.946	^{210m}Bi	3×10^{6} y	58
2.14	^{152}Gd	1.1×10^{14} y	100	5.013	^{231}Pa	3.28×10^{4} y	24
2.23	^{147}Sm	1.06×10^{11} y	100	5.028	^{231}Pa	3.28×10^{4} y	23
2.50	^{174}Hf	2×10^{15} y	100	5.058	^{231}Pa	3.28×10^{4} y	11
3.18	^{190}Pt	7×10^{11} y	100	5.105	^{239}Pu	24,411y	12
3.95	^{232}Th	1.40×10^{10} y	24	5.123	^{240}Pu	6560y	24
4.011	^{232}Th	1.40×10^{10} y	76	5.143	^{239}Pu	24,411y	15
4.147	^{238}U	4.468×10^{9} y	25	5.156	^{239}Pu	24,411y	73
4.196	^{238}U	4.468×10^{9} y	75	5.168	^{240}Pu	6560y	76
4.366	^{235}U	7.04×10^{8} y	18	5.234	^{243}Am	7.37×10^{3} y	11
4.396	^{235}U	7.04×10^{8} y	57	5.276	^{243}Am	7.37×10^{3} y	88
4.415	^{235}U	7.04×10^{8} y	4	5.305	^{210}Po	138.4d	100
4.556	^{235}U	7.04×10^{8} y	4	5.306	^{245}Cm	8.5×10^{3} y	7
4.57	^{210m}Bi	3×10^{6} y	6	5.342	^{246}Cm	4.78×10^{3} y	19
4.597	^{235}U	7.04×10^{8} y	5	5.344	^{228}Th	1.913y	28
4.599	^{226}Ra	1600y	6	5.362	^{245}Cm	8.5×10^{3} y	80
4.617	^{230}Th	7.54×10^{4} y	24	5.386	^{246}Cm	4.78×10^{3} y	81
4.684	^{230}Th	7.54×10^{4} y	76	5.42	^{249}Bk	320d	0.0015
4.722	^{234}U	2.45×10^{5} y	28	5.427	^{228}Th	1.913y	71
4.733	^{231}Pa	3.28×10^{4} y	11	5.443	^{241}Am	432y	13
4.765	^{237}Np	2.14×10^{6} y	17	5.447	^{224}Ra	3.66d	6
4.770	^{237}Np	2.14×10^{6} y	19	5.448	^{214}Bi	19.8m	0.012
4.773	^{234}U	2.45×10^{5} y	72	5.456	^{238}Pu	87.74y	28
4.782	^{226}Ra	1600y	95	5.486	^{241}Am	432y	86
4.787	^{237}Np	2.14×10^{6} y	51	5.490	^{222}Rn	3.823d	100
4.863	^{242}Pu	3.76×10^{5} y	24	5.499	^{238}Pu	87.74y	72
4.896	^{241}Pu	14.35y	0.002	5.512	^{214}Bi	19.8m	0.008
4.903	^{242}Pu	3.76×10^{5} y	76	5.52	^{247}Bk	1.4×10^{3} y	58
4.909	^{210m}Bi	3×10^{6} y	36	5.537	^{223}Ra	11.43d	9
4.95	^{227}Ac	21.773y	1.2	5.605	^{223}Ra	11.43d	26
4.951	^{231}Pa	3.26×10^{4} y	22	5.666	^{251}Cf	900y	55

(Continued)

TABLE 2.17-1 (Continued)

Energy (MeV)	Source	Half-life	Yield[b] (%)	Energy (MeV)	Source	Half-life	Yield[b] (%)
5.68	^{247}Bk	1.4×10^3 y	37	6.115	^{242}Cm	163d	74
5.684	^{224}Ra	3.66d	94	6.119	^{252}Cf	2.64y	84
5.707	^{227}Th	18.72d	8	6.278	^{211}Bi	2.14m	16
5.714	^{223}Ra	11.43d	54	6.28	^{219}At	0.9m	97
5.742	^{243}Cm	28.5y	12	6.287	^{220}Rn	55.6s	100
5.745	^{223}Ra	11.43d	9	6.424	^{219}Rn	3.96s	8
5.755	^{227}Th	18.72d	20	6.551	^{219}Rn	3.96s	11
5.763	^{244}Cm	18.11y	23	6.56	^{222}Ra	38s	96
5.806	^{244}Cm	18.11y	77	6.622	^{211}Bi	2.14m	84
5.812	^{249}Cf	351y	84	6.65	^{218}At	2s	6
5.846	^{251}Cf	900y	45	6.70	^{218}At	2s	94
5.976	^{227}Th	18.72d	23	6.777	^{216}Po	0.15s	100
5.987	^{250}Cf	13.1y	17	6.818	^{219}Rn	3.96s	81
6.002	^{218}Po	3.11m	100	7.28	^{211m}Po	25.2s	91
6.031	^{250}Cf	13.1y	83	7.384	^{215}Po	1.78ms	100
6.037	^{227}Th	18.72d	24	7.448	^{211}Po	0.516s	99
6.051	^{212}Bi	60.6m	25	7.687	^{214}Po	164μs	100
6.071	^{242}Cm	163d	26	8.785	^{212}Po	0.30μs	100
6.076	^{252}Cf	2.64y	15	8.88	^{211m}Po	25.2s	7
6.090	^{212}Bi	60.6m	10	11.65	^{212m}Po	45s	97

a. Note: Values given in this table are subject to minor revision from time to time. Persons requiring the best and most accurate values currently available should consult an authoritative source such as the Oak Ridge Nuclear Data Project, Oak Ridge National Laboratories, Oak Ridge TN.

b. Percentage of the total decay events. Isotopes having multiple alpha emissions are listed once for each alpha-energy. If a yield of less than 100% is listed for any isotope, it indicates that other alpha emissions of that isotope also occur; when combined, the various yields total 100%.

Source: U.S. Bureau of Radiological Health [16] and *Chart of the Nuclides,* 13th ed. [17].

In the cases of alpha emission and spontaneous fission, the emission of the fragment imparts enough kinetic energy to the remaining nucleus to cause significant recoil motion. Atoms of heavy elements that undergo alpha decay are typically transported in recoil for distances of 0.1 to 0.2 mm in air at atmospheric pressure. In solids and liquids, the recoil distances are much smaller, and atoms in the bulk of such condensed phases are not sufficiently affected to produce a significant change in the system. However, atoms lying near an interface (e.g., between a host mineral and a gaseous or liquid environment such as air or water) may be transported far enough by recoil to allow them to escape from a medium in which they are relatively immobile to another in which their mobility is much greater [6]. Recoil transport distances are generally larger with spontaneous fission, because the recoil energies of fission fragments are greater and the residual nuclei are smaller [3].

These alternate recoil pathways are shown schematically in Figure 2.17-4, with 226-radium used as an example. Two very fine spherical grains of solid material, 2-μm diameter, are shown in contact near B. Water is present in the stippled zone of the pore, and the open zone at the right is air-filled. Atoms of ^{226}Ra decay in the upper grain, each yielding an alpha particle (as shown at A) and a ^{222}Rn atom.

- Atom A lies at greater depth within the grain than the recoil range, R, so atom A′ is contained within the grain.
- Atom B′ escapes from the upper grain but buries itself in the lower grain.
- After escape from the upper grain, atom C′ loses the remainder of its recoil energy in the water and is free to diffuse through the pores.
- Atom D′ loses little of its recoil energy in the air and buries itself in the lower grain.

Atoms B′ and D′ may escape from their recoil pockets by diffusion before condensation of the excited atoms of the grain (indirect-recoil effect). Within the mean life[4] of ^{222}Rn (5.5 days), the diffusion distance of ^{222}Rn atoms in the solid grain is less than the width of any line in the figure.

By the recoil-transport mechanism described above, radioactive isotopes like ^{222}Ra and ^{222}Rn can leave mineral grains and enter a mobile liquid or gaseous phase, creating a significant environmental hazard. This effect has been suggested as the cause of anomalously high radon emanation from soils and minerals [14,15]. The amount of ^{222}Rn that may escape from the host soil or mineral is highly variable, as shown in Figure 2.17-5.

Recent studies of environmental radiation health hazards have emphasized the importance attached by health physicists to alpha-emitting decay products of ^{222}Rn, which are a demonstrated cause of cancer of the pulmonary system. Control and amelioration of this hazard is a topic of concern to such diverse industries as underground mining, building construction and agriculture.

4. The "mean life" of a radionuclide is defined as the average time for the number of nuclei in a specified state to decrease (i.e., decay) by a factor of e (2.718 . . .).

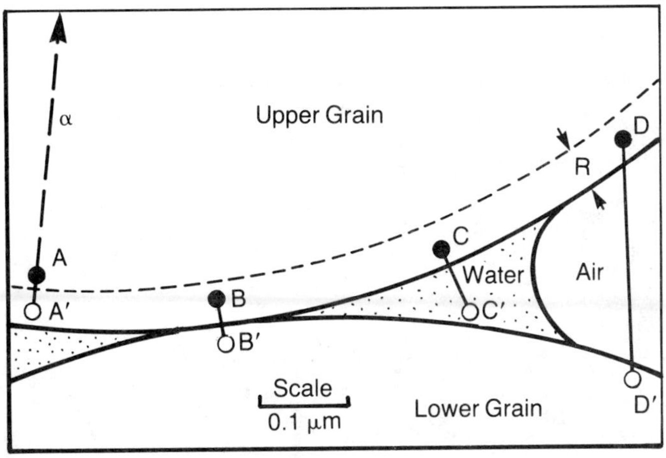

Legend: ● = ^{226}Ra atom, ○ = ^{222}Rn atom, R = recoil range.
(Circles greatly exaggerate size of atoms). See text for discussion.

Source: Adapted from Tanner [15]

FIGURE 2.17-4 Schematic Diagram of Radon Emanation Processes

While radon emanates from rocks and soil almost everywhere, the rate at which this occurs is influenced to a remarkable extent by various environmental conditions such as atmospheric pressure and rain or snowfall [8]. The effect of such temporal influences on the true emanation rate (i.e., the process by which recoil transport of the radon nucleus across a solid-liquid or a solid-vapor interface occurs) is not well understood. As Tanner [14,15] has reported, the production rate of radon in the atmosphere over a radium-bearing soil or rocky earth appears to be significantly influenced in an otherwise unpredictable or "anomalous" manner by, among other things, the presence of an adsorbed liquid layer on the solid.

The evidence of anomalous excess radon emanation from mineral host substances is based on experimental determination of the equilibrium concentrations of ^{222}Rn from samples of crystalline rock in which (a) concentrations of the parent radioisotope ^{222}Ra have been accurately measured by special alpha-radiation detectors, (b) physical properties such as surface area and grain size are known, and (c) the homogeneity of radium distribution has been established. The migration rate of radon atoms in host crystalline substances has been determined both from basic principles and by use of parameters such as atomic radius and lattice dimensions. When all these data are combined, the predicted emanation rate of ^{222}Rn from the solid phase into the surrounding air is reduced by as much as a hundred-fold, compared with the actual production rate within the solid due to decay of ^{226}Ra. The predicted losses are attributable to various mechanisms that would be expected to account for physical entrapment of radon in the solid phase [2,8].

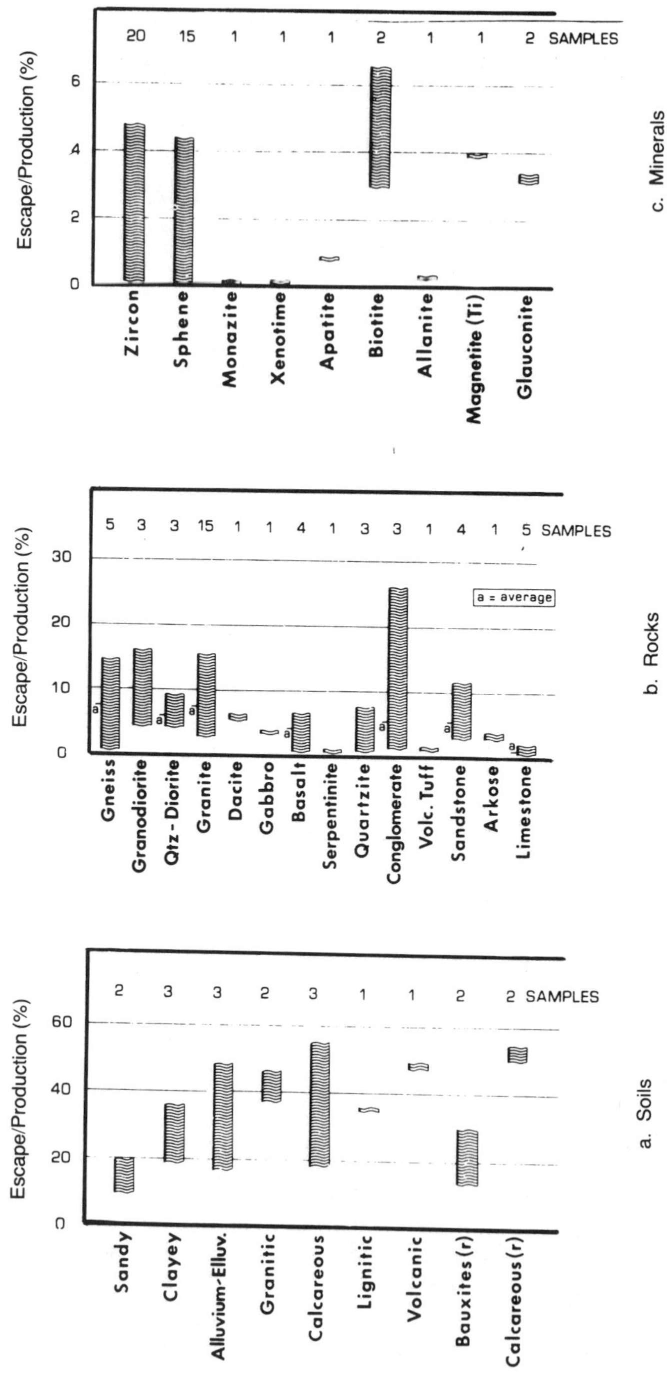

Source: Adapted from Barretto *et al.* [2].

FIGURE 2.17-5 Percentage Escape of 222-Radon from Soils and Minerals

In minerals of density typical of that of rocks and soil, the range of the recoiling atom is on the order of 20-70 nm; thus, only the atoms within that distance from the grain surface have any chance to escape. Moreover, in a microporous host mineral such as sandstone, most escape surfaces are close to the surfaces of adjacent grains, greatly reducing the chance of escape by "direct recoil." Experimental measurements have generally shown, however, that some compensating mechanism actually tends to restore the true emanation yield to a value closer to the actual production rate of ^{222}Rn *in situ*. Reported values for emanating fractions of ^{222}Rn have ranged from 1% to nearly 100% of the theoretical maximum, suggesting that something such as the "indirect recoil" mechanism postulated by Tanner [14,15] is operative to help explain the variation between prediction and experience. The wide range of measured values suggests that experimental variables have not been understood or controlled well enough, and also that existing theoretical models are not adequate to account for observed anomalies. A better method for estimating the emanation rate is clearly needed; this would be useful in establishing standards for the management of uranium mill tailings and other radium-bearing wastes that are important sources of 222-radon.

Recoil or alpha-emission from radioactive decay can also cause physical transport by a form of "internal sputtering" — the ejection of neighboring atoms from a solid surface into an adjacent vapor phase. This process has had significant implications for environmental pollution in the case of 210-polonium, an alpha-emitting isotope that occurs in nature primarily as a sequential decay product of 226-radon [1]. When fabricated into metallic films for industrial purposes, ^{210}Po exhibits instability that has been attributed to such "internal sputtering," resulting in anomalously high diffusion rates through solids, escape from the surface of these solids into the air, and contamination of nearby surfaces. Even such a chemically stable alpha-emitter as 241-americium can create a contamination hazard due to "internal sputtering" and direct alpha-particle damage when incorporated at high concentrations in metal foils and ribbons.

The energy of any spontaneous alpha emission is generally between 2 and 9 MeV (Table 2.17-1), values far greater than the energies of chemical bonds. Thus, recoil transport, combined with the massive ionization and heating of the medium resulting both from recoil and from the alpha particles themselves, is likely to produce some local disruption in any medium that contains alpha-emitters, and potential release to the environment.

2.17.3 Special Properties of "Tracers"

When radioactive substances were discovered, investigators were able to study their chemical behavior only by virtue of the radiation that they produced, because typical concentrations were many orders of magnitude smaller than could otherwise be detected. This principle was later applied in the "tracer" technique, whereby small amounts of radioactive isotopes are incorporated in materials of identical or similar chemical composition so that the "tagged" substance can be identified and measured *in situ* without difficult and laborious separations. These isotopically labeled atoms, ions or molecules intermingle randomly with the matrix material but have essentially no effect on chemical equilibria or on any macroscopic properties such as vapor pressure, solubility or melting point.

A radioactive species may occasionally exist in isolation, unaccompanied by stable isotopes ("carriers"). Even though the quantity may be extremely small, its radioactivity can make it important — not only as a useful experimental tool, but also because of the potential health hazard. The amounts of radiation considered dangerous to human health are typically measured in picocuries. Because one picocurie (pCi) is approximately equivalent to the radioactivity of 10^{-12} gram of 226-radium, the amounts of most radionuclides needed to produce a picocurie are far too small to weigh or to measure by any conventional chemical method.

Until the late 1950s, the handling and processing of radioactive materials at "tracer" concentrations was more an art than a science, because materials in these concentrations often defy conventional mass-action principles of physics and chemistry: they adhere selectively and tenaciously to surfaces of containers, form "radiocolloids" instead of remaining freely dissolved in solutions, and respond erratically to chemical reagents that would be expected to produce complexation, precipitation, and various electrochemical and physical reactions. Moreover, no clear dividing line has yet been established between the concentrations or amounts of material whose behavior follows the rules for relatively large quantities and those that are likely to show anomalies.

The forces that govern such "anomalous" behavior are somewhat better understood now, especially since the advent of adsorption chromatography and the development of the extensive theoretical basis on which it rests. Still, the practicing environmental scientist is likely to find the behavior of ultra-trace-level radioactive materials somewhat unpredictable.

Our knowledge concerning the behavior of the decay products of 238-uranium illustrates the kinds of questions that have not been fully resolved. Table 2.17-2 lists these products, which include 222-radon and its four daughters — 218-polonium, 214-lead, 214-bismuth and 214-polonium. Until fairly recently, radon's chemical and physical behavior puzzled scientists. Although recognized to be a noble gas, its occurrence at ultra-trace-level concentrations made it appear to have unique properties: for example, it was believed to be preferentially soluble in certain natural oils. The behavior of this element has now been explained through a better understanding of adsorption science and of the equilibria in partition between immiscible liquids and between liquids and gases, much of which falls under the general category of "separation science." In fact, at very low concentrations, radon's behavior approaches theoretical ideality in sorption/desorption and partition equilibria.

The chemical behavior of the short-lived radon daughters in the environment remains unexplained, however, because present theoretical models are inadequate and laboratory experimental methods that simulate their properties under the conditions in which nature produces them have not yet been designed. The parent radon atoms are

TABLE 2.17-2

Daughter Nuclides of 238-Uranium

Nuclide[a]	Major Radiation Energies (MeV) and Intensities[b]					
	α		β		γ	
$^{238}_{92}U$	4.15 4.20	(25%) (75%)	---		---	
↓ $^{234}_{90}Th$	---		0.103 0.193	(21%) (79%)	0.063 0.093	(3.5%) (4%)
↓ $^{234}_{91}Pa^m$	---		2.29	(98%)	0.765 1.001	(0.30%) (0.60%)
99.87% ↓ 0.13%						
$^{234}_{91}Pa$	---		0.51 1.13	(66%) (13%)	0.100 0.70 0.90	(50%) (24%) (70%)
↓ $^{234}_{92}U$	4.72 4.77	(28%) (72%)	---		0.053	(0.2%)
↓ $^{230}_{90}Th$	4.62 4.68	(24%) (76%)	---		0.068 0.142	(0.6%) (0.07%)
↓ $^{226}_{88}Ra$	4.60 4.78	(6%) (95%)	---		0.186	(4%)
↓ $^{222}_{86}Rn$	5.49	(100%)	---		0.510	(0.07%)
↓ $^{218}_{84}Po$	6.00	(~100%)	0.33	(~0.019%)	---	
99.98% ↓ 0.02%						
$^{214}_{82}Pb$	---		0.65 0.71 0.98	(50%) (40%) (6%)	0.295 0.352	(19%) (36%)
$^{218}_{85}At$	6.65 6.70	(6%) (94%)	?	(~0.1%)	---	
↓ See next page						

(Continued)

TABLE 2.17-2 (Continued)

Nuclide[a]	Major Radiation Energies (MeV) and Intensities[b]					
	α		β		γ	
$^{214}_{83}$Bi	5.45 5.51	(0.012%) (0.008%)	1.0 1.51 3.26	(23%) (40%) (19%)	0.609 1.120 1.764	(47%) (17%) (17%)
99.98% ↓ 0.02%						
$^{214}_{84}$Po	7.69	(100%)	---		0.799	(0.014%)
$^{210}_{81}$Tl	---		1.3 1.9 2.3	(25%) (56%) (19%)	0.296 0.795 1.31	(80%) (100%) (21%)
$^{210}_{82}$Pb	3.72	(.000002%)	0.016 0.061	(85%) (15%)	0.047	(4%)
$^{210}_{83}$Bi	4.65 4.69	(.00007%) (.00005%)	1.161	(~100%)	---	
~100% ↓ .00013%						
$^{210}_{84}$Po	5.305	(100%)	---		0.803	(0.0011%)
$^{206}_{81}$Tl	---		1.571	(100%)	---	
$^{206}_{82}$Pb	---		---		---	

a. See Tables 2.17-1 and 9.2-1 for half-lives of these elements.

b. Intensities refer to percentage of disintegrations of the nuclide itself, not to original parent of series.

Source: U.S. Bureau of Radiological Health [16]

produced one at a time over periods that are long in relation to chemical reaction rates and to the lives of the daughter atoms themselves, so they and the daughters are created in extreme isolation in nature. Furthermore, each daughter atom is created in an initial state of high ionization and high chemical activity — a so-called "hot atom" — which leads to chemical and physical behavior quite unlike that associated with the corresponding element in ordinary "bulk" concentration. For example, daughter ions produced in air and in the presence of condensation nuclei or other solid and liquid surfaces tend to "plate out"; this phenomenon is widely observed but not well understood, and methods to control it are still based as much on folklore as on established scientific theory.

The potential hazards of radionuclides depend on many factors that include, for example, chemical composition and status, concentration, radiations and energies emitted, half-life, availability, etc. The popular press typically associates high environmental hazard with long radioactive half-life, the idea of persistence being assumed to be of greatest concern. To the environmental scientist, however, the anomalous behavior of some short-lived radionuclides may be of greater importance.

2.17.4 Literature Cited

1. Bagnall, K.W., "The Chemistry of Polonium," in *Advances in Inorganic Chemistry and Radiochemistry,* Vol. IV, H.N. Emeleus and A.G. Sharpe (eds.), Academic Press, New York (1962).

2. Barretto, P.M.C., R.B. Clark and J.A.S. Adams, "Physical Characteristics of Radon-222 Emanation from Rocks, Soils and Minerals: Its Relation to Temperature and Alpha Dose," in *The Natural Radiation Environment,* Vol. II (Proc. Second International Symp.), J.S. Adams, W.M. Lowder and T.F. Gesell (eds.), Houston (1972). Available from NTIS as CONF-720805-P2.

3. Evans, R.D., *The Atomic Nucleus*, McGraw-Hill Book Co., New York (1951)

4. Evans, R.D., "Engineers' Guide to the Elementary Behavior of Radon Daughters," *Health Phys.,* **17**, 229-52 (1969).

5. Friedlander, G., J.W. Kennedy, E.S. Macias and J.M. Miller, *Nuclear and Radiochemistry*, 3rd ed., John Wiley & Sons, New York (1981).

6. Kigoshi, K., "Alpha-Recoil Th-234: Dissolution into Water and the U-234/U-238 Disequilibrium in Nature," *Science*, **173**, 47-48 (1971)

7. Kotegov, K.V. et al., "Technetium," in *Advances in Inorganic Chemistry and Radiochemistry*, Vol. XI, H.J. Emeleus and A.G. Sharpe (eds.), Academic Press, New York (1968).

8. Kraner, H.W., G.L. Schroeder and R.D. Evans, "Measurements of the Effects of Atmospheric Variables on Radon-222 Flux and Soil Gas Concentrations," in *The Natural Radiation Environment*, J.S. Adams and W.M. Lowder (eds.), University of Chicago Press, Chicago, 191-215 (1964).

9. National Council on Radiation Protection and Measurements, "Evaluation of Occupational and Environmental Exposure to Radon and Radon Daughters in the United States," NCRP Report No. 78, Bethesda, MD (1984).

10. Rosholt, J.N. and A.J. Bartel, "Uranium, Thorium and Lead Systematics in Granite Mountains, Wyoming," *Earth Planet. Sci. Lett.*, 141-47 (1969).

11. Rosholt, J.N., B.R. Doe and M. Tatsumoto, "Evolution of the Isotopic Composition of Uranium and Thorium in Soil Profiles," *Geol. Soc. Am. Bull.*, **77**, 987-1004 (1966).

12. Rosholt, J.N., R.E. Zartman and I.T. Nkono, "Lead Isotope Systematics and Uranium Depletion in the Granite Mountains, Wyoming," *Geol. Soc. Am. Bull.*, **84** (1973)

13. Stuckless, S.S. and C. Pizes Ferreira, "Labile Uranium in Granite Rocks," in *Exploration for Uranium Ore Deposits*, IAEA-SM-208/17, Vienna (1976).

14. Tanner, A.B., "Radon Migration in the Ground: A Review," in *The Natural Radiation Environment*, J.S. Adams and W.M. Lowder (eds.), University of Chicago Press, Chicago (1964).

15. Tanner, A.B., "Radon Migration in the Ground: A Supplementary Review," *Proc. 3rd Int'l. Symposium on the Natural Radiation Environment*, Houston (1978).

16. U.S. Bureau of Radiological Health, *Radiological Health Handbook*, International Atomic Energy Agency, Vienna, Austria (1970).

17. Walker, F.W., D.G. Miller and F. Feiner, *Chart of the Nuclides*, 13th ed., General Electric Co., San Jose, CA (1984).

3. KINETICS OF SELECTED PROCESSES

Itamar Bodek, Abraham Lerman

Page

LIST OF TABLES

LIGAND EXCHANGE KINETICS

KINETICS OF ELECTRON TRANSFER (REDOX) REACTIONS

KINETICS OF DISSOLUTION, PRECIPITATION AND CRYSTALLIZATION

LIST OF FIGURES

LIGAND EXCHANGE KINETICS

KINETICS OF ELECTRON TRANSFER (REDOX) REACTIONS

KINETICS OF DISSOLUTION, PRECIPITATION AND CRYSTALLIZATION

3.1 INTRODUCTION

The kinetics or rates of reactions are an important aspect of environmental chemistry, especially with regard to understanding the speciation of chemicals. An understanding of the relative rates of environmental processes allows calculations of speciation based upon an equilibrium model, a non-reaction model or an intermediate model where kinetics must be considered. A simple answer with regard to the kinetics (e.g., that it is either instantaneous or very slow) is useful in determining which way to proceed. Detailed kinetic information is necessary only when the rates of the processes are of the same time frame of interest in the particular environmental system. In that case, a calculation based upon equilibrium conditions may indicate limits to the range of product concentrations. However, as discussed in the following section, relative rates can sometimes affect product identity.

The kinetics of processes is discussed in several locations in this report. These include § 2.3 (Photolysis in Air), § 2.5 (Atmospheric Deposition), § 2.14 (Photolysis in Water), § 2.16 (Diffusion Coefficients) and § 2.17 (Radioactive Processes). In addition, rates are presented in other sections on particular reactions. This chapter focuses on kinetics of ligand exchange (§ 3.2), oxidation-reduction (§ 3.3), and dissolution-precipitation (§ 3.4) reactions. The information presented in this chapter, in general, concerns the following topics:

- Environmental relevance
- Mathematical representation
- Typical values of the rate, or rate constants, and correlations
- The effect of different environments on the reaction rate
- Mechanisms of the particular reaction
- Methods of estimating rates and rate constants.

Sections 3.2 and 3.3 address homogeneous reaction kinetics (i.e., between dissolved ions and dissolved gases), while § 3.4 addresses the rates of heterogeneous reactions such as dissolution, nucleation and recrystallization. The equilibrium conditions for the reactions described in this chapter are given in § 2.9 (Complexation), § 2.10 (Electron Transfer Reactions), and § 2.11 (Solubility and Precipitation Equilibria).

3.2 LIGAND EXCHANGE KINETICS

3.2.1 Introduction

This section addresses rates of substitution of coordinated ligands, especially water molecules, in the primary coordination sphere of metal ions. The principal topics considered are:

- Environmental relevance of the rate of ligand substitution;
- Mathematical representation of reaction rates and their use in estimating the time required for completion of the reaction;
- Mechanism (sequence of steps) of substitution reactions;
- Variation in rates and rate constants due to differences in metal ions, ligands and aqueous environments;
- Typical values of rate constants for substitution in a variety of complex ions; and
- Methods for estimating substitution rates from observed correlations with various properties.

Additional information on complexation can be found in section 2.9. The reading of that section prior to this one may help clarify some topics discussed below. Throughout this section a small k is used to represent a rate constant while a large K is used to represent an equilibrium constant.

3.2.2 Related Environmental Scenarios

An important aspect of environmental chemistry, especially with regard to the speciation of dissolved inorganics, is the determination of the rate and mechanism by which complexation reactions approach equilibrium. When complexation is "fast" (compared with an arbitrary or particularly relevant time frame), equilibrium conditions can be used as a basis for calculation of speciation. If the reaction is very "slow," the reactant will remain in its original form, and little or no reaction occurs. However, if the reaction is "intermediate" in rate, one must consider the reaction process in great detail to estimate the quantities of reactants and products present at various times.

The ease with which a particular complex ion (e.g., $M_{aq}{}^{+x}$) enters into reactions that entail the replacement of one or more of the existing ligands (typically H_2O in environmental systems) in the primary coordination sphere provides a means of assessing reaction rates involving such ions.[1] Taube [23] has characterized metal ions as kinetically *labile* ("fast" substitution rate occurring within time of mixing) or kinetically *inert* ("slow" or "intermediate" substitution rate).

---

1. The term "free" or "uncomplexed" is often used in the literature to describe the aquo complexed metal ion (i.e., $M(H_2O)_y{}^{+x}$ or $M_{aq}{}^{+}$); in reality, the free or uncomplexed ion does not exist in solution.

3.2.3 Mathematical Representation

HALF-LIFE AND OBSERVED RATE

The ligand substitution reaction can be represented as follows:

$$\text{M-X} + \text{Y} \longrightarrow \text{M-Y} + \text{X} \qquad (1)$$

where X is the leaving ligand and Y is the entering ligand on metal ion M.

The *overall* rate of this reaction is generally expressed as

$$\text{Rate} = \frac{d[\text{M-Y}]}{dt} = \frac{d[\text{M-X}]}{dt} = k_{obs}\,[\text{M-X}] \qquad (2)$$

where k_{obs} is the *pseudo* first-order observed rate constant. These reactions are typically first-order in M-X and are conveniently studied under conditions where other reactants are essentially at constant concentrations relative to [M-X]. Equation 2 can also be written in either of the following forms:

$$-\frac{d\,[\text{M-X}]}{[\text{M-X}]} = -d\ \ln\,[\text{M-X}] = k_{obs}dt \qquad (3)$$

or

$$\ln\,[\text{M-X}]_0 - \ln\,[\text{M-X}]_t = k_{obs}\,(t - t_0) \qquad (4)$$

where $[\text{M-X}]_t$ and $[\text{M-X}]_0$ are the concentrations of M-X species at time t and time zero, respectively.

Thus, plots of $\ln\,[\text{M-X}]_0 - \ln\,[\text{M-X}]_t$ versus $(t - t_0)$ should yield a straight line with a slope of k_{obs}. The half-life of an irreversible chemical reaction, which represents the time for reactant M-X to reach half of its initial concentration, is given by

$$t_{1/2} = \frac{0.69315}{k_{obs}} \qquad (5)$$

If homogeneous reaction conditions are assumed, the percentage of original M-X complex remaining after 1, 2, 3, and 4 half-lives have passed is 50%, 25%, 12.5%, and 6.25%, respectively. A first-order or *pseudo* first-order reaction that proceeds for over four half-lives can therefore be considered nearly complete.

To indicate what values of k_{obs} may be environmentally important, calculated half-lives corresponding to various k_{obs} values are listed in Table 3.2-1. The k_{obs} value that corresponds to $t_{1/2}$ values that begin to be important from an environmental perspective (say time of mixing) will thus be on the order of $1-10\ \text{min}^{-1}(t_{1/2} = 4-40\ \text{s})$.

The values listed in Table 3.2-1 are not the *fundamental* rate constants commonly given in the literature, which refer to a single reaction step; *observed* rate constants

TABLE 3.2-1

Half-Life ($t_{1/2}$) for Various Observed First-order Rate Constants (k_{obs})
(calculated by equation 5)

k_{obs} (min^{-1})	$t_{1/2}$ (min)
6.0×10^{-7}	1.2×10^{6}
6.0×10^{-6}	1.2×10^{5}
6.0×10^{-5}	1.2×10^{4}
6.0×10^{-4}	1.2×10^{3}
6.0×10^{-3}	1.2×10^{2}
6.0×10^{-2}	1.2×10^{1}
6.0×10^{-1}	1.2
6.0	1.2×10^{-1}
6.0×10^{1}	1.2×10^{-2}

represent the sum of many possible reaction steps and are therefore related to the overall reaction rates, as expressed in equation 2. These values depend on the reaction mechanism, fundamental rate constants, incoming ligand concentration, pH, temperature, and other factors, as discussed later in this section. Due to the complexity of substitution mechanisms, it is easier to consider the overall reaction in terms of k_{obs} than to calculate the separate contributions of individual steps.

MECHANISMS AND MATHEMATICAL REPRESENTATION

Most ligand substitution reactions proceed as follows:

$$\text{M-X} + \text{Y} \overset{K_{1os}}{\rightleftharpoons} (\text{M-X, Y}) \overset{k_s}{\longrightarrow} (\text{M-Y, X}) \overset{K'_{2os}}{\rightleftharpoons} \text{M-Y} + \text{X} \qquad (6)$$

where M-X is the metal ion with ligand X coordinated in the inner sphere, Y is the incoming ligand, and (M-X, Y) and (M-Y, X) are outer-sphere complexes or ion pairs. (The comma separates the two species in the ion pair.) K_{1os} is the ion pair formation equilibrium constant of (M-X, Y), k_s is the substitution rate constant for the conversion of X from outer to inner sphere within the ion pair, and $K'_{2os} = 1/K_{2os}$, where K_{2os} is the ion pair equilibrium formation constant of (M-Y, X).

The reaction step (M-X, Y)$\rightarrow$(M-Y, X) involves substitution via interchange of ligands in the outer and inner spheres of M and is thus termed an *interchange* mechanism (I). Interchange mechanisms of ligand exchange are generally divided into two categories:

- Dissociative (I_d), where an M-X bond is partially broken before the M-Y bond is formed.

- Associative (I_a), where an M-Y bond is partially formed while the M-X bond is still present.

The situation where bond breaking occurs to form an intermediate of reduced coordination number (e.g., five from a six-coordinate complex) is referred to as a dissociative (D) mechanism. An intermediate of increased coordination (e.g., seven from a six-coordinate reactant) is formed by an associative (A) mechanism. For an interchange (I) mechanism, an intermediate of modified coordination number does not exist.[2]

The mechanism of the substitution reaction determines the mathematical relationship between k_{obs} and the fundamental rate constants available in the literature. As a rule, substitution of cationic complex ions occurs via an interchange mechanism [21]. The generally accepted view [20] is that M^{+2} ions react via I_d and M^{+3} ions via I_a mechanisms.[3]

Using the assumption that k_s is rate limiting, the rate law for equation 6 (I mechanisms) is as follows [24]:

$$\text{Rate} = \frac{d \ln[\text{M-Y}]}{dt} = \frac{k_s\, K_{1os}\, [Y]}{1 + K_{1os}\, [Y]} \tag{7}$$

When K_{1os} (Y) << 1, equation 7 can be approximated by

$$\text{Rate} = \frac{d \ln[\text{M-Y}]}{dt} = k_s K_{1os}\, [Y] \tag{8}$$

Comparison of equations 8 and 3 then gives

$$k_{obs} = k_s\, K_{1os}\, [Y] \tag{9}$$

Thus, for an I mechanism, equation 9 allows the calculation of k_{obs} from data on ion pair formation constants and rates of substitution.

Example 1 Calculate the half-life of reaction 10, given the following information and assumptions:

$$Ni(H_2O)_6^{+2} + SO_4^{-2} \longrightarrow Ni(H_2O)_4SO_4^{0} \tag{10}$$

- Initial $Ni(H_2O)_6{}^{+2}$ concentration is 1×10^{-5} mol/l (0.59 ppm).
- Initial $SO_4{}^{-2}$ concentration is 1×10^{-3} mol/l (96 ppm).
- Ionic strength is 0.5 *M*.

---

2. A, I_a, I_d and D can be roughly conceptualized by the following terminology, which is commonly used in reaction chemistry [1]:

   $SN_2(\text{lim}) \approx A \quad SN_2 \approx I_a \quad SN_1 \approx I_d \quad SN_1(\text{lim}) \approx D$

3. Some notable exceptions are Mn^{+2}, Co^{+3}, and Al^{+3} aquo ions. The few examples of a D mechanism involve neutral or anionic complexes, e.g., $[Co(CN)_5OH_2]^{-2}$. [20,21]

- Rate of water exchange (k_s) on $Ni(H_2O)_6{}^{+2}$ is 3×10^4 s^{-1}.
- Assume reaction proceeds via an I_a mechanism.
- Assume that $Ni(H_2O)_6{}^{+2}$ is the only complex that is reacting.

To estimate $t_{1/2}$ for this reaction, we first estimate the observed rate constant (k_{obs}) under the stated conditions. This requires the following:

(1) Calculation of the outer-sphere association constant (K_{1os}) for $[Ni(H_2O)_6, SO_4]^0$
(2) Value of the rate of water exchange on $Ni(H_2O)_6{}^{+2}$
(3) Verification that the concentration of $SO_4{}^{-2}$([Y]) is approximately constant during the reaction (pseudo first-order conditions.)
(4) Verification that $K_{1os}[SO_4{}^{-2}] = K_{1os}$ [Y] << 1 so that equation 9 can be used rather than the more complex equation 7.

We first calculate K_{1os} for the $[Ni(H_2O)_6, SO_4]^0$ ion pair by the procedure given in section 2.9, using the following parameters in equation 11 of that section:

$$I = 0.5$$
$$Z_1 = +2$$
$$Z_2 = -2$$
$$a = 9.5 \text{Å (Table 2.9-5)} = 9.5 \times 10^{-8} \text{ cm}$$

This leads to a value of $K_{1os} = 5.5\ M^{-1}$.

To approximate k_s for $Ni(H_2O)_6{}^{+2}$, we can use the value of k_{ex} listed in Table 3.2-2 for Ni^{+2}, which is 3×10^4 s^{-1}. As discussed below, k_s is usually greater than k_{ex} in this situation, but they are of the same order of magnitude.

We then compare the value of $[SO_4{}^{-2}]$ with that of $[Ni(H_2O)_6{}^{+2}]$ to verify that the reaction would not markedly affect $[SO_4{}^{-2}]$. Since complete reaction of 1×10^{-5} mol/l of $Ni(H_2O)_6{}^{+2}$ would remove an identical amount of $SO_4{}^{-2}$ and this amount is much smaller than the $[SO_4{}^{-2}]$, which is 1×10^{-3} mol/l, the concentration of $SO_4{}^{-2}$ in solution would be essentially constant. This leads to pseudo first-order conditions.

The above values of K_{1os} and $[SO_4{}^{-2}]$ are next multiplied together to verify that their product is much less than 1:

$$(5.5\ M^{-1})(1 \times 10^{-3}\ M) = 5.5 \times 10^{-3} << 1$$

Having assumed an I_a mechanism and no other reaction paths, we can now use equation 9:

$$k_{obs} = k_s K_{1os}[SO_4^{-2}] = (3 \times 10^4\ s^{-1})(5.5\ M^{-1})(1 \times 10^{-3}\ M) = 1.65 \times 10^2\ s^{-1}$$

In accordance with equation 5, the half-life of the reaction is

$$t_{1/2} = \frac{0.693}{1.65 \times 10^2 s^{-1}} = 4.2 \times 10^{-3}\ s$$

The reaction is thus very rapid. Note that a difference of a factor of 10 in the value of k_s would not significantly alter this conclusion.

The kinetic time frame of ligand substitution reactions in general is frequently paralleled by the rate of water exchange for the aquo complexed ion species, $M(H_2O)_n{}^{+x}$, shown below.

$$[M(H_2O)_n]^{+x} + H_2O^* \rightleftharpoons [M(H_2O)_{n-1}\ H_2O^*)^{+x} + H_2O \qquad k_{ex} \qquad (11)$$

where k_{ex} is the exchange rate constant and H_2O^* is an isotopically labeled (e.g., ^{18}O) water molecule. The water exchange reaction occurs continuously in aqueous media, even though the product and reactants are the same. Isotopic labeling studies allow measurement of the rate of exchange.

Values of k_{ex} for water exchange by a variety of metal ions are given in Table 3.2-2. Rates of exchange for acid dissociated species (e.g., reaction 12) are very different from those for aquo ions (reaction 11).

$$[M(H_2O)_{n-1}\ OH]^{x-1} + H_2O^* \rightleftharpoons [M(H_2O)_{n-2}\ H_2O^*\ OH]^{x-1} + H_2O \qquad (12)$$

TABLE 3.2-2

Water Exchange Rate Constants for a Variety of Metal Ions
(see equation 11)

Ion	k_{ex} (s^{-1})	Ion	k_{ex} (s^{-1})
Cr^{+2}	7×10^9	Hg^{+2}	$>10^4$
Mn^{+2}	3×10^7	Al^{+3}	~ 1
Fe^{+2}	3×10^6	Fe^{+3}	3×10^3
Co^{+2}	1×10^6	Ga^{+3}	$<10^4$
Ni^{+2}	3×10^4	Gd^{+3}	2×10^9
Cu^{+2}	8×10^9	Bi^{+3}	$>10^4$
Be^{+2}	3×10^4	Cr^{+3}	3×10^{-6}
Mg^{+2}	$>10^4$	Rh^{+3}	4×10^{-8}
Ba^{+2}	$>10^4$	VO^{+2}	5.2×10^2
Sn^{+2}	$>10^4$		

Source: Saito and Sasaki [20]; Basolo and Pearson [2].

The relationship between k_s (equation 6) and k_{ex} (equation 11) depends on the mechanism of ligand substitution and the nature of the metal ion and incoming ligand. The following general patterns have been observed [22]:

- For an I_a mechanism, when Y is more nucleophilic than H_2O toward the metal (i.e., M is a "soft" metal and Y is a "soft" ligand[4]), $k_s > k_{ex}$;
- For an I_a mechanism, when Y is less nucleophilic than H_2O toward the metal (i.e., M is a "hard" metal), $k_s \leqslant k_{ex}$; and
- For an I_d mechanism, $k_s \leqslant k_{ex}$, and k_s should be similar for a variety of different Y ligands.

Characteristic rates of substitution for a variety of metal ions are shown in Figure 3.2-1. The similarity of these data to those of Table 3.2-2 suggest that rates of water removal from the inner coordination sphere is often the rate-determining step.

For complexation reactions where an aquo complexed metal ion (i.e., $M(H_2O)_n{}^{+x}$) and a monodentate[5] ligand, L^{-y}, react, the following dissociative interchange mechanism (I_d) has been found generally applicable [8,18]:

$$M(H_2O)_n{}^{+x} + L^{-y} \rightleftharpoons (M(H_2O)_n, L)^{x-y} \qquad K_{os} \qquad (13)$$

$$(M(H_2O)_n, L)^{x-y} \rightarrow (M(H_2O)_{n-1}, L)^{x-y} + H_2O \qquad k_{H_2O} \qquad (14)$$

$$(M(H_2O)_{n-1}, L)^{x-y} \rightarrow ML(H_2O)_{n-1}{}^{x-y} \qquad k_1 \qquad (15)$$

The constants K_{os}, k_{H_2O} and k_1 refer to the equilibrium constant for outer-sphere complex formation, rate-limiting water loss and rate of conversion of outer-sphere to inner-sphere complex, respectively. The constant k_{H_2O} is a first-order rate constant for unaided metal-water bond breaking in the aquo complexed ion, and k_1 is much faster than k_{H_2O}.

Equations 13-15 can be summarized as:

$$M(H_2O)_n{}^{+x} + L^{-y} \rightarrow M(H_2O)\,L^{x-y} + H_2O \qquad k_{ML} \qquad (16)$$

where k_{ML} is the *overall* rate constant for complex formation. The value of k_{ML} is related to K_{os} and k_{H_2O} as follows:

$$k_{ML} = K_{os}\,k_{H_2O} \approx K_{os}\,k_{ex} \qquad (17)$$

Equation 17 is related to equation 9 with Y = H_2O and provides a method for estimating k_{ML} from data for k_{ex}. (See "Use of Water Exchange Rates" in §3.2.6.)

4. For definitions and classifications of "hard" and "soft," see Basolo and Pearson [2].

5. Monodentate ligands are defined in section 2.9.

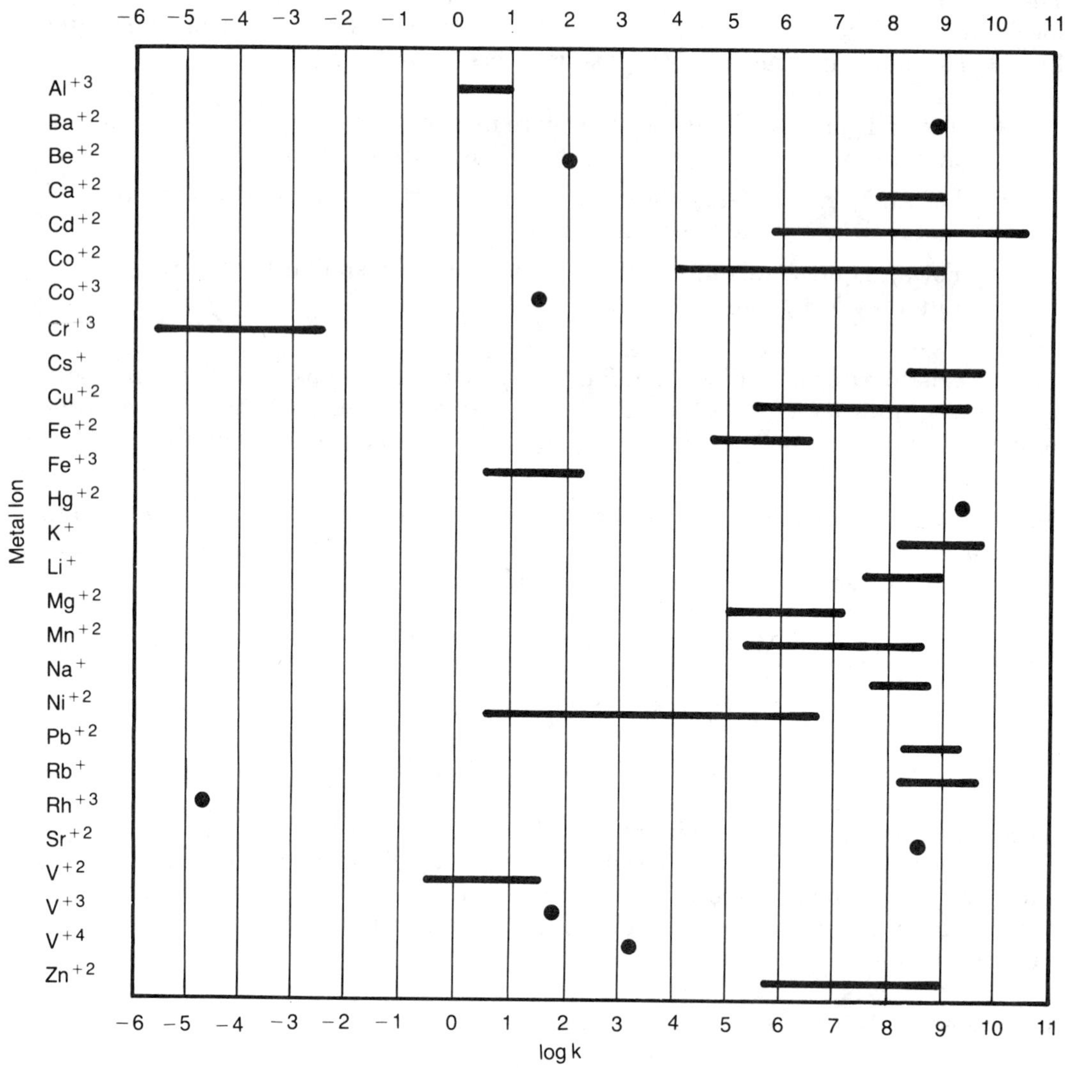

Note: k is expressed in $M^{-1}s^{-1}$. Bars indicate range of values for a variety of inorganic and organic ligands, while dots give single values.

Source: Based on data in Table 1 of Eigen and Wilkins [8]

FIGURE 3.2-1 Second-order Rate Constants for Formation of Metal Complexes in Aqueous Solution for Aquo Ions at 25°C

3.2.4 Variables that Affect Magnitude of Observed Rate Constant (k_{obs})

INTRODUCTION

The utility of k_{obs} in estimating the rate of substitution was described above. To estimate k_{obs}, it is useful to understand what factors affect its value. This section discusses the following factors and their impact on the relative magnitude of k_{obs}:

- Identity and oxidation state of central metal ion,
- Identity of ligands coordinated to the metal ion,
- Identity of incoming ligand,
- Chemical composition of the aqueous environment,
- The mechanism and rate law for the substitution reaction,
- Temperature of the aqueous environment, and
- Ionic strength of the solution.

IDENTITY AND OXIDATION STATE OF CENTRAL METAL ION

The magnitude of k_{obs} is a function of k_s, the rate of water substitution on the metal ion. Significant relationships have been observed between the value of k_s and the identity and oxidation state of the central metal ion:

- High reactivity ($k_s > 10^5\ s^{-1}$) for alkali metals, alkali earth metals, rare earths (+3 oxidation state) and first transition metals of +2 oxidation state in the order Cu(II) ≈ Cr(II) > Zn(II) > Mn(II) > Fe(II);
- Lesser reactivity ($k_s < 10^5\ s^{-1}$) for Ni(II), V(II), Fe(III), Be(II), and Al(III);
- Low reactivity ($k_s < 1\ s^{-1}$) for transition metals of +3 and +4 oxidation state with large crystal field stabilization energies [2], such as Cr(III), Co(III) and Rh(III); see Table 3.2-3.

Metal ions have also been classified as being either "kinetically inert" (reaction occurring longer than one minute at room temperature for a 0.1 *M* solution) or "kinetically labile" (reactions essentially complete within one minute under the same conditions) [2]. The constitution of the electronic shell of the metal and stereochemistry of the ligands (e.g., octahedral, tetrahedral, or square planar) play an important role in determining these reaction rates. For example, octahedral complexes such as Cr(III), Co(III), Fe(III), Mo(III) and Au(III) — all of which have electronic configurations of d^3, d^4 (low spin), d^5 (low spin), or d^6 (low spin) — are essentially inert [2]. Of the metal species that have these electronic configurations, Cr(III) is particularly important from an environmental standpoint. Table 3.2-3 classifies a large number of metal ions into labile and inert categories.

Because of the electrostatic nature of the interaction of metals and ligands, substitution reactions are affected by the size of metal ions and their charge.

TABLE 3.2-3

Inert and Labile Six-coordinate Metal Centers and Complexes

Labile[a]			Inert[a]		
Mn(II)	Y(III)	Mo(V)	Co(III)[b]	V(II)	$Cr(bipy)^{+2}$
Fe(II)	Ti(IV)	W(V)	Rh(III)	Re(III)	$Mn(CN)_6^{-3}$
Fe(III)	Zr(IV)	Re(VI)	Ir(III)	Ru(IV)	$Cr(bipy)_3^{+3}$
Co(II)	Hf(IV)	Ti(II)	Ni(IV)	Os(V)	$Mn(CN)_6^{-4}$
Zn(II)	Ce(IV)	V(III)	Pd(IV)	Re(II)	$Fe(CN)_6^{-3}$
Cd(II)	Th(IV)	Nb(III)	Pt(IV)	Ru(III)	$Fe(phen)_3^{+3}$
Hg(II)	Nb(V)	Ta(III)	Cr(III)	Os(III)	$Fe(bipy)_3^{+3}$
Al(III)	Ta(V)	Mo(IV)	Mo(III)	Ir(IV)	$Mn(CN)_6^{-4}$
Ga(III)	Mo(VI)	W(IV)	W(III)	Ru(II)	$Fe(CN)_6^{-4}$
Tl(III)	W(VI)	Re(V)	Mn(IV)	Os(II)	$Fe(phen)_3^{+2}$
In(III)	Ti(III)	Ru(VI)	Re(IV)	$Cr(CN)_6^{-4}$	$Fe(bipy)_3^{+2}$
Sc(III)	V(IV)	Rare Earths (III)			

a. As defined in the text. No consideration has been given to the composition of the primary coordination spheres in listing these metal oxidation states. biby = 2,2-bibyridyl, phen = o-phenanthrodine.
b. Except CoF_6^{-3}

Source: Basolo and Pearson [2], pp. 143-44.

Table 3.2-4(A) shows how changes in the central metal ion's size and charge can be expected to affect the rate of substitution reactions via three different mechanisms. Since the interaction is not always purely electrostatic, these generalizations should be used with caution.

LIGANDS COORDINATED TO METAL ION

As with the central metal ion, variations in the electrostatic properties of ligands in the first coordination shell also affect the rate of substitution reactions; some useful generalizations are given in Table 3.2-4(B). Since the electrostatic model used is a simplistic representation of the interactions that occur, these generalizations again must be used cautiously.

TABLE 3.2-4

Effect of Changes in Central Atom and Coordinated Ligands on Rate of Substitution Reactions[a]

Change	Mechanism[b] D	I	A
A. Central Metal Atom			
Increase in positive charge	Decrease	c	Increase
Increase in size	Increase	Increase	Increase
B. Coordinated Ligands			
Increase size of leaving group	Increase	c	Decrease
Increase negative charge of leaving group	Decrease	Decrease	Decrease
Increase negative charge of non-labile ligand	Increase	c	Decrease
Increase size of non-labile ligand	Increase	c	Decrease

a. Charge and size effects only; type of bonding assumed to be unchanged.
b. D = dissociative, I = interchange, A = associative
c. Multiple effects oppose each other; therefore, the rate may be either increased or decreased to some extent.

Sources: Benson [4]; Basolo and Pearson [2]

The chemical nature and position of the other coordinated ligands around the metal ion, especially those in the *trans* position,[6] can also influence the rate of substitution. In complexes that contain a "strong" trans labilizing ligand, substitution may occur a million times faster than with a "weak" trans ligand. This effect has been largely observed in square planar complexes such as Pt(II) and Ni(II), but some instances involving reactions of octahedral complexes have also been observed [2].

In Pt(II) complexes, the strength of the *trans* labilizing effect decreases as follows [4]:

$$CN^- \approx C_2H_4 \approx CO \approx NO > R_3P \approx HSO_3^- \approx SC(NH_2)_2 > CH_3^- > C_6H_5^- > SCN^- > NO_2^- > I^- > Br^- > Cl^- > C_5H_5N > NH_3 > OH^- > H_2O$$

These effects occur when bond breaking (i.e., I_d or D mechanisms are in effect. Thus, for example, substitution occurs more rapidly if a ligand is *trans* to a CN^- rather than

6. "*Trans*" and "*cis*" locations with respect to ligand A are at B and C, respectively, in the following schematics for square planar (1) and pseudo octahedral (2) stereochemistries:

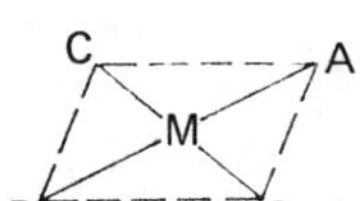

(1)

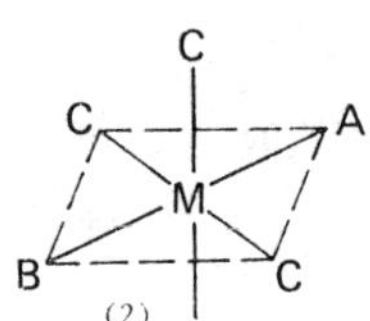

(2)

a Cl^-. The strength of the *trans* effect is enhanced by the bonding ability of the ligand [1]. Ligands that are *cis* to the leaving group in a square planar complex have comparatively little effect on the rate of replacement, unless steric effects are important [2].

The position of a *trans* labilizing ligand in the above series is greatly dependent on the chemical nature of the particular metal complex. Thus, for example, in complexes of *trans*-$Co(en)_2ACl^+$ (where en is ethylenediamine and the ligand A is *trans* to Cl^- in the octahedral complex), the rate of Cl^- loss via an I_d or D mechanisms suggests that the effect of A decreases as follows:

$$NH_2^- >> OH^- > NO_2^- \approx CN^- \approx Cl^- > NCS^-$$

This order is very different from that observed for square planar Pt(II) complexes. Clearly, a change in the chemical composition of the primary coordination sphere (e.g., from $M(H_2O)_6{}^{+3}$ to $M(NH_3)_5H_2O^{+3}$) can strongly affect the rate of substitution of a leaving ligand such as H_2O.

IDENTITY OF INCOMING LIGAND

In a coordination complex, the central metal ion is referred to as the *electrophilic* reagent (acceptor of e^-) and the ligand as the *nucleophilic* agent (donor of e^-). The ease with which a ligand acts as a nucleophilic agent in SN_2 mechanisms (I, I_a or A) is determined by factors such as its Lewis basicity, polarizability, and the "alpha effect"[7] [7]. The relative effect of these ligand properties on the rate of substitution reaction depends on the metal or atom center with which they react. The importance of polarizability of the ligand increases in a series in which the number of outer electrons increases (e.g., I^- vs. Br^-) or the net positive charge decreases, as in Fe(III) to Fe(II) and S(VI) to S(-II). The "alpha effect" appears to be general.

However, as a mixed contribution from each of these properties is generally observed, a relative ranking of ligands can be made only for a particular metal. Table 3.2-5 provides examples of the ranking of ligands with regard to overall nucleophile strength for particular metals or atoms.

A later subsection (§3.2.6) presents a relative ranking of basicity and polarizability for a variety of species as values of H and E_n, respectively. The "Edwards Relationship," a method for estimating the influence of these two parameters on the rate of substitution, is also described in that subsection.

Table 3.2-6 shows the general effect of changes in the size and charge of the entering ligand on the rate of substitution, considering only electrostatic interactions. Again, these generalizations should be used cautiously.

---

7. An effect attributed to the presence of unshared pairs of electrons on the atom adjacent to the donor atom (bonded to the metal).

TABLE 3.2-5

Ranking of Nucleophile Strength for Substitution Reactions

Reactant Atom	Relative Nucleophile Strength	Important Property[a]
Neutral four-coordinate phosphorus	$OOH^- > OH^- \approx OCl^- > NH_2OH > NO_2^- > N_3^- > H_2O$	Basicity >> polarizability
Trigonal boron[b]	$OH^- > OR^- > NH_3 > R_2NH \approx SR^-$	Basicity >> polarizability
Tetrahedral boron[c]	$OH^- > F^- > H_2O$	Basicity >> polarizability
Tetrahedral sulfur	$OH^- \approx CH_3O^- > C_6H_5O^- > RNH_2 > C_6H_5NH_2 > C_6H_5S^-$	Basicity and polarizability
Bivalent sulfur	$RS^- > R_3P > C_6H_5S \approx CN^- > SO_3^{-2} > OH^- > S_2O_3^{-2} > SC(NH_2)_2 > SCN^- > Br^- > Cl^-$	Polarizability and basicity
Trivalent nitrogen[d]	Phosphines > amines > oxygen bases	Polarizability > basicity
Oxygen (in peroxide)	$SO_3^{-2} > S_2O_3^{-2} > SC(NH_2)_2 > I^- > CN^- > SCN^- > NO_2^- > OH^- > N_3^- > BR^- > NH_3 > Cl^- \gtrsim C_5H_5N > H_2O$	Polarizability >> basicity
Square planar platinum (II)	$R_3P \approx$ thiourea $\approx SCN^- \approx I^- > N_3^- > NO_2 >$ Py > aniline > olefin $\approx NH_3 \approx Br^- > Cl^- >$ glycine $\approx OH^- \approx H_2O \approx F^-$	Polarizability >> basicity

a. Extent is highly compound-specific.
b. Based on R_2BX
c. Based on BF_3NH_3
d. Based on NH_2X

Source: Edwards and Pearson [7]

TABLE 3.2-6

Effect of Changes in Entering Ligand on Rate of Substitution for Various Mechanisms[a]

Change in Entering Ligand	Mechanism[b] D	Mechanism[b] I	Mechanism[b] A
Increase negative charge	No effect	Increase	Increase
Increase size	No effect	Decrease	Decrease

a. Charge and size effects only; type of bonding assumed to be unchanged.
b. D = dissociative, I = interchange, A = associative.

Source: Benson [4].

CHEMICAL COMPOSITION OF THE AQUEOUS ENVIRONMENT

The chemical composition of the aqueous environment can influence rates of substitution by (1) altering the speciation of the metal ion and (2) introducing other ions that affect the reaction. Both of these actions allow the overall substitution process to proceed via different reaction paths, which have their own characteristic rates.

The effect of ionic strength on reaction rate is discussed later in this subsection. One of the major determinants of substitution rate is the pH of the solution. Since many complexes, especially aquo complexed metal ions, participate in acid dissociation equilibria (see section 2.7), the pH can affect the relative amounts of $M(H_2O)_n{}^{+x}$ and $M(H_2O)_{n-y}(OH)_y{}^{x-y}$ species. The latter (especially when y = 1) often reacts much faster with ligands than does the completely aquo complexed metal ion [9].

The magnitude of this effect is determined by the individual aquo-hydroxo-complex reaction rate constant and pH. The following examples illustrate the extent to which the rate constant can vary [8,9]:

Reaction	Rate Constant ($M^{-1}\ s^{-1}$)
$Cr(H_2O)_6{}^{+3} + Cl^-$	3.0×10^{-8}
$Cr(H_2O)_5OH^{+2} + Cl^-$	2.3×10^{-5}
$Fe(H_2O)_6{}^{+3} + SCN^-$	1.3×10^{2}
$Fe(H_2O)_5OH^{+2} + SCN^-$	1×10^{4}

Since the speciation of ligands can also be affected by pH, the rate of substitution can be modified. Ligands and other chelating agents containing oxygen and nitrogen donor atoms, such as EDTA and ethylenediamine, are especially affected through acid/base equilibria. For example [8]:

Reaction	Rate Constant ($M^{-1}\ s^{-1}$)
$Ni^{+2} + C_2O_4{}^{-2}$	7.9×10^{4}
$Ni^{+2} + HC_2O_4{}^-$	5.0×10^{3}

The presence of particular metal ions can promote the substitution of coordinated ligands; Hg(II) is by far the most effective in enhancing rates of removal of ligands such as Cl^-, CN^-, $N_3{}^-$ and NCS^- from complexes and replacement by H_2O. Ag(I) and Tl(III) are also effective, but much less so than Hg(II). Metal ions can also accelerate the rate of removal of chelated ligands by coordinating to the freed donor atoms of the chelating ligand [24]:

$$M_1\ (\text{chelate}) + M_2 \rightleftharpoons M_2\ (\text{chelate}) + M_1 \qquad (18)$$

REACTION MECHANISM AND RATE LAW

An understanding of the reaction mechanism and its mathematical representation (rate law) allows the rate of substitution to be defined as a function of the important variables. As previously mentioned, the interchange (I) substitution mechanism is most commonly observed for aquo complexed metal ions.

First-row transition metals with divalent oxidation states generally undergo substitution via I_d; the I_a mechanism is more typical for cationic octahedral complexes of trivalent metals such as V(III), Mo(III), Ru(III), Cr(III), Ir(III) and Re(III), but not Co(III) [21].

Dissociative mechanisms have been proposed for oxo-metal ions of Ti(IV), V(V) and Cr(VI) (e.g., for CrO_4^{-2}), while associative mechanisms appear to operate for W(VI) and Mo(VI) (e.g., MoO_4^{-2}). An increase in rate with increasing concentration of H^+ is observed in the formation of oxo-metal complexes of Cr(VI) and Ti(IV) [20].

The mathematical representation (rate law) is determined by individual steps in the reaction mechanism. Even though the reaction may proceed predominantly along one path, the mechanism can be quite complicated if it completely describes all possible paths for substitution. The rate laws of mechanisms described previously (e.g., I_a, I_d) apply to one path for one form of the metal (e.g., aquo complexed metal ion); in reality, however, substitution can proceed through the hydroxo form of the aquo ion as described above or through catalyzed processes involving H^+ or other ions. A term in the empirically observed rate law which has an *inverse* dependence on hydrogen ion concentration (e.g., $[H^+]^{-1}$) suggests a path via hydroxo species, while a term with *direct* dependence on the concentration of hydrogen or some other ion (e.g., $[H^+]$ or $[Hg^{+2}]$) indicates a path catalyzed by that ion.

Examples of some rate laws for substitution reactions are given in Table 3.2-7. These rate expressions show the strong dependence of the observed rate and observed rate constant on the magnitude of the individual step rate constants and concentrations of other species in solution.

TEMPERATURE

Temperature nearly always affects the magnitude of the rate constants. The rate of a chemical reaction generally varies with temperature according to the empirical Arrhenius expression:

$$k = A\, e^{-E_a/RT} \tag{19}$$

where
- k = rate constant (s^{-1})
- A = Arrhenius pre-exponential factor (s^{-1})
- E_a = activation energy (cal/mole)
- R = gas constant (1.99 cal/deg-mole)
- T = temperature (K)

Equation 19 can be rearranged to give

$$\log k = \log A - \frac{E_a}{2.303\, RT} \tag{20}$$

TABLE 3.2-7

Rate Laws for Some Substitution Reactions[a]

Reaction	Rate Law	Source
$Cr^{+3} + Br^- \rightleftharpoons CrBr^{+2} + H_2O$	$V = [(3.0 \times 10^{-8})(Br^-) + 3.6 \times 10^{-9}(Br^-)(H^+)^{-1}](Cr^{+3})$	[24]
$Fe^{+2} + 3\ bipy \rightleftharpoons Fe(bipy)_3^{+2}$	$V = [(1.4 \times 10^{13})(bipy)^3](Fe^{+2})$	[24]
$CrSO_4^+ \rightleftharpoons Cr^{+3} + SO_4^{-2}$	$V = [(6.1 \times 10^{-7}) + (1.1 \times 10^{-7})(H^+)](CrSO_4^+)$	[24]
$Fe^{+3} + Cl^- \rightleftharpoons FeCl^{+2}$	$V = [(9.4)(Cl^-) + (1.1 \times 10^4)(Cl^-)(H^+)^{-1}](Fe^{+3})$	[24]
$HCrO_4^- + X^{+n} \rightleftharpoons HCrO_3X^{-1+n}$	$V = [(10^5)(X)(H^+)](HCrO_4^-)$	[20]
$MoO_4^{-2} + HC_2O_4^- \rightleftharpoons MoO_4HC_2O_4^{-3}$	$V = [(4.1 \times 10^2)(HC_2O_4^-)](MoO_4^{-2})$	[20]
$VO^{+2} + SO_4^{-2} \rightleftharpoons VOSO_4^0$	$V^b = [(1.5 \times 10^3)(SO_4^{-2})](VO^{+2})$	[20]

a. Data at 25°C unless noted. V refers to rate for product formation (e.g., $dCrBr^{+2}/dt$ for the first reaction) in M/s. Term in [ ] in the rate law refers to k_{obs}. X^{+n} is a variety of Lewis bases. bipy = 2,2-bipyridyl.
b. Rate given is for outer to inner sphere complex conversion.

Most studies of the effect of temperature on rate derive values of E_a from graphical representation of log k *vs.* 1/T. These plots exhibit a linear dependence of log k on 1/T (as expected from eq. 20) over a finite temperature range with a slope of $-E_a/2.303R$ and an intercept of log A. Values of E_a for some substitution reactions are summarized in Table 3.2-8. With equation 21, the value of the rate constant can be extrapolated from one temperature to another given values of E_a and k at some point in the temperature domain for which linear dependence is valid.

$$\log k_{T_1} - \log k_{T_2} = \frac{E_a}{2.3R}\left(\frac{1}{T_1} - \frac{1}{T_2}\right) \tag{21}$$

where k_{T_1} and k_{T_2} are the rate constants at temperatures T_1 and T_2 respectively (in K).

TABLE 3.2-8

Activation Energies for Formation of Metal Complexes from Aquo Complexes in Aqueous Solutions at 25°C

Metal Ion	Ligand	E_a(kcal/mole)
V^{+2}	bipy	15
Mn^{+2}	SO_4^{-2}	6
	Cl^-	9
Fe^{+2}	H_2O^*	8
Co^{+2}	SO_4^{-2}	6
Ni^{+2}	SO_4^{-2}	8
	$C_2O_4^{-2}$	15
	SCN^-	9
	NH_3	10
	en	12
	bipy	14
Cu^{+2}	$HEDTA^{-3}$	8
Cr^{+3}	SO_4^{-2}	28
	SCN^-	26
	H_2O^*	27
Fe^{+3}	Cl^-	17
	SCN^-	13
$FeOH^{+2}$	N_3^-	1
$Co(EDTA)H_2O^-$	H_2O^*	25

bipy = 2,2-bipyridyl
H_2O^* = labeled water
en = ethylenediamine
HEDTA = monoprotonated ethylenediaminetetraacetate

Source: Eigen and Wilkins [8]

Example 2 To illustrate the effect of temperature on reaction rate and the use of equation 21, consider the rate of reaction of Cr^{+3} with SO_4^{-2} to give $CrSO_4^+$ at 15°C. The rate constant is 2.0×10^{-6} M^{-1} s^{-1} at 25°C.

The following values are substituted in equation 21:

T_1 = 298K
T_2 = 288K
E_a = 28,000 cal/mole (Table 3.2-8)
R = 1.99 cal/deg-mole
$\log k_{T_1}$ = $\log (2.0 \times 10^{-6}) = -5.7$

We now calculate the rate constant at 288°K to be:

$$-5.7 - \log k_{288} = \frac{-28{,}000 \text{ cal/mole}}{(2.3)(1.99 \text{ cal/deg-mole})} \left(\frac{1}{298} - \frac{1}{288}\right)$$

$$\log k_{288} = -6.4$$
$$k_{288} = 4.0 \times 10^{-7}\ M^{-1}\ s^{-1}$$

The literature often provides activation energies in the form of enthalpies of activation, $\Delta H^{\ddagger}$ [3], which are related to E_a by equation 22.

$$\Delta H^{\ddagger} = E_a - RT \tag{22}$$

The entropy of activation, $\Delta S^{\ddagger}$, is related to A by equation 23.

$$\Delta S^{\ddagger} = 2.303\ R\ [\log A - \log (k_B T/h)] - R \tag{23}$$

where
R = gas constant (1.99 cal $deg^{-1} mole^{-1}$)
A = Arrhenius pre-exponential factor (s^{-1})
k_B = Boltzmann constant (1.38×10^{-16} erg/deg)
h = Planck constant (6.62×10^{-27} erg-s)
T = temperature (K)

Data for $\Delta H^{\ddagger}$ and $\Delta S^{\ddagger}$ provide an additional source for calculating E_a and evaluating the effect of temperature on the rates.

EFFECT OF IONIC STRENGTH ON RATE CONSTANT

Rate constants are often provided at zero ionic strength or at values of ionic strength different from those encountered in the environmental system of concern. Since ionic strength affects many reaction rate constants, corrections are usually necessary. The effect of ionic strength on the magnitude of a biomolecular rate constant can be estimated by the following equation [24]:

$$\log k = \log k_0 + \frac{2\,\alpha\, Z_A Z_B \sqrt{I}}{1 + \sqrt{I}} \tag{24}$$

where
k = rate constant at ionic strength I
k_0 = rate constant at zero ionic strength
Z_A = charge on reactant A
Z_B = charge on reactant B
α = constant (0.52 at 25°C)
I = ionic strength ($= \frac{1}{2}\Sigma Z_i^2 M_i$, where Z_i = charge of ion i and M_i = molar concentration of i)

This equation has been used quite successfully to correlate the effect of salt concentrations on rate constants for reactions involving ions [12,14]. In contrast to the variables discussed earlier, it is not necessary to know the chemical identity of the species involved in order to model the effect of ionic strength on rates. The limitations of this equation are similar to those expected for the Guntelberg and Davies equations (§ 2.6) for calculating activity coefficients.

According to equation 24, a plot of $\log (k/k_0)$ vs. $\sqrt{I}/(1 + \sqrt{I})$ would be linear with an intercept of zero. Such a plot for a variety of $Z_A Z_B$ product values is given in Figure 3.2-2. The equation predicts an increase in rate over that at $I=0$ with an increase in ionic strength when reactants are of the same charge sign; a decrease in rate is predicted as ionic strength increases if the charges are of opposite sign.

Example 3 What is the effect of increasing ionic strength from zero (infinite dilution) to 0.1 *M* on the rate of reaction of $Co(NH_3)_5\,Br^{+2}$ with OH^- at 25°C?

The following parameter values apply:

α = 0.52
Z_A = +2
Z_B = −1
I = 0.1 *M*
k = rate constant at I = 0.1 *M*
k_0 = rate constant at I = 0 *M*

Substituting these values in equation 24, we obtain

$$\log (k/k_0) = \frac{2\,\alpha\, Z_A Z_B \sqrt{I}}{1 + \sqrt{I}} = \frac{2(0.52)(+2)(-1)\sqrt{0.1}}{1 + \sqrt{0.1}} = -0.5$$

Thus, $\log k = \log k_0 - 0.5$, and $k = 0.316\, k_0$.

Alternatively, we can use Figure 3.2-2 as follows:

Since $I = 0.1$, $\sqrt{I}/(1 + \sqrt{I}) = 0.24\ M$
$Z_A Z_B = (2)(-1) = -2$
A vertical line drawn from 0.24 on the abscissa intersects the [−2] line at $\log (k/k_0) = -0.5$

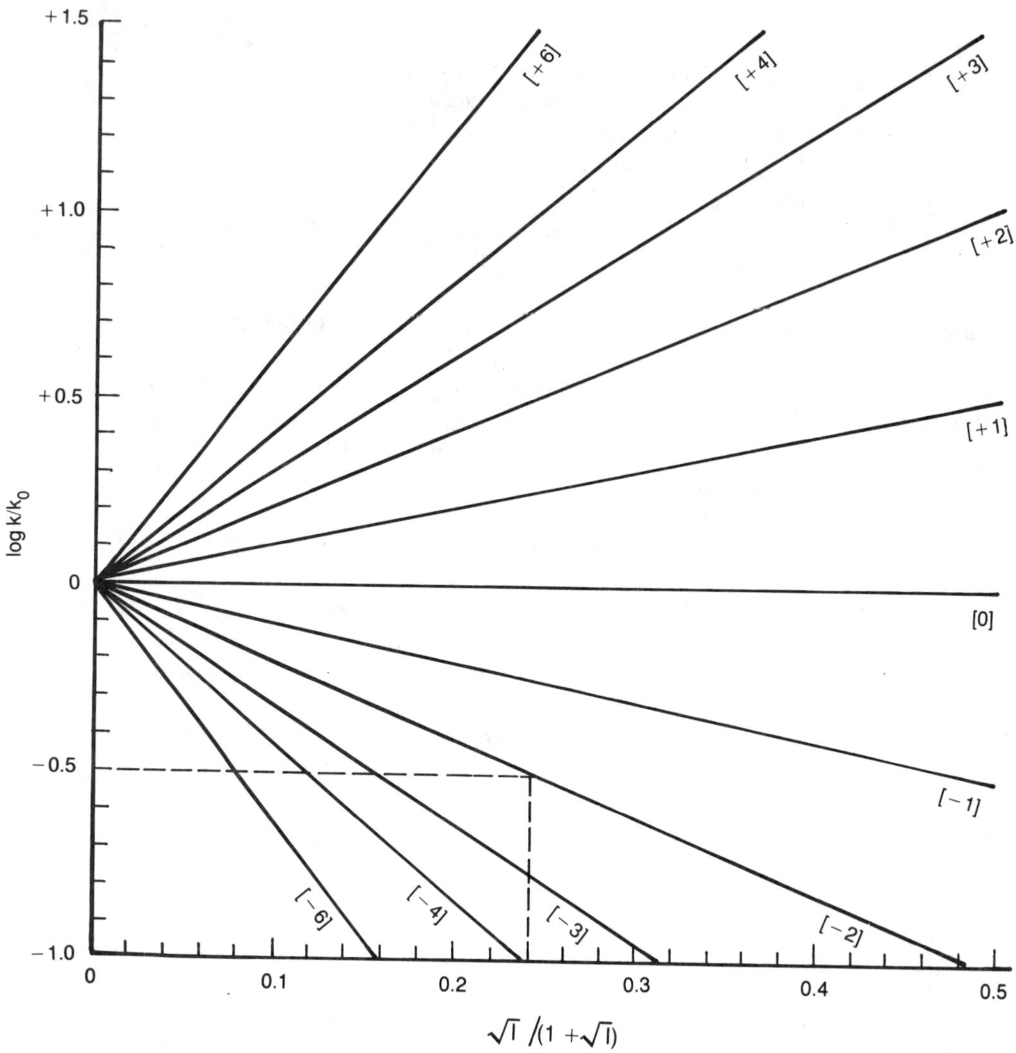

k_0 is the rate constant at I = O; k is the rate constant at I. Numbers in brackets are products of reactants' ionic charges $[Z_A Z_B]$.

Source: Equation 24

FIGURE 3.2-2 Effect of Ionic Strength on Rate Constants

Thus, the rate constant at I = 0.1 is estimated to be about one third that at I = 0, illustrating that even for ions of relatively low charge, the effect can be significant.

3.2.5 Sources of Data on Rates of Ligand Substitution

Kinetic data for ligand substitution reactions of metal complexes can be found in books dealing with mechanisms of inorganic reactions [2,4,24], and numerous review articles have described the reaction chemistry of a variety of metals and types of metal ion complexes [1,8,22]. In addition, journal articles that discuss reactions of particular species provide a source of specific rate constants.

The range of rates of substitution for reactions of aquo metal ions with common ligands (e.g., Cl^-, SO_4^{-2}, EDTA) was displayed in Figure 3.2-1; Figure 3.2-3 is a similar diagram for reactions of various metal ion complexes with other common ligands. As these figures show, the rates of substitution vary over a large range and are highly dependent on the chemical characteristics of the complex ion.

3.2.6 Estimation of Substitution Rates

INTRODUCTION

Estimation of the rates of ligand substitution reactions is complicated by the diverse chemical behavior of metal ions, the presence of several complexes for a particular metal (speciation), the diverse properties of the incoming ligands and the variety of mechanisms by which substitution reactions can proceed. Even for a particular metal ion, generalizations are tenuous, in part because ligand substitution mechanisms can differ greatly for different complexes with the same central metal.

Nevertheless, as discussed in §3.2.2 through §3.2.5, general trends have been observed that allow qualitative estimates of rates of ligand substitution for a variety of species. The initial determination (semi-quantitative) of whether a metal ion is "inert" or "labile" was also discussed earlier; this determination allows an estimate of the need for further consideration of its behavior in a particular scenario (i.e., time frame of importance in the chemical environment of interest).

The rate constant for substitution is very specific to the particular species present under the environmental conditions, and approaches for its estimation are limited. In this section, four useful quantitative methods of estimation are described:

(1) Use of water exchange rates (for dissociative mechanisms),
(2) Use of the Edwards relationship (for associative reactions),
(3) Use of derived molecular orbital bond energies, and
(4) Use of linear free energy relationships (for an I_d ligand substitution mechanism).

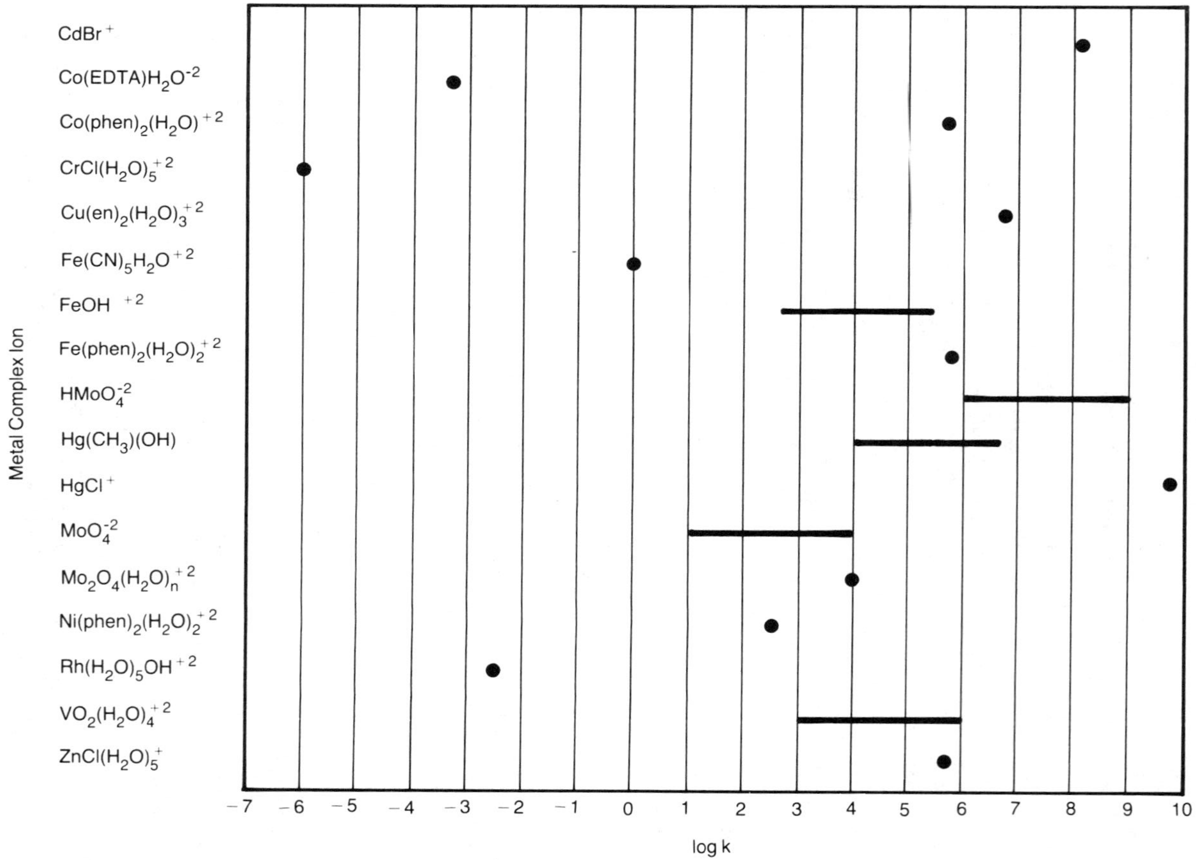

Note: k is expressed in $M^{-1}s^{-1}$. Bars indicate the range of values for a variety of inorganic ligands, while dots represent a single value.

Sources: Eigen and Wilkins [8]; Saito and Sasaki [20]

FIGURE 3.2-3 Kinetic Data for Reactions of Common Ligands with Various Metal Complexes at 25°C

USE OF WATER EXCHANGE RATES

When loss of a coordinated water molecule is the rate-determining step in substitution (via the I_d mechanism, equations 13-17), the rate of formation of the product can be expressed as:

$$\frac{d[ML]}{dt} = k_{ML}\,[M]\,[L] \tag{25}$$

where

k_{ML} = $K_{os}\,k_{H_2O}$ (equation 17)

K_{os} = outer-sphere equilibrium constant for [M, L] ion pair association (M^{-1})

k_{H_2O} = rate constant for metal-water bond rupture (s^{-1})

This relationship between k_{H_2O} and k_{ML} suggests that the mechanism of substitution for a series of ligands is dissociative; it can also be used as a predictive tool if the mechanism is expected to be dissociative.

Values of k_{H_2O} (which is approximately equal to k_{ex} as defined in equation 11) are listed in Table 3.2-2 for a variety of metal ions. A method for estimating K_{os} is given in section 2.9.

Example 4 Rorabacher *et al.* [19] used equation 17 to estimate the rate of reaction of Ni^{+2} with Hen^{+} (protonated ethylenediamine) at 25°C and I = 0.1 *M* from the rate of water exchange on Ni^{+2} as follows:

The reaction of interest is

$$Ni^{+2} + Hen^{+} \longrightarrow NiHen^{+3} \qquad k_{Hen^{+}}$$

The reaction that forms the intermediate ion pair (outer-sphere complex) is

$$Ni^{+2} + Hen^{+} \rightleftharpoons [Ni, Hen]^{+3} \qquad K_{os}$$

By means of an equation similar to equation 11 in section 2.9, the value of K_{os} was calculated to be 0.038 M^{-1}. ($Z_A = +1$, $Z_B = +2$, a = 4.8 Å [from Table 2.9-5], I = 0.1 *M*.)

From Table 3.2-2, the value for k_{H_2O} on Ni^{+2} is $3 \times 10^4\ s^{-1}$. Thus,

$$k_{Hen^{+}} = K_{os}k_{H_2O} = (0.038\ M^{-1})\ (3 \times 10^4\ s^{-1}) = 1.1 \times 10^3\ M^{-1}\ s^{-1}$$

The observed experimental value is 6.3×10^2 [19].

USE OF THE EDWARDS RELATIONSHIP

Edwards [5,6] derived an equation based upon properties of metal ions and ligands which is useful for correlation of rate data for I- and I_a-type inorganic and organic reactions. The equation is:

$$\log k - \log k_o = \alpha E_n + \beta H \tag{26}$$

where

- k = rate constant for substitution reaction of nucleophile (e.g., ligand) with electrophile (e.g., metal), $M^{-1}\ s^{-1}$
- k_o = rate of substitution reaction of H_2O with electrophile (e.g., metal), $M^{-1}\ s^{-1}$
- α,β = empirical parameters specific to the electrophile (e.g., metal)
- E_n = electron donor constant for a specific nucleophile (e.g., ligand)
- H = basicity constant for a specific nucleophile (e.g., ligand)

Values of E_n and H for a variety of ligands and metals are given in Table 3.2-9. Thus, rates of substitution can be estimated by equation 26 based on the electrophilic and nucleophilic properties of the ligand and metal complex.

The values of H are related to pK_a as follows:

$$H = pK_a + 1.74 \tag{27}$$

and the values of E_n are related to E_o by

$$E_n = E_o + 2.70 \tag{28}$$

In the above equations, K_a is the acid dissociation constant of the protonated form of the ligand, and E_o is the electrode potential for reduction of the ligand. If values of K_a and E_o are not available, H and E_n can be estimated by empirical fitting of available kinetic data (equation 26).

Values of α and β in the Edwards relationship for electrophiles are empirically derived by fitting rate data (values of k and k_o) for reaction of a particular electrophile with a variety of nucleophiles (various values of E_n and H) to equation 26. Equation 29 is a convenient form of equation 26 that makes it easier to obtain values of α and β from a variety of rate data for different ligands.

$$(\log k - \log k_o)/E_n = \alpha + \beta(H/E_n) \tag{29}$$

TABLE 3.2-9

Electron Donor Constants and Basicity Parameters for Use in Equation 26[a]

Species	E_n	H	Species	E_n	H
H_2O	0	0	$C_6H_5NH_2$	1.78	6.28
NO_3^-	0.29	0.4	SCN^-	1.83	1.00
SO_4^{-2}	0.59	3.74	NH_3	1.84	11.22
$ClCH_2COO^-$	0.79	4.54	$(CH_3O)_2POS^-$	2.04	4.00
CO_3^{-2}	0.91	12.1	$C_2H_5SO_2S^-$	2.06	−5.00
CH_3COO^-	0.95	6.46	I^-	2.06	−9.00
C_5H_5N	1.20	7.04	$(C_2H_5O)_2POS^-$	2.07	4.00
Cl^-	1.24	−3.00	$CH_3C_6H_4SO_2S^-$	2.11	−6.00
$C_6H_5O^-$	1.46	11.74	$SC(NH_2)_2$	2.18	0.80
Br^-	1.51	−6.00	$S_2O_3^{-2}$	2.52	3.60
N_3^-	1.58	6.46	SO_3^{-2}	2.57	9.00
OH^-	1.65	17.48	CH^-	2.79	10.88
NO_2^-	1.73	5.09	S^{-2}	3.08	14.66

a. Many values are empirically estimated, not calculated by equations 27 and 28.

Source: Edwards [5,6]

Example 5 Estimate the rate of reaction of Fe^{+3} with SO_4^{-2} using equation 26 with available data for reaction of Cl^-, SCN^-, and N_3^-.

Since the values of α and β for Fe^{+3} are not known for these reactions, they must be calculated from the following data, which is partly from Basalo and Pearson [2] and partly from Table 3.2-9:

Ligand	log k (M^{-1} s^{-1})	E_n	H	(log k-log k_o)/E_n	H/E_n
H_2O	2.44[8]	0	0	—	—
Cl^-	0.97	1.24	−3.00	−1.20	−2.4
SCN^-	2.10	1.83	1.00	−0.19	0.55
N_3^-	5.20	1.58	6.46	1.75	4.1

8. The value of k_{H_2O} (s^{-1}) has been divided by 55.5 *M* (concentration of H_2O) to convert to a second-order rate constant in units of $M^{-1}s^{-1}$, like those for the other ligands.

A least-squares fit of the data for a plot of (log k-log k_o)/E_n vs. H/E_n (eq. 29) gives values of $\alpha = -0.222$ and $\beta = 0.456$ for Fe^{+3}, with a correlation coefficient of 0.992.

We can now use these values of α and β, together with the values of E_n and H for SO_4^{-2} (Table 3.2-9), to estimate the rate of reaction of Fe^{+3} with SO_4^{-2} as follows:

$$\begin{aligned}\log k &= \alpha E_n + \beta H + \log k_o \\ &= (-0.222)(0.59) + (0.456)(3.74) + 2.44 \\ &= 4.01\ M^{-1}\,s^{-1}\end{aligned}$$

The experimental value is 3.80 M^{-1} s^{-1} [2].

USE OF DERIVED MOLECULAR ORBITAL BOND ENERGIES

Using a calculated value[9] for the binding energy of water molecules in the first solvation sphere of cations [E(I)], Rode *et al.* [17] obtained the following linear correlation (with a correlation coefficient of 0.991) between the free energy of activation ($\Delta G^{\ddagger}$) for rates of water exchange (k_{ex}) and values of E(I) for 12 metals (see Table 3.2-10):

$$\Delta G^{\ddagger} = 0.118\ E(I) + 1.3 \tag{30}$$

This approach illustrates the ability to calculate binding energies and, thus, rates of exchange of water.

TABLE 3.2-10

Binding Energies, E(I), per Water Molecule in First Hydration Shell of Cations[a]
(kcal/mole)

Ion	Tetrahedral Coordination	Octahedral Coordination	Ion	Tetrahedral Coordination	Octahedral Coordination
Li^+	25.7	22.1	Mn^{+2}	—	63.8
Na^+	17.3	15.4	Fe^{+2}	—	69.9
K^+	14.7	14.2	Co^{+2}	—	75.2
Be^{+2}	116.1	105.4	Ni^{+2}	—	77.8
Mg^{+2}	59.4	50.5	Zn^{+2}	—	71.0
Ca^{+2}	47.1	46.9	Al^{+3}	—	134.3

a. Calculated (see footnote 9)

Source: Rode *et al.* [17]

9. Obtained by MESQUAC (Mixed Electrostatic Quantum Chemical) molecular orbital calculations [10,15-17].

Activation energy and rate are related by the "Eyring relationship":

$$k_{ex} = (k_B T/h)\exp(-\Delta G^{\ddagger}/RT) \tag{31}$$

Assuming that the rate of water loss is rate-determining, we can relate k_{ex} to E(I) as follows:

$$k_{ex} = (k_B T/h)\exp\left(\frac{-[0.118E(I) + 1.3]}{RT}\right) \tag{32}$$

where

k_B = Boltzmann constant (1.380×10^{-16} erg/deg)
T = temperature (K)
h = Planck's constant (6.62×10^{-27} erg-s)
R = gas constant (1.987 cal/deg-mole)

Thus, at 298°K,

$$k_{ex} = (6.21 \times 10^{12})e^{-[1.99 \times 10^{-4} E(I) + 0.0022]}\ s^{-1}$$

Example 6 The value of k_{ex} for an octahedrally coordinated Al^{+3} ion at 298K can be derived using the calculated value of the binding energy of the first coordination water molecule (134.3 kcal/mole) from Table 3.2-10.

Using equation 33, we substitute 134,300 cal/mole for E(I). Thus,

$$\begin{aligned} k_{ex} &= 6.21 \times 10^{12} e^{-[1.99 \times 10^{-4}(134{,}300) + 0.0022]} \\ &= 6.21 \times 10^{12} e^{-(26.73 + 0.0022)} \\ &= 6.21 \times 10^{12}(2.46 \times 10^{-12}) = 14.97\ s^{-1} \end{aligned}$$

The experimental value is approximately 1 s^{-1} (see Table 3.2-2).

USE OF LINEAR FREE ENERGY RELATIONSHIPS

The reaction path profile (see, for example, Wilkins [24]) for an I_d ligand substitution mechanism is typified by reactions such as

$$L_5MY + X \underset{k_{-1}}{\overset{k_1}{\rightleftharpoons}} L_5MX + Y \qquad \Delta G^{\ddagger}, \Delta G^0 \tag{34}$$

where $\Delta G^{\ddagger}$ is the activation energy and ΔG^0 the overall free energy change for the reaction. Equation 34 suggests that for a series of different X ligands substituting the same Y ligand in L_5MY complex ion, a linear free energy relationship (LFER) applies [21], so that

$$\Delta G^{\ddagger} = -a\Delta G^0 + \text{constant} \tag{35}$$

where "a" is the slope of the LFER. Swaddle [21] notes that while values other than 1 have been observed for this slope, it is theoretically equal to 1 for an I_d mechanism.

Equation 35 can be used to calculate activation energies and, thus, rates for a reaction in a series if the overall free energy, ΔG^0, is known. The consequence of a = 1 in this equation is that the *inverse* of reaction 34 proceeds at the same rate (k_{-1}), independent of the nature of Y. (See, for example, Haim [11].)

Equation 35 is more conveniently written as shown below, using the relationships between $\Delta G^\ddagger$ and k_1 and between ΔG^0 and K_1 (and recognizing that $K_0 = k_1/k_{-1}$).

$$\log k_1 = -\log K_0 + B \tag{36}$$

where

k_1 = rate constant for reaction 34
K_0 = equilibrium constant for reaction 34
B = constant (equal to $\log k_{-1}$ for I_d mechanism)

This relationship is illustrated in the series of reactions of $Co(NH_3)_5H_2O$ with various ligands [11,13]:

$$[Co(NH_3)_5H_2O]^{+3} + X^- \rightleftharpoons [Co(NH_3)_5\,X]^{+2} + H_2O \tag{37}$$

where

$$X^- = NO_3^-,\ I^-,\ Br^-,\ Cl^-,\ H_2PO_4^-,\ F^-,\ NCS^-,\ N_3^- \text{ or } HCrO_4^-$$

Example 7 With equation 36, the rate constant for the reaction of I^- with $[Co(NH_3)_5H_2O]^{+3}$ can be calculated using equilibrium constants and rate data for reaction of the cobalt complex with $HC_2O_4^-$.

The reaction of interest is

$$[Co(NH_3)_5H_2O]^{+3} + I^- \rightleftharpoons [Co(NH_3)_5I]^{+2} \qquad \text{k(iodide), K(iodide)}$$

with K(iodide) = $0.12 M^{-1}$ [11] and the rate constant for reaction with iodide being k(iodide).

The reference reaction is

$$[Co(NH_3)_5H_2O]^{+3} + HC_2O_4^- \rightleftharpoons [Co(NH_3)_4C_2O_4H]^{+2} \qquad \text{k(oxalate), K(oxalate)}$$

with k(oxalate) = $2.2 \times 10^{-8}\ s^{-1}$ and K(oxalate) = $35\ M^{-1}$ [11].

Equation 36 can be modified to calculate k(iodide) as follows:[10]

$$\log k(\text{iodide}) = \log k(\text{oxalate}) - \log K(\text{iodide}) + \log K(\text{oxalate})$$

From the values given above,

$\log k(\text{oxalate}) = -7.66$
$\log K(\text{oxalate}) = 1.54$
$\log K(\text{iodide}) = -0.92$

Thus, $\log k(\text{iodide}) = -7.66-(-0.92) + (1.54) = -5.2$

The observed value is -5.08 [11].

3.2.7 Literature Cited

1. Bamford, C.H. and C.F.H. Tipper (eds.), *Reactions of Metallic Salts and Complexes, and Organometallic Compounds*, Vol. VII of *Comprehensive Chemical Kinetics*, Elsevier Publishing Co., New York, 13 (1972).
2. Basolo, F. and R. Pearson, *Mechanisms of Inorganic Reactions*, John Wiley & Sons, New York (1967).
3. Bender, M.L. and L.J. Brubacher, *Catalysis and Enzyme Action*, McGraw-Hill Book Co., New York (1973).
4. Benson, D., *Mechanisms of Inorganic Reactions in Solution*, McGraw-Hill Book Co., London (1968).
5. Edwards, J.O., "Correlation of Relative Rates and Equilibria with Double Basicity Scale," *J. Am. Chem. Soc.*, **76**, 1540 (1954).
6. Edwards, J.O., "Polarizability, Basicity, and Nucleophilic Character," *J. Am. Chem. Soc.*, **78**, 1819 (1956).
7. Edwards, J.O. and R.G. Pearson, "The Factors Determining Nucleophilic Reactivities," *J. Am. Chem. Soc.*, **84**, 16 (1962).
8. Eigen, M. and R.G. Wilkins, "The Kinetics and Mechanism of Formation of Metal Complexes," in *Mechanisms of Inorganic Reactions*, R.K. Murman, R.T.M. Fraser and J. Bauman (eds.), Advances in Chemistry Series 49, American Chemical Society, Washington, D.C. (1965).
9. Espenson, J.E., "Formation Rates of Monosubstituted Chromium(III) Complexes in Aqueous Solution," *Inorg. Chem.*, **8**, 1554 (1969).
10. Fujiwara, S. and B.M. Rode, *Bull. Chem. Soc. Japan*, **52**, 321 (1979).
11. Haim, A., "Some Comments on the Mechanism of Substitution Reactions of Cobalt(III) Complexes of the Pentaammine Class," *Inorg. Chem.*, **9**, 426 (1970).

10. The modified equation is derived by writing equation 36 for the iodide reaction and then for the oxalate reaction, subtracting the second equation from the first, and rearranging assuming the value of B is the same for both reactions.

12. Lamer, V.K., "Reaction Velocity in Ionic Systems," *Chem. Rev.*, **10**, 179 (1932).

13. Langford, C.H., "On the Acid Hydrolysis of $[Co(NH_3)_5X]^{+2}$ Ions and the Mechanism of Interchange," *Inorg. Chem.*, **4**, 265 (1965).

14. Livingston, R., "An Introduction to Chemical Catalysis in Homogeneous Systems," *J. Chem. Ed.*, **7**, 2887 (1930).

15. Rode, B.M. and G.J. Reibnegger, "MESQUAC: Mixed Electrostatic-Quantum Chemical Approach to the Description of Large Complexes," *J. Chem. Soc. Faraday Trans. 2*, **75**, 178 (1979).

16. Rode, B.M. and G.J. Reibnegger, *Monatsh. Chem.*, **110**, 813 (1979).

17. Rode, B.M., G.J. Reibnegger and S. Fujiwara, "Transition State Model for Water Exchange in the First Solvation Shell of Hydrated Cations According to Quantum Calculations," *J. Chem. Soc. Faraday Trans. 2*, **76**, 1268 (1980).

18. Rorabacher, D.B. and D.B. Moss, "Steric Effects in Chelation Kinetics: II. Role of Oxygen as a Donor Atom and the Internal Conjugate Base Effect in Cobalt(II)-Poly(amino alcohol) Reactions," *Inorg. Chem.*, **9**, 1314 (1970).

19. Rorabacher, D.B., T.S. Turan, J.A. Defever and W.G. Nickels, "Steric Effects in Chelation Kinetics: The Formation and Dissocation of Nickel (II) Complexes with Branched Poly(amino alcohols) Related to Ethylenediaminetetraacetic Acid," *Inorg. Chem.*, **8**, 1498 (1969).

20. Saito, K. and Y. Sasaki, *Substitution Reactions of Oxo Metal Complexes*, Vol. I of *Advances in Inorganic and Bioinorganic Mechanisms*, A.G. Sykes (ed.), Academic Press, New York, 182 (1982).

21. Swaddle, T.W., "Activation Parameters and Reaction Mechanism in Octahedral Substitution," *Coord. Chem. Rev.*, **14**, 217-68 (1974).

22. Swaddle, T.W., *Substitution Reactions of Divalent and Trivalent Metal Ions*, Vol. II of *Advances in Inorganic and Bioinorganic Mechanisms*, A.G. Sykes (ed.), Academic Press, New York (1983).

23. Taube, H., "Rates and Mechanisms of Substitution in Inorganic Complexes in Solution," *Chem. Rev.*, **50**, 69 (1952).

24. Wilkins, R.G., *The Study of Kinetics and Mechanism of Reactions of Transition Metal Complexes*, Allyn and Bacon, Boston (1974).

3.3 KINETICS OF ELECTRON TRANSFER (REDOX) REACTIONS

3.3.1 Introduction

The rate of electron transfer between oxidizing and reducing species is important in assessing speciation of inorganic materials in the environment. Assessing how fast a redox reaction will occur is complicated, requiring specific knowledge of the chemistry of the oxidant and reductant and the reaction mechanism, and consideration of all possible effects of the physical and chemical properties of the reaction medium. Accordingly, this section discusses the following topics:

- Exemplar scenarios involving redox reactions;
- Inner- and outer-sphere mechanisms of redox reactions and their mathematical representation;
- Exemplar rates and rate laws for some environmentally important redox reactions and species;
- Chemical and environmental factors affecting redox rates;
- Correlation and estimation of rates for particular mechanisms; and
- Sources of kinetic data.

Additional information on redox reactions, including a discussion of distribution between oxidized and reduced forms and redox mechanisms, is given in section 2.10.

3.3.2 Environmental Relevance

Differences in redox reaction rates can lead to products different from those predicted by equilibrium calculations. For example, although a particular reaction between species A and B may be thermodynamically favorable, the kinetics may be so slow that no reaction occurs in the time frame of interest. If a third species, C, is introduced into the system, reacts rapidly with A, and depletes it, the final distribution of products can be very different from that predicted by the thermodynamic calculation. In the real environment, this is illustrated by the disequilibrium conditions that are observed for many species, as discussed in section 2.10.

COMMON REDOX SCENARIOS

The following are the most common redox situations in the environment:

a. Reactions between two dissolved components (homogeneous),
b. Reactions between a component originally in the gas phase and a dissolved species (heterogeneous), and
c. Reactions between a solid and a dissolved species (heterogeneous).

Examples of reactions that fall under these classifications are:

a. Dissolved iron (II) with dissolved chromium (VI),
b. Gaseous oxygen with dissolved iron (II), and
c. Manganese dioxide (in soil) with dissolved chromium (III).

Although the sources of the two reactants may be in different phases (solid, liquid or gaseous), electron transfer occurs in a single phase. However, unlike reaction in a homogeneous phase, the presence of the source of the reactant as a separate phase can affect the reaction kinetics. For example, in the oxidation of sulfite by oxygen, the reaction occurs via the dissolved oxygen species, but the presence of a constant partial pressure of oxygen above the solution (together with other factors, such as the extent of turbulence) affects the level of dissolved oxygen in solution and thus the rate of the reaction. In the absence of the external source of oxygen and other dissolved ions, the reaction rate is predictable from the concentration of oxygen and sulfite present in solution. Similarly, continuous dissolution of iron (II) from a soil will affect its dissolved level and thus affect the rate of depletion of the oxidant reacting with the iron (II).

RATE-LIMITING STEPS

Various steps in redox mechanisms may control the overall reaction rate. These can be broadly characterized as relating to the availability of one of the reacting species (physical factors) or to the chemical behavior of species (chemical factors). The *physical factors* are exemplified by the following:

- Mode of contact of a solution with a gaseous reactant (such as the atmosphere);
- Availability and rate of dissolution of a reactant from a solid into a dissolved state (e.g., from wastes or soils); and
- Diffusion of dissolved reactants.

The *chemical factors* are exemplified by the following:

- Identity of the reactants [e.g., Fe (II) vs Mn (II)] as well as composition of the coordination sphere [e.g., $Fe(H_2O)_6^{+2}$ vs $Fe(CN)_6^{-4}$];
- Effects of the reaction medium on speciation of the reactants; and
- Mechanism of the formation of the activated complexes (i.e., the precursors to product formation, as discussed below).

An example of the balance between the chemical and physical factors that may limit rate is illustrated in the common mechanism for oxidation of dissolved metal ions, M, by oxygen as follows:

a. Dissolution of oxygen in water

$$O_2\,(g) \underset{k_{-1}}{\overset{k_1}{\rightleftharpoons}} O_2^0 \qquad (1)$$

b. Inner-sphere complex formation and subsequent electron transfer

$$M + O_2^0 \xrightarrow{k_2} [M\text{-}O_2] \xrightarrow{\text{fast}} \text{products} \qquad (2)$$

When $k_1\,pO_2 >> k_2[O_2]\,[M]$, chemical kinetics control the rate; when this relationship is reversed, on the other hand, diffusion of oxygen controls the rate, and the physical geometry of the system (i.e., mode of contact) is of great importance in determining the rate.

In the following sections a greater emphasis is placed on the discussion of chemical factors and homogeneous redox reactions. The discussion is also useful for heterogeneous reactions, provided that some assumptions are made relative to the physical factors and mechanism of the reaction (e.g., that physical factors are not rate-limiting).

3.3.3 Mathematical Representation

Most commonly, redox reaction rates are characterized in the following ways:

- Half-life of the reaction ($t_{1/2}$)
- Observed rate constant (k_{obsd})
- Rate of change of concentration of a reactant or product (d[C]/dt)

The *half-life* of a reaction is defined as the time required for a reactant to reach half of its initial concentration. Half-life is usually used when the concentrations of all other species on which the rate depends are in large excess over that of the reactant of concern, so that pseudo first-order conditions apply, or when the reaction is actually first-order. Half-life is defined for higher order reactions but is not commonly used in those cases. In a first-order reaction, the half-life and the first-order rate constant are related by the equation $k = 0.693\,t_{1/2}$ (See § 3.2 for an expanded discussion of this relationship.) A redox reaction that proceeds for over five half-lives can be considered essentially complete, as only about 3% of the original reactant remains after this time.

The *observed rate constant* is ordinarily measured under conditions in which all species that affect the rate of the redox reaction of a particular species R are defined and their concentrations are kept constant.[1] The value of k_{obsd} is usually determined from a plot of $\ln[R]_{eq} - \ln[R]_t$ versus time, where $[R]_{eq}$ and $[R]_t$ are the concentrations of R at equilibrium and at time t, respectively. If the reaction is run under such conditions (termed pseudo-first-order), the relationship between the rate of the reaction expressed as dR/dt and the observed rate constant, k_{obsd}, is

$$d[R]_t/dt = k_{obsd}[R]_t \tag{3}$$

To determine the effect of different variables on observed rate constants, subsequent experiments such as the following can be performed:

- Variation in the concentration of the oxidant, [OX], at values above 10[R]. A plot of the value of k_{obsd} versus [OX] then yields, for example, a function of the form $k_{obsd} = A\,[OX]$ when other variables are held constant.
- Variation in the concentration of $[H^+]$. A plot of the value of k_{obsd} versus $[H^+]$ could yield, for example, a function of the form: $k_{obsd} = B + C/[H^+]$ when other variables are held constant.
- Variation in the concentration of other ions in solution (such as chloride or sulfate) to determine the effect on the observed rate. A plot of the value of the observed rate constant versus chloride, for example, produces a function of the form: $k_{obsd} = D[Cl^-]$ when other variables, including acidity and concentration of oxidant, are held constant.
- Variation in the temperature to obtain activation energies; this allows calculation of the reaction rates at different temperatures and also helps to define the mechanism of the reaction.

The overall rate law can then be expressed as a function of these variables; for example,

$$d[R]_t/dt = E[OX][R]_t[Cl^-] + (F[OX][R]_t[Cl^-])/[H^+] \tag{4}$$

where the values of E and F are calculated from A, B, C and D given above.

The expression for the rate law (equation 4) suggests the composition of the activated complex by which the reaction proceeds and thus is useful in postulating a reaction mechanism consistent with this rate law. Thus, equation 4 suggests that two types of activated complexes are important in determining the rate of the reaction. They are derived from each of the expressions in the rate law. As the first term in the expression contains R, OX and Cl^-, a structure consisting of all these in the activated complex is indicated, and can be represented as $[R\text{–}Cl\text{--}OX]^{\ddagger}$.

1. This is usually accomplished by using a buffer system or keeping the concentration of the reactant at least 10 times that of R, so that their concentrations undergo minimal change during the progress of the reaction.

Similarly, the second term in equation 4 suggests that a deprotonated form (minus the H^+) of the above structure is also important. For example, if R is an aquo complexed metal ion, the deprotonated form is R–OH and the activated complex would be $[OH\text{–}R\text{–}Cl\text{--}OX]^{\ddagger}$

These structures imply the reactions that must occur before the formation of the activated complex and allow us to write a stepwise mechanism for the redox reaction. For example, reactions to form R–Cl, OH–R–Cl, and R–OH are needed. Since the reactions to form the activated complex are rate-determining, we can assume that all subsequent reactions are faster.

The overall mechanism that fits the above rate law is as follows:

$$R + Cl^- \rightleftharpoons RCl \qquad K_1,\ \text{fast} \qquad (5)$$

$$R \rightleftharpoons ROH + H^+ \qquad K_2,\ \text{fast} \qquad (6)$$

$$ROH + Cl \rightleftharpoons ROH(Cl) \qquad K_3,\ \text{fast} \qquad (7)$$

$$ROH(Cl) + OX \xrightarrow{k_3} ROH(Cl)(OX) \xrightarrow{\text{fast}} \text{Products} \qquad (8)$$

$$RCl + OX \xrightarrow{k_4} RCl(OX) \xrightarrow{\text{fast}} \text{Products} \qquad (9)$$

A proposed mechanism is said to be consistent with the observed kinetic rate law and kinetic data rather than proven, because only the structure of the activated complex is suggested and not the details of how that structure is derived from any prior reactions. Previous and subsequent reactions are generally postulated from experience.

The empirical parameters E and F in equation 4 can be expressed as a function of the rate constants in equations 5–9. For example, assuming that k_3 and k_4 are the rate-determining constants,

$$E = K_1 k_4 \quad \text{and} \quad F = K_2 K_3 k_3$$

As indicated earlier, the mathematical representation of the rate as

$$d[R]_t/dt = f(\text{concentrations})$$

can be very complex. It can be a function of the concentration of many species, including H^+, complexing ions, reactants, intermediates and even products (where reactions are reversible), as well as physical factors such as temperature (discussed in § 3.4) and ionic strength (§ 3.2). The paths that may generate equation 4 include, for example, some involving the intermediate chloro and chloro-hydroxo complexes of the reactant R. More typically, these steps can include aquo ions, other complexes of the oxidant or reductant, products of acid or base dissociation and catalyst species.

It is noteworthy that the rate laws for one class of redox reactions are similar to those found for ligand substitution reactions (§ 3.2). This is due to the presence of a redox mechanism that is limited by the rate of substitution of one of the reactants, as discussed in § 3.3.4 and § 3.3.7.

3.3.4 Mechanisms and Rate Laws for Homogeneous Redox Reactions

INNER- AND OUTER-SPHERE REDOX MECHANISMS

To understand and predict the rate of a homogeneous redox reaction (not including one catalyzed by various ions), one should make a distinction as to whether the reaction proceeds via an inner-sphere or an outer-sphere process. (These processes are described in § 2.10.3 and illustrated in Figures 2.10-2, -3 and -4.) The particular path of electron transfer for a pair of reactants is determined by the relative ease of these two processes: if the outer-sphere path is too slow, redox will proceed via an inner-sphere process, and vice versa. In rare instances, the two processes are comparable in rate and both occur.

The general scheme proposed for a solution-phase redox reaction between a complexed metal ion (M) and a reductant (R) is as follows [7]:

(a) Ion pair formation

$$M + R \rightleftharpoons [M,R] \qquad K_0,\ \text{fast} \tag{10}$$

(b) Activation of the ion pair

$$[M,R] \underset{k_{-1}}{\overset{k_1}{\rightleftharpoons}} [M,R]^{\ddagger} \qquad K_1 \tag{11}$$

(c) Outer-sphere reaction of the activated ion pair

$$[M,R]^{\ddagger} \xrightarrow{k_2} \text{Products} \tag{12}$$

(d) Alternatively, conversion of the activated ion pair to an activated inner-sphere complex

$$[M,R]^{\ddagger} \underset{k_{-3}}{\overset{k_3}{\rightleftharpoons}} [M\text{-}R]^{\ddagger} \qquad K_3 \tag{13}$$

(e) Electron transfer within inner-sphere activated complex and subsequent breakdown to products

$$[M\text{-}R]^{\ddagger} \xrightarrow{k_4} \text{Products} \qquad \text{(inner sphere)} \tag{14}$$

Step *a* represents a rapid equilibrium between the original reactants and an ion pair, which is activated (rearranged with respect to the metal-to-ligand bond distance) in step *b* prior to electron transfer and product formation in step *c* for an outer-sphere process. Alternatively, the activated complex can form an inner-sphere complex in step *d*, followed by electron transfer and product formation in step *e*. An essential feature of the outer-sphere mechanism is that the first coordination spheres of both reactants and products remain intact throughout the process.

Two limiting rate laws for the outer-sphere process can be derived from the above mechanism:

(1) If $k_{-1} >> k_2$, the electron transfer is rate-limiting, and the rate law becomes:

$$d[M]/dt = K_0K_1k_2 [M][R] = k_{obsd} [M][R] \tag{15}$$

(2) If $k_2 >> k_{-1}$, activation of the ion pair is rate-limiting, and the rate law becomes:

$$d[M]/dt = K_0k_1 [M][R] \tag{16}$$

In (1) above, the observed rate constant (k_{obsd}), and thus k_2, is influenced by the overall free energy change (or equilibrium constant) of the redox reaction, and correlations between k_2 and free energy change represented in the aggregate by equations 10, 11 and 12 can be made (see § 3.3.7).

If electron transfer through an outer-sphere precursor (step *c*) is slow, the reaction may proceed through steps *a*, *b*, *d* and *e*, which represent an inner-sphere mechanism. This mechanism is possible only if one of the reactants has a ligand (e.g., Cl^- or OH^-) that can act as a bridge for electron transfer and if the accepting species is sufficiently labile. These and other aspects of the inner-sphere reaction are discussed later in this section.

Two limiting forms can be derived for this mechanism:

(3) If $k_{-3} >> k_4$ (indicating that electron transfer in the inner-sphere complex is rate-limiting), the rate law is

$$d[M]/dt = K_0K_1K_3k_4 [M][R] \tag{17}$$

(4) If $k_4 >> k_{-3}$ and the rate of conversion of $[M\text{-}R]^\ddagger$ to products is faster than dissociation of the complex, substitution can be limited by the rate of complex formation, and the rate law is

$$d[M]/dt = K_0K_1k_3 [M][R] \tag{18}$$

The limit corresponding to equation 18 is characterized by rate constants and activation parameters for the redox reaction with widely different reductants covering a small range of values. In addition, these rates and activation parameters are similar to those observed for substitution at the metal center. This limit is sometimes called a *substitution-controlled* redox reaction.

The rate laws given above are simplified forms of the generally observed rate laws, which typically include other terms, owing to the presence of competitive paths from species in equilibrium with the reactants; however, they are useful for understanding individual terms in rate laws. Table 3.3-1 summarizes the rate equations commonly encountered in inner-sphere redox processes for limiting situations.

EVIDENCE AND REQUIREMENTS FOR INNER-SPHERE REDOX REACTIONS

The preceding discussion suggests that one must classify the mechanism of a redox reaction in order to estimate its rate. The rates of inner-sphere reactions are chiefly a function of substitution (ligand exchange) rates, whereas rates of outer-sphere reactions are more sensitive to the overall free energy change of the reaction (e.g., differences in redox potentials between oxidant and reductant couples).

Redox reactions have been classified as *inner-sphere* processes on the basis of the following evidence [12]:

- Transfer of a ligand from one reactant to the other (see Figures 2.10-3 and -4);
- A redox reaction rate less than or equal to the rate of ligand replacement;
- The rate law is independent (i.e., zero order) in one important reactant (e.g., [R] is missing from equation 18);
- The form of the rate law for the redox reaction is similar to that of substitution;
- Values of the activation energies for the redox reaction and those of substitution are similar; and/or
- Intermediate complexes are detected where the two reactants are bonded (e.g., M-R).

In addition, particular ligand properties are needed for an inner-sphere reaction to occur [14]. In the formation of the precursor to the electron transfer, a ligand bound to one of the reacting species becomes the bridge in the electron transfer step. The presence of a basic unshared pair of electrons on the bridging ligand is a necessary condition for the inner-sphere process. In general, the bridging ligand will have one or two sigma bonds and a lone pair available for electron transfer. Such is the case for M-$\ddot{\underset{..}{O}}$-H and M-$\ddot{\underset{..}{S}}$-CN. Each of these can act as a bridge for electron transfer. The unshared electron pair must be sufficiently basic to form a bond to the other reactant; thus, although the coordinated water molecule (M-H_2O) has a lone pair, it is not a good bridging ligand, because its basicity is too low. The coordinated ammonia ligand has no free lone pair and thus cannot act as a bridge. Similarly, carbon-bonded species such as M-CN have no lone pairs available on the carbon

TABLE 3.3-1

Rate Equations and Mechanisms for Inner-sphere Reactions

Mechanism	Steps	Comments	Rate Law	
1	$MA_x \underset{k_2}{\overset{k_1}{\rightleftharpoons}} MA_{x-1} + A$	Inner-sphere mechanism	For k_1 rate-limiting:	$\frac{d[Prod]}{dt} = k_1[MA_x]$
	$MA_{x-1} + B \underset{k_4}{\overset{k_3}{\rightleftharpoons}} MA_{x-1}B$	Dissociation of complex takes place prior to redox reaction	For k_3 rate-limiting:	$\frac{d[Prod]}{dt} = \frac{k_1k_3}{k_2}\frac{[MA_x][B]}{[A]}$
	$MA_{x-1}B \overset{k_5}{\rightarrow} Products$		For k_5 rate-limiting:	$\frac{d[Prod]}{dt} = \frac{k_1k_3k_5}{k_2k_4}\frac{[MA_x][B]}{[A]}$
2	$B + MA_x \overset{k_1'}{\rightarrow} MA_{x-1}B + A$	Inner sphere	For k_1' rate-limiting:	$\frac{d[Prod]}{dt} = k_1'[MA_x][B]$
	$MA_{x-1}B \overset{k_5}{\rightarrow} Products$	Reducing agent is good nucleophile; aids in replacement of A ligand		
3	$A + B \overset{K_c}{\rightleftharpoons} C$	Rapid precursor [C] formation	For $[B] >> [A]$: and $[A]_0 = [A] + [C]$	$\frac{-d[A_0]}{dt} = \frac{k_3K_c[A]_0[B]}{1 + K_c[B]}$
	$C \overset{k_3}{\rightarrow} C^{\ddagger} \rightarrow$ Products	Rate-limiting activation of precursor prior to electron transfer		
4	$MO_x^{-n} + 2H^+ \overset{K_e}{\rightleftharpoons} H_2MO_x^{2-n}$	Acid/base equilibrium of reactant	For K_e fast and k_1 rate-limiting	$-d[MO_x^{-n}] = K_ek_1[MO_x^{-n}][Y^-][H^+]^2$
	$H_2MO_x^{2-n} + Y^- \overset{k_1}{\rightarrow} YMO_{x-1}^{-(n+1)} + H_2O$	Rate-determining substitution of reactant		

Source: Edwards [12]

for direct bridging via an adjacent atom. A bridge can occur in either of the following ways:

M-A-B-R	M-A-B | R
I. Remote coordination	II. Adjacent coordination

Bridge formation can occur if the atom producing the bridge (e.g., B in I and A in II) has a lone pair available for bonding. However, an effective electron transfer bridge will occur only if the bridging structure is part of a conjugated system. For example, in structure III (below) remote coordination does not lead to an effective electron transfer bridge, since there is lack of conjugation between the oxygen and nitrogen after the other reactant is coordinated to the nitrogen.

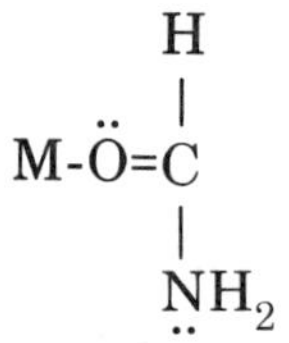

(III)

In structure IV, a remote bridge is possible and effective, because conjugation is retained even after use of an electron pair on the oxygen for coordination to the other reactant.

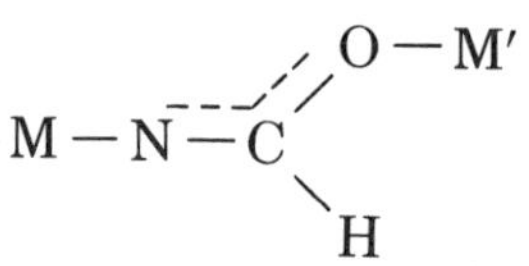

(IV)

Although the presence of a basic lone pair and conjugation are needed for inner-sphere redox, they do not necessarily guarantee it. For example, where the orbital symmetry of the donor and acceptor orbitals [14] is the same but different from that of the bridge system, conjugation is not sufficient; in this case, a stepwise process may occur in which the bridging group acquires an electron and is thus reduced, followed by transfer of the electron from the bridging group to the metal. The ability of the bridging ligand system to be reduced (or oxidized) then becomes an important parameter in determining the rate of reaction.

Some examples of known or suspected inner-sphere redox reactions are given in Table 3.3-2.

TABLE 3.3-2

Examples of Known or Suspected Inner-sphere Redox Reactions

Oxidant	Reductant(s)[a]	Oxidant	Reductant(s)[a]
$Co(C_2O_4)_3^{-3}$	Fe^{+2}	SeO_3^{-2}	I^-, ketones
$P_2O_8^{-4}$	$Fe(terpy)_3^{+2}$	Mn(III)	Various ($C_2O_4^{-2}$, N_2H_4, HN_3, HNO_2)
$P_2O_8^{-4}$	$Fe(phen)_3^{+2}$	Fe(III)	Various (I^-, $S_2O_3^{-2}$, N_2H_4)
$P_2O_8^{-4}$	$Fe(bipy)_3^{+2}$	$S_2O_8^{-2}$	Fe^{+2}, Ag^+
$P_2O_8^{-4}$	VO^{+2}, Ag^+, Fe^{+2}	HNO_2	I^-, N_3^-, $N_2H_5^+$, HSO_3^-
XO_m^{-n} (X=Cr, S, Br, Cl, N)	Various reducing species (e.g., I^-, Br^-, Cl^-, $As(OH)_3$, SO_3^{-2})	Tl^{+3}	Br^-
$HCrO_4^-$	Various (I^-, Br^-, N_2H_4, $As(OH)_3$, H_3PO_2, H_3PO_3)	H_3AsO_4	I^-
TeO_4^{-2}	Glycol	Ce(IV)	Br^-, H_3PO_2, Cl^-, glycols
IO_4^{-2}	Glycol	I_2	$S_2O_3^{-2}$
H_2O_2	HNO_2	Br_2	N_3^-

a. terpy = terpyridine; phen = 1,10-orthophenanthroline; bipy = bipyridyl

Source: Benson [6] and Edwards [12]

EVIDENCE FOR AN OUTER-SPHERE PROCESS

An outer-sphere redox mechanism is indicated on the following bases [6]:

- The ligands that are bonded to both reacting species and that constitute the first coordination shells are substantially inert to substitution, and the rate of the redox reaction is faster than the substitution rate on each of the species.
- The coordination shell of one reactant is rather kinetically inert to substitution, and no ligands capable of forming a bridge are present on this inert metal center.
- The rate law is first-order in each reactant (without modification of the ligands bonded to it).
- The rates of a series of related reactions are very dependent on the difference in electrode potentials between the reacting couples.

Some examples of known or suspected outer-sphere redox reactions are given in Table 3.3-3. It should be noted that the high substitution rates for many aquo metal ion complexes normally preclude assignment of mechanisms based on the rates of ligand exchange. The reactions between persulfate ($S_2O_8^{-2}$) and iron (II) complexes of N–(2–pyridylmethylene) aniline are examples of oxidations that proceed by parallel inner- and outer-sphere mechanisms [12].

TABLE 3.3-3

Examples of Known or Suspected Outer-Sphere Redox Reactions

Oxidant[a]	Reductant[a]
Co^{+3}	Fe^{+2}
Ce(IV)	$Fe(CN)_6^{-4}$
$Mo(CN)_8^{-3}$	$Fe(CN)_6^{-4}$
$IrCl_6^{-2}$	$Mo(CN)_6^{-4}$
$Co(NH_3)_6^{+3}$	$Ru(NH_3)_6^{+2}$
$Co(NH_3)_6^{+3}$	Cr^{+2}
$S_2O_8^{-2}$	$Fe(terpy)_3^{+2}$
$S_2O_8^{-2}$	$Fe(phen)_3^{+2}$
$S_2O_8^{-2}$	$Fe(bipy)_3^{+2}$
Haloamine complexes of Pt(IV)	V^{+2}

a. phen = 1,10-o-phenanthroline; terpy = terpyridyl; bipy = bipyridyl.

Source: Benson [6] and Lappin [19].

OTHER REDOX MECHANISMS

Among the less frequently encountered mechanisms of redox reactions are direct group transfer [18] and free-radical oxidation [4]. Because these are of lesser importance in transition-metal chemistry, and since methods for estimating the rates of these processes were not readily found, they are not discussed here.

3.3.5 Examples of Redox Rates, Rate Laws and Trends

TRENDS IN OXYANION REDOX REACTIONS

Oxyanion species (XO_m^{-n}) often react via substitution of the oxygen atom. Thus, the rates of redox reactions of the various oxyanions are related to the rate of substitution of the oxygen at the central atom. A few parallels in the rate of redox by a variety of oxyanions are generally observed, and these are useful in estimating relative rates [12].

- The rate of oxidation by halates generally follows the order of:

$$IO_3^- > BrO_3^- > ClO_3^-$$

and oxychlor species,

$$OCl^- > ClO_2 > ClO_3^- >> ClO_4^-$$

- Oxidations by perchloric, sulfuric, selenic and nitric acids in aqueous systems are generally slow, while those by chlorous, sulfurous, selenous and nitrous acids are faster.

- Within a periodic table group, for the same oxidation state, the oxyanion whose central atom has the higher atomic number generally reacts faster. For example, arsenic acid reacts more rapidly than phosphoric acid, periodate more rapidly than perchlorate, and telluric more rapidly than either sulfuric or selenic.
- Redox reactions for a large number of oxyanions follow one of the following rate laws:

(a) $\text{Rate} = k\,[XO_m^{-n}]\,[Y^{-x}]\,[H^+]^2$

where

$$XO_m^{-n} = CrO_4^{-2}, SO_4^{-2}, OCl^-, OBr^-, OI^-, ClO_2^-, BrO_2^-, IO_3^-, BrO_3^-, ClO_3^-, NO_2^-, NO_3^-$$

and

$$Y^{-x} = I^-, Br^-, Cl^-, S_2O_3^{-2}, SO_3^{-2}, As(OH)_3^0, C_2O_4^{-2}, HCO_2^-$$

(b) Rate $= k\,[XO_m^{-n}]\,[Y^{-x}]^2\,[H^+]^2$

for reactions between the following pairs of species:

IO_3^-, I^-; BrO_3^-, Br^-; ClO_3^-, Cl^-; IO_3^-, $S_2O_3^{-2}$;
IO_3^-, SO_3^{-2}; $HCrO_4^-$, I^-; $HSeO_3^-$, $S_2O_3^{-2}$.

- Oxygen exchange reactions generally follow one of the following rate laws:

(a) Rate $= k\,[XO_m^{-n}]\,[H^+]^2$

This equation applies, for example, to sulfate, nitrate, chlorite, bromite, carbonate and nitrite ions.

(b) Rate $= k\,[XO_m^{-n}]\,[H^+]$

This equation applies, for example, to iodite, rhenate, hypobromite and hypochlorite ions.

There are exceptions to these trends. For example, relative rates of reduction of chromate by a variety of two-equivalent reducing agents vary widely for no obvious reason [5]. Some correlation between the rate of oxidation and the reduction potential of the reducing species has been noted for chromate (see Figure 3.3-1), but several exceptions do occur.

Of the oxyanions discussed above, chromate is of major environmental concern. A variety of redox processes for conversion of chromium (VI) to chromium (III) and vice versa have been noted [22, 25, 26]. These include oxidation of chromium (III) by oxygen (with or without promotion by uv light) and by manganese dioxide. Reduction of chromium (VI) by iron (II), sulfides, and sulfhydryl groups in organic matter have similarly been noted. The rates of these and other environmentally relevant reactions of chromium are summarized in Table 3.3-4.

OXIDATION OF IRON(II)

Oxidation of iron(II) by oxygen at a pH of 6 – 7.5 and in the absence of catalytic or other contaminant effects obeys the following rate law [27]:

$$\text{Rate} = d[\text{Fe(II)}]/dt = k\,[\text{Fe(II)}]\,[\text{OH}^-]^2 P_{O_2} \tag{19}$$

The value of k at 20.5°C is 1.5×10^{13} l^2 mol^{-2} atm^{-1} min^{-1}. Thus, for a pH of 7.35, a partial pressure of oxygen of 0.195 atm, and a temperature of 20.5°C, the observed first-order rate constant [$k_{obsd} = k \times [OH^-]^2 P_{O_2}$] is equal to 0.147 min^{-1}, which corresponds to a half-life of 4.7 minutes. The oxidation of iron(II) by oxygen is, however, efficiently catalyzed by a variety of species, including copper(II), cobalt(II), manganese(II), manganese dioxide, zeolite and dihydrogen phosphate

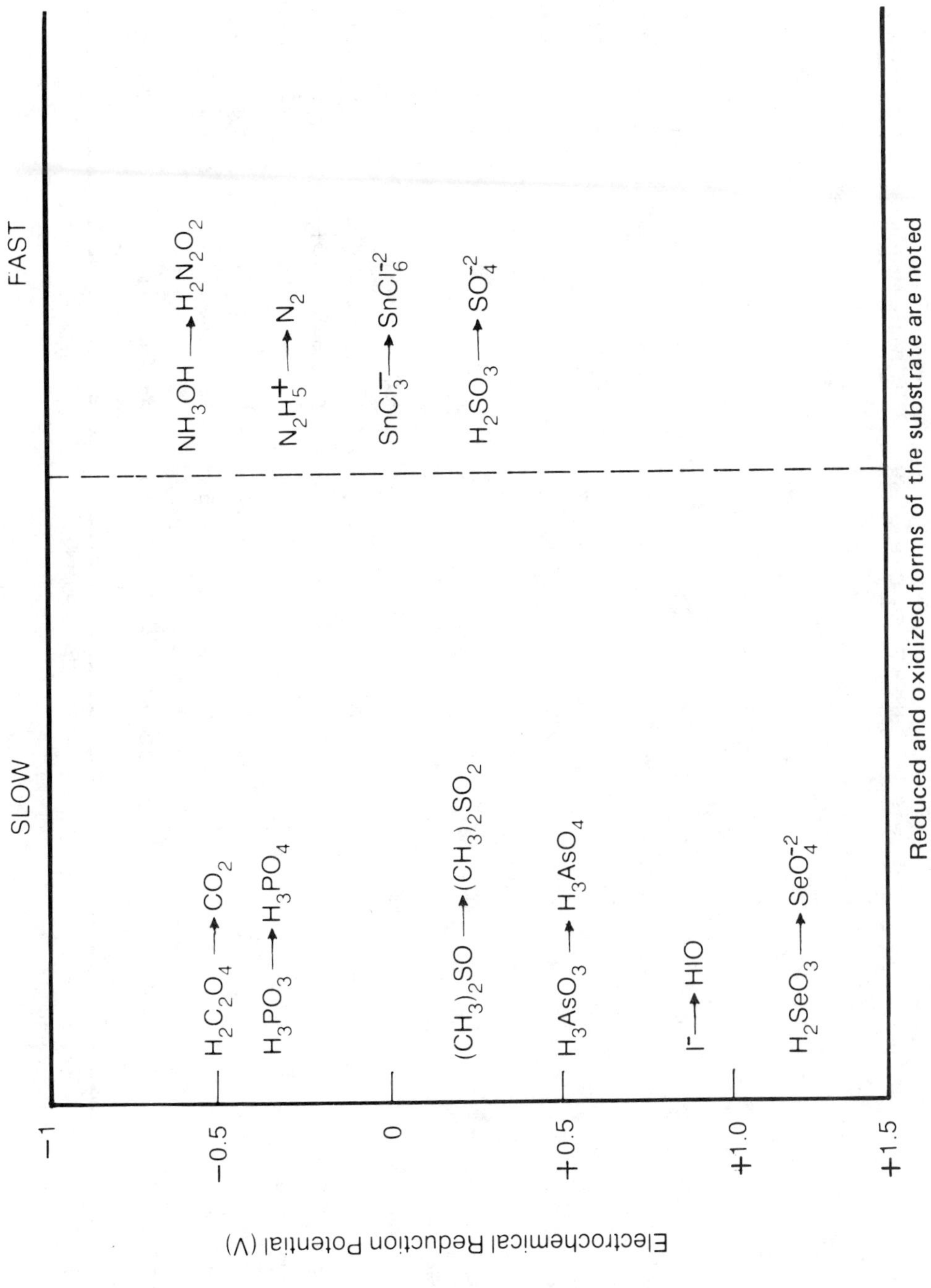

Source: Beattie and Haight [5]

FIGURE 3.3-1 Relative Rates of $HCrO_4^-$ Oxidations of Various Species

TABLE 3.3-4

Kinetic Data for Redox Reactions of Chromium

Reaction	[Cr]	[B][a]	Temp. (°C)	pH	Buffer/Medium	Rate or Half-life	Ref.
Cr(III) + O_2	125 μg/l		25	8.6	10^{-2} *M* $KHCO_3$	0.09 μg/l/d	[25]
			35			1.2 μg/l/d	[25]
Cr(III) + O_2	10^{-5} *M*	Bubbling	25	8.1	Seawater/0.02 *M* borate	$t_{1/2} >> 12$ d	[22]
Cr(III) + MnO_2	125 μg/l	25 mg/l	25	8.6	10^{-2} *M* $KHCO_3$	$t_{1/2}$ = 42 min	[25]
		250 mg/l	25			$t_{1/2}$ ~ 3 min	[25]
Cr(III) + γ-MnOOH (with 10^{-3} citric acid)	10^{-5} *M*	500 mg/l	25	8.1	Seawater/0.02 *M* borate	$t_{1/2} >> 12$ d	[22]
Cr(III) + γ-MnOOH	10^{-5} *M*	30 mg/l	25	8.1	Seawater/0.02 *M* borate	5% in 2 d	[22]
Cr(IV) + Fe(II)	100 μg/l	0.4 mg/l	N.C.	7.5	10^{-2} *M*	$t_{1/2}$ ~ 6 min	[25]
		1.2 mg/l	N.C.	7.1	10^{-2} *M*	$t_{1/2}$ ~ 2 min	[25]
Cr(VI) + S^{-2}	100 μg/l	10^{-3} *M*	N.C.	9.1	N.C.	$t_{1/2} <$ 5 min	[25]
Cr(VI) + Cysteine	100 μg/l	10^{-3} *M*	N.C.	N.C.	N.C.	$t_{1/2}$ ~ 20 min	[25]
Cr(VI) + Mercapto-succinic	100 μg/l	10^{-3} *M*	N.C.	N.C.	N.C.	$t_{1/2} <$ ~ 5 min	[25]
Cr(VI) + Alanine	100 μg/l	10^{-3} *M*	N.C.	N.C.	N.C.	$t_{1/2} >>$ 1 d	[25]
Cr(VI) + Succinic acid	100 μg/l	10^{-3} *M*	N.C.	N.C.	N.C.	$t_{1/2} >>$ 1 d	[25]
Cr(VI) + Ascorbic acid	10^{-5} *M*	10^{-6} *M*	25	8.0-12	Seawater/0.02 *M* borate	$t_{1/2} <$ 72 hr	[22]
Cr(VI) + Ascorbic acid	10^{-5} *M*	10^{-6} *M*	25	2.0- 8	Seawater/0.02 *M* borate	$t_{1/2} >>$ 72 hr	[22]
Cr(VI) + Humic acid	10^{-5} *M*	10^{-6} *M*	25	7.5-12	N.C.	$t_{1/2} >>$ 72 hr	[22]
Cr(VI) + Humic acid	10^{-5} *M*	10^{-6} *M*	25	2.0- 7	N.C.	$t_{1/2} <$ 72 hr	[22]
Cr(VI) + Fulvic acid	1.9×10^{-6} *M*	DOC at 5.8 mg/l	25	5.9	Groundwater	$t_{1/2}$ ~ 5 d	[26]
Cr(VI) + H^{+} [b,c]	1.9×10^{-6} *M*	0.04*M*	25	< 2.0	Aqueous acid	1-10% loss per 29 days	[26]

a. [B] = concentration of second reactant.

b. $4HCrO_4^- + 16H^+ = 4Cr^{+3} + 3O_2 + 10H_2O$

c. H^+ added as HCl, H_2SO_4 or HNO_3.

N.C. = Not cited.

ion. Also, ions that form strong complexes with iron(III) and metal ions that can react directly (via electron transfer) with iron(II) can apparently catalyze the reaction [27].

Both oxidation states of iron can be stabilized by the presence of complexing organic matter, and this can change the oxidation rate. The effect of organic species (Org) on the rate of iron(II) oxidation can be interpreted in terms of the occurrence of two parallel routes for oxidation:

$$Fe^{+2} + O_2^0 \xrightarrow{k_1} Fe^{+3} \qquad (20)$$

$$Fe^{+2} + Org \underset{k_3}{\overset{k_2}{\rightleftharpoons}} Fe(II)\text{-}Org \xrightarrow{k_4} Fe(III) \qquad (21)$$

where k_2/k_3 = K, the formation constant for the Fe(II)–Org complex.

Limiting cases of the above mechanism (see Figure 3.3-2) explain the observed effects of a variety of organic substances on the rate of oxidation. For example, for vanillic acid, phenol, syringic acid and histidine, the value of K is small and $k_4 \ll k_1$; this results in a very small amount of iron(II) being complexed and little effect on the rate of oxidation (curve 1). When K is large and $k_4 \ll k_1$, an oxidation-resistant complex is formed; depending on the value of k_4, total or partial inhibition of oxidation (curves 2 and 2′) occurs. This condition is observed in the presence of tannic acid and glutamine. When $k_4 \gg k_1$, the complex formed is oxidized more rapidly than aquo iron(II), and an enhanced rate (curve 3) is observed.

Rates and rate laws of other redox reactions of iron(II) are given in Table 3.3-5. Reactions with mercury(II), thallium(III) and nitrite are relatively slow, while those with chlorine, cerium(IV) and vanadium(V) are fast.

OXIDATION OF MANGANESE(II) BY OXYGEN

The kinetics of oxidation of manganese(II) by oxygen in waters resembling those of natural systems has been studied [15]. At pH 9, the oxidation to Mn(IV) and subsequent precipitation is complete within minutes; the half-life is approximately 1.5 min. At a lower pH, the reaction is much slower ($t_{1/2} \approx 30$ min at pH 8.5). The reaction rate is inhibited by species that form strong complexes with manganese(II). The presence of 2000 ppm of sulfate inhibits the rate at pH 9, and a half-life of 80 min is observed. Similarly, the presence of 100 ppm of bicarbonate ions increases the half-life to 60 min at pH 9. Catalysis of the reaction by copper(II) has been reported; like the reaction of iron(II) with oxygen, this reaction is probably also catalyzed by a variety of species.

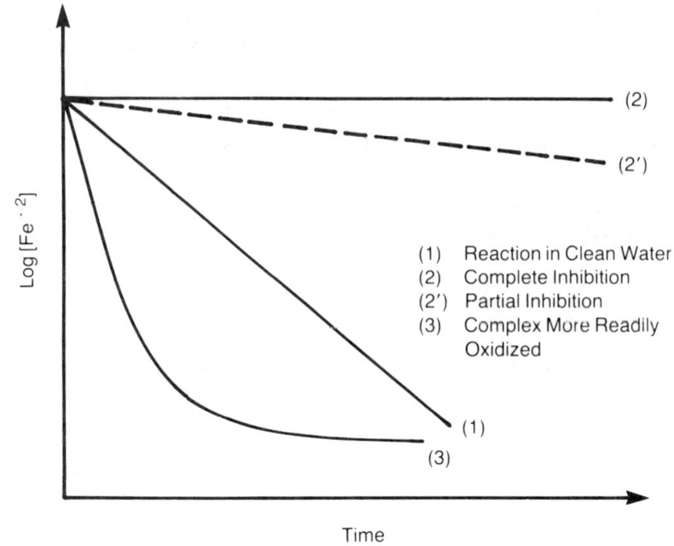

Source: Adapted from Figure 5 in Theis and Singer [29]

FIGURE 3.3-2 Effect of Organic Complexation of Fe^{+2} on Reaction with Oxygen

TABLE 3.3-5

Some Redox Reactions of Fe(II)

Oxidant	Rate = d [Fe(II)]/dt	Rate Constant[a]	Ionic Strength (*M*)	Temperature (°C)
Cl_2	$2k_1$ [Fe(II)] [Cl_2]	9.1×10^2	1	30
Ce(IV)	k_1 [Fe(II)] [Ce(IV)]	5865	2	0.3
Hg(II)	k_1 [Fe(II)] [Hg(II)]	2.0×10^{-6}	c	80
Mn(III)	k_1 [Fe(II)] [Mn(III)]	1.67×10^4	1	25
NO_2^-	k_1 [Fe(II)] [NO_2^-]	7.8×10^{-3}	c	25
	k_1 [H^+] [Fe(II)] [NO_2^-]	0.227[b]	c	25
V(V)	k_1 [VO_2^+] [Fe(II)]	60	c	c
	k_1[H^+] [VO_2^{+2}] [Fe(II)]	3400[b]	c	c

a. Value of k_1 in $l M^{-1} s^{-1}$ unless noted

b. Units are $l^2 M^{-2} s^{-2}$

c. Not available

Source: Banford and Tipper [3].

OXIDATION OF SULFIDE BY OXYGEN

The rate of oxidation of sulfide has been determined in both freshwater and seawater media [1,2,8,23]. These rates are summarized in Table 3.3-6. The reaction is relatively slow. At a sulfide concentration of greater than 50 μmol/l the reaction obeys the following rate law[1]:

$$-d[S(-II)]/dt = k_1\ [S(-II)]\ [O_2^0] = k_2\ [S(-II)] \tag{22}$$

where S (–II) is the sum of sulfide, hydrosulfide and dissolved hydrogen sulfide.

The value of the pseudo-first-order rate constant k_2 at room temperature and pH 8 in seawater saturated with oxygen is 3×10^{-3} min^{-1}.

TABLE 3.3-6

Rates of Oxidation of Sulfide by Oxygen in Natural Systems[a]

Aqueous Medium	$[O_2]$ (μM)	$[S(-II)]_0$ (μM)	pH	Order with Respect to $[S^{-2}]$	$t_{1/2}$ (min)	Ref.
Fresh water	5	100	> 11.0	1.0	130	[2]
Fresh water	800	100	9.4	1.34	3000	[8]
Seawater	200	50	8.0	1.0	180	[1]
Seawater	200	200	8.0	1.0	280	[1]
Seawater	200	40	8.2	1.0	24	[23]

a. Temp. = 23-25°C.

ENVIRONMENTAL REACTIONS OF ARSENIC(III) AND ARSENIC(V)

Laboratory data from experiments using redox agents typically found in natural waters, such as oxygen, hydrogen sulfide and iron(III), suggest that their reactions with dissolved arsenic species occur slowly [9] and that the rates should be considered in speciation calculations. The change in concentration of arsenic(III) and arsenic(V) with time by reaction with these species is shown in Figure 3.3-3.

OXIDATION BY OZONE

The aqueous oxidation of many species by ozone has been studied, and their rates were recently reported [17]. The reactions generally obey the following rate law:

$$\text{Rate} = k_{O3}\,[O_3]\,[\text{Reducing Agent}] \tag{23}$$

Values of k_{O3} for a number of inorganic reducing agents are given in Table 3.3-7.

3.3.6 Chemical and Environmental Factors Affecting Rates

ONE- AND TWO-EQUIVALENT REDOX AGENTS

Oxidizing and reducing agents can be classified as one-equivalent, two-equivalent and one-two-equivalent. These refer to the following reactions:

(a) X(N) ⟶ X(N+1); one-equivalent process
(b) X(N+1) ⟶ X(N+2); one-equivalent process
(c) X(N) ⟶ X(N+2); two-equivalent process

where X(N) is species X in the Nth oxidation state.

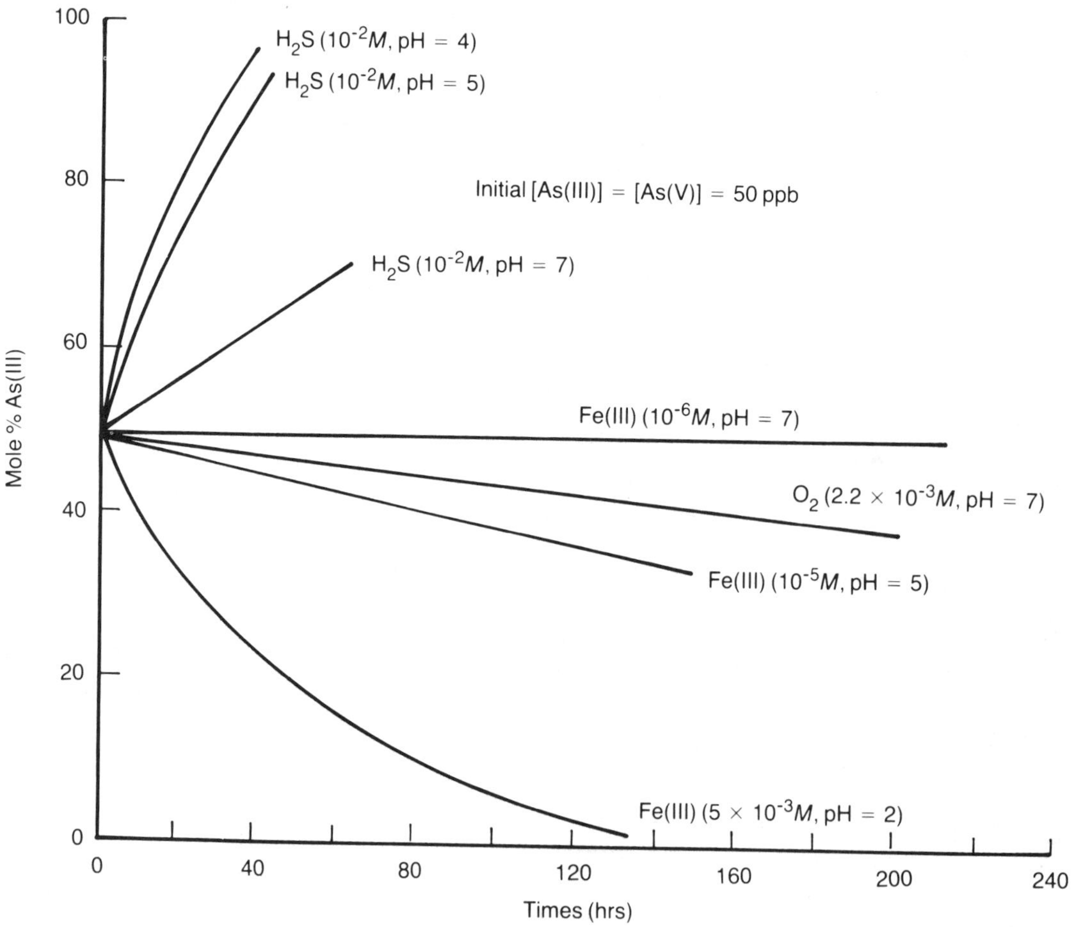

Source: Cherry, Shaik, Tallman and Nicholson [9]

FIGURE 3.3-3 Rates of Conversion of As(III) and As(V) by Different Reactants

TABLE 3.3-7

Rate Constants for Aqueous Oxidations by Ozone[a]

Substrate	k_{O_3} ($M^{-1}s^{-1}$)	Substrate	k_{O_3} ($M^{-1}s^{-1}$)
Cl^-	~0.003	CN^-	$10^3 - 10^5$
Br^-	160	OCN^-	$<10^{-2}$
I^-	$>10^6$	HOCN	$<10^{-2}$
HOCl	<0.002	HS^-	3×10^9
OCl^-	120	H_2S	3×10^4
HOBr	<0.01	H_2SO_3	2×10^4
OBr^-	530	HSO_3^-	3.2×10^5
ClO_3^-	$<10^{-4}$	SO_3^{-2}	1×10^9
BrO_3^-	$<10^{-3}$	Fe^{+2}	$>5 \times 10^5$
IO_3^-	$<10^{-4}$	H_2O_2	$<10^{-2}$
IO_4^-	$<10^{-2}$	HO_2^-	5.5×10^6
NH_3	20	Co^{+2}	0.6

a. Rate = k_{O_3} [O_3] [substrate]. Temperature = 20-23°C.

Source: Hoigné [17]

Species that can react via both (a) and (b) are termed one-two-equivalent agents, while those that react only via (a) are one-equivalent and only via (c) are two-equivalent agents, respectively. For example, Fe(III) produces Fe(II) when it gains one electron. Since Fe(II) is not further reduced by additional electrons, the Fe(III) is only a one-equivalent agent. The process of reducing I_2 to $2I^-$ can occur by simultaneous gain of two electrons, so I_2 is a two-equivalent agent. One-two-equivalent agents are exemplified by transition metal ions that form stable intermediate oxidation states when gaining electrons. For example, V(V) can form V(IV) by gain of one electron or V(III) by gain of two electrons and is thus a one-two-equivalent agent. Examples of other species in each of these categories are shown in Table 3.3-8.

The following general observation regarding the rates of redox reaction between species in these categories has been noted [4]:

- Redox reactions between noncomplementary reactants (i.e., a one- and a two-equivalent) are often slow compared with those between complementary reactants (i.e., a one- and a one-equivalent *or* a two- and a two-equivalent).

TABLE 3.3-8

One-, Two-, and One-Two-Equivalent Oxidizing Agents

One-Equivalent	Two-Equivalent	One-Two-Equivalent
Ce(IV)	I_2	$Cr_2O_7^{-2}$
Co(III)	Br_2	MnO_4^- [a]
Fe(III)	Cl_2	$PtCl_6^{-2}$
$Fe(CN)_6^{-3}$	HBrO	VO_2^+
	HClO	
	IO_3^-	
	BrO_3^-	
	H_2O_2	
	Tl(III)	

a. Complex situations can occur with manganese due to possibility of a variety of oxidation states (e.g., +7, +6, +5, +4, +3, +2)

Source: Higginson and Marshall [16]

This behavior is based on the lower probability of collision of three species and is not an inviolate rule.

It is useful to note the following observations as to what type of species undergo particular reactions [16]:

- Redox reactions between two complex metal ions usually occur in one-equivalent steps.
- Redox reactions between two species derived from non-transition elements (metals and nonmetals) usually occur in two-equivalent steps.
- Redox reactions between one species with a transition metal center and another having a non-transition element as the center may occur in one- or two-equivalent steps. The one-equivalent appears to be more common.
- For free radicals, irrespective of their origin, a one-equivalent step is more probable.

A one-two-equivalent agent can produce unstable intermediate oxidation states (e.g., X(N+1) radicals), which can lead to products such as $\{X(N+1)\}_2$ in addition to the more stable X(N+2) oxidation state [16].

CHARACTERISTICS OF LIGANDS AND METAL IONS

Several characteristics of ligands and metal ions affect the rates of redox reactions. These characteristics include the electronic configuration of the metal or central ion, its lability, the properties of the complexing ions in the medium, and the difference in electrode potential between the reactants (which affects the driving force for the redox reaction). The impact of these and other variables on the rate is discussed below.

Electronic Configuration

Activation of the ion-pair precursor to redox (e.g., equation 11) often involves changing the bond length between the metal and the ligand centers to bring it closer to that of the products. Adding or removing an electron from non-bonding orbitals is easier than from bonding or anti-bonding orbitals, since in the former case bond strength (and thus distance) is not affected. For example, for an octahedral complex (six equivalent ligands around a metal center) it is easier to remove, or add an electron to the t_{2g} orbital than the e_g orbital: the exchange reaction between ruthenium(II) (six electrons in t_{2g}) and ruthenium(III) (five electrons in t_{2g}) is much faster than that between cobalt(II) ($t_{2g}^5\ e_g^2$) and cobalt(III) (t_{2g}^6) [31].

In general, the more similar the molecular geometries of the oxidized and reduced forms of the reactants, the less energy is required to bring about the activated state and the faster the reaction [4].

Substitution-Inert Complexes

In the reaction of two metal complex ions containing ligands that are difficult to replace or that cannot act as bridges, the reaction is likely to proceed in an outer-sphere process whose rate is faster than that based on dissociation or atom transfer [4]. This is true, for example, for reactions of species such as hexacyanoiron (II) with hexacyanoiron (III), tris (o-phenanthroline) iron (II) with tris (o-phenanthroline) iron (III), hexacyanomanganese (II) with hexacyanomanganese (III), with tris (bipyridyl) chromium (II) with tris (bipyridyl) chromium (III). The electronic configuration of metal ions that results in substitution-inert complexes has been discussed in section 3.2.

Presence of Non-Complexing and Complexing Ions in Solution

The rate of a redox reaction is sensitive to the presence of other ions in solution [4]. Usually, the greatest effects are created by anions when both reactants are cationic, and by cations when both reactants are anionic. Anions enhance the rates between cations, usually by forming bridges, if substitutionally labile metal ions are involved. Other ways the rates may be affected include decreases in rate due to stabilization of the reactant oxidation state by complexation and an increase in rate due to reduced electrostatic repulsion by complexing of the reactants and formation of lower-charged reactants (when both reactants are of same charged sign).

Some bridging ligands that can act as multidentate complexing agents enhance the rates over those that act only as monodentate ligands [28]. For example, as illustrated below, ligands of the alpha-hydroxy acid class complexed to a cobalt (III) center can chelate the reducing agent, R, and increase the redox rate over that of a ligand that can form only one bond to the other reactant.

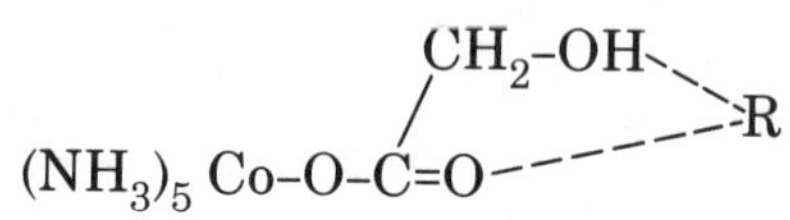

Redox Potential

The greater the difference between the redox potentials for the half-reactions between the two reactants, the greater the probability that the reaction will proceed in an outer-sphere path. This also indicates that for a series of outer-sphere reactions, the larger the difference in electrode potentials, the greater the rate. A quantitative discussion of this relationship is provided in section 3.3.7. Table 3.3-9 lists half-reaction (reduction) potentials for some metal ion species.

ENVIRONMENTAL FACTORS

Factors such as temperature, ionic strength and the presence of other ions in solution affect the rates of redox reactions. The relationships between the rate, the temperature and the ionic strength were discussed in section 3.2. Where inner-sphere processes are involved, temperature dependence is related to the effect of temperature on the rate of substitution. As previously mentioned, a similarity of activation energies between a redox and corresponding substitution reaction is one indication that the redox reaction is proceeding via an inner-sphere mechanism. The presence of other ions may lead to stabilization of the reactants or products; in addition, they can alter the rate by introducing a new, lower-energy path (i.e., catalysis) or by altering the speciation of the reactants.

3.3.7 Methods for Estimating the Rates of Redox Reactions

To estimate the rate of a homogeneous redox reaction, one must know the mechanism by which the reaction proceeds. The estimation methods described below can be classified as those based on the Marcus equation for outer-sphere processes [21], those related to the Marcus or Edwards equations for inner-sphere processes [10, 24], and those where correlations between rates of redox of other species and the particular species of interest have been made.

TABLE 3.3-9

Values of Exchange Rate Constants and Reduction Potentials[a]

Oxidant/Reductant Couple[b]	k_{11}(or k_{22}) ($M \cdot s^{-1}$)	Reduction Potential E^o (V)
Fe(III)/Fe(II)	4.0	+0.73
Cr(III)/Cr(II)	$\leq 2.0 \times 10^{-5}$	−0.41
Eu(III)/Eu(II)	$\leq 1.0 \times 10^{-4}$	−0.43
V(III)/V(II)	1.0×10^{-2}	−0.21
$Fe(CN)_6^{-3}/Fe(CN)_6^{-4}$	3.0×10^{2}	+0.69
$W(CN)_8^{-3}/W(CN)_8^{-4}$	7.0×10^{4}	+0.57
$Mo(CN)_8^{-3}/Mo(CN)_8^{-4}$	3.0×10^{4}	+0.54
$Fe(phen)_3^{+3}/Fe(phen)_3^{+2}$	10^{5}	+1.07
$Ru(NH_3)_6^{+3}/Ru(NH_3)_6^{+2}$	3.0	−0.21
Ce(IV)/Ce(III)	4.4	+1.44
Co(III)/Co(II)	~5.0	+1.82
$IrCl_6^{-3}/IrCl_6^{-2}$	2.0×10^{5}	+1.017
$Co(NH_3)_6^{+3}/Co(NH_3)_6^{+2}$	$<3.3 \times 10^{-12}$	+0.10

a. For use in equation 24.

b. phen = 1,10-o-phenanthroline

Source: Benson [6], Basolo and Pearson [4], Banford and Tipper [3], Latimer [20], Endicott and Taube [13].

The following quantitative methods are described:

- Estimation of the rate of an outer-sphere redox reaction using the Marcus equation or related equations;
- Estimation of the rate of an inner-sphere redox reaction using a Marcus-type equation;
- Estimation of the rate of a substitution-limited inner-sphere redox reaction using an Edwards-type equation; and
- Estimation of the rate of a substitution-controlled inner-sphere redox reaction from ligand exchange data.

When a reaction is known to proceed by an outer-sphere process, the Marcus equation can provide a good initial estimate of rate. For inner-sphere reactions, depending on the available data and detailed mechanism, any of the three methods given above may be applicable. When the mechanism is unknown, the approximate range can be calculated by any of several methods.

MARCUS EQUATION FOR OUTER-SPHERE REACTIONS

Outer-sphere redox reactions are simpler to treat theoretically than inner-sphere reactions because of the complexities of estimating energies for bond breaking. Marcus [21] derived the following relationship for calculating the rate of an outer-sphere electron transfer reaction:

$$k_{12} = (k_{11}\,k_{22}\,K_{12}\,f)^{1/2} \tag{24}$$

where

k_{12} = rate constant for equation 25
K_{12} = equilibrium constant for equation 25
k_{11} = rate constant for equation 26
k_{22} = rate constant for equation 27
f is defined by equation 28

$$OX_1 + RED_2 \overset{k_{12}}{\rightleftharpoons} RED_1 + OX_2 \qquad K_{12} \tag{25}$$

(OX_1 and OX_2 are the oxidized forms of species 1 and 2 respectively, and RED_1 and RED_2 are the reduced forms.)

$$RED_1^* + OX_1 \xrightarrow{k_{11}} OX_1^* + RED_1 \tag{26}$$

$$RED_2^* + OX_2 \xrightarrow{k_{22}} OX_2^* + RED_2 \tag{27}$$

$$\log f = \frac{(\log K_{12})^2}{4\log(k_{11}k_{22}/Z^2)} \tag{28}$$

($Z \approx 10^{11}\ mol^{-1} \bullet l \bullet s^{-1}$)

Equations 26 and 27 are termed the *related isotopic electron exchange reactions*, and the isotopic species are designated by asterisks.

Equation 24 may also be written in terms of the free energies of activation for reactions 26 and 27 and the overall free energy change for reaction 25 as follows:

$$\Delta G^{\ddagger}_{12} = 0.5\,\Delta G^{\ddagger}_{11} + 0.5\,\Delta G^{\ddagger}_{22} + \Delta G^{0}_{12} - 1.15\,RT\log f \tag{29}$$

Equation 29 predicts that a plot of $[\Delta G^{\ddagger}_{12} + 1.15\,RT\log f]$ versus ΔG^{0}_{12} is linear with a slope of 0.5 and an intercept of $0.5\,[\Delta G^{\ddagger}_{11} + \Delta G^{\ddagger}_{22}]$.

For a series of outer-sphere reactions of reactant 1 with several similar reactants 2, the value of the free energy of activation for the exchange reactions is relatively constant; thus, the intercept term would be constant.

When species OX_1 and OX_2 have similar reduction potentials, the value of the f term is about 1 [31]. This leads to simplification of equation 24 to the following:

$$k_{12} = (k_{11}\,k_{22}\,K_{12})^{1/2} \tag{30}$$

The agreement between calculated and observed k_{12} values has been used as evidence that a reaction proceeds by an outer-sphere mechanism. These equations are also useful in predicting the rates if the mechanism is considered to be outer-sphere. Their utility is limited by the available data for the exchange reactions of interest. In some cases the rates of exchange can be back-calculated by these equations if the other needed information is available.

Example 1 Estimate the rate of the outer-sphere redox reaction between hexacyano-iron (II) and octacyanotungsten (VI).

The overall reaction of interest is

$$Fe(CN)_6^{-3} + W(CN)_8^{-4} \longrightarrow Fe(CN)_6^{-4} + W(CN)_8^{-3} \qquad k_{12}, K_{12}, E_{12}^0 \tag{31}$$

The individual exchange reactions are

$$Fe^*(CN)_6^{-3} + Fe(CN)_6^{-4} \longrightarrow Fe^*(CN)_6^{-4} + Fe(CN)_6^{-3} \qquad k_{11} \tag{32}$$

$$W^*(CN)_8^{-4} + W(CN)_8^{-3} \longrightarrow W^*(CN)_8^{-3} + W(CN)_8^{-4} \qquad k_{22} \tag{33}$$

From Table 3.3-9, the reduction potential for the iron–cyanide complex (E_1^0) is +0.69 V. Similarly, the reduction potential for the tungsten-cyanide complex (E_2^0) is +0.57 V. Using these reduction potentials, we can calculate the equilibrium constant for reaction 31 as follows:

$$E_{12}^0 = E_1^0 - E_2^0 = 0.69 - 0.57 = 0.12\text{V}$$

Using equations 7 and 8 in section 2.10 at 25°C, we calculate the equilibrium constant for reaction 31 as

$$\begin{aligned} \log K_{12} &= n\,(23.06)\,E_{12}^0/1.36 \\ &= (1)\,(16.96)\,(0.12) \\ &= 2.04 \end{aligned}$$

This corresponds to an equilibrium constant (K_{12}) of 1.08×10^2.

Using equation 28 with the following parameters, we calculate the value of f:

$$k_{11} = 3 \times 10^2 \text{ mol} \bullet l^{-1} \bullet s^{-1} \text{ (Table 3.3-9)}$$

$$k_{22} = 7 \times 10^4 \text{ mol} \bullet l^{-1} \bullet s^{-1} \text{ (Table 3.3-9)}$$

$$K_{12} = 108 \qquad \log K_{12} = 2.04$$

$$Z = 10^{11} \qquad Z^2 = 10^{22}$$

$$\log f = \frac{(2.04)^2}{4 \log (300 \times 70{,}000/10^{22})}$$

$$= 4.16/(-58.71) = -0.071$$

$$f = 0.85$$

The rate constant for reaction 31 is now calculated from equation 24 with the above value of f:

$$k_{12} = (300 \times 70{,}000 \times 108 \times 0.85)^{1/2}$$

$$= 4.4 \times 10^4 \text{ mol} \bullet l^{-1} \bullet s^{-1}$$

The observed rate constant [6] is 4.3×10^4 mol • l^{-1} • s^{-1}.

The close agreement shown in the example is not representative of other calculations performed using the Marcus equation, as illustrated in Table 3.3-10. The calculated values are sensitive to the equilibrium constant and rates of exchange used. Table 3.3-11 illustrates the effect on the calculated values of using slightly different electrode potentials for the reacting species. Different electrode potentials are commonly observed for species in different reaction media.

TABLE 3.3-10

Examples of Calculated and Observed Rate Constants

Reaction	Calculated[a] ($M^{-1} \bullet s^{-1}$)	Observed[b] ($M^{-1} \bullet s^{-1}$)
Ce(IV) + Fe(II)	5.9×10^6	1.3×10^6
Fe(III) + Cr(II)	$> 6.7 \times 10^5$	$\sim 8.0 \times 10^3$
Fe(III) + V(II)	1.0×10^6	$> 10^5$
Co(III) + Cr(II)	$< 1.0 \times 10^{10}$	$> 3.0 \times 10^2$
$IrCl_6^{-2}$ + $Fe(CN)_6^{-4}$	1.9×10^6	3.8×10^5
$Co(NH_3)_6^{+3}$ + V(II)	6.2×10^{-5}	3.3×10^{-3}
$Co(NH_3)_6^{+3}$ + $Ru(NH_3)_6^{+3}$	$< 3.6 \times 10^{-7}$	1.0×10^{-2}

a. Using equation 24 and data in Table 3.3-9

b. **Source:** Banford and Tipper [3]

TABLE 3.3-11

Sensitivity Analysis of Marcus Equation

Reaction	E_1(V)	E_2(V)	K_{12} (calc)	k_{12} ($M^{-1} \cdot s^{-1}$) Calculated[a]	Observed[b]
$Co(NH_3)_6^{+3} + Ru(NH_3)_6^{+2}$	+0.10	-0.21	1.4×10^{-2}	3.6×10^{-7}	1.0×10^{-2}
		-0.20	2.0×10^{-2}	4.4×10^{-7}	
		-0.10	1.0	3.1×10^{-6}	
Ce(IV) + Fe(II)	+1.44	-0.73	1.1×10^{12}	5.9×10^{5}	1.3×10^{6}
	+1.54	-0.73	5.4×10^{13}	2.3×10^{6}	
	+1.64	-0.73	2.7×10^{15}	8.0×10^{6}	
$IrCl_6^{-2} + Fe(CN)_6^{-4}$	1.0	-0.69	1.8×10^{5}	1.9×10^{6}	3.8×10^{5}
	1.1	-0.69	8.9×10^{6}	8.7×10^{6}	
	1.0	-0.79	3.6×10^{3}	3.6×10^{5}	

a. Calculated by equation 24.

b. **Source:** Banford and Tipper [3].

Relationships between rates for a particular oxidant (A) and a series of reductants (B) and those for the same series with another oxidant (C) have been observed [30]. These relationships are simpler than the Marcus-type equation for outer-sphere reactions; they occur under conditions that simplify the form of the equation, such as when the ion pair equilibrium constants are very similar in the two series. Examples of such relationships are given below.

- For outer-sphere reactions of the type:

$$Co(NH_3)_6^{+3} + B \longrightarrow \text{Products} \qquad k\ (\text{hexaammine})$$

$$Co(NH_3)_5H_2O^{+3} + B \longrightarrow \text{Products} \qquad k\ (\text{pentaammine})$$

$$B\ (\text{oxidized}) + e^- \longrightarrow B\ (\text{reduced}) \qquad E_B^0$$

where k (hexaammine) and k (pentaammine) are the rate constants for the two oxidant series and E_B^0 is the reduction potential for the series of reducing agents. The observed relationship is

$$\log k\ (\text{hexaammine}) - \log k\ (\text{pentaammine}) = 1.3\ E_B^0 + 2.3 \qquad (34)$$

- Similarly, for a series of outer-sphere (one-electron rate-determining step) reactions of platinum(IV) haloamine complexes,

$$\text{Pt(IV) complex} + V^{+2} \longrightarrow \text{Products} \qquad k\ (\text{vanadium})$$
$$\text{Pt(IV) complex} + Ru(NH_3)_6^{+2} \longrightarrow \text{Products} \qquad k\ (\text{ruthenium})$$

the following relationship is observed:

$$\log k\ (\text{vanadium}) = 0.89 \log k\ (\text{ruthenium}) - 1.68 \qquad (35)$$

MODIFIED MARCUS EQUATION FOR INNER-SPHERE REACTIONS

Patel and Endicott [24] estimated the rates of an inner-sphere redox reaction of an oxidant A-L with a reductant C using data for outer-sphere reactions of the same oxidant. They used the relationship shown in equation 36, which is based on modification of that derived by Marcus (see previous section):[2]

$$\log k_{AC} = 0.5 \log F_{AL} + 0.5 \log k_{CC} + 8.45\, E^0_{CC} \qquad (36)$$

where k_{AC} is the rate constant for the following inner-sphere reaction:

$$\text{A-L (oxidized)} + \text{C (reduced)} \xrightarrow{\text{inner sphere}} \text{Products} \qquad k_{AC} \qquad (37)$$

and where A-L is a metal ion with ligand L coordinated to it. E^0_{CC} is the reduction potential for reaction 38:

$$\text{C (reduced)} \longrightarrow \text{C–L (oxidized)} + ne^- \qquad E^0_{CC} \qquad (38)$$

where C-L is the reductant C with ligand L coordinated to it; k_{CC} is the rate of the electron exchange reaction between the reduced and oxidized forms {C (reduced) and C–L (oxidized)}, and F_{AL} is a parameter defined by the following equation:

$$\log F_{AL} = 2 \log k_{AB} - \log k_{BB} - 16.9\, E^0_{BB} \qquad (39)$$

2. This equation is a variation of the following equation found elsewhere [31]:

$$k_{AB}/k_{AC} = (k_{BB}\, K_{AB}/k_{CC}\, K_{AC})^{1/2}$$

k_{AB} = rate constant for the reference outer-sphere reaction (40)

k_{BB} = rate constant for the electron exchange reaction of the reduced and oxidized forms of the outer-sphere reducing agent B

E^0_{BB} = reduction potential for the outer-sphere reducing agent B in reaction 41

$$\text{A-L (oxidized)} + \text{B (reduced)} \xrightarrow{\text{outer sphere}} \text{Products} \qquad k_{AB} \tag{40}$$

$$\text{B (reduced)} \longrightarrow \text{B (oxidized)} + ne^- \qquad E^0_{BB} \tag{41}$$

The observation that a Marcus-type equation predicts the rates for some inner-sphere reactions implies that in certain cases the activation of the oxidant center in both inner- and outer-sphere processes is similar. As this is by no means generally true, the utility of the equation is limited.

Example 2 Estimate the rate of the inner-sphere redox reaction between $Co(NH_3)_5Cl^{+2}$ and Fe^{+2}, using data available for the outer-sphere redox reaction of the cobalt complex with $Ru(NH_3)_6^{+2}$.

The following information is needed [24]:

The reference outer-sphere reaction is

$$Co(NH_3)_5Cl^{+2} + Ru(NH_3)_6^{+2} \rightarrow \text{Products} \qquad k_{AB} = 2.6 \times 10^2\ mol^{-1} \bullet l \bullet s^{-1} \text{ or}$$

$$\log k_{AB} = 2.41$$

The rate constant for the exchange reaction of the ruthenium (II) and ruthenium (III) complexes and the half-reaction potential are

$$Ru(NH_3)_6^{+2} \rightarrow Ru(NH_3)_6^{+3} + e^- \qquad k_{BB} = 1 \times 10^3\ mol^{-1} \bullet l \bullet s^{-1} \text{ or } \log k_{BB} = 3.0$$

$$E^0_{BB} = -0.10\ V$$

The value of $\log F_{AL}$ for the cobalt complex is calculated by use of equation 39:

$$\begin{aligned} \log F_{AL} &= 2 \log k_{AB} - \log k_{BB} - 16.9\, E^0_{BB} \\ &= 2\,(2.41) - 3.0 + 1.69 \\ &= 3.51 \end{aligned}$$

The inner-sphere reaction of interest is

$$Co(NH_3)_5Cl^{+2} + Fe^{+2} \rightarrow Co^{+2} + 5\,NH_3 + FeCl^{+2} \qquad k_{AC} = \text{unknown}$$

Here, chloride is the bridging ligand, and the iron (II) chloride complex is the initial product of the reaction. In aqueous media, the latter complex eventually dissociates to form aquo iron (II) and free chloride ions.

The appropriate exchange reaction for the reducing agent is thus [24],

$$FeCl^{+} \rightarrow FeCl^{+2} + e^{-} \qquad k_{CC} = 22.8 \text{ or } \log k_{CC} = 1.36$$
$$E^0_{CC} = -0.69V$$

(Note that values of E^0 for complexed species can be calculated from those for the aquo ions and related stability constants as given in § 2.10.)

We can now calculate the value of k_{AC} using equation 36.

$$\begin{aligned} \log k_{AC} &= 0.5 \log F_{AL} + 0.5 \log k_{CC} + 8.45\, E^0_{CC} \\ &= 0.5\,(3.51) + 0.5\,(1.36) + 8.45\,(-0.69) \\ &= -3.39 \end{aligned}$$

The observed value [24] is $\log k_{AC} = -2.89$

EDWARDS EQUATION FOR INNER-SPHERE REACTIONS

Edwards [10] developed a relationship for the rates of oxidation of several substrates by hydrogen peroxide. All of the related reactions appear to require initial replacement of oxygen on the hydrogen peroxide center by the reductant. The relationship is

$$\log (k_x/k_o) = \alpha\, E_n + \beta\, H \qquad (42)$$

where

k_x = rate of reaction with species x
k_o = rate of reaction with water (oxygen exchange or substitution)
α, β = empirically determined oxidant constants
E_n = nucleophilic constant for electron donor (reductant)
H = basicity constant for electron donor (reductant)

Table 3.3-12 lists values of E_n and H for some species. Values of α and β are determined from data for a series of related reactions; in the case of hydrogen peroxide, for example, Edwards [10] found that $\alpha = 6.31$, $\beta = -0.394$, and $\log k_o = -16.33$.

Values of k_o, α and β were obtained from empirical fit of the available kinetic data for several reducing agents to equation 42 [10].

Example 3 Estimate the rate of reaction of hydrogen peroxide with chloride, assuming that the reaction proceeds via replacement in a manner similar to that for bromide, iodide and cyanide.

From Table 3.3-12 and values derived by Edwards [10, 11] from the related equations,

$$\alpha = 6.31$$
$$\beta = -0.394$$
$$E_n = 1.24$$
$$H = -3.00$$
$$\log k_o = -16.33$$

The rate is now calculated from equation 42:

$$\log k_x - (-16.33) = (6.31)(1.24) + (-0.394)(-3.00)$$
$$\log k_x = 7.82 + 1.18 - 16.33$$
$$= -7.33$$

The value observed for reaction with chloride is $\log k_x = -6.96$.

TABLE 3.3-12

Donor Constants for Use with Edwards Equation

N	E_n	H	N	E_n	H
NO_3^-	0.29	0.40	SCN^-	1.83	1.00
SO_4^{-2}	0.59	3.74	NH_3	1.84	11.22
$ClCH_2COO^-$	0.79	4.54	$(CH_3O)_2POS^-$	2.04	4.00
CH_3COO^-	0.95	6.46	$C_2H_5SO_2S^-$	2.06	-5.00
C_5H_5N	1.20	7.04	I^-	2.06	-9.00
Cl^-	1.24	-3.00	$(C_2H_5O)_2POS^-$	2.07	4.00
$C_6H_5O^-$	1.46	11.74	$CH_3C_6H_4SO_2S^-$	2.11	-6.00
Br^-	1.51	-6.00	$SC(NH_2)_2$	2.18	0.80
N_3^-	1.58	6.46	$S_2O_3^{-2}$	2.52	3.66
OH^-	1.65	17.48	SO_3^{-2}	2.57	9.00
NO_2^-	1.73	5.09	CN^-	2.79	10.88
$C_6H_5NH_2$	1.78	6.28	S^{-2}	3.08	14.66

Source: Edwards [10]

INNER-SPHERE SUBSTITUTION-CONTROLLED REACTIONS

When the rate of redox reaction is limited by the substitution of a ligand at the metal center (one of the limits described in § 3.3.4), the rate can be estimated using the following equation:

$$k_{obs} = K_{ip} k_w \tag{43}$$

where the rate and equilibrium constants refer to the following reactions:

$$\text{A(oxidant)} + \text{B(reductant)} \rightleftharpoons \text{(A, B)} \qquad K_{ip} \tag{44}$$

$$\text{A(oxidant)} + H_2O \longrightarrow \text{Substitution product} \qquad k_w \tag{45}$$

$$\text{A(oxidant)} + \text{B(reductant)} \longrightarrow \text{Products} \qquad k_{obs} \tag{46}$$

Example 4 Estimate the rate of the substitution-controlled redox reaction between the triaquohydroxodiaminecobalt (III) complex and chromium (II).

The reaction of interest is:

$$Co(NH_3)_2\,(H_2O)_3\,OH^{+3} + Cr^{+2} \longrightarrow Cr^{+3} + Co^{+2} + \text{other products} \qquad k_{obs} \tag{47}$$

The ion pair formation reaction is:

$$Co(NH_3)_2\,(H_2O)_3\,OH^{+3} + Cr^{+2} \longrightarrow [Co(NH_3)_2\,(H_2O)_3\,OH,\,Cr]^{+4} \qquad K_{ip} \tag{48}$$

The substitution rate constant, k_w, to be used in this case is that for the chromium (II) reductant, since substitution at the cobalt (II) center is much slower than at the chromium (II) center (see § 2.9). The value of K_{ip} is calculated by use of the Eigen-Fuoss equation, as described in § 2.9.4; if both ionic charges are +2, $K_{ip} = 1.6 \times 10^{-3}\,M^{-1}$. The first-order rate constant for substitution at the chromium (II) center is about 10^9 s^{-1} [Table 3.2-2]; this indicates a second-order rate constant of 55.5×10^9 mol • l^{-1}, using a value of 55.5 mol • l^{-1} for the concentration of water in an aqueous system. The rate constant for redox reaction is then estimated using equation 43:

$$k_{obs} = (1.6 \times 10^{-3})\,(55.5 \times 10^9)$$
$$= 8.9 \times 10^7\ s^{-1}$$

3.3.8 Sources of Data

The best data on the kinetics of redox reactions are usually found in journal articles that describe kinetic studies of particular reactions. Several books dealing with the mechanisms of inorganic redox reactions in inorganic chemistry include limited compilations of data; however, most of these are not directly relevant to environmental conditions, because the studies were usually done at ionic strengths and in media unlike those found in the environment. Moreover, the conditions seldom resemble those found in the environment with regard to the other species present. Nevertheless, such data help to show whether a given reaction is likely to be fast or slow and, thus, if an understanding of the kinetics is important in that particular case. Data of this kind can be found in references 3, 4, 6, 12 and 31.

3.3.9 Literature Cited

1. Almgran, T.A. and I. Hagstrom, "The Oxidation Rate of Sulfide in Sea Water," *Water Res.*, **8**, 398 (1974).
2. Avrahami, M. and R.M. Goldberg, "The Oxidation of Sulfite at Very Low Concentrations in Aqueous Solution," *J. Chem. Soc. A*, 647 (1968).
3. Banford, C.H. and C.F.H. Tipper, *Comprehensive Chemical Kinetics*, Vol. 7, Elsevier Publishing Co., New York (1972).
4. Basolo, F. and R.G. Pearson, *Mechanisms of Inorganic Reactions*, John Wiley & Sons, New York (1967).
5. Beattie, J.K. and G.P. Haight, "Chromium (VI) Oxidations of Inorganic Substrates," in *Inorganic Reaction Mechanisms*, Part II, J.O. Edwards (ed.), Interscience Publishers, New York (1972).
6. Benson, D., *Mechanisms of Inorganic Reactions in Solution*, McGraw-Hill Book Co., London (1968).
7. Bodek, I. and G. Davies, "Phenomenological Model for Redox Reactions in Solution," *Coord. Chem. Rev.*, **14**, 269 (1974).
8. Chen, K.Y. and J.C. Morris, "Kinetics of Oxidation of Aqueous Sulfide by Oxygen," *Environ. Sci. Technol.*, **6**, 529 (1972).
9. Cherry, J.A., A.U. Shaik, D.E. Tallman and R.V. Nicholson, "Arsenic as an Indicator of Redox Conditions in Groundwater," *J. Hydrol.*, **43**, 373 (1979).
10. Edwards, J.O., "Correlation of Relative Rates and Equilibria with a Double Basicity Scale," *J. Am. Chem. Soc.*, **76**, 1540 (1954).
11. Edwards, J.O., "On the Reactions of Hydrogen Peroxide with Donor Particles," *J. Phys. Chem.*, **56**, 279 (1952).
12. Edwards, J.O. (ed.), "Inorganic Reaction Mechanisms," in *Progress in Inorganic Chemistry*, Vol. 13, S. Lippard (ed.), Interscience Publishers, New York (1970).
13. Endicott, J. and H. Taube, "Studies on Oxidation-Reduction of Ruthenium Ammines," *Inorg. Chem.*, **4**, 437 (1965).

14. Haim, A., "Mechanism of Electron Transfer Reactions," in *Progress in Inorganic Chemistry*, Vol. 30, S. Lippard (ed.), Interscience Publishers, New York (1983).
15. Hem, J.D., "Chemical Equilibria and Rates of Manganese Oxidation," *U.S. Geol. Surv. Water-Supply Pap.*, 1667-A (1963).
16. Higginson, W.C.E. and J.W. Marshall, "Equivalence Changes in Oxidation Reduction Reactions in Solution: Some Aspects of Oxidation of Sulphurous Acid," *J. Chem. Soc.*, 447 (1957).
17. Hoigné, J., H. Bader, W.R. Haag and J. Staehelin, "Rate Constants of Reactions of Ozone with Organic and Inorganic Compounds in Water-III," *Water Res.*, **19**, 993 (1985).
18. Housé, D.A., "Kinetics and Mechanism of Oxidations by Peroxydisulfate," *Chem. Rev.*, **62**, 185 (1962).
19. Lappin, A.G., "Electron Transfer Reactions," in *Mechanisms of Inorganic and Organometallic Chemistry*, Vol. 1, M.V. Twigg (ed.), Plenum Press, New York (1983).
20. Latimer, W.M., *Oxidation Potentials*, 2nd ed., Prentice-Hall, New York (1952).
21. Marcus, R.A., "On the Theory of Oxidation-Reduction Reactions Involving Electron Transfer. V: Comparison and Properties of Electrochemical and Chemical Rate Constants," *J. Phys. Chem.*, **67**, 853 (1963).
22. Nakayama, E., T. Kuwamoto, S. Tsurabo and T. Fujinage, "Chemical Speciation of Chromium in Sea Water," *Anal. Chim. Acta.*, **130**, 401 (1981).
23. Ostland, G.H. and J. Alexander, "Oxidation Rate of Sulfide in Sea Water," *J. Geophys. Res.*, **68**, 3995 (1963).
24. Patel, R.C. and J.F. Endicott, "Electron Transfer Reactions of Cobalt (III): Relative Rate Comparison and Free Energy Relations," *J. Am. Chem. Soc.*, **90**, 6364 (1968).
25. Schroeder, D.C. and G.F. Lee, "Potential Transformations of Chromium in Natural Waters," *Water Air Soil Pollut.*, **4**, 355 (1975).
26. Stollewerk, K.G. and D.B. Grove, "Reduction of Hexavalent Chromium in Water Samples Acidified for Preservation," *J. Environ. Qual.*, **14**, 301 (1985).
27. Stumm, W. and G.F. Lee, "Oxygenation of Ferrous Iron," *Ind. Eng. Chem.*, **53**, 143 (1961).
28. Taube, H., "Bridging Groups in Electron Transfer Reactions," in *Mechanisms of Inorganic Reactions, Adv. Chem. Ser.* No. 49, American Chemical Society, Washington, D.C. (1965).
29. Theis, T.L. and P.C. Singer, "Complexation of Iron (II) by Organic Matter and Its Effect on Iron (II) Oxygenation," *Environ. Sci. Technol.*, **8**, 569 (1974).
30. Twigg, M.V. (ed.), *Mechanisms of Inorganic and Organometallic Reactions*, Vol. 1, Plenum Press, New York (1983).
31. Wilkins, R.G., *The Study of Kinetics and Mechanism of Reactions of Transition Metal Complexes*, Allyn and Bacon, Boston (1974).

3.4 KINETICS OF DISSOLUTION, PRECIPITATION AND CRYSTALLIZATION

3.4.1 Introduction

The goal of this section is to provide the reader with the tools and methods for answering questions on the kinetics of dissolution, nucleation and precipitation of materials of environmental concern. Within this broad category of chemical kinetic processes, the more important and straightforward questions can be stated as follows:

- How fast does a solid material dissolve?
- How fast does the chemical composition or the quality of water change during dissolution?
- How fast can a state of saturation or chemical equilibrium be obtained?
- How fast does a solid precipitate and/or recrystallize?

Each of these questions can arise in an investigation or in monitoring of a great variety of systems containing solids and water. Natural systems made of rocks, soils and running waters, and any number of man-made systems ranging from simple water conduits to complex waste treatment and disposal systems, all have in common one feature: they all contain water and solids that can react with water.

The presentation in this section deals with variables needed in making environmental decisions and obtainable from experimental work, from measurements in natural situations, or from literature data.

Environmental conditions affect to a variable extent the rates of dissolution of solids in undersaturated solutions and the rates of precipitation in supersaturated solutions. As outlined below, the environmental conditions may be divided into two groups — physical and chemical.

3.4.2 Key Variables, Information Needs and Data Sources

ENVIRONMENTAL CONDITIONS

Physical Conditions

The physical conditions of a system that affect the rates of dissolution or precipitation are mainly the following:

- Temperature
- Volume of solution
- Homogeneity of the liquid phase
- Surface area of the solid

Most reactions increase in rate with an increasing temperature. The environmental temperature range of several tens of degrees centigrade makes this effect important for many reactions. For pressure, however, its range of variation near the earth's surface is too small to be significant. Temperature also has indirect effects on dissolution or precipitation processes, such as temperature-induced changes in viscosity of solutions or changes in the rate of water evaporation from open systems.

The volume of water in a system and the surface area of the solid phase are the two physical parameters that control how much solid material will be transferred into a given volume of solution in a given time.

One speaks of a solution as homogeneous if it is well mixed. Mechanical stirring of a solution often increases the rate of dissolution of a solid because of the faster transport of material from the solid surface. In nature, water turbulence plays the role of mechanical stirring. Although methods exist to quantify degrees of turbulence in natural waters or, analogously, degrees of mixing in a stirred volume of water, their treatment usually falls in the domain of dissolution or precipitation processes in engineered systems. Such systems, where the geometric shape of the system (such as the cross-section of a pipe, a reactor, or a body of water) affects the rates of dissolution or precipitation, are outside the scope of this section.

Chemical Conditions

Rates of dissolution or precipitation are, to a large extent, determined by the chemical nature of the components of the system. For example, the rate of dissolution of a carbonate phase, such as $CaCO_3$, depends on the amount of dissolved CO_2 and on the pH of the solution. In this case, the chemical composition of the solution phase includes hydrogen-ion concentration and CO_2. In a case of dissolution of calcium sulfate, $CaSO_4$, the concentration of CO_2 in water may be inconsequential and would not need to be considered. However, the presence of the SO_4^{-2} ion in solution from another dissolved salt, such as Na_2SO_4, might control $CaSO_4$ dissolution by making the solution more nearly saturated with calcium sulfate.

Ionic strength, a parameter based on the concentrations of all the ionic components of the solution (section 2.6), can also affect the rates of dissolution or precipitation.

INFORMATION NEEDED FOR DISSOLUTION AND PRECIPITATION RATES

Each of the four questions in the Introduction refers to a mass of a solid or dissolved material transferred from one phase to another in a unit of time. Questions as to how fast a solid dissolves and how fast solution composition changes are answerable quantitatively, usually in such units as mass per unit time, mass per unit volume per unit time, and mass per unit area per unit time.

Basic data for computation of the rates of dissolution and of the kinetic parameters that enter in the dissolution rate equations are as follows (see also preceding section on Environmental Conditions):

Physical conditions of the system:

- Temperature
- Solution volume (in a system without flow)
- Solution volume flow rate (in a system with flow)
- Surface area of the dissolving solid

Chemical conditions of the system:

- Chemical composition of the solid, or concentrations of the components of interest in the dissolving solid
- Chemical composition of the solution, both initially and as a function of time during dissolution
- Concentrations and chemical states of other components of the system that may affect the rate of dissolution (such as dissolved gases, complexing agents or common ions).

Estimation of the rates of precipitation is often based on monitoring changes in concentration of the precipitating component in solution. In both natural and man-made systems, it is often easier to monitor a change in concentration of the solution than to measure the mass of solid precipitated over a period of time. Therefore, most of the data on rates of precipitation and crystal growth (in closed systems) are derived from measurements of rates of decrease in concentration. To relate the mass of a material removed from solution to the rate of growth of the solid crystals, it is necessary to know the surface area of the solid; this is obtainable either by direct measurements (the gas-adsorption method of solid surface-area determination) or by measuring the numbers and sizes of the crystals formed.

Recrystallization is a process taking place when precipitates age and, more generally, when phase transformations occur in solids under certain environmental conditions. A glassy or amorphous solid may sooner or later become crystalline, a very slow process under environmental conditions but one that speeds up at high temperatures. Amorphous and gelatinous precipitates often contain large proportions of water; removal of this water may result in dry amorphous or crystalline solids.

Another type of recrystallization is dehydration: a solid containing H_2O as part of its stoichiometric chemical formula and its crystal structure recrystallizes to a solid containing fewer H_2O molecules or to an anhydrous form containing none. Such reactions take place under conditions of low humidity or somewhat elevated temperatures. The basic data from which the kinetic parameters of such recrystallization reactions are derived consist of measurements of the volume of crystals of a new phase forming at the expense of the original phase.

LITERATURE SOURCES OF KINETIC DATA

Kinetic data for dissolution and precipitation of solids in aqueous solutions are, in general, more difficult to find than those on the kinetics of reactions in homogeneous or two-fluid systems. While the latter are compiled in reference volumes, one must search the current scientific and engineering literature for the water and solids of environmental concern. Some suggestions on sources of published kinetic data of environmental interest follow.

For systems containing natural waters and mineral solids, articles dealing with the kinetics of dissolution and precipitation reactions, as studied either in a laboratory or in the field, appear in such journals as *Geochimica et Cosmochimica Acta, Limnology and Oceanography, American Journal of Science,* and *Marine Chemistry.*

More specifically oriented toward soils and waters (including the kinetics of ion exchange, leaching and dissolution processes) are the journals *Proceedings of the Soil Science Society of America, Soil Science, Journal of Soil Science,* and *Water, Air and Soil Pollution.*

Kinetics of various chemical reactions of environmental concern is frequently discussed in articles in *Environmental Science and Technology.*

Experimental studies of crystal growth in systems of environmental and medical interest may be found in *Journal of Crystal Growth* and sporadically in the periodical publications of the chemical and mineralogical societies of various countries.

Professional societies, such as the American Chemical Society and the Mineralogical Society of America, occasionally publish proceedings of symposia devoted to particular aspects of the kinetics of solid-water systems.

In recent years, voluminous experimental data on adsorption and leaching kinetics of trace elements and radionuclides have been produced by concerted research supported by the U.S. Department of Energy. Many of these studies were aimed at determining the rates of either uptake or leaching of trace metals and radionuclides in systems consisting of groundwater and rocks, or groundwater and processed nuclear wastes. Experimental determinations of the rates of leaching of components of nuclear fuel waste and of metals in subsurface rocks have been conducted for a variety of materials (glasses, basaltic rocks, clays, volcanic tuffs, and some of their individual mineral components) at temperatures of up to several hundred degrees centigrade and pressures of up to hundreds of atmospheres. Much of the results are available in reports published for the Department of Energy by the National Laboratories (for example, Argonne, Brookhaven, Oak Ridge, and Sandia) and by the major contractors to the D.O.E., such as Battelle-Pacific Northwest Laboratories and Rockwell Hanford Operations, both at Richland, Washington. These publications are not always available from the main clearinghouse of government technical publications in Springfield, Virginia, but can be requested either through the D.O.E. in Washington, D.C., or through the publication and distribution departments of the individual laboratories and agencies.

3.4.3 Dissolution

Dissolution is a chemical reaction that results in transfer of materials from one phase or medium to another. This definition includes a variety of processes involving transfer of chemical species between combinations of solid, liquid and gaseous phases. Thus, one may speak of dissolution of one liquid in another or in a mixture of other liquids, of a gas in other gases or in liquids, and of solids in aqueous solutions. A feature common to all these processes is transfer of the components of one phase — the *dissolving substance* — to another phase, the *solvent*. Chemical species dissolved in a solvent are generally referred to as the *solutes*. A number of examples illustrating different types of dissolution are given below.

DESCRIPTION OF DISSOLUTION PROCESSES[1]

Congruent Dissolution

In congruent dissolution, a solid phase produces dissolved components in the same stoichiometric proportions as in the solid. An example is the dissolution in water of sodium chloride, which occurs in nature as the mineral halite (NaCl):

$$\mathrm{NaCl(s)} \longrightarrow \mathrm{Na^+} + \mathrm{Cl^-} \tag{1}$$

where symbol (s) denotes a solid phase. As each mole of sodium chloride dissolving in water produces 1 mole each of sodium and chloride ions, reaction 1 can be written as a stoichiometrically balanced equality:

$$\mathrm{NaCl(s)} = \mathrm{Na^+} + \mathrm{Cl^-} \tag{2}$$

Here, the equal sign does not necessarily mean that an equilibrium or a state of saturation has been attained; it indicates a stoichiometrically balanced relationship between the reactants and products of a dissolution reaction between halite and water.

Incongruent Dissolution

In the incongruent dissolution of a solid, the proportions of the species in solution differ from those in the solid. Such a stoichiometric imbalance can be achieved only if a new solid forms in a reaction between the original solid and the aqueous solution. An example is the dissolution of kaolinite, a common alumino-silicate mineral; the reaction takes place slowly under certain natural conditions, resulting primarily in dissolution of silica and formation of an aluminum hydroxide mineral, gibbsite:

$$\underset{\text{kaolinite}}{\mathrm{Al_2Si_2O_5(OH)_4(s)}} + \underset{\text{water}}{\mathrm{H_2O}} = \underset{\text{gibbsite}}{\mathrm{2Al(OH)_3(s)}} + \underset{\substack{\text{dissolved}\\ \text{silica}}}{\mathrm{2SiO_2^0}} \tag{3}$$

1. See Appendix A for an explanation of the notation used to designate the states of matter.

In the preceding reaction, the equal sign refers to a stoichiometric balance between the initial reactants (kaolinite and water) and the final products (gibbsite and aqueous SiO_2). The notation does not show any of the intermediate dissolution steps, such as transfer of aluminum species to solution and their precipitation in a new hydroxide phase. Implied in the reaction is conservation of aluminum in the solid and transfer of SiO_2 to solution. Reaction 3 is a process of desilicification; under natural conditions it manifests itself in the development of aluminum hydroxide and oxyhydroxide layers on top of weathered beds of kaolinite.

The shorthand method of writing dissolution reactions as stoichiometrically balanced equations may conceal the occurrence of important compounds that form in the process of dissolution and control it. For example, the net effect of dissolution of Fe_2O_3, occurring in nature as the mineral hematite, can be written as

$$Fe_2O_3(s) + 3H_2O = 2Fe^{+3} + 6OH^- \tag{4}$$

In reaction 4, dissolved Fe^{+3} will be present as a number of hydrated iron species, such as $FeOH^{+2}$, $Fe(OH)_2^+$, and $Fe(OH)_3^0$. However, in the process of dissolution a hydrated mineral may form at the reaction interface. This may be represented by a sequence such as the following:

$$Fe_2O_3(s) + 3H_2O \longrightarrow 2Fe(OH)_3(s) \longrightarrow 2Fe^{+3} + 6OH^- \tag{5}$$

A similarity to a more obvious case of an incongruent dissolution, reaction 3, can be emphasized by a model of dissolution of Fe_2O_3 with concomitant formation of solid iron hydroxide, $Fe(OH)_3$, and aqueous iron and hydroxide species.

Leaching or Dissolution of Minor Components

Leaching is an environmentally important process of dissolution of minor components from a solid matrix. Numerous examples of leaching processes occur in reactions of waters with soils, man-made solid wastes, different types of ash from power plants burning fossil fuels, and — with an eye on the possible future — in reactions of waters with nuclear fuel wastes.

For a trace component X residing in a solid matrix M, one process of dissolution may involve transfer of X from a virtually insoluble matrix. In this case, a solid MX loses component X, becoming solid M in the process:

$$MX(s) \longrightarrow M(s) + X^0 \tag{6}$$

Reaction 6 is analogous to the incongruent dissolution of kaolinite, shown in reaction 3.

If both the matrix component M and the trace component X are soluble, the dissolution reaction is, schematically,

$$MX(s) \longrightarrow M^0 + X^0 \tag{7}$$

Example 1 An example of the latter reaction is dissolution of an impure sodium chloride. Natural NaCl almost always contains minor amounts of bromide, coprecipitated with sodium chloride from the original saline brine of oceanic origin. The composition of such a solid can be written as $NaCl_xBr_{1-x}$. As both NaCl and NaBr are highly soluble in water, the dissolution reaction is

$$\underset{\text{halite with traces of NaBr}}{NaCl_xBr_{1-x}(s)} \longrightarrow Na^+ + xCl^- + (1-x)Br^- \tag{8}$$

Example 2 An example of leaching of a component from a nuclear fuel waste in a form of fused glass is the reaction:

$$(\text{Glass} \cdot \text{Pu})(s) \longrightarrow Pu^{+x} + \text{Glass}(s) \tag{9}$$

In this reaction, only plutonium leached from the glass is being monitored. Other radioactive and nonradioactive components of the glass, as well as SiO_2 dissolving from the glass, are not dealt with in this case. Compare reaction 6.

CRITERIA OF DISSOLUTION

Why do solids dissolve in water? In the language of chemical thermodynamics (see section 2.11), transfer of a component of a solid phase to solution takes place if the chemical potential of this component in the solid is greater than its chemical potential in solution. The chemical potential of a component is the driving force responsible for its transfer from a phase where the chemical potential is higher to a phase where the chemical potential is lower. Dissolution reaction of a commonly occurring form of silica (mineral quartz, SiO_2) can be represented by

$$SiO_2(s) + 2H_2O = H_4SiO_4^0 \tag{10}$$

or, disregarding H_2O in the stoichiometric balance for simplicity, by

$$SiO_2(s) = SiO_2^0 \tag{11}$$

Dissolution takes place if the chemical potential of SiO_2 in the solid is greater than its chemical potential in the aqueous phase, at the same temperature and pressure. The chemical potential of a component in each of the phases is commonly denoted μ, in units of kJ/mol or kcal/mol. Dissolution can occur if the following condition exists:

$$\mu_{SiO_2(s)} > \mu_{SiO_2^0} \tag{12}$$

The case of saturation — that is, no net dissolution or precipitation — corresponds to an equality of the chemical potentials of the SiO_2 component in the two phases:

$$\mu_{SiO_2(s)} = \mu_{SiO_2^0} \tag{13}$$

Precipitation may occur if the chemical potential of SiO_2 in the aqueous phase is greater than in the solid phase, which is the opposite of the case in dissolution:

$$\mu_{SiO_2(s)} < \mu_{SiO_2^0} \tag{14}$$

The inequality of the chemical potentials, defining the conditions of either dissolution or precipitation, does not indicate the speed of either of these processes, or even whether they proceed at all; it indicates only that dissolution or precipitation *may* take place. The following example illustrates the use of the dissolution or precipitation criterion without deriving the mathematical relationships for the chemical potentials of the aqueous and solid species. (Information on chemical potentials and their use is given in section 2.11, which discusses solubilities.)

Example 3 A river water contains 12 mg SiO_2/liter. Will quartz dissolve in this water? Will noncrystalline silica glass dissolve in this river?

To answer these questions, we must establish which of the three criteria, equations 12 through 14, is met within the environmental conditions of the system. At the standard conditions of 25°C and 1 bar total pressure, the following relationship applies to the reaction between quartz and water:

$$\underset{\text{quartz}}{SiO_2(s)} = SiO_2^0$$

The chemical potential is related to the chemical potential in a standard state, μ^0, and the activity, a, of the component by the relationship

$$\mu = \mu^0 + RT \ln a \tag{15}$$

where T is temperature (K) and R is the gas constant, as given under equation 57.

Solid SiO_2 is a pure one-component phase and, by definition, the activity of SiO_2 in it is unity, $a_{SiO_2(s)} = 1$. Therefore, for SiO_2 in quartz, equation 15 reduces to

$$\mu_{qtz} = \mu^0{}_{qtz} \tag{16}$$

For the chemical potential of SiO_2 in quartz at a standard state, one can use the value of the Gibbs standard free energy of formation of quartz [38].

$$\mu^0{}_{qtz} \equiv \Delta G_f^0(qtz) = -856.3 \text{ kJ/mol} \tag{17}$$

For a dilute solution, activity can be replaced by concentration in molal units, giving the following form of the chemical potential:

$$\mu_{SiO_2^0} = \mu^0{}_{SiO_2^0} + RT \ln m_{SiO_2^0} \tag{18}$$

The standard chemical potential of aqueous SiO_2 is

$$\mu^0{}_{SiO_2^0} \equiv \Delta G_f^0(SiO_2^0) = -833.72 \text{ kJ/mol} \tag{19}$$

The concentration of SiO_2 (12 mg/l) is, very nearly,

$$m = \frac{12 \times 10^{-3}\ g/l}{60\ g/mol} \approx 2 \times 10^{-4}\ mol/kg\ or\ molal.$$

Thus, the chemical potential of aqueous SiO_2 in river water is

$$\begin{aligned}\mu^0{}_{SiO_2^0} &= -833.72 + (0.00831)(298)\ \ln\ (2\times10^{-4}) \\ &= -854.0\ kJ/mol\end{aligned} \qquad (20)$$

Comparing equations 20 and 17, we see that the chemical potential of SiO_2 in solution is greater than its chemical potential in the quartz phase ($-854.0 > -856.3$). The conclusion is that quartz will not dissolve in this river water.

For a noncrystalline, amorphous form of SiO_2, the value of the standard chemical potential is

$$\mu^0{}_{SiO_2(am)} = \Delta G_f^0\ (SiO_{2(am)}) = -850.56\ kJ/mol \qquad (21)$$

Since this value is greater than μ_{SiO_2} in the river water,

$$\mu_{SiO_2(am)} > \mu_{SiO_2^0} \qquad (22)$$

The amorphous, glass-like SiO_2 should dissolve in this river water.

DISSOLUTION ON A GLOBAL SCALE

Dissolution occurs on a global scale in the natural process of chemical weathering, which takes place when minerals in rocks react with waters, releasing some of their components to solution and also forming new minerals in the process.

The global rate of chemical weathering can be estimated quite accurately from the measured concentrations of dissolved mineral species in the rivers and groundwaters of the world. The main chemical species in river waters (Table 3.4-1) are the products of reactions between the minerals in the crustal rocks of the earth, and water and atmospheric reactive gases. Ideally, to estimate the global mass of dissolved mineral matter in the rivers of the world, one should have a very comprehensive data base of chemical analyses of the more important rivers, averaged over seasons and years. This global mass, removed from the continents in some unit of time, is a measure of the mean global rate of chemical weathering or, in the language of this section, of a mean global rate of dissolution of the earth's crust. To estimate this rate, we can think either in terms of mass removed per unit area of the land surface per unit of time, or of the thickness of a layer stripped by dissolution per unit of area and per unit of time. The rate of dissolution, R_m, in units of mass, is

$$R_m = \frac{CF}{A}\ (g \cdot cm^{-2} \cdot yr^{-1}) \qquad (23)$$

where C = mean concentration of dissolved mineral species in rivers (g/l)
F = mean water flow rate in rivers (l/yr)
A = area of the continents drained by all the rivers (cm^2)

TABLE 3.4-1

Mean Composition of the Rivers of the World and Rates of Delivery to the Oceans

Dissolved Component	Concentration (mg/kg)				Mass Delivered by Rivers to Oceans[e]
	a	b	c	d	(10^{14} g/yr)
Na^+	4.3	4.8	5.15	7.2	1.93
K^+	2.0	2	1.3	1.4	0.49
Ca^{+2}	15.5	15	13.4	14.7	5.01
Mg^{+2}	3.9	4	3.35	3.65	1.25
Cl^-	3.65	5.7	5.75	8.25	2.15
SO_4^{-2}	11	6.7	8.25	11.5	3.09
HCO_3^-	61[f]	23[g]	52	53	19.5
SiO_2^0	13.3	13	10.4	10.4	3.89

a. Garrels and Mackenzie [17].

b. Holland [19].

c. Meybeck [28] (background or pristine concentrations).

d. Meybeck [28] (values include input from human activity).

e. Values in column (c) multiplied by rate of H_2O delivery to the ocean (3.74×10^{16} kg/yr). From Drever [10].

f. Includes bicarbonate formed from atmospheric CO_2.

g. Refers to bicarbonate species derived from land sources only.

Keep in mind that parts of some continents do not drain into the oceans; regions of Central Asia and Africa have only internal drainage. Similarly, on a smaller scale, internal drainage may occur in the form of a closed basin or lake within a relatively large drainage area. For the purpose of this example, however, the above parameters are [10, 28]:

$$C \approx 100 \text{ mg/l} = 0.1 \text{ g/l}$$
$$F = 3.74 \times 10^{16} \text{ l/yr}$$
$$A = 149{,}000{,}000 \text{ km}^2 = 1.49 \times 10^{18} \text{ cm}^2$$

From these data and equation 23, the mean rate of chemical weathering is

$$R_m \approx 0.003 \text{ g} \cdot \text{cm}^{-2} \cdot \text{yr}^{-1} \tag{24}$$

In detailed studies of erosion rates of watersheds and individual stream basins, variations of as much as a factor of ten have been reported. Such large variations are caused in part by variable rainfall but, significantly, they can also reflect different kinds of rocks in different river basins. Over geologically long periods of time, calcium sulfates, in the form of the minerals gypsum ($CaSO_4 \cdot 2H_2O$) and anhydrite ($CaSO_4$), dissolve in running water faster than crystalline silicate rocks. Thus, a terrain made primarily of calcium sulfate beds weathers more rapidly than a river basin where crystalline rocks are exposed.

The mean rate of dissolution or chemical weathering can also be expressed in terms of a thickness stripped from the earth's surface. In this estimate, the rate of dissolution, R_L, represents the volume of material per unit area dissolved in a unit of time:

$$R_L = \frac{CF}{A\rho} \qquad (cm \cdot yr^{-1}) \tag{25}$$

where ρ is the mean density ($g \cdot cm^{-3}$) of the dissolving materials. From equations 23 and 25, the two estimates of the rate are interrelated:

$$R_L = R_m/\rho \tag{26}$$

A mean density of the crustal materials is

$$\rho = 2.7\ g \cdot cm^{-3}$$

The linear rate of dissolution is therefore

$$R_L = \frac{0.003}{2.7} \approx 0.001\ cm \cdot yr^{-1} \tag{27}$$

Recall that natural erosion of the earth's surface involves two kinds of processes: chemical weathering and physical weathering. The former can be likened to dissolution and the latter to mechanical abrasion. Rivers carry both dissolved and suspended materials, corresponding to the products of the chemical and physical denudation processes. In a later section, the slowness of the global rates will be emphasized by some much faster dissolution rates measured under laboratory conditions.

DIFFERENCES IN KINETICS BETWEEN CLOSED AND OPEN SYSTEMS

In considering the rates of dissolution of natural and environmentally important substances, one must always bear in mind the differences between closed and open systems. A closed system consists of the dissolving phase and water, without flow-through; no material is added by inflow nor removed by outflow. However, precipitation from solution can take place in a closed system, where the mass of the dissolving material is conserved. Open systems, with flow in and out, are common in natural environments, and the kinetics of dissolution in such systems is in certain fundamental aspects always different from the kinetics in closed systems.

Closed Systems

The rate of dissolution in a closed system is a measure of mass of a solid transferred to a volume of solution in a unit of time. As the solute concentration in solution rises when dissolution occurs, the rate of increase in solute concentration is

$$\left(\frac{dC}{dt}\right)_{closed} = R(C, t, \text{environmental parameters}) \tag{28}$$

which indicates that, in a general case, the rate of dissolution may depend on the solute concentration in solution and it may vary with time t; it also depends on such environmental parameters as the temperature, pressure, ionic strength, and viscosity of the solution.

Open Systems

An open system with inflow and outflow may be receiving input of a solute from an external source through the inflow and additional input through dissolution within the system. At the same time, it may be losing solute through outflow. A general form of an equation for the rate of change in concentration within such a system is

$$\left(\frac{dC}{dt}\right)_{open} = \frac{F(t)C_{in}}{V} + R(C, t, \ldots) - \frac{F(t)C}{V} \tag{29}$$

where $F(t)$ = time-dependent volume flow into or out of the system
C_{in} = solute concentration in inflow
V = solution volume in the system
$R(C, t, \ldots)$ = dissolution rate, as defined in equation 28

The flushing or water-residence time of the system — i.e., the length of time needed theoretically to renew the water within the volume of the system — is equal to V/F. In equation 29, concentration in inflow, C_{in}, should not be so great as to suppress the solid dissolution, R.

Solute is removed from the system through outflow. Its concentration in the outflow is the same as throughout the well-mixed volume of the system, C.

An open system that is flushed by solute-free water (that is, no input from outside) is represented by a simpler equation:

$$\left(\frac{dC}{dt}\right)_{open} = R(C, \ldots) - \frac{F(t)}{V}C \tag{30}$$

Equation 30 does not imply that the rate of dissolution in an open system is necessarily smaller than in a closed system (equation 28); the value of dC/dt in either equation depends on R, which in turn depends on concentration C. The remainder of this section will be concerned with dissolution rates in closed systems as represented by the term R in the preceding equations 28 to 30.

DISSOLUTION RATE LAWS

Empirical studies of dissolution demonstrate that the greater the undersaturation of a solution, the faster the concentration increases. Mathematically, this observation is represented by the rate law written as

$$\frac{dC}{dt} = k(C_s - C) \tag{31}$$

where k = dissolution rate parameter ($time^{-1}$)
C_s = concentration at saturation or equilibrium
C = concentration varying with time.

The rate equation is known as the first-order rate law, where the rate term depends on the first power of the concentration terms. If k is a constant under given environmental conditions, dissolution should be faster in an undersaturated solution, where the difference $C_s - C$ is relatively large, than near saturation, when this difference becomes small. For constant k and C_s, Figure 3.4-1 shows how dC/dt changes with concentration C, and Figure 3.4-2 shows how C changes as a function of time.

With constant k and C_s, the integrated form of equation 31 is:

$$C = C_s(1 - e^{-kt}) + C_0e^{-kt} \tag{32}$$

where C_0 is the initial concentration in solution at time t = 0.

The preceding shows that, in order to determine how concentration in solution increases with time, one has to know the value of the rate parameter k, the value of the saturation concentration C_s, and the initial concentration C_0, which may be zero.

Alternatively, to determine the value of k, one must know the saturation concentration C_s (or how it can be obtained, see section 2.11), and a series of data on concentration C as a function of time t.

Intuitively, it is reasonable to suppose that the rate of increase in concentration dC/dt is affected by the volumes of the solid and the solution. Dissolution will in general proceed faster if the surface area of the solid is relatively large. This is reflected in the values of the solid surface area and the solution volume: a value of k in equations 31 and 32 is valid only for a certain set of physical and chemical conditions, among which the surface area of the solid and the solution volume must be known. The dependence of the rate parameter k on these and other conditions will be explained in later sections.

Equation 31 is one of the mathematical forms of dissolution rates. Other forms of dissolution rate equations are given below, and their use is illustrated in a subsequent section.

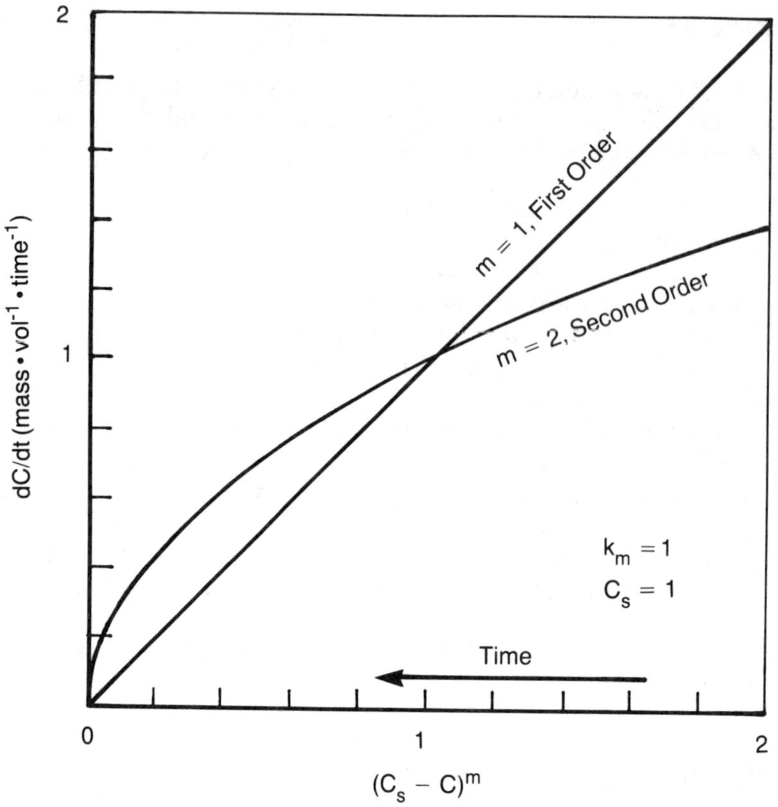

Note: As time progresses, concentration difference C_s − C decreases.

FIGURE 3.4-1 Rate of a Reaction dC/dt as a Function of Distance from Saturation (C_s − C) for First- and Second-order Reactions

Zeroeth Order Law

The zeroeth order represents a constant rate of dissolution:

$$\frac{dC}{dt} = k_0 \qquad (mol \cdot cm^{-3} \cdot s^{-1}) \qquad (33)$$

An integrated form is:

$$C = C_0 + k_0 t \qquad (mol \cdot cm^{-3}) \qquad (34)$$

Dissolution according to this law cannot go on indefinitely, because C cannot increase indefinitely with time. If this law is observed, it can be valid only for some limited period of the dissolution process.

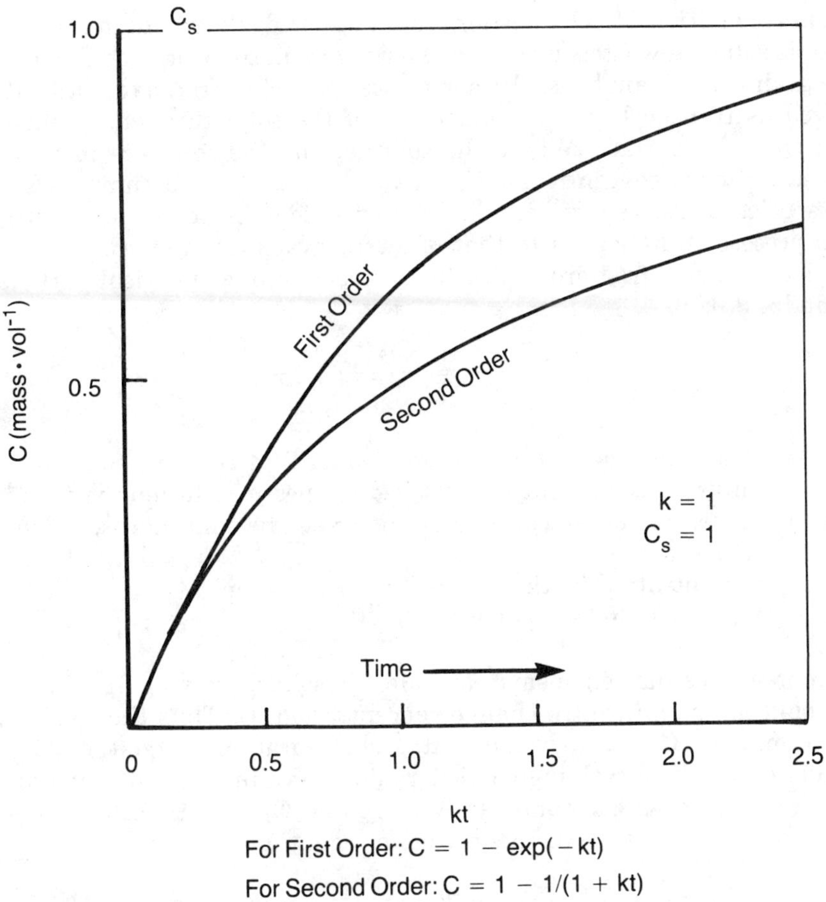

Compare equations 32 and 39

FIGURE 3.4-2 Increase in Concentration in Solution, C, for a First- and Second-order Dissolution Reaction, as a Function of Dimensionless Parameter kt

First Order Law

The following equation, a modified form of equation 31, is a general form of the first-order dissolution process

$$\frac{dC}{dt} = k_1(C_s - C)$$

Many solids in the environment occur in the form of grains or small crystals. It is mathematically convenient to treat dissolution of single grains as if they were spherical particles. One of the problems that may have to be solved is how long the individual grains of a certain initial size can survive during dissolution under given environmental conditions. Such questions may arise in dealing with mixtures of more or less soluble solids in wastes, mixtures of fertilizers in soils, or with the longevities of mineral grains in stream beds. For a spherical particle of radius r, dissolution can be visualized as transport of the components of the solid through a thin layer of solution at the solid surface. Where the solution and the solid are in contact, the solution is saturated at concentration C_s; outward, concentration diminishes, and the dissolving species diffuses away from the solution-solid interface. (The model of such a dissolution process is analogous to that of either evaporation or condensation of a water droplet in air.) A first-order dissolution reaction of a spherical particle is the relationship [5, 33]:

$$-\frac{dr}{dt} = \frac{DV_m(C_s - C)}{r} \tag{35}$$

where V_m = molar volume of the solid ($cm^3 \cdot mol^{-1}$), defined in equation 46
D = diffusion coefficient of dissolved species in solution ($cm^2 \cdot s^{-1}$)
C_s = concentration at equilibrium between the solid and solution ($mol \cdot cm^{-3}$)
C = concentration in the bulk solution ($mol \cdot cm^{-3}$)
r = solid particle radius (cm)

In an environmental situation, a small amount of a solid dissolving in a large volume of water would not significantly change concentration, C. Thus the concentration-difference term, $C_s - C$, is nearly constant, and integration of equation 35 gives an explicit equation for the dissolving particle radius r as a function of time t and other parameters considered as constants (D, V_m, C_s, and C). The time to complete dissolution (t when $r = 0$) is, from integration of equation 35,

$$t = \frac{r_0^2}{2DV_m(C_s - C)} \tag{36}$$

where r_0 is the initial radius of the dissolving particle. Application of this equation is demonstrated in Example 4. The equation may be used to estimate a lower limit of time to dissolution in dilute suspensions. Adsorption of inorganic and organic components, and other reactions on the particle surfaces may slow down the dissolution process. In thicker suspensions and in packed beds, concentration C may not be constant, thereby affecting the dissolution rate as given in equation 35.

Example 4 Estimate the time to dissolution of SiO_2-glass particles, 1 mm in diameter, in an average river water.

The following parameters are needed to obtain t from equation 36:

Initial particle radius given:	$r_0 = 0.05$ cm
Molecular diffusion coefficient of SiO_2^0 in water at 25°C (*see also* § 2.16):	$D = 1 \times 10^{-5}\ cm^2 \cdot s^{-1}$ [21]
Molar volume of SiO_2 glass:	$V_m = 27\ cm^3 \cdot mol^{-1}$ [38]

Solubility of SiO_2 glass at 25°C: $C_s = 1.9 \times 10^{-6}$ mol • cm^{-3} (Example 5)

SiO_2^0 in river waters: $C = 0.17 \times 10^{-6}$ mol • cm^{-3} (Table 3.4-1, column d)

$$t = \frac{0.05^2}{2 \times 1 \times 10^{-5} \times 27(1.9 \times 10^{-6} - 0.17 \times 10^{-6})} = 2.68 \times 10^6 \text{ s}$$

$\cong$ 31 days

The time to dissolution of 31 days is a lower limit estimate for a dilute suspension of silica-glass particles. A related but different case is dealt with in Example 8.

Second and Higher Order Laws

The rate of increase in concentration in solution is some power function of the concentration of the dissolving or reacting species:

$$\frac{dC}{dt} = k_m(C_s - C)^m \tag{37}$$

where m is generally greater than unity.

The integrated form is:

$$C = C_s - \left[\frac{1}{1/(C_s - C_0)^{m-1} + (m-1)k_m t}\right]^{1/(m-1)} \tag{38}$$

In the case of m = 2, a second-order dissolution reaction, the integrated form is

$$\frac{1}{C_s - C} - \frac{1}{C_s - C_0} = k_2 t \tag{39}$$

where $C_s > C > C_0$, and the product k_2t is positive.

Fractional Order Laws

Empirical relationships describing the rates of dissolution of such stoichiometrically complex phases as bone mineral (apatites) and feldspars (sodium and potassium alumino-silicates) give the rates of dissolution in terms of fractional powers of the hydrogen-ion activity, $(a_{H^+})^n$, where the value of n varies over the pH range. For feldspars, some of these relationships are tabulated in a later section (Table 3.4-7).

Time-dependent Laws

Dissolution of many substances and leaching of the matrix components often follow the so-called parabolic rate law, which states that the rate of dissolution is inversely related to the square root of time:

$$\frac{dC}{dt} = k_p t^{-1/2} \tag{40}$$

$$C = C_0 + 2k_p t^{1/2} \tag{41}$$

As in the case of the linear dissolution rate, equation 34, the parabolic rate law cannot apply indefinitely, because C cannot increase indefinitely with time. However, the parabolic rate law is rooted in the mathematical theory of molecular diffusion, and it can be derived for a dissolution process that describes diffusional migration of a component from a solid, when migration takes place through the matrix toward the solid-solution interface. Data on parabolic rates of dissolution of silicates are given in Table 3.4-2.

TABLE 3.4-2

Values of Parabolic Dissolution Rate Parameter (k_{ps}) for Three Silicate Minerals

$k_p = k_{ps}\,S/V$

(C_0 = initial concentration in solution, V = solution volume, S = solid surface area; experimental conditions: 1.0 g solid, 1 liter solution, 25°C)

Mineral	Density ρ ($g \cdot cm^{-3}$)	Specific Surface Area, S_g ($cm^2 \cdot g^{-1}$)	Initial pH	C_0V/S ($10^{-9}\ mol \cdot cm^{-2}$) Mg	Si	$2k_{ps}$ ($10^{-11}\ mol \cdot cm^{-2} \cdot s^{-1/2}$) Mg	Si
Enstatite	3.29	576	3.2	4.9	7.3	7.8	2.4
$Mg_2Si_2O_6$			5.0	2.3	3.0	2.6	2.6
			7.0	0.09	0.2	1.6	1.4
			9.6	1.0	0.57	1.7	2.4
Forsterite	3.13	445	3.2	20.0	5.0	180.0	20.0
Mg_2SiO_4			5.0	9.0	2.0	61.0	3.6
			7.0	2.0	0.7	120.0	2.9
			9.6	4.0	0.6	67.0	0.3
Serpentine	2.62	914	3.2	20.0	0.0	27.0	6.3
$Mg_3Si_2O_5(OH)_4$			5.0	5.8	0.0	2.2	6.8
			7.0	3.4	0.2	1.9	6.8
			9.6	1.0	0.2	1.4	0.7

Source: Luce *et al.* [27]

Dissolution of some solids according to the parabolic rate law may result from the fact that the solid aggregates are made of particles of different sizes, with some of the bigger particles containing overgrowths and broken edges [5]. Washing in water or acid or etching of the solid particles may produce dissolution according to the zeroth order law, equations 33-34.

To obtain a value of the rate parameter k for a parabolic rate dissolution, proceed as follows: To verify the parabolic rate law, concentration C should plot as a straight line function of the square root of time, $t^{1/2}$. The slope of this plot is the value of $2k_p$, given in equation 41. From the numerical data of concentration as a function of time, the rate parameter k_p can be computed by using a concentration value C_1 at time t_1 and C_2 at time t_2:

$$k_p = \frac{C_1 - C_2}{2(t_1^{1/2} - t_2^{1/2})} \quad (g \cdot l^{-1}\ sec^{-1/2}) \tag{42}$$

This value of k_p can be used in equation 41 to compute and plot C as a function of t.

DISSOLUTION RATE PARAMETER, k

In the preceding sections, the dissolution rate parameter k was used in a form that was strictly applicable to homogeneous reactions. In heterogeneous reactions, such as dissolution or precipitation of solids, the reactive surface area of the solid is one of the parameters that determine the reaction rate. The aqueous phase in a thick suspension reaches a state of saturation faster than in a dilute suspension, other factors being equal.

The rate parameter k can be expressed in an internally consistent form for the homogeneous as well as heterogeneous reactions in suspensions, in systems consisting of a solution and a bulk solid, and in packed beds or porous media.

Solid Surface Area and Solution Volume

The importance of these two parameters — the surface area of the solid and the solution volume — and of their ratio will be demonstrated in this section.

Dissolution is a loss of mass by a solid in a time interval dt, which can be written as

$$-dM = k_i\, S\, dt \tag{43}$$

where dM is an increment of mass lost (mol), k_i is some intrinsic dissolution rate constant ($mol \cdot cm^{-2} \cdot s^{-1}$), S is the reactive surface area of the solid (cm^2), and dt is a time interval (s).

Division of both sides of equation 43 by V, the volume of the solution, gives an equation of a form similar to equation 33:

$$\frac{dC}{dt} = k_i \frac{S}{V} \quad (mol \cdot cm^{-3} \cdot s^{-1}) \tag{44}$$

The latter equation shows the effect of the solid surface area on the dissolution rate: the higher the surface/volume ratio in solution, the higher is the reaction rate. This is a zeroeth-order rate equation; for the higher-order rates, the rate constants can also be expressed in terms of the intrinsic rate parameter k, solid surface, and solution volume. For a first-order dissolution reaction,

$$\frac{dC}{dt} = \frac{k_i S V_m}{V}(C_s - C) \qquad (mol \cdot cm^{-3} \cdot s^{-1}) \tag{45}$$

where C_s and C are as defined in equation 31.

V_m, the molar volume of the solid ($cm^3 \cdot mol^{-1}$), is equal to the ratio of gram-formula weight (M, g/mol) and density (ρ, g/cm^3):

$$V_m = \frac{M}{\rho} \tag{46}$$

In equations 45 and 46, the product $k_i V_m$ has the dimensions of velocity ($cm \cdot s^{-1}$); it can be interpreted as a linear rate of dissolution of a flat solid surface receding with a velocity $k_i V_m$. Measured values of such linear velocities of dissolution are listed in Table 3.4-3 for a number of salts. High dissolution rates are generally associated with high values of salt solubility; however, the values of the linear dissolution rates of strongly soluble compounds may be biased by the fact that the reactive surface area of the solid, S, is likely to change substantially during the dissolution of single crystals or agitated suspensions.

Dissolution Rate and Degree of Saturation

The rate of a dissolution reaction that proceeds according to the first-order rate law (equation 31) may be written in the following form:

$$\frac{dC}{dt} = k_1 C_s \left(1 - \frac{C}{C_s}\right) \tag{47}$$

C/C_s is the ratio of concentration of a species in solution to its concentration at equilibrium with its parent solid. This ratio is closely related to the *degree of saturation* of a solution (Ω), as defined in § 2.11.2. For a solid of composition M_pA_q, the degree of saturation of a solution with respect to this solid, written in terms of concentration of the metal-ion species M, is

$$\Omega = \left(\frac{[M]}{[M]_{eq}}\right)^{p+q} \tag{48}$$

where [M] is the concentration of M in solution, $[M]_{eq}$ is its concentration at equilibrium with solid M_pA_q, and p and q are stoichiometric coefficients. We may write C for [M] and C_s for $[M]_{eq}$, obtaining the following form for the degree of saturation:

$$\Omega = \left(\frac{C}{C_s}\right)^{p+q}$$

TABLE 3.4-3

Linear Dissolution Rates and Solubilities of Some Moderately to Strongly Soluble Salts
(Rates of dissolution determined at 25°C, in solutions subjected to rotary stirring at 400 rpm, with the exposed surface of the dissolving solid below the stirrer.)

Solid Phase	Solubility[a] (g/100 g H_2O)	Linear Dissolution Rate ($cm \cdot s^{-1}$)
KI	146.45	3.10×10^{-3}
KBr	67.75	2.85×10^{-3}
KCl	36.62	2.45×10^{-3}
NaCl	35.92	1.75×10^{-3}
$BaCl_2 \cdot 2H_2O$	36.9	1.60×10^{-3}
K_2SO_4	12.04	1.70×10^{-3}
$FeSO_4 \cdot 7H_2O$	29.7	0.80×10^{-3}
$NiSO_4 \cdot 7H_2O$	39.6	0.55×10^{-3}
$CoSO_4 \cdot 7H_2O$	37.8	0.60×10^{-3}
$ZnSO_4 \cdot 7H_2O$	57.9	0.50×10^{-3}
$MgSO_4 \cdot 7H_2O$	38.3	0.50×10^{-3}
$CuSO_4 \cdot 5H_2O$	22.29	0.65×10^{-3}
$CaSO_4 \cdot 2H_2O$[b]	0.21	0.35×10^{-3}

a. For hydrated phases, solubility refers to the anhydrous part of the stoichiometric formula.

b. Selenite (variety of gypsum); dissolution parallel to crystal face (010).

Source: Van Name [42]

or, alternatively, by rearrangement:

$$\frac{C}{C_s} = \Omega^{1/(p+q)} \tag{49}$$

By combining equations 47 and 49, we define the first-order dissolution rate, dC/dt, in terms of the degree of saturation:

$$\frac{dC}{dt} = k_1 C_s \left(1 - \Omega^{1/(p+q)}\right) \tag{50}$$

This expression can be generalized for the rates of dissolution of order m > 1, from equation 37:

$$\frac{dC}{dt} = k_m C_s^{\,m} \left(1 - \Omega^{1/(p+q)}\right)^m \tag{51}$$

The use of an equation of this kind is described in the later section on "Dissolution Rates of Calcite" (see equation 76).

Dissolution Reaction Mechanisms

The chemical kinetic literature draws a clear distinction between "diffusion-controlled" dissolution and "reaction-controlled" dissolution. In the latter, the state of the crustal surface, the distribution of lattice dislocations, "kinks," and "steps" determine the overall rate of the dissolution reaction. In diffusion-controlled dissolution, on the other hand, it is assumed that a thin laminar layer of solution, of the order of 10^{-3}cm in thickness, exists adjacent to the solid surface. At the base of this layer, at the interface between the solid and the solution, the solution is saturated with respect to the solid phase. Dissolved ions or molecules diffuse through the laminar layer into the bulk solution. The process of equilibration at thc base of the layer is attained much faster than the transport of dissolved species away from the solid into the bulk solution. Therefore, the rate of diffusional transport controls the rate of the dissolution reaction.

In surface-controlled (reaction-controlled) dissolution, chemical reactions at the solid surface control the overall dissolution rate. The existence of surface-controlled reactions is corroborated by observations of the dissolving surfaces that show signs of non-uniform corrosion, suggestive of reactions taking place at certain preferred sites. Other evidence is of a negative kind: diffusion-controlled dissolution is believed to be confirmed by a shrinkage of the laminar surface layer when the rate of stirring in solution is increased, and by a temperature dependence of the dissolution rate similar to that of molecular diffusion in water. In practical terms, a thinner surface layer results in a shorter path of transport, such that the rate of dissolution would increase if the layer were made thinner by faster stirring. The temperature dependence of the dissolution rate is commonly expressed as the activation energy of a reaction; for a diffusion-controlled process it is expected to be between 17 and 21 kJ/mol. (A further discussion of activation energies of dissolution is given in a later section.)

Data listed in Table 3.4-4 show that, in general, mineral phases of low solubility dissolve in a reaction-controlled process, but those of higher solubilities dissolve in a diffusion-controlled reaction. (The apparently exceptional case of dissolution of AgCl has been attributed to possible photochemical effects.) Thus, the solubility of a solid is a tentative indication of its dissolution mechanism.

Porous Media

Porous media in the environment are sediments, soils, and packed beds of granular materials. In such systems the pores are generally interconnected, enabling water to pass through them. A characteristic of porous media that bears on solid dissolution rates is a high ratio of the volume of solids to the volume of solution. For example, sediments on the bottom of water reservoirs and lakes typically contain between 10% and 30% by volume of solids, the remainder of the volume being taken up by water. Under heavier loads, deep-seated sands may contain as much as 70% to 80% solids by volume, the remainder consisting of pores filled with either gas or water.

TABLE 3.4-4

Solubilities of Solid Phases and Reaction Mechanisms that Control Dissolution

Solid Phase	Solubility in Pure Water (mol/l)	Reaction Mechanism Controlling Dissolution
$Ca_5(PO_4)_3OH$	2×10^{-8}	Surface-reaction
$KAlSi_3O_8$	3×10^{-7}	Surface-reaction
$NaAlSi_3O_8$	6×10^{-7}	Surface-reaction
$BaSO_4$	1×10^{-5}	Surface-reaction
AgCl	1×10^{-5}	Transport
$SrCO_3$	3×10^{-5}	Surface-reaction
$CaCO_3$	6×10^{-5}	Surface-reaction
Ag_2CrO_4	1×10^{-4}	Surface-reaction
$PbSO_4$	1×10^{-4}	Mixed
$Ba(IO_3)_2$	8×10^{-4}	Transport
$SrSO_4$	9×10^{-4}	Surface-reaction
Opaline SiO_2	2×10^{-3}	Surface-reaction
---------- a		
$CaSO_4 \cdot 2H_2O$	5×10^{-3}	Transport
$Na_2SO_4 \cdot 10H_2O$	2×10^{-1}	Transport
$MgSO_4 \cdot 7H_2O$	3×10^{0}	Transport
$Na_2CO_3 \cdot 10H_2O$	3×10^{0}	Transport
KCl	4×10^{0}	Transport
NaCl	5×10^{0}	Transport
$MgCl_2 \cdot 6H_2O$	5×10^{0}	Transport

a. Note change in mechanism near solubility of 0.002 mol/l.

Source: Berner [5]. *(Copyright 1981, Mineralogical Society of America. Reprinted with permission.)*

In the presence of water, a bulk sediment or a bulk porous medium consists of solids and water. The fraction of a unit bulk volume occupied by water is called *porosity*, denoted by the symbol ϕ. The fraction occupied by the solids is $1-\phi$.

A porous solid filled with an aqueous solution is a heterogeneous medium in which the rates of dissolution of the solid phases must be described by equations that take into account the surface areas of the dissolving solids and the volume of the solution occupying the pore space, analogous to equations 44 and 45. The latter two equations apply in a fairly general way to solids undergoing dissolution. We shall now recast these relatively simple equations in terms of the parameters characteristic of porous media. In essence, this requires that the surface area of the solid (S, cm^2) and the volume of the solution (V, cm^3) be expressed in terms of the surface area and porosity of a porous medium filled with water.

The following reasoning indicates the main relationships between the parameters of a porous medium and those of a more general case of a solid and water.

The volume fraction of the solid in a unit volume of a porous medium, as defined above, is $1-\phi$ ($cm^3 \cdot cm^{-3}$). If the solid has a density of ρ($g \cdot cm^{-3}$), the mass of the solid contained in a unit volume of the porous medium is $\rho(1 - \phi)$ ($g \cdot cm^{-3}$).

If the solid in the porous medium is characterized by a value of the surface area S_g, in units of cm^2 per gram of solid, the surface area of the solid contained in a unit volume of the porous medium is

$$S = S_g\rho(1 - \phi) \qquad (cm^2 \cdot cm^{-3}) \tag{52}$$

The volume of solution, V, filling a unit volume of the pore space of the porous solid is equal to the porosity, as defined earlier:

$$V = \phi \qquad (cm^3 \cdot cm^{-3}) \tag{53}$$

Substitution of the right-hand side of equations 52 and 53 for S and V, respectively, in the dissolution rate equations 44 and 45 gives the following two relationships:

$$\text{Zeroth-order law} \quad \frac{dC}{dt} = k_i \frac{S_g\rho(1 - \phi)}{\phi} = k_0 \tag{54}$$

$$\text{First-order law} \quad \frac{dC}{dt} = k_i V_m \frac{S_g\rho(1 - \phi)}{\phi}(C_s - C) = k_1(C_s - C) \tag{55}$$

The relationships for the rate constants k of dissolution reactions of different orders in porous media and in other solid-solution systems are summarized in Table 3.4-5.

Owing to the dependence of the dissolution rate dC/dt on the ratio S/V (see equations 44, 45, 54, and 55), the rate of reaction varies directly with the surface area of the solid and inversely with the volume of the solution. In porous media, the reaction rate constants k_0 and k_1 in equations 54 and 55 depend on the solid surface area, its

TABLE 3.4-5

Summary of Relationships and Units for Reaction Rate Parameters, k

Reaction Type and Order	k	Units
Homogeneous Reactions		
Zeroth Order	k_0	$mol \cdot cm^{-3} \cdot s^{-1}$
First Order	k_1	s^{-1}
Second Order	k_2	$cm^3 \cdot mol^{-1} \cdot s^{-1}$
Heterogeneous Reactions[a]		
Solid and Solution, Suspensions		
Zeroth Order	$k_0 = k_i S/V$	$mol \cdot cm^{-3} \cdot s^{-1}$
First Order	$k_1 = k_i S V_m/V$	s^{-1}
Porous Media, Packed Beds		
Zeroth Order	$k_0 = k_i S_g \rho(1-\phi)/\phi$	$mol \cdot cm^{-3} \cdot s^{-1}$
First Order	$k_1 = k_i V_m S_g \rho(1-\phi)/\phi$	s^{-1}

a. Definition of symbols:

k_i = rate parameter ($mol \cdot cm^{-2} \cdot s^{-1}$)
S = surface area (cm^2 or $cm^2 \cdot cm^{-3}$)
S_g = specific surface area ($cm^2 \cdot g^{-1}$)
V = solution volume (cm^3 or $cm^3 \cdot cm^{-3}$)
ϕ = pore volume fraction
V_m = solid molar volume ($cm^3 \cdot mol^{-1}$)
ρ = solid density ($g \cdot cm^{-3}$)

density, and the porosity. Some representative values of these parameters are given below to illustrate their effect on reaction rates.

- *Suspension of fine quartz sand at a concentration of 100 mg/l:*

  Surface areas of fine sands commonly range from 1 to a few square meters per gram. Thus, for quartz sand, $S_g = 1\ m^2 \cdot g^{-1} = 10^4\ cm^2 \cdot g^{-1}$, and its concentration in suspension is $10^{-4}\ g \cdot cm^{-3}$. The ratio S/V is

$$S/V = 10^4\ cm^2 \cdot g^{-1} \times 10^{-4}\ g \cdot cm^{-3}$$
$$= 1\ cm^2 \cdot cm^{-3}$$

- *A porous medium of fine quartz sand of the same specific surface area as above and having a porosity value of 40%:*

  The parameters needed for estimation of S/V (from equations 52 and 53 are:

  Surface area: $S_g = 10^4\ cm^2 \cdot g^{-1}$
  Quartz density: $\rho = 2.65\ g \cdot cm^{-3}$
  Porosity: $\phi = 0.4$
  Solid fraction: $1 - \phi = 0.6$

$$S/V = \frac{10^4 \times 2.65 \times 0.6}{0.4}$$

$$= 4 \times 10^4\ cm^2 \cdot cm^{-3}$$

The preceding two estimates of the ratio S/V, one for a fairly dilute suspension and the other for a packed porous medium, indicate a range of values that are encountered in the environment and the possible magnitude of their effect on the rates of dissolution.

EFFECT OF TEMPERATURE ON DISSOLUTION RATES

The dependence of dissolution rate on temperature is most often expressed through a relationship between the reaction rate parameter k and temperature, proposed by Arrhenius in 1889:

$$k = A \exp\left(-\frac{E_a}{RT}\right) \tag{56}$$

$$\ln k = \ln A - \frac{E_a}{RT} \tag{57}$$

where A = pre-exponential factor, in the same units as k
E_a = activation energy for the reaction (kJ/mol)
R = gas constant (8.314×10^{-3} kJ · deg^{-1} · mol^{-1})
T = temperature (K)

The parameters A and E_a are positive quantities, such that k increases with temperature.

Other types of dependence of reaction rates on temperature have been schematically illustrated, but not discussed, by Frost and Pearson [15]. Eyring and Eyring [13] pointed out that among many reactions that do not obey the Arrhenius relationship are certain adsorption reactions and enzyme-catalyzed hydrolysis.

In estimating the values of the rate parameter k at different temperatures, an explicit value of the factor A can be avoided. If k is known at some reference temperature (k_0, T_0) and the activation energy for the reaction, E_a, is also known over some temperature range, then the value of k at some other temperature, T, can be computed from

$$\ln \frac{k}{k_0} = -\frac{E_a}{RT}\left(\frac{1}{T} - \frac{1}{T_0}\right) \tag{58}$$

For a number of dissolution and precipitation reactions, as well as some other environmentally relevant reactions, the values of E_a are given in Tables 3.4-6 and -13.

TABLE 3.4-6

Reaction Rate Parameters, Activation Energies and Equilibrium Constants for Dissolution of Four SiO_2 Solid Phases in Water

Reaction Rate Parameter (k_+, $mol \cdot cm^{-2} \cdot s^{-1}$) as a Function of Temperature (°K)	
Quartz	$\log k_+ = -2.826 - 2.028 \times 10^{-3}T - 4158/T$
α-Cristobalite	$\log k_+ = -4.739 - 3586/T$
β-Cristobalite	$\log k_+ = -4.963 - 3392/T$
Amorphous SiO_2	$\log k_+ = -4.369 - 7.890 \times 10^{-4}T - 3438/T$
Activation Energy (kJ/mol)	
Quartz	$E_a = 72.0 \pm 4.6$
α-Cristobalite	$E_a = 68.7$
β-Cristobalite	$E_a = 65.0$
Amorphous SiO_2	$E_a = 62.9 \pm 2.0$
Equilibrium Constant (K) for Equation 10 as a Function of Temperature (°K)	
Quartz	$\log K = 1.881 - 2.028 \times 10^{-3}T - 1560/T$
α-Cristobalite	$\log K = -0.0321 - 988.2/T$
β-Cristobalite	$\log K = -0.2560 - 793.6/T$
Amorphous SiO_2	$\log K = 0.3380 - 7.889 \times 10^{-4}T - 840.1/T$

Source: Rimstidt and Barnes [37]

RATE OF DISSOLUTION ADVANCE

Half-life of a Reaction

From the first-order reaction rate equations 31 and 32, the half-life can be defined as the time required for concentration in solution to reach a midpoint between the initial C_0 and saturation C_s concentrations. Thus, the time at half-life, $t_{1/2}$, corresponds to the concentration $C = (C_0 + C_s)/2$. Using this value of C and $t_{1/2}$ in equation 32, the half-life is

$$t_{1/2} = \frac{\ln 2}{k_1} = \frac{0.693}{k_1} \tag{59}$$

or, for simplicity and as an approximation,

$$t_{1/2} = \frac{1}{k_1} \tag{60}$$

Similarly, from equation 45, the half-life is

$$t_{1/2} = \frac{V}{k_i S V_m} = \frac{1}{k_1} \qquad (61)$$

where k_1 is defined in Table 3.4-5.

For a second-order dissolution reaction (equation 39), if k_2 is in units of $cm^3 \cdot mol^{-1} \cdot s^{-1}$, the half-life is

$$t_{1/2} = \frac{1}{k_2(C_s - C_0)} \quad (s) \qquad (62)$$

If the rate parameter k' refers to a unit of area and is given as k' in units of $cm^3 \cdot mol^{-1} \cdot cm^{-2}\ S^{-1}$, then k_2 in equation 62 is replaced by the product of k' and the solid surface area S:

$$t_{1/2} = \frac{1}{k'S(C_s - C_0)} \quad (s) \qquad (63)$$

Time to Reach an Equilibrium

In dissolution reactions obeying the first, second or a higher-integer order rate law (equations 32, 38 and 39), the time needed to attain an equilibrium concentration in solution, C_s, is mathematically infinitely long (see Figure 3.4-2). However, a convenient measure of attainment of an equilibrium is some value of concentration C that is very close to an equilibrium concentration C_s. For such a measure, one can use $C = 0.95C_s$; alternatively, one can say that an equilibrium has been effectively reached when a concentration equal to 95% of the difference between the equilibrium and the initial concentrations has been attained:

$$C - C_0 = 0.95(C_s - C_0)$$

The latter condition for attainment of equilibrium, when substituted for C in the first-order dissolution rate equation (32), corresponds to the following value of the dimensionless product kt:

$$k_1 t = 3$$

From the preceding, the time to attain an equilibrium, as defined by the 95% level of concentration, is

$$t = \frac{3}{k_1} \qquad (64)$$

For a second-order reaction (equation 39), the condition of closeness to an equilibrium given above corresponds to the following value of the dimensionless quotient $k_2t(C_s - C_0)$:

$$k_2t(C_s - C_0) = 19$$

Thus, the time to attain saturation in a second-order dissolution reaction is:

$$t = \frac{19}{k_2(C_s - C_0)} \tag{65}$$

The progress of first-order and second-order reactions shown in Figures 3.4-1 and -2 indicates that concentration changes as either a function of the distance from saturation (Figure 3.4-1) or as a function of the product kt (Figure 3.4-2). It should not be generalized from these graphs that the second-order dissolution is slower than the first-order; in Figure 3.4-1, this is the result of an arbitrary choice of the value of $k = 1$ and $C_s = 1$, as noted on the figure.

To summarize the preceding, the time needed to nearly reach an equilibrium in a first-order dissolution reaction can be estimated from equation 64, if the value of the rate parameter k_1 is known. For a second-order reaction, the time to reach an equilibrium is a function of the rate parameter k_2 and of the equilibrium concentration C_s, as given in equation 65. For definitions of the rate parameters k_1 and k_2 in either homogeneous or heterogeneous systems, see Table 3.4-5.

For other dissolution rate laws — the zeroeth-order law, parabolic law, and time-dependent rate laws (equations 34, 41 and 70) — concentrations in solution increase indefinitely with time, and an equilibrium is therefore theoretically unattainable. In practice, this means that such rate laws hold for certain periods of time during the dissolution process but that the approach to equilibrium is controlled by other reaction mechanisms.

EXAMPLES OF DISSOLUTION OF MATERIALS

Dissolution of Silica-containing Minerals

Oxygen, silicon and aluminum are the three most abundant elements in the Earth's crust. In many minerals of which the crust is made, they occur together and are known by the generic names of the silicates and alumino-silicates. The ubiquitous presence of the silicates and alumino-silicates in our environment — in rocks, in soils, in atmospheric dust, in suspension or as sediments in waters, and in man-made ceramic materials — makes it relevant to the context of this section to summarize briefly with a few examples the kinetics of their dissolution in water. The case studies presented in the sections that follow deal with dissolution of pure SiO_2 minerals, a group of alumino-silicate minerals known as feldspars, and a number of other commonly occurring silicate phases.

When silicates crystallize from melts (magma, if the molten rock occurs deep within the Earth's crust, and lava, if the melt is on the surface), individual minerals and groups of structurally similar minerals form within certain ranges of temperature. The sequences of crystallization of the solids from melts depend, in general, on the chemical composition of the melt and the initial and final temperatures of the cooling process.

A qualitative relationship has been observed between the temperature of formation of a silicate mineral in a cooling melt and the resistance of the mineral to weathering and dissolution under the Earth's surface conditions. In general, the higher the temperature of formation of a silicate mineral in the melt, the poorer are its survival chances in contact with waters on the surface. Over long periods of time, high-temperature minerals that become exposed to surface waters may dissolve or react to become more stable forms, whereas the minerals formed at lower temperatures in melts are often preserved in sediments deposited in water for times comparable to the age of the Earth.

As an example of this differential survival of silicates on the Earth's surface, one may consider a beach made of quartz sand. (Quartz is one of the lower-temperature minerals crystallizing from metal-silicate melts.) Such beaches are very common, but those consisting of the green-brown grains of olivine minerals are very rare. Olivines are magnesium and iron silicates, of general composition $(Mg, Fe)_2SiO_4$, that are common in basalts. Olivines form at the higher temperatures in a cooling melt, and they break down faster than quartz through weathering when exposed to water. Beaches made of fine fragments of basalt containing olivines can form in proximity to big masses of basalt or other olivine-containing rocks, such as in Hawaii, where the source rocks are geologically young and close to the shore.

Dissolution of Pure SiO_2 Solids. — The most abundant mineral form of SiO_2 in nature is quartz. The SiO_2 family of minerals, in addition to quartz, includes several polymorphs:[2] tridymite, cristobalite (each of the two represented by an α and a β phase), coesite, stishovite, and silica glass or amorphous silica. α-Quartz is the polymorph that is stable in our environment. Tridymite and cristobalite form at higher temperatures in volcanic and metamorphic rocks. Coesite and stishovite are dense phases that form at high pressures, such as in impact craters of meteorites. Silica glass or structurally amorphous silica occurs as precipitates from water, in skeletal shells secreted by planktonic organisms, and in volcanic rocks.

---

2. Polymorphs are different crystal structures of the same chemical composition. Transformation of one polymorph into another — for example, by the phase transition $CaCO_3$(calcite) $\rightarrow$ $CaCO_3$(aragonite) — is accomplished by changes of temperature and/or pressure. At any given temperature and pressure, only one of the polymorphs is thermodynamically stable, unless the particular temperature and pressure values happen to permit an equilibrium between the two polymorphs.

Chalcedony, agate, onyx, chert, and jasper are different microcrystalline forms of SiO_2 that contain various metal-oxide impurities. A peculiar form of SiO_2 in nature, known by the name of lechatelierite, occurs in quartz sandstones struck by lightning and in meteorite-impact craters; this fused SiO_2 underwent recrystallization at the Earth's surface and sometimes preserves the tube-shaped or bore-shaped irregular channel formed in the sandstone by a lightning discharge. Occurrences of the SiO_2 polymorphs tridymite, cristobalite and glass on the Earth's surface indicate that such phases can metastably coexist without converting to the stable form — quartz — for geologically long periods of time.

A dissolution reaction of SiO_2 in water was presented earlier as

$$SiO_2(s) + 2H_2O = H_4SiO_4^0 \qquad (10)$$

where $SiO_2(s)$ represents any one of its polymorphs. The rate of increase in concentration of dissolved silica, H_4SiO_4, can be represented by the following kinetic equation:

$$\frac{dC}{dt} = \frac{S}{V} k_+ (1 - \frac{C}{K}) \qquad (66)$$

where C = concentration of H_4SiO_4 in a dilute solution (mol/l or mol/kg H_2O)
S = surface area of the solid (cm^2)
V = solution volume or mass of water in it (l or kg H_2O)
k_+ = forward reaction rate parameter ($mol \cdot cm^{-2} \cdot s^{-1}$)
K = equilibrium constant for equation 10.

In its physical units, k_+ is identical to k_i in equation 44. In agreement with the units of the parameters in equation 66, the equilibrium constant K is defined as the solubility of pure SiO_2-solid in a dilute solution where the activity of water is unity, $a_{H_2O} \approx 1$:

$$K = [H_4SiO_4^0] = C_s \text{ (mol/l or mol/kg } H_2O) \qquad (67)$$

Equation 66 is a simplified, practical version of a kinetic equation that was derived in great detail by Rimstidt and Barnes [37], whose data for k_+, K, and the activation energy (E_a) of the dissolution reaction of four SiO_2-solids are summarized in Table 3.4-6. The kinetic equation is valid for dilute solutions, when the differences between the molal and molar concentration scales can be neglected, and the activity coefficient values of the species in solution may be taken as approximately unity.

The quotient C/K in equation 66 expresses the degree of saturation of the solution with respect to the solid. (See section 2.11.2 for the definition of degree of saturation.) If $C/K < 1$, the solid should be dissolving; if $C/K > 1$, precipitation may be taking place, a process that corresponds to a negative value of dC/dt; and at equilibrium,

$C = K$. The equilibrium constant K can also be defined as the ratio of the forward and backward reaction rate constants, k_+ and k_-:

$$K = \frac{k_+}{k_-} \tag{68}$$

Because K, k_+ and k_- are interrelated, only two of these parameters are needed in equation 66 and Table 3.4-6 to compute either dissolution or precipitation reaction rates.

Example 5 Which of two solids — quartz grains or shells of diatoms made of amorphous SiO_2 —should dissolve faster in water containing dissolved silica at a concentration of 4 mg SiO_2/l at 25°C?

From Table 3.4-6, using T = 298 K,

$$\begin{array}{llll} \text{Quartz:} & K = 1.1 \times 10^{-4}, & \log k_+ = -17.38 \\ \text{Glass:} & K = 1.92 \times 10^{-3}, & \log k_+ = -16.14 \end{array}$$

Concentration of silica in solution is

$$C = \frac{4 \times 10^{-3}\ \text{g/l}}{60\ \text{g/mol}} = 6.67 \times 10^{-5}\ \text{mol/liter}$$

In the absence of other information on the surface areas of the two solids, assume that they are the same, making a comparison easier. The ratio of the dissolution rates, from equation 66, is as follows:

$$\frac{(dC/dt)_{gl}}{(dC/dt)_{qtz}} = \frac{k_+\ (\text{amorphous } SiO_2)}{k_+\ (\text{quartz})} \tag{69}$$

$$= 10^{-16.14}/10^{-17.38} = 10^{1.24} = 17$$

Thus, under the conditions stipulated, amorphous SiO_2 is expected to dissolve about 17 times faster than quartz.

Two corollaries can be drawn from the above result:

(1) the concentration of silica in solution during the dissolution processes will effectively be driven by the faster-dissolving phase, and

(2) the concentration in solution may exceed the solubility values of quartz while the diatom shells continue to dissolve, although this does not necessarily indicate that quartz would be precipitating from its supersaturated solution.

Dissolution of Feldspars. — Feldspars are alumino-silicate minerals of major abundance in crustal rocks and sediments. The general formula of their chemical composition is $M^{+n}Al_nSi_{4-n}O_8$, with n = 1 or 2. Potassium feldspar, $KAlSi_3O_8$, sodium feldspar, $NaAlSi_3O_8$, and solid solutions of composition ranging from $NaAlSi_3O_8$ (albite) to $CaAl_2Si_2O_8$ (anorthite) belong to the main rock-forming minerals. Under environmental conditions, it has been observed that the Ca-feldspar and Ca-rich members of the solid solution series dissolve faster than the Na and K-feldspars. Ca-feldspars generally occur in igneous rocks formed at temperatures higher than those with Na-feldspar and K-feldspar, and their faster dissolution correlates with the general phenomenon of the relative rates of weathering of alumino-silicate minerals.

In a study of the rates of dissolution of feldspars in water, at 25°C and 1 atm pressure of CO_2, four stages of dissolution have been identified [8]:

(1) Rapid surface exchange,
(2) Concentration increases as some power function of time,
(3) Parabolic dissolution rate, and
(4) Concentration increases linearly with time (steady dissolution rate).

The time-dependence of concentrations of cations and silica in solution and the approximate duration of each of the four dissolution stages are as follows:

First stage: Surface solution, H^+-cation exchange (<3 min)

Second stage: Concentration rises as a low-power function of time; the stage lasts up to 50 hours,

$$C = k_e t^n \tag{70}$$

Third stage: Increase in concentration obeys the parabolic rate law; duration up to 20 days following the second stage,

$$C = C_0 + 2k_p t^{1/2} \tag{71}$$

The latter equation is identical to equation 41.

Fourth stage: Concentration increases linearly with time; duration up to 30 days following the third stage,

$$C = C_0 + k_0 t \tag{72}$$

Rate parameters (k_{es}, $2k_{ps}$ and k_{0s}) for individual minerals in the second through fourth dissolution stages are listed in Table 3.4-7; these values must be multiplied by S/V to obtain the corresponding rate constants. During the first two stages, more of cations than silica is released to solution, resulting in a cation-deficient layer at the

TABLE 3.4-7

Dissolution Rate Parameters for Feldspars in Water
(at 25°C and 1 atm total CO_2 pressure)

	Power Law (Eq. 65) k_{es} (10^{-10} mol • cm^{-2} • s^{-n})				Parabolic (Eq. 66) $2k_{ps}$ (10^{-13} mol • cm^{-2} • s$^{-1/2}$)				Linear (Eq. 67) k_{0s} (10^{-16} mol • cm^{-2} • s^{-1})			
Mineral[a]	**Na**	**Ca**	**K**	**Si**	**Na**	**Ca**	**K**	**Si**	**Na**	**Ca**	**K**	**Si**
Albite	11.8			0.34	2.98			9.25	25.4			4.41
n =	0.059			0.223								
Oligoclase	6.18	3.90		0.76	2.81	1.10		5.83	1.67	0.89		2.60
n =	0.059	0.053		0.153								
Andesine	2.41	—		0.30	1.14	—		4.05	0.86	—		1.40
n =	0.030			0.199								
Labradorite	3.87	3.58		0.72	1.02	1.29		3.27	0.72	1.09		1.26
n =	0.039	0.057		0.157								
Bytownite	2.31	4.41		0.84	1.34	3.38		3.47	0.54	1.29		1.07
n =	0.061	0.064		0.158								
Anorthite	1.43	3.63		0.96	0.34	3.36		4.01	0.32	2.18		1.32
n =	0.027	0.093		0.173								
Orthoclase	1.49		5.43	0.30	1.49		4.19	3.78	0.66		2.70	1.68
n =	0.087		0.066	0.134								
Microcline	1.85		5.25	0.26	1.16		2.55	3.34	0.55		1.27	1.52
n =	0.067		0.075	0.164								

a. See Table 3.4-8 for composition and specific surface areas of these minerals.

Source: Busenberg and Clemency [8]

feldspar surface. Within the albite-anorthite series, the rate constants for the release of sodium and calcium increase with an increasing mole fraction of each cation in the solid. The faster release of cations continues during the third (parabolic-rate dissolution) stage. When dissolution is continuous, the value of C_0 at the beginning of each dissolution stage is the concentration value attained at the end of the preceding stage. A computational example of silica release during dissolution of oligoclase is given below.

Example 6 Compute SiO_2 concentration in solution in a system closed to water flow and in contact with oligoclase having a solid concentration of 50 $g \cdot l^{-1}$.

After the stage of rapid surface exchange, the solubility of silica is given by equations 70-72. For the dissolution process described by equation 70, Table 3.4-7 lists the value of k_{es} as 0.76×10^{-10} $mol \cdot cm^{-2} \cdot s^{-n}$. To calculate the rate parameter k_e, k_{es} must be multiplied by S/V. From Table 3.4-8, $S_g = 1\ m^2 \cdot g^{-1} = 10^4\ cm^2 \cdot g^{-1}$; therefore,

$$\begin{aligned} k_e &= (0.76 \times 10^{-10}\ mol \cdot cm^{-2} \cdot s^{-n})\ (10^4\ cm^2 \cdot g^{-1})\ (50\ g \cdot l^{-1}) \\ &= 3.8 \times 10^{-5}\ mol \cdot l^{-1} \cdot s^{-n} \end{aligned}$$

If this dissolution stage lasts 30 hr ($t = 1.08 \times 10^5$ s), the concentration of SiO_2 at the end of this time is, by equation 70 and with the power exponent value $n = 0.153$,

$$C = 3.8 \times 10^{-5} \times (1.08 \times 10^5)^{0.153} = 2.23 \times 10^{-4}\ mol \cdot l^{-1} = 13.4\ mg \cdot l^{-1}$$

This C value of 2.23×10^{-4} $mol \cdot l^{-1}$ becomes the new C_0 (initial concentration) for the next dissolution stage (parabolic-rate dissolution), which is characterized by the rate constant value of $2k_{ps} = 5.83 \times 10^{-13}$ $mol \cdot cm^{-2} \cdot s^{-1/2}$. As in the preceding computation, the rate constant k_p is

$$2k_p = 5.83 \times 10^{-13} \times 10^4 \times 50 = 2.92 \times 10^{-7}\ mol \cdot l^{-1}\ s^{-1/2}$$

Allowing the parabolic dissolution stage to last 20 days ($t = 17.28 \times 10^5$ s), we use equation 71 with the value of initial concentration $C = 2.23 \times 10^{-4}$ $mol \cdot l^{-1}$, attained at the end of the preceding stage. Thus,

$$C = 2.23 \times 10^{-4} + 2.92 \times 10^{-7} \times (17.28 \times 10^5)^{1/2}$$

$$= 6.07 \times 10^{-4}\ mol \cdot l^{-1} = 36.4\ mg \cdot l^{-1}$$

This C value becomes the new C_0 (initial concentration) for the next stage of dissolution.

In the next (linear) stage, the dissolution rate constant in Table 3.4-7 is $k_{0s} = 2.60 \times 10^{-16}$ $mol \cdot cm^{-2} \cdot s^{-1}$, and the rate constant k_0 is

$$k_0 = 2.60 \times 10^{-16} \times 10^4 \times 50 = 1.30 \times 10^{-10}\ mol \cdot l^{-1} \cdot s^{-1}$$

Allowing the linear-rate dissolution to continue for 30 days (25.92×10^5 s), we use equation 72 with the initial concentration $C_0 = 6.10 \times 10^{-4}$ $mol \cdot l^{-1}$, from the end of the preceding stage. The new concentration value is

$$\begin{aligned} C &= 6.07 \times 10^{-4} + 1.30 \times 10^{-10} \times 25.92 \times 10^5 \\ &= 9.44 \times 10^{-4}\ mol \cdot l^{-1} \\ &= 56.6\ mg \cdot l^{-1} \end{aligned}$$

TABLE 3.4-8

Composition and Specific Surface Area of Feldspars in the Albite-Anorthite Series

Mineral	Composition (mol %)			Specific Surface Area, S_g
	$NaAlSi_3O_8$	$CaAl_2Si_2O_8$	$KAlSi_3O_8$	$(m^2 \cdot g^{-1})$
Albite	98.9	0.1	1.0	0.83
Oligoclase	69.6	24.0	6.4	1.01
Andesine	49.5	42.7	7.8	1.49
Labradorite	44.5	53.0	2.5	1.04
Bytownite	22.2	77.4	0.4	1.14
Anorthite	5.8	94.0	0.2	1.84
Orthoclase	20.3	0.2	79.5	1.52
Microcline	24.8	0	75.2	1.07

Source: Busenberg and Clemency [8]

The main cation in oligoclase is sodium. For the oligoclase of composition given in Table 3.4-7, the atomic ratio Na/Si is approximately 1/4 = 0.25. The differences between the dissolution rates of the individual components of feldspars are responsible for the ratios in solution that differ from those of the bulk solid. For Na, using the computation procedure shown above, with the data from Table 3.4-7, the Na concentrations at the end of each dissolution stage are consecutively 6.12×10^{-4}, 7.97×10^{-4}, and 10.13×10^{-4} $mol \cdot l^{-1}$. Thus, the Na/Si ratios in solution are 2.7 (after 30 hr), 1.3 (after 21 days), and 1.07 (after 51 days). The higher ratios show that sodium is removed from the solid preferentially to silicon; although the Na/Si ratio decreases with time, even after almost two months of dissolution it is higher than in the bulk solid, reflecting the faster removal of Na from the solid surface.

Example 7 How thick is the leached surface layer of a dissolving feldspar? This question can be answered using the results of the preceding computation. Initially, 50 g of oligoclase was in contact with 1 liter of water; 50 days later, about 100 mg of dissolved solids are in solution. (The amounts of 57 mg SiO_2 and 23 mg Na were computed in Example 6, such that the mass of 100 g of total solids is an orientational figure.) Thus, the fraction of the original solid that dissolved is 0.1 g/50 g = 0.002 or 0.2%. A mass balance between the material in solution, M(soln), in the original solid, M(orig), and in the residual solid, M(res), is

$$M(\text{soln}) = M(\text{orig}) - M(\text{res})$$

$$\frac{M(\text{soln})}{M(\text{orig})} = 1 - \frac{M(\text{res})}{M(\text{orig})} = 0.002$$

For spherical particles dissolving at the surface, the ratio of the remaining to the original mass is equal to the ratio of the cubes of the radii: M(res)/M(orig) = r^3/r_0^3. Thus where r is the radius of the dissolving particle, and r_0 is the initial radius.

$$r^3/r_0^3 = 0.998$$

The fractional decrease in the radius of the fresh feldspar grains is

$$1 - \frac{r}{r_0} = 0.0007 \text{ or } 0.07\%$$

For particles of radius $r_0 = 10\ \mu m$, the decrease translates into a 70Å thickness of the leached layer.

The role of hydrogen-ion concentration in the dissolution of Na-feldspar was considered in detail for the region of linear dissolution, equation 72, by Wollast and Chou [45]. Their results, shown in Figure 3.4-3, indicate that the bulk rate of dissolution goes through a minimum in the neutral pH range, increasing by a factor of 100 under acidic or alkaline conditions. An explanation of this behavior has been given in terms of the transition state theory and activated complexes forming on the mineral surface. The following empirical equation describes the rate of congruent dissolution of sodium feldspar (albite), where R is in $mol \cdot cm^{-2} \cdot s^{-1}$:

$$R = 10^{-13.69} a_{H^+}^{\ 0.49} + 10^{-16.15} + 10^{-18.15}\ a_{H^+}^{\ -0.30} \tag{73}$$

Example 8 In Example 4, the speed with which a silicate particle would dissolve in a diffusion-controlled dissolution reaction was calculated. For a spherical grain of the same diameter (1 mm), the rate of dissolution is many orders of magnitude slower when dissolution proceeds in a surface-controlled reaction. From Figure 3.4-3, the rate of dissolution near neutral pH is about $R = 2 \times 10^{-16}\ mol \cdot cm^{-2} \cdot s^{-1}$. For a feldspar grain of radius $r = 0.05$ cm and molar volume $V_m = 100\ cm^3 \cdot mol^{-1}$, we have

$$\text{Initial dissolution rate} = RV_m = 2 \times 10^{-14}\ cm \cdot s^{-1} = 6.2 \times 10^{-7}\ cm \cdot yr^{-1}.$$

In terms of a time scale measurable in years, the mineral grain is virtually insoluble. The rate of dissolution near pH = 7 translates into a linear rate (in the direction perpendicular to the grain surface) of about 60 Å per year, in the absence of any catalytic or reaction-accelerating process. The time to complete dissolution of a spherical grain, under surface-controlled reaction conditions [21], is:

$$t = \frac{r}{RV_m} = \frac{5 \times 10^{-2}\ cm}{6.2 \times 10^{-7}\ cm/yr} = 80{,}000 \text{ years}$$

A grain ten times smaller in diameter (0.1 mm) will survive under the same conditions one-tenth as long, or 8,000 years. The order of magnitude agrees with the times of renewal or mean age of soils, 10^3 to 10^4 years [21].

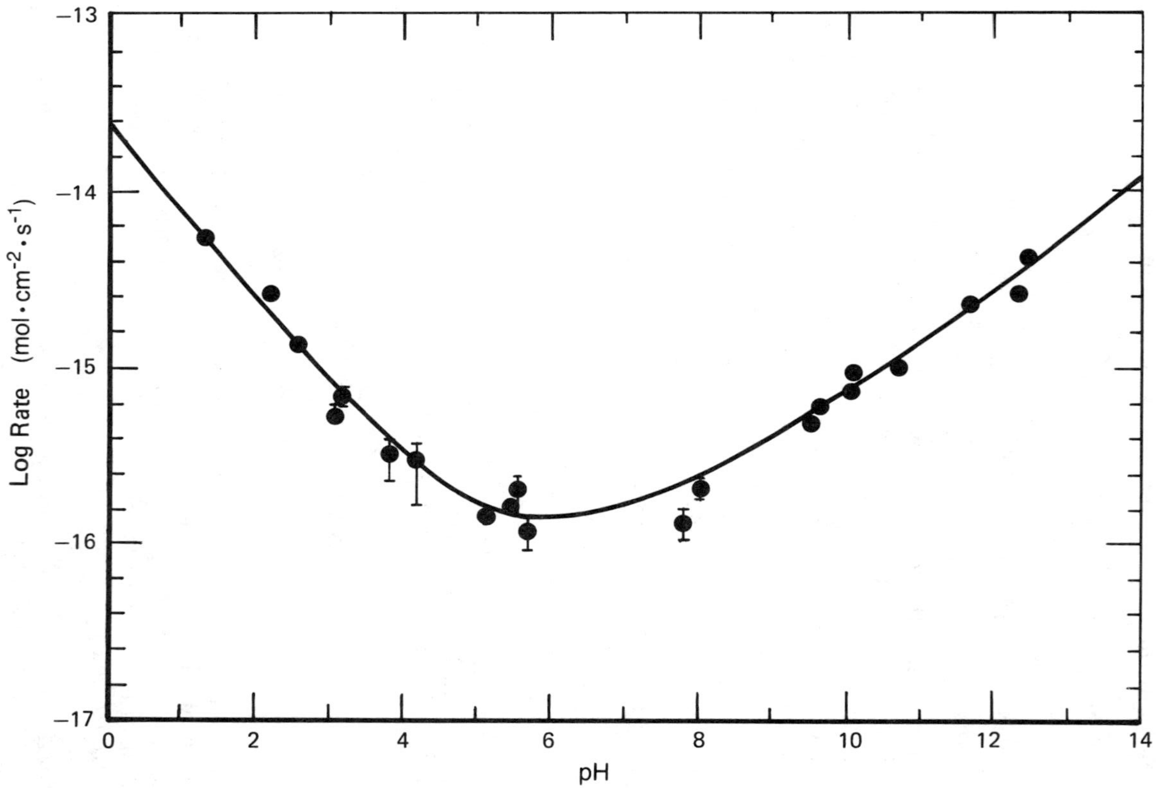

Source: Wollast and Chou [45]. *(Copyright 1985, Reidel Publishing Co. Reprinted with permission.)*

FIGURE 3.4-3 Rate of Dissolution of Sodium-Feldspar (Albite) as a Function of pH, at a Constant Rate of Albite Dissolution

Dissolution Rates of Other Silicate Minerals. — The preceding two sections concerned minerals made of pure SiO_2 and several members of the feldspar group. The variety of the silicate minerals extends to many other groups, such as olivines, pyroxenes, amphiboles, and micas, all of which are characterized by specific crystal structures and chemical compositions of varying complexity.

In Figure 3.4-4 are plotted the rates of dissolution at 25°C of 16 minerals, expressed as log R, against their standard Gibbs free energies of formation. The data for this plot

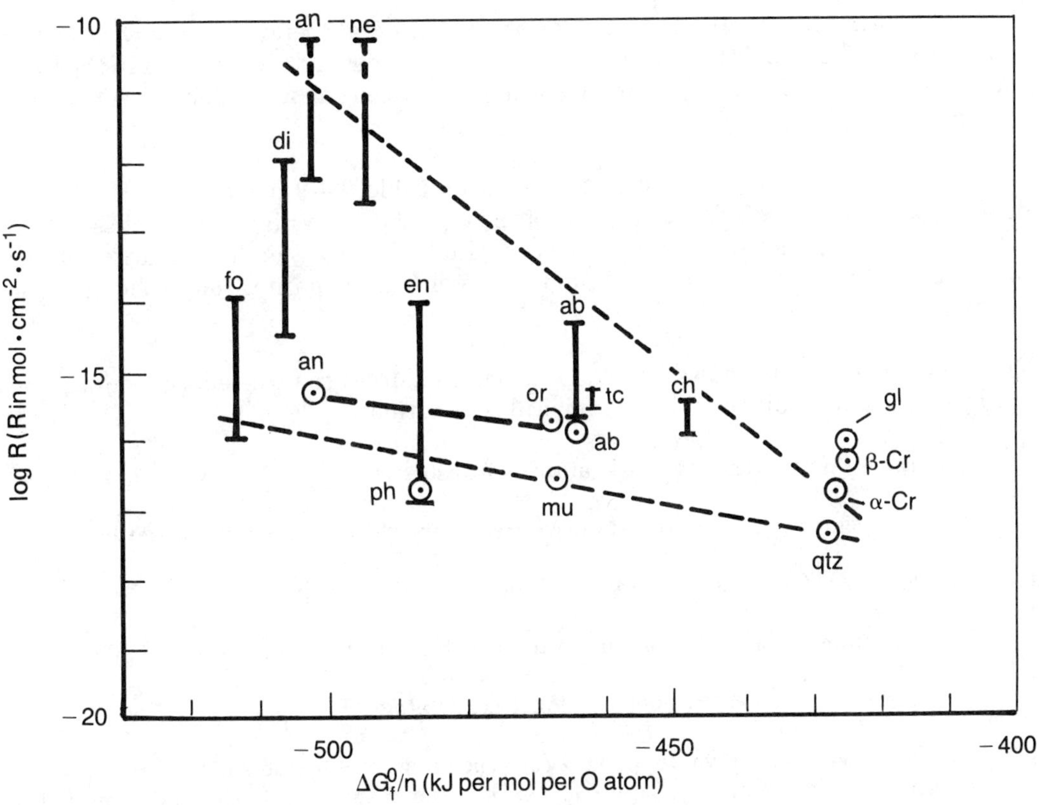

Notations for different solids are given in Table 3.4-9. Dashed lines bracket the range of solubility rates of the common silicate minerals.

Source: See Table 3.4-9 for explanation of symbols.

FIGURE 3.4-4 Rates of Dissolution of Some Silicates, log R, Plotted Against Their Gibbs Free Energies of Formation (25°C), Normalized to the Number of the Oxygen Atoms (n) in the Stoichiometric Formula of the Silicate

are summarized in Table 3.4-9, which also lists the chemical composition of each mineral and the identifying symbols used in Figure 3.4-4. The tabulated values of $\Delta G_f^0/n$ are the free energies of formation per mole, divided by the number of oxygen atoms in the stoichiometric formula of the mineral; this division helps to normalize the ΔG_f^0 values, which vary considerably with the solid compositions and complexity of the stoichiometric formulas. Figure 3.4-4 shows that the solubility rate R increases as ΔG_f^0 becomes more negative.[3]

The range of the log R values plotted for several minerals reflects in many cases the dependence of the dissolution rate on pH. Some of this dependence is noted in the final column of Table 3.4-9, and the subject was also discussed earlier under "Dissolution of Feldspars."

At least some of the alumino-silicates listed in Table 3.4-9 are characterized by a dissolution behavior as a function of pH of the type shown in Figure 3.4-3: dissolution rates are high in the acidic range, then decline to a more or less clearly defined minimum in the range $5 < pH < 7$, and finally increase again through the alkaline pH range.

The following example shows how log k is obtained from the free energy values and indicates the uncertainty of this empirical estimation:

Example 9 What is the solubility rate of white asbestos (chrysotile) in water at 25°C?

Chemical composition and the standard free energy of formation of chrysotile are:

$$Mg_3Si_2O_5(OH)_4, \ \Delta G_f^0 = -4034 \text{ kJ/mol}$$

The total number of oxygen atoms in the stoichiometric formula is n = 9. Thus,

$$\Delta G_f^0/n = -4034/9 = -448 \text{ kcal/mol per O atom}$$

From Figure 3.4-4, the value of log R corresponding to −448 on the horizontal axis is within the range from −15.2 to −17 between the dashed lines. The midpoint of this range is log R ≈ −16. Experimentally determined dissolution rates of silica in chrysotile at 25°C are in the range of log R ≈ −16 to −15.5 over the pH range from 11 to 8 [1].

In the temperature range from about 100°C to 700°C, rates of dissolution of many silicate and alumino-silicate minerals in water indicate increasing solubility with

3. The more negative values of ΔG_f^0 do not necessarily indicate that the minerals are relatively more stable in an aquatic environment. The relative stability of a solid with respect to dissolution — or, in broader terms, its relative resistance to weathering — is determined by the free energy change for a dissolution reaction, not by the ΔG_f^0 values of the solid alone.

TABLE 3.4-9

Dissolution Rates of Selected Silicate Minerals in Water at 25°C, and Other Data for Correlating Dissolution Rates with $\Delta G_f^0/n$

Mineral	Formula	Symbol in Figure 3.4-4	Number of O Atoms n	Free Energy of Formation[a] per O atom $\Delta G_f^0/n$ (kJ/mol per O atom)	Log Dissolution Rate[b] log R (k in $mol \cdot cm^{-2} \cdot s^{-1}$)	Source	pH at Low k	pH at High k
Quartz	SiO_2	qtz	2	−428	−17.4	[37]		
α-Cristobalite	SiO_2	α-Cr	2	−427	−16.8	[37]		
β-Cristobalite	SiO_2	β-Cr	2	−425	−16.3	[37]		
SiO_2 glass	SiO_2	gl	2	−425	−16.1	[37]		
Nepheline	$NaAlSiO_4$	ne	4	−494	−12.6 to −10.3	[41]	7	3
Forsterite	Mg_2SiO_4	fo	4	−513	−16 to −14	[1][c]	5	3
Enstatite	$Mg_2Si_2O_6$	en	6	−487	−16.5 to −14	[1][c]	6	2
Diopside	$CaMgSi_2O_6$	di	6	−506	−14.5 to −12	[1][c]	6	2
K-feldspar	$KAlSi_3O_8$	or	8	−468	−15.8	[8]		
Albite	$NaAlSi_3O_8$	ab	8	−464	−15.8 to −14.3	[45]	6	1.5, 12
Anorthite	$CaAl_2Si_2O_8$	an	8	−502	−12.3 to −10.3	[14]	5.6	2
Albite-Anorthite series		(ab to an)	8	(−464 to −502)	−15.9 to −15.4	[8]		
Chrysotile	$Mg_3Si_2O_5(OH)_4$	ch	9	−448	−16.0 to −15.5	[1][c]	11	8
Talc	$Mg_3Si_4O_{10}(OH)_2$	tc	12	−461	−15.4 to −15.2	[23]		
Muscovite	$KAl_3Si_3O_{10}(OH)_2$	mu	12	−467	−16.6	[24]		
Phlogopite	$KMg_3AlSi_3O_{10}(OH)_2$	ph	12	−487	−16.4	[25]		

a. From Robie *et al.* [38]; phlogopite from Drever [10].
b. Dissolution rate k refers to release of Si, when individual elemental components of solids are reported as having different k-values.
c. Plus other sources and own data on chrysotile.

increasing temperature. Wood and Walther [46] demonstrated this trend by application of the Arrhenius relationship (equation 57) to experimental data and summarized it in the following equation:

$$\log R = -6.85 - \frac{2900}{T} \tag{74}$$

where T is temperature (K) and the dissolution rate R is in units of gram-atoms of oxygen per square centimeter per second (g-at $O \cdot cm^{-2} \cdot s^{-1}$). Compared with some experimental values of R measured at approximately 25°C, the results of this equation are low by about a factor of 10, an uncertainty margin comparable to that of the data plotted in Figure 3.4-4.

Dissolution Rates of Calcite. — Experimental study of the dissolution of calcite ($CaCO_3$) is a subject of considerable environmental importance, owing to the abundance of $CaCO_3$ as the minerals calcite and aragonite, in fresh sediments, in old sedimentary rocks, and in the deposits that form in various engineered structures in contact with water.

The rates of dissolution of calcite in natural waters show, in general, great variability that is related to the physical state of the solid, to the chemical composition of the aqueous phase, and to the pH and the partial pressure of CO_2 of the solution.

Plummer and Wigley [35], some of whose results on the dissolution rate of calcite in water are reproduced in Figure 3.4-5, have cited extensive evidence from several investigators that trace amounts of metals and other species in solution can inhibit the rate of calcite dissolution. The inhibiting effects on calcite dissolution in fresh and ocean waters have been known for trace concentrations of the ferric and chromic ions, scandium, copper, and phosphate; this is presumably due to attachment of the tracer ions to the more reactive surface sites on the solid. In natural waters, particularly in those flowing slowly through groundwater aquifers in limestones, the inhibiting effects of tracer impurities may keep the groundwater below saturation with respect to $CaCO_3$ for many centuries.

The main conclusions of the dissolution rate experiments of Plummer and Wigley are the following:

(1) When the pH of the solution is below 4, dissolution is probably a diffusion-controlled process;

(2) At a pH of about 5 and higher, dissolution is controlled by surface reactions;

(3) Up to pH $\approx$ 5.94, the rate of reaction of the Ca^{+2}-release and the consumption of H^+ ions is very nearly second-order (Figure 3.4-5);

(4) Above pH $\approx$ 5.94, the reaction rate is slower, tending to follow a fourth-order rate law.

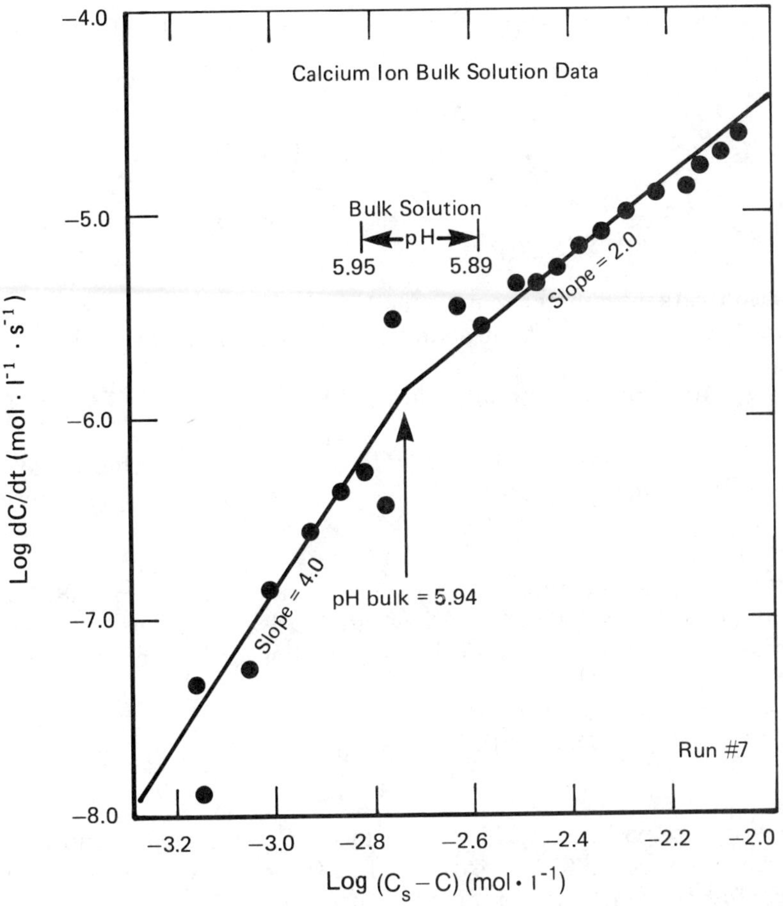

Conditions: 25°C, CO_2 saturated solution at 1 atm, pH range approximately 4 to 6.

Source: Plummer and Wigley [35]. *(Copyright 1976, Pergamon Press, Inc. Reprinted with permission.)*

FIGURE 3.4-5 Rate of Calcite Dissolution as a Function of Distance from Saturation

In the region of the second-order reaction, the rate of release of the Ca^{+2} ions into solution is:

$$\frac{dC}{dt} = \frac{S}{V} k (C_s - C)^2 \quad (mol \cdot l^{-1} \cdot s^{-1}) \tag{75}$$

where S = surface area of the solid (cm^2)
V = solution volume (cm^3)
k = reaction rate parameter ($cm \cdot l \cdot mol^{-1} \cdot s^{-1}$)
C_s = concentration of Ca^{+2} at the solid surface, at equilibrium with the solid (mol/l)
C = concentration of Ca^{+2} in solution (mol/l).

The rate parameter k is pH-dependent; for calcium-ion release, the value of k increases about linearly from 0.012 at pH = 5.70 to 0.034 at pH = 5.94. (The units of k are defined above.)

In an experimental study of the rate of dissolution of calcite at pH = 9.8, and in solutions containing other dissolved electrolytes, Rickard and Sjöberg [36] showed that the rate of dissolution is a function of the degree of saturation of solution with respect to calcite:

$$\frac{dC}{dt} = k_2 S (1 - \Omega^{1/2}) \tag{76}$$

where k_2 is a reaction rate parameter, S is the surface area of the solid, and Ω is the degree of saturation (see section 2.11.2), expressed as the ratio of the ion-concentration product (ICP) and the solubility product (K_{sp}):

$$\Omega = \frac{ICP}{K_{sp}} = \frac{[Ca^{+2}][CO_3^{-2}]}{K_{sp}} \tag{77}$$

Equation 76 for calcite is similar in form to equation 66 for the dissolution rates of SiO_2 minerals.

As a whole, laboratory experiments such as those cited in this section provide information on the reaction mechanisms of the calcite dissolution process, but they do not (and are not claimed to) reproduce the dissolution rates of natural environments, which are often much slower than those achieved in the laboratory.

3.4.4 Precipitation, Nucleation, Growth, and Recrystallization

This section deals with the kinetics of the processes that are the reverse of dissolution — removal of materials from an aqueous to a solid phase.

Although a process of precipitation must begin with nucleation — formation of nuclei and submicroscopic crystals in solution — the theory and the mathematical treatment of the two are very different. Precipitation can be observed and recorded, and its rate can be measured by direct determination of a decrease of some reactant concentration in solution. The treatment and computations based on the results when applied to rates of precipitation are fairly similar to those discussed in the preceding sections on rates of dissolution; for this reason, a case study of precipitation rate will be dealt with before other processes, such as nucleation, crystal growth, and recrystallization.

RATES AND EXAMPLE OF PRECIPITATION

Precipitation of Fe(III)-hydroxide

Ferric hydroxide of composition $Fe(OH)_3$ is a commonly occurring substance in the environment. In aqueous solutions, oxidation of ferrous iron by dissolved oxygen leads to precipitation of ferric hydroxide according to the reaction:

$$Fe^{+2} + 1/4\ O_2^0 + 5/2\ H_2O \longrightarrow Fe(OH)_3(s) + 2H^+$$

The rate of removal of Fe^{+2} from solution is a measure of ferric hydroxide precipitation. The rate of removal of the ferrous iron can be written as [39]:

$$-\frac{d[Fe^{+2}]}{dt} = k\,[Fe^{+2}]\,pO_2\,[OH^-]^2 \qquad (78)$$

where pO_2 is the partial pressure of oxygen (atm), and brackets denote concentrations in mol/liter.

The foregoing rate equation stipulates a dependence on the first power of O_2, whereas a fractional stoichiometric coefficient (1/4) appears in the reaction that precedes it. The 4:1 stoichiometric ratio for Fe/O_2 has been confirmed by White and Yee [44].

At a constant partial pressure of oxygen and a constant pH, the rate equation reduces to the form of a first-order kinetic law:

$$-\frac{d[Fe^{+2}]}{dt} = k_1[Fe^{+2}] \qquad (79)$$

Stumm and Morgan [39] give the value for the reaction rate constant in equation 78, at 20°C, as

$$k = (8.0 \pm 2.5) \times 10^{13}\ l^2 \cdot mol^{-2} \cdot atm^{-1} \cdot min^{-1}$$

From the results of many investigators who have studied Fe^{+2} oxidation in various natural waters, Davison and Seed [9] concluded that the best values of k are slightly lower:

$$k = (2 \pm 1) \times 10^{13}\ l^2 \cdot mol^{-2} \cdot atm^{-1} \cdot min^{-1}$$

The variation of the rate parameter k as a function of pH is shown in Figure 3.4-6.

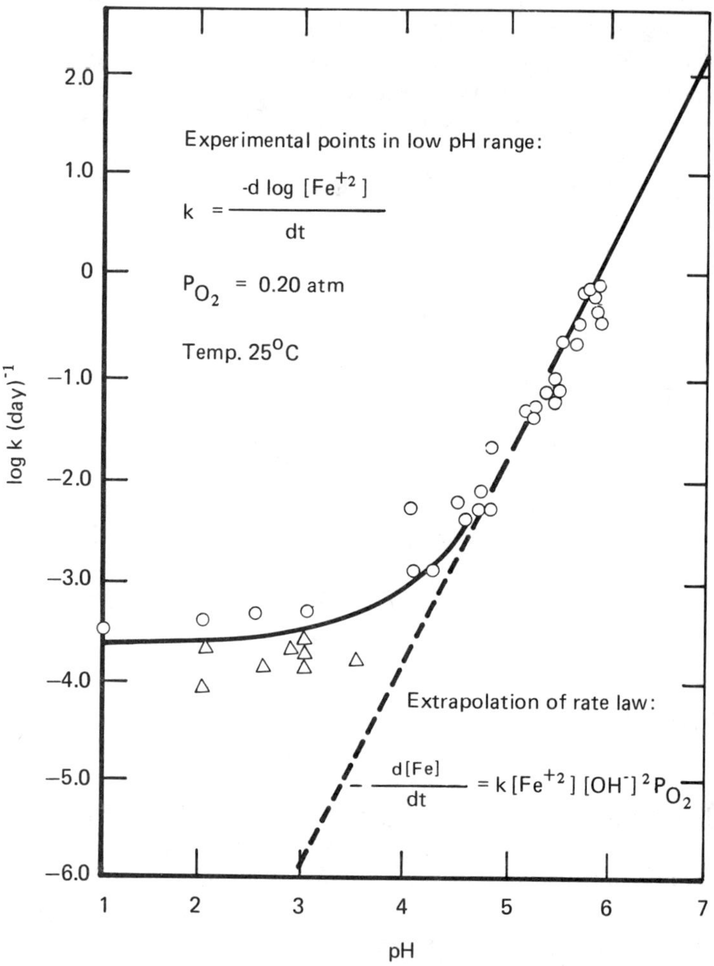

Source: Stumm and Morgan [39]. *(Copyright 1981, John Wiley & Sons. Reprinted with permission.)*

FIGURE 3.4-6 Observed Rate Constant for Oxidation of Fe^{+2} in Water

Dependence of k on Temperature and Solution Composition

For the oxidation reaction of Fe^{+2}, as given in the preceding section, the dependence of k on temperature in the range from 5° to 30°C is reflected in the value of the activation energy parameter E = 96 kJ/mol [39]. This value can be used in equations 56-58 to compute the rate parameter k at temperatures below or above the value for 20°C.

Salts in solution tend to slow down the Fe(II) oxidation reaction [40] according to the relationship

$$\log k = 13.76 - 2.06\, I^{1/2} \tag{80}$$

where I is the ionic strength of the solution. (For definition of I, see section 2.6.) It has also been reported [18] that in solutions of relatively high alkalinity, above an equivalent of 340 mg $CaCO_3$/l, k increases with alkalinity.

Sources of Fe(II)

Reduced iron occurs in many rock-forming minerals, either as the only main metal constituent or as a companion to the oxidized ferric iron, Fe(III). Purely ferrous iron minerals include both common and less abundant species, such as:

FeS_2	—	pyrite, marcasite
$FeCO_3$	—	siderite
$Fe_3(PO_4)_2 \cdot 8H_2O$	—	vivianite
Fe_2SiO_4	—	fayalite, an olivine
Pyroxenes	—	for example, augite

Minerals with iron present in the lattice in two oxidation states are such commonly occurring species as

Fe_3O_4	—	magnetite ($FeO \cdot Fe_2O_3$)
Biotite	—	a mica

Ferric iron occurs in a variety of alumino-silicate minerals and in the three abundant oxides and oxy-hydroxides:

Fe_2O_3	—	hematite
α-FeOOH	—	goethite
γ-FeOOH	—	lepidocrocite
$Fe(OH)_3$	—	amorphous ferric hydroxide

The solubilities of Fe(III) oxy-hydroxides are very low at near-neutral pH values, and the oxidized Fe(II) is generally removed from solution very efficiently.

In the deeper groundwaters and in the deeper parts of some lakes, the decay of organic matter consumes all the oxygen. This promotes reduction of Fe(III) to Fe(II), which (depending on the local environmental conditions) can either react with sulfide in water, forming FeS or FeS_2, or migrate into the water column. Transport of dissolved Fe(II) toward the upper oxygenated layers ultimately results in its oxidation, whereupon the iron returns to the lake bottom as precipitated $Fe(OH)_3$.

Rates of Oxygen Consumption

In a homogeneous Fe(II) solution, the rate of removal of Fe(II), as shown in Figure 3.4-6, translates into a rate of oxygen consumption by the Fe(II) sink.

Below pH $\approx$ 3, the reaction is evidently independent of hydrogen-ion concentration and takes place slowly, with a half-life of about $t_{1/2} = 1/k = 3000$ days. From equation 78, at the atmospheric value of the oxygen partial pressure $pO_2 = 0.2$ atm and for an arbitrarily chosen pH of 6.5, the reaction rate parameter k_1 is

$$k_1 = 2 \times 10^{13} \times 0.2 \times (10^{-7.5})^2 = 0.004 \text{ min}^{-1}$$

The reaction half-life is then

$$t_{1/2} = 1/k_1 = 250 \text{ min}$$

Half-lives of comparable magnitude, 10^2 to 10^3 min, have been reported for the oxidation of iron in pyrite [39].

The oxidation reaction may be retarded through aqueous complex formation with organic materials and, as mentioned earlier, through an increase in the salt concentration in solution.

A study of the oxidation of Fe(II) on the surfaces of silicate minerals (biotite, hornblende and augite) in contact with water has shown that the rate of oxygen consumption increases with acidity: at a pH of 1.5-3.5, the rates of oxidation are significantly higher [44]. Overall, the reported half-lives of the oxygen consumption reactions are about 12 days.

NUCLEATION AND CRYSTAL GROWTH

Crystal nuclei form in supersaturated solutions by processes known as homogeneous and heterogeneous nucleation.

Homogeneous nucleation is the formation of clusters of atoms or molecules of a certain critical size from a supersaturated solution, where they continue to grow as long as the solution remains supersaturated.

In *heterogeneous nucleation*, clusters of atoms or molecules form on an existing substrate. In a supersaturated solution this substrate can consist of crystals of the same phase, such as seeds of NaCl in a solution supersaturated with respect to this mineral, or it can consist of particles of a different phase, such as flakes of mica or practically any other poorly soluble solid.

The more commonly observable nucleation events in the environment are nucleation of ice, nucleation of sulfur and other minerals from the vapor phase emanating from volcanic vents, and nucleation of some commonly occurring minerals in natural waters. In the latter group of processes, fine-grained $CaCO_3$ can nucleate in surface layers of freshwater lakes and oceans, forming characteristic patches of white-colored water ("whitenings") during periods of strong biological productivity. Seasonally strong growth of photosynthetic organisms reduces concentrations of dissolved CO_2 in water, which increases the pH and causes $CaCO_3$ crystals to appear in the surface water layers. Similarly, sodium chloride crystals can nucleate and grow in the surface layer of saline brine lakes, where evaporation may cause local supersaturation with respect to halite. These crystals may grow in a pattern of thin snowflakes and remain on the brine surface for some length of time.

Alkaline brine lakes that are rich in bicarbonate and carbonate, in arid climates, contain relatively high concentrations of dissolved SiO_2. During rainy periods, freshwater runoff overflows the brine, causing a decrease in the pH and subsequent nucleation and precipitation of amorphous SiO_2 gel [11, 12].

Homogeneous Nucleation

The universal presence of suspended solids in natural waters makes it difficult to distinguish between homogeneous and heterogeneous nucleation. In effect, a statement that homogeneous nucleation is an important mechanism of mineral phase formation in natural waters requires proof of its correctness.

A classical theory of homogeneous nucleation in a supersaturated aqueous solution is based on a concept of the free energy change required for the formation of a molecular cluster and the formation of a surface in a supersaturated solution. Supersaturation, relative to the solubility of the bulk phase, provides the driving force needed for the formation of the cluster and the solid surface. Based on the theory of the free energy change in the process of a nucleus formation, the rate of formation of new nuclei (J, number of nuclei per unit volume of solution per unit of time) is represented as a function of temperature, volume of the molecules, interfacial energy between the crystal and solution, and the degree of supersaturation of the solution, as follows [33, 43]:

$$J = A \exp\left(-\frac{\Delta G^*}{RT}\right) \tag{81}$$

where J = rate of nucleation ($cm^{-3} \cdot s^{-1}$)
A = pre-exponential factor, ranging from 10^{25} to 10^{30}
ΔG^* = free energy of formation of a nucleus of critical size
T = temperature (K)
R = gas constant

An explicit form of the free energy term ΔG^* gives the following relationship for the nucleation rate:

$$J = A \exp\left[-\frac{\beta v^2 \sigma^3}{k^3 T^3 (n \ln s)^2}\right] \tag{82}$$

where β = nucleus shape factor (Table 3.4-10)
v = volume of the cluster-forming molecules (cm^3)
σ = interfacial energy ($erg \cdot cm^{-2}$, (as given in Table 3.4-11)
n = number of atoms making up a molecule in the cluster
s = degree of supersaturation, defined as (concentration in solution)/(concentration at saturation with bulk phase)
k = Boltzmann constant (1.38×10^{-16} $erg \cdot deg^{-1}$)

For the rate of appearance of one nucleus, $J = 1\ cm^{-3} \cdot s^{-1}$, supersaturation can be defined as critical supersaturation, s^*. The value of s^* for this event can be derived from equation 82:

$$\ln s^* = \left(\frac{\beta v^2 \sigma^3}{k^3 T^3 n^2 \ln A}\right)^{1/2} = \frac{v\sigma}{kTn}\left(\frac{\beta\sigma}{kT\ln A}\right)^{1/2} \tag{83}$$

It is worth examining the effects of the individual parameters in equation 83 on the theoretical degree of supersaturation needed to induce homogeneous nucleation.

The molecular volume, v, is easily derived from the tabulations of densities (ρ, $g \cdot cm^{-3}$) of molar volumes (V_m, $cm^3 \cdot mol^{-1}$):

$$\begin{aligned} V &= gfw/(\rho N) \quad (cm^3) \\ &= V_m/N \quad (cm^3) \end{aligned} \tag{84}$$

where gfw is the gram-formula weight of the solid and N is the Avogadro number.

For most simple oxides of metals, and for simple anhydrous halides, carbonates, and sulfates, the molar volumes fall in the range from 10 to 80 $cm^3 \cdot mol^{-1}$. This eightfold variation is also reflected in the molecular volumes v.

The cluster shape factor β, given in Table 3.4-10, varies by more than a factor of 10, depending on the cluster shape. Even for such shapes as the sphere and the cube, which are commonly used in theoretical computations of nucleation rates, there is a difference of a factor of two. The dependence of critical supersaturation s^* on the shape factor β is somewhat reduced by the power 1/2.

The values of the interfacial energy, σ, as given in Table 3.4-11, differ from one compound to another by a factor of about two. Because of the 3/2 power, the dependence of s^* on σ is not insignificant. The same type of dependence applies to temperature, but the range of the environmental temperatures from about 10°C to 40°C (283 to 313 K) is fairly small.

TABLE 3.4-10

Geometric Shape Factor, β, in the Homogeneous Nucleation Equation (82)

Body Shape	β
Sphere	16.7552
Icosahedron	20.2162
Dodecahedron	22.2012
Octahedron	27.7128
Cube	32.0000
Rectangular parallelepiped ($4\ell \times 2\ell \times \ell$)	50.8148
Tetrahedron	55.4256
Rod ($10\ell \times \ell \times \ell$)	109.7600
Plate ($10\ell \times 10\ell \times \ell$)	204.8000

Source: Nielsen [33]

TABLE 3.4-11

Interfacial Energy of Crystal Nuclei Against Aqueous Solutions for Spherical and Cubic Nuclei, Critical Supersaturation Ratio for Homogeneous Nucleation, and Critical Diameter of Nucleus

		Interfacial Energy, σ ($erg \cdot cm^{-2}$)		
Solid Phase	Critical Supersaturation Ratio, s*	Spherical Nuclei	Cubic Nuclei	Critical Diameter, d* (Å)
$BaSO_4$	1000; 500	116	151	11
$PbSO_4$	28; 40	74	119	13
$SrSO_4$	39	81	70	12
$PbCO_3$	106	105		11
$SrCO_3$	30	86		12
CaF_2	80	140	140	9
MgF_2	30	129		9
AgCl	5.5	72		15
AgBr	3.7	56		15
Ag_2SO_4	19	62		14

Source: Walton [43]

The uncertainty or variation in the magnitude of the pre-exponential factor A is small, because its natural logarithm appears in equation 83: ln A varies, depending on the value of A, from 57.6 to 69.1, and the difference of 20% is reduced to 10% by the power 1/2.

Some values of critical supersaturation for homogeneous nucleation of a number of inorganic phases are listed in Table 3.4-11.

Heterogeneous Nucleation and Crystal Growth

Mathematical treatment of the initial stages of heterogeneous nucleation is fairly complex, because numerous parameters concerning the structure and energy states of the substrate surface must be known. A process of crystal growth, following the early stage of nucleation, is schematically shown in Figure 3.4-7. The rate of crystal growth in solution is amenable to a much simpler mathematical treatment if one considers growth on existing nuclei. In supersaturated solutions seeded with small crystals of the same composition, relief of the supersaturation is attained through the process of growth, either entirely or at least partly caused by heterogeneous nucleation.

An empirical relationship that applies in general to growth of crystals in solution has the form

$$-\frac{dC}{dt} = k_m S\,(C_s - C)^m \tag{85}$$

where C = concentration in a supersaturated solution
C_s = concentration at saturation with the bulk phase ($C_s < C$)
S = surface area of the growing crystals
k_m = rate parameter for crystal growth
m = the order of the reaction

The order of the crystal growth reaction, m, is often equal to the total number of ions in the stoichiometric formula of the crystalline phase. The reaction order values are summarized in Table 3.4-12. An exception to the rule is $Mg(OH)_2$, whose rate of growth in solution has been described as following the first-order rate law. The mechanism of growth in the early stage of nucleation and growth is considered to be surface-reaction controlled or interface controlled [30].

Temperature Dependence of Growth Rate

The Arrhenius type of dependence of the reaction rate parameter k_2 for second-order crystal growth reactions is written as

$$\ln k_2 = \ln A - E_a/RT \tag{57}$$

where A is a pre-exponential factor, E_a is the activation energy for new growth, and R is the gas constant.

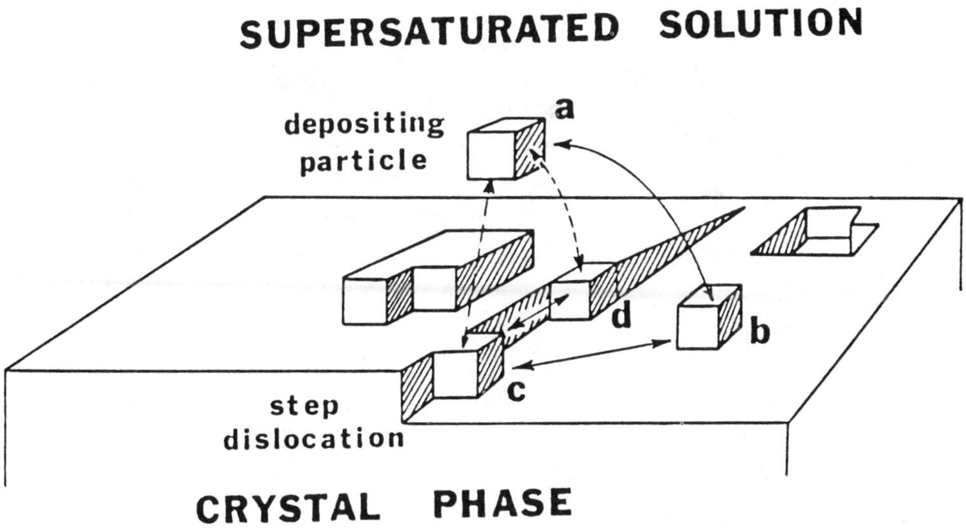

Dissolved species (a) attaches to the crystal surface (b), and migrates by a process of surface diffusion to the face of crystal step (c). At c, the added material is bonded stronger to the crystal than at b, owing to a greater portion of its surface being in contact with the crystal faces. Diffusion from c to d stabilizes the added material further (greater area of contact), enabling it to become incorporated in the crystal lattice.

Source: Nancollas and Reddy [30]. *(Copyright 1974, Ann Arbor Science Publishers. Reprinted with permission from Butterworth Publishers, current holders of the copyright.)*

FIGURE 3.4-7 Schematic Diagram of Crystal Growth in Solution

The values of the activation energy listed in Table 3.4-13 may be compared with the value of $E_a \approx 20$ kJ/mol expected for diffusion-controlled reactions. The values for AgCl and $BaSO_4$ are similar to those for diffusion-controlled growth, but it is not a proof of a diffusion-controlled reaction. Other values are substantially below or above 20 kJ/mol.

Nucleation of Calcite and Inhibition of Growth

In the section on dissolution of calcite ($CaCO_3$), it was mentioned that inhibition of dissolution has been observed for a number of major and minor constituents of natural waters. Inhibition of the oxidation of Fe(II) in water has also been cited in §6.8.

Inhibition of crystal growth is a process of considerable importance in water treatment, in chemical and pharmaceutical industries, and in natural water environ-

TABLE 3.4-12

Order of Precipitation and Crystal Growth Reactions for Equation 85

Solid Phase	Reaction Order m
AgCl	2
$AgIO_3$	2
$BaSO_4$	2
$SrSO_4$	2
$PbSO_4$	2
$CaSO_4 \cdot 2H_2O$	2
$CaC_2O_4 \cdot H_2O$	2
$MgC_2O_4 \cdot 2H_2O$	2
Ag_2CrO_4	3
$Mg(OH)_2$	1
$SrC_2O_4 \cdot H_2O$	1
$CdC_2O_4 \cdot 3H_2O$	1

Sources: References 16, 20,22, 26, 29, 32, 43

TABLE 3.4-13

Activation Energies for Heterogeneous Nucleation in Aqueous Solutions

Solid Phase	E_a ($kJ \cdot mol^{-1}$)	Source
$MgC_2O_4 \cdot 2H_2O$	12.5	[22]
AgCl	20.9	[20]
$BaSO_4$	20.9 ± 2.1	[32]
$CaHPO_4 \cdot 2H_2O$	43.9	[30]
$CaC_2O_4 \cdot H_2O$	49.0 ± 4.2	[29]
$CaSO_4 \cdot 2H_2O$	62.8 ± 2.1	[26]

ments. A systematic study of the inhibition of crystal growth of calcite, magnesium hydroxide, calcium sulfate dihydrate, and dicalcium phosphate dihydrate has been done by Nancollas and Reddy [30], whose data on calcite are given in Table 3.4-14.

The values of k_2 in Table 3.4-14 apply to the reaction rate parameter in the second-order reaction 85. Concentration in solution, C, as a function of time is given by an integrated form of equation 85:

$$\frac{1}{C_s - C} - \frac{1}{C_s - C_0} = k_2 S t \tag{86}$$

where C_0 is the concentration of calcium at the start of the experimental measurements. In equation 86, the surface area of the growing crystals is assumed to be constant. The accuracy of this assumption can generally be either confirmed or disproved only by measurements of the surface area before and after the experiment. Concentrations of calcium (C in equation 86) obtained by measurement in the course of crystal growth over a period of about one hour are used to compute the left-hand side of equation 86, when the initial and equilibrium concentrations must also be known. A plot of the values of the left-hand side of equation 86 against time produces a straight line, indicating that the second-order reaction is obeyed.

TABLE 3.4-14

Effect of Inhibiting Additives on Rates of Calcite Crystallization in a Supersaturated Solution at 25°C

Precipitation Rate Constant,[a] k_s ($cm^6 \cdot mol^{-1} \cdot cm^{-2} \cdot s^{-1}$)	Additive Concentration mg/l	Additive Concentration μ mol/l	Additive[b]
1.62 ± 0.35	0	0	—
1.10	0.05	0.12	ENTMP
0.70	0.08	0.18	ENTMP
0.28	0.12	0.28	ENTMP
0.15	0.50	1.2	ENTMP
0.12	2.5	5.8	ENTMP
0.028	0.5	1.0	TENTMP
0.009	2.5	5.1	TENTMP

a. $k_s = k_2 V/S$, where k_2 is the second-order reaction rate constant ($cm^3 \cdot mol^{-1} \cdot s^{-1}$), V is the solution volume (cm^3), and S is the solid surface area (cm^2).

b. ENTMP = N,N,N′,N′-ethylenediaminetetra (methylenephosphonic) acid

TENTMP = N,N,N′,N′-triethylenetetraaminetetra (methylenephosphonic) acid

Source: Nancollas and Reddy [30]

The inhibiting effect of the phosphonate ions, as shown in Table 3.4-14, is noticeable at low concentrations, of the order of 10^{-7} to 10^{-6} mol/l, in the presence of much higher calcium concentration in the supersaturated solution, of the order of 10^{-4} mol/l. A decrease in the growth rate by a factor of ten is accomplished by a tenfold increase in concentration of the inhibiting agent, in a low concentration range.

KINETICS OF RECRYSTALLIZATION

In the early part of this section, several types of recrystallization and precipitate aging processes were mentioned. For one type of recrystallization — dehydration and hydration reactions of solids — the data base seems to be more developed than for other recrystallization reactions; for this reason, the kinetics of dehydration and hydration are presented here, along with some case studies.

The Nature and Driving Forces of Dehydration

Dehydration is the removal of water molecules that form part of a crystal lattice of a solid. As such, the process clearly differs from drying or desiccation, which is the removal of water that is adsorbed on the surfaces of solids or that fills the capillary and interstitial spaces within solid materials.

A dehydration reaction can be either complete or partial, depending on whether an anhydrous phase or a hydrate with fewer water molecules in the structure is the reaction product. A general reaction balance is

$$A \cdot nH_2O(s) \longrightarrow A(s) + nH_2O\ (g\ \text{or}\ l) \tag{87}$$

$$A \cdot nH_2O(s) \longrightarrow A \cdot (n\text{-}m)H_2O(s) + mH_2O\ (g\ \text{or}\ l) \tag{88}$$

Decomposition of solids into other solids and evolving gases is a chemical problem of considerable industrial and technological concern, particularly in processes at elevated temperatures, where thermal stresses become the main driving forces of decomposition reactions.

In reactions 87 and 88, the partial pressure of water vapor or its activity (that is, relative humidity of the atmosphere) is one of the factors that determine whether dehydration will take place at the given temperature and total pressure. (For more details on the activity of water as a driving force in such reactions, see section 2.7.)

Occurrences of Hydration and Dehydration Reactions

Clay minerals and sheet-layer structured silicates of the mica group contain hydroxyls (OH) in their lattices. At elevated temperatures (see Table 3.4-16), dehydroxylation reactions take place, resulting in evolution of H_2O from the minerals. Among the hydroxides and oxy-hydroxides of simpler stoichiometric composition, dehydration or dehydroxylation reactions can be represented by the following sequences:

$$Mg(OH)_2 \longrightarrow MgO$$
$$Fe(OH)_3 \longrightarrow FeOOH \longrightarrow Fe_2O_3$$
$$Al(OH)_3 \longrightarrow AlOOH \underset{\text{high-T process}}{\longrightarrow} Al_2O_3$$

Hydrated minerals, in which H_2O forms part of the crystal structure, abound in salt deposits [6]. They are formed during the evaporation of ocean and lake waters and are sometimes preserved in the subsurface, which can be mined for rocksalt and various K-, Mg- and Ca-containing minerals.

The following list gives the chemical compositions and mineral names for some hydrated phases and their anhydrous counterparts. (Those in parentheses are either rare or do not occur in nature.)

$CaSO_4 \cdot 2H_2O$	gypsum
($CaSO_4 \cdot 1/2\,H_2O$)	hemihydrate
$CaSO_4$	anhydrite
$MgSO_4 \cdot 7H_2O$	epsomite
$MgSO_4 \cdot 6H_2O$	hexahydrite
$MgSO_4 \cdot 5H_2O$	pentahydrite
($MgSO_4 \cdot 4H_2O$)	
($MgSO_4 \cdot 2H_2O$)	
$MgSO_4 \cdot H_2O$	kieserite
$CaHPO_4 \cdot 2H_2O$	brushite
$CaHPO_4$	monetite
$CaCO_3$	calcite, aragonite
($CaCO_3 \cdot 6H_2O$)	
$Na_2SO_4 \cdot 10H_2O$	mirabilite
Na_2SO_4	thenardite
$NaCl \cdot 2H_2O$	hydrohalite
$NaCl$	halite

Under appropriate conditions, hydrated and anhydrous phases can form independently; they are not necessarily produced by the hydration or dehydration of each other.

Hydrous sulfates of heavy metals, such as $CuSO_4 \cdot 5H_2O$, and those of Ni, Zn, and Co have often been used in laboratory studies of the rates and mechanisms of dehydration reactions. Their occurrences in nature suggest that they are formed by the oxidation of heavy-metal sulfides that are exposed to the surface or near-surface environment.

Dehydration reactions in nature often take place in saline brines, where the activity of water and, hence, its partial pressure in the vapor phase are low. A classic example of such a dehydration reaction is the transformation of gypsum (calcium sulfate dihydrate) to anhydrite (anhydrous calcium sulfate).

Mechanisms and Rates of Dehydration

During dehydration, water molecules escape from the surface of the solid and subsequently from the deeper lattice sites. There are three basic mechanisms of dehydration:

- In **nucleation**, nuclei of the anhydrous phase form according to a pattern that may vary from one solid to another and may also depend on the conditions of the experiment. Instances of spontaneous one-, two-, and three-dimensional nucleation and nucleation along certain preferred crystallographic directions have been reported.
- In **dehydration by a boundary-controlled reaction**, a chemical reaction takes place at a boundary between the hydrated and anhydrous phases. In the course of the reaction, the boundary moves deeper into the solid, as a front of an advancing anhydrous phase forming at the expense of the hydrated solid.
- In a **diffusion-controlled process**, usually associated with dehydration at higher temperatures, the progress of the reaction can be described by a parabolic kinetic law (see equations 40 and 41). Here again, diffusion in one, two or three dimensions, and either isotropic or anisotropic diffusion, may be responsible for different interpretations of the reaction mechanism and mathematically different formulations of the dehydration rates.

For determining the rates of dehydration, the concentration of a dehydrated phase in a volume containing both the hydrated reactant and dehydrated product is not a convenient parameter; instead, the fraction of the hydrated phase reacted, denoted α, is conventionally used in the mathematical formulations of dehydration reactions. For example, the parabolic rate law for a diffusion-controlled dehydration reaction is, in the notation of a reacted fraction:

$$\frac{d\alpha}{dt} = \frac{k'}{2} t^{-1/2} \tag{89}$$

and the integrated form of the kinetic equation is

$$\alpha^2 = kt \tag{90}$$

with $k = (k')^2$.

Several rate equations, with explanatory notes, are given in Table 3.4-15 for different mechanisms of dehydration reactions. Figure 3.4-8 shows the progress of dehydration reactions with time for four dehydration mechanisms.

At room temperatures and atmospheric pressure, even in dry air, dehydration reactions are very slow. Therefore, dehydration rates are commonly measured under partial vacuum and elevated temperatures. The progress of the reaction can be observed in the solid and by monitoring the pressure of the water vapor formed.

TABLE 3.4-15

Rate Equations for Dehydration Reactions

For nucleation and boundary-reaction control:

Avrami-Erofeëv equations:

a. $[-\ln(1-\alpha)]^{1/2} = kt$

b. $[-\ln(1-\alpha)]^{1/3} = kt$

c. $[-\ln(1-\alpha)]^{1/4} = kt$

d. Contracting area: $1-(1-\alpha)^{1/2} = kt$

e. Contracting volume: $1-(1-\alpha)^{1/3} = kt$

For diffusion control:

f. One-dimensional (parabolic) diffusion: $\alpha^2 = kt$

g. Two-dimensional diffusion: $(1-\alpha)\ln(1-\alpha) + \alpha = kt$

h. Three-dimensional diffusion: $[1-(1-\alpha)^{1/3}]^2 = kt$

Explanation of symbols:

α = volume fraction of hydrate phase reacted

t = time

k = reaction rate parameter (differs for each equation)

Source: Brown *et al.* [7]

The rate of dehydration reactions, in a temperature range from about 40°C to 120°C, is of the order of 10^{-7} to 10^{-5} cm $\cdot$ s^{-1}. This represents the rate of advance of a dehydrated phase replacing the hydrated reactant. The two orders of magnitude range corresponds to the difference between 1 cm per day and 1 cm per year.

In general, it is doubtful that rates of dehydration reactions obtained at elevated temperatures in near vacuum (pressures of water vapor in the range of $P_{H_2O} < 1$ torr to over 100 torr, or from less than 0.001 atm to over 0.1 atm) are applicable to dehydration processes in the inorganic or organic environment.

Dehydration of $CaSO_4$-Hydrates

Reactions in the system $CaSO_4$ — H_2O (vapor) involve three hydrated solids and two anhydrous products: $CaSO_4 \cdot 2H_2O$ (gypsum), α-$CaSO_4 \cdot 1/2\ H_2O$, β-$CaSO_4 \cdot 1/2\ H_2O$, γ-$CaSO_4$ (known as soluble anhydrite), and β-$CaSO_4$ (anhydrite). Studies of the dehydration reactions in this system [2-4] have shown the following:

Dehydration of $CaSO_4 \cdot 2H_2O$ proceeds directly to the anhydrous phase when water-vapor pressure is very low, $P_{H_2O} = 10^{-5}$ torr (1 torr = 1 mm Hg = 1/760 atm). At higher water-vapor pressures, dihydrate goes to hemihydrate, which then dehydrates further to the anhydrous phase.

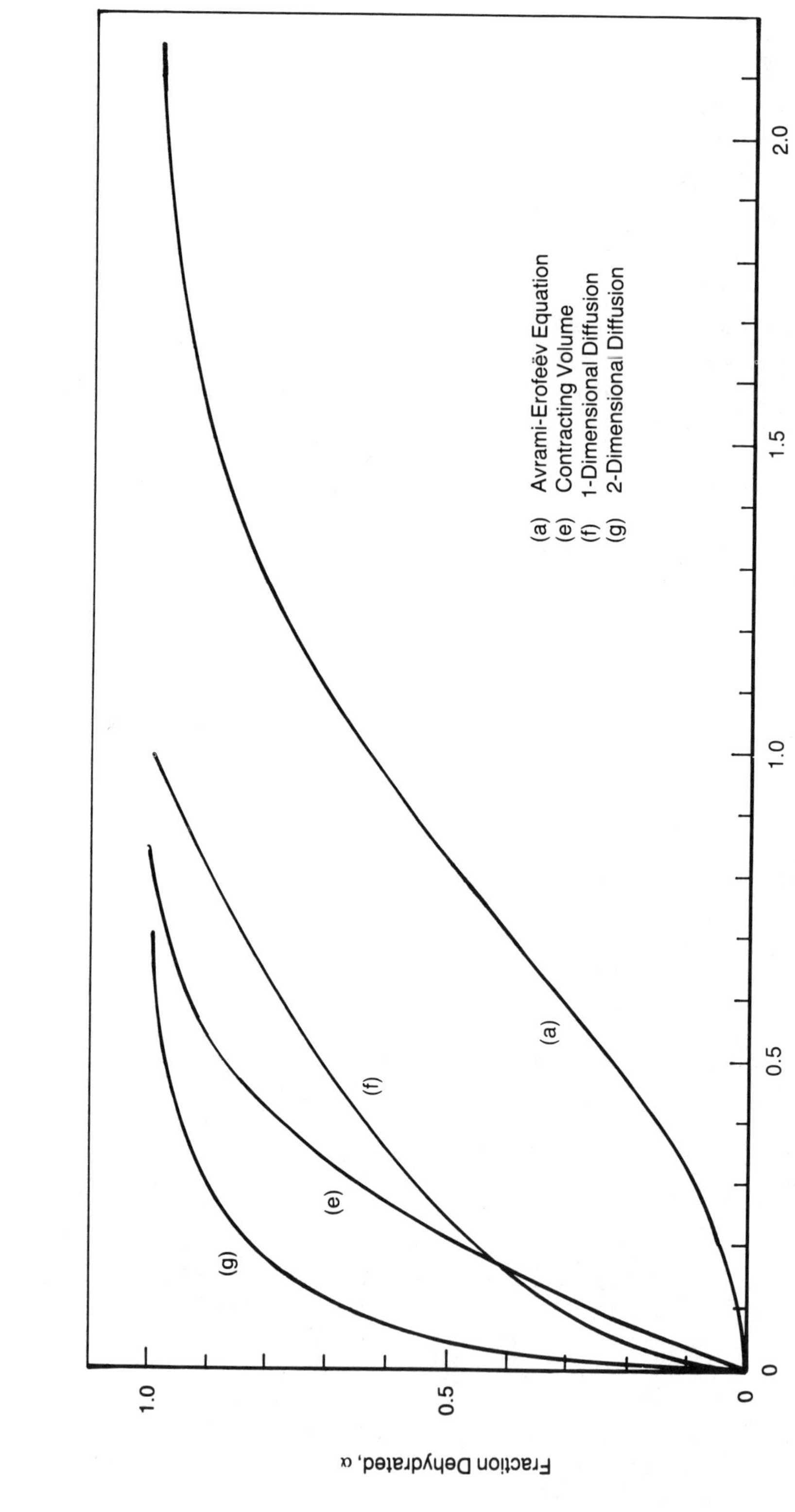

FIGURE 3.4-8 Progress of Dehydration Reactions Computed from Equations a, e, f, and g in Table 3.4-15

Below 110°C, dehydration reactions in $CaSO_4 \cdot 2H_2O$ and in $CaSO_4 \cdot 1/2\ H_2O$ are controlled by a mixed mechanism of nucleation and boundary reaction. Above 110°C, the mechanism is diffusion-controlled. The activation energy of dehydration shows a dependence on the water-vapor pressure; in general, activation energies are higher for a nucleation-controlled dehydration and lower for a diffusion-controlled stage. The values of the activation energy and the rate constants for the different reactions in the calcium sulfate — water vapor system are summarized in Table 3.4-16.

Other Hydrates

There are numerous reports in the literature on the rates of dehydration of copper sulfate hydrates, $CuSO_4 \cdot 5H_2O$ and $CuSO_4 \cdot 3H_2O$. Some of the data on these reactions are shown in Table 3.4-16.

A transformation reaction for calcium oxalate trihydrate to a monohydrate in solution has been studied by Gardner [16]:

$$CaC_2O_4 \cdot 3H_2O\ (s) \longrightarrow CaC_2O_4 \cdot H_2O\ (s) + H_2O\ (l)$$

Although this can be formally considered a dehydration reaction, it involves dissolution of one phase and nucleation of another in solution (see "Nucleation and Crystal Growth" in §3.4.4). The same can be said of the dehydration reaction of $CaSO_4 \cdot 2H_2O$ in a saline brine: the crystal structures of gypsum and anhydrite are different; removal of H_2O from a hydrate phase probably involves dissolution of the reactant, and nucleation and precipitation of the anhydrous product.

TABLE 3.4-16

Kinetic Parameters for Some Dehydration Reactions of the Type *Solid 1* → *Solid 2 + Water Vapor*

Dehydration Reaction	Temperature Range (°C)	Water Vapor Pressure (torr or mm Hg)	E_a (kJ/mol)	log A	$k = A\exp(-E_a/RT)$ (s^{-1}, unless otherwise noted)	Source
$CuSO_4 \cdot 5H_2O \rightarrow CuSO_4 \cdot 3H_2O + 2H_2O$	47-63		104±10	12	$10^{12}\exp(-1.25 \times 10^4/T)$	[31]
$CuSO_4 \cdot 3H_2O \rightarrow CuSO_4 \cdot H_2O + 2H_2O$	70.5-86		134±3	15.5	$10^{15.5}\exp(-1.61 \times 10^4/T)$	[31]
$BaCl_2 \cdot 2H_2O \rightarrow BaCl_2 + 2H_2O$	40-43		142±8		0.5×10^{-6} to 32×10^{-6} $cm \cdot s^{-1}$, for different crystal faces	[34]
$CaSO_4 \cdot 2H_2O \rightarrow CaSO_4 \cdot ½H_2O + 1½H_2O$	80-152					[2]
	<110	4.6	247 }		Range from 1.6×10^{-3}	[2]
	<110	17	146 }		to 1.6×10^{-4}	[2]
	<110		92[a]			[2]
	>110		46±4[b]			[2]
α-$CaSO_4 \cdot ½H_2O \rightarrow \gamma$-$CaSO_4 + ½H_2O$	68-270	10^{-5}	65.3	7.3	$10^{7.3}\exp(-7854/T)$	[4]
		44	151	17.7	$10^{17.7}\exp(-1.82 \times 10^4/T)$	[4]
β-$CaSO_4 \cdot ½H_2O \rightarrow \gamma$-$CaSO_4 + ½H_2O$	115-140	10^{-5}	26.6		3.5×10^{-4} (388°K)	[3]
		24	84.5		1.1×10^{-4} (388°K)	[3]

a. Boundary control
b. Diffusion control

3.4.5 Literature Cited

1. Bales, R.C. and J.J. Morgan, "Dissolution Kinetics of Chrysotile at pH 7 to 10," *Geochim. Cosmochim. Acta,* **49**, 2281 (1985).
2. Ball, M.C. and L.S. Norwood, "Studies in the System Calcium Sulphate—Water. Part I: Kinetics of Dehydration of Calcium Sulphate Dihydrate," *J. Chem. Soc. A,* 1633-37 (1969).
3. Ball, M.C. and L.S. Norwood, "Studies in the System Calcium Sulphate—Water. Part II: The Kinetics of Dehydration of β-$CaSO_4 \cdot 1/2H_2O$," *J. Chem. Soc. A,* 528-30 (1970).
4. Ball, M.C. and L.S. Norwood, "Studies in the System Calcium Sulphate—Water. Part III: Kinetics of Dehydration of α-Calcium Sulphate Hemihydrate," *J. Chem. Soc. A,* 1476-79 (1970).
5. Berner, R.A., "Kinetics of Weathering and Diagenesis," in Lasaga, A.C. and R.J. Kirkpatrick (eds.), "Kinetics of Chemical Processes," *Rev. Mineral.,* **8,** 111-34 (1981).
6. Braitsch, O., *Salt Deposits,* Springer-Verlag, New York (1971).
7. Brown, M.E., D. Dollimore and A.K. Galwey, "Reactions in the Solid State," in Bamford, C.H. and C.F.H. Tipper (eds.), *Comprehensive Chemical Kinetics,* Vol. 22, Elsevier, New York (1980).
8. Busenberg, E. and C.V. Clemency, "The Dissolution Kinetics of Feldspars at 25°C and 1 atm CO_2 Partial Pressure," *Geochim. Cosmochim. Acta,* **40,** 41-49 (1976).
9. Davison, W. and G. Seed, "The Kinetics of the Oxidation of Ferrous Iron in Synthetic and Natural Waters," *Geochim. Cosmochim. Acta,* **47,** 67-80 (1983).
10. Drever, J.I., *The Geochemistry of Natural Waters,* Prentice-Hall, Englewood Cliffs, N.J. (1982).
11. Eugster, H.P., "Hydrous Sodium Silicates from Lake Magadi, Kenya: Precursors of Bedded Chert," *Science,* **157,** 1177-80 (1967).
12. Eugster, H.P. and L.A. Hardie, "Saline Lakes," in Lerman, A. (ed.), *Lakes: Chemistry, Geology, Physics,"* Springer-Verlag, New York, 237-93 (1978).
13. Eyring, H. and E.M. Eyring, *Modern Chemical Kinetics,* Chapman and Hall, London (1963).
14. Fleer, V.M., *The Dissolution Kinetics of Anorthite ($CaAl_2Si_2O_8$) and Synthetic Strontium Feldspar ($SrAl_2Si_2O_8$) in Aqueous Solution at Temperatures Below 100°C,* Ph.D. Dissertation, Pennsylvania State University, University Park, Pa. (1982).
15. Frost, A.A. and R.G. Pearson, *Kinetics and Mechanism,* John Wiley & Sons, New York (1961).
16. Gardner, G.L., "Kinetics of the Dehydration of Calcium Oxalate Trihydrate Crystals in Aqueous Solution," *J. Colloid Interface Sci.,* **54,** 298-310 (1976).
17. Garrels, R.M. and F.T. Mackenzie, *Evolution of Sedimentary Rocks,* Norton, New York (1971).

18. Ghosh, M.M., "Oxygenation of Ferrous Iron (II) in Highly Buffered Waters," in Rubin, A.J. (ed.), *Aqueous Chemistry of Metals,* Ann Arbor Science Publishers, Ann Arbor, Mich. (1974), p. 193.

19. Holland, H.D., *The Chemistry of the Atmosphere and Oceans,* John Wiley & Sons, New York (1978).

20. Howard, J.R., G.H. Nancollas and N. Purdie, "The Precipitation of Silver Chloride from Aqueous Solutions. Part 6: Kinetics of Dissolution of Seed Crystals," *Trans. Faraday Soc.,* **56,** 278-83 (1960).

21. Lerman, A., *Geochemical Processes — Water and Sediment Environments,* John Wiley & Sons, New York (1979).

22. Lichstein, B. and F. Brescia, "The Mechanism of Precipitation of Magnesium Oxalate from Supersaturated Solutions. II: The Heat and Entropy of Activation," *J. Am. Chem. Soc.,* **79,** 1591-92 (1957).

23. Lin, F.-C. and C.V. Clemency, "The Dissolution Kinetics of Brucite, Antigorite, Talc, and Phlogopite at Room Temperature and Pressure," *Am. Mineral.,* **66,** 801 (1981).

24. Lin, F.-C. and C.V. Clemency, "The Kinetics of Dissolution of Muscovites at 25°C and 1 atm CO_2 Partial Pressure," *Geochim. Cosmochim. Acta,* **45,** 571 (1981).

25. Lin, F.-C. and C.V. Clemency, "Dissolution Kinetics of Phlogopite in a Closed System," *Clays Clay Miner.,* **29,** 101 (1981).

26. Liu, S.-T. and G.H. Nancollas, "The Kinetics of Crystal Growth of Calcium Sulfate Dihydrate," *J. Cryst. Growth,* **6,** 281-89 (1970).

27. Luce, R.W., R.W. Bartlett and G.A. Parks, "Dissolution Kinetics of Magnesium Silicates," *Geochim. Cosmochim. Acta,* **36,** 35-50 (1972).

28. Meybeck, M., "Concentrations des Eaux Fluviales en Éléments Majeurs et Apports en Solution aux Océans," *Rev. Geol. Dyn. Geogr. Phys.,* **21,** 215-46 (1979).

29. Nancollas, G.H. and G.L. Gardner, "Kinetics of Crystal Growth of Calcium Oxalate Monohydrate," *J. Cryst. Growth,* **21,** 267-76 (1974).

30. Nancollas, G.H. and M.M. Reddy, "Crystal Growth Kinetics of Minerals Encountered in Water Treatment Processes," in Rubin, A.J. (ed.), *Aqueous-Environmental Chemistry of Metals*, Ann Arbor Science Publishers, 219-53 (1974).

31. Ng, W.-L., C.-C. Ho and S.-K. Ng, "Isothermal Dehydration of Copper Sulphate Pentahydrate and Trihydrate," *J. Inorg. Nucl. Chem.,* **34,** 459-62 (1978).

32. Nielsen, A.E., "The Kinetics of Crystal Growth in Barium Sulfate Precipitation. II: Temperature Dependence and Mechanism," *Acta Chem. Scand.,* **13,** 784-802 (1959).

33. Nielsen, A.E., *Kinetics of Precipitation*, Macmillan, New York (1964).

34. Osterheld, R.K. and P.R. Bloom, "Dehydration Kinetics for Barium Chloride Dihydrate Single Crystals," *J. Phys. Chem.,* **82,** 1591-96 (1978).

35. Plummer, L.N. and T.M.L. Wigley, "The Dissolution of Calcite in Co_2-Saturated Solutions at 25°C and 1 Atmosphere Total Pressure," *Geochim. Cosmochim. Acta,* **40,** 191-202 (1976).

36. Rickard, D. and E.L. Sjöberg, "Mixed Kinetic Control of Calcite Dissolution Rates," *Am. J. Sci.,* **283,** 815 (1983).

37. Rimstidt, J.D. and H.L. Barnes, "The Kinetics of Silica-Water Reactions," *Geochim. Cosmochim. Acta,* **44,** 1683-1700 (1980).

38. Robie, R.A., B.S. Hemingway and J.R. Fisher, *Thermodynamic Properties of Minerals and Related Substances at 298.15 K and 1 bar (10^5 Pascals) Pressure and at Higher Temperatures,* Bulletin 1452, U.S. Geological Survey, 1-456 (1978).

39. Stumm, W. and J.J. Morgan, *Aquatic Chemistry,* John Wiley & Sons, New York (1981).

40. Sung, W. and J.J. Morgan, "Kinetics and Product of Ferrous Iron Oxygenation in Aqueous Systems," *Environ. Sci. Technol.,* **14,** 561-68 (1980).

41. Tole, M.P., A.C. Lasaga, C. Pantano and W.B. White, "Factors Controlling the Kinetics of Nepheline Dissolution," *Geochim. Cosmochim. Acta,* **50,** 379-92 (1986).

42. Van Name, R.G., "The Velocity of Dissolving of Crystals in Liquids," in *International Critical Tables,* Vol. 5, McGraw-Hill Book Co., New York, 55-60 (1929).

43. Walton, A.G., *The Formation and Properties of Precipitates,* John Wiley & Sons, New York (1967).

44. White, A.F. and A. Yee, "Aqueous Oxidation-Reduction Kinetics Associated with Coupled Electron-Cation Transfer from Iron-Containing Silicates at 25°C," *Geochim. Cosmochim. Acta,* **49,** 1263-76 (1985).

45. Wollast, R. and L. Chou, "Kinetic Study of the Dissolution of Albite with a Continuous Flow-through Fluidized Bed Reactor," in Drever, J.I. (ed.), *The Chemistry of Weathering,* Reidel Publishing Company, Boston, 75-96 (1985).

46. Wood, B.J. and J.V. Walther, "Rates of Hydrothermal Reactions," *Science,* **222,** 413-15 (1983).

4. UPTAKE BY BIOTA

Sara E. Bysshe

LIST OF TABLES

BIOLOGICAL UPTAKE PROCESSES AND PROBLEMS

UPTAKE BY TERRESTRIAL VEGETATION

UPTAKE BY MARINE BIOTA

UPTAKE BY FRESHWATER BIOTA

LIST OF FIGURES

BIOLOGICAL UPTAKE PROCESSES AND PROBLEMS

UPTAKE BY TERRESTRIAL VEGETATION

UPTAKE BY MARINE BIOTA

UPTAKE BY FRESHWATER BIOTA

4.1 INTRODUCTION

Investigation of inorganic species in the environment often leads to questions concerning uptake by biological organisms and the potential for accumulation. Concern over accumulation derives from the historical consequences of excessive levels of inorganics such as lead, cadmium and mercury in environmental and food sources, which have led to debilitating and even lethal effects in humans [32, 42, 47].[1]

This chapter is designed to provide an overview of biological uptake of inorganics. It is intended to assist a reader facing a potential or uncertain uptake problem, or a reader seeking an introductory understanding of the subject. The discussion is divided between (a) general biological uptake processes and problems, and (b) issues specifically associated with uptake phenomena in terrestrial, marine and freshwater environments.

The first sections are intended to provide the reader with several key steps that should be followed to establish whether or not specific circumstances will result in risk due to uptake of particular inorganics. In this context, the "Decision Process" presented in § 4.2.4 may help the reader choose an appropriate focus for more detailed review and use of the scientific literature, or it may help him to structure an appropriately inclusive field measurement or laboratory experimental design for investigation of a potential uptake problem.

The latter portions of the chapter explore combinations of circumstances and environments that are likely to result in increases in availability and subsequent uptake of inorganics that could ultimately harm the environment or human consumers. The circumstances emphasized are those in the various environmental compartments that are presently understood to be at potential risk from inorganic uptake. Most of the discussion concerns uptake of trace metals, because of the inherent hazard represented by their toxicity and the fact that they have been the subject of much research. The chapter focuses on agricultural species in the terrestrial environment and on finfish and shellfish in the aquatic environment, again because so much of the concern and research has been in these areas.

Most of the experimental data reviewed for this chapter came from earlier summaries on the subject of uptake. The objective here is to indicate the range of uptake values observed, not to provide a comprehensive collection of data for the elements cited. Since a review of all of the most recently reported uptake values, summarized in this manner, would not provide a better answer for a specific problem in a specific environment, no attempt has been made to provide detailed information of this kind. Several of the references listed at the end of this chapter will be helpful to readers who wish to examine the data and further understand key processes.

1. All references cited in this chapter are grouped in section 4.6.

4.2 BIOLOGICAL UPTAKE PROCESSES AND PROBLEMS

4.2.1 General Description of Uptake Processes

In the most general sense, the uptake of inorganic constituents by plants and animals, both terrestrial and aquatic, is a part of natural cycling processes. Living organisms play a major role in the cycling of elements such as C, O, H and N but a more minor or even incidental role with respect to many other elements. Figure 4.2-1 portrays some of these cycling processes. It can generally be assumed that uptake from soil solution through roots is the predominant uptake pathway for plants. Uptake of gaseous species from the atmosphere through the leaves is a major pathway for a few major constituents (CO_2, O_2, etc.) and may be a common but less widely investigated uptake pathway for other chemicals. A major exposure pathway for animals, especially aquatic ones, is that of chemicals in aqueous solution. Food chain transfer (eating other organisms) can also be an important route. Animals may take up substances through inhalation or by consuming non-food particles, but these are generally considered less significant ingestion routes for aquatic organisms. Ingestion of soil by livestock and humans can be a major transfer pathway for elements such as Pb, F, As, and Hg, among others.

Loss of inorganic constituents from tissues can result from shifts in the equilibria that hold those specific chemicals. Loss of inorganic constituents from living organisms must take place across respiratory membranes such as gills or plant surfaces, or across some other exchange surface that is part of a simple or complex organic system (e.g., digestive system, urinary system). One or many systems may be involved. The rate of loss appears to be quite variable, depending not only on the inorganic constituent involved but also probably other factors such as temperature and reproductive season.

Contributions of inorganics from industrial and agricultural activities have altered equilibria in some ecosystems. While new equilibria are ultimately established among compartments, changes in the availability of some constituents can create problems.

Figure 4.2-2 further illustrates the relationship between uptake and organism requirements. Many inorganics in the environment are required nutrients; examples include Cu, Zn, Cl, Na, and Fe, as well as O, C, N, and H. As the figure illustrates, biological productivity is diminished by deficiencies in these nutrients. On the other hand, even nutrients can be present in such excessive concentrations as to cause chronic toxicity and ultimately death. Some inorganics (e.g., Hg, Cd and Pb) are not essential for growth; at sufficiently low levels, these are not believed to be harmful, but at higher levels they can have toxic effects. Thus, uptake processes contribute directly to both normal life-cycle requirements and possible toxicity.

The term "bioconcentration" or "bioaccumulation" refers to the degree to which an organism accumulates a specific chemical from the environment. Many researchers limit the use of "bioconcentration" to uptake from water, and use "bioaccumulation"

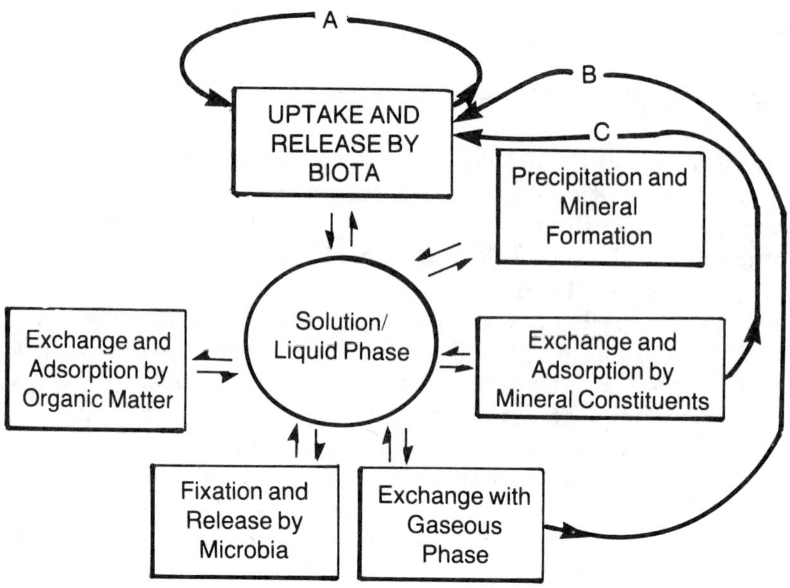

A. Consumption of plants or animals that have concentrated inorganic species.

B. Uptake through inhalation.

C. Uptake through mouthing ingestion of non-food particles.

Source: Adapted from Kabata-Penias and Penias [20]. *(Copyright 1984, CRC Press, Inc. Reproduced with permission.)*

FIGURE 4.2-1 Summary of Cycling/Equilibria Processes Among Environmental Components Controlling Availability of Inorganic Substances for Biological Uptake

to refer to uptake from all possible sources. The terminology is not standardized, and numerous exceptions can be found in the literature. Bioaccumulation is usually not applied to macronutrients but to micronutrients and other substances that tend to accumulate from the environment to adverse levels. The "concentration factor," a further refinement of this terminology, may generally be defined as the degree (factor) to which an organism has concentrated a substance above environmental concentrations [32]:

$$\text{Concentration Factor} = \frac{\text{Concentration of Substance in Organism (wt/wt)}}{\text{Concentration of Substance in Soil or Water (wt/wt or wt/vol)}}$$

Organisms take up a substance at one rate and lose it at a different rate, usually a slower one. These rates are normally affected by the ambient environmental concentrations. Some time is required for a plateau or equilibrium concentration to be reached; at this point, concentrations in the organism are believed to be relatively constant for that environmental concentration and, by definition, uptake and loss

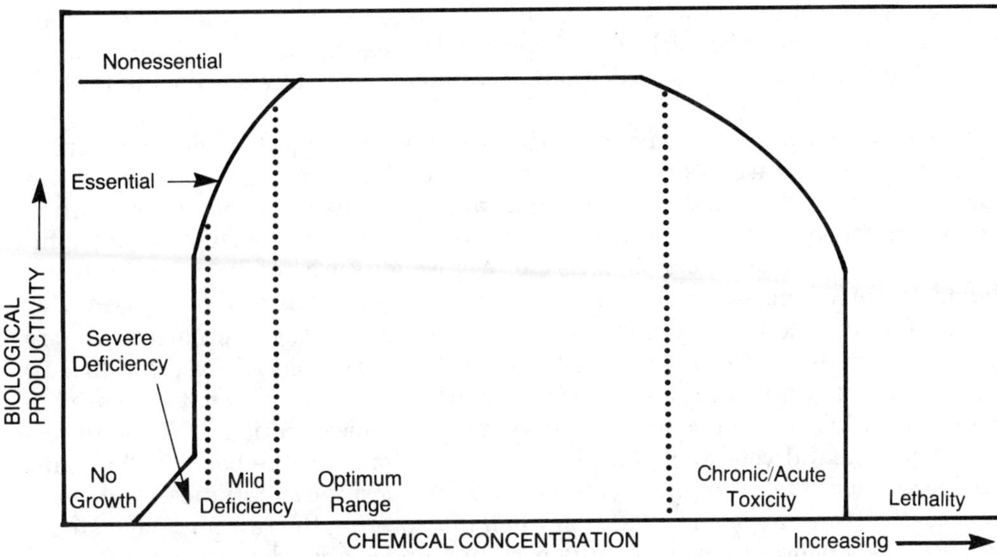

Source: Adapted from Overcash and Pal [28]

FIGURE 4.2-2 Productivity in Biological Populations or Organisms as a Function of Uptake of Essential and Nonessential Substances

rates are the same. It is at this so-called equilibrium concentration that measurements of concentration factor are considered most useful. For example, as of this writing a minimum of 28 days is considered necessary for equilibrium in laboratory-based tests to establish concentration factors in aquatic organisms [32]. In practice, as discussed below, the time required for equilibrium to be achieved has proven quite variable.

4.2.2 Methods for Investigation of Uptake

Uptake, accumulation and concentration factors are determined in various ways. For terrestrial plants, laboratory-type studies might include the addition of a measured concentration of a trace element from a soluble salt (e.g., copper sulfate, lead nitrate) to soil or other growth medium. Research has shown that results from soluble metal salts have no relation to results from tests using actual environmental media [10].

Thus, uptake studies have been developed where inorganics such as metals from contaminated soils, sludges, wastes, or as metal oxides, are introduced into the growth medium (e.g., soil), where test species are grown. These studies are considered to be more environmentally representative of exposure. Even though the latter may result in lower uptake and lower toxicity over the time of the test, and the chemical speciation may be only partially known, they are nearly always considered more realistic and thus more useful [30].

Tests may also be conducted with plants in greenhouse or field studies. The former approach permits greater control and is typically less expensive, but it also tends to result in higher uptake and toxicity than do comparable field experiments [30].

Plants are grown for a certain amount of time, with tests stopped at some particular growth stage. Concentrations of one or more forms of the added trace constituent are measured in plant parts (usually food chain tissues) or in the whole plant. Concentrations of a constituent may be measured and/or reported on a wet weight, dry weight, or ash weight basis of a specified plant tissue. Wet weights are highly irregular, being dependent on such factors as moisture content of the soil, and are not often used. Dry or ash weights can be converted to wet weights by use of the appropriate factors (moisture content, etc.). Wet weight/dry weight ratios vary from 1.1 for grain to 5-10 for roots and tubers, to 10-25 for leaf crops such as lettuce [10]. These data are used to calculate the ratios of plant concentration to medium concentration. Clearly, differences in experimental conditions (such as exposure time and the parts of the plant that are examined) may complicate comparison among different studies. And these differences illustrate the need for caution when interpreting the results from a single test, as well as when comparing results obtained under different test conditions.

The Army Corps of Engineers has developed a standard plant bioassay for uptake of soil toxic elements using yellow nutsedge *(Cyperus esculentus)*, *Spartina altenaflora*, or *Disticichilis spicata* in an experimental test chamber [22]. In such studies, plants are observed for phytotoxic effects, harvested for yield measurement after growth, and then plant material is analyzed for trace metal concentrations.

Aquatic tests are typically done in a tank with commonly used test species (such as *Daphnia* or bluegills). The inorganic chemical is added in a specified form to the water and is typically noted as the concentration of the ion. In such cases, the medium may be from a natural water source (ocean, lake, stream, etc.), or prepared in accordance with specified standards and free of contaminants. (Tests with marine organisms typically use filtered piped-in water.) In both cases, pH, salinity, or concentrations and characteristics of known dissolved ions of importance to uptake, such as carbonate, are usually identified. Laboratory-produced test media do not contain natural organics or natural sediments that might affect availability by forming complexes with test substances. Piped-in water from a natural source, by contrast, would not be free of dissolved organic matter.

The inorganics of concern for uptake may be added in a specific concentration, metered in at a constant rate and/or verified through monitoring for concentration or ionic activity. These different measuring procedures can make the exact levels of testing for a constituent quite variable, even when reported test concentrations are similar. Measurement and calculation of results for concentration factors are similar to the procedure described above for plants. Alternatively, field measurements can be made of uptake in naturally occurring or "caged" organisms in a contaminated aquatic environment, but the calculation of concentration factors is more difficult because exposure is not as well controlled.

The duration of the tests also varies, but regulatory preference (e.g., for a 28-day test or equilibrium conditions) helps to encourage standardization. Extended tests present difficulties, as even controls may be difficult to keep under confined conditions for 28 days or longer. Further, organisms already stressed by confined conditions may not survive (or respond typically to) experimentally altered conditions. Low control mortality is generally the only measurable indicator of success in this type of test.

Some tests require the examination of specific parts of test species if the substance in question preferentially accumulates in certain tissues. (For example, some trace elements tend to concentrate in plant roots or animal liver rather than in leaves or muscle tissue.) Examination of concentrations of an element in a whole organism thus yields an average value and may not be the optimum approach in a given circumstance. The optimum approach should be determined by the intended use of the data. For example, the average concentration may be sufficient if the entire organism is consumed, while specific tissue concentrations may be important if only those tissues are eaten. Either or both approaches may be appropriate for measuring uptake as long as care is taken to avoid inappropriate combination or comparison of results.

Field measurements of the concentrations of numerous elements in plants and animals in the various ecosystem types discussed here are increasingly common. As with laboratory tests, the results may be reported as concentrations per wet, dry, or ashed weight. The concentrations of a number of metals in agricultural crops and aquatic organisms used for human consumption are readily available from sources including: "Market Basket" studies conducted by the FDA that examine nutrient and toxic elements in food; more recent surveys by the USEPA and USDA designed to examine background levels of elements in selected crops [58,59]; a compilation of literature data on aquatic organisms uptake by Eisler [13]; a compilation of monitoring data for trace metals in food chain organisms by Jenkins [18]. Tables 4.2-1, -2, and -3 summarize some of the findings from Jenkins' review [18]. Data on uptake by plants from soil have been reviewed by Baes *et al.* [1]; some of their findings are presented in section 4.3.

However, ambient water/soil concentrations associated with the measured organism values are often unavailable, or the measurements may not be useful. For example, aquatic organisms are not stationary and have variable exposure histories. They may obtain a specific constituent from water, food, or sediments, or all three. Single or a few measurements of a constituent by grab sample at one location may or may not reflect important organism exposure concentrations. In addition to organism movement, seasonal or episodic variations in ambient water quality (e.g., storm flows, snow melt or lake stratification) can dominate the availability and uptake of inorganics.

4.2.3 Estimation of Uptake of Inorganic Chemicals

There are no generally applicable mathematical techniques for estimating the extent of biological concentrations of inorganics, even for combinations of chemicals and biological species known to result in bioaccumulation. This is primarily because the

TABLE 4.2-1

Typical Relative Bioaccumulation and Bioconcentration of Toxic Trace Metals

	Mammals, Birds and Fish	Molluscs, Crustacea, Lower Animals	Higher Plants	Mosses, Lichens and Algae
Antimony	x	x	x	x
Arsenic	xx	xxx	xxx	xx
Beryllium	x	x	x	x
Boron	x	xx	xxx	x
Cadmium	xxx	xxx	xxx	xxx
Chromium	xx	xxx	xxx	xxx
Cobalt	x	x	xxx	x
Copper	xxx	xxx	xxx	xxx
Lead	xxx	xxx	xxx	xxx
Mercury	xxx	xxx	x	x
Nickel	xx	xxx	xxx	xxx
Selenium	xx	x	xxx	x
Tin	xx	x	xxx	x
Vanadium	x	xxx	xx	xx

Bioaccumulation Scale:

x - low or limited
xx - moderate
xxx - high to very high

Source: Jenkins [18]

biological uptake of inorganics is entirely situation-specific, depending on combinations of many factors that affect environmental availability and fate within the organism. In contrast to organics, for example, no single "internal fate" factor such as lipid solubility can serve as the foundation for a quantitative estimation technique for inorganics as a general group. Also, as indicated in sections 4.2.2 above and 4.3 below, previous investigation in this area has been more helpful in defining how to conduct future studies than in providing a data base for mathematical estimation techniques. Partly as a result, laboratory uptake results and field observations often disagree.

It would be easier to calculate the potential uptake of inorganics if we knew what forms or chemical species of a substance were available for uptake by biological organisms. The lack of this knowledge reflects our limited ability to identify and then measure the concentrations of the biologically available forms. This problem is exacerbated by the many chemical reactions and environmental conditions that affect chemical forms and the equilibria among them. If the available and toxic forms of inorganics in various media could be accurately measured and compared with

TABLE 4.2-2

Relative Importance of Biomagnification and Biominification[a] in Food Chains

	Terrestrial			Aquatic		
	Plants	Herbivores	Carnivores	Plants	Herbivores	Carnivores
Antimony	−	+ −	−	B	+	M
Arsenic	M	+ −	M	B	+ 0 M	+ 0 M
Beryllium	+ −					
Boron	B −	−				
Cadmium	B M	B 0 M	+ 0 −	B M	B	B 0 M
Chromium	+ M	+ M	0	B		−
Cobalt	M	0		B	M	M
Copper	M	+ −		B −	M	0 M
Lead	B 0 M	B 0 M	0 M	B −	−	M
Mercury	+ M	+ 0	B 0 −	B 0	B 0	B 0
Nickel	B M	−		B	−	0 −
Selenium	B M	+ M		B		−
Tin	+ 0 M	−		+		
Vanadium	+ M	M		B	B*	

a. Biominification is defined by Jenkins [18] as the decrease in concentration of a trace element in organisms with increasing higher trophic levels in food chains.

b. Symbols used in Table are defined as follows:

- B — biomagnification reported two or more times
- \+ — biomagnification reported once
- M — biominification reported two or more times
- − — biominification reported once
- 0 — no change in trace metal level
- * — tunicates

Source: Jenkins [18]

TABLE 4.2-3

Typical Concentration Ratios of Trace Metals in Edible Tissues of Invertebrates and Fish ($\times 10^{-3}$)[a]

	Marine		Freshwater	
	Invertebrates	Fish	Invertebrates	Fish
Antimony	0.005	0.04	0.01	0.001
Arsenic	0.333	0.333	0.333	0.333
Beryllium	0.2	0.2	0.01	0.002
Cadmium	50.0	3.0	2.0	0.2
Chromium	2.0	0.4	0.04	0.04
Cobalt	1.0	0.1	0.2	0.02
Copper	1.67	0.667	1.0	0.2
Lead	1.0	0.3	0.1	0.3
Mercury	33.3	1.67	100.0	1.0
Nickel	0.25	0.1	0.1	0.1
Selenium	1.0	4.0	0.167	0.167
Tin	1.0	3.0	1.0	3.0
Vanadium	0.05	0.01	3.0	0.01

a. For example, for arsenic in fish, C(fish)/C(water) = 333.

Source: Vaughn *et al.* [56], as adapted by Jenkins [18].

observed uptake, reliable correlations and predictions could be made more readily. Research is also being directed toward verifying fundamental models for equilibria and soil and plant uptake processes in an indirect approach to this problem [10].

Much research effort has been focused on trace metals because of their observed toxicity. Nonetheless, direct measurement of available chemical forms has been hampered by the uncertainty as to which forms should be the basis for calculating uptake concentrations and the proper way to *measure* these forms. Even the measurement of soil properties known to have an effect on the availability of trace metals for uptake is complicated by the variety of approaches and standards used by different agencies [20]. For example, soil pH measurements can give different values depending on the test method used; typical differences between common methods — relative to the simple use of a 1:1 (by volume) soil-water mix — are: (1) use of 0.01 *M* calcium chloride, which gives a pH of about 0.5 units lower; and (2) use of 1 N KCl, which gives a pH of about 1.0 unit lower [10].

However, the standardization of tests for these properties would not permit consistent approximations of the chemical forms available to a plant or their reliable correlation with observed uptake. For example, even so-called direct measurements of a metal in soil do not provide such information; at least six extraction procedures are used to measure plant-available metals in the soil, in addition to measurements of the "total" amount of an element [28]. It is generally conceded that a specific method can give reliable information for a specific soil-plant system [20] and can therefore be used for monitoring at a specific site, but it cannot be used universally and may not be useful in predictions. Attempts are now being made to find more universally applicable approaches for measuring plant-available metals in the soil. (See § 4.3.2.)

In aquatic systems, the free metal ion and some hydrolyzed species are considered the ones available for uptake and toxicity [27]. According to recent versions of some of the EPA Water Quality Criteria documents, there are no ideal analytical methods for determining available toxic concentrations of many trace elements (e.g., lead) in water. This is largely due to the variety of forms in a water system and the lack of definitive information on the uptake and toxicity of each. The EPA's present approach for some metals is to measure the "active element," defined as the amount passing through a 0.45-μm membrane filter following acidification to pH 4 with nitric acid; like other extractions with a specified pH "end point," however, this method does not cover all potentially toxic forms. Methylmercury is a particularly important example [50]. Recent research (e.g., Cowan *et al.* [12]) is making use of modeled theoretical predictions of trace metal speciation in specific, measured water quality circumstances, in combination with examples of uptake under the latter circumstances, to develop correlations between specific predicted chemical species and observed uptake. This may result in a useful tool for approximating uptake potential, at least in controlled situations.

Bioavailability from sediments is not well understood from either a theoretical standpoint or the measurement of actual concentrations. Few states have developed criteria for sediments defining limits for inorganics that bioaccumulate. In California, criteria for determining the hazard of a waste of any material has been applied to contaminated sediments [39]. California (Table 4.2-4) defines the criteria threshold limits of listed inorganics as a combination of the measured water-soluble quantity and the amount available from a leaching test, in this case using sodium citrate. Bioavailability from sediments is expected to be inversely proportional to the sediment organic carbon content for several trace metals (e.g., Cu, Cd, Pb and possibly Hg) [19]; this is probably due to complexation of the metals by the acid functional groups on humic and fulvic acids in the organic matter. Redox reactions and sulfide activity can also affect availability, through the precipitation of a number of trace metal inorganics.

It is clear from the available data base that the uptake of some inorganics is likely to be more problematic than others with regard to:

- Concentration in the environment,
- Environmental bioavailability, and
- Toxicity to organisms in the food chain.

Two reviews have produced rankings of inorganics that have historically caused health problems when consumed by humans:

(1) Table 4.2-5 compares the toxicities of some key metals and their observed tendency to accumulate in aquatic organisms. These qualitative comparisons, which are based on the data reviewed and the apparent nature of uptake for individual metal species, provide a useful indication of risk. It should be cautioned that a lower hazard rating in no way implies that such chemical species may not be present at concentrations that could be harmful.

(2) Table 4.2-6 compares uptake levels in plants that are toxic to the plant with levels that are toxic to grazing animals. In some cases, because plants die first, accumulated metals are not passed along to consumers in high amounts, and the upper ends of the food chain are protected; in other cases, plants are more tolerant, and toxic levels can be transferred to consumers such as livestock.

While the above tables (including Tables 4.2-1, -2 and -3) are possible sources of information for assigning priority to inorganics of environmental concern, there are many site-specific exceptions. Thus, the inorganic environmental problem at a specific site can be reliably defined only through a knowledge of the uptake potential and hazard at that particular site. The following illustrate exceptions to the priorities that one might derive from the majority of evidence:

- Lead is not readily taken up by aquatic organisms except shellfish and is generally concentrated only in plant roots, yet lead poisoning has been implicated in the loss of endangered California condors due to the ingestion of lead bullets in prey [21]. A lead level of 4 mg/kg in waterfowl livers is used as a guideline by the U.S. Fish and Wildlife Service in considering restriction of use of lead shot in specific hunting areas [54]. High levels of lead in urban children have been blamed on soils contaminated by old, flaked paint and by automobile exhaust [47].

- High selenium levels in the Kesterson reservoir and associated wetlands in California have reportedly had adverse effects on wildlife and cattle. The circumstances include the release of selenium previously bound in seleniferous agricultural soils in an alkaline environment. Irrigation waters, which are drained in this area to avoid salt buildup, leach selenium from the soil. The irrigation drainage flows to the reservoir, where biological uptake has led to waterfowl mortalities and birth defects [39]. Human Se toxicity occurred in China when ash from seleniferous coal was used to fertilize cropland [60].

- The availability and subsequent uptake of heavy metals by plants from contaminated dredge material is believed to be controlled to a large extent by the degree of soil oxidation in combination with pH. Plant

TABLE 4.2-4
Criteria for Concentration of Persistent and Bioaccumulative Inorganic Toxic Substances in a Hazardous Waste or Material
(From Section 66699, Title 22 of California Administrative Code)

Substance[a]	STLC[b] (mg/l)	TTLC (wet weight mg/kg)
Antimony and/or antimony compounds	15	500
Arsenic and/or arsenic compounds	5.0	500
Asbestos[c]	—	1.0 (as %)
Barium and/or barium compounds (excluding barite)	100	10,000
Beryllium and/or beryllium compounds	0.75	75
Cadmium and/or cadmium compounds	1.0	100
Chromium (+6) compounds	5	500
Chromium and/or chromium (+3) compounds	560	2,500
Cobalt and/or cobalt compounds	80	8,000
Copper and/or copper compounds	25	2,500
Fluoride salts	180	18,000
Lead and/or lead compounds	5.0	1,000
Mercury and/or mercury compounds	0.2	20
Molybdenum and/or molybdenum compounds	350	3,500
Nickel and/or nickel compounds	20	2,000
Selenium and/or selenium compounds	1.0	100
Silver and/or silver compounds	5	500
Thallium and/or thallium compounds	7.0	700
Vanadium and/or vanadium compounds	24	2,400
Zinc and/or zinc compounds	250	5,000

STLC = soluble threshold limit concentration; TTLC = total threshold limit concentration

a. STLC and TTLC values are calculated on the concentrations of the elements, not the compounds. Values apply to asbestos and elemental metals only if they are in a friable, powdered, or finely divided state.

b. STLC includes both the water-soluble and the leachable quantities using sodium citrate.

c. Includes chrysotile, amosite, crocidolite, tremolite, anthophyllite and actinolite.

Source: California Administrative Code [9]

TABLE 4.2-5

Relative Hazard to Humans from Metals Accumulated in Edible Portions of Fish and Shellfish

Metal	Toxicity to Humans from Oral Ingestion		Bioaccumulative Tendency: Freshwater Fish Muscle		Bioaccumulative Tendency: Marine Fish Muscle		Bioaccumulative Tendency: Marine Shellfish or Crustaceans		Human Hazard Rating		
	low	high	low	high	low	high	low	high	low	med.	high
Aluminum	x			x	x		c		x		
Arsenic		x	x			x		x		x	
Beryllium	x		x		c		c		x		
Boron	x		x		x		x		x		
Cadmium		x	x		x			x		x	
Cesium[a]		x		x	x		x			x	
Chromium	x		x		x		x		x		
Cobalt[a]		x	x		x		x		x		
Copper	x		x		x			x	x		
Iron	x			x		x		x	x		
Lead		x	x		x			x		x	
Manganese	x		x		x		x		x		
Molybdenum	x		x		x		c		x		
Mercury[b]		x		x		x		x			x
Nickel	x		x		x		x		x		
Plutonium[a]		x	c		x		x		x		
Ruthenium[a]		x	x		x		x		x		
Selenium		x	x		x		x		x		
Silver	x		x		x			x		x	
Strontium[a]		x	x		x		x		x		
Zinc	x		x		x			x	x		

a. Radioisotope is primary form of concern.

b. Methylmercury is the species of concern.

c. Insufficient information available.

Source: Phillips and Russo [32]

TABLE 4.2-6

Examples of Criteria for Selected Trace Elements in Soils and Plants
(all values based on dry weight)

Element	Range of Soil Background Levels[a] (ppm)	Range of Plant Background Levels[a] (ppm)	Plant Concentration Known to Be Deficient (ppm)	Excessive Level/ Toxic Threshold in Plants[a] (ppm)	Maximum Chronic Tolerated Level in Grazing Animal Diet (mg/kg)
As	0.1-50	0.01-1.7	—	3-20	50
B	0.2-130	10-200	<5-30	50-200	150
Cd	0.01-1	0.1-0.8	—	5-700	0.05
Cr	5-3000	0.01-1	—	5-30	3000
Cu	2-100	4-30	<2-5	20-50	25-300
Fe	7,000-550,000	3-300	—	—	500-3000
Hg	0.01-0.8	—	—	1-2	—
Mn	100-4000	15-300	<15-25	300-2000	400-2000
Mo	0.02-5	0.2-5	<0.1-0.3	10-100	10-100
Ni	5-5000	0.1-5	—	50-100	50-300
Pb	2-200	0.1-10	—	30-300	30
Se	0.1-38	0.01-2	—	5-100	2
Zn	10-300	8-150	<10-20	>300-1500	300-1000

a. Extreme range of values measured. Concentrations generally apply to forage crop species rather than hyperaccumulators.

Sources: Leeper [24], p. 40
Overcash and Pal [28], p. 394
Kabata-Pendias and Pendias [20], p. 57
Chaney *in* Parr, Marsh and Kla [30], p. 162
Chaney [10]

bioassays have shown that cadmium and zinc availability were lowered in a flooded (reduced) environment, but increased in upland (oxidizing) conditions, which also lowered pH. The reverse was true for the plant uptake of arsenic from contaminated sediments placed in the same two types of experimental environments [22].

The point of these examples is that waste outfalls, waste disposal sites and chemical spills are not the only sources of problems related to uptake of inorganics. As described below, those who must make decisions about inorganic uptake phenomena should employ procedures sufficiently flexible to allow for unconventional sources and a range of potential influencing factors.

4.2.4 Decision Process for Analyzing Problems Related to Uptake of Inorganic Chemicals

The uptake and accumulation of inorganic species are inherently variable processes, and the data base for accurate quantitative estimation of uptake/accumulation phenomena is inadequate. Therefore, related problems are often best approached by a combination of selective literature review and original data collection or experimentation. Figure 4.2-3 illustrates in simplified form a decision process based on such a combination. This approach is modified from decision processes developed for evaluating problems in managing large-volume waste disposal [36].

Until usable approximation techniques become available, the investigator or decision-maker faced with a possible uptake/bioaccumulation problem should begin by obtaining initial information on the identity of the inorganic chemical of interest and on the characteristics of the receiving environment. If available, the following characteristics of the environment should be listed at this stage:

(a) *Media of exposure* (air, surface water, ground water, soil, aquatic sediments, biota)

(b) *Biological target species* (potentially exposed species that are important to man or linked to important species by the food web)

(c) *Exposure history* (geography and chronology of any known sources and sinks of higher-than-background concentrations of the inorganic chemicals of interest, as well as any data on background concentrations)

With the above information, the investigator may conduct a preliminary literature review to determine whether a more extensive review of literature and extrapolation from historical data will adequately describe the scope and magnitude of the problem. Literature appropriate for review at this stage of the investigation would include summary texts such as this one and the references cited in section 4.6. The merits of a more comprehensive review would be uncertain until a preliminary review is completed and indicates one of the following:

(1) The inorganic chemical has a history of causing uptake/bioaccumulation-related problems under similar circumstances;

(2) There is conclusive evidence that the chemical cannot cause an uptake/bioaccumulation-related problem under the given circumstances; or

(3) More information is needed to reach a defensible conclusion.

If either of the first two statements applies, it is reasonable to assume that analysis of factors a,b and c above based on a more detailed literature review will adequately define the problem and identify appropriate mitigation or response options. However, the third conclusion may well prevail if uptake of the particular inorganic in question has not been evaluated previously under sufficiently similar circumstances.

In any case, as indicated in Figure 4.2-3, the next step would be a focused search and review of the above-listed factors in the literature, extending to original research publications on field and laboratory studies of similar conditions. Numerous on-line services accessible by remote terminals provide rapid access to abstracts of articles in relevant journals.

The search and review should focus on defining the problem and finding precedents for its resolution if adequate data are already available. If, on the other hand, there is considerable uncertainty and insufficient data, the review should focus on precedents for the design and interpretation of a meaningful, valid field and/or laboratory data collection program.

If additional data collection is needed, the problem definition effort should first focus on accurate measurement and interpretation of the field situation, to determine whether those conditions represent a reasonable "worst case" for exposure and uptake of the inorganic(s) of concern. "Typical" or "boundary" conditions of interest may be defined later, as appropriate. The keys to achieving the objectives of the field research are:

1. Synoptic (simultaneous) measurement of the levels of the contaminant(s) and other controlling variables in the potentially important exposure media and exposed organisms; and
2. Adequate geographic coverage and duration to determine contamination gradients (including "control" conditions) and important periodic (e.g., seasonal) variations. In particular, monitoring should consider seasonal changes that control contaminant bioavailability, such as the temperature, redox and pH stratification of many temperate lakes or the seasonal water and sediment fluxes of runoff-dominated rivers or agricultural lands.

If the measurement of existing conditions suggests that a "worst case" equilibrium has not yet been achieved, it may be useful to simulate this equilibrium through further laboratory and/or field experimentation in order to assess the ultimate seriousness of the uptake-related problem. However, accurate simulation is extremely difficult and potentially costly because of the need to control the factors mentioned above. The frequently observed discrepancies between laboratory and field results for uptake of inorganics point up the difficulties of laboratory simulation; limited field-scale tests should receive at least preliminary emphasis whenever practical.

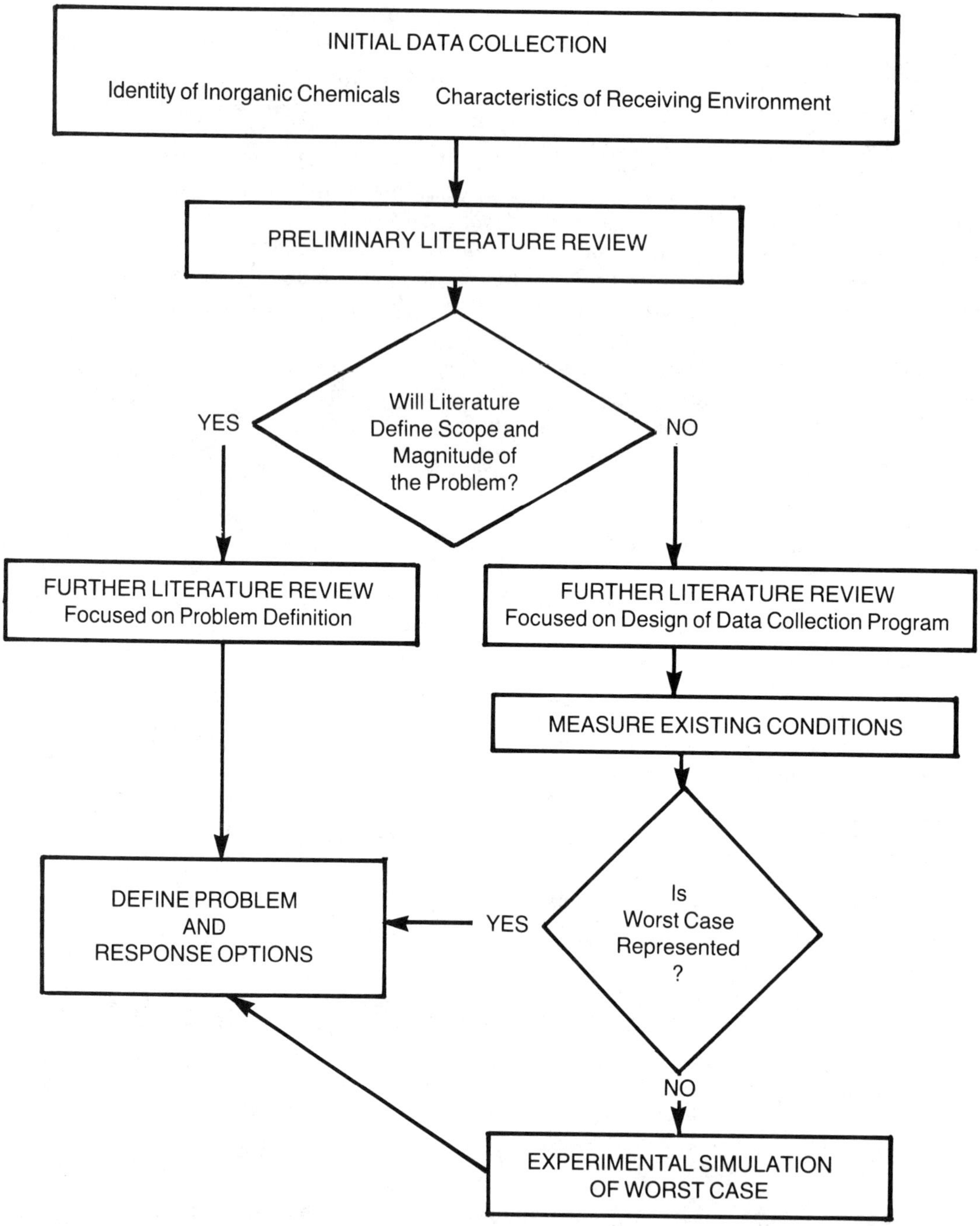

Source: C.B. Cooper, Arthur D. Little, Inc.

FIGURE 4.2-3 Decision Process for Analyzing Problems Related to Uptake of Inorganic Chemicals

4.3 UPTAKE BY TERRESTRIAL VEGETATION

4.3.1 Uptake and Loss Processes in Plants

Plants obtain nutrients through uptake of material by roots from the surrounding soil environment and, to a lesser extent, from the air through their leaves.

The soil environment may be broadly viewed as consisting of mineral matter, organic matter, soil solution, and slowly mixing gases. The relative proportions of these general components, as well as their physical and chemical makeup, are a result of geologic history, recent history and climatic influences. The site-specific characterization of soils is an important input for understanding the material available for uptake by local plant species; for example, soil pH, clay content, and organic matter strongly influence the concentrations of many elements in soil solution.

Inorganic elements that are most readily available for root uptake are those dissolved in soil solutions. They may be in the form of ions, soluble chelates or other complexed forms. Weakly adsorbed ions or species may also be bioavailable. This availability can vary with changes in soil solution chemical characteristics: cations, for example, are more available at low pH, becoming more strongly bound at high pH. The reverse is generally true for anions [20, 28, 30]. The variables that affect adsorption are further discussed in section 2.12.

Root uptake can be either passive, as in diffusion, or active. Active transport processes include transport across membranes by chelating agents or carriers [20, 28]. Such uptake applies to the movement of both macronutrients and micronutrients into a plant; it also applies to the mechanisms for uptake of non-nutrient elements and excessive accumulations of micronutrients.

Another avenue for material entry into plants is through the leaves. Aerosol materials may be deposited on leaf surfaces. Gases such as O_2 and CO_2, or even SO_2 or Hg^0, may diffuse into leaves, where they can become dissolved. Uptake from leaf surfaces is believed to be a combination of passive (diffusion) and active transport [20]. Of course, irrespective of whether airborne materials penetrate leaf surfaces, they may be available to the food chain as the result of grazing.

Phosphorus, nitrogen, calcium and other macronutrients that enter a plant are utilized in plant metabolism. Micronutrients — typically trace element species (anions or cations) — perform a wide variety of functions within a plant system. As their name implies, they are required in small amounts. Key functions include incorporation in enzyme systems and structural materials. Many, however, are simply attached to other large molecules. Their function is unknown, although their essential nature has been established for at least some plant species [20]. (See also Mengel and Kirby [25].)

Micronutrients may be translocated from their entry points to places where they can be utilized, stored, immobilized, or accumulated. The same appears to be true for the uptake of trace elements that have not been shown to be essential. Many such elements remain or are concentrated in fibrous root tissues. Fewer are translocated into leaves, and fewer still into tubers, fruits and seeds. Transport is believed to take place through the plant's vascular tissue (xylem and phloem) and is affected by transpiration (evapotranspiration) intensity [20]. Movement of an element into vascular tissue appears to hinge on the nature and location of chelates formed when it enters the root, and whether or not such chelates are translocated [30].

Loss processes (other than metabolic by-products) are fairly limited. Few trace elements are lost directly at leaf or root surfaces. Roots and old leaves appear to be common sites for accumulation and/or immobilization of excess or non-essential trace elements; thus, trace elements are lost through the dropping of leaves. While fewer trace elements are stored or immobilized in fruits or seeds, the seasonal loss of these structures is another loss process for the plant. If storage is in the roots, trace elements would not be "removed" from the plant until it is decomposed.

The most common (and most complex) mechanisms for removal of excess/toxic materials appear to be within the plant itself, through chemical fixation and/or physical separation within cell structures. Species of plants that are noted accumulators of certain trace elements make extensive use of these mechanisms; hence, while some chemicals may appear to remain within the plant, they are prevented from having a toxic effect on its life processes. There are also tolerant species that appear capable of excluding high levels of a microelement from uptake, a different protective mechanism against excess internal levels [30].

4.3.2 Environmental Factors Influencing Uptake

In any study of uptake in plants and the accumulation of inorganics that are considered potentially toxic, one of the major tasks is to describe the situations or circumstances that result in such adverse effects. The preceding discussion has referred to the variability inherent in interactions between soil and plant systems; some characteristics of this variability will now be outlined.

Plant species differ greatly in their need and uptake potential for various inorganic species. Studies aimed at maximizing agricultural yields have examined macronutrient requirements, largely from the perspective of deficiency. Differences in phosphorus and nitrogen removal from soil of up to an order of magnitude have been observed among a broad range of forage, field and forest vegetative cover [28]. Aboveground portions of field and forage crops have exhibited somewhat less than order-of-magnitude differences in levels of a number of macronutrients; for example, under normal conditions a crop such as coastal Bermuda grass removes from soil and contains relatively high phosphorus levels, while sugar cane may contain less than a third as much of this nutrient.

High levels of some macronutrients appear to cause a number of potentially adverse conditions, one of which is an imbalance in required nutrients. For example, excess potassium can produce an imbalance in the K to (Ca + Mg) ratio in plants in some environmental circumstances. This, in turn, may result in a deficiency of magnesium and a potentially lethal disease in grazing cattle and sheep [30]. High phosphate levels in soil reduce the uptake of many trace elements that are either necessary or potentially toxic. Very high levels of a number of added macronutrients, where soil and plant assimilative/uptake capacity is exceeded, may be transported overland by surface runoff to surface water or leached into groundwater. While these are potentially significant environmental concerns, they are somewhat removed from problems directly resulting from uptake in biota.

Plant species also vary in their uptake of trace elements, both essential and nonessential, even when other environmental factors affecting availability are thought to be controlled. Species and cultivar (genetic strain) variability appears to be of two types: (1) effectiveness in excluding excessive levels of soil trace constituents, and (2) adaptations for excessive plant concentrations of trace metal species. The result is diversity in sensitivity to changes in soil concentrations of these elements. Chaney (see Table 4.3-1) has classified a number of crop species for sensitivity to heavy metals; he suggests that these crops have varying abilities to exclude the increasingly available concentrations of metals found in sludge.

In addition, however, some species can adapt to natural soils that are extremely high in one or more cations, which may or may not be micronutrients. These "hyperaccumulators" are able to adsorb extremely high concentrations, typically of one element only. Such species have been identified for Ni, Zn, Cu, Co, and Cr and Se [30]. For example, while typical species exhibit phytotoxic reactions with foliar concentrations of nickel in the 50-100 ppm range, hyperaccumulator species grow well with foliar concentrations of up to 19,000 ppm.

Another key variable in plant reactions to micronutrients is the manner and degree to which specific trace elements are translocated to other parts of the plant once they are taken up by the roots. Table 4.3-2 summarizes the results of research on trace metal translocation. Root cell sap contains constituents that have an affinity for, or chelate, trace metals to varying degrees [30].

Plant species also vary in terms of whether trace metals are translocated to seeds and fruits as well as shoots and leaves. It has been reported that some species (e.g., corn, beans, fruit trees) exclude excess concentrations from fruits and seeds, while other species (e.g., wheat, oats, soybean, and root crops) exhibit similar microelement concentrations in storage tissues and leaves [30].

Soil composition clearly plays a vital role in determining the availability of trace elements for uptake. Soil pH, organic matter, and clay content all have a strong influence in this regard. The amount of soluble zinc in the soil, for example, decreases with higher values of pH and C.E.C.; since the ability of the soil to adsorb zinc cations is thereby increased, more zinc can be safely added under these conditions (Table 4.3-3). The availability of trace elements can also be altered by chemical reactions such as precipitation with iron oxides or phosphates, microbial activity, and redox conditions.

TABLE 4.3-1

Relative Sensitivity of Crops to Sludge-Applied Heavy Metals

Very Sensitive[a]	Sensitive[b]	Tolerant[c]	Very Tolerant[d]
Chard	Mustard	Cauliflower	Corn
Lettuce	Kale	Cucumber	Sudangrass
Redbeet	Spinach	Zucchini squash	Smooth bromegrass
Carrot	Broccoli	Flatpea	'Merlin' red fescue
Turnip	Radish	Oat	
Peanut	Tomato	Orchardgrass	
Ladino clover	Marigold	Japanese bromegrass	
Alsike clover	Zigzag clover	Switchgrass	
Crown vetch	Red kura clover	Red top	
'Arc' alfalfa	Crimson clover	Buffelgrass	
White sweetclover	Alfalfa	Tall fescue	
Yellow sweetclover	Korean lespedeza	Red fescue	
Weeping lovegrass	Sericea lespedeza	Kentucky bluegrass	
Lehman lovegrass	Blue lupine		
Deertongue	Birdsfoot trefoil		
	Hairy vetch		
	Soybean		
	Snapbean		
	Timothy		
	Colonial bentgrass		
	Perennial ryegrass		
	Creeping bentgrass		

a. Injured at 10% of a high metal sludge at pH 6.5 and at pH 5.5.

b. Injured at 10% of a high metal sludge at pH 5.5, but not at pH 6.5.

c. Injured at 25% high metal sludge at pH 5.5, but not at pH 6.5, and not at 10% sludge at pH 5.5 or 6.5.

d. Not injured even at 25% sludge, pH 5.5

Source: Chaney in Parr [30, pp 58-59]. *(Copyright 1983, Noyes Data Corp. Reprinted with permission.)*

TABLE 4.3-2

Trace Metal Translocation (Root to Shoot) Capabilities

Weakly Chelated/ Easily Translocated to Shoots and Leaves[a]	Moderately Chelated/ Moderate Translocation	Strongly Held/ Little Translocation
Zn	Cu	Pb
Cd	Ni	Cr
Mn	Co	Hg
B		
Se		
Mo		

a. Root cell sap contains chemicals that chelate or hold different microelements to differing degrees.

Source: Chaney in Parr [30, pp 56 and 57]

TABLE 4.3-3

Suggested Safe Limits of Zinc for Sensitive Crops in Three Kinds of Soil Managed at Three pH Levels

Soil Texture	Cation Exchange Capacity	pH	Percentage of C.E.C. Safely Occupied with Zinc	Added Zinc (lb/acre)
Average	15	6.0	5	490
		6.5	10	980
		7.0	20	1,960
Clayey	30	6.0	5	980
		6.5	10	1,960
		7.0	20	3,920
Sand	3	6.0	5	100
		6.5	10	200
		7.0	20	390

Source: Leeper [24]. *(Copyright 1978, Marcel Dekker. Reprinted with permission.)*

4.3.3 Data and Criteria on Uptake Potential

The concern over vegetative uptake of inorganics derives from circumstances where such uptake:

- reduces crop productivity or choice of crop;
- limits soil cover, resulting in erosion or loss of topsoil or water; or
- has an undesirable effect on the food chain.

The interaction between concentrations of a constituent and yield (growth) were illustrated earlier (Figure 4.2-2). With regard to nutrients, concentrations below a threshold value inhibit growth. Above this value, each constituent exhibits a range of concentrations that are optimal for growth. If these concentrations are exceeded, however, yields and even survival are jeopardized. In this model, non-essential constituents are taken up without any effect on growth until a similar toxicity threshold is reached. When plant toxicity occurs, crop loss, soil loss and surface water/ground water effects may become environmental issues. Another issue, important whether or not plant productivity is reduced, is the potential for adverse effects of contaminated vegetation on humans or animals.

The interaction of plants with water, macronutrients, salts, ions and trace metal species all follow the general model shown in Figure 4.2-2 and can result in any of the listed environmental effects. However, the latter are not always the direct result of excessive uptake, as discussed below.

MACRONUTRIENTS

The macronutrients include species of elements such as N, P, K, Na, S, Ca and Mg. As mentioned earlier, a major problem with excess levels of these constituents is the possibility of chemical imbalance in the soil. This, in turn, can alter the proper ratio of other macronutrients within the plant and induce or (more likely) inhibit the uptake of other essential and non-essential nutrients. Long-term severe effects can be of two types: (1) food-chain impacts, resulting from a particular nutrient excess, especially on grazing animals, and (2) plant death, resulting in decreased plant cover and increased potential runoff. Even in less severe cases, many macronutrients are in highly soluble form (e.g., nitrates), and excesses tend to be leached from surface soils (to groundwater, for example).

WATER

Different soil systems have differing capacities for water assimilation before they become hydraulically overloaded. Plants require and take up water. However, when soils become waterlogged for any length of time, various adverse effects can occur: anaerobic conditions may result which, if sustained, may destroy plants and some soil macro- and microbiota; anaerobic conditions may cause chemical changes that increase soluble concentrations of some trace elements; and leaching or runoff from such sites may add contaminants to usable water in other locations.

ACIDS, BASES, SALTS

The most common soluble ions in soils are potassium, sodium, calcium, magnesium, nitrate, chloride and sulfate. Sodium and chloride, in elevated amounts, do not result in increased uptake; rather, they reduce water uptake, increase plant dehydration, and may even cause a breakdown of soil aggregation and its arable potential [28, 30]. Ultimately, the ability of soils to support vegetative cover may be reduced. Although there is a wide variability in plant salt tolerance, the surviving species might not be considered useful.

Changes in soil pH can alter the availability of numerous other constituents because of shifts in pH-dependent equilibria and chemical associations. These reactions are especially important with respect to trace metal species. The following section cites specific examples of the effects of pH on uptake.

UPTAKE OF TRACE METAL SPECIES

Agricultural concern about trace metal species is the result of lowered productivity from nutrient deficiencies as well as the more recently recognized effects of excesses. Much has been written about trace element uptake; the subjects of research cover the full range of deficiency diseases through toxicity and include the effects of many variables on such uptake. Because uptake potential is so dependent on the characteristics of the site and the species, it is impractical to present data that are other than general or explanatory in nature.

Thus, the approach here is to provide examples of the type of information that would be useful in searching out more detail, to help explain that information, and to provide sufficient overview for understanding the complexities of the soil/plant system that control uptake and its impacts. Appendix D (Pollutant Criteria and Standards) has been used as a source of reasonably conservative maximum levels; however, these criteria constantly change as new information becomes available, so those used here may already be somewhat dated.

Some ranges of values for soil and plant concentrations of selected trace elements were listed in Table 4.2-6. A recently completed survey summarizes the levels of Cd, Pb, Zn, Cu and Ni in agricultural soils in the United States [17]. The percentile distributions of metal concentrations is a useful perspective for usual soil metal levels. It should be kept in mind that concentrations within the background-level range may, under some circumstances, have deleterious effects on plants or animals through uptake. Similarly, typical plant levels may be excessive for very sensitive species.

The tabulated ranges of concentrations that are toxic to plants and animals indicate that plants provide foraging animals with some measure of protection from toxic constituents, since Zn, Cu, Ni, Mn, As, B and some others can kill plants at levels that are generally not toxic to animals. There are some notable exceptions, particularly among rare hyperaccumulator plant species. Only elements such as Zn, Cd, Mo and Se are readily available to grazing animals, because they are easily translocated to the exposed portions of plants. Many trace elements are simply held too tightly in the soil or bound in the roots (e.g., species of Fe, Pb, Hg, Al, Ti, Cr^{+3}, Ag, Au, Sn, Hg, Si and Zr). Caution should be used in interpreting these generalizations, as grazing animals can also obtain heavy metals from plant surfaces and direct ingestion of soil.

Much of the research in this area has been devoted to understanding the specific circumstances that inhibit or encourage uptake of individual trace elements. Table 4.3-4 lists some of the findings for 12 selected trace-metal residues; it is not intended as a summary of this information but, rather, a collection of examples of important factors that influence uptake. The table is especially useful for pointing out what factors lead one to expect high or low concentrations of constituents in plants in a particular environment. It also indicates the importance of soil environmental factors, such as the role of pH, clay and organic matter, cation exchange capacity and other conditions that are known to help immobilize or make available the trace metal species. Recent studies have shown that roots of grass species (*Graminae*) excrete chelating substances, "phytosiderophores," that solubilize Fe for uptake [34]. Other species appear to acidify the immediate root environment to solubilize Fe for uptake [10, 11]. It is believed that these mechanisms cannot avoid other metals in the environment. Calcium, as the major soil cation, is believed to play an important role in the control of uptake of divalent metal ions.

Documented interactions among macro and micro soil elements, without reference to levels and the nature of interaction, are summarized in Table 4.3-5. Such interactions may help to explain uptake phenomena of interest or concern, and may also shed light on some otherwise inexplicable uptake patterns that require further inquiry by the reader.

TABLE 4.3-4

Uptake and Accumulation of Selected Trace Elements in Plants

Element	Uptake Potential and Destination	Form	Factors Affecting Uptake	Other Fate
Arsenic (non-essential element)	Uptake seems to be passive, with water. Translocated to many parts of plants; most found in old leaves and roots.	Soluble forms. The very toxic sodium arsenate and arsenic dioxide are extreme examples, but there are many anionic forms of both As^{+3} and As^{+5} that are soluble, the latter being less toxic.	*Promoted by*: Reduction processes that liberate soluble forms from Al and Fe oxides. *Restrained by*: Ferrous sulfate, calcium carbonate, oxidation. *Uncertain*: Phosphate reduces toxicity.	Strong sorption by clays, hydroxides, Fe and Al oxides. Organic matter reduces mobility. Enrichment evident in surface soils. Microbial reactions release volatile forms from soil. Some forms behave in soil in fashion similar to phosphate. Alkylarsine can be formed under anaerobic conditions.
Boron (essential nutrient)	Deficient in some soils. Highest concentration in old leaves; low in seeds and grains. Passive sorption.	Boric acid and other soluble species.	*Promoted by*: Alkaline soil and high soil salinity. *Restrained by*: Gypsum and calcium hypophosphate; pH<7. *Uncertain*: Antagonistic interaction with Cu, Cr, Mo, Mn, Fe, Si.	Can occur naturally in hazardous amounts (especially arid regions).
Cadmium (non-essential element)	Uptake by roots passive or metabolic. Highly mobile in plants — highest in roots, leafy portions, low in seeds. Soluble species highly available; can disturb enzyme activity.	Valence state in soil is Cd^{+2}. Forms complex ions and organic chelates. Absorbed by clay and organic matter.	*Promoted by*: Low soil pH (<6). *Restrained by*: Moderately alkaline soil (above pH 7). *Uncertain*: Organic matter may reduce availability in acid soil, but forms precipitates in alkaline soils.	Stays in zone of incorporation, causing build-up in soil. Zn, Cu, and Se seem to reduce Cd uptake or toxicity.
Chromium (essential element to animals at least)	Concentrates in roots; very slow movement to other plant parts.	Cr^{+6} highly available and toxic, but rapidly reduced to Cr^{+3} in all soils. Little Cr^{+3} is absorbed by roots.	*Promoted by*: Difficult oxidation to Cr^{+6}, which is very mobile and unstable. *Restrained by*: Reduction of Cr^{+6} to Cr^{+3} (which is mobile only in very acid soils; almost all precipitates at pH 5.5). Reduction stimulated by organic matter, lime, and phosphate but also occurs naturally in most soils.	Precipitates in soil

(Continued)

TABLE 4.3-4 (Continued)

Element	Uptake Potential and Destination	Form	Factors Affecting Uptake	Other Fate
Copper (essential element)	Deficiency or toxicity, depending on soil concentration of available form. Uptake by roots. Uptake likely active and also passive (at least in toxic range of soluble species). Movement to leaves related to concentration (supply) in roots.	Soluble forms in soil 99% chelated; organic chelates. Products of hydrolysis, anionic hydroxy compounds or carbonates are inorganic soluble forms (some pH-dependent).	*Promoted by*: Low pH *Restrained by*: Most ions have lowest solubility at soil pH of 7 to 8. Lime, organic matter and phosphate additions to soil reduce phytotoxicity. *Variable/uncertain*: Available species in soil. Interactions with Zn, Fe, Mo, Cd, Cr.	Tends to be immobile in soil because it readily interacts with minerals and also precipitates easily with soil anions.
Iron (essential nutrient in plants and animals)	Deficiency not uncommon. Uptake metabolically controlled. Transport regulated by plant to control shoot Fe.	Soluble Fe, including soluble organic complexes (small fraction of total levels).	*Promoted by*: Acid, waterlogged soils, mobile organic complexes, and chelates. *Restrained by*: Alkaline, well-aerated soils (may become deficient). *Uncertain*: Great number of nutritional and other soil factors affect uptake (e.g., antagonism by many heavy metals, P, etc.).	Iron readily precipitates as oxides, hydroxides and can be in chelated forms where organic matter is high.
Lead (non-essential element)	Absorbed by roots and concentrated there; probably passive. Little translocated to other plant parts; airborne deposition believed to contribute to leafy concentrations.	Ionic forms in soil solution. Airborne sources.	*Promoted by*: Flooding and other anaerobic conditions, low pH, low CEC, low organic levels, phosphate deficient soils (e.g., mine waste). *Restrained by*: pH above 6, lime, phosphate, organic matter. *Variable/uncertain*: Toxicity of soil or plant levels to plant.	Largely immobile in soils due to soil clays, phosphates, sulfates, carbonates, hydroxides, sesquioxides and organic matter. More toxic to rest of food chain. Accumulates in surface soils (soil half-life has been estimated in the 700 to 6000 year range). High soil levels may inhibit microbial processes and reduce decomposition (especially low-CEC soils)
Mercury (non-essential element)	Transfer to plants nearly totally as elemental vapor.	Retained in soil as slightly mobile organo-complexes.	*Promoted by*: Methylation; other organo-compounds also available, but to lesser degree. Extent of methylation in soil uncertain. *Uncertain*: Liming, sulfur compounds, or rock phosphates may reduce uptake; the effect of organic matter.	Migration processes are limited; soil build-up. Reduction to elemental form results in toxic mercury vapor.

(Continued)

TABLE 4.3-4 (Continued)

Element	Uptake Potential and Destination	Form	Factors Affecting Uptake	Other Fate
Molybdenum (essential nutrient)	Found in most tissues	Soluble forms; molybdate; adsorbed or Fe in soil.	*Promoted by*: High pH and saturated soils. *Restrained by*: pH<5.5, humic acid (although organic soils can release it slowly); iron oxides. *Uncertain*: Availability limited by Cu, Mn, Fe.	Least soluble in acid soils. Toxic to grazing animals at levels not toxic to plants.
Nickel (essential element for some plant species)	Soluble forms readily absorbed by roots; mobile, and also accumulates in leaves and seeds. Required by some N-fixing species; may be essential at very low levels.	Possibly soluble organic chelates and possibly also adsorbed on oxides of Fe and Mn.	*Promoted by*: pH below 6.0. *Restrained by*: Lime, Fe, phosphate reduce availability to plants. *Variable*: Addition of organic matter can sometimes reduce availability.	Excess can cause iron deficiency in plants; coprecipitates with Fe and Mn in soil. Adsorbed readily by organic matter.
Selenium (uncertain role in plants; micro-nutrient in animals)	Uptake of water-soluble fraction by roots. Distributed to growing tips and seeds; concentration in roots.	Water-soluble fraction in soils is not fixed by Fe oxides, so readily available. Se-amino acids from plant residue increase available Se in soil.	*Promoted by*: Alkaline, well-oxidized soils. *Restrained by*: Acid and neutral conditions where selenides and selenites (as Fe hydroxides or oxides) are not formed.	Micronutrient in animals. Levels toxic to livestock are lower than toxic levels for plants. Methylated derivatives of microbial activity volatilize. Some plants are hyperaccumulators.
Zinc (essential element)	Deficiency in soil and plants not uncommon; high uptake potential and probability in plants if available; found in various plant parts.	Readily available as hydrated Zn^{+2}, Zn^{+2} with some organic chelates, and adsorbed on Fe, Mn oxides.	*Promoted by*: Low soil pH. *Restrained by*: High soil pH, high clay content, high cation exchange capacity, high phosphate. *Uncertain*: Possible antagonistic effects with Cu, Fe, Cd, As; possible Zn/Cd synergism, in some proportions.	Very soluble in soil solution compared with other metals. Very mobile in acid soil.

Sources: Overcash and Pal [28] (all elements); Kabata-Pendias and Pendias [20] (all elements except Mo); Ryan *et al.* [35] (Cd only); Chaney [10].

TABLE 4.3-5

Interactions Between Major Elements and Trace Elements in Plants

Major Element	Antagonistic Elements	Synergistic Elements
Ca	Al, B, Ba, Be, Cd, Co, Cr, Cs, Cu, F, Fe, Li, Mn, Ni, Pb, Sr, Zn	Cu, Mn, Zn
Mg	Al, Be, Ba, Cr, Mn, F, Zn, Ni[a], Co[a], Cu[a], Fe[a]	Al, Zn
P	Al, As, B, Be, Cd, Cr, Cu, F, Fe, Hg, Mo, Mn, Ni, Pb, Rb, Se, Si, Sr, Zn	Al, B, Cu, F, Fe, Mo, Mn, Zn
K	Al, B, Hg, Cd, Cr, F, Mo, Mn, Rb	—
S	As, Ba, Fe, Mo, Pb, Se	F[b], Fe
N	B, F, Cu	B, Cu, Fe, Mo
Cl	Br, I	—

a. Reported for microorganisms.

b. Mutual pollution causes significant injury.

Source: Adapted from Kabata-Penias and Penias [20]. *(Copyright 1984, CRC Press, Inc. Reprinted with permission.)*

4.3.4 Models and Predictive Techniques

There are no models of plant uptake that are consistently useful for predicting concentrations of trace elements under a broad range of circumstances. Over the years, however, agronomists have produced much data on "optimal" doses of chemical fertilizers (particularly macronutrients) to maximize crop yields. This represents a large data base. A recent publication by Barber [2] summarizes a number of predictive models for plant uptake and soil-plant interactions affecting uptake of metals by plants.

Figure 4.3-1 presents estimates of soil-to-plant concentration factors for most of the stable elements. The two values given for each element, B_v and B_r, are defined as follows:

$$B_v = C_v/C_s \tag{1}$$

$$B_r = C_r/C_s \tag{2}$$

where

B_v = soil-to-plant elemental transfer coefficient for *vegetative* portions of food crops and feed plants,

B_r = soil-to-plant elemental transfer coefficient for *nonvegetative (reproductive)* portions of food crops and feed plants,

	I A	II A	III B	IV B	V B	VI B	VII B	VIII	VIII	VIII	I B	II B	III A	IV A	V A	VI A	VII A
II	Li 0.025 4.0×10^{-3}	Be 0.010 1.5×10^{-3}											B 4.0 2.0		N 30 30		F 0.060 6.0×10^{-3}
III	Na 0.075 0.055	Mg 1.0 0.55											Al 4.0×10^{-3} 6.5×10^{-4}	Si 0.35 0.070	P 3.5 3.5	S 1.5 1.5	Cl 70 70
IV	K 1.0 0.55	Ca 3.5 0.35	Sc 6.0×10^{-3} 1.0×10^{-3}	Tl 5.5×10^{-3} 3.0×10^{-3}	V 5.5×10^{-3} 3.0×10^{-3}	Cr 7.5×10^{-3} 4.5×10^{-3}	Mn 0.25 0.050	Fe 4.0×10^{-3} 1.0×10^{-3}	Co 0.020 7.0×10^{-3}	Ni 0.060 0.060	Cu 0.40 0.25	Zn 15 0.90	Ga 4.0×10^{-3} 4.0×10^{-4}	Ge 0.40 0.080	As 0.040 6.0×10^{-3}	Se 0.025 0.025	Br 1.5 1.5
V	Rb 0.15 0.070	Sr 2.5 0.25	Y 0.015 6.0×10^{-3}	Zr 2.0×10^{-3} 5.0×10^{-4}	Nb 0.020 5.0×10^{-3}	Mo 0.25 0.060	Tc 9.5 1.5	Ru 0.075 0.020	Rh 0.15 0.040	Pd 0.15 0.040	Ag 0.40 0.10	Cd 0.55 0.15	Ir 4.0×10^{-3} 4.0×10^{-4}	Sn 0.030 6.0×10^{-3}	Sb 0.20 0.030	Te 0.025 6.0×10^{-3}	I 0.15 0.050
VI	Cs 0.080 0.030	Ba 0.15 0.015		Hf 3.5×10^{-3} 8.5×10^{-4}	Ta 0.010 2.5×10^{-3}	W 0.045 0.010	Re 1.5 0.35	Os 0.015 3.5×10^{-3}	Ir 0.055 0.015	Pt 0.095 0.025	Au 0.40 0.10	Hg 0.90 0.20	Tl 4.0×10^{-3} 4.0×10^{-4}	Pb 0.045 9.0×10^{-3}	Bi 0.035 5.0×10^{-3}	Po 2.5×10^{-3} 4.0×10^{-4}	Ar 1.0 0.15
VII	Fr 0.030 8.0×10^{-3}	Ra 0.015 1.5×10^{-3}															

Lanthanides	La 0.010 4.0×10^{-3}	Ce 0.010 4.0×10^{-3}	Pr 0.010 4.0×10^{-3}	Nd 0.010 4.0×10^{-3}	Pm 0.010 4.0×10^{-3}	Sm 0.010 4.0×10^{-3}	Eu 0.010 4.0×10^{-3}	Gd 0.010 4.0×10^{-3}	Tb 0.010 4.0×10^{-3}	Dy 0.010 4.0×10^{-3}	Ho 0.010 4.0×10^{-3}	Er 0.010 4.0×10^{-3}	Tm 0.010 4.0×10^{-3}	Yb 0.010 4.0×10^{-3}	Lu 0.010 4.0×10^{-3}
Actinides	Ac 3.5×10^{-3} 3.5×10^{-4}	Th 8.5×10^{-4} 8.5×10^{-5}	Pa 2.5×10^{-3} 2.5×10^{-4}	U 8.5×10^{-3} 4.0×10^{-3}	Np 0.10 0.010	Pu 4.5×10^{-4} 4.5×10^{-5}	Am 5.5×10^{-3} 2.5×10^{-4}	Cm 8.5×10^{-4} 1.5×10^{-3}							

Key:

Li 0.025 4.0×10^{-3}	— Element Symbol — Transfer Coefficient, B_v (see text) — Transfer Coefficient, B_r (see text)

Source: Baes *et al.* [1]

FIGURE 4.3-1 Values of the Soil-to-Plant Concentration Factors B_v and B_r

C_v = elemental concentration in *vegetative* portions of food crops and feed plants (dry weight) at edible maturity,

C_r = elemental concentration in *nonvegetative (reproductive)* portions of food crops and feed plants (dry weight) at edible maturity, and

C_s = elemental concentration in root zone soil (dry weight).

The estimates were generated by Baes *et al.* [1] for use in a computerized model (TERRA) of the terrestrial transport of radionuclides. All estimates of B_v and B_r were based on any combination of: (1) analysis of literature references; (2) correlations with other parameters; (3) elemental statistics; or (4) comparisons of observed and predicted elemental concentrations in foods. In general, no *a priori* bases or protocols were used to produce conservative values. The final values selected were estimated to two significant digits and then rounded off to the nearest 0.5 decimal place. Because of the significant variability in uptake ratios caused by environmental variables, it should be clear that the single values in Figure 4.3-1 must be used with extreme caution.

Acknowledging problems with the usefulness of any one extraction procedure in determining soil trace metal availability for plant uptake, Gerritse *et al.* [15] have found high correlations between concentrations of some trace metals (e.g., Zn, Cd and Cu, but not Pb) measured in soil solutions and uptake from a broad range of soil types. While measurement of soil solution concentrations is not necessarily a simple analytical exercise and the extraction method used is subject to choice, correlations with uptake have been an improvement over soil-extracted metal concentrations. This type of approach, which is aimed at developing generalizable principles, may ultimately lead to useful prediction techniques. The investigator faced with a potential problem of inorganic uptake in a terrestrial system may follow the steps outlined in § 4.2.4 when no better approach is available.

4.4 UPTAKE BY MARINE BIOTA

4.4.1 Introduction

The marine environment serves as the major aquatic source of food for man. It has also become a place for waste discharge and disposal. The understanding of plant uptake processes described in the section above was derived in part from agricultural science. In that instance, cause and effect relationships could sometimes be directly observed, even under field conditions. Observational understanding of inorganic uptake in the marine environment, especially in natural settings, is more difficult.

Like the terrestrial environment, the marine environment is not uniform. At the very simplest level, three general types of marine environments must be considered:

- *Estuaries*, which are characterized by a range of salinities from fresh to salt water that varies seasonally and even daily in some locations, and a degree of seasonal and even daily variation in temperature, sometimes pH, and large seasonal changes in primary productivity with associated dissolved organic matter.
- *Nearshore Areas*, which tend to have less salinity variation than estuaries but are characterized by variability in depth, and hence temperature, by major water exchanges due to seasonal phenomena, storms, and the tides.
- *Offshore Areas*, which usually vary in temperature with depth. Even with the partial exceptions of the euphotic zone near the surface and occasional periods and areas of upwelling, temperature and salinity changes tend to occur less in offshore areas than in the estuaries and nearshore areas above.

While these are simplistic descriptions, they are useful in understanding general observations about marine uptake processes.

A large number of marine species spend some or all of their life cycle in nearshore coastal or estuarine environments, primarily because the shallower and warmer waters, increased light penetration, and energy subsidies such as tides combine to generate an enormous food production capability. However, when we attempt to examine the contamination exposure history of a particular organism, the problem can be far more complex than studying the exposure of fish in a tank. For example, bivalves such as clams and oysters are essentially bottom dwellers in close contact with sediments, but they also spend their very early life floating with other plankton in nearshore/estuarine habitats. Striped bass, sometimes a nearshore fish, spawn in estuaries; during their embryo-to-juvenile life stages, they remain in the estuaries, slowly migrating downstream from brackish to salt water, and the adults move freely in and out of the estuarine and nearshore environments. Salmon migrate through estuaries to freshwater headwaters to spawn, and juveniles migrate downstream and out to sea months later. Species such as the mummychog remain in estuaries for their entire life cycles. Planktonic species, plant and animal, are moved with tides and currents but essentially remain within one ecological regime. Rooted aquatics and marsh plants are clearly stationary while alive, but their decomposition products can be transported to nearshore or offshore environments. Birds and other terrestrial species may feed from the nearshore and estuarine areas without being directly part of them. The complexity of this exposure history is compounded by feeding behavior that may change over the course of development in some species: juveniles of a given fish species may feed on plankton, while the adults feed on other fish.

Many coastal areas, particularly the lands bordering major estuaries, are urbanized and industrialized, causing marine ecosystems to be exposed to large volumes of waste. Inland areas also contribute to this waste load as rivers emptying into estuaries carry portions of effluents and surface runoff from upstream. It is significant that these same waters, out to the edge of the continental shelf, have supported more than 90% of the American commercial seafood catch and virtually all of the recreational fishing [18]. Even fish caught offshore are dependent to some degree on the shallower estuaries and nearshore waters.

Attempts to understand uptake in marine organisms must take these facts into account. Poisonous levels of contaminants (e.g., Cd, Hg) in seafood have directly affected humans and fish-eating birds. Awareness of the potential for adverse effects from this major food resource has led to regulation of the levels of many chemicals in seafood as well as to research into uptake and loss processes.

4.4.2 Uptake Processes

Uptake of inorganic constituents varies with the nature of the organism and the constituent in question. There are a number of obvious differences between algae and fish, as well as similarities. Many trace metal species, including some of the heavy metals, are required micronutrients; they typically play a role in enzyme-mediated processes or form integral components of cell structure.

There are three major routes of uptake, only one or two of which are important to each specific group of organisms. These routes are:

- Direct absorption from water of chemically available forms of an inorganic constituent across cellular or respiratory surfaces;
- Direct absorption of chemically available constituents from sediments through filter feeding or simple contact; and
- Uptake from food containing inorganic constituents.

In contrast to lipophilic organics, where direct uptake from water is usually the most important route, uptake of inorganics may well be directly from the environment *and* from food. Unfortunately, most laboratory experiments have examined only one exposure route at a time — most often, uptake from water.

As with terrestrial plants, inorganic uptake by marine biota can be a source of required nutrients as well as of non-essential elements, and there appears to be no optimum range and/or tolerance for tissue concentrations of a particular trace element, whether it is a nutrient or not. Once certain concentrations are exceeded, however, toxic effects become evident. Figure 4.4-1 shows the relationships between observed uptake and adverse effects in aquatic species. Since the damage threshold for some species can be much lower than for others, their sensitivity to different inorganic constituents varies. Research, especially laboratory studies, has indicated that other factors such as age, size of the individual, and even the season of the year may play a role in observed variations in tolerance [4,13]. This discussion of toxicity, however, is limited to their relationship to uptake processes and effects.

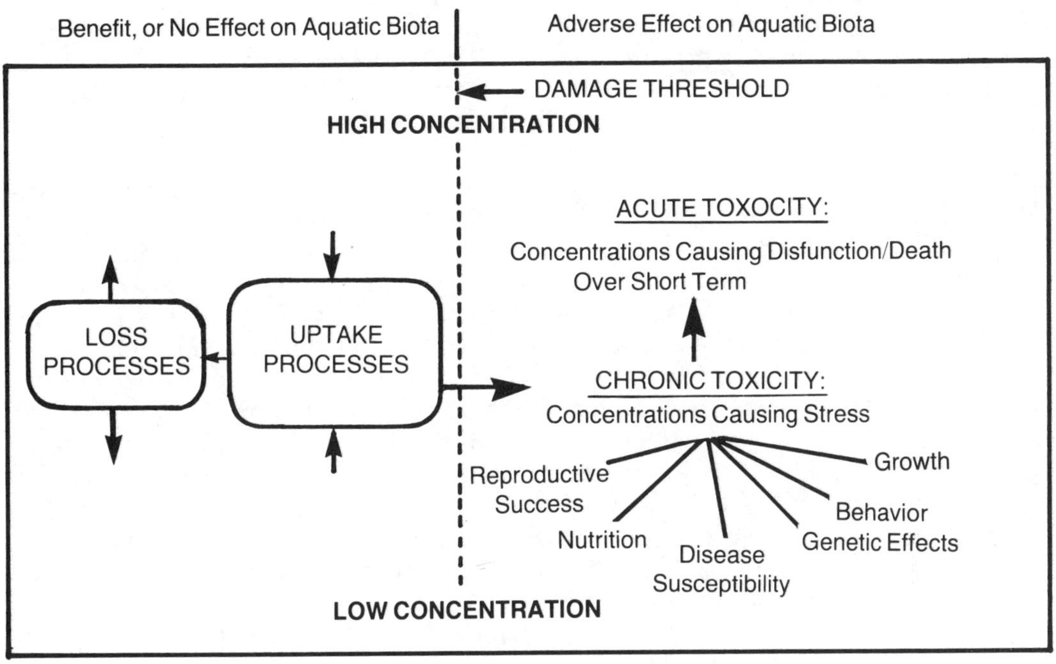

FIGURE 4.4-1 Effects of Introduced Contaminants

Recent investigations of several West Coast marine species have provided direct chemical evidence to explain the mechanisms of observed damage-threshold phenomena. For some time, researchers had noted that some trace metals measured in tissues were associated with molecules in a specific weight range [13]. Brown *et al.* [7], at the Southern California Coastal Water Research Project (SCCWRP), devised a method for examining the location and association of several heavy metal species with specific molecules in fish tissue. They identified three distinct tissue chemical pools into which specific contaminants may become partitioned:

- [ENZ] A high-molecular-weight pool that contains cellular enzyme systems and is the site of toxic action by both heavy metal species and organics.

- [MT] A medium-molecular-weight pool that has been shown to contain a metal-binding protein, metallothionein, and is considered the site of trace metal species detoxification.
- [GSH] A low-molecular-weight pool of compounds such as glutathione, considered the site of organic metabolite detoxification.

As expected, micronutrients such as Cu and Zn, which are components of enzymes (metalloenzymes), were found in the [ENZ] pool of tissues. The authors theorized that such species were not toxic until they were found in the [ENZ] pool in excess of requirements [7]. When tissues of fish from contaminated areas were examined, elevated levels of Cu and Zn were found in the metallothionein-containing fraction as well as the enzyme fraction. The non-essential element, cadmium, was almost entirely bound in the metallothionein-containing fraction. It was theorized that non-essential metal species are partitioned into the [MT] pool and, hence, are not harmful unless assimilative capacity is exceeded. Excesses of essential metal species are also partitioned into the [MT] pool and similarly rendered non-toxic.

Several additional observations concerning uptake phenomena have been made during the course of this research. Fish from Southern California offshore areas heavily polluted by outfall discharges of both heavy metal species and chlorinated hydrocarbons showed reduced metal uptake where chlorinated hydrocarbon uptake was high. Greater uptake of metal species was observed in areas less polluted by hydrocarbons, and with *lower* metal concentrations as well. Subsequent examination of this phenomenon led SCCWRP researchers to hypothesize, with evidence from biochemical examination, that oxygenated metabolites of chlorinated hydrocarbons may interfere with metal uptake in fish; this might cause adverse effects due to deficiencies in uptake of required metal species or allowing some metals to escape detoxification by "spillover" into the [ENZ] pool [5]. This latter hypothesis has come under criticism. The area studied is also heavily influenced by sewage discharges, resulting in high levels of particulate and dissolved organic matter as well as the above noted chlorinated hydrocarbons. Thus, the observed reduction in trace metal uptake could be due to the levels of organic matter in the water decreasing metal bioavailability [23].

Questions remain concerning what combined levels of metal species and organics lead to toxicity, which tissues are most sensitive to excesses, and how such observations may be related to environmental factors such as seasonal changes. Questions also remain concerning the general applicability of these observations across the broad range of marine organisms (beyond the fish, mollusks, and sea urchins studied) and across the range of inorganics other than Cu, Zn and Cd. It is essential that our understanding be advanced to the point at which we can relate such observations to damage thresholds and identifiable environmental causes.

4.4.3 Loss Processes

Keys to the ultimate tissue concentrations of a substance in marine biota are the length of time organisms are exposed to the contaminant, the environmental concentration of the contaminant, the rates of uptake and loss, and often organism size. Sedentary organisms may be in relatively constant contact with contaminated sediments, while seasonal visitors may spend a short time in an area and return to cleaner environs for long periods. For many compounds, depuration rates decline with the size of the organisms: larger fish contain higher levels of a contaminant than smaller ones of the same species and exposure history.

Further, loss of an inorganic contaminant has been shown in research to be represented by a two compartment loss curve. Thus, there is an initial, relatively fast rate and a longer-term slower rate. For regulatory purposes the first order rate loss is considered an adequate initial assumption and is referred to as the "half life" of the chemical in an organism [23].

Finally, mercury may be lost from different marine species at different rates: a biological half-life for mercury in predatory fish, feeding on contaminated food, of greater than 1000 days has been reported [13], while mercury in prey fish in the same experiment exhibited a biological half-life of less than 60 days. These are potentially important contributing factors in the differences often seen in tissue concentrations of organisms taken from a similar environment.

4.4.4 Data and Criteria on Uptake Potential

A fairly large body of data exists on observed concentrations of inorganic contaminants in marine organisms, but the data base on uptake potential/accumulation factors is far smaller. For certain contaminants — mercury, cadmium, chromium, copper, zinc, and a few others — there is much information on uptake concentrations from both field and laboratory data collections; there are significantly less data on trace metal species and other inorganics that have been less problematic or are less well-known.

Two general types of data are available on uptake in marine organisms. The first is from laboratory experiments. As a rule, organisms are placed in contact with one or more target elements for a given length of time prior to sacrifice for measurement of tissue levels of the given contaminant. Some test organisms may also be placed in clean water following exposure, to determine their degree of retention of contaminants. Laboratory experiments typically use a specific chemical form of an element, measure the duration of exposure, and may report species, size, water conditions and water concentrations of the contaminant.

Field collections of organisms may report species, size or age, and existing water or sediment conditions for a specific area. However, it is very difficult to know the exposure history that produced any observed uptake by the organisms collected. Relatively sedentary organisms, including rooted aquatics, mussels, clams and other bivalves, and a broad group of sediment dwellers such as worms or tunicates do have a reasonably constant exposure to sediments. However, they may also principally receive dissolved inorganics from overlying water concentrations that are not necessarily so constant. There is clearly even less certainty with free-swimming community members, whose exposure changes continually.

Another important variable in marine uptake literature is the tissue examined for contaminant concentrations. (See earlier discussion of this factor in § 4.2.2.) Muscle tissue levels and concentration factors for fish or crustacea that are used for human consumption are useful in assessing risk potential for humans. However, muscle tissue is frequently *not* the site of highest concentration of trace metal contaminants that are not organo-metallics. (Mercury, as an organo-mercury compound, concentrates in fish flesh; technically, this is not an exception but often appears so if it is analyzed as total mercury.)

Recent research indicates that organisms have specific sites for detoxifying at least some contaminants. Toxicity to the organism occurs when detoxification capacity is exceeded *at these sites*. Organs important to detoxification may or may not be the ones of highest concentration, but they appear to be locations where accumulation is high. Sites of threshold toxic action may also be quite specific. These latter points are not yet well defined by ongoing research, but it is clear that data on whole body concentrations and/or on tissues that represent low accumulation may be misleading if used as indicators for threshold toxicity. Unfortunately, the linkages between environmental exposure, documented uptake and accumulation, toxicity, and risk to the food chain are not yet well quantified. The particular tissues that should be used for indicators of accumulation in various species are also not yet firmly established.

Selected data on uptake of inorganics in marine organisms have been summarized for specific contaminants. Table 4.4-1 emphasizes information on the general uptake process for several phyla of marine organisms consumed by humans. The field data are taken from Eisler's extensive compilation of such numbers [10], in which the methodology is not specified. One striking generalization can be inferred from these data: uptake and toxicity increase at higher temperatures and lower salinities for all the trace metal species of toxicological concern. These conditions are most common in the nearshore areas and in estuaries; thus, the same areas required for reproductive success for many economically and ecologically important species are the marine environments with potentially highest contaminant loading and uptake. All concentration factors listed are intended to illustrate *relative* uptake potential, since uptake will vary with exposure, environmental availability and, especially, chemical speciation.

Concentration factors measured in the laboratory appear to be generally lower than those from field samples. Possible causes for this include: longer exposure duration in the field, undocumented exposure to higher than measured ambient concentrations, complexation, salinity variation, handling techniques and analysis, analysis of previously deceased rather than freshly sacrificed individuals, and both water and food as sources for uptake. Appendix D (in Part II) lists USEPA and other nationally applicable marine uptake/bioaccumulation (residue) criteria for inorganics.

4.4.5 Models and Predictive Techniques

The general constraints in § 4.2.3 apply particularly to uptake by marine biota. No generally applicable models or estimation techniques are available. One special

TABLE 4.4-1

Some General Characteristics of Trace Metal Uptake in Selected Groups of Marine Organisms

Chemical	Algae/Macrophytes	Molluscs	Crustaceans	Finfish
Arsenic (non-essential)	• Arsenate appears more absorbed, and with adverse effect. • Lab uptake measured, reported in 200x to 3000x range (highest in lipid fractions). • Source of organo arsenic compounds in marine food-web.	• High field bivalve concentration factors reported as 1500x to 18,000x, up to 46,000x (data from Greece). • Short-term lab studies do *not* show significant uptake magnification. • Concentrated in lipid extracts, mostly in non-toxic form. • Loss is in two phases, with half times of 3 days and 32 days. • Increased uptake at lower salinity.	• Concentrated in lipid fractions as organic arsenic likely less toxic to organism. • Salinity, temperature modify uptake rates. • Food chain accumulation.	• High tissue levels (especially in lipid fractions) compared with fresh-water and terrestrial organisms. • Inorganic forms converted to organic forms not toxic to organism or to terrestrial species (man). • Some evidence for reformation of toxic arsenite in fish flesh after death, becoming more toxic to consumer.
Cadmium (non-essential)	• Concentration factors, in field samples, found to be as high as 4200x to 11,000x. • Lab study concentration factors ranged from 11x to 8300x. • Release rates, in clean water, appear to be slow. • Cations in seawater (Ca, Mg, Sr) compete for uptake sites with Cd, Cu, Zn. Availability reduced with organic chelates.	• Concentration factors for field samples of 10,000x were not uncommon. Highest levels in digestive gland, kidney. • Lab studies show concentration factors from 2x or 3x to 20,000x (dry weight). • Higher uptake at lower salinity. • Proteins similar to metallothioneins believed involved in detoxification. • Very slow depuration of body burden of cadmium in clean water.	• Highest levels found in viscera, lowest in muscle. • Uptake values in lab were in 100x to 600x range. Half-life for loss in clean water, up to one year. • Most efficient uptake is through gills at higher concentrations, food at low concentrations. • Increased uptake at lower salinity, higher temperature. • Proteins isolated in midgut that bind to cadmium (metallothioneines).	• Highest levels found in liver and other viscera; whole body range similar to molluscs-crustaceans. • Increasing age, lower temperature, high salinity reduce uptake. • Metallothionein-cadmium bonding interactions are a detoxification mechanism. Calcium in sea-water may also interfere with cadmium effects. • Diet and water are indicated as important exposure pathways. • Significance of lower levels and intermittent exposure (common in ocean) not certain.
Chromium (essential element)	• Concentration factors of up to 8600x reported. • Short-term Cr studies show uptake of 18x. • Little evidence for food chain biomagnification.	• High field accumulations reported at 16,000x to 260,000x (high levels from Greek waters). • Less accumulation by heavier individuals and at high salinities. • Lab studies showed accumulation up to about 100x and 500x. • Uptake from water likely to be greater in lab; uptake from food greater in natural environment.	• Digestive glands and gills contained highest levels. • Uptake equilibrium shown to take up to 5 days. • Some studies show accumulation up to 100x.	• 1000-fold variation in accumulation among species from same location, same collection method. • Laboratory uptake observed in the 1x to 100x range, with higher concentration in liver, low in muscle.

(Continued)

TABLE 4.4-1 (Continued)

Chemical	Algae/Macrophytes	Molluscs	Crustaceans	Finfish
Copper (essential)	• Accumulation range from 2000x to 20,000x in field collections. • Hypothesized variation in species tolerance to very phytotoxic ion. • Dead algae release soluble copper -available for other uptake. • Uptake increases with temperature, decreases with salinity. Chelation complexation with organic matter, iron, citrate found to reduce availability.	• High biomagnification in field collections from 4100x to 15,000x (high levels in Greek waters). • Seasonal, sex, age differences. Uptake increases with increased temperature, decreases at higher salinity. • Highest accumulation in gills, digestive tract, kidney. • Accumulations in long-term lab experiments up to 20,000x.	• High in blood, as a component of respiratory pigments in this and other groups. Muscle contains lowest concentration. Hepatopancreas stores copper; chitin of some species absorbs copper. • Small metal-binding proteins have been identified in digestive tract gland of some species. • Age/developmental stage and season affect uptake. Uptake and/or toxicity greater at lower salinity and higher temperature.	• Concentrations in marine vertebrates lowest among marine biota. • Highest fish concentrations in liver and gills, viscera. Very low in muscle. • Copper-binding proteins have been isolated from some species. • Age, sex, size, season may affect uptake, sensitivity. • Free cupric iron is form affecting toxicity. • More toxic at lower salinity. • Probable increase in uptake following death. • Lab study showed 5000x to 7000x concentration for viscera.
Lead (non-essential)	• Field collections illustrate accumulation ranges from 1200x to 82,000x. • Examples of lab tests show concentrations at 124x for short-term tests.	• Long-term lab studies show accumulations of 200x to more than 20,000x dry weight in various soft tissues. • Highest accumulations in kidney, gills, and digestive gland. • Younger and smaller organisms exhibit faster uptake, slower loss rate. • Higher uptake levels with lower salinity. • Uptake of lead in sediments by some species can be controlled by higher iron levels. • Uptake may be from food or from water.	• Radioisotope studies show uptake of about 100x in whole organism (ranged from 10x in muscle to 1 million x in hepatopancreas). • Localized in exoskeleton and lost with molt. • Uptake and/or toxicity greater at lower salinity, also higher temperature.	• Some fish shown to accumulate from water by 100x. • Lead accumulates in bone, so whole fish levels relatively higher than in muscle or viscera.
Mercury (non-essential)	• Short-term lab experiments showed uptake in the 10,000x to 75,000x range. • Organomercury compounds more readily taken up; greater uptake at lower salinity, high pH. • Great species differences in sensitivity.	• Short-term lab uptake of 100x from water. Long-term lab uptake of 2800x to about 30,000x. Uptake of inorganic forms about ¼ that of organic forms. • Methylmercury more toxic and more readily taken up. Biological processes convert a number of other forms of mercury to organic mercury compounds. Uptake greater in low salinity. Less uptake in presence of selenium.	• Field collections do not show exceedingly high levels *except* in very polluted areas. • Laboratory studies have illustrated uptake factors in the 500x to more than 10,000x range in different tissues over different duration exposures. • Concentrations high in gills, digestive glands. • Relatively lower in muscle and blood. • Bioconcentration from food demonstrated but is faster and higher from solution.	• Field collections show accumulation in *edible flesh*. Most in flesh in organic form and higher in large, upper-level carnivores than species with short food chain. Some uptake believed to represent natural background conditions. • Concentration from food sources believed by some to be greater than from water but is still argued. • Fish appear to have high tolerance for mercury without obvious adverse effects.

(Continued)

TABLE 4.4-1 (Continued)

Chemical	Algae/Macrophytes	Molluscs	Crustaceans	Finfish
Mercury (non-essential) (continued)		• Loss processes are extremely slow over 6 months and up to 3 years in clean water environments. • Uptake highest among organisms with direct contact with sediments.	• Organomercury compounds are more toxic and more readily taken up than inorganic salts. Toxicity/uptake greater at lower salinity — for inorganic salts.	• Concentration factors in long-term lab studies show levels from 450x to 2000x, depending upon life stage and form of mercury. • Depuration rates reported from weeks to years, with loss from muscle relatively slow.
Nickel (possibly essential)	• Reported accumulations in field samples range from 146x to above 1000x.	• High variability in field concentrations; one extraordinary uptake value of 120,000x. • Concentration varies with season of collection. Tissue of highest concentration varies with species. • Long and short-term lab studies show uptake potential from 20x to more than 500x.	• High variability in field collection concentrations. • Lowest concentrations found in edible flesh, highest in exoskeleton.	• Low levels found in field collections of these species. • Not accumulated in amounts of concern to human consumers.
Zinc (essential)	• Field sample concentrations were at least several orders of magnitude above ambient water levels. • Uptake inversely related to organic pesticide uptake, pH and salinity, and directly related to temperature and O_2 levels. High variability in uptake among species (even from same area). • Required element. Macrophytes (especially in marshes) play role in cycling of element.	• Lowest concentration factor from field collections greater than 1500x. • Concentration factors as high as 1.7 to 4 million x in kidney, as high as 26,000x to 100,000x in soft parts. • Filter feeders are high accumulators. • Highest in kidney and digestive gland. • Uptake varies in some species with weight and sex; much can be bound by proteins or in organelles. • Higher temperature and lower salinity increased zinc levels. • Loss rates may be rapid or slow. Reported half-life of 255 days in oysters.	• Zn apparently not limiting in marine environment. Also accumulates beyond "needs," likely bound to an enzyme-organic fraction. • Food considered more important pathway for uptake than water. Losses of internal Zn measured in months at fastest. • High degree of variability in uptake with life stage of species. • Uptake higher at higher temperature and lower salinity.	• Field concentrations generally low. Highest levels in viscera and gonad, low in muscle. • Food greater source of metal than water. • Age and sex of fish affect uptake. • Zn-binding proteins have been identified. • Uptake affected by organo-chlorine pesticides, Cd and Cu as well as temperature, salinity, and others. • Uptake may increase after death.

Note: Concentration factors, where given, represent the only numbers from a small data base or upper-range, "worst case" numbers from a large data base.

Sources: Eisler [13]; Phillips and Russo [32]

circumstance where modeling is used to trigger an "early warning" system is through the "Mussel Watch" program. Mussels are used as indicators of contamination, and are collected and analyzed on a regular basis. Its major benefit is to provide an early warning, but it works well only if samples are collected where contaminant concentrations are high. Because of the lack of uniformity typical of marine water concentrations, it is possible for pulses that are toxic to some organisms to be missed by this method. However, it remains a significant monitoring tool, even if it is not as useful for broader predictive purposes.

The Mussel Watch model, as reported by Goldberg *et al.* [16], is:

$$dM/dt = KM_{sw} - TM$$

where

M	=	metal concentration in bivalve tissue
t	=	age of organism
K	=	constant
T	=	biological half-life of metal
M_{sw}	=	concentration of metal in seawater

The principal limitation of this model is the assumption that M_{sw}, remains constant for any appreciable length of time. In fact, major pulses of different concentrations are common in field exposure situations. A secondary limitation may be the assumption that the constant K can be based on linearity in the availability/uptake relationship; while linearity may prevail under some of the conditions of uptake observed to date, the extent of appropriate generalization from these observations is unclear [13].

4.5 UPTAKE BY FRESHWATER BIOTA

4.5.1 Introduction

The freshwater environment includes a wide range of water characteristics and habitats whose variations affect uptake processes. Uptake in these environments can become problematic because of their heavy use for water supply (e.g., crop irrigation) and recreational harvest of edible finfish and waterfowl. Some freshwater habitats are large, relatively deep lakes, characterized by seasonal changes in temperature, oxygen concentration and pH stratification, with subsequent turnover and mixing. Other lakes that are too small or too shallow to stratify have limited water exchange and are particularly vulnerable to contaminant buildup. Major rivers are characterized by large-volume flows and sediment loads, even with seasonal variations in input. Many small streams and headwaters regularly experience great fluctuations in flow volume, sediment load, temperature, pH and oxygen as the result of seasonal changes and storms. These characteristics can strongly influence exposure to and uptake of inorganics by adding or removing inorganics and changing their availability.

Some less readily observed characteristics can also influence the uptake of inorganic substances in fresh water:

- The nature and degree of watershed development (roads, deforestation, houses, etc.) can influence the quality and quantity of water that enters a surface freshwater system. Inputs may include organic matter such as leaf litter, or nutrients that stimulate the production of organic matter, which in turn affects uptake. Disturbed areas can contribute increased concentrations of leachable substances such as organic acids from natural soils [38]. Watersheds with agricultural and/or urban development may contribute a wide range of organic as well as inorganic substances. Such contributions may increase or decrease the potential for uptake of inorganics that are already present, as well as contributing substances directly for uptake.
- Freshwater systems show great variation in pH, water hardness and alkalinity, and the presence and concentration of a number of solid materials and dissolved constituents (e.g., phosphates and clays or silts) that can form immobilizing complexes with inorganic constituents otherwise available for uptake.

In each freshwater environment, some or all of these factors affect the equilibria among the various chemical forms of an introduced or newly mobilized inorganic substance and, hence, the amounts and chemical species available to biological organisms.

4.5.2 Uptake and Loss Processes in Freshwater Organisms

In general, the discussion of uptake and loss processes for marine biota in § 4.4.2 and § 4.4.3 applies equally to uptake by freshwater biota.

Evidence exists for binding of some heavy metals by specific metallothionein-like proteins in some freshwater fish. Such proteins apparently increase with exposure of the biota to mercury, zinc, or copper [33] or to cadmium [32]. Copper- and cadmium-specific metallothioneins have been identified in freshwater species [33]. Zinc may associate with more than one type of binding protein [32]. The data presented to date for freshwater species (*Salmonids*) do not yet indicate support for the hypothesis (discussed in § 4.4.2) that toxicity occurs when the detoxification capacity of the binding proteins is exceeded; however, they do support the potential corollary hypothesis of protection to those organisms that are preexposed, produce the detoxifying protein, and are then reexposed to otherwise toxic concentrations [33]. Numerical relationships between accumulation levels, protein binding, and toxicity to the organism are not yet defined as well by these data as they have been in the research on marine species.

CHEMICAL ENVIRONMENTAL VARIABLES

There are additional explanations for variations in inorganic metal uptake, primarily involving water chemistry. For a substance to be toxic, it must be available for uptake. A number of inorganic constituents, including many of the trace metals, exist in more than one valence state and are complexed and/or in equilibrium with other substances in the water. Several of these factors, which are known to alter the types or concentrations of ions present, have significant effects on the toxicity of many inorganic species. The ones most frequently cited in demonstrable correlations with toxicity effects are water hardness/alkalinity, pH, organic matter, and (sometimes) temperature. Because they can have such a dramatic effect on toxicity, the EPA has stipulated varying criteria for the allowed concentrations of some elements in surface waters. Figure 4.5-1 shows, for example, how the EPA's freshwater aquatic life criteria increase with increasing hardness of the water.

In fresh-water systems as a whole, these factors vary considerably from one water body to the next; thus, it is important to examine the role that they may play in the uptake of specific constituents. One should also be aware that each of these water quality factors may affect the availability/uptake of only some inorganics, and may affect them in differing degrees.

Hardness and alkalinity are related measures of water quality (carbonate and bicarbonate). They are often similarly high or low. Highly alkaline and hard waters tend to have a high pH; soft water usually has a lower pH, but not always. Where there is an effect on substance toxicity/uptake, each of these factors may act in combination or separately to influence the availability of a specific inorganic species. Examples of modes of interaction include competition for binding sites, complexation to reduce availability, and shifting of equilibria among the forms of an inorganic

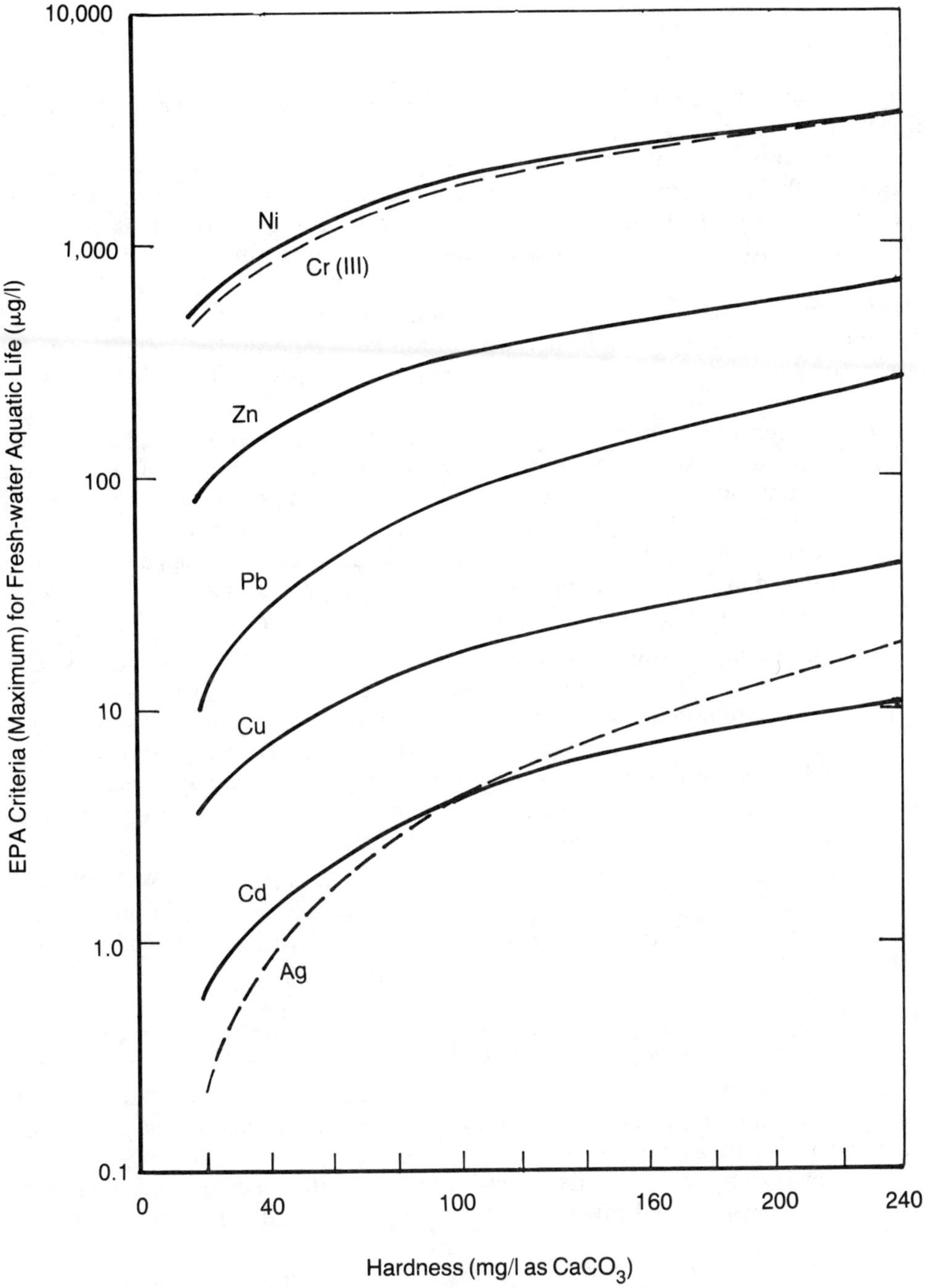

Note: Curves were derived from equations given in Table D-3 of Appendix D. For Ag, Ni, and Zn, criteria are maximum concentrations not to be exceeded at any one time. For Cd, Cr, Cu and Pb, criteria are one-hour average concentrations, not to be exceeded more than once every three years on the average.

FIGURE 4.5-1 EPA Criteria for Fresh-water Aquatic Life as a Function of Water Hardness

substance present. Sometimes the interaction among these factors is understood; at other times it can only be deduced. Unfortunately, the historic body of literature has not always noted what are now considered key water quality parameters (e.g., hardness/alkalinity, pH, organic matter, temperature). Because such parameters appear to affect the availability of toxic forms of a constituent, it must also be assumed that there is some effect on uptake and subsequent accumulation.

Temperature changes affect the metabolism of aquatic biota, and there is evidence that the uptake of trace metals increases at higher temperatures [32]. Several of these interactions are described in the examples below (and are also listed in Table 4.5-1).

- Copper occurs in many forms in natural waters. Soluble cupric ions are believed to be the toxic form as well as the form most available for accumulation. Calcium ions are believed to compete for binding sites on biological surfaces, indirectly interfering with copper uptake [29]. Carbonate ions and increasing pH are thought to shift copper equilibria toward less available soluble forms; thus, copper uptake is believed to be reduced by higher alkalinity/hardness/pH [46,57].
- Cadmium, another inorganic ion that can be highly mobile and has been toxic through food-web accumulation, is precipitated from solution by carbonate and adsorbed by a number of naturally occurring substances. As pH increases, these processes become more effective, and cadmium may precipitate from the water column [42]. In a very recent study, increased calcium levels in water have been shown to reduce the uptake and accumulation of cadmium in fish [31].
- The uptake and toxicity of chromium VI was judged by the EPA to be strongly influenced by water hardness [43]. Uptake of chromate by fish increased with a shift in pH from 7.8 to 6.5, which purportedly caused greater oxidizing action by chromate ions [55].

Mathematical relationships between observed toxicity and water hardness, in those cases where regression correlations are plausible, have been formulated by the EPA for use in approximating safe water levels for some substances (such as Cd, Cu, and Zn; see Appendix D). Such correlations have not been tested/observed for accumulation of these same constituents. A qualitative relationship, similar to that observed for toxicity, is believed to occur; in fact, as noted above, cadmium uptake and accumulation by perfused gills was reduced by increased water hardness.

An additional factor that is believed to affect the uptake of some inorganic species is the presence of organic matter in the water. Loosely defined, this may include humic substances and other suspended organic matter, as well as dissolved amino acids. Different forms of a number of inorganic species found in water are adsorbed or complexed by such material. Most often, this associated material is less toxic or less available for uptake, and may often be precipitated and become associated with the sediments [8, 57]. These factors are most important in the measurement of water and sediment concentrations of a specific inorganic. For example, measurement of "total" levels of a trace metal could include chelated or adsorbed forms that may not be available for uptake by biota. A capability for defining and then measuring available forms has remained largely elusive.

There are exceptions to the above broad generalizations. Methylation renders some inorganics more soluble and more readily available for uptake, and the resulting organometallic species may be more or less toxic than the prior form. The classic example of a more toxic, available form that bioaccumulates is methylmercury. Any mercurial compound introduced into water can apparently be converted to methylmercury. (See section 2.15, Microbial Transformations.) The conversion of methylmercury to elemental mercury can also occur but evidently is not significant unless large amounts of methylmercury exist [32]. Significantly, methylmercury does not behave like its inorganic form; it is more readily taken up, especially by fish, and is lost very slowly. Its half-life is probably 1-3 years. (See Table 4.5-1.) Further, unlike many other inorganics of potential concern to human consumers, methylmercury concentrates in muscle tissue as well as in other organs [32].

Arsenic is another example of an inorganic element that can be methylated by bacteria. In this case, however, the highly toxic methyl and dimethyl products are both highly volatile and readily oxidized. Their bioaccumulative significance is unknown [32].

Suspended or deposited sediments, including clays, can act in a similar fashion to reduce the availability of trace inorganics for uptake in freshwater as well as in terrestrial systems. While the presence of these substances has been shown to reduce the toxicity of some heavy metal species, predictive or quantifiable relationships have not been demonstrated. In view of the variety of potential complexing agents, this is not surprising.

PHYSICAL ENVIRONMENTAL VARIABLES

The above discussion points up the need to understand the environmental circumstances that are coupled with any examination of the fate of inorganic constituents. Streams, for example, usually have small dilution capacity; however, if they flow through an area that contributes large amounts of leaf litter and other organics or a fairly heavy sediment load, the uptake/toxicity potential of inorganics such as some trace metal species can be reduced. Large lakes represent another example: many become stratified during warm months, and the temperature barriers prevent mixing of water between upper and lower levels. The resultant anaerobic condition in the lower level can cause a seasonal decrease in pH, which disturbs adsorption and speciation equilibria and causes some trace metal species to be released from the sediments. During the fall turnover, when surface waters become colder and sink, there is a remixing of the water column and a resuspension of material, increasing potential uptake.

Uptake processes for inorganics are also affected by other watershed and waterbody characteristics. For example, water in areas of igneous or metamorphic bedrock tend to have low alkalinity, while areas of sedimentary or calcareous bedrock geology and accompanying higher weathering rates tend to produce waters with higher alkalinity. (See Figure 4.5-2.) When the former circumstances are accompanied by watershed soils that are acidic and/or characterized by low buffering capacity, lakes in the region are potentially vulnerable to damage from acid precipitation [38].

Patterns of acid precipitation for the eastern United States are illustrated in Figure 4.5-3. The sufficient lowering of pH can have an adverse effect in and of itself. Moreover, such circumstances also lead to the release of otherwise bound inorganic constituents such as aluminum species through the shifting of pH-related equilibria in soils and sediments. More to the point of this discussion, low pH sometimes appears to enhance uptake and toxicity of harmful inorganics, including many trace metal species.

Finally, it is important to note that even when existing circumstances suggest low uptake potential, they may change and alter uptake risk. Annual turnover in large lakes, agricultural drainage modifications, and acid precipitation effects are examples of short- and long-term changes in environmental circumstances that alter uptake potential.

4.5.3 Data and Criteria on Uptake Potential

In general, the data available on uptake of inorganics by freshwater organisms are similar to that for marine organisms (§ 4.4.4.) The major difference is that a far greater volume of data exists for freshwater organisms, but, perhaps as a consequence, it has not been compiled as comprehensively or systematically as that for marine biota or terrestrial vegetation.

As in the two earlier sections, very general summaries concerning the behavior of a number of inorganics in fresh water are given in Table 4.5-1. Note that many trace elements become more available in soft, non-alkaline water but that a few are more available in alkaline situations. The high degree of environmental variability in freshwater systems makes it particularly difficult to generalize uptake rates.

Appendix D lists USEPA, FDA and other nationally applicable freshwater bioaccumulation criteria for inorganics. Table 4.2-4 in this chapter lists criteria developed by the State of California based on combined availability from dissolved and leachable sediment sources.

4.5.4 Models and Predictive Techniques

As for the other environments, no generally applicable models or predictive techniques are available for quantitative estimation of uptake or accumulation of inorganics by freshwater biota. The degree of water hardness, more specifically measured as calcium carbonate concentration, shares a demonstrated relationship with the toxicity of some of the species of heavy metals such as Cu and Zn. For these metals, there has been sufficient laboratory-demonstrated change in toxicity with

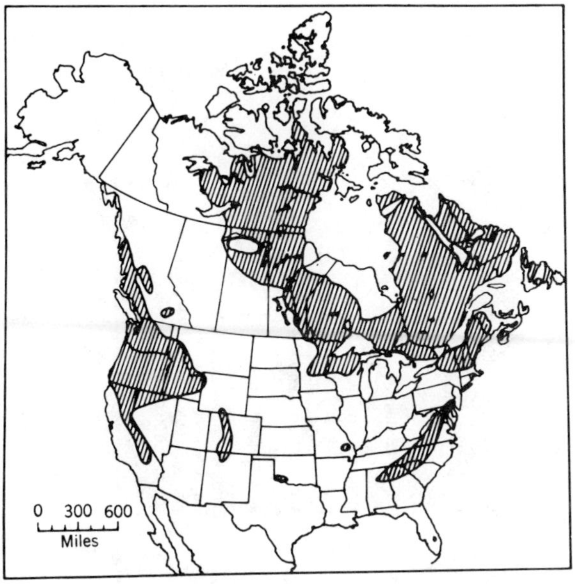

Shaded areas have igneous or metamorphic bedrock geology; unshaded areas have calcareous or sedimentary bedrock geology.

Because of the slower geological weathering rates in areas of metamorphic and igneous bedrock geology, lakes in these areas are expected to be dilute and to have low alkalinities (<0.5 meq HCO_3^- per liter). This hypothesis was tested by compiling lake alkalinity data from several sources; regions having lakes with low alkalinity coincided with the shaded areas.

Source: Galloway and Cowling [11]. *(Copyright 1978, Air Pollution Control Association. Reprinted with permission.)*

FIGURE 4.5-2 Regions in North America Containing Lakes That Are Sensitive to Acidification by Acid Precipitation

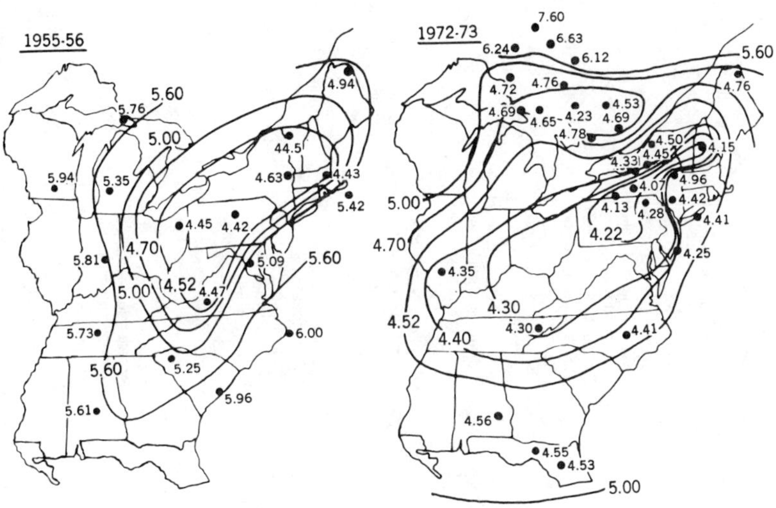

Source: Galloway and Cowling [11]. *(Copyright 1978, Air Pollution Control Association. Reprinted with permission.)*

FIGURE 4.5-3 Weighted Annual Average pH of Precipitation in the Eastern United States

TABLE 4.5-1

Some General Characteristics of Trace Metal Uptake in Selected Groups of Freshwater Organisms

Chemical	General Information	Uptake in Plants	Uptake in Invertebrates	Uptake in Aquatic Vertebrates	Sources
Arsenic (non-essential)	• May be present in trivalent, pentavalent or organic complexes and other organic forms in water. • Wide range of *toxicity* among species tested with inorganic trivalent and pentavalent forms. • Water hardness seems to have little effect on toxicity. • Laboratory data base limited; reflects several orders of magnitude differences in active toxicity among species tested.	• Data not available for freshwater aquatic species.	• Appears to be little difference in uptake of inorganic and several organic forms. • Uptake is low at levels that are not also toxic. • Less than 20x for freshwater species with trivalent, pentavalent and some organic arsenic forms.	• Studies show low bioconcentration and short half-life for concentrations that are not also toxic. • Less than 20x for freshwater species tested with trivalent, pentavalent and some organic arsenic forms. • Fats contain more than other tissue. Does accumulate in muscles.	[32, 40, 41]
Cadmium (non-essential)	• Probably occurs in many forms in fresh water. • Calcium carbonate (hardness) reduces toxicity, but effect of hardness on uptake not demonstrated. • High organic, high pH reduce toxicity, but effect on uptake not demonstrated. • Laboratory acute toxicity tests reflect over 4 orders of magnitude across species tested. Test water hardness is partly responsible for wide range.	• Over long time, concentration factors may be extremely high, and certainly similar to any high uptake potential in animals. (Up to and greater than 1000x.)	• Steady-state concentrations appear to be reached quickly in freshwater crustaceans. • Data from marine molluscs indicate that some invertebrates are very high accumulators and that a long time is required for a steady state to be reached. • Existence of cadmium-binding protein demonstrated in marine molluscs (up to 3500x in freshwater invertebrates.)	• Very broad range of uptake values reported. Very low concentration factor shown for muscle. Highest levels from whole body analysis and viscera. • Apparently takes more than 28 days for steady-state concentrations to be reached. • Elimination of cadmium accumulated from water is slower than elimination of accumulations from food. • Binds with metallothionein-like protein in liver and kidney (up to about 12,000x in muscle).	[32, 42]

(Continued)

TABLE 4.5-1 (Continued)

Chemical	General Information	Uptake in Plants	Uptake in Invertebrates	Uptake in Aquatic Vertebrates	Sources
Chromium (essential element)	• Hexavalent forms vary somewhat in fresh water, and toxicity does not seem affected by hardness. • Trivalent forms form stable complexes with negatively charged species, so less soluble/toxic. • Toxicities of the two are not widely different. Toxicity covers three orders of magnitude across species.	• Uptake of hexavalent form by algae community reported at 8500x. • Data unavailable for trivalent form.	• Data not available for freshwater species; uptake in marine forms in the 80x to 200x range (not extensive data base.)	• Uptake in one tested species less than 5x. • Relatively little accumulates in edible tissues.	[32, 43, 44]
Copper (essential element)	• Required micronutrient • Soluble ions readily complexed by organic matter and calcium carbonate, and absorbed by suspended solids. Higher pH decreases availability. • Only water hardness presently reflected in regulatory evaluation. Other factors above reduce toxicity as well, but effect on uptake less certain. • Significance of tissue residues uncertain.	• A value of 2000x for one species (marine values higher).	• Single value of about 200x for insect larvae. (Marine values higher.)	• Value of 300x for one fish only, 1x for fish muscle in another test. • Uptake appears to begin at levels that also result in chronic effects.	[32, 45]
Lead (non-essential)	• Many problems encountered in analytical measurement. • Precipitation of divalent cation by hydroxide, carbonate and organic matter common. Less toxic in hard water and higher pH. • Up to three orders of magnitude in laboratory acute toxicity — with water hardness and species differences credited. Effect on uptake not noted.	• Data unavailable for freshwater species. (One saltwater value more than 700x.)	• Values reported from 500x to 1700x for whole-body levels. • Highest concentrations in viscera.	• Whole-body concentration factors around 50x. • Mucus concentrates lead to levels much higher than organism. May "contaminate" samples.	[32, 47, 48]

(Continued)

TABLE 4.5-1 (Continued)

Chemical	General Information	Uptake in Plants	Uptake in Invertebrates	Uptake in Aquatic Vertebrates	Sources
Mercury (non-essential)	• Elemental *and* divalent forms can be methylated. • Methylmercury is more water-soluble, many times more toxic, and has higher uptake/bioconcentration potential than many inorganic forms. • Toxicity and uptake increase with increased temperature; higher uptake at lower pH. • Bioconcentration is high because uptake rates are fast, loss rates slow. • Depuration of methylmercury so slow that this form predominates in tissue even with greater exposure to inorganic forms.	• Short-term non-steady-state concentration factor of 29,000x reported for inorganic mercury. • Short-term concentration of up to 2100x for methylmercury.	• Steady-state data unavailable for fresh-water species. • Up to 40,000x for methylmercury for saltwater bivalves. • 10,000x for inorganics in saltwater at steady state. • Short-term methylmercury uptake up to the 8000x range.	• Half-life of methylmercury in fish tissue 2 to 3 years. • Mercury moves from absorbing surface to internal organs via blood, then kidney/gall bladder for recycle or elimination, and muscle for long-term storage. • Uptake via gills *and* intestine. • Selenium antagonism. • Muscle and other tissue concentration similar. • 5000x long-term uptake factor in fish for inorganic forms. Up to 64,000x long-term uptake factor for methylmercury.	[32, 49, 50]
Nickel (may be essential element)	• Toxicity reduced with increase in hardness. $CaCO_3$ likely reduces availability.	• Value of about 10x reported for fresh-water algae.	• Value of 100x reported for fresh-water crustacean (saltwater bivalve values 3 to 4 times higher).	• Value of less than 100x reported for one fresh-water fish species.	[32, 51]
Selenium (essential element)	• Selenates (+6) may be less toxic than selenites (+4) and favored by alkaline and oxidizing conditions. • Essential element in animal nutrition. • Toxicity data base mostly for selenites; indicates great biological species differences. • Data on effects of pH, hardness, etc. insufficient. • Freshwater data unavailable for organic forms which can be formed from inorganic forms in environment.	• None reported for fresh-water species.	• None reported for fresh-water species.	• Uptake/concentration values for selenite in fresh-water fish up to 78x for whole body and 18x for muscle. • Half-life in fish about 2 months.	[32, 52]

(Continued)

TABLE 4.5-1 (Continued)

Chemical	General Information	Uptake in Plants	Uptake in Invertebrates	Uptake in Aquatic Vertebrates	Sources
Zinc (essential element)	• Toxicity reduced with increasing hardness. • Required trace element in most organisms. • Occurs in many forms — dissolved, suspended, etc.	• None reported for fresh-water species. (Up to 16,000x reported for marine species.)	• Uptake values up to 1130x reported (up to 1600x for marine mollusc).	• Uptake values up to 432x reported for fresh-water species.	[32, 53]

(Concentration factors, where noted, represent the only numbers for a small data base *OR* the upper range, "worst case" numbers from a large data base.)

measured changes in calcium carbonate hardness to permit significant correlations by regression analyses. These, in turn, have been used to predict safe and potentially toxic concentrations of the associated metals in waters of varying hardness [42, 45, 53]. Unfortunately, data to date have not allowed comparably demonstrable relationships for varying uptake potential and water hardness. Similarly, reliably predictive relationships have not been broadly developed between toxicity or uptake potential and any other of the important water quality parameters, such as pH, sediment load, organic matter (other than that described by JRB [19]), redox state or temperature. For the time being, they must be discussed ad hoc or qualitatively for most inorganic chemicals. Exceptions include ammonia, where combinations of temperature and pH can be reliably used to predict the percentage of the un-ionized (toxic) form, but not necessarily its uptake pattern.

Some models, such as the EPA's EXAMS, use simplified uptake algorithms based on phenomena such as adsorption (not really uptake) by phytoplankton. These have little predictive value in problem definition for higher species and are mainly used in descriptions of the interim fate of contaminants.

Thus, the type of decision process outlined in § 4.2.4 above is recommended as the basis for identifying and resolving potential uptake-related problems in the freshwater environment.

4.6 LITERATURE CITED

1. Baes, C.F. III, R.D. Sharp, A.L. Sjoreen and R.W. Shor, "A Review and Analysis of Parameters for Assessing Transport of Environmentally Released Radionuclides Through Agriculture," Report No. ORNL-5786 prepared by Oak Ridge National Laboratory for the U.S. Environmental Protection Agency (1984). (NTIS: DE85-000287.)
2. Barber, S.A., *Soil Nutrient Bioavailability — A Mechanistic Approach*, John Wiley & Sons, New York (1984).
3. Bascom, W. (ed.), *Biennial Report, 1981-82*, Southern California Coastal Water Research Project, Long Beach, CA (1982).
4. Bascom, W. (ed.), *Biennial Report, 1983-84*, Southern California Coastal Water Research Project, Long Beach, CA (1984).
5. Brown, D.A., S.M. Bay, P. Szalay, G.P. Hershelman, C.F. Ward, A.M. Westcott and D.J. Greenstein, "Metal and Organic Detoxification/Toxification in Fish Livers and Gonads," in Bascom [4], 195-209.
6. Brown, D.A., R.W. Gossett, G.P. Hershelman, C.F. Ward and J.N. Cross, "Metal and Organic Contaminants in Sediments and Animals," in Bascom [4], 179-93.
7. Brown, D.A., K.D. Jenkins, E.M. Perkins, R.W. Gossett and G.P. Hershelman, "Detoxification of Metals and Organic Compounds in White Croakers," in Bascom [1], 157-71.
8. Brown, V.M., T.L. Shaw and D.G. Shurben, "Aspects of Water Quality and the Toxicity of Copper to Rainbow Trout," *Water Res.*, **8**, 797-803 (1974).
9. California Administrative Code; Title 22, Division 4: Environmental Health, Article 11: "Criteria for Identification of Hazardous and Extremely Hazardous Wastes," distributed by State of California, Documents Section, P.O. Box 1015, North Highlands, CA 95660.
10. Chaney, R.L. (Beltsville, MD), personal communication to W. Lyman (Arthur D. Little, Inc.), June 1986.
11. Chaney, R.L., J.C. Smith, Jr., D.E. Baker, R. Bruins, D. Cole and R.F. Korcals, "Transfer of Sludge Applied Trace Elements to the Food Chain," draft report (unpublished) (1986).
12. Cowan, C.E., E.A. Jenne and R.R. Kinnison, *Methodology for Determining the Relationship between Toxicity and the Aqueous Speciation of a Metal*, PNL-SA-12471, CONF-8504147-1, NTIS No. DE85012442/WEP, Battelle Pacific Northwest Labs., Richland, Wash. (1985).
13. Eisler, R., *Trace Metal Concentrations in Marine Organisms*, Pergamon Press, New York (1981).
14. Galloway, J.N. and E.B. Cowling, "The Effects of Precipitation on Aquatic and Terrestrial Ecosystems: A Proposed Precipitation Chemistry Network," *J. Air Pollut. Control Assoc.*, **28**, 229-35 (1978).
15. Gerritse, R.G., W. Van Driel, K.W. Smilde and B. Van Luit, "Uptake of Heavy Metals by Crops in Relation to Their Concentration in the Soil Solution," *Plant Soil*, **75**, 393-404 (1983).

16. Goldberg, E.D., V.T. Bowen, J.W. Farrington, G. Harvey, J.H. Martin, P.L. Parker, R.W. Rosebrough, W. Robertson, E. Schneider and E. Gamble, "The Mussel Watch," *Envir. Conserv.*, **5**, 101-25 (1978), in Eisler [13].

17. Holmgren, G.G.S., M.W. Meyer, R.B. Daniels, R.L. Chaney and J. Kubota, "Cadmium, Lead, Zinc, Copper and Nickel in Agricultural Soils of the United States," submitted to *J. Environ. Qual.* (1986).

18. Jenkins, D.W., "Biological Monitoring of Toxic Trace Metals: Volume I. Biological Monitoring and Surveillance," Report No. EPA-600/3-80-089, U.S. Environmental Protection Agency, Las Vegas, NV (1980).

19. JRB Associates (Belleview, Wash.), *Initial Evaluation of Alternatives for Development of Sediment-related Criteria for Contaminants in Marine Waters (Puget Sound). Phase II: Development and Testing of the Sediment-Water Equilibrium Partition Approach*, EPA Report 910/9-83-117, U.S. Environmental Protection Agency, Washington, D.C. (1984).

20. Kabata-Pendias, A. and H. Pendias, *Trace Elements in Soils and Plants*, CRC Press, Boca Raton, Fla. (1984).

21. Ketwig, K., "Lost Condors Spark Debate; Lead Poisoning Takes Two," *Audubon Action,* **3** (Aug. 1985).

22. Lee, C.R., B.L. Folsom, Jr., and R.M. Engler, "Availability and Plant Uptake of Heavy Metals from Contaminated Dredged Material Placed in Flooded and Upland Disposal Environments," *Environ. Internat.*, **7**, 65-71 (1982).

23. Lee, H. (U.S. Environmental Protection Agency, Newport, OR), personal communication to W. Lyman (Arthur D. Little, Inc.) (May 1986).

24. Leeper, G.W., *Managing the Heavy Metals on the Land*, Marcel Dekker, New York (1978).

25. Mengel, K. and E.A. Kirkly, *Principles of Plant Nutrition*, 3rd edition, International Potash Institute, Bern, Switzerland (1982).

26. National Academy of Sciences and National Academy of Engineering, *Water Quality Criteria, 1972*, EPA R3-73-033, U.S. Environmental Protection Agency, Washington, D.C. (1973).

27. O'Donnel, J.R., B.M. Kaplan and H.E. Allen, "Bioavailability of Trace Metals in Natural Waters," *Aquatic Toxicology and Hazard Assessment: Seventh Symposium*, ASTM STP 854, R.D. Cardwell, R. Purdy and R.C. Bahner (eds.), American Society for Testing and Materials, Philadelphia, 485-501 (1985).

28. Overcash, M.R. and D. Pal, *Design of Land Treatment Systems for Industrial Wastes — Theory and Practice*, Ann Arbor Science Publishers, Ann Arbor, Mich. (1979).

29. Pagenkopf, G.K., "Gill Surface Interaction Model for Trace-Metal Toxicity to Fishes: Role of Complexation, pH, and Water Hardness," *Environ. Sci. Technol.*, **17**, 342-47 (1983).

30. Parr, J.F., P. B. Marsh and M. Kla, *Land Treatment of Hazardous Wastes*, Noyes Data Corp., Park Ridge, N.J. (1983).

31. Pärt, P., O. Svanberg and A. Kiessling, "The Availability of Cadmium to Perfused Rainbow Trout Gills in Different Water Qualities," *Water Res.*, **19**, 427-34 (1985).

32. Phillips, G.R. and R.C. Russo, *Metal Bioaccumulation in Fishes and Aquatic Invertebrates: A Literature Review*, EPA-600/3-78-103, U.S. Environmental Protection Agency, Environmental Research Laboratory, Duluth, Minn. (1978).

33. Roch, M., J.A. McCarter, A.T. Matheson, M.J.R. Clark and R.W. Olafson, "Hepatic Metallothionein in Rainbow Trout (*salmo gairdneri*) as an Indicator of Metal Pollution in the Campbell River System," *Can. J. Fish. Aquat. Sci.*, **39**, 1596-1601 (1982).

34. Römheld, V. and H. Marschner, "Evidence for a Specific Uptake System for Iron Phytosiderophores in Roots of Grasses," *Plant Physiol.*, **80**, 175-80 (1986).

35. Ryan, J.A., R.L. Chaney, G. Prince, A.D. Otte and J.M. Walker, "Review of the Soil Factors and the Concept of Cumulative Application Rate of Cd on Cd Content of Crops," Background Document from J.A. Ryan, USEPA, WRD, MERL to A.A. Peter, Jr., EPA, OSW (May 13, 1980).

36. Santhanam, C.J., A. Balasco, I. Bodek and C.B. Cooper, *Full-scale Field Evaluation of Waste Disposal from Coal-Fired Electric Generating Plants, Vol. II*, Report EPA-600/7-85-028b, U.S. Environmental Protection Agency, Office of Research and Development, Washington, D.C. (1985).

37. Takagi, S., K. Nomoto and T. Takemoto, "Physiological Aspect of Mugineic Acid, a Possible Phytosiderophore of Craminaceous Plants," *J. Plant Nutrition*, **7**, 469-77 (1984).

38. Talbot, R.W. and A.W. Eizerman, "Acidification of Southern Appalachian Lakes," *Environ. Sci. Technol.*, **19**, 552-57 (1985).

39. U.S. Department of the Interior, *Kesterton Reservoir Closure and Cleanup Plan*, Bureau of Reclamation, Mid-Pacific Regional Office, Sacramento, Calif. (1985).

40. U.S. Environmental Protection Agency, *Ambient Water Quality Criteria for Arsenic*, EPA 440/5-80-021, Office of Water Regulations and Standards, Criteria and Standards Division, Washington, D.C. (1980).

41. U.S. Environmental Protection Agency, "Revised Section B of *Ambient Water Quality Criteria for Arsenic*," Draft (Aug. 19, 1983).

42. U.S. Environmental Protection Agency, *Ambient Water Quality Criteria for Cadmium*, EPA 440/5-80-025, Office of Water Regulations and Standards, Criteria and Standards Division, Washington, D.C. (1980).

43. U.S. Environmental Protection Agency, *Ambient Water Quality Criteria for Chromium*, EPA 440/5-80-005, Office of Water Regulations and Standards, Criteria and Standards Division, Washington, D.C. (1980).

44. U.S. Environmental Protection Agency, "Revised Section B of *Ambient Water Quality Criteria for Chromium*," Draft (Aug. 19, 1983).

45. U.S. Environmental Protection Agency, *Ambient Water Quality Criteria for Copper*, EPA 440/5-80-036, Office of Water Regulations and Standards, Criteria and Standards Division, Washington, D.C. (1980).

46. U.S. Environmental Protection Agency, "Revised Section B of *Ambient Water Quality Criteria for Copper — Aquatic Toxicology*," Draft (Aug. 19, 1983).

47. U.S. Environmental Protection Agency, *Ambient Water Quality Criteria for Lead*, EPA 440/5-80-057, Office of Water Regulations and Standards, Criteria and Standards Division, Washington, D.C. (1980).

48. U.S. Environmental Protection Agency, "Revised Section B of *Ambient Water Quality Criteria for Lead*," Draft (Aug. 19, 1983).

49. U.S. Environmental Protection Agency, *Ambient Water Quality Criteria for Mercury*, EPA 440/5-80-058, Office of Water Regulations and Standards, Criteria and Standards Division, Washington, D.C. (1980).

50. U.S. Environmental Protection Agency, "Revised Section B of *Ambient Water Quality Criteria for Mercury*," Draft (Aug. 19, 1983).

51. U.S. Environmental Protection Agency, *Ambient Water Quality Criteria for Nickel*, EPA 440/5-80-060, Office of Water Regulations and Standards, Criteria and Standards Division, Washington, D.C. (1980).

52. U.S. Environmental Protection Agency, *Ambient Water Quality Criteria for Selenium*, EPA 440/5-80-070, Office of Water Regulations and Standards, Criteria and Standards Division, Washington, D.C. (1980).

53. U.S. Environmental Protection Agency, *Ambient Water Quality Criteria for Zinc*, EPA 440/5-80-079, Office of Water Regulations and Standards, Criteria and Standards Division, Washington, D.C. (1980).

54. U.S. Fish and Wildlife Service, *Fed. Regist.*, **50**, 30849-53 (July 30, 1985).

55. Van Der Putte, I., J. Lubbers and Z. Kolar, "Effect of pH on Uptake, Tissue Distribution and Retention of Hexavalent Chromium in Rainbow Trout (*salmo gairdneri*)," *Aquat. Toxicol.*, **1**, 3-18 (1981).

56. Vaughan, B.E., K.H. Abel, D.A. Cataldo, J.M. Hales, C.E. Hane, L.A. Rancitelli, R.C. Roufson, R.E. Wildung and E.G. Wolf, "Review of Potential Impact on Health and Environmental Quality from Metals Entering the Environment as a Result of Coal Utilization," Battelle Energy Prog. Rpt., Pacific Northwest Laboratory (1975).

57. Winner, R.W., "Bioaccumulation and Toxicity of Copper as Affected by Interactions between Humic Acid and Water Hardness," *Water Res.*, **19**, 449-55 (1985).

58. Wolnik, K.A., F.L. Fricke, S.G. Capar, M.W. Meyer, R.D. Satzger, E. Bonnin and C.M. Gaston, "Elements in Major Raw Agricultural Crops in the United States, 3. Cadmium, Lead and Eleven other Elements in Carrots, Field Corn, Onions, Rice, Spinach, and Tomatoes," *J. Agric. Food Chem.*, **33**, 807-11 (1985).

59. Wolnik, K.A., F.L. Fricke, S.G. Capar, G.L. Braude, M.W. Meyer, R.D. Salzger and E. Bonnin, "Elements in Major Raw Agricultural Crops in the United States, 1. Cadmium and Lead in Lettuce, Peanuts, Potatoes, Soybeans, Sweet Corn and Wheat," *J. Agric. Food Chem.*, **31**, 1240-49 (1983).

60. Yang, G., S. Wang, R. Zhov and S. Sun, "Endemic Selenium Intoxication of Humans in China," *Am. J. Clinical Nutrition*, **37**, 872-81 (1983).

4.7 SOME USEFUL SOURCES

In addition to the sources listed above, the following will also be found useful for background information:

61. Cardwell, R.D., R. Purdy and R.C. Bahner (eds.), *Aquatic Toxicology and Hazard Assessment: Seventh Symposium*, Report No. STP 854, American Society for Testing and Materials, Philadelphia, PA (1985).

62. Solomons, W. and U. Forstner, *Metals in the Hydrocycle*, Springer-Verlag, New York (1984).

Particularly useful sources for *assessment approaches* are given in references 2, 28 and 61.

Particularly useful sources of *data* are references 13, 17, 18, 20, 28, 58 and 59.

5. MATHEMATICAL ENVIRONMENTAL FATE MODELING

Marc Bonazountas, Aviva Brecher, Robert G. Vranka

LIST OF TABLES

AIR MODELING

TERRESTRIAL MODELLING

AQUATIC MODELING

RADIONUCLIDE MIGRATION MODELING

LIST OF FIGURES

BACKGROUND

AIR MODELING

TERRESTRIAL MODELLING

AQUATIC MODELING

5. MATHEMATICAL ENVIRONMENTAL FATE MODELING

5.1 INTRODUCTION

This chapter can be read as a separate document; it describes concepts, uses, and limitations of state-of-the-art mathematical environmental pollutant fate modeling for use in both environmental studies and analyses of environmental quality or human exposure.

The purpose of this chapter is to help readers understand modeling complexities, identify specific modeling packages, and select a documented mathematical model.

The text that follows is divided into five major sections:

- Background (Section 5.2),
- Air Modeling (Section 5.3),
- Terrestrial Modeling (Section 5.4),
- Aquatic Modeling (Section 5.5), and
- Radionuclide Modeling (Section 5.6).

The models have been categorized on the basis of the types of environmental systems they include, *not* the environmental processes involved. Aquatic and terrestrial models differ significantly in the data base (minerals considered) and mode of transport.

Each section of this chapter is fairly self-contained; it includes an introduction, a discussion of key physical and chemical issues related to each modeling category, a discussion of important environmental interactions of inorganic pollutants with the specific environment, an outline of model requirements for input data and model applications, and examples of model applications. References for each modeling category are presented at the end of the subsection for that category.

Readers of this chapter who are not familiar with a particular modeling category should first read section 5.2 and then the relevant section of this chapter (e.g., Terrestrial Modeling). The Background (section 5.2) not only refers to other chapters but also summarizes information on the use of models, general modeling concepts (e.g., multimedia modeling), pathways of pollutants in the environment, exposure assessments, states of inorganic materials, and environmental interactions and properties of materials in modeling. This information is important to a better understanding of modeling concepts, features and models, as well as the power and limitations of environmental models presented in the subsequent sections.

5.2 BACKGROUND

5.2.1 Legal/Regulatory Issues

Models are applied in a variety of ways to assist in decision-making and will be used to a greater extent in the future. Specific statutes or regulations require the use of models in certain situations. Additionally, provisions of the National Environmental Policy Act of 1969 (NEPA), as well as judical decisions implementing NEPA and other environmental statutes, should facilitate the increased use of mathematical models [3].

There are a number of reasons for the increasingly widespread involvement of mathematical and computer models in environmental decisions. For example, Congress mandated the employment of computer models in the 1977 amendments to the Clean Air Act. Under the 1977 Amendments, models must be used in connection with the Prevention of Significant Deterioration (PSD) of air quality and for designation of nonattainment areas. Under the PSD program enacted in the 1977 Amendments, a major emitting facility in an area subject to PSD regulations must obtain a permit prior to commencing construction; this must be preceded by an analysis of air quality impacts projected for the area as a result of growth associated with such facility. The analysis must be based on air quality models specified by regulations promulgated by EPA. The 1977 Amendments also require conferences on air quality modeling every three years, to ensure that air quality models used in the PSD program reflect the current state-of-the-art in modeling [3].

The increasing use of quantitative models, particularly computer models, has placed a new burden upon the courts in their review of environmental decisions based on those models. This burden is a part of the "new era" in environmental decision-making and reflects the increasing involvement of scientific and technical issues in legal decisions. Problems in judicial review arising from the use of computer models and other quantitative methodologies in environmental decision-making are described by Case [3], whose publication is the source of a considerable amount of the information presented here.

The problems arising from the use of an environmental model in regulatory practice are twofold: first, it may actually increase the likelihood that a substantively incorrect decision will be reached. This greater probability of error generally can be traced to the inability of environmental decision-makers to deal with certain aspects of modeling in making such decisions. Second, the use of a model increases the danger that wrong environmental decisions may not be detected and corrected by the reviewing courts. Additionally, certain institutional aspects of the environmental agencies, the courts, and their mutual relationship further contribute to the difficulties of judicial review in cases relying on environmental models. Such institutional problems include the self-perceived lack of scientific expertise on the part of judges, the lack of judicial access to technical resources to assist in analyzing the issues involved, the limits on the court's ability to supplement or go outside the record, and the traditional deference that the courts give to administrative decisions.

Nevertheless, model development and application will increase in the future. Beyond any legal mandates, increasingly complicated and intractable environmental problems will compel the greater use of quantitative models by environmental decision-makers. Many experts believe in environmental fate, exposure and risk modeling, since models contribute to the scientific understanding of environmental quality. (See, for example, Swann and Eschenroeder [4].)

5.2.2 General Concepts and Pathways

INFORMATION REQUIREMENTS

Environmental pathway analysis provides the link between quantification of source emissions and assessments of receptor exposure (e.g., human exposure) through estimation of the ambient concentrations of contaminants in the various environmental media. A successful mathematical modeling effort must quantify fairly accurately the relationship between chemical releases into the environment and actual amounts of these chemicals to which the air, soil and water, as well as humans and other biota, are exposed.

Whether the concern is for human health or environmental impact, concentrations of the chemical compounds at user-specified receptors or media of concern must be estimated. This estimation is designated as "mathematical environmental fate modeling."

Mathematical environmental fate modeling generally requires a knowledge of (1) the distribution of the releases of the material into the natural environment, (2) the environmental conditions influencing the fate (transport, transformation) of the chemical compounds, (3) the physical and chemical properties of the material, and (4) techniques (models) for analyzing the information gathered, as schematically shown in Figure 5.2-1.

Two basic techniques can be employed to investigate environmental pathways: analytic sampling programs and mathematical fate modeling. Sampling programs are costly to design and implement; therefore, computerized mathematical models of environmental processes are frequently used to generate information unavailable by other means, or to estimate data (i.e., environmental concentrations) that would otherwise be costly to obtain.

Optimal use of environmental mathematical models necessitates knowledge of the following:

- Entry and dynamics of chemicals in the environment,
- Potential pathways of pollutants,
- Exposure pathways when confronted with human and biotic exposure,
- Mathematical modeling concepts, and
- Model application, input data, monitoring and validation issues.

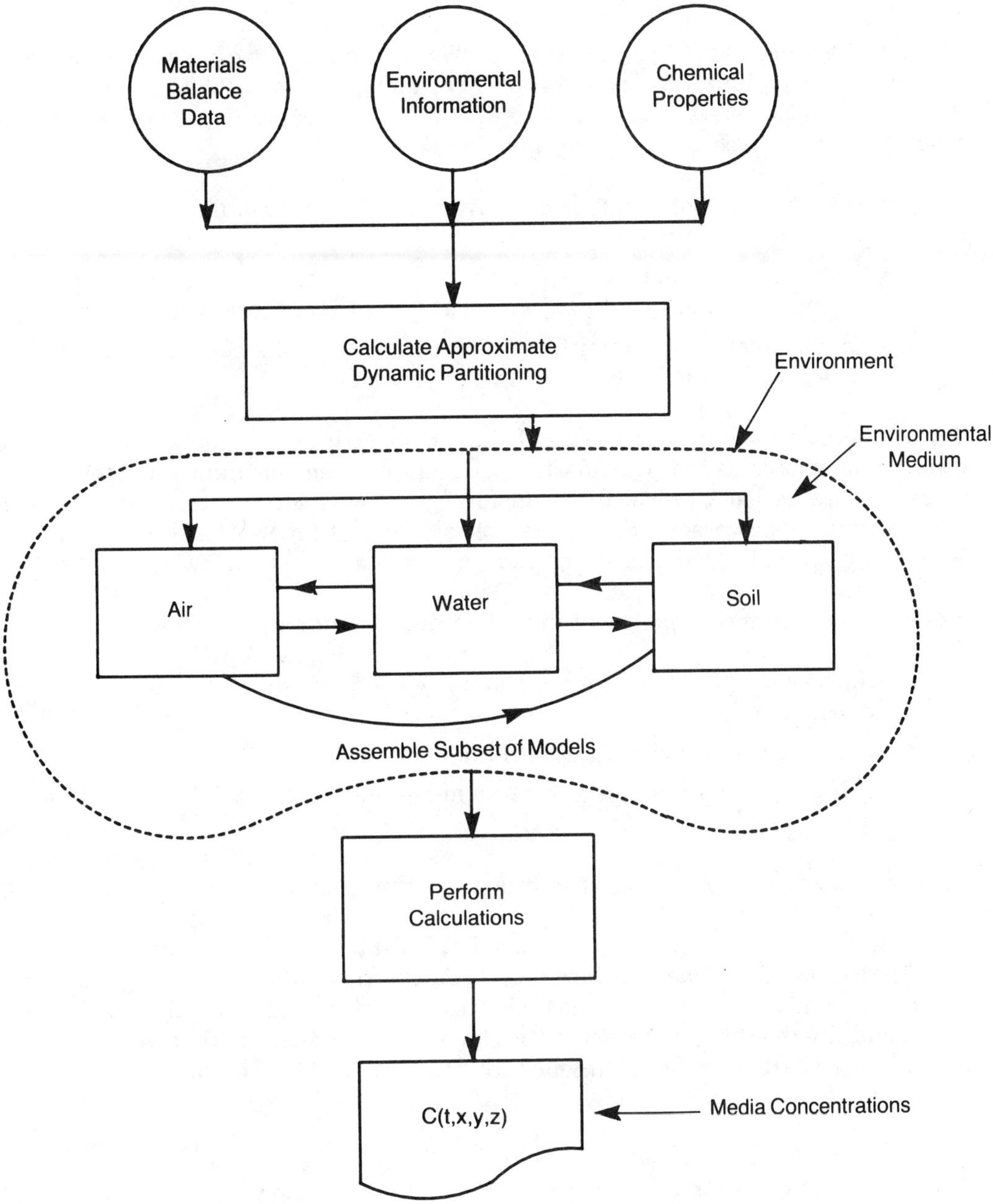

Source: Bonazountas and Fiksel [2]

FIGURE 5.2-1 Schematic for Environmental Modeling

ENTRY AND DYNAMICS OF CHEMICALS IN THE ENVIRONMENT

When identifying pathways and, hence, choosing models, we must consider what becomes of the pollutant as it enters the environmental media. Within any medium, three types of process (defined here as intramedia processes) govern the pollutant concentration at each point at any given time:

- *Advection* — pollutant mass movement in the medium carrying the pollutant.
- *Diffusion* — movement or spread of the pollutant, relative to the mass of the medium, and as driven by molecular or turbulence-scale dynamics.
- *Transformation* — production or consumption of the pollutant, usually driven by chemical reactions in the medium.

Superimposed on these mechanisms of change operating in the bulk volume of each medium are processes that transfer the pollutant from one medium to another. These processes are called "intermedia transfers" and, as such, are distinguished from transformation processes. (Some conceptual model frameworks lump intermedia transfers with embedded transformation processes.)

Examples of intermedia pollutant transfers are as follows:

- *Surface deposition* — rainout (liquid→solid), washout, fallout and dry deposition
- *Evaporation* — codistillation (liquid→gas) or volatilization
- *Adsorption* — liquid (or gas)→solid and *desorption* (solid→liquid (or gas))

These processes are further discussed later in this chapter.

In locating and choosing a model, one can simplify fate assessment efforts by delineating (1) the source release patterns, and (2) the dominant dynamic processes as explained here. Taking the intramedia processes first, one can address model criteria by considering the ratio of characteristic times [2]. The advection time is the principal length scale of the domain, L, divided by the average flow speed, u:

$$t_a \approx L/u \tag{1}$$

Typically, L may be stream reach distance and u the flow velocity.

The diffusion time is approximated by the random walk hypothesis and is given by:

$$t_d \approx \Delta^2/2D \tag{2}$$

where Δ is the characteristic transverse direction (e.g., stream depth) and D is the transverse diffusivity, be it turbulent or molecular.

Finally, the transformation time is approximated by:

$$t_t \approx C/\Delta C_t \tag{3}$$

where ΔC_t is the average rate of concentration change due only to transformation (typically a chemical reaction rate) and C is the average concentration in the domain.

If $t_t << t_a$ and $t_t << t_d$, there is rapid chemical change before any movement occurs. If $t_t >> t_a$ and $t_d << t_a$, there is little chemical change; diffusion spreads the pollutant rapidly, making the mixture homogeneous. If $t_t \approx t_d \approx t_a$, all processes act simultaneously. Taking these cases in order, we see that the first case does not require a model (except possibly a reacting plume in the near field), the second case is approximated by a non-reactive box model, and the third calls for a full reactive diffusion model.

Source geometry, interphase transfer and time dependencies must be superimposed on the above features. For example, the advection distances would be different for point and area sources. Also, the significance of source location must be considered in light of interphase transfer efficiency. For example, a water discharge containing a pollutant of high volatility and low solubility transfers the problem immediately from one of water modeling to one of air modeling. Implicit in these environmental dynamic considerations are three general principles:

- Intramedia processes are largely assessed on the basis of environmental scenarios.
- Intermedia transfers are largely determined by the pollutant's fate properties.
- Chemical transformations can be involved in both of the above.

There are, of course, exceptions to these rules. For example, molecular diffusivity is a pollutant fate property but may control an intramedia process; similarly, rainfall history is an environmental scenario characteristic but may control an intermedia transfer.

An experienced model user can do a simplified systems-level analysis prior to model selection, based on entry characteristics and environmental dynamics of the pollutant. Experienced analysts would advise that it is better to rely on intuition and a few calculations than to construct a formal, logical decision tree for guiding this process. Inexperienced scientists must analyze source characteristics, environmental dynamics, and pollutant fate properties more carefully before proceeding with model selection and application.

There are no specific rules or procedures for approaching model selection based on entry characteristics and environmental dynamics of the pollutant. The only criteria are the model features described in the user's manual, but these are often not objectively reported. Characterization of the sources, the environment and the fate properties is prerequisite to any such procedure.

PATHWAYS OF POLLUTANTS

The two major stages of an environmental modeling pathway analysis are:

(I) Examination of background information relevant to the environment; and

(II) The quantitative analysis.

Stage I may amplify an initial "scan" effort to identify key factors, such as monitoring data evaluations and geographic setting evaluation, that may indicate which media (air, soil, water) must be modeled, and which pathways and media are likely to be of secondary importance. Principally, Stage I aims at (a) identifying potential (probable) pathways of species, (b) evaluating available data on the specific region or site, (c) identifying what receptors in the area might be affected or are of importance for further consideration, and (d) identifying candidate models for estimating media concentrations.

Stage II, the quantitative analysis, involves (a) collection of monitoring and other site data, (b) prioritization of the important pathways, (c) selection of models to simulate these pathways, (d) compilation of input data for the models, (e) performance of the simulations, (f) analysis of model output/results, (g) output validation (using monitoring data) whenever feasible, and (h) design and/or evaluation of control strategies (future actions) for environmental protection which may involve additional model runs.

By carefully considering the fate properties and potential receptor effects during Stage I, the user will gain an understanding of the critical environmental pathways, enabling him to establish priorities for Stage II. For example, for a chemical whose main effect is toxicity to benthic organisms, the pathway leading from the source through the air into the water and to the receptor provides the logic for selecting a model that considers the relevant pathways. Considering relative contributions of multiple release modes, amounts or dominant fate properties, certain pathways will be preferred to others among the rather large array of possible combinations. Flow charts or diagrams are useful tools for identifying and establishing the pathway connections between sources and receptors and thereby defining fate modeling approaches.

Figure 5.2-2 is a typical environmental pathway chart designed for a regional pollutant fate modeling application [2]. It shows the three major environmental media — air, soil and water — and the intermedia and intramedia pathways of pollutants originating from a source (point, area, line, other). In reference to this chart, section 5.3 describes models and pathways relevant to the air compartment of the chart, section 5.4 describes models relevant to the terrestrial (soil) compartment, section 5.5 describes models and pathways related to the aquatic (water) compartment, and section 5.6 describes radionuclide models that are related to all compartments.

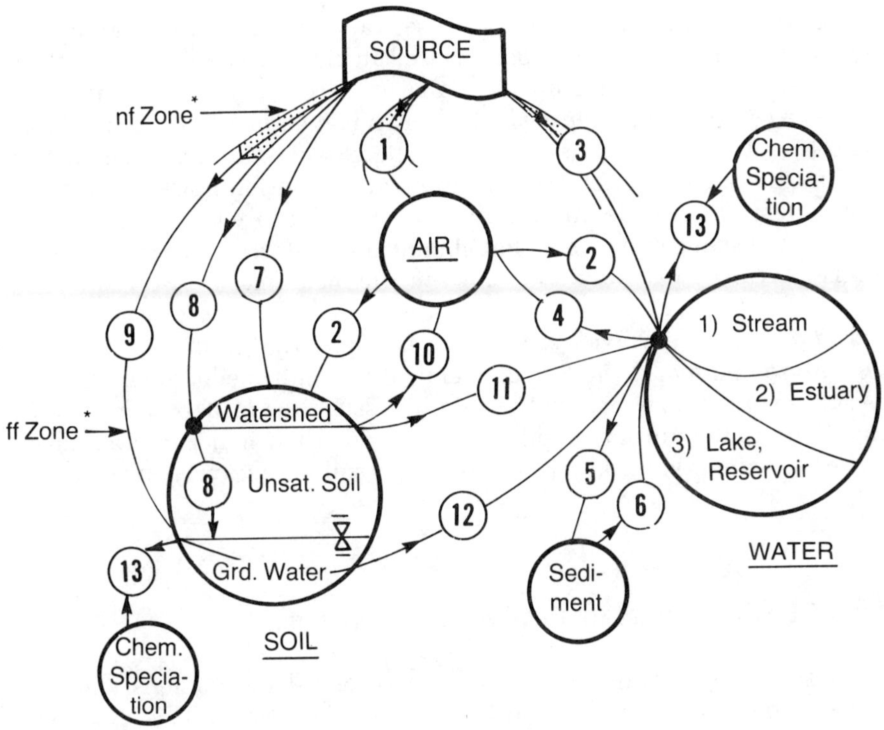

Notes

*	Paths 1,3,9 have near-field (nf) and far-field (ff) zones
**	Degradation, transformation, and out-of/into basin transfer pathways are not shown in this chart.

#1:	Source-to-Air
#2:	Air-to-Surface (i.e., wet/dry depositions)
#3:	Source-to-Water (stream, estuary, lake/reservoir; nf, ff models; overall water quality; pollutant-specific models)
#4:	Water-to-Air (e.g., volatilization, codistillation; pollutant-specific models)
#5:	Water-to-Sediment (e.g., adsorption, diffusion, pollutant-specific models)
#6:	Sediment-to-Water (e.g., desorption, dissolution, dispersion; pollutant-specific models)
#7:	Not a real pathway; rather, pollutant input modes to watershed/soil models
#8:	Source-to-Soil-to-Groundwater
#9:	Source-to-Groundwater (nf, ff models)
#10:	Land-to-Air (e.g., volitilization, dust pollutant particle resuspension)
#11:	Land-to-Water (e.g., via runoff or sedimentation)
#12:	Groundwater-to-Water (e.g., via advection, diffusion)
#13:	Aquatic Equilibria (or chemical speciation for water or soil)

Source: Bonazountas and Fiksel [2]

FIGURE 5.2-2 Environmental Pathways Chart**

Figure 5.2-2 shows, for example, that a model user can proceed to either (a) separately consider each medium and select the appropriate calculation technique or model that characterizes the behavior in a medium (e.g., air), or (b) consider transfer mechanisms to other media (e.g., water via deposition) from that medium. Experience is a key factor in selecting the appropriate time and space scales for the modeling task. Preliminary half-life estimates for transfer and transformation processes and a comparison among them can provide initial indication of which fate properties are important to consider. A possible outcome is that chemical transformations or dilution may effectively terminate the pathway within a specific medium.

Aravamudan *et al.* [1] present five hypothetical examples that illustrate the thinking required to analyze and model the fate of chemicals in multimedia environments. In this chapter, however, we shall deal with modeling issues and models of only the three single medium environments of air, soil, and water and their major interactions with the adjacent environments via key processes (e.g., soil-air *via* volatilization).

5.2.3 Routes of Environmental Interaction and Properties

This section describes important routes of environmental interactions and characteristics of inorganic species with regard to these interactions. This discussion is aimed at summarizing important information about mathematical modeling, processes considered by models, and the input parameters required.

The various paths and interactions are shown schematically in Figure 5.2-3. Major mechanisms discussed below are: volatilization, solubility, fast aqueous reactions, slow aqueous reactions, speciation, soil interaction, and bioaccumulation.

VOLATILIZATION

The evaporative loss of a chemical (volatilization) depends upon the vapor pressure of the chemical and on environmental conditions that influence diffusion from a surface. Volatilization is an important source of material for airborne transport and may lead to the distribution of a chemical over wide areas far removed from the point of release. Vapor pressure values indicate the tendency of pure materials to vaporize in an unperturbed situation. Vapor pressure data combined with solubility or, more precisely, activity coefficient data permit estimation of rates of evaporation of dissolved species from water, using Henry's Law constants and other parameters.

Chemicals with relatively low vapor pressure, high adsorptivity onto solids, or high solubility are less likely to volatilize and become airborne. However, many chemicals with very low vapor pressures can volatilize at surprisingly high rates, because of their low solubilities in water (i.e., high activity coefficients) or low adsorptivity to solids.

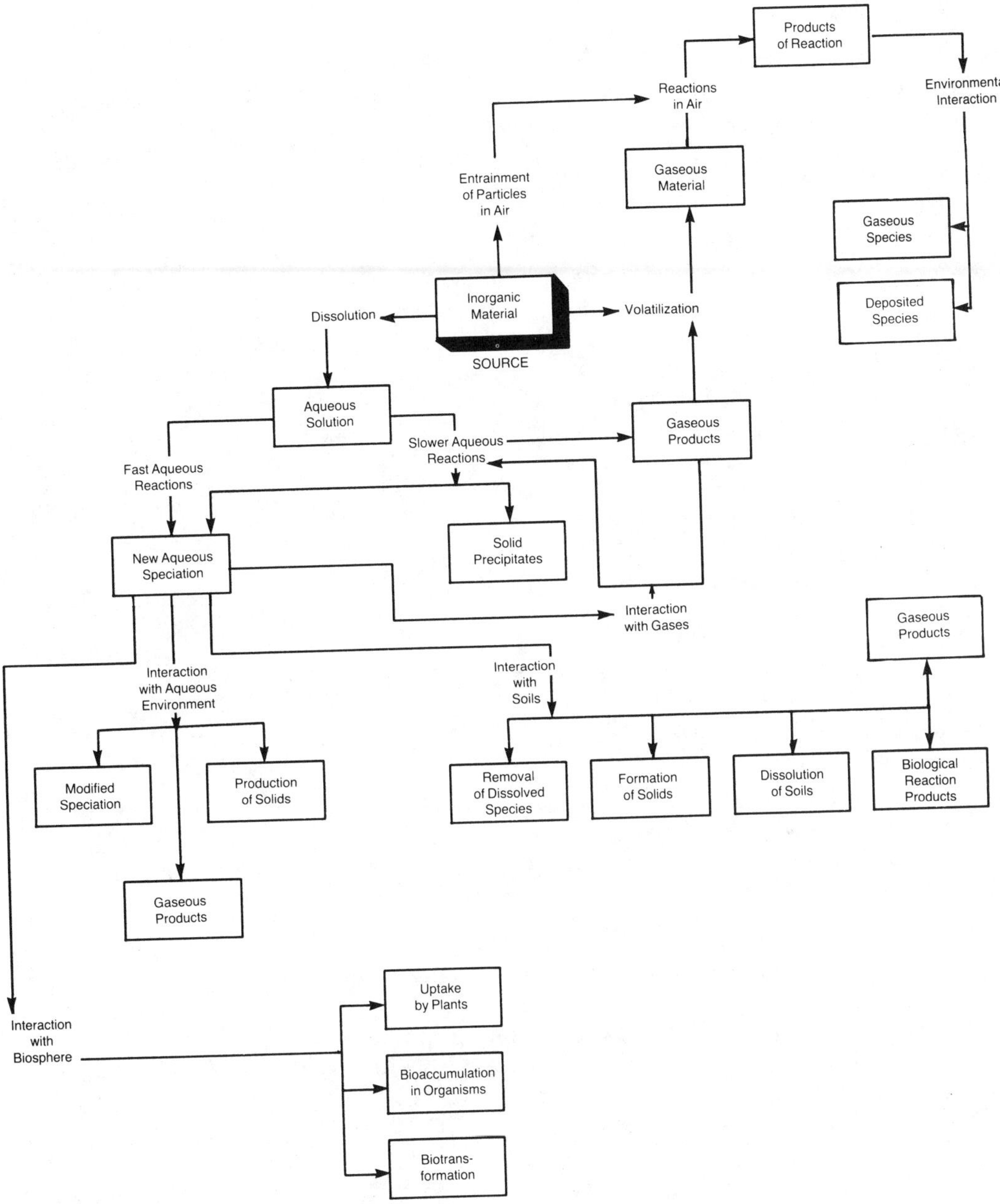

FIGURE 5.2-3 Important Mechanisms of Environmental Interactions of Inorganic Materials

SOLUBILITY

The equilibrium relationships for dissolution and dissociation of inorganic salts in aqueous solution, and other reactions, govern the chemical solubility. Solubility product constants for many species are tabulated in the literature; those that are not available may be calculated from Gibbs free energies of formation (which are tabulated in the literature) as described in section 2.11.

The aqueous solubility of a salt in the presence of a common ion (i.e., one of the ions into which the salt dissociates) may be either depressed or increased. An increase can occur when a charged complex is formed from a neutral complex by additional association, i.e., complexation (e.g., $AgCl + Cl^- \rightarrow AgCl_2^-$). When this process is not favored, the common ion effect suppresses solubility, as when sodium sulfate is dissolved in a saturated solution of calcium sulfate, causing the latter to precipitate. Complexes can also form between cations and other species (ligands). The extent to which metals form complexes varies significantly; alkali and alkaline earth metals form very weak complexes (i.e., ion pairs), while other metals can form highly stable complexes (e.g., copper ammonia complexes).

The effect of complexation (see section 2.9) on solubility can be taken into account mathematically by employing a series of equilibrium relationships written in terms of complexation equilibrium constants (stability constants). The above factors are frequently taken into account by mathematical modeling packages.

FAST REACTIONS IN AQUEOUS MEDIA

Fast reactions of inorganic dissolved species in aqueous media can be classified under the following general categories:

- Reactions with dissociation (of solvent molecules)
- Substitution reactions (with solvent or dissolved species)
- Redox reactions (with dissolved gases or other ions)

These reactions can lead to formation of new complex ions (those with different ligands in the coordination sphere and those with different oxidation state of the metal centers) and/or other ions. For details, the reader is referred to appropriate sections of Chapters 2 and 3.

The fast reactions of metal-containing species are important, because the medium in which the particular inorganic substance is dissolved largely determines the speciation of the particular metal ion. Since the extent of environmental interaction (e.g., soil attenuation, uptake by plants, toxicity) will depend on the chemical form (species) of the metal, knowledge of the behavior with regard to rapid reactions is essential.

In addition, when solutions of these metal-containing species are mixed with the environment (e.g., when leachate mixes with groundwater), chemical modification of the species will occur initially via thermodynamically favored rapid reactions and subsequently by the possible slower reactions. Such knowledge is important when selecting and applying the aqueous equilibria or speciation models described in section 5.5.5.

SLOWER REACTIONS IN AQUEOUS MEDIA

In the category of slower reactions in aqueous media, one may consider the following:

- Ligand substitution reactions of kinetically rather inert ions;
- Electron transfer reactions (e.g., inner-sphere mechanisms for kinetically inert ions and some outer-sphere reactions);
- Reactions with dissolved gaseous species (e.g., oxidation by dissolved oxygen, H_2S formed by bacterial reduction);
- Precipitation of solids by formation of insoluble species via substitution; and
- Oxidation reactions (e.g., formation of metal hydroxides and metal sulfides).

SPECIATION

These slower reactions affect the speciation and concentration of inorganic substances in the aqueous phase and determine the type and extent of further interactions in the same manner as the fast reactions, although they take place over a different time frame. (See Chapter 3.) For the relatively inert metal centers such as Co(III) and Cr(III), immediate modification of the speciation of the ion may *not* occur upon mixing with other aqueous phases; the reaction may not proceed until the mixture has reacted with soil components.

Thus, the time frame for some reactions may indicate that characterizing the speciation in the original aqueous phase (e.g., leachate) will dictate the chemical behavior of the compound with the environment. In other cases, it will be necessary to understand the speciation under the exact conditions of environmental interaction, such as adsorption. Information on chemical speciation is provided in sections 2.6-2.13.

SOIL AND SEDIMENT INTERACTIONS

Dissolved aqueous inorganic species interact with soils. The more important mechanisms of interaction include (a) formation of precipitates, (b) adsorption of components onto soil surfaces, (c) modification of speciation by soil constituents (solid and liquid phases), and (d) reactions induced by bacteria.

The surface properties of soil particles that dictate adsorption behavior can be modified by the acid/base buffer system in contact with the soil. (See section 2.12.) Most minerals display amphoteric properties toward solutions — i.e., they behave as weak acid or bases. The buffer capacity of soil, such as that related to its $CaCO_3$ content, affects the potential of soil water pH to fluctuate. Microorganisms can cause important changes in the solubility of soil minerals and other precipitates by altering solution conditions and catalyzing reactions.

Conceptual models for the adsorption of species onto soils have been developed to incorporate factors such as electrostatic energy and chemical bonding effects (e.g., van der Waals, dipole effects and covalent bonding). In particular, Langmuir adsorption isotherms consistent with this conceptual model have been generated. As previously mentioned, solution chemical equilibria modify the adsorptive behavior of the soil as well as the species present in solution and their tendencies to be adsorbed; therefore, simplified adsorption, attenuation or speciation models must be used with caution. Supported computerized packages are preferable to the simpler models.

BIOACCUMULATION

Much information is available on the effects of individual inorganic constituents on a large variety of organisms in specific environments. However, such information has generally been difficult to use in developing techniques for the estimation of effects, and no such methods relating to the effect of inorganic compounds on organisms in aquatic or terrestrial environments have yet been developed. Chapter 4 discusses this subject further.

TERRESTRIAL PLANTS

If there are exceptions to the general lack of estimation techniques and models relating inorganic chemical concentrations, biological effects and fate, they are likely to be found in agricultural theory and modeling. The application of soil amendments (fertilizers) as macronutrients (e.g., phosphates, nitrates) or micronutrients (e.g., Mn), or for improving the soil chemical environment to release nutrients (liming), is a relatively well-developed science. At the least, there is an understanding of optimum ranges of concentrations of agriculturally related chemicals for the growth of crop species. This stands in contrast to the paucity of information regarding uptake of organic soil contaminants by plants.

5.2.4 Literature Cited

1. Aravamudan, K., M. Bonazountas, A. Eschenroeder, D. Gilbert, L. Nelken, K. Scow, R. Thomas, W. Tucker and C. Unger, *An Environmental Partitioning Model for Risk Assessment of Chemicals: Part 1. Narrative Descriptions of the Methodology*, Arthur D. Little, Inc. Report to U.S. Environmental Protection Agency, Office of Toxic Substances, Contract 68-01-0387 (1979).
2. Bonazountas, M. and J. Fiksel, "Environmental Mathematical Pollutant Fate Modeling Handbook/Catalogue," Draft Report prepared for U.S. Environmental Protection Agency, Office of Policy and Resource Management, Washington, D.C. (1982).

3. Case, C.D., "Problems in Judicial Review Arising from the Use of Computer Models and Other Quantitative Methodologies in Environmental Decision Making," *Environ. Aff.*, **10**, 251 (1983).

4. Swann, R.L. and A. Eschenroeder, *Fate of Chemicals in the Environment: Compartmental and Multimedia Models for Prediction*, ACS Symposium Series 225, American Chemical Society, Washington, D.C. (1983).

5.3 AIR MODELING

This section describes principles of air-quality modeling and summarizes well-documented or Government-approved computerized models. Information and references on relationships between ambient concentrations and emissions for assessing the health risks of airborne carcinogens are presented by Eschenroeder *et al.* [20]. General information on mathematical modeling packages is presented by Bonazountas and Fiksel [4], Drake and Laulainan [15], Bass and Benkley [3], the U.S. Environmental Protection Agency [41], Aravamudan *et al.* [1], and others.

5.3.1 Background

Atmospheric fate modeling generally considers the transport and diffusion of pollutants from sources to receptors by the air pathway. Receptors can be defined as locations at which exposure to pollutants can take place.

Atmospheric models characterize the dispersion and deposition of pollutants during transport [37]. Generally, the output of air quality models consists of air concentrations of pollutants of interest in time and space as well as the amount of material deposited on the earth's surface by wet and dry deposition processes. The time and space aspect of these quantities depends on the characteristics of the model and how it is applied.

Models for estimating exposure levels are used to predict short-term average (acute) and long-term average (chronic) exposure to air pollutants. They vary in complexity, required input data, and form of output. If the exposure to be determined includes contact with a pollutant through inhalation, average ambient concentrations over a given time at a receptor must be combined with the inhalation rate to give an integrated exposure level. Estimation of exposure to pollutants through dermal contact requires the ambient air concentration of the pollutant at the receptor along with the rate of deposition.

The selection of a model for a given application must consider the following:

- Definition of emissions with regard to the space and time in question;
- Wind flow patterns as a function of location, topography, urban influences, surface roughness and large-scale weather patterns;
- Physical processes and chemical reactions of pollutants as they are transported and dispersed in the atmosphere; and
- Transport of pollutants from other regions that create background levels for the region in question;
- Use of data bases that are sufficient to calibrate a model and validate it for predictive purposes.

Present atmospheric models range from simple empirical kinds to very complex, time-dependent ones [18]. Simple empirical models are based on a deterministic analysis of air quality, emissions and meteorological data [10,12]. Time-dependent models are derived from basic physical and chemical principles related to the processes of transport, diffusion, chemical transformation and removal [24,35]. Some semi-empirical models, an intermediate class, share the simplicity of empirical models but approach the predictive ability of complex, time-dependent models [28,40]; such models, which are based on the Gaussian plume approach, are considered in greater detail in this review.

Semi-empirical models provide sufficient flexibility to address exposure to inorganic pollutants in the atmosphere. Some that differ with source emission type, geographic extent and averaging time are included in the list of approved models of the EPA Users Network of Applied Models for Air Pollution (UNAMAP) [42]. Details on the use of a given model in the UNAMAP series are given in the related user's manual. Information on several of these models is reported in section 5.3.4.

5.3.2 Types of Pollutant Sources

A number of different types of pollutant sources may have to be analyzed when atmospheric pollutant concentrations and human exposure levels are calculated. A model can be selected on the basis of the type of source and receptor to be considered. Generally, the calculation of exposure levels must consider point, area and line sources, as described below.

POINT SOURCES

Point sources are the most commonly encountered release modes in air quality assessments. For an exhaust stack, the following specifications are required: stack height, gas exit velocity, exhaust temperature and emission rate of each pollutant [11,21,38,40]. Many of these parameters can be measured or can be estimated from data on a similar stack.

Emission rates can also be calculated by using either a mass balance approach or emission factors related to the process being modeled. For complex industrial facilities, a number of stacks may have to be considered. As a simple worst-case screening approach, these multiple sources can be treated either as (1) a single source, with all of the emissions assumed to originate from one stack, or (2) individually, with the relative location of each source being included in the modeling analysis [23]. Complex sources may also include low stacks and roof-top vents from which pollutants can be entrained in building wakes before being transported downwind. For some analyses, these emission characteristics must be specified, and models used to simulate their unique transport mechanisms.

AREA SOURCES

When emissions from a number of sources are spread more or less uniformly over a given area [16], a different characterization is required for modeling exposure assessment. This type of source would include emissions from residences or an extended area of small industrial facilities. In this case, emissions are usually treated as uniform pollutant releases at a given height in units of mass per unit area. Models designed to simulate the transport of pollutants from an area source should be utilized [9,22,33,34,35].

LINE SOURCES

Transportation activities can produce a number of pollutants of concern. These releases can be treated as line sources for a given roadway; for combined roadways in a larger area, they can be treated as area sources. Emission rates are usually based on the number of vehicle miles traveled on the road segment that is modeled. Line source models should be used to characterize the transport and exposure to pollutants [8,32].

5.3.3 Physical and Chemical Factors/Problems

Transformation and removal mechanisms in the atmosphere must be considered [20] for many pollutants to appropriately characterize exposure levels. Depletion mechanisms in the atmosphere include photochemical reactions [14,31], wet deposition, dry deposition and gravitational settling.

Chemical reactions usually are treated as first-order decay with half-life terms [2,7]. Gravitational settling and deposition caused by washout or turbulent transfer to the surface can be simulated by specifying settling velocities, deposition velocities and scavenging ratios for the given pollutants [29,36,39]. A number of the models described below can simulate these processes.

5.3.4 Modeling Issues/Approaches, Input

Since most models were formulated to study dispersion of stack exhausts, this section will focus on modeling this aspect. Five processes (steps) can be of importance in mathematical air modeling [13,18,19]:

(1) Plume rise from the stack exit,

(2) Atmospheric turbulence and dispersion mechanisms,

(3) Chemical pollutant transformation,

(4) Loss to surface by deposition or other mechanisms, and

(5) Near-field dispersion versus far-field (long range) transport.

Current models and computer codes address these processes in varying degrees of completeness and detail; many do not treat one or more of these processes at all. A thorough discussion of the processes is given by Eschenroeder *et al.* [20], and Figure 5.3-1 illustrates where they occur in a typical plume.

The most used and most reliable atmospheric models involve the Gaussian plume concept (Figure 5.3-2), in which emissions and transport are assumed to be steady-state or pseudo steady-state. These models have the following properties:

- Concentrations from a continuous, steady source vary in direct proportion to the source strength or rate of emission.

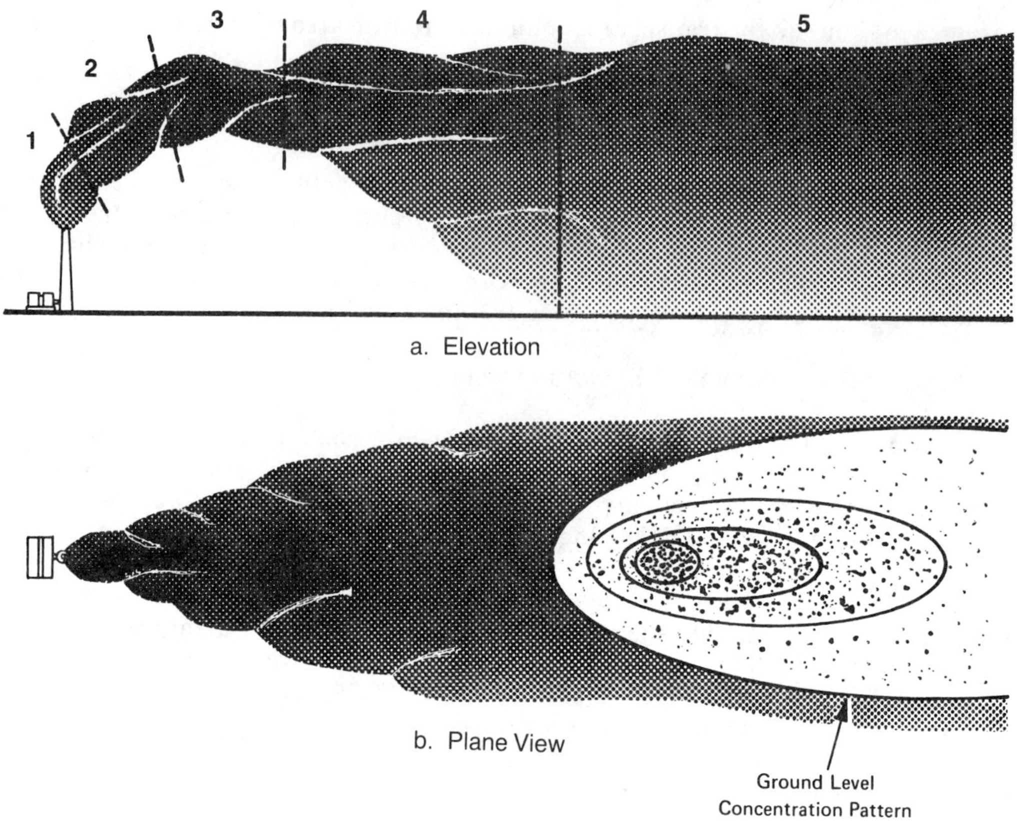

a. Elevation

b. Plane View

Source: Eschenroeder *et al.* [20]. *(Copyright 1985, Electric Power Research Institute. Reprinted with permission.)*

FIGURE 5.3-1 Typical Plume Behavior and Ground-level Concentrations in a Neutrally Stratified Atmosphere

- Concentrations vary inversely with the mean wind speed at the source.
- The distance from the source to the receptor and intensity of atmospheric turbulence determine the height and width of the plume at the receptor.
- Concentrations may decrease because of chemical reactions, radioactive decay and removal.
- Practical time scales are 1 to 2 hours, with adaptation for daily and monthly time averages; maximum space scales are 10 to 30 km.
- Extra provisions are made to account for complex terrain and different atmospheric stability conditions.

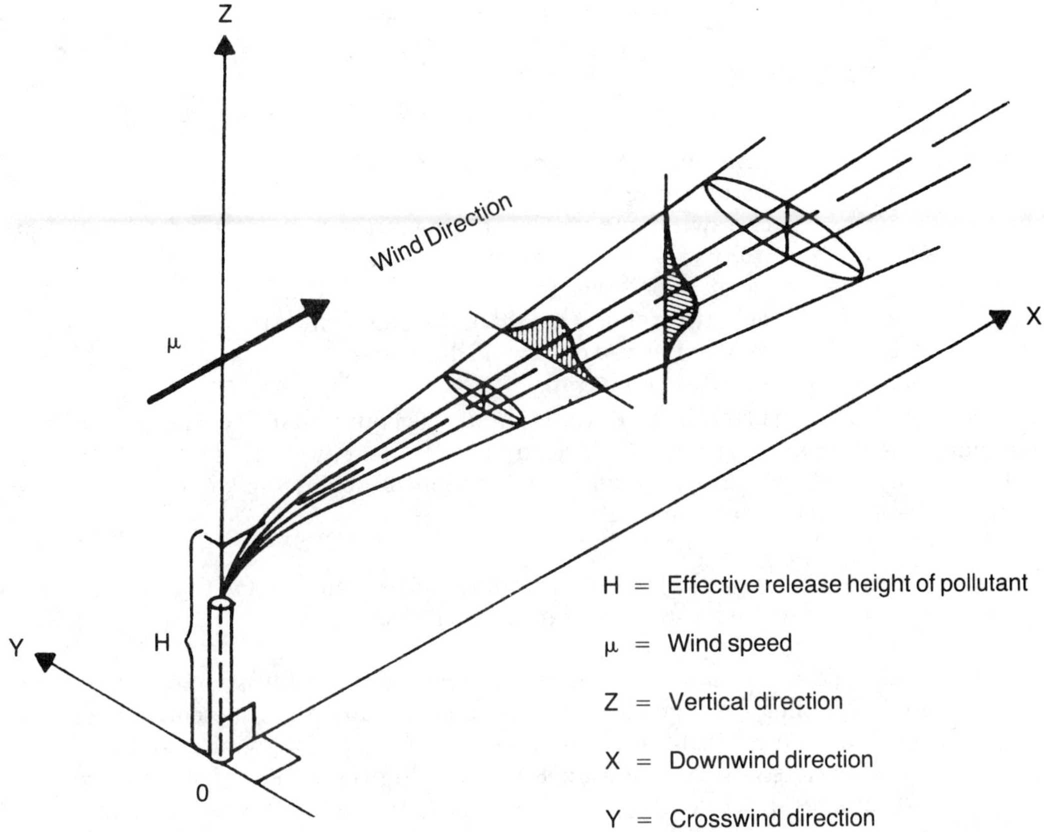

Source: Turner [40]

FIGURE 5.3-2 Schematic of a Standard Gaussian Plume, Showing Plume Rise and Vertical and Crosswind Diffusion Patterns

Dispersion models of this kind can provide reasonable estimates of exposure levels without great investment of time and money [2]. The basic equation for ground-level concentrations from an elevated point source is [40]:

$$X\,(x,y,z = 0) = \frac{Q}{\pi\sigma_y\,\sigma_z\,\bar{\mu}}\ \exp\left[-\left(\frac{H^2}{2\sigma_z^{\,2}}\right) - \left(\frac{y^2}{2\sigma_y^{\,2}}\right)\right] \tag{4}$$

where
- X = atmospheric concentration (g/m^3)
- Q = pollutant release rate (g/s)
- σ_y, σ_z = crosswind and vertical plume concentration standard deviation (m)
- $\bar{\mu}$ = mean wind speed (m/s)
- H = effective release height of the pollutant (m)
- x,y = downwind and crosswind distances (m)
- z = vertical height (m)

Most of the models reported in a later section utilize equation 4 or derivatives of it to simulate transport and to calculate exposure levels. The parameters σ_y and σ_3 can be obtained from plots of values given as a function of downwind distance from the source (x) and other parameters [40].

When using models, one must specify a number of parameters for the simulations. These parameters fall into the following three categories:

- *Meteorological Inputs*: To simulate the transport and diffusion of pollutants, an atmospheric model requires information on the meteorological variables that control these processes — wind direction, wind speed, and atmospheric stability (a parameter describing the extent of turbulence and changes in wind direction). These data, related to the specific simulation, are generally derived from measurements made at a site that is similar to the one being simulated.
- *Source Input Parameters*: In an evaluation of the exposure to a specific pollutant, the estimated emission rate or release rate can be the most critical but least known parameter. Information on emission rates for less common pollutants are generally not well documented; however, failure to properly characterize the release rate may result in large discrepancies between predicted and actual exposure levels. The most accurate data for point source emissions come from isokinetic stack sampling to measure the concentrations of gas and particulates.

  Emission estimates based on empirical studies of comparable sources should be used with caution, because the conditions associated with the latter can be vastly different from those that apply to the studied source. Not only should release rates be properly characterized, but the time of day of a given release should be incorporated, because it can drastically affect the magnitude of a predicted exposure level.
- *Receptor Characteristics*: The locations of the key receptors relative to the sources to be evaluated must be specified when estimating exposure

levels. This includes specification of not only the horizontal distances and directions from the source but also the elevation of the receptors relative to the source. This is particularly important when evaluating exposure to pollutants in areas of hilly terrain. Certain models identified below have algorithms that can simulate pollutant transport in complex or hilly terrain.

The propagation of uncertainty is a major concern in air modeling estimates [30]. Consistent with the objective of assessing the accuracy of concentration estimates made by the model, it is necessary that explicit and rational treatment be given to the sources and propagation of uncertainty in those estimates. An analysis of uncertainty has been made by Eschenroeder *et al.* [20]. To the extent possible, this treatment employs standard methods of analysis of propagated error and divides the total propagated uncertainty into two components: (1) a model or method error contribution, originating from uncertainty in the algorithms and (2) errors due to uncertainty in one or more data inputs [20].

5.3.5 Survey of Air Models

The models contained in the UNAMAP Series address the wide variety of the problems identified above and have some of the analytical capabilities required to simulate pollutant transport phenomena for assessing human exposure levels. These models and their capabilities are identified in Table 5.3-1. A user's guide for each model can be obtained from the National Technical Information Service (NTIS). Typical minimum data required for using the models are: wind direction, wind speed, atmospheric stability, atmospheric mixing height data, ambient temperature, precipitation, stack characteristics, sources and receptors, terrain data, and pollutant data.

Each model identified in Table 5.3-1 can be modified to fulfill a specific, limited use. The differences between the models involve such terms as the spreading of plumes as a function of atmospheric stability [27,28], vertical mixing [26], plume rise [5,6], source configuration [25], terrain considerations [17], acute or chronic exposure, etc. Most of the models are quite different with regard to specific capabilities, and for almost any need one can find an appropriate model.

5.3.6 Model Application

A typical screening approach for evaluating peak ground-level concentrations resulting from stack emissions would include the use of the model PTPLU. This model of the EPA-UNAMAP series can calculate the maximum ground-level concentrations under a number of meteorological conditions. For a given stack with identified operating parameters, the model will cycle through a number of combinations of wind speeds and atmospheric stabilities. It will identify the maximum concentration for each wind speed/stability category and indicate the downwind distance from the stack to where the maximum occurs. The model's output covers the typical range of meteorological conditions, and it indicates the expected short-term (up to one hour) average acute exposure levels that a receptor may encounter.

TABLE 5.3-1

Features of Selected Air Quality Models

MODEL FUNCTION	APRAC-1	CDM	CRSTER	HIWAY-2	ISC	MPTER	PAL	PTMAX PTDIS PTMTP PTPLU	RAM	TCM	TEM	VALLEY	COMPLEX I, II
Point Source		●	●		●	●	●	●	●	●	●	●	●
Area Source	●	●			●		●		●	●	●	●	
Line Source	●			●			●						
Volume Source					●								
Short-Term	●		●	●	●	●	●	●	●		●	●	●
Long-Term		●	●		●				●	●		●	●
Plume Trapping	●	●	●	●	●	●	●	●	●		●	●	
Half-Life		●			●	●			●	●	●	●	●
Gravitational Settling					●	●[c]	●[c]						
Aerodynamic Effects		●		●	●		●						
Buoyant Plume Rise		●	●		●	●	●	●	●	●	●	●	
Momentum Plume Rise					●	●			●	●	●		
Stack Downwash					●	●		●	●		●		
Turbulent Deposition						●[c]	●[c]						
Terrain Adjustment			●		●	●						●	●
Complex Terrain						●						●	●
Calibration		●								●			
Urban/Rural Dispersion Coefficient[a]	U	U	●	R	●	●	R	R	●	●	●	●	●
Background Calculated	●												
Street Canyon Flows	●												
Wind Speed Extrapolation		●	●		●	●	●	●	●	●	●		●
Highway Generated Turbulence				●	●		●						
Emission Rates Function of Met. Conditions					●								
Emission Rate Function of Time of Day	●				●	●			●				●

(Continued)

TABLE 5.3-1 (Continued)

MODEL FUNCTION	APRAC-1	CDM	CRSTER	HIWAY-2	ISC	MPTER	PAL	PTMAX PTDIS PTMTP PTPLU	RAM	TCM	TEM	VALLEY	COMPLEX I, II
Buoyancy Induced Dispersion						●		●					●
σ_y Function of Averaging Time and Stability											●		
Inversion Penetration Factors											●		
Particle Resuspension													
Washout													
No. of Sources (Max.)[d]	1200	200 P 2500 A	19	24	100[b]	250	180	25 (PTMTP)	250 P 100 A	Un- limited	300 P 50 A	50	250
No. of Receptors (Max.)	625	Un- limited	180	50	400[b]	180	30	30 (PTMTP)	150	2500	2500	112	180

a. U = Urban Only, R = Rural Only

b. The maximum number of sources and receptors are interrelated: These values are reasonable estimates.

c. These capabilities are available on a version which is not part of the UNAMAP series of models.

d. P = Point, A = Area

Source: Bonazountas and Fiksel [4]

This screening analysis is useful for quickly estimating peak short-term average levels, and it can be used to determine whether a problem may exist. It generally is geared to calculate a conservative worst case. More accurate levels can be determined by using other models that utilize site-specific meteorological data and can better simulate atmospheric reactivity and atmospheric transport mechanisms such as gravitational settling, deposition and washout.

5.3.7 Literature Cited

1. Aravamudan, K., M. Bonazountas, A. Eschenroeder, D. Gilbert, L. Nelken, K. Scow, R. Thomas, W. Tucker and C. Unger, *An Environmental Partitioning Model for Risk Assessment of Chemicals: Part 1. Narrative Descriptions of the Methodology*, Arthur D. Little, Inc. Report to U.S. Environmental Protection Agency, Office of Toxic Substances, Contract 68-01-0387 (1979).
2. *Air Quality Display Model*, TRW Systems Group Manual for NAPCA, HEW, Contract No. PH-22-68-60, Washington, D.C. (1969).
3. Bass, A. and C.W. Benkley, *Development of Mesoscale Air Quality Models: Vol. 2, User's Guide to MESOPLUME (Mesoscale Plume Segment) Model*, EPA-600/7/80-059, U.S. Environmental Protection Agency, Research Triangle Park, N.C. (1980).
4. Bonazountas, M. and J. Fiksel, "Environmental Mathematical Pollutant Fate Modeling Handbook/Catalogue," draft report prepared for the U.S. Environmental Protection Agency, Office of Policy and Resource Management, Washington, D.C. (1982).
5. Briggs, G.A., "Plume Rise," USAEC Critical Review Series, TID-25075, Clearinghouse for Fed. Sci. and Tech. Info. (1969).
6. Briggs, G.A., "Plume Rise Predictions," Lectures on Air Pollution and Environmental Impact Analysis, American Meteorological Society, Boston (1975), pp. 59-111.
7. Busse, A.D. and J.R. Zimmerman, *User's Guide for the Climatological Dispersion Model*, U.S. Environmental Protection Agency, EPA-R4-73-024, Research Triangle Park, N.C. (1973).
8. Calder, K.L., "On Estimating Air Pollution Concentrations from a Highway in an Oblique Wind," *Atmos. Environ.*, **7**, 863-68 (1973).
9. Calder, K.L., "Multiple-Source Plume Models of Urban Air Pollution — Their General Structure," *Atmos. Environ.*, **11**, 403-14 (1977).
10. Calder, K.L. and W.S. Meisel, "Source-oriented Empirical Air Quality Models," *Proc. Conf. Environ. Modeling and Simulation*, Cincinnati, April 19-22, 1976 (1976).
11. Carson, J.E. and H. Moses, "The Validity of Several Plume Rise Formulas," *J. Air Pollut. Control Assoc.*, **19**, 862-66 (1969).
12. Chang, T.Y. and B. Weinstack, "Generalized Rollback Modeling for Urban Air Pollution Control," *J. Air Pollut. Control Assoc.*, **25**, 1033-37 (1975).

13. Committee on Atmospheric Turbulence and Diffusion, "Accuracy of Dispersion Models," *Bull. Am. Meteorol. Soc.*, **59**, 1025-26 (1978).

14. Demerjian, K.L., A.J. Kerr and J.G. Calvet, "The Mechanism of Photochemical Smog Formation," in *Advances in Environmental Science and Technology*, J.N. Pots and R.L. Metcalf (eds.), John Wiley & Sons, New York (1974).

15. Drake, R.L. and N.S. Laulainan, *Mathematical Models for Atmospheric Pollutants*, EPRI Report No. EA-1131 (1979).

16. Egan, B.A. and J.R. Mahoney, "Numerical Modeling of Advection and Diffusion of Urban Area Source Pollutants," *J. Appl. Meteorol.*, **11**, 312-22 (1972).

17. Egan, B.A., "Turbulent Diffusion in Complex Terrain," Lectures on Air Pollution and Environmental Impact Analysis, American Meteorological Society, Boston, 112-35 (1975).

18. Environmental Protection Agency, *Guideline on Air Quality Models (Revised)*, OAQPS Guideline Series, Research Triangle Park, N.C. (1986).

19. Environmental Protection Agency, *Summaries and Proposed Recommendations Concerning Air Quality Models*, Research Triangle Park, N.C. (1980).

20. Eschenroeder, A.Q., G.C. Magil, C.R. Woodruff, "Assessing the Health Risks of Airborne Carcinogens," Final Report EPRI EA-4021, Electric Power Research Institute, Palo Alto, Calif. (1985).

21. Gibson, L.V. and L.K. Peters, "A Short-term Air Quality Model for Sulfur Dioxide in Louisville, Kentucky," *J. Air Pollut. Control Assoc.*, **27**, 218-23 (1977).

22. Gifford, F.A., "Atmospheric Transport and Dispersion over Cities," *Nucl. Saf.*, **13**, 391-402 (1972).

23. Guzewich, D.C. and W.J.B. Pringle, "Validation of the EPA-PTMTP Short-term Gaussian Dispersion Model," *J. Air Pollut. Control Assoc.*, **27**, 540-42 (1977).

24. Liu, M.K. and J.H. Seinfeld, "On the Validity of Grid and Trajectory Models of Urban Air Pollution," *Atmos. Environ.*, **9**, 555-74 (1975).

25. Pasquill, F., "The Estimation of Windborne Material," *Meteorol. Mag.*, **90**, 33-39 (1961).

26. Pasquill, F., *The "Gaussian-Plume" Model with Limited Vertical Mixing*, U.S. Environmental Protection Agency, EPA-600/4-76-042, Research Triangle Park, N.C. (1976).

27. Pasquill, F., *Atmospheric Dispersion Parameters in Gaussian Plume Modeling: Part II. Possible Requirements for Change in the Turner Workbook Values*, U.S. Environmental Protection Agency, EPA-600/4-76-030b, Research Triangle Park, N.C. (1976).

28. Pasquill, F., *Atmospheric Diffusion*, 2nd ed., John Wiley & Sons, New York (1974).

29. Peterson, K.R., T.V. Crawford and L.A. Lawson, *CPS: A Continuous-Point-Source Computer Code for Plume Dispersion and Deposition Calculations*, UCRL-52049, Lawrence Livermore Laboratory, Livermore, Calif. (1976).

30. Ragland, K.W., "Worst-case Ambient Air Concentrations from Point Sources Using the Gaussian Plume Model," *Atmos. Environ.*, **10**, 371-74 (1976).
31. Reynolds, S.D., P.M. Roth and J.H. Seinfeld, "Mathematical Modeling of Photochemical Air Pollution: I. Formulation of the Mode," *Atmos. Environ.*, **7**, 1033-61 (1973).
32. Sharma, V. and L.O. Myrup, "Diffusion from a Line Source in an Urban Atmosphere," *Atmos. Environ.*, **9**, 907-22 (1975).
33. Sharma, V., "An Area-source Model for Urban Air-pollution Applications," *Atmos. Environ.*, **10**, 1027-32 (1976).
34. Shieh, L.J. and P.K. Halper, *Numerical Comparison of Various Model Representations for a Continuous Area Source*, IBM Scientific Center, Palo Alto, Calif. (1971).
35. Shir, C.E., "Numerical Investigation of the Atmospheric Dispersion of Stack Effluents," *IBM J. Res. Dev.*, **16**, 171-79 (1972).
36. Slade, D.H., "Deposition of Particles and Gases," in *Meteorology and Atomic Energy*, Section 5-3, U.S. Atomic Energy Commission (1977).
37. Slade, D.H. (ed.), *Meteorology and Atomic Energy*, TID-24190, Atomic Energy Commission, Washington, D.C. (1968).
38. Smith, M.E., "Recommended Guide for the Prediction of the Dispersion of Airborne Effluents," American Society of Mechanical Engineers (1968).
39. Sullivan, D.A. and B.A. Woodcock, "Gravitational Settling and Dispersion Modeling of Fugitive Dust Sources," presented at the 75th Annual Meeting of the Air Pollution Control Association, New Orleans, La. (1982).
40. Turner, D.B., "Workbook of Atmospheric Dispersion Estimates," U.S. Environmental Protection Agency, Office of Air Programs, Research Triangle Park, N.C. (1974).
41. U.S. Environmental Protection Agency, "Guideline on Air Quality Models (Revised)," Office of Air Quality Planning and Standards, Research Triangle Park, N.C. (1986). Report No. EPA/450/2-78/027R.
42. "User's Network for Applied Modeling of Air Pollution (Version 4)," National Technical Information Service (NTIS), U.S. Department of Commerce, Springfield, Va. (1982).

5.4 TERRESTRIAL MODELING

5.4.1 Introduction

This section describes principles of state-of-the-art mathematical terrestrial (land, soil, groundwater) modeling for use primarily in environmental quality studies and secondarily in environmental decision-making. Terrestrial modeling of inorganic species is complex and must deal with many uncertainties; therefore, it has not been extensively employed in decision-making.

Subsequent paragraphs provide a short background on terrestrial modeling (section 5.4.2), describe the principal processes considered by models (section 5.4.3), summarize the differences between organic and inorganic pollutant computer FORTRAN codes (section 5.4.4), list speciation codes (section 5.4.5), and illustrate applications of speciation models (section 5.4.6).

Readers should also review section 5.5, Aquatic Modeling, because of the chemical and mathematical similarities between the two modeling categories [2].

5.4.2 Background

Terrestrial chemical fate modeling has traditionally been performed for three distinct subcompartments: (1) the land surface (or watershed), (2) the unsaturated soil (or "soil") zone, and (3) the saturated (or "groundwater") zone of a region. In general, the mathematical simulation is structured around two major processes, the *hydrologic cycle* and the *pollutant cycle*, each of which is associated with a number of physicochemical processes. Land surface models also account for a third cycle, *sedimentation*.

Land surface models describe pollutant fate on land (known as watershed), the unsaturated soil zone of the region, and pollutant contribution to the water body of the area. Unsaturated soil zone models simulate both (1) soil moisture movement and (2) soil moisture and soil solid quality conditions of a soil zone profile extending between the ground surface and the groundwater table. Groundwater models describe the fate of pollutants in aquifers [3].

When used properly and with an understanding of their limitations, mathematical models can greatly assist decision-makers in determining the importance of pollutant pathways in the terrestrial environment [9,41]. For this reason, the use of models has grown dramatically over the past decade. Although the number of terrestrial model types is very large, there are only a few fundamental modeling concepts.

Soil zone modeling is highly complex, because the physical and chemical dynamics of a soil subcompartment — in contrast to those of a water or air subcompartment — are governed by external ("out-compartmental") forces such as precipitation, air temperature, and solar radiation. Water and air modeling are generally simpler, because the dynamics of these compartments are governed by "in-compartmental" forces.

Chemical modeling in soil systems provides information on the distribution of elements (e.g., metal species) within a soil matrix consisting of soil-solids, soil-moisture (in the soil zone) or soil-water (in groundwater), and soil-air (in the soil zone), as schematically shown in Figure 5.4-1 for the soil zone of a region [8]. The objective is to determine the amount of pollutants in the solid phase, aqueous phase and/or gaseous phase at a given point and time.

5.4.3 Principal Processes

An evaluation of the fate of inorganic compounds in soil and groundwater requires a detailed consideration of the physical, chemical and biological processes and reactions involved, such as complexation, adsorption, precipitation, oxidation-reduction [29], chemical speciation and biological reactions (Figure 5.4-2) to determine (for metallic species) the free metal concentration in soil solutions. These processes can affect such characteristics as species solubility, availability for biological uptake, physical transport, and corrosion potential [28,44].

To describe the complex interactions involved [3], various kinds of models have been developed. These are, for example, "adsorption" models (which utilize mathematics involved in specific adsorption, surface complex formation and ion exchange), surface complexation models, constant-capacitance models, cation-exchange models, and overall fate-modeling packages that take into account the effects of one or more geochemical processes. One category ("equilibrium" or "speciation" models) is designed to determine the distribution of inorganic species in the soil water. Chemically based computer models of soluble trace metal speciation are being employed increasingly in decision-making, such as in studies related to sewage and effluents applied to agricultural land [45].

The transport of particular chemical species in terrestrial systems is of interest to a variety of scientists [10,11], since measurement or reporting of "total" concentration of a particular inorganic compound [15] in the soil may be misleading in many environmental management situations [13]. Toxic effects of trace metals, for example, may be affected more by their chemical form than by their total concentration [16,2]. Therefore, mathematical computer models capable of simulating the distribution of inorganic pollutant species in soil and groundwater systems are valuable tools for analyzing contaminant pathways.

5.4.4 Models

At this stage of intensive research on terrestrial chemical modeling, we can group the prevailing concepts into two major categories: (1) *geochemical* or, more appropriately, *pollutant fate* models, and (2) *speciation equilibria* models. This terminology is not standard but is employed here for convenience.

Geochemical models may be applicable for simplified overall inorganic pollutant fate estimates via adsorption processes (routine) or for many chemical and biological processes that govern the fate of the total pollutant mass from dissolved or partially

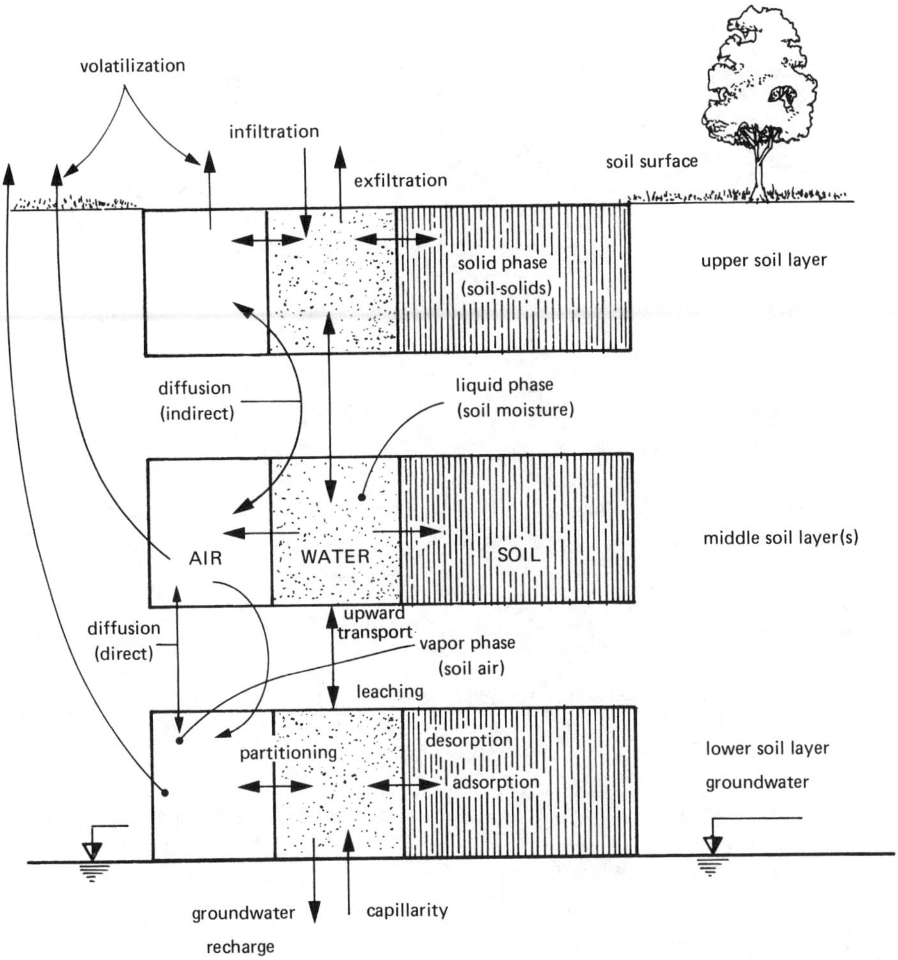

Source: Bonazountas and Wagner [7]

FIGURE 5.4-1 Schematic of Phases in a Soil System

dissolved (immiscible) fluids. Speciation models estimate the distribution of metals in various forms, but certain "speciation" codes can simulate the fate of individual dissolved pollutant species and the total mass of dissolved pollutants, since they combine fate and speciation in one package.

The variety of geochemical (pollutant fate) models has dramatically increased during the last decade. These models can be employed for both organic and inorganic pollutants, since the chemical processes are simulated via their overall equilibrium coefficients (e.g., adsorption coefficients). Information on such models is beyond the scope of this chapter and is not presented here, but some (mainly for organic pollutant models) is given by Bonazountas [9]. (See Table 5.4-1.) A study on the use and limitations of the three EPA unsaturated soil zone models SESOIL, PRZM and PESTAN of Table 5.4-1 has been conducted by Hern and Melancon [17a,17b] and others [33a].

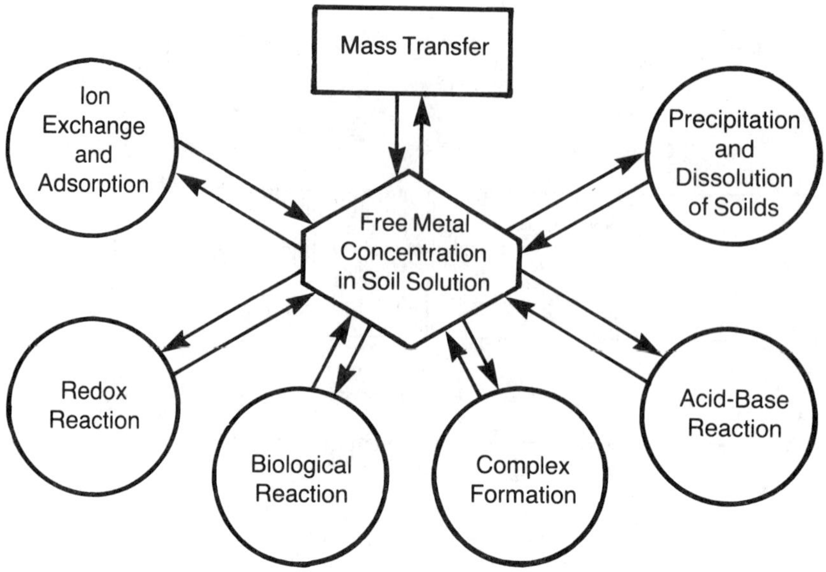

Source: Adapted from Mattigod *et al.* [32]. *(Copyright 1981, American Society of Agronomy. Reprinted with permission.)*

FIGURE 5.4-2 Principal Controls on Free Trace Metal Concentrations in Soil Solutions

The status of geochemical speciation equilibria models for inorganic pollutants in soils is described below. Note that speciation computer codes account for additional processes — redox reaction [26], adsorption [25], complexation [23,24], and others (e.g., [18,27,37]). Speciation equilibria models are based on chemical thermodynamic principles. Some of the computer codes available for speciation calculations in soil and groundwater aqueous systems are described below.

Speciation calculations and procedures are described in sections 2.9-2.13. Steps of speciation calculations are not of concern in the following section, which focuses on integrated computerized modeling codes.

Excellent state-of-knowledge reviews of chemical equilibria codes (models) of inorganic pollutants in soils are presented by numerous investigators [19,21,22,31,34,38,42,49,52]. The following discussion has drawn on these reviews, particularly that of Sposito [49]. Readers interested in details should refer to the original publications.

TABLE 5.4-1

Some Well-documented Pollutant Fate Models and Their Features

Model Acronym	Model Type[a]		Model Formulation[b]					Mathematics[c]		Chemistry Issues[d]					User Concerns[e]				Source
	Unsaturated Zone	Groundwater	Flow Module	Solute Module	TDE Approach	Compartmental	Statistical, Other	Analytical	Numerical	Organics	Inorganics	Metals	Gaseous Phase	Increased Chemistry	Input Data Req'd.	Calibration	Level of Effort	Application Study	
PESTAN	•			•	•			•		•					L	L	L	•	[15]
PRZM	•		•	•		•			•	•					M	M	M	•	[10a]
SCRAM	•		•	•	•				•	•					H	H	H	•	[1]
SESOIL	•		•	•		•	•	•	•	•	•	•	•	•	M	M	M	•	[7]
AT123D		•		•	•			•		•	•	•			M	L	M		[56,57]
PATHS		•		•	•		•	•	•	•	•	•			L	M	M	•	[36]
MMT/VVT		•	•	•	•				•	•					H	H	H	•	[11]
FEMWASTE		•		•					•	•					H	H	H	•	[57]
R. WALK		•		•	•				•	•					H	H	H	•	[44]
USGS Models	•	•	•	•	•			•	•	•	•	•	•		H	H	H	•	[4]
Other Models																			[22]

a. The use of complex models (e.g., a numerical soil and groundwater package) that can handle more than one compartment is not always desirable, since generalized packages tend to be cumbersome unless especially designed.

b. The most representative characteristics are given. The Traditional Differential Equation (TDE) approach applies to the flow and solute module. "Other" includes linear analytic system solutions, for example.

c. The most representative characteristics are given.

d. Almost all models can simulate organic, inorganic and metal fate, assuming that a careful calibration via an adsorption coefficient may alter the model output to predict measured/monitored values. However, not all models have by design increased chemistry capabilities (e.g., cation exchange capacity; complexation); therefore, the most representative capabilities are indicated.

e. L = low, M = medium, H = high input data requirements. In general, numerical models have higher input data requirements and calibration needs and therefore may better represent spatial resolution of a domain. Compartment models provide an optimal compromise. The level of effort is intuitively defined here.

Source: Adapted from Bonazountas [9]. *(Copyright 1983, American Chemical Society. Reprinted with permission.)*

5.4.5 Speciation Codes/Models

Although geochemical speciation modeling was developed relatively recently, more than a dozen comprehensive computer codes are already available. Kincaid *et al.* [22] provide an overview of the geochemical code history (up to 1983) in which they group models into four major families according to their evolutionary stage (Figure 5.4-3).

Speciation models can be grouped today into two categories: (1) "speciation" codes, which account for speciation equilibria of inorganic pollutants for a terrestrial water compartment (e.g., codes of Figure 5.4-3), and (2) "coupled speciation" codes, which can simulate both speciation equilibria and solute transport of individual species in the terrestrial (soil, groundwater) environment in both time and space.

Coupled codes are formulated by two mathematical methods, one described by Miller and Benson [34] and the other by Jennings *et al.* [21]. The former interfaces the computer code for equilibrium distributions of species (e.g., code of Figure 5.4-3) with the code for transport and performs calculations in two steps: (1) species estimation and (2) species transport. This procedure is repeated in time. Jennings' method consists of solving simultaneously a system of equations describing chemical reactions, advective-dispersive transport of reacting contaminants, and interphase mass transfer. The method of solution is irrelevant here.

Coupled speciation codes such as FIESTA [23,52] and CHEMTRN [34] are the subject of current research, as are the extension of the speciation codes. Although coupled codes appear to be powerful tools, they are also large and require extensive input data; as a result, they are complex and frequently inefficient (in terms of input data requirements and effort) for obtaining practical solutions to problems.

Coupled codes are not further analyzed in this report; rather, emphasis is placed on speciation equilibria (geochemical simple compartment) codes, because these models have been more extensively applied and the activity coefficients have been validated by laboratory and field studies.

The significant advances made in aquatic chemistry during the past decade [50] have yielded an approach to the soil solution that is parallel to that taken for other natural waters [48]. In that respect, there are many similarities between models presented in this section and the one that follows (see § 5.5); for example, the model GEOCHEM [50] has been developed from the water bodies speciation model REDEQL2, a program created originally by Morel and Morgan and their co-workers at Caltech [33].

The basic geochemical and aquatic equilibria processes modeled by existing computer codes include adsorption/desorption, precipitation/dissolution, reduction/oxidation, ion exchange, and aqueous speciation in soil solution. Various mathematical expressions are used to model these processes.

For aqueous speciation in soil solutions, Sposito [49] reports that chemical modeling can be accomplished either by a "specific interaction" approach or an "ion-association"

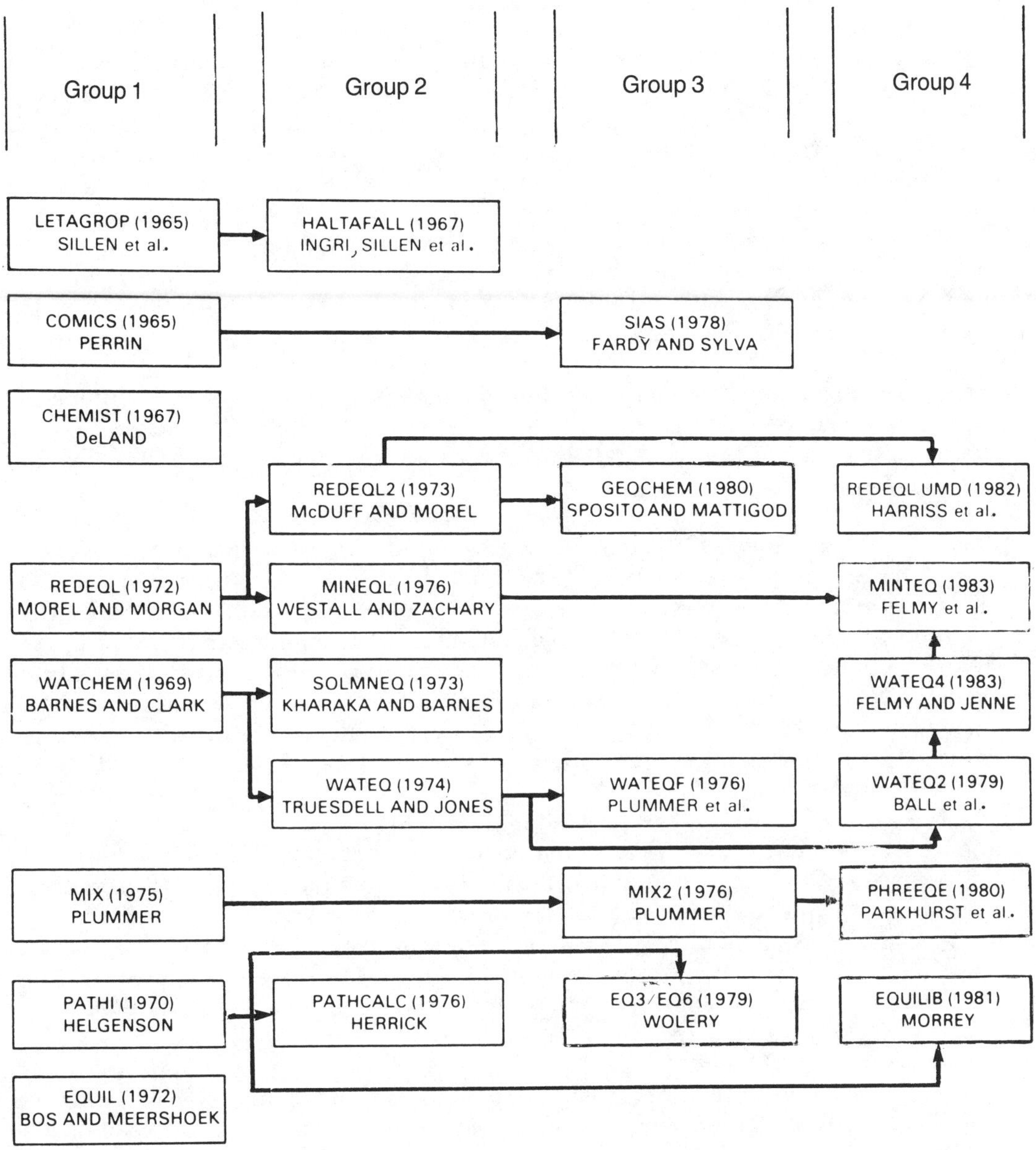

Source: Adapted from Kincaid *et al*. [22]. *(Copyright 1984, Electric Power Research Institute. Reprinted with permission.)*

FIGURE 5.4-3 Evolution of Geochemical Codes

approach. In the former, the composition of an electrolyte solution is described in terms of the total number of molalities of the stoichiometric components of neutral solutes, while the thermodynamic properties of the solution are expressed with "mean ionic" activity coefficients for neutral solutes introduced into standard chemical thermodynamic equations. In the ion-association approach, the composition of an electrolyte solution is described in terms of the molalities of molecular species presumed to exist in the solution, and the thermodynamic properties are expressed with "single-species" activity coefficients for the assumed molecular constituents. The latter approach appears to be used universally by soil chemists, despite its more tenuous relationship with rigorous thermodynamics. For additional details, the reader is referred to the original work of Sposito [49].

The geochemical data bases used are the most important features of a computer code. Data bases can provide typical values of kinetic and adsorption rates at typical temperatures. Reasonably good data bases are available for GEOCHEM and MINTEQ, among others.

Table 5.4-2 lists features of computer codes that are suitable for geochemical modeling. Of the nine codes summarized, the first six listed are very capable in the areas of aqueous speciation, adsorption/desorption and precipitation/dissolution [22]. In addition, all these codes are documented, available in the public domain in the United States, recently updated, and frequently accompanied with a sizeable data base. CHEMTRN features are not presented in Table 5.4-2 because of the limited information available during the production of this report. Table 5.4-3 provides user-related information for these codes. Many other codes in the literature (e.g., BALANCE [40]) are not reported here.

Model selection is a major issue in mathematical geochemical modeling. Factors of importance are the study objectives, the features of the model, input requirements, monitoring data available to validate prediction, model documentation, and cost for an application. As it is difficult to recommend a universal computer code that can handle all of the above factors, decision-makers must first review a number of codes that may meet project requirements and then select one that best suits their study objectives.

Two of the best known geochemical speciation codes are GEOCHEM and MINTEQ. CHEMTRN, a coupled geochemical/speciation and transport code, is also receiving increased attention. These codes are briefly described below.

- GEOCHEM is based on the computer program REDEQL2 for calculating equilibrium speciation of chemical elements in a soil solution. The component species are identified as uncomplexed metal cations, the free proton, uncomplexed ligands, and the free electron. Single species activity coefficients are calculated in the program. The model contains critical thermodynamic data for soils, a method for calculating cation exchange, and a correcting method for non-zero ionic strength up to 3 molal. Currently, the model stores thermodynamic data for 36 metals and 69 ligands, and it can handle more than 200 soluble complexes and

TABLE 5.4-2

Summary of Geochemical Code Capability, Adaptability and Availability

Criterion	GEOCHEM	REDEQL. UMD	MINTEQ	PHREEQE	EQUILIB	EQ3/EQ6	SIAS	SOLMNEQ	CHEMIST
Code Classification	III	III	III	IIIR	II	IIR	I	II	I
Number of Elements	36 (94)	3	32	19 (26)	26		10	25	
Aqueous Speciation									
Number of ligands/species	69/ –	65/ –	16/373	8/120	186	~20	10	162	(169)
Number of redox species	48 (60)	20	22	6	20	6	(200)	10	Variable
Activity-coefficient correction	DA,HN	DA,SHM	DA	ECH	HN	HN	None	HN	None
Calculation of pH	Yes	Yes	Yes	Yes	Yes	Yes	No	No	No
Method of iteration	N-R	N-R	N-F	N-R,C-F	Pred.	N-R	B-S	B-S	N-R
Adsorption									
Model	J-H	Swiss	Three mdls.	Two mdls.	None	None	None	None	None
Number of species	10	20	No limit	Variable	None	None	None	None	None
Number of surfaces	6	5	3	NA	None	None	None	None	None
Number of species	No limit	None		None	None	None	None	None	None
Precipitation/Dissolution									
Number of minerals	~500	<500	238	24	200	250	None	158	None
Quantitative mass transfer	Yes	Yes	Yes	Yes	Yes	Yes	None	No	No
Automatic selection of mineral	Yes	Yes	Yes	No	Yes	Yes	None	No	No
Capability for solid solutions	No	No	No	No	None	Yes	None	No	No
Gas Generation									
Ammonia	No	No	No	No	Yes	No	No	No	No
Oxygen	Yes	No	Yes	Yes	Yes	Yes	No	No	No
Hydrogen	No	No	No	Yes	Yes	Yes	No	No	No
Carbon dioxide	Yes	Yes	Yes	Yes	Yes	Yes	No	Yes	No
Code Structure									
Size (32 bit words, K)	434	62	68.1	>64	20.5	363	34	30	25
Modularity	Yes	Yes	Yes	Yes	Yes	Yes	No	No	No
Language	FORTRAN IVG	FORTRAN IV	FORTRAN IV	FORTRAN IVH	FORTRAN IV	FORTRAN 4.6	FORTRAN H	PL/1	FORTRAN

(Continued)

TABLE 5.4-2 (Continued)

	GEOCHEM	REDEQL. UMD	MINTEQ	PHREEQE	EQUILIB	EQ3/EQ6	SIAS	SOLMNEQ	CHEMIST
System	IBM 4314 VAX	CDC CYBER 171	UNIVAC 1144 PDP 11/70 VAX	Amdahl DEC VAX	CDC 7600 PDP VAX	CDC 7600 CDC 6600 VAX	IBM/360/65	IBM/360	IBM/360
Other Criteria									
Latest documentation date	1980	1982	1982	1980	1978	1979	1978	1973	1968
Data Base									
Temperature range	No	No	Yes	Yes	Yes	Yes	No	Yes	No
Easily modified	Yes	Yes	Yes	Yes	Yes	Yes	U-S	No	U-S

Legend: ( ) = maximum, U-S = user supplied, J-H = James-Healy, N-R = Newton-Raphson, B-S = back substitution, Pred. = predictor/back substitution, DA = Davies, HN = Helgeson-Nigrini, SHM = Sun-Harriss-Mattigod, ECH = extended Debye-Hückel, I = chemical speciation, II = I + mass transfer by precipitation or dissolution, III = II + adsorption and ion exchange, IIR and IIIR = pseudo kinetics.

Source: Kincaid *et al.* [22]. *(Copyright 1984, Electric Power Research Institute. Reprinted with permission.)*

TABLE 5.4-3

Advantages and Disadvantages of Chemical Speciation Models

Code	Characteristics
GEOCHEM	Available in public domain. Documentation marginal. Recently updated. Models important processes. Includes adsorption. Data base probably the largest available. Modularity not yet evaluated.
REDEQL.UMD	Should be available in the public domain soon. Well documented, although not in final form. Recently updated. Models important processes. Data base and modularity not yet evaluated.
MINTEQ	Available in the public domain. Well documented. Models important processes. Includes adsorption. Modular construction. Data base is the best documented; one of largest available.
PHREEQE	Available in the public domain. Documented. Models precipitation/ dissolution. Includes adsorption. Data base supplied by user. Modular construction.
EQUILIB	EPRI proprietary code. Well documented. Models precipitation, not adsorption. Data base quite extensive, reasonably well documented. Can use other data bases. Construction is modular. Method of solution involves unique elements.
EQ3/EQ6	Publicly available. Documentation available for EQ3 only. Updated version of PATHI construction. Contains precipitation, no adsorption but structured for inclusion. Contains capability to model paths of chemical changes.
SOLMNEQ	Publicly available. Documentation old. Does not contain unique characteristics. Data base not easily modifiable.
CHEMIST	Publicly available. Old and sketchy documentation. No precipitation/adsorption. No unique modeling characteristics.
CHEMCSMP	Publicly available. Old and very inadequate documentation. May have unique kinetic characteristics but tied to IBM system.
SIAS	Update of an old program. Documentation sketchy. SIAS capabilities covered in better documented, more inclusive codes.

Source: Kincaid *et al*. [22]. *(Copyright 1984, Electric Power Research Institute. Reprinted with permission.)*

solids. Adsorbed metal species are described in the model, and adsorbed ligand species will be considered in the future. A new version of GEOCHEM is being prepared [49]; therefore, interested scientists should contact its developers [50].

- MINTEQ is similar in structure to MINEQL, and both of these are similar in overall structure to GEOCHEM, since they originate from REDEQL. As shown in Figure 5.4-3, MINTEQ was formed from MINEQL and the data base of WATEQ [5,43]. The model includes ion speciation, redox equilibria, calculation of activity coefficients, solubility, adsorption, and mass transfer. The model and its large data bases are well documented, the latter containing data on more than 35 metals and 60 ligands.

As reported in a comparative analysis by Sposito [49], a principal difference between MINEQL (and consequently MINTEQ) and GEOCHEM is that MINEQL can accept the concentration of any free ionic species, soluble complex, or dissolved gas as input data to be held fixed during a calculation, whereas GEOCHEM can do this only for the activities of H^+, e^-, CO_2(g), and N_2(g). Therefore, if desired, the concentration of Cu^{+2} can be specified as a fixed input datum in MINEQL, a feature that gives the code user great flexibility in speciation calculations.

- CHEMTRN is a one-dimensional geochemical transport/speciation (coupled) model for solutes in a saturated porous medium. The model includes dispersion/diffusion, advection, ion exchange, formation of complexes and speciation in the aqueous phase, and the dissociation of water. The mass action, transport, and site constraint equations are expressed in differential/algebraic form and are solved simultaneously. This coupled model is in the development stage; therefore, no data base accompanies the computer code.

5.4.6 Model Applications

Reports in the literature on the chemical forms of heavy metals in soil solutions are limited [13,29]. One reason for this is the large number of chemical forms in which metals exist and the analytical problems associated with their determination [39]; another reason is that the modeling of metal species in soil solutions has only recently become a matter of concern.

The use of chemical equilibrium models in decision-making has been described by many researchers [6,12,13,14,20,21,24,30,31,34,45,46,47,52,53,54,55] who have employed and validated models such as GEOCHEM with data from field (e.g., [6]) and laboratory [13] analyses.

5.4.7 Example of Model Application

This section illustrates the use of a model to estimate the chemical speciation of metals in a soil system. The modeling experiment was conducted by the Tennessee Valley Authority [6] and relates to land-treated sewage sludge. It is presented in four sections:

(1) Statement of problem,
(2) Field monitoring,
(3) Model application, and
(4) Discussion of findings.

STATEMENT OF PROBLEM

Anaerobically digested, air-dried sewage sludge was applied (in 1971) in plots of 3.6 by 15.0 meters and was incorporated into the soil to a depth of 15 cm. The application rates were 50, 100 and 200 tons/ha.

Cultivation of the plots began after 12 months with sweet corn (*Zea Mays L. cv.* Silver Queen). At that time, the plots were subdivided into areas of 3.6 by 7.5 meters. Half of each plot received three additional sludge applications — in that year, 12 months later, and 24 months later — at the rate used in the initial application. The other half of each plot received no additional sludge, so that residual effects from the original applications could be monitored. The sludge contained both organic and inorganic pollutants.

FIELD MONITORING

Before application, the sludge was analyzed for inorganic species and total organic carbon. A field monitoring program was conducted in 1971. The soil of the area is a Sango silt loam soil (Glossic Fragiudult) with a pH of 4.9 and a cation exchange capacity of 7 mol (NH_4^+) kg^{-1}.

The field monitoring program (collection and analysis of contaminated soils) was repeated in 1977. Soil samples were collected, air-dried, and crushed to less than 2 mm. Water was added to 150-g samples to bring the soil water potential (see Appendix C) to -0.33 bar, and the samples were then incubated for 7 days at 25°C. Soil solutions were recovered by centrifugation [17] and analyzed for total Cd, Zn, Mn, Fe, Al, Cu, Ni, Pb, Ca, Mg, Na, K, P, Cl, sulfate, organic carbon, conductivity and pH. The chemical composition of soil solutions from untreated and sludge-treated soils is shown in Table 5.4-4.

Concentrations of Zn, Cd, Mn, Ca, Mg, Na, K, PO_4-P, SO_4-S, and Cl in the soil solution tended to increase with sludge application, particularly in plots that received multiple sludge applications. Concentrations of Cu, Ni, Pb, and Fe were generally low (less than 10^{-6} *M*) at all sludge application rates, but soluble Cu and Ni concentrations in soil treated with sludge at 800 t ha^{-1} were markedly higher than those in soil receiving lesser amounts of sludge.

MODEL APPLICATION

An attempt was made to duplicate the above conditions and estimate species of chemicals in the soil systems via a mathematical model. The GEOCHEM model was used to predict the equilibria of the metals in solution because its data base can adequately support simulations of the species presented.

To facilitate both model use and interpretation of the data obtained, GEOCHEM calculations were performed using three sets of conditions: (a) metal-organic complexes were ignored; (b) metal-organic complexes were approximated by a mixture

TABLE 5.4-4

Chemical Composition of Soil Solutions from Untreated and Sludge-treated Soils

		Sludge applied, t ha^{-1}							
		A[a]				*B*[a]			
Parameter	**Units**	**0**	**50**	**100**	**200**	**0**	**200**	**400**	**800**
Zn	$\times 10^{-5} M$	0.35	7.94	10.72	16.98	0.40	11.75	16.60	36.31
Cd	$\times 10^{-8} M$	0.98	2.24	3.09	79.4	1.26	3.02	58.9	144.5
Cu	$\times 10^{-6} M$	<1.0	<1.0	<1.0	3.16	<1.0	<1.0	<1.0	6.31
Ni	$\times 10^{-6} M$	<1.0	<1.0	<1.0	<1.0	<1.0	<1.0	<1.0	6.76
Pb	$\times 10^{-6} M$	<1.0	<1.0	<1.0	<1.0	<1.0	<1.0	<1.0	<1.0
Mn	$\times 10^{-4} M$	3.72	3.09	3.89	3.55	3.63	5.37	6.46	7.41
Fe	$\times 10^{-6} M$	<1.0	<1.0	<1.0	<1.0	<1.0	<1.0	<1.0	<1.0
Al	$\times 10^{-4} M$	1.86	1.86	1.86	<1.0	1.86	1.86	<1.0	<1.0
Ca	$\times 10^{-3} M$	1.95	3.63	3.72	4.07	2.19	3.89	4.68	9.12
Mg	$\times 10^{-4} M$	1.20	5.89	4.07	4.37	0.95	4.37	4.68	6.31
K	$\times 10^{-4} M$	7.59	19.95	10.23	9.55	8.13	10.0	10.72	11.75
Na	$\times 10^{-3} M$	3.31	3.89	3.89	3.98	2.51	8.32	8.51	9.12
PO_4-P	$\times 10^{-6} M$	2.6	2.9	4.2	15.49	2.57	9.33	26.92	60.26
SO_4-S	$\times 10^{-3} M$	0.26	0.85	0.98	1.15	0.44	0.91	1.66	3.39
Cl^-	$\times 10^{-1} M$	0.60	1.55	1.29	1.23	0.51	0.72	1.07	1.35
Organic C (g m^{-3}		160	130	160	190	170	150	180	250
Conductivity (dS m^{-1})		1.54	2.25	2.14	2.08	1.56	2.03	2.38	3.88
pH		4.7	4.9	5.1	5.6	4.6	5.0	5.2	5.5

a. *A* denotes a single application in 1971; *B* denotes repeat applications of initial rate in 1972, 1973, and 1974.

Source: Behel *et al.* [6]. *(Copyright 1983, American Society of Agronomy. Reprinted with permission.)*

model [31]; and (c) metal-organic complexes were modeled using stability constants obtained experimentally [48] for a sludge fulvic acid (fulvate model). It is almost always necessary to make assumptions of this kind when using a model.

The mixture sub-model of GEOCHEM predicted that less than 16% of the total (t) Cd_t Zn_t, and Mn_t in solution would be complexed with organic and inorganic ligands. Therefore, it was not surprising that exclusion of the mixture model from GEOCHEM has a minor effect on the speciation of Cd, Zn, and Mn (Table 5.4-5). The results of the overall simulation are shown in Table 5.4-6.

The predominant species of Zn in the soil solutions was Zn^{+2} (Table 5.4-7). Even though sludge applications increased total soluble Zn (Zn_t) from 4×10^{-6} *M* to more than 1×10^{-4} *M*, 87-97% of the Zn_t was present as Zn^{+2}. For most sludge treatments, SO_4 complexes accounted for 5-8% of Zn_t. Fulvate complexes (mixture model) with Zn tended to increase with sludge applied, but the amount of Zn_t associated with soluble organic C was always less than 2%. The distribution of Zn species was essentially the same after application of 200 t ha^{-1} in a single treatment or in four annual 50-t ha^{-1} increments. Soluble Zn_t increased to a greater extent with the single 200-t ha^{-1} sludge application ($1.7 \times 10^4 M$) when compared with four increments of 50 t ha^{-1} ($1.2 \times 10^4 M$).

The activity of Zn^{+2} in the soil solutions was similar to that supported by several Zn solid phases. The activity of Zn^{+2} ranged from approximately 10^{-5} to 10^{-3} *M* in solutions of pH 5 to 6. Based on stability diagrams [27], the Zn^{+2} in soil solution could be controlled by one or more of the following solids: (a) $ZnFe_2O_4$ in equilibrium with soil Fe; (b) soil Zn; and (c) Zn_2SiO_4 in equilibrium with amorphous Si. At similar pH values, Zn^{+2} activities obtained from this study were greater than those previously reported for soils incubated with sewage sludge [13]. The pH range of the sludge-treated soils was too narrow to evaluate the pH dependence of Zn^{+2}, as predicted by the solubility of known Zn solid phases.

DISCUSSION OF FINDINGS

According to Behel *et al.* [6], GEOCHEM calculations were close to monitored values and indicated that Zn^{+2} and Cd^{+2} are the predominant species in the soil solution of acid soils treated with sludges of metal additions. The amount approached the maximum recommended as safe for growth of agronomic crops.

Computer modeling of trace metal equilibria can provide a useful framework for understanding the nature of the multitude of reactions that take place in a soil system. However, a field monitoring program is needed to guide model application with data and calibration parameters.

Model output validation is essential to any soil modeling effort, although this term has a broad meaning in the literature. For the purpose of this section, we can define validation as "the process that analyzes the validity of final model output," namely, the validity of the predicted pollutant concentrations or mass in the soil column (or in

TABLE 5.4-5

Effect of Excluding Organic Ligands from Consideration by GEOCHEM on the Percentage of Free Metals in Soil Solution

Sludge Applied[a] ($t\ ha^{-1}$) A	B	Organic Ligands Included[b]	Cd A	Cd B	Zn A	Zn B	Cu A	Cu B	Ni A	Ni B	Mn A	Mn B
			(calculated % of total present as free metal ion)									
0	0	Fulvate model	67	67	79	80	—[c]	—	—	—	69	73
		Mixture model	92	91	97	96	—	—	—	—	98	97
		None	94	93	98	96	—	—	—	—	98	97
50	200	Fulvate model	69	72	84	84	—	—	—	—	80	81
		Mixture model	85	89	94	94	—	—	—	—	95	95
		None	86	90	95	95	—	—	—	—	96	96
100	400	Fulvate model	66	67	80	81	—	—	—	—	74	78
		Mixture model	84	83	93	91	—	—	—	—	95	84
		None	86	86	94	92	—	—	—	—	95	93
200	800	Fulvate model	64	63	79	78	69	69	—	72	74	76
		Mixture model	82	79	91	87	19	19	—	87	77	80
		None	86	82	93	88	93	88	—	90	94	90

a. *A* denotes 1 application in 1971; *B* denotes repeat applications of initial rate in 1972, 1973, and 1974.

b. Fulvate model — calculations based on equilibrium constants for fulvic acid extracted from sewage sludge [48]; Mixture model — mixture model for soluble fulvic acid [31].

c. Total concentrations below detection limits of analytical methods used.

Source: Behel *et al.* [6]. *(Copyright 1983, American Society of Agronomy. Reprinted with permission.)*

groundwater), as compared to available knowledge of measured pollutant concentrations from monitoring data (field sampling).

Disagreement in the absolute levels of concentration (predicted versus measured) does not necessarily indicate that either method of obtaining data (modeling, field sampling) is incorrect or that either data set should be revised. Laboratory analysis of field samples is difficult and uncertain. Field sampling approaches and modeling approaches produce two different perspectives of the same situation [9].

Field information, data, sampling and analysis are subject to temporal and spatial uncertainties. Model results are based on assumptions and utilize deterministic approaches; therefore, a disagreement of concentrations is to be expected.

TABLE 5.4-6

Concentration and Speciation of Mn, Cu, and Ni in the Soil Solution As Affected by Rate and Frequency of Sludge Application

Metal	Sludge Applied ($t\ ha^{-1}\ y^{-1}$)	No. of Applications[a]	Total Concentration ($\times 10^{-5}M$)	Free Metal	Metal Complexes SO_4	PO_4	MM[b]
				------ % of total metal ------			
Mn	0	0	37	98	1.8	<0.1	0.3
		0	36	97	2.8	<0.1	0.3
	50	1	31	95	4.2	<0.1	0.3
		4	54	95	4.0	<0.1	0.4
	100	1	39	95	4.9	<0.1	0.5
		4	65	84	6.0	<0.1	9.5
	200	1	36	77	4.5	<0.1	17.9
		4	74	80	8.3	0.2	11.4
Cu	200	1	0.32	19	1.4	<0.1	79.8
	200	4	0.63	19	2.5	<0.1	78.3
Ni	200	4	0.68	87	9.0	0.1	3.3

a. Single application in 1971; repeat applications of initial rate in 1972, 1973, and 1974.

b. MM denotes mixture model for fulvate [31].

Source: Behel *et al.* [6]. *(Copyright 1983, American Society of Agronomy. Reprinted with permission.)*

TABLE 5.4-7

Concentration and Speciation of Zn and Cd in the Soil Solution As Affected by Sludge Rate and Frequency of Application

Sludge Applied	No. of	Total	Free	Complexed Zn[b]		Total	Free	Complexed Cd[b]		
		Zn	Zn^{+2}	SO_4	MM[c]	Cd	Cd^{+2}	SO_4	Cl	MM[c]
($t\ ha^{-1}y^{-1}$)	Applications[a]	($\times 10^{-5}M$)	------ % of total ------			($\times 10^{-8}M$)	------ % of total ------			
0	0	0.35	97	2.2	0.5	0.98	92	2.7	3.6	1.7
	0	0.40	96	3.5	0.4	1.3	91	4.2	2.9	1.5
50	1	7.9	94	5.3	0.4	2.2	85	6.0	7.5	1.5
	4	11.8	94	5.0	0.7	3.0	89	5.9	3.3	1.9
100	1	10.7	93	6.0	0.9	3.1	84	6.9	6.2	2.4
	4	16.6	91	8.1	1.0	59	83	9.4	4.4	2.8
200	1	17.0	91	6.7	1.7	79	82	7.7	5.6	4.6
	4	36.3	87	11.3	1.3	144	79	13.0	4.6	3.6

a. A single application in 1971; repeat applications of initial rate in 1972, 1973, and 1974.

b. Complexes not given represented $< 0.1\%$ of total soluble metal.

c. MM denotes mixture model for fulvate [31].

Source: Behel *et al.* [6]. *(Copyright 1983, American Society of Agronomy. Reprinted with permission.)*

5.4.8 Literature Cited

1. Adams, R.T. and Kurisu, F.M., *Simulation of Pesticide Movement in Small Agricultural Watersheds*, Final Report, Environmental Research Laboratory, U.S. Environmental Protection Agency, Athens, Ga. (1976).
2. Allen, H.E., R.H. Hall and T.D. Brishin, "Metal Speciation: Effects on Aquatic Toxicity," *Environ. Sci. Technol.*, **14**, 441-43 (1980).
3. Anderson, M.P., "Movement of Contaminants in Groundwater: Chemical Processes," in *Groundwater*, National Academy Press, Washington, D.C. (1984).
4. Appel, C.A. and J.D. Bredehoeft, *Status of Groundwater Modeling in the U.S. Geological Survey*, U.S. Department of the Interior, Washington, D.C. (1978).
5. Ball, J.W., E.A. Jenne and D.K. Nordstrom, "WATEQ2 — A Computerized Chemical Model for Trace and Major Element Speciation and Mineral Equilibria in Natural Waters," in *Chemical Modeling in Aqueous Systems: Speciation, Sorption, Solubility, and Kinetics*, E.A. Jenne (ed.), American Chemical Society, ACS Symposium Series 93, Washington, D.C. (1979).
6. Behel, D., Jr., D.W. Nelson and L.E. Sommers, "Assessment of Heavy Metal Equilibria in Sewage Sludge-treated Soil," *J. Environ. Qual.*, **12**, 181 (1983).
7. Bonazountas, M. and J. Wagner, *SESOIL: A Seasonal Soil Compartment Model*, Draft Report, EPA Contract No. 68-01-6271, Office of Toxic Substances, U.S. Environmental Protection Agency, Washington, D.C. (1984, 1981).
8. Bonazountas, M., J. Wagner, and B. Goodwin, *Evaluation of Seasonal Soil/Groundwater Pollutant Pathways via SESOIL*, Final Report, Office of Water Regulations and Standards, U.S. Environmental Protection Agency, Washington, D.C. (1981).
9. Bonazountas, M., "Soil and Groundwater Fate Modeling: A Review," in *Fate of Chemicals in the Environment: Compartmental and Multimedia Models for Predictions*, R.L. Swann and A. Eschenroeder (eds.), ACS Symposium Series No. 225, American Chemical Society, Washington, D.C. (1983).
10. Bower, C.A. and L.V. Wilcox, "Soluble Salts," in *Methods of Soil Analysis*, C.A. Black (ed.), American Society of Agronomy, Madison, Wis., 947-48 (1965).

10a. Carsel, R.F., C.N. Smith, L.A. Mulkey, J.D. Dean, and P.P. Jowise, "User's Manual for the Pesticide Root Zone Model (PRZM): Release 1," Report No. EPA/600/3-84/109, U.S. Environmental Protection Agency, Athens, GA (1984).

11. Cole, C.R., "Information on the MMT/VVT Model to be Obtained by the Author," Battelle Pacific Northwest Laboratory, Richland, Wash. 99352 (1980).
12. Dowdy, R.H. and V.V. Volk, "Movement of Heavy Metals in Soils," Chapter 15 in Nelson *et al.*, Special Publication No. 11, Soil Science Society of America, American Society of Agronomy, Madison, Wis. (1983).
13. Emmerich, W.E., L.J. Lund, A.L. Page and A.C. Chang, "Predicted Solution Phase Forms of Heavy Metals in Sewage Sludge-treated Soils," *J. Environ. Qual.*, **11**, 182-86 (1982).

14. Emmerich, W.E., "Chemical Forms of Heavy Metals in Sewage Sludge-amended Soils as They Relate to Movement through Soils," Ph.D. Dissertation, University of California, Riverside (1980).

15. Enfield, C.G., R.F. Carsel, S.Z. Cohen, T. Phan and D.M. Walters, "Approximating Pollutant Transport to Groundwater," USEPA, RSKERL, Ada, Okla. (unpublished paper) (1980).

16. Florence, T.M., "Trace Metal Species in Fresh Waters," *Water Res.*, **11**, 681-87 (1977).

17. Gillman, G.P., "A Centrifuge Method for Obtaining Soil Solution," Division of Soils Report No. 16, Commonwealth Scientific and Industrial Research Organization, Melbourne, Australia (1976).

17a. Hern, S.C. and S.M. Melancon, "Guidelines for Field Testing Soil Fate and Transport Models," Report No. EPA/600/4-86/020, U.S. Environmental Protection Agency, Las Vegas, NV (1986).

17b. Hern, S.C. and S.M. Melancon, *Vadoze Zone Modeling of Organic Pollutants*, Lewis Publishers, Inc., Chelsea, Michigan (1986).

18. James, R.V. and J. Rubin, "Applicabilities of Local Equilibrium Assumption to Transport through Soils of Solutes Affected by Ion Exchange," in *Chemical Modeling in Aqueous Systems*, E.A. Jenne (ed.), Symposium Series No. 93, American Chemical Society, Washington, D.C. (1979).

19. Jenne, E.A., "Chemical Modeling — Goals, Problems, Approaches and Priorities," in *Chemical Modeling in Aqueous Systems*, E.A. Jenne (ed.), Symposium Series No. 93, American Chemical Society, Washington, D.C. (1979).

20. Jennings, A.A. and D.J. Kirkner, "Instantaneous Equilibrium Approximation Analysis," *J. Hydraul. Div. Am. Soc. Civ. Eng.* (1984).

21. Jennings, A.A., D.J. Kirkner and T.L. Theis, "Multicomponent Equilibrium Chemistry in Groundwater Quality Models," *Water Resour. Res.*, **18**, 1089-96 (1982).

22. Kincaid, C.T., J.R. Morrey and J.E. Rogers, *Geochemical Models for Solute Migration*, Vol. 1, "Process Description and Computer Code Selection," Report EA-3417, Electric Power Research Institute, Palo Alto, Calif. (1984).

23. Kirkner, D.J., H.W. Reeves and A.A. Jennings, "Finite Element Analysis of Multicomponent Contaminant Transport Including Precipitation-dissolution Reactions," Laible *et al.* (eds.), Proc. 5th International Conference, Burlington, Vt., Springer-Verlag, New York (1984).

24. Kirkner, D.J., T.L. Theis and A.A. Jennings, "Multicomponent Solute Transport with Sorption and Soluble Complexation," *Adv. Water Resour.*, **7**, 120 (1984).

25. Langmuir, D., "Techniques in Estimating Thermodynamic Properties for Some Aqueous Complexes of Geochemical Interest," in *Chemical Modeling in Aqueous Systems*, E.A. Jenne (ed.), Symposium Series No. 93, American Chemical Society, Washington, D.C. (1979).

26. Lindberg, R.D. and D.D. Runnells, "Groundwater Redox Reactions: An Analysis of Equilibrium State Applied to Eh — Measurements and Geochemical Modeling," *Science*, **225**, 925 (1984).

27. Lindsay, W.L., *Chemical Equilibria in Soils*, John Wiley & Sons, New York (1979).

28. Lyman, W.J., W.F. Reehl and D.H. Rosenblatt, *Handbook of Chemical Property Estimation Methods*, McGraw-Hill Book Co., New York (1982).

29. Mahier, R.J., F.T. Bingham, G. Sposito and A.L. Page, "Cadmium-enriched Sewage Sludge Application to Acid and Calcareous Soils: Relation Between Treatment, Cadmium in Saturation Extracts, and Cadmium Uptake," *J. Environ. Qual.*, **9**, 359-64 (1980).

30. Matthews, P.J., "Control of Metal Applications Rates from Sewage Sludge Utilization in Agriculture," *CRC Crit. Rev. Environ. Control*, **14**, (1984).

31. Mattigod, S.V. and G. Sposito, Chemical Modeling of Trace Metal Equilibria in Contaminated Soil Solutions Using the Computer Program GEOCHEM," in *Chemical Modeling in Aqueous Systems*, E.A. Jenne (ed.), Symposium Series No. 93, American Chemical Society, Washington, D.C. 837-56 (1979).

32. Mattigod, S.V., G. Sposito and A.L. Page, "Factors Affecting the Solubilities of Trace Metals in Soils," in *Chemistry in the Environment*, American Society of Agronomy, Soil Science Society of America, ASA Special Publication No. 40 (1981).

33. McDuff, R.E. and F.M.M. Morel, "Description and Use of the Chemical Equilibrium Program REDEQL-2," Tech. Rep. EQ-73-02, W.M. Keck Laboratory, California Institute of Technology, Pasadena, Calif. (1974).

33a. Melancon, S.M., J.E. Pollard and S.C. Hern, "Evaluation of SESOIL, PRZM and PESTAN in a Laboratory Column Leaching Experiment," *Environ. Toxicol. Chem.*, **5**, 865-878 (1986).

34. Miller, C.W. and L.V. Benson, "Simulation of Solute Transport in a Chemically Reactive Heterogeneous System: Model Development and Application," *Water Resour. Res.*, **19**, 381-91 (1983).

35. Murarka, L., "Planning Workshop on Solute Migration from Utility Solid Waste," Publication EA-2415, Electric Power Research Institute, Palo Alto, Calif. (1982).

36. Nelson, R.W. and J.A. Schur, "Assessment of Effectiveness of Geologic Oscillation Systems: PATHS Groundwater Hydrologic Model," Battelle Pacific Northwest Laboratory, Richland, Wash. (1980).

37. Nelson, D.W., D.E. Elrick and K.K. Tanji (eds.), "Chemical Mobility and Reactivity in Soil Systems," Special Publication No. 11, Soil Science Society of America, American Society of Agronomy, Madison, Wis. (1983).

38. Nordstrom, D.K., L.N. Plummer, T.M.L. Wigley, T.T. Wolery, J.W. Ball, E.A. Jenne, R.L. Bassett, D.A. Crerar, T.M. Florence, B. Fritz, M. Hoffman, G.R. Holdren, G.M. Lafon, S.V. Mattigod, R.E. McDuff, F. Morel, M.M. Reddy,

G. Sposito and J. Thrailkill, "A Comparison of Computerized Chemical Models for Equilibrium Calculations in Aqueous Systems," in *Chemical Modeling in Aqueous Systems*, E.A. Jenne (ed.), Symposium Series No. 93, American Chemical Society, Washington, D.C. (1979).

39. Norvell, W.A., "Equilibria of Metal Chelates in Soil Solution," in *Micronutrients in Agriculture*, J.J. Mortvedt, P.M. Giordano, and W.L. Lindsay (eds.), Soil Science Society of America, Madison, Wis. (1972).

40. Parkhurst, D.L., L.N. Plummer and D.C. Thorstenson, *BALANCE — A Computer Program for Calculating Mass Transfer for Geochemical Reactions in Ground Water*, Report USGS/WRI-82-14, U.S. Geological Survey, Reston, Va. (1982).

41. Parkhurst, D.L., D.C. Thorstenson and L.M. Plummer, "PHREEQE: A Computer Program for Geochemical Calculations," *Water-Resour. Invest. U.S. Geol. Surv.*, pp. 80-96 (1980). (NTIS Association No. PB81 167801.)

42. Plummer, L.M., D.L. Parkhurst and D.R. Kouir, "MIX2: A Computer Program for Modeling Chemical Reactions in Natural Waters," *Water-Resour. Invest. U.S. Geol. Surv.*, pp. 61-75 (1975).

43. Plummer, L.M., B.F. Jones and A.H. Truesdell, "WATEQF — A FORTRAN IV Version of WATEQ, A Computer Program for Calculating Equilibrium of Natural Water," *Water-Resour. Invest. U.S. Geol. Surv.*, **76** (1976).

44. Prickett, T.A., T.G. Naymik and C.G. Lonnquist, *A 'Random-Walk' Solute Transport Model for Selected Groundwater Quality Evaluations*, Bulletin 65, Illinois State Water Survey, Champaign, Ill. (1981).

45. Senesi, N. and G. Sposito, "Residual Copper(II) Complexes in Purified Soil and Sewage Sludge Fulvic Acids: Electron Spin Resonance Study," *Soil Sci. Am. J.*, **48**: 1247-53 (1984).

46. Sposito, G., F.T. Bingham, S.S. Yadav and C.A. Inouye, "Trace Metal Complexation by Fulvic Acid Extracted from Sewage Sludge: II. Development of Chemical Models," *Soil Sci. Soc. Am. J.*, **46**, 51-56 (1982).

47. Sposito, G., K.M. Holtzclaw and C.S. LeVesque-Madore, "Trace Metal Complexation by Fulvic Acid Extracted from Sewage Sludge: I. Determination of Stability Constants and Linear Correlation Analysis," *Soil Sci. Soc. Am. J.*, **45**, 465-68 (1981).

48. Sposito, G., *The Thermodynamics of Soil Solutions*, Clarendon Press, Oxford, England (1981).

49. Sposito, G., "Chemical Models of Inorganic Pollutants in Soils," *CRC Crit. Rev. Environ. Control*, **15**, No. 1 (1985).

50. Sposito, G. and S.V. Mattigod, *GEOCHEM: A Computer Program for the Calculation of Chemical Equilibria in Soil Solutions and Other Natural Water Systems*, Dept. of Soil and Environmental Sciences, University of California, Riverside, Calif. (1980).

51. Stumm, W. and J.J. Morgan, *Aquatic Chemistry*, 2nd ed., John Wiley & Sons, New York (1981).

52. Theis, D.L., D.J. Kirkner and A.A. Jennings, *Multi-Solute Subsurface Transport Modeling for Energy and Solid Wastes*, Dept. of Civil Engineering, University of Notre Dame, Notre Dame, Ind. (1982).

53. Truesdell, A.H. and B.F. Jones, "WATEQ — A Computer Program for Calculating Chemical Equilibria of Natural Water," *J. Res. U.S. Geol. Surv.*, **2** (1974).

54. Westall, J. and H. Hohl, "A Comparison of Electrostatic Models for the Oxide/Solution Interface," *Adv. Colloid Interface Sci.*, **12**, 265-94 (1980).

55. Wolery, T.J., *Calculation of Chemical Equilibrium Between Aqueous Solution and Mineral: The EQ3/6 Software Package*, UCRL-52658, Lawrence Livermore Laboratory, Livermore, Calif. (1979).

56. Yeh, G.T. and D.S. Ward, *FEMWASTE: A Finite-element Model of Waste Transport through Saturated-Unsaturated Porous Media*, Publication No. 1462, ORNL-5601, Oak Ridge National Laboratory, Oak Ridge, Tenn. (1981).

57. Yeh, G.T. and D.S. Ward, *FEMWATER: A Finite-Element Model of Water Flow through Saturated-Unsaturated Porous Media*, ORNL-5567, Oak Ridge National Laboratory, Oak Ridge, Tenn. (1980).

5.5 AQUATIC MODELING

Until recently, virtually all modeling studies directed at examining the migration and fate of metals have neglected important chemical interactions that control their behavior in aquatic systems. For example, a study of Pb, Cd, Zn, Cu and S movement through Crooked Creek Watershed in Missouri conducted by Munro *et al.* [31] considered only metal adsorption through the use of an equilibrium partitioning coefficient [7].

The EPA is now giving states the latitude to establish site-specific standards, in recognition of the major impacts to water quality of metal species likely to harm an ecosystem. As a result, there is a growing need for more thorough investigation of the chemical reactions of inorganic species in aquatic systems and models for calculating the distribution and activities of species in aqueous regimes.

An evaluation of the fate of trace metals in surface waters (stream, lake, estuary, coastal waters) requires the consideration of speciation, adsorption, precipitation, and other chemical reactions or processes. These processes can affect the solubility, bioavailability, toxicity, physical transport, and corrosion potential of metals. Various computerized models of the complex interactions have been developed for specific applications. This section briefly describes several metal speciation models and important limitations to their usefulness.

5.5.1 Background

Computer modeling of metal speciation in aquatic environments has developed only in the last decade, and the number of available models is small. Some are quite narrow in applicability, in that their use is restricted to certain environments — e.g., anoxic sediments, seawater, and waters overlying carbonate mineralization. The historical development of terrestrial water models for metals was shown in Figure 5.4-3.

The models that consider the speciation of trace metals in waters are equilibrium models. For natural water systems, the model must be able to establish the following for various ranges of pH, pe, ionic strength and temperature:

- Equilibrium concentrations of trace metals complexed with organic and inorganic ligands,
- Adsorption and desorption equilibria,
- Precipitation and dissolution of solids, and
- Redox equilibria.

Equilibrium constants for organometallic interactions in natural systems [41,42] are not well defined; models often incorporate organic ligands such as amino acids rather than humic and fulvic acids, which in real systems may act as the major sources of complexes of organic ligands. With regard to adsorption and desorption equilibria, the type of adsorption surface (i.e., clays or hydrous metal oxides) and extent of surface area are important aspects of the modeling of trace metal behavior in water. Computer code flexibility (e.g., pH, pe) allows more representative computation of the equilibria in well-characterized natural waters.

We shall not deal here with details of the specific reactions included in aqueous chemical modeling, since these are treated in Chapter 2 of this report. Felmy *et al.* [6] have prepared a primer on key concepts for aqueous chemistry, with special emphasis on speciation, activity, adsorption, and solid-phase reaction as modeled by a recent EPA computer code designated MEXAMS (Metals Exposure Analysis Modeling System; see Section 5.5.4).

5.5.2 Modeling Issues

A discussion of computerized chemical models for equilibrium calculations in aqueous systems is given by Nordstrom *et al.* [32], and Jenne [17] has written an excellent essay on the goals, problems, approaches, and priorities of chemical modeling. Information from the work of these researchers is included in the following paragraphs.

There are several ways to formulate aqueous models, such as the Bjerrum ion association theory [4], the Fuoss ion association theory [9] and the Reilly, Wood and Robinson mixed electrolyte theory [37]. Each theory carries a set of assumptions and restrictions.

When using an ion association theory, which is by far the most common method, one can formulate the species distribution in two distinct but thermodynamically related ways: (1) the equilibrium constant approach and (2) the Gibbs free-energy approach. As described by Nordstrom *et al.* [32], both approaches are subject to the conditions of *mass balance* and *chemical equilibrium.*

The mass balance condition requires that the computed sum of the free and derived species be equal to the given total concentration. Chemical equilibrium requires that the most stable arrangement for a given system be found, as defined by the equilibrium constants for all mass action expressions of the system, or through the use of Gibbs free energies for all of the components and derived species. In the equilibrium constant approach, the mass action expressions are substituted into the mass balance conditions, resulting in a set of nonlinear equations that must be solved simultaneously. The Gibbs free-energy approach is simply a transformation of variables through the thermodynamic relation, which allows a different numerical approach.

Most models use the equilibrium constant approach, which utilizes measured equilibrium constants for all mass action expressions of the system. The Gibbs approach uses free-energy minimization techniques, and Gibbs free-energy values are required. In both cases, the most stable condition is sought, and a solution to a set of nonlinear equations is required. The solution involves interactive procedures, several of which are discussed by Nordstrom *et al.* [32]. The use of the Gibbs free-energy minimization approach is primarily useful for simple systems because of the limited availability of free-energy values. The equilibrium constant approach has the advantage of a larger data base and is therefore generally preferable.

The inputs required for these models include an estimated concentration of all the species considered (usually the total concentration), partial pressures of any gases to be considered, information regarding adsorbing surfaces, the solids allowed to precipitate, the pH (which can be calculated) and the redox potential, Eh (or pe, if redox equilibria are considered). Data base requirements are further discussed in section 5.5.6.

As most of the models function similarly, their outputs are generally of a similar type. They provide the final concentration of the free ion, as well as the concentrations of complexes. The extent of adsorption, cation exchange and precipitation is also indicated, if allowed in the particular run or model. Specific outputs are discussed in section 5.5.7.

Although several computerized codes are now available and are widely applied, all have significant limitations. The general limitations are described in the next section, followed by reference to specific modeling codes.

5.5.3 Modeling Limitations

Nordstrom *et al.* [32] and Jenne [17] cite a number of major factors that limit the effectiveness of models. One is that all the models assume equilibrium conditions; that is, the kinetics of precipitation, oxidation-reduction and adsorption are ignored. This may not be a valid assumption in some cases, because of kinetically unfavorable chemical processes, biological transformation, and physical transport. However, it provides a good approximation of some environments, such as the hypolimnion of a thermally stratified lake, where most of the above chemical reactions are assumed to be kinetically rapid for modeling purposes.

As the outputs of these models are strongly related to their inputs, the characterization of the situation to be modeled will directly affect the results obtained. In this respect, models are more dependably used in a descriptive way and "after the fact" (i.e., to reproduce an existing situation); they may have limited applicability if used in a predictive way (with default generalized data) for a specific future environment. In some cases, the input data are also limited by the state-of-the-art, such as in the characterization of organic matter in natural systems.

For all models/codes there can be large variation in the stability constants for the same reaction. These differences can dramatically affect the results, as shown by Nordstrom *et al.* [32] for 15 models applied to two test cases (a river water and a seawater). The authors attributed the largest source of differences between predicted and observed values to the fact that the thermodynamic data (equilibrium constants) used by the various codes varied by up to three orders of magnitude. Concerted efforts are now being made to develop consistent and documented data bases.

Most equilibrium constants are measured at 25°C, and most models are applicable only at that temperature. Other temperatures can be used in some models, but the activation energy data base required to adjust equilibrium constants is somewhat limited [13].

The calculations are valid over a specific range of ionic strength; approximations become more inexact as ionic strength increases. At high ionic strength, different approaches are needed, and various forms of the Debye-Hückel or Davies equations are used [17]. In addition, assumptions about redox conditions may not be realistic. (The user must choose which redox reactions to consider and the dominant redox potential.) Nordstrom *et al.* [32] recommend that redox reactions be analyzed separately in some cases. The total number of complexes considered by an aqueous model affects the overall output.

The modeling of adsorption is not well developed, owing to a general lack of understanding of the pheonomenon. In addition, surface modifications may cause solids in natural systems to behave quite differently from pure phases [5,13]. Modeling of the adsorption process is accomplished via equilibrium constants, which are frequently estimated in the laboratory and later adjusted in real work applications.

Harriss *et al.* [13] report that at very low concentrations, which may be of particular interest to the user, inaccuracies may result from computational limitations. In addition, the validity of models at high or low concentrations and in real-world applications is largely untested because of inadequate analytical capabilities for measuring dissolved species. Monitoring and model validation programs have recently been established by the EPA for the MEXAMS model [7].

Other important restrictions on the usefulness of chemical models are the limited capability to characterize organic ligands in natural waters, the scarcity of thermodynamic data, limitations in scientists' training, inadequate literature reviews, and the lack of integrated studies.

There is no general-purpose model that can be used for all applications. While there are numerous limitations to the use of equilibrium models, the general consensus is that they can provide useful results if applied correctly and with an understanding of the differences between simulated and real systems.

Users tend to employ models that are well documented, have an integrated or thorough data base, and are supported by an organization. Remarkable support for water quality modeling in the United States has been provided by the EPA's Environmental Research Laboratory in Athens, Georgia. The recently developed model MEXAMS, for example, was supported by this organization.

5.5.4 Model Codes

Nordstrom *et al.* [32] have surveyed over 30 computer codes and described the features of 14 of them. Therefore, this information is not duplicated here. Some of these codes were specifically developed for "water quality" models; others were developed to be "geochemical" (soil moisture or groundwater quality) codes.

Although certain terrestrial (geochemical) codes (e.g., GEOCHEM) can be used to simulate the speciation and fate of metals in a water body by "tailoring" the size of a model compartment (e.g., sediment) to reflect the size of the corresponding compartment of the water body, codes specifically developed for aquatic modeling should be used.

The models REDEQL2, WATEQ2, MINEQL and MEXAMS are now available and widely accepted, applied, supported, and validated. Various versions of these models exist in the literature also (e.g., [13,38]). Table 5.5-1 summarizes some major features of the four models discussed below; interested readers should refer to the original works for details.

REDEQL2

REDEQL2 is an offspring of REDEQL (see Figure 5.4-3) developed by Morel and Morgan at the California Institute of Technology [28]. The University of Minnesota in Duluth undertook modifications and improvements to the program in 1978-1982.

General Features

REDEQL2 is a multiphase, multicomponent aqueous equilibrium program that can impose metal precipitation and dissolution and simulate adsorptive behavior. It is currently able to model systems at 25°C and has an ethalpy data base to broaden this aspect of the program. According to model developers, REDEQL2 can be used for both laboratory and natural systems. Its application may handle metal speciation in algal culture media, degradation of NTA in natural waters, metal speciation in sewage oxidized and diluted by seawater, zinc speciation with regard to toxicity to fish, and metal speciation relative to changes in the Great Lakes [35].

As of May 1985, the data base for REDEQL2 included 36 metals, 65 ligands, 212 solids, 2239 complexes (of which 113 are ternary) and 22 redox reactions. To model water quality, the chemical equilibrium and redox equilibrium constants are needed at infinite dilution (I = 0) at 25°C. Constants are adjusted by the model for various ambient temperatures. Since most data are also measured at a higher ionic strength, the program makes certain adjustments. The University of Minnesota researchers added an ability to handle ternary complexes ($M_iL_jL_k$ or $M_iM_jL_k$) in which H^+ and OH^- are not M or L, respectively. Some equilibrium constants for these complexes are not well known and must be measured. The REDEQL2 program can accommodate redox equilibria in which the reaction causes the ion to reverse charge, such as Cr^{+3} going to CrO_4^{-2}.

TABLE 5.5-1

Descriptive Features of REDEQL2, WATEQ2, MINEQL and MEXAMS

Parameter	REDEQL2[a]	WATEQ2	MINEQL	MEXAMS[a]
Model type	Equilibrium constant	Free energy minimization	Equilibrium constant	Equilibrium constant
Metals	36	26	34	7
Ligands	65	80	57	More than 60
Organics	Amino acids	Fulvate and humates	14 Amino acids	Amino acids
Complexes	2239	220	Almost unlimited	Almost unlimited
Ternary	113	—	No	No
Minerals	$Pb_5(SO_4)_3Cl$, Chlorite, illite, Na-montmorillonite, dolomite, microcline	309-all types	Allowed as inputs	Allowed as inputs
Gases	CO_2, N_2, O_2	CO_2, O_2, CH_4	Allowed as inputs	Allowed as inputs
Adsorption surfaces	MnO_2, $Fe(OH)_3$, TiO_2, SiO_2	No	No	Six algorithms (various surfaces)
Redox reactions	22	12	18	18
Activity coefficient	Magnuson or Davies	Extended Debye-Hückel or Davies	Davies	Extended Debye-Hückel
Temperature	25°C	0-100°C	25°C	25°C
Inorganic carbon	Controlled by pCO_2, HCO_3-Alk or input [ ]	Calc. from HCO_3-Alk	Controlled by pCO_2, HCO_3-Alk or input [ ]	Controlled by pCO_2, HCO_3-Alk, or input [ ]
pH	Fixed or variable	Field sample pH	Fixed or variable	Fixed or variable
Solids not allowed to precipitate	Yes	No	Yes	Yes
Iterations	Less than 2000 (Newton-Raphson)	Less than 40 (back substitution)	Less than 2000 (Newton-Raphson)	Less than 2000 (Newton-Raphson)
Inputs	[ ], Redox rxns, ionic strength, solids	Complete chemical analysis of water, trace elements	Same as REDEQL2	Chemical analysis of water, pH, dissolved O_2

a. [ ] = "concentration(s)"

Adsorption in REDEQL2 is calculated using the James-Healy approach; the dielectric constants of the solids are important surface characteristics. The free energy of adsorption is used as the fitting parameter. SiO_2, MnO_2, $Fe(OH)_3$ and TiO_2 were chosen as sorption surfaces, because (1) with the exception of TiO_2, all are thought to be commonly occurring sorption surfaces in soils, (2) they are available and used for laboratory work, and (3) data are available for these species.

Input parameters of REDEQL2 include the concentration or activity of each metal, ligand, and solid; the ionic strength, if it is to remain fixed; the pH, also if fixed; the redox reactions; the solids prohibited from precipitating; the adsorption surface area and adsorption surfaces; and (optionally), the pe, CO_2, and N_2. Some options available to the user are detailed below.

Multicase Runs

The program can handle a set number of ligands and metals per case run. Up to ten cases may be run at the same time; however, the free-ion concentrations resulting from the previous run must be used as input to the next run. To prevent dissimilar parameters between case runs, the types of metals and ligands are fixed, and non-precipitable solids, redox reactions and charge-balance ions are not permitted to change. The pe, total concentrations, gaseous partial pressure, and pH are allowed to vary.

Charge Balance at Fixed pH

The overall charge of the final solution calculated for a non-fixed pH is balanced by adjusting the hydrogen and hydroxide ion concentrations to achieve a charge-balanced solution. If the pH is kept constant, the program continues to adjust the concentrations of one or more chosen ligands until the charge of the solution is less than 0.1% of the ionic strength.

pH Calculation

The pH of the final solution may be calculated with each interaction or remain fixed throughout. The user is cautioned against the former option for waters that cannot be exactly characterized, such as natural waters. Since the pH is determined by balancing the charge of all the metals, ligands, and electrons, the exact concentration of these components must be known to obtain a meaningful pH; therefore, the calculated pH option is best used for laboratory waters only.

Inorganic Carbon

The partial pressure of CO_2, carbonate alkalinity and total inorganic carbon are the three options available to the user to control the inorganic carbon (H_2CO_3, HCO_3^-, and CO_3^{-2}) distribution in the program. The alkalinity case is usable only for options in which the pH is fixed and in sample runs containing no carbonate solids. The CO_2 option determines the amount of inorganic carbon dissolved in solution at a fixed pH.

One drawback of this option is that CO_2 will be drawn from the atmosphere if the solution is undersaturated with respect to CO_3^{-2}. To avoid "draining" the atmosphere of CO_2, the concentration of CO_3^{-2} is increased by the program until the situation is alleviated.

Ionic Strength

The ionic strength of the solution, if not fixed, is calculated at the end of the program for the new concentrations of metals and ligands. If the ionic strength has changed more than 0.5% from the initial input value, the program will resume calculations using thermodynamic data adjusted to the new ionic strength.

Redox Reactions at Fixed Potential

REDEQL2 calculates the redox reactions for the specified set of reactions from potential data input as dissolved O_2, pe, or Eh (volts). A fixed pH must accompany a fixed potential, since chemical speciation is dependent on both parameters. Again, knowledge of possible redox reactions is required in selecting redox reactions to be included in the program; typically, the program can accommodate only two or three reactions of this type, so it is best to avoid including one that is not dynamic for the given pe and pH. Accordingly, if the concentration of an ion involved in a redox reaction is less than 10^{-10} *M* at the pH and pe of the system, or if the redox species form a ratio of less than 10^{-5} or greater than 10^5, the reaction should be excluded from the program.

Redox Reactions with Changing Potential

This option is usable only at a variable pH, since some reactions cause changes in both the pH and the potential. The initial redox state and concentrations of the participating redox species must be specified in order to use this option.

Minerals

REDEQL2 includes data in its thermodynamic data file for six alumino-silicate clays, which can be present in either the suspended or solid state [25].

WATEQ2

WATEQ2 is a model of the WATEQ-series [40] that evolved from WATCHEM (Figure 5.4-3) and is designed to accept water analysis with on-site values for pH, Eh, and temperature [2,3]. It is essentially a water-rock equilibrium model in which almost every conceivable ore and mineral containing a trace metal is included.

WATEQ2 is a Gibbs free-energy minimization program in which the anion concentrations are juggled until the difference in Gibbs energy between calculation steps is within 0.1% of the input concentrations. The thermodynamic data base is of substantial size and concerns solid forms of trace elements, polysulfides, metastable solids, sparingly soluble salts and complexes of major ions. Organic ligands are incorporated as groups of molecular weight 6550 g/mole or 2000 g/mole to simulate fulvate and humate ligands. Twelve redox reactions are included.

The required inputs are standard water parameters such as conductivity, dissolved solids, dissolved organic carbon, salinity, and the following species: Fe (II and III), Mn (II), Cr (II), Zn (II), Cd (II), Pb (II), Ni (II), Ag (I), As (III, V, total), Cs (I), Rb (I), Li (I), Ba (II), Sr (II), I, Br, and dissolved O_2. The outputs are the concentration in ppm, molality, activity, and the activity coefficient. The program calculates a standard deviation from errors propagated from analytical and thermophysical data.

MINEQL

MINEQL [43] evolved from REDEQL [28], aiming to be a more compact program of greater clarity and flexibility. At present, it cannot accommodate adsorption equilibria, but a subroutine is being devised [35]. The input ligands and metals are classified according to phase and solubility, and each ion or complex is placed in one of the following classifications:

Type I — Soluble components (e.g., metals)
Type II — Soluble complexes (e.g., CO_3^{-2})
Type III — Precipitated solids not allowed to dissolve (includes gases)
Type IV — Precipitated solids that dissolve

Presently, MINEQL has 50 cations and 198 neutral and anionic ligands in its thermochemical data bank. Creation of new species or modifications of formulation constraints indicate the flexibility of the program, in that only two of the three needed parameters (species, equilibrium constant and the type of category) will prompt these changes.

MEXAMS

MEXAMS, the Metals Exposure Analysis Modeling System [6], provides an enhanced capability for assessing the impact of priority pollutant metals on aquatic systems. It allows the user to consider the complex chemistry affecting the behavior of metals in conjunction with the transport processes that affect their migration and fate. This is accomplished by linking MINTEQ, a geochemical model, with EXAMS, an aquatic exposure assessment model.

General Features

MINTEQ is a thermodynamic equilibrium model that computes aqueous speciation, adsorption, and precipitation/dissolution of solid phases [6,7]. It has a well-documented thermodynamic data base that contains equilibrium constants and other accessory data for seven priority pollutant metals: As, Cd, Cu, Pb, Ni, Ag and Zn. The model was developed by combining the best features of MINEQL and WATEQ4. EXAMS, which was designed for the rapid evaluation of organic pollutants in water bodies, was modified to achieve linkage to MINTEQ.

Applicability

MEXAMS was developed to provide EPA with a predictive tool capable of performing screening-level analyses, but it can also be used on a more site-specific basis to investigate potential impacts of different metal sources such as industrial discharges or mine drainage. An example of the former is a study by Morel *et al.* on the fate of trace metals discharged from a Los Angeles County treatment plant [7].

According to the model manual, other applications of MEXAMS relate to improving the information available from bioassays and involve the use of the model to identify what is and what is not known about the behavior of priority pollutant metals in aquatic systems.

Limitations

Despite the many capabilities that both MINTEQ and EXAMS offer, model developers report several limitations that MEXAMS users should be aware of: (1) MINTEQ's thermodynamic data base contains equilibrium constants and accessory data for only seven pollutant metals (As, Cd, Cu, Pb, Ni, Ag, Zn); (2) the data base contains no data for processes of inorganic complexation; (3) only equilibrium processes that govern the kinetics of precipitation/dissolution, oxidation/reduction, and adsorption are considered; (4) MEXAMS has been tested on only relatively simple problems.

Guidelines for Use

MEXAMS is available in an interactive mode; in fact, the user can separately employ MINTEQ or EXAMS via an interactive process. In MINTEQ, species are assigned to one of four categories: species complex, fixed species, dissolved solids subject to precipitation, and species not considered. MINTEQ contains six algorithms for calculating adsorption, and the user must provide data and algorithm selection.

The more data the user has on the water chemistry of the system, the more accurate will be the predicted results. This does not mean that the user must have data for all the components to be simulated, since the default data base may substantially facilitate operations. However, certain parameters are of greater importance and should be given special attention; these are: pH, Eh (pe), temperature, ionic strength, major anions, major cations, trace constituents (H_2S, PO_4^{-3}, F^-, Fe (II), Mg (II), Al (III), Ba (II), Sr (II), other). The MEXAMS and MINTEQ manuals provide detailed information on the above topics.

5.5.5 Availability/Suitability of Codes

Of the codes previously discussed, MEXAMS receives the strongest support (from EPA's Environmental Research Laboratory). REDEQL2 appears to have the strongest data base and to be suited for many purposes. WATEQ2 is restricted to calculating metal-ligand equilibria for overlying rock formations, while MINEQL2 appears to offer great flexibility and compactness. There is no general guidance as to which code should be used for a specific situation; the user must either review reported codes and select the one that is suitable to the relevant metals and ligands, or he must consult an organization that supports a model and request assistance for a site-specific application.

5.5.6 Data Availability for Models

As previously mentioned, the types of reactions considered in REDEQL2 and equivalent models include metal ion hydrolysis, ligands, protonation, complexation, precipitation, redox and adsorption. Each such reaction can be described quantitatively by a stability or formation constant that relates the concentrations of reactants and products. At the computed equilibrium, all equilibrium constant expressions are satisfied simultaneously, and the sum of the individual species of metals and ligands is equal to a fixed total concentration. The input data required for this computation include: (1) total concentrations of all metals and ligands to be considered, (2) equilibrium constants for the types of reactions described above, (3) ionic strength, which may be calculated or specified, (4) pH, which also may be calculated or specified, and (5) redox potential in the case of redox reactions.

The fact that models have sometimes been used quite successfully indicates that data are available for some cases. However, most of the waters that may have to be modeled are not well characterized, making the use of models difficult in terms of calibration and validation. The following sections discuss some of the problems that may be encountered.

5.5.7 Issues Related to Modeling

TOTAL CONCENTRATION OF METALS AND LIGANDS

For any given system that the model is intended to represent, the concentrations of all ligands present in the system which can significantly complex any or all of the trace metals present must be known. A considerable body of knowledge on thermodynamic data and/or the complexes that are formed has been built up through the study of simple systems containing one metal and one ligand at concentrations higher than those ordinarily found in natural water.

However, natural systems are considerably more complex in terms of the diversity of trace metal species that may exist and the extremely low concentrations of individual species. Until recently, most workers used stability constants singly or in pairs to enable them to calculate the equilibrium concentration of a particular species. In recent reviews of trace metal speciation, however, researchers have noted that this procedure is invalid, since any natural water contains many metals capable of reacting with each ligand and many ligands that can react with each metal. The correct equilibrium concentration of any species can be computed only by simultaneously taking into account all the competing equilibria.

STABILITY CONSTANTS

The extent of complexation of trace metals in solution can be calculated if the concentration of ligands and the stability constants of the appropriate reactions are known. Unfortunately, the stability constants of all possible complexes have not been measured. Sibley and Morgan [39] cite the example of borate complexes, for which stability constants are not available, and chloro-complexes of copper and lead, which have several reported constants for the same complex.

IONIC STRENGTH

The equilibrium quotient is not constant for all concentrations when analytically measured values are used, because the behavior (concentration vs. activity) of dissolved participants in chemical reactions is not "ideal." (See section 2.6.)

Some measurement techniques serve to determine directly the activity of a particular ion. Ion-selective electrodes, for example, are available and can measure the activities of a large number of cations and anions of interest in natural water. The first of these electrodes to be developed, and still the best known, is the glass electrode, which has been used for many years to determine the hydrogen ion activity (pH) of solutions.

If direct measurement of the activity of an ion is not possible, it can be calculated by multiplying the activity coefficient by the analytically determined concentration of the ionic species. (See section 2.6.) The method commonly employed for this calculation uses some form of the Debye-Hückel equation, which involves, in turn, determination of a quantity defined as the ionic strength. The importance of this fact to the investigator concerned with modeling natural or laboratory water systems is that the required equilibrium constant data are usually specified for solutions of zero ionic strength.

In solutions containing less than 50 mg/l of dissolved ions, activity coefficients for most ions are 0.95 or more. In such dilute solutions, activity values are equal to measured concentrations (within ordinary analytical error). If concentrations are near 500 mg/l of dissolved solids, the value of the activity coefficient may be as low as 0.70 for divalent ions. At the maximum ionic strength where the Debye-Hückel equation can be accurately used, the activity coefficients for some divalent ions may be less than 0.50 [14]. For the concentration ranges of principal interest in natural water chemistry, the Debye-Hückel equation and equilibrium constants applicable at zero ionic strength provide sufficient accuracy.

However, investigators working with more highly concentrated solutions who elect to use the models in the mode in which ionic strength is specified have no choice but to use constants that are directly applicable to those solutions. As noted in the user's guide for REDEQL, the usual approach is to allow the program to calculate ionic strength (unless one is working with a solution of very constant (and well known) ionic strength, such as seawater), because ionic strength is difficult to estimate accurately without knowing the concentrations of all the species present.

pH

Like ionic strength, the pH of the solution under consideration can be either specified or calculated by the program. Also, as with ionic strength, if the pH value is specified, it must be carefully interpreted. As noted by Hem, the pH of a water sample represents the interrelated results of a number of chemical equilibria, which may be altered by the sampling operation: "A pH measurement taken at the moment of sampling may represent the original equilibrium conditions . . . satisfactorily, but if the water is put into a sample bottle and the pH is not determined until the sample is taken out for analysis some days, weeks, or months later, the measured pH may be influenced by other reactions such as oxidation of ferrous iron, and the laboratory pH can be a full unit different from the value at the time of sampling. A laboratory determination of pH can be considered as applicable only to the sample solution in the sample bottle at the time the determination is made." [14]

pe

In computations in which redox reactions are to be considered, a value for pe (the negative logarithm of the electron activity) must be specified. Field and laboratory measurements of the redox potential, Eh, from which pe is calculated are subject to misinterpretation in much the same way as are pH measurements. The redox potential of sample solutions exposed to air can be affected by atmospheric oxygen. Measurements on reducing systems, such as groundwater containing a mixture of ferrous and ferric iron, can be made to give meaningful answers only if oxygen is very carefully excluded from all parts of the sampling and measuring system [14].

SUSPENDED SOLIDS

Inorganic species such as common clay minerals, hydrous oxides of iron and manganese, silica, and sulfides present as suspended particulate matter in natural waters [4] have been shown in both laboratory [19] and field [21] studies to be able to adsorb significant quantities of trace elements from solution. Models that consider adsorption on suspended solids may allow only for simple precipitation, which does not involve occlusion or adsorption of other trace metals by the precipitate.

In some of their modeling efforts, Sibley and Morgan [39] described adsorption in terms of the James-Healey model for aqueous metals and oxide surfaces. In that approach, adsorbing surfaces are treated as ligands, and the adsorption process is described as a metal-ligand equilibrium. They concluded that significant quantities of certain metals could occur in fresh waters as adsorbed species, and that trace metal speciation is, in fact, determined primarily by competition between adsorption and complexation.

Most reports in the recent literature agree with the conclusion of Florence and Batley [8]: "In the past, the presence of soluble, stable metal-organic complexes in natural waters has been overemphasized at the expense of appreciation of the influence of colloidal particles, both organic and inorganic . . . almost nothing is known about the concentration of organic and inorganic colloidal particles or their metal-binding ability. For these reasons the results from even the most sophisticated computer model of trace metals speciation in natural waters bear little, if any, relation to reality."

While the above statement may be overly harsh, the representation of adsorption clearly presents a stumbling block to the modeling of at least some waters.

ORGANIC SPECIES

It is now well known that a number of organic materials of both natural and anthropogenic origin are capable of complexing with several heavy metals (e.g., copper and iron) and thereby affecting environmental reactions such as bioavailability. A number of studies have investigated this phenomenon.

A statement by Jenne [17] aptly summarizes the applicability of currently available data to chemical modeling of heavy metal speciation in aquatic systems: "The information being developed in these studies has not been shown to be particularly useful to quantitative chemical modeling . . . Complexation capacity values have served to identify and dramatize the potentially important role of dissolved organic compounds as ligands capable of promoting the solubilization and hydrologic transport of trace elements. However, there seems little point in continuing these determinations, since their magnitude is now known and it is not apparent how they can be used in quantitative chemical modeling."

A similar conclusion was reached by McCrady and Chapman [24] after a study of the copper-complexing capacity of several natural waters: "The theoretical equilibrium approach for predicting free cupric ion concentrations in natural waters, utilizing REDEQL, should be used with caution, realizing that at best, quantitative results might be obtained at the higher copper concentrations. At the present time, it is not possible to quantitatively predict and model the effect of a vast array of organic interactions in natural water."

What is required for modeling efforts are measurements of the molar concentrations of the complexing organic species (in different surface waters and at different times), their stability constants for the trace metals of concern, and the effects of pH, ionic strength and temperature on these stability constants. The situation will be much simplified if it is found that there are only a few important natural organic ligands for each metal.

A number of stability constants for metal-natural organic complexes have been reported for a few metals (see below); however, the values for a particular metal can vary widely and could be expected to account only qualitatively for the complexation phenomena in a modeling effort.

- *Organic Chemicals in Surface Waters* — In surface waters free from significant amounts of man-made pollution, a large majority of the dissolved organic carbon (DOC) is of natural origin; fulvic acid, humic acid, tannins and lignins are usually the major components [20]. The total concentration of these species (DOC) in natural waters may range from several parts per billion to several hundred parts per million [18], though a more common range is probably 2-20 mg/l. Particulate organic carbon (POC) will also contain chemicals capable of complexing or adsorbing metals. POC concentrations may in some cases be higher than DOC concentrations, but the effective surface area and adsorption capacity are lower.

Humic and fulvic acids are amorphous, yellow-brown or black, hydrophylic, polydisperse substances of wide-ranging molecular weight (<10,000 for fulvic acid, 10,000-30,000 for humic acid) [20]. Tannins and lignins are high-molecular-weight polycyclic aromatic compounds and are highly resistant to biological degradation [20]. A significant amount of this material is probably present in macromolecular or colloidal form [10]. In one study of pond waters, approximately 70% of the total organic carbon was in the filtrate fraction that passed a 0.0032 μm (nominal pore diameter) filter but was retained on a 0.0009 μm filter [11].

In surface waters without strong vertical mixing, these organics — with their complexed metals and other adsorbed pollutants — may be concentrated in the surface microlayer. Some anthropogenic (organic) pollutants may also act to bind heavy metals in waters, but no literature data support this conjecture.

- *The Complexation Process* — The most probable reaction mechanism between humic compounds and metal ions is the formation of complex bonds with carboxyl and phenolic groups [18]. Information on this subject is presented in section 2.9.

- *Stability Constants* — The literature and the data bases of models list a number of stability constants (in terms of log K) for various metal ions with humic (HA) and fulvic (FA) acids. It should be noted that many of these values were measured under acidic conditions that are seldom found in natural waters. The values of K are a function of pH and ionic strength. In general, K increases with increasing pH and decreases with increasing ionic strength [18]. Also, the values of K obtained for humic and fulvic acids are usually much smaller than those associated with man-made chelating agents such as EDTA, NTA, NEDTA and CDTA; thus the latter values should not be used as surrogates in chemical modeling.

The preceding paragraphs have described some of the limitations associated with modeling aquatic systems. While these are important considerations, they should not prevent the use and continued development of such models.

5.5.8 Model Applications

The first equilibrium models were developed for use in seawater, but their use has now expanded to many applications in fresh waters, both groundwaters and surface waters. These are briefly discussed below.

MODELING OF LABORATORY AND NATURAL WATERS

Computer equilibrium models have been used to predict the speciation of metals in natural waters in order to estimate the toxic effects of contaminants on water quality and biota (e.g., refs. 10,22). For example, Pagenkopf [33,34] used the COMICS program to determine zinc speciation in laboratory test waters; similarly, Andrew *et al.* [1] used REDEQL to determine speciation of copper in Lake Superior water.

Magnuson *et al.* [23] reevaluated the data used by Andrew *et al.* with REDEQL2 and found that the inclusion of new species, especially $Cu(OH)_2^0$, greatly changed the predicted speciation, since this species then became dominant. These authors also performed a sensitivity analysis using a three-metal, three-ligand system; by varying pK by 0.05 unit, they produced concentration changes of 6-30%. Both of these studies used measured dissolved copper as the input, and no adsorption, organic complexing, or precipitation was allowed. In addition, no analytical work was conducted in conjunction with the modeling to verify the results.

Glass [12] also used REDEQL2 to model Lake Superior water, choosing citrate as a representative ligand in place of organic carbon. EDTA, NTA and cysteine were included to illustrate trace quantities of organic ligands. The author stated that these results (Table 5.5-2) agree generally with what is known about chemical speciation in Lake Superior; for example, less than 1% of total copper is measured as the free ion, as predicted by the model. Precipitation and adsorption were not considered in this study.

McCrady and Chapman [24] used REDEQL2 to compare copper speciation in reconstituted water, well water and natural river waters. No adsorption, precipitation, or complexing with organics was allowed. A cupric ion electrode was used to determine the concentration of that species analytically. These authors found that the model predicted the concentration of the cupric ion quite well in the reconstituted water and the well water, but the concentrations predicted in river water showed poor agreement with the measured concentrations at low levels (0.1-0.5 m moles/liter): the ratio of measured copper ion to computed copper ion ranged from 0.02 to 0.26 at these levels. Agreement improved at higher copper concentrations and was better in some rivers than in others.

Hoffman and Eisenreich [15] used a modified version of REDEQL2 to determine the source of variation of iron and manganese in the hypolimnion of Lake Mendota. They included three adsorbing surfaces (SiO_2, MnO_2, and Fe_3O_4) and organic complexes with citrate, sulfosalicylate, glycine, phthalate, and dipyridyl. The concentrations of these organics were based on one-tenth of the DOC concentration according to the equivalents of carbon for each ligand. Numerous solids were allowed to precipitate.

TABLE 5.5-2

Calculated Distribution of 18 Metals and 12 Ligands in Lake Superior Water[a]

	Ligand:	H_2O	OH	CO_3	Cl	SO_4	SiO_3	PO_4	NO_3	CIT	EDTA	NTA	CYST
	Total →	−1.74	−1.74	3.058	4.442	4.438	4.350	7.490	4.784	6.00	25.0	25.0	25.0
Metal	↓	Free → ↓	6.17	5.41	4.44	4.46	11.36	12.23	4.78	16.13	36.88	29.77	29.11
Ca	3.487	3.50	7.93	5.19(2)	b	5.82	b	9.18	b	6.08(3)	28.08(2)	25.29	b
Mg	3.959	3.97	7.40	5.63(2)	b	6.20	b	9.15	b	7.36(2)	32.06(2)	26.77	29.04
Na	4.252	4.25	b	8.55	b	8.11	b	b	b	b	38.50	31.74	b
K	4.889	4.89	b	b	b	8.34	b	b	b	b	b	b	b
Sr	6.75	6.75	11.58	b	b	b	b	12.74(2)	b	10.13	32.93(2)	29.84	b
Ba	7.0	7.00	12.03	b	b	b	b	b	b	26.78	33.98(2)	30.50	b
Fe(III)	7.268	18.61	9.30(3)	b	21.77(3)	19.20(2)	14.45	18.27	b	7.27(3)	26.80(3)	26.15(3)	43.52
Cu(II)	7.796	11.23	11.35(5)	10.09(4)	12.83(5)	13.56	b	15.51(2)	b	7.80(3)	27.22(3)	26.72(2)	b
Zn	7.959	8.18	9.38(5)	8.40(2)	11.22(5)	10.51	b	13.26	b	11.35	26.37(2)	25.78	26.45(2)
Mn(II)	8.26	8.29	10.62(2)	9.55	11.72(3)	10.61	b	12.97	b	11.07(2)	28.97(2)	29.08	32.86
Cr(III)	8.420	15.82	8.42(3)	b	19.58(2)	17.81	b	14.67	b	b	26.01(3)	b	b
Ni	8.47	8.90	9.83	8.91(2)	12.93(2)	11.22	b	14.27	b	9.13(3)	25.08(3)	25.79(2)	47.37
Co(II)	8.47	8.73	10.14(3)	8.84(2)	11.86(2)	10.85	b	14.01	b	26.11(2)	27.21(2)	26.32	27.80(2)
Pb	9.015	10.88	10.79	9.03	13.58	12.80	b	b	b	11.83	27.56	27.27	26.74
Cd	9.745	10.00	12.23(6)	10.12(2)	12.30(5)	12.32	b	18.56	14.17	11.57(3)	28.18(3)	28.09	b
Ag	10.0	10.02	14.22(2)	b	11.41(4)	13.47	b	b	b	19.39	38.62(2)	b	b
Hg	10.30	19.67	10.31(5)	13.94(2)	11.81(5)	21.90(2)	b	b	b	37.96	32.65(3)	33.01(2)	32.85
CHg	11.30	12.35	22.65(2)	b	11.34	b	b	b	b	b	b	b	25.45
Protonated ligand			−1.74	3.06(2)	b	10.13	4.35(2)	7.51	b	7.23(4)	33.46(3)	27.18(3)	25.22(3)

a. Values are the negative logarithm of the concentration (moles/liter). Values in parentheses are numbers of different complexes that contain the ligand in the total metal-ligand concentration. CIT = citrate, a representative organic ligand in place of fulvic acids; EDTA = ethylenediaminetetraacetate; NTA = nitrilotriacetate; CYST = cysteine; CHg = methylmercuric ion; pH = 7.8.

b. Stability constants for these complexes were not available.

Source: Glass [12]. *(Copyright 1977, New York Academy of Sciences. Reprinted with permission.)*

The calculated manganese concentrations agreed well with the trend in observed values. The authors concluded that the variations in total manganese concentration were due to pH-dependent dissolution of solid $MnCO_3$. The pH-dependent desorption from oxide surfaces was apparently of secondary importance.

OTHER USES OF EQUILIBRIUM MODELS

There have been a few, somewhat peripherally related uses of water body models. Some of these applications deal with soil quality (e.g., speciation in groundwater). Potter *et al.* [36] used SOLMNEQ and WATEQ to predict *in situ* leaching of uranium ore bodies. The models were not able to predict uranium concentrations very well, primarily due to the lack of consideration of kinetics and adsorption.

Minear and Rose [27] used WATEQF, MINEQL, and REDEQL2 to calculate metal speciation in coal mining spoil bank groundwater. Numerous solid phases were included in this analysis. The authors concluded that well samples were undersaturated with respect to the solid phases considered, implying that the system was not controlled by dissolution equilibrium.

Richter and Theis [38] used REDEQL2 to determine speciation and adsorption of heavy metals in fly-ash leachate, using silicon, iron, and manganese oxides as the adsorbing surfaces present. The soluble metal concentrations measured at the wells agreed favorably (within one-half log unit) with the calculated values.

Mouvet and Bourg [30] used the European model ADSORP to study adsorption and speciation of trace metals on sediment of the Meuse river; Pb, Ni, Ca, Mg, Cu and Zn were investigated. Calculated values of the adsorbed/dissolved distribution agreed well with observed values after some realistic data manipulation. This work indicates that dissolved trace metal concentration in the river is controlled by adsorption, not by a precipitation mechanism, and that the relationship between organic matter and suspended matter greatly influences the adsorption of such metals as Cu and Pb.

Finally, an application of the MEXAMS model for metal speciation and fate is being conducted for the Eight Mile River, Connecticut. The program is sponsored by the EPA and is executed by the state of Connecticut (Department of Natural Resources), under the support of the New England Interstate Water Pollution Control Commission (NEIWPCC, Boston). Model simulation results are expected by the end of 1986.

5.5.9 Example of Model Application

This example deals with the use of a model to estimate chemical speciation of trace metals in seawater. The metals originate in a wastewater treatment plant. The example is adapted from a study by Morel *et al.* [29] for Los Angeles County. It is presented in four sections:

(1) Problem statement,

(2) Field monitoring,

(3) Model application, and
(4) Discussion of results.

PROBLEM STATEMENT

The Joint Water Pollution Control Project (JWPCP) of the Los Angeles County Sanitation District discharges approximately 370 mgd of wastewater into the sea. The sewage consists of both domestic and industrial wastes containing high metal concentrations, whose effects on aquatic life are dramatic and easily measured. For example, chromium and lead are highly concentrated in both the sewage and the seawater at a ratio of C (sewage)/C (seawater) of about 4,000 and 6,000 respectively. These high values initiated a project aimed at mathematical modeling of the speciation of trace metals discharged.

The sludge from the JWPCP sewage treatment is fairly typical of that produced by primary treatment techniques. The sedimented sludge is digested anaerobically and is then centrifuged. The centrifugate is remixed with the primary effluent, contributing much of its particulate — and metal — load.

The primary sewage effluent is discharged via two submarine outfalls (a 90-inch "Y" and a 120-in. "dog leg") at a depth of 60 meters at the edge of the continental shelf. The rise of the plume as well as the dilution of sewage by seawater were well studied during the field monitoring program.

FIELD MONITORING

The sewage composition and aquatic environment were monitored at the plant, and a study of the metal concentrations in the nearby sediments was made; thus, an experimental data base was available for model application and validation.

Table 5.5-3 gives the average concentration and mass emission rates in the treated effluent of 11 metals selected for the modeling application.

MODEL APPLICATION

REDEQL, a general program for the computation of chemical equilibrium in aqueous systems [26], was selected for the modeling effort. Model selection was based on both the experience of the researchers with the particular package and the adequacy of data in the model's database.

Since the model did not account for the ligands of primary interest, a set of representative organic ligands and an absorbing surface have been added (Table 5.5-4). Addition of data to the database of a speciation model is to be expected in such modeling efforts; these data must be compiled and put into the system, with the assistance of the model developer or the agency that supports the model (e.g., the EPA).

TABLE 5.5-3

Trace Metals in JWPCP Sewage Outfall

	Estimated Concentration (μg/l)			Mass Emission Rates to So. Calif. Bight (tons/yr)		
	JWPCP[a] Sewage	So. Calif.[b] Seawater	Sewage-seawater Ratio	JWPCP[c]	Total[d]	JWPCP/ Total Ratio
Fe	9900	10.0	1000	5300	20,000	0.27
Zn	2400	10.0	250	1300	2,400	0.54
Cr	860	0.2	4000	460	700	0.66
Cu	560	3.0	200	300	1,000	0.30
Pb	250	0.04	6000	140	550	0.25
Ni	240	6.0	40	130	400	0.33
Mn	130	2.0	70	70	350	0.20
Cd	30	0.1	300	16	65	0.25
Ag	20	0.3	70	11	20	0.55
Co	10	0.2	50	5	25	0.20
Hg	1	0.03	30	0.5	10	0.05

a. Data from Southern California Coastal Water Resources Project (SCCWRP) 1971 averages. Redox level (pe) not experimentally measured but was deduced to be about −3.6.

b. Selected from SCCWRP tabulation of literature data. These data show great variability. This is particularly true for the less abundant metals, the estimates of which are probably too high.

c. Mass emission rates obtained by multiplying column 1 by 540 × 10^9 l/yr (1971 flow).

d. Sum of wastewater discharges, surface runoff, vessel coating, ocean dumping, rainfall, and aerial fallout as estimated by critical examination of SCCWRP data.

Source: Morel *et al.* [29]. *(Copyright 1975, American Chemical Society. Reprinted with permission.)*

The inputs to REDEQL include the analytical concentrations of trace metals found in the sewage, the analytical concentrations of major metals and ligands, and the measured pH of 7.7, as reported in Table 5.5-4 (column 1).

Model application is not a straightforward process. For example, because the redox level (pe) is not an experimentally measured quantity, the user must attempt to find a range of pe where the calculated partition of metals between the particulate and soluble phases matches the experimentally determined filterable/non-filterable fractionation. For this particular example, the pe is best described as a function of the bacterially mediated reduction of sulfate,

$$SO_4^{-2} + 8H^+ + 8e^- = S^{-2} + 4\ H_2O$$

coupled with the oxidation of organic carbon.

Table 5.5-4, column 2, summarizes the first output (first simulation) of metal species estimated by the model for the JWPCP. To make the model more realistic, organic ligands were included. Model users selected a set of organics that represented various functional groups and complexing behaviors for the organic fraction of various wastewaters. Acetate, tartrate, glycine, salicylate, glutamate, and phthalate were added, each at an equivalent carbon concentration of 10^{-3} M. The results are presented in the third column of Table 5.5-4.

To make the model even more realistic, an adsorbing surface was added and given characteristics similar to those of metal oxides. The computations of absorption were then made according to the model of James-Healy [16]. The results are presented in the last column of Table 5.5-4.

Figure 5.5-1 represents the result of a large number of computations for various mixtures of sewage and seawater at various redox potentials. The pH imposed at each dilution was calculated from the carbonate alkalinity and total carbonate of the corresponding sewage-seawater mixture. The ionic strength was approximated by the appropriate dilution of sewage (I = 0.01) with seawater (I = 0.5). The lines in Figure 5.5-1 indicate the domain of pe and dilution in which various insoluble forms of the trace metals are stable. (For simplicity, a detailed description of soluble speciation has been omitted from the figure.)

Figure 5.5-1 shows that, in general, the solubilization of metals is affected to a much greater extent by oxidation than by dilution. For a detailed discussion of this subject and the modeling results, the reader is referred to the original work of Morel *et al.* [29].

DISCUSSION OF RESULTS

Mathematical speciation modeling can be a useful tool if the necessary data are available. In their work, Morel *et al.* argued that "overall, an equilibrium model for sewage can be made to closely follow the existing experimental data, but such an effort is not warranted until more detailed analytical data are obtained." This statement applies to almost all water modeling studies.

For the particular example described, Morel *et al.* indicate that much insight has been gained into the factors that govern the fate of trace metals by using chemical equilibrium models. However, model applications are not always straightforward and often present difficulties that are not easily overcome.

5.5.10 Conclusions

The following conclusions can be drawn from the discussion in the previous sections:

(1) Over a dozen computerized models are presently used for investigating metal speciation in aqueous systems. The MEXAMS, REDEQL2, WATEQ2 and MINEQL models appear to be well developed. Of these,

TABLE 5.5-4
Metal Distribution Among Various Species in Three Equilibrium Models of JWPCP Sewage

− Log Total Concn.	Inorganic Model[a] (%)	Inorganic Model with Addition of Organics[a,b] (%)	Inorganic Model with Addition of Organics and Adsorbing Surface[a,b,c] (%)
Fe, 3.7	Fe_3O_4(s), 100	Fe_3O_4(s), 100	Fe_3O_4(s), 100
Cr, 4.8	$Cr(OH)_3$(s), 97	$Cr(OH)_3$(s), 97	$Cr(OH)_3$(s), 52
	$Cr(OH)_4^-$, 3	$Cr(OH)_4^-$, 3	$Cr(OH)_4^-$, 3
			ADS, 45
Cu, 5.0	CuS(s), 100	CuS(s), 100	CuS(s), 100
Cd, 6.5	CdS(s), 100	CdS(s), 100	CdS(s), 100
Pb, 6.0	PbS(s), 100	PbS(s), 100	PbS(s), 100
Zn, 4.5	ZnS(s), 100	ZnS(s), 99	ZnS(s), 99
Ag, 6.7	Ag_2S(s), 100	Ag_2S(s), 100	Ag_2S(s), 100
Hg, 8.3	HgS(s), 100	HgS(s), 100	HgS(s), 100
Ni, 5.4	NiS(s), 42	NiS(s), 22	NiS(s), 14
	Ni^{+2}, 2	Ni^{+2}, 2	Ni^{+2}, 2
	$Ni(CN)_4^{-2}$, 56	$Ni(CN)_4^{-2}$, 56	$Ni(CN)_4^{-2}$, 56
		GLY, 10	GLY, 10
		GLU, 10	GLU, 10
			ADS, 8
Co, 6.8	CoS(s), 97	CoS(s), 95	CoS(s), 93
	Co^{+2}, 2	Co^{+2}, 2	Co^{+2}, 2
		GLU, 2	GLU, 2
			ADS, 2
Mn, 5.6	Mn^{+2}, 50	Mn^{+2}, 50	Mn^{+2}, 45
	$MnHCO_3^+$, 24	$MnHCO_3^+$, 24	$MnHCO_3^+$, 21
	$MnSO_4$, 19	$MnSO_4$, 19	$MnSO_4$, 17
	$MnCl^+$, 6	$MnCl^+$, 6	$MnCl^+$, 6
			ADS, 9

a. Other inputs to the model: pH = 7.7, pe = −3.6, pCa = 2.75, pMg = 3.0, pBa = 5.0, pAl = 4.5, pCO_3 = 2.0, pSO_4 = 2.3, pCl = 1.8, pF = 4.0, pNH_3 = 2.2, pPO_4 = 3.3, pCN = 5.0. Ionic strength = 0.01, T = 25°C.

b. Organic ligands added: pAcetate = 3.3, pGlycine = 3.8, pTartrate = 3.6, pGlutamate = 3.7, pSalicylate = 3.8, pPhthalate = 3.9.

c. Adsorbing surface: 10 $m^2 l^{-1}$, constant potential. The computation is performed according to the James-Healy model which considers coulombic, solvation, and specific chemical energy interactions. The chemical energy term is essentially a fitting parameter and has been chosen arbitrarily in the model to represent high chemical affinity of the adsorbing surface for all metals (about 10 kcal/mol).

Source: Morel *et al.* [29]. *(Copyright 1975, American Chemical Society. Reprinted with permission.)*

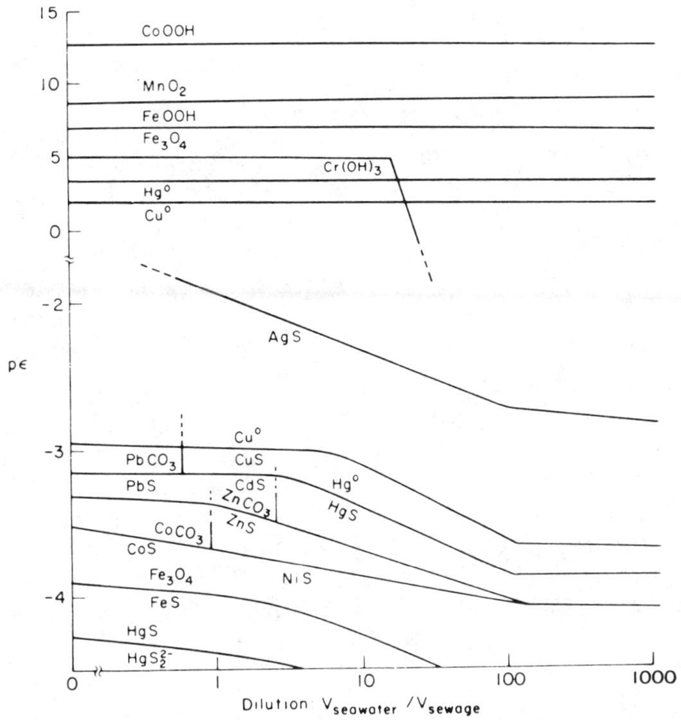

Composition of diluting seawater:
pH = 8.2, pCa = 2.0, pMg = 1.2, pBa = 6.7, pFe = 6.8, pZn = 6.8, pCr = 8.4,pCu = 7.3, pPb = 9.7, pNi = 7.0. pMn = 7.4, pCd = 9.0, pAg = 8.5, pCo = 8.3, pHg = 9.8, pAl = 7.0, pCO_3 = 2.6, pSO_4 = 1.5, pCl = 3.6, pF = 4.2, pNH_3 = 6.0, pPO_4 = 5.7, pCN = 9.0. Ionic strength = 0.5. Redox level (pe) set at −3.6.

Source: Morel *et al.* [29]. *(Copyright 1975, American Chemical Society. Reprinted with permission.)*

FIGURE 5.5-1 Metal Solids as a Function of Dilution and Oxidation

MEXAMS appears to have the greatest support and REDEQL2 to be the most advanced code.

(2) The waters to be modeled should be well characterized with respect to Eh, pH, redox potential, ionic strength, concentrations of major cations and anions, concentrations of trace metals and organic ligands, and concentration and nature of suspended solids. While this is not commonly the case, enough is known about water chemistry to permit

reasonable estimates of input values for these data in many cases. However, the sensitivity of the models to each of these parameters varies, and the results of model runs using estimated water quality data may have to be interpreted with caution.

(3) There may be a large degree of uncertainty within the model itself, with a large portion relating to uncertainty in the stability constants. Some stability constants measured in the laboratory and used in modeling of natural waters may apply only to the time period immediately following the formation of the complex and thus do not reflect long-term stability. The error or uncertainty ranges associated with measured values are seldom reported; some values may be in error by as much as an order of magnitude.

(4) The representation of organic complexation and adsorption can be a major problem in estimating the speciation of metals. Both the concentration and nature of the organic ligands (e.g., humic and fulvic acids) and the suspended solids must be known before reasonable model parameters can be chosen. Uncertainties in the stability constants and adsorption coefficients selected can easily be an order of magnitude. Reconstituted and well waters — both low in organics and suspended solids — will be easier to model in this regard than those with higher levels.

(5) Previous investigators have found that computer models can fairly well represent speciation of some reconstituted and well waters. However, validation has been accomplished only by measurement of the free ion (versus total metal concentration), such as Cu^{+2}; thus, the relative importance of other species cannot be confirmed. When organic complexing has been ignored in the modeling of speciation in river waters, the results have been poor — especially when metal concentrations were low.

5.5.11 Literature Cited

1. Andrew, R.W., K.E. Biesinger and G.E. Glass, "Effects of Inorganic Complexing on the Toxicity of Copper to *Daphnia Magna*," *Water Res.*, **11**, 309-15 (1977).
2. Ball, J.W., D.K. Nordstrom and E.A. Jenne, "Additional and Revised Thermochemical Data and Computer Code for WATEQ2 — A Computerized Chemical Model for Trace and Major Element Speciation and Mineral Equilibria of Natural Waters," Water Resources Report WRI-78-116, U.S. Geological Survey (January 1980).
3. Ball, J.W., E.A. Jenne and D.K. Nordstrom, "WATEQ2 — A Computerized Chemical Model for Trace and Major Element Speciation and Mineral Equilibria of Natural Waters," in *Chemical Modeling in Aqueous Systems*, E.A. Jenne (ed.), Symposium Series No. 93, American Chemical Society, 815-35 (1979).

4. Bjerrum, N. "Ionic Association. I: Influence of Ionic Association on the Activity of Ions at Moderate Degrees of Association," *Kyl. Danske Videnskab. Math-fys. Medd.*, **7**, 1-48 (1926).
5. Dempsey, B.A. and P.C. Singer, "Role of Adsorption in Controlling the Distribution of Metals — The Great Lakes," presented at ACS, Environmental Chemistry division, Honolulu (April, 1978).
6. Felmy, A.R., D.C. Girvin and E.A. Jenne, *MINTEQ — A Computer Program for Calculating Aqueous Geochemical Equilibria*, Final Report, EPA-600/3-84-032 (1984).
7. Felmy, A.R., S.M. Brown, Y. Onishi, S.B. Yabusaki and R.S. Argo, *MEXAMS — The Metals Exposure Analysis Modeling System*, Final Report, EPA-600/3-84-031 (1985).
8. Florence, T.M. and G.E. Batley, "Chemical Speciation in Natural Waters," *CRC Crit. Rev. Anal. Chem.*, **9**, 219 (1980).
9. Fuoss, R.M., "Properties of Electrolyte Solutions," *Chem. Rev.*, **17**, 27-42 (1935).
10. Gachter, R., J.S. Davis and A. Mares, "Regulation of Copper Availability to Phytoplankton by Macromolecules in Lake Water," *Environ. Sci. Technol.*, **12**, 1416-21 (1978).
11. Giesy, J.P., Jr., L.A. Briese and G.J. Leversee, "Metal Binding Capacity of Selected Marine Surface Waters," *Environ. Geol.*, **2**, 257-68 (1978).
12. Glass, G.E., "Identification and Distribution of Inorganic Components in Water: What to Measure?" *Ann. N.Y. Acad. Sci.*, **298**, 31-46 (1977).
13. Harriss, D.K., V.R. Magnuson, D.K. Taylor and M.S. Sur, "REDEQL2: Program and Data Base Modifications," presented to the Division of Environmental Chemistry, American Chemical Society, Miami Beach, Fla. (September 1978).
14. Hem, J.D., *Study and Interpretation of the Chemical Characteristics of Natural Water*, 2nd ed., Geological Survey Water-Supply Paper 1473, U.S. Government Printing Office, Washington, D.C. (1970).
15. Hoffman, M.R. and S.J. Eisenreich, "A Chemical Equilibrium Model for the Variation of Iron and Manganese in the Hypolimnion of Lake Mendota," Paper presented before the Division of Environmental Chemistry, American Chemical Society, Miami, Fla. (September 1978).
16. James, R.O. and T.W. Healy, *J. Colloid Interface Sci.*, **40**, 42 (1972).
17. Jenne, E.A., "Chemical Modeling — Goals, Problems, Approaches, and Priorities," in *Chemical Modeling in Aqueous Systems*, E.A. Jenne (ed.), Symposium Series No. 93, American Chemical Society, Washington, D.C. (1979).
18. Khalid, R.A., R.P. Gambrell, M.G. Verloo and W.H. Patrick, Jr., "Transformations of Heavy Metals and Plant Nutrients in Dredged Sediments as Affected by Oxidation Reduction Potential and pH. Volume I: Literature Review," Contract Report D-77-4, Dredged Material Research Program, U.S. Army Engineer Waterways Experiment Station, Vicksburg, Miss. (1977).
19. Kinniburgh *et al.* (1975): cited by Perwak *et al.* [35].

20. Lawrence, J., "Semi-quantitative Determination of Fulvic Acid Tannin and Lignin in Natural Waters," *Water Res.*, **14**, 373-77 (1980).

21. Leland (1975): cited by Perwak *et al.* [35].

22. Lion, L.W. and J.O. Leckie, "Copper in Marine Microlayers: Accumulation, Speciation and Transport," in *Abstracts of the 178th National Meeting*, American Chemical Society, Division of Environmental Chemistry, Washington, D.C., 88-91 (1979).

23. Magnuson, V.R., D.K. Harriss, M.S. Sun, D.K. Taylor and G.E. Glass, "Relationships of Activities of Metal-ligand Species to Aquatic Toxicity," in *Chemical Modeling in Aqueous Systems*, E.A. Jenne (ed.), Symposium Series No. 93, American Chemical Society, Washington, D.C. (1979).

24. McCrady, J.K. and G.A. Chapman, "Determination of Copper Complexing Capacity of Natural River Water, Well Water and Artificially Reconstituted Water," *Water Res.*, **13**, 143-50 (1979).

25. McDuff, M.E. and F.M. Morel, "Description and Use of the Chemical Equilibrium Program REDEQL2," Tech. Rept. EQ-72-02, California Institute of Technology, Pasadena, Calif. (1973, updated July 1975).

26. McDuff, R.E. and F. Morel, "REDEQL, A General Program for the Computation of Chemical Equilibrium in Aqueous Systems," Tech. Rept. EQ-72-01, Keck Laboratories, California Institute of Technology, Pasadena, Calif. (1972).

27. Minear, R.A. and R.R. Rose, "Application of Various Chemical Models to the Saturation Zone of a Coal Mining Spoil Bank System," Paper presented before the Division of Environmental Chemistry, American Chemical Society, Miami, Fla. (September 1978).

28. Morel, F.M. and J.J. Morgan, "A Numerical Method for Computing Equilibria in Aqueous Chemical Systems," *Environ. Sci. Technol.*, **6**, 58-67 (1972).

29. Morel, F.M., J.C. Westall, C.R. O'Melia and J.J. Morgan, "Fate of Trace Metals in Los Angeles County Wastewater Discharge," *Environ. Sci. Technol.*, **9**, 756 (1975).

30. Mouvet, C. and A.C.M. Bourg, "Speciation of Copper, Lead, Nickel and Zinc in the Meuse River," *Water Res.*, **17**, 641-49 (1983).

31. Munro *et al.* (1976): cited by Felmy *et al.* [6].

32. Nordstrom, D.K., *et al.*, "A Comparison of Computerized Chemical Models for Equilibrium Calculations in Aqueous Systems," in *Chemical Modeling in Aqueous Systems*, E.A. Jenne (ed.), Symposium Series No. 93, American Chemical Society, Washington, D.C., 857-92 (1979).

33. Pagenkopf, G.K., "Metal-Ion Transport Mediated by Humic and Fulvic Acids," in *Organometals and Organometalloids — Occurrence and Fate in the Environment*, F.E. Brinkman and J.M. Bellama (eds.), Symposium Series No. 82, American Chemical Society, Washington, D.C. (1978).

34. Pagenkopf, G.K., "Zinc Speciation and Toxicity to Fishes," in *Toxicity to Biota of Metal Forms in Natural Water*, Proceedings of a workshop, October 7-8, 1975, Great Lakes Research Advisory Board (1976).

35. Perwak, J., B. Goodwin, W. Lyman, L. Nelken and J. Stevens, "Chemical Modeling of Metals in Water," Memorandum Report to Environmental Protection Agency, Contract No. 68-01-5959, Arthur D. Little, Inc., Cambridge, Mass. (1980).

36. Potter, R.W. II, J.M. Thompson, M.A. Clynne and V.L. Thurmond, "Equilibrium, Kinetic, and Chromatographic Controls of the Solution Composition Obtained During the *in situ* Leaching of an Uranium Ore Body," Paper presented before the Division of Environmental Chemistry, American Chemical Society, Miami, Fla. (September 1978).

37. Reilly, P.J., R.H. Wood and R.A. Robinson, "Prediction of Osmotic and Activity Coefficients in Mixed-Electrolyte Solutions," *J. Phys. Chem*, **75**, 1305-15 (1971).

38. Richter, R.O. and T.L. Theis, "Application of Chemical Modeling Using 'REDEQL2' to the Speciation and Adsorption of Heavy Metals in Fly Ash Leachates," Paper presented before the Division of Environmental Chemistry, American Chemical Society, Miami, Fla. (September 1978).

39. Sibley and Morgan (1975): cited by Perwak *et al.* [35].

40. Truesdell, A.H. and B.F. Jones, "WATEQ, A Computer Program for Calculating Chemical Equilibria of Natural Waters," *J. Res. U.S. Geol. Surv.*, **2**, 233-48 (1974).

41. Van denBerg, C.M.G. and J.R. Kramer, "Conditional Stability Constants for Copper Ions with Ligands in Natural Waters," in *Chemical Modeling in Aqueous Systems*, E.A. Jenne (ed.), Symposium Series No. 93, American Chemical Society, Washington, D.C. (1979).

42. Vuceta, J. and J. Morgan, "Chemical Modeling of Trace Metals in Role of Complexation and Adsorption," *Environ. Sci. Technol.*, **12**, 1302-8 (1979).

43. Westall, J.C., J.L. Zachary and F.M. Morel, "MINEQL — A Computer Program for the Calculation of Chemical Equilibrium Composition of Aqueous Systems," Tech. Note No. 18, EPA Grant R-803738, Massachusetts Institute of Technology (1976).

5.6 RADIONUCLIDE MIGRATION MODELING

This section briefly summarizes the modeling issues and approaches adopted by radionuclide transport models. These models have been developed primarily for assessing performance and risks of deep-mined nuclear waste repositories sited in salt, basalt or granite geologic media.

5.6.1 Introduction

To analyze the long-term performance of a geologic nuclear waste disposal system, separate models are needed for (a) migration of formation fluids (e.g., oxic or anoxic brines in salt media, vadose-zone pore water in tuff) and groundwater to the waste; (b) dissolution (leaching) of the nuclear wastes by the formation fluids; (c) migration of radionuclides via groundwater to the accessible environment (air, water, soil); and (d) exposure of biota to radiation. Mills and Vogt [13] discuss the use of various radiological assessment models and review the available codes. A publication of the National Research Council [14] describes the overall geologic isolation system concept and related criteria.

A number of recent reviews prepared for the Department of Energy (DOE), the Nuclear Regulatory Commission (NRC) and the U.S. Environmental Protection Agency (EPA) have focused on the radionuclide migration models. These reviews [14,15,18] contain both listings of codes and detailed discussions of their features, capabilities and modeling approaches. A synoptic overview of processes modeled, solution procedures, and other features of existing models is presented in Table 5.6-1. A directory of selected radionuclide migration computer codes is presented in section 5.6.3.

5.6.2 Principal Processes

Radionuclide migration models are one-, two-, or three-dimensional groundwater flow models that have been interfaced to geo/radio-chemistry models (subroutines). The various modeling concepts are described by Bonazountas [2].

A typical radionuclide transport model combines a source model, a groundwater flow model, and a contaminant transport (e.g., dispersion, advection, adsorption) model into a single code. For example, GETOUT [7], which has been widely used for assessing the performance of deep geologic repositories, includes each of these three models with the following assumptions:

(1) *Source term*: constant rate of release of radionuclides into groundwater, infinite solubility (of radionuclides).

(2) *Groundwater flow*: uniform one-dimensional flow path, steady-state fluid flow.

(3) *Radionuclide transport*: decay chains with up to three members, axial dispersion, and sorption based on linear isotherm and instantaneous equilibrium.

TABLE 5.6-1

Synopsis of Selected Radionuclide Transport Models for Nuclear Waste Repository Performance Assessment

Model	Solution Procedure	Dimensions	Media	Process Modeled	Reference
BRINEMIG	FD	1	G	HC	1
CCC	IFDM	3	G	SF,HT,GM	4,18
CFEST	FE	2,3	—	U,HC,ST	1,4
CHAINT	FE	2	F	ST	4,18
DNET	PN	2	G,F	SF,DS	18
DPCT	FE,PT	2	G	SF,ST	18
EQ3/EQ6	—	—	—	GC	1,18
FEMWASTE	FE	2	G	ST	4,15,18
FEMWATER	FE	2	G	SF,U	4,15,18
FTRANS	FE	2	F,G,DP	ST,U	4,6,14
GETOUT	A	1	G,F	ST,PT	1,7,18
KONBRED (MOC)	FD,MC	2	G	SF,ST	4,18
LAYFLO	SA	1	G	PT,ST,GC	1
MAGNUM2D	FF	2	G,F,DP	SF,HT	4,18
MMT	PT	2	G	ST	1,8,15,18
NUTRAN	PN,N1	1,2	G,F,DP	PT,GC,DS	4,18
NWFT/DVM	PN/PT	2	G,F	SF,ST	18
PATHS	A	2	G	SF,ST	18
PHREEQE	—	—	—	GC	4,18
SEGOL	A	2,3	G	U,ST	4,15,18
SHAFT79	IFDM	3	G	SF,HT	4,18
SHALT	FE	2	G	SF,HT,ST,HC	18
SWENT	FD	3	G	SF,HT,HC	1,9
SWIFT	FD	3	G	SF,HT,ST,HC	4,18
TRACR3D	IFDM	3D	F	U,PT,ST	4,15,18
TRANSONE	FE	1	G	ST,U,SF	4,15,18
TRIPM	FE	2	G	SF,U,GC	1,15
TRUST	IFDM	3	G	SF,U,GM	4,15,18
UNSAT2	FE	2	G	SF,U	4,15,18
VS2D	FD	2	G	U,SF,ST	4,15,18
V3	FD	2	F	SF	18
WATEQ	—	—	—	GC	18

(Continued)

TABLE 5.6-1 (Continued)

Abbreviations:

A	— Analytical Solution	HT	— Heat Transport
DP	— Double Porosity Media	IFDM	— Integrated Finite Difference Method
DS	— Dissolution	MC	— Method of Characteristics
F	— Fractured Porous Media	NI	— Numerical Integration
FD	— Finite Difference	PN	— Pipe Network
FE	— Finite Element	PT	— Particle Tracker
G	— Granular Type Porous Media	SA	— Semi-Analytical
GC	— Geochemistry	SF	— Saturated Flow
GM	— Geomechanical	ST	— Solute Transport
HC	— Coupled Heat and Solute Transport	U	— Unsaturated Flow

The sophistication and capabilities of available codes vary widely. Source-term models can range in complexity from a simple specified concentration or flux boundary condition in a transport model, to the complex modeling of equilibria among various species in aqueous solution surrounding radioactive waste (e.g., PROTOCOL [11,17]). Flow models, besides dimensional capabilities, may model flow only in media that are porous or can be represented as porous, while others are capable of modeling fractured or dual-porosity media. Nuclide transport code capabilities range from those that can model only a single decaying species (e.g., CFEST, KONBRED) to those that can model decaying chains of arbitrary length (e.g., SWIFT and CHAINT). Besides decay, most transport codes consider one or more of the following solute transport processes: dispersion, diffusion, and retardation (including various adsorption processes). Special-purpose brine migration models (e.g., BRINEMIG) have been developed for the particularly impermeable domal salt repository medium, where brine inclusions slowly migrate toward the emplaced waste (thermal gradient) prior to waste dissolution.

Bedrock radionuclide models are medium-specific, because most of the flow in hard-rock (basalt and granite) geologic media is through joints or fractures. Two approaches are available for fractured flow modeling: (1) the discrete-fracture (DF) approach, which models individual fractures and treats the rock matrix as a porous medium, and (2) the dual-porosity (DP) approach, which superimposes a high-porosity medium (fractures) and a low-porosity medium (rock matrix) with a transfer function relating fluid flow between them. Typical solute transport codes for fractured media (e.g., FTRANS and TRUMP) assume saturated flow conditions and consider adsorption of radionuclides onto the rock matrix.

For a shallow tuff repository, unsaturated flow conditions would apply for modeling radionuclide transport [15]. Models developed for this purpose are FEMWASTE, TRANSONE and SEGOL [4,15].

5.6.3 Radionuclide Codes

This section considers only generalized, documented models capable of simulating advection, adsorption, decay and speciation of radionuclide chains in groundwater. Of particular interest are codes noted for their computational efficiency, user-friendliness, and validated history by a number of field tests (cf. [1] and [10]).

Table 5.6-1 provided an overview of available models. Table 5.6-2 lists widely used, well-documented radionuclide transport codes and their main features. These codes have been developed to various levels of sophistication, approximation, accuracy and flexibility; they can be used as stand-alone packages or as end-to-end codes with other models.

Model development status and confidence in model predictions are often related. Although partial validation of selected radionuclide transport models (e.g., SWENT) has been claimed and documented [9], this amounts merely to satisfactory agreement between model predictions and field observations for very specific conditions (e.g., short-term leachate migration from a landfill, or uranium-front movement in an aquifer). In most cases, such as in the random-walk 1-D multicomponent mass-transport (MMT) code, numerical code verification for a test case has been performed, but field validation has not yet been accomplished.

A number of analytical solutions to radionuclide transport equations have also been developed and are noted briefly here, in light of their value and frequent use in verifying the accuracy of numerical codes.

Lester *et al.* [12] published analytical solutions for the transport of a three-component chain and considered both impulse releases and decaying band releases. Their solutions were limited, however, by the assumptions of a constant fluid velocity in a one-dimensional flow and a single retardation factor applicable to all chain members. Nevertheless, the solutions have served as benchmarks for subsequent numerical solutions to the chain transport problem.

Extensive work on this problem has been carried out by a group at the University of California/Berkeley and Lawrence Berkeley Laboratory. This group has published a set of analytical solutions to a variety of chain transport problems (e.g., Harada *et al.* [5] and Chambre *et al.* [3]). These solutions include a variety of waste form dissolution and local transport problems, one-dimensional flow and transport, one-dimensional flow with three-dimensional transport, two-dimensional flow and transport, and one-dimensional flow and transport of a single radionuclide in fractured media.

5.6.4 Model Applications

It is difficult to assess the suitability of a general-purpose code, because the applications of the code to a variety of siting assessments and contaminant flow analyses are often waste-specific, site-specific, and different in both spatial and temporal resolution. However, several well-documented, developed and maintained

TABLE 5.6-2

Selected Radionuclide Migration Models

A. Saturated Flow of Single Species

CFEST	Couples fluid energy and solute transport (CFEST): A general-purpose, three-dimensional transport code that simulates seasonal energy storage (nonisothermal) in underground confined aquifers. Can handle single-phase Darcy (laminar, transient) flows in a horizontal plane, a vertical plane, or in three dimensions. Both transient and steady-state single-species transport problems can be treated. This is an extension of the FE3DGW code.
KONBRED (GSMOC)	A one- and two-dimensional model of groundwater flow and solute transport, to compute the concentration of a dissolved chemical species in a saturated aquifer at any specified place and time. Computes changes in convective transport, hydrodynamic dispersion, and mixing from fluid sources. The groundwater flow equation is solved by a finite-difference technique and the solute transport equation by the method of characteristics. A modified version accounts for decay of a single species but does not consider formation of radioactive daughter products.

B. Saturated Flow of Multiple Species

CHAINT	A one- and two-dimensional, finite-element model for the simulation of multicomponent nuclide transport in saturated fractured or porous media. Processes modeled include advection, dispersion, diffusion, sorption, chain-decay, and mass release. Principal input required is the groundwater flow field, which can be obtained from a groundwater flow model.
GETOUT	Developed for the transport of radionuclide chains along a 1-D path. Uses analytical solutions for 1-, 2- or 3-member chains and approximate solutions for longer chains. Release may be either an impulse or a band. Assumes infinite solubility of elements and equilibrium adsorption. Model is widely applicable to steady-state problems.
LAYFLO	1-D, semianalytical model for the migration of a 3-member decay chain in a multilayered geological medium. It accounts for longitudinal dispersion, decay, sorption and desorption, in fully saturated media with steady flow.
MMT	A discrete-parcel, random-walk, 1-D *M*ulticomponent *M*ass *T*ransport code for radionuclides in groundwater. It accounts for transport, decay, and sorption-desorption of radionuclides in a steady/uniform simplified groundwater aquifer.
NUTRAN	Calculates the dose-to-man from radioactivity transported from waste repositories via groundwater pathways. Evaluates the combined effect of systems of natural and engineering barriers. The program considers leaching and dissolution of waste and calculates the migration of radionuclides from the repository as a function of time. The time-dependent concentrations of various radionuclides are calculated along the flow paths, modeled as a one-dimensional network.

(Continued)

TABLE 5.6-2 (Continued)

NWFT/DVM	The Network Flow and Transport Model/Distributed Velocity Model (NWFT/DVM) is a 1-D flow code, which simulates a flow system as a 1-D (pipe) network of segments, but also treats leach- and solubility-limited sources, and handles decay chains subject to adsorption, advection and dispersion in transport.
PATHS	A 2-D numerical model for contaminant transport. Treats both steady and transient saturated flow systems, for single-component transport in homogeneous geology. The model is based on analytical solutions for the groundwater potential. Geochemical retardation is considered, and dispersion is ignored. This is an oversimplified worst-case engineering analysis tool for releases from waste storage facilities, with no field validation to date.
SWENT	A one-, two- or three-dimensional finite difference, transient or steady-state groundwater-flow, nuclide-transport, and heat-transport model, applicable to complex geometries. Saturated flow in an isothermal or non-isothermal porous medium, as well as sorption and desorption mechanisms, can be modeled. For nuclide decay, the code considers conservation of dissolved contaminants, energy and total liquid mass. The code is fairly general and contains many options for geometry, processes, and boundary conditions. Based upon the nuclide inventory and the biosphere transport, the model can calculate the potential hazard to people from release to the biosphere.
TRIPM	A two-dimensional FE model for the simultaneous transport of water and reacting solutes through saturated and unsaturated porous media. Predicts transport and decay.
C. Codes for Unsaturated Flow Conditions	
FEMWASTE	A two-dimensional finite element code for waste transport through a porous medium with varying degrees of saturation. The transport mechanisms include advection, hydrodynamic dispersion, chemical adsorption, and first-order decay. Computes the velocity flow field required for the solute transport subroutine.
FEMWATER	A two-dimensional finite element model of water flow through porous media with varying degrees of saturation. An aquifer may be modeled with anisotropic parameters. Nonlinearities such as the dependence of the hydraulic conductivity and storage on the pressure head are considered. Output includes a continuous pressure distribution and velocity field. Output is formatted for use as input to the companion code, FEMWASTE, which simulates contaminant transport in the medium.
SEGOL	A two- or three-dimensional code designed for solute transport in unsaturated flow systems. Moisture flow is based on Darcy's Law. The transport mechanisms include advection, dispersion, sorption and decay.
TRASONE	A one-dimensional finite element code for mass transport through unsaturated and saturated porous media.

(Continued)

TABLE 5.6-2 (Continued)

TRUST	An integrated finite difference algorithm for one-, two- or three-dimensional fluid flow problems in unsaturated, deformable porous media. The code uses a mixed explicit-implicit approach in setting up the matrix equations and internally calculates the time-step size. The model considers pressure-dependent density variations of water [16].
UNSAT2	A two-dimensional finite element code for flow through unsaturated porous media. Has features similar to FEMWATER and can also model axisymmetric problems. Each material may have local anisotropy, with the principal hydraulic conductivities oriented at any desired angle with respect to the coordinates.
VS2D	A two-dimensional finite difference model for flow in saturated-unsaturated media. Can handle nonlinear problems caused by infiltration into dry soils, and anisotropy and discontinuities in permeabilities and porosities.
D. Codes for Fractured Media	
FTRANS	A two-dimensional finite element code for groundwater flow and single- or multiple-species radionuclide transport in fractured porous media. Fractured-flow systems may be modeled using either the dual-porosity or the discrete-fracture approach, or a combination of the two.
TRACR3D	A three-dimensional, integrated, finite-difference program for solving two-phase fluid flow problems in fractured media. Special features include one- or two-phase flow with tracers in either phase, Langmuir sorption, radioactive decay and capillary effects.
E. Coupled Flow and Thermal Codes	
CCC	This code models heat and fluid flow by conduction and convection in two- or three-dimensional saturated porous media. Can also calculate one-dimensional subsidence. An integrated finite difference method is employed to solve mass-, momentum-, and energy-balance equations.
MAGNUM2D	This finite-element model can simulate groundwater flow and heat transport in fractured porous rock systems. Both dual-porosity and discrete-fracture models can be used.
F. Special-purpose Codes	
BRINEMIG	One-dimensional, finite difference code for modeling intergranular brine migration to nuclear waste, and, possibly, *from* the waste, up the thermal gradient, through isotropic and homogeneous salt. This is a special-purpose, non-Darcian flow code limited in applicability to salt repository sites.

Sources: Battelle [1], Curtis *et al*. [4], Oster [15], Thomas *et al*. [18].

TABLE 5.6-3

SWENT Model Validation Studies

Site Selected	Description of Problem	Variables Observed in Field and Simulated by Model with Good Agreement	Processes Validated
Canadian Air Force Base Borden Landfill, Ontario, Canada	Leachate migration from landfill	1. Piezometric surface 2. Chloride ion concentration profiles in waste region and in plume after 36 years	Fluid flow and trace component transport
Thermal Energy Storage Experiment at Mobile, Alabama	Injection, storage, and recovery of heat from a confined aquifer	Temperature profiles in the aquifer (3 dimensional) at end of each cycle, with injection period, storage period, and recovery period	Coupled fluid and heat flow
Carrizo Aquifer, Wintergarden, S.W. Texas	Uranium front movement in the aquifer	Concentrations of ^{14}C, ^{234}U and ^{238}U in the aquifer	Fluid flow and trace component transport

Source: INTERA Environmental Consultants, Inc. [9].

codes in the public domain (e.g., SWENT) have been successfully applied to analyses of injection of industrial wastes into saline aquifers, *in situ* solution mining, migration of contaminants from landfills, and saltwater intrusion in coastal regions, as well as to aquifer characterization. Table 5.6-3 summarizes the characteristics of three field-validation studies of the SWENT model; these results indicate that SWENT is generally applicable for three-dimensional radionuclide transport in multilayered aquifer systems [9].

5.6.5 Literature Cited

1. Battelle Memorial Institute, Office of Nuclear Waste Isolation, "Performance Assessment Plans and Methods for a Salt Repository Project," DOE-BMI/ONWI-545, Columbus, Ohio (August 1984).
2. Bonazountas, M., "Soil and Groundwater Fate Modeling," in *Fate of Chemicals in the Environment*, R. Swann and A. Eschenroeder (eds.), American Chemical Society, ACS Symposium Series No. 225, Washington, D.C. (1983).
3. Chambre, P.L., T.H. Pigford, A. Fujita, T. Kanki, A. Kobayaski, H. Leung, D. Ting, Y. Sato and S.J. Zavoshy, *Analytical Performance Models for Geologic Repositories*, Vols. I and II, LBL-14842, UC-70, report to the U.S. Department of Energy by the University of California, Lawrence Berkeley Laboratory (1982).
4. Curtis, R.H., R.J. Wart and E.L. Skiba, Acres American, Inc. and Teknekron Research, Inc., "A Summary of Repository Design Models," NUREG/CR-3450, prepared for U.S. Nuclear Regulatory Commission (October 1983).
5. Harada, M., P.L. Chambre, M. Foglia, K. Higashi, F. Iwamoto, D. Leung, T.H. Pigford and D. Ting, "Migration of Radionuclides through Sorbing Media. Analytical Solution — I," LBL-10500, UC-70, Draft Report to the U.S. Department of Energy, University of California, Lawrence Berkeley Laboratory (1980).
6. INTERA Environmental Consultants, Inc., "FTRANS: A Two-Dimensional Code for Simulating Fluid Flow and Transport of Radioactive Nuclides in Fractured Rock for Repository Performance Assessment," ONWI-426, prepared for Office of Nuclear Waste Isolation, Battelle Memorial Institute, Columbus, Ohio (April 1983).
7. INTERA Environmental Consultants, Inc., "GETOUT: A Computer Code for One-Dimensional Analytical Solution for Radionuclide Transport," ONWI-433, prepared for Office of Nuclear Waste Isolation, Battelle Memorial Institute, Columbus, Ohio (April 1983).
8. INTERA Environmental Consultants, Inc., "MMT: A Random-Walk One-Dimensional Multicomponent Mass Transport Code," ONWI-432, prepared for Office of Nuclear Waste Isolation, Battelle Memorial Institute, Columbus, Ohio (April 1983).
9. INTERA Environmental Consultants, Inc., "SWENT: A Three-Dimensional Finite Difference Code for the Simulation of Fluid, Energy and Solute Radionuclide Transport," ONWI-457, prepared for Office of Nuclear Waste Isolation, Battelle Memorial Institute, Columbus, Ohio (April 1983).

10. "INTRACOIN-International Nuclide Transport Code Intercomparison Study," Final Report, Level 1 (Code Verification), SKI 84:3 (September 1984).

11. Jackson, D.D., "PROTOCOL, a Numerical Simulator for the Dissolution of Inorganic Solids in Aqueous Solutions," UCRL-91631, University of California (September 1984).

12. Lester, D.H., G. Jansen and H.C. Burkholder, "Migration of Radionuclide Chains through an Adsorbing Medium," AIChE Symposium Series 71, pp. 152-202 (1975).

13. Mills, M. and D. Vogt, Teknekron Research, Inc., "A Summary of Computer Codes for Radiological Assessment," NUREG/CR-3209, prepared for U.S. Nuclear Regulatory Commission (March 1983).

14. National Research Council, Waste Isolation Systems Panel, "A Study of the Isolation System for Geologic Disposal of Radioactive Wastes," National Academy Press (1983).

15. Oster, C.A., "Review of Groundwater Flow and Transport Models in the Unsaturated Zone," NUREG/CR-2917, prepared for U.S. Nuclear Regulatory Commission (1982).

16. Pacific Northwest Laboratories, "Uranium Recovery Research Sponsored by the NRC at PNL: Quarterly Progress Reports for 1984," PNL-5015-1 through -4 (1985).

17. Pickrell, G. and D. Jackson, "User's Guide to PROTOCOL, A Numerical Simulator for the Dissolution Reactions of Inorganic Solids in Aqueous Solutions," UCID-20236, University of California (October 1984).

18. Thomas, S.D., B. Ross and J.W. Mercer, GeoTrans Inc. and Teknekron Research, Inc., "A Summary of Repository Siting Models," NUREG/CR-2784, prepared for U.S. Nuclear Regulatory Commission (July 1982).

6. MATRIX ELEMENTS AND LIGANDS

Deborah L. Cerundolo, Susan F. Coons, Elizabeth D. Gibson, Christopher P. Loreti

Page

LIST OF TABLES

SODIUM (Na)

POTASSIUM (K)

CALCIUM (Ca)

MAGNESIUM (Mg)

SILICON (Si)

ALUMINUM (Al)

IRON (Fe)

HYDROGEN AND HYDROXIDE ION (H^+, OH^-)

CHLORIDE (Cl^-)

SULFATE (SO_4^{-2})

CARBONATES (HCO_3^-, CO_3^{-2}, etc.)

LIST OF FIGURES

SODIUM (Na)

POTASSIUM (K)

CALCIUM (Ca)

MAGNESIUM (Mg)

SILICON (Si)

ALUMINUM (Al)

IRON (Fe)

CHLORIDE (Cl^-)

SULFATE (SO_4^{-2})

CARBONATES (HCO_3^-, CO_3^{-2}, etc.)

6.1 INTRODUCTION

This chapter briefly discusses the important environmental chemistry of seven of the most common elements in the earth's crust (sodium, potassium, calcium, magnesium, silicon, aluminum, and iron) and the major ligands typically associated with these elements in natural systems (OH^-, Cl^-, SO_4^{-2}, CO_3^{-2}, and HCO_3^-). Silicon, aluminum and iron, along with oxygen, account for almost 90% of the igneous rock mass; the four elements that make up the remaining 10% —calcium, sodium, potassium, and magnesium — provide the major cation content of aqueous systems.

The elemental composition of sedimentary rocks, which are produced during weathering of igneous rocks, is generally similar to the parent igneous rock with the addition of some CO_2, H_2O, and HCl. However, the mineralogy of igneous and sedimentary rocks is quite different. Silicates and silicate-rich aluminosilicates (feldspars) are characteristic igneous rocks; carbonates and silicate-poor aluminosilicates (clays) are abundant sedimentary materials. The major constituents of rocks — oxides of aluminum, silicon, and iron — are relatively insoluble. The net effect of weathering, therefore, is largely the progressive stripping of the four major cations (Ca^{+2}, Na^+, K^+, Mg^{+2}) from the aluminosilicates. In the weathering process, natural waters acquire a part of their dissolved load, most of which originates in the dissolution of carbonates, sulfates, chlorides and sulfides (e.g., $CaCO_3$, $MgCO_3$, $CaMg(CO_3)_2$, $CaSO_4$, $MgSO_4$, NaCl, KCl, and FeS_2).

The major or matrix elements discussed in this chapter were chosen specifically because of their predominance in the environment. Information on their environmental properties and processes in soil/water systems is provided in order to further understanding of the fate and transport of trace elements or pollutants that may be complexed (or otherwise affected) by them. In most natural waters, the major dissolved components are sodium, potassium, calcium, and magnesium ions, and the following ligands: Cl^-, HCO_3^-, CO_3^{-2}, SO_4^{-2}, and OH^-. The major cations are predominantly present as free hydrated ions, while a significant fraction of the anions are associated with metal ions. Silicon, aluminum, and iron, in combination with oxygen, constitute the major portion of the mineral fraction of natural soils.

Information on each major/matrix element or ligand is provided in a separate section. For the seven elements, information is provided on: uses, natural occurrence, concentration and predominant speciation in natural waters, and properties and speciation at the solid/solution interface. For the ligands, the available information is generally organized into: use and occurrence (sources), environmental transport, speciation in natural waters or soils, and complexation/solubility. The section on carbonates also provides a discussion of alkalinity/acidity. The hydrogen/hydroxide ions section is generally a discussion of pH in soils and natural waters and the formation of metal hydroxides. References are presented at the end of each section.

6.2 SODIUM (Na)

6.2.1 Production and Uses

The sodium compounds produced in major quantities in the United States are the chloride, hydroxide, carbonate, sulfate, and bicarbonate. The principal uses of these compounds include applications in chemical industries, pulp and paper manufacture, glass manufacture, food production and gasoline production, and for deicing of roadways.

6.2.2 Natural Sources

Sodium is an alkali metal occurring as a major component of the earth's crust. Its concentration in the lithosphere and igneous rocks is 2.8% and 2.5% respectively [9,10]. The lower reported concentrations in sedimentary rock and soils (0.07-0.7%) are due to weathering, which leaches ions from the surface of alumino-silicates and chlorides; this process accounts for the levels of dissolved sodium in natural aqueous systems.

Feldspars (anhydrous alumino-silicates of potassium, sodium, calcium, and occasionally of other cations such as barium) are the predominant sodium-containing minerals in igneous rock. In sedimentary rock, the principal minerals are montmorillonite (layer silicates with a typical interlayer spacing of 120-180 nm) and halite (NaCl). When feldspars are exposed to aqueous solutions, there is a rapid initial exchange of cations for hydrogen ions, followed by a slow dissolution in which the framework elements (e.g., aluminum and silicon) are released. Adsorption of ligands (e.g., oxalate, fluoride) to the surface of soil minerals modifies the surface charge and has been reported to increase the rate of weathering [12].

Table 6.2-1 lists the logarithms of equilibrium constants reported for reactions describing the dissolution of sodium from feldspars, montmorillonite and halite. These equilibrium constants can be used to establish stability relations among the various minerals in the environment. Stumm and Morgan [12] used the stability constants for reactions 1, 3, 4, and 5 to develop a predominance diagram for the sodium-containing minerals (Figure 6.2-1). The activities of sodium relative to hydrogen, $(Na^+)/(H^+)$, and of silicic acid, (H_4SiO_4), are plotted as $pH - pNa$ and $-\log (H_4SiO_4)$, respectively. The dissolution of feldspars and release of cations into bulk solution proceeds by a series of congruent and incongruent reactions [2,4,10]. The diffusional transport of material out from the reactant mineral is the rate limiting step in the transfer between feldspar and aqueous solutions. The initial reaction is the rapid exchange of surface cations for hydrogen ions; this is followed by the slower release of large amounts of cations and silicic acid and the appearance of gibbsite [$Al(OH)_3$] as an intermediate reaction product. The precipitated gibbsite then reacts with the feldspar to produce kaolinite [$Al_2Si_2O_5(OH)_4$]. After destruction of all gibbsites, kaolinite continues to precipitate at the expense of the feldspar as the activity of silicic acid again increases.

TABLE 6.2-1

Equilibrium Constants for Dissolution Reactions of Selected Minerals

Reaction	Log K (25°C)	Ref.
Sodium Feldspars		
1. $NaAlSi_3O_8$ (s) + H^+ + 4½ H_2O ⇌ ½[$Al_2Si_2O_5(OH)_4$] (s) + 2 H_4SiO_4 + Na^+	−1.9	[12]
2. ½[$Na_2O \cdot Al_2O_3 \cdot 6\,SiO_2$] (s) + H^+ + 2½ H_2O ⇌ ½[$Al_2Si_2O_5(OH)_4$] (s) + 2 H_2SiO_3 + Na^+	−0.68	[10]
Sodium Montmorillonite		
3. $NaAl_7Si_{11}O_{30}(OH)_6$ (s) + H^+ + 11½ H_2O ⇌ 3½[$Al_2Si_2O_5(OH)_4$] (s) + 4 H_4SiO_4 + Na^+	−9.1	[12]
Kaolinite		
4. $Al_2Si_2O_5(OH)_4$ (s) + 5 H_2O ⇌ $Al_2O_3 \cdot 3H_2O$ (s) + 2 H_4SiO_4	−9.4	[12]
Amorphous Silica		
5. SiO_2(s) + 2 H_2O ⇌ H_4SiO_4 −2.7 [12]	−2.7	[12]
Halite		
6. NaCl (s) ⇌ Na^+ + Cl^- 1.54 [10]	1.54	[10]

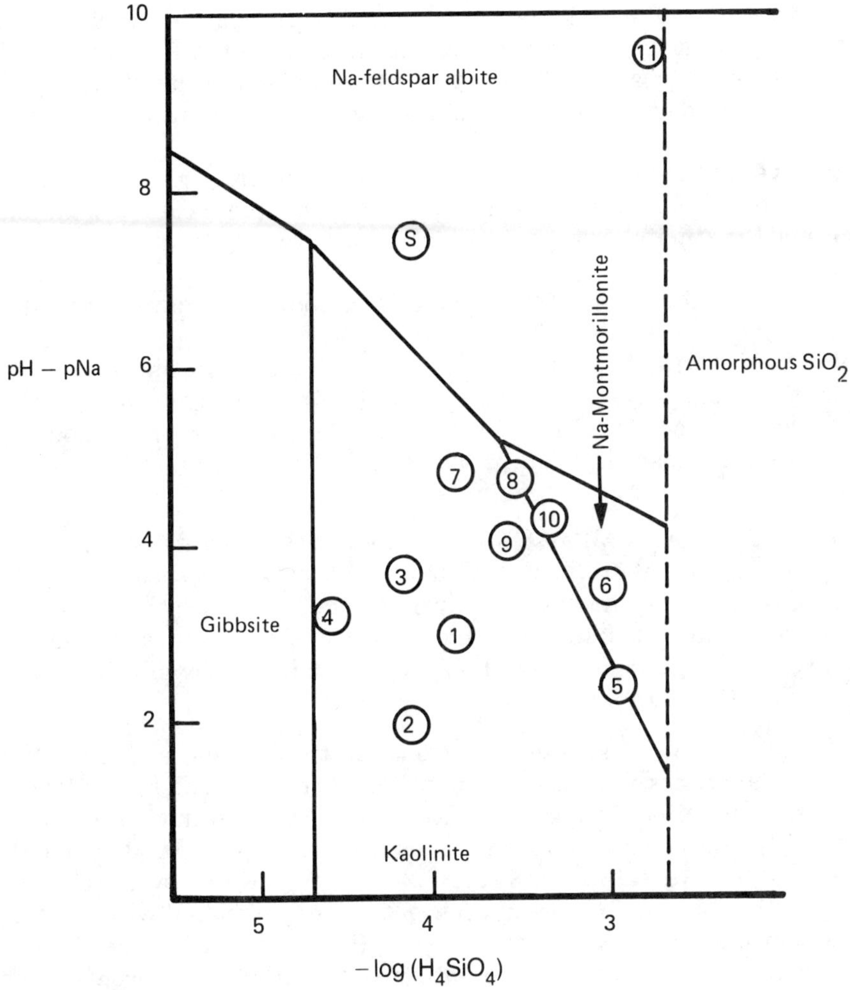

Legend:

1 – Stream/granite
2 – Stream/quartzite
3 – Stream/sandstone
4 – Lake Erie
5 – Groundwater/granite
6 – Groundwater/plagioclase
7 – Groundwater/sandstone
8 – Groundwater/shale
9 – Groundwater/limestone
10 – Groundwater/dolomite
11 – Closed basin – soda lake
S – Seawater

Source: Stumm and Morgan [12]. *(Copyright 1981, John Wiley & Sons. Reprinted with permission.)*

FIGURE 6.2-1 Stability Relationship of Sodium-Containing Minerals

The measured values of pH, pNa, and p(H_4SiO_4) for 11 natural aqueous systems and seawater are plotted in Figure 6.2-1. Most of the points fall within the stability field of kaolinite, suggesting that the waters would be in equilibrium with kaolinite; feldspars would be unstable and would be degraded under these conditions. The measurements for seawater and a closed-basin lake suggest that equilibrium with feldspars may be achieved in the presence of high sodium concentrations.

On the average, approximately 35% of the sodium ion in river water originates from the weathering of alumino-silicates [12]. The remainder is due to weathering of halite (45%) and the atmospheric deposition of sodium transported from the oceans (20%).

6.2.3 Concentration and Predominant Speciation in Natural Systems

Table 6.2-2 illustrates the range of aqueous sodium concentrations in natural waters. Where the information is available, the type of rock being drained by the freshwater system is indicated. Concentrations in natural freshwaters range from 0.6 to 60 mg/l; in contrast, the concentration in seawater is 10,750 mg/l.

Sodium exists in natural systems in a single valence state of +1. In most natural waters, sodium ion is sufficiently abundant such that its speciation is not generally affected by variations in the concentrations of other minor and trace ions. However, because of its natural abundance, changes in sodium concentration may be significant enough to alter the ionic strength of the solution, thereby affecting the activities of other ionic species in solution.[1]

Almost all the salts of sodium are strong electrolytes and are highly dissociated in most natural waters. It is possible, however, for sodium and ligand concentrations to exceed the solubility products in concentrated brines formed on evaporation and/or freezing [7]. The major sodium ion species present in natural waters and their relative concentrations are shown in Table 6.2-3. In freshwater systems, the sodium ion complexes are present at very low concentrations, compared with that of the uncomplexed aqueous ion. In seawater, however, the concentration of sodium sulfate complex [$NaSO_4^-$] is significant, representing 1-2% of the sodium content. Formation of $NaSO_4^-$ has been shown to vary inversely with both pressure and temperature [8]. Organic sodium complexes (e.g., NaEDTA) are much less stable in aqueous solution than are the organic trace-metal complexes.[2]

The activities of environmental sodium complexes (NaCl, $NaCO_3^-$, $NaHCO_3$, NaOH, and $NaSO_4^-$) at equilibrium with representative soil pore water concentrations[3] of the complexing ions have been examined as a function of pH [9]. In well-drained soils, uncomplexed sodium ion is the only important sodium species up to pH 10, at which point sodium carbonate complexes may become important.

1. The extent to which ionic strength affects the activity coefficients of ions is further discussed in section 2.6, which also lists calculated sodium activity coefficients.
2. Equilibrium constants for selected sodium complexes are given in section 2.9.
3. Representative soil pore water concentrations are: 10^{-3} *M* Na^+, 10^{-4} *M* to 10^{-2} *M* Cl^-, 10^{-4} *M* to 10^{-2} *M* SO^{-2} and $10^{-3.5}$ *M* to $10^{-2.5}$ *M* CO_2(g).

TABLE 6.2-2

Measured Sodium Concentrations in Natural Waters

Aqueous System[a]	Sodium Conc. (mg/l)	Ref.
Stream (granite)	2.3	[12]
Stream (quartzite)	0.6	[12]
Freshwater (halite)	3.0	[10]
Freshwater (feldspars)	3.9	[10]
Groundwater (shale)	57	[12]
Groundwater (granite)	9.0	[12]
Groundwater (feldspars)	23	[12]
Brine (halite)	120,000	[5]
Pore Water (well-drained soil)	23	[9]
River	5.1[b]	[3]
River	7.1[c]	[3]
River	6.0(avg)	[10]
Rainwater	<1	[5]
Rainwater	2.1	[10]
Public Water Supply in 100 U.S. Cities	12 (median) 1.1-200 (range)	[13] [13]
Seawater	10,750	[10]

a. The type of rock being drained is given in parentheses.

b. Natural (unpolluted) component of river.

c. River water including contribution from human activities.

TABLE 6.2-3

Predominant Sodium Complexes in Freshwater and Seawater

	Freshwater [12] ($\sim$10mg/l Na)	Seawater [7, 12] (10,750 mg/l Na)
Na^+ (free ion)	100%	98-99%
$NaHCO_3$	0.1%	$<0.1\%$
$NaCO_3^-$	0.005%	$<0.1\%$
$NaSO_4^-$	0.08%	1-2%

Competition for available ligands and significant formation of sodium complexes may occur when sodium concentrations are very high. However, to permit successful competition with the trace elements, the ratio of the molar concentration of sodium to the molar concentration of other ions (major or trace elements) would have to be on the order of the inverse ratio of the stability constants for the complexes in question. For example, the stability constants for $NaCO_3^-$ and $CaCO_3$ ion association in natural waters at zero ionic strength, 25°C and 1 atmosphere are $10^{0.77}$ and $10^{3.15}$, respectively [10]. These data suggest that with constant CO_3^{-2} concentration and a sodium ion molar concentration equivalent to 240 ($10^{3.15}/10^{0.77}$) times the calcium ion molar concentration, formation of $NaCO_3^-$ complex is no longer negligible in comparison with $CaCO_3$ complex formation.

6.2.4 Solid-Solution Interface

Precipitation reactions do not generally remove significant amounts of sodium from solution under environmental conditions in non-arid regions. However, ion exchange onto soils and sediments in contact with aqueous solutions occurs rapidly and is a major mechanism for removing sodium ions from solution; in effect, it controls the attenuation of sodium in the soil-water system and the mobility of sodium in the subsurface environment. Once sodium is introduced, the soil/sediment quickly adsorbs some of it, retarding the mobility of this ion and affecting other equilibria.

The pH_{zpc} for natural materials such as feldspars, montmorillonite, albite, and kaolinite ranges from 2 to 4.6.[4] This indicates that in natural waters at pH $\approx$7 the surface will adsorb available cations to neutralize the surface charge. Appendix C summarizes the exchange properties, including cation exchange capacity (CEC), of selected natural materials.

The affinity of an ion exchanger for sodium ion, relative to other cations, is a function of its charge, hydrated radius, and concentration in the contacting solution. In general, relative to an ion of higher charge, sodium is easier to remove from clay surfaces and has less tendency to replace ions bound on the surface. For example, calcium (Ca^{+2}) would be expected to bind more strongly than Na^+. (H^+ is an exception, in that it tends to behave as a divalent or trivalent ion and has a stronger affinity for exchange sites than do ions of higher charge [1].)

The relative affinity of ion exchange surfaces for ions of the alkali metal series, all of which exist in the same valence state, generally increases with decreasing size of the hydrated ionic radii, which is just the reverse of the order of the crystal ionic radii [6]. The relative affinity of the alkali series is $Cs^+ > Rb^+ > K^+ > Na^+ > Li^+$.

Because of its charge and hydrated radius size, sodium binds fairly weakly to exchange sites. In natural waters, ions such as Cs^+ or even $CuOH^+$ may preferentially occupy ion exchange sites at the surface, due to the fact that Na^+ has a higher degree of hydration and a larger hydrated radius [12].

4. The zero point of charge (pH_{zpc}) of a solid material corresponds to the pH at which the surface is uncharged. In the absence of adsorbable ions other than H^+ and OH^-, it is equivalent to the isoelectric point of the material [12].

Another factor influencing the exchange of sodium on natural ion exchange materials is its ionic concentration, a function of the laws of mass action. For example, ion exchangers exhibit a stronger affinity for Ca^{+2} than for Na^{+} in dilute solutions, but this selectivity decreases with increasing ionic strength [12]. Data reported by Sayles and Mangelsdorf [11] and summarized in Table 6.2-4 illustrate that clay materials initially in equilibrium with freshwater systems contain primarily Ca^{+2}. However, when these materials are equilibrated with seawater having a higher ionic strength, the adsorbed ions are primarily Na^{+}. Since the seawater concentration of sodium is higher than that of calcium, sodium ions successfully compete with calcium for available surface sites.

The physical properties (e.g., permeability) of soils are also altered by increased levels of adsorbed sodium. Different interlayer spacings may be observed in the ion exchange framework, depending on the extent of hydration of the adsorbed ions and the elasticity of the material itself. (Organic exchangers may be relatively elastic, while zeolites are quite inelastic.) Stumm and Morgan [12] describe a composite balance among the electrostatic, Van der Waals and osmotic forces influencing the swelling pressure in the ion exchanger; the extent of swelling increases with the extent to which the adsorbed ion is hydrated. Thus, for example, if strongly hydrated Na^{+} replaces less strongly hydrated Ca^{+2}, twice as many ions are required to neutralize the ion exchange charge. The greater degree of hydration results in swelling and reduces soil permeability. Therefore, in waters of high sodium concentration, the bottom sediments are less permeable to water, which could result in reduced drainage in the aqueous system.

TABLE 6.2-4

Average Exchange Composition (Equivalent Fraction) of Clays Equilibrated Initially with River Water and Subsequently with Seawater

	Average Equivalent Fraction[a]		
	In River Water	**In Seawater**	**Change[b]**
Na^{+}	0.04	0.47	+0.43
K^{+}	0.01	0.06	+0.05
Ca^{+2}	0.60	0.16	−0.44
Mg^{+2}	0.25	0.32	+0.07
H^{+}	0.1	0	−0.10

a. The fraction of total exchange sites occupied by the cation.

b. A (+) sign indicates uptake by the clay; a (−) sign indicates release to the surrounding waters.

Source: Sayles and Mangelsdorf [11]. *(Copyright 1982, Prentice-Hall, Inc. Reprinted with permission.)*

6.2.5 Literature Cited

1. Bear, F.E. (ed.), *Chemistry of the Soil*, Reinhold Publishing Corp., New York (1964).
2. Busenberg, E. and C. Clemency, "The Dissolution Kinetics of Feldspars at 25°C and 1 atm CO_2 Partial Pressure," *Geochim. Cosmochim. Acta*, **40**, 41-49 (1976).
3. Drever, J.I., *The Geochemistry of Natural Waters*, Prentice-Hall, Inc., Englewood Cliffs, NJ (1982).
4. Helgeson, H.C., "Kinetics of Mass Transfer Among Silicates and Aqueous Solutions," *Geochim. Cosmochim. Acta*, **35**, 421-69 (1971).
5. Hem, J.D., "Study and Interpretation of Chemical Characteristics of Natural Waters," Water Supply Paper No. 1473, 2nd ed., U.S. Geological Survey, Washington DC (1970).
6. Horne, R.A., *The Chemistry of Our Environment*, John Wiley & Sons, New York (1978).
7. Horne, R.A., *Marine Chemistry*, John Wiley & Sons, New York (1969).
8. Kester, D.R. and R.M. Pytkowicz, "Effect of Temperature and Pressure on Sulfate Ion Association in Seawater," *Geochim. Cosmochim. Acta*, **34**, 1039-51 (1970).
9. Lindsay, W.L., *Chemical Equilibria in Soils*, John Wiley & Sons, New York (1979).
10. Morel, F.M.M., *Principles of Aquatic Chemistry*, John Wiley & Sons, New York (1983).
11. Sayles, F.L. and P.C. Mangelsdorf, "Cation Exchange Characteristics of Amazon River Suspended Sediment and Its Reaction with Seawater," *Geochim. Cosmochim. Acta*, **43**, 767-79 (1977) (as cited in Drever [3]).
12. Stumm, W. and J.J. Morgan, *Aquatic Chemistry*, 2nd ed., John Wiley & Sons, New York (1981).
13. Todd, D.K. (ed.), *The Water Encyclopedia*, Water Information Center, Inc., New York (1970).

6.3 POTASSIUM (K)

6.3.1 Production and Uses

Commercial production of potassium is generally limited to the extraction of ores from underground deposits of soluble potassium salts, generally referred to as potash. Approximately 98% of the potassium recovered is potassium chloride (muriate of potash); the remaining 2% consists mostly of potassium sulfates. The soluble minerals that are processed include sylvite (KCl), carnallite ($KCl \cdot MgCl_2 \cdot 6H_2O$), kainite ($KCl \cdot MgSO_4 \cdot 2.75H_2O$), and langbeinite ($K_2SO_4 \cdot 2MgSO_4$) [9].

Fertilizers account for about 95% of the U.S. potassium consumption. About 3% of this is the sulfate (K_2SO_4 or $K_2SO_4 \cdot 2MgSO_4$); the remainder is KCl. Of the domestically consumed KCl, about 3% is converted by aqueous electrolysis into the hydroxide (KOH) for use in the manufacture of chemicals for the chemicals and ceramics industries [14].

6.3.2 Natural Sources

Potassium is an alkali metal that forms a major component of the earth's crust. It shares with sodium the position of seventh most abundant element. Common minerals such as alums, feldspars, micas and evaporites (e.g., sylvite) contain potassium. Its average concentration in igneous rock is 2.7%; in sedimentary rock and soils, it comprises 2.6% and 0.16-1.67%, respectively [10,11]. The slightly lower concentration in sedimentary rocks and the markedly lower concentration in soils is the result of weathering. Potassium is also an essential nutrient for plants and animals: it is the fourth most abundant element in plants and the seventh most abundant in animals [7].

Weathering — in this case, the leaching of ions from the surface of alumino-silicates and chlorides — accounts for the levels of dissolved potassium in natural aqueous systems. According to Stumm and Morgan [12], approximately 71% of the potassium in river water originates from the weathering of alumino-silicates; the remainder is due to weathering of sylvite (14%) and to atmospheric deposition (<14%). The element is more difficult to liberate from silicate minerals than is sodium and is more readily reincorporated into solid weathering products, especially certain clay minerals. Consequently, the concentration of dissolved potassium in most natural water systems is considerably lower than that of sodium.

The principal potassium minerals of igneous rocks include feldspars (orthoclase and microcline), micas (biotite, phlogopite, muscovite) and a feldspathoid (leucite). Potassium is present in sediments as unaltered feldspar and mica particles, or as illite ($3K_2O \cdot 2MgO \cdot 9Al_2O_3 \cdot 28SiO_2 \cdot 2H_2O$) and other clay minerals. Evaporite rocks (sylvite, acranite, carnallite) include beds of potassium salts, which are responsible for the high potassium concentrations in brines [6,10].

Table 6.3-1 lists the equilibrium constants (log K) reported for reactions that influence the dissolution of potassium from feldspars, micas, and sylvite. The stability relationships of the predominant potassium mineral phases are illustrated in Figure 6.3-1; in this diagram the activities of potassium relative to hydrogen, $(K^+)/(H^+)$, and of silicic acid are plotted as pH − pK and − log (H_4SiO_4), respectively. The dissolution of feldspars and release of cations into bulk solution proceeds by a series of congruent and incongruent reactions [2,5,12]. The diffusional transport of material out from the reactant mineral is the rate limiting step in the transfer between feldspars and aqueous solutions. The initial reaction is the rapid exchange of surface cations for hydrogen ions; this is followed by the slower release of large amounts of cations and silicic acid and the appearance of gibbsite $(Al(OH)_3)$ as an intermediate reaction product. The precipitated gibbsite then reacts with the feldspar to produce kaolinite $(Al_2Si_2O_5(OH)_4)$. After destruction of all the gibbsite, kaolinite continues to precipitate at the expense of the feldspar as the activity of silicic acid again increases.

Figure 6.3-1 also indicates measured values of pH − pK and − log(H_4SiO_4) for 11 natural aqueous systems and seawater. Most of the aqueous systems fall within the stability field of kaolinite; feldspars and micas would be unstable and, therefore, degraded under these conditions. The measurements for seawater (S) indicate that, in the presence of high potassium concentration, equilibrium with micas may be achieved [12].

6.3.3 Concentration and Predominant Speciation in Natural Waters

Table 6.3-2 illustrates the range of aqueous potassium concentrations observed in natural waters. With the exception of brine in a closed-basin lake (soda lake), concentrations in natural freshwater systems (rivers, streams and groundwater) range from 0.31 to 7.8 mg/l; in seawater, the average concentration is 390 mg/l.

Potassium exists in natural systems in a single valence state of +1. In environmental systems, changes in solution redox conditions affect potassium only indirectly; i.e., they affect other aqueous components with which it may react. In general, potassium forms fairly weak complexes; the common ligands (e.g., Cl^-, OH^-) do not form appreciable amounts of potassium complexes.[1]

Most potassium salts are strong electrolytes and highly dissociated in natural waters. In freshwater systems, potassium sulfate represents 0.13% of the dissolved potassium, and unassociated potassium ion accounts for the remainder [12]; in seawater, potassium sulfate represents 1% of the potassium, while free potassium ion accounts for the remaining 99% [8]. No additional potassium complexes are normally present in significant concentrations.

The activities of environmental potassium complexes (KSO_4^-, KOH^0, $K_2CO_3^0$) in equilibrium with representative concentrations[2] of the complexing ions in soil pore

1. Equilibrium constants for selected potassium complexes are tabulated in Section 2.9.
2. Representative soil pore water concentrations are: 10^{-3} M K^+, 10^{-4} M to 10^{-2} M SO_4^{-2}, and $10^{-3.5}$ M to $10^{-2.5}$ M $CO_2(g)$.

TABLE 6.3-1

Equilibrium Constants for Weathering Reactions of Selected Potassium Minerals

Reaction	Log K (25°C)
Potassium Feldspars ($KAlSi_3O_8$)	
$3KAlSi_3O_8(s) + 2H^+ + 12H_2O \rightleftharpoons KAl_3Si_3O_{10}(OH)_2(s) + 6H_4SiO_4 + 2K^+$	−4.9[a]
$2KAlSi_3O_8(s) + 2H^+ + 9H_2O \rightleftharpoons H_4Al_2Si_2O_9(s) + 4H_4SiO_4 + 2K^+$	−1.0[a]
$KAlSi_3O_8(s) + H^+ + 5/2\ H_2O \rightleftharpoons ½(H_4Al_2Si_2O_9)(s) + 2H_2SiO_3 + K^+$	−3.54[b]
$KAlSi_3O_8$ (K-glass) $+ 4H^+ + 4H_2O \rightleftharpoons Al^{+3} + 3H_4SiO_4 + K^+$	7.87[c]
$KAlSi_3O_8$ (high sanidine) $+ 4H^+ + 4H_2O \rightleftharpoons Al^{+3} + 3H_4SiO_4 + K^+$	1.40[c]
$KAlSi_3O_8$ (microcline) $+ 4H^+ + 4H_2O \rightleftharpoons Al^{+3} + 3H_4SiO_4 + K^+$	1.00[c]
Potassium Micas [$KAl_3Si_3O_{10}(OH)_2$]	
$2KAl_3Si_3O_{10}(OH)_2(s) + 2H^+ + 3H_2O \rightleftharpoons 3H_4Al_2Si_2O_9(s) + 2K^+$	6.5[a]
$KAl_3Si_3O_{10}(OH)_2(s) + H^+ + 3/2H_2O \rightleftharpoons 3/2\ (H_4Al_2Si_2O_9)(s) + K^+$ (muscovite)	3.51[b]
$KAl_3Si_3O_{10}(OH)_2\ (s) + 10H^+ \rightleftharpoons 3Al^{+3} + 3H_4SiO_4 + K^+$ (muscovite)	13.44[c]
$2KAl_3Si_3O_{10}(OH)_2(s) + 2H^+ + 18H_2O \rightleftharpoons 3(Al_2O_3 \cdot 3H_2O)(s) + 6H_4SiO_4 + 2K^+$ (gibbsite)	−7.6[a]
Kaliophilite	
$KAlSiO_4(s) + 4H^+ \rightleftharpoons Al^{3+} + H_4SiO_4 + K^+$	13.05[c]
$KAlSi_2O_6(s) + 4H^+ + 2H_2O \rightleftharpoons Al^{+3} + 2H_4SiO_4 + K^+$	6.72[c]
Illite (clay)	
$K_{0.6}Mg_{0.25}Al_{2.3}Si_{3.5}O_{10}(OH)_2 + 8H^+ + 2H_2O \rightleftharpoons 0.25\ Mg^{+2} + 2.3Al^{+3} + 3.5H_4SiO_4 + 0.6K^+$	10.35[c]
Sylvite	
$KCl(s) \rightleftharpoons K^+ + Cl^-$	0.98[b]

a. Garrels and Christ [4]
b. Morel [11]
c. Lindsay [10]

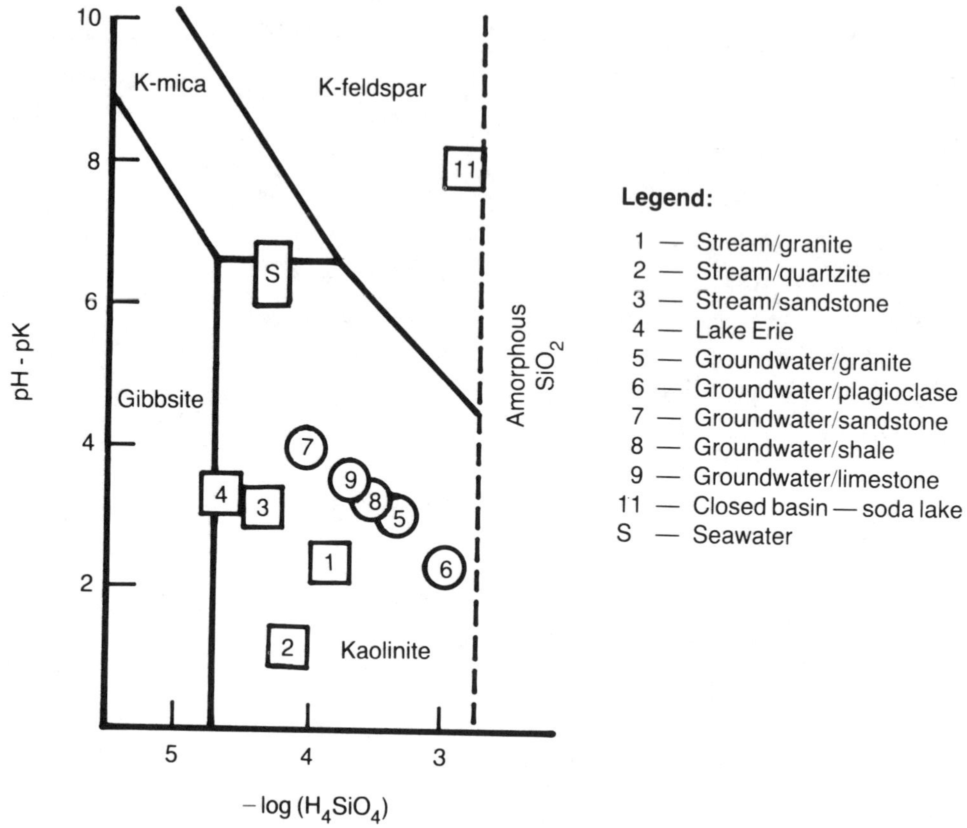

Source: Stumm and Morgan [12]. *(Copyright 1981, John Wiley & Sons. Reprinted with permission.)*

FIGURE 6.3-1 Stability Relationship of Potassium-containing Minerals

water have been examined as a function of pH [10]. In well-drained soils, the uncomplexed ion is expected to be the predominant potassium species from pH 4 to pH 10; dissolved potassium complexes are of only minor importance.

6.3.4 Solid-Solution Interface

Precipitation reactions do not generally remove significant amounts of potassium from solution under environmental conditions; ion exchange onto soils and sediments in contact with aqueous solutions is the primary removal process. In effect, ion exchange controls the attenuation of potassium in the soil-water system and the mobility of potassium in the subsurface environment. Once potassium is introduced, the soil/sediment quickly adsorbs some of it, retarding the mobility of this ion and affecting other equilibria [3].

TABLE 6.3-2

Measured Potassium Concentrations in Natural Waters

Aqueous System[a]	Potassium Conc. (mg/l)	Source
Hydrothermal Fluid	585	[11]
Seawater	390	[11]
Freshwater (feldspars)	1.17	[11]
Stream (granite)	0.78	[12]
Stream (quartzite)	0.31	[12]
Stream (sandstone)	0.62	[12]
Groundwater (granite)	3.9	[12]
Groundwater (plagioclase)	1.2	[12]
Groundwater (sandstone)	3.9	[12]
Groundwater (shale)	2.5	[12]
Groundwater (limestone)	7.8	[12]
Closed-basin Lake	778	[12]
River	2.9 (Mississippi)	[6]
River	2.3 (mean-worldwide)	[6]
Rainwater	0.42 (average)	[6]
Snow	0.6 (Nevada, 1958)	[6]
Public Water Supply (100 U.S. cities)	1.6 (median) 0-30 (range)	[13]

a. The type of rock being drained is given in parentheses.

The affinity of an ion exchanger for potassium ion, relative to other cations, is a function of charge, hydrated radius, and concentration of potassium in the contacting solution. In general, the affinity of the exchanger for multivalent ions is greater than that for monovalent ions. For example, calcium (Ca^{+2}) would be expected to bind preferentially to K^+. This selectivity for ions of higher valence decreases with increasing ionic strength of the solution. (H^+ is an exception, in that it tends to behave like divalent or trivalent ions and has a stronger affinity for exchange sites than do other ions [1].)

The relative affinity of ion exchange sites for the metals in the alkali series is $Cs^+ > Rb^+ > K^+ > Na^+ > Li^+$ [12]. For ions of the alkali metal series, all of which exist in the same valence state, the relative affinity generally increases with decreasing size of the hydrated ionic radii; in contrast, the relative affinity decreases with increasing size of the crystal ionic radii [7]. The preference of clays and most other natural ion exchangers for K^+ over Na^+ helps to explain why Na^+/K^+ ratios observed in natural waters and especially in seawater are larger than unity, even though potassium and sodium are almost equally abundant in igneous rocks. Ion-exchange processes continuously remove K^+ from solution and return it to the solid phase [12].

A third factor influencing the exchange of potassium on natural ion exchange materials is ionic concentration, in accordance with laws of mass action. For example, when nutrient solutions and fertilizers are applied to increase the ionic strength and potassium concentration of soil solutions, they minimize the valence effect that can inhibit the vital uptake of potassium by plants.

6.3.5 Literature Cited

1. Bear, F.E. (ed.), *Chemistry of the Soil,* Reinhold Publishing Corp., New York (1964).
2. Busenberg, E. and C. Clemency, "The Dissolution Kinetics of Feldspars at 25°C and 1 atm CO_2 Partial Pressure," *Geochem. Cosmochim. Acta,* **40**, 41-49 (1976).
3. Drever, J.I., *The Geochemistry of Natural Waters*, Prentice-Hall, Inc., Englewood Cliffs, NJ (1982).
4. Garrels, R.M. and C.L. Christ, *Solutions, Minerals, and Equilibria*, Freeman, Cooper & Co., San Francisco (1965).
5. Helgeson, H.C., "Kinetics of Mass Transfer Among Silicates and Aqueous Solutions," *Geochim. Cosmochim. Acta*, **35**, 421-69 (1971).
6. Hem, J.D., "Study and Interpretation of Chemical Characteristics of Natural Waters," Water Supply Paper No. 1473, 2nd ed., U.S. Geological Survey, Washington DC (1970).
7. Horne, R.A., *The Chemistry of Our Environment*, John Wiley & Sons, New York (1978).
8. Horne, R.A., *Marine Chemistry*, John Wiley & Sons, New York (1969).
9. *Kirk-Othmer Encyclopedia of Chemical Technology,* 3rd ed., John Wiley & Sons, New York (1978).
10. Lindsay, W.L., *Chemical Equilibria in Soils*, John Wiley & Sons, New York (1979).
11. Morel, F.M.M., *Principles of Aquatic Chemistry*, John Wiley & Sons, New York (1983).
12. Stumm, W. and J.J. Morgan, *Aquatic Chemistry*, 2nd ed., John Wiley & Sons, New York (1981).
13. Todd, D.K. (ed.), *The Water Encyclopedia*, Water Information Center, Inc., New York (1970).
14. U.S. Department of the Interior, Bureau of Mines, "Mineral Facts and Problems," U.S. Government Printing Office, Washington, DC (1980).

6.4 CALCIUM (Ca)

6.4.1 Production and Uses

The major calcium compounds produced commercially in the United States are the carbonate, chloride, and sulfate. Calcium carbonate, the major constituent of limestone, is primarily used in the manufacture of paper; other markets include the plastics, paint, and rubber industries, the food and drug industries, and the treatment of water supplies and industrial wastes. Major markets for calcium chloride are in deicing, dust control, road stabilization, and production of concrete products. Calcium sulfate is used in the construction industry, in the production of heavy chemicals, and in the food and pharmaceutical industries [5].

6.4.2 Natural Sources

Calcium, a metallic element, is fifth in abundance in the earth's crust, of which it forms approximately 3.6% [5]; its concentrations in igneous and sedimentary rocks are 3.5% and 3.3%, respectively [7]. The average calcium concentration in soils is reported to be approximately 1.4%; it varies primarily because of differences in the parent materials and the level of rainfall [6]. Soils that have developed from calcareous parent materials (i.e., those containing calcium carbonate) often have calcium carbonate still present in the soil profile. However, with advanced weathering and high rainfall, calcium carbonate and most other calcium minerals are depleted [6].

Calcium is widely distributed in the common minerals of rocks and soil. The chain silicates (i.e., pyroxenes[1] and amphiboles[2]) and alumino-silicates (i.e., plagioclase feldspars[3]) are the principal calcium-containing igneous rocks. Calcium also occurs in other silicate minerals which are produced during metamorphism. The most common forms of calcium in sedimentary rocks are the carbonates — calcite ($CaCO_3$), aragonite ($CaCO_3$), and dolomite ($CaMg(CO_3)_2$); other calcium-containing minerals in sediments include the sulfates — gypsum ($CaSO_4 \cdot 2H_2O$) and anhydrite ($CaSO_4$) — and the fluoride, fluorite (CaF_2). Calcium is also a component of montmorillonite and some zeolites. Calcium carbonate is present in sandstone and detrital rock as a cement between particles or a partial filling of interstices [3].

Natural water systems acquire their chemical characteristics by dissolution and by chemical reactions with solids, liquids, and gases with which they come in contact during the hydrological cycle. Weathering of rocks (dissolution) and formation of soils/sediments (precipitation) are the chemical processes that regulate the composi-

---

1. Single-chain silicates; common calcium-containing pyroxenes include wollastonite ($CaO \cdot SiO_2$), hedenbergite ($CaO \cdot FeO \cdot 2SiO_2$), and diopside ($CaO \cdot MgO \cdot 2SiO_2$).
2. Double-chain silicates; common calcium-containing amphiboles include tremolite ($2CaO \cdot 5MgO \cdot 8SiO_2 \cdot H_2O$) and iron tremolite ($2CaO \cdot 5FeO \cdot 8SiO_2 \cdot H_2O$).
3. Feldspars are alumino-silicates of Na, K, Ca, or occasionally other cations. The most common calcium-containing feldspar is anorthite ($CaO \cdot Al_2O_3 \cdot 2SiO_3$). Plagioclase feldspars are a series consisting of a "solid solution" of $CaAl_2Si_2O_8$ – $NaAlSi_3O_8$.

tion of natural waters. Weathering reactions, which involve the interaction of water and the atmosphere with the earth's crust and subsequent removal of the major cations, occur because the minerals contained in the uppermost layer of the lithosphere are thermodynamically unstable in the presence of water and the acidity of the atmosphere. The solubilities of the various calcium minerals govern the levels of calcium in solution.

The stabilities of various calcium-containing minerals, as represented by their equilibrium constants, are given in Table 6.4-1; the overall reactions are also presented. Precise thermochemical data are not available for many of the minerals.

TABLE 6.4-1

Equilibrium Constants for Weathering Reactions of Selected Calcium Minerals

Reaction	log K (25°C)
Silicates	
$CaO \cdot FeO \cdot 2SiO_2$ (hedenbergite) + $4H^+ \rightleftharpoons Ca^{+2} + Fe^{+2} + 2H_2SiO_3^0$	4.60[a]
$2CaO \cdot 5MgO \cdot 8SiO_2 \cdot H_2O$ (tremolite) + $14H^+ \rightleftharpoons 2Ca^{+2} + 5Mg^{+2} + 8H_2SiO_3^0$	4.46[a]
$CaSiO_3$ (wallastonite) + $2H^+ + H_2O \rightleftharpoons Ca^{+2} + H_4SiO_4^0$	13.27[c]
$CaMg(SiO_3)_2$ (diopside) + $4H^+ + 2H_2O \rightleftharpoons Ca^{+2} + Mg^{+2} + 2H_4SiO_4^0$	21.16[c]
Alumino-silicates	
$CaAl_2Si_2O_8$ (anorthite) + $2H^+ + H_2O \rightleftharpoons Al_2Si_2O_5(OH)_4$ (kaolinite) + Ca^{+2}	14.4[b]
$CaAl_2Si_2O_8$ (anorthite) + $8H^+ \rightleftharpoons Ca^{+2} + 2Al^{+3} + 2H_4SiO_4^0$	23.33[c]
Montmorillonite	
$CaAl_{14}Si_{22}O_{60}(OH)_{12}(s) + 2H^+ + 23H_2O \rightleftharpoons 7\,Al_2Si_2O_5(OH)_4(s) + 8H_4SiO_4^0 + Ca^{+2}$	−15.4[b]
Carbonates	
$CaCO_3$ (calcite) + $2H^+ \rightleftharpoons Ca^{+2} + CO_2(g) + H_2O$	9.74[c]
or $CaCO_3$ (calcite) $\rightleftharpoons Ca^{+2} + CO_3^{-2}$	−8.30[b]
$CaCO_3$ (aragonite) + $2H^+ \rightleftharpoons Ca^{+2} + CO_2(g) + H_2O$	9.97[c]
$CaCO_3 \cdot 6H_2O$ (ikaite) + $2H^+ \rightleftharpoons Ca^{+2} + CO_2(g) + 7H_2O$	11.78[c]
$CaMg(CO_3)_2$ (dolomite) + $4H^+ \rightleftharpoons Ca^{+2} + Mg^{+2} + 2CO_2(g) + 2H_2O$	18.46[c]
Sulfates	
$CaSO_4 \cdot 2H_2O$ (gypsum) $\rightleftharpoons Ca^{+2} + SO_4^{-2} + 2H_2O$	−4.62[a]
$CaSO_4$ (anhydrite) $\rightleftharpoons Ca^{+2} + SO_4^{-2}$	−4.41[c]
Oxides, Hydroxides	
CaO (lime) + $2H^+ \rightleftharpoons Ca^{+2} + H_2O$	32.95[c]
$Ca(OH)_2$ (portlandite) + $2H^+ \rightleftharpoons Ca^{+2} + 2H_2O$	22.80[c]
Fluorides	
CaF_2 (fluorite) $\rightleftharpoons Ca^{+2} + 2F^-$	−10.41[c]

a. Morel [7] (solubility constants)
b. Stumm and Morgan [8]
c. Lindsay [6]

Predominance diagrams have been produced by Stumm and Morgan [8] to graphically illustrate the stability relationships of several mineral phases. The weathering process causes the appearance or disappearance of stable or metastable solid phases, and the redistribution of aqueous species. As for natural water systems, these diagrams are based on the concept of partial equilibrium, in which a system is in equilibrium with respect to one process but not to others. Stumm and Morgan [8] plotted the concentration levels of the selected parameters measured in several natural water systems on these graphs to illustrate whether, in the specific systems, calcium was still being dissolved from minerals or if the levels of calcium ion were being limited by solubility.

The reaction progress (Figure 6.4-1a) for the dissolution of calcium feldspars by CO_2-enriched water illustrates the complexity of the dissolution process. The dissolution causes the initial release of calcium ion, aluminum ion, and silicic acid, and the consumption of hydrogen ions:

$$CaAl_2Si_2O_8\ \text{(anorthite)} + 8H^+ \longrightarrow 2Al^{+3} + Ca^{+2} + 2H_4SiO_4^0 \qquad (1)$$

After an initial congruent dissolution[4], the released $[Al^{+3}]$ precipitates as $Al(OH)_3$ (gibbsite):

$$\text{Anorthite} + 2H^+ \longrightarrow 2Al(OH)_3(s) + Ca^{+2} + 2H_4SiO_4^0 \qquad (2)$$

From then on, the dissolution of feldspars is incongruent; with increasing $[H_4SiO_4^0]$, the gibbsite/kaolinite boundary is reached.

$$Al(OH)_3(s) + H_4SiO_4^0 \longrightarrow \tfrac{1}{2}Al_2Si_2O_5(OH)_4\ \text{(kaolinite)} + 5H_2O \qquad (3)$$

After all gibbsite has been consumed, the solution composition leaves the gibbsite/kaolinite boundary and traverses the kaolinite field in the direction of montmorillonite, $Ca_{0.33}Al_{4.67}Si_{7.33}O_{20}(OH)_4$.

$$7\ \text{kaolinite} + 8H_4SiO_4 + Ca^{+2} \longrightarrow 3\ \text{montmorillonite} + 2H^+ + 23H_2O \qquad (4)$$

Eventually, even calcium carbonate (calcite) might precipitate. The initial concentration of hydrogen ion is governed by the equilibration of the CO_2 level in the atmosphere with the water system.

The measured values of pH, pCa, $p(HCO_3)$, $-\log[H_4SiO_4]$ and P_{CO_2} (partial pressure of CO_2) for eleven natural aqueous systems and seawater are plotted in Figure 6.4-1. In Figure 6.4-1a most of the points for natural water systems fall within the stability field of kaolinite, indicating that most waters are in equilibrium with kaolinite; feldspars would be unstable and would degrade under these conditions. The measure-

4. Congruent dissolution occurs when the entire solid dissolves; incongruent dissolution occurs when part of the solid dissolves, leaving a solid phase different in composition from the original [2].

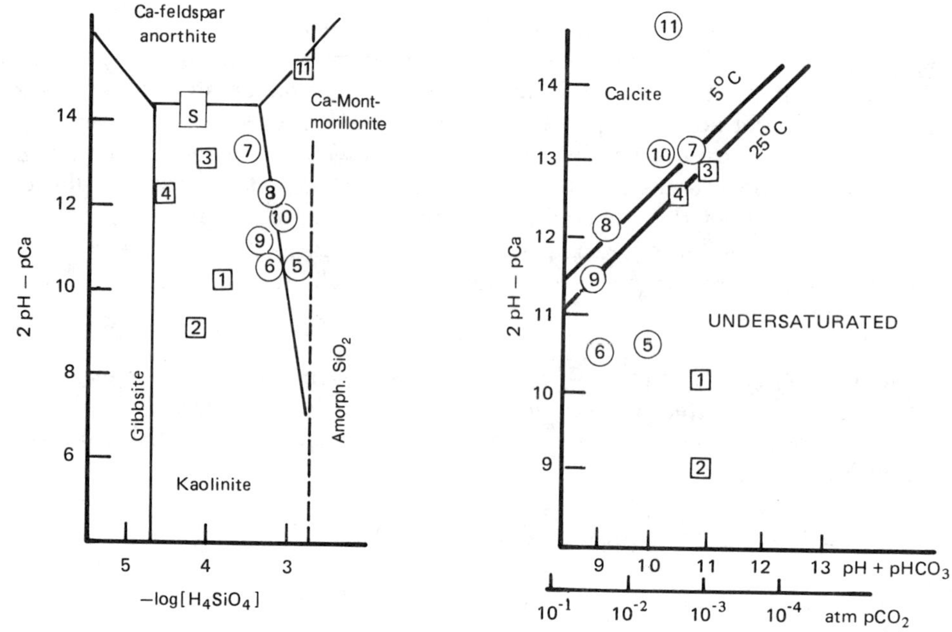

a. Dissolution of Feldspars, Intermediate Stage

b. $CaCO_3$ Solubility Equilibrium

Legend

□ — Surface waters
○ — Groundwaters
1 — Stream/granite
2 — Stream/quartzite
3 — Stream/sandstone
4 — Lake Erie
5 — Groundwater/granite
6 — Groundwater/plagioclase
7 — Groundwater/sandstone
8 — Groundwater/shale
9 — Groundwater/limestone
10 — Groundwater/dolomite
11 — Closed basin (soda lake)
S — Seawater

Source: Adapted from Stumm and Morgan [8]. *(Copyright 1981, John Wiley & Sons. Reprinted with permission.)*

FIGURE 6.4-1 Stability Relationship of Calcium-containing Minerals

ments for seawater (S) and a closed-basin lake (11) suggest that equilibrium with feldspars may eventually be attained if the calcium content is high enough, since these points fall directly on the boundary line.

The dissolution of calcite ($CaCO_3$) by a CO_2-enriched water at a constant partial pressure of CO_2 (Figure 6.4-1b) illustrates the limiting effect that solubility plays in natural water systems. The homogeneous equilibrium

$$CO_2(g) + H_2O \rightleftharpoons HCO_3^- + H^+ \qquad (5)$$

is always assumed to be obtained during the $CaCO_3$ dissolution process. The $CaCO_3$ dissolution process can then be characterized by:

$$CaCO_3(s) + 2H^+ \rightleftharpoons Ca^{+2} + CO_2(g) + H_2O \quad (6)$$

or

$$CaCO_3(s) + 2H^+ \rightleftharpoons Ca^{+2} + H^+ + HCO_3^- \quad (7)$$

Therefore, $[Ca^{+2}]/[H^+]^2$ can be plotted as a function of either $(P_{CO_2})^{-1}$ or $([H^+][HCO_3^-])^{-1}$; the approach to calcite saturation can be represented by a vertical line.

Most of the natural water systems are undersaturated with respect to calcite, as seen in Figure 6.4-1b; however, the groundwater/dolomite (10) and closed-basin lake (11) are oversaturated, and calcite may actually precipitate.

The levels of calcium ion in aqueous systems are largely controlled by the precipitation of calcite. Although the addition of Ca^{+2} will also affect the activities of other species by the formation of insoluble species (e.g., CaF_2, $CaPO_4$), the concentration of these species is generally so low that the net effect of the Ca^{+2} concentration is minimal.

6.4.3 Concentration and Predominant Speciation in Natural Waters

Table 6.4-2 illustrates the range of aqueous calcium concentrations in natural waters. Where the information is available, the rock type being drained by the fresh water system is indicated. Concentrations in natural fresh waters (rivers, streams, and groundwaters) range from 1 to 126 mg/l; the concentration in seawater is 412 mg/l.

In general, calcium salts are not as strong electrolytes as those of sodium or potassium (i.e., they are less completely dissociated at comparable concentrations), and some calcium complexes are stable in aqueous systems.[5] The major calcium complexes in fresh water and seawater and their relative concentrations are presented in Table 6.4-3. In both fresh water and seawater systems, the calcium ion complexes are present at low concentrations compared with that of the uncomplexed aqueous ion. In fresh water, the concentrations of both the calcium sulfate ($CaSO_4^0$) and calcium bicarbonate ($CaHCO_3^+$) complexes are significant, representing 2.3% and 1.9% of the total calcium content, respectively [8]; in seawater, the concentration of $CaSO_4^0$ represents 8% of the total calcium content [4].

Carbonate and hydroxide ion pairs of calcium have been reported, but their concentrations are generally significant only in strongly alkaline waters. Calcium does form stronger complexes with organic ligands, as well as ion pairs (e.g., with phosphate); however, the influence of these ligands and ion pairs on the concentration of calcium in natural waters is not likely to be significant in most natural water systems, as the concentration of Ca^{+2} is considerably higher than that of the complexing species [3].

5. Equilibrium constants for selected calcium complexes are tabulated in section 2.9.

TABLE 6.4-2

Measured Calcium Concentrations in Natural Waters

Aqueous System[a]	Calcium Concentration (mg/l)	Source
Hydrothermal Fluid	1,000	[7]
Seawater	412	[7]
Groundwater (granite)	13	[8]
Groundwater (plagioclase)	32	[8]
Groundwater (sandstone)	40	[8]
Groundwater (shale)	126	[8]
Groundwater (limestone)	80	[8]
Groundwater (dolomite)	63	[8]
Closed-basin Lake (soda lake)	1	[8]
Stream (granite)	4	[8]
Stream (quartzite)	2	[8]
Stream (sandstone)	32	[8]
River[b]	13.4	[2]
River[c]	14.7	[2]
Riverwater (mean-worldwide)	15	[3]
Rainwater (average)	1.4	[3]
Public Water Supply (100 U.S. cities)	26 (median)	[9]
	0-145 (range)	[9]

a. The type of rock being drained is given in parentheses.
b. Natural (unpolluted) component of river.
c. Including contribution from human activities.

TABLE 6.4-3

Predominant Calcium Complexes in Fresh Water and Seawater

Complex	Fresh Water (~28 mg/l Ca)	Seawater (412 mg/l Ca)
Ca^{+2}	95.0%	91.0%
$CaHCO_3^+$	1.9%	1.0%
$CaCO_3^0$	0.9%	0.2%
$CaSO_4^0$	2.3%	8.0%

Percentages do not total exactly 100 because of rounding.

Sources: Fresh water — Stumm and Morgan [8]
Seawater — Horne [4]

The activities of several environmental calcium complexes in equilibrium with soil-Ca or calcite are plotted in Figure 6.4-2 as a function of $CO_2(g)$, anion activities, and pH.[6] Figure 6.4-2a shows that the chloride complexes of calcium contribute very little to the total soluble calcium. The $CaHCO_3^+$ ion is insignificant in acid soils but becomes more significant in submerged soils, where the pH approaches neutrality and the partial pressure of $CO_2(g)$ increases. Calcite supports the neutral ion pair, $CaCO_3^0$, at a concentration of $10^{-5.27}M$ independent of $CO_2(g)$ and pH; however, this ion pair is not important in acidic soil.

Figure 6.4-2b shows that $CaSO_4^0$ is significant at sulfate concentrations greater than $10^{-4}M$. None of the phosphate complexes ($CaH_2PO_4^+$, $CaHPO_4^0$, and $CaPO_4^-$) are significant in acid soils; however, in neutral and calcareous soils, $CaHPO_4^0$ and $CaPO_4^-$ are significant. The hydrolysis species ($CaOH^+$ and $Ca(OH)_2^0$) and nitrate complexes ($CaNO_3^+$ and $Ca(NO_3)_2^0$) are not important in the pH range of most soils [6].

6.4.4 Solid-Solution Interface

Ion exchange of calcium with a variety of surface sites in soils governs the concentration of calcium ions in soil water and the soil itself. Colloid fractions of clays and hydrous oxides especially contribute to these surface sites. Much of the calcium present in soils is held onto the colloidal surface as adsorbed cations. However, calcium ions may be released to the soil solution by exchange with other positively charged ions and subsequently readsorbed. Thus, during liming, when soluble calcium ions are added to the soil, the colloidal surface can adsorb large quantities of calcium ions. The following reaction, which involves the exchange of protons (from carbonic acid) for adsorbed calcium ions, is an example of the general cation exchange reaction:

$$\underset{\text{Adsorbed Calcium}}{\text{Colloid-Ca}} + \underset{\text{Carbonic Acid}}{2H_2CO_3} \longrightarrow \underset{\text{Adsorbed Hydrogen}}{\text{Colloid}<^{H}_{H}} + \underset{\text{Soluble Bicarbonate}}{Ca(HCO_3)_2} \qquad (8)$$

Although all cations may be adsorbed by soil colloids, certain cations, such as calcium, are favored under natural conditions. In humid regions, the order of adsorption of cations is H^+, Al^{+3}, and Ca^{+2} first, Mg^{+2} second, and K^+ and Na^+ third. In well-drained arid and semiarid regions, the order is usually Ca^{+2} and Mg^{+2} first, Na^+ and K^+ second, and H^+ last. Table 6.4-4 illustrates the predominance of calcium in selected surface soils [1].

6. Representative soil pore-water concentrations are: $10^{-2.5}$ M Ca^{+2}, 10^{-4} M to 10^{-2} M Cl^-, 10^{-4} M to 10^{-2} M SO_4^{-2}, 10^{-6} M to 10^{-4} M $H_2PO_4^-$, 10^{-4} M to 10^{-2} M NO_3^-, and $10^{-3.5}$ M to $10^{-2.5}$ M $CO_2(g)$.

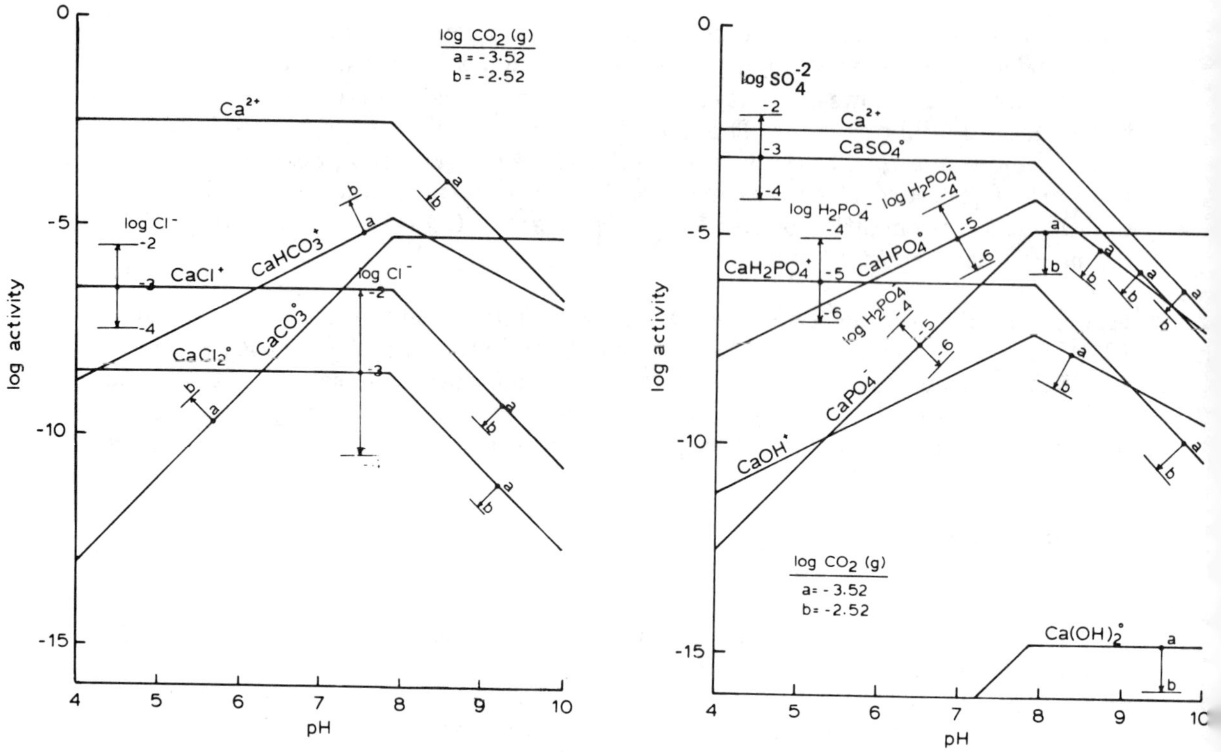

a. Chloride, Nitrate and Bicarbonate Complexes

b. Sulfate and Phosphate Complexes

Source: Lindsay [6]. *(Copyright 1979, John Wiley & Sons. Reprinted with permission.)*

FIGURE 6.4-2 Complexes of Ca^{+2} in Equilibrium with Soil-Ca or Calcite as Affected by CO_2(g), Anion Activities and pH for a Theoretical System

TABLE 6.4-4

Relative Proportion of Adsorbed Metallic Cations Present in Certain Surface Soils of the United States[a]

Soil	Ca	Mg	K	Na
Penn loam (N.J.)	60.8	15.8	19.0	4.4
Mardin silt loam (N.Y.)	90.7	5.0	3.1	1.2
Webster series soil (Iowa)	76.8	20.4	1.2	1.6
Sweeney clay loam (Calif.)	76.1	21.3	1.3	1.3
Red River Valley soil (Minn.)	73.9	21.5	4.2	0.4
Keith silt loam (Nebr.)	77.1	13.3	7.1	2.5
Holdrege silt loam (Nebr.)	66.5	20.9	11.1	1.5

a. The percentage figures in each case are based on the sum of the metallic cations taken as 100. Note the geographic distribution of the soil samples.

Source: Brady [1]. *(Copyright 1974, Macmillan Publishing Co. Reprinted with permission.)*

6.4.5 Literature Cited

1. Brady, N.C., *The Nature and Properties of Soils,* Macmillan Publishing Co., New York (1974).
2. Drever, J.I., *The Geochemistry of Natural Waters,* Prentice-Hall, Inc., Englewood Cliffs, N.J. (1982).
3. Hem, J.D., "Study and Interpretation of Chemical Characteristics of Natural Waters," Water Supply Paper No. 1473, 2nd ed., U.S. Geological Survey, Washington, D.C. (1970).
4. Horne, R.A., *Marine Chemistry,* John Wiley & Sons, New York (1969).
5. *Kirk-Othmer Encyclopedia of Chemical Technology,* 3rd ed., John Wiley & Sons, New York (1979).
6. Lindsay, W.L., *Chemical Equilibria in Soils,* John Wiley & Sons, New York (1979).
7. Morel, F.M.M., *Principles of Aquatic Chemistry,* John Wiley & Sons, New York (1983).
8. Stumm, W. and J.J. Morgan, *Aquatic Chemistry,* 2nd ed., John Wiley & Sons, New York (1981).
9. Todd, D.K. (ed.), *The Water Encyclopedia,* Water Information Center, Inc., New York (1970).

6.5 MAGNESIUM (Mg)

6.5.1 Production and Uses

Magnesium occurs widely in nature, primarily in the ores dolomite, magnesite and carnallite, and (as the chloride) in seawater, underground natural brines and salt deposits. Magnesium metal, which is produced by electrolysis of molten magnesium chloride or thermal reduction of magnesium oxide, is mostly used in the manufacture of aluminum and magnesium alloys. Most of the magnesium compounds produced (chiefly the carbonate, chloride and sulfate) are used to manufacture a variety of refractory minerals. The remainder is used in the chemical processing and pharmaceutical industries [4,9].

6.5.2 Natural Sources

Magnesium, a metallic element, is eighth in abundance in the earth's crust, of which it forms approximately 2.06% [9]. Its concentration in igneous and sedimentary rocks is 1.8% and 1.6%, respectively [6]. In igneous rocks, magnesium is typically a constituent of the dark-colored ferromagnesium minerals (olivine[1], pyroxenes[2], amphiboles[3] and dark-colored micas), along with a variety of less common species. In altered rocks (e.g., metamorphic rocks), magnesium mineral species such as chlorite, montmorillonite and serpentine occur. Sedimentary rocks of magnesium include the carbonates such as magnesite and hydromagnesite, the hydroxide (brucite), and mixtures of magnesium and calcium carbonates (e.g., dolomite, a specific structure with magnesium and calcium present in equal amounts) [2]. Structural formulas and common names of magnesium minerals are included in Table 6.5-1.

The average magnesium content of soils is only 0.5%, because the element is removed from soils by weathering. The solubility relationships of the various magnesium minerals determine which are stable in soils and which are most likely to control the activity of Mg^{+2} and magnesium complexes in natural water systems. In general, the silicates and alumino-silicates undergo a rapid surface exchange of H^+ for Mg^+, followed by a slower dissolution [1]. The stabilities of various magnesium-containing minerals, as represented by their equilibrium constants, are listed in Table 6.5-1; the overall reactions are also presented. Thermochemical data for many of these minerals are not known with any precision.

The stability and weathering of the magnesium silicates, magnesium alumino-silicates, and other magnesium minerals are addressed separately in the following discussions. The solubilities of several magnesium silicates are presented graphically in Figure 6.5-1 in terms of log $[Mg^{+2}]$ versus pH. For purposes of illustration,

1. Class of nesosilicates in which the SiO_4 tetrahedra are isolated and bound to each other only by ionic bonds from interstitial cations; common magnesium-containing olivines include olivine, $(Mg, Fe)_2SiO_4$, and forsterite, Mg_2SiO_4.
2. Single-chain silicates; common magnesium-containing pyroxenes include enstatite, $MgO \cdot SiO_2$, and diopside, $CaO \cdot MgO \cdot 2SiO_2$.
3. Double-chain silicates; a common magnesium-containing amphibole is tremolite, $2CaO \cdot 5MgO \cdot SiO_2 \cdot H_2O$.

TABLE 6.5-1

Equilibrium Constants for Weathering Reactions of Selected Magnesium Minerals

Reaction	log K (25°C)[a]
Silicates	
$Mg_{1.6}Fe(II)_{0.4}SiO_4$ (olivine) $+ 4H^+ \rightleftharpoons 1.6\,Mg^{+2} + 0.4\,Fe^{+2} + H_4SiO_4^0$	26.18
Mg_2SiO_4 (forsterite) $+ 4H^+ \rightleftharpoons 2Mg^{+2} + H_4SiO_4^0$	28.87
Mg_2SiO_4 (forsterite) $+ 4H^+ \rightleftharpoons 2Mg^{+2} + H_2SiO_3^0 + H_2O$	7.11[b]
$MgSiO_3$ (clinoenstatite) $+ 2H^+ + H_2O \rightleftharpoons Mg^{+2} + H_4SiO_4^0$	11.42
$MgSiO_3$ (enstatite) $+ 2H^+ \rightleftharpoons Mg^{+2} + H_2SiO_3^0$	5.82[b]
$Mg_3Si_4O_{10}(OH)_2 \cdot 2H_2O$ (vermiculite) $+ 2H_2O + 6H^+ \rightleftharpoons 3Mg^{+2} + 4H_4SiO_4^0$	30.39
$MgCa(SiO_3)_2$ (diopside) $+ 4H^+ + 2H_2O \rightleftharpoons Mg^{+2} + Ca^{+2} + 2H_4SiO_4^0$	21.16
$MgCa(SiO_3)_2$ (diopside) $+ 4H^+ \rightleftharpoons Mg^{+2} + Ca^{+2} + 2H_2SiO_3^0$	5.30[b]
$Mg_3Si_2O_5(OH)_4$ (chrysotolite) $+ 6H^+ \rightleftharpoons 3Mg^{+2} + 2H_4SiO_4^0 + H_2O$	32.87
$Mg_2Si_3O_6(OH)_4$ (sepiolite) $+ 2H_2O + 4H^+ \rightleftharpoons 2Mg^{+2} + 3H_4SiO_4^0$	15.89
$Mg_6Si_4O_{10}(OH)_8$ (serpentine) $+ 12H^+ \rightleftharpoons 6Mg^{+2} + 4H_4SiO_4^0 + 2H_2O$	61.75
$Mg_3Si_4O_{10}(OH)_2$ (talc) $+ 4H_2O + 6H^+ \rightleftharpoons 3Mg^{+2} + 4H_4SiO_4^0$	22.26
$2CaO \cdot 5MgO \cdot 8SiO_2 \cdot H_2O$ (tremolite) $+ 14H^+ \rightleftharpoons 2Ca^{+2} + 5Mg^{+2} + 8H_2SiO_3^0$	4.46[b]
Alumino-silicates	
$Mg_5Al_2Si_3O_{10}(OH)_8$ (chlorite) $+ 16H^+ \rightleftharpoons 5Mg^{+2} + 2Al^{+3} + 3H_4SiO_4^0 + 6H_2O$	60.12
$Mg_2Al_4Si_5O_{18}$ (cordierite) $+ 2H_2O + 16H^+ \rightleftharpoons 2Mg^{+2} + 4Al^{+3} + 5H_4SiO_4^0$	45.46
$K_{0.6}Mg_{0.25}Al_{2.3}Si_{3.5}O_{10}(OH)_2$ (illite) $+ 8H^+ + 2H_2O \rightleftharpoons 0.6K^+ + 0.25Mg^{+2} + 2.3Al^{+3} + 3.5H_4SiO_4^0$	10.34
$Mg_{0.2}(Si_{3.81}Al_{1.71}Fe(III)_{0.22}Mg_{0.29})O_{10}(OH)_2$ (Mg-montmorillonite) $+ 6.76H^+ + 3.24H_2O \rightleftharpoons$ $0.49Mg^{+2} + 1.71Al^{+3} + 0.22Fe^{+3} + 3.81H_4SiO_4^0$	2.67
$Mg_{2.71}Fe(II)_{0.02}Fe(III)_{0.46}Ca_{0.06}K_{0.1}(Si_{2.91}Al_{1.14})O_{10}(OH)_2$ (vermiculite) $+ 10.36H^+ \rightleftharpoons$ $2.71Mg^{+2} + 0.02Fe^{+2} + 0.46Fe^{+3} + 0.06Ca^{+2} + 0.1K^+ + 1.14Al^{+3} + 2.91H_4SiO_4^0 + 0.36H_2O$	38.05

(Continued)

TABLE 6.5-1 (Continued)

Reaction	log K (25°C)[a]
Oxides	
MgO (periclase) + $2H^{-} \rightleftharpoons Mg^{-2} + H_2O$	21.74
Hydroxides	
$Mg(OH)_2$ (brucite) + $2H^{+} \rightleftharpoons Mg^{+2} + 2H_2O$	16.84
$Mg(OH)_2$ (brucite) $\rightleftharpoons Mg^{+2} + 2OH^{-}$	-11.1[b]
$MgOH^{+} \rightleftharpoons Mg^{+2} + OH^{-}$	-2.6[b]
Carbonates	
$MgCO_3$ (magnesite) + $2H^{+} \rightleftharpoons Mg^{+2} + CO_2(g) + H_2O$	10.69
$MgCO_3$ (magnesite) $\rightleftharpoons Mg^{+2} + CO_3^{-2}$	7.46[b]
$MgCO_3 \cdot 3H_2O$ (nesquehonite) + $2H^{+} \rightleftharpoons Mg^{+2} + CO_2(g) + 4H_2O$	13.49
$MgCO_3 \cdot 3H_2O$ (nesquehonite) $\rightleftharpoons Mg^{+2} + CO_3^{-2} + 3H_2O$	-4.67[b]
$MgCO_3 \cdot 5H_2O$ (lansfordite) + $2H^{+} \rightleftharpoons Mg^{+2} + CO_2(g) + 6H_2O$	13.62
$MgCa(CO_3)_2$ (dolomite) + $4H^{+} \rightleftharpoons Mg^{+2} + Ca^{+2} + 2CO_2(g) + 2H_2O$	18.46
Sulfates	
$MgSO_4 \rightleftharpoons Mg^{+2} + SO_4^{-2}$	8.18
Phosphates	
$MgHPO_4 \cdot 3H_2O$ (newberyite) $\rightleftharpoons Mg^{+2} + HPO_4^{-2} + 3H_2O$	-5.8[b]
$Mg_3(PO_4)_2 \cdot 8H_2O$ (bobierrite) $\rightleftharpoons 3Mg^{+2} + 2PO_4^{-3} + 8H_2O$	-25.2[b]

a. Lindsay [5], except as noted
b. Morel [6], solubility constants

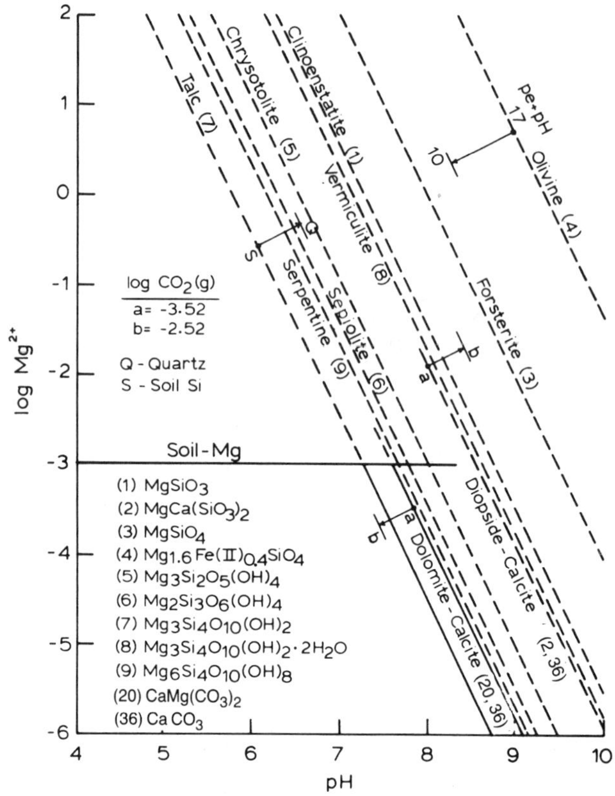

Source: Lindsay [5]. *(Copyright 1979, John Wiley & Sons. Reprinted with permission.)*

FIGURE 6.5-1 Solubility of Several Magnesium Silicates in Equilibrium with Soil-Si and Soil-Fe, with Indicated Changes for CO_2(g), Quartz, and Redox

$[H_4SiO_4^0]$ was fixed by soil-SiO_2 (amorphous); this represents the early stages of weathering, where most primary magnesium silicates (i.e., unaltered silicates) are weathering. The solubility of magnesium silicates decreases in the following order:

> olivine > forsterite > clinoenstatite > vermiculite > diopside, in equilibrium with calcite (calcium carbonate) > chrysotolite > sepiolite > serpentine > talc.

Dolomite, in equilibrium with calcite, is included in Figure 6.5-1 for reference.

The stabilities of the silicate minerals shift with changes in $H_4SiO_4^0$ activity; as soils weather, $[H_4SiO_4^0]$ declines and magnesium silicates become more soluble [shift from solubility control due to an amorphous soil-SiO_2 concentration (S) to control due to quartz (Q)]. Also, an increase in CO_2 levels will increase the solubility of the magnesium silicates while decreasing the solubility of dolomite. In submerged soils, where CO_2(g) levels are high, dolomite is the most stable mineral phase. Below pH 7.5, most magnesium minerals are too soluble to persist in soils. However, in alkaline soils that are also high in soluble $H_4SiO_4^0$, some magnesium silicates may actually form (i.e., talc, serpentine, sepiolite and chrysotolite).

The solubilities of several magnesium alumino-silicates are depicted in Figure 6.5-2 in terms of log [Mg^{+2}] versus pH. To develop this diagram, conditions typical of the intermediate stages of weathering were assumed, i.e., where secondary magnesium alumino-silicates would most likely be weathering: [Al^{+3}] and [$H_4SiO_4^0$] were fixed by kaolinite and quartz, [Fe^{+3}] by soil-Fe, [Ca^{+2}] by soil-Ca or calcite, and [K^+] by soil-K. Mg-montmorillonite, chlorite and dolomite-calcite are much more stable than illite, cordierite and vermiculite. The more soluble minerals disappear readily, whereas the more stable ones will remain until they are removed in the later stages of weathering, as soils become more acidic. Again, the stability of these minerals is greatly affected by changes in the $H_4SiO_4^0$ activity; as [$H_4SiO_4^0$] drops during the advanced stages of weathering and kaolinite and gibbsite coexist (Q to G), Mg-montmorillonite becomes highly unstable.

The solubilities of several additional magnesium minerals are plotted as a function of pH in Figure 6.5-3. Periclase is much too soluble to persist in soils; it will generally hydrolyze to brucite, which itself is not stable, or precipitate in the form of a less soluble mineral. The carbonates (lansfordite, nesquehonite and magnesite) are also very soluble and will not remain in soils, except under highly alkaline conditions. $MgSO_4$ is highly soluble (solubility product = 8.15) and will not form in well-drained soils. Only dolomite in equilibrium with calcite is stable enough to be present in the soil structure.

Natural water systems acquire their chemical composition through weathering reactions, which involve the interaction of water and the atmosphere with the earth's crust and subsequent removal of the major cations. The genealogy of most fresh water systems can generally be attributed to the dissolution of the underlying rock. In seawater, however, a variety of removal processes control the concentration of the major cations, including magnesium. These processes involve not only simple equilibration with solid phases in the sediment, but also ion exchange in detrital clays, high- and low-temperature reactions with volcanic material at oceanic ridges, and biological activity.

Incorporation into sediment is an important removal process for magnesium, as well as for the other major species. A small amount of magnesium is ion-exchanged for calcium on clay minerals of ocean sediment. Also, a small amount of magnesium carbonate is deposited with calcite in seawater; however, carbonate formation in general only removes about 6% of the magnesium supplied by rivers. There is a significant uptake of magnesium (about 24% of the river input of magnesium) by sediment in which sulfate reduction is taking place, although the actual mechanism is not known. In addition, about 4% of the river input of magnesium is buried in interstitial water. High-temperature alteration of basalts at hydrothermal vents and (to a more limited extent) low-temperature alteration apparently constitute the most important sink for magnesium in seawater [1].

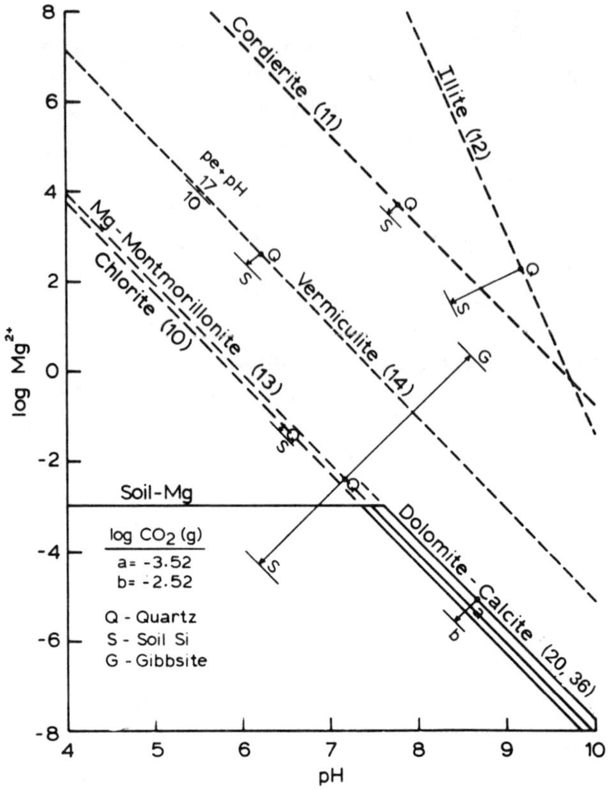

(10) $Mg_5Al_2Si_3O_{10}(OH)_8$
(11) $Mg_2Al_4Si_5O_{18}$
(12) $K_{0.6}Mg_{0.25}Al_{2.3}Si_{3.5}O_{10}(OH)_2$
(13) $Mg_{0.2}(Si_{3.81}Al_{1.71}Fe(III)_{0.22}Mg_{0.29})O_{10}(OH)_2$
(14) $Mg_{2.71}Fe(II)_{0.02}Fe(III)_{0.46}Ca_{0.06}K_{0.1}(Si_{2.91}Al_{1.14})O_{10}(OH)_2$
(20) $CaMg(CO_3)_2$
(33) $CaCO_3$

Source: Lindsay [5]. *(Copyright 1979, John Wiley & Sons. Reprinted with permission.)*

FIGURE 6.5-2 Solubility of Magnesium Aluminosilicates in Equilibrium with Kaolinite, Quartz, Ca^{+2} at $10^{-2.3}$ *M*, K^+ at 10^{-3} *M*, and Soil-Fe

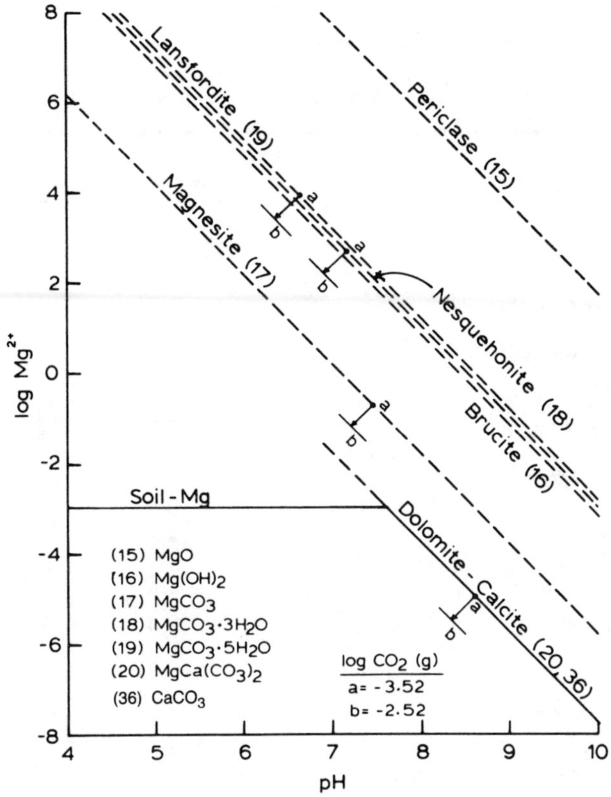

Source: Lindsay [5]. *(Copyright 1979, John Wiley & Sons. Reprinted with permission.)*

FIGURE 6.5-3 Solubilities of Oxides, Hydroxides, and Carbonates of Magnesium

6.5.3 Concentration and Predominant Speciation in Natural Waters

Table 6.5-2 illustrates the range of aqueous magnesium concentrations in natural waters. Where available, the rock type being drained by the water system is indicated. Concentrations in natural fresh waters (rivers, streams and groundwaters) range from 0.2 mg/l for a stream over quartzite to 76.8 mg/l for groundwater over shale; the reported concentration in seawater is 1300 mg/l.

In general, the magnesium ion, Mg^{+2}, is the predominant form of dissolved magnesium in natural waters; however, some magnesium complexes do form.[4] The major magnesium complexes in fresh water and seawater and their relative concentrations are outlined in Table 6.5-3. In both fresh water and seawater, the magnesium sulfate

4. Equilibrium constants for selected magnesium complexes are tabulated in section 2.9.

TABLE 6.5-2

Measured Magnesium Concentrations in Natural Waters

Aqueous System[a]	Concentration (mg/l)	Source
Seawater	1,300	[6]
Groundwater (granite)	3.9	[7]
Groundwater (plagioclase)	15.3	[7]
Groundwater (sandstone)	7.7	[7]
Groundwater (shale)	76.8	[7]
Groundwater (limestone)	9.7	[7]
Groundwater (dolomite)	38.5	[7]
Closed-basin Lake (soda lake)	0.6	[7]
Stream (granite)	0.6	[7]
Stream (quartzite)	0.2	[7]
Stream (sandstone)	2.4	[7]
River[b]	3.35	[1]
River[c]	3.65	[1]
River Water (mean-worldwide)	4.1	[2]
Public Water Supply (100 U.S. cities)	6.25 (median)	[8]
	0-120 (range)	[8]

a. Parentheses indicate type of rock being drained.
b. Natural (unpolluted) component of river.
c. Including contribution from human activities.

TABLE 6.5-3

Predominant Magnesium Complexes in Fresh Water and Seawater

	Fresh Water ($\sim$7.3 mg/l Mg)	Seawater ($\sim$1312 mg/l Mg)
Mg^{+2}	96.3%	87%
$MgHCO_3^+$	0.4%	1%
$MgCO_3$	0.5%	0.3%
$MgSO_4$	2.6%	11%

Sources: Fresh water — Stumm and Morgan [7]
Seawater — Horne [3]

(Percentages do not total 100 because of rounding.)

ion pair complex ($MgSO_4^0$) is the most significant complex present, representing 2.6% and 11% of the total magnesium content, respectively. The concentrations of the bicarbonate and carbonate complexes are significant but considerably less than that of the sulfate complex.

6.5.4 Speciation in Soil Solution

Figure 6.5-4 shows the activities of several magnesium complexes of environmental concern at equilibrium with soil-Mg or dolomite-calcite, plotted as a function of pH, pCO_2 and anion activity.[5] The $MgSO_4^0$ ion pair contributes significantly to the total magnesium in solution, especially as the SO_4^{-2} increases above 10^{-4} *M*. The $MgHPO_4^0$ ion pair also becomes significant as pH and $H_2PO_4^-$ activity increase. The $MgHCO_3^+$ and $MgCO_3^0$ species contribute little to the total magnesium content in acid soils but become important in slightly alkaline soils and submerged soils, where CO_2 levels are generally higher. In the presence of dolomite or calcite, the $MgCO_3^0$ ion pair is independent of pH and pCO_2 and has a fixed activity of $10^{-6.2}$ *M*. Complexes of $Mg(NO_3)_2^0$ and $MgCl_2^0$ are insignificant at the normal levels of NO_3^- and Cl^- found in soils. Also, the $MgOH^+$ and $Mg(OH)_2^0$ species are insignificant and contribute very little to the total soluble magnesium content [5].

6.5.5 Literature Cited

1. Drever, J.I., *The Geochemistry of Natural Waters,* Prentice-Hall, Inc., Englewood Cliffs, N.J. (1982).
2. Hem, J.D., *Study and Interpretation of the Chemical Characteristics of Natural Waters,* Water Supply Paper No. 1473, 2nd ed., U.S. Geological Survey, Washington, D.C. (1970).
3. Horne, R.A., *Marine Chemistry,* John Wiley & Sons, New York (1969).
4. *Kirk-Othmer Encyclopedia of Chemical Technology,* 3rd ed., John Wiley & Sons, New York (1978).
5. Lindsay, W.L., *Chemical Equilibria in Soils,* John Wiley & Sons, New York (1979).
6. Morel, F.M.M., *Principles of Aquatic Chemistry,* John Wiley & Sons, New York (1983).
7. Stumm, W. and J.J. Morgan, *Aquatic Chemistry,* 2nd ed., John Wiley & Sons, New York (1981).
8. Todd, D.K. (ed.), *The Water Encyclopedia,* Water Information Center, Inc., New York (1970).
9. U.S. Department of the Interior, Bureau of Mines, "Mineral Facts and Problems," U.S. Government Printing Office, Washington, D.C. (1980).

5. Representative soil pore-water concentrations for acid and near-neutral soils are: 10^{-3} *M* Mg^{+2}, 10^{-4} *M* to 10^{-2} *M* SO_4^{-2}, 10^{-4} *M* to 10^{-2} *M* NO_3^- or Cl^-, 10^{-6} *M* to 10^{-4} *M* $H_2PO_4^-$, and $10^{-3.5}$ *M* to $10^{-1.5}$ *M* $CO_2(g)$.

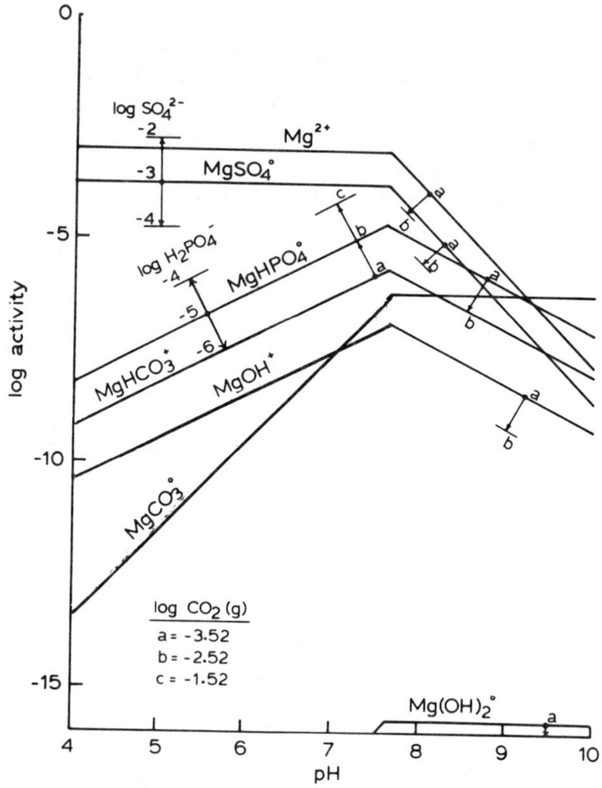

Source: Lindsay [5]. *(Copyright 1979, John Wiley & Sons. Reprinted with permission.)*

FIGURE 6.5-4 Stability of Potentially Significant Magnesium Complexes in Equilibrium with Soil-Mg or Dolomite-Calcite

6.6 SILICON (Si)

6.6.1 Production and Uses

Silicon is used for deoxidizing and as a strengthening alloy in the production of iron, steel, and nonferrous metals. Silicon metal is the starting material for intermediate products such as silanes, which are used to formulate silicone resins, lubricants, elastomers, antifoaming agents, and water-repellent compounds. High-purity silicon is the base material for semiconductor devices and is used in almost all integrated circuits. Other applications include photovoltaic power systems, infrared optical instruments, abrasives, heating elements, and heat- and corrosion-resisting hardware.

In the form of sand and clay, silicon-bearing minerals are used to make concrete and brick. Silicon compounds are useful refractory materials for high-temperature work, and the silicates are used to make pottery and enamels. As sand, silica is a principal ingredient of glass.

Important sources of silicon are quartz, quartzite, and sandstone. Silicon metal and ferrosilicon are produced by smelting processes. Silicon metal is refined and further processed to yield the highly purified, single-crystal silicon required for use in semiconductors.

6.6.2 Occurrence

Silicon comprises 25.7% of the earth's crust, by weight, and is second only to oxygen as the most abundant element. It does not occur free in nature but combined with oxygen and other elements to form oxides and silicate minerals. The oxide appears in many forms, among them sand, quartz, rock crystal, amethyst, agate, flint, jasper, and opal. Silicate minerals include granite, asbestos, feldspar, clay, and mica.

The Si^{+4} ion does not occur as such, but is typically surrounded by four oxygen atoms. The SiO_4^{-4} tetrahedron is a fundamental building unit of most minerals in igneous and metamorphic rocks and is present in most other rocks and soils, as well as in natural water at low concentrations.

Silicon accounts for 29.7% of igneous rock and 27.4% of sedimentary rock [9]. In igneous rock, it occurs as crystalline SiO_2, or quartz. Quartz also constitutes the bulk of the grains in most sandstones and is a common mineral in metamorphic rock.

6.6.3 Properties and Speciation in Water

The term "silica," meaning the oxide SiO_2, is used to refer to dissolved silicon in natural water, although the actual form is hydrated and is generally represented as $H_4SiO_4^0$ or $Si(OH)_4^0$ [4].

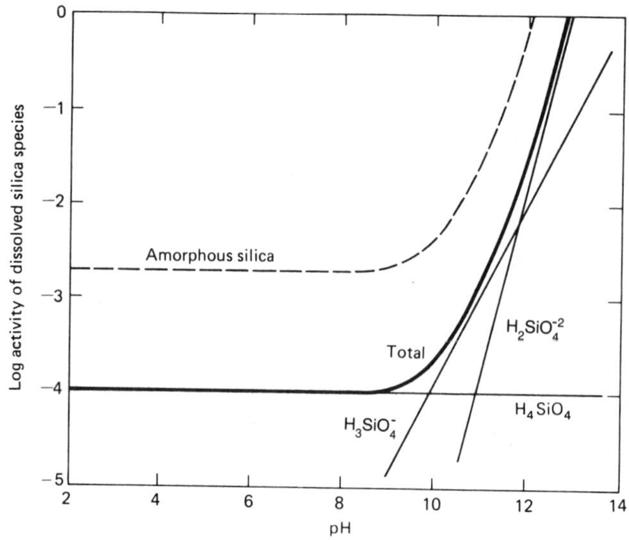

Note: Dashed line is the corresponding sum for equilibrium with amorphous silica. Heavy line represents the sum of the activities of individual species.

Source: Drever [3]. *(Copyright 1982, Prentice-Hall, Inc. Reprinted with permission.)*

FIGURE 6.6-1 Activities of Dissolved Silica Species in Equilibrium with Quartz at 25°C

Most of the dissolved silica in natural waters results from the chemical breakdown of silicate minerals in weathering processes. The solubility of SiO_2 can be characterized by the following equilibria [13]:

$$SiO_2(s, quartz) + 2H_2O = Si(OH)_4^0 \qquad \log K = -3.7 \qquad (1)$$

$$SiO_2(s, amorphous) + 2H_2O = Si(OH)_4^0 \qquad \log K = -2.7 \qquad (2)$$

$$Si(OH)_4^0 = SiO(OH)_3^- + H^+ \qquad \log K = -9.46 \qquad (3)$$

$$SiO(OH)_3^- = SiO_2(OH)_2^{-2} + H^+ \qquad \log K = -12.56 \qquad (4)$$

$$4Si(OH)_4^0 = Si_4O_6(OH)_6^{-2} + 2H^+ + 4H_2O \qquad \log K = -12.57 \qquad (5)$$

The equilibrium constants are valid at 25°C. In these equations, silicic acid is written as $Si(OH)_4^0$ (rather than as $H_4SiO_4^0$), to emphasize that silicon tends to coordinate with hydroxy and oxy ligands and form multinuclear species [13].

Figure 6.6-1 shows the activities of dissolved silica species in equilibrium with quartz and amorphous silica at 25°C. Natural waters normally have more dissolved silica than the corresponding quartz equilibrium value, but less than the values shown for amorphous silica, which is the likely upper equilibrium limit [4]. In natural water with a pH <9, the most probable form of aqueous silica is monomeric silicic acid,

$Si(OH)_4^0$. Above pH 9, aqueous silica becomes ionized, forming silicate ions [10]. The total dissolved silica concentration increases dramatically at high pH values.

The dissolution of a silicate can be typified by the reaction of the sodium feldspar albite with water containing H^+ to form kaolinite and soluble products [4]:

$$2NaAlSi_3O_8(s) + 2H^+ + 9H_2O = Al_2Si_2O_5(OH)_4(s) + 4H_4SiO_4^0 + 2Na^+ \quad (6)$$

The kaolinite produced by this reaction may dissolve reversibly:

$$Al_2Si_2O_5(OH)_4(s) + 6H^+ \rightleftharpoons 2Al^{+3} + 2H_4SiO_4^0 + H_2O \quad (7)$$

At sufficiently high aqueous concentrations, silica may be precipitated:

$$H_4SiO_4^0 = SiO_2(s) + 2H_2O \quad (8)$$

Concentrations of silica in natural waters commonly range from 1 to around 30 mg/l. However, higher concentrations (29 to 363 mg/l) have been found in groundwaters and are related to rock type and the temperature of the water [4]. Waters in equilibrium with amorphous silica contain 100-140 mg/l silica; those in equilibrium with quartz contain about 6-12 mg/l silica; and waters in equilibrium with clay minerals, feldspars, mica, and other siliceous rock-forming minerals contain intermediate concentrations. Weathering of rock minerals in contact with water not only dissolves silica in relatively large amounts but also produces siliceous weathering products, such as clay minerals [10].

Dissolved silica concentrations are generally higher in water seeping through soils (subsurface flow) than in surface waters or groundwater aquifers. For instance, during stream rise, dissolved silica initially decreases while overland flow comprises much of the streamflow and then increases as subsurface flow becomes the major component of streamflow. With decreasing discharge, groundwater becomes an increasing proportion of streamflow, and silica concentration slowly decreases [7].

Concentrations in seawater vary widely. Silicon is found both as dissolved silicate ion and as suspended silica. Dissolved silicon exists mostly as monomeric silicic acid, $Si(OH)_4^0$ [11]. Silicon is an important nutrient in the oceans, and the silicon content of seawater, both dissolved and particulate, is closely linked to the growth and decay of organisms. Seawater near the surface is very low in silica (often less than 1 mg/l), apparently because marine organisms extract and use it in their shells and skeletons. This depletion of silica contributes to an increase in pH. Similar depletion effects are noticeable in the surface layers of some lakes and reservoirs [4]. The major oceans of the world seem to have very different silicon profiles; the deep waters of the Pacific and Indian oceans are much richer in silicon than those of the Atlantic [5].

Colloidal hydroxides of silicon are known to form at the pH of seawater and to be potent scavengers of polyvalent elements [2].

6.6.4 Properties and Speciation at the Solid/Solution Interface

FORMS IN SOIL SOLIDS

Soil contains approximately 32% silicon by weight [8]. The element is present in primary silicate minerals, secondary alumino-silicates, and various forms of $SiO_2(s)$. Quartz, the most stable SiO_2 mineral, is also the most common mineral in soils, accounting for up to 90% of the solid soil phase [6]. Feldspars, anhydrous alumino-silicates of primarily K, Na, and Ca, account for about 60% by weight of igneous rocks and large fractions of sedimentary rocks [1].

Table 6.6-1 shows the dissolution reactions of selected silicon-bearing minerals. The equilibrium constants shown can be used to establish stability relations among the

TABLE 6.6-1

Selected Equilibrium Reactions and Activity Constants of Silicate Minerals[a]

Reaction	Log K^{0b}
SiO_2 Minerals	
SiO_2(silica glass) + $2H_2O \rightleftharpoons H_4SiO_4^0$	−2.71
SiO_2(amorp) + $2H_2O \rightleftharpoons H_4SiO_4^0$	−2.74
SiO_2(coesite) + $2H_2O \rightleftharpoons H_4SiO_4^0$	−3.05
SiO_2(soil) + $2H_2O \rightleftharpoons H_4SiO_4^0$	−3.10
α-SiO_2(tridymite) + $2H_2O \rightleftharpoons H_4SiO_4^0$	−3.76
α-SiO_2(cristobalite) + $2H_2O \rightleftharpoons H_4SiO_4^0$	−3.94
α-SiO_2(quartz) + $2H_2O \rightleftharpoons H_4SiO_4^0$	−4.00
Alumino-silicates	
$Al_2Si_2O_5(OH)_4$(kaolinite) + $6H^+ \rightleftharpoons 2Al^{+3} + 2H_4SiO_4^0 + H_2O$	5.45
$Al_2Si_4O_{10}(OH)_2$(pyrophyllite) + $6H^+ + 4H_2O \rightleftharpoons 2Al^{+3} + 4H_4SiO_4^0$	−1.92
$NaAlSi_3O_8$(high albite[c]) + $4H^+ + 4H_2O \rightleftharpoons Na^+ + Al^{+3} + 3H_4SiO_4^0$	3.67
$NaAlSi_3O_8$(low albite[c]) + $4H^+ + 4H_2O \rightleftharpoons Na^+ + Al^{+3} + 3H_4SiO_4^0$	2.74
$KAl_2(AlSi_3O_{10})(OH)_2$(muscovite) + $10H^+ \rightleftharpoons K^+ + 3Al^{+3} + 3H_4SiO_4^0$	13.44
$K_{0.6}Mg_{0.25}Al_{2.3}Si_{3.5}O_{10}(OH)_2$(illite) + $8H^+ + 2H_2O \rightleftharpoons 0.6K^+ + 0.25Mg^{+2} + 2.3Al^{+3} + 3.5H_4SiO_4^0$	10.35
$Mg_{0.2}(Si_{3.81}Al_{1.71}Fe(III)_{0.22}Mg_{0.29})O_{10}(OH)_2$ (Mg-montmorillonite) + $6.76H^+ + 3.24H_2O \rightleftharpoons 0.49Mg^{+2} + 1.71Al^{+3} + 0.22Fe^{+3} + 3.81H_4SiO_4^0$	2.68

a. Equilibrium reactions for various silicate species at 25°C.
b. K^0 is the equilibrium constant in which all terms are expressed in activities; e.g., for kaolinite, $K^0 = (Al^{+3})^2(H_4SiO_4^0)^2/(H^+)^6$
c. High albite is a more soluble, less stable sodium alumino-silicate mineral than low albite.
Source: Lindsay [8]. *(Copyright 1979, John Wiley & Sons. Reprinted with permission.)*

SPECIES IN SOIL SOLUTION

Silicon is released rapidly from minerals into the soil solution. The concentration of Si in soil solutions and drainage waters is very dependent on several soil and climatic factors, especially pH. Silicon is usually more mobile in highly alkaline soils; at pH >9.5, dissolved concentrations rise sharply.

Since the solubility of soil SiO_2 is intermediate between that of quartz and amorphous silica (Table 6.6-1), it is used as a reference solubility level of silica in soils [8]. In the normal pH range of soils, $H_4SiO_4^0$ comprises the major silicate species in solution [8]. Concentrations range from 1 to about 200 mg soluble Si per liter of soil solution [6]. Only at pH values above 8.5 do the ionic silicates contribute significantly to total silica in solution.

various minerals. Prolonged weathering of primary minerals leads to the deposition of sedimentary minerals, such as the alumino-silicates kaolinite and montmorillonite [8]. Free SiO_2(s) may become depleted in highly weathered soils, leaving the sesquioxides of iron and aluminum as the major residual minerals. In such soils the solubility of $H_4SiO_4^0$ drops below that of quartz [8].

Because of the complexity of silicon chemistry, equilibrium conditions predicted from solubility relationships are not often encountered. Surface adsorption of silicic acid frequently prevents these equilibrium conditions from being reached [12].

Figure 6.6-2 shows the stability relationships of gibbsite and several alumino-silicate minerals that are fairly common and stable in soils. The stability lines of these minerals are plotted in terms of the activity of $H_4SiO_4^0$. Mg-montmorillonite is most stable in the presence of high $H_4SiO_4^0$ concentrations, which typically occur during the weathering of silica-rich primary minerals. As pH increases, Mg-montmorillonite becomes more stable at lower $H_4SiO_4^0$ activities. Pyrophyllite is only slightly less stable. At pH 7, if $H_4SiO_4^0$ drops below $10^{-3.7}$ *M*, kaolinite becomes more stable than Mg-montmorillonite. When $H_4SiO_4^0$ drops below $10^{-5.31}$ *M*, gibbsite becomes the most stable mineral. In the advanced stages of soil weathering, the alumino-silicates disappear, leaving mainly gibbsite and iron oxides [8].

In soils, amorphous silicates contribute to anion adsorption processes, and silicate and phosphate ions may compete for sites on mineral soil particles [6]. Displacement of silicate from soil colloids by phosphate has been reported. As phosphate fixation on soil colloids increases, so does the amount of silica in solution [1]. The resulting presence of monosilicic acid in solution increases the sorption of heavy metal cations, such as Co, Ni, and Zn, by clays [6]. (Sorption constants for various species on silica-type materials are given in the sections of this report that cover the individual species.)

In acid soils, silicate and phosphate ions form insoluble precipitates that may fix several other cations, such as Fe and Al oxides. Significant amounts of organic matter in flooded soil induce Si mobility, probably because monosilicic acid is released by the reduction of Fe hydrous oxides [6].

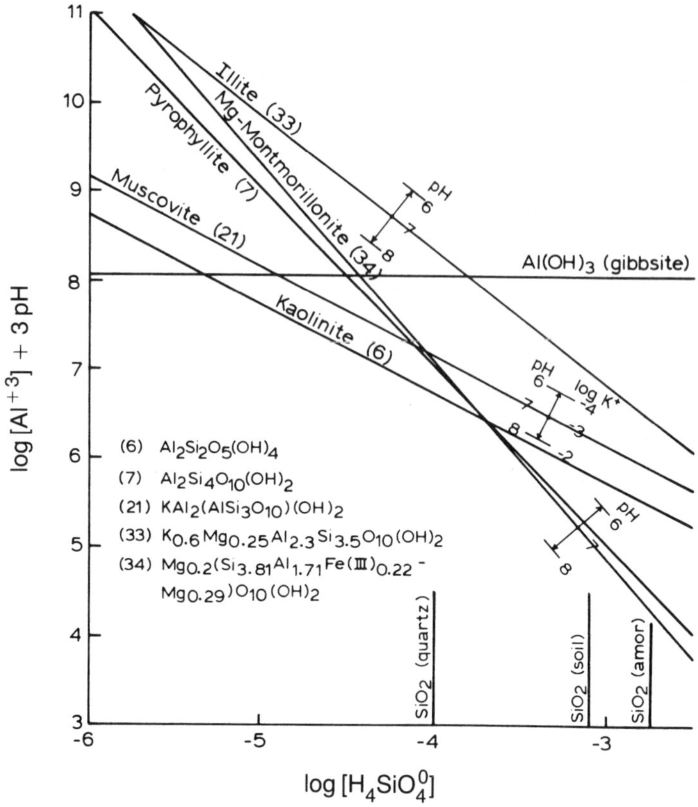

Source: Lindsay [8]. *(Copyright 1979, John Wiley & Sons. Reprinted with permission.)*

FIGURE 6.6-2 Summary Diagram of Several Mixed Alumino-silicates in Equilibrium with 10^{-3} *M* K^+, 10^{-3} *M* Mg^{+2}, and Soil-Fe with Indicated Changes for pH and $[K^+]$

6.6.5 Literature Cited

1. Bear, F.E. (ed.), *Chemistry of the Soil*, Reinhold Publishing Corp., New York (1964).
2. Bowen H.J.M., *Environmental Chemistry of the Elements*, Academic Press, London (1979).
3. Drever, J.I., *The Geochemistry of Natural Waters*, Prentice-Hall, Inc., Englewood Cliffs, N.J. (1982).
4. Hem, J.D., "Study and Interpretation of Chemical Characteristics of Natural Waters," Water Supply Paper No. 1473, 2nd ed., U.S. Geological Survey, Washington, D.C. (1970).
5. Horne, R.A. *Marine Chemistry: The Structure of Water and the Chemistry of the Hydrosphere*, John Wiley & Sons, New York (1969).

6. Kabata-Pendias, A. and H. Pendias, *Trace Elements in Soils and Plants*, CRC Press, Inc., Boca Raton, Fla. (1984).
7. Kennedy, V.C., "Silica Variation in Stream Water with Time and Discharge," in Hem, J.D. (chairman), *Nonequilibrium Systems in Natural Water Chemistry*, American Chemical Society, Washington, D.C. (1971).
8. Lindsay, W.L., *Chemical Equilibria in Soils*, John Wiley & Sons, New York (1979).
9. Morel, F.M.M., *Principles of Aquatic Chemistry*, John Wiley & Sons, New York (1983).
10. Polzer, W.L., "Geochemical Control of Solubility of Aqueous Silica," in Faust, S.D. and J.V. Hunter (eds.), *Principles and Applications of Water Chemistry*, Proc. 4th Rudolfs Conference, John Wiley & Sons, New York (1967).
11. Sillén, L.G., "The Physical Chemistry of Sea Water," in M. Sears (ed.), *Oceanography,* Publication No. 67, American Association for the Advancement of Science (1961), cited in Horne [5].
12. Strober, W., "Formation of Silicic Acid in Aqueous Suspensions of Different Silica Modifications," in *Equilibrium Concepts in Natural Water Systems*, R.F. Gould (ed.), *Adv. Chem. Ser.*, **67**, 161-82 (1967), as cited in Lindsay [8].
13. Stumm, W. and J.J. Morgan, *Aquatic Chemistry*, 2nd ed., John Wiley & Sons, New York (1981).

6.7 ALUMINUM (Al)

6.7.1 Production and Uses

Aluminum is the most widely used nonferrous metal because of its many important features — low density, high electrical and thermal conductivity, high reflectivity, and corrosion resistance. The pure metal is soft and weak and is typically combined with other metals. Alloying and heat treatment can yield products that are light, strong, and readily formable.

Frequent applications are for sheet, castings, and wrought materials, especially when weight and corrosion resistance are important. The major industry categories in which aluminum is used are transportation, construction, electrical, containers and packaging, consumer durables, and mechanical equipment.

Bauxite, an impure hydrated oxide ore, is the main source of aluminum. The bauxite is purified and hydrolyzed to produce alumina (aluminum oxide, Al_2O_3), which is then electrolytically reduced to aluminum metal.

6.7.2 Occurrence

Aluminum is the most abundant metallic element in the earth's crust (8.1 wt %). However, because it is very reactive, the free metal is not found in nature. A group III element, aluminum occurs only in the trivalent state. The aluminum ion is capable of fourfold coordination with oxygen and forms many of the same kinds of compounds as silicon.

Aluminum occurs in many silicate rock minerals, such as the feldspars, feldspathoids, micas, and many amphiboles. Concentrations in rocks commonly range from 0.45% to 10% [9].

Table 6.7-1 shows the dissolution reactions of selected aluminum-bearing minerals. The equilibrium constants shown can be used to establish stability relations among the various minerals. Prolonged weathering of primary minerals leads to the deposition of sedimentary clay minerals, such as the alumino-silicates kaolinite and montmorillonite. As soils weather, silicon is lost more rapidly then aluminum, and the aluminum precipitates as hydrated aluminum oxides, such as gibbsite and boehmite. These are constituents of bauxites and laterites. In sulfate-rich environments, aluminum forms sulfate minerals such as alunite [$KAl_3(SO_4)_2(OH)_6$] and basaluminite [$Al_4(OH)_{10}SO_4 \cdot 5H_2O$].

The activity of Al^{+3} in soil and groundwaters is largely controlled by precipitation/dissolution reactions and can be estimated from pH, total Al, and activities of F^- and SO_4^{-2}. Gibbsite [$Al(OH)_3$] is a common control on solubility of aluminum minerals in many geochemical environments; at lower pH values and higher SO_4^{-2} concentrations, alunite and basaluminite are important solubility controls. Ion exchange is an important retention mechanism for aluminum in acid-to-neutral pH regimes [6].

TABLE 6.7-1

Selected Equilibrium Reactions and Activity Constants of Aluminum Minerals[a]

Reaction	Log K^0
Oxides and Hydroxides	
$Al(OH)_3$(amorp) + $3H^+ \rightleftharpoons Al^{+3} + 3H_2O$	9.66
α-$Al(OH)_3$(bayerite) + $3H^+ \rightleftharpoons Al^{+3} + 3H_2O$	8.51
γ-AlOOH(boehmite) + $3H^+ \rightleftharpoons Al^{+3} + 2H_2O$	8.13
$Al(OH)_3$(norstrandite) + $3H^+ \rightleftharpoons Al^{+3} + 3H_2O$	8.13
γ-$Al(OH)_3$(gibbsite) + $3H^+ \rightleftharpoons Al^{+3} + 3H_2O$	8.04
α-AlOOH(diaspore) + $3H^+ \rightleftharpoons Al^{+3} + 2H_2O$	7.92
Sulfates	
$KAl_3(SO_4)_2(OH)_6$(alunite) + $6H^+ \rightleftharpoons K^+ + 3Al^{+3} + 2SO_4^{-2} + 6H_2O$	3.04
Alumino-silicates — See Table 6.6-1	
Aluminum Phosphates	
$AlPO_4$(berlinite) + $2H^+ \rightleftharpoons Al^{+3} + H_2PO_4^-$	0.50
$AlPO_4 \cdot 2H_2O$(variscite) + $2H^+ \rightleftharpoons Al^{+3} + H_2PO_4^- + 2H_2O$	−2.50
$H_6K_3Al_5(PO_4)_8 \cdot 18H_2O$(K-taranakite) + $10H^+ \rightleftharpoons 3K^+ + 5Al^{+3} + 8H_2PO_4^- + 18H_2O$	−22.30
$H_6(NH_4)_3Al_5(PO_4)_8 \cdot 18H_2O$($NH_4$-taranakite) + $10H^+ \rightleftharpoons 3NH_4^+ + 5Al^{+3} + 8H_2PO_4^- + 18H_2O$	−19.10

a. Equilibrium reactions at 25°C

Source: Lindsay [11]. *(Copyright 1979, John Wiley & Sons. Reprinted with permission.)*

6.7.3 Properties and Speciation in Water

The concentration of aluminum in natural waters depends primarily on pH and on the presence of complexing agents. Although it is an abundant element, aluminum rarely occurs in natural waters in concentrations greater than a few tenths of a milligram per liter, as long as the pH is nearly neutral and concentrations of strong complexing agents are low [12]. The typical concentration in seawater has been reported as 0.01 mg/l, probably present as $Al(OH)_4^-$, AlF_2^+, and AlF^{+2} [7]. In U.S. surface waters, the mean concentrations of suspended and dissolved aluminum have been measured at 3860 μg/l and 74 μg/l, respectively. The suspended form was observed in 97% of the samples, while dissolved aluminum was detected in only 31% of the samples [14].

The concentration of aluminum in river waters can vary significantly with flow. It readily precipitates in natural waters to form particulate or colloidal hydroxides; however, under the turbulent conditions of high flow, it can be dissolved from suspended minerals, perhaps aided by complex formation [14].

Aluminum is readily soluble at pH <4. High concentrations have been found in acid mine waters, acid sulfate soil water, acid geothermal waters, and poorly buffered lakes, streams, and groundwaters that receive large inputs of acid runoff [13].

In aqueous solutions, Al^{+3} does not occur as the free ion but is surrounded by six molecules of water to form $Al(H_2O)_6^{+3}$. As the pH increases, protons are removed from the coordinated waters to give the hydrolysis products shown in Table 6.7-2. Progressive hydrolysis leads to the formation of colloidal aluminum hydroxide.

TABLE 6.7-2

Hydrolysis Reactions and Activity Constants of Aluminum[a]

Reaction[b]	Log K^0
$Al^{+3} + H_2O \rightleftharpoons AlOH^{+2} + H^+$	−5.02
$Al^{+3} + 2H_2O \rightleftharpoons Al(OH)_2^+ + 2H^+$	−9.30
$Al^{+3} + 3H_2O \rightleftharpoons Al(OH)_3^0 + 3H^+$	−14.99
$Al^{+3} + 4H_2O \rightleftharpoons Al(OH)_4^- + 4H^+$	−23.33
$Al^{+3} + 5H_2O \rightleftharpoons Al(OH)_5^{-2} + 5H^+$	−34.24
$2Al^{+3} + 2H_2O \rightleftharpoons Al_2(OH)_2^{+4} + 2H^+$	−7.69

a. Equilibrium reactions at 25°C.

b. Hydrolysis species are written without the water of hydration even though the water is present. For example, the first reaction could also be written as $Al(H_2O)_6^{+3} \rightleftharpoons Al(H_2O)_5OH^{+2} + H^+$

Source: Lindsay [11]. *(Copyright 1979, John Wiley & Sons. Reprinted with permission.)*

The most important forms of dissolved aluminum are Al^{+3}, $Al(OH)_2^+$, and $Al(OH)_4^-$. Each species predominates over a certain pH range. The cation Al^{+3} predominates in many solutions where pH <4.0. At pH values around 5-6, $Al(OH)_2^+$ predominates. Around pH 6.0, aluminum solubility reaches a minimum. Above neutral pH, the predominant dissolved form of aluminum is the anion $Al(OH)_4^-$. Figure 6.7-1 shows the total activity of dissolved aluminum in equilibrium with gibbsite [4]. (The total activity of dissolved aluminum is the sum of the activities of the three dissolved species.)

Aluminum also shows a strong tendency to form polymers. Ultimately, the neutral mineral gibbsite is precipitated [13]. The polymerization of aluminum hydroxide species is affected by the presence of dissolved silica. When enough silica is present, the aluminum is rapidly precipitated as poorly crystallized clay mineral species.

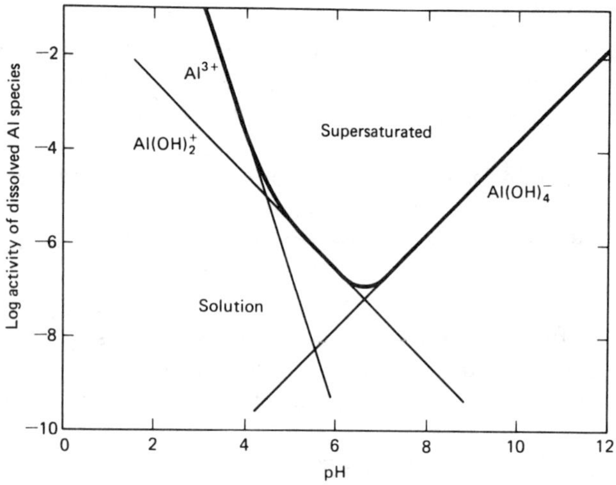

Note: Heavy line is sum of individual activities.

Source: Drever [4]. *(Copyright 1982, Prentice-Hall, Inc. Reprinted with permission.)*

FIGURE 6.7-1 Activities of Dissolved Aluminum Species in Equilibrium with Gibbsite ($Al(OH)_3$) at 25°C

Certain anions (e.g., SO_4^{-2}, F^-) form strong aqueous complexes with aluminum that can dominate solution speciation at higher ligand concentrations [6]. In the presence of fluoride, strong complexes of aluminum and fluoride are formed; AlF_2^+ and AlF_2^+ are most likely in natural water containing fluoride in concentrations ranging from a few tenths mg/l to a few mg/l. Soluble phosphate complexes of aluminum have been reported, and the sulfate complex $AlSO_4^+$ may predominate in acid solutions in the presence of elevated sulfate levels [7]. Table 6.7-3 shows equilibrium constants for selected aluminum complexes.[1]

The influence of phosphate on aluminum is not clear. Polymeric aluminum orthophosphates are known [3]. Again, the stability and degree of aggregation depend primarily on pH. The most stable compound is probably $AlPO_4$, which has a solubility product of 6.3×10^{-19} [13].

When various ligands compete for aluminum, the distribution of aluminum species depends mainly on the concentrations of the ligands and on pH (OH^- is a ligand as well). For instance, in a solution containing fluoride in excess over aluminum, almost all the aluminum would be associated with fluoride ions below neutral pH. In an acidic environment, fluoride is the most important ligand for aluminum (except in waters with high concentrations of sulfate ion). Hydroxy complexes dominate above

1. Equilibrium constants for several other aluminum complexes are presented in section 2.9.

TABLE 6.7-3

Equilibrium Reactions and Activity Constants for Selected Aluminum Complexes

Reaction	Log K^0 (25°C)
$Al^{+3} + F^- \rightleftharpoons AlF^{+2}$	6.98
$Al^{+3} + 2F^- \rightleftharpoons AlF_2^+$	12.60
$Al^{+3} + 3F^- \rightleftharpoons AlF_3^0$	16.65
$Al^{+3} + 4F^- \rightleftharpoons AlF_4^-$	19.03
$Al^{+3} + SO_4^{-2} \rightleftharpoons AlSO_4^+$	3.20
$Al^{+3} + 2SO_4^{-2} \rightleftharpoons Al(SO_4)_2^-$	1.90

Source: Lindsay [11]. *(Copyright 1979, John Wiley & Sons. Reprinted with permission.)*

neutral pH. The complexing power of some natural organic compounds is comparable to that of strongly complexing inorganic ligands such as fluoride; under suitable conditions, organic compounds may increase aluminum solubility by 100-fold [10].

The presence of silica at activities above 10^{-4} decreases the equilibrium activities of dissolved aluminum at all pH values. Different silica activities influence the solubility lines for gibbsite, kaolinite, and pyrophyllite (see, for example, Figure 6.6-2). At any particular silica activity, the most stable aluminum silicate is generally that with the lowest equilibrium concentration of dissolved aluminum [4].

6.7.4 Properties and Speciation at the Solid/Solution Interface

SOILS

The median content of aluminum in soils has been reported as 71,000 mg/kg soil, with a range of 10,000 to 300,000 mg/kg [2]. Many solid aluminum compounds have been identified in soils; the most stable of these typically include the hydroxide (gibbsite) and the clay minerals kaolinite and montmorillonite. At pH <4 and sulfate activities greater than 10^{-3}, alumino-sulfate minerals such as alunite and basaluminite are more stable than gibbsite and form in soils and sediments [11].

The first section of Table 6.7-1 shows the solubilities of aluminum oxides and hydroxides. The most soluble form of aluminum hydroxide expected in soils is amorphous $Al(OH)_3$. The other hydroxides have slightly lower solubilities. The activity of Al^{+3} in equilibrium with any of these minerals depends on pH and decreases 1000-fold for each unit increase in pH [11].

Other important forms of aluminum in soil include the fluoride, sulfate, and phosphate species. Fluoride complexes (Table 6.7-3) can be extremely important in soils: for example, they increase total soluble aluminum by approximately 60-fold in a soil at pH 6 in equilibrium with gibbsite [11]. Alunite (Table 6.7-1) has been reported in acid sulfate soils. Aluminum also forms insoluble phosphate minerals in soils. (See final section of Table 6.7-1.)

Because gibbsite is one of the most stable hydroxides and occurs in many soils and sediments, it is generally used to estimate Al^{+3} activities. However, many alumino-silicates can depress the activity of Al^{+3} below that of gibbsite. For example, when the activity of $H_4SiO_4{}^0$ is greater than $10^{-5.31}$, this species controls Al^{+3} activity by the formation of kaolinite and/or Mg-montmorillonite, which depresses the activity of Al^{+3} considerably below that of gibbsite [11].

The stability relationships of gibbsite [$Al(OH)_3$] and several alumino-silicate minerals that are fairly common and stable in soils was shown previously in Figure 6.6-2. The stability lines of the minerals are plotted in terms of the activity of $H_4SiO_4^0$. Mg-montmorillonite is most stable under high $H_4SiO_4^0$ concentrations, which typically occur during the weathering of silica-rich primary minerals. Increasing pH makes this mineral more stable. Pyrophyllite is only slightly less stable than Mg-montmorillonite. At pH 7, if $H_4SiO_4^0$ drops below $10^{-3.7}$ *M*, kaolinite becomes more stable than Mg-montmorillonite. When $H_4SiO_4^0$ drops below $10^{-5.31}$ *M*, gibbsite becomes the stable mineral. In the advanced stages of soil weathering, soluble silica is largely depleted and the alumino-silicates disappear, leaving mainly gibbsite and iron oxides [11].

The adsorption of aluminum by soils and soil clays, which is affected by soil acidity, determines the toxicity of this element in acid soils. For example, organic matter present in soils lowers the pH of the soil solution below neutral, making elements such as aluminum more soluble. Fulvic and humic acids act as strong chelating agents for aluminum. Table 6.7-4 gives some Langmuir adsorption constants for aluminum on clay minerals and silica.

The principal determinant of soil acidity appears to be the relative amounts of exchangeable and nonexchangeable Al on soil cation exchange sites, mainly those associated with layer lattice silicates. The buildup of exchangeable Al leads to soil acidity. Below pH 5.5, Al becomes much more mobile and actively competes with other cations for exchange sites. Solutions of neutral soils contain about 400 μg/l Al, but concentrations as high as 5,700 μg/l have been reported in acid soil solutions (pH 4.4) [9].

SEDIMENTS

The mean elemental composition of major sediments ranges from 9,000 (in limestone sediments) to 94,000 (marine clays) mg Al per kg sediment [2]. Estuaries have been observed to trap elements such as aluminum, which is usually present in greater concentrations in river water than in sea water. Sediments near the mouths of rivers contain large amounts of aluminum [2].

TABLE 6.7-4

Langmuir Adsorption Constants[a] for Aluminum

Adsorbent	Analyte Conc. (*M*)	pH	A_m	K_L
Clay Minerals				
Montmorillonite				
Untreated	0-0.002	4.0	660-720[b]	5.8-6.1[b]
Ca-saturated	0-0.002	4.0	890	4.7
Silica	$10^{-6.7} - 10^{-5.4}$	5.00	100	4.6
		5.25	560	4.8

a. Explanation of symbols for Langmuir equation (see section 2.12):

A_m = Langmuir adsorption maximum ($\mu mol \cdot g^{-1}$)
K_L = Langmuir constant (log M^{-1})

b. Estimated values

Source: EPRI [6]

Clays, the most common sedimentary aluminum-bearing minerals, typically consist of alternating layers of silicon and aluminum. Representative clay minerals in ocean sediments are kaolinite in the tropics, illite in temperate latitudes, and montmorillonite in regions of volcanic activity. Both illite and montmorillonite strongly incorporate potassium into their lattices and may be the main sinks for potassium in the ocean. As colloids, they also readily bind ions having multiple charges [2].

6.7.5 Toxicity and the Effects of Acid Precipitation

The toxicity of aluminum depends on its solubility and speciation, both of which are controlled primarily by the pH of the environment. The toxicity of aluminum in acid environments has become a topic of increasing concern. The element is an important buffer in soils and waters, but the elevated concentrations that accompany decreases in pH have been found to have toxic effects. Acid precipitation exacerbates the problem, because it increases the mobility of aluminum.

Aluminum levels in soil solution affected by acid precipitation have been shown to be orders of magnitude higher than the levels in a soil solution whose chemistry is dominated by carbonic acid. The higher levels of aluminum are caused by the lower pH value of acid precipitation and by complexes formed with inorganic and organic ligands. Percolation of acid precipitation through the soil tends to dissolve the least stable soil minerals and raise the levels of aluminum significantly in the subsurface runoff, which ultimately finds its way into the channel system of a watershed [15].

At pH values of 4.0-5.2, aluminum can be acutely toxic to fish at concentrations as low as 0.1 mg/l [5]. The toxic effects peak at pH 5.0-5.2 [1]. Complexing agents are important, because they affect the concentration and availability of aluminum to fish. Calcium interferes antagonistically with aluminum toxicity. Survival times in aluminum solutions at low pH have been shown to increase when complexing agents such as fluoride and citrate were added. As organic ligands, such as humic acids, also form complexes with aluminum, some rivers of low pH with a high organic load may be able to support healthy fish populations at abnormally high aluminum concentrations [13].

Seasonal variations in aluminum concentrations have been observed. These patterns are important, because fish species are at sensitive life stages only at certain times of the year. Major short-term increases in aluminum concentrations have been observed during periods of acidic runoff, often after snowmelt; however, a recent study found the highest concentrations of aluminum during the midwinter thaw, not during the spring melt, which could result from the depletion of a labile pool of aluminum in soil [8]. These major short-term increases in the concentration of aluminum, combined with a reduced pH level, produce conditions acutely toxic to fish.

Aluminum toxicity is also often reported for plants grown in acid soils. A high availability of aluminum in acid soils is a limiting factor in the production of most field crops [9].

6.7.6 Literature Cited

1. Baker, J.P. and C.L. Schofield, "Aluminum Toxicity to Fish in Acidic Waters," *Water Air Soil Pollut.*, **18**, 289-309 (1982), as cited in Odonnell *et al.* [13].
2. Bowen, H.J.M., *Environmental Chemistry of the Elements*, Academic Press, London (1979).
3. Callis, C.F., J.R. van Wazer and P.G. Arvan, "The Inorganic Phosphates as Polyelectrolytes," *Chem. Rev.*, **54**, 777 (1954), as cited in Odonnell *et al.* [13].
4. Drever, J.I., *The Geochemistry of Natural Waters*, Prentice-Hall, Inc., Englewood Cliffs, N.J. (1982).
5. Driscoll, C.T., J.P. Baker, J.J. Bisogni and C.L. Schofield, "Effect of Aluminum Speciation on Fish in Dilute Acidified Waters," *Nature*, **284**, 161-64 (1980), as cited in Odonnell *et al.* [13].
6. Electric Power Research Institute, *Chemical Attenuation Rates, Coefficients, and Constants in Leachate Migration. Volume I: A Critical Review*, EPRI EA-3356 (February 1984).
7. Hem, J.D., *Study and Interpretation of the Chemical Characteristics of Natural Water*, 2nd ed., Water Supply Paper No. 1473, U.S. Geological Survey, Washington, D.C. (1970).
8. Hooper, R.P. and C.A. Shoemaker, "Aluminum Mobilization in an Acidic Headwater Stream: Temporal Variation and Mineral Dissolution Disequilibria," *Science*, **229**, 463-65 (2 August 1985).

9. Kabata-Pendias, A. and H. Pendias, *Trace Elements in Soils and Plants*, CRC Press, Inc., Boca Raton, Fla. (1984).
10. Lind, C.J. and J.D. Hem, *Effect of Organic Solutes on Chemical Reactions of Aluminum*, Geological Survey Water-Supply Paper 1927-G, U.S. Government Printing Office, Washington, D.C. (1975), as cited in Odonnell *et al.* [13].
11. Lindsay, W.L., *Chemical Equilibria in Soils*, John Wiley & Sons, New York (1979).
12. Nordstrom, D.K., "The Effect of Sulphate on Aluminum Concentrations in Natural Waters: Some Stability Relations in the System $Al_2O_3 - SO_3 - H_2O$ at 298 K," *Geochim. Cosmochim. Acta*, **46**, 681-92 (1982), as cited in Odonnell *et al.* [13].
13. Odonnell, A.R., G. Mance and R. Norton, *A Review of the Toxicity of Aluminum in Fresh Water*, Technical Report TR 197, Water Research Centre, Great Britain (July 1984).
14. Singer, P.C., *Trace Metals and Metal-Organic Interactions in Natural Waters*, Ann Arbor Science Publishers, Ann Arbor, Mich. (1974).
15. University of California, Riverside, *Effects of Acid Precipitation on Soil Leachate Quality*, EPA-600/3-80-015 (January 1980).

6.8 IRON (Fe)

6.8.1 Production and Uses

The iron oxide minerals magnetite, hematite, and goethite are important components of iron ore, which is used primarily to produce commercial-grade iron and steel. The most widely used metallic materials in the world today, iron and steel are ubiquitous in transportation, construction, and machinery as structural members, coverings and containers, and mechanical elements.

6.8.2 Occurrence

Iron accounts for 5% (by weight) of the earth's crust. In nature, it exists in the Fe(II) (ferrous) and Fe(III) (ferric) oxidation states. Igneous rock contains 4.6% (weight) iron, typically in the ferrous form, while sedimentary rock contains 4.4% [13].

The reactions of iron in weathering processes depend on the Eh-pH system of the environment and on the oxidation state of its compounds. Low pH and redox potential levels favor the ferrous form; high pH and/or redox levels create a stable environment for ferric iron. The electrode potential for the Fe(II)/Fe(III) couple is -0.771 V.

Table 6.8-1 shows the dissolution reactions of selected iron-bearing minerals. The listed equilibrium constants can be used to establish stability relations among the various minerals in the environment [13]. Iron released from primary minerals precipitates as Fe(III) oxides and hydroxides [11]. These compounds are important iron-bearing minerals in natural waters and soils. The hydroxide complexes of Fe(III) account for most of the soluble iron concentration throughout the pH range of natural waters [13]. Table 6.8-2 presents solubility product constants for selected iron compounds.

Significant amounts of iron are also released into the environment as a result of coal mining. During these operations, pyrite (FeS_2), found in coal seams, is exposed to air and water and is oxidized to sulfate by bacterial and chemical reactions. This process releases dissolved ferrous iron and acidity into the mine drainage. The ferrous iron then reacts with oxygen to form ferric iron, which hydrolyzes to form insoluble ferric hydroxide ($Fe(OH)_3$). The precipitated ferric hydroxide deposited in the mine and streams acts as a reservoir for soluble ferric iron which then reacts with pyrite to generate soluble ferrous iron [14]. (See section 6.9.3 for further discussion.)

6.8.3 Properties and Speciation in Water

In aqueous solution, ferric and ferrous iron hydrolyze to give various hydrolysis species (Table 6.8-3). The reactions are highly dependent on pH. Figure 6.8-1 shows how the hydrolysis species of Fe(III) are distributed as a function of pH.

TABLE 6.8-1

Selected Equilibrium Reactions and Activity Constants for Ferric and Ferrous Minerals

Reaction	Log K^0 (25°C)
Fe(III) Oxides and Hydroxides	
$Fe(OH)_3$ (amorph) + $3H^+ \rightleftharpoons Fe^{+3} + 3H_2O$	3.54
$Fe(OH)_3$ (soil) + $3H^+ \rightleftharpoons Fe^{+3} + 3H_2O$	2.70
$1/2\ \gamma\text{-}Fe_2O_3$ (maghemite) + $3H^+ \rightleftharpoons Fe^{+3} + 3/2\ H_2O$	1.59
γ-FeOOH (lepidocrocite) + $3H^+ \rightleftharpoons Fe^{+3} + 2H_2O$	1.39
$1/2\ \alpha\text{-}Fe_2O_3$ (hematite) + $3H^+ \rightleftharpoons Fe^{+3} + 3/2\ H_2O$	0.09
α-FeOOH (goethite) + $3H^+ \rightleftharpoons Fe^{+3} + 2H_2O$	−0.02
Other Fe(III) Minerals	
$FeCl_3$ (molysite) $\rightleftharpoons Fe^{+3} + 3Cl^-$	13.25
$Fe_2(SO_4)_3 \rightleftharpoons 2Fe^{+3} + 3SO_4^{-2}$	2.89
$KFe_3(SO_4)_2(OH)_6$ (jarosite) + $6H^+ \rightleftharpoons K^+ + 3Fe^{+3} + 2SO_4^{-2} + 6H_2O$	−12.51
Fe(II) Minerals	
$FeO + 2H^+ \rightleftharpoons Fe^{+2} + H_2O$	13.48
$Fe(OH)_2 + 2H^+ \rightleftharpoons Fe^{+2} + 2H_2O$	12.90
$FeCl_2$ (laurencite) $\rightleftharpoons Fe^{+2} + 2Cl^-$	9.00
$FeCO_3$ (siderite) + $2H^+ \rightleftharpoons Fe^{+2} + CO_2(g) + H_2O$	7.92
$FeSO_4 \rightleftharpoons Fe^{+2} + SO_4^{-2}$	2.65
Redox Reactions	
Fe_3O_4 (magnetite) + $8H^+ + 2e^- \rightleftharpoons 3Fe^{+2} + 4H_2O$	35.69
Fe_3O_4 (magnetite) + $8H^+ \rightleftharpoons 3Fe^{+3} + e^- + 4H_2O$	−3.42
$Fe_3(OH)_8$ (ferrosic oxide) + $8H^+ + 2e^- \rightleftharpoons 3Fe^{+2} + 8H_2O$	43.75

Source: Lindsay [11]. *(Copyright 1979, John Wiley & Sons. Reprinted with permission.)*

In natural waters, iron can be present as several species, forming complexes with both inorganic and organic materials. With organic substances, ferric iron forms stronger complexes than does ferrous iron and also forms complexes with chloride, fluoride, sulfate, and phosphate.[1]

In natural streams, iron occurs as particulate ferric hydroxide or as some form of organic complex. In lakes and rivers, water near the bottom may be depleted of oxygen and attain a low Eh; such waters may contain dissolved ferrous (reduced) iron concentrations up to several mg/l. Typical total iron concentrations in groundwater range from 1 to 10 mg/l. Ferrous iron is the most common form of dissolved iron in

1. See section 2.9 for equilibrium constants for selected iron complexes.

TABLE 6.8-2

Solubility Product Constants for Selected Solid Iron Compounds

Compound	pK_{sp} (25°C)
$FeCO_3$	10.68
$Fe(OH)_2$	15.1
FeS	18.1
$Fe_3(PO_4)_2$	26.4
$Fe(OH)_3$	38.8
α-FeOOH	41.5
FeOOH (amorph)	39.5

Source: Kotrlý and Šůcha [10]. *(Copyright 1985, John Wiley & Sons. Reprinted with permission.)*

TABLE 6.8-3

Hydrolysis Reactions and Equilibrium Activity Constants of Iron

Reaction	Log K^0 (25°C)
Fe(III) Hydrolysis	
$Fe^{+3} + H_2O \rightleftharpoons FeOH^{+2} + H^+$	−2.19
$Fe^{+3} + 2H_2O \rightleftharpoons Fe(OH)_2^{+} + 2H^+$	−5.69
$Fe^{+3} + 3H_2O \rightleftharpoons Fe(OH)_3^{0} + 3H^+$	−13.09
$Fe^{+3} + 4H_2O \rightleftharpoons Fe(OH)_4^{-} + 4H^+$	−21.59
$2Fe^{+3} + 2H_2O \rightleftharpoons Fe_2(OH)_2^{+4} + 2H^+$	−2.90
Fe(II) Hydrolysis	
$Fe^{+2} + H_2O \rightleftharpoons FeOH^{+} + H^+$	−6.74
$Fe^{+2} + 2H_2O \rightleftharpoons Fe(OH)_2^{0} + 2H^+$	−16.04
$Fe^{+2} + 3H_2O \rightleftharpoons Fe(OH)_3^{-} + 3H^+$	−31.99
$Fe^{+2} + 4H_2O \rightleftharpoons Fe(OH)_4^{-2} + 4H^+$	−46.38
$3Fe^{+2} + 4H_2O \rightleftharpoons Fe_3(OH)_4^{+2} + 4H^+$	−45.39

Source: Lindsay [11]. *(Copyright 1979, John Wiley & Sons. Reprinted with permission.)*

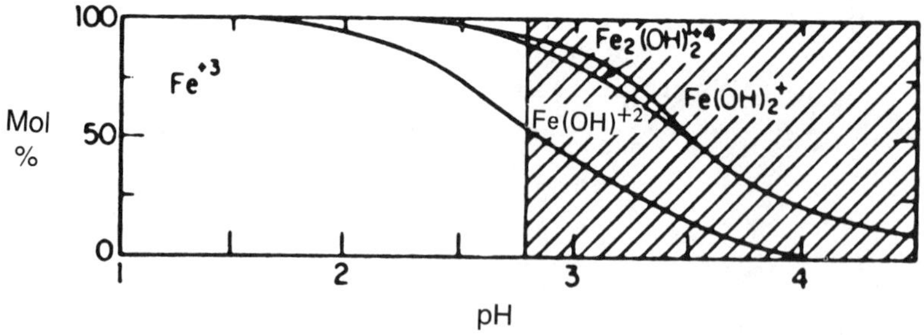

$[Fe(III)_{total}] = 10^{-4}$ *M*. Shaded area indicates conditions under which solution is oversaturated with respect to $Fe(OH)_3(s)$ ($K_{sp} = 10^{-38}$).

Source: Stumm and Morgan [14]. *(Copyright 1981, John Wiley & Sons. Reprinted with permission.)*

FIGURE 6.8-1 Distribution of Hydrolysis Species of Fe(III) as a Function of pH

groundwater. Colloidal ferric hydroxide is typically present in surface waters. Water of very low pH, which may result from industrial waste discharge and mine drainage, can contain high concentrations of ferric and ferrous iron. Brines may contain more than 1,000 mg/l of ferrous iron [5].

Iron is an important minor constituent of seawater, where its average concentration is 0.01 mg/l and it occurs primarily in the Fe(III) oxidation state [6]. Since dissolved iron occurs at low levels, the speciation is highly dependent on solution composition.

The main variables affecting the solubility of iron include pH, redox potential (Eh), and concentrations of the dissolved carbon dioxide and sulfur species. Figure 6.8-2 is an Eh-pH diagram showing the stability conditions of the solid phases $Fe(OH)_3$, $FeCO_3$, FeS_2 and FeS. Pyrite (FeS_2) is stable under strongly reducing conditions over a wide pH range. In moderate oxidizing conditions above pH 5, ferric hydroxide forms. Between these two regions iron is fairly soluble, especially at low pH. This diagram indicates that, under strongly alkaline and oxidizing conditions, Fe(VI) in the form of the ferrate ion (FeO_4^{-2}) is stable. However, this species is unlikely to be found in natural environmental systems due to its strong oxidizing ability [1].

When pyrite is exposed to oxygenated water or ferric hydroxide is in contact with reducing materials, iron tends to go into solution [5]. Formation of solid ferrosic hydroxide, $Fe_3(OH)_8$, may control Fe^{+2} levels in groundwaters [3]. The relationships shown in Figure 6.8-2 are further illustrated by an example of the freshwater chemistry of iron, described by Horne [6] for a small coastal pond (Figure 6.8-3). Suspended iron sulfides are carried upward by diffusion or other mixing processes and are oxidized when they reach the oxygen-containing waters. Subsequent hydrolysis of Fe(III) forms amorphous ferric hydroxides, which may be colloidal and remain sus-

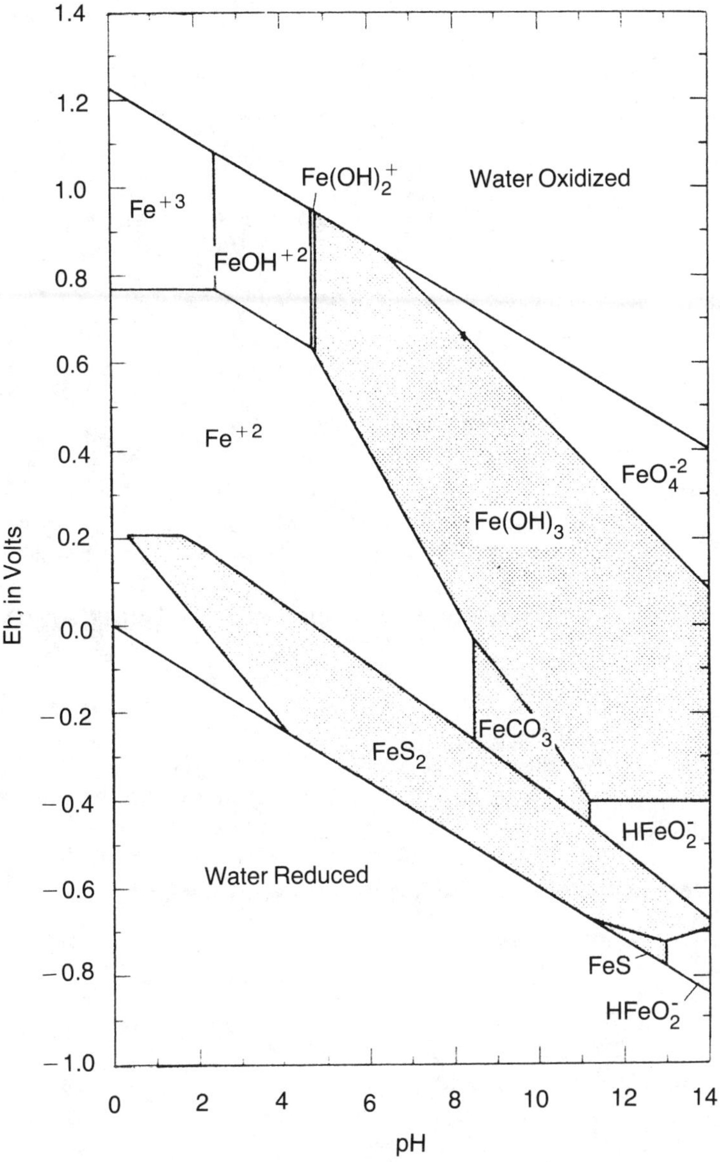

Shaded areas are stability regions for solids; predominant dissolved species are shown in unshaded areas. Activity of sulfur species = 96 mg/l as SO_4^{-2}, CO_2 species = 1,000 mg/l as HCO_3^-, and total dissolved iron activity = 0.0056 mg/l.

Source: Hem [5]

FIGURE 6.8-2 Fields of Stability for Solid and Dissolved Forms of Iron as a Function of Eh and pH at 25°C and 1 Atmosphere

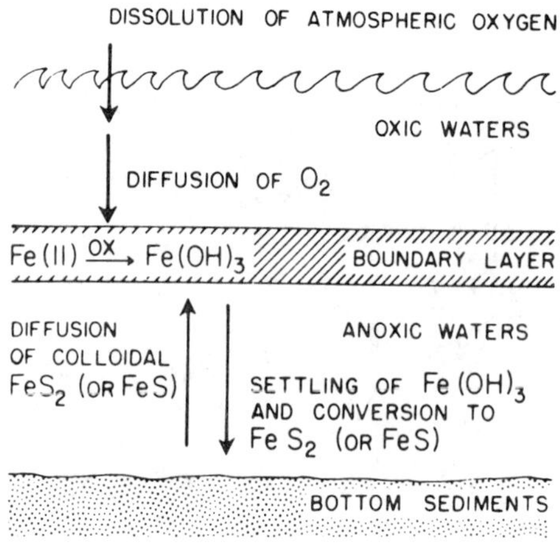

Source: Horne [6]. *(Copyright 1978, John Wiley & Sons. Reprinted with permission.)*

FIGURE 6.8-3 Iron Chemistry and Transport in Oyster Pond, Cape Cod

pended at the boundary layer between the reduced water and the surface oxygenated water. With further coagulation, the larger particles settle into the anoxic waters, are converted to sulfides, and return to the bottom sediments.

6.8.4 Properties and Speciation at the Solid/Solution Interface

SOILS

The observed iron content of soils is the result of chemical and physical processes that have acted upon the iron originally present in rocks. The most common range of iron in soils is 0.5-5% [7]; the average content of iron in soils is estimated at 3.8% [11].

Iron occurs in soils mainly as Fe(III) oxides and hydroxides in the form of small, discrete grains of oxide mineral or amorphous coatings on other minerals. Goethite is the predominant mineral form [7]. Jarosite, another Fe(III) mineral, is often found in acid sulfate soils containing pyrite (FeS_2) [11]. The Fe(II) minerals are generally too soluble to persist in normal aerobic soils. As they dissolve, they release Fe(II), which is oxidized and precipitates as Fe(III) oxides [12].

The formation of iron oxides is affected by organic matter and bacteria: iron forms metal chelates with organic matter, and some bacteria are involved in iron cycling, as explained elsewhere in this report. In fact, the distribution of extractable iron in soil largely depends on the presence of organic matter. Soil humic acid strongly adsorbs or complexes with iron at pH >3 [8].

Other soil constituents that adsorb Fe(III) are crystalline oxides, clay minerals, and hydrous oxides of iron and manganese. Table 6.8-4 gives some adsorption constants for iron on clay minerals, silica, and organic matter. (See section 2.12 for the equations in which the listed parameters are used.)

TABLE 6.8-4

Exemplar Adsorption Constants for Iron

Adsorbent	Analyte Conc.(*M*)	pH	A_m [a]	K_L [b]	A [c]
Clay Minerals					
Kaolinite	$10^{-5.5} - 10^{-5}$	3	4.3	5.5	—
Ca-saturated	$10^{-5.5} - 10^{-5}$	4	14.9	6.0	—
Na form	0 – 0.015	3	1430	2.6	—
Bentonite, Na form	0 – 0.015	3	6180	2.8	—
Illite, Na form	0 – 0.015	3	4920	2.9	—
Montmorillonite	$10^{-2.5}$	>0.5	—	—	220
Na form	$10^{-2.5}$	>1.0	—	—	260
		>1.6			280
		>2.1			280
H form	$10^{-2.5} - 10^{-2.2}$	—	460	4.2	—
Silica					
SiO_2	$10^{-3.9}$	2.7	—	—	4.0
SiO_2	$0 - 10^{-3.4}$	2.5	0.5	6.4	—
		3.0	1.0	6.7	—
Organic Matter					
Humic acid (soil)	$10^{-4.3} - 10^{-3.3}$	2.4	1760	3.2	—

a. A_m = Langmuir adsorption maximum (μmol/g)
b. K_L = Langmuir constant (log M^{-1})
c. A = adsorption (μmol/g)

Source: Electric Power Research Institute [3]. *(Copyright 1984, Electric Power Research Institute. Reprinted with permission.)*

The solubility of iron in soil depends primarily on the solubility of the hydrous Fe(III) oxides [4,11]. Hydrolysis, complexation and redox processes are also important. In addition, Fe solubilities may be significantly affected by the formation of other iron compounds, such as phosphates, sulfides, and carbonates [7].

The solubilities of Fe(III) oxides commonly found in soils are shown on the first six lines of Table 6.8-1. The solubility of Fe(III) is usually controlled by the most soluble oxide present; thus, soil-$Fe(OH)_3$ generally controls the activity of Fe(III) in soils (Figure 6.8-4) [11]. In this discussion, soil-$Fe(OH)_3$ is defined as an amorphous phase having a greater degree of structural order than freshly precipitated $Fe(OH)_3$ (amorp).

Soil-$Fe(OH)_3$, magnetite, or siderite can control the solubility of Fe(II) in soils, depending on redox and CO_2(g). Under oxidizing conditions (pe + pH $>$ 11.5), soil-$Fe(OH)_3$ (which is intermediate in solubility to amorphous hydroxide and crystalline oxide) controls the solubility; below 11.5, magnetite (Fe_3O_4) is the stable phase until siderite ($FeCO_3$) forms, as determined by the partial pressure of CO_2(g) [11].

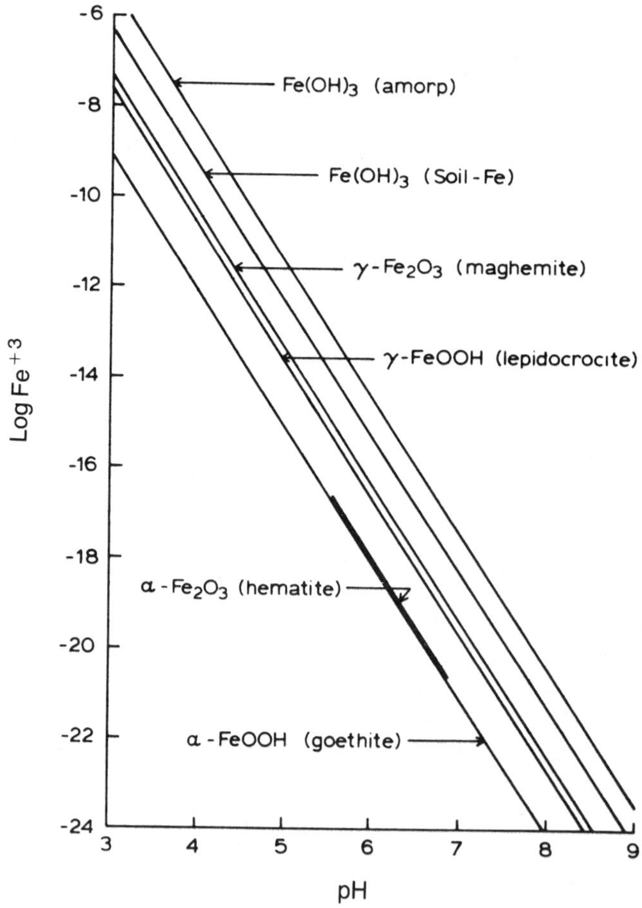

Source: Lindsay [11]. *(Copyright 1979, John Wiley & Sons. Reprinted with permission.)*

FIGURE 6.8-4 Activity of Fe^{+3} Maintained by Fe(III) Oxides and Soil-Fe

SEDIMENTS

Iron is a major element of freshwater and oceanic sediments. Concentrations range from 1% to 9% [6,9]. The form of iron varies, depending on the redox conditions. Surface sediments may be somewhat oxidized and often contain ferric iron. Soluble (reduced) iron content increases with sediment depth. Ferrous sulfide (FeS), the predominant inorganic form of iron [9], may in turn be transformed to pyrite (FeS_2), a common iron sulfide in sediments [2].

In sediments containing enough organic matter, anaerobic bacteria reduce sulfate in the interstitial water to form sulfide species, which react with iron-containing minerals in the sediment to form a variety of iron sulfides.

Hydroxides of iron (e.g., FeOOH) can precipitate from natural waters to form colloidal particles that aggregate and settle on the bottom. The particles adhere to surfaces and aggregate to form coatings, granules, and nodules.

6.8.5 Effect of Iron on Concentrations of Other Metals

Although iron itself is not considered toxic, it is environmentally significant because of its interaction with metals that are toxic. Iron oxides adsorb many elements and participate in the attenuation of most trace and heavy metals. In soils and sediments where conditions are not strongly reducing, adsorption by solid iron oxides is one of the most important controls on the distribution of trace elements. In an aerobic sediment or water system, the metal bonding may be almost irreversible. If the oxides are dissolved by reduction, the adsorbed metals may be released and appear in solution, or they may be precipitated as some other phase, such as a sulfide [2,9]. (However, if ferrous iron is present in large quantities, it can tie up the sulfide to the extent that insufficient amounts are available to precipitate heavy metals [9].)

Precipitation of iron oxides during dredging operations has been reported to decrease dissolved concentrations of cadmium, copper, lead and zinc [15]. Iron (and manganese) oxides are used as scavengers in wastewater treatment and may be very effective in retarding the migration of pollutants in the subsurface [2]. Deep-sea nodules, concretions of manganese and iron oxides growing on the ocean floor, are known to concentrate trace elements, such as Co, Ni, Zn and Pb, from seawater. As the oxides aggregate and settle, they act as scavengers for trace elements. (Data on adsorption of trace elements by iron oxides are given in the various chapters on individual elements, and a general discussion is presented in section 2.12.)

In soils, various trace elements are concentrated by iron oxides, including Zn, Pb, Mn, Ni, Cu, Co, V, Mo, and Cr. The most widely observed sorption capacities of iron oxides are those for phosphates, molybdates, and selenites. The adsorption is dependent on pH and is greatest for various ions on iron oxides at pH 4-5. Iron oxides also react with carbonates in soil systems. Fe(III) can be incorporated in hydrated phosphates, and Fe(II) reacts with sulfur to form the stable minerals pyrite and jarosite [7].

6.8.6 Literature Cited

1. Cotton, F.A. and G. Wilkinson, *Advanced Inorganic Chemistry*, John Wiley & Sons, New York (1972).
2. Drever, J.I., *The Geochemistry of Natural Waters*, Prentice-Hall, Englewood Cliffs, N.J. (1982).
3. Electric Power Research Institute, *Chemical Attenuation Rates, Coefficients, and Constants in Leachate Migration. Volume I: A Critical Review*, EPRI EA-3356 (February 1984).
4. Fuller, W.H., *Movement of Selected Metals, Asbestos, and Cyanide in Soil: Applications to Waste Disposal Problems*, EPA-600/2-77-020, NTIS PB-266 905 (April 1977).
5. Hem, J.D., *Study and Interpretation of the Chemical Characteristics of Natural Water*, 2nd ed., Water Supply Paper No. 1473, U.S. Geological Survey, Washington D.C. (1970).
6. Horne, R.A., *The Chemistry of Our Environment*, John Wiley & Sons, New York (1978).
7. Kabata-Pendias, A. and H. Pendias, *Trace Elements in Soils and Plants*, CRC Press, Boca Raton, Fla. (1984).
8. Kerndoff, H. and M. Schnitzer, "Sorption of Metals on Humic Acid," *Geochim. Cosmochim. Acta*, **44**, 1701-08 (1980).
9. Khalid, R.A., R.P. Gambrell, M.G. Verloo and W.H. Patrick, Jr., *Transformations of Heavy Metals and Plant Nutrients in Dredged Sediments as Affected by Oxidation Reduction Potential and pH*, U.S. Army Contract No. DACW39-74-C-0076 (May 1977).
10. Kotrlý, S., and L. Šůcha, *Handbook of Chemical Equilibria in Analytical Chemistry*, John Wiley & Sons, New York (1985).
11. Lindsay, W.L., *Chemical Equilibria in Soils*, John Wiley & Sons, New York (1979).
12. Lindsay, W.L., "Inorganic Phase Equilibria of Micronutrients in Soil," in *Micronutrients in Agriculture*, W.L. Lindsay (ed.), Soil Science Society of America, Madison, Wis. (1972), as cited in [4].
13. Morel, F.M.M., *Principles of Aquatic Chemistry*, John Wiley & Sons, New York (1983).
14. Stumm, W. and J.J. Morgan, *Aquatic Chemistry*, 2nd ed., John Wiley & Sons, New York (1981).
15. Windom, H.L., *Investigations of Changes in Heavy Metals Concentrations Resulting from Maintenance Dredging of Mobile Bay Ship Channel, Mobile Bay, Alabama*, Mobile District, U.S. Army Corps of Engineers, Mobile, Ala. (1973), as cited in [9].

6.9 HYDROGEN AND HYDROXIDE IONS (H^+, OH^-)

Hydrogen and hydroxide ions are fundamental to the chemistry of aqueous systems. Both ions are present in all water samples and generally are not considered pollutants except at concentrations far greater than those found in nature. For example, the International Joint Commission [4] established an objective for the pH of the Great Lakes of between 6.5 and 9, a range of more than 300-fold in the concentration of H^+. The hydrogen and hydroxide ions are considered in this section because of their importance in determining the presence and form of other chemical species in aqueous solutions.

6.9.1 pH and the Dissociation of Water

Equation 1 represents the autoprotolysis of water.

$$H_2O + H_2O \rightleftharpoons H_3O^+ + OH^- \quad (1)$$

The equilibrium constant for this equation, K_w, expressed in terms of activities is referred to as the ion product of water. In dilute solutions, the activity of water equals one, and by convention H^+ is used in place of H_3O^+ [10]. Thus K_w represents the product of (H^+) and (OH^-). Its value at 25°C is 10^{-14}. A solution is said to be neutral when the activities of H^+ and OH^- are equal, i.e., at a pH of one-half of pK_w. As Table 6.9-1 shows, the value of K_w, and hence the pH of neutrality, is strongly dependent on temperature.

The values of K_w given in this table for pure water are not valid for seawater, because K_w is affected by high ionic strength. A value of $10^{-13.2}$ has been given for seawater of 3.5% salinity at 25°C [3].

Originally, pH was defined as the negative logarithm of the hydrogen ion concentration. Today, however, it is defined and measured as the negative logarithm of the hydrogen ion activity. The hydrogen ion activity is most often measured using a glass electrode. The glass electrode develops a potential relative to the reference electrode that varies with the activity of H^+ in solution. The measured potential is compared to that obtained for reference buffer values. Procedures for measuring pH in water can be found, for example, in reference [9], and the measurement of soil solution pH is described in references [1] and [6].

The pH scale used by the National Bureau of Standards and endorsed by IUPAC[1] is based on an infinitely dilute reference state; pH is determined relative to a standard buffer whose pH is known in terms of H^+ activity [10].

The NBS method does not permit accurate measurement of the pH of seawater, which has a high ionic strength, but the results can be useful for comparative purposes if the method is used consistently. No single pH scale for seawater has yet become standard [10].

1. International Union of Pure and Applied Chemistry (Oxford, U.K.)

TABLE 6.9-1

Ion Product of Water

°C	K_w	pK_w	pH of "Neutral" Solution
0	0.12×10^{-14}	14.93	7.47
15	0.45×10^{-14}	14.35	7.18
20	0.68×10^{-14}	14.17	7.08
25	1.01×10^{-14}	14.00	7.00
30	1.47×10^{-14}	13.83	6.92
40	2.95×10^{-14}	13.53	6.76

Source: Snoeyink and Jenkins [8] and Harned and Owen [5]. *(Copyright 1980, John Wiley & Sons. Reprinted with permission.)*

6.9.2 Soil pH

Soil pH is generally measured with a glass electrode [6], often after the soil has been slurried and equilibrated with water. The ratio of soil solids to added water affects the measured pH, which may decrease by as much as 1 pH unit or more as the proportion of soil increases. The pH also normally decreases with increasing equilibrium salt concentration [1]. For a given soil, other important factors affecting the pH include the dryness of the sample, its exposure to CO_2, and the particle size fraction of the slurry used to represent the soil. These factors should be considered in the comparison of samples; particular attention should be given to moisture content — i.e., whether the samples have been air-dried or are field-moist [6].

Some workers have noted the existence of a junction (extraneous) potential in soil pH measurements, especially in very concentrated soil colloid suspensions [6], leading to inaccurate measurements. The junction potential can be virtually eliminated, however, by measuring the pH in salt solutions of 0.01 *M* or greater [1]. Allowing the tip of the calomel reference electrode used in the pH measurement to contact only the supernatant above the colloidal phase, while suspending the glass pH electrode within the colloid also minimizes the junction potential [1].

The relationship between soil pH and cation exchange capacity as well as the influence of the former on mineral availability are discussed in Appendix C. Soil nomenclature as it relates to pH is also described.

6.9.3 The pH of Natural Waters

The pH of natural waters is affected by various physical and biological processes, both natural and man-made. In any water exposed to the atmosphere, the effects of gaseous carbon dioxide must be considered. When carbon dioxide from the atmosphere dissolves in water, carbonic acid is formed. The quantity of carbon dioxide in solution is further influenced by aquatic biota, which produce CO_2 during respiration and consume it through photosynthesis. The entire carbonate system is considered in greater detail in the carbonate/bicarbonate section (§ 6.12). Appendix B (Table B-9) lists pH values for several natural waters.

One of the major anthropogenic influences on the pH of surface waters is "acid mine drainage," which is caused by the oxidation of pyrite (FeS_2) produced as a byproduct of coal and ore mining. According to Stumm and Morgan [10], the formation of H^+ ions can be characterized by the following sequence of reactions:

$$2FeS_2(s) + 7O_2 + 2H_2O \rightleftharpoons 2Fe^{+2} + 4SO_4^{-2} + 4H^+ \quad (2)$$
$$4Fe^{+2} + O_2 + 4H^+ \rightleftharpoons 4Fe^{+3} + 2H_2O \quad (3)$$
$$Fe^{+3} + 3H_2O \rightleftharpoons Fe(OH)_3(s) + 3H^+ \quad (4)$$
$$FeS_2(s) + 14Fe^{+3} + 8H_2O \rightleftharpoons 15Fe^{+2} + 2SO_4^{-2} + 16H^+ \quad (5)$$

Equation 3 is the rate-determining step of this series. At pH <3.5, it is independent of pH, but it can be catalyzed by *Ferrobacillus* and *Thiobacillus* microorganisms, increasing the reaction rate by several orders of magnitude [11].

While pH values as low as 3 are found in waters of coal-mining areas [10], neutral or higher pH values are not proof of the absence of mine drainage. Wentz [11] found a poor correlation between pH and metal concentration in streams in mining areas, probably because H^+ ions are neutralized by the alkalinity of the receiving water without the removal of the accompanying metal ion.

A sample of pure water would be expected to have a pH of about 5.6 when in equilibrium with a natural (clean) atmosphere (see § 6.12.2). The pH of snow and rainwater, however, is reduced by atmospheric pollutants, primarily sulfur dioxide (SO_2) and oxides of nitrogen (NO_x). The transformation of SO_2 and NO_x to sulfuric and nitric acids in the atmosphere is a complex process involving a number of reaction pathways. While the direct oxidation of SO_2 and NO_x with gaseous oxygen is favored thermodynamically, these reactions are extremely slow. The gas-phase oxidations of SO_2, NO and NO_2 are believed to occur primarily through reactions with highly reactive intermediates, excited molecules, atoms and free radicals [2].

Sulfuric acid can be formed by several gas-phase reactions of SO_2 with reactive intermediates. The most important of these is probably the reaction with the hydroxy radical ($HO\cdot$), which proceeds as follows [2]:

$$HO\cdot + SO_2 \rightleftharpoons HOSO_2\cdot \quad (6)$$
$$HOSO_2\cdot + O_2 \rightleftharpoons HO_2\cdot + SO_3 \quad (7)$$
$$SO_3 + H_2O \rightleftharpoons H_2SO_4 \quad (8)$$

Hydroxy radicals can be formed in the atmosphere by a number of processes, including the reaction of an excited oxygen atom (formed by the photodissociation of ozone) with an H_2O molecule [2]. Because the HO_2 generated by the oxidation of $HOSO_2$ reacts in turn with NO ($HO_2 + NO \rightleftharpoons HO + NO_2$) to form HO, the net effect is that the hydroxy radical is not consumed in the reaction sequence.

There is little question that aqueous-phase as well as gas-phase processes contribute significantly to the formation of H_2SO_4 in the atmosphere [2]. Upon dissolution in water, SO_2 forms the hydrate $SO_2 \cdot H_2O$ (or sulfurous acid, H_2SO_3), which dissociates rapidly to form the bisulfite ion (HSO_3^-), which in turn is oxidized to sulfate. The oxidation of bisulfite with aqueous oxygen is very slow if catalysts such as trace metals are not present; ozone (O_3) and hydrogen peroxide (H_2O_2) are thought to be primarily responsible for sulfate formation [2]. The aqueous-phase oxidation of SO_2 is thought to occur mainly in the following way:

$$SO_2(g) + H_2O(liq) \rightleftharpoons H_2SO_3^0 \quad (9)$$
$$H_2SO_3^0 \rightleftharpoons H^+ + HSO_3^- \quad (10)$$
$$HSO_3^- \rightleftharpoons H^+ + SO_3^{-2} \quad (11)$$
$$HSO_3^- + H_2O_2 \rightleftharpoons SO_4^{-2} + H^+ + H_2O \quad (12)$$
$$HSO_3^- + O_3 \rightleftharpoons SO_4^{-2} + H^+ + O_2 \quad (13)$$

At low pH, reaction 12 predominates over reaction 13.

The gas-phase oxidation of NO to NO_2 occurs principally with ozone and peroxy radicals, while NO_2 is oxidized to HNO_3 by $HO\cdot$ in a reaction similar to that with SO_2 [2]:

$$HO\cdot + NO_2 \rightleftharpoons HNO_3 \quad (14)$$

The aqueous-phase chemistry resulting in the formation of nitric acid is not well understood. However, it is known that dinitrogen pentoxide (N_2O_5) can be effectively scavenged by water droplets to form HNO_3 directly [2]. Dinitrogen pentoxide is formed by the following two gas-phase reactions:

$$NO_2 + O_3 \rightleftharpoons O_2 + NO_3 \quad (15)$$
$$NO_3 + NO_2 \rightleftharpoons N_2O_5 \quad (16)$$

Dinitrogen pentoxide then reacts with water to form nitric acid:

$$N_2O_5 + H_2O(liq) \rightleftharpoons 2H^+ + 2NO_3^- \quad (17)$$

The contribution of this mechanism to the total nitric acid content of the atmosphere has not yet been evaluated [2].

6.9.4 Metal Hydroxides

The preceding discussion has focused on the definition of pH and on forms of pollution that result in the production of H^+ ions. Table 6.9-2 lists the acidity constants of some inorganic acids found in nature; others are listed in Table 2.7-3 (Part I).

TABLE 6.9-2

Acidity Constants of Some Common Acids
(in order of decreasing acid strength)

	Reaction	– Log Acidity Constant, pK (approx.) at 25°C
Perchloric acid	$HClO_4 \rightleftharpoons ClO_4^- + H^+$	–7
Hydrogen chloride	$HCl \rightleftharpoons Cl^- + H^+$	~–3
Sulfuric acid	$H_2SO_4 \rightleftharpoons HSO_4^- + H^+$	~–3
Nitric acid	$HNO_3 \rightleftharpoons NO_3^- + H^+$	–1
Hydronium ion	$H_3O^+ \rightleftharpoons H_2O + H^+$	0
Bisulfate	$HSO_4^- \rightleftharpoons SO_4^{-2} + H^+$	1.9
Acetic acid	$CH_3COOH \rightleftharpoons CH_3COO^- + H^+$	4.7
Hypochlorous acid	$HOCl \rightleftharpoons OCl^- + H^+$	7.6
Ammonium ion	$NH_4^+ \rightleftharpoons NH_3 + H^+$	9.3
o-Silicic acid	$Si(OH)_4 \rightleftharpoons SiO(OH)_3^- + H^+$	9.5
Hydrogen peroxide	$H_2O_2 \rightleftharpoons HO_2^- + H^+$	—
Silicate	$SiO(OH)_3^- \rightleftharpoons SiO_2(OH)_2^{-2} + H^+$	12.6
Bisulfide	$HS^- \rightleftharpoons S^{-2} + H^+$	14

Source: Adapted from Stumm and Morgan [10]

In aqueous solutions, metal ions complex with the hydroxide ligand. Table 6.9-3 illustrates the type and variety of monomeric hydroxide complexes that can be formed. The relative distribution and concentration of these complexes is dependent upon a variety of factors, including the metal, its pK values, and other reactions (e.g., polymerization, solubility equilibria).

The complexation of ferric ion with OH^- can be written as follows:

$$Fe^{+3} + OH^- \rightleftharpoons FeOH^{+2} \quad (18)$$

TABLE 6.9-3

Hydroxide Complex Ions of Some Common Metals

Metal	Species
Ag(I)	$AgOH^0$, $Ag(OH)_2^-$, $Ag(OH)_3^{-2}$
Al(III)	$AlOH^{+2}$, $Al(OH)_2^+$, $Al(OH)_3^0$, $Al(OH)_4^-$
Ba(II)	$BaOH^+$
Ca(II)	$CaOH^+$
Cd(II)	$CdOH^+$, $Cd(OH)_2^0$, $Cd(OH)_3^-$, $Cd(OH)_4^{-2}$
Co(II)	$CoOH^+$, $Co(OH)_2^0$, $Co(OH)_3^-$
Cr(III)	$CrOH^{+2}$, $Cr(OH)_2^+$, $Cr(OH)_3^0$, $Cr(OH)_4^-$
Cu(II)	$CuOH^+$, $Cu(OH)_2^0$, $Cu(OH)_3^-$, $Cu(OH)_4^{-2}$, $Cu_2(OH)_2^{+2}$
Fe(II)	$FeOH^+$, $Fe(OH)_2^0$, $Fe(OH)_3^-$, $Fe(OH)_4^{-2}$
Fe(III)	$FeOH^{+2}$, $Fe(OH)_2^+$, $Fe(OH)_3^0$, $Fe(OH)_4^-$
Hg_2(I)	Hg_2OH^+
Hg(II)	$HgOH^+$, $Hg(OH)_2^0$, $Hg(OH)_3^-$
Mg(II)	$MgOH^+$
Mn(II)	$MnOH^+$, $Mn(OH)_2^0$, $Mn(OH)_3^-$
Ni(II)	$NiOH^+$, $Ni(OH)_2^0$, $Ni(OH)_3^-$
Pb(II)	$PbOH^+$, $Pb(OH)_2^0$, $Pb(OH)_3^-$
Zn(II)	$ZnOH^+$, $Zn(OH)_2^0$, $Zn(OH)_3^-$, $Zn(OH)_4^{-2}$

Source: Pagenkopf [7]. *(Copyright 1978, Marcel Dekker. Reprinted with permission.)*

When this reaction is added to the dissociation of water (equation 19), we see that the reaction of an aqueous metal ion to form a metal hydroxide complex results in the release of a proton; i.e., the metal ion acts as an acid.[2]

$$\begin{array}{l} Fe^{+3} + OH^- \rightleftharpoons FeOH^{+2} \\ \qquad\quad H_2O \rightleftharpoons H^+ + OH^- \\ \hline Fe^{+3} + H_2O \rightleftharpoons FeOH^{+2} + H^+ \end{array} \tag{19}$$

The equilibrium constant for reaction 19 is included in Table 2.7-3 along with the constants for several other metals. To illustrate the strength of metal ions as acids, the pKa for the formation of $FeOH^{+2}$ and H^+ from Fe^{+3} ($pK_a = 2.6$) is of the same order of magnitude as the first pK_a for phosphoric acid (2.0).

2. While outwardly different from equation 10 of section 2.7, which describes the behavior of trace metals in aqueous systems, equation 19 is just a simpler way of expressing the same reaction. For the sake of convenience, the six waters of hydration surrounding the aquated Fe^{+3} ion are not shown here; if they were (and if equation 10 of section 2.7 showed $Fe(OH_2)_6^{+3}$ reacting with H_2O), the two equations would be identical.

The acid dissociation constants given in Table 2.7-3 can be used to calculate the formation constants for metal hydroxide complexes such as those listed in Table 6.9-3 (the formation of $FeOH^{+2}$ as expressed in equation 19, for example). The relationship between acid dissociation constants and formation constants of hydroxide complexes is described in section 2.7 and illustrated in Example 6 of that section.

Table 6.9-4 lists the equilibrium constants for reactions of solid hydroxides and oxides with H^+; these reactions are especially important, because they promote the release of metals from the solid phase to the aqueous phase. The release of metal ions occurs as the free H^+ ions react with the hydroxide or oxide and are thereby removed. This can create a buffering effect and keep pH relatively constant.

6.9.5 Literature Cited

1. Bohn, H.L., B.L. McNeal and G.A. O'Connor, *Soil Chemistry*, John Wiley & Sons, New York (1979).
2. Committee on Atmospheric Transport and Chemical Transformation in Acid Precipitation, National Research Council, *Acid Deposition Atmospheric Processes in Eastern North America*, National Academy Press, Washington, D.C. (1983).
3. Culberson, C. and R.M. Pytkowicz, "Ionization of Water in Seawater," *Mar. Chem.*, **1**, 309-16 (1973).
4. *Great Lakes Water Quality Agreement of 1978*, International Joint Commission, Canada and the United States, Ottawa, Ontario (1978).
5. Harned, H.S. and B.B. Owen, *The Physical Chemistry of Electrolyte Solutions*, 3rd ed., Reinhold Publishing Co., New York (1958), as cited by Snoeyink and Jenkins [8].
6. Jackson, M.L., *Soil Chemical Analysis*, Prentice-Hall, Englewood Cliffs, N.J. (1958).
7. Pagenkopf, G.K., *Introduction to Natural Water Chemistry*, Marcel Dekker, New York (1978).
8. Snoeyink, V.L. and D. Jenkins, *Water Chemistry*, John Wiley & Sons, New York (1980).
9. *Standard Methods for the Examination of Water and Wastewater*, 16th ed., American Public Health Association, Washington, D.C. (1985).
10. Stumm, W. and J.J. Morgan, *Aquatic Chemistry*, 2nd ed., John Wiley & Sons, New York (1981).
11. Wentz, D.A., "Stream Quality in Relation to Mine Drainage in Colorado," in *Water Resource Problems Related to Mining*, R.F. Hadley and D.T. Snow (eds.), American Water Resources Association, Minneapolis (1974).

TABLE 6.9-4

Equilibrium Constants for Reactions Involving H^+/OH^- with Metal Hydroxides and Oxides

Reaction	log K at 25°C	I
$H_2O(liq) \rightleftharpoons H^+ + OH^-$	−14.00	0
	−13.79	1 *M* $NaClO_4$
$Fe(OH)_3(amorph) \rightleftharpoons Fe^{+3} + 3OH^-$	−38.7	3 *M* $NaClO_4$
$Fe(OH)_3(amorph) \rightleftharpoons FeOH^{+2} + 2OH^-$	−27.5	3 *M* $NaClO_4$
$Fe(OH)_3(amorph) \rightleftharpoons Fe(OH)_2{}^+ + OH^-$	−16.6	3 *M* $NaClO_4$
$Fe(OH)_3(amorph) + OH^- \rightleftharpoons Fe(OH)_4{}^-$	−4.5	3 *M* $NaClO_4$
$2Fe(OH)_3(amorph) \rightleftharpoons Fe_2(OH)_2{}^{+4} + 4OH^-$	−51.9	3 *M* $NaClO_4$
$FeOOH(amorph) + 3H^+ \rightleftharpoons Fe^{+3} + 2H_2O$	3.55	3 *M* $NaClO_4$
α-$FeOOH(s) + 3H^+ \rightleftharpoons Fe^{+3} + 2H_2O$	1.6	3 *M* $NaClO_4$
$2Fe^{+3} + 2H_2O \rightleftharpoons Fe_2(OH)_2{}^{+4} + 2H^+$	−2.95	0
α-$FeOOH(s) + 3H^+ \rightleftharpoons Fe^{+3} + 2H_2O$	0.5	0
$FeOOH(amorph) + 3H^+ \rightleftharpoons Fe^{+3} + 2H_2O$	2.5	0
$Fe(OH)_2(s)(active) + 2H^+ \rightleftharpoons Fe^{+2} + 2H_2O$	12.85	0
$2Al^{+3} + 2H_2O \rightleftharpoons Al_2(OH)_2{}^{+4} + 2H^+$	−7.7	0
$3Al^{+3} + 4H_2O \rightleftharpoons Al_3(OH)_4{}^{+5} + 4H^+$	−13.9	0
$13Al^{+3} + 28H_2O \rightleftharpoons Al_{13}O_4(OH)_{24}{}^{+7} + 32H^+$	−98.7	0
α-$Al(OH)_3(s) + 3H^+ \rightleftharpoons Al^{+3} + 3H_2O$	8.5	0
$Al(OH)_3(amorph) + 3H^+ \rightleftharpoons Al^{+3} + 3H_2O$	10.8	0
$CuO(s) + 2H^+ \rightleftharpoons Cu^{+2} + H_2O$	7.65	0
$2Cu^{+2} + 2OH^- \rightleftharpoons Cu_2(OH)_2{}^{+2}$	17.0(18°C)	0
$ZnO + 2H^+ \rightleftharpoons Zn^{+2} + H_2O$	11.14	0
$Zn(OH)_2(amorph) + 2H^+ \rightleftharpoons Zn^{+2} + 2H_2O$	12.45	0
β-$Cd(OH)_2(s) + 2H^+ \rightleftharpoons Cd^{+2} + 2H_2O$	13.65	0
$Mn(OH)_2(s)(active) + 2H^+ \rightleftharpoons Mn^{+2} + H_2O$	15.2	0
$HgO + 2H^+ \rightleftharpoons Hg^{+2} + H_2O$	2.56	0
$Mg^{+2} + H_2O \rightleftharpoons MgOH^+ + H^+$	−11.44	0
$Mg(OH)_2(s)(brucite) + 2H^+ \rightleftharpoons Mg^{+2} + 2H_2O$	16.84	0
$Ca^{+2} + H_2O \rightleftharpoons CaOH^+ + H^+$	−12.85	0
$Ca(OH)_2(s) + 2H^+ \rightleftharpoons Ca^{+2} + 2H_2O$	22.8	0

(Acid dissociation constants for a variety of aqueous metal ions are given in Table 2.7-3).

Source: Stumm and Morgan [10]. *(Copyright 1981, John Wiley & Sons. Reprinted with permission.)*

6.10 CHLORIDE (Cl^-)

6.10.1 Occurrence and Uses

Most of the chlorine on earth exists in the oceans in the form of the chloride ion (Cl^-). Chloride is the predominant ion in seawater on both a mass and molar basis, having a concentration of 19.35 g/kg (0.55*M*) versus 10.77 g/kg (0.47*M*) for sodium, the next most abundant ion [13]. The abundance of chloride in the oceans is generally explained by the volatility of elemental chlorine, which became separated from crustal rock at an early stage of the geochemical history of the earth [8]. As common chlorides are highly soluble in water, chloride minerals and chloride deposited from the atmosphere readily dissolve from soils, enter rivers or groundwater and eventually flow into the sea.

About 8.5 million metric tons of deicing salts (mostly sodium chloride with some calcium chloride) were used on U.S. roads in 1970 [4]. The use of such salts can have pronounced localized effects. In addition to damaging vegetation along the roadway, the salts also influence receiving waters. The use of deicing salts in Rochester, N.Y. was found to increase the chloride concentration in a nearby bay more than fivefold; in at least one year, the runoff was sufficiently saline to prevent complete vertical mixing of the bay during the spring and to create a density gradient that prolonged the period of summer stratification by a month [4]. It has been suggested that mercury and other heavy metals may be released from sediments in waters receiving runoff from salted roads as the salt cations compete for exchange sites on the sediments and the heavy metals form complexes with chloride [5], as described below.

6.10.2 Environmental Importance

Chloride is not considered toxic except at very high concentrations, and it is essential to mammals and probably all organisms [3]. The chloride ion should not be confused with other forms of chlorine in more oxidized form, such as aqueous chlorine, $Cl_2(aq)$, hypochlorous acid, HOCl, and hypochlorite ion, OCl^-, all of which are highly toxic [3]. The World Health Organization [14] recommends a maximum chloride concentration in drinking water of 250 mg/l; this limit is not based on adverse health effects but, rather, on the unpleasant taste imparted to water at concentrations above 200-300 mg/l.

The major stress that chloride and other salts cause on plants is osmotic [2]. High salt concentrations in soils make it difficult for plants to obtain water. Plants vary in their ability to exclude chloride from their above-ground parts. In plants less able to exclude chloride (and not adapted to a saline environment), chloride toxicity manifests itself in excessive accumulations near the end of the plant's transpiration stream, leading to the necrosis and burning of leaf tips and margins, and to eventual death [2].

6.10.3 Chloride Transport through the Environment

The chlorine transported via rivers to the ocean is estimated at 260 million metric tons annually [3]. This is by far the largest flux of chlorine in the environment; mining and combustion account for annual transfers of 81 and 1.5 million metric tons of chlorine respectively. In contrast, the weathering of rocks is thought to release only about 0.52 million metric tons (as Cl^-) per year. A small fraction of oceanic chlorine returns to land in aerosols and rain: the annual deposition of chlorine near seacoasts may amount to 100 kg/ha, decreasing rapidly to 1 or 2 kg/ha in the interior [2].

6.10.4 Chloride Chemistry

The reactions of chloride in the environment are confined largely to dissolution-precipitation and complexation. Its acid-base chemistry is insignificant, because HCl is such a strong acid ($pK_a \approx -3$) [12]; the conjugate base (the chloride ion) is extremely weak:

$$Cl^- + H_2O \rightleftharpoons OH^- + HCl \qquad K_b = 10^{-17}$$

Under typical environmental conditions, the chloride ion undergoes neither oxidation nor reduction. Molecular chlorine, Cl_2, is a strong oxidizing agent:

$$\tfrac{1}{2}\,Cl_2(aq) + e^- = Cl^- \qquad \log K = 23.6 \qquad E^0 = 1.40V$$

Hence, chlorine exists in solution in its reduced form as chloride. Figure 6.10-1 is an Eh-pH diagram for a chlorine-water system of 0.04M total chlorine concentration at 25°C and zero ionic strength. At equilibrium, Cl^- is the stable chlorine species within the stability limits (dashed lines) of water. In dilute solutions, $Cl_2(aq)$ exists only at low pH. Cl_2, OCl^- and HOCl are all unstable or metastable in water.

Chloride salts dissolve readily in water. Table 6.10-1 lists the equilibrium constants of some chloride minerals. Of these, halite (NaCl) and sylvite (KCl) are the most common; their solubilities are orders of magnitude greater than those of other minerals, such as gypsum and calcite, both of which precipitate before NaCl during the evaporation of seawater [3].

Despite the high solubility of halite and sylvite, chloride is the dominant anion in groundwater systems only at depths where contact with salt strata may occur or in waters that have moved long distances [6]. This phenomenon can be explained by a lack of halite or sylvite along the flow path of the groundwater. When groundwater comes in contact with chloride salts early in its flow path, chloride becomes the dominant anion directly.

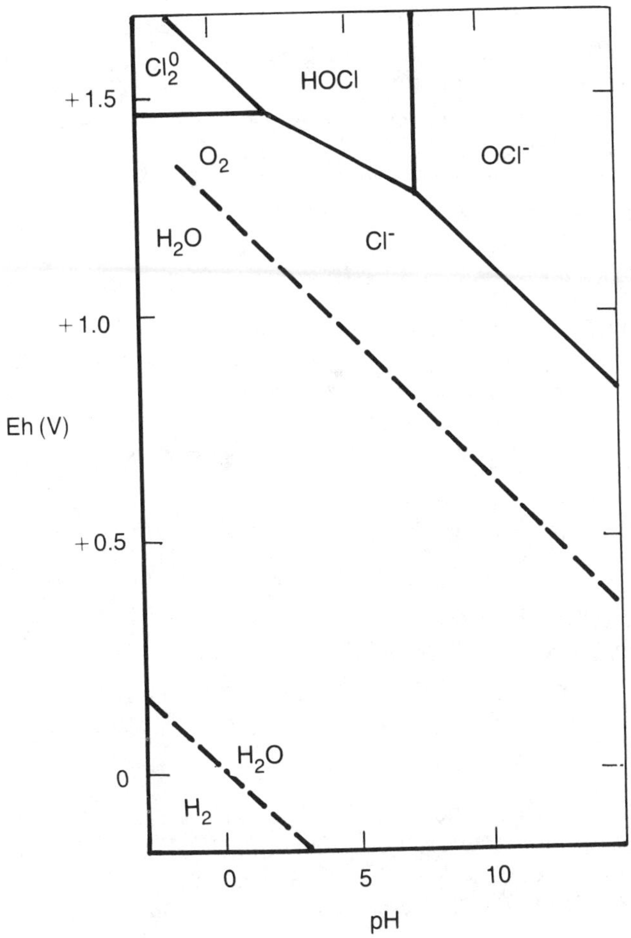

Cl_T = 0.04 *M*; unit atomic ratios for oxidants and reductants at the boundaries. Dashed lines represent stability limits of water.

Source: Stumm and Morgan [13]. *(Copyright 1981, John Wiley & Sons. Reprinted with permission.)*

FIGURE 6.10-1 Eh-pH Diagram for the Chlorine System

Complexation of the chloride ion is of interest in environmental chemistry because of the effect on speciation and mobility of the metals that complex strongly with it. The tendency of heavy metals such as cadmium, lead and mercury to complex with chloride increases their solubility, leading to higher total concentrations in solution than would be expected from the solubility product alone. Besides increasing the concentration of metal chlorides, the complexation of the metal ion reduces its tendency to form precipitates with other ions such as hydroxide and phosphate.

TABLE 6.10-1

Solubility Product Constants of Selected Chlorides
(at 25°C and zero ionic strength)

Solid	Dissolution Reaction	pK_{sp}
AgCl	$AgCl \rightleftharpoons Ag^+ + Cl^-$	9.752
Hg_2Cl_2	$Hg_2Cl_2 \rightleftharpoons Hg_2^{+2} + 2Cl^-$	17.91
KCl	$KCl \rightleftharpoons K^+ + Cl^-$	−0.932[a]
NaCl	$NaCl \rightleftharpoons Na^+ + Cl^-$	−1.553[a]
$PbCl_2$	$PbCl_2 \rightleftharpoons Pb^{+2} + 2Cl^-$	4.78
TlCl	$TlCl \rightleftharpoons Tl^+ + Cl^-$	3.74

a. From Sillen and Martell [11]

Source: Martell and Smith [9], except as indicated

Table 6.10-2 lists the major metal-chloride complexes found in natural waters and the log equilibrium constants for their formation. Chloride complexation causes a significant fraction of some metals to exist in complexed form. In seawater, evidence suggests that $NaCl^0$, KCl^0, $MgCl^+$ and $CaCl^+$ account for 13%, 17%, 43% and 47% of the respective total metal concentrations [3].

The behavior of heavy metals with chlorides has been investigated for both aqueous and soil systems. Hahne and Kroontje [7] calculated that in a 1.0*M* chloride solution, the conditional solubility of $Zn(OH)_2$ would be more than double that based on the solubility product, while the solubility of $Hg(OH)_2$ would be several orders of magnitude greater.

6.10.5 Chloride in Soils

The chloride ion is extremely mobile in soils. Almost no soil retention occurs; thus, chloride in solution moves through soil at virtually the same speed as the displacing solution itself [2]. Because of its high mobility, data on chloride sorption to soils and sediments are rare and none are presented here.

Chloride is typically the predominant anion in saline soils and the second most abundant anion after carbonate/bicarbonate in sodic soils [2]. The effect of chloride on the behavior of heavy metals in soils is similar to that in aqueous solutions. In one study, for a given cadmium application, the total concentration in the soil extract increased over tenfold with increasing chloride additions [1]. However, the uptake of cadmium by Swiss chard grown in these soils was found to be correlated primarily with the free Cd^{+2} concentration. The authors of the study [1] suggested that the assessment of metal availability based on total metal concentration be modified to account for metal speciation.

TABLE 6.10-2

Equilibrium Constants for Chloride Complexes
(for aqueous solutions at 25°C and zero ionic strength unless otherwise noted)

Reaction	log K	Reaction	log K
$Ag^{+} + Cl^{-} \rightleftharpoons AgCl^{0}$	3.31	$Hg^{+2} + 3Cl^{-} \rightleftharpoons HgCl_3^{-}$	15.1[b]
$Ag^{+} + 2Cl^{-} \rightleftharpoons AgCl_2^{-}$	5.25	$Hg^{+2} + 4Cl^{-} \rightleftharpoons HgCl_4^{-2}$	15.4[b]
$Bi^{+3} + Cl^{-} \rightleftharpoons BiCl^{+2}$	3.5	$Mg^{+2} + Cl^{-} \rightleftharpoons MgCl^{+}$	−0.10[a]
$Ca^{+2} + Cl^{-} \rightleftharpoons CaCl^{+}$	−0.11[a]	$Mn^{+2} + Cl^{-} \rightleftharpoons MnCl^{+}$	0.6[b]
$Cd^{+2} + Cl^{-} \rightleftharpoons CdCl^{+}$	1.98	$Ni^{+2} + Cl^{-} \rightleftharpoons NiCl^{+}$	0.6[b]
$Cd^{+2} + 2Cl^{-} \rightleftharpoons CdCl_2^{0}$	2.6	$Pb^{+2} + Cl^{-} \rightleftharpoons PbCl^{+}$	1.59
$Cd^{+2} + 3Cl^{-} \rightleftharpoons CdCl_3^{-}$	2.4	$Pb^{+2} + 2Cl^{-} \rightleftharpoons PbCl_2^{0}$	1.8
$Cd^{+2} + 4Cl^{-} \rightleftharpoons CdCl_4^{-2}$	1.7	$Pb^{+2} + 3Cl^{-} \rightleftharpoons PbCl_3^{-}$	1.7
$Co^{+2} + Cl^{-} \rightleftharpoons CoCl^{+}$	0.5[b]	$Pb^{+2} + 4Cl^{-} \rightleftharpoons PbCl_4^{-2}$	1.4
$Cr^{+3} + Cl^{-} \rightleftharpoons CrCl^{+2}$	0.1[b]	$Sn^{+2} + Cl^{-} \rightleftharpoons SnCl^{+}$	1.64
$Cu^{+2} + Cl^{-} \rightleftharpoons CuCl^{+}$	0.4	$Sn^{+2} + 2Cl^{-} \rightleftharpoons SnCl_2^{0}$	2.43
$Cu^{+} + 2Cl^{-} \rightleftharpoons CuCl_2^{-}$	5.5[c]	$Tl^{+3} + Cl^{-} \rightleftharpoons TlCl^{+2}$	7.72
$Fe^{+3} + Cl^{-} \rightleftharpoons FeCl^{+2}$	1.48	$Tl^{+} + Cl^{-} \rightleftharpoons TlCl^{0}$	0.49
$Fe^{+3} + 2Cl^{-} \rightleftharpoons FeCl_2^{+}$	2.13	$Zn^{+2} + Cl^{-} \rightleftharpoons ZnCl^{+}$	0.43
$Hg^{+2} + Cl^{-} \rightleftharpoons HgCl^{+}$	7.2[b]	$Zn^{+2} + 2Cl^{-} \rightleftharpoons ZnCl_2^{0}$	0.61
$Hg^{+2} + 2Cl^{-} \rightleftharpoons HgCl_2^{0}$	14.0[b]	$Zn^{+2} + 3Cl^{-} \rightleftharpoons ZnCl_3^{-}$	0.5

a. I = 1.0

b. From Morel [10]

c. 20°C

Source: Martell and Smith [9], except as indicated

6.10.6 Literature Cited

1. Bingham, F.T., G. Sposito and J.E. Strong, "The Effect of Chloride on the Availability of Cadmium," *J. Environ. Qual.*, **13**, 71-74 (1984).
2. Bohn, H.L., B.L. McNeal and G.A. O'Connor, *Soil Chemistry*, John Wiley & Sons, New York (1979).
3. Bowen, H.J.M., *Environmental Chemistry of the Elements*, Academic Press, London (1979).
4. Bubeck, R.C., W.H. Diment, B.L. Deck, A.L. Baldwin and S.D. Lipton, "Runoff of Deicing Salt: Effect on Irondequoit Bay, Rochester, New York," *Science*, **172**, 1128-32 (1972).
5. Feick, G., R.A. Horne and D. Yeaple, "Release of Mercury from Contaminated Freshwater Sediments by the Runoff of Road Deicing Salt," *Science*, **175**, 1142-43 (1972).
6. Freeze, R.A. and J.A. Cherry, *Groundwater*, Prentice-Hall, Englewood Cliffs, N.J. (1979).
7. Hahne, H.C.H. and W. Kroontje, "Significance of pH and Chloride on Behavior of Heavy Metal Pollutants: Mercury (II), Cadmium (II), Zinc (II), and Lead (II)," *J. Environ. Qual.*, **2**, 444-50 (1973).
8. Hem, J.D., *Study and Interpretation of the Chemical Characteristics of Natural Water*, Water Supply Paper No. 1473, 2nd ed., Geological Survey, U.S. Department of the Interior, Washington, D.C. (1970).
9. Martell, A.E. and R.M. Smith, *Critical Stability Constants,* Vol. 4, *Inorganic Complexes* and Vol. 5, *First Supplement,* Plenum Press, New York (1976 and 1982).
10. Morel, F.M.M., *Principles of Aquatic Chemistry*, John Wiley & Sons, New York (1983).
11. Sillén, L.G. and A.E. Martell, *Stability Constants of Metal-Ion Complexes*, The Chemical Society, London, Special Publications No. 17 (1964) and No. 25 (1971).
12. Snoeyink, V.L. and D. Jenkins, *Water Chemistry*, John Wiley & Sons, New York (1980).
13. Stumm, W. and J.J. Morgan, *Aquatic Chemistry*, 2nd ed., John Wiley & Sons, New York (1981).
14. World Health Organization, *Guidelines for Drinking-Water Quality. Vol. 1: Recommendations*, Geneva (1984).

6.11 SULFATE (SO_4^{-2})

6.11.1 Occurrence and Uses

Sulfur accounts for 0.1% of the earth's crust on average [4], being widely distributed in reduced form (S^{-2}) as metallic sulfides in both igneous and sedimentary rocks. Sulfate (SO_4^{-2}) occurs primarily in evaporite sediments and, to a much lesser extent, in certain igneous-rock minerals of the feldspathoid group [14]. Gypsum ($CaSO_4 \cdot 2H_2O$) and anhydrite ($CaSO_4$) are the major sulfate minerals; barium and strontium sulfates are relatively rare, and sodium sulfate occurs in some closed basins [14]. In seawater, sulfate is the second most abundant anion (next to Cl^-), with a concentration of 2.7 g/kg [11]; in free or complexed form, it accounts for virtually all the sulfur in the oceans.

Two metal sulfates play important roles in the purification of drinking waters and wastewaters. Aluminum sulfate, $Al_2(SO_4)_3 \cdot 14.3\ H_2O$, and ferrous sulfate (copperas), $FeSO_4 \cdot 7H_2O$, are the most common coagulants used in water treatment [7]. Both react to form hydroxide flocs, which remove turbidity from the water through coagulation and flocculation.

6.11.2 Environmental Importance

Sulfate is generally not considered toxic to plants or animals. In man, high concentrations have a laxative effect, which is enhanced when sulfate is associated with magnesium [24]. A solution of 1000 mg/l of SO_4^{-2} has a purgative effect on adults, and lower concentrations may affect children. While sulfates generally have less influence on the taste of drinking water than do chlorides and carbonates, the taste threshold varies according to the accompanying cation from 200 to 500 mg SO_4^{-2}/l. *Guidelines for Drinking-Water Quality*, published by The World Health Organization [24], recommends a sulfate concentration limit of 400 mg/l SO_4^{-2} based on taste considerations, while the U.S. EPA recommends the somewhat lower maximum of 250 mg/l [8].

Plants obtain sulfur, which is essential to all organisms [3], primarily through their roots in the form of sulfate. Although sulfate is not toxic to plants, the acidity that often accompanies its production (during the oxidation of pyrite, for example) may be. The production of sulfate and hydrogen ions by the oxidation of pyrite is described in § 6.9.3, and other sulfur oxidation processes that have the effect of reducing the pH of soil are discussed below. The drainage of marine floodplains containing sediments with high concentrations of sulfides may also lead to low-pH soils, because H^+ ions are produced when sulfides are oxidized to sulfate [2].

6.11.3 Sulfate Transport through the Environment

The transport of sulfate through the environment is considered in more detail in section 8.2, which examines the entire sulfur cycle. The oxidation of gaseous sulfur dioxide in the atmosphere, which leads to the formation of sulfuric acid, is described in § 6.9.3.

6.11.4 Sulfate Chemistry

ACID-BASE CHEMISTRY

Sulfuric acid, H_2SO_4, is a strong acid and dissociates completely in dilute aqueous solution to form the bisulfate ion, HSO_4^-:

$$H_2SO_4 \rightleftharpoons H^+ + HSO_4^- \quad K_{a,1} = 10^3 \text{ [23]}$$

The bisulfate ion is a much weaker acid:

$$HSO_4^- \rightleftharpoons SO_4^{-2} + H^+ \quad K_{a,2} = 10^{-1.9} \text{ [23]}$$

but is still stronger than many common acids, such as phosphoric and acetic. Thus, the sulfate ion, the conjugate base of bisulfate, is a relatively weak base, having a pK_b of 12.1 [23]. Except in strongly acid waters, virtually all bisulfate will dissociate. Bisulfate predominates only below a pH of about 1.8; at pH 3.9, only about 1% of the sulfate exists as HSO_4^- [14]. In most natural waters the acid-base behavior of SO_4^{-2} is of no concern.

REDOX REACTIONS

The oxidation-reduction reactions of sulfate are more complex than its acid-base reactions, because sulfur can exist in oxidation states ranging from -2 to $+6$. As Figure 6.11-1 shows, sulfate (in which sulfur exists in its fully oxidized hexavalent form) is the predominant form of sulfur over most of the pH and Eh range encountered in natural waters. This diagram refers to a solution of 96 mg/l SO_4^{-2}; with decreasing or increasing concentrations, the boundaries of the solid sulfur region would expand or contract respectively. The S^{-2} ion occurs only at very high pH, while HSO_4^-, as previously discussed, dominates only in very acidic ($pH \leq 2$) solutions.

Although it is not evident from Figure 6.11-1, the slowness of the sulfur redox reactions allows non-equilibrium forms to persist for significant periods [14]. For example, in waters that contain dissolved oxygen, equilibrium calculations indicate that sulfide could not exist; however, data collected in South San Francisco Bay showed that sulfide could be detected in waters with up to 3 mg/l of dissolved oxygen [22].

Microorganisms are important in mediating redox reactions involving sulfate that otherwise would be extremely slow. Unless certain species of bacteria with a suitable food or energy supply are present, very little sulfate reduction occurs in natural systems [14]. However, nearly all anaerobic, organically rich environments — marine, estuarine and fresh-water sediments and waterlogged soils such as marshes and paddy fields — contain such bacteria [5]. *Desulfovibrio desulfuricans* is the most common sulfate reducer in waterlogged soils and fresh-water sediments and, like all other species of sulfate-reducing bacteria, is an obligate anaerobe [5]. The pH range in which these bacteria can grow is narrow, and no growth occurs below pH 5.5 [1]. The reduction of sulfate can have profound effects on an ecosystem, because the hydrogen sulfide produced is highly toxic to most life forms other than some tolerant bacteria and algae [6,9].

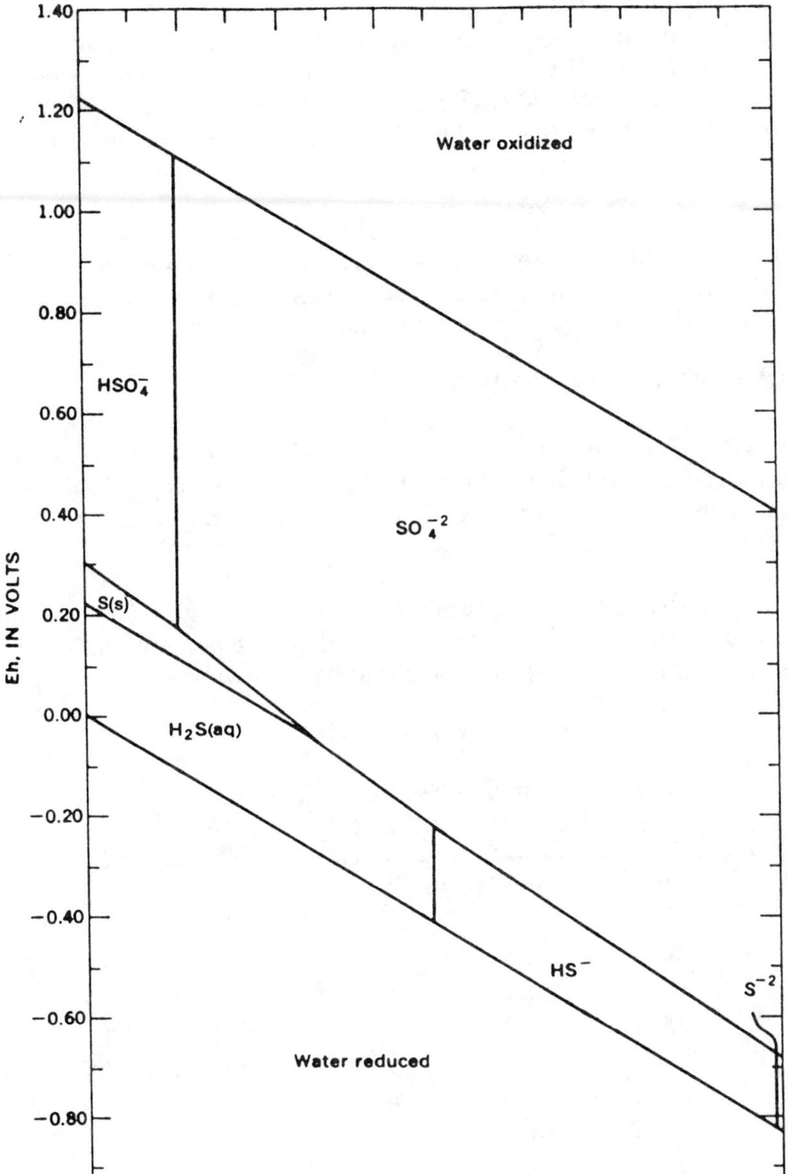

Total dissolved sulfur activity = 96 mg/l as SO_4^{-2}

Source: Hem [14]

FIGURE 6.11-1 Fields of Dominance of Sulfur Species at Equilibrium at 25°C and 1 Atmosphere

Microbes also play an important role in reactions of elemental sulfur and reduced sulfur compounds to form sulfate; if they are not present, the rates of such reactions are very slow. For example, the oxidation of pyrite (FeS_2) is catalyzed by the presence of certain autotrophic bacteria [23]. Those of the genus *Thiobacillus*, which are the best known sulfur-oxidizing bacteria, exist in small numbers in most soils, but they proliferate when elemental sulfur or reduced sulfur species are added [1]. The pH optima for Thiobacilli bacteria range from 2 to 3.5 for *T. thiooxidans* and *T. ferrooxidans* to neutral or slightly alkaline for *T. denitrificans*, *T. thioparus* and *T. novellus* [5].

Brown [5] has noted that the production of sulfate by bacteria is important not only in making sulfur available to plants but also in reducing the pH of the soil. The reduced pH may lead to the release of otherwise insoluble forms of PO_4^{-3}, Mg^{+2}, Ca^{+2}, Mn^{+2}, Fe^{+2}, Fe^{+3} and Al^{+3}, sometimes to toxic concentrations [1].

COMPLEXATION AND SOLUBILITY

Sulfate tends to associate with cations. These associations generally occur as ion pairs, such as $NaSO_4^-$ and $CaSO_4^0$, rather than inner-sphere complexes. As sulfate concentrations in solution increase, a greater proportion of sulfate ions becomes associated [14].

The extent of ion association in natural waters is difficult to measure, and reported values vary widely. Table 6.11-1 compares the calculated percentages of total aqueous sulfate associated in various forms for simulated seawaters and fresh water. The

TABLE 6.11-1

Sulfate Ion Association
(percentage of total SO_4^{-2} at 25°C)[a]

	Free SO_4^{-2}	$NaSO_4^-$	KSO_4^-	$MgSO_4^0$	$CaSO_4^0$	Source
Seawater						
	15.9	61.8	1.6	17.3	3.5	[16][b]
	39.0	37.1	0.4	19.5	4.0	[20]
	40	37	1	19	3	[18]
	62.9	16.4	0.5	17.4	2.8	[13]
	35	26	1	33	6	[22][c]
	54	21	1	21.5	3	[10]
Fresh Water[d]						
	90	—	—	3	6	[18]
	90	—	—	4	4	[22][c]
	88	—	—	4	8	[23][c]

a. Percentages may not total 100% due to rounding.
b. Considers Cl^- ion associations in system, unlike other sources.
c. Adapted.
d. Composition varies from source to source.

largest discrepancies in seawater speciation between the various investigators occur in the fraction of free sulfate and that associated with sodium. These discrepancies can be explained by the use of differing K values, the assumption of different salt concentrations, and the consideration of different systems of complexes. Part of the difference between the first row of values and the others may be explained by the fact that it is the only chemical system to consider chloride complexation. Much less sulfate association occurs in fresh water, as shown by the consistently higher percentages of free sulfate in that medium.

The equilibrium constants for associated sulfate species are listed in Table 6.11-2. As Morel has noted [18], the primarily electrostatic binding of the sulfate ion is indicated by the narrow range of equilibrium constants (~log K = 2.2-2.8) for sulfate associations with divalent metal ions.

TABLE 6.11-2

Sulfate Stability Constants
(at 25°C and zero ionic strength unless otherwise indicated)

Complex	log K or log β	Complex	log K or log β
$AgSO_4^-$	1.3	$HgSO_4^0$	2.5[a]
$BaSO_4^0$	2.7	$Hg(SO_4)_2^{-2}$	3.6[a]
$CaSO_4^0$	2.31	KSO_4^-	0.85
$CdSO_4^0$	2.46	$MgSO_4^0$	2.23
$Cd(SO_4)_2^{-2}$	3.2[a]	$MnSO_4^0$	2.26
$Cd(SO_4)_3^{-4}$	2.7[a]	$NH_4SO_4^-$	1.11[b]
$CoSO_4^0$	2.36	$NaSO_4^-$	0.70
$CrSO_4^+$	3.0[a]	$NiSO_4^0$	2.32
$CuSO_4^0$	2.36	$PbSO_4^0$	2.75
$FeSO_4^0$	2.2	$SrSO_4^0$	2.55
$FeSO_4^+$	4.04	$ZnSO_4^0$	2.38
$Fe(SO_4)_2^-$	5.38	$Zn(SO_4)_2^{-2}$	3.1[a]

a. From Morel [18]

b. 18°C

Source: Martell and Smith [17] except as indicated

Solubility products of some sulfates are listed in Table 6.11-3. The solubilities (on a mass of sulfate per mass of solution basis) vary widely; gypsum, the most widely occurring sulfate mineral, is of moderate solubility — i.e., greater than calcite ($CaCO_3$) and dolomite ($CaMg(CO_3)_2$) but less than halite (NaCl). The limit of sulfate solubility in natural waters is commonly related to the solubility of gypsum [19], due to the relative abundance of calcium in natural waters and the low solubility of gypsum as compared with sodium and magnesium sulfates.

TABLE 6.11-3

Sulfate Solubility in Water
(at 25°C and zero ionic strength)

Solid Phase	log K
Ag_2SO_4	−4.83
$BaSO_4$ (barite)	−9.96
$CaSO_4$ (anhydrite)	−4.62
$HgSO_4$	−6.13
$PbSO_4$	−7.79
$SrSO_4$ (celestite)	−6.50

Source: Martell and Smith [17]

6.11.5 Sulfate in Soil

Soil sulfates can be divided into the following categories [12]: easily soluble sulfate, adsorbed sulfate, and insoluble sulfate (e.g., co-precipitated with $CaCO_3$). Easily soluble sulfate usually accounts for 1-10% of total soil sulfur, but its concentration varies with depth, being low in sandy subsoils and very high in calcareous, poorly drained lower horizons [5]. In humic soils containing appreciable amounts of hydrated aluminum or ferric oxides, adsorbed sulfate is important; in calcareous soils, co-precipitation with $CaCO_3$ may make sulfate unavailable to plants [5].

Similarly, Hue, Adams and Evans [15] state that sulfate retention in acid soils has been shown to increase with decreasing pH or organic matter content, or with increasing solution SO_4^{-2}, kaolinite, Fe oxide or Al oxide content of the soil. They note, however, that acid (pH<5) horizons of forest soils have been reported to show weak or no correlation between SO_4^{-2} retention and soil pH, clay content, or extractable Al fraction, but these soils did show a strong negative correlation between sulfate retention and organic matter content.

In general, soils retain sulfate weakly [21]. Like HCO_3^- and CO_3^{-2}, sulfate is less mobile than Cl^- and NO_3^-, which move through soils at virtually the same speed as the wetting front or displacing solution [2], but it is more mobile than strongly retained anions such as phosphate. Sulfate sorption isotherm data for several soils and soil minerals are given in Table 6.11-4.

TABLE 6.11-4

Langmuir Constants for Sulfate Sorption on Several Soils and Minerals

Soil or Mineral	Sorbate Conc. (*M*)	Electrolyte			Langmuir Constants[a]	
		Identity	Conc. (*M*)	pH	A_m	K_L
Clay Minerals:						
Allophane (synthetic)	$0\text{-}10^{-2.3}$	NaCl	0.01	5.0	33	3.2
Fe Oxides:						
Goethite (α-FeOOH)	$10^{-3.4}$	NaCl	0.01	5.0	29.4[b]	
Amorphous Fe Oxide	$10^{-5}\text{-}10^{-3}$	$NaNO_3$	0.1	5.0	1160	4.5
Soil 1 (Robertson)	$10^{-4.1}\text{-}10^{-2.8}$	$CaCl_2$	0.005	4.2[c]	11.1	3.6
Soil 2 (Wungong)	$10^{-4.1}\text{-}10^{-2.8}$	$CaCl_2$	0.005	5.2[c]	3.2	3.4
Soil 3 (Manjimup)	$10^{-4.1}\text{-}10^{-2.8}$	$CaCl_2$	0.005	6.2[c]	4.4	3.5
Soil 4	$0\text{-}10^{-2.6}$	KCl	0.02	6.0[c]	3.75	4.04
Soil 5	$0\text{-}10^{-2.6}$	KCl	0.02	5.4[c]	0.54	3.43
Soil 6	$0\text{-}10^{-2.6}$	KCl	0.02	5.9	1.88	3.94

Soil Description:

	Free Fe(%)	Free Al (%)	CEC (meq/100 g)
1 (Robertson)	16.1	5.2	1.2
2 (Wungong)	10.3	0.8	0.3
3 (Manjimup)	26.3	2.1	0.5

	Fe_2O_3 (%)	Gibbsite (%)	Kaolinite (%)	Organic Carbon (%)
4	18	33	11	0.7
5	16	8	33	2.3
6	17	10	43	1.0

a. A_m = Langmuir sorption maximum (μmol/g); K_L = Langmuir constant (log M^{-1})

b. Quantity of sulfate sorbed under the given conditions (μmol/g)

c. Soil pH

Source: Rai *et al*. [21] *(Copyright 1984, Electric Power Research Institute. Reprinted with permission.)*

6.11.6 Literature Cited

1. Alexander, M., *Introduction to Soil Microbiology*, John Wiley & Sons, New York (1977), as cited in Brown [5].
2. Bohn, H.L., B.L. McNeal and G.A. O'Connor, *Soil Chemistry*, John Wiley & Sons, New York (1979).
3. Bowen, H.J.M., *Environmental Chemistry of the Elements*, Academic Press, London (1979).
4. Brasted, R.C., "Sulfur, Selenium, Tellurium, Polonium and Oxygen" in *Comprehensive Inorganic Chemistry*, Vol. 8, D. Van Nostrand, New York (1961), as cited in Brown [5].
5. Brown, K.A., "Sulfur in the Environment: A Review," *Environ. Pollut.* (Series B) **3**, 47-80 (1982).
6. Castenholz, R.W., "The Effect of Sulphide on Blue-green Algae of Hot Springs. I: New Zealand-Iceland," *J. Phycol.*, **12**, 54-68 (1976), as cited in Brown [5].
7. Clark, J.W., W. Viessman, Jr. and M.J. Hammer, *Water Supply and Pollution Control*, 3rd ed., Harper and Row, New York (1977).
8. *Code of Federal Regulations*, 40CFR143 (July 1, 1985).
9. Czurda, V., "Hydrogen Sulfide as a Factor for the Growth of Green Algae," *Zentralbl. Bakteriol. Parasitenkd.* (II) **103**, 285-311 (1941), as cited in Brown [5].
10. Garrels, R.M. and M. Thompson, *Am. J. Sci.*, **260**, 57 (1962), as given in Stumm and Morgan [23].
11. Goldberg, E.D., "Chemistry — The Oceans as a Chemical System," in *Composition of Sea Water, Comparative and Descriptive Oceanography*, H.M. Hill (ed.), Vol. 2 of *The Sea*, Wiley-Interscience, New York, 3-25 (1963), as given in Snoeyink and Jenkins [22].
12. Halstead, R.L. and P.J. Rennie, "The Effects of Sulfur on Soils in Canada" in *Sulfur and Its Inorganic Derivatives in the Canadian Environment*, National Research Council of Canada Report No. 15015, 181-220 (1977), as cited in Brown [5].
13. Hanor, J.S., *Geochim. Cosmochim. Acta*, **33**, 894 (1969), as given in Riley, J.P. and R. Chester, *Introduction to Marine Chemistry*, Academic Press, London (1971).
14. Hem, J.D., *Study and Interpretation of the Chemical Characteristics of Natural Water*, 2nd ed., Water Supply Paper 1473, U.S. Geological Survey, Washington, D.C. (1970).
15. Hue, N.V., F. Adams and C.E. Evans, "Sulfate Retention by an Acid BE Horizon of an Ultisol," *Soil. Sci. Soc. Am. J.*, **49**, 1196-1200 (1985).
16. Johnson, J.S. and R.M. Pytkowicz, "Ion Association of Chloride and Sulfate with Sodium, Magnesium and Calcium in Seawater at 25°C," *Mar. Chem.*, **8**, 87-93 (1979).

17. Martell, A.E., and R.M. Smith, *Critical Stability Constants,* Vol. 4, *Inorganic Complexes* and Vol. 5, *First Supplement,* Plenum Press, New York (1976 and 1982).
18. Morel, F.M.M., *Principles of Aquatic Chemistry*, John Wiley & Sons, New York (1983).
19. Nriagu, J.O. and J.D. Hem, "Chemistry of Pollutant Sulfur in Natural Waters," in *Sulfur in the Environment. Part II: Ecological Impacts*, J.O. Nriagu (ed.), John Wiley & Sons, New York, 211-70 (1978).
20. Pytkowicz, R.M. and J.E. Hawley, "Bicarbonate and Carbonate Ion Pairs and a Model of Seawater at 25°C," *Limnol. Oceanogr.*, **19**, 223-34 (1974), as given in Johnson and Pytkowicz [16].
21. Rai, D., J. Zachara, A. Schwab, R. Schmidt, D. Girvin and J. Rogers, *Chemical Attenuation Rates, Coefficients and Constants in Leachate Migration. Vol. I: A Critical Review*, Report EA-3356 to EPRI by Pacific Northwest Laboratories (Battelle Institute), Richland, Wash. (1984).
22. Snoeyink, V.L. and D. Jenkins, *Water Chemistry*, John Wiley & Sons, New York (1980).
23. Stumm, W. and J.J. Morgan, *Aquatic Chemistry*, 2nd ed., John Wiley & Sons, New York (1981).
24. World Health Organization, *Guidelines for Drinking-Water Quality. Vol. I: Recommendations*, Geneva (1984).

6.12 CARBONATES

6.12.1 Sources and Concentrations in Water and Soil

Carbonate minerals, primarily calcite ($CaCO_3$) and dolomite ($CaMg(CO_3)_2$), have been estimated to contain more than 83% of the carbon on earth; most of the remaining 17% is organic carbon in sediments [14]. Calcium carbonate sediments form in shallow waters of the oceans as marine organisms with $CaCO_3$ supporting structures die and accumulate on the sea bottom. Organisms that utilize and secrete calcium carbonate are usually found in warm or shallow waters, which are often supersaturated with $CaCO_3$. These organisms are, in effect, catalysts for its precipitation [6], and include several blue-green, green, yellow-green and brown algae, a group of red algae, most echinoderms and mollusks, and some barnacles [2]. However, organisms may also build their skeletal structures in waters undersaturated with $CaCO_3$ [3]. Deep ocean waters, unlike shallow ones, tend to be undersaturated and thus dissolve solid calcium carbonate sinking from above.

While carbonate precipitates can also result from chemical reactions in seawater, this is not common except in very warm and shallow bays. Calcite is the first salt to precipitate as seawater is evaporated [2]. Calcite precipitates may be converted to dolomite during diagenesis (the transformation of sediments into sedimentary rocks) by ion exchange with Mg^{+2}.

In soil systems, inorganic carbon exists mostly as $CaCO_3$. Carbonate species (e.g., H_2CO_3, HCO_3^-, CO_3^{-2}) naturally present in rain tend to accumulate in soils of arid areas. Where evapotranspiration exceeds precipitation, the flux of water into the soil is not sufficient to remove sparingly soluble salts like $CaCO_3$ from the soil [1]. Another source of calcium carbonate in soils is groundwater: as groundwater nears the surface, evapotranspiration and CO_2 loss to the atmosphere (which decreases $CaCO_3$ solubility) may cause $CaCO_3$ to precipitate [1]. Dolomite formation might be expected in calcite-containing soils where concentrations of Mg^{+2} in the solid phase exceed 5 mole %, the maximum isomorphic substitution possible in calcite; however, this process seems to be rare, apparently requiring a pH of 9.5 or higher [1].

The content of CO_2 plays an important role in soil chemistry. The amount of CO_2 in the soil solution is a function of microbial and root activity and the rate of upward diffusion of CO_2. In turn, the CO_2 and calcium concentrations and the soil solution pH determine the speciation of inorganic carbon in the soil. Few metal carbonates can form in acid soils, since the concentration of the carbonate ion in solution is small at low pH, being converted to HCO_3^- and H_2CO_3 (see § 6.10.2). In alkaline soils, carbonate minerals become important and may determine the limits of the solubilities of many metal ions [8].

The partial pressure of CO_2 in well-drained agricultural soils ranges from 10 to 100 times the partial pressure of CO_2 in the atmosphere (0.0003 atm) [12]; 0.01 atm is often considered typical, although a lesser value might be more appropriate for drier and warmer soils [1]. Because the rate of CO_2 diffusion in aqueous solutions is several orders of magnitude less than that in air, flooded soils in which respiration (biological oxidation of organic matter) is taking place may develop P_{CO_2} concentrations close to the partial pressure of O_2 in the atmosphere (0.209 atm) [12] as O_2 is converted to CO_2 [1]. Such a high CO_2 concentration can lower the pH substantially. The change in CO_2 concentration (and hence pH and speciation) can occur within a few hours [1], leading to large differences between field and laboratory measurements. Redox conditions do not affect carbonate [1] but may affect the oxidation state of the associated cation (e.g., Fe^{+2} vs. Fe^{+3}) present in the system.

6.12.2 Speciation of CO_2 in the Aqueous Environment

The carbonate system in water is important because of the ubiquity of carbon dioxide and carbonate-bearing minerals in the environment. Carbon dioxide gas, after dissolving in water, is hydrolyzed to form carbonic acid, and the dissolution of carbonate minerals such as calcite results in the formation of carbonate and a cation. The major reactions of the aqueous carbonate system can be represented by the following equilibria:

$$CO_2(g) \rightleftharpoons CO_2^0 \qquad pK_H \qquad (1)$$

$$CO_2^0 + H_2O \rightleftharpoons H_2CO_3^0 \qquad K_{eq} \qquad (2)$$

$$H_2CO_3^0 \rightleftharpoons H^+ + HCO_3^- \qquad pK'_{a,1} \qquad (3)$$

$$HCO_3^- \rightleftharpoons H^+ + CO_3^{-2} \qquad pK_{a,2} \qquad (4)$$

$$CaCO_3(s) \rightleftharpoons Ca^{+2} + CO_3^{-2} \qquad pK_{so} \qquad (5)$$

Since it is difficult to distinguish between CO_2^0 and $H_2CO_3^0$ by chemical measurement, the conventional approach is to assume the existence of an $H_2CO_3{}^*$ species whose concentration is the sum of the $H_2CO_3^0$ and CO_2^0 concentrations [12]:

$$[H_2CO_3{}^*] = [CO_2^0] + [H_2CO_3^0] \qquad (6)$$

Consequently, the dissociation of carbonic acid is defined as:

$$H_2CO_3{}^* \rightleftharpoons H^+ + HCO_3^- \qquad pK_{a,1} \qquad (7)$$

In actuality, most of the hypothetical $H_2CO_3{}^*$ exists as CO_2^0. Reported values for the equilibrium constant for equation 2 ($K_{eq} = [H_2CO_3^0]/[CO_2^0]$ range from 1.01×10^{-3} to 2.86×10^{-3} at 25°C [14]; an intermediate value of $10^{-2.8}$ is typically used. Since K_{eq} is small, $[CO_2^0]$ is usually assumed equal to $[H_2CO_3{}^*]$ for purposes of calculation.

Table 6.12-1 shows the effect of temperature on the equilibrium constants of reactions 1, 4, 5 and 7.

TABLE 6.12-1

Temperature Dependence of Some Important Carbonate Equilibrium Constants
(at zero ionic strength)

Reaction	Equilibrium Constant, pK					
	5°C	10°C	15°C	20°C	25°C	40°C
$CO_2(g) \rightleftharpoons CO_2^0$; pK_H	1.20	1.27	1.34	1.41	1.47	1.64
$HCO_3^- \rightleftharpoons CO_3^{-2} + H^+$; $pK_{a,2}$	10.56	10.49	10.43	10.38	10.33	10.22
$CaCO_3(s) \rightleftharpoons Ca^{+2} + CO_3^{-2}$; pK_{so}	8.09	8.15	8.22	8.28	8.34	8.51
$H_2CO_3^* \rightleftharpoons HCO_3^- + H^+$; $pK_{a,1}$	6.52	6.46	6.42	6.38	6.35	6.30
$CaCO_3(s) + H^+ \rightleftharpoons Ca^{+2} + HCO_3^-$; $p(K_{so}/K_{a,2})$	−2.47	−2.34	−2.21	−2.10	−1.99	−1.71

Source: Larson and Buswell [7]. *(Copyright 1980, John Wiley & Sons. Reprinted with permission.)*

THE CLOSED AQUEOUS SYSTEM

In closed aqueous carbonate systems (e.g., deep oceans, some groundwater systems, rapid titration of samples), CO_2 exchange with the surrounding environment is minimal or very slow. This system (at 25°C) can be described by the equilibrium expressions of equations 4 and 7 above, along with the ion product of water and the sum of the inorganic carbon species, C_T, that make up the carbonate system, as follows:

$$[H^+][OH^-] = 10^{-14} \tag{8}$$

$$\frac{[H^+][HCO_3^-]}{[H_2CO_3^*]} = 10^{-6.35} \tag{9}$$

$$\frac{[H^+][CO_2^{-2}]}{[HCO_3^-]} = 10^{10.33} \tag{10}$$

$$C_T = [H_2CO_3^*] + [HCO_3^-] + [CO_3^{-2}] \tag{11}$$

Equation 9 expresses the equilibrium constant for the dissociation of $H_2CO_3^*$ (reaction 7).

The concentrations of the various carbonate species and OH^- can be plotted as a function of H^+ concentration for a fixed C_T concentration. This has been done in Figure 6.12-1 in terms of negative logarithms for a $C_T = 10^{-3}$ *M* solution, using constants valid at I = 0 and 25°C. When we introduce the possibility of solid $CaCO_3$ formation in the system, we can plot the maximum calcium concentration possible before the precipitate forms (dashed line). From Table 6.12-1 at 25°C,

$$[Ca^{+2}][CO_3^{-2}] = 10^{-8.3} = K_{so} \tag{12}$$

Thus,

$$[Ca^{+2}]_{max} = 10^{-8.3}/[CO_3^{-2}] \tag{13}$$

$$pCa^{+2}{}_{max} = 8.3 - pCO_3^{-2}$$

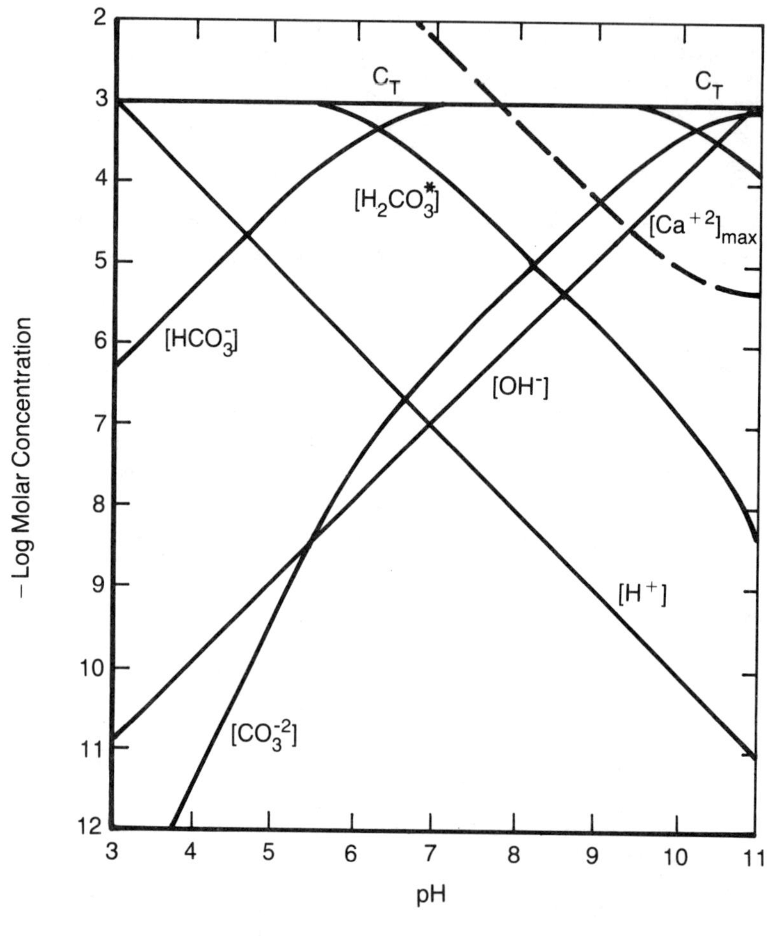

Dashed line represents maximum Ca^{+2} concentration before $CaCO_3$ precipitates.

Source: Adapted from Snoeyink and Jenkins [12]

FIGURE 6.12-1 Negative Log Concentration versus pH for the Various Species Present in a Closed Carbonate System at 25°C with $C_T = 10^{-3}$ *M* and I = 0

By an analogous procedure, similar plots can be made for other carbonate salts.

Note that the equilibrium values used to plot the curves in Figure 6.12-1 are for an ionic strength of zero and that the formation of calcium carbonate and calcium bicarbonate ion pairs is not considered in this analysis. At high ionic strengths (e.g., $>10^{-2}$ *M*, see § 2.6), the accuracy of the plots is diminished significantly.

The curves can be used to determine the concentrations of carbonate species and $[Ca^{+2}]_{max}$ when C_T differs from 10^{-3} *M*, because the concentrations of these species can be expressed as multiples of C_T. The curves are simply adjusted up or down by the appropriate amount. For example, for $C_T = 10^{-4}$ *M*, the various carbonate lines would be shifted down one log concentration unit, and that for $[Ca^{+2}]_{max}$ would be shifted up by the same amount.

THE OPEN SYSTEM

In an open system — i.e., where the carbonate solution is exposed to the atmosphere — the total carbonate concentration varies with the amount of atmospheric CO_2 that dissolves in the solution. Since the partial pressure of CO_2 is constant under normal atmospheric conditions, the concentration of aqueous CO_2 is also constant once equilibrium is attained. As noted earlier, almost all H_2CO_3* exists as CO_2(aq); therefore, in accordance with Henry's Law,

$$[H_2CO_3^*] = [CO_2^0] = K_H P_{CO_2(g)} \tag{14}$$

By replacing equation 11 for a closed system with equation 14 for an open system, one can plot the theoretical concentrations of various carbonate species for any desired partial pressure of CO_2 above the solution. In Figure 6.12-2, a P_{CO_2} of $10^{-3.5}$ atm (that of the normal atmosphere) and a temperature of 25°C are assumed, resulting in a H_2CO_3* concentration of 10^{-5} *M*. Again, the maximum Ca^{+2} concentration is plotted as a function of pH by using equation 13, and again the caveat regarding ionic strength effects and formation of ion pairs applies.

Figure 6.12-2 can be used to determine the pH of a solution of pure water exposed to the atmosphere. If the only carbonate species present are those resulting from the dissolution of atmospheric CO_2 in the sample, then a pH of about 5.6, corresponding to the point where $[H^+]$ equals $[HCO_3^-]$, results. The hydrogen and bicarbonate concentrations must be equal, because in any solution the total anion and cation charges must be equal. Where $[H^+] = [HCO_3^-]$, the concentrations of OH^- and CO_3^{-2} are negligible and can be ignored in the charge balance.

In a closed system, C_T is constant; in an open system exposed to a fixed P_{CO_2}, $[H_2CO_3^*]$ is constant, and each of the aqueous carbonate species is expressible as a multiple of P_{CO_2}. Hence, the plot can be used to determine the speciation in situations where P_{CO_2} is much higher than in the open atmosphere (e.g., soil waters). To find the concentration of various carbonate species at different partial pressures, the lines are shifted up or down in a way analogous to the shift based on C_T in a closed system. For example, for a 50-fold increase in P_{CO_2}, the plots for the carbonate species in Figure 6.12-2 would be shifted up by 1.7 units (log 50) and the $[Ca^{+2}]_{max}$ line would be shifted down by 1.7 units.

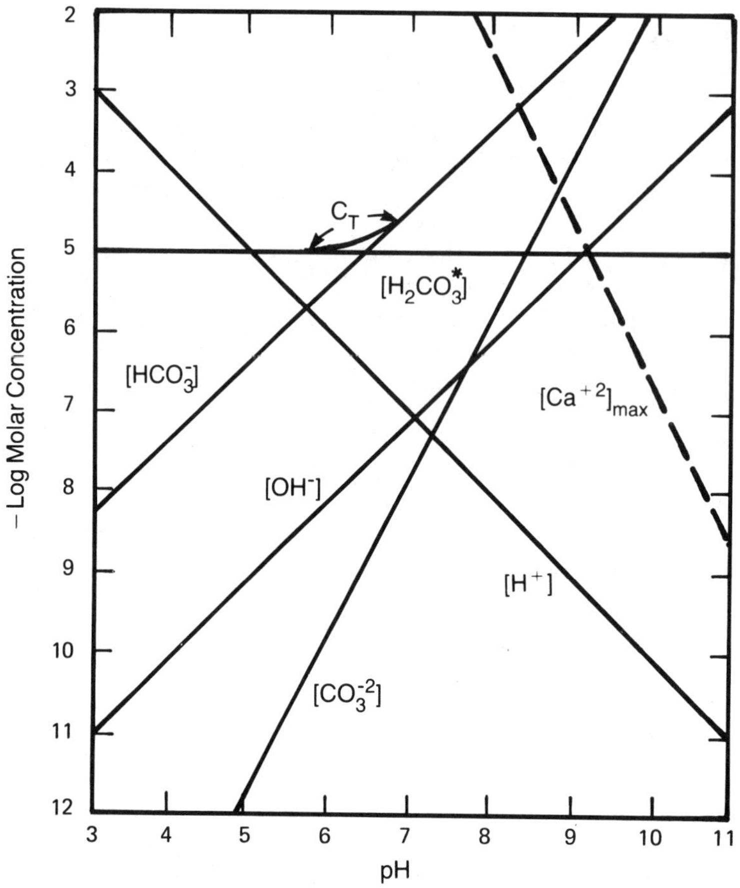

Dashed line represents maximum Ca^{+2} concentration before $CaCO_3$ precipitates.

Source: Adapted from Snoeyink and Jenkins [12]

FIGURE 6.12-2 Negative Log Concentration versus pH for the Various Species Present in an Open Carbonate System Exposed to $10^{-3.5}$ atm CO_2 at 25°C and I = 0

6.12.3 Alkalinity

Alkalinity is a quantitative measure of the ability to react with H^+ ions: the higher the alkalinity of a solution, the greater its capacity to react with H^+. Two kinds of alkalinity are in common use — total and carbonate. These are discussed below and summarized in Table 6.12-2.

TABLE 6.12-2

Alkalinity and Acidity Equations and Equivalence Points for the Carbonate System

Equivalence Point	Endpoint pH[a] (approx.)	Endpoint Indicator	Definition (Alkalinity and Acidity in eq/liter)
pH_{CO_2}	4.5	Methyl Orange	Total Alkalinity = $[HCO_3^-] + 2[CO_3^{-2}] + [OH^-] - [H^+]$ Mineral Acidity = $[H^+] - [HCO_3^-] - 2[CO_3^{-2}] - [OH^-]$
$pH_{HCO_3^-}$	8.3	Phenolphthalein	Carbonate Alkalinity = $[CO_3^{-2}] + [OH^-] - [H_2CO_3^*] - [H^+]$ Carbon Dioxide Acidity = $[H^+] + [H_2CO_3^*] - [CO_3^{-2}] - [OH^-]$
$pH_{CO_3^{-2}}$	~10.8	None[b]	Caustic Alkalinity = $[OH^-] - [HCO_3^-] - 2[H_2CO_3^*] - [H^+]$ Total Acidity = $[H^+] + [HCO_3^-] + 2[H_2CO_3^*] - [OH^-]$

a. True equivalence-point pH varies somewhat with total carbonate carbon, C_T.
b. Endpoint is poorly defined because of the buffering effect of water.

Source: Adapted from Snoeyink and Jenkins [12].

The alkalinity of fresh waters that are relatively unpolluted — i.e., free of weak acids other than H_2CO_3, such as H_3PO_4, H_4SiO_4, H_3BO_3 and organic acids — is closely related to the concentrations of their carbonate species. The *total alkalinity* is defined as [14]:

$$\text{Total Alk} = [HCO_3^-] + 2[CO_3^{-2}] + [OH^-] - [H^+] \quad (15)$$

and is measured (as meq/l) by the stoichiometric amount of strong acid required to convert the HCO_3^- and CO_3^{-2} to $H_2CO_3^*$ and OH^- to H_2O. Typical values of the total alkalinity of natural waters are listed in Table B-9 (Appendix B). Values for rivers are roughly 100 mg/l as $CaCO_3$.[1]

Carbonate alkalinity corresponds to the amount of acid needed to convert the carbonate to bicarbonate and OH^- to H_2O. It is defined as follows:

$$\text{Carbonate Alk} = [CO_3^{-2}] + [OH^-] - [H_2CO_3^*] - [H^+] \quad (16)$$

A third form of alkalinity, *caustic alkalinity*, corresponds to the amount of strong acid required to convert the OH^- to H_2O and is defined as:

$$\text{Caustic Alk} = [OH^-] - [HCO_3^-] - 2[H_2CO_3^*] - [H^+] \quad (17)$$

Caustic alkalinity is rarely used, owing to the difficulty of its measurement.

Total alkalinity is determined by titration to pH $\approx$ 4.5 (methyl orange indicator). Carbonate alkalinity is determined by titration to pH $\approx$ 8.3 (phenolphthalein indicator). The endpoint for the titration to determine caustic alkalinity is poorly defined, because of the buffering effect of water [12].

1. To convert from alkalinity measured in meq/l to mg/l as $CaCO_3$, multiply by 50 mg/l as $CaCO_3$ per meq/l.

In most natural waters, carbonate species are the predominant form of alkalinity. Other weak acids such as silicic, phosphoric, and boric (in seawater) as well as organic acids and a wide variety of dissolved or suspended materials may also contribute to the alkalinity of a water [12]. Where species other than carbonates contribute to alkalinity, the terms "carbonate" and "total" alkalinity should not be used, since they are based on the carbonate system; equations 15 and 16 no longer fully describe the ability of these systems to neutralize acid. The alkalinity is then determined simply as the amount of acid required to reach a particular pH.[2] For wastewaters in which weak acids and bases other than those of the carbonate system are present, an endpoint pH of 3.7 has been suggested [13].

For systems where carbonate species are solely or primarily responsible for alkalinity, Deffeyes [4] developed a series of plots relating total alkalinity to total carbonate carbon, C_T. Figure 6.12-3 is one of these plots for various values of pH at 25°C and one atmosphere pressure. If the pH and total carbonate carbon ($[H_2CO_3{}^*]$ + $[HCO_3^-]$ + $[CO_3^{-2}]$) are known, the alkalinity can be read off the ordinate.

Figure 6.12-3 can also be used to determine the effect of the addition of various chemical species to the system. The addition of a strong base or acid results in moves vertically upward or downward respectively by an amount equal to the milliequivalents of acid or base added. The addition of CO_2 increases the total carbonate carbon equivalently, moving the point parallel to the x-axis while the total alkalinity remains unchanged and the pH decreases. The addition of a soluble carbonate salt results in a 1:2 increase in total carbonate carbon vs. alkalinity, and the addition of a soluble bicarbonate salt results in a 1:1 increase. Dilution causes a proportional decrease in total carbonate carbon and total alkalinity.

The direction of the vectors for the addition of various species should be apparent from equation 15. Adding a strong base or acid is equivalent to adding OH^- or H^+ respectively. Bicarbonate increases the alkalinity and C_T at a 1:1 ratio and carbonate at a 2:1 ratio, as indicated by their coefficients in the total alkalinity equation. Dilution decreases alkalinity and total carbonate carbon by the same proportion, but not necessarily by the same absolute amount. Increasing CO_2^0 does not affect the alkalinity. (CO_2^0 does not appear in the total alkalinity equation, but it does increase C_T.)

To illustrate the use of Figure 6.12-3, let us assume that 0.5 millimole of $NaHCO_3$ and 0.5 millimole of Na_2CO_3 are added to one liter of water with a pH of 6 and a total carbonate carbon of 1 millimole/liter (point A). What is the pH and alkalinity of the resulting solution? The addition of 0.5 m*M* of $NaHCO_3$ increases the C_T to 1.5 m*M* and the alkalinity by 0.5 meq/l, from 0.3 to 0.8. The resultant pH is 6.4 (point B). Then the addition of 0.5 m*M* Na_2CO_3 brings C_T to 2.0 m*M* and increases the alkalinity by 1.0 meq/l to 1.8 meq/l, resulting in a pH of about 7.3 (point C). If this solution is then diluted with an equal quantity of pure water of pH 7, both the C_T and the alkalinity would be halved to 1.0 m*M* and 0.9 meq/l respectively, while the pH would remain about 7.3 (point D).

---

2. The endpoint pH is determined by the pK values of the other bases present.

Points A, B, C and D refer to the example explained in the text.

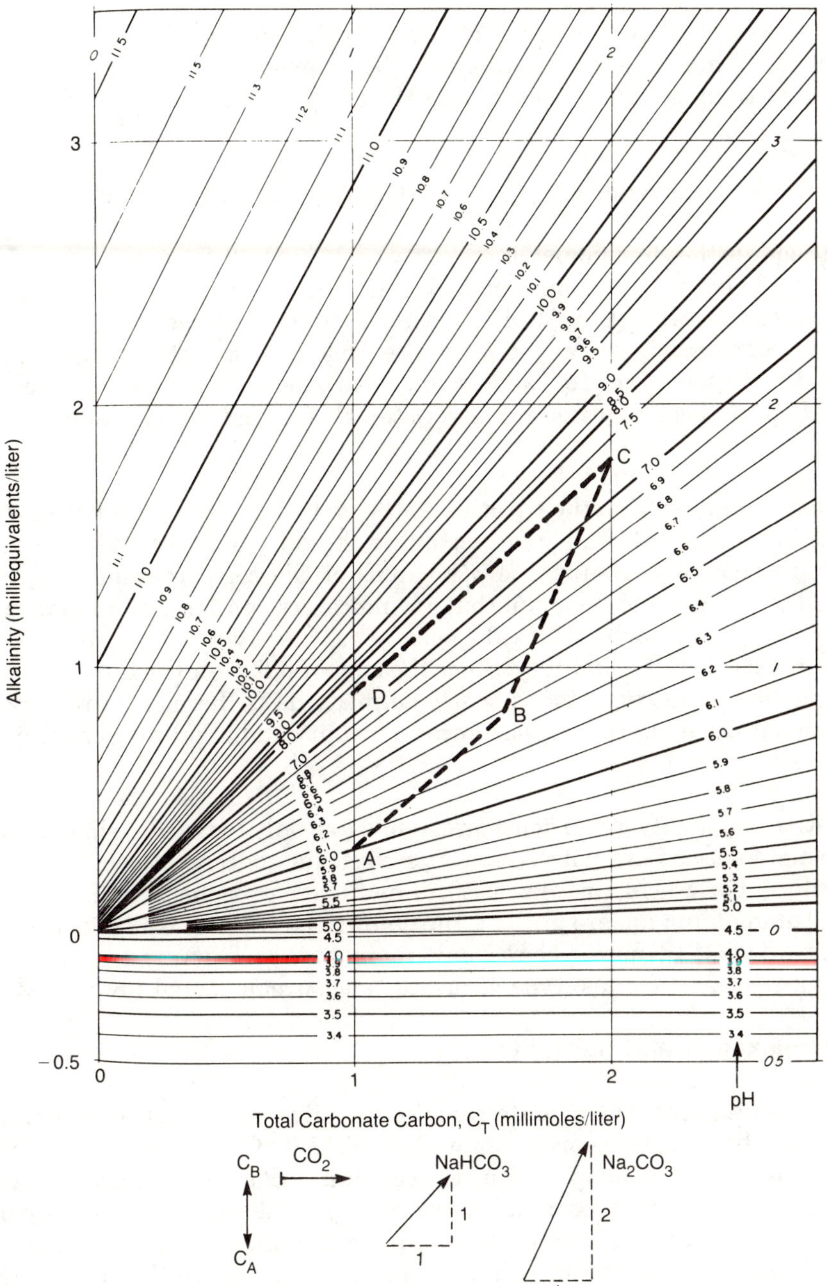

Source: Adapted from Deffeyes [4]. *(Copyright 1965, American Society of Limnology and Oceanography. Reprinted with permission.)*

FIGURE 6.12-3 Total Alkalinity versus Total Carbonate Carbon for a Series of pH Values at 25°C

6.12.4 Acidity

Acidity is a quantitative measure of the capacity to react with basic ions. The commonly used acidities are mineral and carbon dioxide, as shown in Table 6.12-2. If no acids other than carbonate species and H^+ are present, *mineral acidity* is defined as:

$$\text{Mineral Acidity} = [H^+] - [HCO_3^-] - 2[CO_3^{-2}] - [OH^-] \quad (18)$$

This value equals the stoichiometric quantity of strong base required to convert H^+ to H_2O, $H_2CO_3^*$ being the predominant carbonate species at this point. The amount of base needed to convert $H_2CO_3^*$ to HCO_3^- and H^+ to H_2O is known as *carbon dioxide acidity* and corresponds to the point where HCO_3^- is the predominant carbonate species. A third type of acidity, *total acidity*, is sometimes calculated; it equals the quantity of strong base required to convert the HCO_3^- and $H_2CO_3^*$ to CO_3^{-2} and H^+ to H_2O.

The solutions obtained from the titration to the endpoints for mineral acidity, carbon dioxide acidity, and total acidity correspond to those for total, carbonate and caustic alkalinity, respectively. Hence, the pH values of the corresponding endpoints are equal. Methyl orange and phenolphthalein indicators can be used to determine the mineral acidity and carbon dioxide acidity (respectively) as well as the total alkalinity and carbonate alkalinity. Since the endpoint for total acidity corresponds to that for caustic alkalinity, it too is poorly defined and is therefore rarely measured. The relationships among the various forms of acidity and alkalinity are summarized in Table 6.12-2.

The definitions of acidity given above are applicable only to waters where bases other than carbonate species are not present at significant concentrations. Where species such as silica, boric acid (in seawater), dihydrogen phosphate ($H_2PO_4^-$), hydrogen sulfide and ammonium ion are present, the definitions of total acidity (and alkalinity) must be modified [12]. These species would increase the acidity by an amount related to the number of acidic sites, type of acidity measurement and pK values.

6.12.5 Complexation and Solubility

Aqueous carbonate and bicarbonate complex readily with metal ions. Values for a number of stability constants are given in Table 6.12-3. (Other sources give somewhat different values.) Sodium, magnesium, and calcium carbonate complexes form in both fresh and salt waters. In fresh water, however, only small concentrations of carbonate and bicarbonate complexes form; using stability constants for major inorganic ligand-metal ion complexes, Stumm and Morgan [14] have calculated that 99% of the total carbonate carbon exists as free ions in a simulated fresh water.

Much larger proportions of CO_3^{-2} and HCO_3^- exist as complexes in seawater. In various models of seawater, 70-80% of HCO_3^- has been found to exist as free ions, 8-11% as $NaHCO_3^0$, 7-19% as $MgHCO_3^+$, 3-7% as $CaHCO_3^+$ and 0-0.4% as $KHCO_3^0$.

TABLE 6.12-3

Carbonate and Bicarbonate Stability Constants
(Complexation constants for M + CO_3 = MCO_3 or M + HCO_3 = $MHCO_3$ at 25°C and zero ionic strength)

Complex	log K or log β	Complex	log K or log β
$BaCO_3^0$	2.78	$CaHCO_3^+$	1.1
$CaCO_3^0$	3.15	$CdHCO_3^+$	1.8[a,c]
$CdCO_3^0$	4.4[a]	$CoHCO_3^+$	2.4[a,c]
$CoCO_3^0$	6.1[a,b]	$CuHCO_3^+$	2.7[a]
$CuCO_3^0$	6.75	$FeHCO_3^+$	1.10
$Cu(CO_3)_2^{-2}$	9.92	$HgHCO_3^+$	2.5[a,c]
$HgCO_3^0$	6.3[a,b]	$MgHCO_3^+$	1.01
$MgCO_3^0$	2.92	$MnHCO_3^+$	1.27
$MnCO_3^0$	5.0[a,b]	$NaHCO_3^+$	−0.25
$NaCO_3^-$	1.27	$NiHCO_3^+$	2.5[a,c]
$NiCO_3^0$	6.2[a,b]	$ZnHCO_3^+$	1.50
$ZnCO_3^0$	4.8[a]		

a. From Mantoura, Dickson and Riley [9]; temperature not specified but probably 25°C

b. Estimated from plot of log β versus electronegativity

c. Estimated from the relation log β_{MCO_3}/log β_{MHCO_3} = 2.5

Source: Martell and Smith [10], except as indicated

Even greater proportions of CO_3^{-2} are believed to exist in complexed form: the same models show roughly 8-10% of CO_3^{-2} existing as free ions, 16-19% as $NaCO_3^-$, 44-67% as $MgCO_3^0$, and 7-21% as $CaCO_3^0$ [14].

The solubilities of carbonate minerals are intermediate between those of highly soluble chlorides and sulfates (e.g., NaCl, $MgSO_4$) and highly insoluble phosphates (e.g., $Ca_3(PO_4)_2$). Equilibrium constants for the dissolution of carbonate compounds are listed in Table 6.12-4. Like the stability constants for complex formation, these values vary somewhat from source to source.

As described above, the concentrations of HCO_3^- and CO_3^{-2} in natural waters depend on the P_{CO_2} to which the solutions are exposed. Thus, bicarbonate and carbonate solubilities (mass/volume basis) vary with P_{CO_2}. Seidell [11] reports carbonate solubilities for a variety of conditions. Carbonate solubility may also be strongly affected by ionic strength. Barium carbonate, for example, is several times more soluble in water of ionic strength 0.1 (K_{sp} = 3.2×10^{-8}) than in water of zero ionic strength (K_{sp} = 4.9×10^{-9}) [5].

TABLE 6.12-4

Equilibrium Constants for Dissolution of Carbonate Compounds
(at 25°C and zero ionic strength)

	Reaction	log K
	$Ag_2CO_3(s) \rightleftharpoons 2Ag^+ + CO_3^{-2}$	−11.09
	$BaCO_3(s) \rightleftharpoons Ba^{+2} + CO_3^{-2}$	−8.30
Aragonite	$CaCO_3(s) \rightleftharpoons Ca^{+2} + CO_3^{-2}$	−8.22
Calcite	$CaCO_3(s) \rightleftharpoons Ca^{+2} + CO_3^{-2}$	−8.35
Dolomite	$CaMg(CO_3)_2(s) \rightleftharpoons Ca^{+2} + Mg^{+2} + 2CO_3^{-2}$	−16.7[a]
	$CdCO_3(s) \rightleftharpoons Cd^{+2} + CO_3^{-2}$	−13.74
	$CoCO_3(s) \rightleftharpoons Co^{+2} + CO_3^{-2}$	−9.98
	$CuCO_3(s) \rightleftharpoons Cu^{+2} + CO_3^{-2}$	−9.63
Malachite	$CuOH(CO_3)_{0.5}(s) + 2H^+ \rightleftharpoons Cu^{+2} + 3/2H_2O + 1/2CO_2(g)$	6.49[a]
Malachite	$Cu_2CO_3(OH)_2(s) \rightleftharpoons 2Cu^{+2} + CO_3^{-2} + 2OH^-$	−33.78
Azurite	$Cu_3(CO_3)_2(OH)_2(s) \rightleftharpoons 3Cu^{+2} + 2CO_3^{-2} + 2OH^-$	−45.96
Azurite	$Cu(OH)_{0.67}(CO_3)_{0.67}(s) + 2H^+ \rightleftharpoons Cu^{+2} + 4/3H_2O + 2/3CO_2(g)$	6.47[a]
Siderite	$FeCO_3(s) \rightleftharpoons Fe^{+2} + CO_3^{-2}$	−10.68
	$Hg_2CO_3(s) \rightleftharpoons Hg_2^{+2} + CO_3^{-2}$	−16.05
Magnesite	$MgCO_3(s) \rightleftharpoons Mg^{+2} + CO_3^{-2}$	−7.46
Hydromagnesite	$Mg_4(CO_3)_3(OH)_2 \cdot 3H_2O(s) \rightleftharpoons 4Mg^{+2} + 3CO_3^{-2} + 2OH^- + 3H_2O$	−29.5[a]
Nesquehonite	$MgCO_3 \cdot 3H_2O(s) \rightleftharpoons Mg^{+2} + CO_3^{-2} + 3H_2O$	−5.19[a]
Rhodochrosite	$MnCO_3(s) \rightleftharpoons Mn^{+2} + CO_3^{-2}$	−9.30
	$NiCO_3(s) \rightleftharpoons Ni^{+2} + CO_3^{-2}$	−6.87
Cerussite	$PbCO_3(s) \rightleftharpoons Pb^{+2} + CO_3^{-2}$	−13.13
Strontianite	$SrCO_3(s) \rightleftharpoons Sr^{+2} + CO_3^{-2}$	−9.03
	$ZnCO_3(s) \rightleftharpoons Zn^{+2} + CO_3^{-2}$	−10.00
	$Zn(OH)_{1.2}(CO_3)_{0.4}(s) + 2H^+ \rightleftharpoons Zn^{+2} + 1.6H_2O + 0.4CO_2(g)$	9.8[a]

a. From Stumm and Morgan [14]

Source: Martell and Smith [10], except as indicated

6.12.6 Literature Cited

1. Bohn, H.L., B.L. McNeal and G.A. O'Connor, *Soil Chemistry*, John Wiley & Sons, New York (1979).
2. Bowen, H.J.M., *Environmental Chemistry of the Elements*, Academic Press, London (1979).
3. Cloud, P.E., Jr., "Carbonate Precipitation and Dissolution," in *Chemical Oceanography*, Vol. 2, J.P. Riley and G. Skirrows (eds.), Academic Press, London (1965), as cited by Horne [6].
4. Deffeyes, K.S., "Carbonate Equilibria: A Graphic and Algebraic Approach," *Limnol. Oceanogr.*, **10**, 412-26 (1965).
5. Fritz, J.S. and G.H. Schenk, Jr., *Quantitative Analytical Chemistry*, Allyn and Bacon, Boston (1966).
6. Horne, R.S., *Marine Chemistry*, John Wiley & Sons, New York (1969).
7. Larson, T.E. and A.M. Buswell, "Calcium Carbonate Saturation Index and Alkalinity Interpretations," *J. Am. Water Works Assoc.*, **34**, 1664 (1942), as cited by Snoeyink and Jenkins [12].
8. Lindsay, W.J., *Chemical Equilibria in Soils*, John Wiley & Sons, New York (1979).
9. Mantoura, R.F.C., A. Dickson and J.P. Riley, "The Complexation of Metals with Humic Materials in Natural Water," *Estuarine Coastal Mar. Sci.*, **6**, 387-408 (1978).
10. Martell, A.E. and R.M. Smith, *Critical Stability Constants*, Vol. 4, *Inorganic Complexes* and Vol. 5, *First Supplement*, Plenum Press, New York (1976 and 1982).
11. Seidell, A., *Solubilities of Inorganic and Metal Organic Compounds*, 4th ed., American Chemical Society, Washington, D.C. (1970).
12. Snoeyink, V.L. and D. Jenkins, *Water Chemistry*, John Wiley & Sons, New York (1980).
13. *Standard Methods for the Examination of Water and Wastewater*, 13th ed., M.J. Taras, A.E. Greenberg, R.D. Hoak and M.C. Rand (eds.), American Public Health Association, Washington, D.C. (1971).
14. Stumm, W. and J.J. Morgan, *Aquatic Chemistry*, 2nd ed., John Wiley & Sons, New York (1981).

7. TRACE METALS

Madeline F. McComish, Joo Hooi Ong

LIST OF TABLES

ARSENIC (As)

BARIUM (Ba)

BERYLLIUM (Be)

CADMIUM (Ce)

CHROMIUM (Cr)

COPPER (Cu)

LEAD (Pb)

MANGANESE (Mn)

MERCURY (Hg)

NICKEL (Ni)

SELENIUM (Se)

THALLIUM (Tl)

ZINC (Zn)

LIST OF FIGURES

ANTIMONY (Sb)

ARSENIC (As)

BARIUM (Ba)

BERYLLIUM (Be)

CADMIUM (Cd)

CHROMIUM (Cr)

COPPER (Cu)

LEAD (Pb)

MANGANESE (Mn)

MERCURY (Hg)

NICKEL (Ni)

SELENIUM (Se)

SILVER (Ag)

THALLIUM (Tl)

ZINC (Zn)

INTRODUCTION

This chapter briefly discusses the important environmental chemistry of 15 trace metals/metalloids: antimony, arsenic, barium, beryllium, cadmium, chromium, copper, lead, manganese, mercury, nickel, selenium, silver, thallium, and zinc. These 15 elements were chosen for specific discussion primarily because of their environmental and health significance. Table D-1 in Appendix D shows the standards and criteria applicable to many inorganics; the metals/metalloids covered here are those most frequently mentioned in environmental and health standards and criteria.

The major/matrix metals were discussed in Chapter 6. Environmentally important organometallic compounds of the trace metals are discussed briefly in each metal subsection as applicable. Organometallics as a group are discussed in section 10.1. The occurrence of trace metals/metalloids in the environment is described in Appendix B. Information on uptake by biota, primarily terrestrial plants and freshwater and saltwater fishes, is provided in Chapter 4.

Information on each trace metal is provided in a separate subsection, in alphabetical order by element name. For each metal, information is provided on:

- Occurrence and uses
- Speciation reactions in water:
  - acid-base dissociation
  - complexation
  - redox reactions
- Solubility/precipitation
- Sorption on soils and sediments

Where appropriate, information is also provided on:

- Biotransformation
- Volatilization

The references are given at the end of the subsection on each metal. Additional sources of information that may be of interest to the reader are listed after the references.

7.1 ANTIMONY (Sb)

7.1.1 Occurrence and Uses

Antimony occurs in nature mainly as sulfides or oxides [1]. The principal antimony minerals are the native element (Sb), stibnite (Sb_2S_3), kermesite (Sb_2S_2O), senarmontite (Sb_2O_3), jamesonite ($2PbS \cdot Sb_2O_3$) and boulangerite ($5PbS \cdot 2Sb_2S_3$) [2].

The most important metallurgical use for antimony is as an alloying ingredient in lead and other metals, to increase hardness and stiffness. Antimonial lead is used principally as a grid metal alloy in lead-acid storage batteries. Other uses of antimony alloys include type metal, bearing metal, cable covering, and sheets and pipes. High-purity antimony (>99.999%) has limited application in the manufacture of semiconductor devices [5].

The chemical properties of antimony are similar to those of arsenic in many ways. It can exhibit a number of oxidation states: −3, 0, +3 and +5 [1]. There is, however, comparatively little information regarding its behavior in the natural environment.

7.1.2 Speciation Reactions in Water

Sb(III) and Sb(V) exist as hydroxo-complexed species in water under typical environmental conditions. Both strongly coordinate with oxygen. "Antimonous acid" is the name given to $Sb(OH)_3$, and "antimonic acid" is the name given to $Sb(OH)_5$. These formulas are just representations of Sb(III) and Sb(V) in water; the actual structures of the species in solution are not known [1]. The existence of Sb^{+3} or Sb^{+5} ions is improbable [2]. Cotton and Wilkinson [4] suggested that only Sb(III) has definite cationic chemistry. In the following discussion, the formulas $Sb(OH)_3$ and $Sb(OH)_5$ are used to represent these acids in solution.

Figure 7.1-1 shows the hydrolysis of Sb(III) in equilibrium with Sb_2O_3. Over the pH range expected in typical natural environments, $Sb(OH)_3^0$ is the dominant species. Figure 7.1-2 shows the hydrolysis of Sb(V) at a concentration of 10^{-5} *M* in solution. Over the pH range of typical natural environments, $Sb(OH)_6^-$ is the dominant species. The polynuclear hydroxo complexes $Sb_{12}(OH)_{64}^{-4}$ and $Sb_{12}(OH)_{65}^{-5}$ have been reported [6], but they are present only when total Sb concentrations are much higher (>10^{-3} *M*); under environmental conditions, Sb concentrations are too low to generate important levels of these complexes.

Figure 7.1-3 shows the speciation of antimony with varying pH and redox conditions in the Sb-H_2O system. In oxidizing environments, $Sb(OH)_6^-$ is dominant at pH values greater than about 3. $Sb(OH)_3^0$ is dominant in relatively reducing environments over a wide pH range (1.5-12) [6].

There is little information on the complexation of antimony with ligands other than OH^-. The Sb(III) complex with S^{-2}, $Sb_2S_4^{-2}$, has been suggested by Rai *et al.*; this compound would be important in reducing environments when S^{-2} concentrations

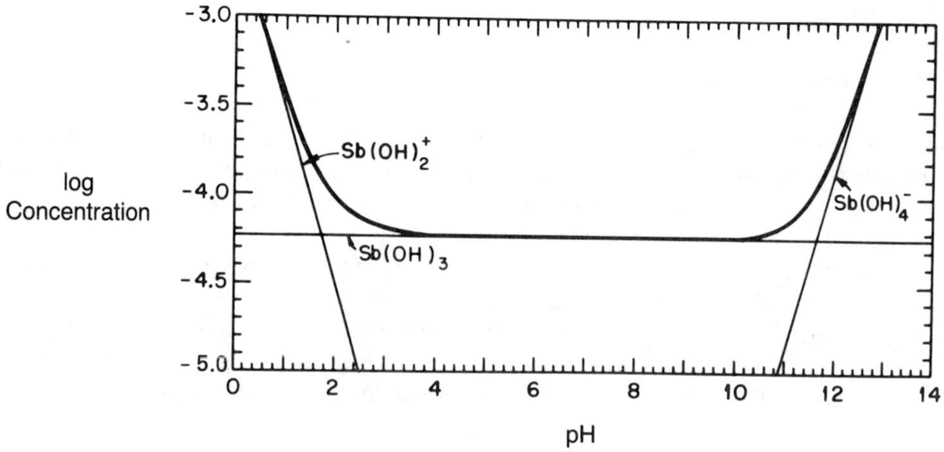

Source: Baes and Mesmer [1]. *(Copyright 1976, John Wiley & Sons. Reprinted with permission.)*

FIGURE 7.1-1 Calculated Hydrolysis Products of Sb(III) and Total Concentration of Dissolved Sb(III) (Heavy Curve) in Equilibrium with Sb_2O_3 at 25°C and I = 1 *M*

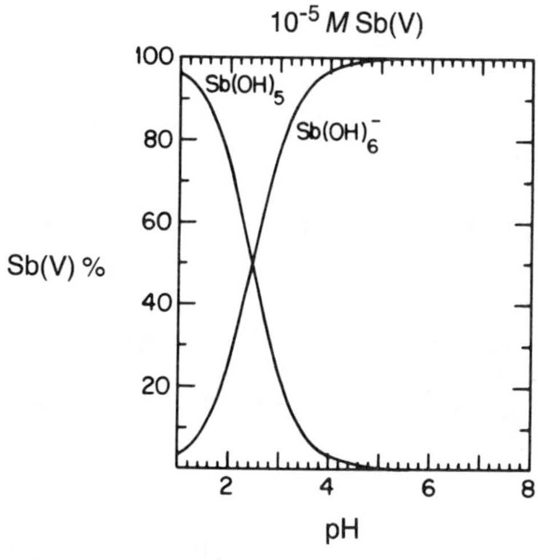

Source: Baes and Mesmer [1]. *(Copyright 1976, John Wiley & Sons. Reprinted with permission.)*

FIGURE 7.1-2 Calculated Distribution of Hydrolysis Species of Sb(V) at 10^{-5} *M* in 0.5 *M* $(CH_3)_4$ NCl at 25°C

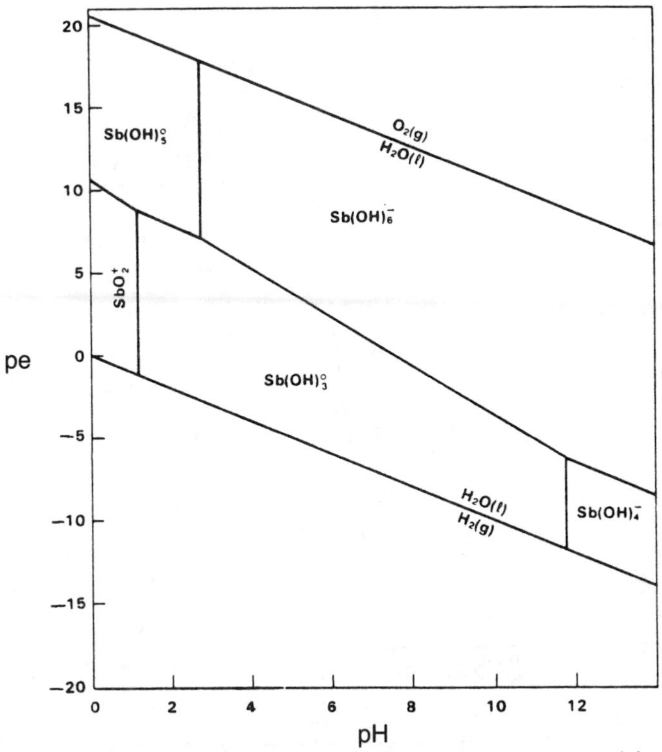

Source: Rai *et al.* [6]. *(Copyright 1984, Electric Power Research Institute. Reprinted with permission.)*

FIGURE 7.1-3 pe-pH Diagram for Sb-H_2O System at 25°C

exceed 10^{-6} *M* [6]. Insoluble antimony compounds in sediments can be converted to gaseous SbH_3 under reducing conditions. This mobile form of antimony can then react with water under oxidizing conditions to form soluble hydroxo-complexes, which could then be mobilized in the water system [2].

7.1.3 Solubility/Precipitation

The known Sb(V) and Sb(III) oxides, Sb_2O_5 and Sb_2O_3, are highly soluble. Figure 7.1-4 shows the dominant solution species in equilibrium with the least soluble solid phase with varying pe + pH. The concentration of total Sb under most environmental conditions is too low to cause the precipitation of Sb_2O_3 or Sb_2O_5. In the presence of sulfur, Sb_2S_3 (stibnite) may be the most likely precipitate; however, because of stable $Sb_2S_4^{-2}$ complex formation, Sb_2S_3(s) is not likely to form [6].

7.1.4 Sorption on Soils and Sediments

Callahan *et al.* [2] state that antimony may have an affinity for clay and other mineral surfaces. They also suggest that co-precipitation of antimony with hydrous

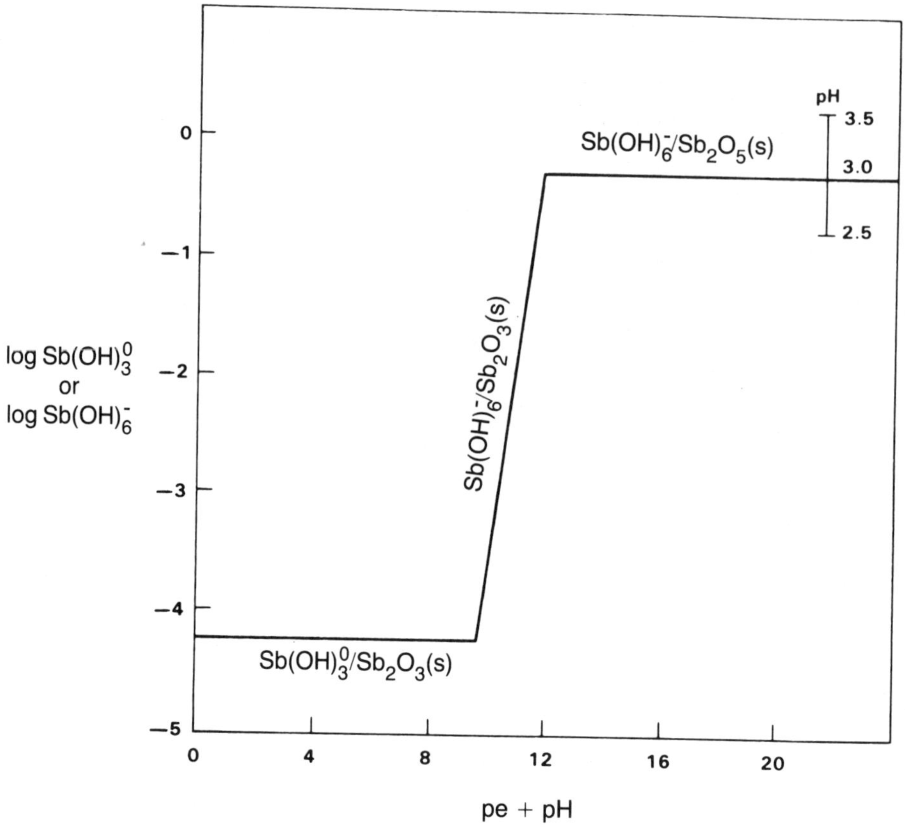

Source: Rai *et al.* [6]. *(Copyright 1984, Electric Power Research Institute. Reprinted with permission.)*

FIGURE 7.1-4 Solubility of Sb_2O_3(s) and Sb_2O_5 with Varying Redox Potential

Fe, Mn and Al oxides is a mechanism for controlling the mobility of Sb in the environment. They consider complexation with natural organics to be of little importance. Quantitative data on sorption of antimony was not found.

7.1.5 Biotransformation

Alkylated forms of antimony have been found in natural waters [7]. Biomethylation of antimony in the natural environment has not been clearly demonstrated but is considered likely [3] because of antimony's similarity and proximity to Sn, Pb, As, Se and Te in the periodic table. All of the latter elements have been shown to be biomethylated.

7.1.6 Volatilization

Stibine, SbH_3, and its methylated derivatives may be volatilized from aquatic systems. Since stibine is rapidly oxidized in water, the remobilization of these compounds as oxides in water is more likely than volatilization and oxidation in the atmosphere [2].

7.1.7 Literature Cited

1. Baes, C.F., Jr. and R.E. Mesmer, *The Hydrolysis of Cations*, John Wiley & Sons, New York (1976).
2. Callahan, M., M. Slimak, N. Gabel, I. May, C. Fowler, J. Freed, P. Jennings, R. Durfee, F. Whitmore, B. Maestri, W. Mabey, B. Holt and C. Gould, *Water-Related Environmental Fate of 129 Priority Pollutants. Volume I: Introduction and Technical Background, Metals and Inorganics, Pesticides and PCBs*, EPA-440/4-79-029a, EPA Contracts 68-01-3852 and 68-01-3867, Office of Water Planning and Standards, U.S. Environmental Protection Agency, Washington, D.C. (1979).
3. Challenger, F., "Biosynthesis of Organometallic and Organometalloidal Compounds," in *Organometals and Organometalloids*, F.E. Brinckman and J.M. Bellama (eds.), Symposium Series 82, American Chemical Society, Washington, D.C., 1-22 (1978).
4. Cotton, F.A. and G. Wilkinson, *Advanced Inorganic Chemistry*, Interscience Publishers, New York (1972).
5. *Kirk-Othmer Encyclopedia of Chemical Technology*, 3rd ed., Vol. 3, John Wiley & Sons, New York (1978).
6. Rai, D., J. Zachara, A. Schwab, R. Schmidt, D. Girvin and J. Rogers, *Chemical Attenuation Rates, Coefficients, and Constants in Leachate Migration. Vol I: A Critical Review*, Report EA-3356 to EPRI by Pacific Northwest Laboratories (Battelle Institute), Richland, Wash. (1984).
7. Turner, D.R., "Relationships Between Biological Availability and Chemical Measurements," in *Metal Ions in Biological Systems*, Vol. 18, H. Sigel (ed.), Marcel Dekker, New York, 137-64 (1984).

7.1.8 Additional Sources of Information

Doak, G.D. and L.D. Freedman, *Organometallic Compounds of Arsenic, Antimony, and Bismuth*, John Wiley & Sons, New York (1970).

McAuliffe, C.A., *Transition Metal Complexes of Phosphorus, Arsenic, and Antimony Ligand*, Krieger Publishers, Melbourne, Fla. (1973).

7.2 ARSENIC (As)

7.2.1 Occurrence and Uses

In its natural state, arsenic is usually associated with sulfide ores. Over 100 minerals and ores contain arsenic. The principal arsenic-bearing minerals include: arsenopyrite (FeAsS), niccolite (NiAsS), cobaltite (CoAsS), tennantite ($Cu_{12}As_4S_{13}$), enargite (Cu_3AsS_4), and native arsenic [5]. The principal arsenic compounds produced are arsenic trioxide (As_2O_3) and arsenic metal, from which other compounds are made. About 70% of all arsenic used is in pesticides, principally the following:

- Monosodium methanearsenate (MSMA) — $HAsO_3CH_3Na$
- Disodium methanearsenate (DSMA) — $Na_2AsO_3CH_3$
- Arsenic acid — H_3AsO_4
- Dimethylarsinic acid (cacodylic acid) — $(CH_3)_2AsO_2H$

Other uses of arsenic and arsenic compounds are in wood preservatives, glass manufacture, alloys, electronics, catalysts, feed additives, and veterinary chemicals [23].

There are many arsenic forms of environmental significance, including arsenious acids (H_3AsO_3, $H_2AsO_3^-$, $HAsO_3^{-2}$), arsenic acids (H_3AsO_4, $H_2AsO_4^-$, $HAsO_4^{-2}$), arsenites, arsenates, methylarsenic acid ($H_2AsO_3CH_3$), dimethylarsinic acid ($(CH_3)_2AsO_2H$), arsine (AsH_3), dimethylarsine ($HAs(CH_3)_2$) and trimethylarsine ($As(CH_3)_3$). These forms illustrate the various oxidation states that arsenic commonly exhibits (−III, 0, III, V) and the resulting complexity of its chemistry in the environment.

7.2.2 Speciation Reactions in Water

Arsenic(V) chemistry resembles that of phosphorus(V). In aqueous systems, it exhibits anionic behavior. In aerobic waters, arsenic acid predominates only at extremely low pH (<2); within a pH range of 2 to 11, it is replaced by $H_2AsO_4^-$ and $HAsO_4^{-2}$. Arsenious acid appears at low pH and under mildly reduced conditions, but is replaced by $H_2AsO_3^-$ as the pH increases. Only when the pH exceeds 12 does $HAsO_3^{-2}$ appear. At low pH in the presence of sulfide, $HAsS_2$ can form; arsine, arsine derivatives and arsenic metal can occur under extreme reducing conditions [24]. Figure 7.2-1 shows the speciation of arsenic under varying pH and redox conditions. Figures 7.2-2 and -3 show the composition of arsenic in aerobic and reduced systems, respectively.

Since it forms anions in solution, arsenic does not form complexes with simple anions like Cl^- and SO_4^{-2}, as do cationic metals. Rather, anionic arsenic complexes behave like ligands in water. Arsenic forms bonds with organic sulfur, nitrogen and carbon. Arsenic (+III) reacts with sulfur and sulfhydryl groups such as cystine, organic dithiols, proteins and enzymes, but it does not react with amine groups or organics with reduced nitrogen constituents. On the other hand, arsenic (+V) reacts with

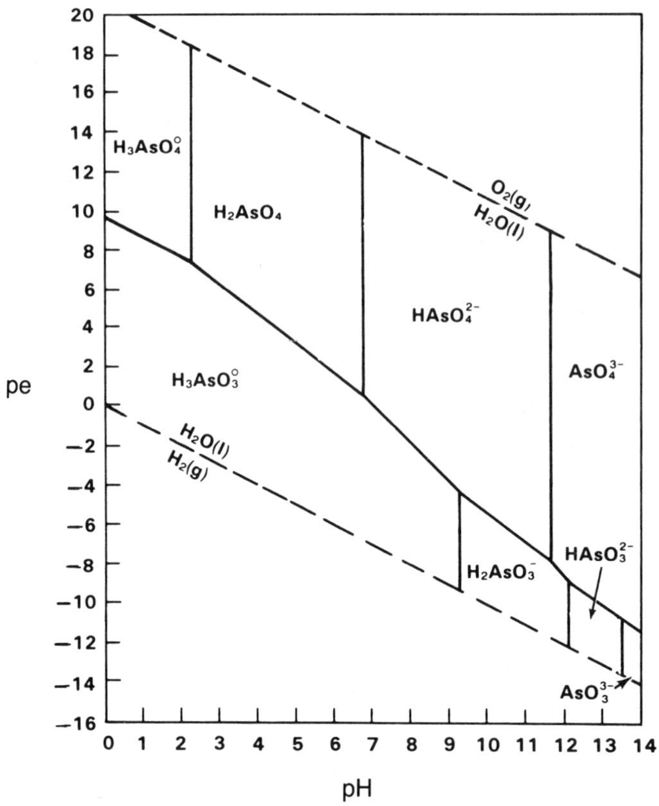

Source: Rai *et al.* [22]. *(Copyright 1984, Electric Power Research Institute. Reprinted with permission.)*

FIGURE 7.2-1 The pe-pH Diagram for Arsenic in Water at 25°C

reduced nitrogen groups such as amines but not sulfhydryl groups. Carbon forms organoarsenicals with both the trivalent and pentavalent forms [23]. The complexation of arsenic (+III and +V) by dissolved organic matter in natural environments prevents sorption and coprecipitation with solid-phase organics and inorganics; essentially, it increases the mobility of arsenic in aquatic systems and in the soil [5].

7.2.3 Solubility/Precipitation

The behavior of arsenic as a ligand suggests that the presence of metal cations could result in precipitation of arsenates and arsenites. Wagemann [25] studied barium, chromium, iron and calcium at typical freshwater concentrations as possible factors controlling the concentration level of total dissolved arsenic in solution. He postulated that at typical freshwater concentrations, Ba is the most likely controlling metal. Figure 7.2-4 shows the theoretical conditions for the precipitation of $Ba_3(AsO_4)_2$.

AsS and/or As_2S_3 may form in very anoxic environments where sulfur is present [8]. Some authors [5,8,22] suggest that $FeAsO_4$ may control As levels at high redox

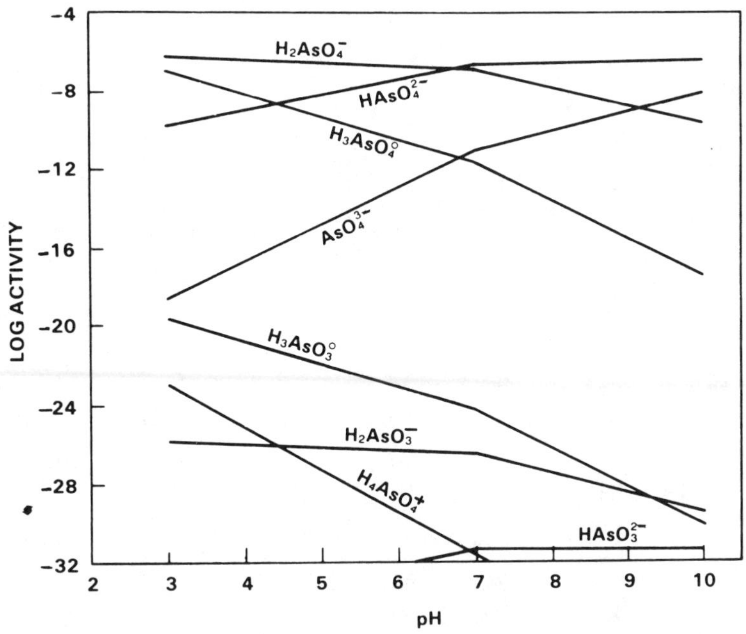

Total concentration of soluble As assumed to be 10^{-6} *M*.

Source: Rai *et al.* [22]. *(Copyright 1984, Electric Power Research Institute. Reprinted with permission.)*

FIGURE 7.2-2 Activities of As Aqueous Species Under Aerobic Conditions (pe + pH = 16)

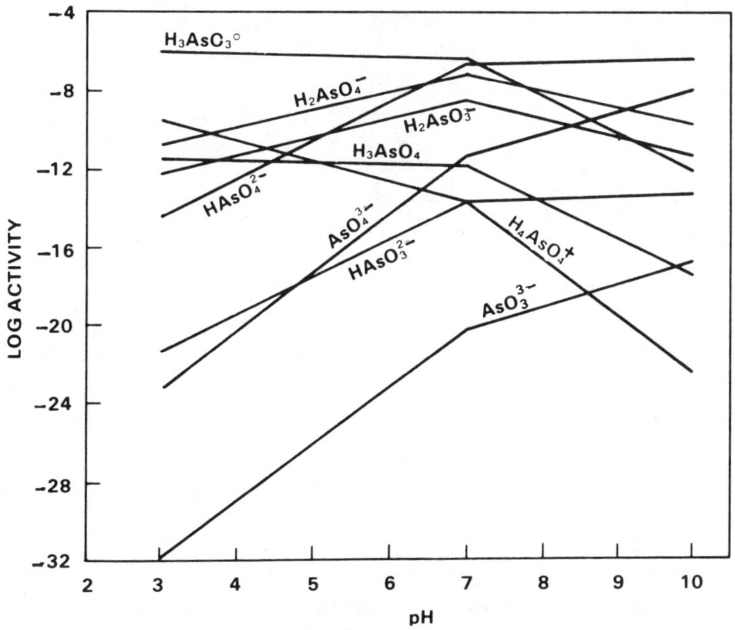

Total concentration of soluble As assumed to be 10^{-6} *M*.

Source: Rai *et al.* [22]. *(Copyright 1984, Electric Power Research Institute. Reprinted with permission.)*

FIGURE 7.2-3 Activities of As Aqueous Species Under Reduced Conditions (pe + pH = 7)

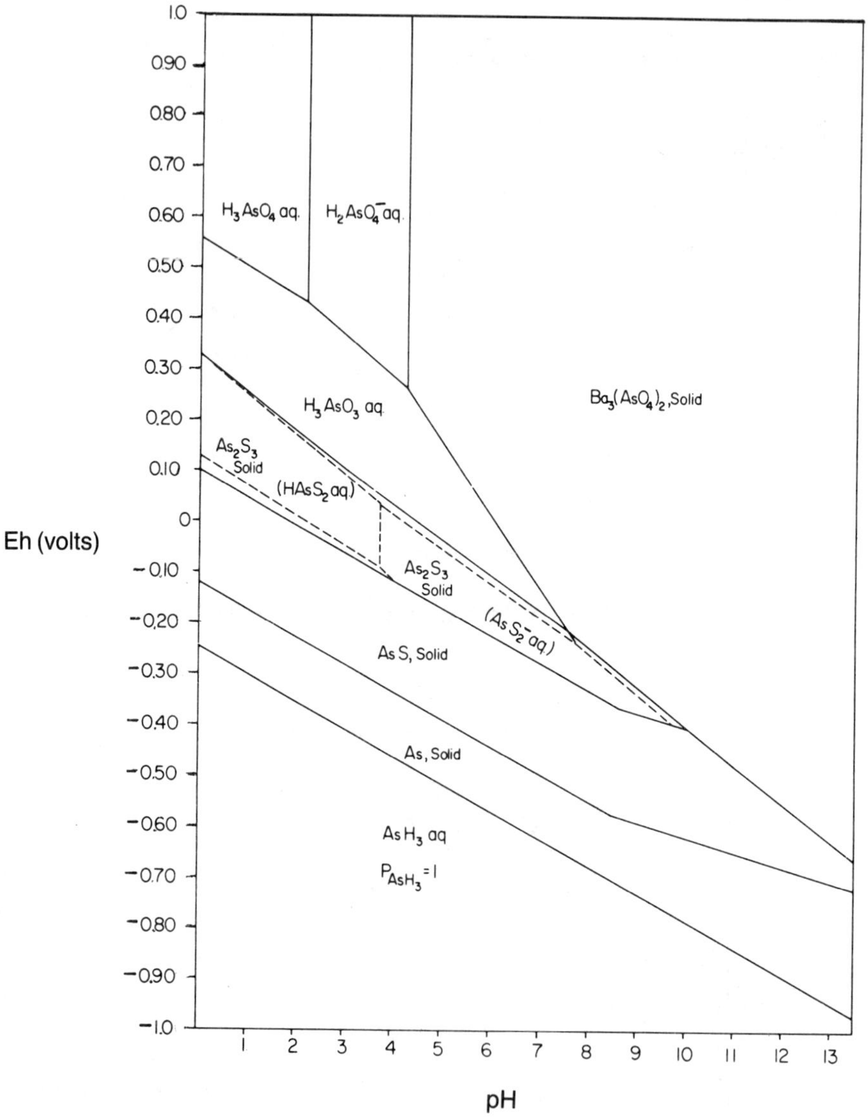

Source: Wagemann [25]. *(Copyright 1978, Pergamon Press, Inc. Reprinted with permission.)*

FIGURE 7.2-4 Fields of Stability for Important Arsenic Species at 25°C and 1 Atm in the Presence of 10^{-5} *M* Total Arsenic, 10^{-3} *M* Total Sulfur and 2.2×10^{-7} *M* Total Barium

(pe>12.5) and acidic pH (pH<2.3). $Pb_3(AsO_4)_2$ and $Mn_3(AsO_4)_2$ have also been suggested as possible solids in natural environments [14], but it has not been established whether the mechanism is precipitation or sorption onto solid surfaces.

The precipitation of arsenic depends largely on redox potential and pH. However, the presence of other minerals is also important; co-precipitation of arsenic with sulfides and iron oxides is a significant removal mechanism for arsenic in aqueous systems, as described below.

7.2.4 Sorption on Soils and Sediments

Arsenic is strongly sorbed onto soils and sediments, and sorption is one of the principal means by which arsenic is removed from solution. The process actually comprises three mechanisms: (1) specific adsorption onto Fe and Al hydrous oxides, clays, and carbonates; (2) co-precipitation with Fe oxides; and (3) isomorphic substitution of arsenic with phosphorus in minerals [24]. Co-precipitation with hydrous Fe oxides is probably the most common mechanism under most environmental conditions [5]. Ferguson and Gavis [5,8] state that co-precipitation with manganese oxides can also occur.

Each of the arsenic species possesses different chemical properties that affect its sorption behavior. In anaerobic sediments, Holm *et al.* [16,22] reported that arsenate is more strongly sorbed than methylarsenic acid. Cacodylic acid (dimethylarsinic acid) is less strongly sorbed than arsenate or methylarsenic acid. Wauchope [27] found that the same order applied in aerobic sediments.

Ferguson and Anderson [7] stated that the sorption of arsenate on iron and aluminum hydroxide followed the Langmuir isotherm, while arsenite was less strongly sorbed on the same materials and seemed to vary linearly with concentration. Iron oxide surfaces are positively charged at particular pH conditions, and arsenate and arsenite may be sorbed as negatively charged ions. Gulens, Champ and Jackson [12] speculated that both As(V) and As(III) anions form complexes with Fe(III) in solution; the difference in As(V) and As(III) sorption could then be explained by the greater solubility of the Fe(III)-As(III) complex. Whatever the mechanism, As(III) appears to be less strongly sorbed than As(V).

The factors affecting arsenic sorption on soils and sediments include redox conditions, pH, the presence of certain competing anions and complexing ions, salinity, clay content, and hydrous oxide content. Sorbed As(V) in sediments may be remobilized if conditions become sufficiently reduced for As(III) to form. Arsenic also appears to be more mobile under alkaline conditions. The maximum sorption of As(V) on kaolinite and montmorillonite is at pH 5 [9]. The sorption of As(III) increases beyond this pH; at pH 8, more As(III) is sorbed than As(V). Gupta and Chen [13] showed that sorption of As(III) onto alumina, bauxite and carbon decreases at pH values above 9, while for As(V), the decrease occurs when the pH is above 7. Similar pH dependencies have been shown for sorption on Al and Fe oxides [1,21].

The rate and extent of arsenic sorption onto alumina, bauxite and carbon decreases with increased salinity [13]. The presence of other ions also affects arsenic sorption, because they compete for sorption sites or complex with arsenic. Because of its similarity to arsenate, phosphate competes strongly with arsenate on iron oxides, soils and sediments [4,15,16,18]. The anions that have little or no effect include Cl^- and SO_4^{-2} [17,18]. Organics that complex with arsenic presumably would increase arsenic mobility and decrease its sorption.

The literature provides sorption constants for arsenic either as a particular arsenic species or for total arsenic sorption onto various sorbents, soils and sediments. Different types of sorption constants are also reported: sorption capacities at equilibrium, Freundlich constants, Langmuir constants, and sorption for specific experimental conditions. A brief summary of each kind of constant as measured for arsenic is given below. The reader is encouraged to refer to the original documents for details on experimental conditions, because values vary greatly depending on the particular conditions of each measurement.

Fuller [10,11] found the mobility of arsenic in clay soils to be low to moderate but much higher for loamy and sandy soils (6-10 cm/day for loamy sand). Using leachates of varying concentrations from landfills and soils, Arthur D. Little, Inc. [3] measured sorption capacity by means of batch tests, as shown in Table 7.2-1.

TABLE 7.2-1

Sorption Capacity for Total Arsenic at Equilibrium for Soils

Soil Type	Solution Conc. at Equilibrium (ppb)	Sorption Capacity (μg total As/g soil)
Alluvial material	<0.2-420	1.1-215
Residual soil	<0.2-225	1.1-28.9
Silty fine sand with little clay	0.4-483	1.0-252
Gravelly, well-graded silty sand	14-477	1.0-121
Brown, clayey sand	2.2-495	1.0- 80
Fine sand	10-514	1.1-7.9

Source: Arthur D. Little, Inc. [3].

Elkhatib *et al.* [6] reported values for arsenic (III) sorption by various soils in West Virginia. Freundlich isotherms were used to model the sorption for Lily (fine loamy siliceous), Chavies (coarse loamy mixed), and Upshur (fine mixed) soils. The results are shown in Table 7.2-2.

Sorption of arsenic(V) has been described by the Langmuir isotherm. For kaolinite and montmorillonite, A_m was reported [9] to be 7.19 and 9.9 μmol/g respectively, and the corresponding K_L values were 3.54 and 3.57 (log M^{-1}).[1] For soils, A_m (total) was reported by Wangan *et al.* [26][1] to vary between 3.31 and 4.47.

TABLE 7.2-2

Freundlich Isotherm Constants[a] for Some West Virginia Soils

Soil Type	Lily		Chavies		Gilpin		Pope		Upshur	
Horizon	A	Bt	A	Bt	A	Bt	A	Bw	A	Bt
K_F	28.3	22.7	77.4	93.8	30.2	19.2	53.9	102.0	33.6	21.5
1/N	0.399	0.428	0.437	0.450	0.674	0.958	0.571	0.659	0.545	0.636

a. K_F, 1/N = Freundlich constants for $S = K_F C^{1/N}$, where S = mg As (III)/kg and C = mg As (III)/liter.

Source: Elkhatib *et al.* [6]

Wauchope and McDowell [28] measured sorption, S (μmol/g), for various arsenic species onto sediments from five lakes and one creek in the Mississippi River alluvial flood plain. Their results are shown in Table 7.2-3.

TABLE 7.2-3

Sorption (S) of Arsenic Species Onto Sediments After Two Hours

	Arsenate $H_2AsO_4^-$		Methanearsenate $HAsO_3CH_3^-$		Cacodylate $HAsO_2(CH_3)_2$	
Sorbate Conc. (μ*M*)	0.37	3.97	0.35	3.53	0.40	3.99
Sorption[a] (μmol/g)	−0.023-0.063	0.35-0.71	−0.008-0.057	−0.016-0.58	−0.011-0.051	0.017-0.48

a. Negative values indicate desorption from the sediments.

Source: Wauchope and McDowell [28]

1. A_m = Langmuir sorption maximum, μmol/g; K_L = Langmuir constant, log M^{-1}; A_m (total) is the sum of $A_m(i)$ when more than one type of site, i, is present.

7.2.5 Biotransformation

The biological cycling of arsenic in the environment is described in section 2.15, which deals with the microbial transformation of inorganic pollutants. Methylation of arsenic is important because of the extremely toxic products that result. Also, this process transfers arsenic from sediments back to the water column in aquatic systems, increasing arsenic mobility in the environment. Biotransformation of arsenic can produce highly volatile compounds such as arsine (AsH_3), dimethylarsine ($HAs(CH_3)_2$) and trimethylarsine ($As(CH_3)_3$).

The processes and conditions involved in arsenic cycling in water were illustrated in section 2.15 (Figure 2.15-3). This cycle has been found to occur in both freshwater and saltwater systems [2,5,8]. The dominant processes in the arsenic cycle of an environmental system is determined by the environmental conditions (aerated or reduced, pH, microbial population, etc.) in the system.

7.2.6 Volatilization

In extremely anoxic environments, arsine may be produced. Methylation of arsenic can also result in highly volatile methylated arsine derivatives. Arsine is probably rapidly oxidized under aerobic conditions or in the atmosphere [5,19,20], but dimethylarsine and trimethylarsine may be more persistent because of their lower rates of oxidation [5].

7.2.7 Literature Cited

1. Anderson, M.A., J.F. Ferguson and J. Gavis, "Arsenate Adsorption on Amorphous Aluminum Hydroxides," *J. Colloid and Interface Sci.,* **54,** 391-99 (1976).

2. Andreae, M.O., "Distribution and Speciation of Arsenic in Natural Waters and Some Marine Algae," *Deep Sea Research,* **25,** 391-402 (1978).

3. Arthur D. Little, Inc., *Full-Scale Field Evaluation of Waste Disposal from Coal-Fired Electric Generating Plants. Appendix F (Part 4),* EPA Contract No. 68-02-3167, Final Report to U.S. EPA, Office of Research and Development, Washington, D.C. (1984).

4. Barrow, N.J., "On the Displacement of Adsorbed Anions from Soil: II. Displacement of Phosphate by Arsenate," *Soil Sci.,* **117,** 28-33 (1974), as cited by Rai *et al.* [22].

5. Callahan, M., M. Slimak, N. Gabel, I. May, C. Fowler, J. Freed, P. Jennings, R. Durfee, F. Whitmore, B. Maestri, W. Mabey, B. Holt and C. Gould, *Water-Related Environmental Fate of 129 Priority Pollutants. Volume I: Introduction and Technical Background, Metals and Inorganics, Pesticides and PCBs,* EPA-440/4-79-029a, EPA Contracts 68-01-3852 and 68-01-3867, Office of Water Planning and Standards, U.S. Environmental Protection Agency, Washington, D.C. (1979).

6. Elkhatib, E.A., O.L. Bennett and R.J. Wright, "Arsenite Sorption and Desorption in Soils," *Soil Sci. Soc. Am. J.,* **48,** 1025-30 (1984).
7. Ferguson, J.F. and M.A. Anderson, "Chemical Forms of Arsenic in Water Supplies and Their Removal," in *Chemistry of Water Supply, Treatment, and Distribution,* A.J. Rubin (ed.), Ann Arbor Science Publishers, Ann Arbor, Mich. 137-58 (1974).
8. Ferguson, J.F. and J. Gavis, "A Review of the Arsenic Cycle in Natural Waters," *Water Res.,* **6,** 1259-74 (1972).
9. Frost, R.R. and R.A. Griffin, "Effect of pH on Adsorption of Arsenic and Selenium from Landfill Leachate by Clay Minerals," *Soil Sci. Soc. Am. J.,* **41,** 53-57 (1977), as cited by Rai *et al.* [22].
10. Fuller, W.H., *Investigation of Landfill Leachate Pollutant Attenuation by Soils,* EPA 600/2-78-158, NTIS PB 286 995, Municipal Env. Res. Lab., Cincinnati, Ohio (1978).
11. Fuller, W.H., *Movement of Selected Metals, Asbestos, and Cyanide in Soil: Applications to Waste Disposal Problems,* EPA 600/2-77-020, PB 266 905, Municipal Environmental Research Laboratory, Cincinnati, Ohio (1977).
12. Gulens, J., D.R. Champ, and R.E. Jackson, "Influence of Redox Environments on the Mobility of Arsenic in Ground Water," in *Chemical Modeling in Aqueous Systems,* E.A. Jenne (ed.), American Chemical Society, Washington, DC, 81-95 (1979).
13. Gupta, S.K. and K.Y. Chen, "Arsenic Removal by Adsorption," *J. Water Pollut. Control Fed.,* **50,** 493-506 (1978).
14. Hess, R.E. and R.W. Blancher, "Dissolution of Arsenic from Waterlogged and Aerated Soils," *Soil Sci. Soc. Am. J.,* **41,** 861-65 (1977).
15. Hingston, F.J., A.M. Posner and J.P. Quirk, "Competitive Adsorption of Negatively Charged Ligands on Oxide Surfaces," *Disc. Faraday Soc.,* **52,** 234-342 (1971), as cited by Rai *et al.* [22].
16. Holm, T.R., M.A. Anderson, D.G. Iverson, and R.S. Stanforth, "Heterogeneous Interactions of Arsenic in Aquatic Systems," in *Chemical Modeling in Aqueous Systems,* E.A. Jenne (ed.), Symposium Series 93, American Chemical Society, Washington, D.C., 711-36 (1979), as cited by Rai *et al.* [22].
17. Leckie, J.O., M.M. Benjamin, K. Hayes, G. Kaufman and S. Altman, *Adsorption/Coprecipitation of Trace Elements from Water with Iron Oxyhydroxide,* EPRI-RP-910, Electric Power Research Institute, Palo Alto, Calif. (1980), as cited by Rai *et al.* [22].
18. Livesey, N.T. and P.M. Huang, "Adsorption of Arsenate by Soil and Its Relation to Chemical Properties and Anions," *Soil Sci.,* **131,** 88-94 (1981), as cited by Rai *et al.* [22].
19. Parris, G.E. and F.E. Brinckman, "Reactions which Relate to the Environmental Mobility of Arsenic and Antimony: I. Quarternization of Trimethylarsine and Trimethylstibine," *J. Org. Chem.,* **40,** 3801-03 (1975), as cited by Callahan *et al.* [5].

20. Parris, G.E. and F.E. Brinckman, "Reactions which Relate to the Environmental Mobility of Arsenic and Antimony: II. Oxidation of Trimethylarsine and Trimethylstibine," *Environ. Sci. Technol.*, **10,** 1128-34 (1976), as cited by Callahan *et al.* [5].

21. Pierce, M.L. and C.B. Moore, "Adsorption of Arsenite on Amorphous Iron Hydroxide from Dilute Aqueous Solution," *Environ. Sci. Technol.*, **14,** 214-16 (1980).

22. Rai, D., J. Zachara, A. Schwab, R. Schmidt, D. Girvin and J. Rogers, *Chemical Attenuation Rates, Coefficients, and Constants in Leachate Migration. Vol. I: A Critical Review,* Report EA-3356 to EPRI by Pacific Northwest Laboratories (Battelle Institute), Richland, Wash. (1984).

23. Reimers, R.S., "Disposal Problems and Mitigation Techniques for Hazardous Wastes," Enviro-Med Laboratories, Ruston, La. (Undated).

24. Scow, K., M. Byrne, M. Goyer, L. Nelken, J. Perwak, M. Wood and S. Young, *An Exposure and Risk Assessment for Arsenic,* Final Draft Report to U.S. Environmental Protection Agency, EPA Contract 68-01-6160, Monitoring and Data Support Division, Office of Water Regulations and Standards, Washington, D.C. (1981).

25. Wagemann, R., "Some Theoretical Aspects of Stability and Solubility of Inorganic Arsenic in the Freshwater Environment," *Water Res.*, **12,** 139-45 (1978).

26. Wangan, L.E., E.A. Stallings and R.D. Walker, *The Transport of Contaminants from Energy Process Waste Leachates through Subsurface Soils and Soil Components: Laboratory Experiments,* LA-9406-MS, Los Alamos National Laboratory, Los Alamos, N. Mex. (1982).

27. Wauchope, R.D., "Fixation of Arsenical Herbicides, Phosphate and Arsenate in Alluvial Soils," *J. Environ. Qual.*, **4,** 355-58 (1975).

28. Wauchope, R.D. and L.L. McDowell, "Adsorption of Phosphate, Arsenate, Methanearsonate, and Cacodylate by Lake and Stream Sediments: Comparisons with Soils," *J. Environ. Qual.*, **13,** 499-504 (1984).

7.2.8 Additional Sources of Information

Blackwood, T.R., S.R. Archer, and T.K. Corwin, *Status Assessment of Toxic Chemicals: Arsenic,* EPA-600/2-79-210B, NTIS PB80-146319 (1979).

Braman, R.S., "Arsenic in the Environment," in *Arsenical Pesticides,* E.A. Woolson (ed.), Symposium Series No. 7, American Chemical Society, Washington, D.C., 108-23 (1975).

Hawkins, D.B., "Environmental Fate of Arsenic in Groundwater," IWR-77, Institute of Water Resources, Fairbanks, Alaska (1976).

National Academy of Sciences, *Arsenic,* Washington, D.C. (1977).

Onishi, H. and E.B. Sandell, "Geochemistry of Arsenic," *Geochim. Cosmochim. Acta,* **7,** 1-33 (1955).

Penrose, W.R., "Arsenic in the Marine and Aquatic Environments: Analysis, Occurrence and Significance," *CRC Crit. Rev. Environ. Control,* **4,** 465-82 (1974).

Walsh, L.M., M.E. Sumner and D.R. Keeney, "Occurrence and Distribution of Arsenic in Soils and Plants," *Environ. Health Perspec.,* **19,** 57-71 (1977).

7.3 BARIUM (Ba)

7.3.1 Occurrence and Uses

Barium, one of the alkaline-earth metals, occurs in nature primarily as the sulfate (barite) and the carbonate (witherite). About 85% of the barium sulfate mined annually is used in oil- and gas-well drilling; the remainder is used in the paint, glass and rubber industries and for the production of other barium compounds [2]. Barium carbonate is no longer mined commercially [5]. Barium chemicals such as the carbonate, chloride, hydroxide and nitrate are used in the brick, ceramic, photographic and chemical manufacturing industries and as lubricating oil additives, rodenticides, insecticides and stabilizers for plastics. The metal and its alloys are used to remove gases from vacuum tubes [2,5].

7.3.2 Speciation Reactions in Water

The only oxidation state of importance under environmental conditions is Ba(II).

Barium does not hydrolyze appreciably except in highly alkaline environments ($pH \geqslant 10$)[1]. The acid dissociation constant of barium is listed in Part I of this report (Table 2.7-3).

In aqueous environments containing Cl^-, SO_4^{-2}, NO_3^- and CO_3^{-2} ions at $pH \leqslant 9.3$, Ba^{+2} is the dominant species. In waters with high sulfate concentration, appreciable levels of $BaSO_4^0$ occur. In the presence of carbonate at $pH > 9.3$, $BaCO_3^0$ is an important species [8,9]. Figure 7.3-1 shows the effect of pH on the activities of various aqueous barium species in equilibrium with barite ($BaSO_4$) and witherite ($BaCO_3$).

As Ba^{+2} forms relatively weak complexes with Cl^-, OH^- and NO_3^-, these complexes do not contribute significantly to the total barium concentration [9].

Table 2.9-1 (Part I) lists some stability constants of barium complexes with inorganic ions. Barium forms complexes with natural organics to a limited extent [8,9].

7.3.3 Solubility/Precipitation

The chloride, hydroxide and nitrate of barium are water-soluble; their solubility products range from $\sim 10^{0.5}$ to $10^{-2.3}$ [2,3,5]. The fluoride, oxalate, carbonate, sulfate and phosphate salts of barium have very low solubility [5]. Barite ($BaSO_4$) and witherite ($BaCO_3$) exhibit fast precipitation kinetics and are the compounds that may limit barium concentration in natural waters [9].

Barium sulfate, which has a solubility product of $10^{-9.96}$ at 25°C and zero ionic strength [7], may control barium concentrations in natural water at $pH \leqslant 9.3$. At $pH > 9.3$, barium carbonate (whose solubility product is $10^{-8.3}$ at 25°C and zero ionic strength [7]) may control the concentration [4,6,9].

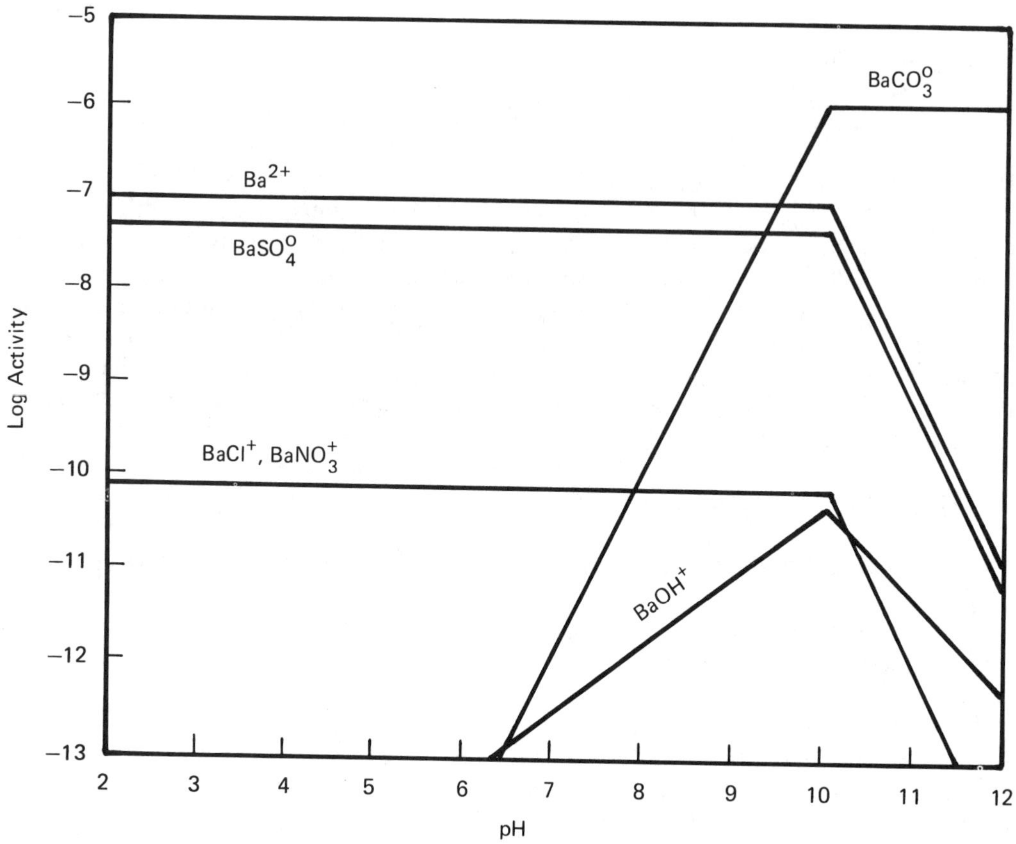

$[Cl^-] = [SO_4^{-2}] = 10^{-3}\,M$, $[NO_3^-] = 10^{-4}\,M$, $[CO_2(g)] = 10^{-3.52}$ atm

Source: Rai *et al.* [9]. *(Copyright 1984, Electric Power Research Institute. Reprinted with permission.)*

FIGURE 7.3-1 Activities of Barium Species When Ba^{+2} Activity Is Controlled by Barite ($BaSO_4$) or Witherite ($BaCO_3$)

Solid barium arsenate may control the concentration of total dissolved arsenic in fresh water. (See § 7.2.3 and Figure 7.2-4.)

7.3.4 Sorption on Soils and Sediments

Both specific and nonspecific adsorption of barium onto oxides and soils have been observed.[1] Specific sorption occurs onto metal oxides and hydroxides [6,9]. Adsorption

1. *Specific adsorption* refers to chemical reaction of the adsorbed substance with the surface [10]. *Nonspecific adsorption* involves electrostatic or coulombic interactions [9].

onto metal oxides probably acts as a control over the concentration of barium in natural waters. Barium is reported to occur in deep-sea manganese nodules and also in freshwater manganese deposits [6]. Barium displaces other sorbed alkaline earth metals from MnO_2, SiO_2 and TiO_2 under typical environmental conditions, but other alkaline earth metals displace barium from Al_2O_3 [9].

Electrostatic forces account for a large fraction of the nonspecific sorption of barium on soil and subsoil. The retention of barium, like that of other alkaline earth cations, is largely controlled by the cation exchange capacity of the sorbent. Complexation by soil organic material occurs to a limited extent [9]. Little data are available on barium sorption in soil; Table 7.3-1 lists some sorption constants for this element.

TABLE 7.3-1

Sorption Constants for Barium

	Al Oxides	Fe Oxides	Mn Oxides	Sediment
A_m(μmol/g)	17	17,60	2050	1.0
Log K_L(log M^{-1})	—	—	4.6	—
K_d(ml/g)	—	—	—	530, 2800

Source: Rai *et al.* [9]
Definitions (see also section 2.12):
A_m and K_L are Langmuir constants

$$S = \frac{K_L A_m C}{1 + K_L C}$$

where

S = μmoles sorbed at equilibrium per gram of solid
A_m = maximum adsorption capacity of solid (μmol/g)
K_L = sorption constant (related to binding energy of sorbate) in l/mol
C = total sorbate concentration in solution at equilibrium (mol/l)
K_d = ratio of the quantity of sorbate sorbed per gram of solid (μmol/g) to the concentration of the sorbate remaining in solution at equilibrium (μmol/l)

7.3.5 Literature Cited

1. Baes, C.R. and R.E. Mesmer, *The Hydrolysis of Cations,* John Wiley & Sons, New York (1976).
2. Clayton, G.D. and F.E. Clayton, *Patty's Industrial Hygiene and Toxicology,* 3rd rev. ed., Volume 2A, John Wiley & Sons, New York (1981).
3. Dean, J.A. (ed.), *Lange's Handbook of Chemistry,* 12th ed., McGraw-Hill Book Company, New York (1979).
4. Faust, S.D. and O.M. Aly, *Chemistry of Natural Waters,* Ann Arbor Science Publishers, Ann Arbor, Mich. (1981).
5. Grayson, M. and D. Eckroth (eds.), *Kirk-Othmer Encyclopedia of Chemical Technology,* Vol. 3, John Wiley & Sons, New York (1978).
6. Hem, J.D., "Study and Interpretation of Chemical Characteristics of Natural Waters," Water Supply Paper No. 1473, 2nd ed., U.S. Geological Survey, Washington, D.C. (1970).
7. Koltrý, S. and L. Šůcha, *Handbook of Chemical Equilibria in Analytical Chemistry,* Halsted Press, New York (1985).
8. Morel, F.M.M., *Principles of Aquatic Chemistry,* John Wiley & Sons, New York (1983).
9. Rai, D., J. Zachara, A. Schwab, R. Schmidt, D. Girvin and J. Rogers, *Chemical Attenuation Rates, Coefficients, and Constants in Leachate Migration. Vol I: A Critical Review,* Report EA-3356 to EPRI by Pacific Northwest Laboratories (Battelle Institute), Richland, Wash. (1984).
10. Salomons, W. and U. Förstner, *Metals in the Hydrocycle,* Springer-Verlag, New York (1984).

7.3.6 Other Sources of Information

Broadbent, F.E. and J.B. Ott, "Soil Organic Matter Metal Complexes: 1. Factors Affecting Retention of Various Cations," *Soil Sci.,* **83,** 419-27 (1957).

Kinniburgh, D.G., M.L. Jackson and J.K. Syers, "Adsorption of Alkaline Earth, Transition, and Heavy Metal Cations by Hydrous Oxide Gels of Iron and Aluminum," *Soil Sci. Soc. Am. J.,* **40,** 796-99 (1976).

Murray, J.W., "The Interaction of Metal Ions at the Manganese Dioxide — Solution Interface," *Geochim. Cosmochim. Acta,* **39,** 505-19 (1975).

7.4 BERYLLIUM (Be)

7.4.1 Occurrence and Uses

Unlike some other trace metals (e.g., copper and mercury) that concentrate in sulfide minerals, beryllium is concentrated in silicates. In fact, beryl ($Be_3Al_2Si_6O_{18}$) is the only important beryllium mineral. Beryllium also occurs in the minerals bromellite (BeO), chrysoberyl ($BeAl_2O_4$) and beryllonite ($NaBePO_4$) [2]. Trace amounts of beryllium are added to other metals such as copper to increase strength, hardness and resistance to corrosion [4,11]. Beryllium is used in X-ray tubes because of its ability to transmit low-energy X-rays, in nuclear weapons and nuclear reactors. It is also used as a heat sink material in low-weight, high-performance aircraft brakes. Other uses are in inertial guidance components, space optics, and microelectronics [5].

7.4.2 Speciation Reactions in Water

ACID-BASE DISSOCIATION

In aqueous solution, beryllium exhibits only the +2 oxidation state [1]. Table 2.7-3 (Part I) lists pK_a values for beryllium hydroxides. Figure 7.4-1 shows the hydrolysis products of beryllium; however, the solubility of $Be(OH)_2$ is very low (solubility product $\approx 10^{-21}$) at the pH range of most natural systems [2,8,9].

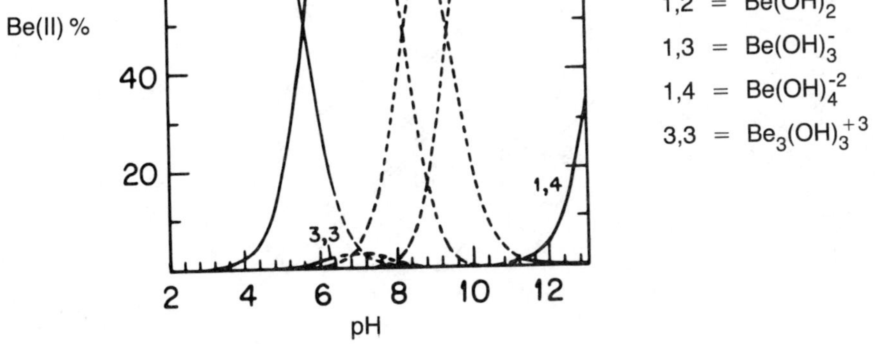

Dashed curves denote supersaturation with respect to $\alpha - Be(OH)_2$

Source: Baes and Mesmer [1]. *(Copyright 1976, John Wiley & Sons. Reprinted with permission.)*

FIGURE 7.4-1 Calculated Distribution of Hydrolysis Products with I = 1 *M*, T = 25°C and 10^{-5} *M* Be(II)

Several authors have reported the existence of polynuclear hydroxo-beryllium compounds such as Be_2OH^{+3}, $Be_3(OH)_3^{+3}$ and $Be_6(OH)_8^{+4}$ [1,6,9], but significant amounts of these species are unlikely at the low beryllium concentrations and the pH ranges normally encountered in the environment, as illustrated in Figure 7.4-2.

COMPLEXATION

Be(II) forms stronger complexes with OH^- and F^- than with Cl^-, SO_4^{-2} and NO_3^-, as shown in Figure 7.4-2. (Note the lower concentration of F^-.) In aquatic systems under typical environmental conditions, where very little or no F^- is present, the hydroxo-complexes $BeOH^+$, $Be(OH)_2^0$ and Be^{+2} should be the dominant dissolved species. In a system with F^-, BeF^+ and BeF_2^0 would also be significant. Rai *et al.* [9] concluded from available data that Cl^-, NO_3^-, SO_4^{-2} and polynuclear hydroxo-complexes of beryllium do not contribute significantly to total soluble Be.

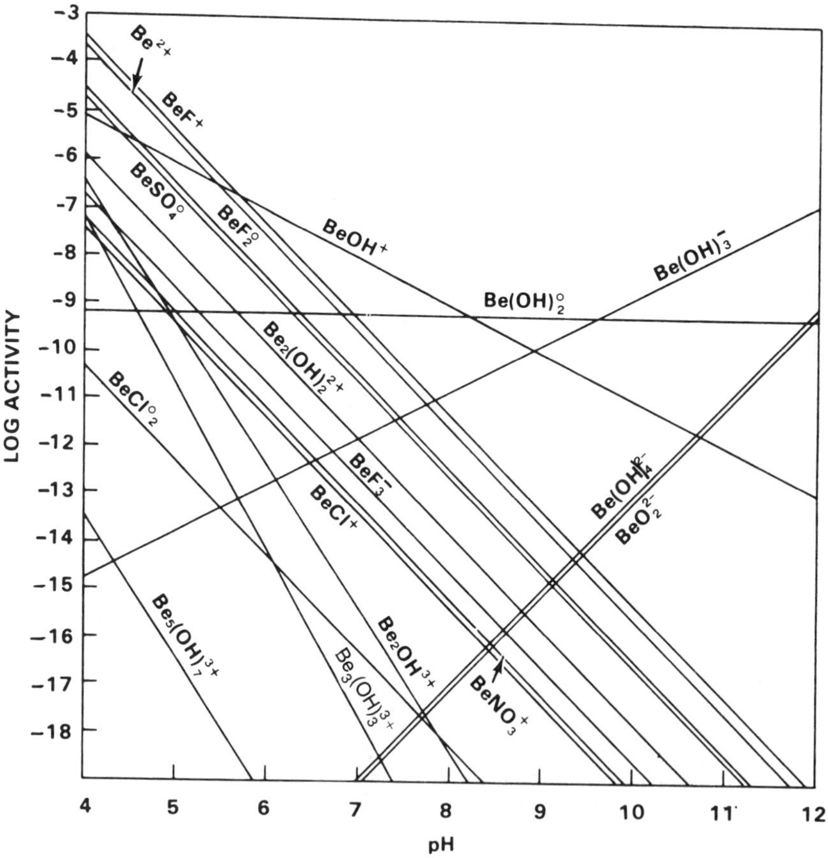

$SO_4^{-2} = Cl^- = NO_3^- = 10^{-3.0}$ *M* and $F^- = 10^{-5.0}$ *M*

Source: Rai *et al.* [9]. *(Copyright 1984, Electric Power Research Institute. Reprinted with permission.)*

FIGURE 7.4-2 Calculated Beryllium(II) Speciation in an Aquatic System in Equilibrium with Be_2SiO_4 and SiO_2(am) at 25°C

Data on stability constants for beryllium complexes are not as prevalent as for most other trace metals. The stability constants of Be(II) complexes with organics such as citrate^{-4} and salicylate^{-2} indicate that Be(II) does not form strong complexes with organic ligands. Table 7.4-1 shows values of stability constants. In typical natural environments containing only trace concentrations of beryllium, there would probably be little Be(II)-organic complexes present.

TABLE 7.4-1

Stability Constants for Beryllium(II) Complexes[a]
(Expressed as logarithm of overall formation constant, β)

OH^-	F^-	Cl^-
BeL^+ 8.6 (25°C, 0 *M*)	BeL^+ 4.71 (25°C, 0.5 *M*)	BeL^+ −0.8 (25°C, 0 *M*)
BeL_2 14.4 (25°C, 0 *M*)	BeL_2 8.32 (25°C, 0.5 *M*)	BeL_2 −0.7 (25°C, 0 *M*)
BeL_3^- 18.8 (25°C, 0 *M*)	BeL_3^- 11.1 (25°C, 0.5 *M*)	
BeL_4^{-2} 18.6 (25°C, 0 *M*)	BeL_4^{-2} 13.1 (25°C, 1.0 *M*)	
BeL_3^{+3} 33.1 (25°C, 0 *M*)		
SO_4^{-2}	**Citrate^{-4}**	**Salicylate^{-2}**
BeL 1.95 (25°C, 0 *M*)	BeH_2L 1.4 (34°C, 0.15 *M*)	BeL_2^{+2} 22.02 (20°C, 0.1 *M*)
BeL_2^{-2} 1.78 (25°C, 1.0 *M*)	$BeHL^-$ 2.2 (34°C, 0.15 *M*)	BeL 12.37 (20°C, 0.1 *M*)
BeL_3^{-4} 2.08 (25°C, 1.0 *M*)		

a. Concentrations in parentheses indicate ionic strength of solution.

Source: Martell and Smith [7]

REDOX REACTIONS

The only oxidation state of importance under environmental conditions is Be(II). Changes in solution redox potential, Eh, do not affect the oxidation state. The standard electrode potential for the Be^{+2}/Be^0 couple is −1.85V (25°C), which indicates that the reduction of Be(II) to Be(s) does not occur in nature; water would decompose under the conditions needed for this reduction to take place [12]. Since sulfide solids of beryllium are not environmentally important, Eh changes do not affect the solubility of beryllium.

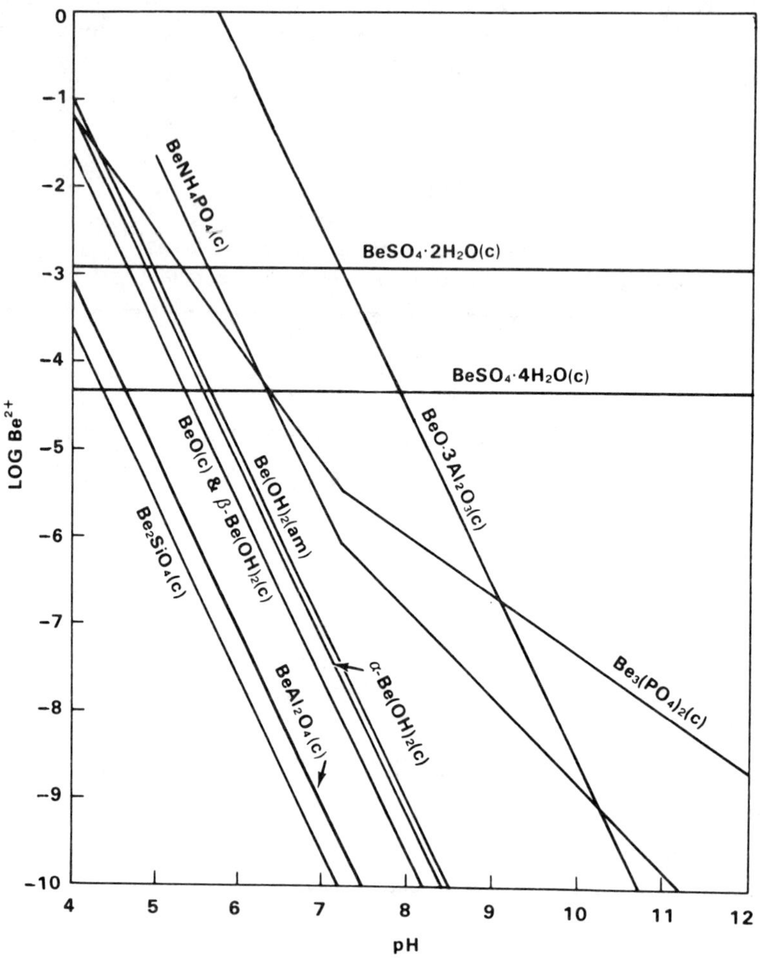

$SO_4^{-2} = 10^{-3.0}$, $H_2PO_4^- = 10^{-5.5}$ (at pH < 7.2) and $HPO_4^{-2} = 10^{-5.5}$ *M.*

Al activity is controlled by $\gamma Al(OH)_3(c)$ and Si activity by $SiO_2(am)$. Concentration of NH_4^+ not known.

Source: Rai *et al.* [9]. *(Copyright 1984, Electric Power Research Institute. Reprinted with permission.)*

FIGURE 7.4-3 Relative Solubility of Beryllium Solid Phases at 25°C

7.4.3 Solubility/Precipitation

As mentioned above, $Be(OH)_2$ has a low solubility (solubility product of $10^{-21.31}$ for α-$Be(OH)_2$ and $10^{-21.7}$ for β-$Be(OH)_2$) and is likely to precipitate under typical environmental conditions [7]. The phosphate, $Be_3(PO_4)_2$, is also relatively insoluble: its solubility product is about 10^{-38} [12]. Figure 7.4-3 shows that the silicate (Be_2SiO_4) and aluminate ($BeAl_2O_4$) are relatively insoluble. The sulfates and chlorides are soluble, with solubility products of $10^{5.34}$ (α-$BeSO_4$) and around 10^{34} (β-$BeCl_2$) [6]. In terms of solubility in water, 42.5 g of $BeSO_4 \cdot 4H_2O$ will dissolve in 100 ml of water [13]. Figure 7.4-3 shows the relative solubility of various beryllium compounds with varying pH. Rai *et al.* [9] suggested that β-$Be(OH)_2$ may be an important concentration-limiting solid phase.

7.4.4 Sorption on Soils and Sediments

There is little information on the sorption of Be(II) by soils, sediments or specific sorbents. Romney and Childress [10] found that ions of the ^7Be isotope were strongly sorbed by soils containing montmorillonite and illite-like clay minerals but not by kaolinite. The sorption mechanisms are not known with certainty [3,9]. Quantitative data on sorption constants were not found.

7.4.5 Literature Cited

1. Baes, C.F., Jr. and R.E. Mesmer, *The Hydrolysis of Cations,* John Wiley & Sons, New York (1976).
2. Callahan, M., M. Slimak, N. Gabel, I. May, C. Fowler, J. Freed, P. Jennings, R. Durfee, F. Whitmore, B. Maestri, W. Mabey, B. Holt and C. Gould, *Water-Related Environmental Fate of 129 Priority Pollutants. Volume I: Introduction and Technical Background, Metals and Inorganics, Pesticides and PCBs,* EPA-440/4-79-029a, EPA Contracts 68-01-3852 and 68-01-3867, Office of Water Planning and Standards, U.S. Environmental Protection Agency, Washington, D.C. (1979).
3. Dreesen, D.R., "Review of Literature on the Environmental Quality Aspects of Beryllium: Implications to the U.S. Army Yuma Proving Ground Test Program," Draft, Los Alamos National Laboratory, N. Mex. (1984).
4. Fuller, W.H., *Movement of Selected Metals, Asbestos, and Cyanide in Soil: Applications to Waste Disposal Problems,* EPA 600/2-77-020, PB 266 905, Municipal Environmental Research Laboratory, Cincinnati, Ohio (1977).
5. *Kirk-Othmer Encyclopedia of Chemical Technology,* 3rd ed., Vol. III, John Wiley & Sons, New York (1978).
6. Kragten, J., *Atlas of Metal-Ligand Equilibria in Aqueous Solution,* Ellis Horwood Ltd. (Wiley), New York (1978).
7. Martell, A.E. and R.M. Smith, *Critical Stability Constants,* Vol. 3, *Other Organic Ligands* and Vol. 4, *Inorganic Complexes,* Plenum Press, New York (1977 and 1976).

8. Parker, V.B., D.D. Wagman and W.H. Evans, "Selected Values of Chemical Thermodynamic Properties," Technical Note 270-6, National Bureau of Standards, U.S. Govt. Printing Office, Washington, D.C. (1971), as cited in Rai *et al.* [9].

9. Rai, D., J. Zachara, A. Schwab, R. Schmidt, D. Girvin and J. Rogers, *Chemical Attenuation Rates, Coefficients, and Constants in Leachate Migration. Vol I: A Critical Review,* Report EA-3356 to EPRI by Pacific Northwest Laboratories (Battelle Institute), Richland, Wash. (1984).

10. Romney, E.M. and J.D. Childress, "Effects of Beryllium in Plants and Soil," *Soil Sci.*, **100**, 210-17 (1965), as cited in Fuller [4].

11. Sienko, M.J. and R.A. Plane, *Chemistry: Principles and Properties,* McGraw-Hill Book Co., New York (1956).

12. Sillen, L.G. and A.E. Martell, *Stability Constants of Metal-Ion Complexes*, Special Publication No. 7, The Chemical Society, London (1964).

13. Weast, R.C. (ed.), *Handbook of Chemistry and Physics*, 52nd ed., The Chemical Rubber Co., Cleveland (1972).

7.4.6 Additional Sources of Information

World Health Organization, "Beryllium and Beryllium Compounds," *International Agency for Research on Cancer Monographs*, **23**, 143-204 (1980).

Drury, J.S., C.R. Shriver, E.B. Lewis, L.E. Touill and A.S. Hammons, *Reviews of the Environmental Effects of Pollutants: VI. Beryllium*, Oak Ridge National Laboratory, Oak Ridge, Tenn. (1978).

7.5 CADMIUM (Cd)

7.5.1 Occurrence and Uses

Cadmium is the second member of the Group IIB triad (Zn, Cd, Hg) in the periodic table. Environmental concern over cadmium stems from its widespread occurrence as a result of industrial use and disposal practices, its high toxicity, and its bioavailability. Its toxicity is due to a strong affinity for sulfhydryl groups, which causes it to compete with zinc for binding in enzymes [1,16]. Cadmium is accumulated and retained in the human body, causing erothrocyte destruction, renal degradation, chronic pulmonary problems and skeletal deformity [5,6,12].

Cadmium is commonly associated with zinc in carbonate and sulfide ores and is also a byproduct of the refining of other metals. It is used principally in electroplating, batteries, pigments, plastic stabilizers, photovoltaic devices and alloys [16].

7.5.2 Speciation Reactions in Water

ACID-BASE DISSOCIATION

The Cd^{+2} ion has little tendency to hydrolyze at pH < 8, but at values above 11, nearly all the cadmium is present in hydroxo-complex form [1]. The pK_a values of various cadmium species are given in Part I, Table 2.7-3. Figure 7.5-1 shows the distribution of cadmium hydrolysis products at different pH values; the data were calculated by use of the acid dissociation constant at zero ionic strength (see Part I, section 2.7).

COMPLEXATION

Salinity and water hardness, along with pH, determine the speciation of cadmium in natural waters [3]. Although its speciation is similar to that of zinc, cadmium is more highly complexed by Cl^-. Therefore, in a transition from fresh water to seawater, such as when a river enters an ocean, cadmium speciation changes dramatically [14]. In fresh water, at typical environmental pH values of 6-8, Cd^{+2} is the predominant species. In seawater, the chloro-complexes of Cd ($CdCl^+$, $CdCl_2$) predominate. $CdCl_3^-$ and Cd^{+2} are present to a lesser extent (1-10% of total cadmium). Tables 2.9-1 and 2.9-2 (Part I) contain complexation equilibrium constants for a variety of ligand complexes of cadmium. Figure 7.5-2 shows the chemical speciation of cadmium in unpolluted, organic-free aquatic systems as a function of pH.

The speciation of cadmium is not affected by the levels of organic ligands normally found in natural waters [17,23]. However, in waters polluted with high concentrations of organic matter, organic materials can affect the form in which cadmium is present [2]. Shephard *et al.* [21] reported that in one part of a contaminated freshwater lake, organic binding of cadmium accounted for 53% of the total cadmium; this high percentage of organic complexing was related to the production of organic matter by aquatic macrophytes and phytoplankton. The conditional stability constants (K_o)

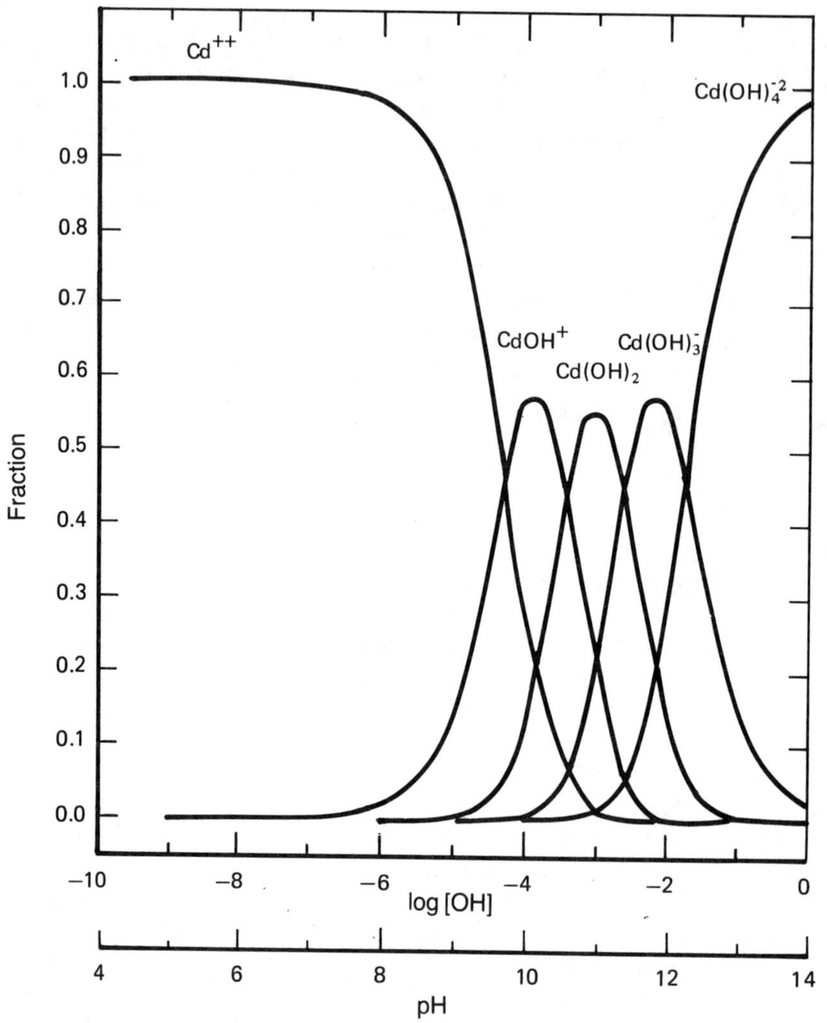

Source: Hahne and Kroontje [7]. *(Copyright 1973, American Society of Agronomy, Inc. Reprinted with permission.)*

FIGURE 7.5-1 Calculated Distribution of Molecular and Ionic Species of Divalent Cadmium (Assuming Zero Ionic Strength) in Pure Water at Various pH Values

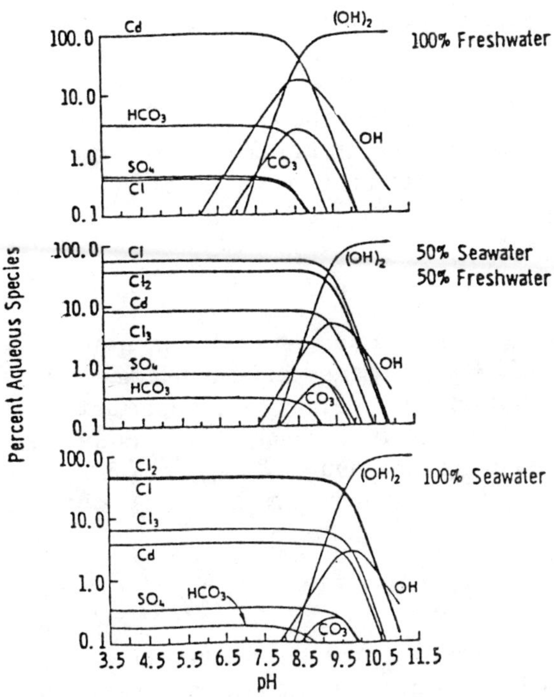

Legend:

Cd = Cd^{+2}

Cl = $CdCl^{+}$

Cl_2 = $CdCl_2$

Cl_3 = $CdCl_3^{-}$

OH = $CdOH^{+}$

OH_2 = $Cd(OH)_2$

HCO_3 = $CdHCO_3^{+}$

CO_3 = $CdCO_3$

SO_4 = $CdSO_4$

Molal Concentrations of Chemical End Members Used in the Calculations

Component	Freshwater	Seawater
Calcium	0.000375	0.0104
Magnesium	0.000168	0.054
Sodium	0.000274	0.4752
Potassium	0.000059	0.01
Chloride	0.00022	0.554
Alkalinity	0.000955	0.00234
Sulfate	0.000955	0.0284
Ionic strength[a]	0.0021	0.661

a. Effective molal ionic strength (after computation of species distribution).

Source: Long and Angino [14]. *(Copyright 1977, Pergamon Press, Inc. Reprinted with permission.)*

FIGURE 7.5-2 Chemical Speciation of Cadmium(II) in Seawater-Fresh Water Mixtures with Total Cadmium Concentration <10 ppm

of cadmium complexes with humic materials from various natural environments have been determined at 20°C, pH = 8 and I = 0.02; the values of log K_o range from 4.57 to 5.08 [15].[1]

Cadmium is considered a kinetically labile metal. The characteristic second-order rate constant for exchange of one water molecule in the primary solvation shell of Cd(II) ion is $9 \times 10^7\ M^{-1}\ s^{-1}$ [17]. Cadmium reacts about ten times faster than zinc with the corresponding ligand [25].

REDOX REACTIONS

The only oxidation state of importance for cadmium under environmental conditions is +2. Cadmium sulfide is formed over a wide pH range under reducing conditions in the presence of sulfide ions [5]. Precipitation of sulfide may be an important control on the mobility of cadmium in reducing environments. The mobility of cadmium is much less in reducing conditions than in oxidizing conditions [2,3]. Figure 7.5-3 shows the fields of stability of solid and dissolved cadmium species in the cadmium-carbon dioxide-sulfur-water system as a function of Eh and pH.

7.5.3 Solubility/Precipitation

Constants for solubility equilibria of naturally occurring cadmium-containing solids are given in Table 7.5-1. Figure 7.5-4 illustrates the solubility of several cadmium minerals in water with varying pH. (See also Figure 7.5-3 for fields of stability of solid and dissolved cadmium species in water.) To illustrate the use of Figure 7.5-4, suppose that one wishes to determine the concentration of Cd^{+2} in a solution of $Cd_3(PO_4)_2$ with a CO_2 level of 0.0003 atm and a pH of 8, where phosphate levels are fixed by TCP-calcite. A line is first drawn parallel to the TCP-calcite line through b to represent CO_2 = 0.0003 atm. A vertical line drawn from pH = 8 intersects this line at log $[Cd^{+2}] = -5.8$.

Since the solubility of a mineral containing other species besides cadmium is determined by the concentration of these other species (e.g., Ca, Si, CO_2), the mineral lines are a function of the concentration of these species in solution. Thus, the $Cd_3(PO_4)_2$ line is plotted for several conditions, depending on whether phosphate levels are controlled by $FePO_4 \cdot 2H_2O$ (strengite) – $Fe(OH)_3$ (soil); $CaHPO_4 \cdot 2H_2O$ (DCPD); $Ca_3(PO_4)_2$ (TCP); or TCP-calcite ($CaCO_3$) – CO_2 (at two CO_2 levels). The $CdSiO_3$ line is plotted with soil-silica, amorphous silica, or quartz as the mineral controlling silica levels, and the $CdCO_3$ line is plotted at two $CO_2(g)$ levels.

Compounds of cadmium with the common ligands in surface waters are more soluble in neutral and acidic solution than in alkaline solution; therefore, cadmium is more readily transported at pH ≤7 [2,5]. Dissolved carbonate also reduces the solubility of cadmium: for example, at pH = 8.3 when $Cd(OH)_2$ is the solid phase and no carbonate

---

1. See section 7.7.2 for further discussion of K_o.

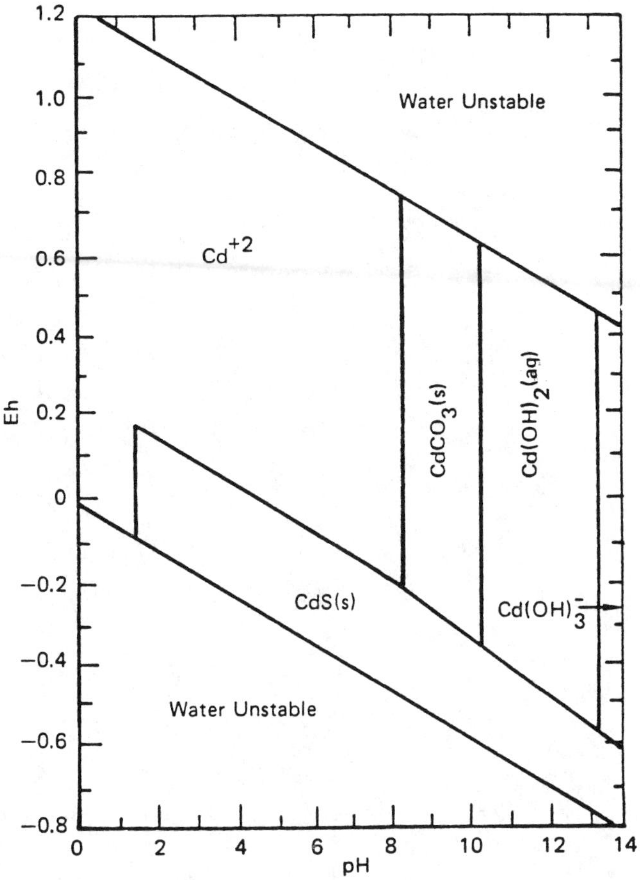

Cd Concentration = $10^{-7.05}$ *M*, Dissolved CO_2 = 10^{-3} *M*,
Dissolved S Species = 10^{-3} *M*

Source: Hem [8]. *(Copyright 1972. Reprinted with permission.)*

FIGURE 7.5-3 Fields of Stability of Solid and Dissolved Cadmium Species in the System Cd + CO_2 + S + H_2O at 25°C, 1 atm, in Water

TABLE 7.5-1

Constants for Solubility Equilibria for Common Cadmium Minerals and Solids
(at 25°C and I = 0 unless noted)

Ore/Solids	Chemical Formula and Reaction	Log K	Source
Greenockite	$CdS(s) \rightleftharpoons Cd^{+2} + S^{-2}$	– 27.0	[23]
	$CdS(s) + 2H^{+} \rightleftharpoons Cd^{+2} + H_2S(aq)$	– 5.73[a]	[10]
Cadmium Hydroxide	$\beta\text{-}Cd(OH)_2(s) + 2H^{+} \rightleftharpoons Cd^{+2} + 2H_2O$	13.61	[22]
	$Cd(OH)_2(s) \rightleftharpoons Cd^{+2} + 2OH^{-}$	– 14.39	[22]
Monteponite	$CdO + 2H^{+} \rightleftharpoons Cd^{+2} + H_2O$	15.14[a]	[13]
	$CdO(s) + H_2O \rightleftharpoons Cd^{+2} + 2OH^{-}$	– 12.62[a]	[22]
Octavite	$CdCO_3(s) + 2H^{+} \rightleftharpoons Cd^{+2} + H_2O + CO_2(g)$	6.16[a]	[13]
	$CdCO_3(s) \rightleftharpoons Cd^{+2} + CO_3^{-2}$	– 13.74[a]	[6]
Cadmium Phosphate	$Cd_3(PO_4)_2(s) + 4H^{+} \rightleftharpoons 3Cd^{+2} + 2H_2PO_4^{-}$	1.00[a]	[13]
(tribasic)	$Cd_3(PO_4)_2(s) \rightleftharpoons 3Cd^{+2} + 2PO_4^{-3}$	– 32.6[a,b]	[22]
Cadmium Silicate	$CdSiO_3(s) + 2H^{+} + H_2O \rightleftharpoons Cd^{+2} + H_2SiO_4$	7.63[a]	[13]
Cadmium Sulfate	$CdSO_4(s) \rightleftharpoons Cd^{+2} + SO_4^{-2}$	– 0.04[a]	[13]
	$CdSO_4 \cdot H_2O(s) \rightleftharpoons Cd^{+2} + SO_4^{-2} + H_2O$	– 1.59[a]	[13]
	$CdSO_4 \cdot 2Cd(OH)_2(s) + 4H^{+} \rightleftharpoons 3Cd^{+2} + SO_4^{-2} + 4H_2O$	22.65[a]	[13]
	$2CdSO_4 \cdot Cd(OH)_2(s) + 2H^{+} \rightleftharpoons 3Cd^{+2} + 2SO_4^{-2} + 2H_2O$	6.73[a]	[13]
	$Cd_4(OH)_6SO_4(s) \rightleftharpoons 4Cd^{+2} + 6OH^{-} + SO_4^{-2}$	– 50[a]	[10]

a. Ionic strength not specified
b. 20°C

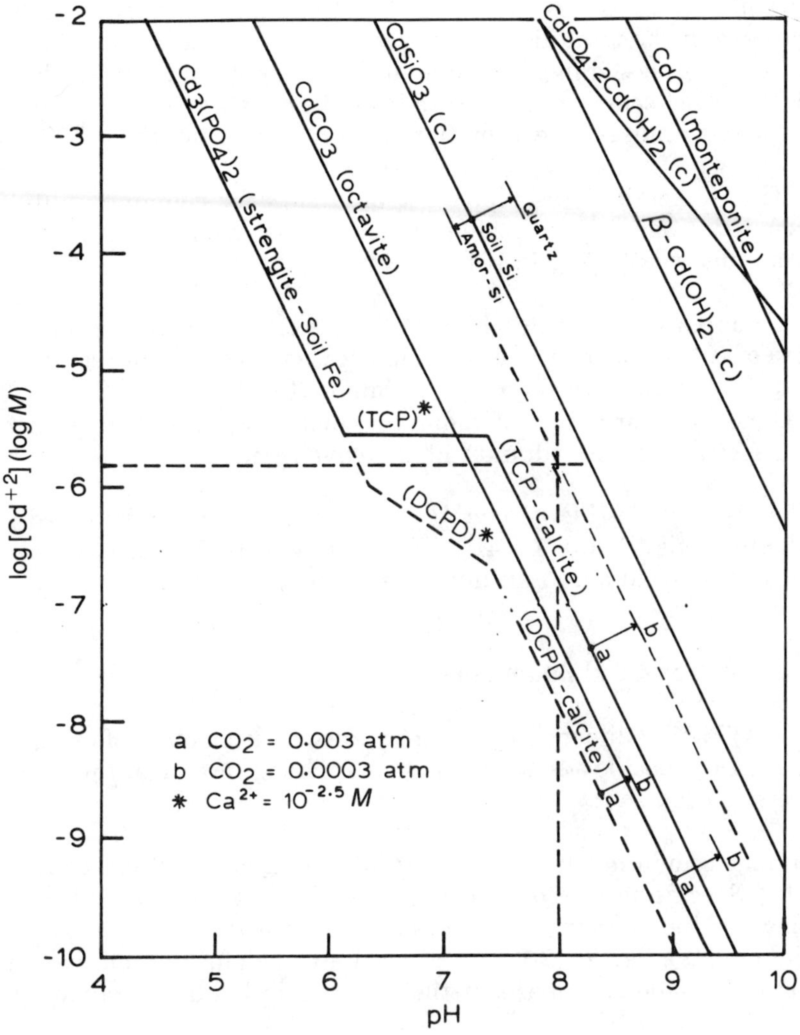

Note: See text for explanation of TCP and DCPD lines.

Source: Adapted from Lindsay [13]. *(Copyright 1979, John Wiley & Sons. Reprinted with permission.)*

FIGURE 7.5-4 Solubility of Several Cadmium Minerals

is present, the calculated solubility of cadmium is 637 mg/l, while at a total carbonate concentration of 5×10^{-4} *M* the solubility is 0.11 mg/l [6]. This occurs because of the low solubility of $CdCO_3$ ($K_{sp} = 1.8 \times 10^{-14}$). Also, under oxidizing (aerobic) conditions, an increase in the total dissolved carbon dioxide species by one log unit (e.g., from 10^{-3} to 10^{-2} *M*) results in a decrease in the solubility of cadmium by the same amount [6]. Dissolved cadmium levels are lower in hard waters (>120 mg/l $CaCO_3$) than in soft waters because of the formation of the relatively insoluble $CdCO_3$ [1,3,6].

Bottom muds in lakes and reservoirs act as a sink for cadmium sulfide, which precipitates in reducing (anaerobic) environments. This precipitate is highly resistant to carbonation and weathering [5,12].

Significant amounts of cadmium chloride complexes begin to form at a chloride concentration of 10^{-3} *M* (35 ppm). These complexes, which are highly soluble, compete with the sparingly soluble precipitates of cadmium ($Cd(OH)_2$, CdO, $CdCO_3$, CdS, and $Cd_3(PO_4)_2$. Hence, the formation of cadmium chloride complexes may increase the mobility and distribution of cadmium in the environment [7,16].

Octavite ($CdCO_3$) and $Cd_3(PO_4)_2$ have been reported to control dissolved cadmium levels in alkaline soils (Figure 7.5-4). At pH values below 7.5, Cd^{+2} concentrations are controlled by soil-cadmium, an undefined complex [13,19].

7.5.4 Sorption on Soils and Sediments

Adsorption on soil and sediments significantly affects the mobility of cadmium in the environment [2,24]. Specific adsorption and ion exchange are important mechanisms in this process [19,20].

The adsorption of cadmium often correlates with the cation exchange capacity (CEC) of the soil [19]. Specific adsorption to calcite ($CaCO_3$) and hydrous oxides of iron and aluminum may be the most important mechanism at low environmental concentrations of cadmium [19]. Section 2.12 (Part I) contains helpful information on general adsorption processes and describes a method for calculating the mobility of cadmium in soil.

Clay minerals, carbonate minerals, oxides and, to a lesser extent, organic matter have been noted as adsorbents that control the behavior of cadmium in soil and sediments [9,16]. Conflicting data have been reported regarding the relative importance of cadmium adsorption by organic material in soils and sediments [2,5,9,16,18,19].

Removal of cadmium from solution increases greatly as the pH increases through a critical range of 6-8 [2,10]. In an experiment measuring the removal of cadmium from solution by two hydrous oxides and two Delaware soils (Metapeak, a silt loam, and Evesboro, a loamy sand), Huang *et al.* [10] found that 10^{-3} *M* cadmium was completely removed when the pH exceeded 9. It is unclear whether precipitation or adsorption was responsible for the removal. Below pH 6, little or no cadmium was removed.

The presence of anions also affects cadmium adsorption. Sulfate and chloride ions, for example, reduce adsorption by amorphous $Fe(OH)_3$ [4,19,20], and chloride reduces adsorption by clay minerals, lepidocrocite (α-FeOOH), silica and alumina [19]. On the other hand, ligands such as thiosulfate, humic acid, nitriloacetate, glycine, tartrate and phosphate increase cadmium adsorption [10,11,19,20]. Another influence on the adsorption of cadmium is competition by other metal cations: the presence of Ca^{+2}, Mg^{+2} and trace metal cations reduces adsorption by soils, clay minerals, Fe and Mn oxides, and alumina [19].

Adsorption coefficients for cadmium on various soils and sediments are available in the literature. The values vary widely with conditions and with the equation used to describe the phenomena. Rai *et al.* [19] has compiled adsorption data for cadmium from many sources; the ranges of these values are given in Table 7.5-2.

TABLE 7.5-2

Typical Ranges of Sorption Constants for Cadmium

Sorption Constant	Clay	Fe Oxides	Mn Oxides	Organic Matter	Soils	Sediment
A_m (μmol/g)	0.1-460	5[a]	17-2000[b]	5.6-190	0.01-109	2.0-173
Log K_L (log M^{-1})	1.9-7.8	6[a]	4.0-6.3[b]	2.6-5.0	3.0-7.5	4.0-6.0
K_d (ml/g)	1.5-900	80-180[b]	4[a]			1.2-2.5

a. Only one value given
b. Only two values given

Source: Adapted from Rai *et al.* [19]. *(Copyright 1984, Electric Power Research Institute. Reprinted with permission.)*

Definitions (see also section 2.12):

A_m and K_L are Langmuir constants

$$S = \frac{K_L A_m C}{1 + K_L C}$$

where

S = μ moles sorbed at equilibrium per gram of solid
A_m = maximum adsorption capacity of solid (μ mol/g)
K_L = sorption constant related to binding energy of sorbate in 1 mol^{-1}
C = total sorbate concentration in solution at equilibrium (mol l^{-1})

K_d = ratio of the quantity of sorbate sorbed (μ mol/g) per gram of solid to the concentration of the sorbate remaining in solution at equilibrium (μ mol/ml)

Sorption constants are a function of several variables, such as the nature of the different solids present in the sorbent and the extent of modification of the surface sites by acid/base equilibria (see Section 2.12). The Langmuir adsorption maximum (A_m) and K_d values in Table 7.5-2 were measured at pH 5-8.5 and range over approximately two orders of magnitude for a given sorbent. Log K_L values are less variable but nevertheless range up to a factor of approximately 3 for a given sorbent. Sorption constants for specific experimental conditions can be found in Rai *et al.* [19] and other sources in the literature.

7.5.5 Literature Cited

1. Baes, C.F. and R.E. Mesmer, *The Hydrolysis of Cations*, John Wiley & Sons, New York (1976).
2. Callahan, M., M. Slimak, N. Gabel, I. May, C. Fowler, J. Freed, P. Jennings, R. Durfee, F. Whitmore, B. Maestri, W. Mabey, B. Holt and C. Gould, *Water-Related Environmental Fate of 129 Priority Pollutants. Volume I: Introduction and Technical Background, Metals and Inorganics, Pesticides and PCBs*, EPA-440/4-79-029a, EPA Contracts 68-01-3852 and 68-01-3867, Office of Water Planning and Standards, U.S. Environmental Protection Agency, Washington, D.C. (1979).
3. Cooper, C.B., S. Bysshe, L. Nelken, K. Scow and W. Tucker, *Risk Assessments of Priority Pollutants: Cadmium* (Rough Draft), EPA Contract 68-010-3857, Task Order No. 11 (April, 1979).
4. Davies-Colley, R.J., P.O. Nelson and K.J. Williamson, "Copper and Cadmium Uptake by Estuarine Sedimentary Phases," *Environ. Sci. Technol.*, **18**, 491-99 (1984).
5. Faust, S.D. and O.S. Aly, *Chemistry of Natural Waters*, Butterworth, Woburn, Mass. (1981).
6. Fergusson, J.E., *Inorganic Chemistry and the Earth*, Pergamon Press, Elmsford, N.Y. (1982).
7. Hahne, H.C.H. and W. Kroontje, "Significance of pH and Chloride Concentration on Behavior of Heavy Metal Pollutants — Mercury (II), Cadmium (II), Zinc (II) and Lead (II)," *J. Environ. Qual.*, **2**, 444-50 (1973).
8. Hem, J.D., "Chemistry and Occurrence of Cadmium and Zinc in Surface and Ground Waters," *Water Resour. Res.*, **8**, 661-71 (1972).
9. Hickey, M.G. and J.A. Kittrick, "Chemical Partitioning of Cadmium, Copper, Nickel and Zinc in Soil and Sediments Containing High Levels of Heavy Metals," *J. Environ. Qual.*, **13**, 372-76 (1984).
10. Huang, C.P., H.A. Elliot and R.M. Ashmead, "Interfacial Reactions and the Fate of Heavy Metals in Soil-Water Systems," *J. Water Pollut. Control Fed.*, **49**, 745-56 (1977).
11. Kuo, S. and B.L. McNeal, "Effects of pH and Phosphate on Cadmium Sorption by a Hydrous Ferric Oxide," *Soil Sci. Soc. Am. J.*, **48**, 1040-44 (1984).

12. Lee, F.Y. and J.A. Kittrick, "Elements Associated with the Cadmium Phase in a Harbor Sediment as Determined with the Electron Beam Microprobe," *J. Environ. Qual.*, **13**, 337-40 (1984).
13. Lindsay, W.L., *Chemical Equilibria in Soils*, John Wiley & Sons, New York (1979).
14. Long, D.T. and E.E. Angino, "Chemical Speciation of Cd, Cu, Pb, and Zn in Mixed Freshwater, Seawater and Brine Solutions," *Geochim. Cosmochim. Acta*, **41**, 1183-91 (1977).
15. Mantoura, R.F.C., A. Dickson and J.P. Riley, "The Complexation of Metals with Humic Materials in Natural Waters," *Estuarine Coastal Mar. Sci.*,**6**, 387-408 (1978).
16. Moore, J.W. and S. Ramamoorthy, *Heavy Metals in Natural Waters*, Springer-Verlag, New York (1984).
17. Morel, F.M.M., *Principles of Aquatic Chemistry*, John Wiley & Sons, New York (1983).
18. Nriagu, J.O. and R.D. Coker, "Trace Metals in Humic and Fulvic Acids from Lake Ontario Sediments," *Environ. Sci. Technol.*, **14**, 443-46 (1980).
19. Rai, D., J. Zachara, A. Schwab, R. Schmidt, D. Girvin and J. Rogers, *Chemical Attenuation Rates, Coefficients, and Constants in Leachate Migration. Vol.I: A Critical Review*, Report EA-3356 to EPRI by Pacific Northwest Laboratories (Battelle Institute), Richland, Wash. (1984).
20. Salomons, W. and U. Förstner, *Metals in the Hydrocycle*, Springer-Verlag, New York (1984).
21. Shephard, B.K., A.W. McIntosh, G.K. Atchison and D.W. Nelson, "Aspects of the Aquatic Chemistry of Cadmium and Zinc in a Heavy Metal Contaminated Lake," *Water Res.*, **14**, 1061-66 (1980).
22. Sillén, L.G., and A.E. Martell, *Stability Constants of Metal-Ion Complexes*, Special Publication No. 17, The Chemical Society, Burlington House, London (1964).
23. Stumm, W. and J.J. Morgan, *Aquatic Chemistry*, 2nd ed., John Wiley & Sons, New York (1981).
24. U.S. Environmental Protection Agency, *Ambient Water Quality Criteria for Cadmium*, Report No. EPA-440/5-80-025 (October 1980).
25. Wilkins, R.G., *The Study of Kinetics and Mechanism of Reactions of Transition Metal Complexes*, Allyn and Bacon, Boston (1974).

7.5.6 Other Sources of Information

Arthur D. Little, Inc., *Full-Scale Evaluation of Waste Disposal from Coal-Fired Generating Plants. Appendix F (Part 4)*, EPA Contract No. 68-02-3167, Final Report to U.S. EPA, Office of Research and Development, Washington, D.C. (1984)

Cavallaro, N. and M.B. McBride, "Copper and Cadmium Adsorption Characteristics of Selected Acid and Calcareous Soils," *Soil Sci. Soc. Am. J.*, **42**, 550-56 (1978).

Fuller, W.H., *Movement of Selected Metals, Asbestos, and Cyanide in Soil: Applications to Waste Disposal Problems*, EPA 660/2-77-020, PB 266 905, Municipal Environmental Research Laboratory, Cincinnati, Ohio (1977).

Garcia-Miragaya, J., "Specific Sorption of Trace Amounts of Cadmium by Soils," *Commun. Soil Sci. Plant Anal.*, **11**, 1157-66 (1980).

Gardiner, J., "The Chemistry of Cadmium in Natural Waters — I. A Study of Cadmium Complex Formation Using the Cadmium Specific-Ion Electrode," *Water Res.*, **8**, 23-30 (1974).

Gardiner, J., "The Chemistry of Cadmium in Natural Waters — II. The Adsorption of Cadmium on River Muds and Naturally Occurring Solids," *Water Res.*, **8**, 157-64 (1974).

Guy, R.D. and C.L. Chakrabarti, "Studies of Metal-Organic Interactions on Model Systems Pertaining to Natural Waters," *Can. J. Chem.*, **54**, 2600-11 (1976).

John, M.K., "Cadmium Adsorption Maxima of Soils as Measured by the Langmuir Isotherm," *Can. J. Soil Sci.*, **52**, 343-50 (1975).

Kinkade, M.L. and H.E. Erdman, "The Influence of Hardness Components (Ca^{+2} and Mg^{+2}) in Water on the Uptake and Concentration of Cadmium in a Simulated Fresh Water Environment," *Environ. Res.*, **10**, 308-13 (1975).

Levi-Minzi, R., G.F. Soldatini and R. Riffaldi, "Cadmium Adsorption by Soils," *J. Soil Sci.*, **27**, 10-15 (1976).

Patterson, J.W., H.E. Allen and J.J. Scala, "Carbonate Precipitation for Heavy Metal Pollutants," *J. Water Pollut. Control Fed.*, **49**, 2397-2410 (1977).

Petruzelli, G., G. Guidi and L. Lubrano, "Organic Matter as an Influencing Factor on Copper and Cadmium Adsorption by Soils," *Water Air Soil Pollut.*, **9**, 263-68 (1978).

Ramamoorthy, S. and B.R. Rust, "Heavy Metal Exchange Processes in Sediment-Water Systems," *Environ. Geol.*, **2**, 165-72 (1978).

Silviera, D.J. and L.E. Sommers, "Extractability of Copper, Zinc, Cadmium, and Lead in Soils Incubated with Sewage Sludge," *J. Environ. Qual.*, **6**, 47-52 (1977).

Singh, B. and G.S. Sekhon, "Adsorption, Desorption, and Solubility Relationships of Lead and Cadmium in Some Alkaline Soils," *J. Soil Sci.*, **28**, 271-75 (1977).

Sommers, L.E. and W.L. Lindsay, "Effect of pH and Redox on Predicted Heavy Metal-Chelate Equilibria in Soils," *Soil Sci. Soc. Am. J.*, **43**, 39-47 (1979).

Weber, W.H. and H.S. Posselt, "Equilibrium Models and Precipitation Reactions for Cadmium (II)," in *Aqueous Environmental Chemistry of Metals*, A.J. Rubin (ed.), Ann Arbor Science Publishers, Ann Arbor, Mich. (1974).

7.6 CHROMIUM (Cr)

7.6.1 Occurrence and Uses

Chromium is a member of the transition metal series of elements in the periodic table. It is widely distributed in the earth's crust and is an important industrial metal. Chromite, $(Fe, Mg)O(Cr, Al, Fe)_2O_3$, is the only commercially important ore. Chromite belongs to the spinel group of minerals of a general composition $M(II)M(III)O_4$, where Cr(III) and other metals may occupy the M(III) sites in different proportions. The highest grade of this ore contains 52-56% of chromic oxide and 10-26% ferrous oxide. The theoretical end member, $FeOCr_2O_3$, would contain 68% chromic oxide and 32% ferrous oxide [5, 10, 14].

Large quantities of chromite are used in the steel industry to produce ferroalloys that are resistant to corrosion, oxidation and stress. Chromium is also used in the production of stainless steel, heat-resistant and chemically resistant (refractory) bricks, pigments, electroplating, and in the dye, tanning, textile and chemical industries [10, 14, 27].

Hexavalent chromium in the workplace is a suspected carcinogen. Trivalent chromium is considered an essential trace element in humans [7, 27].

7.6.2 Speciation Reactions in Water

Chromium occurs commonly in the +2, +3, and +6 oxidation states [2]; the latter two are the most important and stable aqueous species.

ACID-BASE DISSOCIATION

Chromous ion (Cr^{+2}) is unstable in water, and little is known about its hydrolysis. $Cr(OH)_2$ precipitates near pH 6 and is easily oxidized [2].

The hydrolysis behavior of Cr^{+3} is complicated by the slow kinetics of its polymerization reactions [2,6]. Chromic ion (Cr^{+3}) produces the mononuclear species $CrOH^{+2}$, $Cr(OH)_2^+$, $Cr(OH)_4^-$ and the neutral species $Cr(OH)_3^0$. The polynuclear species $Cr_2(OH)_2$ and $Cr_3(OH)_4^{+5}$ form slowly at 25°C [2]. Figure 7.6-1 shows the predicted distribution of Cr^{+3} hydrolysis species when no other complexing agents are present.

Because Cr^{+6} hydrolyzes extensively, only neutral or anionic (hydroxo or oxo) species occur in water [2, 13]. The predicted predominance of Cr^{+6} species in various pH regions is shown in Figure 7.6-2. All stable Cr^{+6} complexes are anionic oxyspecies; the anionic form varies according to pH and may be chromate (CrO_4^{-2}), hydrochromate ($HCrO_4^-$), or dichromate ($Cr_2O_7^{-2}$). At low pH and high total chromium (VI) concentration, the dichromate ion ($Cr_2O_7^{-2}$) predominates. Dichromate would become important only at very high Cr concentrations ($>10^{-2}$ *M* Cr), which might occur in cases of heavy industrial pollution. At pH >6.5, Cr(VI) is in the form of the chromate ion [2, 5, 13]. The predominant Cr(VI) species for concentrations of Cr

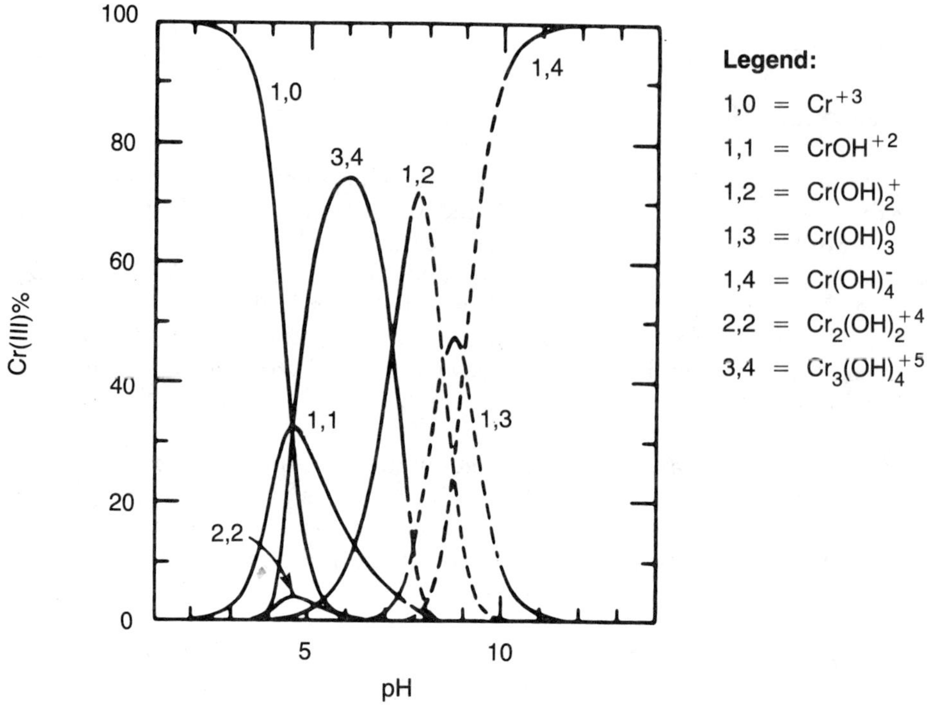

I = 1 *M*, T = 25°C, 10^{-5} *M* Cr(III)

Dashed curves denote regions supersaturated with respect to $Cr(OH)_3$

Source: Baes and Mesmer [2]. *(Copyright 1976, John Wiley & Sons. Reprinted with permission.)*

FIGURE 7.6-1 Predicted Distribution of Cr(III) Species at Varying pH

that might exist in natural water are hydrochromate and chromate [13]. Table 2.7-3 (Part I) lists pK values for chromium (III and VI) aquo species.

COMPLEXATION

The distribution of chromium among its aqueous inorganic species in the environment is, according to thermodynamic considerations, highly dependent upon pH and Eh and, to a lesser extent for Cr(IV), upon the presence of complexing ligands [20].

The chromium (III) center forms six coordinate octahedral complexes with a variety of ligands and can exist as cationic and anionic complexes, depending on the number and charge of ligands attached. The chromium (III) metal center complexes with many species, including water, fluoride, ammonia, cyanide, thiocyanate, oxalate, sulfate, and numerous organic ligands [5, 14, 20, 23, 27]. In neutral or basic solutions,

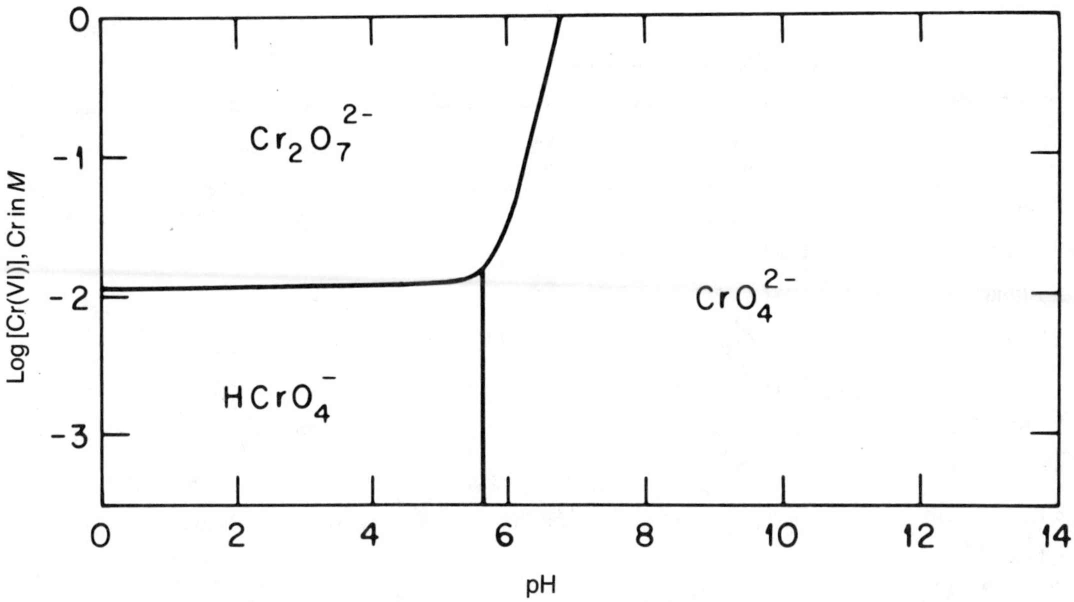

Solid lines represent conditions of equal concentration of the respective Cr(VI) species in the neighboring regions of the diagram.

Source: Baes and Mesmer [2]. *(Copyright 1976, John Wiley & Sons. Reprinted with permission.)*

FIGURE 7.6-2 Calculated Predominance Diagram for Cr(VI) – OH^- Species at 25°C and I = 1 *M*

chromium (III) forms polynuclear complexes in which adjacent chromium atoms are linked through hydroxo- or oxo-bridges [5, 8, 14]. As noted in section 3.2, Cr(III) tends to form complexes whose ligand rates of exchange are low [10, 23]; therefore, the vast majority of substitution reactions of Cr(III) complexes take place slowly and are easily measured [28]. Because of their kinetic inertness, many of these complexes persist for relatively long periods in solution, even under conditions that cause them to become thermodynamically unstable [8, 10, 14], and many Cr(III) complexes can be isolated as solids [14]. The characteristic second-order rate constant for exchange of one water molecule in the primary solvation shell of Cr(III) ion is 3.6×10^{-5} M^{-1} s^{-1} [15].

Cr(III) forms relatively strong complexes with oxygen donor ligands and neutral ligands [14]. These complexes are probably ultimately converted to $Cr(OH)_3$ [5, 10]. The formation of Cr-organic complexes has a major influence on the distribution of chromium species in seawater [20]. Table 2.9-1 (in Part I) lists complexation equilibrium constants for a variety of inorganic complexes of chromium (III); stability constants for various organic complexes are given in Table 7.6-1.

TABLE 7.6-1

Stability Constants of Cr^{+3} Organic Complexes

Ligand	Formula	Donor Atoms	Log β_1[a]	Log β_2[b]	Log β_3[c]
Oxalic Acid	$C_2H_2O_4$	O	5.34	10.51	15.44
Glycine	$C_2H_5O_2N$	O, N	8.62	16.27	—
α-Alanine	$C_3H_7O_2N$	O, N	8.53	15.97	—
Serine	$C_3H_7O_3N$	O, N	8.0	14.2	19.4
Succinic Acid	$C_4H_6O_4$	O	6.42	10.99	13.85
Phthalic Acid	$C_8H_6O_4$	O, O	5.52	10.00	12.48
Asparagine	$C_4H_8O_3N_2$	O, N	7.7	13.6	18.5
Ethylenediamine	$C_2H_8N_2$	N, N	16.5	30.5	—
Ethylenediamine tetra-acetic acid	$C_{10}H_{10}O_8N_2$	O, N	23.40	—	—

a. Refers to reaction $Cr^{+3} + L \rightleftharpoons ML$
b. Refers to reaction $Cr^{+3} + 2L \rightleftharpoons ML_2$
c. Refers to reaction $Cr^{+3} + 3L \rightleftharpoons ML_3$

Source: Sillén and Martell [24], as reported by Moore and Ramamoorthy [14]. *(Copright 1984, Springer-Verlag. Reprinted with permission.)*

In municipal waste waters in southern California, 97-99% of the dissolved chromium was found to be in the Cr(III) form. The more oxidized form, Cr(VI), constituted less than 1% of the total chromium. Chlorination did not increase the concentrations of Cr(VI). Clean coastal waters contained predominantly Cr(VI) [11, 14, 20, 21]. The distribution of chromium among inorganic Cr(III), inorganic Cr(VI), and organically bound Cr(III) species depends on the presence of organic materials, manganese oxide in suspended particulates and H_2S released from sediments [20]. Figure 7.6-3 depicts the speciation of chromium in freshwater in the absence of organic substances at pH 6. Figure 7.6-4 shows the corresponding information for seawater at pH 8.

REDOX REACTIONS

The two important oxidation states of chromium in natural waters are +3 and +6. Under the conditions of pH and pe found in most oxygenated waters, Cr(VI) is the thermodynamically stable species [13, 20]. However Cr(III), being kinetically stable, could persist if it was bound to naturally occurring solids [14]. Laboratory studies have indicated that Cr(III) and Cr(VI) are interconvertible under natural conditions [5]. Cr(III) is oxidized in natural waters by a large excess of MnO_2; oxidation by

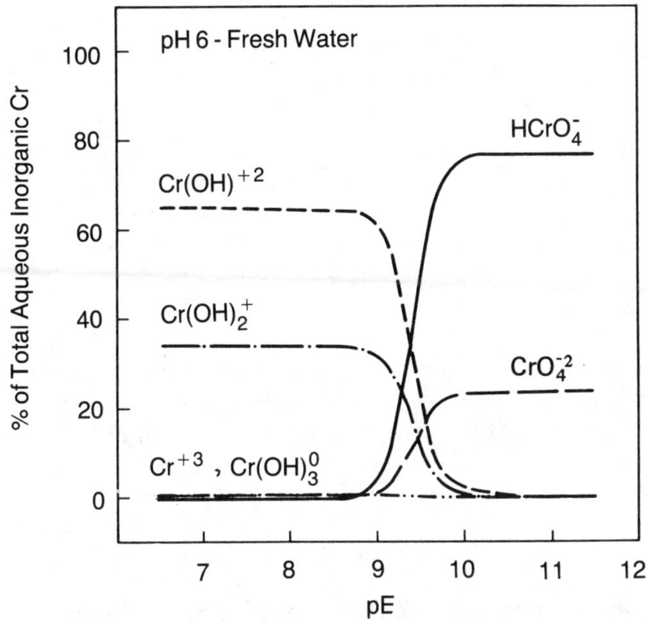

Source: Schmidt [20]

FIGURE 7.6-3 Distribution of Aqueous Inorganic Chromium Species in Fresh Water at pH 6

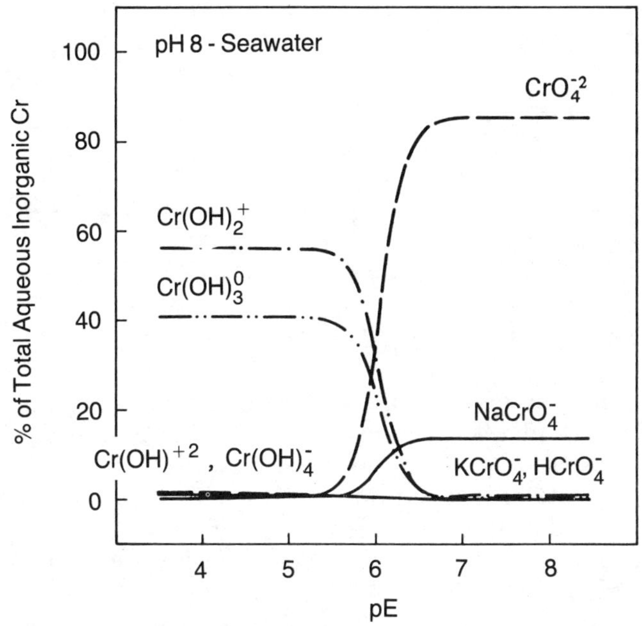

Source: Schmidt [20]

FIGURE 7.6-4 Distribution of Aqueous Inorganic Chromium Species in Seawater at pH 8

oxygen occurs slowly under natural water conditions [5, 14, 16, 20, 21]. Ultraviolet light may promote oxidation of Cr(III) to Cr(VI). Complexed Cr(III) can have different redox behavior from that of the aquochromium (III) ion [22].

Cr(VI) is a strong oxidizing agent in acidic waters [5, 23, 25, 27] but is relatively stable in most natural waters [27]. Species such as $CrO_3(s)$, CrO_4^{-2} and $CrO_2Cl_3^-$ strongly oxidize organic matter on contact [5]. In natural waters containing organic matter, Cr(VI) can be reduced to Cr(III) [10, 16, 20, 22, 25], which is then precipitated as $Cr(OH)_3$ [10]. Cr(VI) is also easily reduced by Fe^{+2}, H_2S, dissolved sulfides, cysteine and mercaptosuccinic acid [5, 10, 14].

Table 7.6-2 lists reduction potentials and logarithms of equilibrium constants for some reactions of chromium species. Figure 7.6-5 illustrates an Eh-pH diagram for the important species of chromium in a system containing 10^{-4} *M* chromium in water at 25°C.

TABLE 7.6-2

Reduction Potentials and Logarithms of Equilibrium Constants of Some Chromium Species
(at 25°C and zero ionic strength)

Reaction	E^0 (V)	Log K
$1/2\ Cr_2O_7^{-2} + 7\ H^+ + 3e^- = Cr^{+3} + 7/2\ H_2O$	+1.33	67.4
$HCrO_4^- + 7\ H^+ + 3e^- = Cr^{+3} + 4\ H_2O$	+1.20	60.8
$Cr^{+3} + e^- = Cr^{+2}$	−0.41	−6.9
$Cr^{+3} + 3e^- = Cr(s)$	−0.74	−37.5
$Cr^{+2} + 2e^- = Cr(s)$	−0.86	−29.0

Source: Kotrlý and Šůcha [12]

7.6.3 Solubility/Precipitation

The concentration of chromium in some natural waters is limited by the low solubility of Cr^{+3} oxides [9, 10]. $Cr(OH)_3$ and Cr(III) coprecipitated with Fe oxides are thought to be the Cr(III) solubility controlling solids [19]. Precipitation of $Cr(OH)_3$ at pH >5 is thought to be the dominant fate of chromium in natural waters [5]. As the pH rises above 5, Cr(III) becomes less soluble; the solubility product for the hydroxide at pH 8.5 is 2.9×10^{-29} [22]. Among all the Cr(III) solid phases for which data are available, $FeCr_2O_4$ is the least soluble at pE + pH $\leq$14 [19]. Chromium (III) solubility varies greatly with water characteristics such as hardness, alkalinity and salinity. Cr^{+3} has low solubility in seawater [10, 27]. The low solubility of chromic oxide under neutral or basic conditions limits its rate of reaction (e.g., with oxygen); thus, precipitated

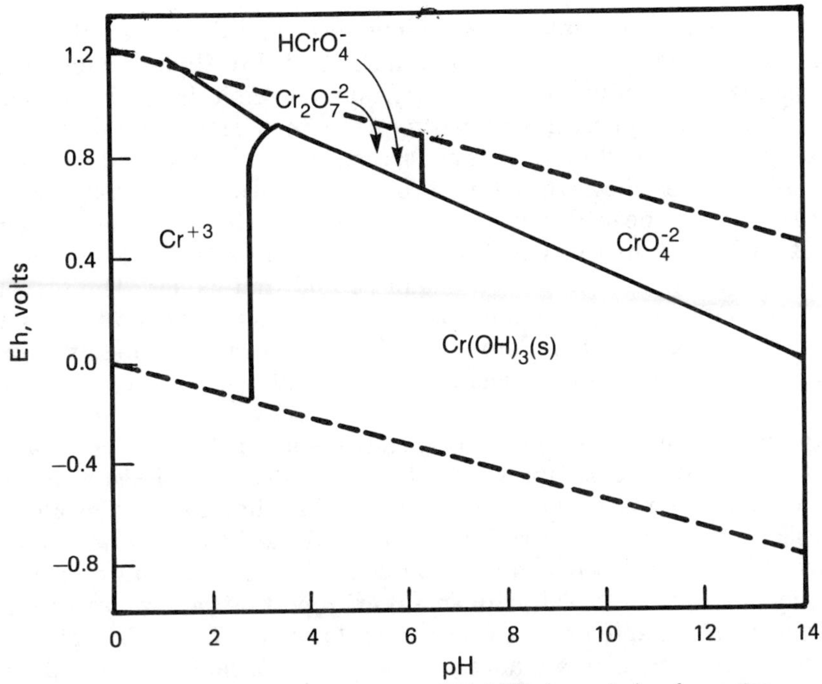

Dashed lines represent stability boundaries for water.

$[Cr_T] = 10^{-4}\ M$

Source: Adapted from Pourbaix [18]. *(Copyright 1966, Pergamon Press. Reprinted with permission.)*

FIGURE 7.6-5 Eh-pH Diagram for the System Chromium-water at 25°C

chromic oxide can persist under conditions in which it is thermodynamically unstable. Complexation of Cr(III) can prevent the precipitation of the hydrous oxide at pH values at which it is likely to occur [27].

$BaCrO_4$, $PbCrO_4$, and $AgCrO_4$ readily precipitate from solution [19]. Log K_s values at zero ionic strength and 25°C are -9.67 (Ba^+), -13.75 (Pb^{+2}), and -11.92 (Ag^{+2}) [12]. Lead chromate is the species most likely to control Cr^{+6} solubility in environmental conditions, since levels of other metals are typically low [19].

Speciation studies of municipal waste waters of California, which discharge 600-700 metric tons of chromium annually into coastal waters, showed that chromium occurred principally (67-98%) in the particulate state [14].

7.6.4 Sorption on Soils and Sediments

The adsorption of Cr(III) and Cr(VI) is complicated by redox changes that can occur in soil environments. Organic matter in soils can bring about the spontaneous reduction of Cr(VI) to Cr(III), even under alkaline conditions [4]. Manganese oxides in soil oxidize Cr(III) to Cr(VI) [3].

Chromate (CrO_4^{-2}) and dichromate ($Cr_2O_7^{-2}$) ions form many soluble salts and thus represent the mobile and potentially toxic forms in soils [5]. Under anaerobic conditions, these species are rapidly reduced to Cr(III) and precipitated as hydroxides and oxides. As soil conditions approach neutrality (pH 7), the formation of dichromates is minimized [21]. Hexavalent chromium is not sorbed to any significant degree by clays or hydrous metal oxides; however, it is strongly sorbed by activated carbon and therefore may have some affinity for organic matter in the environment [5]. Adsorption of Cr(VI) by hydrous metal oxides [13] and soil [5] decreases as pH increases. Some adsorption occurs on manganese and iron oxides. The presence of other anions (e.g., sulfate and phosphate) significantly affects the extent of adsorption by competing for adsorption sites. Formation of ion pairs, such as dissolved calcium chromate, can also reduce the extent of adsorption [19].

In contrast to Cr(VI), the adsorption of Cr(III) increases as pH increases. In a laboratory study of adsorption of chromium onto clay minerals, under fixed conditions, it was found that 30 to 300 times more Cr(III) was adsorbed than Cr(VI). The amounts of Cr(III) adsorbed correspond to adsorption of the hydrolyzed Cr(III) species on cation exchange sites [5]. Table 7.6-3 lists sorption constants for chromium on two clay species. Sorption of Cr(III) onto bed sediments is an important removal mechanism [5, 14]. At low concentration and low pH, where the solubility of Cr(III) is higher, strong specific adsorption on iron and manganese oxides occurs. Formation of soluble organic complexes reduces the degree of adsorption [19].

Organic material may also be an important adsorbent in soil [19]. Slight enrichment of chromium occurs in the humic acid fraction of sediments; the enrichment factor (defined as the ratio of metal concentration in the humic acid fraction to its total concentration in the associated sediments) is 0.83-1.9, compared with 0.46-0.81 for zinc and 21-40 for copper [17].

The chromium (VI) sorption capacity of various soils has been measured by batch tests with a field-collected leachate from a utility coal fly-ash disposal pond. Tests were conducted by adding variable amounts of soil (0.05-0.25 g) to 50 ml of leachate solution. Since this leachate did not contain significant amounts of chromium, Cr(VI) in the form of chromate ion was added to give a final concentration of 278 ppb Cr(VI) [1]. The results are given in Table 7.6-4 and suggest that Cr(VI) is not strongly adsorbed by these materials.

7.6.5 Bioaccumulation

As an essential nutrient, chromium is bioaccumulated by a variety of aquatic organisms [5, 9, 10, 14, 26]. There is a great range of sensitivity to chromium among different species of aquatic life. Hexavalent chromium is more soluble and generally more toxic than the trivalent form. Water hardness, salinity, temperature and (to some extent) pH affect the bioaccumulation of chromium by aquatic organisms [9, 10].

TABLE 7.6-3

Sorption Constants of Chromium (III) and Chromium (VI)

Sorbent	Oxidation State	pH	Sorption Constants[a] (A_m)
Montmorillonite Clay	Cr(III)	2.5	345
	Cr(III)	3.0	632, 649
	Cr(III)	4.0	2622, 2689
	Cr(VI)	3.0	3.64, 12.8
	Cr(VI)	4.0	2.50, 10.3
	Cr(VI)	5.0	2.22, 8.03
	Cr(VI)	7.0	0.98, 8.03
Kaolinite Clay	Cr(III)	2.5	63.5
	Cr(III)	3.0	96.3
	Cr(III)	4.0	206, 283
	Cr(VI)	3.0	1.79, 3.64
	Cr(VI)	4.0	0.85, 2.50
	Cr(VI)	5.0	0.62, 2.22
	Cr(VI)	7.0	0.29, 0.98

a. A_m (μmol/g) = Langmuir adsorption maximum (see Section 2.12). Values of K_L not reported.

Source: Rai *et al.* [19]. *(Copyright 1984, Electric Power Research Institute. Reprinted with permission.)*

TABLE 7.6-4

Sorption Capacity for Chromium(VI)[a] at Equilibrium for Soils

Soil Type	Sorption Capacity (μg Cr(VI)/g soil)
Alluvial material	0.47-0.96
Residual soil (silty sand)	0.47-1.45
Silty fine sand with little clay	0.40-0.54

a. Field-collected leachate, pH 9.3, containing 278 ppb Cr(VI) as chromate

Source: Arthur D. Little, Inc. [1]

7.6.6 Literature Cited

1. Arthur D. Little, Inc., *Full-Scale Evaluation of Waste Disposal from Coal-Fired Generating Plants. Appendix F (Part 4),* EPA Contract No. 68-02-3167, Final Report to U.S. EPA, Office of Research and Development, Washington, D.C. (1984).
2. Baes, C.F. and R.E. Mesmer, *The Hydrolysis of Cations,* John Wiley & Sons, New York (1976).
3. Bartlett, R. and B. James, "Behavior of Chromium in Soils: III. Oxidation," *J. Environ. Qual.*, **8**, 31-35 (1979).
4. Bartlett, R.J. and J.M. Kimble, "Behavior of Chromium in Soils: II. Hexavalent Forms," *J. Environ. Qual.*, **5**, 383-85 (1976).
5. Callahan, M., M. Slimak, N. Gabel, I. May, C. Fowler, J. Freed, P. Jennings, R. Durfee, F. Whitmore, B. Maestri, W. Mabey, B. Holt and C. Gould, *Water-Related Environmental Fate of 129 Priority Pollutants. Vol. 1: Introduction and Technical Background, Metals and Inorganics, Pesticides and PCBs,* EPA-440/4-79-029a, EPA Contracts 68-01-3852 and 68-01-3867, Office of Water Planning and Standards, U.S. Environmental Protection Agency, Washington, D.C. (1979).
6. Carlin, R.L., *Transition Metal Chemistry, Vol. 1,* Edward Arnold Ltd., London (1965).
7. Clayton, G.D. and F.E. Clayton, *Patty's Industrial Hygiene and Toxicology,* 3rd rev. ed., *Vol. 2A,* John Wiley & Sons, New York (1981).
8. Cotton, R.A. and G. Wilkinson, *Advanced Inorganic Chemistry,* 3rd ed., Interscience Publishers, Div. John Wiley & Sons, New York (1972).
9. Ecological Analysts, Inc., *The Sources, Chemistry, Fate, and Effects of Chromium in Aquatic Environments,* American Petroleum Institute, Washington, D.C. (1981).
10. Faust, S.D. and O.S. Aly, *Chemistry of Natural Waters,* Butterworth, Woburn, Mass.(1981).
11. Jan, T.K. and D.R. Young, "Chromium Speciation in Municipal Wastewaters and Seawater," *J. Water Pollut. Control Fed.*, **50**, 2327-36 (1978).
12. Kotrlý, S. and L. Šůcha, *Handbook of Chemical Equilibria in Analytical Chemistry,* Halstead Press, Div. John Wiley & Sons, New York (1985).
13. MacNaughton, M.G., "Adsorption of Chromium(VI) at the Oxide-Water Interface," in *Biological Implications of Metals in the Environment,* H. Drucker and R.R. Wildung (eds.), CONF-750929, Springfield, Va., NTIS, 240-53 (1977).
14. Moore, J.W. and S. Ramamoorthy, *Heavy Metals in Natural Waters,* Springer-Verlag, New York (1984).
15. Morel, F.M.M., *Principles of Aquatic Chemistry,* John Wiley & Sons, New York (1983).

16. Nakayama, E., T. Kuwamoto, S. Tsurubo and T. Fujinaga, "Chemical Speciation of Chromium in Seawater. Part 2: Effects of Manganese Oxides and Reducible Organic Materials on the Redox Processes of Chromium," *Anal. Chim. Acta*, **130**, 401-4 (1981).
17. Nriagu, J.O. and R.D. Coker, "Trace Metals in Humic and Fulvic Acids from Lake Ontario Sediments," *Environ. Sci. Technol.*, **14**, 443-46 (1980).
18. Pourbaix, M., *Atlas of Electrochemical Equilibria in Aqueous Solutions*, Pergamon Press, Elmsford, N.Y. (1966).
19. Rai, D., J. Zachara, A. Schwab, R. Schmidt, D. Girvin and J. Rogers, *Chemical Attenuation Rates, Coefficients and Constants in Leachate Migration. Volume 1: A Critical Review,* Report EA-3356 to EPRI by Pacific Northwest Laboratories (Battelle Institute), Richland, Wash. (1984).
20. Schmidt, R.L., *Thermodynamic Properties and Environmental Chemistry of Chromium,* DOE Contract DE-AC06-76RLO 1830, Pacific Northwest Laboratories (Battelle Institute), Richland, Wash. (1984).
21. Schroeder, D.C. and G.F. Lee, "Potential Transformations of Chromium in Natural Waters," *Water Air Soil Pollut.*, **4**, 355-65 (1975).
22. Shivas, S.A.J., "Factors Affecting the Oxidation State of Chromium Disposed in Tannery Waters," *J. Am. Leather Chem. Assoc.,* **75**, 42-48 (1980).
23. Sienko, M.J. and R.A. Plane, *Chemistry*, 5th ed., McGraw-Hill Book Co., New York (1976).
24. Sillén, L.G. and A.E. Martell, *Stability Constants of Metal-Ion Complexes, Supplement No. 1*, Special Publication No. 25, The Chemical Society, Burlington House, London (1971).
25. Stollenwerk, K.G. and D.B. Grove, "Reduction of Hexavalent Chromium in Water Samples Acidified for Preservation," *J. Environ. Qual.*, **14**, 396-99 (1985).
26. Stumm, W. and J.J. Morgan, *Aquatic Chemistry*, 2nd ed., John Wiley & Sons, New York (1981).
27. U.S. Environmental Protection Agency, *Ambient Water Criteria for Chromium*, Report No. EPA-440/5-80-035, Washington, D.C. (October 1980).
28. Wilkins, R.G., *The Study of Kinetics and Mechanism of Reactions of Transition Metal Complexes*, Allyn and Bacon, Inc., Boston (1974).

7.6.7 Other Sources of Information

Bray, D.J., "Toxicity of Chromium Compounds Formed in Refractories," *Am. Ceram. Soc. Bull.*, **64**, 1012-16 (1985).

Cranston, R.E. and J.H. Murray, "The Determination of Chromium Species in Natural Waters," *Anal. Chim. Acta*, **99**, 275 (1978).

Emerson, S., R.E. Cranston and P.S. Liss, "Redox Species in a Reducing Fjord: Equilibrium and Kinetic Considerations," *Deep-Sea Res.*, **26A**, 859 (1979).

Espenson, J.H., "Formation Rates of Monosubstituted Chromium (III) Complexes in Aqueous Solution," *Inorg. Chem.*, **8**, 1554-56 (1969).

Fuller, W.H., *Movement of Selected Metals, Asbestos, and Cyanide in Soil: Applications to Waste Disposal Problems,* EPA 600/2-77-020, PB 266 905, Municipal Environmental Research Laboratory, Cincinnati (1977).

James, B.R. and R.J. Bartlett, "Behavior of Chromium in Soils: V. Fate of Organically Complexed Cr(III) Added to Soil," *J. Environ. Qual.*, **12**, 169-72 (1983).

James, B.R. and R.J. Bartlett, "Behavior of Chromium in Soils: VI. Interactions Between Oxidation-Reduction and Organic Complexation," *J. Environ. Qual.*, **12**, 173-76 (1983).

Lollar, R.M., "Bases for Delisting Solid Wastes Containing Trivalent Chromium," Paper Presented to Session 41: *Hazard Waste Delisting — Case Studies*, American Institute of Chemical Engineers, Cleveland (1982).

Nakayama, E., T. Kuwamoto, S. Tsurubo, H. Tokoro and T. Fujinaga, "Chemical Speciation of Chromium in Seawater. Part 1: Effect of Naturally Occurring Organic Materials on the Complex Formation of Chromium (III)," *Anal. Chim. Acta*, **130**, 289 (1981).

Nakayama, E., T. Kuwamoto, S. Tsurubo, H. Tokoro and T. Fujinaga, "Chemical Speciation of Chromium in Seawater. Part 3: The Determination of Chromium Species," *Anal. Chim. Acta*, **130**, 247 (1981).

Proceedings of the Symposium on "Health Aspects of Chromium-containing Materials" (September 15, 1977, Baltimore, Md.), Industrial Health Foundation, Inc., Pittsburgh, Pa. (1978).

7.7 COPPER (Cu)

7.7.1 Occurrence and Uses

Copper occurs in nature mostly in the form of oxides and sulfides. The most abundant form is chalcopyrite ($CuFeS_2$). Copper metal and alloys and copper sulfate are the main forms of copper produced. The principal uses of copper are in wire, electrical and electronic equipment, brass, electroplating, and construction. Copper sulfate and chromated copper arsenate are used in agriculture [9].

7.7.2 Speciation Reactions in Water

ACID-BASE DISSOCIATION

Aquo-copper ions in water act as acids. The pK_a values of aquo- and hydroxo-copper ions are shown in Table 2.7-3 in Part I of this report. An important Cu(II) species present in aquatic systems is $Cu_2(OH)_2^{+2}$, which is formed by the dimerization of $CuOH^+$. The formation of polynuclear hydroxo compounds like $Cu_2(OH)_2^{+2}$ is a much slower process than hydrolysis. $Cu_2(OH)_2^{+2}$ is not stable thermodynamically under normal environmental conditions and is generally formed only when high concentrations of $Cu(OH)_2^0$ exist in solution, such as by oversaturation with respect to $Cu(OH)_2(s)$. Once formed, however, it may exist as a metastable species for years [12]. Figure 7.7-1 shows the percentage of different hydrolyzed Cu(II) species with varying pH under conditions where no complexing agents are present.

COMPLEXATION

Copper forms complexes with both inorganic and organic ligands in aquatic systems. The complexes with OH^- and CO_3^{-2} are strong, but those with Cl^- and SO_4^{-2} are relatively weak. Figure 7.7-2 shows the chemical speciation of copper as a function of pH in unpolluted, organic-free aquatic systems. In acidic environments ($pH<7$), Cu^{+2} is the dominant species. In a typical environment (pH = 6-8), the predominant species are Cu^{+2} , $Cu(OH)_2^0$, $CuHCO_3^+$, $CuCO_3^0$, and $CuOH^+$. In aquatic systems polluted by industrial reagents and natural organic ligands, the speciation of copper is greatly influenced by the extent of formation of copper complexes with these ligands. Because of the great stability of its complexes with organic matter, Cu(II) will tie up organic ligands and may affect the complexation of other metals with organics [12]. Even though major cations (Ca^{+2}, Mg^{+2}) form much weaker complexes with organic species than does copper, the presence of these cations in natural environments at concentrations orders of magnitude larger than copper concentrations reduces the availability of free organic ligands for complexation with copper.

Stability constants for Cu(II) complexes with some inorganic and organic ligands are shown in Tables 2.9-1 and -2 in Part I of this report. Stability constants of Cu(II)-humate complexes vary with the source of humic material and aquatic composition, particularly pH. For freshwater lakes and rivers, logarithms of the overall

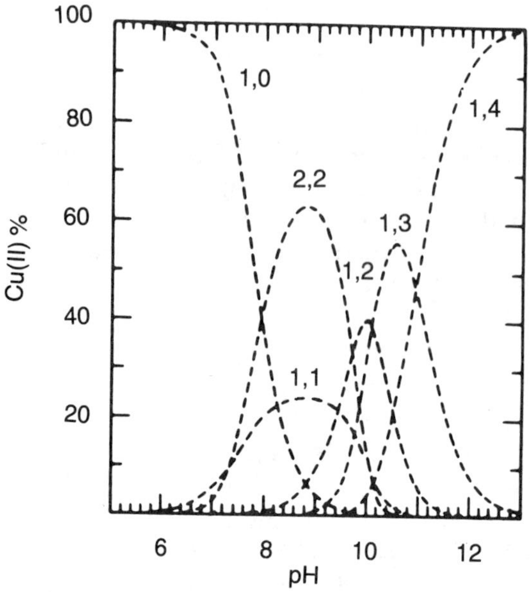

Legend:

1,0 = Cu^{+2}

1,1 = $CuOH^{+}$

1,2 = $Cu(OH)_2^0$

1,3 = $Cu(OH)_3^-$

1,4 = $Cu(OH)_4^{-2}$

2,2 = $Cu_2(OH)_2^{+2}$

I = 1 *M*, T = 25°C, 10^{-5} *M* Cu(II)

Dashed curves denote supersaturation with respect to CuO.

Source: Baes and Mesmer [1]. *(Copyright 1976, John Wiley & Sons. Reprinted with permission.)*

FIGURE 7.7-1 Hydrolyzed Cu(II) Species at Varying pH

humate conditional stability constants[1] ranging from 7.2 to 9.5 have been found at pH values between 6.6 and 8.8 [13].

Complexation of Cu(II) by ligands occurs rapidly: most of these complexes are formed at a second-order rate constant of about 10^9 $M^{-1}s^{-1}$ [8]. This implies that, for example, if the Cu(II) concentration in an aquatic system is $10^{-5}M$ and the Cl^- concentration is $10^{-3}M$, the half life for formation of the $CuCl^+$ complex is on the order of 10^{-6} second.

---

1. The overall conditional stability constant, K_o is defined by assuming an aggregate site-type as follows [13]:

$$K_o = \frac{[ML]}{[M^{+2}]\,[L_T - ML]}$$

where
$[ML]$ = total complexed Cu(II)
$[M^{+2}]$ = total free aqueous Cu(II)
$[L_T]$ = total ligand concentration
$[L_T - ML]$ = total uncomplexed ligand

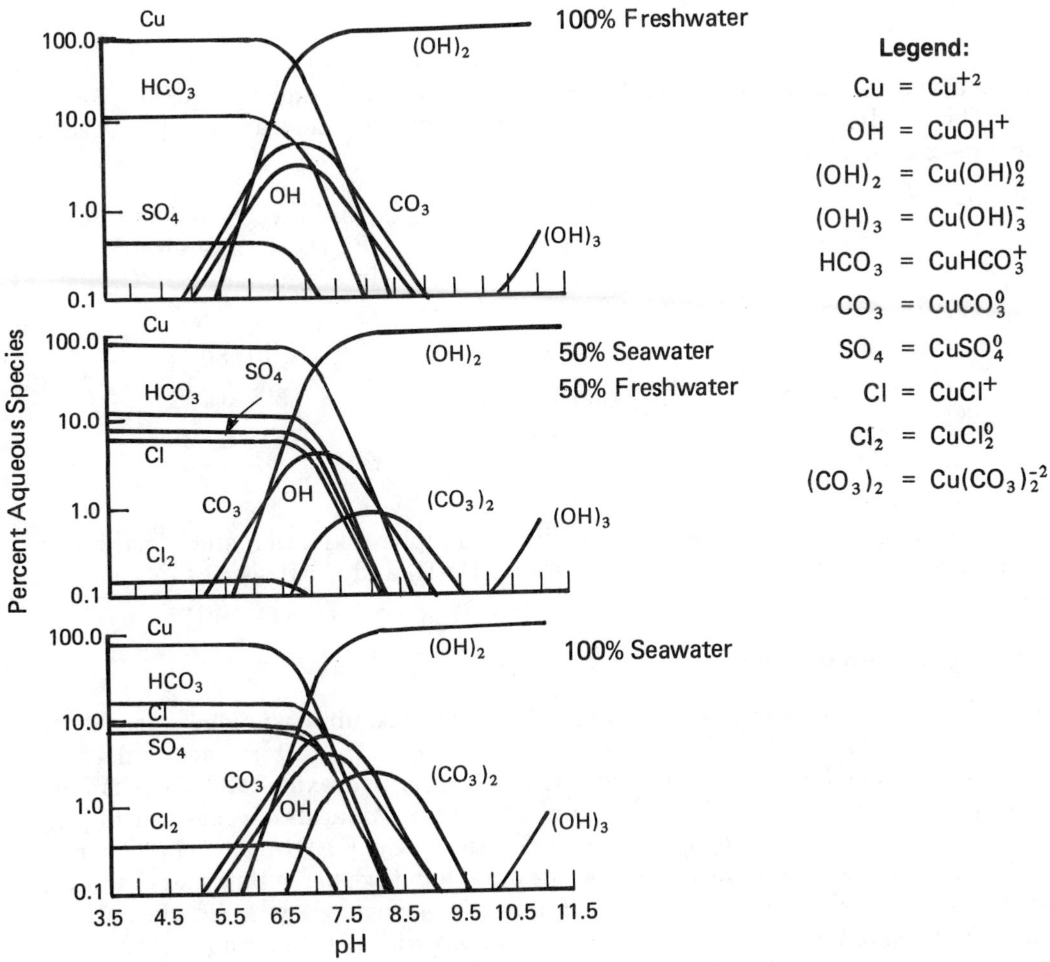

Composition of Systems (Molal Concentrations)

Species	Freshwater	Seawater
Ca	0.000375	0.0104
Mg	0.000168	0.054
Na	0.000274	0.4752
K	0.000059	0.01
Cl	0.00022	0.554
SO_4	0.000955	0.0284
Total Alkalinity	0.000955	0.00234
Ionic Strength	0.0021	0.661

Source: Long and Angino [6]. *(Copyright 1977, Pergamon Press, Inc. Reprinted with permission.)*

FIGURE 7.7-2 Chemical Speciation of Cu(II) in Unpolluted, Organic-free Aquatic Systems with Total Copper Concentration < 10 ppm

REDOX REACTIONS

Copper exists as Cu(II) under aerobic conditions and Cu(I) under anaerobic conditions. The equilibrium constants and standard electrode potentials of copper redox reactions are as follows [11]:

	Log K at 25° C	Standard Electrode Potential (Pt, H_2/H^+ reference) International V, 25°C
$Cu^{+2} + e^- \rightleftharpoons Cu^+$	2.68	+0.1586
$Cu^{+2} + 2e^- \rightleftharpoons Cu(s)$	11.38-11.44	+0.337-0.338
$Cu^+ + e^- \rightleftharpoons Cu(s)$	8.76-8.80	+0.518-0.521

The presence of complexing ligands stabilizes the higher oxidation state, making it less amenable to reduction. Figure 7.7-3 shows Cu(II) and Cu(I) species in a simple system.

7.7.3 Solubility/Precipitation

Copper oxides and sulfides are sparsely soluble; typical solubility products are $10^{-48.5}$ for chalcocite (Cu_2S), $10^{-45.96}$ for azurite ($Cu_3(CO_3)_2(OH)_2$), $10^{-33.78}$ for malachite ($Cu_2CO_3(OH)_2$), and $10^{-36.1}$ for covellite (CuS). Cupric hydroxide ($Cu(OH)_2$) is also sparsely soluble with a solubility product of $10^{-19.32}$ [7]. Minerals of copper with iron, e.g., cupric ferrite (α-$CuFe_2O_4$), cuprous ferrite (α-$Cu_2Fe_2O_4$), and chalcopyrite ($CuFeS_2$), are also virtually insoluble. On the other hand, copper sulfates, oxysulfates, and phosphates are very soluble [5,8]. Figure 7.7-4 shows how a simple natural system of dissolved and solid copper species changes with pH. In aerobic waters, the solid species typically found are the oxide and hydroxide. In anaerobic systems containing sulfur, copper sulfide, cuprous oxide and metallic copper are present [9].

7.7.4 Sorption on Soils and Sediments

Copper is strongly sorbed by hydrous iron and manganese oxides, clay minerals (e.g., kaolinite) and organic matter. It exceeds zinc, lead and cadmium in its tendency to be sorbed to solids. Sorption is probably the most important controlling mechanism in determining copper mobility in the environment. Organic complexation, specific adsorption, precipitation, and ion exchange are important mechanisms for sorption of copper onto soils and sediments [10]. Several investigators have demonstrated that the sorption of copper on hydrous iron and manganese oxides is due to co-precipitation of copper in the iron and manganese oxide lattice [2].

The sorption process is strongly dependent on pH. This can be explained by the following:

- $CuOH^+$ is the preferred sorbed species rather than Cu^{+2} [10].

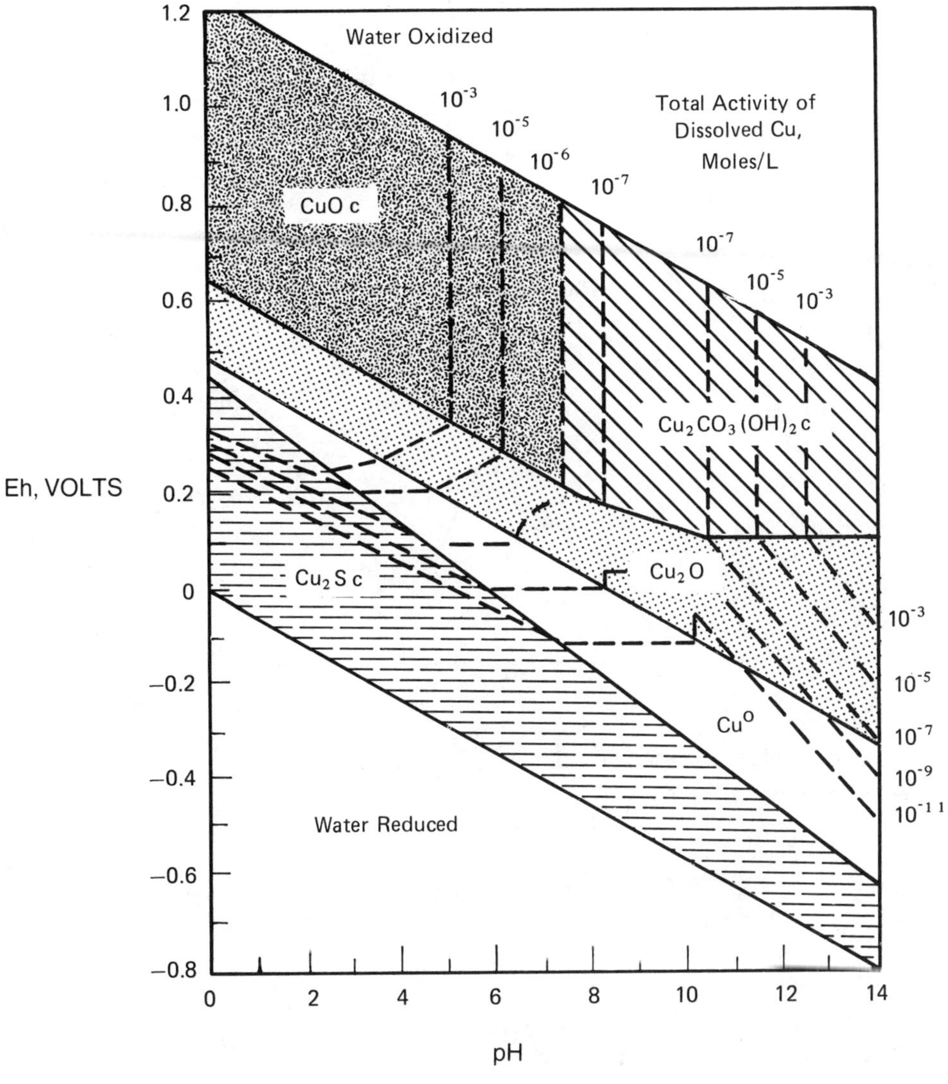

Source: Hem [3]. *(Copyright 1975, Pergamon Press. Reprinted with permission.)*

FIGURE 7.7-3 Copper Speciation in Oxidizing and Reducing Systems at 25°C, 1 atm in the System Containing Cu-H_2O-C-S with Total Dissolved C = 10^{-3} *M* and Total Dissolved S = 10^{-4} *M*

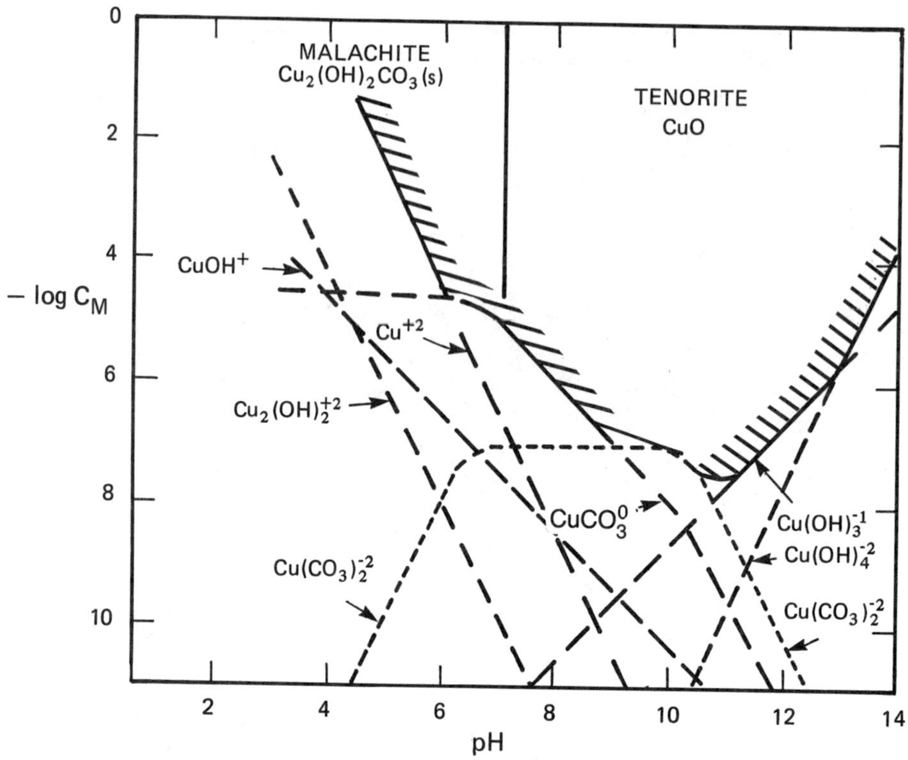

Note: Total carbon maintained at 10^{-2} M throughout.

Source: Stumm and Morgan [12]. *(Copyright 1981, John Wiley & Sons. Reprinted with permission.)*

FIGURE 7.7-4 Solubility and Speciation of Cu(II) in a Natural Carbonate-bearing Water

- Precipitation and co-precipitation are important sorption mechanisms.
- pH affects the surface charge on oxides and, thus, their sorptive capacity for cations.

The presence of anions also appears to influence the sorption of copper. Complexing ligands may remobilize the copper. However, some anions may increase sorption by the formation of copper-ligand bonds that increase free-electron sharing with the surface ions. In an experiment on the effect of pH and anions on the sorption of copper by Metapeak soil, Huang *et al.* [4] found that sorption is very low below pH 4 and increases rapidly above pH 6, coming to a maximum at pH 7. The anions used (phosphate, glycine/tartrate, nitriloacetate, and humic acid) all increase sorption significantly at the lower pH (pH $\leq$ 6) but have no effect above pH 6.

Langmuir isotherm sorption constants for copper are listed in Table 7.7-1. Note that these are typical values; those reported by Rai *et al.* vary over an extremely large range, because they are affected by experimental conditions, type of sorbent, concen-

TABLE 7.7-1

Typical Values of Sorption Constants for Copper

Sorption Constant[a]	Clay	Fe Oxides	Mn Oxides	Organic Matter	Soils	Sediment
A_m (μmol/g)	6	130	1670	200	28	34
Log K_L (log M^{-1})	5	5	4	5	5	5
K_d (ml/g)	3	3600	8000			

a. Definitions (see also section 2.12):

A_m and K_L are Langmuir constants

$$S = \frac{K_L A_m C}{1 + K_L C}$$

where

S = moles sorbed at equilibrium per gram of solid
A_m = maximum adsorption capacity of solid
K_L = sorption constant (related to binding energy of sorbate)
C = total sorbate concentration in solution at equilibrium

K_d = ratio of the quantity of sorbate sorbed per gram of solid to the amount of the sorbate remaining in solution at equilibrium

Source: Adapted from Rai *et al.* [10].

trations of solute, other substances present in the soil or sediment, and many other factors. In the cited source the Langmuir maximum sorption capacity (A_m) varied by factors of between 2 and 100 for the same sorbent. The Langmuir constant, K_L, was more uniform, varying by up to a factor of 3. K_d values varied by up to a factor of 30. Extreme caution should be exercised when using these typical values, and the sensitivity of the environmental system to variations in these sorption constants should also be assessed.

7.7.5 Literature Cited

1. Baes, C.F., Jr. and R.E. Mesmer, *The Hydrolysis of Cations,* John Wiley & Sons, New York (1976).
2. Callahan, M., M. Slimak, N. Gabel, I. May, C. Fowler, J. Freed, P. Jennings, R. Durfee, F. Whitmore, B. Maestri, W. Mabey, B. Holt and C. Gould, *Water-Related Environmental Fate of 129 Priority Pollutants. Volume I: Introduction*

and Technical Background, Metals and Inorganics, Pesticides and PCBs, EPA-440/4-79-029a, EPA Contracts 68-01-3852 and 68-01-3867, Office of Water Planning and Standards, U.S. Environmental Protection Agency, Washington, D.C. (1979).

3. Hem, J.D., Discussion of "Role of Hydrous Metal Oxides in the Transport of Heavy Metals in the Environment," by G.F. Lee in *Heavy Metals in the Aquatic Environment*, P.A. Krenkel (ed.), Pergamon Press, Oxford, England, 149-53 (1975).
4. Huang, C.P., H.A. Elliot and R.M. Ashmead, "Interfacial Reactions and the Fate of Heavy Metals in Soil-Water Systems," *J. Water Pollut. Control Fed.*, **49**, 745-56 (1977).
5. Lindsay, W.L., *Chemical Equilibria in Soils,* John Wiley & Sons, New York (1979).
6. Long, D.T. and E.E. Angino, "Chemical Speciation of Cd, Cu, Pb, and Zn in Mixed Freshwater, Seawater, and Brine Solutions," *Geochim. Cosmochim. Acta*, **41**, 1183-91 (1977).
7. Martell, A.E. and R.M. Smith, *Critical Stability Constants*, Vol. 4, *Inorganic Complexes*, Plenum Press, New York (1976).
8. Morel, F.M.M., *Principles of Aquatic Chemistry*, John Wiley & Sons, New York (1983).
9. Perwak, J., S. Bysshe, M. Goyer, L. Nelken, K. Scow, P. Walker and D. Wallace, *An Exposure and Risk Assessment for Copper*, Final Draft Report to U.S. Environmental Protection Agency, EPA Contract 68-01-3857, Monitoring and Data Support Division, Office of Water Regulations and Standards, Washington, D.C. (1980).
10. Rai, D., J. Zachara, A. Schwab, R. Schmidt, D. Girvin and J. Rogers, *Chemical Attenuation Rates, Coefficients, and Constants in Leachate Migration. Vol I: A Critical Review*, Report EA-3356 to EPRI by Pacific Northwest Laboratories (Battelle Institute), Richland, Wash. (1984).
11. Sillén, L.G. and A.E. Martell, *Stability Constants of Metal-Ion Complexes*, Special Publication No. 7, The Chemical Society, London (1964).
12. Stumm, W. and J.J. Morgan, *Aquatic Chemistry*, 2nd ed., John Wiley & Sons, New York (1981).
13. Van den Berg, C.M.G. and J.R. Kramer, "Conditional Stability Constants for Copper Ions with Ligands in Natural Waters," in *Chemical Modeling in Aqueous Systems*, E.A. Jenne (ed.), Symposium Series 93, American Chemical Society, Washington, D.C., 115-32 (1979).

7.7.6 Additional Sources of Information

Lopez, J.M. and G.F. Lee, "Environmental Chemistry of Copper in Torch Lake, Michigan," *Water Air Soil Pollut.*, **8**, 373-85 (1977).

National Research Council, *Copper*, PB-262 425, Washington, D.C. (1977).

Sunda, W.G. and P.J. Hanson, "Chemical Speciation of Copper in River Water," in *Chemical Modeling in Aqueous Systems*, E.A. Jenne (ed.), Symposium Series 93, American Chemical Society, Washington, D.C., 147-80 (1979).

Sylva, R.N., "The Environmental Chemistry of Copper (II) in Aquatic Systems," *Water Res.*, **10**, 792-98 (1976).

7.8 LEAD (Pb)

7.8.1 Occurrence and Uses

Lead is the most abundant of the natural heavy elements. The primary form of lead in nature is galena (PbS), a relatively insoluble ore. Lead also occurs as plattnerite (PbO_2), cerussite ($PbCO_3$) and anglesite ($PbSO_4$) [17]. The main uses of lead are in lead acid storage batteries and antiknock gasoline additives. In batteries, the main form of lead used is lead metal and lead oxides. Tetraethyl and tetramethyl lead are also used in antiknock gasoline additives. Lead metal and lead alloys are used in red lead and litharge, ammunition, solder, weights and ballast, pigments, cable covering, brass and bronze, sheet lead, bearing metals, caulking metals, pipes, and type metal [12].

7.8.2 Speciation Reactions in Water

ACID-BASE DISSOCIATION

Aquo-lead ions act as acids in water. The pK_a values of aquo-lead species are shown in Table 2.7-3 (Part I). Figure 7.8-1 shows how the fractions of various hydrolyzed Pb(II) species change with pH under conditions where no other complexing agents are present. The total dissolved Pb(II) in the system shown is 10^{-9} *M*. The predominant species under acidic conditions is Pb^{+2}, but $PbOH^{+}$ dominates in the typical pH range found in the environment.

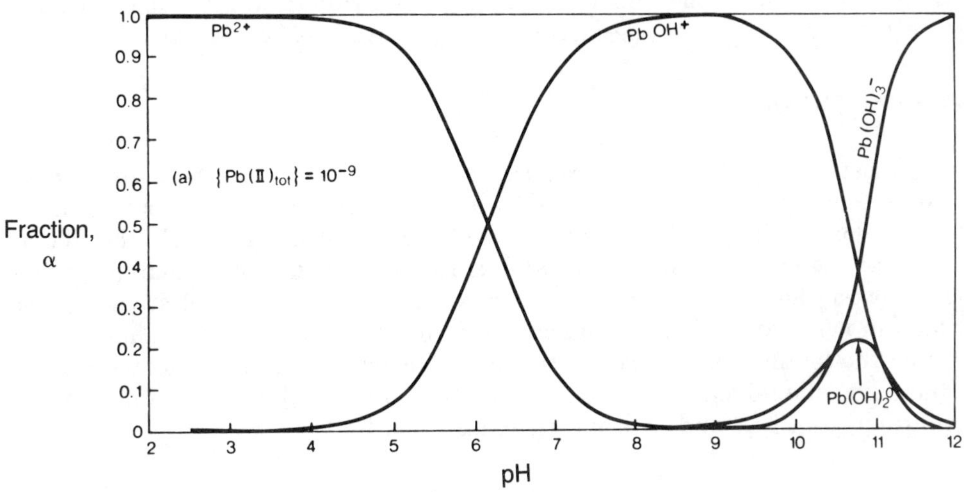

Total lead(II) activity = 10^{-9} *M*

Source: Rickard and Nriagu [14]. *(Copyright 1978, Elsevier/North-Holland Biomedical Press. Reprinted with permission.)*

FIGURE 7.8-1 Calculated Distribution of Lead(II) Hydroxy Species in Terms of pH and Fraction of Total Dissolved Lead(II)

Polynuclear hydroxo complexes are formed at higher concentrations of total dissolved Pb(II). Several sources [1, 14, 16] have reported the existence of $Pb_3(OH)_4^{+2}$, $Pb_4(OH)_4^{+4}$, and $Pb_6(OH)_8^{+4}$. However, as the concentration of Pb(II) expected in natural systems is low ($< 10^{-6}$ *M*) and other complexing agents are also present, significant amounts of these complexes are unlikely.

COMPLEXATION

Lead is complexed by both inorganic (e.g., OH^-, CO_3^{-2}, Cl^-) and organic (e.g., humic and fulvic acid) ligands present in aquatic systems. Some stability constants for Pb(II) complexes are shown in Table 2.9-1 (Part I); these indicate that Pb(II) forms strong complexes with OH^- and CO_3^{-2} and relatively weak complexes with Cl^- and other common inorganic ligands. However, in systems with very high Cl^- concentrations, such as estuaries and seawater, $PbCl^+$ is the dominant species.

Figure 7.8-2 shows the chemical speciation of Pb(II) in model unpolluted, organic-free aquatic systems. The free aquated Pb^{+2} ion is predominant in freshwater systems for much of the pH range found in the natural environment. The other significant species are $PbHCO_3^+$, $PbCO_3^0$, and $PbOH^+$. In seawater, $PbCl^+$ is dominant over the normal pH range. Pb^{+2}, $PbCl_2^0$, and $PbCl_3^-$ are present in significant concentrations as well. At a higher pH, carbonate and hydroxide complexes prevail.

Humic matter in natural environments also forms complexes with Pb(II). The data on stability constants of Pb-humate complexes are limited. With "Aldrich" synthetic humic material at pH 6.8, values of K_1 and K_2 were found to be 6.53 and 5.30 respectively[1] [6]. Saar and Weber [15] found conditional stability constants of lead (II)-fulvic acid complexes ranging from around 4 to 6.3 (pH between 4 and 6).

REDOX REACTIONS

The oxidation states 0, +2, and +4 are possible for lead. However, Pb(IV) exists only in extremely oxidizing conditions not usually found within the ranges of environmental pH and redox conditions [5, 13]. Figure 7.8-3 shows calculated Pb(II) and Pb(IV) species in a simple system of lead and water containing no other complexing species. As this diagram shows, PbO_2 and Pb^0 have very narrow regions of stability, the former only in very oxidizing environments and the other under very reduced conditions. Under the conditions as shown in the figure, several possible lead oxides are not stable, including the mixed valence oxide, Pb_3O_4, and PbO [7]. Pb(II) is stable under most conditions as Pb^{+2} and hydroxy complexes of Pb(II).

1. K_1 and K_2 are conditional constants for binding sites of type i:

K_i = $(ML)/[(M^{+2})(L_{iT} - ML_i)]$
ML = complexed Pb(II)
M^{+2} = free aquated Pb(II)
L_{iT} = total sites of type i
$L_{iT} - ML_i$ = total uncomplexed sites of type i

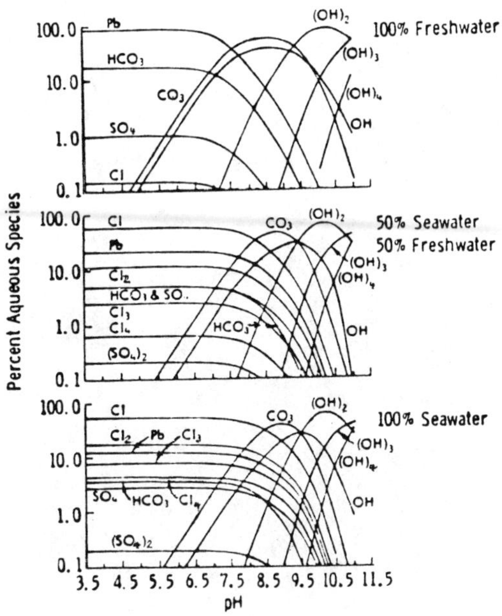

Legend:

Pb = Pb^{+2} (free aq.)

OH = $PbOH^{+}$

$(OH)_2$ = $Pb(OH)_2^0$

$(OH)_3$ = $Pb(OH)_3^-$

$(OH)_4$ = $Pb(OH)_4^{-2}$

HCO_3 = $PbHCO_3^+$

CO_3 = $PbCO_3^0$

Cl = $PbCl^+$

Cl_2 = $PbCl_2^0$

Cl_3 = $PbCl_3^-$

Cl_4 = $PbCl_4^{-2}$

SO_4 = $PbSO_4^0$

$(SO_4)_2$ = $Pb(SO_4)_2^{-2}$

Molal concentrations of the chemical end-members used in the calculations are the same as those listed under the speciation diagram for cadmium shown earlier (Figure 7.5-2).

Source: Long and Angino [9]. *(Copyright 1977, Pergamon Press, Inc. Reprinted with permission.)*

FIGURE 7.8-2 Chemical Speciation of Lead(II) in Unpolluted, Organic-free Aquatic Systems

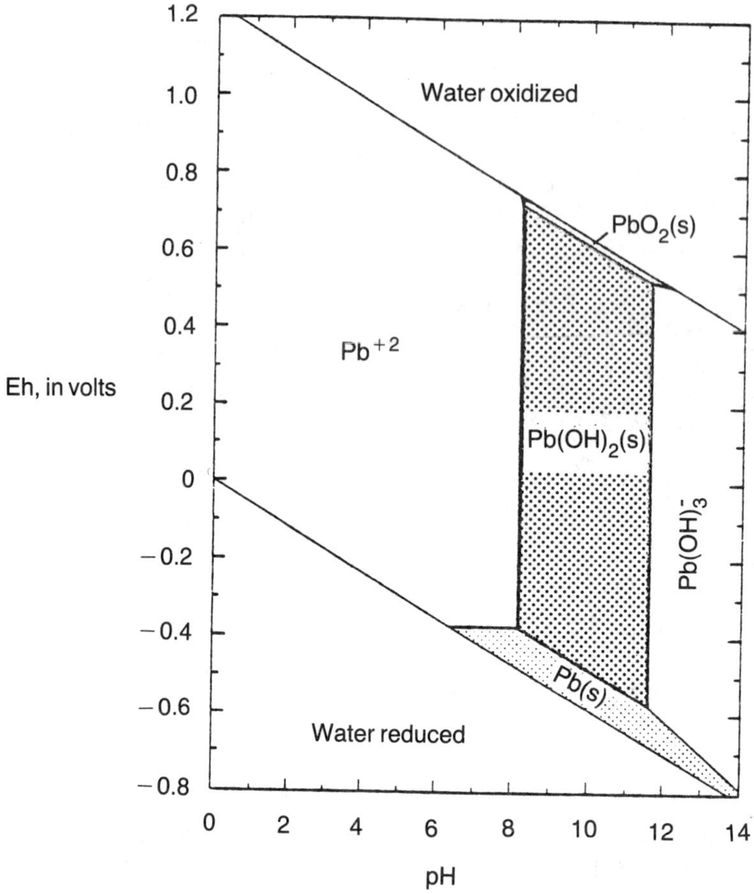

At 25°C and 1 atm; lead activity = $10^{-8.3}$ *M*

Source: Hem [7]

FIGURE 7.8-3 Calculated Lead Speciation in Oxidizing and Reducing Systems Containing Lead and Water

Redox conditions influence the type of Pb(II) precipitate formed in natural systems. Since PbS is extremely insoluble (see below) in the pH range of 6-9, it is likely to precipitate whenever Pb^{+2} and S^{-2} occur together — i.e., under redox conditions in which S^{-2} species are stable [14].

7.8.3 Solubility/Precipitation

Pb(II) forms compounds of low solubility with several major anions that occur in the natural environment, e.g., CO_3^{-2}, OH^-, S^{-2}, SO_4^{-2}, and PO_4^{-3}. The solubility products of some solid species are $10^{-27.5}$ for PbS [10], $10^{-13.1}$ for $PbCO_3$ [10], $10^{-7.79}$ for $PbSO_4$ [10], $10^{-43.5}$ for $Pb_3(PO_4)_2$ [10] and $10^{-84.4}$ for $Pb_5(PO_4)_3Cl$ [14]. Depending on the pH, redox conditions and ions present, any of these low-solubility solids may control the total dissolved Pb(II) in the system. Figure 7.8-4 shows the calculated stability fields for some common lead minerals.

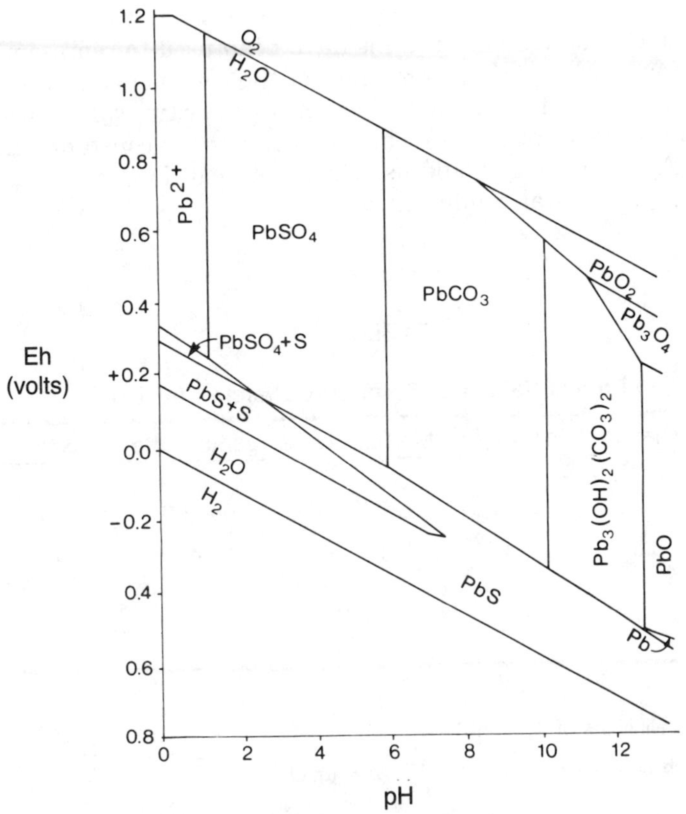

Total dissolved sulfur = $10^{-1.5}$ *M*, total dissolved carbonate = 10^{-1} *M*. Field boundaries pertain to dissolved species concentration of 10^{-6}*M*.

Source: Rickard and Nriagu [14]. *(Copyright 1978, Elsevier/North-Holland Biomedical Press. Reprinted with permission.)*

FIGURE 7.8-4 Stability Fields for Common Lead Minerals

7.8.4 Sorption on Soils and Sediments

The reported mechanisms for sorption of lead onto soils and sediments are ion exchange, specific adsorption, co-precipitation with hydrous oxides, and incorporation into cationic lattice sites in crystalline sediments [2, 13]. Organic matter, clay minerals and oxides of manganese and iron are important sorbents of lead in soils, and the extent of sorption appears to increase with increasing pH [13].

Both Langmuir and Freundlich isotherms have been used to represent the sorption of lead onto various sorbents. Typical values of the sorption constants are listed in Table 7.8-1. (These are medians except where only one or two values are available, in which case mean values are given.) The ranges of these constants are very wide, particularly for A_m; thus, one should use them with caution and check the original sources for the experimental conditions.

TABLE 7.8-1

Typical Values of Sorption Constants for Lead

	Clays	Fe Oxides	Mn Oxides	Organic Matter	Soils	Sediments
A_m(μmol/g)	~40	2400	2700	300-500	31	20
Log K_L(log M^{-1})	3.5	~4	3.8	~4	4.5	4.9
K_F(l/g)					~15	~2.2
1/N					~0.2	1.08

See Table 7.7-1 for definitions of A_m and K_L.

K_F, 1/N are Freundlich constants for $S = K_F C^{1/N}$, A = μmol/g
C = μ *M*

Source: Adapted from Rai *et al.* [13]

Soil systems contain many anions that may influence the sorption of lead. Huang *et al.* [8] showed that the addition of humic acids increased the sorption of lead by Metapeak soil (Figure 7.8-5). These authors hypothesized that the anions accumulate in the double layer of a positively charged particle and that complexes or chelates are subsequently formed between the lead ions and the anionic ligands already sorbed. Another explanation offered by Huang *et al.* is that lead-ligand associates are formed first, followed by sorption through specific chemical bonds such as the sharing of free electrons.

7.8.5 Biotransformation, Volatilization and Photolysis

Laboratory experiments have demonstrated lead methylation in lake sediments, but evidence of its occurrence in the natural environment is lacking [3]. Although microorganisms appear to be responsible for lead methylation, the mechanism has not been determined. The methylation of lead nitrate and lead chloride to tetramethyllead in some sediments may be due to the presence of ligand systems that form stable monomethyllead complexes, which then undergo a series of methylation steps leading to tetramethyllead [3]. Tetraalkyllead compounds have been detected in fish tissue [4, 11].

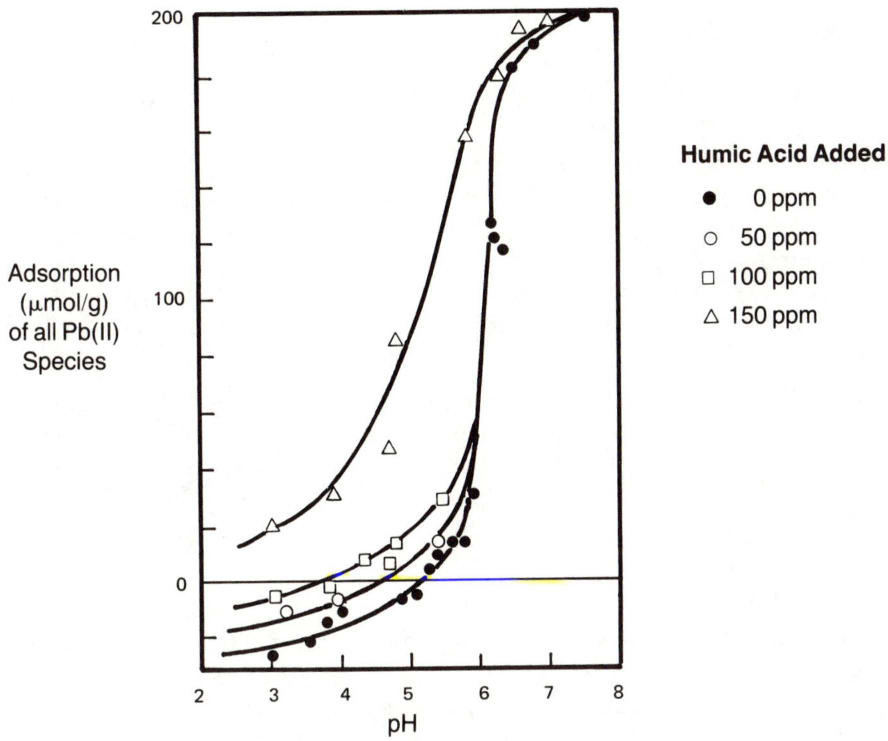

5 g/l soil, 10^{-3} *M* Pb (from chloride), I = 0.1 *M* (NaCl), 24 hr mixing.

Source: Huang *et al.* [8]. *(Copyright 1977, Water Pollution Control Federation. Reprinted with permission.)*

FIGURE 7.8-5 Effect of Humic Acid on the Sorption of Lead(II) by Metapeak Soil at Various pH Values

The chemical identity of the initial species of lead appears to affect the natural amount of tetramethyllead formed. In laboratory experiments with natural sediments, conversions of an organic lead (IV) species, Me_3PbOAc, due to *Aeromonas* species cultures were found to be higher and more consistent than those of an inorganic lead (II) species, $Pb(NO_3)_2$. Conversion of the latter was sporadic [3]. Most of the lead released to the environment consists of inorganic lead(II) compounds, and it is unclear whether their methylation occurs under natural conditions. Temperature, pH, and microbial population and activity all have an effect on this process.

Tetramethyllead, $(CH_3)_4Pb$, is quite volatile. Once formed, however, it can be rapidly oxidized in aerobic waters. The PbClBr produced by automobile engines is believed to be photolyzed in the atmosphere to PbO and halogens when exposed to sunlight [2].

7.8.6 Literature Cited

1. Baes, C.F., Jr. and R.E. Mesmer, *The Hydrolysis of Cations,* John Wiley & Sons, New York (1976).
2. Callahan, M., M. Slimak, N. Gabel, I. May, C. Fowler, J. Freed, P. Aennings, R. Durfee, F. Whitmore, B. Maestri, W. Mabey, B. Holt and C. Gould, *Water-Related Environmental Fate of 129 Priority Pollutants. Volume I: Introduction and Technical Background, Metals and Inorganics, Pesticides and PCBs,* EPA-440/4-79-029a, EPA Contracts 68-01-3852 and 68-01-3867, Office of Water Planning and Standards, U.S. Environmental Protection Agency, Washington, D.C. (1979).
3. Chau, Y.K. and P.T.S. Wong, "Occurrence of Biological Methylation of Elements in the Environment," in *Organometals and Organometalloids: Occurrence and Fate in the Environment,* F.E. Brinckman and J.M. Bellama (eds.), Symposium Series 82, American Chemical Society, Washington, D.C., 39-53 (1978).
4. Chau, Y.K., P.T.S. Wong, Ct. A. Bengert and O. Kramar, "Determination of Tetraalkyllead Compounds in Water, Sediment, and Fish Samples," *Anal. Chem.,* **51**, 186-88 (1979).
5. Cotton, R.A. and G. Wilkinson, *Advanced Inorganic Chemistry,* 3rd ed., Interscience Publishers, New York (1972).
6. Guy, R.D. and C.L. Chakrabarti, *Can. J. Chem.,* **54**, 2600-11 (1976), as cited in Morel, F., *Principles of Aquatic Chemistry,* John Wiley & Sons, New York (1983).
7. Hem, J.D., "Inorganic Chemistry of Lead in Water," in *Lead in the Environment,* T.G. Lovering (ed.), Professional Paper 957, U.S. Geological Survey, Washington, D.C., 5-11 (1976).
8. Huang, C.P., H.A. Elliot and R.M. Ashmead, "Interfacial Reactions and the Fate of Heavy Metals in Soil-Water Systems," *J. Water Pollut. Control Fed.,* **49**, 745-56 (1977).
9. Long, D.T. and E.E. Angino, "Chemical Speciation of Cd, Cu, Pb, and Zn in Mixed Freshwater, Seawater, and Brine Solutions," *Geochim. Cosmochim. Acta,* **41**, 1183-91 (1977).

10. Martell, A.E. and R.M. Smith, *Critical Stability Constants*, Vol. 4, *Inorganic Complexes*, Plenum Press, New York (1976).

11. Mor, E.D. and A.M. Beccaria, "A Dehydration Method to Avoid Loss of Trace Elements in Biological Samples," *Proceedings of International Experts' Discussion on Lead Occurrence, Fate and Pollution in the Marine Environment*, Rovinj, Yugoslavia (1977), as cited in Perwak *et al.* [12].

12. Perwak, J., M. Goyer, L. Nelken, E. Payne and D. Wallace, *An Exposure and Risk Assessment for Lead,* U.S. EPA Contract 68-01-5949, Final Draft Report to Monitoring and Data Support Division, Office of Water Regulations and Standards, U.S. Environmental Protection Agency, Washington, D.C. (1980).

13. Rai, D., J. Zachara, A. Schwab, R. Schmidt, D. Girvin and J. Rogers, *Chemical Attenuation Rates, Coefficients, and Constants in Leachate Migration. Vol I: A Critical Review,* Report EA-3356 to EPRI by Pacific Northwest Laboratories (Battelle Institute), Richland, Wash. (1984).

14. Rickard, D.T. and J.O. Nriagu, "Aqueous Environmental Chemistry of Lead," in *The Biogeochemistry of Lead in the Environment*, J.O. Nriagu (ed.), Part A, Elsevier/North-Holland Biomedical Press, New York, 219-84 (1978).

15. Saar, R.A. and J.H. Weber, "Lead(II)-Fulvic Acid Complexes. Conditional Stability Constants, Solubility, and Implications for Lead(II) Mobility," *Environ. Sci. Technol.*, **14**, 877-80 (1980).

16. Stumm, W. and J.J. Morgan, *Aquatic Chemistry,* 2nd ed., John Wiley & Sons, New York (1981).

17. Zimdahl, R.L. and J.J. Hassett, "Lead in Soil," in *Lead in the Environment*, W.R. Boggess (ed.), Prepared for the National Science Foundation, NSF/RA-770214, Washington, D.C., 93-98 (1977).

7.8.7 Additional Sources of Information

Besides the references listed above, reviews of lead behavior in the environment can be found in:

Bock, K.J., K.A. Daum, E. Merian, L.W. Newland, C.R. Pearson, H. Stache and M. Zander, "Anthropogenic Compounds," Vol. 3, Part B of *The Handbook of Environmental Chemistry*, O. Hutzinger (ed.), Springer-Verlag, Berlin (1982).

Hepple, P. (ed.), *Lead in the Environment*, Applied Sciences Publishers Ltd. (Elsevier), New York (1973).

Konrad, Z. and M. Branica (eds.), "Lead in the Marine Environment," *Proceedings of International Experts' Discussion on Lead Occurrence, Fate and Pollution in the Marine Environment*, Rovinj, Yugoslavia, 18-22 October 1977, Pergamon Press, New York (1980).

Lyman, D.R. and L. Piantanida (eds.), *Environmental Lead: Proceedings of the 2nd International Symposium on Environmental Lead Research*, Ecotoxicology and Environmental Quality Series, Academic Press, New York (1981).

7.9 MANGANESE (Mn)

7.9.1 Occurrence and Uses

Manganese, an essential nutritional element, is found in every kind of plant and animal. It also occurs in a great variety of minerals widely scattered over the earth and on the floor of the Pacific, Atlantic and Indian oceans in manganese "nodules." The most important commercial source is pyrolusite (β-MnO_2). Other important minerals are manganite (γ-$MnOOH$), hausmannite (Mn_3O_4) and rhodochrosite ($MnCO_3$). The principal uses of this metal are in steel making, in non-ferrous alloys, as MnO_2 in the manufacture of dry cell batteries and in the chemical industry [3,10,14,15].

7.9.2 Speciation Reactions in Water

Manganese is a multi-valent element; it can exist in the +2, +3, +4, +6, and +7 oxidation states. Manganous ion (Mn^{+2}) is the most thermodynamically stable aqueous oxidation state [2,13].

ACID-BASE DISSOCIATION

Mn^{+2} does not readily hydrolyze. At ordinary concentrations, hydrolysis is just detectable before precipitation of $Mn(OH)_2$ at pH >8. The solid $Mn(OH)_2$ is only slightly amphoteric [2]. Figure 7.9-1 shows how pH affects the percentage of various hydrolyzed Mn(II) species when no complexing agents are present. Figure 7.9-2 shows the distribution of Mn(II) hydrolysis products at varying pH in solutions saturated with $Mn(OH)_2$.

The other oxidation states of manganese do not have significant aqueous chemistry. Mn^{+3} hydrolyzes in strong acid but disproportionates in natural waters, to give Mn^{+2} and MnO_2 (see "Redox Reactions" below) [2]. Mn^{+4} has no significant aqueous chemistry, as this oxidation state will either rapidly disproportionate or form the extremely insoluble MnO_2 [2]. Mn(VI) (MnO_4^{-2}) is stable only under very basic conditions. Mn(VII) (MnO_4^-) is unstable with respect to liberation of oxygen from water; however, this reaction proceeds slowly [2,17]. Table 2.7-3 (Part I) lists pK values for manganese aquo species.

COMPLEXATION

Redox conditions and pH, along with salinity, affect the speciation of manganese in natural waters. Figure 7.9-3 depicts the chemical speciation of Mn^{+2} during the transition from typical river water (pH ~7.5) to seawater (pH ~8.1); the concentrations of Mn^{+2} and humic ligand were assumed to be constant. In fresh water, the Mn^{+2} ion represents over 90% of the species present [8, 9, 17]. In seawater, Mn^{+2} remains the major species, but $MnCl^+$ and $MnSO_4^0$ are also present in significant amounts [8,9,13,17]. $MnCO_3^0$ and $MnHCO_3^+$ are found in both fresh water and seawater, accounting for 1-10% of the available Mn^{+2} [8].

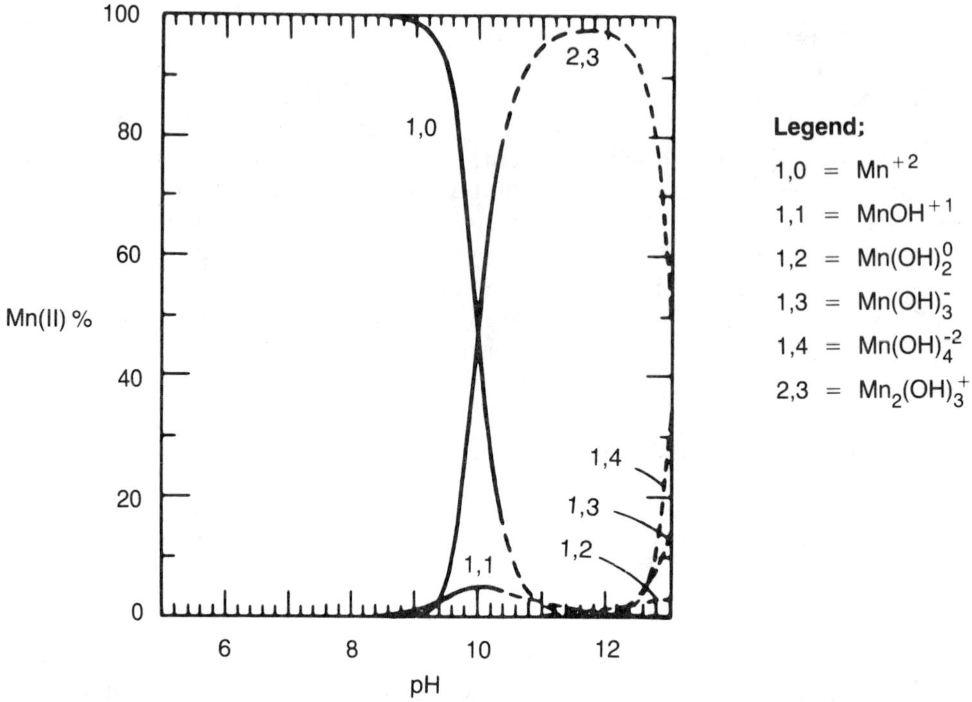

I = *M*, T = 25°C, 10^{-5} *M* Mn(II)

Dashed curves denote regions supersaturated with respect to $Mn(OH)_2$.

Source: Baes and Mesmer [2]. *(Copyright 1976, John Wiley & Sons. Reprinted with permission.)*

FIGURE 7.9-1 Hydrolyzed Mn(II) Species at Varying pH in Aquatic Systems Not Containing Complexing Agents

Manganese(II) does not form strong complexes with the organic ligands (humic and fulvic acids) typically found in aqueous environments. Overall conditional stability constants (K_0) for manganese complexes with humic materials from various natural environments have been determined at 20°C, pH = 8 and I = 0.02; the values of log K_0 range from 3.7 to 4.85 [8].[1] Tables 2.9-1 and -2 (Part I) list complexation equilibrium constants for a variety of inorganic and organic ligand complexes of Mn^{+2}.

Mn^{+2} is considered a kinetically labile metal ion [19]. The characteristic second-order rate constant for exchange of one water molecule in its primary solvation shell is $3.4 \times 10^6 M^{-1} s^{-1}$ [9]. The reaction rates of Mn(II) ion with ligands commonly found in the aquatic environment are similar to those of Zn(II).

1. See § 7.7.2 for further discussion of K_0.

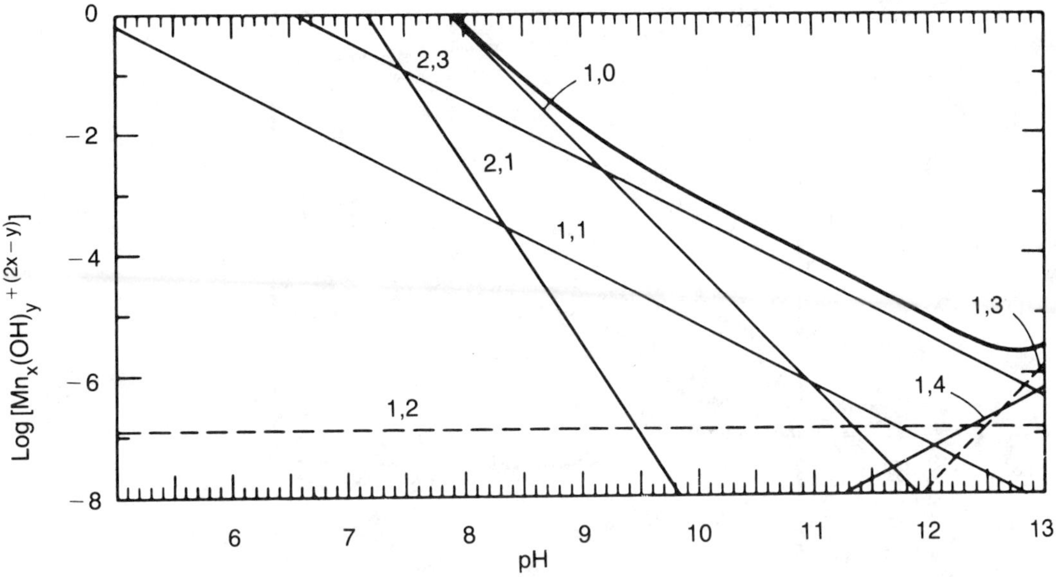

Legend:

1,0 = Mn^{+2}	1,2 = $Mn(OH)_2^0$	1,4 = $Mn(OH)_4^{-2}$
1,1 = $MnOH^{+1}$	1,3 = $Mn(OH)_3^-$	2,1 = $Mn_2(OH)^{+3}$
		2,3 = $Mn_2(OH)_3^+$

I = 1 *M* (sodium sulfate), T = 25°C. Heavy curve is the total concentration of Mn(II)

Source: Baes and Mesmer [2]. *(Copyright 1976, John Wiley & Sons. Reprinted with permission.)*

FIGURE 7.9-2 Distribution of Mn(II) Hydrolysis Products at Varying pH in Solutions Saturated with $Mn(OH)_2$

REDOX REACTIONS

Unlike Cr(II), Mn(II) is a very poor reducing agent and is resistant to oxidation in neutral or acidic solutions. The rate of Mn(II) oxidation increases with increasing pH [10,17]. At higher pH, $Mn(OH)_2$ precipitates and can also be oxidized to Mn(III) (manganic) ion [2,15]. The oxygenation of soluble Mn(II) in simple media is autocatalytic and heterogeneous [10, 17]. The products of Mn(II) oxidation are non-stoichiometric and show average degrees of oxidation ranging from about $MnO_{1.3}$ to $MnO_{1.9}$ (30-90% oxidation to MnO_2) under varying oxidizing conditions [17]. In many natural waters, the oxidation of Mn(II) to MnOOH or to Mn(IV) is catalyzed by certain microorganisms [17].

Mn(III) is a powerful oxidizing agent that is easily reduced to Mn(II) or disproportionates [2,15] according to the following reaction:

$$2Mn^{+3} + 2H_2O \rightleftharpoons Mn^{+2} + MnO_2(c) + 4H^+ \qquad \log K \sim 9 \qquad (1)$$

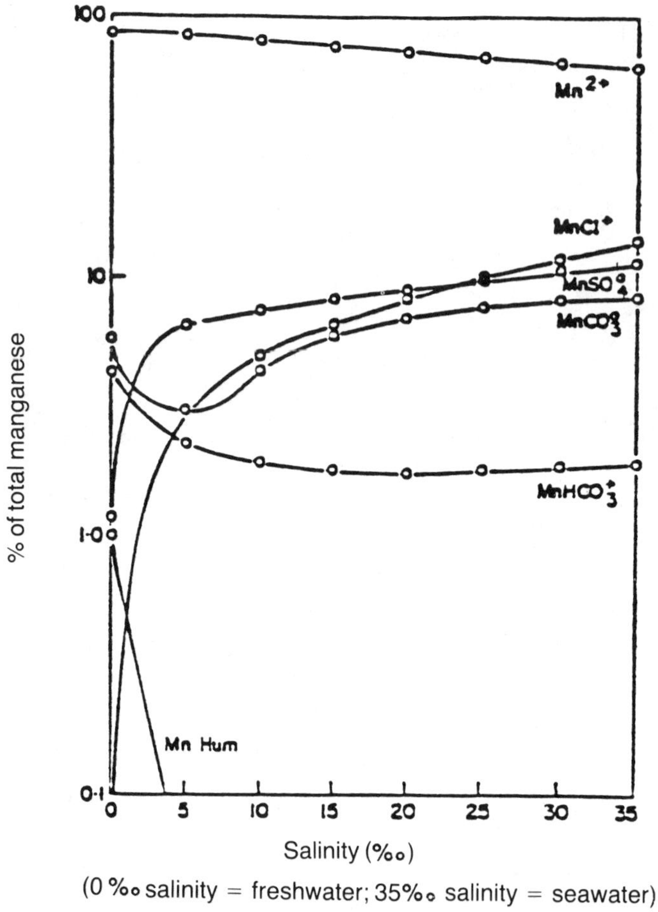

Composition of Waters Used in Construction of Model
(mol/kg solution)

Species	Average River Water	Seawater
Na^+	2.83×10^{-4}	0.46847
K^+	5.9×10^{-5}	0.01020
Mg^{+2}	1.69×10^{-4}	0.05307
Ca^{+2}	3.74×10^{-4}	0.01028
Cl^-	2.20×10^{-4}	0.54590
SO_4^{-2}	1.17×10^{-4}	0.02823
Carbonate Alkalinity	9.54×10^{-4}	0.0024
Total Inorganic Carbon	1.02×10^{-3}	0.0022
Total Mn^{+2}	10^{-8}	10^{-8}
Total Humic Ligand	10^{-6}	10^{-6}
pH	7.5	8.1

Source: Mantoura *et al.* [8]. *(Copyright 1978, Academic Press, Inc., Ltd. Reprinted with permission.)*

FIGURE 7.9-3 Equilibrium Speciation of Manganese(II) as a Function of Salinity Along a Model Estuary

MnO_2, the most stable Mn(IV) compound, is readily attacked by reducing reagents in acid solutions. MnO_4^{-2} [Mn(VI)] and MnO_4^- [Mn(VII)] are thermodynamically unstable in natural waters [2]. Figure 7.9-4 shows the fields of stability of solid and dissolved manganese species in the manganese-carbon dioxide system as a function of pe and pH. Reduction potentials for some manganese species are given in Table 7.9-1; these values show that MnO_4^-, MnO_2, and Mn(III) are all strong oxidants.

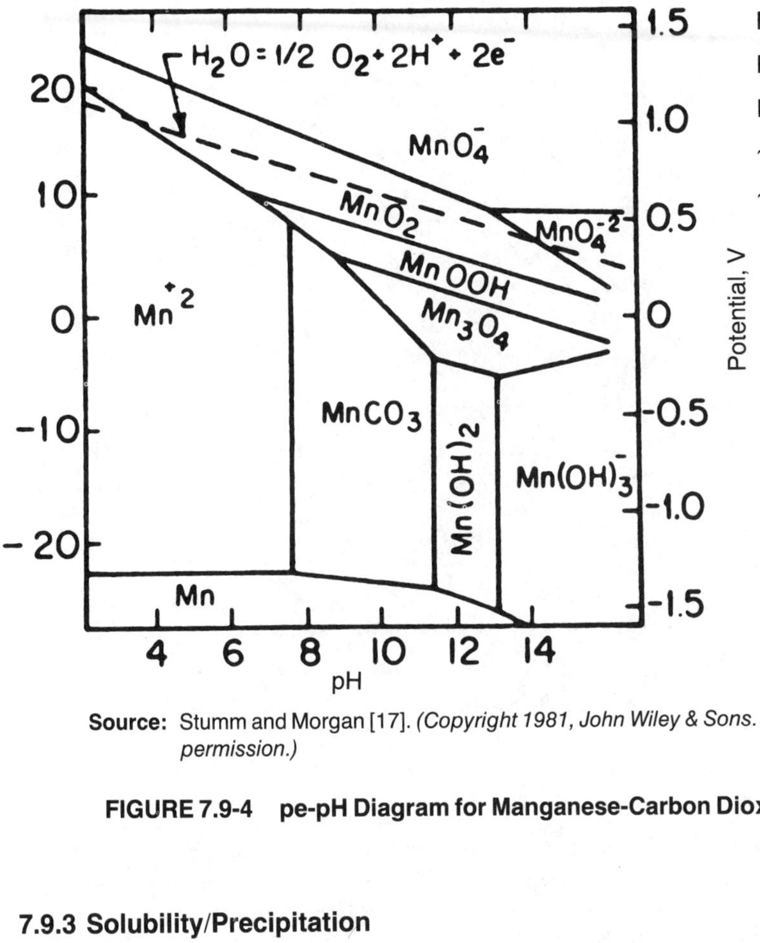

Solid phases considered:

$Mn(OH)_2(s)$	(pyrochroite)
$MnCO_3(s)$	(rhodochrosite)
$Mn_3O_4(s)$	(hausmannite)
γ-MnOOH	(manganite)
γ-MnO_2	(nsutite)

Source: Stumm and Morgan [17]. *(Copyright 1981, John Wiley & Sons. Reprinted with permission.)*

FIGURE 7.9-4 pe-pH Diagram for Manganese-Carbon Dioxide System

7.9.3 Solubility/Precipitation

$MnCO_3$ (rhodochrosite) has been suggested as the solubility-controlling solid for Mn^{+2} under reducing conditions and relatively high pH and CO_2 levels; at higher redox potentials (e.g., pe = 10, pH 7), where oxidation of Mn(II) occurs (see Figure 7.9-4), the controlling solids are γ-MnOOH (manganite), β-MnO_2 (pyrolusite), and mixed-valence Mn oxides [7,10,13,14]. The Mn(III) and Mn(IV) oxides and hydroxides are essentially insoluble under natural aquatic conditions [10]: the solubility product of $Mn(OH)_4$, for example, is $\sim 10^{-56}$ [6]. However, natural organics can cause these oxides to enter solution through the reduction of manganese to the soluble manganous (II) form, which increases the mobility of manganese and its availability to organisms; these reactions are photocatalyzed in seawater and river water [18] and proceed more rapidly at pH>7 [16]. Together with the slow kinetics of the reaction in which Mn(II) is oxidized by dissolved oxygen in natural waters [10], this explains the presence of soluble Mn^{+2} in surface oxic waters [18].

TABLE 7.9-1

Reduction Potentials and Logarithms of Equilibrium Constants of Some Manganese Species
(at 25°C and I = 0 except as noted)

Reaction	E^0(V)	Log K
$MnO_4^- + e^- \rightleftharpoons MnO_4^{-2}$	+0.57	9.6
$MnO_4^- + 4H^+ + 3e^- \rightleftharpoons MnO_2(s,\beta) + 2H_2O$	+1.68	85.1
$MnO_4^- + 8H^+ + 5e^- \rightleftharpoons Mn^{+2} + 4H_2O$	+1.51	127.5
$MnO_4^{-2} + 2H_2O + 2e^- \rightleftharpoons MnO_2(s) + 4OH^-$	+0.511	17.7
$Mn(IV) + e^- \rightleftharpoons Mn(III)$	+1.652	27.93[a]
$MnO_2(s) + 4H^+ + 2e^- \rightleftharpoons Mn^{+2} + 2H_2O$	+1.23	41.6
$Mn_3O_4 + 8H^+ + 2e^- \rightleftharpoons 3Mn^{+2} + 4H_2O$	+1.755	59.34
$Mn(III) + e^- \rightleftharpoons Mn(II)$	+1.488	25.15[a]
$Mn(OH)_3(s) + 3H^+ + e^- \rightleftharpoons Mn^{+2} + 3H_2O$	+1.841	31.12
$Mn(OH)_3(s) + e^- \rightleftharpoons Mn(OH)_2(s) + OH^-$	+0.154	2.6
$Mn^{+2} + 2e^- \rightleftharpoons Mn(s)$	−1.17	−39.5
$Mn(OH)_2(s) + 2e^- \rightleftharpoons Mn(s) + 2OH^-$	−1.55	−52.4

a. Ionic strength = 7.5M H_2SO_4

Source: Kotrlý and Šůcha [5]. *(Copyright 1985, John Wiley & Sons. Reprinted with permission.)*

In natural waters, both MnO_2 and γ-MnOOH(s) are formed by oxidation of Mn^{+2}. MnO_2, which imparts turbidity and stains clothing during laundering processes, makes manganese a nuisance in water supplies [4,10]. In lake waters, γ-MnOOH(s) is the final product of the sequence of Mn(II) oxidation, with Mn_3O_4 as an intermediate [17]. Kotrlý and Šůcha [5] list the following solubility products at zero ionic strength and 25°C: $MnCO_3 = 10^{-9.3}$, $Mn(OH)_2$(active) $= 10^{-12,8}$, MnS(green) $= 10^{-13.5}$, and MnS(pink) $= 10^{-10.5}$.

Manganese minerals in soil are difficult to characterize, possibly because mixed-valence manganese oxides can form in reduced soil as Mn^{+2} becomes oxidized. In the pe + pH range of 15-16, where oxidation of Mn(II) can occur, MnOOH, $MnCO_3$ and Mn(IV) minerals can coexist [7].

7.9.4 Sorption on Soils and Sediments

Sorption of manganese is complicated by redox reactions that produce aqueous compounds of different oxidation states. Specific adsorption, ion exchange, and organic complexation all affect the retention of manganese by soil, but it is not clear which of these processes is most important [13].

Limited studies indicate that at Mn concentrations $<10^{-4}$ *M*, specific adsorption is the dominant retention mechanism. Manganese oxides have a higher specific adsorption for manganese than do iron oxides, possibly because the Mn^{+2} ion is oxidized at the surface of Mn oxides and is incorporated into a new surface solid phase [12, 13]. Chemisorption to $CaCO_3$ occurs in calcareous materials, possibly due to the formation of $MnCO_3$ [7, 13].

Regression analysis of the dependence of Mn(II) adsorption capacity on soil properties suggests that cation exchange capacity is the most important sorption mechanism for this element at manganese solution concentrations $>10^{-4}$ *M*. However, the sorption of Mn is usually far less than the cation exchange capacity of soil. Release of Mn^{+2} from soil can occur when more strongly binding metals undergo ion exchange with adsorbed Mn or replace it in the Mn oxide structure [13].

When Mn^{+2} is adsorbed by clay minerals, iron oxides or manganese oxides, hydrogen ions are usually released. This can result from the following reactions of Mn^{+2} with the solid surfaces of hydroxylated Fe, Al or Mn sorbents [13]:

$$\text{(Sor)-OH} + \text{Mn}^{+2} \rightleftharpoons \text{(Sor)-OMn}^{+} + \text{H}^{+} \quad (2)$$

$$\text{(Sor)-OH} + \text{Mn}^{+2} \rightleftharpoons \text{(Sor)-OMnOH} + 2\text{H}^{+} \quad (3)$$

where (Sor)-OH is hydroxylated Fe, Al or Mn.

Complexation of manganese by organic matter may strongly influence its adsorption at lower concentrations [13]. Mn^{+2} binds loosely to soil humates by forming outer-sphere complexes [14]. The enrichment of manganese in the humic acid fraction of sediment is low; the enrichment factor (defined as the ratio of metal concentration in the humic acid fraction to its total concentration in the associated sediments) is 0.06-0.32, compared with 0.46-0.81 for zinc and 21-40 for copper [11].

The manganese(II) sorption capacity of various soils has been measured by batch tests with a field-collected leachate containing 17-18 ppm of manganese [1]. In some soil samples, the release of manganese was observed. The results are given in Table 7.9-2.

Table 7.9-3 lists ranges of sorption constants for manganese on various sorbents. Clay minerals and hydrous iron oxides adsorb copper, zinc and nickel more strongly than they do manganese [13].

TABLE 7.9-2

Sorption Capacity for Total Manganese, Mn(T), at Equilibrium for Soils

Soil Type	Sorption Capacity (μg Mn(T)/g soil)
Alluvial material	34.1-1,880
Residual soil (silty sand)	a
Gravelly, well-graded, silty sand	29.2-52.0
Brown, clayey sand	>172-336
Fine sand	19.0-26.0
Clay core liner	28.9-59.0
Very fine sand with little silt	a

a. Release of manganese from soil observed

Source: Arthur D. Little, Inc. [1]

High sorption capacity was observed for the soils having highest clay content (alluvial material, 28%; brown, clayey sand, 20%) and highest total organic carbon (brown, clayey sand, 0.88%).

7.9.5 Literature Cited

1. Arthur D. Little, Inc., *Full-Scale Field Evaluation of Waste Disposal from Coal-Fired Electric Generating Plants. Appendix F (Part 4)*, EPA Contract No. 68-02-3167, Final Report to U.S. EPA, Office of Research and Development, Washington, D.C. (1984).
2. Baes, C.F. and R.E. Mesmer, *The Hydrolysis of Cations,* John Wiley & Sons, New York (1976).
3. Clayton, G.D. and F.E. Clayton, *Patty's Industrial Hygiene and Toxicology,* 3rd rev. ed., Vol. 2A, John Wiley & Sons, New York (1981).
4. Faust, S.D. and O.S. Aly, *Chemistry of Natural Waters,* Butterworth, Woburn, Mass. (1981).
5. Kotrlý, S. and L. Šůcha, *Handbook of Chemical Equilibria in Analytical Chemistry,* Halstead Press, Div. of John Wiley & Sons, New York (1985).
6. Lelong, F., Y. Tardy, G. Grandin, J.J. Trescases and B. Boulange, "Pedogenesis Chemical Weathering and Processes of Formation of Some Supergene Ore Deposits," in *Handbook of Stratabound and Stratiform Ore Deposits,* Vol. 3, K.H. Wolf (ed.), Elsevier, Amsterdam (1976), as cited by Salomons and Förstner [14].
7. Lindsay, W.L., *Chemical Equilibria in Soils,* John Wiley & Sons, New York (1979).

TABLE 7.9-3

Typical Values of Sorption Constants for Manganese

Sorption Constant	Clays	Mn Oxides	Fe Oxides	Calcite	Soil
A_m(μmol/g)	1.3-610	1130-3230	—	8.7, 9.8[a]	0.91-79
Log K_L(log M^{-1})	1.9-3.5	1.6-3.6	—	4.3, 5.2[a]	3.4-4.3
K_F	0.011-23	—	—	—	—
1/N	0.18-1.2	—	—	—	—
K_d($l \cdot g^{-1}$)	—	30-50,000	20[b]	—	—

a. Only two values
b. Only one value
Source: Rai *et al.* [13]. *(Copyright 1984, Electric Power Research Institute. Reprinted with permission.)*

Definitions (see also section 2.12):

A_m and K_L are Langmuir constants

$$S = \frac{K_L A_m C}{1 + K_L C}$$

where

S	=	μ moles sorbed at equilibrium per gram of solid
A_m	=	maximum sorption capacity of solid (μ mol/g)
K_L	=	sorption constant (related to binding energy of sorbate) in $l \cdot mol^{-1}$
C	=	total sorbate concentration in solution at equilibrium ($mol \cdot l^{-1}$)
K_F, 1/N	=	Freundlich constants for $A = K_F C^{1/N}$, where a = μmol/g and C = μM
K_d	=	ratio of the quantity of sorbate sorbed per gram of solid (mol/g) to the concentration of the sorbate remaining in solution at equilibrium ($mol \cdot l^{-1}$)

8. Mantoura, R.F.C., A. Dickson and J.P. Riley, "The Complexation of Metals with Humic Materials in Natural Waters," *Estuarine Coastal Mar. Sci.*, **6**, 387-408 (1978).
9. Morel, F.M.M., *Principles of Aquatic Chemistry,* John Wiley & Sons, New York (1983).
10. Morgan, J.J., "Manganese," in *Principles and Applications of Water Chemistry,* S.D. Faust and J.V. Hunter (eds.), John Wiley & Sons, New York (1967).
11. Nriagu, J.O. and R.D. Coker, "Trace Metals in Humic and Fulvic Acids from Lake Ontario Sediments," *Environ. Sci. Technol.*, **14**, 443-46 (1980).
12. Posselt, H.S., F.J. Anderson and W.J. Weber, "Cation Sorption on Colloidal Hydrous Manganese Dioxide," *Environ. Sci. Technol.*, **2**, 1087-93 (1968).

13. Rai, D., J. Zachara, A. Schwab, R. Schmidt, D. Girvin and J. Rogers, *Chemicals Attenuation Rates, Coefficients, and Constants in Leachate Migration. Vol I: A Critical Review,* Report EA-3356 to EPRI by Pacific Northwest Laboratories (Battelle Institute), Richland, Wash. (1984).

14. Salomons, W. and U. Förstner, *Metals in the Hydrocycle,* Springer-Verlag, New York (1984).

15. Sienko, M.J. and R.A. Plane, *Chemistry,* 5th ed., McGraw-Hill Book Co., New York (1976).

16. Stone, A.T. and J.J. Morgan, "Reduction and Dissolution of Manganese(III) and Manganese(IV) Oxides by Organics: 2. Survey of the Reactivity of Organics," *Environ. Sci. Technol.,* **18**, 617-24 (1984).

17. Stumm, W. and J.J. Morgan, *Aquatic Chemistry,* 2nd ed., John Wiley & Sons, New York (1981).

18. Sunda, W.G., S.A. Huntsman and G.R. Harvey, "Photoreduction of Manganese Oxides in Seawater and Its Geochemical and Biological Implications," *Nature,* **301**, 234-36 (1983).

19. Wilkins, R.G., *The Study of Kinetics and Mechanism of Reactions of Transition Metal Complexes,* Allyn and Bacon, Boston (1974).

7.9.6 Other Sources of Information

Bricker, O., "Some Stability Relations in the System Mn-O_2-H_2O at 25°C and One Atmosphere Total Pressure," *Am. Mineral.,* **50**, 1296-1354 (1965).

Hemstock, G.A. and P.F. Low, "Mechanisms Responsible for Retention of Manganese in the Colloidal Fraction of Soil," *Soil Sci.,* **76**, 331-43 (1953).

McBride, M.B., "Chemisorption and Precipitation of Mn^{+2} on $CaCO_3$ Surfaces," *Soil Sci. Soc. Am. J.,* **43**, 693-98 (1979).

McKenzie, R.M., "Proton Release During Adsorption of Heavy Metal Ions by a Hydrous Manganese Dioxide," *Geochim. Cosmochim. Acta,* **43**, 1855-57 (1979).

Murray, J.W., "The Interaction of Metal Ions at the Manganese Dioxide - Solution Interface," *Geochim. Cosmochim. Acta,* **39**, 505-19 (1975).

Ponnamperuma, F.M., T.A. Loy and E.M. Tianco, "Redox Equilibria in Flooded Soils: II. The Manganese Oxide Systems," *Soil Sci.,* **108**, 48-57 (1969).

Schuman, L.M., "Effect of Soil Properties on Manganese Adsorption Isotherms for Four Soils," *Soil Sci.,* **124**, 77-81 (1977).

Schwab, A.P. and W.L. Lindsay, "Effect of Redox on the Solubility and Availability of Iron," *Soil Sci. Soc. Am. J.,* **47**, 217-20 (1983).

Stone, A.T. and J.J. Morgan, "Reduction and Dissolution of Manganese(III) and Manganese(IV) Oxides by Organics: 1. Reaction with Hydroquinone," *Environ. Sci. Technol.,* **18**, 450-56 (1984).

Takematsu, N., "Sorption of Transition Metals on Manganese and Iron Oxide, and Silicate Minerals," *J. Oceanogr. Soc. Japan,* **35**, 36-42 (1979).

7.10 MERCURY (Hg)

7.10.1 Occurrence and Uses

Mercury is generally considered a chalcophilic element, i.e., one that tends to concentrate in sulfides. Cinnabar (α-HgS) is its principal and most common ore. Metacinnabar (β-HgS) is another common mineral, and the metal is also found in the uncombined state. Small amounts occur in other sulfides [12]. Mercury is produced and used primarily in the liquid elemental form and in various amalgams with metals. Its chief uses are in electrical apparatus (~42%) as mercury metal, Zn-Hg amalgam, HgO, and Cd-Hg amalgam. Approximately 25% of the mercury consumed in the United States is used in the liquid elemental form to make chlorine and caustic by the mercury cell process. Other uses are in paint manufacturing (phenylmercuric acetate), industrial and control instruments (liquid Hg), dental preparations (Ag-Hg, Sn-Hg amalgams) and the production of catalysts, fungicides, bactericides, and pharmaceuticals [31].

Besides the metal and amalgams, mercury compounds that are of potential environmental significance include phenylmercury compounds, HgO, and methylated forms of Hg. The chemistry of this element in the environment is complex, not only because of its various oxidation states (0, +1, +2) but also because of biotic and abiotic methylation (see § 7.10.5) and the volatility of several forms of Hg.

7.10.2 Speciation Reactions in Water

ACID-BASE DISSOCIATION

Mercury(II) is hydrolyzed in water, behaving as an acid. From calculations of Hg(II) in a system containing no other ligands, Hahne and Kroontje [14] found that it hydrolyzes between pH 2 and 6. Figure 7.10-1 shows the theoretical hydrolysis of Hg(II). $HgOH^+$ is dominant between pH 2.2 and 3.8, but $Hg(OH)_2^0$ becomes dominant at higher values of pH.

Bondietti *et al.* [4] postulated that $Hg(OH)_3^-$ can also exist at very high pH and that the polynuclear complexes $Hg_2(OH)^{+3}$ and $Hg_3(OH)_3^{+3}$ might also be found. It is unknown whether any of these species would be significant under environmental conditions.

Mercury(I) ions generally form weaker complexes than do mercury(II) ions. In water, Hg(I) hydrolyzes to produce small amounts of Hg_2OH^+ [2]. The stability constant for the formation of this complex is $10^{8.7}$ (defined as $Hg_2^{+2} + OH^- = Hg_2OH^+$ at 25°C for 0.5 *M* [28]). From this, the pK of Hg_2^{+2} is calculated to be 5.3. It is a weaker acid than Hg^{+2}.

The speciation of mercury in natural aquatic systems is complicated by the process of methylation, which forms aquated CH_3Hg^+ ions as described in § 7.10.5. These exist in aqueous systems as $CH_3\text{-}Hg\text{-}OH_2^+$ ions with a covalent bond between Hg and O

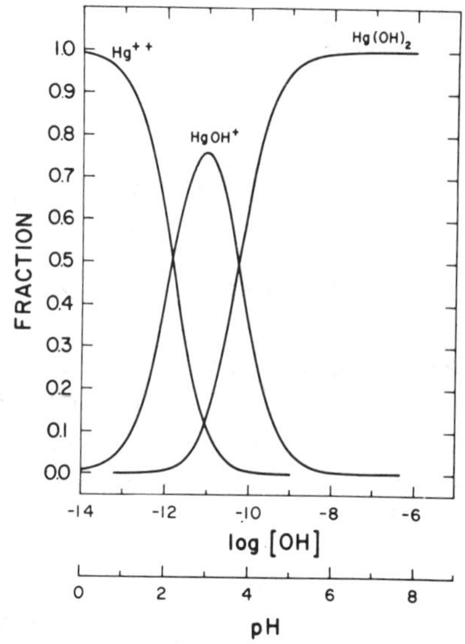

Source: Hahne and Kroontje [14]. *(Copyright 1973, American Society of Agronomy, Inc. Reprinted with permission.)*

FIGURE 7.10-1 Calculated Hydrolysis of Mercury(II) with Varying pH

[40]. As a cation, CH_3-Hg-OH_2^+ behaves like an acid and hydrolyzes to form $(CH_3Hg)_2OH^+$ and CH_3HgOH [4]. The latter is the stable methylmercury species in fresh water [40]. Other alkyl mercury ions may complex with OH^- in a similar manner [4].

COMPLEXATION

Mercury(II) forms complexes with the inorganic and organic ligands in aquatic systems. Table 2.9-1 (see Part I) shows the stability constants for some of these complexes. Those for $HgCl^+$, $HgCl_2^0$, $HgCl_3^-$ and $HgCl_4^{-2}$ are $10^{7.2}$, $10^{14.0}$, $10^{15.1}$ and $10^{15.4}$ respectively (given as formation constants at 25°C and zero ionic strength) [29]. The stability constants for $HgCO_3^0$ and $HgSO_4^0$ are, respectively, $10^{6.3}$ [27] and $10^{2.5}$ [29] at 25°C and I = 0. Hg^{+2} therefore forms relatively strong complexes with Cl^- and CO_3^{-2}. Hahne and Kroontje [14] calculated the complexation of Hg(II) with Cl^- as shown in Figure 7.10-2. The presence of a complex depends highly on the [metal]/[ligand] ratio.

Considering the strong Hg(II) complexation with Cl^- and also the strong hydrolysis of Hg(II) in aquatic systems, Cl^- complexes and hydrolyzed species of Hg(II) are likely to be the predominant species in water under aerobic environmental conditions. Stumm and Morgan [40] suggested that the principal dissolved inorganic species are probably $HgCl_4^{-2}$ in seawater and $Hg(OH)_2^0$ and HgOHCl in fresh water under aerobic conditions. The other inorganic species present under aerobic conditions are

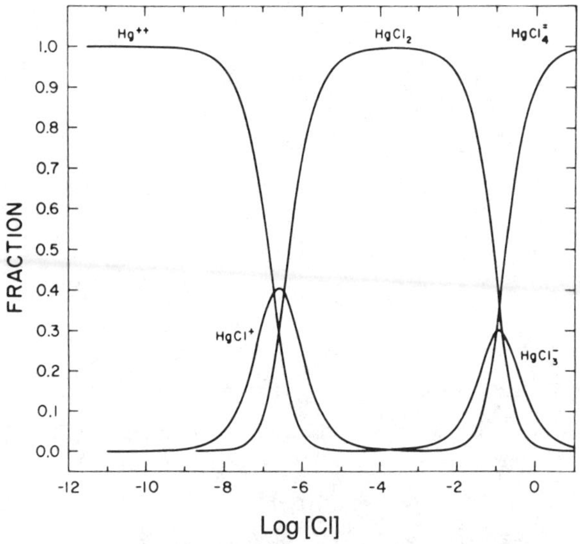

[Cl] is Cl^- concentration in *M*

Source: Hahne and Kroontje [14]. *(Copyright 1973, American Society of Agronomy, Inc. Reprinted with permission.)*

FIGURE 7.10-2 Calculated Complexation of Mercury(II) with Cl^-

Hg^{+2}, $HgCl^+$, $HgCl_2^0$, and $HgCl_3^-$ [29]. Figure 7.10-3 shows the speciation of mercury under aerobic conditions when no organics are present; note the change in predominant species from $Hg(OH)_2$ in fresh water to $HgCl_4^{-2}$ in saline water.

Several authors [6,12,29] report that HgS will precipitate under mildly reducing conditions in the presence of sulfur. Under mildly reducing conditions, the dissolved species in equilibrium with HgS are different from those found in aerobic systems: $[S\text{-}Hg\text{-}S]^{-2}$ and other sulfide complexes predominate (Figure 7.10-4).

Mercury(II) forms strong complexes with organics; the literature [6] indicates that it has a strong affinity for sulfhydryl groups and amino groups in proteins, amino acids and other organics in the environment.

Mantoura, Dickson and Riley [27] measured stability constants by gel complexometry. For fresh-water humic compounds isolated from rivers and lakes, the logarithm of the overall conditional stability constant[1] of Hg(II)-humate complexes ranged from 18.4 to 21.1. For humic compounds from surface water of a sea loch (lake), log K_o was 20.6. Sediment from the same sea loch had a log K_o of 21.3. In seawater, log K_o for Hg(II)-humate complexes was around 18.0. The fulvic acid of peat had a log K_o with Hg(II) of 18.3. The lowest value of log K_o (5.2) was for the humic acid fraction of soil [27].

1. See § 7.7.2 for definition of this constant (K_0).

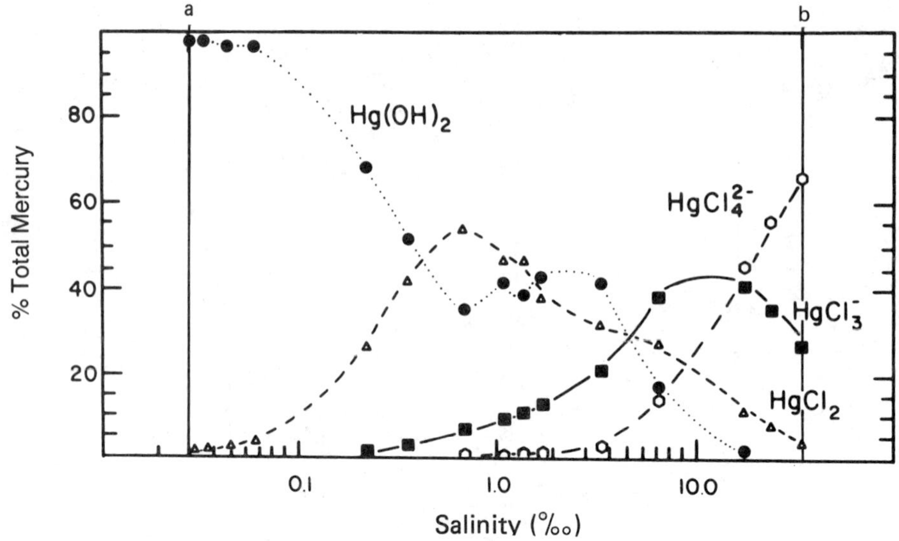

Total mercury concentration = 10^{-9} *M*

Line 'a' corresponds to fresh water:

$Cl^- = 5 \times 10^{-4}$ *M*, pH = 7.19

Line 'b' corresponds to seawater:

Salinity = 36 ‰, pH = 8.29

Salinity (‰) is defined as the total grams of dissolved solids per kilogram of seawater.

Source: Adapted from Morel [29]. *(Copyright 1983, John Wiley & Sons. Reprinted with permission.)*

FIGURE 7.10-3 Calculated Inorganic Speciation of Mercury in an Estuary as a Function of Salinity

Mantoura, Dickson and Riley [27] found that the complexation of mercury(II) and copper with humic compounds is of major importance (>99.9% of metal complexed) in fresh water. In sea and estuarine waters, the competition from Cl^- greatly decreases the proportion of Hg(II) bound with humic components as salinity increases. The stability constants for Hg(II)-Cl complexes are very high, and the concentration of Cl^- in seawater also far exceeds that of natural organics. Figure 7.10-5 shows the complexation of mercury with increasing salinity in the presence of humic compounds.

According to Stumm and Morgan [40], the methylmercury cation CH_3Hg^+ tends to add a single ligand, forming complexes with S, P, O, N, halogens and C-containing ligands. Complexes of this cation with S^{-2} and S-containing ligands are highly stable (log K for CH_3HgS^- = 21.0, log K for $(CH_3Hg)_2S$ = 16.3) [28]. Logarithms of stability constants for CH_3HgCl, $CH_3HgCO_3^-$ and $CH_3HgSO_4^-$ are 5.25, 6.1 and 0.94 respec-

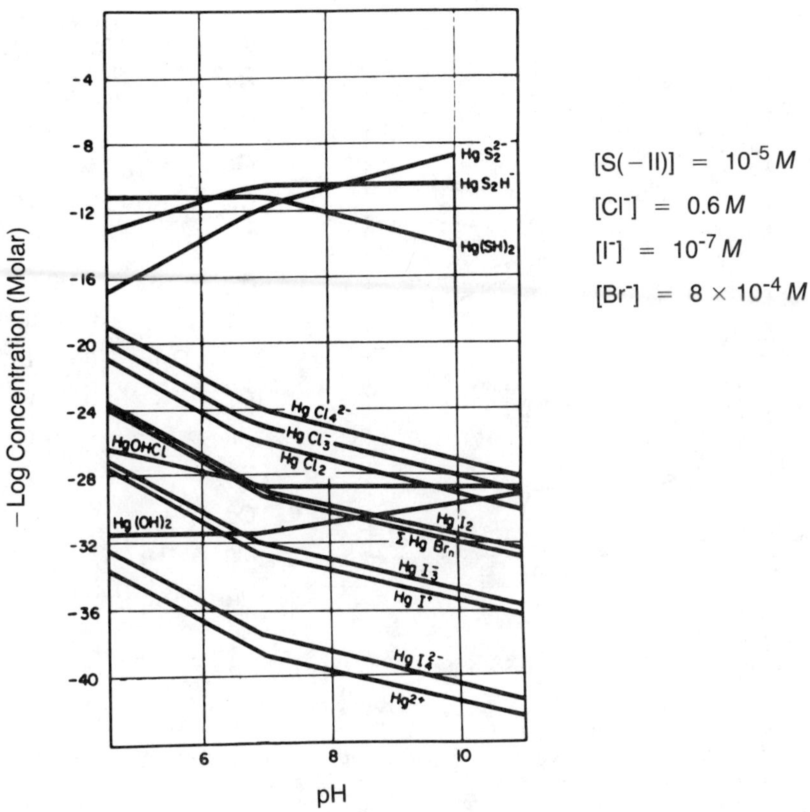

Source: Stumm and Morgan [40]. *(Copyright 1981, John Wiley & Sons. Reprinted with permission.)*

FIGURE 7.10-4 Calculated Hg(II) Speciation in Seawater Where Hg Concentration Is Limited by HgS (Solid)

tively [40]. The rate of complexation is extremely high, and the CH_3Hg^+ unit is kinetically inert toward decomposition (i.e., breaking of the C-Hg bond). Figure 7.10-6 shows the calculated composition of Hg species at equilibrium in aquatic systems when methylation has occurred and methylated mercury species are present. In seawater, methyl chloro-complexes are the most important; in fresh water, both methyl hydroxo- and methyl chloro-complexes are important. Please refer to Section 7.10.5 on biotransformation of mercury, and Section 2.15 in Part I for discussions on biologically mediated reactions.

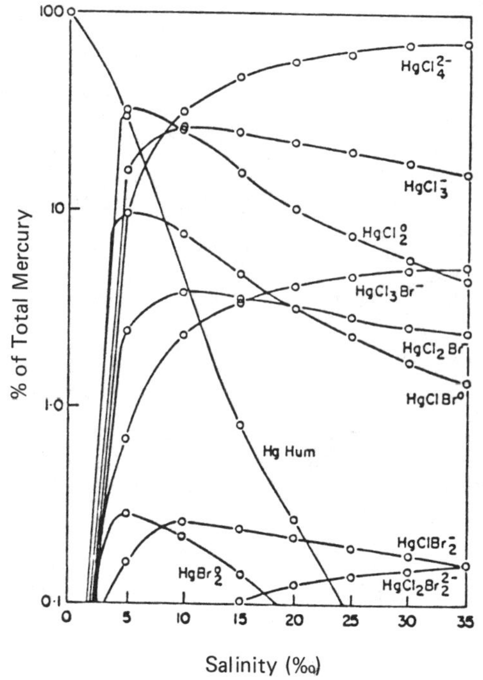

Species	[mol (kg solution)$^{-1}$] Average River Water	Seawater
Na^+	2.83×10^{-4}	0.46847
K^+	5.9×10^{-5}	0.01020
Mg^{+2}	1.69×10^{-4}	0.05307
Ca^{+2}	3.74×10^{-4}	0.01028
Cl^-	2.20×10^{-4}	0.54590
Br^-	0	0.000842
SO_4^{-2}	1.17×10^{-4}	0.02823
Alk (CO_3)	9.54×10^{-4}	0.0024
C_T	1.02×10^{-3}	0.0022
Cu	10^{-8}	
Zn	10^{-8}	
Cd	10^{-8}	
Mn	10^{-8}	
Hg	10^{-9}	
Ni	10^{-8}	
Co	10^{-9}	
Hum	10^{-6}	

Source: Mantoura, Dickson and Riley [27]. *(Copyright 1978, Academic Press, Inc. (London) Ltd. Reprinted with permission.)*

FIGURE 7.10-5 Calculated Equilibrium Speciation of Hg(II) as a Function of Salinity in a Model Estuary in the Presence of Humic Matter, pH ≈ 7-8.2, and T = 20°C

REDOX REACTIONS

Mercury exhibits three oxidation states in the environment: +1 (Hg_2^{+2}), +2 (Hg^{+2}) and zero (Hg(liq)). The equilibrium constants for their redox reactions are as follows:

Reaction	Log K (25°C)	Source
$Hg^{+2} + 2e^- \rightleftharpoons Hg_2^{+2}$	30.68	[39]
$Hg_2^{+2} + 2e^- \rightleftharpoons 2Hg(liq)$	26.79	[39]
$Hg^{+2} + Hg(liq) \rightleftharpoons Hg_2^{+2}$	1.94	[39]
$Hg^{+2} + 2e^- \rightleftharpoons Hg(liq)$	28.86	[25]
$\frac{1}{2}Hg_2^{+2} + e^- \rightleftharpoons Hg^0$	6.93	[25]
$Hg^{+2} + 2e^- \rightleftharpoons Hg^0$	22.34	[25]

The standard electrode potentials for the first two reactions are 0.9075 and 0.7925 international volts respectively (Pt, H_2/H^+ reference) at 25°C [39]. In aqueous solu-

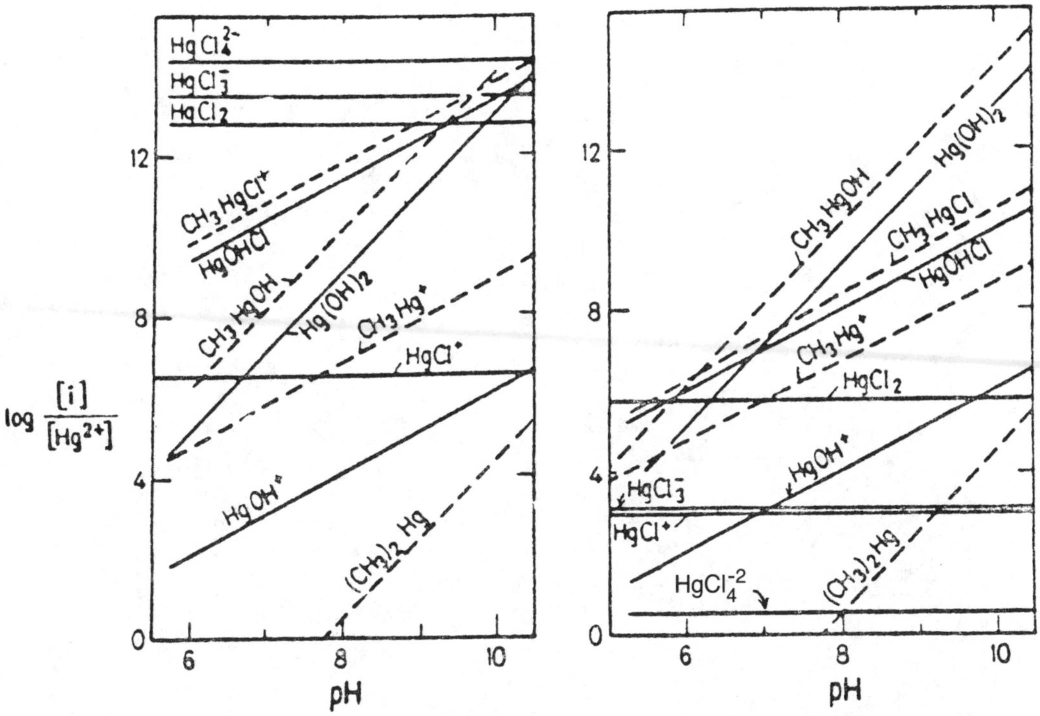

(a) Seawater: 0.6 *M* Cl^-, 10^{-4} *M* $CH_4(aq)$

(b) Fresh Water: 2×10^{-4} *M* Cl^-, 10^{-4} *M* $CH_4(aq)$

Source: Stumm and Morgan [40]. *(Copyright 1981, John Wiley & Sons. Reprinted with permission.)*

FIGURE 7.10-6 Calculated Speciation at Equilibrium for Systems Containing Methyl Mercury at 20°C

tions, Hg(I) actually exists as Hg_2^{+2}, where the Hg(I) species are joined by a covalent bond [38]. Figures 7.10-7 and -8 show fields of stability for solid, liquid and aqueous forms of mercury at various pH and Eh levels. Hg(liq) is stable over a wide range of conditions in normal environments. Under mildly reducing conditions, in the presence of sulfur, HgS is expected to precipitate.

7.10.3 Solubility/Precipitation

Mercury is very soluble in oxidized aquatic systems. The solubility products of the relatively insoluble mercury compounds are $10^{-52.7}$ (black, β-HgS), $10^{-53.3}$ (red, α-HgS), $10^{-28.62}$ (HgI_2), $10^{-25.44}$ ($Hg(OH)_2$), $10^{-54.77}$ (Hg_2S) and $10^{-28.33}$ (Hg_2I_2) [25,28]. A saturated solution of Hg(liq) has been reported as $10^{-6.52}$ *M* Hg^0 [25] or 0.02-0.03 mg/l at 20°C [5]. The solubility of mercury compounds, however, is greatly influenced by its strong complexation behavior with common ligands. Hahne and

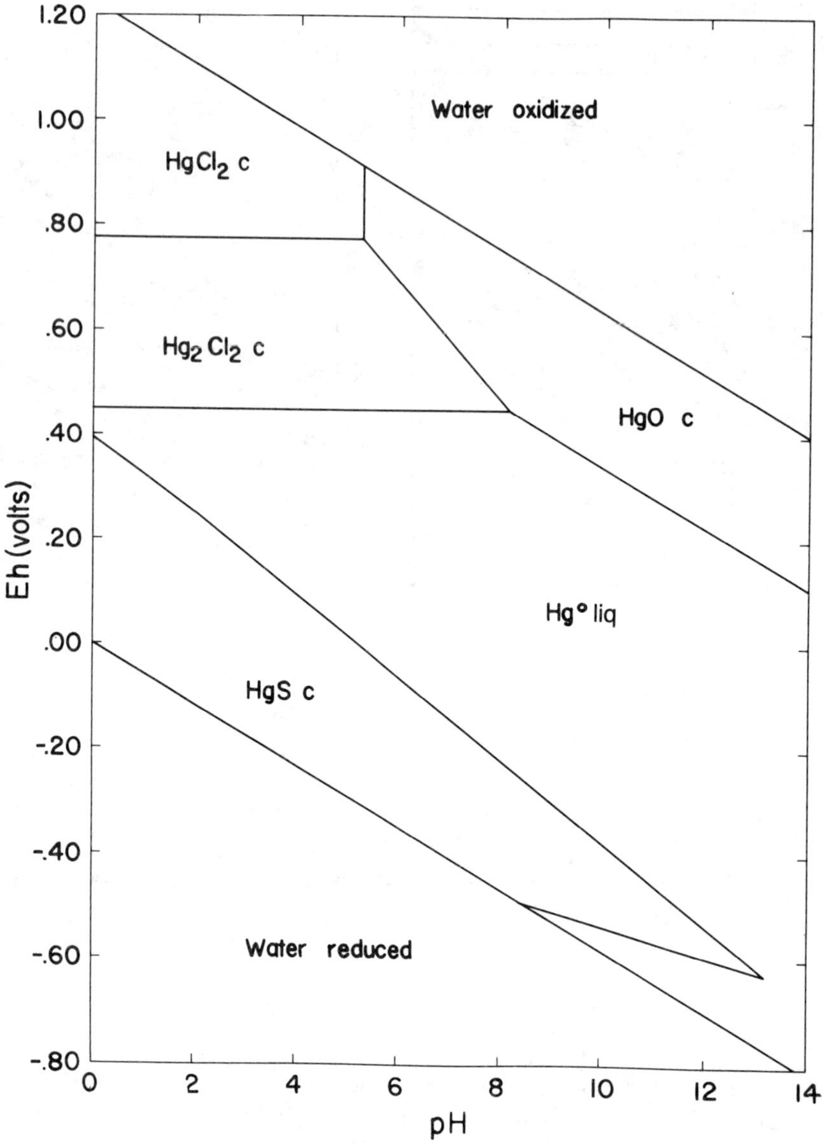

System contains 36 ppm Cl^-, 96 ppm sulfur as SO_4^{-2}.

Source: Hem [15]

FIGURE 7.10-7 Fields of Stability for Solid (c) and Liquid (liq) Mercury Species at 25°C and 1 Atmosphere

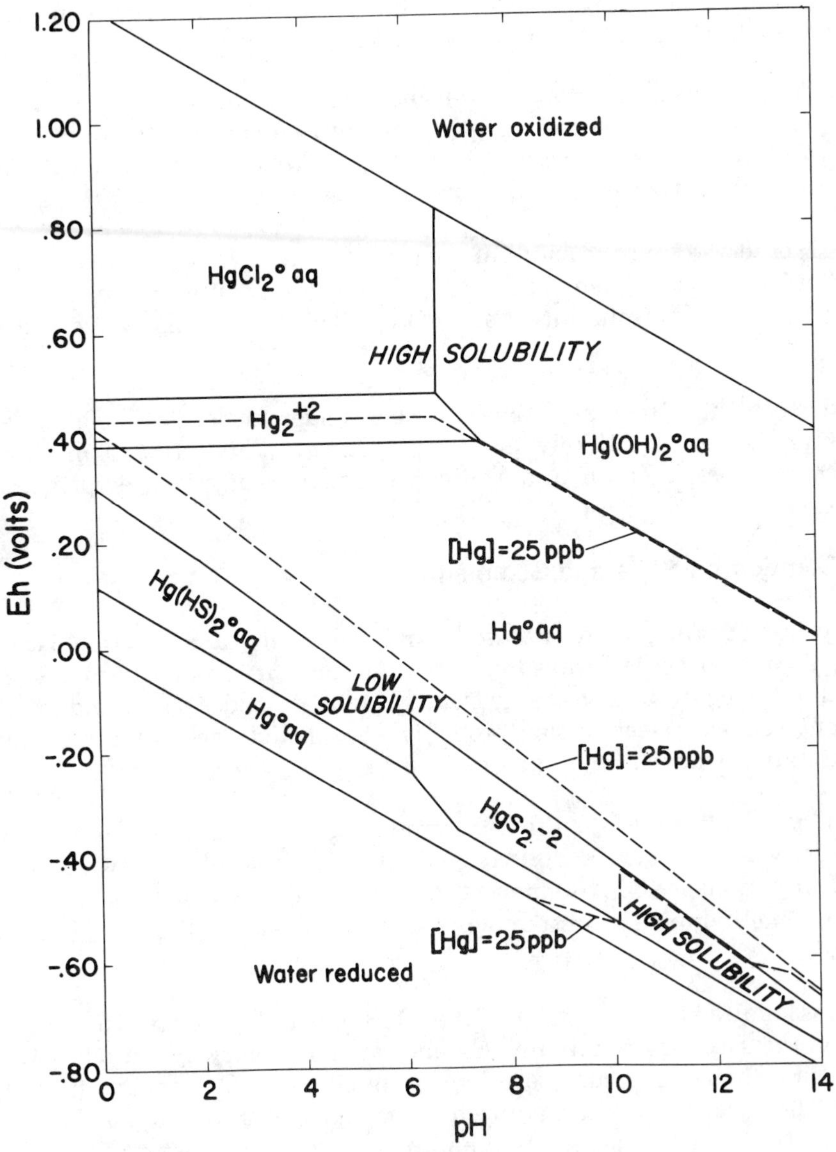

System contains 36 ppm Cl^-, 96 ppm sulfur as SO_4^{-2}.

Dashed line indicates approximate solubility of mercury in this system.

Source: Hem [15]

FIGURE 7.10-8 Fields of Stability for Aqueous Mercury Species at 25°C and 1 Atmosphere

Kroontje [14] calculated that instead of the theoretical 0.039 ppm Hg^{+2} that would be expected in water on the basis of its solubility product, up to 107 ppm of Hg(II) can exist as soluble $Hg(OH)_2^0$ when no other ligands are present.

The effect of Cl^- complexation is even more pronounced: at a Cl^- concentration of 35,460 ppm, the solubilities of $Hg(OH)_2$ and HgS were calculated to be increased by factors of 10^5 and 3.6×10^7 respectively. At a Cl^- concentration of only 3.5 ppm, the solubilities of these two compounds increase by factors of 55 and 408 respectively [14].

Trost [41] observed that the solubility of cinnabar (α-HgS) was increased by 30 times at a pH of 5 — and even more at higher pH — by the addition 850 ppm of humic acid (expressed as C). The solubilities of HgO and $HgCl_2$ were also increased by humic acid.

At sufficiently high mercury concentrations under moderately oxidizing environments, Hg^0 is the most likely insoluble species [6]. $Hg(OH)_2$ is another possible insoluble species [29]. In anaerobic waters, mercury is precipitated as HgS [5,29,31,32].

7.10.4 Sorption on Soils and Sediments

Mercury is strongly sorbed onto soils and sediments, and sorption is one of the important controlling pathways for removal of mercury from solution. The primary sorbents in the environment are hydrous Fe and Mn oxides, clays and organics [31]. Important removal mechanisms probably also include organic complexation and co-precipitation with sulfides such as iron sulfides.

Mercury has an affinity for sulfhydryl groups in organic matter and clays [31]. The amount sorbed has also been found to be most closely related to the organic content of the soil or sediment [33]. Reimers and Krenkel [34] found that the sequence of sorption of methylmercury in sediments was organics >> illite >> montmorillonite >> sand.

Numerous studies have been made of the effect of pH, Cl^-, dissolved oxygen and other factors on the sorption of inorganic Hg and methylmercury by various sorbents. (Eh would also influence mercury sorption by its effect on the speciation of Hg in the system.) The effect of pH does not appear to be consistent: Khalid *et al.* [20] found that Hg^{+2} sorption on Mississippi sediments increased between pH 5 and 8, while Ramamoorthy and Rust [33] stated that sorption of this ion on Ottawa River sediments was little affected by pH. The optimal sorption of $HgCl_2$ on organics occurred at pH <5 [1]. The desorbed Hg at pH >5 is believed to be due to the formation of soluble Hg-humate complexes [1,34].

The presence of Cl^- appears to decrease sorption onto the sorbents that have been tested. In studies with sediments, a decrease of sorption of Hg with increasing Cl^- concentration has been reported [7,34]. Similar results have been obtained with

Hg^{+2} sorption onto peat [11] and Mn oxides [26]. The highly stable Hg-Cl complexes appear to be poorly sorbed [22]. Some investigators believe that $Hg(OH)_2$ is the preferred sorbed species [26,32].

Increasing dissolved oxygen also appears to increase the amount of Hg^{+2} and methylmercury in solution, presumably by decreasing concentrations of sulfide [11]; this implies that the dredging of sediments could solubilize mercury in an aquatic system [35].

As discussed above, inorganic and organic Hg species are strongly sorbed onto soils and sediments. Once this process has occurred, desorption of Hg is negligible [33,35]. The desorption of Hg^{+2} from river sediments was found to be less than 1% after 70 hours of agitation with distilled water [33]. Moreover, sorption is rapid [26]: Rogers *et al.* [36] found that the sorption rate of inorganic Hg is 10^3-10^5 times greater than the rate of desorption.

Median values for distribution coefficients (K_d) for Hg sorption on Fe oxides and bentonite have been reported to be around 1000 l/g and 150 l/g respectively.[2] Constants reported for a Freundlich isotherm of Hg on Fe oxides were 90.8 l/g (K_F) and 0.76 (1/N). A mean value of ~7 μmol/g was found for A_m for two soils. In sediments, median values of A_m and K_L were reported to be ~50 μmol/g and ~6 (log M^{-1}) respectively for several sediments. The range of reported values in the literature is extremely large. The related measurements are very specific to experimental conditions, type of sorbent, concentration of solute, other substances present in the soil or sediment, and many other factors; therefore, the above values should be used with great caution.

Langmuir constants for soils and sediments are shown in Table 7.10-1.

---

2. These values and those in the following two sentences are adapted from Rai *et al.* [32]. The various sorption constants are defined as follows:

K_d = ratio of the quantity of sorbate sorbed per gram of solid to the concentration of the sorbate remaining in solution at equilibrium

A_m,K_L = Langmuir constants, where $S = (K_L A_m C)/(1 + K_L C)$ and S = moles sorbed at equilibrium per gram of solid, A_m = maximum sorption capacity of solid, K_L = sorption constant (related to binding energy of sorbate), and C = total sorbate concentration in solution at equilibrium

K_F,1/N = Freundlich constants for $S = K_F C^{1/N}$, where S is in μmol/g and C is in μ*M* units.

TABLE 7.10-1

Langmuir Constants for Mercury Sorption onto Soils and Sediments

	% Organic Carbon	% Clay	Particle Size (mm)	A_m (μmol/g)	Log K_L (log M^{-1})
Soils [16]					
Asquith	0.79	8.8	—	1.6	—
Oxbow	2.98	45.5	—	12.3	—
Sediments (calculated from [33])	**% Organic Matter**				
	10.78		0.35	463.6	5.9
	1.81	—	0.225	448.7	4.8
	0.42	—	0.240	349.0	4.2
	0.57	—	0.260	219.4	4.3

7.10.5 Biotransformation and Methylation/Demethylation

The biological cycling of mercury in the environment has been described in Section 2.15 on microbial transformation of inorganic pollutants. An excellent review of mercury cycling is also given by Wood [44]. Microorganisms are extremely important in a number of reactions involving Hg speciation. The disproportionation reaction, $Hg_2^{+2} \rightleftharpoons Hg^{+2} + Hg^0$ is catalyzed by microorganisms. Aerobes can solubilize Hg^{+2} from HgS by oxidizing the S^{-2} to sulfate via sulfite. Another important reaction is the enzymatic conversion of Hg^{+2} to Hg^0. Microorganisms also reduce methylmercury to Hg^0 and methane [44], and they degrade phenylmercuric acetate in soil to Hg^0, which then can be lost as mercury vapor [21].

The biological methylation of Hg can occur under anaerobic conditions (and, to a lesser extent, under aerobic conditions) in both water and soil solution. The importance of methylation in sediments is that it can result in remobilization of precipitated or sorbed mercury. The rate of biomethylation depends on a variety of factors, which can include concentration of available Hg^{+2}, the microbial population, pH, temperature, redox potential, and the synergistic or antagonistic effects of other metabolic or chemical processes [44]. The form of Hg that is methylated is Hg^{+2}; however, Hg(liq), phenylmercury, and most other Hg species are available for methylation, since they can be converted to Hg^{+2} upon release into the environment. HgS is least accessible to methylation because of its extreme insolubility [31], but even this species may be solubilized by certain aerobes and subsequently methylated. The higher the concentration of Hg^{+2}, the higher the end concentration of methylated mercury [3]. Similarly, the extent of biomethylation is directly related to the microbial concentration in the system.

Hg^{+2} concentration and pH appear to influence the proportions of monomethyl- and dimethylmercury resulting from biomethylation in a system. D'Itri [9] reported that low initial Hg^{+2} concentrations and pH ⩾7 result in dimethylmercury, while higher initial Hg^{+2} concentrations and acidic pH result in monomethylmercury. The optimum pH for monomethylmercury formation in the laboratory or in sediments is around 4.5 [45].

The production of the dimethylmercury species under basic conditions is thought to be a result of a disproportionation reaction of CH_3Hg^+ to give $(CH_3)_2Hg$ and Hg^{+2} [45]. The absence of strong complexing agents such as S^{-2} appears to favor biomethylation [31].

Abiotic (chemical) methylation has been reported in the literature. An example is the chemical reaction of methyl groups from other compounds (e.g., water-soluble methylsilicon compounds [8] and methyl groups present as contaminants in ethyl-lead compounds [18]) with Hg^{+2} to give methylmercury. Abiotic methylation is reported to be directly proportional to temperature and Hg^{+2} concentration and inversely proportional to pH (at pH >5) [37].

Compared with methylation, demethylation is a much slower process [31]. The CH_3Hg^+ unit, once formed, is kinetically inert toward decomposition [40], which is observed only at elevated levels of methylmercury [10]. Biological demethylation is also described in §2.15. Chemical demethylation (photolytic decomposition) of methylmercury-sulfur complexes has been reported by Lexmond *et al.* [24].

7.10.6 Volatilization and Photolysis

Metallic mercury and dimethylmercury (boiling point 96°C) are volatile under normal environmental conditions [42]. The vapor pressure of liquid mercury is 0.0018 mm Hg at 25°C [38]. Loss of mercury from soils following conversion to volatile species occurs more quickly for the most soluble species such as $HgCl_2$ and $Hg(NO_3)_2$ than for those that are less soluble (e.g., HgO and HgS). The volatilization rate is also a function of the type of soil, being slower in loam than in sand and still slower in clay [37]. Dimethylmercury is very insoluble in water and has a high vapor pressure, resulting in rapid volatilization from aquatic systems [31]. Stumm and Morgan [40] reported a large value of 8.5 atm M^{-1} for the Henry's law constant for Hg^0(aq). At normal temperatures, the Hg^0 solute is readily lost from aqueous systems. The ratio of the mass transfer coefficient[3] of mercury to that of oxygen was measured as $K_L^{Hg}/K_L^{O_2}$ = 0.94 ± 0.08 between 0° and 30°C; this value is higher than that of many common volatiles (e.g., ethylene 0.87; propane 0.72; trichloroethylene 0.57) [30].

---

3. The mass transfer coefficient is defined [30] using the two-film model for mass transfer,

$$N = K_L a(C_L - p/H)V_L$$

where:

N	=	mass transfer rate (mol/s)
K_L	=	overall mass transfer coefficient (m/s)
a	=	interfacial area per unit volume of the aqueous phase (m^{-1})
C_L	=	concentration of contaminant in aqueous phase (mol/m^3)
p	=	partial pressure (Pa)
H	=	Henry's law constant ($Pa\text{-}m^3/mol$)
V_L	=	volume of aqueous phase (m^3)

Wood [44] reported that once dimethylmercury is in the atmosphere, it is photolyzed by UV light, yielding Hg^0 and methyl radicals; the latter can react with hydrogen atoms to form methane or couple to form ethane. Other investigators [17,19,43] have suggested the photolytic breakdown of dimethylmercury to monomethylmercury. The photodecomposition of phenylmercury compounds has also been reported [46].

7.10.7 Literature Cited

1. Anderson, A., "Mercury in Soil," *Grundforbattring*, **20**, 95 (1968), Dept. Sec. State Transl. Bureau No. 5433 (1970), as cited by Reimers *et al.* [35].
2. Baes, C.F. and R.E. Mesmer, *The Hydrolysis of Cations*, John Wiley & Sons, New York (1976).
3. Bisogni, J.J. and A.W. Lawrence, "Kinetics of Mercury Methylation in Aerobic and Anaerobic Aquatic Environments," *J. Water Pollut. Control Fed.*, **47**, 135-52 (1975), as cited by Callahan *et al.* [6].
4. Bondietti, E.A., R.M. Perhac, F.H. Sweeton and T. Tamura, "Toxic Metals in Lake and River Sediments," in *Ecology and Analysis of Trace Contaminants*, Progress Rept. June 1973, ORNL-NST — EATC-1, Oak Ridge National Laboratory, Oak Ridge, Tenn. (1973).
5. Burrows, W.D., R.S. Reimers and E.B. Shin, "Chemistry of Mercury in the Environment," *CRC Crit. Rev. Environ. Control*, **4**, 251-339 (1974).
6. Callahan, M., M. Slimak, N. Gabel, I. May, C. Fowler, J. Freed, P. Jennings, R. Durfee, F. Whitmore, B. Maestri, W. Mabey, B. Holt and C. Gould, *Water-Related Environmental Fate of 129 Priority Pollutants. Volume I: Introduction and Technical Background, Metals and Inorganics, Pesticides and PCBs*, EPA-440/4-79-029a, EPA Contracts 68-01-3852 and 68-01-3867, Office of Water Planning and Standards, U.S. Environmental Protection Agency, Washington, D.C. (1979).
7. DeGroot, A.J., J.J.M. DeGoeij and C. Zegers, "Contents and Behavior of Mercury as Compared with Other Heavy Metals in Sediments from the Rivers Rhine and Ems," *Geol. Mijnbouw*, **50**, 393 (1971), as cited by Reimers *et al.* [35].
8. DeSimmone, R.E., *Chem. Commun.*, **13**, 780 (1972), as cited by Wood [44].
9. D'Itri, F.M., *The Environmental Mercury Problem*, CRC Press, Cleveland (1972), as cited by Perwak *et al.* [31].
10. Fagerstrom, T. and A. Jernelov, "Some Aspects of the Quantitative Ecology of Mercury," *Water Res.*, **6**, 1193-1202 (1972), as cited by Callahan *et al.* [6].
11. Feick, G., E.E. Johanson and D.S. Yeaple, *Control of Mercury Contamination in Fresh Water Sediments*, Environmental Protection Technology Series, EPA-R2-72-077 (1972), as cited by Reimers *et al.* [35].
12. Fleisher, M., "Summary of the Literature on the Inorganic Geochemistry of Mercury," in *Mercury in the Environment*, Professional Paper 713, U.S. Geological Survey, Washington, D.C., 6-13 (1970).

13. Fuller, W.H., *Movement of Selected Metals, Asbestos, and Cyanide in Soil: Applications to Waste Disposal Problems,* EPA 600/2-77-020, PB 266 905, Municipal Environmental Research Laboratory, Cincinnati, Ohio (1977).

14. Hahne, H.C.H. and W. Kroontje, "Significance of pH and Chloride Concentration on Behavior of Heavy Metal Pollutants — Mercury(II), Cadmium(II), Zinc(II), and Lead(II)," *J. Environ. Qual.*, **2**, 444-50 (1973).

15. Hem, J.D., "Chemical Behavior of Mercury in Aqueous Media," in *Mercury in the Environment*, Professional Paper 713, U.S. Geological Survey, Washington, D.C., 19-24 (1970).

16. Hogg, T.J., J.W.B Stewart and J.R. Bettany, "Influence of the Chemical Form of Hg on Its Adsorption and Ability to Leach through Soil," *J. Environ. Qual.*, **7**, 440-45 (1978), as cited by Rai *et al.* [32].

17. Holm, H.W. and M.F. Cox, *Mercury in Aquatic Systems: Methylation, Oxidation-Reduction, and Bioaccumulation*, EPA 660/3-74-021, U.S. EPA, Washington, D.C. (1974), as cited by Callahan *et al.* [6].

18. Jernelov, A., *Report to the Government of Ontario* (1971), as cited by Wood [44].

19. Johnson, D.L. and R.S. Braman, "Distribution of Atmospheric Mercury Near Ground," *Environ. Sci. Technol.*, **8**, 1003-09 (1974), as cited by Callahan *et al.* [6].

20. Khalid, R.A., R.P. Gambrell and W.H. Patrick, Jr., "Sorption and Release of Mercury by Mississippi River Sediment as Affected by pH and Redox Potential," in *Biological Implications of Metals on the Environment*, Proc. 15th Annual Hanford Life Sciences Symposium, Sept. 29-Oct. 1, 1975, Richland, Wash., 297-314 (1975), as cited by Perwak *et al.* [31].

21. Kimura, Y. and V.L. Miller, "The Degradation of Organo-Mercury Fungicides in Soils," *J. Agr. Food Chem.*, **12**, 253-57 (1964), as cited by Fuller [13].

22. Kinniburgh, D.G. and M.L. Jackson, "Adsorption of Mercury(II) by Iron Hydrous Oxide Gel," *Soil Sci. Soc. Am. J.*, **42**, 45-47 (1978), as cited by Rai *et al.* [32].

23. Lee, G.F., "Role of Hydrous Metal Oxides in the Transport of Heavy Metals in the Environment," in *Heavy Metals in the Aquatic Environment*, P.A. Krenkel (ed.), Pergamon Press, Oxford, England, 137-47 (1975).

24. Lexmond, T.M., F.A.M. de Haan and M.J. Frissel, "On the Methylation of Inorganic Mercury and the Decomposition of Organo-Mercury Compounds — A Review," *Neth. J. Agri. Sci.*, **24**, 79-97 (1976), as cited by Perwak *et al.* [31].

25. Lindsay, W.L., *Chemical Equilibria in Soils,* John Wiley & Sons, New York (1979).

26. Lockwood, R.A. and K.Y. Chen, "Adsorption of Hg(II) by Hydrous Manganese Oxides," *Environ. Sci. Technol.*, **11**, 1028-34 (1973), as cited by Lee [23].

27. Mantoura, R.F.C., A. Dickson and J.P. Riley, "The Complexation of Metals with Humic Materials in Natural Waters," *Estuarine Coastal Mar. Sci.*, **6**, 387-408 (1978).

28. Martell, A.E. and R.M. Smith, *Critical Stability Constants*, Vol. 4, *Inorganic Complexes*, Plenum Press, New York (1976).

29. Morel, F.M.M., *Principles of Aquatic Chemistry*, John Wiley & Sons, New York (1983).
30. Okouchi, S. and S. Sasaki, "Volatility of Mercury in Water," *J. Hazard. Mater.*, **8**, 341-48 (1984).
31. Perwak, J., M. Goyer, L. Nelken, K. Scow, M. Wald and D. Wallace, *An Exposure and Risk Assessment for Mercury,* Final Draft Report to U.S. Environmental Protection Agency, EPA Contracts 68-01-3857 and 68-01-5949, Monitoring and Data Support Division, Office of Water Regulations and Standards, Washington, D.C. (1980).
32. Rai, D., J. Zachara, A. Schwab, R. Schmidt, D. Girvin and J. Rogers, *Chemical Attenuation Rates, Coefficients, and Constants in Leachate Migration. Vol I: A Critical Review*, Report EA-3356 to EPRI by Pacific Northwest Laboratories (Battelle Institute), Richland, Wash. (1984).
33. Ramamoorthy, S. and B.R. Rust, "Mercury Sorption and Desorption Characteristics of Some Ottawa River Sediments," *Can. J. Earth Sci.*, **13**, 530-36 (1976).
34. Reimers, R.S. and P.A. Krenkel, "Kinetics of Mercury Adsorption and Desorption in Sediments," *J. Water Pollut. Control Fed.*, **46**, 352-65 (1974), as cited by Reimers *et al.* [35].
35. Reimers, R.S., P.A. Krenkel, M. Eagle and G. Tragitt, "Sorption Phenomenon in the Organics of Bottom Sediments," in *Heavy Metals in the Aquatic Environment*, P.A. Krenkel (ed.), Pergamon Press, Oxford, England, 117-29 (1975).
36. Rogers, J.S., P.M. Huang and W.K. Lian, "Chemical Dynamics of Mercury Sorption and Desorption of Freshwater Sediments," *Verh. Internat. Verein. Limnol.*, **21**, 174-80 (1981), as cited by Rai *et al.* [32].
37. Rogers, R.D., *Volatility of Mercury from Soils Amended with Various Mercury Compounds*, EPA-600/3-78-046, Office of Research and Development, U.S. EPA (1978), as cited by Perwak *et al.* [31].
38. Sienko, M.J. and R.A. Plane, *Chemistry: Principles and Properties*, McGraw-Hill Book Co., New York (1966).
39. Sillen, L.G. and A.E. Martell, *Stability Constants of Metal-Ion Complexes*, Special Publication No. 7, The Chemical Society, London (1964).
40. Stumm, W. and J.J. Morgan, *Aquatic Chemistry*, 2nd ed., John Wiley & Sons, New York (1981).
41. Trost, P.B., "Effects of Humic-Acid-Type Organics on Secondary Dispersion of Mercury," Unpublished Ph.D. Thesis, Colorado School of Mines, Colo. (1970), as cited by Reimers *et al.* [35].
42. Weast, R.C. (ed.), *Handbook of Chemistry and Physics*, 60th Edition, CRC Press, Inc., Boca Raton, Florida (1979).
43. Williston, S.H., "Mercury in the Atmosphere," *J. Geophys. Res.*, **73**, 7051-55 (1968), as cited by Callahan *et al.* [6].
44. Wood, J.M., "Metabolic Cycles for Toxic Elements in the Environment," in *Heavy Metals in the Aquatic Environment*, P.A. Krenkel (ed.), Pergamon Press, Oxford, England, 105-12 (1975).

45. Wood, J.M. and E.L. Lien, *Report ot the Government of Ontario* (1971), as cited by Wood [44].

46. Zepp, R.G., N.L. Wolfe and J.A. Gordon, "Photodecomposition of Phenylmercury Compounds in Sunlight," *Chemosphere*, **3**, 93-99 (1973), as cited by Callahan *et al.* [6].

7.10.8 Additional Sources of Information

Gavis, J. and J.F. Ferguson, "The Cycling of Mercury through the Environment," *Water Res.*, **6**, 989-1008 (1972).

Iverson, W.P., C. Huey, F.E. Brinckman, K.L. Jewett and W. Blair, "Biological and Nonbiological Transformations of Mercury in Aquatic Systems," in *Heavy Metals in the Aquatic Environment*, P.A. Krenkel (ed.), Pergamon Press, Oxford, England, 193-95 (1975).

Klein, D.H. and E.D. Goldberg, "Mercury in the Marine Environment," *Environ. Sci. Technol.*, **4**, 765-876 (1972).

Krenkel, P.A., "Mercury: Environmental Considerations, Part I," *CRC Crit. Rev. Environ. Control*, **3**, 303-73 (1973).

National Academy of Sciences, *An Assessment of Mercury in the Environment*, Washington, D.C. (1978).

Nriagu, J.O. (ed.), *The Biogeochemistry of Mercury in the Environment*, Elsevier/North-Holland Biomedical Press, Amsterdam, The Netherlands (1979).

Tierney, D.R., T.R. Blackwood and T.M. Briggs, *Status Assessment of Toxic Chemicals: Mercury*, PB 80-146384, EPA-600/2-79-2101 (1979).

7.11 NICKEL (Ni)

7.11.1 Occurrence and Uses

Nickel ores occur primarily in the form of the oxides, silicates and sulfides and are usually associated with other sulfide, silicate or arsenide minerals. As the silicate, nickel is found in the lattice of hydrated magnesium-iron minerals such as garnierite. Pentlandite, the main sulfide ore, also contains copper and iron [11]. About 84% of all nickel produced is used in alloys. Nickel metal, nickel(II) sulfate, nickel(II) chloride and nickel(II) cyanide are used in electroplating. Other uses of nickel are in batteries and as catalysts [11].

7.11.2 Speciation Reactions in Water

ACID-BASE DISSOCIATION

Aquo-nickel ions act as acids in water. The pK_a values of nickel are shown in Table 2.7-3 in Part I of this report. Aqueous nickel is not readily hydrolyzed to the hydroxide under typical environmental conditions, where the presence of other ligands would result in other nickel species; significant formation of hydroxide complexes occurs only under basic conditions (pH>7).

COMPLEXATION

Nickel tends to complex with both inorganic and organic ligands in aquatic systems. Figure 7.11-1 shows the chemical speciation of nickel in an organic-free aquatic system as a function of pH. Over most of the pH range, Ni^{+2} and $NiSO_4^0$ are the predominant species. $Ni(OH)_2^0$ and $Ni(OH)_3^-$ appear only when the pH is highly basic. In the pH range of 8-10, $NiOH^+$ and $NiHCO_3^+$ may also be important.

In polluted aquatic systems where industrial reagents and natural organic ligands are present, the speciation of nickel is greatly affected by the extent of formation of nickel complexes with these ligands. Rashid and Leonard [14] found that humic acid solubilized nickel from nickel carbonate. By the same token, the quantity of nickel required to cause precipitation was greatly increased when humic acid was added to solutions containing nickel and inorganic ligands. These results suggest that nickel-humic complexes are important in aquatic systems. Figure 7.11-2 shows the equilibrium speciation of nickel(II) as a function of salinity in the presence of humic acid; note that nickel-humate complexes exist only in systems with low salinity. Overall stability constants for Ni(II) complexes with some inorganic and organic ligands are listed in Tables 2.7-1 and -2 in Part I of this report. For freshwater and marine humic compounds isolated from surface waters, Mantoura, Dickson and Riley [8] found that the logarithms of the overall conditional stability constants, K_0, of Ni(II)-humate

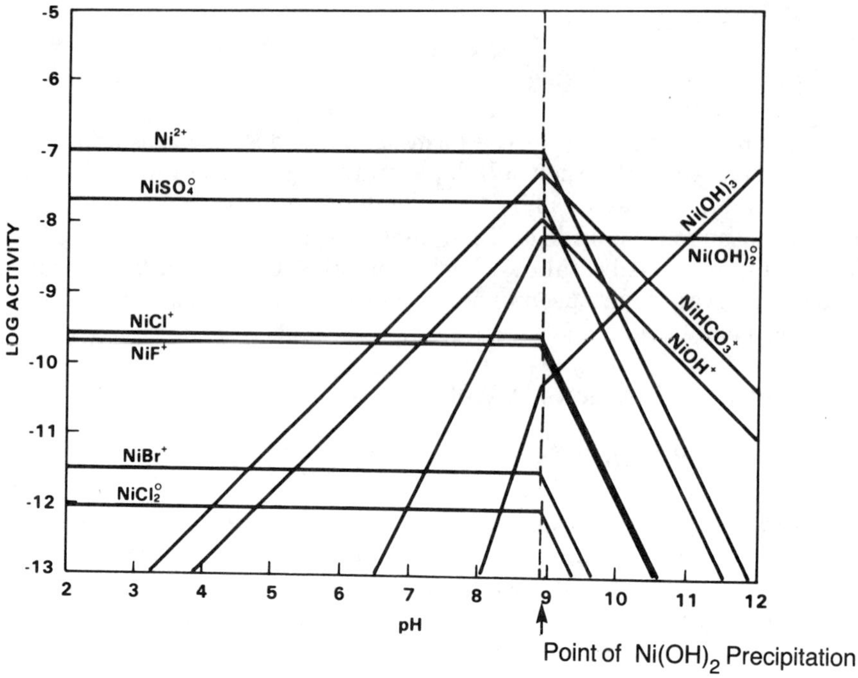

$SO_4^{-2} = Cl^{-} = 10^{-3}\,M$; $F^{-} = 10^{-4}\,M$,

$Br^{-} = 10^{-5}\,M$; $CO_2(g) = 10^{-3.52}$ atm

Source: Rai *et al.* [13]. *(Copyright 1984, Electric Power Research Institute. Reprinted with permission.)*

FIGURE 7.11-1 Activities of Various Ni^{+2} Species When Activities of Ni^{+2} Are Fixed at 10^{-7} *M* or Controlled by $Ni(OH)_2$

complexes ranged from 5.14 to 5.51 at pH 8 and an ionic strength of 0.02 *M*.[1] For the fulvic acid fractions of horticultural peat and soil, the overall conditional stability constants obtained were 4.98 and 4.2-4.35 respectively.

---

1. The overall conditional stability constant, K_0, is defined on the basis of an aggregate site-type:

$$K_0 = \frac{[ML]}{[M^{+2}]\,[L_T - ML]}$$

where

$[ML]$ = total complexed Ni(II)

$[M^{+2}]$ = total free aqueous Ni(II)

$[L_T]$ = total ligand concentration

$[L_T - ML]$ = total uncomplexed ligand

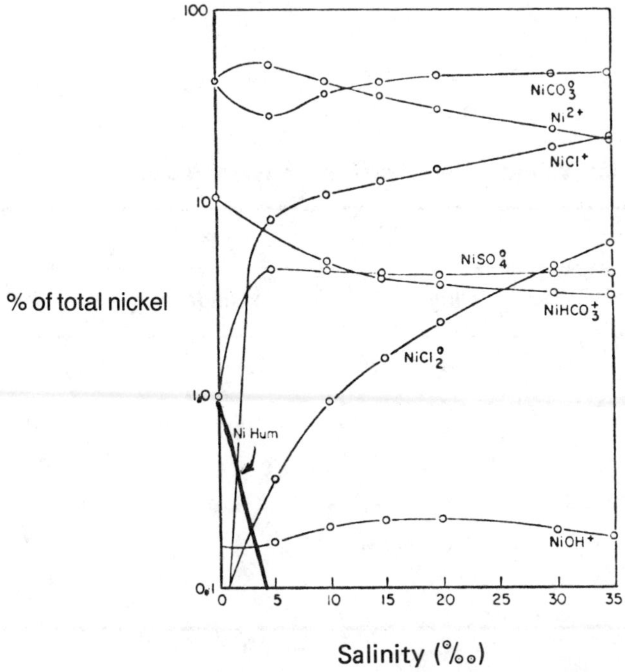

	Concentration (mol/kg solution)		
Species	**Average River Water**		**Seawater**
Na^+	2.83×10^{-4}		0.46847
K^+	5.9×10^{-5}		0.01020
Mg^{+2}	1.69×10^{-4}		0.05307
Ca^{+2}	3.74×10^{-4}		0.01028
Cl^-	2.20×10^{-4}		0.54590
SO_4^{-2}	1.17×10^{-4}		0.02823
Alkalinity (carbonate)	9.54×10^{-4}		0.0024
Total Inorganic Carbon	1.02×10^{-3}		0.0022
Cu		10^{-8}	
Zn		10^{-8}	
Cd		10^{-8}	
Mn		10^{-8}	
Hg		10^{-9}	
Ni		10^{-8}	
Co		10^{-9}	
HUM		10^{-6}	

Source: Mantoura, Dickson and Riley [8]. *(Copyright 1978, Academic Press, Inc., Ltd. Reprinted with permission.)*

FIGURE 7.11-2 Equilibrium Speciation of Nickel(II) as a Function of Salinity in a Model Estuary, pH ≈ 7-8.2

TABLE 7.11-1

Typical Values of Sorption Constants for Nickel

Parameter	Soils		
	Clay	Fe Oxides	Mn Oxides
K_d (l/g)	0.2*	100**	40
S (μmol/g)	40	8.6	5.6
A_m (μmol/g)			600*
Log K_L (log M^{-1})			3.5*
K_F (l/g)			0.27
1/N			0.95

*Only 2 values
**Only 1 value

Source: Adapted from Rai *et al.* [13].

Definition of parameters:

K_d = distribution coefficient (l/g)

S = sorption (μmol/g)

A_m,K_L = Langmuir constants, where $S = (K_L A_m C)/(1 + K_L C)$ and S = moles sorbed at equilibrium per gram of solid, A_m = maximum sorption capacity of solid, K_L = sorption constant related to binding energy of sorbate, and C = total sorbate concentration in solution at equilibrium

K_F,1/N = Freundlich constants for $S = K_F C^{1/N}$, where S = μmol/g and C = μ*M*

TABLE 7.11-2

Freundlich Isotherms for Nickel Sorption on Soils

Freundlich Isotherm[a]	Type of Soil					
	Silt Loam	Sandy Loam	Loamy Sand	Sand	Clay	Clay Loam
K_F (l/g)	0.057-0.44	0.11-0.48	0.005-0.01	0.31-0.50	0.6-0.73	0.51
1/N	0.57-0.96	0.81-0.99	0.87-0.92	1.01-1.18	0.95-1.03	1.02

a. See definitions in Table 7.11-1.

Source: Bowman *et al.* [2], as reported in Rai *et al.* [13].

The rate of complexation of Ni(II) by ligands is slower than that of Cu(II) by a factor of about 10^6 [12]. The second-order rate constant for most Ni(II) complexes is about 2.9×10^3 M^{-1} s^{-1}; this implies that, for example, if the Ni(II) concentration in an aquatic system is 10^{-5} M and the Cl^- concentration is 10^{-3} M, the half-life for formation of $NiCl^+$ complex is on the order of 0.2 second. Complexation of Ni(II) by ligands that normally exist in the environment is therefore very rapid in terms of environmental time frames.

REDOX REACTIONS

Although nickel can have a valence of +1, +3 or +4 in certain compounds, the only oxidation state of importance under environmental conditions is Ni(II). The standard electrode potential for the Ni^{+2}/Ni^0 couple is −0.270V to −0.237V (25°C); thus, in a stability diagram (such as that given in Figure 7.15-3 for zinc), the reduction of Ni(II) to Ni(s) would be below the stability zone for water [15]. In the presence of sulfur and very reducing environments, the extremely insoluble nickel sulfide solid is formed; under aerobic conditions and below pH 9, nickel forms complexes with carbonate, sulfate, hydroxide and other naturally occurring ligands [4].

7.11.3 Solubility/Precipitation

The relatively insoluble nickel compounds that are environmentally important are the sulfide, hydroxide and carbonate. Their approximate solubility products are $10^{-19.4}$ (α-NiS), $10^{-24.9}$ (β-NiS), $10^{-26.6}$ (γ-NiS), $10^{-15.2}$ ($Ni(OH)_2$) and $10^{-6.87}$ ($NiCO_3$) [9]. Rai *et al.* [13] suggest that $NiFeO_4$ and NiS may be the solubility-controlling precipitates under oxidizing and reducing conditions, respectively. Since the solubility product of nickel hydroxide is about 10 orders of magnitude lower than that of nickel carbonate, the latter is highly unlikely to precipitate under typical environmental conditions [13]. Under aerobic conditions and normal values of pH, complexation decreases the activity of the free aqueous nickel ion, thereby discouraging precipitation reactions.

7.11.4 Sorption on Soils and Sediments

Fuller [6] considered nickel, together with chromium and mercury, to be among the most generally mobile of the heavy metals that he studied. The mechanisms for sorption may be specific adsorption, ion exchange and/or co-precipitation. The sorptive behavior of nickel in soil has not been extensively investigated, but iron and manganese oxides, clay minerals and (to some extent) organic matter are believed to be important sorbents of nickel [13].

Nickel sorption on iron and manganese oxides is pH-dependent, probably because $NiOH^+$ is preferentially sorbed and also because the surface charge on sorbents is affected by pH [13]. The presence of other cations may reduce nickel sorption. Sodium, and particularly calcium, are two effective competitors for sorption sites [2,5,10].[1] Other cations — e.g., other heavy metals — are probably also effective in reducing

1. As cited by Rai *et al.* [13].

sorption of nickel. Complexing ligands such as SO_4^{-2} [3,10,16][1] and organic acids [7] reduce the sorption of nickel by sorbents. Remobilization of nickel from solid phases also appears to be possible in the presence of fulvic and humic acids [7]. Thus, sorption of nickel may be moderately effective in reducing the mobility of this element in natural, unpolluted systems but may be minimal where complexation is important (e.g., in organic-rich, polluted systems) [4].

Typical sorption constants for nickel are given in Table 7.11-1. Median values are listed except where only one or two values are available, in which case mean values are given. Because of the sparseness of the data and their applicability only to very specific conditions, such constants should be used with great caution.

Bowman *et al.* reported a number of Freundlich constants for nickel sorption on some types of soils. Their data are summarized in Table 7.11-2.

Amoozegar-Fard, Fuller and Warrick [1] developed a simple, field-oriented regression equation to predict the movement of certain metals in a column of soil. The equation and parameters that they obtained for nickel are described in section 2.12.

7.11.5 Literature Cited

1. Amoozegar-Fard, A., W.H. Fuller and A.W. Warrick, "An Approach to Predicting the Movement of Selected Polluting Metals in Soils, " *J. Environ. Qual.*, **13**, 290-97 (1984).
2. Bowman, R.S., M.E. Essington and G.A. O'Connor, "Soil Sorption of Nickel: Influence of Solution Composition," *Soil Sci. Soc. Am. J.*, **45**, 860-65 (1981).
3. Bowman, R.S. and G.A. O'Connor, "Control of Nickel and Strontium Sorption by Free Metal Activity," *Soil Sci. Soc. Am. J.*, **46**, 933-36 (1982).
4. Callahan, M., M. Slimak, N. Gabel, I. May, C. Fowler, J. Freed, P. Jennings, R. Durfee, F. Whitmore, B. Maestri, W. Mabey, B. Holt and C. Gould, *Water-Related Environmental Fate of 129 Priority Pollutants. Volume I: Introduction and Technical Background, Metals and Inorganics, Pesticides and PCBs*, EPA-440/4-79-029a, EPA Contracts 68-01-3852 and 68-01-3867, Office of Water Planning and Standards, U.S. Environmental Protection Agency, Washington, D.C. (1979).
5. Doner, H.E., J.J. Traina and A. Pukite, "Interaction of Nickel, Copper, and Cadmium with Soil Manganese in Saline Solutions," *Soil Sci.*, **133**, 369-77 (1982).
6. Fuller, W.H., *Movement of Selected Metals, Asbestos, and Cyanide in Soil: Applications to Waste Disposal Problems*, EPA 600/2-77-020, PB 266 905, Municipal Environmental Research Laboratory, Cincinnati, Ohio (1977).
7. Jackson, K.S. and G.B. Skippen, "Geochemical Dispersion of Heavy Metals via Organic Complexing: A Laboratory Study of Copper, Lead, Zinc, and Nickel Behavior at a Simulated Sediment-Water Boundary," *J. Geochem. Explor.*, **10**, 117-38 (1978).

8. Mantoura, R.F.C., A. Dickson and J.P. Riley, "The Complexation of Metals with Humic Materials in Natural Waters," *Estuarine Coastal Mar. Sci.,* **6**, 387-408 (1978).
9. Martell, A.E. and R.M. Smith, *Critical Stability Constants*, Vol. 4, *Inorganic Complexes*, Plenum Press, New York (1976).
10. Mattigod, S.V., A.S. Gibali and A.L. Page, "Effect of Ionic Strength and Ion Pair Formation on the Adsorption of Nickel by Kaolinite," *Clays Clay Miner.,* **27**, 411-16 (1979).
11. McNamara, P., M. Byrne, B. Goodwin, K. Scow, W. Steber, R. Thomas and M. Wood, *An Exposure and Risk Assessment for Nickel,* EPA Contract 68-01-5949, Final Draft Report to Monitoring and Data Support Division, Office of Water Regulations and Standards, U.S. Environmental Protection Agency, Washington, D.C. (1981).
12. Morel, F.M.M., *Principles of Aquatic Chemistry,* John Wiley & Sons, New York (1983).
13. Rai, D., J. Zachara, A. Schwab, R. Schmidt, D. Girvin and J. Rogers, *Chemical Attenuation Rates, Coefficients, and Constants in Leachate Migration. Vol. I: A Critical Review,* Report EA-3356 to EPRI by Pacific Northwest Laboratories (Battelle Institute), Richland, Wash. (1984).
14. Rashid, M.A. and J.D. Leonard, "Modifications in the Solubility and Precipitation Behavior of Various Metals as a Result of Their Interaction with Sedimentary Humic Acid," *Chem. Geol.,* **11**, 89-97 (1973).
15. Sillén, L.G. and A.E. Martell, *Stability Constants of Metal-Ion Complexes,* Special Publication No. 17, The Chemical Society, London (1964).
16. Theis, T.L. and R.O. Richter, "Adsorption Leaching of Nickel Species at Oxide Surfaces," *Adv. Chem. Ser.,* **189**, 73-96 (1980).

7.11.6 Additional Sources of Information

Nickel, National Academy of Sciences, Washington, D.C. (1975).

Versar, Inc., *Preliminary Investigation of Effects on the Environment of Boron, Indium, Nickel, Selenium, Tin, Vanadium and Their Compounds: Volume III — Nickel,* Springfield, Va. (1975).

7.12 SELENIUM (Se)

7.12.1 Occurrence and Uses

The chemistry of selenium is quite similar to that of sulfur, as both elements are in Group VI of the periodic table. Thus, sulfides or natural sulfur deposits often contain selenium. Naturally occurring sulfides of bismuth, iron, mercury, silver, copper, lead, and zinc have been found to contain up to 20% selenium. Jarosite [$KFe_3(SO_4)_2(OH)_6$] and barite ($BaSO_4$), two sulfate-containing minerals, contain low levels of selenium [3,9].

Selenium is produced commercially from copper refinery slimes [6,9,16]. It is used in the electronics industry, chiefly in the form of metal selenides, to make photoelectric cells and rectifiers. Selenium dioxide (SeO_2) is widely used in the production of other selenium compounds [6]. Sodium selenite (Na_2SeO_3) is used as a decolorizer for glass and ceramics; ammonium selenite [$(NH_4)_2SeO_3$] and diarsenic selenide (As_2Se) are also used in the glass industry. As small amounts of the element are essential for animal nutrition, sodium selenate (Na_2SeO_4) and selenite (Na_2SeO_3) are added to animal feeds. Other selenium compounds are used in pigments and in the steel and rubber industries [6,9,10,11].

7.12.2 Speciation Reactions in Water

Selenium is found in nature in the -2, 0, $+4$, and $+6$ oxidation states. The reduced state is represented by the selenide (Se^{-2}) species. Elemental Se exists in several allotropic forms. The higher oxidation states are represented by selenites (SeO_3^{-2}) and selenates (SeO_4^{-2}).

Selenium, like sulfur, forms oxyanions and exhibits anionic chemistry in aqueous solution. The speciation of selenium is determined by the Eh and pH of the solution.

Elemental selenium, Se, which is essentially insoluble in water, is the most stable form of the element over a wide range of redox and pH values and is favored by low pH and reducing conditions [2,3,9,10,11,20]. Its reduction to hydrogen selenide (H_2Se) occurs at an electrode potential ($E_0 = -0.40V$) below that of the H^+/H_2 couple; H_2Se is therefore unstable in water, but its oxidation to elemental selenium is a slow reaction [2,9]. Rai *et al.* [19] report that $H_2Se(aq)$ and HSe^- are the major aqueous species of selenium in the Eh-pH range that the other references cited above designate as the stability field for elemental selenium.

Listed below are logarithms of acid dissociation equilibrium constants for hydrogen selenide (H_2Se), selenious acid (H_2SeO_3), and hydrogen selenate ($HSeO_4^-$) at 25°C and zero ionic strength (unless otherwise noted) [14].

$$H_2Se^0 \rightleftharpoons HSe^- + H^+ \qquad \log K = -3.89 \qquad (1)$$

$$HSe^- \rightleftharpoons Se^{-2} + H^+ \qquad \log K = -15\ (20°C) \qquad (2)$$

$$H_2SeO_3 \rightleftharpoons HSeO_3^- + H^+ \qquad \log K = -2.75 \qquad (3)$$

$$HSeO_3^- \rightleftharpoons SeO_3^{-2} + H^+ \qquad \log K = -8.5 \qquad (4)$$

$$HSeO_4^- \rightleftharpoons SeO_4^- + H^+ \qquad \log K = -1.7 \qquad (5)$$

Selenious acid species occur under the intermediate to slightly oxidizing conditions encountered in aerobic waters [3,9,19,22]. Over a pH range of about 3 to 9, the biselenite ion ($HSeO_3^-$) predominates; at pH >9, selenite (SeO_3^{-2}) is the major species [3,9,20]. The distribution of biselenite and selenite species can be estimated from the buffer equation:

$$\log ([SeO_3^{-2}]/[HSeO_3^-]) = pH - 8.5 \qquad (6)$$

Below pH 7 and under mildly reducing conditions, selenites are rapidly reduced to elemental selenium [3,16,20]; however, the oxidation of Se to SeO_3^{-2} is not readily achieved [9,11].

Selenates (SeO_4^{-2}) are stable under alkaline oxidizing conditions [3,9,11,19]. Their conversion to selenite or elemental selenium in acidic environments appears to be kinetically inhibited [3,9]; the conversion of selenite to selenate is also very slow [9,11]. Selenic acid is not a strong oxidizing agent. It is reduced more rapidly by halide ions than by more powerful reducing agents; no appreciable reduction of selenate occurs with H_2S, SO_2, or Fe^{+2} [15]. Selenate is the form of selenium that has the greatest environmental impact because of its stability in alkaline pH, its solubility and its ready availability to plants [3,9,11].

Reduction potentials for some selenium species are given in Table 7.12-1.

Under moderately reducing conditions, many heavy metals are precipitated as the selenides [3,9]. (See §7.12.3 for solubility products of some heavy metal selenides.) Under oxidizing conditions, selenite and selenate anions can form soluble complexes with many of the trace metals found in natural waters [19]. Table 7.12-2 contains complexation equilibrium constants for reactions of various metal cations with selenite and selenate ligands. Seawater levels of selenium are lower than freshwater levels [20]; this is attributed to adsorption of selenite by hydrous metal oxides [9].

Selenium forms covalent bonds with both carbon and sulfur atoms [7]. For the most part, the organic compounds of selenium that are found in the aquatic environment contain selenium in the −2 oxidation state. As much as 45% of the selenium in seawater has been found to be in organic form [20]. Many of these compounds decompose in the environment to form Se^0 [3].

7.12.3 Solubility/Precipitation

The stability and very low solubility of elemental selenium (Se) in aqueous systems [3,20] indicate that it is a major inert sink for selenium in the aquatic environment;

this is important in determining the overall fate of the element [3,9]. The heavy metal selenides that precipitate under moderately reducing conditions are also extremely insoluble and can have an important effect on the cycling of selenium in soils and the aquatic environment [9]. The following are log K_{sp} values, at 25°C and zero ionic strength, for some heavy metal selenides of environmental interest [14]: -26.0(FeSe), -60.8(Cu_2Se), -48.1(CuSe), -29.4(ZnSe), -35.2(CdSe), -64.5(HgSe), and -42.1(PbSe).

TABLE 7.12-1

Reduction Potentials and Equilibrium Constants of Some Selenium Reactions
(at 25°C and zero ionic strength)

Reaction	E^0 (V)	Log K
$SeO_4^{-2} + 4H^+ + 2e^- \rightleftharpoons H_2SeO_3 + H_2O$	+1.15	38.9
$H_2SeO_3 + 4H^+ + 4e^- \rightleftharpoons Se(s) + 3H_2O$	+0.74	50
$Se(s) + 2H^+ + 2e^- \rightleftharpoons H_2Se(g)$	−0.37	−12.5
$Se(s) + 2H^+ + 2e^- \rightleftharpoons H_2Se$	−0.40	−13.5

Source: Kotrlý and Šůcha [14]

TABLE 7.12-2

Logarithms of Complexation Equilibrium Constants for Metals with Selenite and Selenate Ligands
(at 25°C and zero ionic strength, unless noted)

Metal	Log K		
	$MSeO_3$ [a]	$M(SeO_3)_2$ [b]	$MSeO_4$ [c]
Ag^+	2.4[d]	3.76[d]	—
Cd^{+2}	—	5.1[d]	2.27
Hg^{+2}	—	12.5[d]	—
Mn^{+2}	—	—	2.43
Co^{+2}	—	—	2.70
Ni^{+2}	—	—	2.67
Zn^{+2}	—	—	2.19

a. Refers to reaction $M^{+n} + SeO_3^{-2} \rightleftharpoons MSeO_3^{(n-2)}$

b. Refers to reaction $M^{+n} + 2\,SeO_3^{-2} \rightleftharpoons M(SeO_3)_2^{(n-4)}$

c. Refers to reaction $M^{+n} + SeO_4^{-2} \rightleftharpoons MSeO_4^{(n-2)}$

d. 1 *M* ionic strength

Source: Smith and Martell [21]

The solubility of hydrogen selenide gas in deoxygenated, demineralized water at a partial pressure of 1 atm H_2Se was determined to be 0.083 mol/l at 25.1°C and 0.069 mol/l at 34.7°C. Henry's constant (K_h) as a function of temperature can be obtained from the following equation [8]:

$$\ln K_h = 12.4425 - 1772.35/T \tag{7}$$

Selenite salts are generally less soluble than the corresponding selenates; most selenate salts are similar in solubility to the sulfate salts of the corresponding metals [3,9]. The very low solubilities of ferric selenite and the ferric hydroxy selenite[1] are important to the environmental cycling of selenium [3,9]. Selenites also form stable adsorption complexes with ferric oxides that are even less soluble than the ferric selenites [9,11,19]. The probability that selenite will either form insoluble compounds or adsorbates with ferric oxides or will be reduced to insoluble selenium metal minimizes the possibility for its transport in the environment [3].

Figure 7.12-1 describes the relative solubility of selenium solids at pe + pH = 5 and in equilibrium with amorphous $Fe(OH)_3$ and α-$Cu_2Fe_2O_4$. Under these conditions, $FeSe_2$ is the least soluble of the solid phases reported. Figure 7.12-2 is a similar plot for these solids at pe + pH = 10 and in equilibrium with amorphous $Fe_2(OH)_3$; Se is the least soluble phase under these conditions [19].

7.12.4 Sorption on Soils and Sediments

Limited data are available regarding the sorption of selenium on soils. The process is affected by redox conditions, pH, hydrous oxide content (particularly iron oxides), clay content, organic materials and the presence of competing anions [3,9,10,11,16,19].

Figure 7.12-3 shows Eh as a function of pH for some redox couples that may influence selenium oxidative states in soils. In soils under reducing conditions, elemental Se and heavy metal selenides may occur in the Eh-pH areas between the $HSeO_3^-$/Se and H^+/H_2 lines [11].

Experimental evidence suggests that cryptocrystalline and amorphous forms of silica, alumina and iron (III) oxide control selenium adsorption [3,9,19]. Under acidic conditions, selenite(IV) and selenate(VI) adsorb strongly on goethite, amorphous iron oxyhydroxide and gibbsite; fractional adsorption decreases with increasing pH [3,9,10,11,19]. Selenite adsorption is irreversible on goethite, partially reversible on hydrous alumina, and reversible on gibbsite [19].

Selenate may be adsorbed more strongly by clay materials than is selenite [19], but the opposite is true for hydrous oxides of iron [9,19]. Phosphate and sulfate reduce selenium adsorption on iron oxides and are effective in releasing up to 90% of soil-adsorbed selenite and selenate [19]. The ion exchange behavior of selenate is very similar to that of sulfate [16].

---

1. Faust and Aly [9] give K_s values of $2.0 \pm 1.7 \times 10^{-31}$ for $Fe_2(SeO_3)_3$ and $10^{-61.7}$ for $Fe_2(OH)_4SeO_3$.

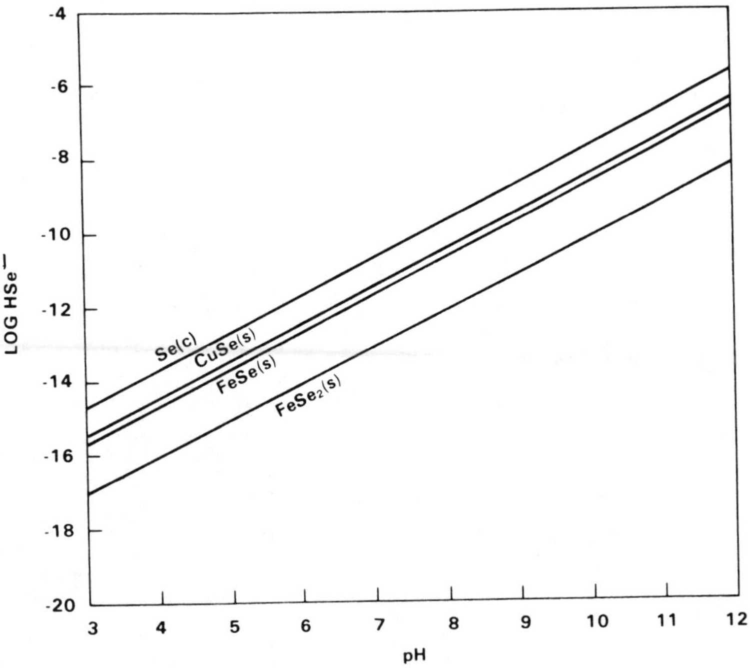

Source: Rai *et al.* [19]. *(Copyright 1984, Electric Power Research Institute. Reproduced with permission.)*

FIGURE 7.12-1 Relative Solubility of Se Solids at pe + pH = 5 and in Equilibrium with Amorphous $Fe(OH)_3$ and α-$Cu_2Fe_2O_4$

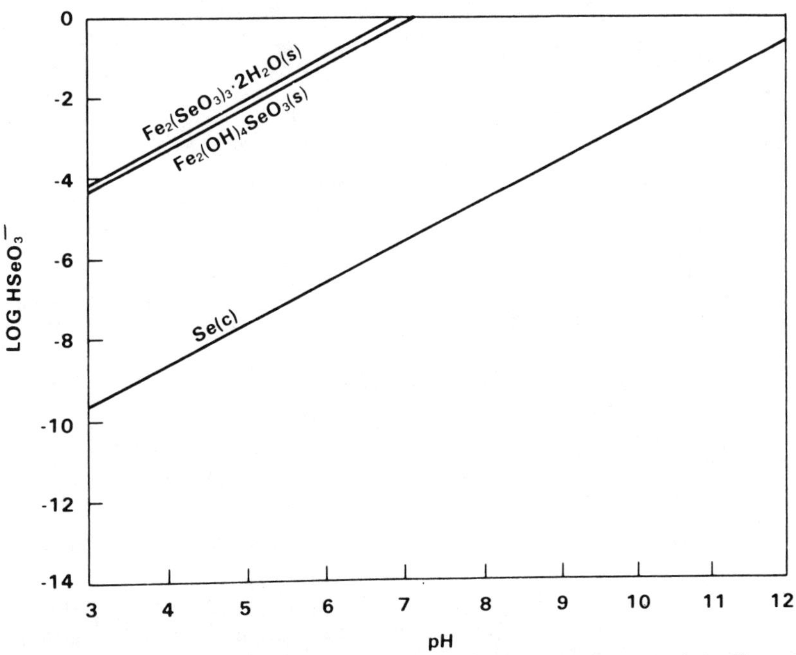

Source: Rai *et al.* [19]. *(Copyright 1984, Electric Power Research Institute. Reproduced with permission.)*

FIGURE 7.12-2 Relative Solubility of Se Solids at pe + pH = 10 and in Equilibrium with Amorphous $Fe(OH)_3$

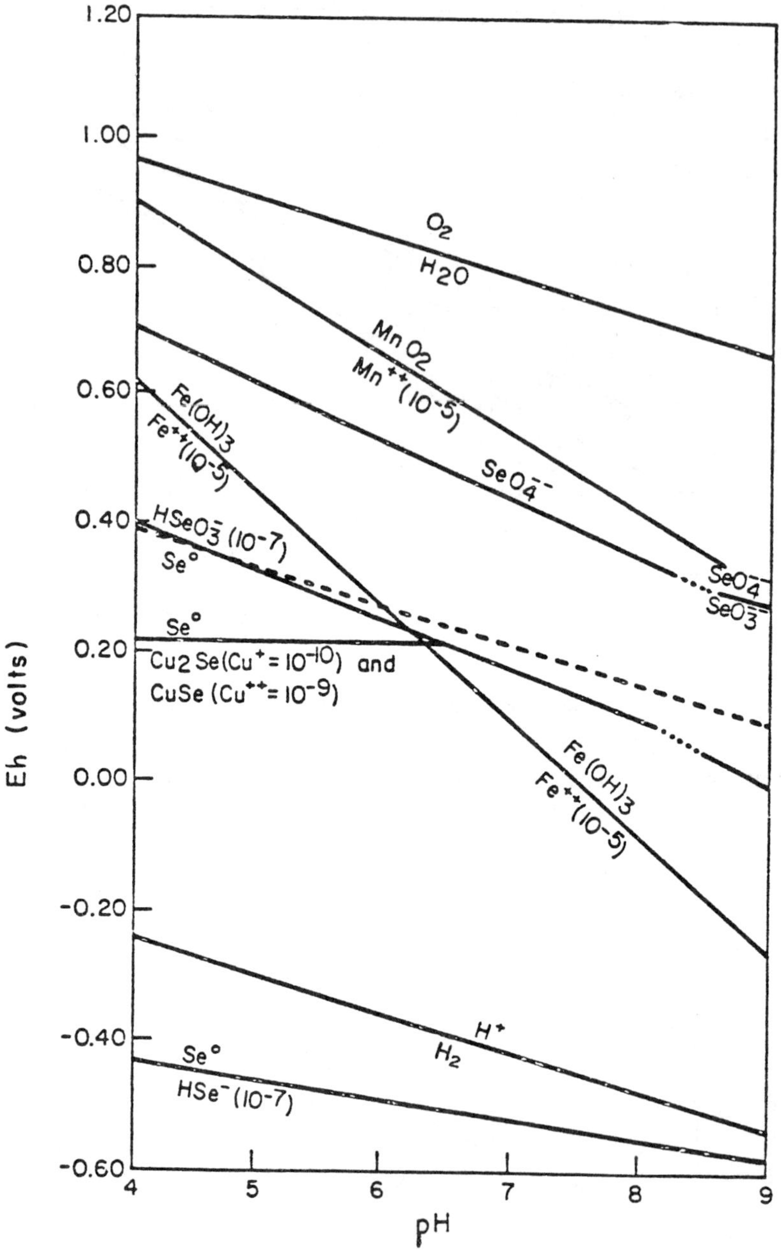

(Dashed line represents Pearsalls [18] dividing line between oxidized and reduced soils. Partial pressure of H_2 = 1 atm; partial pressure of O_2 = 0.2 atm)

Source: Geering *et al.* [12]. *(Copyright 1968, Reprinted with permission.)*

FIGURE 7.12-3 Oxidation-Reduction Potentials of Selenium Couples and Some Redox Couples That Could Affect Oxidation State of Selenium in Soils, as a Function of pH.

Arthur D. Little, Inc. [1] measured the sorption capacity of various soils for selenium using batch tests and an actual leachate containing 125 ppb of selenium(IV). Hingston *et al.* [13] also used batch tests to measure the sorption capacity of goethite for selenium(IV) at various pH values. The results are given in Table 7.12-3.

TABLE 7.12-3

Sorption Capacity of Soils for Total Selenium at Equilibrium

Soil Type	Sorption Capacity (μg Se(IV)/g soil)	Source
Alluvial Material	0.24-12	[1]
Residual Soil (Silty Sand)	0.25-5.8	[1]
Gravelly, Well-graded Silty Sand	0.3-2.4	[1]
Brown Clayey Sand	0.2-1.0	[1]
Fine Sand	0.1-0.2	[1]
Silty Fine Sand with Little Clay	0.25-3	[1]
Goethite (Iron Oxide), pH 7	6700-7100	[13]
Goethite (Iron Oxide), pH 10.5	63-1970	[13]

Both Se(IV) and Se(VI) adsorption on oxides and soils conforms to the Langmuir equation [9,19]. A multi-site Langmuir equation is needed for some soils that contain more than one adsorbing species [19]; for such soils, simultaneous adsorption equilibria are described by multiple Langmuir constants. The Langmuir constants for oxides vary significantly with pH [13]. Table 7.12-4 lists Langmuir constants for the sorption onto iron oxide, clay and soil. The adsorptivity of oxyanions (MO_4^{-2}) of the following elements generally decreases in the order [19]:

$$As > Cr \geqslant Mo \geqslant Se \approx S$$

7.12.5 Bioaccumulation

Selenium is an essential animal nutrient but is toxic at high concentrations. It enters the food chain almost entirely through plants, primarily in the form of selenates [9,10,11]. In acid soils, little selenium is available; in alkaline soils, however, plants can accumulate high levels of the element, even when the amount in the soil is small [9,10]. Some plants (e.g., locoweed) can accumulate levels of selenium that are toxic to livestock [9].

The bioaccumulation of selenium occurs in a number of aquatic organisms and has been correlated with the concentration of mercury that they contain [3,9]. Since selenium reacts with mercury to produce a non-reactive, non-toxic compound, the accumulation of moderate amounts of selenium by these organisms may be beneficial [3]. Fish accumulate selenium primarily from their diet and very little from water [3]. Further discussion of the bioconcentration of selenium in aquatic organisms is given in Chapter 4.

TABLE 7.12-4

Langmuir Constants of Selenium(IV) and Selenium(VI)

Sorbent	Oxidation State	pH	Langmuir Constant[a]
Goethite	Se(IV)	3-11	A_m = 20-120 K_L = 3.5-5.0
Clay	Se(IV)	3	A_m = 4.2, 8.34[b] K_L = 3.23, 3.45[b]
	Se(IV)	5	A_m = 2.95, 6.99[b] K_L = 3.37, 3.49[b]
	Se(IV)	7	A_m = 2.4, 5.25[b] K_L = 3.26, 3.25[b]
Soil	Se(IV)	7.6-8.1	A_m^{I} = 2.5-18 K_L^{I} = 2.0-3.59 A_m^{II} = 51.5-240 K_L^{II} = 1.23-2.24
	Se(IV)	10.5	A_m^{I} = 10[c] K_L^{I} = 2.91[c] A_m^{II} = 79.5[c] K_L^{II} = 7.08[c]
Soil	Se(VI)	7.6-8.1	A_m^{I} = 3.5-16 K_L^{I} = 3.39-4.27 A_m^{II} = 25.5-128 K_L^{II} = 1.88-3.53
	Se(VI)	10.5	A_m^{II} = 100[c] K_L^{II} = 1.94[c]

a. K_L values given = log M^{-1} (i.e., take antilog of value shown to get K_L in units of M^{-1}).
b. Two values given
c. Only value given

Source: Rai *et al.* [19]. *(Copyright 1984, Electric Power Research Institute. Reprinted with permission.)*

See Table 7.5-2 for definitions of A_m and K_L. Superscripts I and II refer to two adsorption sites.

Bioaccumulation of toxic levels of selenium has been caused by drainage of undiluted drainage water from the San Joaquin Valley in California, where irrigation of seleniferous soils concentrates selenium compounds in the underlying clay. Because of the alkaline environment peculiar to this area, the selenium is dissolved in the drainage water, taken up by algae, and bioaccumulated in fish and waterfowl that eat the algae. The drainage of this underground sump water has led to the poisoning of the Kesterton National Wildlife Refuge by selenium and possibly heavy metals [17].

7.12.6 Biotransformation

Section 2.15 (Part I) discusses the microbial transformation of selenium and other inorganic pollutants. Several species of microorganisms can reduce Se(IV) to either Se or Se^{-2}. The selenium(IV) appears to be bound to cell proteins through thiol groups attached to adjoining carbon atoms. The ability of microorganisms to reduce soluble selenium compounds to the insoluble and nontoxic elemental state could have important implications for the environmental cycling of the element [9].

Biological methylation of selenium occurs in plants, mammals and microbes, as well as in soil and the aquatic environment. Inorganic selenium undergoes biological methylation to two gaseous compounds, $(CH_3)_2Se$ (dimethyl selenide) and $(CH_3)_2Se_2$ (dimethyl diselenide) [3,9]. Challenger [4] has suggested the following conversion pathway for selenite to dimethylselenide, where methionine is the methyl-group donor:

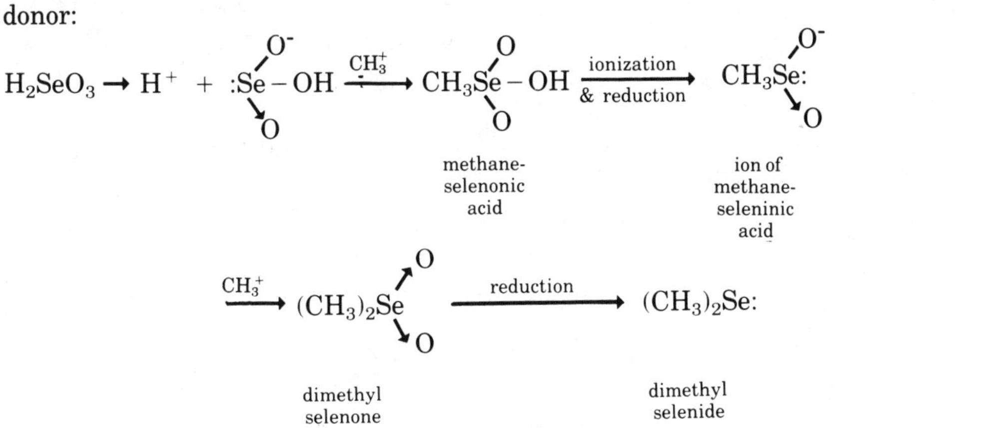

(8)

The conversion of selenium to volatile methylated products in aquatic environments proceeds under both aerobic and anaerobic conditions and is temperature-dependent; the production of volatile selenium decreases at temperatures below 20°C. Selenate, selenite and organic compounds such as selenocystine $[(SeCH_2CH(NH_2)CO_2H)_2]$, selenourea $(H_2NCSeNH_2)$ and seleno-DL-methionine $[CH_2SeCH_2CH_2(NH_2)CO_2H]$ undergo biomethylation to dimethylselenide, dimethyldiselenide and an unidentified volatile selenium compound [5]. Dimethylselenide and dimethyldiselenide, which are volatile and of low solubility, are slowly oxidized by molecular oxygen to stable, water-soluble species [20]. Dimethylselenide is about one five-hundredth as toxic as selenite, and there is no evidence of its accumulation in fish or other levels of the food chain [9].

7.12.7 Volatilization

Biomethylation, with the subsequent volatilization of dimethyl selenide and dimethyldiselenide, may be an important fate process for selenium [3]. Although this is a detoxifying process [9], selenium is released into the environment as a result. Volatile H_2Se can be formed under reducing conditions; however, as previously discussed, this compound is unstable in the aquatic environment.

7.12.8 Literature Cited

1. Arthur D. Little, Inc., *Full-Scale Evaluation of Waste Disposal from Coal-Fired Generating Plants. Appendix F (Part 4)*, EPA Contract 68-02-3167, Final Report to U.S. EPA, Office of Research and Development, Washington, D.C. (1984).
2. Baes, C.F. and R.E. Mesmer, *The Hydrolysis of Cations*, John Wiley & Sons, New York (1976).
3. Callahan, M., M. Slimak, N. Gabel, I. May, C. Fowler, J. Freed, P. Jennings, R. Durfee, F. Whitmore, B. Maestri, W. Mabey, B. Holt and C. Gould, *Water-Related Environmental Fate of 129 Priority Pollutants. Vol. 1: Introduction and Technical Background, Metals and Inorganics, Pesticides and PCBs*, EPA-440/4-79-029a, EPA Contracts 68-01-3852 and 68-01-3867, Office of Water Planning and Standards, U.S. Environmental Protection Agency, Washington, D.C. (1979).
4. Challenger, F., *Adv. Enzymol.*, **12**, 429 (1951), as cited by Faust and Aly [9].
5. Chau, Y.K., P.T.S. Wong, B.A. Silverberg, P.L. Luxon and G.A. Bengert, "Methylation of Selenium in the Aquatic Environment," *Science*, **192**, 1130-31 (1976).
6. Clayton, G.D. and F.E. Clayton, *Patty's Industrial Hygiene and Toxicology*, Vol. 2A, 3rd rev. ed., John Wiley & Sons, New York (1981).
7. Cotton, R.A. and G. Wilkinson, *Advanced Inorganic Chemistry*, 3rd ed., Interscience Publishers, New York (1972).
8. Dubeau, C., J.C. Sisi and N. Ozanne, "Solubility of Hydrogen Selenide Gas in Water," *J. Chem. Eng. Data*, **16**, 78-79 (1971).
9. Faust, S.D. and O.S. Aly, *Chemistry of Natural Waters*, Butterworth, Woburn, Mass. (1981).
10. Fergusson, J.E., *Inorganic Chemistry and the Earth*, Pergamon Press, Elmsford, N.Y. (1982).
11. Fuller, W.H., *Movement of Selected Metals, Asbestos, and Cyanide in Soil: Applications to Waste Disposal Problems*, EPA 600/2-77-020, PB 266 905, Municipal Environmental Research Laboratory, Cincinnati, Ohio (1977).
12. Geering, H.R., E.E. Cary, L.H.P. Jones and W.H. Allaway, "Solubility and Redox Criteria for the Possible Forms of Selenium in Soils," *Soil Sci. Soc. Am. Proc.*, **32**, 35-40 (1968), as cited by Fuller [11].

13. Hingston, F.J., A.M. Posner and J.P. Quirk, "Adsorption of Selenite by Goethite," in *Adsorption from Aqueous Solution*, Advances in Chemistry Series No. 79, American Chemical Society, Washington, D.C., 82-90 (1968).

14. Kotrlý, S. and L. Šůcha, *Handbook of Chemical Equilibria in Analytical Chemistry*, Halstead Press, New York (1985).

15. Latimer, W.M., *The Oxidation States of the Elements and Their Potentials in Aqueous Solution*, 2nd ed., Prentice-Hall, Inc., New York (1952).

16. Maneval, J.E., G. Klein and J. Sinkovic, *Selenium Removal from Drinking Water by Ion Exchange*, EPA/600/2-85/074, Cooperative Agreement CR-810254, Water Engineering Research Laboratory, Office of Research and Development, U.S. Environmental Protection Agency, Cincinnati, Ohio (1985).

17. Marshall, E., "Selenium Poisons Refuge, California Politics," *Science*, **229**, 144-46 (1985).

18. Pearsall, W.H., "Soil Complexes in Relation to Plant Communities, Part I: Oxidation-Reduction Potentials in Soils," *J. Ecol.*, **26**, 181-93 (1938).

19. Rai, D., J. Zachara, A. Schwab, R. Schmidt, D. Girvin and J. Rogers, *Chemical Attenuation Rates, Coefficients and Constants in Leachate Migration. Volume 1: A Critical Review*, Report EA-3356 to EPRI by Pacific Northwest Laboratories (Battelle Institute), Richland, Wash. (1984).

20. Robberecht, H. and R. Van Grieken, "Selenium in Environmental Waters: Determination, Speciation and Concentration Levels," *Talanta*, **29**, 823-44 (1982).

21. Smith, R.M. and A.E. Martell, *Critical Stability Constants. Volume 4: Inorganic Complexes*, Plenum Press, New York (1976).

22. Stumm, W. and J.J. Morgan, *Aquatic Chemistry*, 2nd ed., John Wiley & Sons, New York (1981).

7.12.9 Additional Sources of Information

Benjamin, M.M. and N.S. Bloom, "Effects of Strong Binding of Anionic Adsorbates on Adsorption of Trace Metals on Amorphous Iron Oxyhydroxide," in *Adsorption from Aqueous Solutions*, P.H. Tenari (ed.), Plenum Press, New York, 41-60 (1981).

Chapman, W.H., H.L. Fisher and M.W. Pratt, *Concentration Factors of Chemical Elements in Edible Aquatic Organisms*, Lawrence Radiation Laboratories, Livermore, Calif. (1968), as cited by Callahan *et al.* [3].

Davis, J.A. and J.O. Leckie, "Surface Ionization and Complexation at the Oxide/Water Interface. 3: Adsorption of Anions," *J. Colloid Interface Sci.*, **74**, 32-43 (1980).

Frost, R.R. and R.A. Griffin, "Effect of pH on Adsorption of Arsenic and Selenium from Landfill Leachate by Clay Minerals," *Soil Sci. Soc. Am. J.*, **41**, 53-57 (1977).

Klayman, D.L. and W.H.H. Gunther (eds.), *Organic Selenium Compounds: Their Chemistry and Biology*, Wiley-Interscience Publishers, New York (1973).

Koeman, J.H., W.H.M. Peeters, C.H.M. Koukstaad-Hol, P.S. Tjioe and J.J.M. DeGoeij, "Mercury-Selenium Correlations in Marine Mammals," *Nature*, **245**, 385-86 (1973).

Lakin, H.W., "Selenium in Our Environment," in *Trace Elements in the Environment*, E.L. Hothny (ed.), Advances in Chemistry Series 123, American Chemical Society, Washington, D.C., 96-111 (1973).

National Research Council, "Selenium," in *Medical and Biological Effects of Environmental Pollutants*, PB-251 318, National Academy of Sciences, Washington, D.C., 1-42, 59-72 (1976).

Rajan, S.S.S. and J.H. Watkinson, "Adsorption of Selenite and Phosphate on an Allophane Clay," *Soil Sci. Soc. Am. J.*, **40**, 52-54 (1976).

Rosenfeld, I. and O.A. Beath, *Selenium: Geobotany, Biochemistry, Toxicity, and Nutrition*, Academic Press, New York (1964).

Singh, M., N. Singh and P.S. Relan, "Adsorption and Desorption of Selenite and Selenate Selenium on Different Soils," *Soil Sci.*, **132**, 134-41 (1981).

7.13 SILVER (Ag)

7.13.1 Occurrence and Uses

Silver occurs in nature primarily as the sulfide (Ag_2S) or associated with other sulfides, usually those of copper, lead and arsenic. It is also found in the elemental state. Other silver minerals are cerargyrite (AgCl), proustite ($3Ag_2S \cdot As_2S_3$), pyrargyrite ($3Ag_2S \cdot Sb_2S_3$), and stephanite ($5Ag_2S \cdot Sb_2S_3$)[2]. The primary use of silver (about 40% of the total U.S. supply) is in photographic film and chemicals. Another 20% is used in electrical contacts and conductors. Other uses of the metal, its alloys and compounds include batteries, sterling ware, solder, catalysts, electroplating, jewelry, coins, medallions, dental and medical applications, and cloud seeding [10].

A number of other silver compounds are commonly found in nature or enter the environment from various sources: bromide, fluoride, iodide, cyanide, thiocyanate, oxide, hydroxide, carbonate, nitrate, sulfate, ferrocyanide complexes, and thiosulfate complexes. With the exception of the nitrate, most of the above are relatively insoluble [10].

7.13.2 Speciation Reactions in Water

ACID-BASE DISSOCIATIONS

Silver does not hydrolyze appreciably in aqueous solution. The pK_a values of silver are shown in Table 2.7-3 in Part I of this report. In addition to mono-, di- and trihydroxy complexes, Ag(I) also forms a polynuclear complex, $Ag_3(OH)_4^-$, but its concentration is very low compared with other species (Figure 7.13-1); it forms only when the pH is very high and other ligands are not present. Under typical environmental conditions, hydroxide complexes of silver occur at very low concentrations.

COMPLEXATION

Ag(I) tends to complex with ligands present in aquatic systems. The predominant inorganic species under normal environmental conditions in simple aquatic systems are Ag^+, $AgCl^0$, $AgCl_2^-$, and a variety of sulfur complexes such as AgS^-, $AgHS^0$, $AgHS_2^{-2}$, $Ag(HS)_2^-$ and $Ag(H_2S_3)^{-2}$ [9]. Figure 7.13-2 shows the chemical speciation of silver in an aquatic system that is in equilibrium with Ag_2S solid.

Ag(I) forms strong complexes with sulfide ligands. Weaker complexes are formed with the halide ligands (e.g., Cl^- and Br^-). The speciation of Ag(I) in fresh and marine waters is affected by the concentration of Cl^-. In river water, Jenne *et al.* [5] found that $AgHS^0$ exceeded Ag^+ and $AgCl^0$. In marine water, however, $AgCl_2^-$ was the dominant species, followed by AgHS.

Because other components are common in systems that contain silver, complexes of silver with cyanides and thiosulfates are frequently encountered. There is not much information on Ag-humate complexes, but overall formation constants for Ag(I) with synthetic organics (see Morel [9]) suggest that Ag-organic complexes may be important in natural systems.

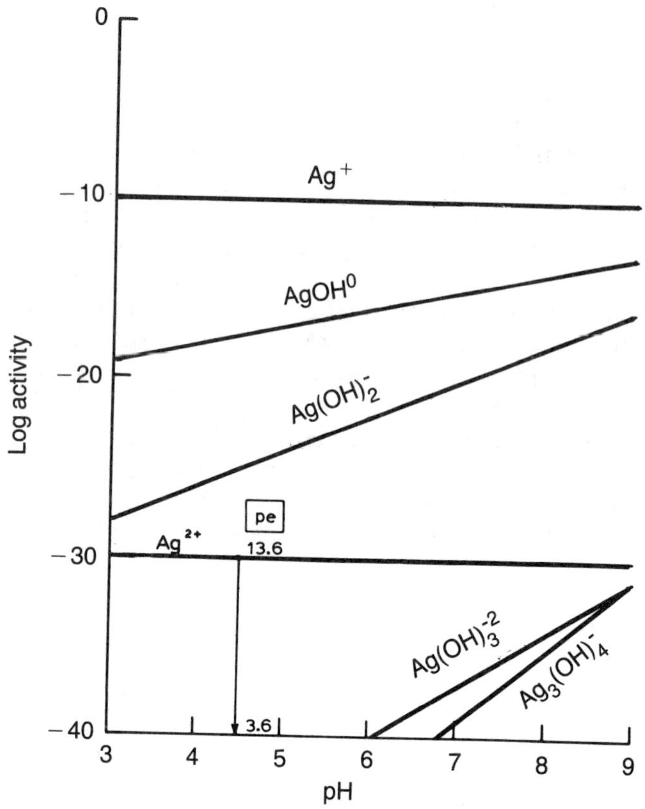

Source: Lindsay [7]. *(Copyright 1979, John Wiley & Sons. Reprinted with permission.)*

FIGURE 7.13-1 Hydrolyzed Ag(I) Species, at Varying pH, at Equilibrium with 10^{-10} *M* Ag^+

REDOX REACTIONS

Although Ag(II) and Ag(III) are known, they are unstable in the natural environment [10]. Silver can exist under typical conditions as Ag(I) or as elemental Ag(s). The equilibrium constant and standard electrode potential for the reaction: $Ag^+ + e^- \rightleftharpoons Ag(s)$, are $10^{13.51}$ and 0.799 International Volts at 25°C respectively [12]. Figure 7.13-3 shows silver speciation with varying pH and redox conditions. Metallic silver is stable over a broad range of pH and Eh and may form in nature. In more reducing conditions and the presence of sulfur, Ag_2S may form [1].

7.13.3 Solubility/Precipitation

The oxides, sulfates, carbonates, and phosphates of silver are relatively soluble; e.g., the solubility product of silver sulfate is $10^{-4.8}$ [8]. These compounds are not likely to

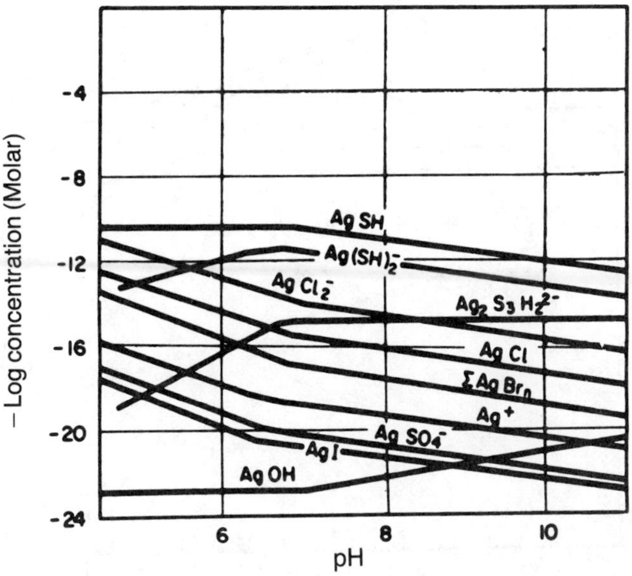

Conditions:

Temperature = 25°C

Pressure = 1 atm

No other metals present

Inorganic anion concentrations:

[S(−II)]	=	10^{-5} *M*
[Cl⁻]	=	0,6 *M*
$[SO_4^{-2}]$	=	0.01 *M*
[I⁻]	=	1×10^{-7} *M*
[Br⁻]	=	8×10^{-4} *M*

Source: Stumm and Morgan [13]. *(Copyright 1981, John Wiley & Sons. Reprinted with permission.)*

FIGURE 7.13-2 Chemical Speciation of Silver in Organic-free Aquatic System at Equilibrium with Solid Ag_2S

remain as solids or to precipitate in the natural environment. Of the common silver compounds, the sulfide has the lowest solubility product ($10^{-50.1}$) [8]. As discussed above, this compound is likely to precipitate under reducing conditions in the presence of sulfur. Under reducing conditions, Ag(s) may also form. Precipitates of silver halides may be found in systems where the halide is present in sufficiently high concentrations. Morel [9] postulates that Ag_2S, AgCl and AgBr are the dominant solid species in aquatic systems.

7.13.4 Sorption on Soils and Sediments

There is little quantitative or qualitative information on sorption of silver, but several authors [3,6] have shown that manganese dioxide has a strong affinity for silver. The order of sorption of silver onto a number of sorbents appears to be $MnO_2 > Fe(OH)_3 >$ clays $> Fe_2O_3$ [6]. However, almost all of the silver is desorbed from

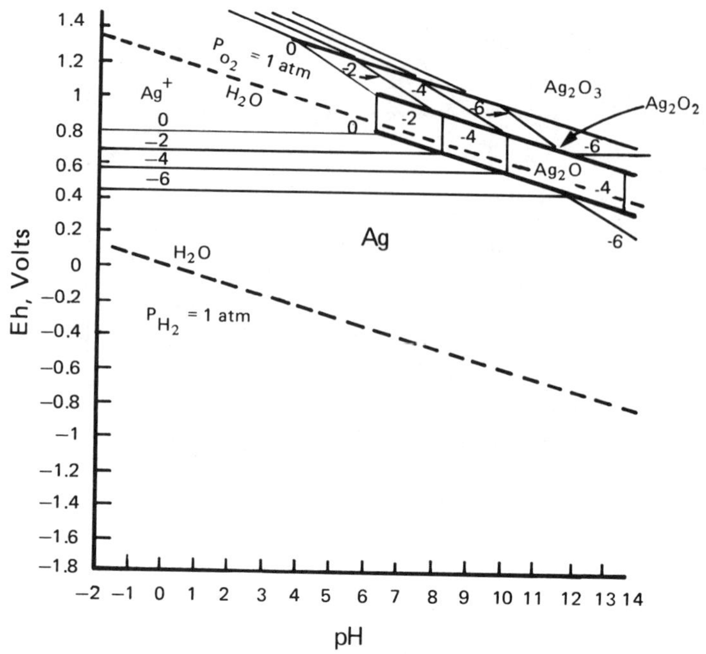

Source: Adapted from Shumilova and Zhutaeva [11]. *(Copyright 1978, Marcel Dekker. Reprinted with permission.)*

FIGURE 7.13-3 Silver Speciation in Water

MnO_2 upon contact with seawater, and significant desorption also occurs with the other sorbents; this may reflect the strong tendency of silver to be mobilized through complexation. Organic matter also appears to sorb silver [4].

7.13.5 Literature Cited

1. Brookins, D.G., "Eh-pH Diagrams for Elements for Z = 40 to Z = 42. Application to the Oklo Natural Reactor, Gabon," *Chem. Geol.*, **23** (1978).
2. Callahan, M., M. Slimak, N. Gabel, I. May, C. Fowler, J. Freed, P. Jennings, R. Durfee, F. Whitmore, B. Maestri, W. Mabey, B. Holt and C. Gould, *Water-Related Environmental Fate of 129 Priority Pollutants. Volume I: Introduction and Technical Background, Metals and Inorganics, Pesticides and PCBs*, EPA-440/4-79-029a, EPA Contracts 68-01-3852 and 68-01-3867, Office of Water Planning and Standards, U.S. Environmental Protection Agency, Washington, D.C. (1979).
3. Chao, T.T. and B.J. Anderson, "The Scavenging of Silver by Manganese and Iron Oxides in Stream Sediments Collected from Two Drainage Areas of Colorado," *Chem. Geol.*, **14,** 159-66 (1974).

4. Freeman, R.A., *The Ecological Kinetics of Silver in an Alpine Lake Ecosystem* (preprint), 2nd ASTM Symposium on Aquatic Toxicology, Cleveland, Ohio (1977).
5. Jenne, E.A., D.C. Girvin, J.W. Ball and J.M. Burchard, "Inorganic Speciation of Silver in Natural Waters — Fresh to Marine," in *Environmental Impacts of Artificial Ice Nucleating Agents,* D.A. Klein (ed.), Dowden, Hutchinson and Ross, Stroudsburg, Pa. (1978), pp. 41-61.
6. Kharkar, D.P., K. Turekian and K. Bertine, "Stream Supply of Dissolved Ag, Mo, Sb, Se, Cr, Co, Rb, and Cs to the Oceans," *Geochim. Cosmochim. Acta,* **32,** 285-98 (1968).
7. Lindsay, W.L., *Chemical Equilibria in Soils,* John Wiley & Sons, New York (1979).
8. Martell, A.E. and R.M. Smith, *Critical Stability Constants,* Vol. 4, *Inorganic Complexes,* Plenum Press, New York (1976).
9. Morel, F.M.M., *Principles of Aquatic Chemistry,* John Wiley & Sons, New York (1983).
10. Scow, K., M. Goyer, L. Nelken, E. Payne, K. Saterson, P. Walker and M. Wood, *An Exposure and Risk Assessment for Silver,* EPA Contracts 68-01-3857 and 68-01-5949, Final Draft Report to Monitoring and Data Support Division, Office of Water Regulations and Standards, U.S. Environmental Protection Agency, Washington, D.C. (1980).
11. Shumilova, N.A. and G.V. Zhutaeva, "Silver" in *Encyclopedia of Electrochemistry of the Elements,* Vol. VIII, A.J. Bard (ed.), Marcel Dekker, Inc., New York (1978).
12. Sillén, L.G. and A.E. Martell, *Stability Constants of Metal-Ion Complexes,* Special Publication No. 17, The Chemical Society, London (1964).
13. Stumm, W. and J.J. Morgan, *Aquatic Chemistry,* 2nd ed., John Wiley & Sons, New York (1981).

7.13.6 Additional Sources of Information

Boyle, R.W., *The Geochemistry of Silver and Its Deposits,* Bulletin 160, Geological Survey of Canada, Ottawa (1968).

Carson, B.L. and I.C. Smith, *Silver and Appraisal of Environmental Exposure,* National Institute of Environmental Health Sciences, Tech. Rept. No. 3, Washington, D.C. (1975).

Lindsay, W.L. and M. Sadiq, "Theoretical Solubility Relationships of Silver in Soils," in *Environmental Impacts of Artificial Ice Nucleating Agents,* D.A. Klein (ed.), Dowden, Hutchinson and Ross, Stroudsburg, Pa. (1978), pp. 25-40.

Thompson, N.R., *Silver: Comprehensive Inorganic Chemistry,* Pergamon Press, New York (1973).

7.14 THALLIUM (Tl)

7.14.1 Occurrence and Uses

Thallium is introduced into the environment principally as waste from the production of other metals, in whose ores it is found as an impurity [2]. The greatest amounts are found in potash minerals that have no commercial significance [8]. Thallium occurs at concentrations of 6-16% in crookesite, $(Cu,Tl,Ag)_2Se$, lorandite, $TlAsS_2$, hutchinsonite, $4(Tl,Pb)S \cdot Ag_2S \cdot 5As_2S_3$, and vrbaite, $Tl(As,Sb)_3S_5$; however, these mineral deposits are too small to be of commercial value at the present time [3,8].

Thallium is obtained commercially as a by-product from the roasting of pyrite ores (FeS_2), from the smelting of lead and zinc, or as a by-product of cadmium production [3,8]. Since 1972, when it was banned as a rodenticide, only a few thousand pounds a year are used in the United States [3]. Thallium metal is used in low-melting alloys and optical glasses; its compounds are used in lenses, plates, and prisms in optical and X-ray systems, in photoelectric cells, and in chemical synthesis.

7.14.2 Speciation Reactions in Water

ACID-BASE DISSOCIATION

Under environmental conditions, thallium occurs in the +1 and +3 oxidation states. The Tl^+ ion does not hydrolyze appreciably under normal environmental conditions [1,8]. The Tl^{+3} ion, which is found in very oxidizing waters, is a relatively strong acid in aqueous solution [8] and will hydrolyze in quite acidic solution to form $Tl(OH)_3$ and the colloidal oxide Tl_2O_3 [4]. The neutral species $Tl(OH)_3$ exhibits a wide range of pH stability [1].

Figure 7.14-1 shows how pH affects the percentage of various hydrolyzed Tl(III) species when no complexing agents are present. Table 2.7-3 in Part I lists pK_a values for Tl^{+3}.

COMPLEXATION

The Tl^+ ion is similar in its chemistry to both the alkali metal ions and the Ag^+ ion [1,8]. The similarity to the alkali metals can be attributed to the fact that these cations have similar atomic radii and the same valence [2,6].

Like the alkali metal hydroxides, TlOH is soluble in aqueous solutions and gives strongly basic solutions [4,8]. Tl^+ is also similar to the alkali metal cations, in that it forms relatively few strong complexes [4]; for example, stability constants with inorganic anions are typically 0 to 1 log unit larger than those with K^+ ion.

Thallium(I) resembles silver, in that it forms an insoluble sulfide and weak halide complexes that are light-sensitive and sparingly soluble [1,8].

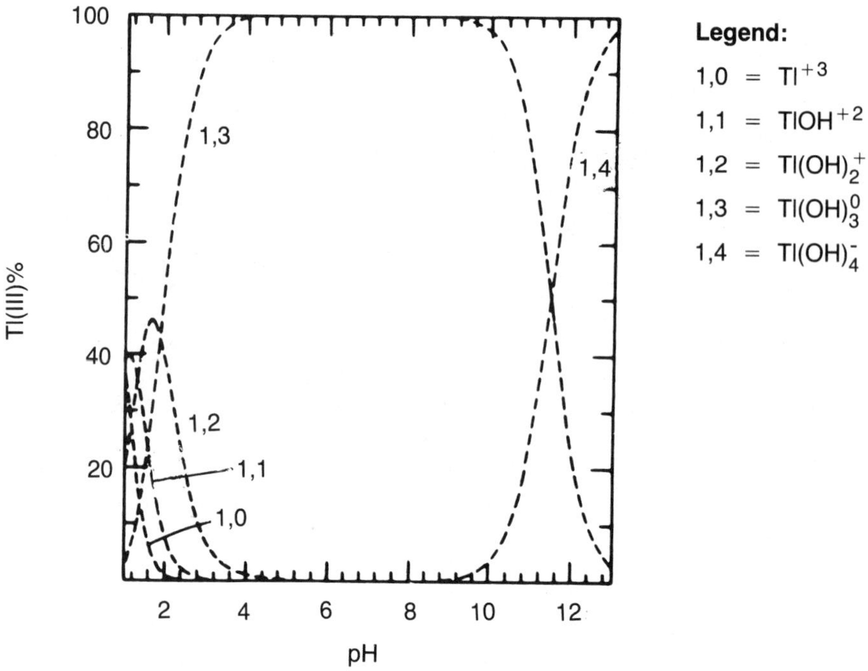

I = 1 *M*, T = 25°C, 10^{-5} *M* Tl(III)

(Dashed curves denote regions supersaturated with respect to Tl_2O_3)

Source: Baes and Mesmer [1]. *(Copyright 1976, John Wiley & Sons. Reprinted with permission.)*

FIGURE 7.14-1 Hydrolyzed Tl(III) Species at Varying pH

The Tl^{+3} cation, which can occur in aerobic waters, forms stable halide complexes and will form mixed chloride-hydroxide complexes over a narrow pH range [1]. Although this ion is easily reduced in aqueous solution (see REDOX below), the higher oxidation state is stabilized by the presence of chloride [4] and thus should be more stable in seawater than fresh water. Figure 7.14-2 is a predominance diagram for $Tl^{+3} - OH^- - Cl^-$ species at varying concentrations of Cl^- with no other complexing agents present. This figure predicts that $TlCl_4^-$ is the predominant form of Tl(III) in seawater (where log [Cl] = 0.25) [13].

At the Eh and pH concentrations normally found in natural waters, Tl^+ should be the predominant species in both fresh water and seawater [2]. In very oxidizing waters, however, Tl^{+3} may be present with $Tl(OH)_3^0$ as the predominant species in fresh water, and $Tl(OH)_3^0$ and $TlCl_4^-$ may predominate in seawater [1,2,4].

Increases in pH decrease thallium-inorganic interactions but cause increased thallium-humic acid interactions. Therefore, thallium-organic interactions may be important in natural waters [2]. Information on the speciation of thallium in natural waters is sparse.

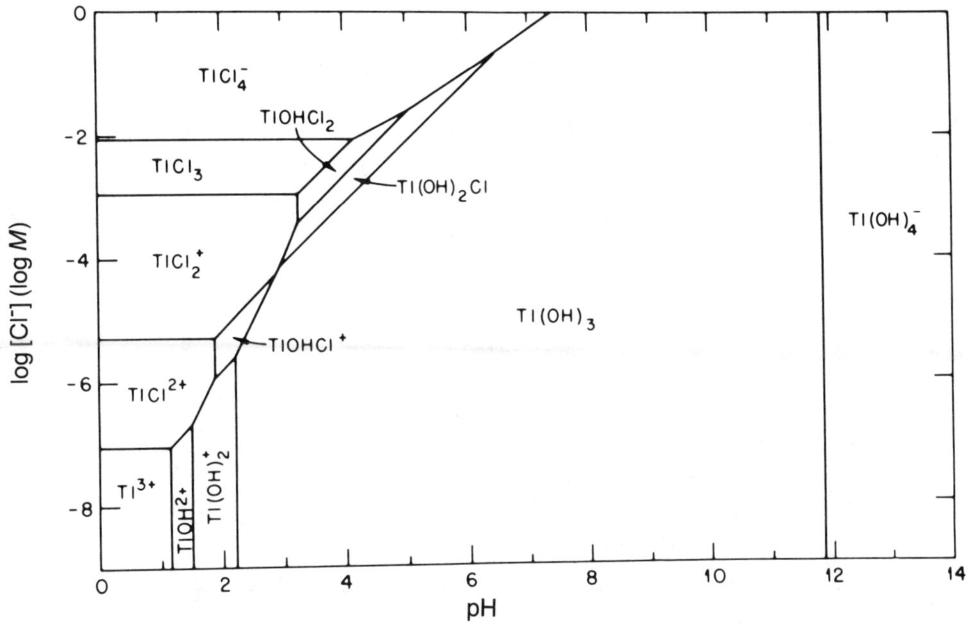

(Lines represent equal concentrations of species in adjacent fields.)

Source: Baes and Mesmer [1]. *(Copyright 1976, John Wiley & Sons. Reprinted with permission.)*

FIGURE 7.14-2 Calculated Predominance Diagram for the $Tl^{+3} - OH^- - Cl^-$ Species at 25°C and Ionic Strength of 3 *M* $NaClO_4$

Table 7.14-1 lists complexation equilibrium constants and stability constants for thallium I and III with commonly occurring ligands. Table 7.14-2 lists complexation equilibrium constants for thallium I and III with some sequestering agents of commercial interest. Section 2.9 (Part I) contains a discussion of the formation of metal complexes.

TABLE 7.14-1

Logarithms of Complexation Equilibrium Constants (K) and Overall Stability Constants (β_2 and β_3) for Thallium (I) and (III) with Commonly Occurring Ligands
(at 25°C and zero ionic strength, unless noted)

Metal	MCl^a	$MCl_2{}^b$	$MCl_3{}^c$	MSO_4	$M(OH)_2{}^a$	$M(OH)_2{}^b$	$M(OH)_3{}^c$
Tl^+	0.49	0.0	—	1.37	0.79	-0.8^d	—
Tl^{+3}	7.72	13.48	16.5	2.27^d	13.4	26.4	38.7

Source: Kotrlý and Šůcha [7]

a. Refers to reaction $M + L \rightleftharpoons ML$ where L = Cl(or OH), K

b. Refers to reaction $M + 2L \rightleftharpoons ML_2$ where L = Cl(or OH), β_2

c. Refers to reaction $M + 3L \rightleftharpoons ML_3$ where L = Cl(or OH), β_3

d. 3 *M* ionic strength

TABLE 7.14-2

Logarithms of Complexation Equilibrium Constants (K) for Thallium I and III with Sequestering Agents of Commercial Interest
(1:1 complex formation at 25°C and 0.10 *M* ionic strength, unless noted)

	Sequestering Agent[a]		
Metal	NTA	EDTA	CIT
Tl^{+}	4.75[b]	6.54[c]	1.04
Tl^{+3}	20.9[c]	35.3	—

Source: Kotrlý and Šůcha [7]

a. NTA = $N(CH_2COOH)_3$
EDTA = $(HOOCCH_2)_2NCH_2CH_2N(CH_2COOH)_2$
CIT = $(HOOCCH_2)(OH)(COOH)C(CH_2COOH)$

b. 20°C

c. Zero ionic strength

REDOX REACTIONS

In very oxidizing waters, thallium can exist as Tl^{+3}. This cation is easily reduced to Tl^{+} via the reaction $Tl^{+3} + 2e \rightleftharpoons Tl^{+}$ ($E^0 = +1.25V$; $K_0 = 42.6$ at 25°C) [6,12]. This reaction is affected by pH and the presence of complexing ions. The presence of Cl^- promotes the formation of complexes that stabilize Tl^{+3}, thereby decreasing the potential ($E = +0.77V$ in 1 *M* HCl and $+1.26V$ in 1 *M* $HClO_4$[4]).

In reducing environments, thallium exists as Tl(I) or may be precipitated as the metal: $Tl^{+} + e \rightleftharpoons Tl(s)$ ($E^o = 0.33V$; $K_o = -5.64$ at 25°C [12]).

7.14.3 Solubility/Precipitation

Thallium(I) is much more soluble than other environmentally important heavy metals [2]. Tl_2O and TlOH are readily soluble in water, and salts such as the sulfate, nitrate, acetate, and carbonate are moderately so [4,6]. The halides, with the exception of the very soluble TlF, are sparingly soluble [4]. The solubility products (at $I=0$ and 25°C) for Tl(I) halides are: $10^{-3.74}$ (TlCl), $10^{-5.44}$ (TlBr) and $10^{-7.23}$ (TlI) [7]. For Tl_2S, the solubility product is $10^{-21.2}$ [7]. In reducing environments, precipitation of Tl_2S may be an important mechanism for the removal of thallium from solution [2].

Thallium(III) forms a relatively insoluble hydroxide (K_{sp} of $Tl(OH)_3 = 10^{-45.2}$) but a very soluble chloride [6,7]. Precipitation of $Tl(OH)_3$, depending on the relative kinetics of reduction compared with hydrolysis, may be an effective mechanism for removing thallium from solution in oxidizing waters by means of the following sequence: thallium(III) precipitates as the oxide or hydroxide and settles to bed sediments, where it is reduced to Tl(I); the latter then reacts with sulfide to form insoluble Tl_2S [2].

7.14.4 Sorption on Soils and Sediments

Thallium is strongly adsorbed by montmorillonite clays. Magorian *et al.* [9] report that a 1-g/l suspension of the montmorillonite clay, hectorite, removed 97% of a 100 μg/l concentration of thallium(I) at pH 8.1; adsorption was less effective at pH 4.0. Thallium is not adsorbed as strongly as Cu or Zn ions under similar conditions [2].

A very strong enrichment of thallium, compared with its crustal abundance, was found in manganese nodules in the Pacific Ocean [11]. An affinity for hydrous metal oxides has been suggested for thallium [2].

Thallium sorption has not been extensively studied. However, available data indicate that sediments actively adsorb thallium and provide an effective sink for this ion in aquatic environments [2].

7.14.5 Bioaccumulation

In non-reducing waters, adsorption and bioaccumulation are the primary mechanisms for removal of thallium from solution [2]. Bioaccumulation is facilitated by the solubility of Tl(I) in most aquatic systems, which makes it readily available to aquatic organisms. Moreover, Tl(I) can replace potassium in various enzymes and thus inhibit enzyme activity [3,4,8]: in some experiments, Tl(I) had an affinity 10 times greater than potassium for the site activated by potassium [8].

The alga *Ulothrix sp.* was able to concentrate thallium by a factor of 127 to 220 within one hour; in comparison, the bioconcentration factors[1] for 2.7 hours were 114 for lead, 30 for cadmium, 80 for zinc, and 313 for copper [9]. In a study on the effect of thallium and other heavy metals on photosynthesis in freshwater algae, ca. 1.5×10^{-5} *M* thallium was found to decrease photosynthesis by 50%. It also appeared to be irreversibly bound to the algae, which would indicate that it is available not only for bioaccumulation but for food chain magnification as well [10].

7.14.6 Biotransformation

Biomethylation of thallium(I) has been observed under laboratory conditions in anaerobic, mixed bacterial cultures [5]. Tl^+ undergoes oxidative methylation — i.e., its oxidation number increases during biomethylation. Me_2Tl^+ was the only methylated thallium species detected, probably because: (1) $MeTlX_2$ compounds (where X is an anion such as chloride or acetate) are generally unstable or tend to decompose in aqueous solution [5,8] and (2) Me_3Tl decomposes instantly in water to form stable Me_2Tl^+ and methane [5]. Me_2Tl^+ compounds are less toxic to bacteria than Tl^+ [5]. Biomethylation of thallium has not yet been verified under natural conditions.

1. The bioconcentration factor relates the concentration of an element in aquatic animals to the concentration in the water in which they live [14]. See Chapter 4, Part I for further information.

7.14.7 Literature Cited

1. Baes, C.F. and R.E. Mesmer, *The Hydrolysis of Cations*, John Wiley & Sons, New York (1976).
2. Callahan, M., M. Slimak, N. Gabel, I. May, C. Fowler, J. Freed, P. Jennings, R. Durfee, F. Whitmore, B. Maestri, W. Mabey, B. Holt and C. Gould, *Water-Related Environmental Fate of 129 Priority Pollutants. Volume I: Introduction and Technical Background, Metals and Inorganics, Pesticides and PCBs*, EPA-440/4-79-029a, EPA Contracts 68-01-3852 and 68-01-3867, Office of Water Planning and Standards, U.S. Environmental Protection Agency, Washington, D.C. (1979).
3. Clayton, G.D. and F.E. Clayton, *Patty's Industrial Hygiene and Toxicology*, 3rd rev. ed., Vol. 2A, John Wiley & Sons, New York (1981).
4. Cotton, R.A. and G. Wilkinson, *Advanced Inorganic Chemistry*, 3rd ed., Interscience Publishers, New York (1972).
5. Huber, F., U. Schmidt and H. Kirchmann, "Aqueous Chemistry of Organolead and Organothallium Compounds in the Presence of Microorganisms," in *Organometals and Organometalloids*, F. Brinckman and J. Bellama(eds.), American Chemical Society, Washington, D.C. (1978).
6. Korenman, I.M., *Analytical Chemistry of Thallium*, Israel Program for Scientific Translations, Jerusalem (1963).
7. Kotrlý, S. and L. Šůcha, *Handbook of Chemical Equilibria in Analytical Chemistry*, Halsted Press, New York (1985).
8. Lee, A.G., *The Chemistry of Thallium*, Elsevier Publishing Company, New York (1971).
9. Magorian, T.R., K.G. Wood, J.G. Michalovic, S.L. Pek and M.M. Van Lier, "Water Pollution by Thallium and Related Metals," NTIS PB 253 333, Springfield, Va. (1974).
10. Overnell, J., "The Effect of Some Heavy Metal Ions on Photosynthesis in a Freshwater Alga," *Pestic. Biochem. Physiol.*, **5**, 19-26 (1975).
11. Salomons, W. and U. Förstner, *Metals in the Hydrocycle*, Springer-Verlag, New York (1984).
12. Sillén, L.G. and A.E. Martell, *Stability Constants of Metal-Ion Complexes*, Special Publication No. 17, The Chemical Society, Burlington House, London (1964).
13. Stumm, W. and J.J. Morgan, *Aquatic Chemistry*, 2nd ed., John Wiley & Sons, New York (1981).
14. U.S. Environmental Protection Agency, *Ambient Water Quality Criteria for Thallium*, Report No. EPA-440/5-80-074 (October 1980).

7.14.8 Other Sources of Information

Britten, J.S. and M. Blank, *Biochem. Biophys. Acta*, **159**, 160 (1968).

Gehring, P.J. and P.B. Hammond, *J. Pharmacol. Exp. Ther.*, **155**, 187 (1967).

Inturrisi, C.E., *Biochem. Biophys. Acta*, **178**, 630 (1969).

Mathis, B.J. and N.R. Kevern, "Distribution of Mercury, Cadmium, Lead, and Thallium in an Eutrophic Lake," *Hydrobiologia,* **46**, 207-22 (1975).

Zitco, V., "Toxicity and Pollution Potential of Thallium," *Sci. Total Environ.*, **4**, 185 (1975).

7.15 ZINC (Zn)

7.15.1 Occurrence and Uses

Zinc is a member of the Group IIB triad (Zn, Cd, Hg) in the periodic table. It occurs in nature primarily as the sulfide, carbonate, silicate and oxide. The metal is readily extracted from numerous minerals, principally sphalerite (ZnS) and smithsonite ($ZnCO_3$). The most important uses of the metal and its alloys are in galvanizing, diecasting and the manufacture of brass. Zinc oxide, zinc sulfide and other compounds are used in plastics, paints, cosmetics and agricultural applications [14,19].

7.15.2 Speciation Reactions in Water

ACID-BASE DISSOCIATION

Zinc hydrolyzes negligibly below pH 6, but significant amounts of aqueous hydrolysis products can form above pH 7. The pK values of zinc aquo species are found in Table 2.7-3, Part I. Figure 7.15-1 shows the distribution of zinc hydrolysis products at different pH values, based on acid dissociation constants at zero ionic strength (see section 2.7). Very small amounts of $Zn(OH)^+$ are formed, because pK_2 is much smaller than pK_1.

COMPLEXATION

Figure 7.15-2 depicts its chemical speciation in freshwater and seawater systems that contain no other trace elements or organic substances. Tables 2.9-1 and -2 (Part I) list complexation equilibrium constants for a variety of inorganic and organic ligand complexes of zinc. At pH = 4-7, the predominant species are Zn^{+2} in fresh water and Zn^{+2} and $ZnCl^+$ in seawater. At pH > 7.5, $Zn(OH)_2^0$ is the principal species in both environments.

The concentration levels of organic ligands that are typically found in aqueous systems do not markedly affect the speciation of zinc [12]. In polluted areas, however, organic materials can affect the chemical form in which zinc is present [1].

Overall conditional stability constants (see §7.7.2) of zinc complexes with humic materials were determined at pH 8 and 20°C. The range of log K_0 for humic material from freshwater lakes and rivers was 5.05-5.41. For humic material from seawater, log K_0 ranged from 5.27 to 5.31 [10].

Complexation of zinc by ligands proceeds rapidly: the rate constants are $\sim 5 \times 10^7\ s^{-1}$ [20]. The characteristic second-order rate constant for exchange of one water molecule in the primary solvation shell of Zn(II) ion is $5 \times 10^6\ M^{-1}\ s^{-1}$ [12]. The calculated half-life for 1 ppm of Zn^{+2} reacting with 50 ppm of NH_3 is on the order of 10^{-5} second.

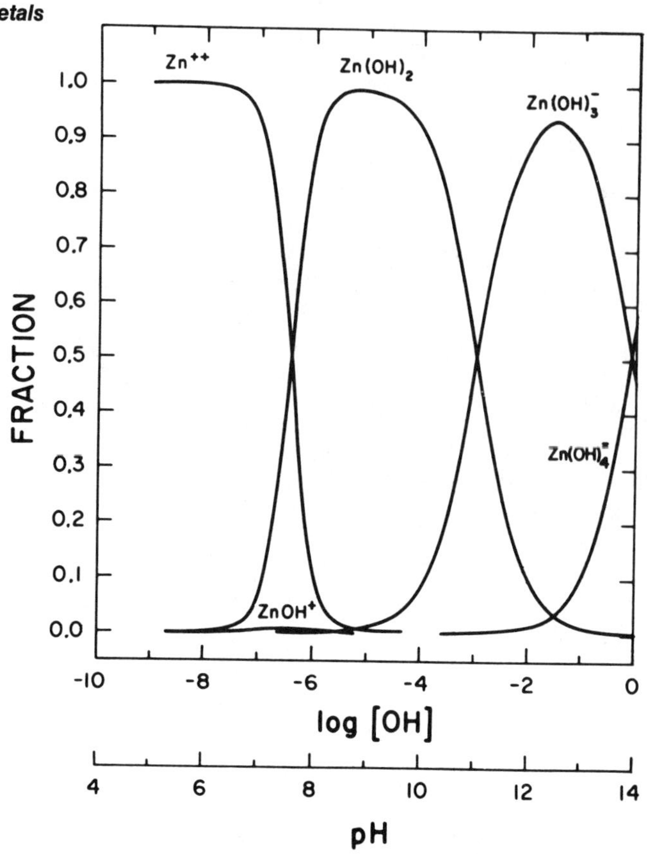

Source: Hahne and Kroontje [4]. *(Copyright 1973, American Society of Agronomy, Inc. Reprinted with permission.)*

FIGURE 7.15-1 Distribution of Molecular and Ionic Species of Divalent Zinc at Different pH Values and I = 0

REDOX REACTIONS

The only oxidation state of importance for zinc under environmental conditions is +2. However, the speciation of ligands that react with Zn^{+2} (e.g., sulfur) can be affected by Eh [1]. Figure 7.15-3 shows the fields of stability of solids and dissolved zinc species in the zinc-sulfur-carbon dioxide-water system as a function of Eh and pH. Three solid phases are possible in this system: the sulfide, carbonate, and hydroxide of zinc. In aerated water, the latter two are stable [16]. Precipitation of the sulfide is an important control on the mobility of zinc in reducing environments [19].

7.15.3 Solubility/Precipitation

Constants for solubility equilibria of naturally occurring zinc-containing solids are given in Table 7.15-1. Figure 7.15-4 illustrates the solubility of several of these minerals in water of varying pH.[1] The Zn^{+2} ion is the only form of dissolved zinc

1. See §7.5.3 for an explanation of this type of solubility graph.

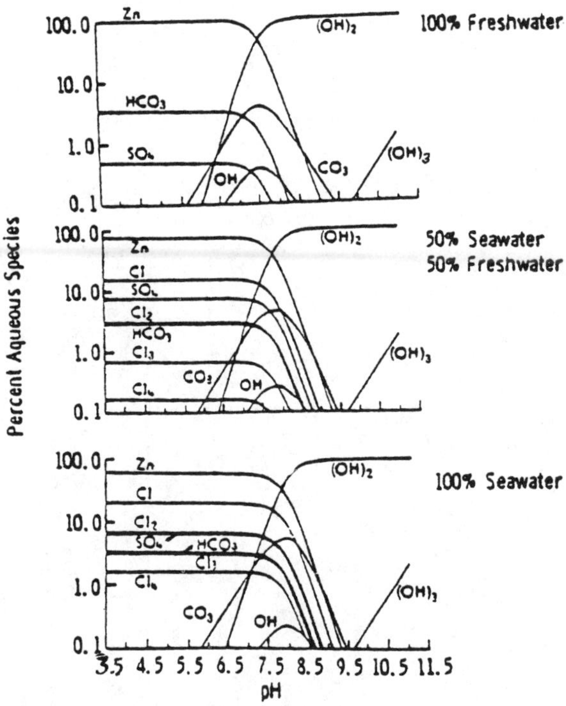

Legend:

Zn = Zn^{+2}

Cl = $ZnCl^{+}$

Cl_2 = $ZnCl_2$

Cl_3 = $ZnCl_3^-$

Cl_4 = $ZnCl_4^{-2}$

OH = $ZnOH^+$

$(OH)_2$ = $Zn(OH)_2$

$(OH)_3$ = $Zn(OH)_3^-$

HCO_3 = $ZnHCO_3^+$

CO_3 = $ZnCO_3$

SO_4 = $ZnSO_4$

Molal Concentrations of Chemical End Members Used in the Calculations

Component	Freshwater	Seawater
Zinc (total)	<10 ppm	<10 ppm
Calcium	0.000375	0.0104
Magnesium	0.000168	0.054
Sodium	0.000274	0.4752
Potassium	0.000059	0.01
Chloride	0.00022	0.554
Alkalinity	0.000955	0.00234
Sulfate	0.000955	0.0284
Ionic strength[a]	0.0021	0.661

a. Effective molal ionic strength (after computation of species distribution).

Source: Long and Angino [9]. *(Copyright 1977, Pergamon Press, Inc. Reprinted with permission.)*

FIGURE 7.15-2 Chemical Speciation of Zinc(II) in Seawater and Fresh Water

considered in this figure. (See also Figure 7.15-3 for fields of stability of solid and dissolved zinc species in water.) The $ZnFe_2O_4$ line is plotted for various conditions, depending on which minerals control iron levels (amorphous-Fe, soil-Fe, maghemite, or goethite). The Zn_2SiO_4 line is plotted with soil-silica, amorphous silica, or quartz as the mineral controlling silica levels, and $ZnCO_3$ is plotted at two CO_2(g) levels.

The solubility of zinc in natural waters is strongly affected by pH, redox potential (Figure 7.15-3), temperature, and the presence of complexing ligands, competing ions and precipitating reagents. As its species formed with the common ligands in surface waters are soluble in neutral and acidic solution, in the absence of strong sorbing material, zinc is readily transported in most natural waters and is one of the most mobile of the heavy metals [19].

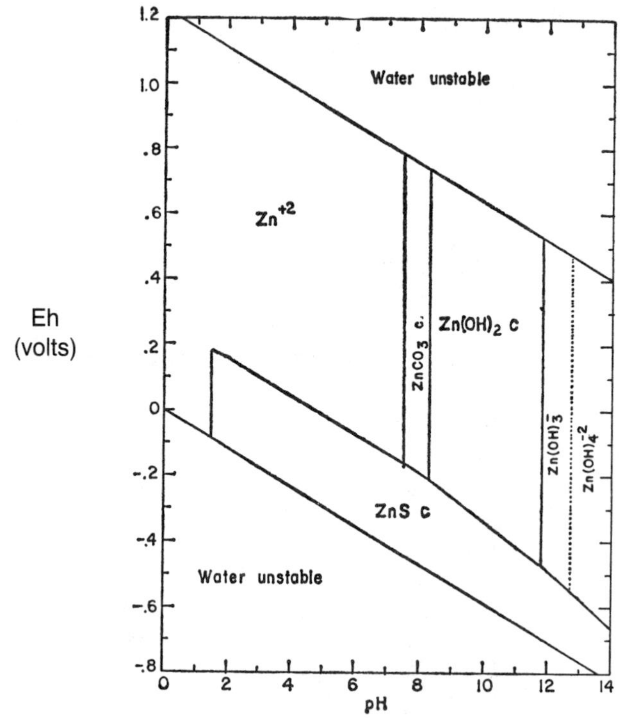

Concentrations: Zn = 10^{-5} *M*

Dissolved CO_2 = 10^{-3} *M*

Dissolved S Species = 10^{-3} *M*

Source: Hem [5]. *(Copyright 1972, Pergamon Press, Inc. Reprinted with permission.)*

FIGURE 7.15-3 Fields of Stability of Solid and Dissolved Zinc Species in the System Zn + CO_2 + S + H_2O at 25°C and 1 atm in Water

TABLE 7.15-1

Constants for Solubility Equilibria for Common Zinc Ores and Solids
(at 25°C and zero ionic strength unless noted)

Ore/Solid	Chemical Formula and Reaction	Log K	Source
Zincite	$ZnO(s) + 2H^+ \rightleftharpoons Zn^{+2} + H_2O$	11.14	[18]
	$ZnO(s) + H_2O \rightleftharpoons Zn^{+2} + 2\,OH^-$	−17	[17]
Zinc Hydroxide	$Zn(OH)_2(amorph) + 2H^+ \rightleftharpoons Zn^{+2} + 2H_2O$	12.45	[18]
	$Zn(OH)_2(s) \rightleftharpoons Zn^{+2} + 2OH^-$	−15.47[a]	[17]
Smithsonite	$ZnCO_3(s) + 2H^+ \rightleftharpoons Zn^{+2} + CO_2(g) + H_2O$	7.95	[18]
	$ZnCO_3(s) \rightleftharpoons Zn^{+2} + CO_3^{-2}$	−10.7[a]	[2]
Sphalerite	$ZnS(s) \rightleftharpoons Zn^{+2} + S^{-2}$	−24.7	[18]
Willemite	$Zn_2SiO_4 + 4H^+ \rightleftharpoons 2Zn^{+2} + H_4SiO_4$	13.15[a]	[8]
Franklinite	$ZnFe_2O_4 + 8H^+ \rightleftharpoons Zn^{+2} + 2Fe^{+3} + 4H_2O$	9.85[a]	[8]
Zincosite	$ZnSO_4 \rightleftharpoons Zn^{+2} + SO_4^{-2}$	3.41[a]	[8]
Hopeite	$Zn_3(PO_4)_2 + 4H_2O + 4H^+ \rightleftharpoons 3Zn^{+2} + 2H_2PO_4^- + 4H_2O$	3.80[a]	[8]
Hydrozincite	$Zn_5(OH)_6(CO_3)_2 \rightleftharpoons 5Zn^{+2} + 2CO_3^{-2} + 6\,OH^-$	−68.2[b]	[18]
	$Zn_5(OH)_6(CO_3)_2 + 10H^+ \rightleftharpoons 5Zn^{+2} + 2CO_2(g) + 8H_2O$	9.8	[18]

a. Ionic strength of system not specified.

b. I = 0.1 *M*

The formation of soluble hydroxy complexes of zinc has a great impact on the solubility of $Zn(OH)_2(s)$. If its solubility is calculated without including the formation of hydroxy or other complexes, a value of 0.28 ppm of Zn^{+2} is obtained ($\log K_{sp} = -15.5$, pH = 8.93, I = 0). Consideration of the magnitude of the formation constants for $ZnOH^+$ and $Zn(OH)_2^0$ shows that the reaction $Zn(OH)_2(s) \rightleftharpoons Zn(OH)_2^0$ would have an equilibrium constant of 0.00245 ($\log K = -2.61$)[2]; this leads to an overall solubility of 160 ppm Zn(II) existing as soluble $Zn(OH)_2^0$. Mobilization of Zn^{+2} can thus occur as $Zn(OH)_2^0$ species at pH values above 7.7 [4].

As mentioned earlier, zinc sulfide precipitates in a reducing environment. In oxidizing waters, precipitation of $Zn(OH)_2$, $ZnCO_3$, or $Zn_4(OH)_6SO_4$ can occur when zinc is present in high concentrations [1,19].

2. This equilibrium constant can be derived from the sum of the following three reactions:

$Zn(OH)_2(s) \rightleftharpoons Zn^{+2} + 2OH^-$, $K_{sp} = 10^{-15.5}$

$Zn^{+2} + OH^- \rightleftharpoons ZnOH^+$, $K_1 = 10^{4.4}$

$ZnOH^+ + OH \rightleftharpoons Zn(OH)_2^0$, $K_2 = 10^{8.49}$

$Zn(OH)_2(s) \rightleftharpoons Zn(OH)_2^0$, $K = K_{sp}K_1K_2 = K_{sp}\beta_2 = 10^{-2.6}$

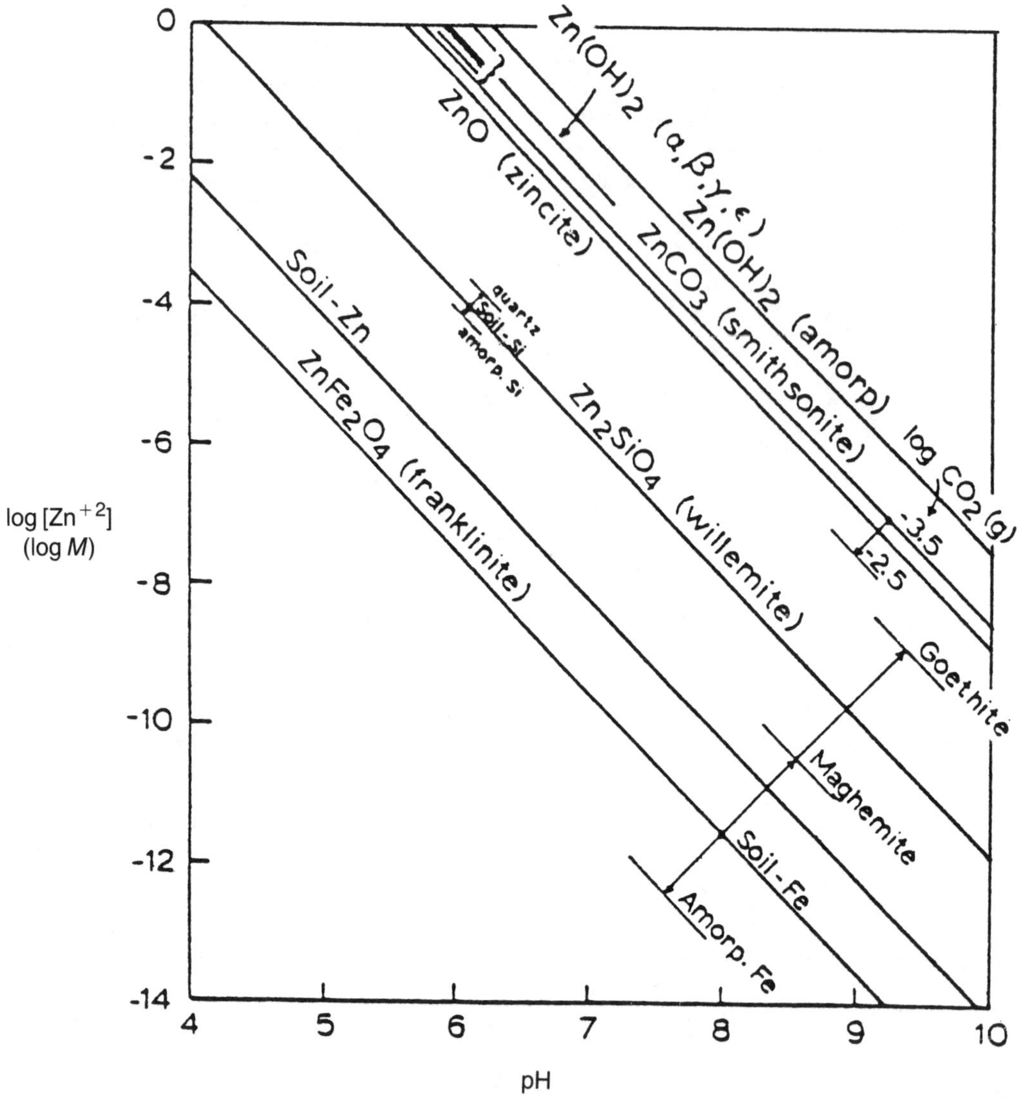

Source: Lindsay [8]. *(Copyright 1979, John Wiley & Sons. Reprinted with permission.)*

FIGURE 7.15-4 Solubility of Several Zinc Minerals Compared with That of Soil-Zinc

The solubility of zinc in soil water appears to be related to soil-Zn adsorption equilibria [8,15]. The solubility of the unidentified form known as soil-Zn is described by the following equation:

$$\text{soil-Zn} + 2H^+ \rightleftharpoons Zn(II) + \text{soil} \begin{matrix} \diagup H \\ \diagdown H \end{matrix} \qquad (1)$$

The equilibrium constant for this reaction is given by:

$$K_0 = \frac{[Zn^{+2}]}{[H^+]^2} \qquad (2)$$

Since $K_0 = 5.8$ [8],

$$\log [Zn^{+2}] = 5.8 - 2pH \qquad (3)$$

The line labeled "Soil-Zn" in Figure 7.15-4 is a plot of equation 3. Equations 1-3, which are derived from experimental measurements using soils of different pH, permit Zn(II) solubility relationships in soil to be compared with those of known minerals [8]. Zn(II) concentration in the aqueous phase of soils is a function of several species. The presence of crystalline iron oxides can lower dissolved Fe(III) levels and thus raise the solubility line for franklinite ($ZnFe_2O_4$) shown in Figure 7.15.4 closer to that of soil-Zn. This has prompted speculation that franklinite in equilibrium with Fe(III) oxides may be the solubility-controlling species referred to as soil-Zn [8].

7.15.4 Sorption on Soils and Sediments

The sorption of zinc on soil and sediments strongly impacts its mobility in the environment [1,19]. For this reason, severe zinc contamination tends to be confined to the region of the source [14]. Hydrous iron and manganese oxides, clay minerals [15], carbonate minerals and, to a lesser extent, organic matter [6,11] have been noted as sorbents that control the behavior of zinc in soil and sediments. In soil samples, the distribution of zinc has been found to be 30-60% associated with Mn and Fe oxides and 20-45% with the clay fraction [3].

The enrichment of zinc in the humic acid fraction of the soil is much less than that of copper: the enrichment factor (defined as the ratio of metal concentration in the humic acid fraction to its total concentration in the associated sediments) is 0.46-0.81 for zinc but 21-40 for copper. Less than 5% of the zinc in sediments is estimated to be bound to organic matter [13,16].

Attenuation due to specific adsorption (see § 7.3.4), co-precipitation and ion exchange plays a role in the sorption of zinc on soils and sediments [15,16,19]. Co-precipitation of zinc during precipitation of manganese and iron hydrous oxides occurs in solutions where originally high levels of reduced manganese and iron were exposed to oxygen; the sorbed zinc is incorporated into the crystal lattice structure of the oxides [1]. There is an inverse correlation between sediment grain size and the concentration of zinc sorbed — i.e., smaller particles adsorb more. Section 2.12 (Part I) contains helpful information on general sorption processes.

The sorption of zinc increases greatly as the pH increases through a critical range of 5 to 7 [1,16]. In an experiment measuring the effect of pH on zinc sorption onto two hydrous oxides and Metapeak and Evesboro soils (see § 7.5.4), Huang *et al.* [7] found that 10^{-3} *M* zinc was completely removed from solution when the pH exceeded 8. At pH 5, little or no zinc was sorbed. Nonspecifically (electrostatically) sorbed, non-complexing monovalent anions (NO_3^-, Cl^-, ClO_4^-) have little effect on zinc sorption. Specifically (chemically) sorbed multivalent anions (e.g., SO_4^{-2}, CrO_4^{-2}, SeO_3^{-2}, AsO_4^{-3}, PO_4^{-3}) may enhance zinc sorption by Fe oxides, possibly by decreasing the net charge on the surface of the particles [15,18].

Sorption coefficients for zinc on a variety of soils and sediments are available in the literature. The values vary widely with conditions and with the equations used to describe the phenomena. Ranges of sorption constants for zinc are presented in Table 7.15.2. Additional information on soil mobility is given in § 2.12.3.

The value ranges for A_m (the Langmuir sorption maximum) and K_d given in Table 7.15-2 are very large. Rai *et al.* [15] and other references should be consulted for sorption constants that are specific to experimental conditions.

TABLE 7.15-2

Typical Values of Sorption Constants for Zinc

	Clay	Fe Oxides	Mn Oxides	Organic Matter	Soils	Sediment
A_m(μmol/g)	6.1-560	5.0-600	34-4650	12,180[b]	0.89-24	47-180
Log K_L(log M^{-1})	2.7-3.3	2.7-6.7	3.4-6.3	3.4,5.0[b]	4.2-7.5	3.8-4.8
K_d(ml/g)	2[a]	1.3-500	20-300	—	—	—

a. Only one value given.

b. Only two values given.

Source: Rai *et al.* [15]

Definitions: A_m,K_L = Langmuir constants, where $S = (K_L A_m C)/(1 + K_L C)$ and S = μmoles sorbed at equilibrium per gram of solid, A_m = maximum sorption capacity of solid (μmol/g), K_L = sorption constant (related to binding energy of sorbate in $l \cdot mol^{-1}$), and C = total sorbate concentration in solution at equilibrium ($mol \cdot l^{-1}$)

K_d = ratio of the quantity of sorbate sorbed per gram of solid (μmol/g) to the concentration of the sorbate remaining in solution at equilibrium (μmol/ml)

7.15.5 Literature Cited

1. Callahan, M., M. Slimak, N. Gabel, I. May, C. Fowler, J. Freed, P. Jennings, R. Durfee, F. Whitmore, B. Maestri, W. Mabey, B. Holt and C. Gould, *Water-Related Environmental Fate of 129 Priority Pollutants. Volume I: Introduction and Technical Background, Metals and Inorganics, Pesticides and PCBs*, EPA-440/4-79-029a, EPA Contracts 68-01-3852 and 68-01-3867, Office of Water Planning and Standards, U.S. Environmental Protection Agency, Washington, D.C. (1979).
2. Faust, S.D. and O.S. Aly, *Chemistry of Water Treatment*, Butterworth, Woburn, MA (1983).
3. Fergusson, J.E., *Inorganic Chemistry and the Earth*, Pergamon Press, Elmsford, N.Y. (1982).
4. Hahne, H.C.H. and W. Kroontje, "Significance of pH and Chloride Concentration on Behavior of Heavy Metal Pollutants — Mercury (II), Cadmium (II), Zinc (II) and Lead (II)," *J. Environ. Qual.*, **2,** 444-50 (1973).
5. Hem, J.D., "Chemistry and Occurrence of Cadmium and Zinc in Surface and Ground Waters," *Water Resour. Res.*, **8,** 661-71 (1972).
6. Hickey, M.G. and J.A. Kittrick, "Chemical Partitioning of Cadmium, Copper, Nickel and Zinc in Soil and Sediments Containing High Levels of Heavy Metals," *J. Environ. Qual.*, **13,** 372-76 (1984).
7. Huang, C.P., H.A. Elliot, and R.M. Ashmead, "Interfacial Reactions and the Fate of Heavy Metals in Soil-Water Systems," *J. Water Pollut. Control Fed.*, **49**, 745-56 (1977).
8. Lindsay, W.L., *Chemical Equilibria in Soils*, John Wiley & Sons, New York (1979).
9. Long, D.T. and E.E. Angino, "Chemical Speciation of Cd, Cu, Pb, and Zn in Mixed Freshwater, Seawater and Brine Solutions," *Geochim. Cosmochim. Acta*, **41,** 1183-91 (1977).
10. Mantoura, R.F.C., A. Dickson and J.P. Riley, "The Complexation of Metals with Humic Materials in Natural Waters," *Estuarine Coastal Mar. Sci.*, **6,** 387-408 (1978).
11. Moore, J.W. and S. Ramamoorthy, *Heavy Metals in Natural Waters*, Springer-Verlag, New York (1984).
12. Morel, F.M.M., *Principles of Aquatic Chemistry*, John Wiley & Sons, New York (1983).
13. Nriagu, J.O. and R.D. Coker, "Trace Metals in Humic and Fulvic Acids from Lake Ontario Sediments," *Environ. Sci. Technol.*, **14,** 443-46 (1980).
14. Perwak, J., M. Goyer, L. Nelken, G. Schimke, K. Scow, P. Walker, D. Wallace and A. Wechsler, *Exposure Assessments of Priority Pollutants: Zinc*, EPA Contract 68-01-3857, Task Order No. 9 (January 1980).

15. Rai, D., J. Zachara, A. Schwab, R. Schmidt, D. Girvin and J. Rogers, *Chemical Attenuation Rates, Coefficients, and Constants in Leachate Migration. Vol. I: A Critical Review*, Report EA-3356 to EPRI by Pacific Northwest Laboratories (Battelle Institute), Richland, Wash. (1984).

16. Salomons, W. and U. Förstner, *Metals in the Hydrocycle*, Springer-Verlag, New York (1984).

17. Sillén, L.G. and A.E. Martell, *Stability Constants of Metal-Ion Complexes,* Special Publication No. 17, The Chemical Society, Burlington House, London (1964).

18. Stumm, W. and J.J. Morgan, *Aquatic Chemistry*, 2nd ed., John Wiley & Sons, New York (1981).

19. U.S. Environmental Protection Agency, *Ambient Water Quality Criteria for Zinc,* Report No. EPA-440/5-80-079 (October 1980).

20. Wilkins, R.G., *The Study of Kinetics and Mechanism of Reactions of Transition Metal Complexes*, Allyn and Bacon, Boston (1974).

7.15.6 Other Sources of Information

Arthur D. Little, Inc., *Full-Scale Field Evaluation of Waste Disposal from Coal-Fired Electric Generating Plants. Appendix F (Part 4)*, EPA Contract No. 68-02-3167, Final Report to U.S. EPA, Office of Research and Development, Washington, D.C. (1984).

Bar-Yosef, B., "pH-Dependent Zinc Adsorption by Soil," *Soil Sci. Soc. Am. J.,* **43,** 1095-99 (1979).

Cavallaro, N. and M.B. McBride, "Zinc and Copper Sorption and Fixation by an Acid Soil Clay: Effect of Selective Dissolutions," *Soil Sci. Soc. Am. J.,* **48,** 1050-54 (1984).

Frost, R.R. and R.A. Griffen, "Effect of pH on Adsorption of Copper, Zinc, and Cadmium from Landfill Leachate by Clay Minerals," *J. Environ. Sci. Health,* **A.12,** 139-56 (1977).

Fuller, W.H., *Movement of Selected Metals, Asbestos, and Cyanide in Soil: Applications to Waste Disposal Problems,* EPA 600/2-77-020, PB 266 905, Municipal Environmental Research Laboratory, Cincinnati, Ohio (1977).

Guy, R.D. and C.L. Chakabarti, "Studies of Metal-Organic Interactions in Model Systems Pertaining to Natural Waters," *Can. J. Chem.,* **54,** 2600-11 (1976).

Montgomery, J.R. and R.J. Santiago, "Zinc and Copper in 'Particulate' Forms and 'Soluble' Complexes with Inorganic or Organic Ligands in the Guanajibo River and Coastal Zone, Puerto Rico," *Estuarine Coastal Mar. Sci.,* **6,** 111-16 (1978).

Nriagu, J.O. (ed.), *Zinc in the Environment,* John Wiley & Sons, New York (1980).

Patterson, J.W., H.E. Allen and J.J. Scala, "Carbonate Precipitation for Heavy Metal Pollutants," *J. Water Pollut. Control Fed.,* **49,** 2397-2410 (1977).

Shuman, L.M., "The Effect of Soil Properties on Zinc Adsorption by Soils," *Soil Sci. Soc. Am. Proc.,* **39,** 454-58 (1975).

Sidhu, A.A., N.S. Randhawa and M.K. Sinha, "Adsorption and Desorption of Zinc in Different Soils," *Soil Sci.,* **124,** 211-18 (1977).

Silviera, D.J. and L.E. Sommers, "Extractability of Copper, Zinc, Cadmium, and Lead in Soils Incubated with Sewage Sludge," *J. Environ. Qual.,* **6,** 47-52 (1977).

Sommers, L.E. and W.L. Lindsay, "Effect of pH and Redox on Predicted Heavy Metal-Chelate Equilibria in Soils," *Soil Sci. Soc. Am. J.,* **43,** 39-47 (1979).

8. POLLUTANT/TRACE LIGANDS

Christopher P. Loreti

Page

LIST OF TABLES

NITROGEN (N)

SULFUR (S)

PHOSPHORUS (P)

FLUORIDE (F)

LIST OF FIGURES

NITROGEN (N)

SULFUR (S)

PHOSPHORUS (P)

FLUORIDE (F)

INTRODUCTION

Important ligands that are present in waters and soils at concentrations generally lower than those of the major/matrix ligands of Chapter 6 are described in the following four subsections. These ligands are of interest as indicators of pollution and for their effects on the environment, rather than as threats to human health. The ligands are defined by their principal elements: nitrogen, sulfur, phosphorus, and fluorine. Because sulfate is discussed in Chapter 6, other sulfur species are covered here. The fluorine subsection includes some information on bromine and iodine, which were not deemed significant enough to merit separate sections.

The information contained in each of the subsections is arranged in the following format:

- Elemental cycle
- Species of environmental importance
- Aqueous chemistry
  - Redox
  - Acid base
  - Complexation and solubility
- Fate in soil

Literature sources are given at the end of each subsection.

8.1 NITROGEN (N)

Nitrogen gas (N_2) is the largest constituent of the atmosphere, accounting for almost 80% by volume of clean air at sea level. Because nitrogen gas is relatively inert, however, the environmental chemistry of nitrogen consists primarily of the reactions and transport of nitrogen species in forms more oxidized or reduced than N_2.

8.1.1 The Nitrogen Cycle

Nitrogen is essential to all organisms and exists in a variety of forms, including gaseous molecular nitrogen, nitrate (NO_3^-), and ammonia (NH_3). As a result, the nitrogen cycle is quite complex. The transformations of nitrogen species, their sources, and pathways through the environment are shown qualitatively in Figure 8.1-1.

The conversion of gaseous nitrogen in the atmosphere to more reactive oxidized and reduced forms is termed "fixation."[1] Biologically mediated fixation is carried out by blue-green algae, some anaerobic bacteria, free-living *Azobacter,* and symbiotic *Rhizobium* bacteria living in nodules on the roots of legumes [1]. During biological fixation, N_2 is reduced to NH_3 or NH_4^+; unlike N_2, these forms of nitrogen are usable by plants. In the manufacture of fertilizers, N_2 in the atmosphere is fixed as ammonia. Atmospheric nitrogen is also fixed as a by-product of combustion; the nitrogen in combustion air (and in fuel) is oxidized to gaseous NO and NO_2 (nitrogen oxides or NO_x). Nitric oxide, NO, is also formed by lightning.

Nitrogen oxides in the atmosphere react with organic compounds to form smog and other pollutants such as peroxyacyl nitrates[2] (PANs), which are eye irritants and toxic to some plants [7]. Nitric acid, HNO_3, is formed upon the further oxidation of NO_x to N_2O_5 and reaction with water; it is then washed out of the atmosphere as rain or deposited as an aerosol. (See §6.9.3 for a description of the chemistry involved.)

Nitrate, nitrite (NO_2^-), and ammonium ions in the soil are available for uptake by plants, a process known as assimilation. The reverse of assimilation is mineralization, the conversion of organic N to ammonia or ammonium (ammonification) and its subsequent oxidation to nitrite and nitrate (nitrification). Ammonification occurs primarily through the action of microorganisms and some animals such as protozoa and earthworms, which excrete ammonium in their wastes [10]. Bacteria of the genus *Nitrosomonas* (and several others) oxidize ammonium ions to nitrite ions, which are then oxidized to nitrate through the action of *Nitrobacter* and *Nitrococcus* bacteria [10]. Under anaerobic conditions, nitrate ions in soils are reduced to gaseous N_2 or N_2O (nitrous oxide) by certain microorganisms; this "denitrification" process facilitates the transport of nitrogen from the terrestrial environment to the atmosphere.

---

1. "Fixation" as used here should not be confused with the fixation of hazardous waste, which means the conversion to a less reactive (or leachable) form.

2. Chemical formula: $R\overset{\overset{\displaystyle O}{\|}}{C}OONO_2$ where, for the simpler peroxyacyl nitrates, R = H, CH_3, C_2H_5, C_6H_5, etc.

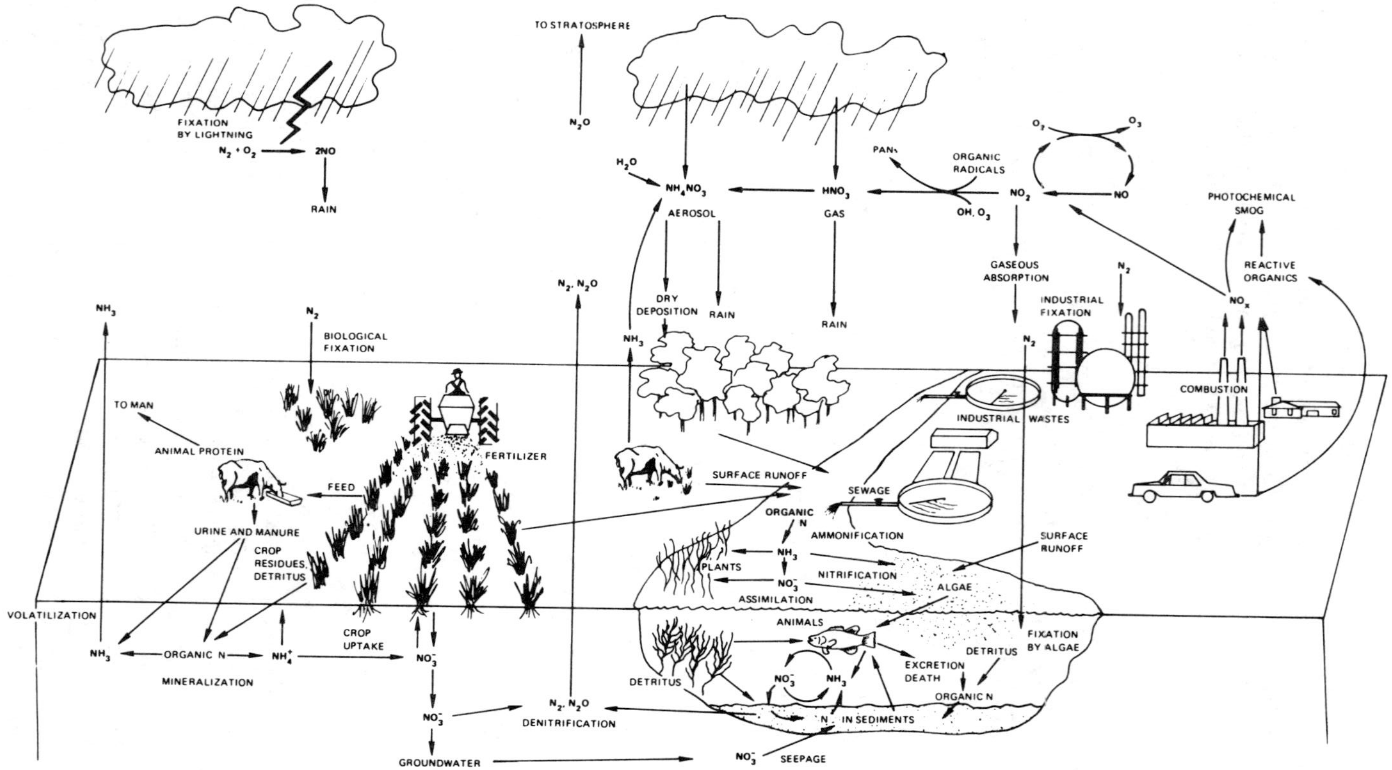

Source: National Research Council [7]

FIGURE 8.1-1 Schematic Representation of the Nitrogen Cycle, Emphasizing Human Activities that Affect the Fluxes of Nitrogen

The magnitude of the nitrogen flows described above and the major sinks for nitrogen in the environment are given in Table 8.1-1 and Figure 8.1-2. This figure represents the pools of mobile nitrogen, not the complete global inventory. Some of the values given, such as those for industrial nitrogen fixation and the size of the atmospheric nitrogen pool, are known with reasonable precision; others, such as the size of the organic nitrogen pool and the rate of nitrogen fixation (and denitrification) in the ocean, are supported by only limited data and are therefore uncertain [12].

TABLE 8.1-1

Rates of Nitrogen Transfer Between Pools Shown in Figure 8.1-2
(10^7 metric tons of N/year)

Transfer	Rate
(a) Nitrogen fixation, land	9.7
(b) Nitrogen fixation, ocean	3.1
(c) Nitrogen fixation, atmospheric	0.75
(d) Nitrogen fixation, industrial	4.0
(e) Nitrogen fixation, combustion	1.8
(f) Weathering processes	0.44
(g) Runoff, organic	2.2
(h) Runoff, inorganic	1.3
(i) Assimilation, land	180
(j) Assimilation, sea	140
(k) Mineralization, land	190
(l) Mineralization, sea	140
(m) Denitrification, land	12
(n) Denitrification, sea	4.0
(o) Ammonium fallout, rainout, and washout, land	6.6
(p) Ammonium fallout, rainout, and washout, sea	1.3
(q) Nitrogen from fossil fuel (largely ammonium)	0.35
(r) Ammonium volatilization, plants and animals	5.3
(s) Ammonium volatilization, soil	2.2
(t) NO_x fallout, rainout, and washout, land	2.5
(u) NO_x fallout, rainout, and washout, sea	0.88
(v) To organic pool, land	190
(w) To organic pool, sea	140

Source: Adapted from Subcommittee on Ammonia, NRC [12].

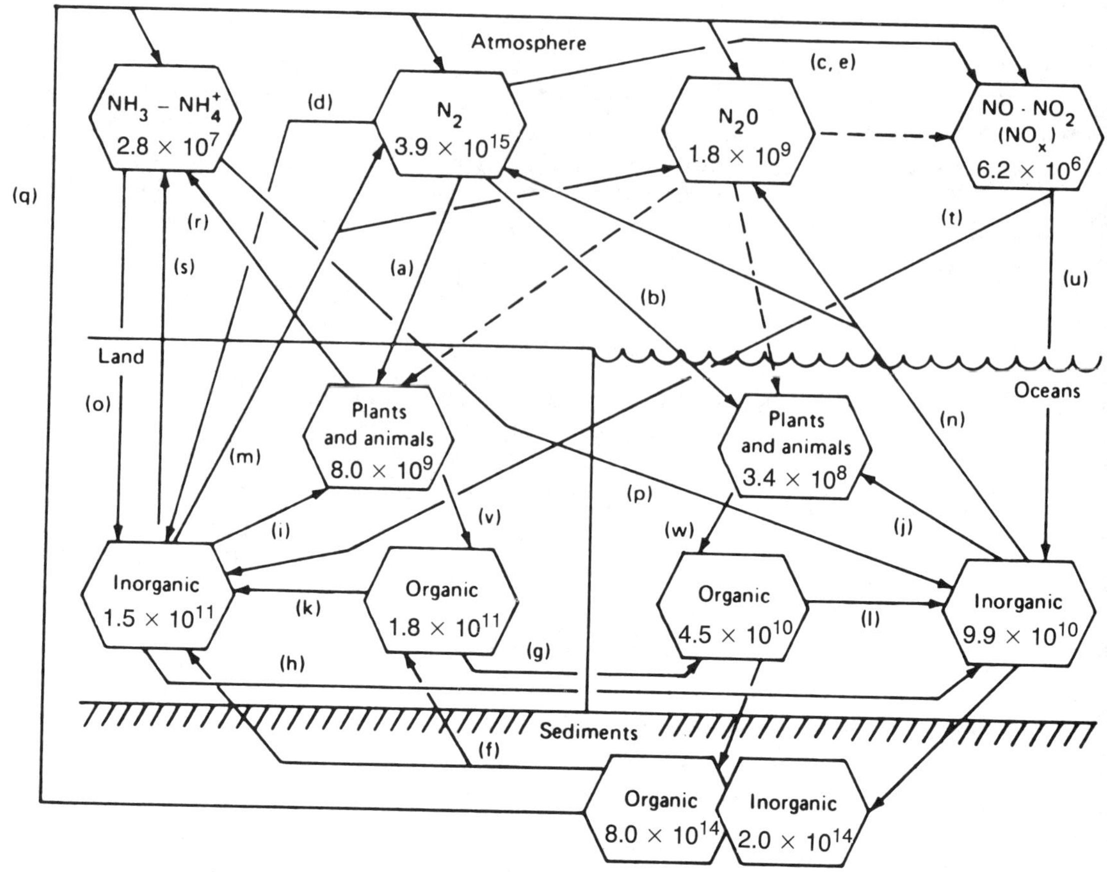

Numbers refer to metric tons of N. See Table 8.1-1 for rates of transfer (arrows).

Source: Adapted from NRC [12]

FIGURE 8.1-2 Sizes of Pools of the Nitrogen Cycle

According to Table 8.1-1, nitrogen is removed from the atmosphere faster than it enters. The atmospheric sink for nitrogen is so large that this loss has a negligible effect on the quantity of N_2 in the atmosphere; however, the net transfer of nitrogen to the terrestrial environment due to the cultivation of legumes and the production of fertilizers has raised concern that this fixed nitrogen be properly managed and not cause the nitrate concentration of surface and groundwaters to reach unacceptable levels [3].

8.1.2 Species of Environmental Importance

Nitrate, nitrite, and ammonia are the nitrogen species of greatest environmental interest because of their interactions with living organisms. Cyanides, CN^-, are toxic in the form of HCN, but they are discussed in § 10.13 rather than here; though important in industry, cyanides play a minor role in the natural environmental chemistry of nitrogen, because their occurrence in plants and animals is as short-lived metabolic intermediates [13].

Nitrates and nitrites are of human health concern primarily because of the effects of nitrites. Under certain conditions, nitrate is reduced to nitrite in the gastrointestinal tract, where it is hazardous to warm-blooded animals and human infants under three months of age [13]. Nitrite reaching the bloodstream reacts with hemoglobin to form methemoglobin, reducing the transport of oxygen through the body. The maximum legally allowable nitrate concentration in drinking water has been set by the U.S. EPA at 10 mg/l nitrate-nitrogen (40CFR:141, 1985), and the recommended maximum nitrite concentration has been proposed to be 1 mg/l nitrite-nitrogen (*Fed. Reg.*, **50**: 46936, 1985).

Nitrate is also important in biological systems, where nitrogen is often a limiting nutrient. Like phosphate, high concentrations of nitrate (several hundred μg/l) in lake surface waters can cause algal blooms [7] if nitrogen is the limiting factor for growth. Because of the rapid assimilation of nitrate by aquatic plants, the concentration of nitrate in surface waters is generally less than that in groundwater [7].

Ammonia is toxic to many aquatic organisms, but its toxicity in solution is highly dependent on pH. Ammonia protonates in solution to form NH_4^+ (see below), which is far less toxic than NH_3. The distribution of NH_3 and NH_4^+ is indicated by the pH. The EPA's "Red Book" recommends a criterion of 0.02 mg/l NH_3 (un-ionized) to protect aquatic life and gives total NH_3 and NH_4^+ concentrations that correspond to this NH_3 concentration as a function of pH and temperature [13].

8.1.3 Aqueous Nitrogen Chemistry

The chemical reactions of nitrogen-containing species in the environment are dominated by biological mediation. Theoretical analyses based on thermodynamics, such as Pourbaix diagrams,[3] can be misleading unless the effects of living organisms are taken into account. Nonetheless, equilibrium calculations provide stability boundaries for the reactions of nitrogen-containing species.

REDOX REACTIONS

Nitrogen exists in oxidation states ranging from −III to +VI in a variety of species [12]. Table 8.1-2 summarizes the important redox reactions of nitrogen and their equilibrium constants. Many of these reactions are very slow and do not reach equilibrium in nature.

One of the reactions listed in Table 8.1-2 is the reduction of NO_3^- to $N_2(g)$. Comparison of the equilibrium constant for this reaction with the redox potential of oxygenated water (§ 2.10) indicates that gaseous nitrogen can be oxidized to nitrate upon

3. Pourbaix diagrams indicate the predominant chemical species expected to be present, at equilibrium, as a function of redox potential (pe) and pH. Examples are given in Chapters 6 and 7.

TABLE 8.1-2

Equilibrium Constants for Reactions Involving Nitrogen
(at 25°C)

Reaction			log K
$NO_2(g)$	⇌	$1/2\ N_2O_4(g)$	0.42
HNO_2^0	⇌	$H^+ + NO_2^-$	−3.15
$NO(g)$	⇌	NO^0	−2.73
$N_2O(g)$	⇌	N_2O^0	0.54
$NH_2(g)$	⇌	$1/2\ N_2H_4(g)$	17.18
$1/2\ N_2H_4(g)$	⇌	$1/2\ N_2H_4^0$	2.74
$1/2\ N_2H_4^0 + 1/2\ H^+$	⇌	$1/2\ N_2H_5^+$	3.98
$NH_3(g)$	⇌	NH_3^0	1.76
$NH_3^0 + H^+$	⇌	NH_4^+	9.28
$NH_4^+ + H_2O$	⇌	$NH_4OH^0 + H^+$	−9.25
NO_3^-	⇌	$NO_3(g) + e^-$	−39.87
$NO_3^- + H^+$	⇌	$1/2\ N_2O_5(g) + 1/2\ H_2O$	−9.08
$NO_3^- + 2H^+ + e^-$	⇌	$NO_2(g) + H_2O$	13.03
$NO_3^- + 2H^+ + 2e^-$	⇌	$NO_2^- + H_2O$	28.64
$NO_3^- + 3H^+ + 2e^-$	⇌	$1/2\ N_2O_3(g) + 3/2\ H_2O$	30.62
$NO_3^- + 4H^+ + 3e^-$	⇌	$NO(g) + 2H_2O$	48.41
$NO_3^- + 5H^+ + 4e^-$	⇌	$1/2\ N_2O^0 + 5/2\ H_2O$	75.52
$NO_3^- + 6H^+ + 5e^-$	⇌	$1/2\ N_2(g) + 3H_2O$	105.15
$NO_3^- + 6H^+ + 16/3\ e^-$	⇌	$1/3\ N_3^- + 3H_2O$	84.80
$NO_3^- + 7H^+ + 6e^-$	⇌	$NH_2OH^0 + 2H_2O$	67.69
$NO_3^- + 7H^+ + 6e^-$	⇌	$NH(g) + 3H_2O$	46.81
$NO_3^- + 8H^+ + 7e^-$	⇌	$NH_2(g) + 3H_2O$	74.02
$NO_3^- + 10H^+ + 8e^-$	⇌	$NH_4^+ + 3H_2O$	119.07

Source: Lindsay [5]. *(Copyright 1979, John Wiley & Sons. Reprinted with permission.)*

dissolution in water. However, nitrate (and nitrite) are minor constituents of natural waters. In seawater, nitrate concentration ranges from 1 to 500 μg/l (as N) and nitrite from 0.1 to 50 μg/l (as N) [8]. Thermodynamically, all the N_2 in solution would be expected to be oxidized to NO_3^-, and all the oxygen in the atmosphere would be exhausted before equilibrium between dissolved N_2 and NO_3^- in the oceans was reached [4]. Without the organisms described above, which mediate and thus enhance the rate of N_2 fixation, the oxidation of N_2 proceeds extremely slowly.

In Figure 8.1-3, the calculated concentrations of nitrogen species in an aqueous solution of 10^{-3} *M* total N are plotted as functions of pe at 25°C and pH 7. Equilibria that are favored thermodynamically but not observed in nature, such as those involving N_2, N_2O, and N_3^-, are ignored. The figure includes the equilibrium with NO, based on the hypothesis that NO may be an intermediate in the denitrification process but is reduced to N_2 or N_2O before reaching the equilibrium shown [5].

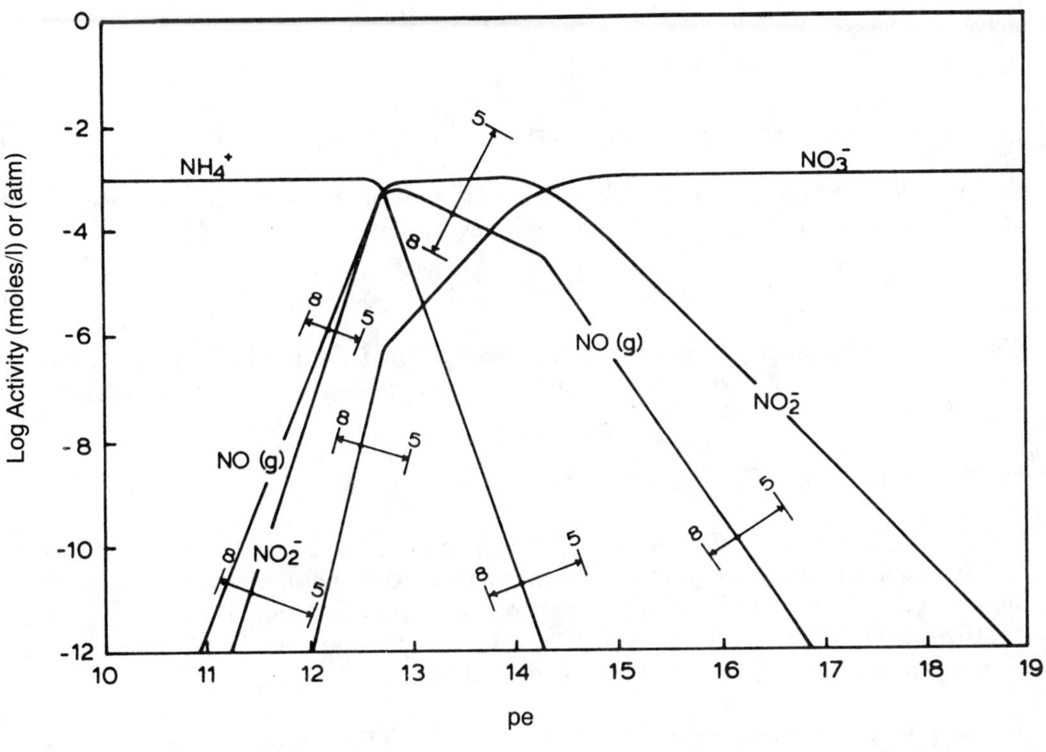

Total nitrogen concentration = 10^{-3} *M*, pH = 7. Bars show effect of changes to pH 5 and 8.

Source: Lindsay [5]. *(Copyright 1979, John Wiley & Sons. Reprinted with permission.)*

FIGURE 8.1-3 Effect of Redox Potential on Activity of Nitrogen Species When Equilibrium with N_2(g), N_2O^0, and N_3^- Is Excluded from Consideration

The nitrate ion acts as a strong oxidizing agent when present at high concentration and elevated temperature. In dilute solutions (<1 *M*) of nitric acid or nitrate salts at normal temperatures, the kinetics of oxidation are so slow that the ion appears to have virtually no oxidizing power. Nitrogen in the form of nitrite can act as a reducing or oxidizing agent. Nitrite oxidizes many substances such as sulfide, sulfite, and iodide [7].

ACID-BASE CHEMISTRY

Nitric acid (HNO_3) is a strong acid ($pK_a = -1$ [11]); hence, the nitrate ion, its conjugate base, has no interesting acid-base chemistry in natural waters. Nitrous acid (HNO_2) is much weaker ($pK_a = 3.36$ [7]), and in acidic waters such as bogs, swamps, and mine drainage waters, as well as in acidic soils, a large fraction can exist in undissociated form [7]. While HNO_2 will be oxidized to HNO_3 in oxygenated waters, it can also react with itself to form nitrous acid anhydride (N_2O_3), which in turn can react with nonionized secondary amines to form nitrosamines [7].

Ammonia acts as a base in solution [12]:

$$NH_3 + H_2O \rightleftharpoons NH_4^+ + OH^- \qquad K_b = 1.77 + 10^{-5}\ (25°C) \tag{1}$$

The pK_b of ammonia (4.75 at 25°C) is similar to that of the borate ion. The conjugate acid of ammonia, the ammonium ion, can act as a very weak acid [12]:

$$NH_4^+ + H_2O \rightleftharpoons NH_3 + H_3O^+ \qquad K_a = 5.64 + 10^{-10}\ (25°C) \tag{2}$$

With a pK_a of 9.25, the ammonium ion is more acidic than the bicarbonate ion ($pK_a = 10.3$ [12]); however, a significant amount (>10%) of total N will occur as NH_3^0 only at a pH greater than 8.2 ($I = 0$, 25°C).

COMPLEXATION AND SOLUBILITY

Nitrate displays little tendency to complex with aqueous metals in natural waters. Nitrite can form strong complexes with many transition metals, but as its concentration in natural waters is usually very low, such complexation is not likely to be significant [7]. Ammonia forms a variety of relatively strong aqueous complexes, the stability constants of which are given in Table 8.1-3.

Nitrate and nitrite salts are highly soluble in water. On a mass per mass basis, both sodium nitrate ($NaNO_3$) and sodium nitrite ($NaNO_2$) are more soluble than sodium chloride in water. Sodium nitrate, for example, has a saturation concentration of over 8 *M* at 22°C [9]. Ammonium salts are also highly soluble: ammonium chloride (NH_4Cl) is slightly more soluble than sodium chloride at room temperature, and ammonium nitrate (NH_4NO_3) is more soluble than sodium nitrate (both comparisons on a mass per mass basis). Liquid ammonia (NH_3) has a high solubility in water, because, like water, it is a polar molecule. (See § 10.6 for information on the solubility and volatility of ammonia in water.)

The concentration of the nitrate ion in surface waters is limited by biological reactions that consume it, rather than by the solubility of nitrate species. Unlike phosphates, which form insoluble precipitates, nitrates are unlikely to precipitate out of solution and accumulate in an inactive form in sediments. Nitrate reaching rivers is flushed into the ocean if it is not assimilated by aquatic organisms.

TABLE 8.1-3

Stability Constants of Ammonia Complexes
(at 25°C and zero ionic strength unless otherwise indicated)

Metal Ion	Complex	log K or log β	Metal Ion	Complex	log K or log β
Ag^+	$AgNH_3^+$	3.31	Cu^{+2}	$CuNH_3^{+2}$	4.04
	$Ag(NH_3)_2^+$	7.22		$Cu(NH_3)_2^{+2}$	7.47
				$Cu(NH_3)_3^{+2}$	10.27
Cd^{+2}	$CdNH_3^{+2}$	2.55		$Cu(NH_3)_4^{+2}$	11.75
	$Cd(NH_3)_2^{+2}$	4.56			
	$Cd(NH_3)_3^{+2}$	5.90	Ni^{+2}	$NiNH_3^{+2}$	2.72
	$Cd(NH_3)_4^{+2}$	6.74		$Ni(NH_3)_2^{+2}$	4.89
				$Ni(NH_3)_3^{+2}$	6.55
Co^{+2}	$CoNH_3^{+2}$	1.99[a]		$Ni(NH_3)_4^{+2}$	7.67
	$Co(NH_3)_2^{+2}$	3.50[a]		$Ni(NH_3)_5^{+2}$	8.34
	$Co(NH_3)_3^{+2}$	4.43[a]			
	$Co(NH_3)_4^{+2}$	5.07[a]	Zn^{+2}	$ZnNH_3^{+2}$	2.21
	$Co(NH_3)_5^{+2}$	5.13[a]		$Zn(NH_3)_2^{+2}$	4.50
	$Co(NH_3)_6^{+2}$	4.39[a]		$Zn(NH_3)_3^{+2}$	6.86
				$Zn(NH_3)_4^{+2}$	8.89

a. 20°C

Source: Martell and Smith [6]

8.1.4 Nitrogen-containing Ligands in Soil

In soils, the nitrate ion is far more mobile than the ammonium ion. Because the ammonium ion bears a positive charge, the cation exchange sites in the soil tend to retain it [12]. Nitrogen is typically leached from soils as nitrate [1]. In most soils of the United States, neither anion sorption nor anion repulsion (by negatively charged soil surfaces) is important; therefore, as a first approximation, the movement of nitrate in soils equals that of the water in which it is contained [7], provided uptake by biota does not occur. Because nitrate sorption by soils is generally insignificant, isotherm data are scant.

The high mobility of nitrate in soils can lead to high nitrate concentrations in groundwater beneath soils that receive large nitrogen loadings. The irrigation of acid soils in which NO_3^- has accumulated over geologic time, the application of large quantities of organic matter in the form of sewage sludge, manure, or wastewater, and over-fertilization with nitrogen-containing fertilizers all can lead to high nitrate concentrations in groundwater [1]. Roughly half of the nitrogen applied as fertilizer is not assimilated by plants [1] but leaves the soil either by leaching, erosion, or denitrification. Since more nitrogen is applied as fertilizer than any other element, on a global basis [2], the potential for nitrates entering surface and groundwaters is great.

8.1.5 Literature Cited

1. Bohn, H.L., B.L. McNeal and G.A. O'Connor, *Soil Chemistry*, John Wiley & Sons, New York (1979).
2. Bowen, H.J.M, *Environmental Chemistry of the Elements*, Academic Press, London (1979).
3. Delwiche, C.C., "The Nitrogen Cycle," *Sci. Am.*, **223**, 137-46 (1970).
4. Hutchinson, G.E., "Chemistry of Lakes," in *A Treatise on Limnology*, Vol. 1, Part 2, John Wiley & Sons, New York (1975).
5. Lindsay, W.J., *Chemical Equilibria in Soils*, John Wiley & Sons, New York (1979).
6. Martell, A.E. and R.M. Smith, *Critical Stability Constants*, Vol. 4, *Inorganic Complexes* and Vol. 5, *First Supplement*, Plenum Press, New York (1976 and 1982).
7. National Research Council, *Nitrates: An Environmental Assessment*, National Academy of Sciences, Washington, D.C. (1978).
8. Riley, J.P. and R. Chester, *Introduction to Marine Chemistry*, Academic Press, London (1971).
9. Seidell, A., *Solubilities of Inorganic and Metal Organic Compounds*, 4th ed., American Chemical Society, Washington, D.C. (1970).
10. Soderlund, R. and T. Rosswall, "The Nitrogen Cycles," in *Environmental Chemistry*, Vol. 1, Part B, O.E. Hutzinger (ed.), Springer-Verlag, Berlin, 61-81 (1985).
11. Stumm, W. and J.J. Morgan, *Aquatic Chemistry*, 2nd ed., John Wiley & Sons, New York (1981).
12. Subcommittee on Ammonia, National Research Council, National Academy of Sciences, *Ammonia*, University Park Press, Baltimore (1979).
13. U.S. Environmental Protection Agency, *Quality Criteria for Water*, Washington, D.C. (1976).

8.2 SULFUR (S)

The sulfate ligand was considered in section 6.11. In this section, the chemistry and environmental significance of other sulfur-containing compounds are described.

8.2.1 The Sulfur Cycle

The cycle of sulfur in the environment is shown qualitatively in Figure 8.2-1. In soils and sediments, sulfur cycling involves the transformation of sulfur between its +6 and −2 oxidation states, reactions largely mediated by microorganisms. Reduction of sulfate to sulfide by microorganisms (anaerobic sulfate reducers such as *desulfovibrio* bacteria) occurs in virtually any anoxic habitat containing organic matter and sulfate [19]. Hydrogen sulfide is oxidized by chemosynthetic bacteria known as colorless sulfur bacteria, which oxidize sulfide with O_2, and by photosynthetic green and purple sulfur bacteria [12], which are obligate anaerobes. The oxidation of H_2S to SO_4^{-2} also occurs through the mediation of thiobacilli bacteria (aerobic sulfide oxidizers), and organic S is oxidized to SO_4^{-2} through the action of aerobic heterotrophic microorganisms [13]. The importance of microbial processes is further emphasized by the fact that no chemical reaction for the reduction of SO_4^{-2} to S^{-2} at normal temperatures and pressures is known [6] and that 50-60% of global S oxidation is estimated to occur microbially [18]. Sulfur-oxidizing bacteria can be found even under extreme environmental conditions; bacteria of the genus *Thiomicrospira* thrive in the sulfide-containing waters of marine hydrothermal vents at depths of 2500 meters, and other (probably) sulfur-oxidizing bacteria have been found inside larger organisms, such as the mouthless large worm that inhabit these waters [6].

Figure 8.2-1 illustrates the close linkage between sulfide transformations and those of other chemical species. The primary geochemical consequence of SO_4^{-2} reduction is the formation of iron sulfides; Fe^{+2}, which is stable under anoxic conditions, combines with S^{-2} to form the black precipitate characteristic of most anoxic environments [6]. As iron sulfides form, phosphate is converted from insoluble to soluble form, linking the sulfur and phosphorus cycles [13].

The sulfur content of various compartments of the environment and the fluxes between them are provided in Figure 8.2-2 and Table 8.2-1. The large contribution of anthropogenic sulfur to the atmosphere is apparent from the comparison of sulfur fluxes from land to air: 65×10^9 kg-S/yr from soil microbes vs. 46×10^9 kg-S/yr for sulfur released by oil and coal burning. The potential effects of anthropogenic sulfur emissions are likely greater than global balances such as Table 8.2-1 suggest, since these emissions tend to come from concentrated sources like coal-burning power plants.

8.2.2 Sulfur Species of Environmental Importance

Hydrogen sulfide is generally the sulfur species of greatest environmental concern, because it is highly toxic to most life forms other than some tolerant bacteria and algae [5]. Hydrogen sulfide can also cause considerable damage to plant roots and is

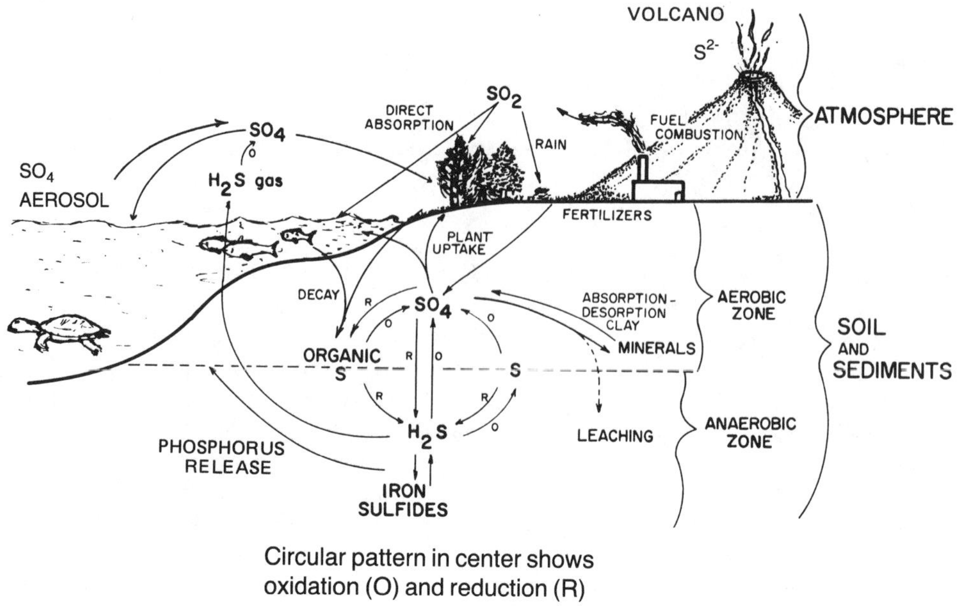

Source: Adapted from Odum [13]. *(Copyright 1971, W.B. Saunders. Reprinted with permission.)*

FIGURE 8.2-1 Sulfur Cycle Linking Air, Water, and Soil

especially toxic to citrus trees and rice [1]. Elemental sulfur (S_8) is very toxic to most bacteria and fungi (but not green algae), as it readily penetrates the cell membrane, and is then reduced to H_2S within the cell [10].

The threshold concentration of gaseous H_2S for toxic respiratory effects in man is about 10 $\mu g/m^3$ [14]. Because H_2S has a strong odor (suggestive of rotten eggs), it can be detected by smell at a concentration of about 3 $\mu g/m^3$, which is lower than the threshold level [16].

The toxicity of sulfides in water is due primarily to H_2S rather than HS^- or S^{-2} ions [16]. At pH 7, H_2S and HS^- concentrations are approximately equal, while at pH 6 about 90% of the sulfide occurs as H_2S. Because H_2S is oxidized to sulfate or elemental sulfur in aerated waters, the toxic effects of H_2S releases on fish and other aquatic life have been minimized. Nonetheless, the U.S. EPA Red Book [16] recommends a criterion of 2 $\mu g/l$ undissociated H_2S to protect fish and aquatic life in fresh and marine waters.

The National Interim Secondary Drinking Water Regulations standard for human consumption is 50 $\mu g/l$ of H_2S [15]. The secondary standards are based on aesthetic considerations; a standard based on health effects in humans does not exist, since at concentrations high enough to be toxic to humans the water would be rendered

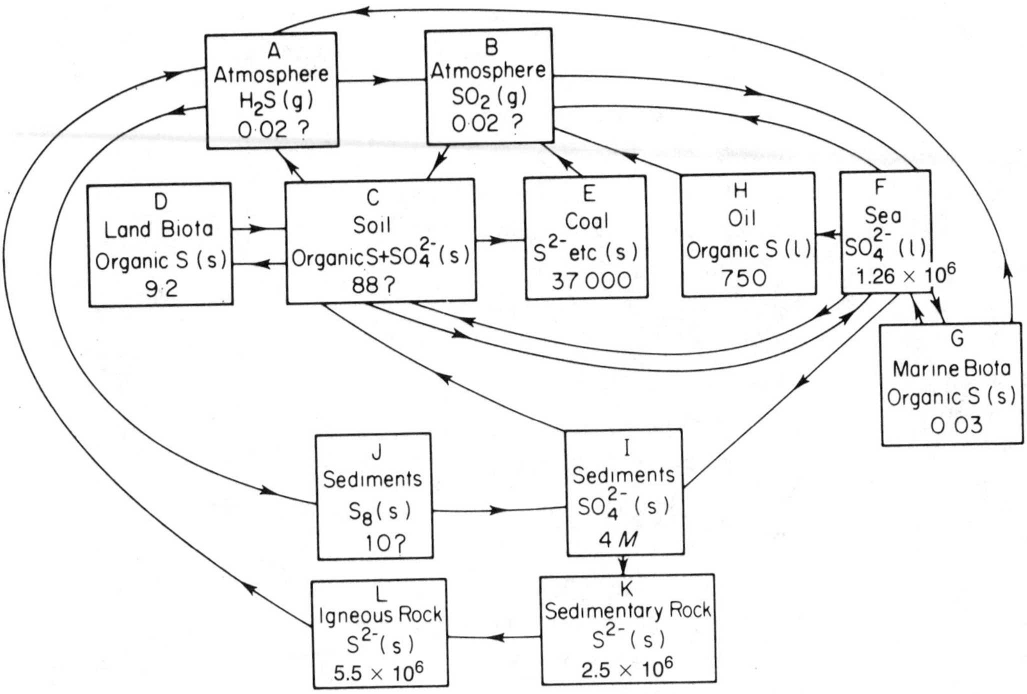

- Figures are masses of S in 10^{12} kg.
- A and B are uncertain, based on 5 μg H_2S and 10 μg SO_2 per m^3 of air.
- C is equally uncertain, assuming C/S is 28.5 in soil; D and G assume C/S is 90 in biota.
- H includes an unknown mass of H_2S trapped with natural gas.

Source: Bowen [3]. *(Copyright 1979, Academic Press. Reprinted with permission.)*

FIGURE 8.2-2 Diagrammatic Sulfur Cycle

TABLE 8.2-1

Transfers of the Sulfur Cycle Shown in Figure 8.2-2

Transfer	Description	Magnitude (10^9 kg S/yr)
A-B	Atmospheric oxidation of H_2S to SO_2	120[a]
A-J	Bacterial conversion of H_2S to S_8	0.00002
B-C	Transfer of SO_2 + SO_4^{-2} from air to soil	116
B-F	Transfer of SO_2 + SO_4^{-2} from air to sea	96
C-A	H_2S formed by soil microbes	65
C-D	Uptake of soil SO_4^{-2} by land plants	5800
C-E	S fixed in coal	0.14?
C-F	River transport of S to sea	137
D-C	S returned to soil by land biota	5800[a]
E-B	SO_2 formed by burning coal	42
F-B	SO_4^{-2} in spray from sea to air	44
F-C	SO_4^{-2} spray transferred to land	4
F-G	Uptake of SO_4^{-2} by marine plants	280
F-H	S fixed as petroleum in sediments	0.0035
F-I	S fixed as sulfates in sediments	55/95[b]
G-A	Release of H_2S by sea biota	55
G-F	Release of organic S to sea by biota	23?[a]
H-B	SO_2 formed by burning oil	4.1
I-C	S released to soil by rock weathering	14/42[b]
	+ S applied as fertilizer to soil	11/26[b]
I-K	Reduction of SO_4^{-2} to S^{-2} in sediments	0-80[a]
J-I	Oxidation of S_8 to form sulfates	10
K-L	Conversion of sediments to igneous rock	7.8?
L-A	Volcanic emission of S(g)	2?

a. Determined by the sum or difference of other values in this table.
b. Multiple values indicate estimates of different authors.

Source: Bowen [3]. *(Copyright 1979, Academic Press. Reprinted with permission.)*

unpalatable by its taste and odor. (For the same reason, sulfide criteria for livestock have not been established [16].) The removal of sulfide from drinking water is also desirable because sulfide is corrosive to metal and concrete.

8.2.3 Aqueous Sulfur Chemistry

REDOX CHEMISTRY

Sulfate is the major sulfur-containing ligand in natural waters, as described in section 6.11 and illustrated by Figure 6.11-1, the pe-pH relationship for equilibrium sulfur species in an aqueous system. However, reduced sulfur species can coexist with oxygen in aqueous solutions for some time. Sulfide species in oxygen-saturated seawater (pH 8) have a lifetime on the order of several hours [18], and as sulfate is reduced or sulfide is oxidized, a number of compounds of intermediate oxidation state can be formed. These include sulfite (SO_3^{-2}), thiosulfate ($S_2O_3^{-2}$), polysulfides (S_n^{-2}) and solid sulfur in colloidal or orthorhombic form (S_8-coll or S_8-orth) [12]. These intermediate sulfur species are generally not abundant, as they usually become oxidized or reduced to a greater degree. Figure 8.2-3 shows the predominance area diagram for a $2 \times 10^{-2}M$ aqueous sulfur system.[1] The formation of polysulfide, which dominates in a small area at the center of the plot, is discussed in Section 2.8.

ACID-BASE CHEMISTRY

Both H_2S and H_2SO_3 act as weak acids in solution. The first and second pK_a's for hydrogen sulfide are 7.02 and 13.9 respectively; for sulfurous acid (H_2SO_3), the corresponding values are 1.91 and 7.18 (all at 25°C and zero ionic strength) [11]. Thiosulfuric acid ($H_2S_2O_3$) has first and second pK_a's of 0.6 and 1.6 respectively ($I = 0$, 25°C) [9], making it a moderately strong acid; however, these acids have little influence on the pH of natural waters. In general, their concentrations are low, and all three are oxidized to sulfate in aerobic environments.

COMPLEXATION AND SOLUBILITY

As the solubility and complexation of sulfides, sulfites, and thiosulfates are discussed in detail in § 10.7, the subject is only briefly covered here. The formation constants for several sulfide, sulfite, and thiosulfate complexes are listed in Table 8.2-2, and Table 8.2-3 gives the solubility products for several sulfides, sulfites and thiosulfates. The metal sulfides listed tend to be highly insoluble, and several metals (e.g., silver and mercury) also tend to form strong sulfide complexes, as shown in Table 8.2-2. The low solubility of metal sulfides is credited with the formation of economically significant metal ore deposits. It is generally considered that bacterial sulfur reduction has made an indirect contribution to the formation of these deposits [6]. The sulfides of the alkali metals and barium, which are not included in Table 8.2-3, are readily soluble in water [8].

---

1. Sulfate formation is not considered in this diagram, since it would cover the sulfite and most of the thiosulfate areas [12]. See Figure 6.11-1 for a diagram that does include sulfate formation.

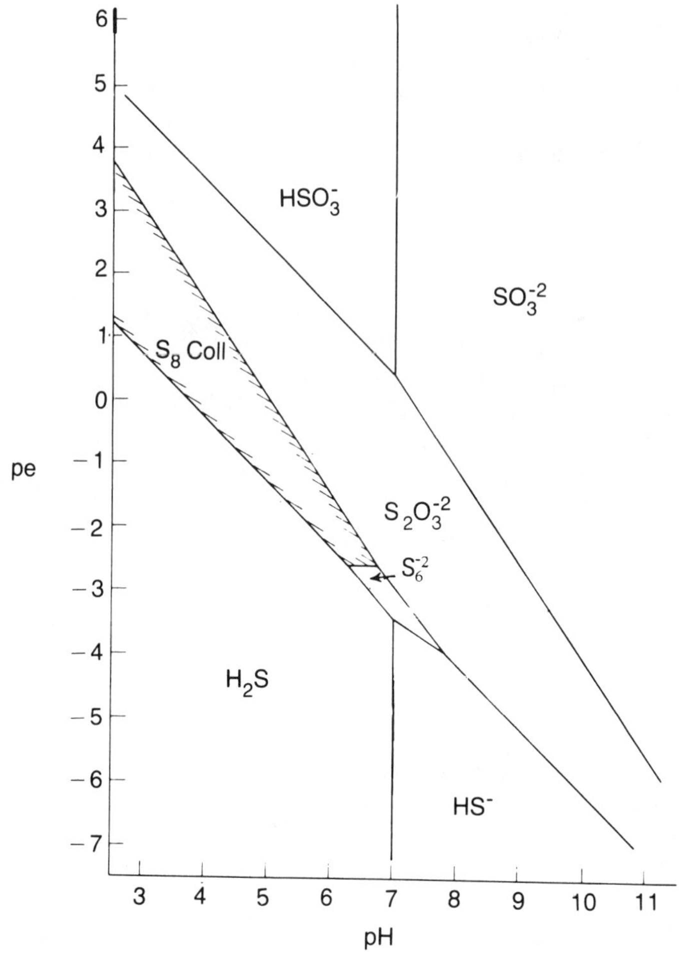

$S_T = 2 \times 10^{-2}$ *M* (25°C). Sulfate formation is not considered.

Source: Morel [12]. *(Copyright 1983, John Wiley & Sons. Reprinted with permission.)*

FIGURE 8.2-3 pe-pH Diagram for an Aqueous Sulfur System

8.2.4 Inorganic Sulfur in Soil

Inorganic sulfur typically accounts for less than 25% of total sulfur in most agricultural soils [7]. Most inorganic sulfur in well-drained soils is sulfate (see § 6.11.5); reduced inorganic forms account for less than 1% of total sulfur [17]. Elemental sulfur and sulfide are the only inorganic forms of sulfur found in soils besides sulfate [2]. The accumulation of elemental S is rare, since it is stable only under acidic and mildly reducing conditions [2], but hydrogen sulfide may be formed by the microbial reduction of sulfate or organic sulfur species. In strip mine spoils, tidal mud flats, and marine, estuarine and fresh-water sediments, sulfide

TABLE 8.2-2

Stability Constants of Thiosulfate, Sulfide and Sulfite Complexes
(at 25°C and zero ionic strength unless otherwise indicated)

Metal	Complex	log K or log β	Metal	Complex	log K or log β
Ag^{+}	$AgS_2O_3^-$	8.82[a]	Fe^{+3}	$FeS_2O_3^+$	3.3[b]
	$Ag(S_2O_3)_2^{-3}$	13.67[a]	Hg^{+2}	HgS^0	7.9[b]
	$Ag(S_2O_3)_3^{-5}$	14.2[a]		HgS_2^{-2}	14.3[b]
	AgS^-	19.2[b]		$Hg(SO_3)_2^{-2}$	24.07[d]
	$AgSO_3^-$	5.60		$Hg(SO_3)_3^{-4}$	25.96[d]
	$Ag(SO_3)_2^{-3}$	8.68		$Hg(S_2O_3)_2^{-2}$	29.23
	$Ag(SO_3)_3^{-5}$	9.00		$Hg(S_2O_3)_3^{-4}$	30.6
Ba^{+2}	$BaS_2O_3^0$	2.27	K^{+}	$KS_2O_3^-$	0.96
Ca^{+2}	$CaS_2O_3^0$	1.98	Mg^{+2}	$MgS_2O_3^0$	1.82
Cd^{+2}	CdS^0	19.5[b]	Mn^{+2}	$MnS_2O_3^0$	1.95
	$CdS_2O_3^0$	3.92	Na^{+}	$NaS_2O_3^-$	0.53
	$Cd(S_2O_3)_2^{-2}$	6.3	Ni^{+2}	$NiS_2O_3^0$	2.06
	$Cd(S_2O_3)_3^{-4}$	6.4[b]	Pb^{+2}	$PbS_2O_3^0$	3.0[b]
	$Cd(S_2O_3)_4^{-6}$	8.2[b]	Zn^{+2}	ZnS^0	16.6[b]
	$Cd_2(S_2O_3)_2^0$	13.0[b,c]		$ZnS_2O_3^0$	2.35
Ce^{+3}	$CeSO_3^+$	8.04		$Zn(S_2O_3)_2^{-2}$	2.5[b]
Co^{+2}	$CoS_2O_3^0$	2.05			

a. 20°C
b. From Morel [12]
c. For complexes consisting of more than one metal atom, the given stability constant is for the overall formation of the ligand from the individual anions and cations, e.g, $2Cd^{+2} + 2S_2O_3^{-2} = Cd_2(S_2O_3)_2^0$
d. 18°C

Source: Martell and Smith [11] unless otherwise indicated.

TABLE 8.2-3

Solubility Products of Sulfides, Sulfites, and Thiosulfates
(at 25°C and zero ionic strength)

Solid	log K_{sp}	Solid	log K_{sp}
Ag_2S	−50.1	HgS (black)	−52.7
Ag_2SO_3	−13.82	HgS (red)	−53.3
BaS_2O_3	− 4.79	MnS (pink)	−10.5
$CaSO_3$	− 6.5	MnS (green)	−13.5
CdS	−27.0	α-NiS	−19.4
α-CoS	−21.3	β-NiS	−24.9
β-CoS	−25.6	γ-NiS	−26.6
CuS	−36.1	PbS	−27.5
Cu_2S	−48.5	α-ZnS	−24.7
FeS	−18.1	β-ZnS	−22.5

Source: Martell and Smith [11]

accumulation is common [4]. (See § 6.9.4 for a discussion of acid mine drainage resulting from pyrite (FeS_2) oxidation.)

Despite the formation of hydrogen sulfide gas in soils, little sulfur volatilizes to the atmosphere [2]. Hydrogen sulfide in soils reacts with iron and other transition metals to form insoluble sulfides, but, where Fe concentrations are low, H_2S may accumulate to toxic concentrations [4].

The mobility of sulfide in soil is extremely low, due to the low solubility of metal sulfides. Sulfide in solution tends to precipitate on the soil as FeS and thus is removed from solution. If the precipitated sulfides are later exposed to an oxidizing environment, they may be converted to sulfate, which is retained only weakly by soils [2]. (See § 6.11.5.)

8.2.5 Literature Cited

1. Alexander, M., *Introduction to Soil Microbiology,* John Wiley & Sons, New York (1977), as cited in Brown [4].
2. Bohn, H.L., B.L. McNeal and G.A. O'Connor, *Soil Chemistry,* John Wiley & Sons, New York (1979).

3. Bowen, H.J.M., *Environmental Chemistry of the Elements,* Academic Press, London (1979).

4. Brown, K.A., "Sulfur in the Environment: A Review," *Environ. Pollut. B,* **3,** 47-80 (1982).

5. Czurda, V., "Hydrogen Sulfide as a Factor for the Growth of Green Algae," *Zentralbl. Bakteriol. Parasitenkd.,* II, **103,** 285-311 (1941) and Castenholtz, R.W., "The Effect of Sulfide on Blue-Green Algae of Hot Springs, I. New Zealand-Iceland," *J. Phycol.,* **12,** 54-68 (1976), both as cited (together) in Brown [4].

6. Grant, W.D. and P.E. Long, "Environmental Microbiology" in *The Natural Environment and the Biogeochemical Cycles,* Vol. 1, Part D of *Environmental Chemistry,* O. Hutzinger (ed.), Springer-Verlag, Berlin, 125-237 (1985).

7. Halstead, R.L. and P.J. Rennie, "The Effects of Sulfur on Soils in Canada," in *Sulfur and Its Inorganic Derivatives in the Canadian Environment,* National Research Council of Canada, Report 15015, 181-220 (1977), as cited in Brown [4].

8. Jellinek, F., "Sulfides," in *Inorganic Sulphur Chemistry,* G. Nickless (ed.), Elsevier Publishing Co., New York, 669-747 (1968).

9. Kotrlý, S. and L. Šůcha, *Handbook of Chemical Equilibria in Analytical Chemistry,* John Wiley & Sons, New York (1985).

10. Lyr, H., *Plant Diseases,* Vol. 1, J.G. Horsfall and E.B. Cowling (eds.), Academic Press, New York, 239 (1977), as cited in Bowen [3].

11. Martell, A.E. and R.M. Smith, *Critical Stability Constants,* Vol. 4, *Inorganic Complexes and Vol. 5, First Supplement,* Plenum Press, New York (1976 and 1982).

12. Morel, F.M.M., *Principles of Aquatic Chemistry,* John Wiley & Sons, New York (1983).

13. Odum, E.P., *Fundamentals of Ecology,* 3rd ed., W.B. Saunders, Philadelphia (1971).

14. U.S. Environmental Protection Agency, "Water Quality Criteria 1972," Report No. EPA-R3,73.033 (1973), as cited in Brown [4].

15. U.S. Environmental Protection Agency, "National Interim Secondary Water Regulations," *Fed. Reg.,* **42,** 62 (March 1977), as cited in Dohnalek, D.A. and J.A. FitzPatrick, "The Chemistry of Reduced Sulfur Species and Their Removal from Groundwater Supplies," *J. Am. Water Works Assoc.,* **75,** 298-308 (1983).

16. U.S. Environmental Protection Agency, "Quality Criteria for Water" (Red Book) (July 1976).

17. Williams, C.H., "Sulfur Deficiency in Australia," *Sulfur Inst. J.,* **8,** 5-8 (1972), as cited in Brown [4].

18. Zehnder, A.J.B. and S.H. Zinder, *Handbook of Environmental Chemistry,* Vol. 1A, O. Hutzinger (ed.), Springer-Verlag, Berlin (1980), as cited by Grant and Long [6].

19. Zinder, S.H. and T.D. Brock, "Microbial Transformations of Sulfur in the Environment" in *Sulfur in the Environment: Part II. Ecological Impacts,* J.O. Nriagu (ed.), John Wiley & Sons, New York, 445-66 (1978).

8.3 PHOSPHORUS (P)

The environmental importance of phosphorus stems not from its abundance — other elements are far more abundant — but rather from its role as a plant nutrient. Since orthophosphate species (PO_4^{-3}, HPO_4^{-2}, $H_2PO_4^{-}$) are the predominant forms of phosphorus in the natural environment, this discussion of phosphorus reactions and transport focuses on the chemistry of orthophosphates.

8.3.1 The Phosphorus Cycle

Phosphorus is an essential nutrient for all organisms [2]. For this reason, biological assimilation and mobilization play a large part in the phosphorus cycle, as shown in Figure 8.3-1. Estimates of the annual quantity of phosphorus mined for agricultural

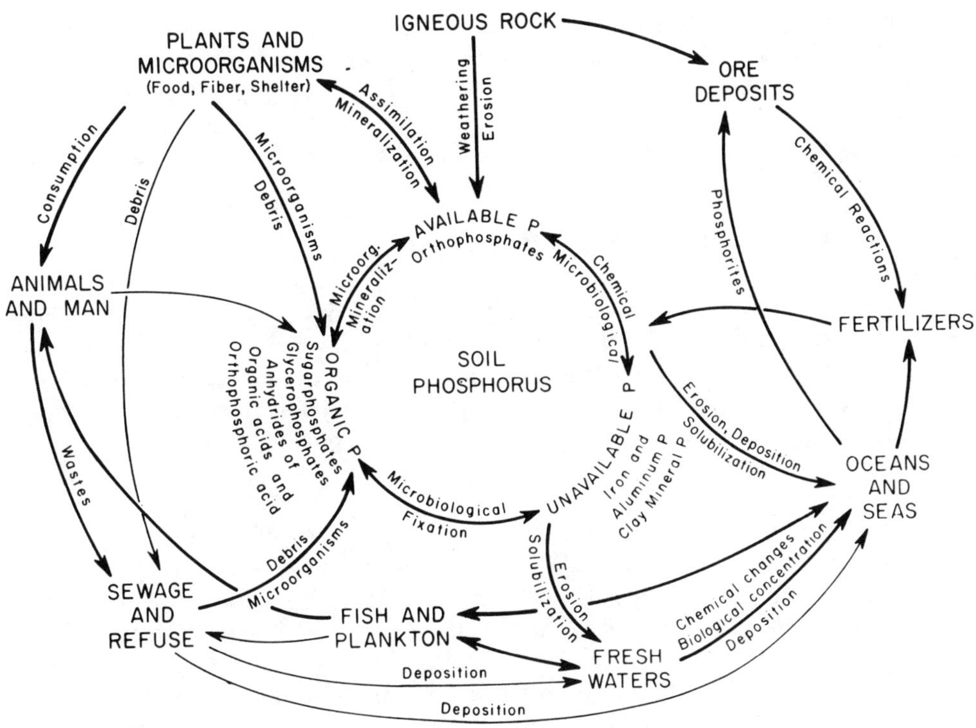

Source: Fuller [5]. *(Copyright 1972, Litton Educational Publishing, Inc. Reprinted with permission of R.W. Fairbridge, current holder of the copyright.)*

FIGURE 8.3-1 The Phosphorus Cycle

use range from 13 million tonnes (metric tons) [18] to 11.2 million tonnes [2], as phosphorus use in fertilizers is exceeded only by that of nitrogen and potassium. By comparison, the annual release of phosphorus through the weathering of rocks and its transfer from rivers to oceans have been estimated by one source to be 4 million and 0.74 million tonnes, respectively [2], while another source has given estimates of 13 and 17 million tonnes, respectively [18].

Part of the phosphorus washed into the oceans is returned to land biologically. The deposit of guano in coastal areas by sea birds has led to the formation of economically significant phosphate accumulations, some amounting to several hundred thousand tonnes before they were mined [14]. Phosphorus is also returned to land through the removal of fish by man (an estimated 55,000 tonnes of phosphorus per year [16]) and other animals. However, fish harvesting and guano deposits offset only a very small fraction of the phosphorus lost to sea, half of which is due to man's activities [18].

Notably absent from the phosphorus cycle in Figure 8.3-1 is an atmospheric component. Worldwide anthropogenic emissions of phosphorus (resulting from combustion) are estimated to be only 36,000 tonnes annually, or roughly 0.04% of the corresponding figure for sulfur. Phosphorus in the atmosphere appears only in dust particles and microbial debris [5], since environmentally stable phosphate compounds are generally not volatile [4].

8.3.2 Species of Environmental Importance

While most inorganic phosphates are considered non-toxic, elemental phosphorus is extremely toxic and is subject to bioaccumulation in much the same way as mercury [23]. For this reason, the EPA recommends that the concentration of elemental phosphorus not exceed 0.10 μg/l in marine or estuarine waters [23]. However, waters containing significant concentrations of elemental phosphorus are rare, as it is readily oxidized in nature [5].

Phosphate salts, on the other hand, are widespread and of considerable environmental concern as nutrients; their fate in aquatic environments has been extensively studied over the past two decades. In surface waters where phosphorus is the limiting factor for growth, inputs of phosphate can result in obnoxious algal blooms, making the water unfit or unattractive for swimming, fishing or drinking. Blooms have been observed in waters with phosphate concentrations as low as 5×10^{-6} M. Where phosphate inputs to a lake are continually elevated (and phosphorus is limiting), eutrophication is accelerated.

The total (organic plus inorganic) phosphate concentration is more important than the dissolved inorganic concentration in determining biological productivity, since algae are able to utilize dissolved organic phosphate (e.g., that released by dead algae) [4]. Concern over the discharge of phosphate from sewage systems has led to the banning of synthetic detergents containing phosphates in many states [4] and to the construction of tertiary treatment systems for phosphorus removal at sewage treatment plants that discharge into sensitive waters.

8.3.3 Phosphorus Chemistry

REDOX REACTIONS

Phosphorus exists in various compounds in oxidation states ranging from −III to V [5]. It is most commonly found in the phosphate (V) state. Elemental phosphorus (oxidation state zero) is extremely reactive [5] and thus short-lived. Phosphine (PH_3), which contains phosphorus in its fully reduced form (−III), has been found in polluted springs and in the hypolimnion of lakes and marshes under highly reducing conditions [5]. Nevertheless, phosphates in inorganic and organic form are the predominant form of phosphorus in nature, and only orthophosphate (PO_4^{-3}) and its protonated forms are likely to be important in the environment [8].

The equilibrium constants for the redox reactions below [11] indicate the expected dominance of phosphorus in its fully oxidized state as phosphate.

Reaction	log K	
$H_2PO_4^- + 2H^+ + 2e^- = H_2PO_3^- + H_2O$	−19.87	(1)
$H_3PO_4^0 + 5H^+ + 5e^- = P\text{ (c, white)} + 4H_2O$	−35.21	(2)
$H_2PO_4^- + 9H^+ + 8e^- = PH_3(g) + 4H_2O$	−36.73	(3)

Phosphate is thus a poor oxidizing agent, and oxidation reactions are rarely encountered under natural conditions.

ACID-BASE CHEMISTRY

When completely deprotonated, phosphoric acid (H_3PO_4) yields the orthophosphate ion. It loses its first proton readily but the succeeding two far less so, as the following acid dissociation constants show [15]:

Reaction	log K	
$H_3PO_4^0 \rightleftharpoons H^+ + H_2PO_4^-$	− 2.15	(4)
$H_2PO_4^- \rightleftharpoons H^+ + HPO_4^{-2}$	− 7.20	(5)
$HPO_4^{-2} \rightleftharpoons H^+ + PO_4^{-3}$	−12.35	(6)

The distribution of phosphate species in fresh water at 25°C is shown as a function of pH in Figure 8.3-2. This diagram applies to waters of low ionic strength (dissolved solids content <1,000 mg/l) [7]. The presence of HPO_4^{-2} in waters of moderate pH (roughly 5-9) will add to the alkalinity of the water, as it is converted to $H_2PO_4^-$ upon the addition of a strong acid. (See § 6.12.3 for a discussion of alkalinity.)

Other acids containing phosphorus, such as phosphorous acid (H_3PO_3) and pyrophosphoric acid ($H_4P_2O_7$), are of little environmental significance due to their environmental instability. The dissociation of H_3PO_3 results in $H_2PO_3^-$, which is readily oxidized to $H_2PO_4^-$, as shown by the equilibrium constant for equation 1.

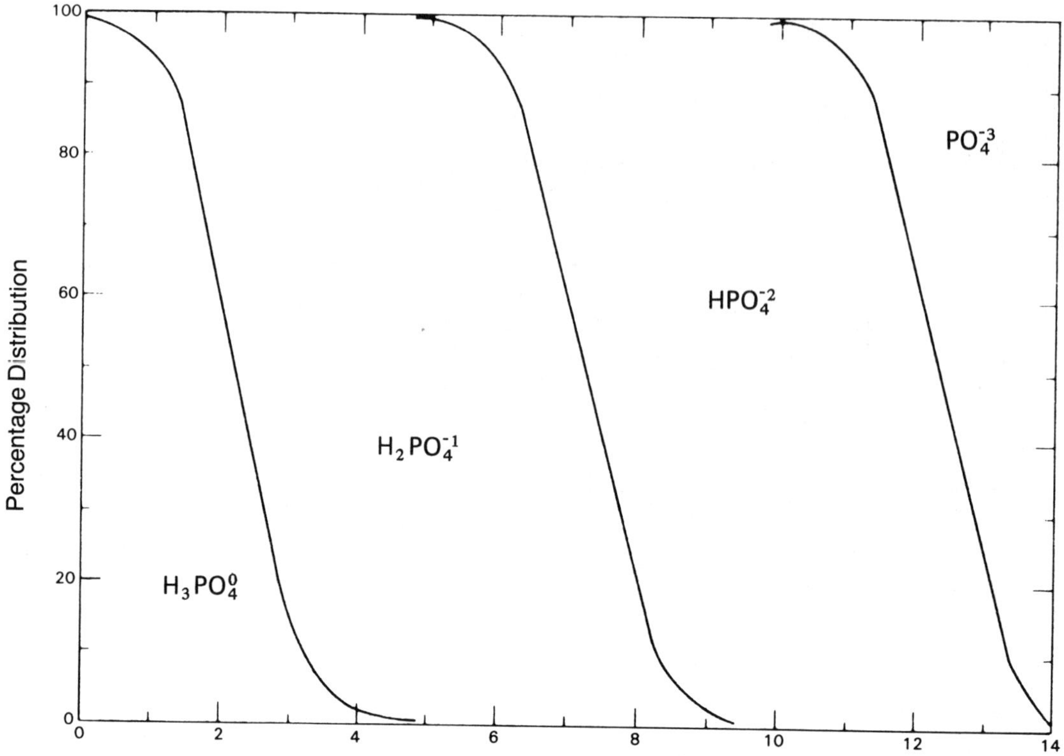

(T = 25°C, P = 1 atm, I = 0)

Source: Hem [7]

FIGURE 8.3-2 Percentage of Total Dissolved Phosphate in Solution as a Function of pH

Pyrophosphoric acid is unstable because pyrophosphate is rapidly hydrolyzed to the phosphate ion by microorganisms [21].

PHOSPHATE COMPLEXATION AND SOLUBILITY

Both orthophosphates and polyphosphates (e.g., $P_2O_7^{-4}$) complex with metal ions. (Polyphosphates are described in § 2.8.) Orthophosphate complexation is important for transition and high-positive-valence ions but less so for alkali and alkaline earth metals. Polyphosphates form relatively strong complexes with all three types of metals [24]. Stability constants for ortho-, pyro-, and tripolyphosphate are given in Table 8.3-1.

The amount of phosphate complexation in seawater varies with the form (degree of protonation) of the phosphate. In seawater of average salinity and pH 8.0 at 20°C, 87% of the inorganic phosphate exists as HPO_4^{-2}, 12% as PO_4^{-3}, and 1% as $H_2PO_4^-$ [9].

TABLE 8.3-1

Logarithms of Complexation Equilibrium Constants (K) and Overall Stability Constants (β) for Metals with Ortho-, Pyro-, and Tripolyphosphate[a]
(at 25°C and zero ionic strength)

Metal	Orthophosphate		Pyrophosphate		Tripolyphosphate	
Ba^{+2}					$BaP_3O_{10}^{-3}$	6.3
					$BaHP_3O_{10}^{-2}$	3.7
Ca^{+2}	$CaPO_4^-$	6.46[b]	$CaP_2O_7^{-2}$	6.8[b]	$CaP_3O_{10}^{-3}$	8.1[b]
	$CaHPO_4^0$	2.74[b]	$CaHP_2O_7^-$	4.0	$CaHP_3O_{10}^{-2}$	4.9
	$CaH_2PO_4^+$	1.4[b]				
Cd^{+2}			$CdP_2O_7^{-2}$	8.7	$CdP_3O_{10}^{-3}$	9.8[b]
					$CdHP_3O_{10}^{-2}$	5.4
Co^{+2}	$CoHPO_4^0$	3.1	$CoP_2O_7^{-2}$	7.9	$CoP_3O_{10}^{-3}$	9.7
Cu^{+2}	$CuHPO_4^0$	4.1	CuP_2O_7	9.8	$CuP_3O_{10}^{-3}$	11.1
	$CuH_2PO_4^+$	1.7	$CuHP_2O_7^-$	6.1	$CuHP_3O_{10}^{-2}$	6.3
			$Cu(P_2O_7)_2^{-4}$	12.5		
			$CuH_2P_2O_7^0$	3.1		
Fe^{+2}	$FeHPO_4^0$	3.6[b]				
	$FeH_2PO_4^+$	2.7[b]				
Fe^{+3}					$FeH_2P_3O_{10}^0$	7.03[c]
					$FeH_3P_3O_{10}^+$	6.37[c]
K^+	$KHPO_4^-$	1.0	$KP_2O_7^{-3}$	2.1[b]	$KP_3O_{10}^{-4}$	2.8
Mg^{+2}	$MgPO_4^-$	4.8	$MgP_2O_7^{-2}$	7.2[b]	$MgP_3O_{10}^{-3}$	8.6[b]
	$MgHPO_4^0$	2.91[b]	$MgHP_2O_7^-$	4.7	$MgHP_3O_{10}^{-2}$	4.9
	$MgH_2PO_4^-$	1.2				
Mn^{+2}					$MnP_3O_{10}^{-3}$	9.9
					$MnHP_3O_{10}^{-2}$	5.6
Na^+	$NaHPO_4^-$	1.1	$NaP_2O_7^{-3}$	2.29[b]	$NaP_3O_{10}^{-4}$	2.7[b]
			$Na_2P_2O_7^{-2}$	4.2	$NaHP_3O_{10}^{-3}$	2.4
			$NaHP_2O_7^{-2}$	1.4[b]		
Ni^{+2}	$NiHPO_4^0$	3.0	$NiP_2O_7^{-2}$	7.7	$NiP_3O_{10}^{-3}$	9.5
			$NiHP_2O_7^-$	5.0	$NiHP_3O_{10}^{-2}$	5.5

(Continued)

TABLE 8.3-1 (Continued)

Metal	Orthophosphate		Pyrophosphate		Tripolyphosphate	
Pb^{+2}	$PbHPO_4^0$	3.1[b]	$PbP_2O_7^{-2}$	9.5		
	$PbH_2PO_4^+$	1.5[b]	$Pb(P_2O_7)_2^{-6}$	10.2		
Sr^{+2}	$SrPO_4^-$	5.5	$SrP_2O_7^{-2}$	5.4	$SrP_3O_{10}^{-3}$	7.2[b]
	$SrHPO_4^0$	2.1			$SrHP_3O_{10}^{-2}$	4.4
	$SrH_2PO_4^+$	0.7				
Zn^{+2}	$ZnHPO_4^0$	3.3	$ZnP_2O_7^{-2}$	8.7[b]	$ZnP_3O_{10}^{-3}$	10.3
	$ZnH_2PO_4^+$	1.6	$Zn(P_2O_7)_2^{-6}$	11.0[b]	$ZnHP_3O_{10}^{-3}$	5.7

a. Values given for complexes involving a protonated form of the phosphate (HPO_4^{-2}, $HP_2O_7^{-3}$, $HP_3O_{10}^{-4}$, etc.) are for the formation of the complex by the metal and the entire ligand (e.g., Na + HPO_4^{-2} = $NaHPO_4^-$, not Na + H^+ + PO_4^{-3} = $NaHPO_4^-$).
b. From Martell and Smith [13].
c. From Kotrlý and Šůcha [10].

Source: Adapted from Morel [15], except as indicated.

Forty-four percent of the HPO_4^{-2} and 99.6% of the PO_4^{-3} are believed to exist in complexed form with metals other than sodium [9].

Figure 8.3-3 shows the distribution of aqueous phosphate in equilibrium with Ca^{+2}, Mg^{+2} and Fe^{+3} species for a soil-water system. At pH 4-6, free $H_2PO_4^-$ is the dominant species, but at higher pH values $CaHPO_4^0$, $MgHPO_4^0$, HPO_4^{-2} and $CaPO_4^-$ become important.

The most notable characteristic of metal phosphates is their low solubility. Table 8.3-2 lists the solubility equilibrium constants for various metal phosphates and phosphate minerals. The low solubility of phosphates strongly affects their behavior in aqueous and soil systems (as described below) and is important in water treatment processes. Phosphate salts are employed in industrial water softening processes to remove calcium [21], and aluminum and iron salts are used to precipitate phosphates from sewage [3].

The concentration of phosphates in natural waters is limited by the formation of insoluble species such as $FePO_4$ and $CaHPO_4$. In the oceans and many lakes, iron(III) oxides in sediments are the primary sink for phosphates [22]. Under most conditions phosphates will remain as part of the sediments; however, if the bottom of a lake becomes anaerobic, resulting in reducing and acidic conditions, phosphate minerals may be converted to soluble forms as insoluble $FePO_4$ is reduced to soluble $Fe_3(PO_4)_2$ and insoluble $CaHPO_4$ is acidified to soluble $Ca(H_2PO_4)_2$ [4].

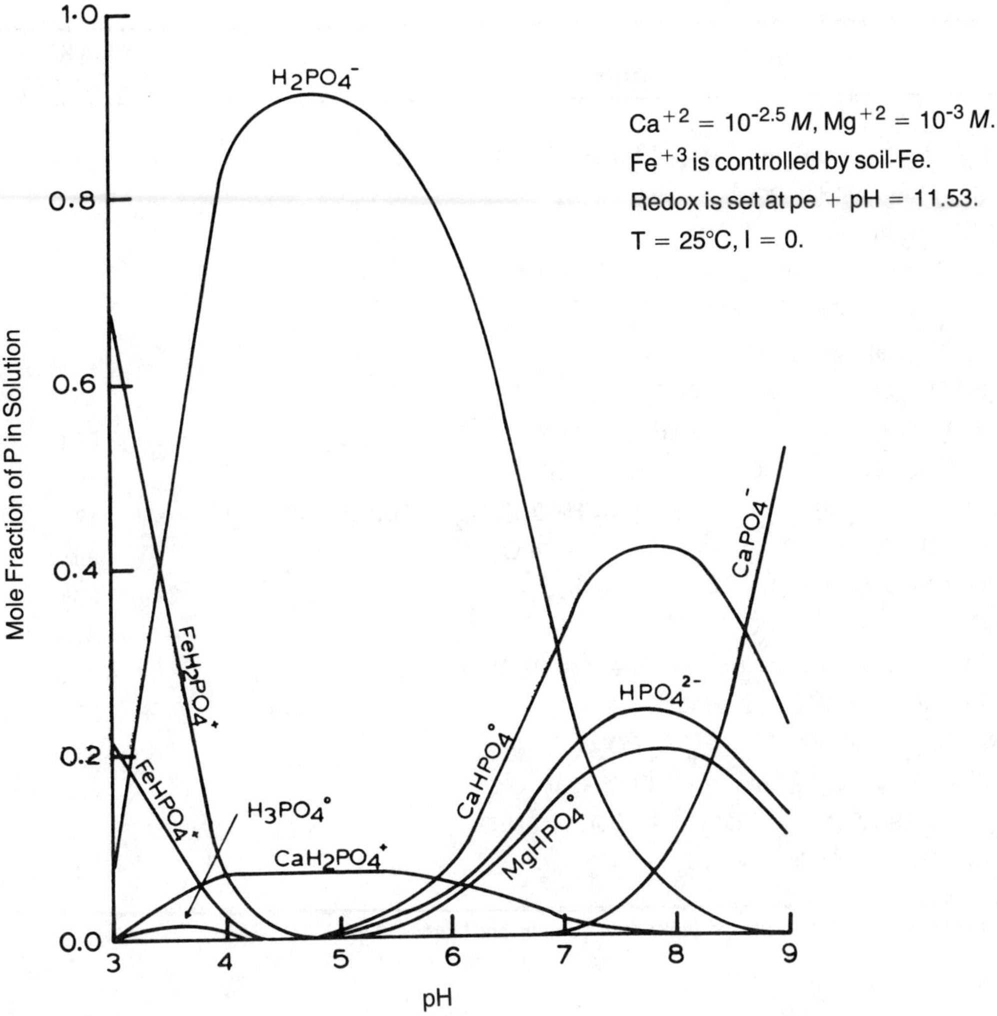

Source: Lindsay [11]. *(Copyright 1979, John Wiley & Sons. Reprinted with permission.)*

FIGURE 8.3-3 Effect of pH on Phosphate Species in Soil Solution

TABLE 8.3-2

Phosphate Solubility Equilibrium Constants

Reaction	log K (25°C, I = 0)
$AlPO_4 \cdot 2H_2O(s)$(variscite) $\rightleftharpoons Al^{+3} + PO_4^{-3} + 2H_2O$	−21
$AlPO_4(s) \rightleftharpoons Al^{+3} + PO_4^{-3}$	−20.6
$Ca(H_2PO_4)_2(s) \rightleftharpoons Ca^{+2} + 2H_2PO_4^-$	−1.14
$CaHPO_4 \cdot 2H_2O(s) \rightleftharpoons Ca^{+2} + HPO_4^{-2} + 2H_2O$	−6.6
$CaHPO_4(s) \rightleftharpoons Ca^{+2} + HPO_4^{-2}$	−6.6
$\beta\text{-}Ca_3(PO_4)_2(s) \rightleftharpoons 3Ca^{+2} + 2PO_4^{-3}$	−24.0
$Ca_4H(PO_4)_3(s) \rightleftharpoons 4Ca^{+2} + 3PO_4^{-3} + H^+$	−46.9
$Ca_5(PO_4)_3OH(s) \rightleftharpoons 5Ca^{+2} + 3PO_4^{-3} + OH^-$	−55.6
$Ca_{10}(PO_4)_6F_2(s) \rightleftharpoons 10Ca^{+2} + 6PO_4^{-3} + 2F^-$	−118
$Ca_{10}(PO_4)_6(OH)_2(s) + 6H_2O \rightleftharpoons 4[Ca_2(HPO_4)(OH)_2] + 2Ca^{+2} + 2HPO_4^{-2}$	−17
$CaHAl(PO_4)_2(s) \rightleftharpoons Ca^{+2} + Al^{+3} + H^+ + 2PO_4^{-3}$	−39
$FePO_4 \cdot 2H_2O(s)$(strengite) $\rightleftharpoons Fe^{+3} + PO_4^{-3} + 2H_2O$	−26.4
$FePO_4(s) \rightleftharpoons Fe^{+3} + PO_4^{-3}$	−21.9
$Fe_3(PO_4)_2 \cdot 8H_2O(s)$(vivianite) $\rightleftharpoons 3Fe^{+2} + 2PO_4^{-3} + 8H_2O$	−36.0
$Fe_3(PO_4)_2(s) \rightleftharpoons 3Fe^{+2} + 2PO_4^{-3}$	~ −32
$FeNH_4PO_4(s) \rightleftharpoons Fe^{+2} + NH_4^+ + PO_4^{-3}$	~ −13
$MgHPO_4 \cdot 3H_2O(s) \rightleftharpoons Mg^{+2} + HPO_4^{-2} + 3H_2O$	−5.8
$Mg_3(PO_4)_2 \cdot 8H_2O(s) \rightleftharpoons 3Mg^{+2} + 2PO_4^{-3} + 8H_2O$	−25.2
$MgNH_4PO_4(s) \rightleftharpoons Mg^{+2} + NH_4^+ + PO_4^{-3}$	−12.6

Sources: Morel [15]; Snoeyink and Jenkins [21]; Stumm and Morgan [22].

The most common phosphate minerals (which are also of very low solubility) are apatites, $Ca_5X(PO_4)_3$, where X is Cl, F or OH, corresponding to chlor-, fluor-, and hydroxy-apatite, respectively. Ninety-five percent of the earth's solid phosphate exists as fluorapatite [19]. However, the form in which phosphates are deposited depends on their environment. While variscite ($AlPO_4 \cdot 2H_2O$) and strengite ($FePO_4 \cdot 2H_2O$) appear to be more common in soil and fresh-water sediments, apatite prevails in certain marine sediments [22].

8.3.4 Fate of Phosphorus Species in Soil

The fate of phosphorus species in soil and water environments is of special concern because of the importance of phosphorus as an essential nutrient. Phosphate added to soil as fertilizer is quickly sorbed and later "fixed" (probably precipitated) into less soluble forms [1]. While the precise mechanism of phosphate sorption is uncertain (see § 2.12), phosphate fixation is appreciable in all but very coarse-textured soils [1]; only about one-fourth of fertilizer phosphate is usable by plants, the rest being lost to the occluded soil fraction [20].

The sorption isotherm equations for three soils are given in Table 8.3-3. Three different types of equations — linear, Freundlich and Langmuir — were used to fit the data, indicating the specificity of the isotherm equations to the soils and the experimental conditions. Thus, these equations should be taken as examples only and not as representative of soil sorption in general.

TABLE 8.3-3

Phosphate Sorption on Soils

Soil	Equation[a]	Equation Type	Solution Composition ($\mu g\ PO_4$-P/ml)	Source
Oldsmar Fine Sand[b]	$S = 17C^{0.345}$	Freundlich	0-100 in 0.01 N $CaCl_2$	[12]
Oldsmar Fine Sand[c]	$S = 0.23C$	Linear		
Loam[d]	$S = \frac{C}{0.0102C + 0.0622}$	Langmuir	1.9-12	[17] (adapted)
Loam[e]	$S = \frac{C}{0.0082C + 0.393}$	Langmuir	1.5-22	
Clay[f]	$S = \frac{C}{0.0040C + 0.011}$	Langmuir	0.8-16	

a. S = Phosphate sorbed onto soil ($\mu g\ PO_4$-P/g soil)
C = Phosphate concentration in solution ($\mu g\ PO_4$-P/ml)

b. 96% sand, 1.5% silt, 2.4% clay, 1.56% organic matter, pH = 4.75, obtained at a depth of 0-10 cm.

c. 97% sand, 1.6% silt, 1.4% clay, 0.00% organic matter, pH = 6.35, obtained at a depth of 30-40 cm.

	Size Fraction (μm)	Percent Clay	pH
d. Quay fine sandy loam	2-0.2	5.1	7.0-7.9
	<0.2	15.8	
e. Owyhee silt loam	2-0.2	11.4	
	<0.2	12.5	
f. Pierre clay	2-0.2	11.9	
	<0.2	32.1	

Organic phosphates may represent the principal form of phosphate in some soils. Almost all of the phosphate taken up by plants is converted to organic form [4], and decomposer microorganisms also produce organic phosphates; three-fourths of all organic phosphates in soils appear to be produced microbially [6]. Unlike insoluble metal phosphates, they are not permanently lost from the soil solution. Being salts or metal complexes of phosphate esters, they release their phosphate slowly by hydrolysis, a process that is greatly accelerated by phosphatase enzymes [4].

The concentration of phosphate in soil waters is on the order of 10^{-5} to 10^{-6} M [1]. In water draining from agricultural soils, it ranges from 10^{-7} M [1] to 2-30 $\times$ 10^{-6} M [21], but this is not considered an important loss from the soil [4]. The additional loss of soluble phosphate due to fertilization is also believed to be negligible. Erosion of soil to which phosphate is "fixed," however, may be responsible for much of the roughly seven million tonnes of phosphorus resulting from man's activities that is drained to the oceans annually [4].

8.3.5 Literature Cited

1. Bohn, H.L., B.L. McNeal and G.A. O'Connor, *Soil Chemistry,* John Wiley & Sons, New York (1979).
2. Bowen, H.J.M, *Environmental Chemistry of Elements,* Academic Press, London (1979).
3. Clark, J.W., W. Viessman, Jr., and M.J. Hammer, *Water Supply and Pollution Control,* 3rd ed., Harper and Row, New York (1977).
4. Emsley, J., "The Phosphorus Cycle" in *The Natural Environment and the Biogeochemical Cycles,* Vol. 1, Part A of *The Handbook of Environmental Chemistry,* O.E. Hutzinger (ed.), Springer-Verlag, Berlin, 147-67 (1980).
5. Fuller, W.H., "Phosphorus: Element and Geochemistry," in *The Encyclopedia of Geochemistry and Environmental Sciences,* R.W. Fairbridge (ed.), Van Nostrand Reinhold, New York, 942-46 (1972).
6. Fuller, W.H. and R.H. Fox, *Soil Sci.,* **97** (1964), as cited in Emsley [4].
7. Hem, J.D., *Study and Interpretation of the Chemical Characteristics of Natural Water,* 2nd ed., Water Supply Paper 1473, U.S. Geological Survey, Washington, D.C. (1970).
8. Hutchinson, G.E., "Chemistry of Lakes," Part 2 of Vol. I in *A Treatise on Limnology,* John Wiley & Sons, New York (1975).
9. Kester, D.R. and R.M. Pytkowicz, *Limnol. Oceanogr.,* **12** (1967), as cited in Stumm and Morgan [22].
10. Kotrlý, S. and L. Šůcha, *Handbook of Chemical Equilibria in Analytical Chemistry,* Halsted Press, New York (1985).
11. Lindsay, W.J., *Chemical Equilibria in Soils,* John Wiley & Sons, New York (1979).

12. Mansell, R.S., H.M. Selim, P. Kanchanasut, J.M. Davidson and J.G.A. Fiskell, "Experimental and Simulated Transport of Phosphorus through Sandy Soils," *Water Resour. Res.* **13,** 189-94 (1977).

13. Martell, A.E. and R.M. Smith, *Critical Stability Constants*, Vol. 4, *Inorganic Complexes* and Vol. 5, *First Supplement*, Plenum Press, New York (1976 and 1982).

14. McKelvey, V.E., "Abundance and Distribution of Phosphorus in the Lithosphere," in *Environmental Phosphorus Handbook,* E.J. Griffith, A. Beeton, J.M. Spences and D.T. Mitchell (eds.), John Wiley & Sons, New York (1973).

15. Morel, F.M.M., *Principles of Aquatic Chemistry,* John Wiley & Sons, New York (1983).

16. Odum, E.P., *Fundamentals of Ecology,* 3rd ed., W.B. Saunders, Philadelphia (1971).

17. Olsen, R.S. and F.S. Watanabe, "A Method to Determine a Phosphorus Adsorption Maximum of Soils as Measured by the Langmuir Isotherm," *Soil Sci. Soc. Am. Proc.,* **21** (1957).

18. Pierrou, U., "Nitrogen, Phosphorus, and Sulphur Global Cycles," SCOPE Report 7, B.H. Svensson and R. Soderlund (eds.), *Ecol. Bull.,* **22** (Stockholm) (1976), as cited in Emsley [4].

19. Rankana, K. and T.G. Sahama, *Geochemistry,* University of Chicago Press, Chicago (1949), as cited in Emsley [4].

20. Russell, E.W., *Soil Conditions and Plant Growth,* 10th ed., Longman, London (1973), as cited in Emsley [4].

21. Snoeyink, V.L. and D. Jenkins, *Water Chemistry,* John Wiley & Sons, New York (1980).

22. Stumm, W. and J.J. Morgan, *Aquatic Chemistry,* 2nd ed., John Wiley & Sons, New York (1981).

23. U.S. Environmental Protection Agency, "Quality Criteria for Water" (July 1976).

24. Van Wazer, J.R., "The Compounds of Phosphorus," in *Environmental Phosphorus Handbook,* E.J. Griffith, A. Beeton, J.M. Spences and D.T. Mitchell (eds.), John Wiley & Sons, New York (1973).

8.4 FLUORINE (F)

Fluorine makes up 0.06-0.09% by weight of the upper layers of the lithosphere [5] and is widely distributed in both the lithosphere and hydrosphere [20]. It is probably an essential element for mammals [4], but at high concentrations it is toxic to both animals and plants.

8.4.1 The Fluorine Cycle

Like the more abundant sulfur and nitrogen, fluorine cycles through the environment. The environmental transport of fluorine species is illustrated in Figure 8.4-1. On a global basis, the annual transfer of fluorine from rivers to oceans has been estimated at 3.7 million tonnes (metric tons), the quantity removed by mining at 2 million tonnes, and the amount released into the atmosphere by combustion at 0.022 million tonnes [4].

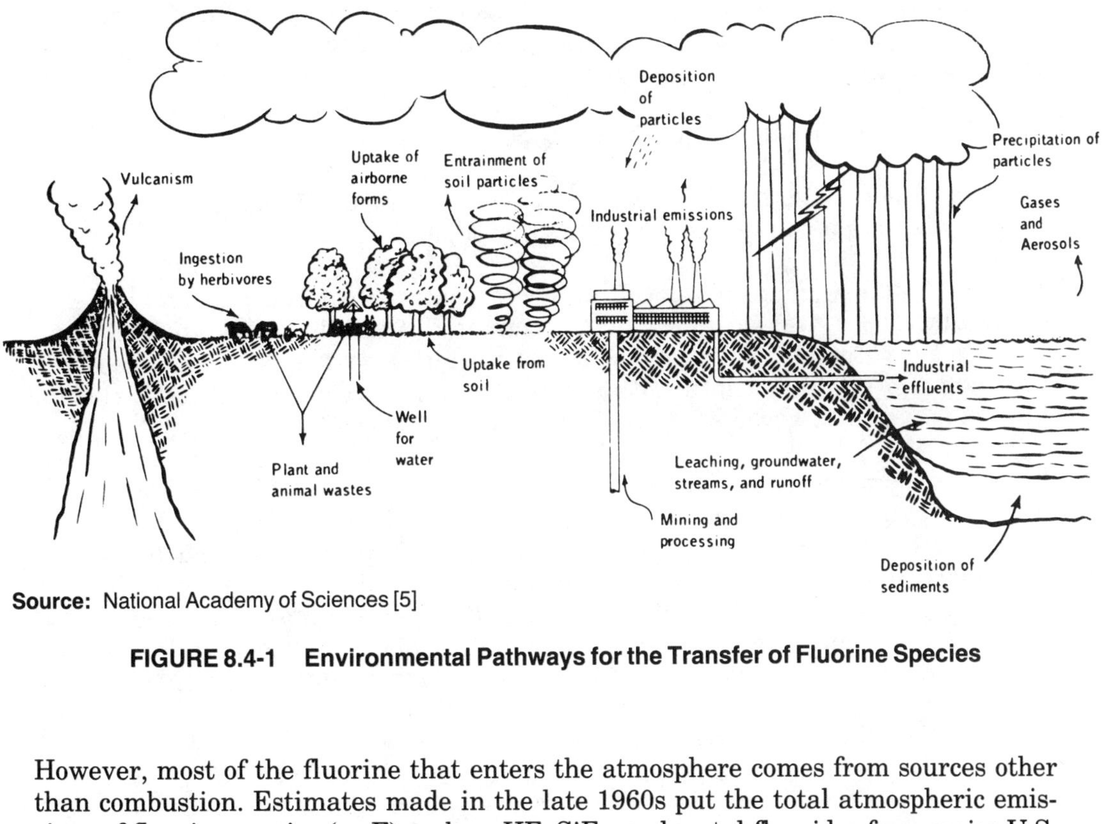

Source: National Academy of Sciences [5]

FIGURE 8.4-1 Environmental Pathways for the Transfer of Fluorine Species

However, most of the fluorine that enters the atmosphere comes from sources other than combustion. Estimates made in the late 1960s put the total atmospheric emissions of fluorine species (as F) such as HF, SiF_4, and metal fluorides from major U.S. industrial sources at 108,000 tonnes, of which only 14,500 came from the combustion of coal [5]. Releases of fluorine-containing species into the atmosphere come from such industries as brick and tile products manufacture, steel and aluminum manufacture, and the production of phosphoric acid, superphosphate fertilizer, and elemental phosphorus. Fluorides entrained in windblown dust and emissions from fumaroles and volcanoes also contribute to the atmospheric content.

Fluoride ions enter surface waters as a result of the natural leaching of soils and from various man-made sources. Many of the industries that release fluoride species into the air also produce aqueous wastes containing dissolved fluorides [8], in part because systems to remove air emissions of fluorides result in aqueous fluoride wastes. Between 5,500 and 27,000 tonnes of fluorides may enter U.S. waterways annually as the result of phosphate mining and processing. Municipal sewerage systems also release fluoride in their effluents. Roughly half the communities in the United States fluoridate their water (as of 1975), and most of the fluoride added to public water supplies ends up in sewage, adding to that contributed by industrial sources. However, secondary sewage treatment may reduce fluoride in the final effluent by more than half. The portion that reaches the oceans may be volatilized (e.g., as HF), enter the atmosphere in aerosols, or be deposited in sediments [8].

Fluoride enters the soil environment through the deposition of atmospheric species and (in agricultural areas) through the use of fluorine-containing pesticides. Fluoride is also added to soils inadvertently as an impurity in phosphate fertilizers. From 0.5 to 4 percent of the weight of these fertilizers may be fluoride [8] since fluorapatite, $CaF_2(PO_4)_6$, is the chief constituent of the phosphate rock used in their manufacture [5].

8.4.2 Species of Environmental Importance

The fluoride ion is the species of greatest environmental concern, since, as described below, it is the predominant form of fluorine under natural conditions.

The ingestion by humans of fluoride at low concentrations has been found to reduce the incidence of dental caries [2]. Many communities add fluoride to their drinking water to a concentration of 1 mg/l as a public health measure. At higher concentrations (>2 mg/l) fluoride can result in the mottling of developing teeth, and skeletal fluorosis may be observed at concentrations of 3.0-6.0 mg/l [9].

Fluoride's effects on livestock are similar to those on humans and in severe cases may lead to crippling. Forage containing fluoride concentrations of 30-40 mg/kg may be extremely toxic to cattle, while sheep, swine, and other species seem to be more tolerant [12]. Many plant species, including forage crops, bioaccumulate fluorides from the air. According to the USDA [12], fluoride has caused more damage to livestock worldwide than any other air pollutant, primarily through the deposition of gaseous and particulate species on forage and their subsequent ingestion.

8.4.3 Aqueous Fluorine Chemistry

REDOX CHEMISTRY

Fluorine is the most electronegative element and one of the most powerful oxidizing agents known [20].

$$F_2 + 2e^- = 2F^- \qquad E^0 = 2.87 \text{ V [22]}$$

In comparison, the corresponding values for Br_2(aq) and I_2(aq) are 1.087 V and 0.535 V, respectively [22]. The great strength of fluorine as an oxidizer means that the element is unlikely to be found in other than its reduced form, F^-. Fluoride is the only oxidation state of fluorine found in natural aqueous systems [10], whereas Br^- (under very acidic conditions) and I^- can be oxidized by O_2 in solution.

ACID-BASE CHEMISTRY

Hydrofluoric acid (HF) is a relatively weak acid ($pK_a = 3.2$) compared with HBr ($pK_a = -9$) and HI ($pK_a = -9.5$) [11]; hence, the fluoride ion, F^-, can act as a base ($pK_b = 10.8$), albeit a weak one. Hydrofluoric acid is known for its ability to dissolve glass, a trait attributed to the formation of fluosilicate ions (SiF_6^{-2}) [19].

Various volatile fluorine compounds are hydrolyzed in the atmosphere to form acids. Gaseous emissions of anhydrous HF, for example, yield hydrofluoric acid, and silicon tetrafluoride reacts to form fluosilicic acid [2]:

$$3SiF_4(g) + 2H_2O \rightarrow SiO_2(s) + 2H_2SiF_6(g)$$

However, ionic particulate fluorides (e.g., CaF_2 and NaF) and some particulate complexes (e.g., Na_3AlF_6 and Na_2SiF_6) do not produce acidic reaction products when dissolved in water [5].

COMPLEXATION AND SOLUBILITY

Fluoride complexes readily with Al^{+3}, Be^{+2}, and Fe^{+3}, as shown by the stability constants in Table 8.4-1. Aluminum fluoride complexes are likely to be found in waters with pH values below neutrality; however, in natural waters the fluoride concentration commonly exceeds that of dissolved aluminum, leading to a lesser proportion of F^- than Al^{+3} in complexed form [10]. In seawater, with a F^- concentration of 8×10^{-5} mole/kg, 51% of F^- has been calculated to exist as the free ion, 47% complexed as MgF^+ and 2% as CaF^+ [21].

Compared with bromide and iodide, fluoride can form either weaker or stronger complexes, depending on the particular metal species. The log stability constants for AgF^0, $AgBr^0$ and AgI^0 are 0.4, 4.68, and 6.55 respectively [11]. However, the corresponding values for ZnF^0, $ZnBr^0$, and ZnI^0 are 1.15, −0.19, and −0.47 respectively [11], showing just the opposite trend.

The common fluoride minerals generally have low solubilities. The most abundant of these is fluorite (CaF_2), which has a $K_{sp}$of $10^{-10.14}$. On a mass basis, both CaI_2 and $CaBr_2$ are more than 30,000 times as soluble as CaF_2. The solubility products of other common fluoride minerals are given in Table 8.4-2. Solutions that contain moderate concentrations of $Na^+$and Al^{+3} may reduce the amount of soluble fluoride by the precipitation of cryolite [2], but few natural waters have sufficient dissolved alumina to make this reaction possible [10].

TABLE 8.4-1

Stability Constants of Fluoride Complexes
(at 25°C and zero ionic strength unless otherwise indicated)

Metal Ion	Complex	log K or log β	Source
Ag^{+}	AgF^{0}	0.4	[15]
Al^{+3}	AlF^{+2}	7.0	[15]
	AlF_2^{+}	12.6	[15]
	AlF_3^{0}	16.7	[15]
	AlF_4^{-}	19.1	[15]
Be^{+2}	BeF^{+}	5.99	[11]
	BeF_2^{0}	8.32[a]	[15]
Ca^{+2}	CaF^{+}	1.1	[15]
Cd^{+2}	CdF^{+}	1.0	[16]
	CdF_2^{0}	1.4	[16]
Co^{+2}	CoF^{+}	1.0	[16]
Cr^{+3}	CrF^{+2}	5.2	[15]
	CrF_2^{+}	9.2	[16]
	CrF_3	12.0	[16]
Cu^{+2}	CuF^{+}	1.5	[16]
Fe^{+2}	FeF^{+}	1.4	[16]
Fe^{+3}	FeF^{+2}	6.0	[15]
	FeF_2^{+}	10.6	[16]
	FeF_3^{0}	13.7	[15]
Hg^{+2}	HgF^{+}	1.6	[15]
Mg^{+2}	MgF^{+}	1.8	[15]
Mn^{+2}	MnF^{+}	1.9	[16]
Na^{+}	NaF^{0}	−0.45	[15]
Ni^{+2}	NiF^{+}	1.1	[16]
Pb^{+2}	PbF^{+}	2.0	[16]
	PbF_2^{0}	3.4	[16]
Zn^{+2}	ZnF^{+}	1.15	[15]

a. I = 0.5 *M*

TABLE 8.4-2

Fluoride Solubility Products
(at various temperatures between 18° and 25°C)

Solid	log K_{sp}	Solid	log K_{sp}
BaF_2	−5.76[a]	$K_2[SiF_6]$	−6.06
$Ba[SiF_6]$	−6	$K_2[ZrF_6]$	−3.3
CaF_2	−10.50[a]	LiF	−2.77[a]
$Ca[SiF_6]$	−3.09	MgF_2	−8.18[a]
CeF_3	−19.1[a]	$Na_3[AlF_6]$	−9.39
$Cs[BF_4]$	−4.7	PbF_2	−7.44[a]
$Cs_2[PtF_6]$	−5.62	$Rb_2[PtF_6]$	−6.12
$Cs_2[SiF_6]$	−4.90	$Rb_2[SiF_6]$	−6.3
CrF_3	−10.18	ScF_3	−17.37
$K_2[PtF_6]$	−4.54	SrF_2	−8.54[a]

a. From Martell and Smith [15]

Source: Dean [6], except as indicated

A study of the Chesapeake Bay concluded that it has no effective sinks for fluoride through precipitation or other processes [2]; all fluoride discharged into the bay apparently flows into the ocean. However, interstitial water may be a sink for small amounts: an extensive survey of coastal and estuarine waters found fluoride enriched in interstitial waters relative to surface waters [23].

8.4.4 Fate of Fluoride in Soil

Over 90% of the natural fluoride content of soils is insoluble or tightly bound to soil particles [14]. Soils tend to have lower concentrations of fluoride at their surface than at a depth of a few feet, indicating that water seeping into the ground may remove soluble fluoride from the surface and that little fluoride is available for uptake by plants; however, research results vary concerning the degree to which fluoride added by pollution or fertilization is available for uptake by plant roots [8].

Several soil characteristics influence the availability of fluoride. It is more readily available in high-clay soils than in sandy ones [7,17], and calcium tends to increase F^- immobilization [13]. The effect of pH is less clear. Fluoride is strongly sorbed onto such soil minerals as goethite (FeOOH) under acid conditions [3], and the sorption of F^- on soils has been reported to increase with decreasing pH [17]; however, others

[2,7] have noted increased fluoride availability with decreasing pH. Sorption data for fluoride on several soils are given in Table 8.4-3. Table 8.4-4 lists the distribution coefficients for iodide sorption on several soil minerals at three pH values.

TABLE 8.4-3

Langmuir Constants for Fluoride Sorption on Several Soils and Minerals

	Sorbate Conc. (*M*)	Electrolyte			Langmuir Constants[a]	
		Indentity	Conc. (*M*)	pH	A_m	K_L
Clay Minerals:						
Kaolinite	$0\text{-}10^{-2.4}$	$MgCl_2$	0.01	6.2	21.9	3.48
Halloysite (expanded)	$0\text{-}10^{-2.4}$	$MgCl_2$	0.01	7.0	175	3.12
Alumina:						
Freshly Precipitated	$0\text{-}10^{-2.4}$	$MgCl_2$	0.01	6.3	1820	4.44
Gibbsite	$0\text{-}10^{-2.4}$	$MgCl_2$	0.01	7.0	11.7	3.68
Soils:						
Loam	$0\text{-}10^{-2.4}$	$MgCl_2$	0.01	7.3[b]	9.0	3.12
Clay	$0\text{-}10^{-2.4}$	$MgCl_2$	0.01	8.0[b]	11.3	3.18
Sandy Loam	$0\text{-}10^{-2.4}$	$MgCl_2$	0.01	7.2[b]	4.6	3.48
Silt Loam	$0\text{-}10^{-2.4}$	$MgCl_2$	0.01	7.7[b]	8.1	3.39
Clay Loam	$0\text{-}10^{-2.4}$	$MgCl_2$	0.01	5.6[b]	69.3	3.68
(1)[c]	$10^{-9}-10^{-2.6}$	NaCl	0.1	5.9[b]	9.9	3.86
(2)[c]	$10^{-9}-10^{-2.6}$	NaCl	0.1	4.8[b]	26.7	3.39
(3)[c]	$10^{-9}-10^{-2.6}$	NaCl	0.1	6.4[b]	32.9	2.93
(4)[c]	$10^{-9}-10^{-2.6}$	NaCl	0.1	7.7[b]	5.3	3.10
(5)[c]	$10^{-9}-10^{-2.6}$	NaCl	0.1	5.9[b]	7.8	3.67

a. A_m = Langmuir sorption maximum (μmol/g);
K_L = Langmuir constant (log M^{-1})

b. Soil pH

c. Composition of soils:

	Clay (%)	Organic Carbon (%)	Al (μg/g)
(1)	20.0	0.18	1000
(2)	43.8	0.53	2300
(3)	25.5	4.31	1200
(4)	19.0	0.79	790
(5)	10.4	0.94	825

Source: Rai *et al.* [18]. *(Copyright 1984, Electric Power Research Institute. Reprinted with permission.)*

TABLE 8.4-4

Distribution Coefficients for Iodide Sorption
(m^3/kg)

	Distribution Coefficient,[a] log K_d		
Mineral	**pH 6**	**pH 7**	**pH 8**
Metal Hydroxides:			
$Fe(OH)_3$	−1.5	−1.7	−1.9
$Al(OH)_3$	−1.4	−1.9	−2.2
Silicate Minerals:			
$(Mg,Fe)_2SiO_4$ (olivine)	−1.5	−2.1	−2.7
$(Al,Mg)_8(Si_4O_{10})_3(OH)_{10}$ (montmorillonite)	−2.4	−3.7	—
SiO_2 (quartz)	−2.7	−2.9	−3.1

a. Determinations based on a 10^{-12} *M* solution of $^{131}I^-$ with the following initial composition: 165 ppm Na^+, 3.9 ppm K^+, 18 ppm Ca^{+2}, 4.3 ppm Mg^{+2}, 123 ppm HCO_3^-, 9.6 ppm SO_4^{-2}, 70 ppm Cl^-, 12 ppm total SiO_2, and pH = 8.2 to simulate groundwater. See § 2.12.6 for a description of the distribution coefficient.

Source: Allard *et al.* [1]

8.4.5 Literature Cited

1. Allard, B., B. Torstenfelt, K. Anderson and J. Rydberg, "Possible Retention of Iodine in the Ground," in *Scientific Basis for Nuclear Waste Management*, Vol. 2, C.J.M. Northrup, Jr. (ed.), Plenum Press, New York (1980).
2. Berkowitz, J.B., M.M. Goyer, J.C. Harris, W.J. Lyman, L.H. Nelken, and D.H. Rosenblatt, "Chemistry, Toxicology and Potential Environmental Effects of Selected Organic Pollutants," Volume III, Final Report to the U.S. Army Toxic and Hazardous Materials Agency (June 1978).
3. Bohn, J.L., B.L. McNeal and G.A. O'Connor, *Soil Chemistry*, John Wiley & Sons, New York (1979).
4. Bowen, H.J.M., *Environmental Chemistry of the Elements*, Academic Press, London (1979).
5. Committee on Biological Effects of Atmospheric Pollutants, *Fluorides*, National Academy of Sciences, Washington, D.C. (1971).
6. Dean, J.A. (ed.), *Lange's Handbook of Chemistry*, 11th ed., McGraw-Hill Book Co., New York (1973).

7. Gisiger, L., "The Solubility of Various Fluorine Compounds in Soil," *Fluoride Q. Rep.*, **1**, 21-26 (1968), as cited in Groth [8].

8. Groth, E., III, "Fluoride Pollution," *Environment*, **17**, 29-38 (1975).

9. *Guidelines for Drinking-Water Quality: Vol. 1 — Recommendations*, World Health Organization, Geneva (1984).

10. Hem, J.D., *Study and Interpretation of the Chemical Characteristics of Natural Water*, 2nd ed., Water Supply Paper 1473, U.S. Geological Survey, Washington, D.C. (1970).

11. Kotrlý, S. and L. Šůcha, *Handbook of Chemical Equilibria in Analytical Chemistry*, John Wiley & Sons, New York (1985).

12. Lillie, R.J., *Air Pollutants Affecting the Performance of Domestic Animals: A Literature Review*, Agricultural Handbook No. 380, Agricultural Research Service, USDA, Washington, D.C. (1970), as cited in Groth [8].

13. MacIntire, W.H., "Air Versus Soil as Channels for Fluoride Contamination of Vegetation in Two Tennessee Locales," in *Proc. U.S. Technical Conference on Air Pollution*, L.C. McCabe (ed.), Interdepartmental Committee on Air Pollution, Washington, D.C., 53-58 (1950), as cited in Groth [8].

14. Marier, J.R. and D. Rose, *Environmental Fluoride*, Pub. No. 12,206, National Research Council of Canada, Ottawa (1971), as cited in Groth [8].

15. Martell, A.E. and R.M. Smith, *Critical Stability Constants*, Vol. 4, *Inorganic Complexes* and Vol. 5, *First Supplement*, Plenum Press, New York (1976 and 1982).

16. Morel, F.M.M., *Principles of Aquatic Chemistry*, John Wiley & Sons, New York (1983).

17. Peek, D.C. and V.V. Volk, "Fluoride Sorption and Desorption in Soils," *Soil Sci. Soc. Am. J.*, **49**, 583-86 (1985).

18. Rai, D., J. Zachara, A. Schwab, R. Schmidt, D. Girvin and J. Rogers, *Chemical Attenuation Rates, Coefficients, and Constants in Leachate Migration. Vol. I: A Critical Review*, Report EA-3356 to EPRI by Pacific Northwest Laboratories (Battelle Institute), Richland, Wash. (1984).

19. Sienko, M.J. and R.A. Plane, *Chemistry: Principles and Properties*, McGraw-Hill Book Co., New York (1966).

20. Simpson, D.R., "Fluorine: Element in Geochemistry," in *Encyclopedia of Geochemistry and Environmental Sciences*, R.W. Fairbridge (ed.), Van Nostrand Reinhold, New York (1972).

21. Stumm, W. and J.J. Morgan, *Aquatic Chemistry*, 2nd ed., John Wiley & Sons, New York (1981).

22. Weast, R.C. (ed.), *Handbook of Chemistry and Physics*, 52nd ed., Chemical Rubber Company, Cleveland (1971).

23. Windom, H.L., "Fluoride Concentration in Coastal and Estuarine Waters of Georgia," *Limnol. Oceanogr.*, **16**, 806-10 (1971); *Chem. Abstr.*, **76**:63046 (1972), as cited in Berkowitz *et al.* [2].

9. RADIONUCLIDES

Donald B. Lindsay

Page

LIST OF TABLES

PROPERTIES OF SELECTED RADIONUCLIDES

9.1 INTRODUCTION: GENERAL CONSIDERATIONS

All of the known elements have radioactive isotopes, no two of which are exactly alike in all respects. An extensive and still-growing body of physical data has been developed to characterize and identify them. Fortunately, from the standpoint of the environmental specialist, only a small number of them are of practical concern; most are only of academic interest.

This chapter, which is addressed primarily to environmental management personnel, concentrates on only a few radionuclides that are expected to have some practical significance. Some of these are found in significant concentrations in the natural environment; the others, which are created artificially, are made in sufficient quantity and are sufficiently long-lived to represent a potential hazard to the environment.

Furthermore, the radionuclides discussed here exhibit chemical and physical properties — apart from radioactivity itself — that are not specified elsewhere in this volume and that would not readily be inferred from descriptions of the nonradioactive form of the element given in other chapters.

Finally, the discussion excludes or, at most, briefly summarizes the properties of certain radioelements such as plutonium; although they satisfy the previous criteria, these elements are deemed to be of interest primarily to those involved in military weapons programs or the nuclear power industry.

Our purpose in including some of the naturally occurring radionuclides, notably 226-radium and its short-lived decay products, is partly to give appropriate emphasis to what is generally acknowledged to be the single most important radioactive environmental pollutant at this time (222-radon) and partly to convey a sense of the fraction of natural and man-made environmental radioactivity that makes up the relatively constant "background" to which any new component would be added.

The introduction of radioactive materials into the human environment brings with it the question of their disposition. Generally speaking, the addition of small quantities of substances that are already present in relatively large amounts and are not subject to human control is thought to be tolerable, provided that some benefit is realized in the process. Thus, the release of small quantities of 14-carbon and 3-hydrogen (tritium) to the environment in gaseous or water-soluble forms is tolerated at levels that are acknowledged to be too small to produce a detectable increase in the existing "background."

Where such rapid dispersal and virtual disappearance into the "background" cannot be assured, other measures must be taken to protect the environment. For instance, deliberate dilution with nonradioactive isotopes, combined with appropriate chemical and/or physical dispersion, may make environmental concentrations acceptably small. Alternatively, radionuclides that are alien to the present terrestrial environment and are produced in quantities large enough to represent an environmental hazard are currently required to be confined so long as their radioactivity persists at significant levels. Examples of such species are found in the transuranic nuclides such as 239-plutonium and the mixed fission-product wastes associated with spent nuclear fuel elements.

The criteria for control of radionuclides in the environment center mostly on predicted adverse effects on human health, balanced against expected benefits to human welfare of uses of radioactive materials and of practices that expose us to ionizing radiation. As a result, these criteria are constantly under review and are subject to considerable variation in different places in the world and at different times. The selection of radionuclide species in the following pages is therefore somewhat arbitrary, being based on current U.S. practice. Table 9.2-1 lists the nuclides that have been selected for discussion, together with some basic data on their radioactive properties, sources and principal areas of occurrence.

9.2 PROPERTIES OF SELECTED RADIONUCLIDES

The 39 radionuclides listed in Table 9.2-1 are culled from the nearly two thousand species listed in the literature [18, 36] and represent those likely to be of most interest to the reader. Nearly three-quarters of this selected group are found in nature. Some of the latter are being generated continuously by cosmic radiation; the remainder are primordial in origin, consisting of surviving fractions of the longest-lived species initially present at the creation of the earth, together with their shorter-lived radioactive progeny, if any. The remaining one-quarter of those listed are created artificially, either as a result of controlled nuclear fission or by use of nuclear particle accelerators.

The radioactive properties of all these nuclides are independent of their origin or means of production. Each nuclide exhibits a set of properties which, although differing sometimes only slightly from others, is nevertheless ultimately unique in one way or another. This fact permits individual species to be identified completely by physical means, usually by detection and measurement of the emitted radiations alone.

Properties other than the radioactive ones are much less varied and tend to resemble those of the elemental species to which individual nuclides belong. Only in rare instances does a nuclide's radioactivity *per se* affect its physical or chemical behavior. (See section 2.17 in Part I, Radioactive Processes and Properties.) These are the exceptions to the general rule that isotopes of an element are chemically indistinguishable from each other. The subsections that follow note these exceptions and also describe the anomalous chemical behavior of extremely small concentrations of matter in which measurable quantities of radionuclides are sometimes found.

9.2.1 Tritium (3-Hydrogen; ^3H)

Tritium is widely present in the environment, being a highly mobile atom, and commingles spontaneously with stable hydrogen. It is produced naturally in the atmosphere by the collisions of cosmic ray protons and neutrons with atoms of nitrogen, oxygen and argon [7]. The most recent estimate of the world inventory of

TABLE 9.2-1

Basic Information on Radionuclides Considered

Isotope	Element	Half-life	Principal Emission and Energy[a,b]	Source[c]	Occurrence
^{3}H	Tritium	12.33 y	β^- 0.0186 (no γ)	Natural (cosmic rays) ^{6}Li (n, α)	Atmosphere, biosphere, hydrosphere, weapons tests, reactor wastes, research and medical wastes
^{14}C	Carbon	5730 y	β^- 0.156 (no γ)	Natural (cosmic rays) ^{14}N (n, p)	Atmosphere, biosphere, hydrosphere, lithosphere, research and medical wastes
^{32}P	Phosphorus	14.28 d	β^- 1.710 (no γ)	Natural (cosmic rays) ^{31}P (n, γ) ^{34}S (d, α) ^{32}S (n, p)	Research and medical wastes
^{35}S	Sulfur	87.2 d	β^- 0.167 (no γ)	Natural (cosmic rays) ^{34}S (n, γ) ^{37}Cl (d, α)	Research wastes
^{40}K	Potassium (0.118% isotopic abundance)	1.25×10^{9} y	β^- 1.33 γ 1.46	Natural (primordial)	Biosphere, hydrosphere, and lithosphere
^{50}V	Vanadium (0.25% isotopic abundance)	3.9×10^{17} y	γ 1.554, 0.793	Natural (primordial)	Biosphere, hydrosphere, and lithosphere
^{51}Cr	Chromium	27.7 d	γ 0.320 e^- 0.315	^{50}Cr (n, γ)	Research and medical wastes
^{87}Rb	Rubidium (27.83% natural isotopic abundance)	4.89×10^{10} y	β^- 0.273 (no γ)	Natural (primordial)	Lithosphere
^{99m}Tc	Technetium	6.01 h	γ 0.142	Fission ^{98}Mo (n, γ), ^{99}Mo (β^-)	Weapons tests, research and medical wastes, reactor wastes
^{99}Tc	Technetium	2.13×10^{5} y	β^- 0.293 (weak γ)	^{99m}Tc (I.T.)	Weapons tests, research and medical wastes, reactor wastes
^{115}In	Indium (95.7% isotopic abundance)	4.4×10^{14} y	β^- 0.48 (no γ)	Natural (primordial)	Lithosphere

(Continued)

TABLE 9.2-1 (Continued)

Isotope	Element	Half-life	Principal Emission and Energy[a,b]	Source[c]	Occurrence
^{125}I	Iodine	59.9 d	γ 0.0355 e^- 0.004, 0.030	^{123}Sb (α, 2n)	Medical and research wastes
^{129}I	Iodine	1.6×10^7 y	β^- 0.150 e^- 0.005, 0.034 γ 0.0396	Fission	Reactor wastes, weapons tests
^{131}I	Iodine	8.040 d	β^- 0.606 γ 0.364 (others)	Fission	Research and medical wastes reactor wastes, weapons tests
^{133}Xe	Xenon	5.25 d	β^- 0.346 e^- 0.045, 0.075 γ 0.081	Fission	Reactor wastes, weapons tests, research and medical wastes
^{137}Cs	Cesium	30.17 y	β^- 0.512 e^- 0.624, 0.656 γ 0.662	Fission	Reactor wastes, weapons tests
^{142}Ce	Cerium (11.08% isotopic abundance)	5×10^{16} y	α 1.5	Natural (primordial)	Lithosphere
^{147}Pm	Promethium	2.62 y	β^- 0.224 (weak γ)	^{146}Nd (n, γ) ^{147}Nd (β^-)	Research wastes
^{154}Eu	Europium	8.5 y	β^- 0.58, 0.26 (others) γ 1.274 (others)	^{153}Eu (n, γ)	Research wastes
^{176}Lu	Lutetium	3.7×10^{10} y	β^- 0.57 γ 0.31, 0.202	Natural (primordial)	Lithosphere
^{187}Re	Rhenium (62.6% natural abundance)	4.5×10^{10} y	β^- 0.0026	Natural (primordial)	Lithosphere
^{201}Tl	Thallium	73.07 h	γ 0.167 (others)	^{203}Tl (p, 3n)	Medical wastes
^{210}Pb	Lead	22.3 y	β^- 0.017, 0.061 γ 0.0465	Natural (primordial; decay product of ^{238}U)	Atmosphere, hydrosphere, lithosphere
^{214}Pb	Lead	26.8 m	β^- 0.67, 0.73, 1.03 γ 0.352 (others)	Natural (primordial; decay product of ^{238}U)	Atmosphere, hydrosphere, lithosphere
^{210}Bi	Bismuth	5.01 d	β^- 1.161 (weak γ)	Natural (primordial) ^{210}Pb (β^-)	Ubiquitous

(Continued)

TABLE 9.2-1 (Continued)

Isotope	Element	Half-life	Principal Emission and Energy[a,b]	Source[c]	Occurrence
^{214}Bi	Bismuth	19.8 m	β^- 3.27, 1.54 γ 0.609 (others)	Natural (primordial) ^{214}Pb (β^-)	Ubiquitous
^{210}Po	Polonium	138.4 d	α 5.304 γ 0.803	Natural (primordial) ^{210}Bi (β^-)	Ubiquitous, industrial wastes
^{214}Po	Polonium	163.7 μs	α 7.687 γ 0.7997	Natural (primordial) ^{214}Bi (β^-)	Ubiquitous
^{218}Po	Polonium	3.11 m	α 6.00	Natural (primordial) ^{222}Rn (α)	Ubiquitous
^{220}Rn	Radon	55.6 s	α 6.288	Natural (primordial; decay product of ^{232}Th)	Atmosphere, lithosphere
^{222}Rn	Radon	3.824 d	α 5.4897	Natural (primordial; decay of ^{226}Ra)	Ubiquitous
^{226}Ra	Radium	1600 y	α 4.7845, 4.598 γ 0.186	Natural (primordial; decay of ^{238}U)	Lithosphere, hydrosphere
^{232}Th	Thorium (100% natural abundance)	1.40×10^{10} y	α 4.01, 3.95 γ 0.059	Natural (primordial)	Lithosphere
^{234}Th	Thorium	24.1 d	β^- 0.19, 0.10 γ 0.24, 0.0924 (others)	Natural (primordial; decay of ^{238}U)	Lithosphere
^{234}Pa	Protactinium	1.17 m	β^- 2.29	Natural (primordial; decay of ^{238}U)	Lithosphere
^{235}U	Uranium (0.72% natural abundance)	7.04×10^{8} y	α 4.16-4.6 γ 0.186	Natural (primordial)	Lithosphere

(Continued)

TABLE 9.2-1 (Continued)

Isotope	Element	Half-life	Principal Emission and Energy[a,b]	Source[c]	Occurrence
^{238}U	Uranium (99.27% natural abundance)	4.468×10^9 y	α 4.20	Natural (primordial)	Lithosphere
^{239}Pu	Plutonium	2.411×10^4 y	α 5.16 (others)	^{238}U (n, γ) ^{239}U (β^-) ^{239}Np (β^-)	Military wastes, reactor wastes
^{241}Am	Americium	432 y	α 5.4857, 5.4430 γ 0.0595, 0.0263 (others)	^{239}Pu (n, γ)	Military wastes, reactor wastes, industrial ionization sources

a. The principal forms of radioactive emission are identified by the conventional symbols α, β^-, e^- and γ, which are defined as follows [24]:

α : Alpha Particle. Charged particle emitted from the nucleus of an atom, having a mass and a charge equal in magnitude to that of a helium nucleus; i.e., two protons and two neutrons.

β^- : Beta Particle. Charged particle emitted from the nucleus of an atom, with a mass and charge equal in magnitude to that of an electron.

e^- : Internal Conversion Electron. An orbital electron ejected from an atom as a result of interaction with a gamma ray (q.v.) emitted from the nucleus. Such ejected electrons are mono-energetic, their energy being equal to the difference between their initial binding energy and that of the gamma ray that causes their ejection. (Note: internal conversion electrons should not be confused with the phenomenon of electron capture, designated either by the symbols "EC" or "ϵ," a process which results in emission of a characteristic x-ray from whichever shell of electrons the captured electron originated. Such x-rays are also produced by ejection of internal-conversion electrons.)

γ : Gamma Ray. Short-wavelength electromagnetic radiation (energy range 0.01 MeV to 9 MeV) emitted from and originating in the nucleus. The term "x-ray" also refers to electromagnetic radiation associated with radioactive decay, but is customarily restricted to photons emitted from the region of the orbiting electrons, not from the nucleus.

b. All energy values are given in millions of electron volts (MeV).

c. Nuclear reaction mechanisms are indicated by conventional symbolic notation:

n = neutron;
p = proton;
d = deuteron;
α = alpha particle;
γ = gamma ray;
I.T. = isomeric transition.

Sources: Lederer and Shirley [18]; Walker *et al.* [36]

this nuclide from natural sources is 1 × 10^{18} Bq (30 MCi) [32]. This continuing inventory, resulting from a balance between a relatively constant production rate and radioactive decay ($T_{1/2}$ = 12.33 y), was greatly perturbed during the 1960s by military weapons tests, conducted in the atmosphere, which added an estimated 4,500 MCi of ^3H to the environment [11]; smaller amounts are constantly being introduced in wastes from nuclear reactors. Over 99% of ^3H from all sources appears as tritiated water (HTO) and is widely distributed in the surface waters of the earth. Because the radiation from ^3H is very weak (beta rays with a maximum energy of 0.0186 MeV), this nuclide is not considered to be a major toxicological hazard.

9.2.2 14-Carbon

This nuclide is produced by cosmic rays in nature and is also created artificially in nuclear fission. About 300 MCi exist in the total terrestrial inventory from cosmogenic sources [17]; much smaller amounts are contributed annually from various research wastes and in reactor fuel wastes. At sufficiently low concentrations, ^{14}C wastes are exempted from treatment as nuclear waste, since ^{14}C is virtually inseparable from stable ^{12}C once the isotopes have been intermingled. No natural chemical or physical processes can re-concentrate the radioisotope with respect to the stable one. The half-life for radioactive decay is 5,730 years, and decay is by beta ray emission (E_{max} = 0.156 MeV). The chemical behavior of carbon (as carbonate) in the environment is discussed in section 6.12 of this volume.

9.2.3 32-Phosphorus

Although ^{32}P is created in small quantities by cosmic rays in the atmosphere, the half-life of the isotope (14.28 days) is too short to permit significant concentrations to occur [16]. The nuclide is made artificially by a number of methods and may occur in the environment as a waste product, but its short half-life, absence of penetrating gamma radiation, and direct decay (by beta-ray emission, E_{max} = 1.710 MeV) to a stable daughter nuclide (^{32}S) make it relatively harmless to the environment. It is generally used in some degree of dilution with stable ^{31}P, with which it acts as a true "tracer." The chemical behavior of phosphorus is discussed in sections 8.3 and 10.8.

9.2.4 35-Sulfur

Like ^{32}P, ^{35}S is made in very small quantities by cosmic ray action but is primarily created artificially [16]. Like ^{32}P, it is generally used as a tracer in combination with its stable isotopes. It decays directly by beta-ray emission (E_{max} = 0.1677 MeV) with a half-life of 87.2 days to a stable isotope, ^{35}Cl. The environmental behavior of sulfur compounds is discussed in sections 8.2 and 10.7.

9.2.5 40-Potassium

This nuclide is of interest because it is primordial in origin; it is one of the few radionuclides of the lighter elements that exhibit a very long half-life (1.25 × 10^9 years). ^{40}K occurs widely in nature, being an essentially universal isotopic component of natural potassium. It accounts for over 90% of the total radioactivity of ocean water [11] and is present in all human bodies because it is contained in most foods. Its penetrating gamma radiation (1.46 MeV) and moderately energetic beta particles (1.32 MeV max.) make it a major contributor to terrestrial radiation background. Chemically, ^{40}K is essentially inseparable from other potassium isotopes, of whose total it constitutes only 0.118%; since it decays directly to stable ^{40}Ca, it presents no special management problems. The chemical behavior of potassium in the environment is discussed in section 6.3.

9.2.6 50-Vanadium

^{50}V is the only radioactive isotope of vanadium. It is a very long-lived (3.9×10^{17} years) remnant of primordial origin and is ubiquitous in deposits of natural uranium [16]. Isotopically, it represents only 0.25% of natural vanadium, and it decays directly either by beta-ray emission to stable ^{50}Cr or, by electron capture, to stable ^{50}Ti. Its long half-life makes it so slightly radioactive, and its low isotopic fraction makes its contribution to environmental radioactivity so slight, that it is only of academic interest.

9.2.7 51-Chromium

This isotope is entirely man-made (by neutron-activation of ^{50}Cr) and is used as a radio-tracer in medical and industrial research [16]. It has a relatively short half-life (27.7 days), decaying directly by electron capture and emission of a 0.32-MeV gamma ray to stable ^{51}V [18]. It is virtually inseparable from stable chromium once the isotopes are commingled, so it presents no unusual problem to environmental management. The environmental behavior of stable chromium is discussed in section 7.6, and that of ^{51}Cr can be taken to be identical to it.

9.2.8 87-Rubidium

This long-lived (4.89×10^{10} years) isotope of rubidium occurs both in nature, where it is a remnant of primordial genesis, and as a product of nuclear fission [16]. It represents a substantial fraction (27.83%) of all rubidium isotopes in nature and, being an alkali element, is dispersed widely, along with ^{40}K. Because of the relative abundance of potassium, however, the gamma radiation contributed by ^{87}Rb is only about 1% of the total in seawater. Its presence in the natural environment is useful for geochronological measurements because of its wide distribution and stability of concentration [11]. It is of interest here only because of its slight contribution to the natural radioactive background.

9.2.9 99m/99-Technetium

Technetium is a rarity among elements with atomic number less than 92, in that it has no stable isotopes and does not occur in nature. The most familiar of its isotopes is $^{99m}Tc/^{99}Tc$, a radionuclide that decays in two distinct steps. The first decay is by internal isomeric transition, emitting only low-energy gamma rays (0.143 MeV) with a half-life of 6.01 hours; the second step is much slower ($T_{1/2} = 2.13 \times 10^5$ y) and involves only beta-ray ^{99}Tc emission, decaying to stable ^{99}Ru [16].

^{99m}Tc is widely used in medical diagnosis as a radiotracer. It exhibits a reliable chemical behavior when oxidized to the +7 state, in which it forms the stable pertechnetate ion (TcO_4^-) in aqueous solutions [26]. Reduction of $^{99m}TcO_4^-$ with large excesses (10^8 - 10^9:1) of $SnCl_2$ in the presence of a variety of organic and inorganic substrates has been found to precipitate, chelate or bind the technetium quite strongly, usually as the reduced Tc^{+4} ion, so that it can be carried through various physiological systems in this strongly chelated or adsorbed condition. The extremely small quantities of ^{99m}Tc encountered in most circumstances (typically below 10^{-9} *M*) makes the determination of chemical speciation rather difficult, but empirical results have been found to be highly consistent. Most of what is known of the aqueous chemistry of technetium has been learned from studies of ^{99}Tc, since its longer half-life permits it to be prepared and used conveniently at somewhat higher concentrations.

In recent years, however, the availability of new methods for separation and analysis of very low concentrations of materials has permitted studies of ^{99m}Tc at tracer levels that have resolved some of the uncertainties previously associated with its chemical properties [22, 28, 29, 35]. These studies have shown that, in general, ^{99m}Tc behaves according to conventional theory but that the presence of such trace-level reactants as dissolved oxygen often produces confusing results when not adequately controlled. Hwang *et al.* [13] showed that ^{99m}Tc exists in several complexes as Tc(V) and that under some aqueous conditions Tc(IV) disproportionates into Tc(III) and Tc(V).

An interesting fact confirmed by Vine and Wahl [35] is that ^{99m}Tc produced by beta-decay from aqueous solutions of ^{99}Mo(VI) occurs almost entirely as Tc(VII), the beta decay apparently causing uniform oxidation of technetium to its highest state. Other isotopes of technetium produced by beta decay of corresponding short-lived molybdenum isotopes resulting from fission of ^{235}U are ordinarily found in lower valence states as well as in (VII). Vine and Wahl speculated that molybdenum isotopes produced in fission, which are known to occur originally in low oxidation states, may be incompletely oxidized by subsequent beta-decay, thus yielding a mixture of technetium oxidation states.

The behavior of technetium in groundwater is highly dependent on its oxidation state. As might be expected, under oxidizing conditions TcO_4 is the predominant species, because it is very soluble and is very slightly sorbed to mineral substrates [14]. Under reducing conditions, technetium precipitates as the very insoluble TcO_2.

Table 9.2-2 lists solubilities and retardation factors[1] for technetium and eight other radionuclides that occur in high-level nuclear wastes and are described in this section. The data for technetium, drawn from the work of Bondietti and Francis [5], show the effect of reducing and oxidizing conditions as well as two different pH conditions on solubility.

1. The retardation factor, K, is the ratio (expressed in ml/g) of the pore velocity of groundwater to the net velocity of transport of a dissolved contaminant, assuming local chemical equilibrium of the contaminant dissolved in the liquid and sorbed on the rock. K is related to the distribution coefficient, K_d, by: $K = 1 + \rho K_d(1 - e)/e$, where e is the porosity, ρ is the density of the solid phase, and K_d is in cm^3/g.

TABLE 9.2-2

Solubilities and Retardation Factors of Some Hazardous Radioactive Elements

Element	Solubility (log ppm)					Retardation Factor ($1 \times 10K_d$)[a]				
	Most Probable	Reducing: Eh = −0.2		Oxidizing: Eh = +0.2						
		pH = 9	pH = 6	pH = 9	pH = 6	Granite	Basalt	Tuff	Clay, Soil, Shale	Salt
Tc	−3	−10	High	High	High	1 5 40	5 100	1 5 100	1 5 20	1 5 20
I	High	High	High	High	High	1 1 1	1 1 50	1 1 1	1 1 1	1 1 1
Cs	High	High	High	High	High	100 1,000 10,000	100 1,000 10,000	60 500 10,000	200 1,000 20,000	1 10 2,000
Pb	−1	−1	0	−1	0	10 50 200	20 50 500	20 50 500	20 50 500	5 20 100
Ra	−2	−3	−1	3	−1	50 500 5,000	50 500 5,000	50 500 5,000	50 500 5,000	5 50 500
Th	−3	−4	−4	−4	−4	500 5,000 10,000	500 5,000 10,000	500 5,000 10,000	500 5,000 50,000	300 1,000 5,000
U	−3	−3	−5	High	High	10 50 500	20 50 1,000	5 40 200	50 200 5,000	10 20 60
Pu	−3	−5	−4	−5	−3	10 200 5,000	100 500 5,000	50 200 5,000	500 1,000 20,000	10 200 10,000
Am	−4(?)	−8	−5	−8	−5	500 3,000 50,000	60 500 50,000	300 1,000 50,000	200 800 50,000	300 1,000 5,000

a. High and low values indicate range; middle value is considered "most likely."

Source: Waste Isolation Systems Panel [37]. Compilation of data attributed to K.B. Krauskopf, Stanford University.

Table 9.2-2 also shows the effect of several different substrates on retardation factor. A rather wide range of retardation factor values is reported for all the elements in the table. This variation is attributable to uncertainties in experimental results. A "most likely" value for each of the nuclides is included (intermediate between the extremes in each instance), reflecting the opinion of K.B. Krauskopf, the researcher who compiled these data [37]. In computing the retardation factor values, a typical value of 10 is assigned to $\rho(1-e)/e$; this leads to a range extending from K = 1 to K = 100 for different rocks and different experiments. (Krauskopf chooses K = 5.)

9.2.10 115-Indium

^{115}In is a very long-lived, naturally occurring radionuclide of primordial origin. Over 95% of the world's supply of indium exists in this one isotopic form, whose half-life is 4.4×10^{14} years and which decays by beta-ray emission (E_{max} = 0.48 MeV) with no associated gamma radiation. Indium has been included in this list, not because it presents a radiation hazard that might require the application of appropriate control technology, but rather to call attention to the fact that it is one of many natural substances in the environment that are slightly radioactive. Because of its long half-life, indium has a low specific radioactivity (disintegrations per second per gram). Its lack of gamma radiation makes its health hazard even smaller, even though the element is almost entirely composed of this one radioactive isotope.

9.2.11 Iodine Isotopes

Three radioactive isotopes of iodine are environmentally significant: ^{125}I, ^{129}I, and ^{131}I. Two of these, ^{125}I and ^{131}I, are important in medical and scientific research and in clinical medicine. The third, ^{129}I, is a long-lived fission product that occurs in reactor wastes (spent fuel elements) and in atmospheric debris from nuclear weapons tests. All iodine radioisotopes are readily susceptible to isotopic dilution by admixture of stable ^{127}I, and the observed chemical behavior of the radioisotopes at tracer concentrations generally corresponds to that of the stable form at higher concentrations. The environmental chemistry of iodine is described in section 8.4.

^{131}I, the most widely used iodine radionuclide in medicine, is made by nuclear fission of ^{235}U and decays to stable xenon by beta-ray and gamma-ray emission (E_γ = 0.364 MeV) with a half-life of 8.040 days. ^{125}I is made by cyclotron bombardment of ^{123}Sb by accelerated alpha particles (helium nuclei) and decays with a half-life of 59.9 days by electron capture to stable tellurium. ^{129}I decays with a relatively long half-life of 1.6×10^7 years, principally by beta-ray emission, yielding stable xenon.

9.2.12 133-Xenon

Interest in ^{133}Xe centers on the fact that it is a noble or "rare" gas and is therefore virtually inert to chemical reactions. The only practical methods of controlling its behavior in the environment are the physical methods (adsorption, condensation, etc.) that are effective for all noble gases, whether stable or radioactive. Because of its lack of chemical reactivity, xenon is not preferentially concentrated or retained in the human body and is therefore not considered to present a major risk to human health when in the atmosphere, even in relatively high concentrations [16]. ^{133}Xe is used in medical diagnostic studies of the pulmonary system and occurs in wastes from nuclear weapons tests and reactor operations. It decays primarily by beta-ray emission (with associated 0.081 MeV gamma radiation and conversion electrons) to stable cesium [36].

A frequent problem in handling and disposing of ^{133}Xe is its rather high solubility in certain polymers and organic liquids. Like krypton, neon, argon, and radon, xenon's solubility in such materials is potentially useful but also potentially troublesome. It has been studied by several investigators, and a considerable body of literature (e.g., [23]) has been published on the subject.

9.2.13 137-Cesium

^{137}Cs is produced in nuclear fission and occurs in atmospheric debris from weapons tests, where its moderately long half-life (30.17 years) and high solubility in water predispose it to concentrating in the hydrosphere. According to Klement [16], "The accumulation of ^{137}Cs deposited from past nuclear tests is the major source of long-lived external gamma radiation from fallout." ^{137}Cs decays by beta-ray, gamma-ray (E_γ = 0.662 MeV) and conversion-electron emission to stable barium.

Chemically, ^{137}Cs in tracer concentrations appears to behave identically to the stable form (^{133}Cs) occurring in higher concentrations. In aqueous solution the major species is Cs^+, which undergoes neither oxidation nor reduction in that medium. As shown in Table 9.2-2, the solubility of Cs^+ is always found to be high, irrespective of pH or Eh, but the reported retardation factors vary widely, depending in part on the nature of the mineral substrate but also on other experimental conditions. Another demonstration of the wide range of these values is shown in Table 9.2-3, which gives some reported values of distribution coefficients (K_d) as a function of substrate; data are shown for both brine and fresh water, but Muller *et al.* consider the differences to be insignificant, given the wide variations reported by different investigators for fresh water. There seems to be general agreement that cesium is strongly sorbed on common rocks from dilute solutions but not from brine; sorption is especially strong on rocks that have layer-silicate minerals (micas and clay minerals) [37].

9.2.14 147-Promethium

All promethium isotopes are radioactive, and none is found in nature. The element was discovered in 1947 by analysis of mixed fission products, using ion-exchange techniques [12, 38]. The only widely used isotope is ^{147}Pm, which decays with a half-life of 2.62 years by beta-ray emission (E_{max} = 0.224 MeV; weak gamma-ray) to stable samarium.

Chemically, promethium is a rare earth. The relatively small amount of research that has been done on the tracer-level chemistry of this element suggests that it should be handled only in the presence of larger (carrier) concentrations of chemically similar

TABLE 9.2-3

Reported Ranges of Cesium Distribution Coefficients, K_d

	log K_d	(ml/g)
Solid Substrate	**Freshwater**	**Brine**
Reference Clays	6.2	2.4
	3.8	0.4
Sandstone,	3.5	—
Alluvium	2.0	—
Dolomite,	1.8	−0.9
Limestone	1.0	—
Granite	3.0	—
	1.0	—
Basalt	4.0	0.5
	2.5	0.2
Shale	3.7	—
	2.5	—
Tuff	4.3	—
	2.0	—
Salt	—	−2.5

Source: Muller *et al.* [21]

substances such as lanthanum or praseodymium. Tracer concentrations are effectively "carried" on both fluoride and oxalate salts of other rare earths, divalent calcium and tetravalent thorium. Hydrous oxides of iron and the rare earths are used to "scavenge" promethium, but these are generally less selective than the fluorides and oxalates [27]. Ion exchange permits the separation, purification and analysis of promethium at even the lowest tracer concentrations. Its behavior in ion-exchange systems can always be inferred from its position between neodymium and samarium. Both cation and anion exchangers can be useful in separating promethium from uranium, thorium, plutonium and various fission products.

All rare earths are strongly adsorbed on cation exchangers from solution in dilute mineral acids and are eluted by concentrated acids. For the best resolution in such separations, however, complex-formers are used, whose stability generally increases with increasing atomic number of the cation. Two types of complex-formers are in general use: alpha-hydroxy-carboxylic acids (citric, glycolic, lactic, etc.), and amino-polycarboxylic acids (EDTA, nitrilotriacetic acid, diethylenetriamine-penta-acetic acid, etc.). Because Pm^{+3} (and other trivalent rare earths) are weakly adsorbed on anion exchangers from mineral acids, separation from more strongly adsorbed ions is possible.

In bulk form, promethium oxide, Pm_2O_3, does not appear to react with atmospheric moisture, presumably because its intense radioactivity makes the surface too warm to permit condensation of H_2O [9]. When immersed in water, however, Pm_2O_3 dissolves rapidly until equilibrium is reached. The solubility product varies inversely with temperature [20], leading to solubilities ranging from 3.5×10^{-6} *M* at 10°C down to about 1.4×10^{-6} *M* at 40°C. Pellets of Pm_2O_3 immersed in water (pH 6.0) showed a dissolution rate (i.e., leachability) of about 4-12 μg/hr-cm^2 of pellet surface; in sea-water (both synthetic and natural) at a pH of 8.4, the rate was about one-tenth as much. However, the reaction of Pm_2O_3 with water, leading to formation of $Pm_2O_3 \cdot 3H_2O$, as shown by x-ray diffraction and thermogravimetric analysis, causes rapid physical disintegration of pellets, with resulting marked increase in surface area.

Promethium-147 is used to a modest extent in the manufacture of solid beta radiation sources but has no present use in medicine. It is therefore only of minor concern as an environmental hazard.

9.2.15 154-Europium

Natural europium consists of two stable isotopes, ^{151}Eu (47.82%) and ^{153}Eu (52.18%). Both have extraordinarily high cross-sections (i.e., probabilities) for capture of thermal neutrons, yielding the short-lived ^{152}Eu and 8.5-year ^{154}Eu respectively. The isotope decays by beta and gamma emission (E_{max} = 1.274 MeV) to stable gadolinium; this property makes detection of europium possible at ultra-trace levels by neutron activation. Past use of this analytical method has contributed much of the present inventory of ^{154}Eu in low-level waste repositories [16].

The chemistry of europium is typical of the rare earths. As the radioactive form is always accompanied by some of the stable isotope and is readily susceptible to dilution by addition of more of the stable material, environmental control of ^{154}Eu is relatively simple and straightforward.

Geochemical studies compiled by Wolfsberg [39] showed tuff, both devitrified and zeolitized, to be an effective scavenger of europium and other radionuclides. Sorption ratios for europium from groundwater ranged from 90 to 7500 ml/g for devitrified tuff and from 1200 to 2500 ml/g for zeolitized tuff; similar values were obtained by both sorption and desorption methods.

9.2.16 176-Lutetium

This long-lived isotope of lutetium ($T_{1/2}$ = 3.7 × 10^{10} years) occurs in nature, being of primordial origin, and is concentrated primarily in the lithosphere. ^{176}Lu decays by beta and gamma emission (E_{γ} = 0.3, 0.202 MeV) to form stable hafnium. The nuclide constitutes only 2.6% of the isotopic composition of natural lutetium and is of interest only as a ubiquitous component of the natural radiation environment [16]. Its chemistry is that of a typical rare earth.

9.2.17 187-Rhenium

^{187}Re is a long-lived ($T_{1/2}$ = 4.5 × 10^{10} years) radionuclide and represents about 63% of the total elemental rhenium in nature. It is of primordial origin and occurs primarily in the lithosphere. It decays by emission of very weak beta rays (E_{max} = 0.0026 MeV) to form stable osmium [18]. Its lack of penetrating radiation makes it relatively innocuous as an environmental hazard. Its chemical behavior is similar to that of manganese and technetium, being highly water-soluble under oxidizing conditions but relatively insoluble and therefore relatively immobile when reduced to the (IV) state.

9.2.18 201-Thallium

The presence of ^{201}Tl in significant quantities in our environment is a quite recent phenomenon, caused by the discovery that, physiologically, thallium mimics potassium in its uptake by cardiac tissue. Because the distribution of potassium (and, therefore, of thallium) in the human heart is important in diagnostic procedures, and because ^{201}Tl offers advantages in practice that are not offered by any radioisotope of potassium, its use in diagnostic medicine has grown rapidly. ^{201}Tl decays with a half-life of 73.07 hours and emits a gamma ray of 0.167 MeV energy. The nuclide is made by bombardment of stable ^{203}Tl with electrically accelerated protons[2], and it decays by electron capture to form stable mercury. Its half life is short enough to permit most of what is made to decay in controlled storage, thus minimizing the environmental hazard of this highly water-soluble nuclide.

In the unlikely event that ^{201}Tl entered the environment, its high water-solubility would predispose it to concentrate initially in groundwater. The water-related properties and likely environmental fate of thallium has been reviewed by Callahan *et al.* [8]. The element is reported to be strongly adsorbed by montmorillonitic clays [19], especially in alkaline solution (pH 8.1), and has been found in relatively high concentrations (compared to those found in water, biota, etc.) in sediments in fresh-water lakes.

The only other important process by which thallium is removed from groundwater is bioaccumulation, although data to support this observation are scanty. Bioconcentration factors reported by various investigators range from 10 to 150,000 for various marine plants and animals [8].

---

2. The nuclear reaction is $^{203}\text{Tl}\,(p, 3n) \longrightarrow {}^{201}\text{Pb} \xrightarrow{E.C.} {}^{201}\text{Tl}$.

9.2.19 210-Lead

^{210}Pb occurs in nature as a late decay product in the long chain of successive radioactive decays beginning with 238-uranium. Because of its moderately long half-life ($T_{½}$ = 22.3 years), ^{210}Pb is usually present in the environment in concentrations large enough to assure that its chemical properties will be those ordinarily associated with its stable isotopes. The fact that ^{210}Pb owes its existence to sequential decay of a number of physically and chemically dissimilar substances, however, leaves open the possibility that the supply equilibrium could be upset by some process of selective transfer, so that the eventual daughter-product ^{210}Pb could occur in true trace-level concentrations. While no published data have been found to support any anomalous chemical behavior of ^{210}Pb at tracer concentrations, the possible occurrence of such behavior cannot now be ruled out. ^{210}Pb decays by beta and gamma emission (E_{γ} = 0.0465 MeV), producing radioactive ^{210}Bi [36].

As indicated in Table 9.2-2, lead is believed to be moderately soluble in groundwater, based mostly on calculations and inferences with respect to Eh, pH and ionic environment. Estimates of retardation due to sorption are said to be no more than guesses, given the present state of knowledge, but are believed to indicate modest sorption on most types of natural rock [37]. Additional data on the environmental chemistry of lead are provided in section 7.8.

9.2.20 214-Lead

See 222-Radon (§ 9.2.27)

9.2.21 210-Bismuth

^{210}Bi is formed in nature by the sequential decay of a long chain of radionuclides, beginning with ^{238}U. The immediate parent of ^{210}Bi is ^{210}Pb, and because of the much shorter (5-day) half-life of ^{210}Bi, the two nuclides are usually found together in virtually "secular" equilibrium [36]. Separation of ^{210}Bi from ^{210}Pb is routinely done by deliberate addition of stable bismuth salts to aqueous solutions in order to prevent unwanted effects due to tracer-concentration anomalies. The actual chemical behavior of bismuth isotopes at tracer concentrations has not been a subject of significant research, and no reliable data are available to describe its behavior in the environment.

9.2.22 214-Bismuth

See 222-Radon (§ 9.2.27)

9.2.23 210-Polonium

All 27 isotopes of polonium, like those of all elements of higher atomic weight, are radioactive. With a radioactive half-life of 138.4 days, ^{210}Po is one of the longest-lived polonium isotopes. Being the last radionuclide in the ^{238}U decay chain, it is the only one to be available in quantities large enough to make it accessible for conventional chemical studies. Even so, as Bonner and Kahn point out [6], an amount of ^{210}Po that is readily measurable by radioassay, 1×10^4 disintegrations per minute (dpm), represents a mass of only 5×10^{-15} gram-atoms of the element. Because ^{210}Po decays by alpha radiation (E_α = 5.3 MeV) with its associated high linear-energy transfer to the matrix, as well as the significant nuclear recoil that results from the emission of the massive alpha particle, weighable quantities of ^{210}Po are susceptible to self-heating, auto-oxidation and general physical and chemical degradation, which makes handling and measurement difficult.

In spite of these difficulties, much effort was made, especially during the first decades following the discovery of nuclear fission, to establish the chemical properties of polonium, largely through experiments with ^{210}Po. A thorough review of the subject is provided by K.W. Bagnell [4].

When effects due to its highly energetic alpha radiation (at high concentrations) and to the competition of various trace-level reactants (at tracer concentrations) are considered and accounted for, polonium's chemistry is found to be predictable, based on its position in the periodic table and its similarity to tellurium. In nature, ^{210}Po is commonly encountered in extremely small quantities and/or at extreme dilution; thus, one must be prepared to encounter what appears to be anomalous chemical behavior, such as that described by Bonner and Kahn [6]. Also, one must not fail to consider the fact that the intense alpha radiation effects exhibited by high concentrations of polonium salts, or more commonly by electroplated films of pure metallic polonium, are known to result in physical and chemical degradation of the product. The difficulties associated with preventing spontaneous contamination of the environment by such commercial products as static-electricity-neutralizing polonium brushes have led to the virtual disappearance of such products from the market and their replacement by more stable radioelements such as 241-americium.

9.2.24 214-Polonium

See 222-Radon (§ 9.2.27)

9.2.25 218-Polonium

See 222-Radon (§ 9.2.27)

9.2.26 220-Radon

^{220}Rn (also known as "thoron" because of its origin in the decay series of 232-thorium and to help distinguish it from the more familiar isotope, ^{222}Rn) is a ubiquitous natural substance occurring widely in rocks and soils that have some thorium content, or which have been contaminated by some intermediate daughter-product of ^{232}Th as a result of leaching or other slow geologic transport. ^{220}Rn, which has a half-life of 55.6 seconds, emits alpha radiation (E_α = 6.288 MeV), decaying to radioactive ^{216}Po and, thence, through several more radionuclides to stable ^{208}Pb [6, 36].

Because radon is a noble gas, it is virtually unreactive chemically. Once allowed to escape from confinement in rocks and soils, its behavior in the environment is limited mostly by its diffusion coefficient in air and its solubility in water. (Radon distributes approximately equally on a volumetric basis between air and water at atmospheric pressure.) Because it has such a short half-life, however, its occurrence in the atmosphere or in surface water is limited to concentrations that are environmentally insignificant, compared with those associated with the longer-lived ^{222}Rn.

9.2.27 222-Radon

^{222}Rn, commonly referred to simply as "radon" because of its predominance compared with other isotopes of the element, is currently considered to be the prime environmental hazard of all radionuclides. It occurs only in nature and is a late decay-product in the series originating with ^{238}U. It is distributed widely throughout the lithosphere and hydrosphere, since it originates in uranium- and/or radium-bearing minerals (it is the first decay-product of ^{226}Ra), and it is moderately soluble in water and a variety of organic liquids [5]. Being an unreactive noble gas and having an adequately long half-life of 3.8 days, ^{222}Rn diffuses rapidly when released into the atmosphere and can be transported widely over the surface of the earth by natural air movement. Once diluted in the air, it becomes essentially unrecoverable for purposes of large-scale decontamination [2]. Its hazard to the environment is primarily due to the fact that it acts as an atmospheric carrier and distributor of its own radioactive daughter products, the first four of which are relatively short-lived and include two alpha-emitting isotopes that have been almost unequivocally identified as inducers of human bronchio-pulmonary cancer. These are commonly recognized "radon daughters": ^{218}Po, ^{214}Pb, ^{214}Bi, and ^{214}Po.

In general, ^{222}Rn behaves physically and chemically as theory would predict. By reaction at elevated pressure and temperatures with vigorous oxidants such as dioxygenyl hexafluoroantimonate and bromine trifluoride, Stein [30, 31] was able to produce highly unstable solid salts of radon, but under ordinary conditions radon is as inert as other noble gases. Among the known methods for separating and concentrating radon [2], probably the most useful is that of adsorption (and thermal desorption) on an active solid substrate such as activated carbon (charcoal). Thomas [33], Kapitanov [15] and others have shown the value of activated carbon as an adsorbent for radon, especially at the very low concentrations found in environmental air. In addition to adsorption, the use of low-permeability barriers to confine radon and prevent its escape into the atmosphere is virtually the only practical means of controlling this environmentally hazardous natural gas.

As indicated previously, the health hazard associated with ^{222}Rn is attributed almost entirely to its short-lived daughters. All of these are isotopes of elements (polonium, bismuth, and lead) that are solid in bulk form and under ordinary conditions, but the way in which they are formed in the natural decay of ^{222}Rn makes their properties radically different from those of the bulk materials.

The science and technology of radon daughter control is still quite primitive and largely empirical, being limited to the behavior of single ions in air, and to their migration, reaction with, attachment to and separation from other single ions, colloidal particles and various solid surfaces. Concentration effects are essentially irrelevant to the study of radon daughters, since they are formed in extreme isolation from others like them; they are controlled primarily by close-order forces of adsorption between individual ions and extremely small particles, or by the recoil imparted to them by their own alpha-decay. Little experimental work has been done to explain the physical and chemical behavior of the radon daughters, leaving almost all knowledge on the subject at an empirical and primitive level. Proceedings of a six-day international conference on "Radiation Hazards in Mining," held at the Colorado School of Mines in October 1981 [25], describe some of the most recent studies on radon daughters but present very little theoretical material. A fairly thorough review of the relevant literature up to the date of publication (1978) was prepared by Arthur D. Little, Inc. [3].

9.2.28 226-Radium

^{226}Ra, commonly known simply as "radium" because of its relative abundance and importance compared with the other 24 isotopes of this element, is the fifth successive daughter product of ^{238}U and the immediate parent of ^{222}Rn. ^{226}Ra decays by alpha-ray emission (E_α = 4.78 MeV) with a half-life of 1600 years. The historical importance of radium is indicated by the fact that until recently the unit of radioactive decay rate was the curie (Ci), named after the element's discoverer. Originally defined as the radioactivity of one gram of ^{226}Ra, the curie was subsequently simplified to an approximate value, 3.7×10^{10} dps. The long-time use of radium as a source of penetrating gamma radiation was facilitated by the encapsulation of a quantity of a solid salt of ^{226}Ra in an impervious container, which prevented escape of the gaseous daughter, ^{222}Rn, and allowed it to decay into the successive daughter nuclides; two of these, ^{214}Pb and ^{214}Bi, are the principal emitters of the penetrating gamma radiation [8].

The chemistry of ^{226}Ra is straightforward in practice, since the relatively long half-life of the nuclide permits operations to involve concentrations that are high enough to obey familiar chemical principles. In the radioactive decay of radium, as suggested by Tanner [32], it is likely that recoil of the parent nucleus (which becomes ^{222}Rn in the process) causes an increased fraction of radon to escape from its host mineral, compared with what would be predicted from intracrystalline migration or diffusion; however, the mechanism involved is still not entirely clear.

When exposed to groundwater, radium is likely to encounter either dissolved sulfate or carbonate anions, which could precipitate the radium as $RaSO_4$ or $RaCO_3$. Although both of these salts are moderately insoluble, a sulfate concentration of 1×10^{-4} *M* would still permit an equilibrium concentration of about 0.1 ppm ($10^{-6.4}$ *M*) of Ra^{+2} to exist in solution. The solubilities listed in Table 9.2-2 are described as estimates based on minimal data [37], and one cannot be confident that the insolubility of either of these species will prevent contamination.

9.2.29 232-Thorium and 234-Thorium

Natural thorium, of primordial origin, began as only one isotope, ^{232}Th. Its other naturally occurring isotopes, including ^{234}Th, exist as a result of radioactive decay of primordial thorium and uranium. ^{232}Th decays by alpha-emission (E_α = 4.01, 3.95 MeV) with a half-life of 1.40 × 10^{10} years [36]. Its chemistry follows conventional principles at all concentration levels of practical concern. The chemical behavior of ^{234}Th is of only academic interest, since its short half-life (24.1 days) and chemical compatibility with its matrix (predominantly atoms of ^{232}Th) give it little opportunity for differential reactions.

In groundwater, assuming pH > 5, thorium exists primarily as $Th(OH)_4$. A conservative solubility of 1 ppb is "most probable" [37]. The latter source observes that "reported retardation factors are uniformly high, ranging from 500 (for brine solutions) to 1,000,000 (for clay), but data are scanty."

9.2.30 234-Protactinium

^{234}Pa is a short-lived ($T_{1/2}$ = 1.17 minutes) daughter of ^{234}Th. It decays by beta-ray emission (E_{max} = 2.29 MeV) to yield ^{234}U [34]. Its very short half-life and chemical and physical compatibility with its natural matrix make its chemical reactions of little significance.

9.2.31 235-Uranium and 238-Uranium

Together, ^{235}U and ^{238}U comprise all the natural uranium of direct primordial origin on earth. (^{234}U, the only other naturally occurring isotope, is a third-generation daughter of ^{238}U.) Only 0.72% of natural uranium occurs as ^{235}U, which decays by alpha emission (E_α = 4.16-4.6 MeV) with a half-life of 7.04 × 10^8 years. It has an unusually high cross-section for fission under the impact of thermal neutrons or fast neutrons, which made it the original material of choice for nuclear reactors and nuclear weapons [18, 36].

^{238}U comprises 99.27% of natural uranium, decays by alpha emission (E_α = 4.20 MeV) with a half-life of 4.47 × 10^9 years, and is susceptible to fission only under the impact of fast neutrons. The chemistry of both ^{235}U and ^{238}U is not significantly affected by their radioactivity, and because of their long half-lives, the problem of tracer-level chemistry is not encountered in practice.

Uranium exists in several valence states ranging from 0 to VI. In nature it occurs almost entirely as U(IV) or U(VI) oxides (UO_2 and U_3O_8) and in ground water as UO_2^{+2} or as complexes such as the carbonate [10]. Under oxidizing conditions, uranium is soluble as UO_2^{+2} , and complexation can increase that solubility under all conditions present in normal groundwater. Carbonate complexation renders UO_2^{+2} soluble even under fairly strong reducing conditions (Eh = −0.1 V). The solubility of solid UO_2 is strongly affected by its degree of crystallinity, since amorphous UO_2 is reported to be as much as five orders of magnitude more soluble than the crystalline form [21]. Table 9.2-2 gives a solubility value of 1 ppb as "most probable" for UO_2, but oxidizing conditions, complexation and degree of crystallinity can all affect this property. Retardation factors reported in the literature vary widely, depending mostly on the nature of the substrate, as shown in Table 9.2-2; the presence of organic matter is reported to raise the retardation factor significantly [37]. Uranium occurs widely in natural rocks, soils and both groundwater and seawater [16], but it is considered relatively unimportant as an environmental hazard because of its very long radioactive half-life and low chemical toxicity.

9.2.32 239-Plutonium

^{239}Pu, like all plutonium isotopes, is virtually nonexistent in nature. It is made artificially by neutron activation of ^{238}U followed by successive beta-decays of the product ^{239}U to ^{239}Np and thence to ^{239}Pu. The chemistry of ^{239}Pu has been extensively investigated and is the subject of much literature, most of which is classified and therefore not generally available. Since ^{239}Pu is the exclusive property of government and military organizations, its environmental management is not a matter of direct concern to the general public. Nevertheless, because of the high degree of toxicity of this element, and the possibility that accidental contamination of the environment could occur in the course of military and/or scientific experiments conducted by government agencies, some of the more important facts about plutonium will be reviewed here.

Plutonium is a member of the actinide group of elements, so-called because of their systematic resemblance to the lanthanide group, and can exist in at least six different oxidation states, Pu(II), Pu(III), Pu(IV), Pu(V), Pu(VI) and Pu(VII) as well as the metal. All but Pu(II), which oxidizes spontaneously in water, can be stabilized in aqueous solution [1]. The chemical properties of plutonium are generally similar to those of uranium, leading to the conclusion that plutonium, if allowed to come in contact with environmental water, would be readily dissolved, readily complexed, and potentially widely disseminated. Plutonium tends to form colloids when present as the thermodynamically stable hydroxide in groundwater, and a knowledge of the behavior of such colloidal particulates is essential to the management of groundwater contamination by this element [1]. Sorption of plutonium on natural rocks and soils is fairly high [36] as shown in Table 9.2-2, although, like the other actinides, this behavior of plutonium may be influenced adversely (i.e., reducing the retention) by complexing, especially with organic ligands.

9.2.33 241-Americium

Neither ^{241}Am nor any other isotope of americium exists in nature. Like ^{239}Pu, it is made artificially by neutron activation of ^{238}U, ^{239}Pu and other actinides, followed by successive beta decays of the products. ^{241}Am decays principally by alpha emission (E_α = 5.4857 MeV, 5.4430 MeV), accompanied by several gamma rays (E_γ = 0.0595 MeV, 0.0263 MeV, others), and has proved useful as a replacement for ^{210}Po in alpha-sources for industry and commerce. A popular model of a domestic smoke alarm employs a small source of ^{241}Am to create ions whose mobility is affected by the presence of smoke particles in the air around the sensor. The decay half-life of ^{241}Am is 432 years, making it more stable over time than ^{210}Po ($T_{1/2}$ = 138 days), and its physical and chemical stability are both greater than those of polonium.

Americium in the (IV) oxidation state forms a relatively soluble oxide, AmO_2, but its precise solubility has not been reported [21, 37]. The only stable oxidation state in solution is generally agreed to be (III), but widely divergent reports as to the solubility of $Am(OH)_3$, whether measured or calculated, are all that is available at present.

Americium is generally reported to be strongly sorbed by all common rocks at ambient pH values [37]. Complexation of Am(III) by suitable ligands is expected to strongly affect adsorption, however, reducing the retention.

9.3 LITERATURE CITED

1. Andelman, J.B. and T.C. Rozell, "Plutonium in the Water Environment," in *Radionuclides in the Environment*, R.F. Gould (ed.), American Chemical Society, Washington, D.C. (1970).
2. Arthur D. Little, Inc., *Advanced Techniques for Radon Gas Removal,* prepared for U.S. Bureau of Mines, Report H0230022 (1975).
3. Arthur D. Little, Inc., *Engineering Evaluation of Radon Daughter Removal Techniques,* prepared for U.S. Bureau of Mines, Report J0265011 (1978).
4. Bagnell, K.W., "The Chemistry of Polonium," in *Advances in Inorganic Chemistry and Radiochemistry,* H.J. Emeleus and A.G. Sharpe (eds.), Academic Press, New York (1962).
5. Bondietti, E.A. and C.W. Francis, "A Reference Analysis on the Use of Engineered Barriers for Isolation of Spent Nuclear Fuel in Granite and Basalt," PNL-3530, Pacific Northwest Laboratory, Battelle Memorial Institute, Richland, Wash. (1979).
6. Bonner, N.A. and M. Kahn, "Behavior of Carrier-Free Tracers," in *Radioactivity Applied to Chemistry,* A.C. Wahl and N.A. Bonner (eds.), John Wiley & Sons, New York (1951).
7. Butler, G.C. and C. Hyslop, "Radioactive Substances," in *Anthropogenic Compounds,* Springer-Verlag, New York (1980).
8. Callahan, M., M. Slimak, N. Gabel, I. May, C. Fowler, J. Freed, P. Jennings, R. Durfee, F. Whitmore, B. Maestri, W. Mabey, B. Holt and C. Gould, "Thallium," in *Water-Related Environmental Fate of 129 Priority Pollutants,* Vol. I, Chap. 18, EPA-440/4-79-029a, EPA Contracts 68-01-3852 and 68-01-3867, Office of Water Planning and Standards, U.S. Environmental Protection Agency, Washington, D.C. (1979).
9. Chikalla, T.D., H.T. Fullam and F.P. Roberts, "Preparation and Properties of Promethium Oxide," in *Promethium Technology,* Chapter 10, E.J. Wheelwright (ed.), American Nuclear Society, Hinsdale, Ill. (1973).

10. Deutsch, W.J. and R.J. Serne, "Uranium Mobility in the Natural Environment," in *Geochemical Behavior of Disposed Radioactive Waste*, G.S. Barney, J.D. Navratil and W.W. Schulz (eds.), American Chemical Society, Washington, D.C. (1984).

11. Fukai, R. and Y. Yokohama, "Natural Radionuclides in the Environment," in *The Natural Environment and the Biogeochemical Cycle,* Springer-Verlag, New York (1982).

12. Garner, C.S., "Radioactivity Applied to the Discovery and Investigation of the Newer Elements," in *Radioactivity Applied to Chemistry,* A.C. Wahl and N.A. Bonner (eds.), John Wiley & Sons, New York (1951).

13. Hwang, L.L.Y., N. Ronca, N.A. Solomon and J. Steigman, "Extraction of Tc(V) from Radiopharmaceuticals," *Int. J. Appl. Radiat. Isot.,* **35**, 825-30 (1984).

14. Johnstone, J.K. and K. Wolfsberg (eds.), *Evaluation of Tuff as a Medium for a Nuclear Waste Repository: Interim Status Report on the Properties of Tuff,* SAND80-1464, Sandia National Laboratories, Albuquerque, N. Mex. (1980).

15. Kapitanov, Y.T., I.V. Pavlov, N.P. Semikin and A.S. Serdyukova, "Adsorption of Radon on Activated Carbon," *Izv. Vyssh. Ucheb. Zaved., Geol. Razvedka,* **10,** 118-25, (1967).

16. Klement, A.W. (ed.), *CRC Handbook of Environmental Radiation,* CRC Press, Boca Raton, Fla. (1982).

17. Lal, D. and E. Peters, "Cosmic Ray Produced Radioactivity on Earth," in *Encyclopedia of Physics,* Springer-Verlag, New York, 600-1 (1967).

18. Lederer, C.M. and V.S. Shirley (eds.), *Table of Isotopes,* 7th ed., John Wiley & Sons, New York (1978).

19. Magorian, T.R., K.G. Wood, J.G. Michalovic, S.L. Pek and M.M. VanLier, *Water Pollution by Thallium and Related Metals,* NTIS PB 253333, Springfield, Va., 145-60 (1974).

20. Meloche, C.C. and F. Vratny, "Solubility Product Relations in the Rare Earth Hydrous Hydroxides," *Anal. Chim. Acta,* **20**, 415-19 (1959).

21. Muller, A.B., N.C. Finley and F.J. Pearson, Jr., *Geochemical Parameters Used in the Bedded Salt Reference Repository Risk Assessment Methodology,* Sandia Laboratories, Albuquerque, N. Mex., prepared for U.S. Nuclear Regulatory Commission, Washington, D.C., NUREG/CR-1996, SAND81-0557 (1981).

22. Pinkerton, T.C., W.R. Heineman and E. Deutsch, "Separation of Technetium Hydroxyethylidene Diphosphonate Complexes by Anion-Exchange High Performance Liquid Chromatography," *Anal. Chem.,* **52**, 1108-10 (1980).

23. Ponto, R.A. and M.K. Loken, "Radioactive Gases: Production, Properties, Handling, and Uses," in *Radiopharmaceuticals,* G. Subramanian, B.A. Rhodes, J.F. Cooper and V.J. Sodd (eds.), Society of Nuclear Medicine, New York (1975).

24. Public Health Service, U.S. Dept. of Health, Education and Welfare, *Radiological Health Handbook,* rev. ed., Rockville, Md. (1970).

25. "Radiation Hazards in Mining: Control, Measurement and Medical Aspects," Colorado School of Mines, Golden, Colo. (Oct. 1981), Sponsored by U.S. Mine Safety & Health Admin. *et al.* (1982).

26. Richards, P. and J. Steigman, "Chemistry of Technetium as Applied to Radiopharmaceuticals," in *Radiopharmaceuticals,* G. Subramanian, B.A. Rhodes, J.F. Cooper and V.S. Sodd (eds.), Society of Nuclear Medicine, New York (1975).

27. Roberts, F.P., "Analytical Chemistry of Promethium," in *Promethium Technology,* Chap. 9, E.J. Wheelwright (ed.), American Nuclear Society, Hinsdale, Ill. (1973).

28. Russell, C.D., "Electrochemistry of Technetium," *Int. J. App. Radiat. Isot.,* **33,** 883-89 (1982).

29. Russell, C.D. and J. Majerik, "Tracer Electrochemistry of Pertechnetate: Chelation of ^{99m}Tc by EDTA After Controlled-Potential Reduction at Mercury and Platinum Cathodes," *Int. J. Appl. Radiat. Isot.,* **29**, 109-14 (1978).

30. Stein, L., "Removal of Radon from Air by Oxidation with Bromine Trifluoride," *J. Inorg. Nucl. Chem.,* **35,** 39-43 (1973).

31. Stein, L., "Removal of Xenon and Radon from Contaminated Atmospheres with Dioxygenyl Hexafluoroantimonate, O_2SbF_6," *Nature,* **243**, 30-2 (May 1973).

32. Tanner, A.B., "Radon Migration in the Ground: A Supplementary Review," in *Natural Radiation Environment III,* Vol. 1, Symposium Proceedings, Houston, Tex., April 1978, T.F. Gessel and W.M. Lowder (eds.), pp. 5-56, USDOE Rept. CONF-780422, NTIS (1978).

33. Thomas, J.W., "Radon Adsorption by Activated Carbon in Uranium Mines," *Noble Gas Symposium Proceedings,* Las Vegas, Nev., U.S. Atomic Energy Commission/HASL (Sept. 1973).

34. U.N. Scientific Committee on the Effects of Atomic Radiation, *Sources and Effects of Ionizing Radiation,* Official Record of the General Assembly, 32nd Session, Suppl. No. 40 (A/32/40), United Nations, New York (1977).

35. Vine, E.N. and A.C. Wahl, "The Chemical Behavior of Tracer-Level Technetium Isotopes," *Int. J. Appl. Radiat. Isot.,* **33**, 861-66 (1982).

36. Walker, F.W., D.G. Miller and F. Feiner, *Chart of the Nuclides,* 13th ed., General Electric Co., San Jose, Calif. (1984).

37. Waste Isolation Systems Panel, *A Study of the Isolation System for Geologic Disposal of Radioactive Wastes,"* National Academy Press, Washington, D.C. (1983).

38. Wheelwright, E.J., "Early Developments in Promethium Chemistry," in *Promethium Technology,* Chap. 1, E.J. Wheelwright (ed.), American Nuclear Society, Hinsdale, Ill. (1973).

39. Wolfsberg, K., "Sorptive Properties of Tuff and Nuclide Transport and Retardation," in Johnstone and Wolfsberg [14], pp. 39-48.

10. CHEMICAL CLASSES

Itamar Bodek, Daniel J. Ehntholt, Anita E. Glazer, Christopher P. Loreti, Warren Lyman

LIST OF TABLES

ORGANOMETALLICS

ACIDS

CYANIDES

HYDRIDES

HYPOCHLORITES AND CHLORATES

PEROXIDES

LIST OF FIGURES

ORGANOMETALLICS

ACIDS

GASES

BORON-CONTAINING COMPOUNDS

SILICON-CONTAINING COMPOUNDS

INTRODUCTION

This chapter discusses the environmental chemistry of 16 chemical classes. These classes are variously defined; some are groups of compounds sharing certain chemical or physical properties (acids, bases, salts, and gases), while others are groups of compounds containing a specific element (sulfur-containing, phosphorus-containing, boron-containing, and silicon-containing compounds) or an element in a specific oxidation state or functional group (oxides, carbonyls, carbides, cyanides, hydrides, hypochlorites and chlorates, and peroxides). The chemical classes included were derived from a review of major industrial inorganic chemicals and of inorganic chemicals regulated by the EPA or OSHA. The classes thus include chemicals that are of commercial importance or that are considered problematic from an environmental or occupational health standpoint.

The class of organometallics (section 10.1) is given special treatment and is covered more extensively than any of the other classes. Although the inclusion of organometallics is somewhat out of keeping with the primary focus of this report on inorganics, it reflects an awareness of their growing commercial and environmental importance, and it provides for the fact that they were not covered in the companion work on organic chemicals.[1]

The discussion of each class includes general data on the types of compounds that represent it, focusing on specific chemicals that have been selected because of their commercial or environmental importance. For these selected chemicals, major commercial uses are listed, along with the physical and chemical properties that govern their fate in the environment. Information is then provided on the reactions and transformations that members of the class are likely to undergo under environmental conditions. Such reactions may include:

- Volatilization
- Reaction with air (e.g., oxidation)
- Dissolution in and/or reaction with water (solubility, acid-base dissociation, hydrolysis)
- Speciation in water (complexation, oxidation/reduction, polymerization)
- Mobility and degradation in soils.

1. Lyman, W.J., W.F. Reehl and D.H. Rosenblatt (eds.), *Handbook of Chemical Property Estimation Methods: Environmental Behavior of Organic Compounds*, McGraw-Hill Book Co., New York (1982).

10.1 ORGANOMETALLICS

10.1.1 Introduction

In the context of this report, organometallic compounds are an important class of chemicals for several reasons. First, they have many commercial applications, such as paint additives, pesticides, and gasoline additives; hundreds of premanufacture notice (PMN) forms have been submitted to the EPA for a wide variety of new organometallic compounds. Second, they are important from a biological standpoint: some play an essential role in plant photosynthesis and oxygen transport in blood, while others (such as methylmercury and other toxic compounds) are formed by microorganisms in soils and sediments. (See section 2.15.) Third, organometallics of various types are believed to play a significant role in the mobility and bioavailability of metals in the environment. (As described in section 2.9, trace metal interaction with humic and fulvic acids is of particular interest.) And finally, organometallics bridge the gap between organic chemicals, which were covered in a previous report [23], and the inorganic chemicals covered in this report.

Unfortunately, few data are available on the environmental chemistry of organometallics and even less on methods for estimating their environmentally important properties. The few exceptions (i.e., chemicals for which some data are available) include the organotins, tetraethyl lead, methylated forms of mercury, and metal complexes with humic and fulvic acids.

In assessing the environmental fate of an organometallic, it is important to know what happens to the chemical when it dissolves in water. Does it remain essentially unchanged in solution, permitting it to be treated (for property estimation and fate assessment purposes) like a neutral organic chemical? Or does it react with water to yield a hydrated metal ion and an organic ion, each of which can be considered separately in an environmental assessment? The next subsection focuses on such questions, describing categories of organometallics and a method for roughly assessing the probable reactivity of the metal-to-organic bonds or other metal-ligand bonds.

The final two subsections describe estimation methods for one bulk property (the boiling point) and review some property-property and structure-property estimation methods that have recently been derived for various solution properties of organotins; both physicochemical and toxicological properties are included.

As noted above, some information on the biologically mediated methylation of metals is presented in section 2.15,[1] and the complexation of metals with humic and fulvic acids is covered in the section on complexation (§2.9).

---

1. Additional information on the environmental fate of methylated metal(loid)s is provided by Brinckman *et al.* [8].

10.1.2 Classification of Organometallics

NEED FOR CLASSIFICATION

The term "organometallic" has been applied to almost any substance containing a metal or metalloid and carbon atoms [10], but it most commonly refers to compounds containing metal-carbon bonds. Because of the wide variety of substances that contain both a metal or metalloid and a carbon atom (for example, chelate complexes, metal organic salts, and metal/metalloid-to-R bonds, where the bond to the organic group (R) may be to carbon, oxygen, nitrogen or sulfur atoms), it is useful to classify organometallics on the basis of the processes that affect their environmental behavior. These processes include dissolution, complexation, and breaking of particular metal- or metalloid-to-carbon or other atom bonds. This method of classification characterizes an organometallic in terms of its initial process of environmental speciation, which is of primary concern in environmental assessment. It also helps to associate the compound with particular processes and estimation methods described in other chapters.

CLASSES AND ENVIRONMENTAL BEHAVIOR

Organometallics have been grouped below into three categories based upon their expected environmental behavior and the coverage of relevant properties in this report. The three categories reflect the compounds that are on various regulatory lists as well as those that have major uses (Table 10.1-1).

- **Category I**: When dissolved in water, the compound yields an initial species with the same metal-to-organic bonds for the metal of interest, as shown in equation 1 for a metal or metalloid (M) and an organic group (R). Subsequent environmental reactions are based upon the reactivity of the particular M-R framework bonds. This category can thus include salts of a metal-organic species ([M-R][X]) where, although the original compound separates into the respective ions in solution, all the metal-organic bonds in the ion are retained upon dissolution, as shown in equation 2. (Charges on the ions are omitted.)

$$\text{M–R} + H_2O \longrightarrow \text{M–R}_{aq} \qquad (1)$$

or

$$[\text{M–R}][\text{X}] + H_2O \longrightarrow \text{M–R}_{aq} + \text{X}_{aq} \qquad (2)$$

where R = carbon-containing organic fragment

M = metal or metalloid

X = counter ion of salt (e.g., Cl^-, Na^+)

aq = designation of aqueous species

TABLE 10.1-1

Categorization of Some Organometallics of Regulatory and Environmental Importance
(See text for explanation of categories)

Category I	Category II	Category III	Source[a]
Ethyl silicate Alkyl mercury compds Alkyl lead compds Alkyl tin compds Nickel carbonyl Iron pentacarbonyl	Sodium fluoroacetate	Zinc stearate	[3]
Diethylarsine Phenylmercuric acetate	Lead acetate Thallium(I) acetate	Ferric dextran Mercury fulminate	[11]
Dibutyltin bis-(isooctylmaleate)	Calcium naphthenate	Cobalt naphthenate	[31]
Alkoxymethyl-polysilazanes	Sodium salt of naphthalenedisulfonic acid	Metal polyisobutenyl-succinate Iron complex of sub-stituted phenol azophenyl-pyrazolone	[31]
Lead alkyls	Calcium carbide		[9]
	Sodium propionate Calcium propionate		[9]
Aluminum tris-o-ethyl phosphonate Disodium methylarsonate[b] Triphenyltin hydroxide Triphenyltin acetate			[30]

a. List of chemicals in which organometallic is included. Category assignment made by authors of this report.

b. Although upon dissolution this compound yields separate dissolved ions, one of these ions (methylarsonate) retains the metalloid-organic bond of the original compound and is thus placed in Category I.

- **Category II**: Salts of organic anions with labile metals ([M][R]) which, upon dissolution in water, rapidly yield solutions containing predominantly the separated components of the salts, i.e., the hydrated metal ions and the organic portions, as shown in equation 3 (charges omitted).

$$[M][R] + H_2O \longrightarrow M_{aq} + R_{aq} \tag{3}$$

- **Category III**: Metal-organic complexes (e.g., chelates), $[M\text{-}R_n]$, of relatively non-labile metals with strongly complexing ligands that, upon dissolution in water, yield metal-organic complexes for which reversible ligand exchange reactions are of importance, as shown in equation 4 (with charges omitted), along with any associated counter ions.

$$\begin{aligned} M\text{-}R_n + H_2O \longrightarrow\ & [M\text{-}R_n]_{aq} \\ & \quad\rightleftharpoons \\ & [M\text{-}R_{n-1}]_{aq} + R_{aq} \\ & \quad\rightleftharpoons \\ & [M\text{-}R_{n-2}]_{aq} + R_{aq} \\ & \quad\rightleftharpoons \\ & \text{etc.} \end{aligned} \tag{4}$$

Table 10.1-1 lists some compounds of regulatory and environmental importance in each of the above categories. These categories are not mutually exclusive, as particular compounds can behave in a manner consistent with one or more categories; a gradation occurs within each category as compounds begin to exhibit properties associated with the next category. However, this method of classification provides a useful indication of the processes that are most important in environmental speciation.

Examples of Category I compounds are tetraethyl lead ($Pb(CH_2CH_3)_4$) and diethylarsine ($(CH_3CH_2)_2AsH$), both of which can be considered to initially produce a dissolved molecular species of the same composition in water. Similarly, the sodium salt of methylarsonate ($Na_2CH_3AsO_3$) produces a methylarsonate anion in solution that retains the metalloid-organic bond. For all these compounds the chemical stability of the metal- or metalloid-to-carbon bond is of paramount importance in estimating persistence and environmental fate. (Acid-base equilibria should also be considered, if they modify the species in question.)

Examples of compounds in Category II are sodium acetate ($NaC_2H_3O_2$) and calcium propionate ($Ca(C_3H_5COO)_2$); upon dissolution in water, these yield the labile aquo metal ions of sodium and calcium and the counter-organic anions of acetate and propionate, respectively. For this category, aqueous environmental chemistry is determined by the properties of the respective ions and not the original molecule.

Examples of Category III compounds include transition metal cyanides, which form distinct coordination complexes in solution and whose subsequent environmental speciation can be based on reactions such as reversible complexation equilibria (see § 2.9 and § 3.2). The distinction between all these categories, especially for Cagetory II and III compounds, is very dependent on the kinetic stability or lability of the metal-to-organic ligand bonds.

The equilibria and kinetics of reactions of inorganic ion complexes are discussed in other sections (e.g., § 2.9 and § 3.2) of this report. The environmental behavior and estimation of properties of various organic species are addressed in reference 23. Thus, the environmental behavior of compounds in Categories II and III can be inferred by consulting other sections of this report and information available for organic species. As such, this section focuses on compounds in Category I, whose behavior is strongly influenced by the specific metal or metalloid and its bonding to the elements that are linked to the organic framework (R).

BONDING IN ORGANOMETALLIC COMPOUNDS

The behavior of organometallic compounds in the environment is strongly affected by the type(s) of bonds that they contain. Knowledge of the variation in bonding within a series of compounds can therefore be very helpful in predicting what they will do under given circumstances.

Theoretical chemists usually classify bonding in terms of sigma, pi and ionic bonds. In sigma bonds, overlapping electronic orbitals are coaxial. The axes of overlapping pi bonds are parallel, as in the second carbon-carbon bond of ethylene; pi bonds may also involve back-donation of electrons from a metal's d-orbitals (carbon has no d-orbitals) to available anti-bonding orbitals of the organic moieties. Unlike sigma and pi bonds, pure ionic bonds exhibit attraction based not on orbital overlapping, but on the force between opposite electrical charges. For the purpose of predicting environmental behavior of Category I compounds, it is practical to consider both sigma and pi bonds as "covalent" — at one extreme of bonding behavior — and to define an intermediate group between this extreme and ionic bonding.

Many compounds contain more than one type of bond and thus cannot be described as having only a particular bond character. However, as a first cut separation for assessing their environmental behavior, we can assign compounds into the following subcategories within Category I:

- Subcategory Ia: Compounds whose nonpolar covalent bonds give them a predominantly "organic" character,
- Subcategory Ib: Compounds of intermediate character that cannot be placed in Ia or Ic, and
- Subcategory Ic: Species that contain mostly ionic bonds and are thus treated as "inorganic" compounds.

Subcategories Ia and Ic represent the extremes of a gradual progression in overall polarizability or ionic character. The degree of ionization or polarity [$M\delta^{+}$---$X\delta^{-}$] of bonds in a compound is particularly important when considering relative rates of reactions such as hydrolysis in water. Since the categories form a continuum, subcategory Ic overlaps Category II.

BOND CHARACTER

As discussed above, the classification of diverse organometallic compounds into separate and discrete categories is not possible. However, trends in behavior can be inferred within a homologous series of compounds — such as those that contain a variable metallic center with identical (or very similar) organic components — on the basis of the relative position of the metallic center in the periodic table. For example, Rochow [29] divided most of the periodic table into three sections, each associated with a different kind of organometallic compound typically formed by the related elements (Figure 10.1-1). The section boundaries do not signify a sharp differentiation in properties of the compounds or character of the bonds that are formed by these elements; they merely reflect general trends within the enclosed group. The properties of an organometallic compound clearly depend both on the nature of the metal or metalloid atom and that of the organic(s) attached to that atom. It is therefore essential to focus on the nature of the bond(s) between them.

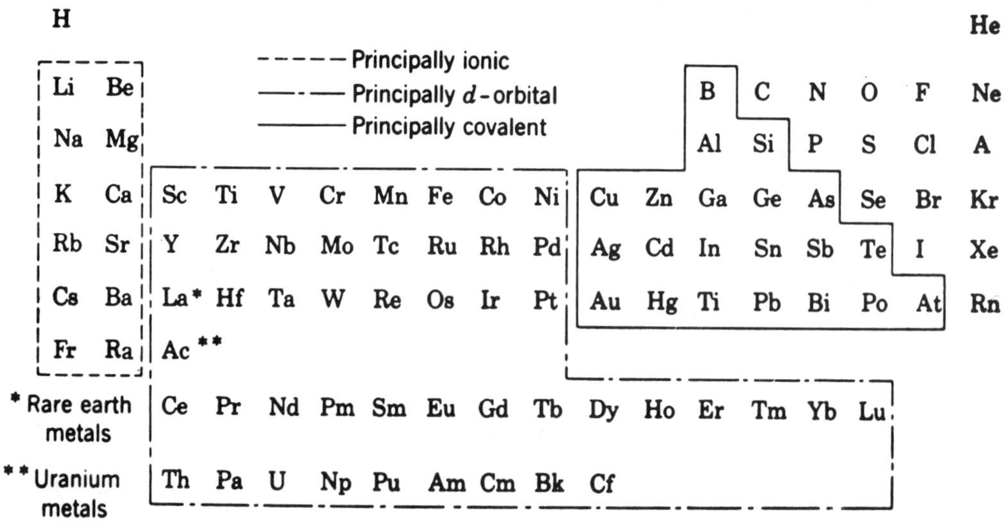

Source: Rochow [29]. *(Copyright 1957, John Wiley & Sons. Reprinted with permission.)*

FIGURE 10.1-1 Types of Bonding in Organometallic Compounds

One approach, a semiquantitative method for determining bond character, utilizes electronegativity values. A summary of values for three scales of electronegativity — the Allred-Rochow [2], Pauling [1], and Mulliken scales — are given in Table 10.1-2. The Allred-Rochow values have the advantage of wide availability for most elements and of empirical consistency. Although most chemists subscribe to the qualitative concept of electronegativity, precise definitions are elusive. Pauling defines it as the ability of an atom in a molecule to attract electrons to itself [3]. Accordingly, the same element in different oxidation states has different electronegativities, i.e., a higher value at higher oxidation state. This is particularly true for metal atoms. The values given in Table 10.1-2 should therefore not be taken as an exact measure of electronegativity in a particular compound but only as rough guides.

Simply on the basis of electronegativity values, the bond polarity should increase as the difference between electronegativity of the bonded atoms increases. However, electronegativity values are empirical parameters that do not take into account effects of molecular structure. Other factors, such as the availability of metal orbitals, steric factors, and mechanisms of particular reactions, also need to be considered in evaluating electronegativity difference as an indicator of bond stability or reactivity in a particular process.

Eisch [14] found general correlations of differences in electronegativity and bond character (which are then related to reactivity properties). For example, metal-alkyl bonds tend to be covalent in character when the difference in electronegativity of the metallic atom and carbon is less than or equal to about one unit. When this difference is less than about 1.5, the organometallic species is expected to remain associated in solution. Differences of greater than about 1.5 are characteristic of highly polar molecules or ionic species in solution. Electronegativity can therefore be used to predict qualitative trends of metallic-organic bond lability (see section 3.2). This process is described below.

BOND CHARACTER CLASSIFICATION

Electronegativity can be used to classify bonds according to the subcategories defined above (Ia, Ib, Ic). The following procedure is used:

Step 1: Draw the structure of the molecule in a form that shows the individual atoms bonded to the metallic center.

Step 2: Identify the different metallic center-to-organic moiety (M-R) bonds.

Step 3: List the electronegativity values (E_M) for the metal or metalloid and for each R (E_R) for all the M-R bonds (Table 10.1-2), and calculate the absolute value of each difference for each bond (ΔE_{MR}) as follows:

$$\Delta E_{MR} = | E_M - E_R | \tag{5}$$

TABLE 10.1-2

Electronegativities of the Elements

(Values in bold type are calculated by using the Allred-Rochow formula; those in italics are estimated by Pauling's method and those in Roman type are calculated by Mulliken's method.)

I[a]	II	III	IV	II	II	II	II	II	II	II	II	III	IV	III	II	I	
H **2.20**																	He
Li **0.97** *0.98* 0.94	Be **1.47** *1.57* 1.46											B **2.01** *2.04* 2.01	C **2.50** *2.55* 2.63	N **3.07** *3.04* 2.33	O **3.50** *3.44* 3.17	F **4.10** *3.98* 3.91	Ne
Na **1.01** *0.93* 0.93	Mg **1.23** *1.31* 1.32											Al **1.47** *1.61* 1.81	Si **1.74** *1.90* 2.44	P **2.06** *2.19* 1.81	S **2.44** *2.58* 2.41	Cl **2.83** *3.16* 3.00	Ar
K **0.91** *0.82* 0.80	Ca **1.04** *1.00*	Sc **1.20** *1.36*	Ti **1.32** *1.54*	V **1.45** *1.63*	Cr **1.56** *1.66*	Mn **1.60** *1.55*	Fe **1.64** *1.83*	Co **1.70** *1.88*	Ni **1.75** *1.91*	Cu **1.75** *1.90* 1.36	Zn **1.66** *1.65* 1.49	Ga **1.82** *1.81* 1.95	Ge **2.02** *2.01*	As **2.20** *2.18* 1.75	Se **2.48** *2.55* 2.23	Br **2.74** *2.96* 2.76	Kr
Rb **0.89** *0.82*	Sr **0.99** *0.95*	Y **1.11** *1.22*	Zr **1.22** *1.33*	Nb **1.23**	Mo **1.30** *2.16*	Tc **1.36**	Ru **1.42**	Rh **1.45** *2.28*	Pd **1.35** *2.20*	Ag **1.42** *1.93* 1.36	Cd **1.46** *1.69* 1.4	In **1.49** *1.78* 1.80	Sn **1.72** *1.96*	Sb **1.82** *2.05* 1.65	Te **2.01** 2.10	I **2.21** *2.66* 2.56	Xe
Cs **0.86** *0.79*	Ba **0.97** *0.89*	*	Hf **1.23**	Ta **1.33**	W **1.40** *2.36*	Re **1.46**	Os **1.52**	Ir **1.55** *2.20*	Pt **1.44** *2.28*	Au *1.42* 2.54	Hg *1.44* 2.00	Tl **1.44** *2.04*	Pb **1.55** *2.33*	Bi **1.67** *2.02*	Po **1.76**	At **1.96**	Rn
Fr **0.86**	Ra **0.97**	**															

*La	Ce	Pr	Nd	Pm	Sm	Eu	Gd	Tb	Dy	Ho	Er	Tm	Yb	Lu
1.08 1.10	**1.06** 1.12	**1.07** 1.13	**1.07** 1.14	**1.07**	**1.07** 1.17	**1.01**	**1.11** 1.20	**1.10**	**1.10** 1.22	**1.10** 1.23	**1.11** 1.24	**1.11** 1.25	**1.06**	**1.14** 1.27
****Ac**	Th	Pa	U	Np	Pu	Am	Cm	Bk	Cf	Es	Fm	Md		
1.00	**1.11**	**1.14**	**1.22** 1.38	**1.22** 1.36	**1.22** 1.28	←		~1.2 (estimated)				→		

a. Roman numeral headings give the oxidation status used for the Pauling-type values.

Source: Cotton and Wilkinson [12]. *(Copyright 1972, Wiley Interscience. Reprinted with permission.)*

Step 4: Compare the value of ΔE_{MR} with those in the following categories and bond characters [14]:

ΔE_{MR}	Bond Character	Subcategory
< 1.0	Covalent	Ia
1.0 to 1.5	Intermediate	Ib
> 1.5	Ionic	Ic

Example 1 Classify the types of bonds in (i) dimethylmercury [$Hg(CH_3)_2$] and in (ii) fluoromethylmercury [$HgF(CH_3)$].

Step 1. (i) CH_3-Hg-CH_3 (ii) CH_3-Hg-F

Step 2. (i) Two Hg-C bonds (ii) One Hg-C and one Hg-F bond

Step 3. Using the Pauling values (italic numbers) in Table 10.1-2,

(i) $E_{Hg} = 1.44$

$E_C = 2.55$

$\Delta E_{Hg\text{-}C} = 1.11$

(ii) $E_F = 3.98$

$\Delta E_{Hg\text{-}F} = 2.54$

Step 4. Hg-C bonds fall into Category Ib and the Hg-F bond into Category Ic. This implies that dimethylmercury (i) is more covalent in character than fluoromethylmercury (ii) and that (i) is expected to exist initially as the molecular species in aqueous solution while (ii) is expected to have a labile Hg-F bond and to initially produce two species (F^- and CH_3Hg^+ ions) in solution. The subsequent chemical reactions of (ii) in water are mainly functions of the properties of these ions, rather than those of the original molecular species. The behavior of (i) in solution is more a function of the composition of the original compound.

Example 2 Rank the following compounds in the expected order of increasing organic-like behavior:

Group 1 — (i) chlorotrimethyltin [$ClSn(CH_3)_3$], (ii) dichlorodimethyltin [$Cl_2Sn(CH_3)_2$], and (iii) trichloromethyltin [Cl_3SnCH_3].

Group 2 — (iv) chlorotriethoxytin [$ClSn(OCH_2CH_3)_3$] and (v) chlorotriethyltin [$ClSn(CH_2CH_3)_2$].

Group 1 — The difference in electronegativity for all the bonds is calculated as shown in example 1 for these structures:

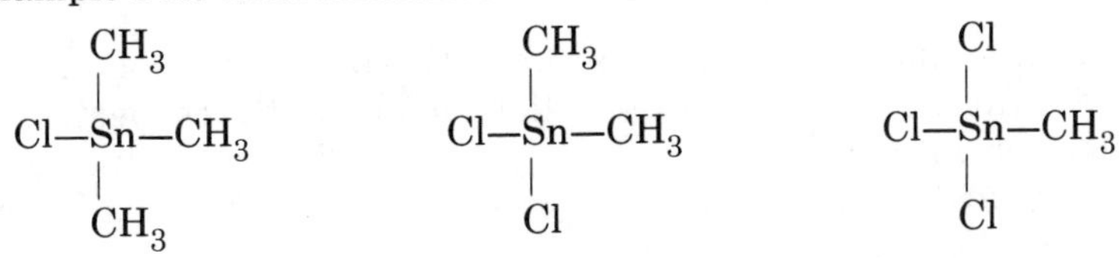

(i) (ii) (iii)

E_{Sn} = 1.96
E_C = 2.55
E_{Cl} = 3.16
For Sn—C the value of $\Delta E_{Sn\text{-}C}$ = 0.59
For Sn—Cl the value of $\Delta E_{Sn\text{-}Cl}$ = 1.20

The sum of the differences in electronegativities for the three compounds are: (i) 2.97, (ii) 3.58, (iii) 4.19. The relative magnitude of these values for bonds at the Sn center is in the order of (iii) > (ii) > (i). Thus, compound (iii) is expected to behave the least like an organic species and compound (i) the most. Note that if CH_3 on structure (iii) were replaced by a larger organic group, this compound would still behave in a more organic-like manner. The Sn-C bonds belong to Subcategory Ia (covalent) and the Sn-Cl bonds to Subcategory Ib (intermediate).

Group 2 — Similarly, for structures (iv) and (v) we have:

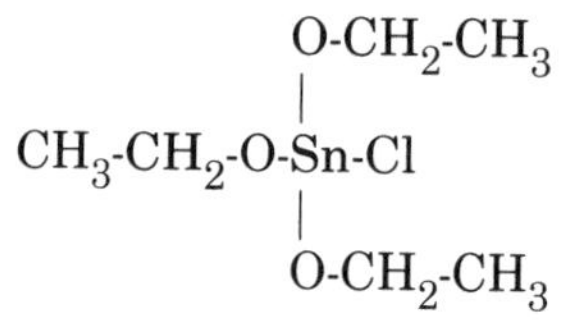
(iv)

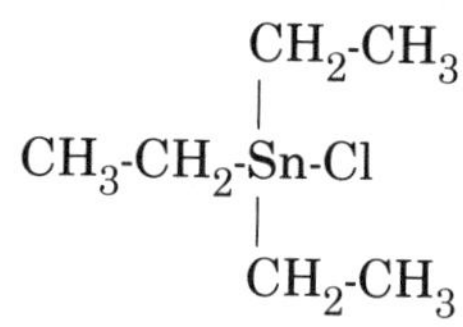
(v)

E_O = 3.44
Thus, $\Delta E_{Sn\text{-}O}$ = 1.48
The Sn-O bonds border on being in Subcategory Ic (ionic).

The sum of the differences for all the tin bonds is 5.64 for compound (iv) and 2.97 for compound (v). Thus, compound (v) is expected to behave more like an organic species than compound (iv).

The prior discussion treats M-R bonds as primarily two-atom interactions. This is clearly not the case. The stability of an organometallic species is also strongly affected by the nature of the entire organic group attached to the metallic atom. The polarity of the M-R bond, where R is the organic group, depends to a great extent on the organic ion stability of the R group under the given polarizing field of the metallic atom M. Trends in organic ion stability can be used to predict the susceptibility of the organometallic compound to reaction with water or oxygen: in general, the more stable the organic ion, the more polar the M-R bond and the more susceptible it is to reaction.

The acid dissociation constants (pK_a values) of the R-H acids are good indicators of organic ion (R^-) stability. Generally, the smaller the pK_a of the respective organic acid, the more electron-attracting the R^- group will be. Typical values of pK_a for various organic acids are given in Table 10.1-3 and illustrate the range of expected stability of the organic anion counterpart.

TABLE 10.1-3

Acid Dissociation Constants for Organic Acids

Conjugate Acid	pK_a	Conjugate Base
Methane	40	$CH_3:^-$
Toluene	35	$C_6H_5CH_2:^-$
Aniline	27	$C_6H_5NH:^-$
t-Butanol	19	$C_4H_9O:^-$
Water	15.7	$HO:^-$
Phenol	10	$C_6H_5O:^-$
RNH_3^+	~10	RNH_2
p-Nitrophenol	7.2	*p*-NO_2-$C_6H_4O:^-$
Pyridinium ion	5.2	Pyridine
Carboxylic acids	4.5 ± 0.5	Carboxylate anions
p-Nitroanilinium	1.0	*p*-Nitroaniline
$CH_3OH_2^+$	-2	Methanol
$C_6H_5OH_2^+$	-6.7	Phenol

Source: Hendrickson, Cram, and Hammond [15]. *(Copyright 1970, McGraw-Hill Book Co. Reprinted with permission.)*

Since organic ion stabilities are a function of several factors including electronic, inductive, resonance and steric effects, they cannot be relied upon as the sole indicator of trends in bond polarity. However, for reactions involving an ionic intermediate, these stabilities can be good indicators of relative rates of reactions that proceed via similar mechanisms.

10.1.3 Estimation of Boiling Points

Many methods for estimating the boiling points of various organometallic compounds have been published [4,5,6,15,22-26]. Most of these are based on the Kinney equation (see Table 10.1-4) and data on boiling point numbers for the metal or metalloid and the organic groups bonded to them [4,15,22-26]. Other empirical equations relating boiling points for specific series of substituted germanium and silicon compounds have also been published [5,6]; these equations are summarized in Table 10.1-4. The fourth column of this table lists sources of boiling point numbers (group contributions) to be used with the Kinney equation for a specific series of compounds. For the series of compounds cited that are based on the Kinney equation, the average error associated with these estimation methods is generally less than 4%; for those based on other empirical equations, the error is about 7%.

Example 3 Estimate the boiling point of $(CH_3)_2SiCl_2$.

Using the data in Table 10.1-11, list and sum all the boiling point numbers (b.p.n.) for components of the molecule.

b.p.n. (Si) = 4.20

b.p.n. (Cl) = 4.17 (for a $=SiX_2$ structure)

b.p.n. (CH_3) = 3.17 (for a $=SiX_2$ structure)

The sum of all b.p.n. values is thus

$$(\text{b.p.n.}) = 4.20 + 2(4.17) + 2(3.17) = 18.88$$

By use of the Kinney equation for alkyl silicon compounds (Table 10.1-4),

$$\text{b.p. } [(CH_3)_2SiCl_2] = 230.14 \times (\,(\text{b.p.n.}))^{1/3} - 543$$

$$= 230.14\,(18.88)^{1/3} - 543$$

$$= 69.8°C$$

The experimental value is 70.2°C [22].

TABLE 10.1-4

Equations Used to Calculate Boiling Points of Organometallic Compounds

Compounds	Equation[a]	Average Error (°C)	Note	Source
Organogermanium hydrides Tetrasubstituted germanes Unsaturated and cyclic organogermanium	b.p. = 230.14 × (b.p.n.)$^{1/3}$–543 (Kinney equation)	± 2	b.p.n. data given in Tables 10.1-5 through 10.1-9	[25]
Organogermanium alcohols Organogermanium esters of organic acids Digermanium compounds	b.p. = 230.14 × (b.p.n.)$^{1/3}$–543	± 2.1	b.p.n. data given in Tables 10.1-5 through 10.1-9	[26]
Alkyl and haloalkyl germanium	b.p. = 230.14 × (b.p.n.)$^{1/3}$–543	± 3.6	b.p.n. data given in Table 10.1-10	[4]
Alkyl and phenyl silicon	b.p. = 230.14 × (b.p.n.)$^{1/3}$–543	± 2.4	b.p.n. data given in Table 10.1-11	[22]
Chlorosilanes	b.p. = 230.14 × (b.p.n.)$^{1/3}$–543	± 0.9	Used modified b.p.n. values of method described in [12]	[24]
Organogermanium carboxylates	[b.p.(A), °K] = n(0.25) [b.p.(B)] + (4-n) (k) [b.p.(C)]	± 6.6	(A) = $R_nGe(OCOR')_{4-n}$ (B) = R_4Ge (C) = R′COOH k = 0.3425 to 0.3767	[5]
Organogermanium mercaptides	[b.p.(A), °K] = 0.750 [b.p. (B)] + 0.527 [b.p. (C)]	±6.6	(A) = R_3GeSR' (B) = R_4Ge (C) = R′SH	[5]
Organosilicon fluoroesters	[b.p.(A), °K] = n(0.25) [b.p.(B)] + (4-n) (k) [b.p.(C)]	NA	(A) = organosilicon ester (B) = R_4Si (C) = RCOOH k = 0.284 to 0.302	[6]

a. b.p. = boiling point (°C, except for the last three entries which are entirely in degrees K); b.p.n. = boiling point number

TABLE 10.1-5

Boiling Point Numbers of Alkyl Groups for Germanes[a,b]

Type of Configuration	CH_3	C_2H_5	n-C_3H_7	i-C_3H_7	n-C_4H_9	i-C_4H_9	n-C_5H_{11}
R_4Ge	3.05	6.1	8.3	8.2	10.2	9.4	12.2
R_3Ge–	3.35	6.2	8.5	8.2	10.8	10.55	13.2
R_2Ge=	3.48	6.4	8.9	8.2	11.5	10.55	14.2
RGe≡	3.8	6.6	9.4	8.2	12.2	11.6	15.0

a. b.p.n. for Ge (for single Ge molecules) = 4.4, H (bonded to Ge) = 1.0

b.p.n for Ge (in a Ge_n chain) = 4.4 + 0.3n, b.p.n. of C = 0.8

b. Additional values are given by Nowak [25] for various unsaturated and cyclic compounds.

Source: Nowak [25]. *(Copyright 1969, Ars Polona. Reprinted with permission.)*

TABLE 10.1-6

Boiling Point Numbers for Halides and Halogenoids of Germanes

Configuration of Compound		F	Cl	Br	I	NCO	NCS	CN
GeX_4		1.55	3.95	6.68	—	7.5	—	—
	R = H	2.5	4.7	6.7	—	—	—	—
$RGeX_3$	R = CH_3	4.43	5.2	7.1	10.3	8.6	14.5	—
	larger R	4.2	5.1	7.5	11.3	8.8	14.2	—
	R = H	4.1	4.7	7.0	—	—	—	—
R_2GeX_2	R = CH_3	5.85	6.5	8.4	11.4	9.4	16.2	—
	larger R	5.1	6.4	8.7	12.0	9.8	15.9	—
	R = H	6.9	7.9	9.9	12.8	11.6	20.0	—
R_3GeX	R = CH_3	4.8	7.4	9.1	11.0	9.8	16.9	12.4
	larger R	4.5	7.3	9.3	12.1	10.2	18.2	12.2

Source: Nowak [26]. *(Copyright 1969, Ars Polona. Reprinted with permission.)*

TABLE 10.1-7

Boiling Point Numbers of Some Alkoxy Groups Bonded to Germanium

Configuration	OCH_3	OC_2H_5	OC_3H_7n	OC_3H_7i	OC_4H_9n	OC_4H_9i	OC_4H_9t
Ge≡	5.6	6.8	8.75	7.65	11.2	9.7	8.3
RGe≡	5.87	6.73	8.73	7.9	11.15	9.65	8.25
$R_2Ge=$	6.1	6.72	8.73	8.1	11.15	9.6	8.2
R_3Ge-	6.4	6.7	8.72	8.2	11.1	9.55	8.1

Source: Nowak [26]. *(Copyright 1969, Ars Polona. Reprinted with permission.)*

TABLE 10.1-8

Boiling Point Numbers of –SR Groups for $Ge(SR)_4$ Compounds

SCH_3	$-SC_2H_5$	$-SC_3H_7n$	$-SC_3H_7i$	$-SC_4H_9n$	$-SC_4H_9i$	$-SC_4H_9t$	$-SC_5H_{11}n$
11.0	12.5	14.0	12.6	16.0	14.7	13.0	18.0

Source: Nowak [26]. *(Copyright 1969, Ars Polona. Reprinted with permission.)*

TABLE 10.1-9

Boiling Point Numbers for Various Functional Groups Bonded to Germanium[a]

Groups	b.p.n.
$R_3Ge(CH_2)_nOH$ (primary alcohol)	b.p.n. (OH) = 12.73 – 0.03Y where Y = sum of atomic weights of all C and H atoms (except H in OH groups) in the molecule
$R_3Ge(CH_2R)_n-CH(OH)R'$ (secondary OH)	b.p.n.(OH) = 10.83 – 0.03Y
$R_3Ge(CH_2)_n-CR'_2OH$ (tertiary OH)	b.p.n.(OH) = 10.63 = 0.03Y
Esters of organic acids	b.p.n.(OCO) = 10.15 – 0.0273X where X = molecular weight of the rest of the molecule (with OCO groups)
Alkyl groups	Calculate by summing values using C = 0.8 and H = 1.0

a. Based on b.p.n. of Ge = 4.4 and H (bonded to Ge) = 0.8.

Source: Nowak [25]. *(Copyright 1969, Ars Polona. Reprinted with permission.)*

TABLE 10.1-10

Boiling Point Numbers of Atoms and Groups for Germanium Compounds[a]

Group	Configuration of Alkyl Groups			
	$RGe\equiv$	$R_2Ge=$	R_3Ge-	R_4Ge
CH_3	2.59	2.80	2.88	2.29
C_2H_5	6.72	5.94	5.63	5.41
n-C_3H_7	8.45	8.30	7.96	7.46
i-C_3H_7	8.83	7.64	7.67	...
n-C_4H_9	12.27	10.88	10.01	9.50
Halogen	**Configuration of Halogen or Isocyanate**			
	$RGeX_3$	R_2GeX_2	R_3GeX	GeX_4
F	2.99	3.73	2.79	0.81
Cl	4.27	5.67	6.54	3.20
Br	6.51	7.50	7.96	6.09
I	10.12	10.91	11.08	...
NCO	7.70	8.65	9.10	6.70

a. Based on Ge = 7.413 and H (bonded to Ge) = 0.085.

Source: Anderson [4]. *(Copyright 1962, American Chemical Society. Reprinted with permission.)*

TABLE 10.1-11

Atomic and Group Boiling Point Numbers for Silicon Compounds

	Si, 4.20; H, 0.60 Type of compound (X = Cl, Br. R = Me, Et, n-Pr, Ph)				
	$\equiv SiX$	$= SiX_2$	$-SiX_3$	SiX_4	R_3SiX
Cl	5.05	4.17	3.59	3.39	4.51
Br	7.27	6.47	6.08	5.88	6.58
	$RSi\equiv$	$R_2Si=$	R_3Si-	R_4Si	
Me	3.43	3.17	3.02	2.74	
Et	6.83	6.17	5.91	5.89	
n-Pr	9.47	8.97	8.33	7.85	
Ph	18.87	18.76	18.41	17.73	

Source: Lewis and Newkirk [22]. *(Copyright 1947, American Chemical Society. Reprinted with permission.)*

10.1.4 Estimation of Other Properties

It is not currently possible to estimate other environmentally important properties of organometallics except in a few limited cases. Most of the available methods are for organotins, and involve correlations between physicochemical and toxicological properties and such molecular properties or parameters as the octanol-water partition coefficient, molecular surface area, chromatographic retention time, acid dissociation constant, or molecular connectivity index [7,19-21,33].

Part of the difficulty in developing property-property or structure-property relationships of the type typically derived for organics [23] is that many organometallics do not yield a single, stable (neutral) molecular species upon dissolution in water; rather, they react fairly easily with water, yielding a variety of degradation products. Degradation products may go on to form complexes with other ions present in solution. The degree of degradation and the type and extent of complexation depend upon a number of environmental variables.

An example of the degradation and speciation of an organometallic in water is given by Laughlin *et al.* [20], who hypothesized the following reactions for Sn(IV) following the dissolution of tributyltin oxide in seawater at pH 8:

$$(Bu_3Sn)_2O + H_2O \rightleftharpoons 2\,Bu_3SnOH$$

$$Bu_3SnOH + H_3O^+ \rightleftharpoons Bu_3SnOH_2^+ + H_2O$$

$$Bu_3SnOH_2^+ + HOCO_2^- \rightleftharpoons Bu_3SnOCO_2^- + H_3O^+$$

$$Bu_3SnOH_2^+ + Cl^- \rightleftharpoons Bu_3SnCl + H_2O$$

$$(Bu = \text{n-butyl} = -CH_2CH_2CH_2CH_3)$$

If it is known, or probable, that an organometallic does not degrade or complex after dissolution in water, it may be possible to treat the compound as an organic chemical and use estimation methods derived for such chemicals [23]. This approach is more likely to be successful if the metal is shielded or surrounded by the organic fragment(s) of the molecule so that it does not interact with water.

Even when some degradation and subsequent complexation occur in aqueous media, it may still be possible — at least for organotins — to derive some correlations for physicochemical and toxicological properties with other chemical properties or molecular descriptors. For example, Table 10.1-12 lists several correlations that use total molecular surface area (TSA) as the independent variable. The parameters in Table 10.1-12 are used in the following equation:

$$\ln \text{ or } \log(\text{Property}) = m\,(\text{TSA}) + \text{constant} \qquad (6)$$

TABLE 10.1-12

Topological Predictors for Organotin Physicochemical and Toxicological Properties

Property and Type of Compound[a]	Measured Units	Parameters for Equation 6[b]			
		m(obsd)	Constant	N	r^2
Aqueous Solubility R_4M (R=alkyl; M=C, Si, Ge, Sn)	–log S molal	0.0224	0.442	9	0.992
Partition Coefficient $R_nR'_{4-n}Sn$, octanol/water	log P $\Sigma\pi$	0.0263	–1.98	12	0.991
Liquid Chromatographic Retention $R_nR'_{4-n}Sn$, HPLC	ln k′	0.0117	–1.94	12	0.995
Gas Chromatographic Retention $R_nR'_{4-n}Sn$ (R=alkyl,vinyl)	RI min	0.0525	3.64	7	0.983
Microbial Uptake R_3SnCl(R=Et,Pr,Bu)	% uptake Δg	0.459	-69.4	3	0.944
Algal Reproduction (Growth) $R_3SnX{\cdot}H_2O$ (X=Cl,Br,CO_3)	ln IC_{50} Sn, mg/l	-0.0365	7.69	5	0.957
Larval Survival (Toxicity) $R_3SnX{\cdot}H_2O$ (X=alkyl,Ph,c-Hx)	ln LD_{50} n*M*	-0.0146	9.18	8	0.938

a. R = alkyl group, M = metal

b. N = number of chemicals in test set used to derive equation.
r = correlation coefficient

Source: Brinckman *et al.* [7]

TSA is a computer-calculated number based on van der Waals' atomic radii and bond angles, and with a subsequent elimination of occluded surfaces [7]. The calculation of TSA is described in a number of publications [7,16,27,32]. The most accurate values will be obtained when the molecule is considered as a whole and the details of the molecular conformation are considered. However, approximate values may be obtained from the fragment constant values given in Tables 10.1-13 and -14.

Example 4 Estimate the water solubility of triphenyltin chloride, Ph_3SnCl. (Dissociation not considered.)

The total surface area (TSA) is first estimated from the fragment constants in Table 10.1-13:

$$\text{TSA} = \underset{3\,\text{Ph}}{3(92.3)} + \underset{\text{Sn}}{17.7} + \underset{\text{Cl}}{27.6}$$

$$= 322.2\ \text{Å}^2$$

The first entry of Table 10.1-12 yields the following form of equation 6:

$$- \log (\text{Solubility}) = 0.0224(\text{TSA}) + 0.442$$
$$= 0.0224\,(322.2) + 0.442$$
$$= 7.66$$

Thus,

$$\text{Solubility} = 2.2 \times 10^{-8} \text{ molar}$$

Another example of a correlation for organotins is given by Vighi and Calamari [33], who measured the 24-hr EC_{50} (effective concentration for 50% of the animals) toxicity to *Daphnia magna*. EC_{50} values for four series of organotins ($RSnX_3$, R_2SnX_2, R_3SnX, R_4Sn) were correlated with their octanol-water partition coefficients (K_{ow}), acid dissociation constants (pK_a) and molecular connectivity indices ($^1\chi$) both individually and in combination.[2]

TABLE 10.1-13

Group and Atom Surface Area (TSA) Values[a]

Organic Group	TSA, Å^2	Standard Deviation
$-CH_2-$	20.7	1.5
Methyl ($-CH_3$)	32.7	0.02
Ethyl ($-CH_2CH_3$)	55.4	0.01
n-Propyl ($-C_3H_7$)	74.4	1.6
i-Propyl ($-C_3H_7$)	75.6	2.2
n-Butyl ($-C_4H_9$)	95.4	1.8
i-Butyl ($-C_4H_9$)	95.6	1.9
cyclo-Hexyl ($-C_6H_{11}$)	117.9	2.2
Phenyl ($-C_6H_5$)	92.3	0.1
Inorganic Group		
Sn	17.7	0.9
Cl	27.6	2.3
CO_3	45.4	3.3
OH	20.2	0.05
OH_2	23.4	2.2

a. Mean TSA derived from both 4- and 5-coordinate R_3Sn structures.

Source: Craig and Brinckman [13] *(Copyright 1986, Longman Group Ltd. Reprinted with permission.)*

2. The molecular connectivity index is a calculated topological index. The calculations are based on graph theory [17,18].

TABLE 10.1-14

Group Surface Areas from Noncyclic Aliphatic Alcohols and Hydrocarbons[a]

Group	Surface Area ($\mathring{A}^2$) Range[b]	Midpoint
Primary OH	19.4-21.0	20.2
Secondary OH	17.9-19.6	18.8
Tertiary OH	15.6-18.9	17.2
Primary CH_3	29.4-33.2	31.3
Secondary CH_3	27.6-31.0	29.3
Tertiary CH_3	26.1-29.3	27.7
CH_2	13.4-22.8	18.1
CH	6.2-10.6	8.4
C	0- 0.7	0.35
Incremental CH_2[c]	18.1	—
Terminal CH_3[c]	33.2	—

a. Based on calculation Method C of Valvani *et al.* [32] in which molecule is considered as a collection of spherical groups and in which the solvent radius is omitted.

b. Group surface area will depend on the degree of substitution surrounding the molecular group. In general, the greater the substitution, the smaller the group surface area and vice versa.

c. In straight chain extended conformation.

Source: Valvani *et al.* [32] *(Copyright 1976, American Chemical Society. Reprinted with permission.)*

The best correlation, excluding only the R_4Sn series of chemicals, used all three parameters:

$$\log \frac{1}{EC_{50}} = 0.207 \log K_{ow} + 0.513 pK_a + 0.206(^1\chi) + 0.824 \tag{7}$$

$$N = 12,\ r = 0.989$$

The acid dissociation constants of a few organometallic acids (with substituted aromatic portions) have been shown to be predictable via the standard Hammett-type equation. The following, for example, are given by Perrin [28] for the structures shown:

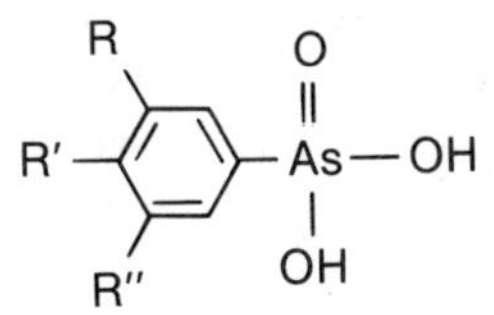

$$pK_a = 3.54 - 1.05 \sum_i \sigma_i \qquad (8)$$

σ_i = substituent constant for each R

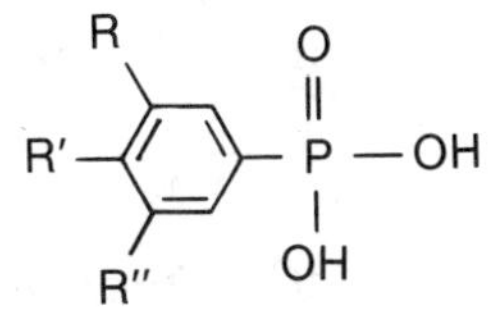

$$pK_a = 1.84 - 0.76 \sum_i \sigma_i \qquad (9)$$

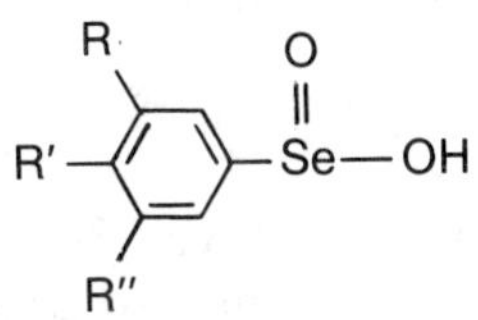

$$pK_a = 4.74 - 0.91 \sum_i \sigma_i \qquad (10)$$

Values of σ for several hundred substituents are given by Perrin [28].

The above comments pertain primarily to solution properties of the chemical. It is quite probable that bulk properties of organometallics (e.g., density, viscosity, heat capacity, and vapor pressure as well as boiling point) could also be estimated via the property-property and structure-property methods used for organic chemicals. To our knowledge, however, this hypothesis has not been tested. Structure-property correlations must await sufficient experimental data from which appropriate fragment constants can be derived.

10.1.5 Literature Cited

1. Allred, A.L., "Electronegativity Values from Thermochemical Data," *J. Inorg. Nucl. Chem.*, **17**, 215-21 (1961).
2. Allred, A.L. and E.G. Rochow, "A Scale of Electronegativity Based on Electrostatic Force," *J. Inorg. Nucl. Chem.*, **5**, 264-68 (1958).
3. American Conference of Governmental Industrial Hygienists, "Threshold Limit Values for Chemical Substances in Work Air," adopted by ACGIH for 1982, Cincinnati, Ohio (1982).
4. Anderson, H.H., "Boiling Points and Boiling Point Numbers of Organogermanium Compounds," *J. Chem. Eng. Data*, **7**, 565-67 (1962).
5. Anderson, H.H., "Boiling Points of Organogermanium Carboxylates and Mercaptides," *J. Chem. Eng. Data*, **12**, 371 (1967).

6. Anderson, H.H. and T.C. Hager, "Organosilicon Fluoroesters," *J. Am. Chem. Soc.*, **81**, 1584 (1959).
7. Brinckman, F.E., G.J. Olson, W.R. Blair and E.J. Parks (National Bureau of Standards, Gaithersburg, Md.), "Implications of Molecular Speciation and Topology of Environmental Metals: Uptake and Toxicity of Organotins," prepared for presentation at *Aquatic Toxicology and Hazard Assessment*, 10th Symposium (May 4-6, 1986), New Orleans, La. (To be published by the American Society for Testing and Materials in 1987.)
8. Brinckman, F.E., G.J. Olson and W.P. Iverson, "The Production and Fate of Volatile Molecular Species in the Environment: Metals and Metalloids," in *Atmospheric Chemistry*, E.D. Goldberg (ed.), Springer-Verlag, New York (1982).
9. "Chemical Products Synopsis," Manville Corp., Cortland, N.Y. (1983).
10. Coates, G.E., M.L.H. Green and K. Wade, *Organometallic Compounds*, 3rd ed., Methuen & Co. Ltd., London (1967).
11. Code of Federal Regulations, **40**, Part 261.32 (July 1, 1984).
12. Cotton, F.A. and G. Wilkinson, *Advanced Inorganic Chemistry*, 3rd ed., Interscience Publishers, New York (1972).
13. Craig, P.J. and F.E. Brinckman, "Currents and Pathways of Organometallic Compounds in the Environment, General Considerations," in *Organometallic Compounds in the Environment, Principles and Reactions*, P. Craig (ed.), Longman Group Ltd., Essex, United Kingdom (1986).
14. Eisch, J.J., *The Chemistry of Organometallic Compounds*, The Macmillan Company, New York (1967).
15. Hendrickson, J.B., D.J. Cram and G.S. Hammond, *Organic Chemistry*, 3rd ed., McGraw-Hill Book Co., New York, 303 (1970).
16. Herman, R.B., "Theory of Hydrophobic Bonding.II: The Correlation of Hydrocarbon Solubility in Water with Solvent Cavity Surface Area," *J. Phys. Chem.*, **76**, 2754-59 (1972).
17. Kier, L.B. and L.H. Hall, *Molecular Connectivity in Chemistry and Drug Research*, Academic Press, New York (1976).
18. Kier, L.B. and L.H. Hall, *Molecular Connectivity in Structure-Activity Analysis*, Research Studies Press, Ltd., Letchworth, Hertfordshire, United Kingdom (1986) (available in the United States from John Wiley & Sons, New York).
19. Laughlin, R.B., Jr., W. French, R.B. Johannesen, H.E. Guard and F.E. Brinckman, "Predicting Toxicity Using Computed Molecular Topologies: The Example of Triorganotin Compounds," *Chemosphere*, **13**, 575-84 (1984).
20. Laughlin, R.B., Jr., H.E. Guard and W.M. Coleman, "Tributyltin in Seawater: Speciation and Octanol-Water Partition Coefficient," *Environ. Sci. Technol.*, **20**, 201 (1986).
21. Laughlin, R.B., Jr., R.B. Johannesen, W. French, H. E. Guard and F.E. Brinckman, "Structure-Activity Relationships for Organotin Compounds," *Environ. Toxicol. Chem.*, **4**, 343-51 (1985).

22. Lewis, R.N. and A.E. Newkirk, "A Method of Calculating the Boiling Points of Silicon Compounds," *J. Am. Chem. Soc.*, **69**, 701 (1947).
23. Lyman, W.J., W.F. Reehl and D.H. Rosenblatt, *Handbook of Chemical Property Estimation Methods. Environmental Behavior of Organic Compounds*, McGraw-Hill Book Company, New York (1982).
24. MacKenzie, C.A., A.P. Mills and J.M. Scott, "Physical Properties of Trichlorosilane and Some of Its Derivatives," *J. Am. Chem. Soc.*, **72**, 2032 (1950).
25. Nowak, M., "Calculations of the Normal Boiling Points of Organogermanium Compounds," *Bull. Acad. Pol. Sci. Ser. Sci. Chim.*, **17**, 425 (1969).
26. Nowak, M., "Calculation of the Normal Boiling Points of Organogermanium Alcohols, Esters of Organic Acids and Compounds Containing Two Ge Atoms in a Molecule," *Bull. Acad. Pol. Sci. Ser. Sci. Chim.*, **17**, 439 (1969).
27. Pearlman, R.S., "SAREA: Van der Waals (and Accessible) Surface Area of Molecules," *Quantum Chem. Prog. Exch. Bull.*, **1**, 15-16 (1981).
28. Perrin, D.D., B. Dempsey and E.P. Serjeant, *pK_a Prediction for Organic Acids and Bases*, Chapman and Hall, New York (1981).
29. Rochow, E.G., D.T. Hurd and R. Lewis, *The Chemistry of Organometallic Compounds*, John Wiley & Sons, New York (1957).
30. Thompson, W.T., *Agricultural Chemicals*, Books 1 to 4, Thompson's Publications, Fresno, Calif. (1983-1984).
31. "Toxic Substances Control Act," 90 Stat. 2003, 15 U.S.C. Sec. 2601 (1976).
32. Valvani, S.C., S.H. Yalkowsky and G.L. Amidon, "Solubility of Nonelectrolytes in Polar Solvents. VI: Refinements in Molecular Surface Area Computations," *J. Phys. Chem.*, **80**, 829-35 (1976).
33. Vighi, M. and D. Calamari, "QSARs for Organotin Compounds on *Daphnia magna*," *Chemosphere*, **14**, 1925-32 (1985).

10.2 ACIDS

10.2.1 Introduction

Acids may be defined as compounds that tend to release protons (H^+ ions), either by reaction with a base or when dissolved in a polar solvent such as water. Such acids are known as protic acids. (According to a more general definition, acids are substances that tend to accept an electron pair from another species; this discussion, however, will be limited to the protic acids.) Acids have a wide variety of commercial uses, some of which are listed in Table 10.2-1.

Many of the inorganic protic acids are commonly sold and used in aqueous solution rather than in pure form. (An important exception is chlorosulfuric acid, HSO_3Cl, which is used as a pure liquid.) These solutions may be formed by dissolution of either the pure acids or other compounds that form acidic solutions in water. For example, As_2O_5, P_4O_{10}, SeO_2, and CrO_3 produce solutions of arsenic acid, phosphoric acid, selenious acid, and chromic acid, respectively, when dissolved in water. In some cases, an acid exists in solution but cannot be easily isolated as a pure substance. For example, nitric acid (HNO_3) is commercially produced as a 50-70% (by weight) solution; the pure liquid tends to decompose, releasing gaseous nitrogen oxides. The solid compound known commercially as chromic acid is actually chromium trioxide, CrO_3; the species H_2CrO_4 exists only in aqueous solution.

The pure acids that can be isolated may be gases (e.g., HF, HCl, HBr, H_2S), liquids (e.g., H_2SO_4) or solids (e.g., H_3AsO_4, H_3PO_4, H_2SeO_3). They vary in solubility from moderately soluble to infinitely miscible with water. The solubilities of some acids are given in Table 10.2-2.

The environmental properties of the gases HBr, HCl, HF, and H_2S are discussed in a later section (10.6). This section deals primarily with the environmental chemistry of aqueous solutions of acids. The major reactions of environmental importance for acids are:

- Volatilization from solution,
- Acid dissociation (production of H^+ ions) and its consequences in aqueous solutions, and
- Reactions of counter ions (anions), such as oxidation/reduction, metal complexation, and precipitation.

10.2.2 Volatilization from Solution

Volatilization of some acids from their aqueous solutions can be considerable. The vapor pressure of the acid over solution generally increases at higher aqueous concentrations and temperatures. However, solutions of acids often deviate significantly from Henry's Law[1] — i.e., as the aqueous concentration increases, the vapor pressure

1. See section 10.6 for a discussion of Henry's Law.

of the acid becomes considerably higher than would be predicted. (This is explained by the fact that as the solution becomes more concentrated, its pH is lowered, thus decreasing its tendency to dissolve more acid.) Vapor pressures of some acids over their aqueous solutions at 20°C are shown in Figure 10.2-1.

TABLE 10.2-1

Commercial Uses of Some Acids

Acid	Major Uses
H_3AsO_4 (arsenic)	Desiccant for defoliation of cotton bolls prior to harvesting; fining agent to remove bubbles in glass manufacture.
HCl (hydrochloric)	Steel pickling; other metal-cleaning processes; ore processing; production of chlorine dioxide from sodium chlorate.
$HClO_4$ (perchloric)	Reagent in research and in analytical chemistry for determination of metallic elements; production of high purity metal perchlorates.
HF (hydrofluoric)	Glass etching and polishing; manufacture of aluminum fluoride and cryolite for aluminum production; manufacture of fluorine-containing organic compounds; uranium processing; petroleum alkylation; stainless steel pickling.
HNO_3 (nitric)	Production of ammonium nitrate for use as fertilizer; manufacture of cyclohexanone and other organic compounds; stainless steel pickling; metal etching; rocket propellant; nuclear-fuel processing.
H_3PO_4 (phosphoric)	Manufacture of phosphate salts and fertilizers; metal treatment; food additives; refractories; catalysis.
H_2SO_4 (sulfuric)	Production of phosphate fertilizers (by converting phosphate rock to phosphoric acid and solid calcium sulfates); production of inorganic pigments, textile fibers, explosives, petroleum products, alcohols, pulp and paper, detergents, and other chemicals; leaching agent for ores; pickling agent for iron and steel; component of lead storage batteries.
HSO_3Cl (chlorosulfuric)	Reagent for organic synthesis of detergents, pharmaceuticals, dyes and pigments, ion-exchange resins, and other products; catalyst for organic reactions; smoke-forming agent in warfare.

Source: *Kirk-Othmer* [4].

TABLE 10.2-2

Solubilities and Acid Dissociation Constants for Some Acids

Acid	Acid Dissociation Constant[a,b] pK_{a1}	pK_{a2}	pK_{a3}	Solubility (g/100g H_2O)[c]
H_3AsO_4	2.24	6.96	11.50	302 (12.5°C)[d] ($H_3AsO_4 \cdot ½H_2O$)
HBr	−9	—	—	193 (25°C)
HCl	−7	—	—	71.9 (20°C)
$HClO_4$	−7.3	—	—	very sol.
H_2CrO_4	−0.20	6.51	—	—
HF	3.17	—	—	very sol.
HNO_3	−1.34	—	—	very sol.
H_3PO_4	2.148	7.199	12.35	very sol.
H_2SO_4	−3.0	1.99	—	very sol.
H_2SeO_3	2.75	8.5	—	166.7 (20°C)

a. As cited in Kotrlý and Šůcha [5]. Values are given for conditions of zero ionic strength (I = 0) and 25°C.

b. Additional pK data are given in section 2.7.

c. **Source**: Dean [2], except as noted.

d. **Source**: Weast [6].

10.2.3 Acid Dissociation

Protic acids dissociate in aqueous solution according to the general reaction

$$HA + H_2O \rightleftharpoons H_3O^+ + A^- \tag{1}$$

or, more simply,

$$HA \rightleftharpoons H^+ + A^- \tag{2}$$

The extent to which this reaction occurs is described by the acid dissociation constant, $K_a = [H^+]\,[A^-]\,/[HA]$. Strong acids are extensively dissociated in solution and therefore have high K_a values, while weak acids are only slightly dissociated and have low K_a values.

Acids with two or three ionizable protons may undergo two or three successive dissociation reactions. For example [5]:

$$H_3AsO_4 \rightleftharpoons H_2AsO_4^- + H^+ \qquad pK_{a1} = 2.24 \tag{3}$$

$$H_2AsO_4^- \rightleftharpoons HAsO_4^{-2} + H^+ \qquad pK_{a2} = 6.96 \tag{4}$$

$$HAsO_4^{-2} \rightleftharpoons AsO_4^{-3} + H^+ \qquad pK_{a3} = 11.50 \tag{5}$$

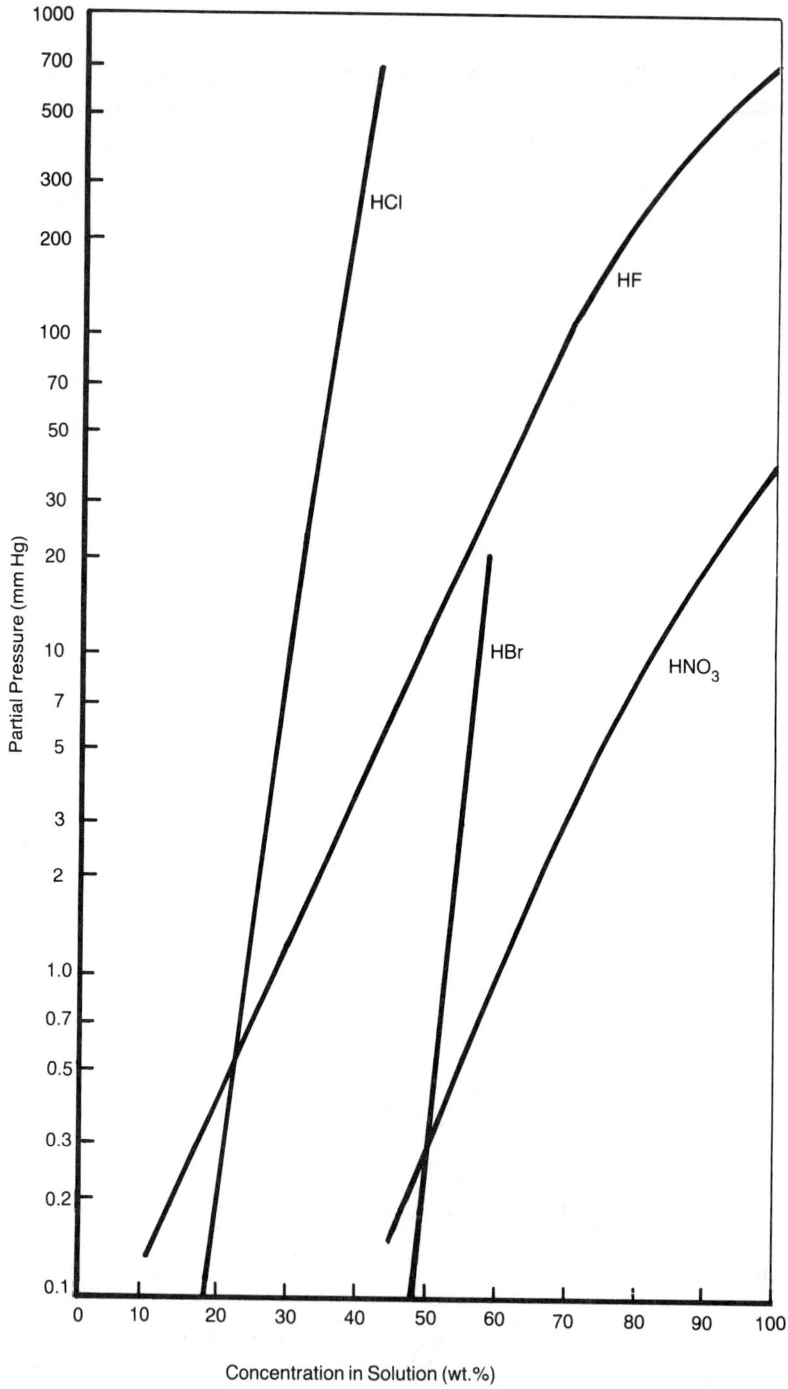

Source of Data: HF — Kirk-Othmer [4], Vol. 10, p. 736
Other acids — Green [3], pp. 3-64, 3-70

FIGURE 10.2-1 Partial Pressures of Some Acids over Their Aqueous Solutions at 20°C

In a solution containing such species, the relative abundance of the protonated and unprotonated forms varies with the pH of the solution.

Acid dissociation constants for several acids are given in Table 10.2-2. For a more extensive discussion of acid dissociation constants and acid/base equilibria, see section 2.7.

Aqueous acid solutions are essentially miscible with water. If a concentrated acid solution is released into a body of water, the primary effect will be to lower the pH. This change in pH will affect the acid/base equilibria of various ions in solution, producing a higher concentration of protonated species. It may also result in the volatilization of gaseous species such as H_2S or HCN, or in the dissolution of basic species such as metal oxides or hydroxides from sediments or suspended particulates.

10.2.4 Reactions of Anions

The anions produced when acids dissociate in water may undergo a variety of reactions. The conjugate bases of acids with more than one ionizable proton participate in acid/base equilibria, as discussed above. Anions may also react with cations in solution to form either solid precipitates or soluble complexes, as illustrated below.

$$S^{-2} + Fe^{+2} \rightleftharpoons FeS(s) \tag{6}$$

$$SO_4^{-2} + Fe^{+3} \rightleftharpoons FeSO_4^{+} \tag{7}$$

Certain anions undergo important oxidation or reduction reactions. For example, As(V) may be reduced to As(III), and sulfate (SO_4^{-2}) to sulfide (S^{-2}); alternatively, Se(IV) may be oxidized to Se(VI), or sulfide may be oxidized to sulfate. The extent of complexation, precipitation, or oxidation/reduction reactions depends on the identity of the anion. These reactions are discussed in the sections of this report dealing with the individual metals and ligands, and in the sections on Complexation (2.9) and Oxidation/Reduction (2.10).

10.2.5 Chlorosulfuric Acid

Chlorosulfuric acid, known commercially as chlorosulfonic acid, warrants a separate discussion because it is the only one of the acids discussed here that is not used in aqueous solution; in fact, it reacts violently with water. This strong acid is a clear, colorless liquid (freezing point −81 to −80°C; boiling point 151-152°C, with decomposition) that is actually a mixture of chlorosulfuric acid in equilibrium with small amounts of hydrogen chloride, sulfur trioxide, and related compounds.

The vapor pressure, P (in Pa), of the liquid acid is given by

$$\log P = 11.496 - 2752/T \tag{8}$$

where T is the temperature in °K. At 25°C, the vapor pressure calculated from this equation is 182 Pa, or 1.37 mm Hg.

When heated, chlorosulfuric acid decomposes, producing sulfuryl chloride (SO_2Cl_2), sulfuric acid, pyrosulfuryl dichloride ($S_2O_5Cl_2$) and pyrosulfuric acid ($H_2S_2O_7$) in equilibrium with the original acid. If heated beyond the boiling point, decomposition results in the formation of sulfur dioxide, chlorine, and water.

The reaction of chlorosulfuric acid with water releases heat and produces dense fumes of hydrochloric and sulfuric acids. Chlorosulfuric acid reacts with most organic compounds, sometimes with charring. Its major uses are in organic synthesis as a sulfating, sulfonating, or chlorosulfonating agent. It is also a fairly strong dehydrating agent [1].

10.2.6 Literature Cited

1. Burrus, H.O., "Chlorosulfuric Acid," in *Kirk-Othmer Encyclopedia of Chemical Technology*, 3rd ed., Vol. 5, John Wiley & Sons, New York (1982).
2. Dean, J.A. (ed.), *Lange's Handbook of Chemistry*, 12th ed., McGraw-Hill Book Co., New York (1979).
3. Green, D.W. (ed.), *Perry's Chemical Engineers' Handbook*, 6th ed., McGraw-Hill Book Co., New York (1984).
4. *Kirk-Othmer Encylopedia of Chemical Technology*, 3rd ed., John Wiley & Sons, New York (1982).
5. Kotrlý, S. and L. Šůcha, *Handbook of Chemical Equilibria in Analytical Chemistry*, Ellis Horwood Ltd., Chichester, England (1985).
6. Weast, R.C. (ed.), *CRC Handbook of Chemistry and Physics*, 54th ed., CRC Press, Cleveland (1973).

10.3 BASES (HYDROXIDES)

10.3.1 Introduction

Bases may generally be defined as compounds that tend to accept a proton. This discussion, however, is limited to a particular class of bases — the hydroxides, compounds containing the hydroxide ion (OH^-) — and focuses on the environmental chemistry of the alkali and alkaline earth metal hydroxides. Several of these compounds have important commercial applications, as shown in Table 10.3-1.

TABLE 10.3-1

Commercial Uses of Hydroxides

Compound	Common Name	Major Uses
NaOH	Caustic soda	Production of organic and inorganic chemicals; refining of kraft pulp; petroleum refining; manufacture of soaps and detergents; food processing; metal processing (e.g., extraction of alumina from bauxite); textile processing; water and acid waste stream treatment.
KOH	Caustic potash	Production of chemicals, fertilizers, and other agricultural products; soaps and detergents; scrubbing and cleaning operations (e.g., industrial gases); dyes and colorants; rubber chemicals.
CsOH	Cesium hydroxide	Removal of sulfur from heavy oils; suggested for use in place of NaOH or KOH in alkaline storage batteries for low-temperature operation (to $-50°C$).
$Ca(OH)_2$ (may contain $Mg(OH)_2$)	Hydrated lime	Component of masonry mortars; soil stabilization; steel fluxing, other metallurgical processes; potable water and wastewater treatment; desulfurization of stack gases; solid waste treatment; chemical manufacture; sugar refining.

Source: *Kirk-Othmer* [4].

The hydroxides of the alkali metals, such as NaOH, KOH, and CsOH, are white, crystalline solids, but they are generally used in aqueous solution rather than as pure solids. A commercially important example of an alkaline earth metal hydroxide is lime. Although the name is sometimes used to refer to limestone ($CaCO_3$ with $MgCO_3$ and various impurities), lime is actually a mixture of calcium and magnesium oxides and hydroxides that is produced when limestone is subjected to high temperature. The primary product of this decomposition of limestone is calcium oxide (CaO), or quick-lime, which reacts with water to form hydrated lime, calcium hydroxide ($Ca(OH)_2$). Hydrated lime is a white, fluffy powder that is slightly soluble in water [1].

Environmentally important reactions of the hydroxides include:

- Dissolution in water and pH effects,
- Reaction with acids, and
- Complexation and precipitation of the ions formed upon dissolution in water.

Each of these reactions is further discussed below.

10.3.2 Dissolution in Water

When hydroxides are dissolved in water, they dissociate to produce free hydroxide ions (thus raising the pH of the solution) and the counter metal cations:

$$NaOH \rightleftharpoons Na^+ + OH^- \quad (1)$$

The hydroxide ion may then react with free H^+ or any acidic species that may be present, forming water:

$$OH^- + H^+ \rightleftharpoons H_2O \qquad K = 10^{14}\ (25°C) \quad (2)$$

The hydroxides of the alkali metals are highly soluble in water, while those of the alkaline earth metals are only slightly soluble. The solubility of the alkaline earth metal hydroxides increases with increasing atomic number of the metal, i.e., in the order $Mg(OH)_2 < Ca(OH)_2 < Sr(OH)_2 < Ba(OH)_2$ [2].

The solubilities of some hydroxides are given in Table 10.3-2. These solubilities are affected by pH, temperature, and the presence of other species in solution. Increased pH causes decreased solubility, because a higher OH^- concentration reduces the amount of solid hydroxide that can dissociate into free metal ions and OH^- ions. The effect of temperature varies: with increasing temperature, the alkali metal hydroxides become more soluble, but calcium hydroxide becomes less soluble. The solubility of $Ca(OH)_2$ and $Mg(OH)_2$ is enhanced by the presence of certain inorganic salts and organic compounds in solution [1].

The dissolution of alkali hydroxides in water is a strongly exothermic process; thus, their solutions generate heat when diluted. Dilution of sodium hydroxide solutions of 40% or greater concentration can generate enough heat to raise the temperature above the boiling point, causing dangerous eruptions of the solution [6].

TABLE 10.3-2

Solubilities of Hydroxides in Water

Compound	Solubility ($g/100g\ H_2O$)[a]	$\log K_{sp}$ (25°C, I = 0)[b]
NaOH	109 (20°C)	—
KOH	112 (20°C)	—
CsOH	386 (15°C)	—
$Ca(OH)_2$	0.173 (20°C)	−5.19
$Mg(OH)_2$	0.0009 (18°C)[c]	−11.15

a. **Source:** Dean [3], except as noted

b. **Source:** Kotrlý and Šůcha [5]

c. **Source:** Weast [8]

Highly soluble hydroxides like NaOH and KOH can produce strongly basic solutions, but solutions of slightly soluble hydroxides have an upper limit on pH, determined by the amount of solid hydroxide that will dissolve. For example, at 25°C, $Ca(OH)_2$ solutions have a maximum pH of about 12.3, and $Mg(OH)_2$ solutions have a maximum pH of about 10.3 (as calculated from K_{sp} values).

10.3.3 Reaction with Acids (Neutralization)

Bases are characterized by their reaction with acids to form neutral salts. The alkali and alkaline earth hydroxides can react both with strong acids such as HCl and H_2SO_4 and with gases that produce weak acids in solutions, such as hydrogen sulfide, sulfur dioxide, and carbon dioxide. For example:

$$Ca(OH)_2 + 2HCl \rightarrow CaCl_2 + 2H_2O \tag{3}$$

$$2NaOH + H_2S \rightarrow Na_2S + H_2O \tag{4}$$

Equilibrium constants for the reactions of $Ca(OH)_2$ and $Mg(OH)_2$ with H^+ ions are given in Table 6.9-4. Hydroxides, both as solids and in aqueous solution, absorb CO_2 readily from the air, reacting to form carbonates [10]; thus, for example, CsOH forms Cs_2CO_3. The reaction of NaOH or KOH with hydrogen sulfide is used commercially for the extraction of H_2S from natural gas [6, 9].

10.3.4 Complexation and Precipitation

An important consequence of the addition of soluble hydroxides to natural waters is the formation of metal complexes and the precipitation of solid metal hydroxides and other species. Since most of the transition metals form sparingly soluble hydroxides, addition of a highly soluble hydroxide such as NaOH to water containing transition metal ions may result in the precipitation of metal hydroxides. This process can be used for metal removal in wastewater treatment [6]. Additional information on the formation of metal hydroxides can be found in section 2.7 (Acid/Base Equilibria).

The metal ion concentration at which a solid hydroxide will precipitate is strongly dependent on the pH of the solution. The concentration of free metal ions in equilibrium with the solid hydroxide decreases with increasing pH (cf. Figure 2.11-9). However, the actual solubility of the metal hydroxide may increase at high pH values, due to the formation of soluble hydroxo complexes [7]. For example, iron (III) may form $Fe_2(OH)_2^{+4}$, $FeOH^{+2}$, $Fe(OH)_2^+$, and $Fe(OH)_4^-$; the relative abundance of these species in equilibrium with solid $Fe(OH)_3$ is a function both of total Fe(III) concentration and of pH (i.e., OH^- concentration).

The formation of polynuclear hydroxo complexes is discussed further in section 2.8, Polymerization. Equilibrium constants for some reactions involving metal hydroxides and hydroxo complexes are given in Table 6.9-4, and Figure 2.11-10 shows the solubility of some metal hydroxides as a function of pH.

The addition of calcium hydroxide to natural waters may also result in the formation and precipitation of solid calcium salts such as $CaCO_3$.

10.3.5 Literature Cited

1. Boynton, R.S., "Lime and Limestone," in *Kirk-Othmer* [4], Vol. 14.
2. Cotton, F.A. and G. Wilkinson, *Advanced Inorganic Chemistry: A Comprehensive Text*, 3rd ed., Interscience Publishers, div. of John Wiley & Sons, New York (1972).
3. Dean, J.A. (ed.), *Lange's Handbook of Chemistry*, 12th ed., McGraw-Hill Book Co., New York (1979).
4. *Kirk-Othmer Encyclopedia of Chemical Technology*, 3rd ed., John Wiley & Sons, New York (1982).
5. Kotrlý, S. and L. Šůcha, *Handbook of Chemical Equilibria in Analytical Chemistry*, Ellis Horwood Ltd., Chichester, England (1985).
6. Leddy, J.J., I.C. Jones, Jr., B.S. Lowry, F.W. Spillers, R.E. Wing and C.D. Binger, "Alkali and Chlorine Products," in *Kirk-Othmer* [4], Vol. 1.
7. Snoeyink, V.L. and D. Jenkins, *Water Chemistry*, John Wiley & Sons, New York (1980).
8. Weast, R.C. (ed.), *CRC Handbook of Chemistry and Physics*, 54th ed., CRC Press, Cleveland (1973).
9. Whaley, T.P., "Sodium, Potassium, Rubidium, Cesium and Francium," in Trotman-Dickenson, A.F. (exec. ed.), *Comprehensive Inorganic Chemistry*, Pergamon Press, Oxford, England (1973).
10. Williams, C.J., "Cesium and Cesium Compounds," in *Kirk-Othmer* [4], Vol. 5.

10.4 SALTS

10.4.1 Introduction

Salts will be defined for the purposes of this discussion as compounds that, when dissolved in water, initially yield the constituent ions of the solid as hydrated cations and anions. For example, calcium carbonate, $CaCO_3$, dissolves to form aqueous Ca^{+2} and CO_3^{-2} ions, and ammonium chloride, NH_4Cl, dissolves to form aqueous NH_4^+ and Cl^- ions. These dissolved ions can subsequently react to form different ionic or non-ionic species under some conditions (e.g., $NH_4^+ \rightleftharpoons NH_3 + H^+$).

The chemistry of salts that is relevant to their behavior in the environment concerns their water solubility and the reactions of the ions produced in aqueous solution. Since dissolution is the first step in any aqueous reaction of a salt, solubility is discussed first, followed by a summary of the reactions of ions in aqueous solution.

10.4.2 Solubility

Solubility in water is a key factor in the environmental behavior of salts. Salts range from highly soluble to nearly insoluble in water. For slightly soluble salts, the extent of dissociation in water is described by the solubility product, K_{sp}. For a salt with the general formula A_xB_y that dissociates in water according to the equation

$$A_xB_y(s) \rightleftharpoons xA^{+y} + yB^{-x}$$

the solubility product is defined as:

$$K_{sp} = \frac{\{A^{+y}\}^x\{B^{-x}\}^y}{\{A_xB_y(s)\}}$$

where $\{A^{+y}\}$, $\{B^{-x}\}$, and $\{A_xB_y(s)\}$ are the activities of the species in solution. Since the activity of the solid phase can be taken as unity, $K_{sp} = \{A^{+y}\}^x\{B^{-x}\}^y$. In solutions with very low ionic strength, the activities of the aqueous species can be approximated by their molar concentrations, so $K_{sp} = [A^{+y}]^x[B^{-x}]^y$. (See section 2.11 for further discussion of solubility products.) Some examples of salts with various degrees of solubility are given in Table 10.4-1.

Many salts can exist either in anhydrous form or as hydrates, i.e., with one or more molecules of water included in the structural formula. The solubilities of hydrates may differ from those of the anhydrous salts. For example, anhydrous sodium sulfate (Na_2SO_4) has a solubility of 19.5 g/100 g H_2O or 1.37 *M* at 20°C, while its heptahydrate ($Na_2SO_4 \cdot 7H_2O$) has a solubility of 44.1 g/100 g H_2O or 1.64 *M* at the same temperature [1].

A property related to solubility is *deliquescence*, the tendency of a crystalline solid to absorb atmospheric moisture until it dissolves into a saturated solution. Certain salts, such as hydrated calcium chloride, $CaCl_2 \cdot 2H_2O$, and zinc chloride, $ZnCl_2$, are

TABLE 10.4-1

Aqueous Solubilities of Some Salts

Very Low Solubility	K_{sp}	Moderate to High Solubility	g/100g H_2O at 20°C
$BaCO_3$	5.1×10^{-9}	Tl_2SO_4	4.87
$CaCO_3$ (calcite)	4.5×10^{-9}	$MgSO_4$	33.7
$CaCO_3$ (aragonite)	6.0×10^{-9}	NaCl	35.9
$CaSO_4$	9.1×10^{-6}	$AlCl_3$	45.8
$SrCO_3$	1.1×10^{-10}	K_2CO_3	111
TlCl	1.7×10^{-4}	NH_4NO_3	192
$Pb_3(AsO_4)_2$	4.0×10^{-36}	CH_3COOK	256
$PbCrO_4$	2.8×10^{-13}	$KMnO_4$	6.34
Tl_2SeO_3	2×10^{-39}	Na_2MoO_4	65.3
		$Na_2Cr_2O_7$	183

Source: Dean [1].

markedly deliquescent, exhibiting this behavior at 30% and 10% relative humidity, respectively. However, all soluble salts will deliquesce in sufficiently humid air. Deliquescence will occur spontaneously when the partial pressure of water in the air exceeds the vapor pressure of water in a saturated solution of the salt. In a closed environment, the process will stop when the water vapor in the atmosphere is so much depleted that its partial pressure equals that of the saturated solution. Factors affecting the rate of the process include crystal size and the rate of diffusion of water vapor into the crystal lattice. In general, more soluble salts have a greater tendency to deliquesce than less soluble salts, because their saturated solutions have lower vapor pressures [2].

10.4.3 Reactions of Ions Produced in Aqueous Solution

Once a salt is dissolved in water, the cations and anions produced may undergo any of several possible reactions. These include:

- Complexation,
- Oxidation or reduction,
- Acid/base equilibria in solution,
- Formation, via acidification, of gaseous species that volatilize from solution, and
- Formation, in combination with other ions, of insoluble salts that precipitate out of solution.

These general types of reactions are described in Part I of this report. The degree to which particular anions or cations participate in these reactions varies considerably. Examples of compounds for which each type of reaction is of greater or lesser importance are given below.

- **Complexation of Cation**

Important for:
Transition metal salts
(e.g., $CuSO_4$, $FeCl_3$, $FeSO_4$, $MnSO_4$, $Pb(CH_3COO)_2$, $PbCrO_4$, $Pb_3(AsO_4)_2$, $ZnCrO_4$, $ZnSO_4$)
Al^{+3} salts
(e.g., $AlCl_3$, AlF_3, $Al_2(SO_4)_3$)
Tl salts
(e.g., TlCl, Tl_2CO_3, $TlNO_3$, Tl_2SO_4, Tl_2SeO_3)

Less important for:
Alkali metal salts
(e.g., NaCl, Na_2CO_3, Na_2SO_4, Na_2CrO_4, $Na_2Cr_2O_7$, Na_2MoO_4, CH_3CH_2COONa, KCl, K_2CO_3, K_2SO_4, $KMnO_4$, CH_3COOK)
Ammonium salts
(e.g., NH_4Cl, NH_4NO_3, $(NH_4)_2MoO_4$, $(NH_4)_2SO_4$)
Alkaline earth metal salts
(e.g., $BaCO_3$, $CaCO_3$, $CaSO_4$, $CaCrO_4$, $MgSO_4$, $SrCO_3$)

- **Chemical Oxidation or Reduction Reactions of Cation**

Important for:
Many transition metal salts
(e.g., salts of Cu, Cr, Fe, Mn)
Tl salts

Not important for:
Alkali metal salts
Ammonium salts
Alkaline earth metal salts
Al^{+3} salts
Zn^{+2} salts

- **Chemical Oxidation or Reduction Reactions of Anion**

Important for:
CrO_4^{-2} salts
$Cr_2O_7^{-2}$ salts
MnO_4^- salts
AsO_4^{-2} salts
SeO_3^{-2} salts

Not important for:
Cl^- salts
F^- salts
CO_3^{-2} salts
NO_3^- salts
SO_4^{-2} salts

- **Acid/Base Equilibria Involving Cation**

Important for:
Salts of metals in oxidation states greater than 1, except for alkaline earth metals (e.g., salts of Al^{+3}, Cu^{+2}, Fe^{+2}, Fe^{+3}, Mn^{+2}, Pb^{+2}, Tl^{+3}, Zn^{+2})

Less important for:
Alkali metal salts
Alkaline earth metal salts
Salts of other metals in oxidation states of 1 (e.g., salts of Cu^{+}, Tl^{+})

- **Acid/Base Equilibria Involving Anion**

Important for:
CO_3^{-2} salts
CrO_4^{-2} salts
$Cr_2O_7^{-2}$ salts
AsO_4^{-3} salts
SeO_3^{-2} salts
Acetate (CH_3COO^-) salts
Propionate ($CH_3CH_2COO^-$) salts
MoO_4^{-2} salts

Moderately important for:
F^- salts
SO_4^{-2} salts

Less important for:
Cl^- salts
NO_3^- salts

- **Formation of Gases via Acidification**
(Gas formed is given in parentheses)

Important for:
CO_3^{-2} salts (CO_2)
F^- salts (HF)
S^{-2} salts (H_2S)

Not important for:
Cl^- salts
NO_3^- salts
SO_4^{-2} salts
SeO_3^{-2} salts
AsO_4^{-3} salts
CrO_4^{-2} salts
$Cr_2O_7^{-2}$ salts
MoO_4^{-2} salts
MnO_4^- salts

- **Formation (and Precipitation) of Insoluble Salts by Cation**

Important:
When anions that tend to form insoluble salts (e.g., CO_3^{-2}, OH^-, S^{-2}, SO_4^{-2}) are present in solution.
For salts of metals in higher oxidation states ($\geqslant +3$), and for certain slightly soluble salts (e.g., $BaCO_3$, $CaCO_3$, $CaSO_4$, $SrCO_3$).

Not important:
When anions present in solution do not form insoluble salts (e.g., NO_3^-, usually Cl^-).
For salts of alkali metals or ammonium.

- **Formation (and Precipitation) of Insoluble Salts by Anion**

Important:
When cations that tend to form insoluble salts (e.g., metals in oxidation states $\geq +3$) are present in solution.
For salts of CO_3^{-2}, SO_4^{-2}, CrO_4^{-2}, MoO_4^{-2}, AsO_4^{-2}

Not important:
When cations present in solution do not form insoluble salts (e.g., Na^+, K^+, NH_4^+)
For salts of NO_3^-, and most salts of Cl^-

Participation of an ion in certain of these reactions may affect the solubility of compounds containing that ion. For example, if an ion forms soluble complexes, the solubility of a salt containing that ion will be enhanced by the presence of appropriate counter ions. Volatilization of a gas formed by an ion or precipitation of an insoluble compound formed by an ion will also effectively enhance the solubility of a salt containing that ion. If an ion participates in acid/base equilibria, the solubility of a salt containing that ion will vary with the pH of the solution.

10.4.4 Literature Cited

1. Dean, J.A. (ed.), *Lange's Handbook of Chemistry*, 12th ed., McGraw-Hill Book Co., New York (1979).
2. Parker, S.P. (ed.), *McGraw-Hill Encyclopedia of Chemistry*, McGraw-Hill Book Co., New York (1983).

10.4.5 Additional Sources of Information

Cotton, F.A. and G. Wilkinson, *Advanced Inorganic Chemistry: A Comprehensive Text*, 3rd ed., Interscience Publishers, div. of John Wiley & Sons, New York (1972).

Martell, A.E. and R.M. Smith, *Inorganic Complexes* and *First Supplement*, Vols. 4 and 5 of *Critical Stability Constants*, Plenum Press, New York (1976 and 1982).

Trotman-Dickenson, A.F. (exec. ed.), *Comprehensive Inorganic Chemistry*, Pergamon Press, Oxford, England (1973).

10.5 OXIDES

10.5.1 Introduction

Oxides are compounds in which oxygen is combined with other elements. Nearly all of the elements (except He, Ne, and Ar) form compounds with oxygen. Many oxides have important commercial applications, especially as catalysts, as pigments, and in ceramics. Others are of environmental importance because they are formed in combustion processes.

This section covers the simple oxides, binary compounds of oxygen. A number of mixed oxides, containing two or more different types of cation, are also known (e.g., $MgAl_2O_4$, $FeTiO_3$, $CaTiO_3$). Some occur naturally as minerals or ores, while others are manufactured for use as catalysts. The properties of some of the naturally-occurring mixed oxides are discussed in sections 2.11, 3.4, and 6.1-6.8 of this report, and discussion of the oxygen-containing species of trace metals is included in Chapter 7. In addition to its occurrence in oxides, oxygen is present in many compounds in the form of oxy-anions (e.g., CO_3^{-2}, SO_4^{-2}). The environmental chemistry of various oxy-anions is discussed in sections 6.11-6.12, 8.1-8.3, and 10.15. Other chemical classes containing oxygen are the carbonyls and the peroxides, discussed in sections 10.11 and 10.16, respectively.

Oxides exhibit a wide range of physical and chemical properties, depending on the element with which oxygen is combined. They may be solids, gases, or (rarely) liquids, ionic or covalent, non-volatile or highly volatile, and water-insoluble or highly soluble. In terms of chemical behavior, oxides are commonly classified according to whether they are acidic, basic, amphoteric, or neutral in aqueous solution. Properties of a number of oxides are summarized in Table 10.5-1.

This discussion will focus on the following environmentally significant reactions of oxides:

- Volatilization of solid oxides
- Dissolution in water
- Acidic or basic behavior in aqueous solution

10.5.2 Non-Metal Oxides

The oxides of virtually all non-metals are covalent. They generally form discrete molecules rather than infinite lattice structures; the exceptions include B_2O_3 (see section 10.9) and SiO_2 (see Chapter 6 and section 10.10), which form polymeric substances. Many of the non-metal oxides are gases. For a discussion of the properties of the gaseous oxides CO, CO_2, ClO_2, N_2O, NO, NO_2, OF_2 and SO_2, see section 10.6.

TABLE 10.5-1

Properties of Some Solid Oxides[a]

Compound	Melting Point (°C)[b]	Solubility (g/100 ml H_2O)[c]	pK_{sp}[d]	Acid/Base Character of Aqueous Solution[e]
Al_2O_3	2047 ± 8	0.000098 (29°) (α-alumina)		amphoteric[f]
As_4O_6	313 (subl.)	3.7 (20°C)	0.68 (cubic form)	acidic[f]
As_2O_5	>827	65.8 (20°)[g]		acidic[f]
CaO	~2587	0.131 (10°) dec.		basic[f]
CdO	826	insol.		
Cr_2O_3	2300	insol.		amphoteric
CrO_2	427 (dec.)	insol.		
CrO_3	187	166 (15°)[g]		acidic
Cu_2O	1230	insol.	14.7	
CuO	1336	insol.	20.35	
FeO	1368	insol.		
Fe_3O_4	1538 (dec.)	insol.		
Fe_2O_3	1562	insol.	42.7	
MgO	2800 ± 13	0.0086 (30°)		basic
Mn_3O_4	1590 (β)	insol.		
MnO_2	847	insol.		
MoO_3	795	0.138 (20°)[g]		acidic
Na_2O	920 (dec.)	dec. to NaOH		basic[f]
OsO_4	40 (white) 41 (yellow)	6.23 (25°)		acidic
P_4O_6	24	dec. to H_3PO_3		acidic[f]
P_4O_{10}	566 (triple point, stable form)	dec. to H_3PO_4		acidic[f]
Sb_4O_6	655	v. sl. sol.		amphoteric[f]

(Continued)

TABLE 10.5-1 (Continued)

Compound	Melting Point (°C)[b]	Solubility (g/100 ml H_2O)[c]	pK_{sp}[d]	Acid/Base Character of Aqueous Solution[e]
Sb_2O_5	380 (loses O_2)	v. sl. sol.		acidic
SeO_2	340-390	38.4 (14°)		acidic
TiO_2	1855	insol.		
Tl_2O_3	717 (decomp. starts at ~90°C)	insol.	45.2	
VO	2077	insol.		
V_2O_5	670	0.8 (20°)	13.34	acidic
ZnO	1975 (dec.)		16.66	amphoteric
ZrO_2	2690 (β)	insol.[e]	54.1	

a. For properties of gaseous oxides, see Table 10.6-1.
b. **Source:** Samsonov [5]
c. **Source:** Weast [6], except as noted
d. **Source:** Kotrlý and Šůcha [4]
e. **Source:** Cotton and Wilkinson [1], except as noted
f. **Source:** Ebsworth, Connor and Turner [3]
g. **Source:** Dean [2]

Other non-metal oxides, such as P_4O_{10}, As_4O_6 and SeO_2, are volatile molecular solids. Vapor pressures of some of these oxides are listed in Table 10.5-2. They are sometimes called acidic oxides or acid anhydrides, because they produce acidic solutions when dissolved in water:

$$SeO_2(s) + H_2O \longrightarrow H_2SeO_3^0 \quad (1)$$

$$As_4O_6(s) + 6H_2O \longrightarrow 4H_3AsO_3^0 \quad (2)$$

In some cases, acidic oxides are referred to simply as acids; for example, chromium trioxide, CrO_3, may be called chromic acid.

10.5.3 Metal Oxides

The metal oxides are solid compounds. Their bonds range from ionic to covalent. The vast majority are ionic compounds with infinite lattice structures, and they have negligible vapor pressures at environmental temperatures. However, OsO_4, RuO_4 and a few others have covalent bonding and are volatile, molecular solids [3]. (The vapor pressure of OsO_4 is given in Table 10.5-2.)

TABLE 10.5-2

Vapor Pressures of Some Solid Oxides

Compound	Vapor Pressure at 25°C (Nt/m²)[a]
As_2O_3 (claudetite)	7.94×10^{-5} (As_4O_6 vapor)
As_2O_3 (arsenolite)	2.51×10^{-7} (As_4O_6 vapor)
OsO_4 (tetragonal)	1.45×10^{4}
OsO_4 (cubic)	69.2
P_2O_5	9.77×10^{-14} (α-P_4O_{10} vapor)
P_2O_5 (metastable)	3.72×10^{4} (P_4O_{10} vapor)
SeO_2	2.21×10^{-5}

a. 1 Nt/m² = 1 Pascal (Pa) = 7.5×10^{-3} mm Hg = 9.9×10^{-6} atm.

Source: Adapted from Samsonov [5]

The oxides that are completely ionic, including those of the alkali and alkaline earth metals (Groups I and II), dissolve readily in water, forming the oxide ion (O^{-2}) and the counter metal cation. The oxide ion is hydrolyzed to yield hydroxide [1]:

$$O^{-2} + H_2O \longrightarrow 2OH^- \quad (3)$$

As this reaction produces a basic solution, these oxides are classified as basic oxides, or basic anhydrides. Some of the metals form ionic oxides that are insoluble in water but usually dissolve in dilute acids [1]; for example,

$$MgO\ (s) + 2H^{+} \longrightarrow Mg^{+2} + H_2O \tag{4}$$

Other metal oxides are amphoteric — that is, they behave as bases toward strong acids and as acids toward strong bases [1]:

$$ZnO(s) + 2H^{+} \longrightarrow Zn^{+2} + H_2O \tag{5}$$

$$ZnO(s) + 2OH^{-} + H_2O \longrightarrow Zn(OH)_4^{-2} \tag{6}$$

The amphoteric oxides are intermediate between ionic and covalent in character. A few other metals, generally the least electropositive ones, form oxides that may be considered acidic. Though insoluble in water, they dissolve in bases as follows [1]:

$$Sb_2O_5\ (s) + 2OH^{-} + 5H_2O \longrightarrow 2\ Sb(OH)_6^{-} \tag{7}$$

10.5.4 Trends in Acid-Base Character of Oxides

Some non-metals and transition metals form several oxides, ranging in character from acidic to basic for a given element, depending on the element's oxidation state. The oxide with the element in the highest oxidation state is generally the most acidic of such a group, as in the series CrO (basic), Cr_2O_3 (amphoteric), CrO_3 (acidic) [1,3].

Among the non-transition-metal elements, the most electronegative generally form the most acidic oxides. Thus, moving from left to right across the periodic table, the oxides of a given row of elements progress from basic to acidic. Within a column (group) of elements, the acidic character of the oxides M_xO_y decreases as the atomic number increases, so long as x and y remain constant [3].

10.5.5 Literature Cited

1. Cotton, F.A. and G. Wilkinson, *Advanced Inorganic Chemistry: A Comprehensive Text*, 3rd ed., Interscience Publishers, New York (1972).
2. Dean, J.A. (ed.), *Lange's Handbook of Chemistry*, 11th ed., McGraw-Hill Book Co., New York (1973).
3. Ebsworth, E.A.V., J.A. Connor and J.J. Turner, "Oxygen," in *Comprehensive Inorganic Chemistry*, Vol. 2, A.F. Trotman-Dickenson (exec. ed.), Pergamon Press, Oxford, England (1973).
4. Kotrlý, S. and L. Šůcha, *Handbook of Chemical Equilibria in Analytical Chemistry*, Halsted Press, a div. of John Wiley & Sons, New York (1985).
5. Samsonov, G.V. (ed.), *The Oxide Handbook*, IFI/Plenum Data Corp., New York (1973).
6. Weast, R.C. (ed.), *CRC Handbook of Chemistry and Physics*, 54th ed., CRC Press, Cleveland (1973).

10.6 GASES

10.6.1 Introduction

The gases covered in this section, while sharing the same physical state at room temperature, are quite diverse from a chemical standpoint. The discussion focuses on the properties and reactions that are environmentally important; the latter include:

- Dissolution in water and volatilization from solution,
- Reactions with water,
- Aqueous oxidation-reduction reactions,
- Oxidation of gases in the atmosphere, and
- Adsorption onto solids.

The commercial uses of some environmentally important gases are given in Table 10.6-1, and Table 10.6-2 lists molecular weights, boiling points and solubilities of a number of gases. Additional data on various gases are given elsewhere in this report, such as in §10.2, §10.9, §10.11, §10.13 and §10.14.

10.6.2 Dissolution in Water/Volatilization from Solution

The solubilities of gases in water vary widely. For example, at 20°C and 1 atm pressure, the solubility of CO is 0.0028 g/100 g H_2O, while that of HBr is 198 g/100 g H_2O [7]. In addition, the solubility of a given gas may vary with temperature, pressure, and pH of the solution.

In general, gases become less soluble in water as the temperature increases. (This is the reverse of the trend for most liquids and solids.) For example, the solubility of carbon dioxide (at 1 atm pressure), in grams CO_2 per 100 grams H_2O, is 0.335 at 0°C, 0.126 at 30°C, and 0.058 at 60°C [7]. Equations showing the variation of solubility with temperature for several gases are given in Table 10.6-3.

The relationship between pressure and solubility is useful for determining the partial pressure of a gas over a dilute aqueous solution — that is, the tendency for a gas to volatilize from water. In general, the concentration of a gas in air above a dilute solution is proportional to its concentration in the solution. This relationship can be expressed by Henry's Law:

$$P_A = H_A N_A \qquad (1)$$

where P_A is the vapor pressure of the solute A in the gas phase above the solution, N_A is the mole fraction of the solute A in the aqueous phase (moles of A/total moles), and H_A is the Henry's Law constant for the solute at the given temperature.

Henry's Law is applicable only to dilute solutions. At high concentrations, the relationship between vapor pressure and aqueous concentration becomes nonlinear. Even at low concentrations, however, this equation must be used with caution: the Henry's Law "constant" for several inorganic gases (particularly those that react with water

TABLE 10.6-1

Uses of Gases

Gas	Uses
CO	Fuel; metallurgical reducing agent; manufacture of chemicals
CO_2	Manufacture of chemicals (mainly urea and methanol); secondary oil recovery (added to old oil wells to increase production); refrigeration; beverage carbonation
Cl_2	Production of inorganic and organic chemicals (especially vinyl chloride, propylene oxide, and methylene chloride); water sanitization and disinfection; sewage treatment; bleach for textiles, pulp and paper
ClO_2	Bleaching of wood pulp; water treatment; textile bleaching; odor control
HBr	Manufacture of inorganic and organic bromides; alkylation catalyst in petroleum industry; catalyst for organic reactions
HCl	Steel pickling; other metallurgical processes; chemical reagent and intermediate; food processing; and many other uses
HF	Manufacture of aluminum fluoride and cryolite; manufacture of chlorofluorocarbons (pressurizing gases, refrigerants, and polymers); and many other uses
H_2S	Production of inorganic sulfides (for dyes, rubber chemicals, pesticides, polymers, plastics additives, leather and drugs); kraft pulping; and other uses
NF_3	Fluorine source for hydrogen fluoride and deuterium fluoride (HF/DF) high-energy chemical lasers
NH_3	Nitrogen source for fertilizers; production of explosives; source of nitrogen for fiber and plastics industry
N_2	As an inert gas atmosphere to preclude oxidation during metallurgical, chemical and food handling processes; as a pressure source for pneumatic instruments, hydraulic accumulators, and other applications; in liquid form, as an expendable refrigerant for freezing food or blood and for forming metals and plastics
N_2O	Anesthetic; propellant for whipped cream; in liquid form, medium for freezing food
NO, NO_2	Intermediates in nitric acid production. (These gases, referred to collectively as NO_x, are air pollutants created by the combustion of hydrocarbon fuels.)
OF_2	Mainly a laboratory chemical; proposed as an oxidizer for rocket applications
O_2	Purification of liquid iron in steelmaking; enrichment of fuel-air flame in smelting non-ferrous metals; oxidizing agent for large-volume chemical manufacture; wastewater treatment; in breathing apparatus for medical, high-altitude, and underwater use; and other uses
O_3	Water treatment
SO_2	Intermediate in sulfuric acid production; sulfite pulping in pulp and paper industry; food processing; water treatment; and many other uses

Source: *Kirk-Othmer* [8]

TABLE 10.6-2

Environmental Properties of Some Gases

Gas	Molecular Weight[a]	Boiling Point at 1 atm (°C)[a]	Solubility in H_2O at 1 atm, 20°C[b] (g/100g H_2O)	Henry's Law Constant, H, at 20°C[c] (mmHg/mole fraction)	Products of Reaction with H_2O
CO	28.01	−191.5	0.002838	4.07×10^7	—
CO_2	44.01	−78.5 (sub)	0.1688	1.08×10^6	H_2CO_3, HCO_3^-, CO_3^-, H^+
$(CN)_2$	52.04	−20.7	450cc/100cc H_2O	—	CN^-, NO_x, other cmpds
Cl_2	70.91	−34.6	0.7293	5.72×10^5[(j)]	HOCl, Cl^-, H^+
ClCN	61.04	12.66	2500cc/100cc H_2O	—	Cl^-, CN^-, NO_x
ClO_2[g]	67.45	10.9[b]	8.70 (15°C)	—	$HClO_3$, HCl
HBr	80.92	−67.0	198[d,h]	—	Br^-, H^+
HCl	36.46	−84.9	71.9[h]	—	Cl^-, H^+
HF	20.01	19.54	100 (773 mm Hg)[d,h]	—	F^-, H^+
H_2S	34.08	−60.7	0.3846	3.67×10^5[(f)]	HS^-, S^{-2}, H^+
NF_3	71.00	−128.8	0.0055 (25°C)	—	—
NH_3	17.03	−33.35	52.9[i]	2.10×10^3	NH_4^+, OH^-
N_2	28.01	−195.8	0.001901	6.11×10^7	—
N_2O	44.01	−88.5	0.1211[e]	1.50×10^6	—
NO	30.01	−151.8	0.006173	2.01×10^7	Converted to NO_2 on contact with O_2
NO_2[k]	46.01	21.2	—	—	HNO_3, HNO_2

(Continued)

TABLE 10.6-2 (Continued)

Gas	Molecular Weight[a]	Boiling Point at 1 atm (°C)[a]	Solubility in H_2O at 1 atm, 20°C[b] (g/100g H_2O)	Henry's Law Constant, H, at 20°C[c] (mmHg/mole fraction)	Products of Reaction with H_2O
OF_2	54.00	−144.8	0.016 (0°C)	—	HF, O_2
O_2	32.00	−182.96	0.004339	3.04×10^7	—
O_3	48.00	−111.9	0.0021	2.86×10^6	—
SO_2	64.06	−10	11.28[i]	2.62×10^4	$SO_2 \cdot H_2O$, HSO_3^-, SO_3^{-2}, H^+

Sources: a. Weast [19], b. Dean [4], c. Washburn [18], d. Dean [3], e. *Kirk-Othmer* [8], f. Strauss and Kaufmann [15].

g. Decomposes in sunlight [12].

h. See also Figure 10.2-1.

i. See also Figure 10.6-1.

j. Henry's Law constant for molecular chlorine, not total chlorine. Determined at 1 atm pressure of pure chlorine. Value from ref. 1.

k. Exists as a mixture of NO_2 and N_2O_4 at environmental temperatures; liquid at boiling point is 0.1% NO_2, vapor at 140°C is 100% NO_2.

TABLE 10.6-3

Variation of Aqueous Solubility of Gases with Temperature

		Coefficients in the Equation:[a] $R \ln x = A + B/T + C \ln T + DT$			
Gas	**Temp Range K**	**A** cal K^{-1} mol^{-1}	**B** cal mol^{-1}	**C** cal K^{-1} mol^{-1}	**D** cal K^{-2} mol^{-1}
Helium	273-334	−233.163	8737.84	32.2652	−0.0119726
Neon	273-339	−310.827	12766.8	43.6185	−0.0127534
Argon	274-347	−336.760	16170.1	46.2117	−0.00608793
Krypton	274-348	−270.967	15992.9	33.2892	0.0260485
Xenon	285-345	−360.119	18744.6	49.0332	−0.00311323
Radon	276-370	−499.309	25864.2	69.3241	0.00101227
Hydrogen	274-339	−357.802	13897.5	52.2871	−0.0298936
Nitrogen	273-346	−327.850	16757.6	42.8400	0.0167645
Oxygen	274-348	−286.942	15450.6	36.5593	0.0187662
Ozone[b]	277-293	*−29.7374*	*3905.44*		
Carbon monoxide	273-353	−341.325	16487.3	46.3757	
Carbon dioxide	273-353	−317.658	17371.2	43.0607	−0.00219107
NH_3	273-373	−162.446	2179.59	32.9085	−0.119722
HN_3	273-323	222.522	−4868.54	−36.7903	
NF_3	283-323	−503.754	24595.5	70.0480	
N_2F_4	288-318	−682.444	33572.0	96.1500	
N_2O	273-313	−180.950	13205.8	20.0399	0.0238544
NO	273-353	−333.515	16358.8	45.3253	−0.0519354
H_2S	273-333	−297.158	16347.7	40.2024	0.00257153
SO_2	283-386	−29.8850	5709.15	0.601884	
SF_6	276-323	−877.854	42051.0	125.018	
Cl_2	283-313	215.390	−4826.15	−38.1252	0.0177270
Cl_2O	273-293	*−14.3490*	*3574.66*		
ClO_2	283-333	112.751	284.523	−21.3532	
H_2Se	298-343	−147.799	9197.72	18.2608	
PH_3	298-323	−309.240	16364.9	41.5653	
AsH_3	273-299	−286.171	15437.9	38.0934	
Air	273-373	−319.323	15492.6	43.0259	0.0196194

a. R is the gas constant (1.9872 cal·K^{-1}·mol^{-1}), T is the temperature in K, and x is the mole fraction solubility at 1 atm partial pressure of the gas. Italicized entries are for the equation $RT \ln x = B + AT$.

b. Ozone decomposes slowly in aqueous solutions. It has been suggested that this behavior might be in part responsible for the divergence of reported solubility data.

Source: Wilhelm, Battino and Wilcock [20]. *(Copyright 1977, American Chemical Society. Reprinted with permission.)*

to produce acidic or basic solutions) varies with concentration. Thus, when using Henry's Law to estimate the solubility or vapor pressure of a gas, it is advisable to use a value of H_A determined at a concentration comparable to the one in question.

The variation in units used for Henry's Law constants should also be noted. For the form of Henry's Law given above (equation 1), H_A would be in units of pressure over mole fraction. However, Henry's Law is sometimes expressed as

$$P_A = H'_A \, x_A \tag{2}$$

where x_A is the mole *ratio* of the solute in water (moles of A/mole of water) [5,16]. Since the mole ratio x_A is equal to $N_A/1\text{-}N_A$, a Henry's Law constant H'_A calculated by equation 2 is not equivalent to H_A as defined by equation 1, except for very dilute solutions ($N_A << 1$), for which the mole ratio and mole fraction are essentially interchangeable. In addition, Henry's Law constants may be given using concentration units of moles solute per volume of solution (proportional to N_A) or mass of solute per mass of water (proportional to x_A.) Thus, it is important to note the units when using H_A values from the literature.

Example 1 The Henry's Law constant for hydrogen sulfide at 20°C is given as 3.67×10^5 mm Hg per mole fraction unit (Table 10.6-2). What is the vapor pressure at 20°C of H_2S above water in which 2 μg H_2S per liter has been dissolved?

Since H_2S dissociates to H^+ and HS^- in aqueous solution, we must first calculate the concentration of undissociated H_2S in solution. According to Table 2.7-2, the equilibrium constant at 20°C for the reaction

$$H^+ + HS^- \rightleftharpoons H_2S$$

is given by log K = 7.11, or $K = 1.29 \times 10^7$. Thus,

$$\frac{[H_2S]}{[H^+][HS^-]} = 1.29 \times 10^7$$

At a pH of 7, $[H^+] = 10^{-7}$, so $[H_2S]/[HS^-] = 1.29$, or $[HS^-] = [H_2S]/1.29$.

The total concentration of H_2S dissolved is

$$2 \times 10^{-6} \text{ g/l} \div 34.1 \text{ g/mol} = 5.87 \times 10^{-8} \text{ mol/l}$$

Assuming that all of the H_2S dissolved is present as H_2S or HS^- (i.e., neglecting other species, such as S^{-2}), we have

$$[H_2S] + [HS^-] = 5.87 \times 10^{-8} \text{ mol/l}$$

Using the equilibrium relationship described above to express $[HS^-]$ in terms of $[H_2S]$ gives

$$[H_2S] + [H_2S]/1.29 = 5.87 \times 10^{-8} \text{ mol/l}$$

Thus,

$$[H_2S] = 3.30 \times 10^{-8} \text{ mol/l}$$

In order to convert the concentration to a mole fraction, we assume that for such a dilute solution, the total number of moles in one liter of solution is essentially the same as in 1000g of water; thus,

$$N_A \approx x_A = \frac{3.3 \times 10^{-8} \text{ mol } H_2S}{10^3 \text{g } H_2O \,/\, 18 \text{ g/mol } H_2O} = 5.94 \times 10^{-10}$$

We can now use equation 1:

$$\begin{aligned} P_A &= H_A N_A \\ &= (3.67 \times 10^5 \text{ mm Hg/mole fraction})(5.94 \times 10^{-10}) \\ &= 2.2 \times 10^{-4} \text{ mm Hg} \end{aligned}$$

Table 10.6-2 gives Henry's Law constants for several gases. Figure 10.6-1 shows the variation of vapor pressure with aqueous concentration for SO_2 and NH_3, and Figure 10.2-1 shows similar data for HF, HCl, HBr and HNO_3.

The solubility of gases is also of importance in determining the "deposition velocity" of pollutant gases in the atmosphere. The amount of a gas that dissolves in water droplets in the atmosphere affects the rate of "wet deposition" of that gas — that is, the rate at which it is brought to earth by precipitation. (See § 2.5.)

10.6.3 Oxygen Solubility

The solubility of oxygen in natural waters is of particular importance, both because oxygen is essential for aquatic life and because dissolved oxygen is frequently an indicator of water quality. The presence of dissolved oxygen in surface waters generally (but not always) implies the absence of large quantities of oxygen-demanding organic wastes. Dissolved oxygen helps to decrease the amount of certain inorganic materials, such as iron and manganese, either by precipitating the oxidized form or by preventing the reduction and leaching of these ions from sediments; it also promotes the biological oxidation of ammonia to nitrate. In addition, dissolved oxygen prevents the anaerobic decomposition of organic materials, a process that may cause the formation of hydrogen sulfide, carbon dioxide and methane [17].

The solubility of oxygen (like that of other gases) decreases with increasing temperature. Gas solubility also decreases with increasing salt concentration, because under these conditions (e.g., in seawater) much of the H_2O is associated with ions as water of hydration and therefore dissolves less O_2 [14]. The solubility of oxygen in fresh and salt water at various temperatures is presented in Table 10.6-4.

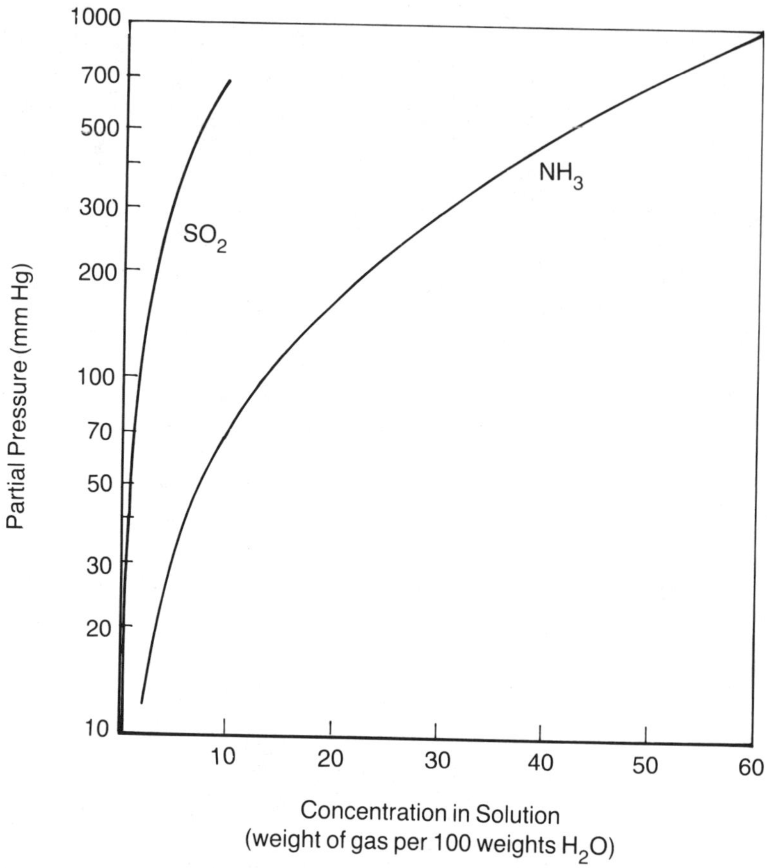

Source of Data: Green [7], pp. 3-101, 3-103

FIGURE 10.6-1 Partial Pressures of SO_2 and NH_3 over Their Aqueous Solutions at 20°C

TABLE 10.6-4

Solubility of Oxygen in Water
(from an atmosphere of 20.94% O_2 and 100% R.H.)

Temperature (°C)	O_2 Solubility (mg/l)[a] Distilled Water (chlorinity =0 ‰)	Sea Water (chlorinity = 20 ‰)[b]	Henry's Law Constant (atm/mole fraction)[c]
1	14.2	11.1	2.55×10^4 (0°)
10	11.3	9.0	3.27×10^4
20	9.1	7.3	4.01×10^4
30	7.5	6.2	4.75×10^4

a. **Source**: Riley and Chester [13], p. 107. (Units of solubility have been converted from ml/l.)

b. Chlorinity was originally defined as the chlorine equivalent of the total halide concentration in g/kg seawater. Because chlorinity is determined by the titration of seawater with $AgNO_3$, it is now defined as the mass in grams of Ag necessary to precipitate the halogens (Cl^- and Br^-) in 328.5233 g of seawater. (Since the relative concentration of the major components of seawater is nearly constant, by measuring one constituent, e.g., chlorinity, the composition of other components can be characterized.) Most seawater has a chlorinity between 18‰ and 21‰ [16].

c. **Source**: Green [7].

10.6.4 Reactions with Water

A number of gases react with water to form an acidic or basic solution. The products of these reactions are listed in Table 10.6-2. The acidic gases HF, HCl, HBr and H_2S react with water to produce aqueous protons; for example,

$$HF(g) + H_2O \rightleftharpoons HF^0 + H_2O \rightleftharpoons H_3O^+ + F^- \quad (3)$$

The extent of this reaction — and, thus, the amount of gaseous HF dissolved — is affected by the pH of the solution. A higher pH allows more aqueous HF to dissociate, thereby increasing the solubility of HF gas. The same trend applies to the other protic acids mentioned above. For further discussion of this process, see §10.2.

Several of the gaseous oxides (e.g., CO_2, ClO_2, NO_2, OF_2, SO_2) also form acidic solutions when they dissolve in water, by reactions such as the following:

$$CO_2(g) + H_2O \rightleftharpoons H_2CO_3^0 \rightleftharpoons HCO_3^- + H^+ \rightleftharpoons CO_3^{-2} + 2H^+ \quad (4)$$

$$2NO_2(g) + H_2O \rightleftharpoons HNO_3^0 + HNO_2^0 \quad (5)$$

The extent of these reactions is affected by pH; thus, these gases, like the protic acids, are more soluble at higher pH. Although the majority of the gaseous oxides form acidic solutions, CO and N_2O do not react with water and therefore produce neutral solutions.

Ammonia, on the other hand, reacts with water to produce a basic solution:

$$NH_3^0 + H_2O \rightleftharpoons NH_4^+ + OH^- \quad (6)$$

Accordingly ammonia becomes more soluble as the pH *decreases.*

As a result of the reactions described above, various anions or cations are released into solution (e.g., F^- from HF, CO_3^{-2} from CO_2, NH_4^+ from NH_3). Subsequent reactions of these ions are discussed in other sections of this report.

Some gases form solid hydrates in aqueous solution. For example, Cl_2 forms chlorine hydrate crystals $Cl_2 \cdot 8H_2O$ at temperatures below 9.6°C [10], and ClO_2 forms a solid hydrate below 18.2°C [6].

10.6.5 Aqueous Oxidation-Reduction Reactions

Oxidation-reduction reactions constitute an important part of the aqueous chemistry of several gases. SO_2 is a good reducing agent, while O_2, O_3, OF_2, ClO_2, and NO_2 are all oxidants. Oxidation potentials for some relevant half-reactions are given in Table 10.6-5. (For discussion of redox reactions and oxidation potentials, see § 2.10 and § 3.3.)

Ozone is one of the strongest oxidizing agents; it is second only to fluorine in this respect. In aqueous solution, ozone oxidizes bromide and iodide to bromine and iodine, respectively, and oxidizes transition metal ions to their highest oxidation states. Because the hydroxides of transition metal ions are typically less soluble for the higher oxidation states (e.g., Fe^{+3} *vs.* Fe^{+2}), treatment with ozone followed by filtration is a common method for removing metals, most frequently iron and manganese, from polluted waters. For example, the oxidation of iron proceeds as follows:

$$2Fe^{+2} + O_3 + H_2O \longrightarrow 2Fe^{+3} + O_2 + 2OH^- \quad (7)$$

$$Fe^{+3} + 3H_2O \longrightarrow Fe(OH)_3(s) + 3H^+ \quad (8)$$

Ozone can also oxidize aqueous solutions of ammonia, nitrogen oxides, and the nitrogenous anions nitrite, cyanide, cyanate and thiocyanate [11].

Oxygen is not as strong an oxidant as ozone but will oxidize certain metal ions, e.g., Cr^{+2} and Fe^{+2}, in aqueous solution [2]. (See § 2.10 and § 3.3.)

Chlorine dioxide is a strong oxidizing agent, reacting with both organic and inorganic materials. In aqueous solution, however, it may be either oxidized (to ClO_3^-) or reduced (to ClO_2^- or Cl^-), depending on what other species are present. In alkaline solution, ClO_2 undergoes a rapid disproportionation, forming both chlorate and chlorite [6,12]:

$$2ClO_2^0 + 2OH^- \longrightarrow ClO_3^- + ClO_2^- + H_2O \quad (9)$$

TABLE 10.6-5

Redox Potentials for Some Gases

Half-Reaction	E^0(V)
CO, CO_2	
$CO_2(g) + 2H^+ + 2e^- = CO(g) + H_2O$	−0.103[a]
ClO_2	
$ClO_2 + e^- = ClO_2^-$	+1.15
$ClO_2\ (aq) + e^- = ClO_2^-$	+0.954
$ClO_2(g) + H^+ + e^- = HClO_2$	+1.27
$ClO_3^- + 2H^+ + e^- = ClO_2 + H_2O$	+1.15
N_2O, NO, NO_2/N_2O_4	
$2NO_3^- + 4H^+ + 2e^- = N_2O_4(g) + 2H_2O$	+0.81
$NO_3^- + 4H^+ + 3e^- = NO(g) + 2H_2O$	+0.96
$2NO_3^- + 2H_2O + 2e^- = N_2O_4 + 4OH^-$	−0.85
$N_2O_4 + 2e^- = 2NO_2^-$	+0.88
$N_2O_4 + 2H^+ + 2e^- = 2HNO_2$	+1.07
$N_2O_4 + 4H^+ + 4e^- = 2NO + 2H_2O$	+1.03
$HNO_2 + H^+ + e^- = NO(g) + H_2O$	+0.99
$2HNO_2 + 4H^+ + 4e^- = N_2O + 3H_2O$	+1.27
$NO_2^- + H_2O + e^- = NO + 2OH^-$	−0.46
$2NO_2^- + 3H_2O + 4e^- = N_2O + 6OH^-$	+0.15
$2NO(g) + 2H^+ + 2e^- = N_2O(g) + H_2O$	+1.59
$2NO + H_2O + 2e^- = N_2O + 2OH^-$	+0.76
O_2, O_3	
$O_2 + 4H^+ + 4e^- = 2H_2O$	+1.229
$O_2 + 2H_2O + 4e^- = 4OH^-$	+0.401
$O_3 + 2H^+ + 2e^- = O_2 + H_2O$	+2.07
$O_3 + H_2O + 2e^- = O_2 + 2OH^-$	+1.24
SO_2	
$SO_4^{-2} + 4H^+ + 2e^- = H_2SO_3 + H_2O$	+0.20
$2H_2SO_3 + H^+ + 2e^- = HS_2O_4^- + 2H_2O$	−0.08
$2H_2SO_3 + 2H^+ + 4e^- = S_2O_3^{-2} + 3H_2O$	+0.40[a]
$H_2SO_3 + 4H^+ + 4e^- = S(s) + 3H_2O$	+0.45

Source: Weast [19], except as noted

a. Kotrlý and Šůcha [9]

10.6.6 Oxidation of Gases in the Atmosphere

Sulfur dioxide (SO_2) and the oxides of nitrogen (NO_x) can be oxidized to sulfuric and nitric acids in the atmosphere, resulting in the phenomenon of "acid rain." The mechanisms by which these transformations occur are discussed in §6.9.3.

10.6.7 Adsorption onto Solids

The extent to which gases adsorb onto solids is of importance in determining their environmental fate and persistence. The extent of adsorption of a gas onto a solid surface such as soil affects the rate and quantity of its release into the atmosphere and subsequent dissipation. In addition, the extent to which gases are adsorbed on atmospheric particles helps determine the rate at which they are brought to earth by the process of dry deposition.

Gases may adsorb onto solid surfaces by either physical or chemical means, but for natural surfaces at environmental temperatures, physical adsorption is the more important process. From data on the extent of adsorption of various gases onto clean charcoal, stronger adsorption has been correlated with higher normal boiling points or with lower saturated vapor pressures. However, the extent of adsorption onto natural surfaces may vary widely, as it is affected by physical factors such as the unit surface area and water content of the soil, by the pH and chemical reactivity of minerals present, and by microbial action. For further discussion and data on this process, see §2.4.

10.6.8 Literature Cited

1. Aieta, E.M. and P.V. Roberts, "Henry Constant of Molecular Chlorine in Aqueous Solution," *J. Chem. Eng. Data*, **31**, 51-53 (1986).
2. Cotton, F.A. and G. Wilkinson, *Advanced Inorganic Chemistry: A Comprehensive Text*, 3rd ed., Interscience Publishers, div. of John Wiley & Sons, New York (1972).
3. Dean, J.A. (ed.), *Lange's Handbook of Chemistry*, 11th ed., McGraw-Hill Book Co., New York (1973).
4. Dean, J.A. (ed.), *Lange's Handbook of Chemistry*, 12th ed., McGraw-Hill Book Co., New York (1979).
5. Gerrard, W., *Solubility of Gases and Liquids: A Graphic Approach*, Plenum Press, New York (1976).
6. Gordon, G., R.G. Kieffer and D.H. Rosenblatt, "The Chemistry of Chlorine Dioxide," *Prog. Inorg. Chem.*, **15**, 201-86 (1972).
7. Green, D.W. (ed.), *Perry's Chemical Engineers' Handbook*, 6th ed., McGraw-Hill Book Co., New York (1984).
8. *Kirk-Othmer Encyclopedia of Chemical Technology*, 3rd ed., John Wiley & Sons, New York (1982).

9. Kotrlý, S. and L. Šůcha, *Handbook of Chemical Equilibria in Analytical Chemistry*, Halsted Press, a div. of John Wiley & Sons, New York (1985).
10. Leddy, J.J., I.C. Jones, Jr., B.S. Lowry, F.W. Spillers, R.E. Wing and C.D. Binger, "Alkali and Chlorine Products," in *Kirk-Othmer* [8], Vol. 1.
11. Nebel, C., "Ozone," in *Kirk-Othmer* [8], Vol. 16.
12. Noack, M.G. and R.L. Doerr, "Chlorine Oxygen Acids and Salts (ClO_2, $HClO_2$)," in *Kirk-Othmer* [8], Vol. 5.
13. Riley, J.P. and R. Chester, *Introduction to Marine Chemistry*, Academic Press, London (1971).
14. Snoeyink, V.L. and D. Jenkins, *Water Chemistry*, John Wiley & Sons, New York (1980).
15. Strauss, H.J. and M. Kaufmann (eds.), *Handbook for Chemical Technicians*, McGraw-Hill Book Co., New York (1976).
16. Stumm, W. and J.J. Morgan, *Aquatic Chemistry*, 2nd ed., John Wiley & Sons, New York (1981).
17. U.S. Environmental Protection Agency, *Quality Criteria for Water*, USEPA, Washington, D.C. (July 1976).
18. Washburn, E.W. (ed.), *International Critical Tables of Numerical Data, Physics, Chemistry and Technology*, Vol. 3, McGraw-Hill Book Co., New York (1928).
19. Weast, R.C. (ed.), *CRC Handbook of Chemistry and Physics*, 54th ed., CRC Press, Cleveland (1973).
20. Wilhelm, E., R. Battino and R.J. Wilcock, "Low-Pressure Solubility of Gases in Liquid Water," *Chem. Rev.*, **77**, 219-62 (1977).

10.7 SULFUR-CONTAINING COMPOUNDS

10.7.1 Introduction

Sulfur forms a wide variety of compounds, both naturally occurring and man-made. Since sulfur can exist in oxidation states ranging from +6 to −2, stable sufur compounds can be found over a wide range of environmental (oxidizing or reducing) conditions. The most stable forms of sulfur are sulfate (SO_4^{-2}), the most oxidized form, and sulfide (S^{-2}), the most reduced form [9].

Sulfur compounds described elsewhere in this report include sulfuric acid (section 10.2, Acids), sulfur dioxide (section 10.6, Gases), sulfates (section 10.4, Salts), and peroxosulfates (section 10.16, Peroxides). In addition, the sulfate ligand is discussed in section 6.11, and other sulfur-containing ligands are discussed in section 8.2. This section focuses on the environmental chemistry of commercially produced sulfur compounds, which include sulfides, sulfur-halogen compounds, sulfur oxides (SO_2 and SO_3), and salts of sulfur oxo anions (e.g., sulfate SO_4^{-2}, sulfite SO_3^{-2}, metabisulfite $S_2O_5^{-2}$, and dithionite $S_2O_4^{-2}$). Examples of these compounds and their commercial uses are given in Table 10.7-1.

10.7.2 Sulfides

The sulfides are compounds in which sulfur has a formal oxidation state of −2. Not all of the sulfides are ionic, however; both the nature of the bonding and the chemical and physical properties of the sulfides vary widely. The sulfides of the non-metals (e.g., CS_2, P_2S_5, As_2S_3) are covalent, molecular solids and liquids with stable vapor phases [4]. Carbon disulfide, CS_2, is a volatile, flammable liquid with a normal boiling point of 46.25°C and a vapor pressure of 39.66 kPa at 20°C. It is slightly soluble in water: 2.10 g CS_2 dissolve in 1 kg of solution at 20°C [12].

In contrast, the sulfides of the metals are mainly ionic solids and may be water-soluble. They usually decompose to the elements upon vaporization [4]. Their dissolution in water yields hydrated metal cations and sulfide (S^{-2}) ions. The sulfide ion, a strong base, may subsequently be hydrolyzed to HS^- and in part to H_2S:

$$S^{-2} + H_2O \rightleftharpoons HS^- + OH^- \qquad \log K = -0.1 \text{ [7]} \tag{1}$$

$$HS^- + H_2O \rightleftharpoons H_2S + OH^- \qquad \log K = -7.0 \text{ [7]} \tag{2}$$

The relative amounts of S^{-2}, HS^-, and H_2S present depend on the pH of the solution. As more S^{-2} is hydrolyzed, more of the metal sulfide can dissolve; thus, the solubility of the sulfide increases with decreasing pH [4]. In addition, H_2S may volatilize from solution, causing a further shift in the S^{-2}—HS^-—H_2S equilibrium. (For a discussion of gas volatilization from solution, see section 10.6, Gases.) However, except in shallow marshes, H_2S volatilization is not believed to be a major pathway for sulfur in natural environments [9].

TABLE 10.7-1

Examples of Sulfur Compounds and Their Uses

Compound	Name	Principal Uses
Sulfur Oxides		
SO_2	Sulfur dioxide	Intermediate in sulfuric acid production; sulfite pulping in pulp and paper industry; food processing; water treatment
SO_3	Sulfur trioxide	Organic sulfonating agent for production of detergents; raw material for chlorosulfuric acid and oleum
Salts of Oxyacids		
Na_2SO_3	Sodium sulfite	Pulp manufacture; water treatment; photography (film preservative)
$NaHSO_3$	Sodium bisulfite	(Not stable as a solid; name sometimes used to refer to $Na_2S_2O_5$)
$Na_2S_2O_5$	Sodium metabisulfite (sodium pyrosulfite; sodium disulfite)	Food preservative; bleach; photography; water treatment; leather tanning; organic synthesis
$Na_2S_2O_4$	Sodium dithionite (sodium hydrosulfite; sodium sulfoxylate)	Dyeing (as a reductant for dyes); pulp bleaching; other bleaching and textile uses
$Na_2S_2O_3$	Sodium thiosulfate	Fixative in photography; antichlor in paper and textile bleaching; extraction of silver from ore; manufacture of matches; soap preservative; organic industry[a]
Sulfides		
CS_2	Carbon disulfide	Manufacture of viscose rayon, cellophane film, and carbon tetrachloride
NiS	Nickel sulfide	Catalyst for hydrogenation of petroleum distillates
SeS_2	Selenium disulfide	Active ingredient of antidandruff shampoos
NaHS	Sodium hydrosulfide	Leather industry; ore flotation; paper industry; oil refining; regeneration of lead sulfite; organic industry[a]

(Continued)

TABLE 10.7-1 (Continued)

Compound	Name	Principal Uses
Halides and Oxyhalides		
S_2Cl_2	Sulfur monochloride	Raw material for manufacture of lubricant additives and rubber vulcanizing agents
SF_4	Sulfur tetrafluoride	Limited commercial application; fluorinating agent for compounds containing carbonyl groups
SF_6	Sulfur hexafluoride	Insulating material in circuit breakers, high voltage coaxial lines, transformers, and other electrical equipment
SO_2F_2	Sulfuryl fluoride	Insecticide/fumigant for structure-damaging pests

a. From Meyer [8]

Source: *Kirk-Othmer* [5], unless otherwise noted

If metal cations are present together with sulfide in aquatic environments, poorly soluble metal sulfides may form and precipitate. Of particular importance, because of the abundance of iron, is the formation of iron sulfides [9]. The reverse process is also of environmental concern: when sulfide minerals dissolve, the sulfide ions may be oxidized to sulfate, releasing hydrogen ions (H^+) into the solution. This can result in the acidification of natural waters near mining operations. (See § 6.9.3 for a discussion of FeS_2 oxidation and the problem of acid mine drainage.) Another possible fate of dissolved sulfide is the formation of polysulfides S_x^{-2} from H_2S and elemental sulfur, which is discussed in section 2.8.

In general, the sulfides of the main-group elements show decreasing solubility with increasing positive charge of the metal cation. Those with low positive charges on the cation (e.g., Na^+, K^+, Mg^{+2}, Ca^{+2}, Al^{+3}) are hydrolyzed by either water or moist air, while those with high positive charges (e.g., Mo^{+6}, W^{+6}) are almost insoluble, and some form thioanions (e.g., MoS_4^{-2}, WS_4^{-2}) that are stable in aqueous solution [4].

Sulfides of the transition metals are often alloy-like or semimetallic [1], and they may be semiconductors or metallic conductors [4]. They are generally less soluble than the sulfides of the main group elements. Among the transition metals, the solubility of the sulfides increases as one proceeds to the right or down on the periodic table [4].

The solubility of metal sulfides may be considerably enhanced by complexation. That is, although a metal sulfide may have a low solubility product, suggesting a low concentration of the aquo metal cation in solution, the cation may form a variety of thio complexes whose presence in solution allows a much greater concentration of solid sulfide to dissolve. Generally, the metal ions that form insoluble sulfides tend to form thio complexes. For example, although silver sulfide, Ag_2S, has a solubility product of about 10^{-50}, which would indicate an aquo Ag^+ concentration of only about 10^{-20} *M* at pH 7, the actual total solubility of Ag_2S is about 10^{-7} *M*, due to the formation of the complexes AgSH, $Ag(SH)_2^-$, and $Ag_2S_3H_2^{-2}$ [11].

Among the transition-metal ions, the tendency to coordinate with S^{-2} increases in the order $Mn^{+2} < Fe^{+2} < Co^{+2} \approx Zn^{+2} < Ni^{+2} < Cu^{+2}$. The tendency to form thio complexes is especially strong with the "B-type" metal cations, or "soft acids": Cu^+, Ag^+, Au^+, Tl^+, Ga^+, Cd^{+2}, Sn^{+2}, Tl^{+3}, Au^{+3}, and In^{+3} [11]. Stability constants for sulfide complexes are given in Table 8.2-2, and solubility products for several metal sulfides are given in Table 8.2-3.

10.7.3 Sulfur Trioxide

Sulfur trioxide (SO_3), as available commercially, is a liquid to which stabilizers are added to prevent its polymerization. It is also commonly available as "fuming sulfuric acid" or "oleum," a solution of 25%-65% SO_3 in sulfuric acid [10]. Sulfur trioxide is a strong organic sulfonating and dehydrating agent, used primarily for the production of detergents and as a raw material for chlorosulfuric acid and oleum [2].

At room temperature, SO_3 without added stabilizers must be very pure to remain a liquid. In the presence of only trace amounts of water, it polymerizes to a solid form known as beta-SO_3 or asbestos-like SO_3 [10], which has a melting point of 32.5°C [2]. Two other solid forms of SO_3 are known: gamma-SO_3 or ice-like SO_3, formed by cooling of the pure liquid to 16.8°C, and alpha-SO_3, a cross-linked polymeric form with a melting point of 62.3°C. All three solid modifications yield liquid gamma-SO_3 when melted. (However, heating of the solid polymeric forms can cause explosions, as vapor pressures greater than atmospheric pressure develop before melting occurs [2].) The liquid has a vapor pressure of 265 mm Hg at 25°C [10] and a normal boiling point of 44.8°C [2].

Sulfur trioxide is a highly reactive compound. It combines vigorously with water to form sulfuric acid, and it is a powerful oxidizing agent [1].

10.7.4 Sulfur Dioxide, Sulfites and Related Compounds

Sulfur dioxide, SO_2, is a common pollutant gas resulting primarily from the combustion of coal and oil. It is quite soluble in water, 36 volumes dissolving in one volume of water at 20°C [14]. (More solubility data for SO_2 are given in section 10.6, Gases.) When SO_2 is dissolved in water, an acidic solution is produced. This is frequently referred to as a solution of sulfurous acid, H_2SO_3, but free H_2SO_3 is not actually observed; rather, the predominant species is the bisulfite ion, HSO_3^-, which exists in equilibrium with hydrated sulfur dioxide and the sulfite ion, SO_3^{-2}, according to the following equations [1,6]:

$$SO_2 \cdot xH_2O \rightleftharpoons HSO_3^- + H_3O^+ + (x-2)H_2O \qquad K = 1.3 \times 10^{-2} \tag{3}$$

$$HSO_3^- + H_2O \rightleftharpoons SO_3^{-2} + H_3O^+ \qquad K = 1.0 \times 10^{-7} \tag{4}$$

The equilibrium constant for equation 3, which may be referred to as the K_a for "sulfurous acid," can be formulated as [1]:

$$K = \frac{[HSO_3^-][H^+]}{[\text{total dissolved } SO_2] - [HSO_3^-] - [SO_3^{-2}]}$$

At high concentrations, the metabisulfite ion $S_2O_5^{-2}$ (sometimes called disulfite or pyrosulfite) is also formed:

$$2HSO_3^- \rightleftharpoons S_2O_5^{-2} + H_2O \tag{5}$$

Sulfite solutions may be oxidized by atmospheric oxygen, yielding sulfate and other oxidized species. In the absence of catalysts, the reaction proceeds slowly, via a radical chain mechanism. The rate of the reaction increases rapidly with increasing pH [9] and is greatly enhanced in the presence of a variety of metal ions. A similar oxidation process is responsible for the formation of H_2SO_4 in acid rain, discussed in § 6.9.3.

Salts of both sulfite (SO_3^{-2}) and metabisulfite ($S_2O_5^{-2}$) with the alkali metals are used as commercial products. (The solid salt known commercially as sodium bisulfite is actually sodium metabisulfite, $Na_2S_2O_5$.) Both Na_2SO_3 and $Na_2S_2O_5$ are solids that are oxidized by moist air (the sulfite more readily than the metabisulfite), yielding sodium sulfate, Na_2SO_4. They dissolve readily in water to produce a sulfite-bisulfite solution, with the relative proportions of sulfite, bisulfite, and metabisulfite depending on pH. When pure sodium sulfite is dissolved in water, it yields an alkaline solution, while a solution of pure sodium metabisulfite has an acidic pH [14]; both compounds are used as reducing agents in a variety of commercial applications, some of which are listed in Table 10.7-1.

A related compound is sodium dithionite, $Na_2S_2O_4$ (also known as sodium hydrosulfite). In solid form, this compound decomposes exothermically when exposed to heat, moisture, or air. When the dry salt is heated above 190°C, it releases SO_2 violently, and in moist air it can ignite spontaneously. In alkaline solution, it is used as a reducing agent, especially for dyes. A stronger reducing agent than sulfite, it is capable of reducing many metal ions (e.g., Cu^+, Ag^+, Pb^{+2}, Sb^{+3}, Bi^{+3}) to the metals, and it also reduces iodine, oxygen, peroxides, and ferric salts [14]. However, dithionite decomposes rapidly if the solution is acidified or heated above 50°C, according to the following equation [14]:

$$2S_2O_4^{-2} + H_2O \rightarrow 2HSO_3^- + S_2O_3^{-2} \tag{6}$$

Sodium thiosulfate, $Na_2S_2O_3$, is used primarily as a fixative in photography, due to its ability to complex silver ions from the silver bromide in photographic emulsions. The thiosulfate ion also forms complexes with several other transition metal ions [1].

Both sulfite and thiosulfate are good complexing agents. Formation constants for sulfite and thiosulfate complexes are given in Table 8.2-2.

10.7.5 Halogenated Sulfur Compounds

Several compounds of sulfur with the halogens have commercial importance, as shown in Table 10.7-1; their chemical and physical properties are summarized in Table 10.7-2. They are gases and liquids with widely varying properties. While sulfur hexafluoride and sulfuryl fluoride are chemically inert under environmental conditions, sulfur tetrafluoride is highly reactive, fuming in moist air and instantly hydrolyzed by water [1,3]. Sulfur monochloride, although generally stable at environmental temperatures, is readily hydrolyzed by water [14].

10.7.6 Literature Cited

1. Cotton, F.A. and G. Wilkinson, *Advanced Inorganic Chemistry, A Comprehensive Text*, 3rd ed., Interscience Publishers, div. of John Wiley & Sons, New York (1972).
2. Donovan, J.R. and J.M. Salamone, "Sulfuric Acid and Sulfur Trioxide," in *Kirk-Othmer* [5], Vol. 22.
3. Eibeck, R.E. and W. Mears, "Sulfur — Sulfur Fluorides," under "Fluorine Compounds, Inorganic," in *Kirk-Othmer* [5], Vol. 10.
4. Jellinek, F., "Sulphides," in G. Nickless (ed.), *Inorganic Sulphur Chemistry*, Elsevier, Amsterdam (1968).
5. *Kirk-Othmer Encyclopedia of Chemical Technology*, 3rd ed., John Wiley & Sons, New York (1982).
6. Lyons, D. and G. Nickless, "The Lower Oxy-Acids of Sulphur," in G. Nickless (ed.), *Inorganic Sulphur Chemistry*, Elsevier, Amsterdam (1968).
7. Martell, A.E. and R.M. Smith, *Inorganic Complexes*, Vol. 4 of *Critical Stability Constants*, Plenum Press, New York (1976).
8. Meyer, B., *Sulfur, Energy and Environment*, Elsevier Scientific, Amsterdam (1977).
9. Nriagu, J.O. and J.D. Hem, "Chemistry of Pollutant Sulfur in Natural Waters," in J.O. Nriagu (ed.), *Sulfur in the Environment. Part II: Ecological Impacts*, John Wiley & Sons, New York (1978).
10. Schmidt, M. and W. Siebert, "Sulphur," in A.F. Trotman-Dickenson (exec. ed.), *Comprehensive Inorganic Chemistry*, Vol. 2, Pergamon Press, Oxford, England (1973).

TABLE 10.7-2

Environmental Properties of Some Sulfur-Halogen Compounds

Compound	Name	Physical State at Room Temp.	Boiling Point (°C)	Reaction in Water	Other Properties of Interest
S_2Cl_2	Sulfur monochloride	Liquid	135.6	Hydrolyzed at moderate rate at environmental temperature, faster at higher temperature[a]	
SF_4	Sulfur tetrafluoride	Gas	−40	Instantly hydrolyzed ($SF_4 + 2H_2O \rightarrow SO_2 + 4HF$)[b]	Highly reactive; fumes in moist air[b]
SF_6	Sulfur hexafluoride	Gas	(−63.8 sublimes)	No reaction; slightly soluble	Chemically inert; high dielectric strength[b]
SO_2F_2	Sulfuryl fluoride	Gas	−55.4	No reaction; solubility 10 g/100 cc H_2O at 9°C	Chemically inert; slowly hydrolyzed by strong aqueous alkali[c]

a. From Weil [14]

b. From Eibeck and Mears [3]

c. From Cotton and Wilkinson [1]

Source: Weast [13], unless otherwise noted.

11. Stumm, W. and J.J. Morgan, *Aquatic Chemistry: An Introduction Emphasizing Chemical Equilibria in Natural Waters*, 2nd ed., John Wiley & Sons, New York (1981).
12. Timmerman, R.W., "Carbon Disulfide," in *Kirk-Othmer* [5], Vol. 4.
13. Weast, R.C. (ed.), *CRC Handbook of Chemistry and Physics*, 54th ed., CRC Press, div. of Chemical Rubber Co., Cleveland (1973).
14. Weil, E.D., "Sulfur Compounds," in *Kirk-Othmer* [5], Vol. 22.

10.8 PHOSPHORUS-CONTAINING COMPOUNDS

The discussion below is limited to phosphides, hypophosphite, phosphorus sulfides, and phosphorus halides and oxyhalides. The environmentally important properties of phosphates are covered in section 8.3, "Phosphorus-containing Ligands," while phosphine is included in section 10.14, "Hydrides," and phosphorus pentoxide, in section 10.5, "Oxides."

10.8.1 Phosphides

Phosphides are binary compounds of phosphorus. They may be divided into two classes: those that are reactive and easily undergo hydrolysis, and those that are metallic in nature and not easily hydrolyzed [3]. Phosphides of the alkali and alkaline earth metals react rapidly with water to form hydrogen phosphides (predominantly phosphine, PH_3), but phosphides of group IIIA, IVA, and VA elements are much less reactive [4].

Aluminum phosphide, AlP, has a melting point of over 1700°C [4]. It is stable in dry air, but when highly purified, it reacts readily with water to form phosphine, PH_3; this reaction is accelerated in acid or alkaline solutions [7]. Aluminum phosphide is widely used as a grain fumigant, because the phosphine gas produced upon its contact with moisture is highly toxic to pests but does not have long-lasting deleterious effects on the environment [5].

Several metals form more than one phosphide. Zinc, for example, forms both ZnP_2 and Zn_3P_2, which are stable under ordinary conditions. (Note that ZnP_2 is one of the phosphides that does not obey the classical valence rules.) The melting point of ZnP_2 is about 700°C, but when heated to this temperature in a vacuum, it decomposes to form Zn_3P_2 [4]. Neither ZnP_2 nor Zn_3P_2 reacts with water. ZnP_2 reacts slowly with acids to form phosphine, while Zn_3P_2 is stable in cold, non-oxidizing acids [4]. The latter phosphide is used as a rodenticide [5].

10.8.2 Phosphorus Halides and Oxyhalides

The phosphorus halides may be solids, liquids or gases at 25°C. Physical properties of several trihalides (PX_3), phosphorus oxychloride ($POCl_3$) and the pentachloride (PCl_5) are given in Table 10.8-1.

The phosphorus halides react with liquid water or water vapor at varying rates: phosphorus triiodide reacts the fastest, phosphorus trifluoride the slowest, and phosphorus trichloride at an intermediate rate [4].

TABLE 10.8-1

Properties of Some Phosphorus Halides

Compound	Melting Point (°C)	Boiling Point (°C)	Vapor Pressure, log P (torr @ T °C)[a]
PBr_3	−40.5	173.2	$6.9155 - \frac{1590.5}{(T + 221.0)}$
PCl_3	−91	75	$6.8267 - \frac{1196}{(T + 227.0)}$
$POCl_3$	2	105	$6.8658 - \frac{1297.2}{(T + 220.0)}$
PCl_5	162 (sublimes)[b]	162 (sublimes)[b]	$10.2068 - \frac{2903.1}{(T + 237.0)}$
PF_3	−151.30	−101.38	$6.8604 - \frac{620.22}{(T + 257.0)}$

a. Equation is applicable over temperature range at which halide is liquid. In the case of PCl_5, it applies up to 160°C.

b. Weast [6]

Source: Dean [2], except as indicated.

Phosphorus trichloride forms various products when it hydrolyzes, depending on the ratio of H_2O to PCl_3. At H_2O/PCl_3 mole ratios greater than 3, phosphonic acid (H_3PO_3, also referred to as phosphorous acid) is produced [5]:

$$PCl_3 + 3H_2O \rightleftharpoons H\underset{\underset{OH}{|}}{\overset{\overset{O}{\|}}{P}}OH + 3HCl \qquad (1)$$

At mole ratios between 2.5 and 3, a mixture of pyrophosphonic ($H_4P_2O_5$) and phosphonic acid results:

$$3PCl_3 + 8H_2O \rightleftharpoons 9HCl + H\underset{\underset{HO}{|}}{\overset{\overset{O}{\|}}{P}}O\underset{\underset{OH}{|}}{\overset{\overset{O}{\|}}{P}}H + H\underset{\underset{OH}{|}}{\overset{\overset{O}{\|}}{P}}-OH \qquad (2)$$

At ratios less than 2.5, an additional compound of indeterminate structure is formed; this is believed to be a polymer of randomly disposed P-O-P and P-P linkages with other groups along the chain consisting of $-OH$, $=O$, $-H$ and sometimes $-Cl$ or other halides [5].

The phosphorus halides are easily oxidized. Phosphorus trichloride, trifluoride and tribromide readily add chlorine, fluorine and sometimes bromine to form pentahalides [4]. Phosphorus pentachloride exists in equilibrium with PCl_3 and Cl_2 in the gas phase:

$$PCl_5(g) \rightleftharpoons PCl_3(g) + Cl_2(g) \tag{3}$$

The equilibrium constant for reaction 3 has been expressed as a function of temperature (between 150 and 230°C) as follows [4]:

$$\log K = -20{,}000/4.57T + 1.75 \log T + 6.66 \tag{4}$$

The equilibrium constant, K, is given in mm Hg, and T is the absolute (Kelvin) temperature. Equation 4 means that at 160°C, the approximate sublimation point, 13.5% of the PCl_5 is dissociated [4].

Pure PCl_3 reacts rapidly with O_2 to form $POCl_3$; however, impurities such as iron halides, sulfur compounds, and rubber inhibit the reaction [4]. Phosphorus oxychloride can also be formed by the controlled hydrolysis of PCl_5. (Uncontrolled, PCl_5 reacts violently with water to form HCl and H_3PO_4 [3].) The oxyhalides of fluorine and bromine, but not iodine, are also known to exist [4].

Phosphorus trichloride is the most important phosphorus halide industrially because of its use as an intermediate in the production of phosphorus oxychloride (which is used in the synthesis of phosphate esters [3]), phosphorus sulfochloride ($PSCl_3$), phosphonic acid, dialkyl phosphonates (R_2HPO_3), and phosphorus pentachloride; the latter is used in the manufacture of chlorophosphazenes $(NPCl_2)_n$ and as a catalyst and chlorinating agent in inorganic syntheses [1].

10.8.3 Phosphorus Sulfides

Phosphorus combines with sulfur to form four well-characterized crystalline compounds, P_4S_7, P_4S_5, P_4S_3, and P_4S_{10}; the latter two are produced commercially [3]. Tetraphosphorus decasulfide, P_4S_{10} (also referred to as phosphorus pentasulfide, P_2S_5) is a solid under environmental conditions, with a melting point of 286-290°C and a boiling point of 514°C [6]. The vapor pressure of P_4S_{10} can be approximated by the following equation [4]:

$$\log P = -4940/T + 9.17 \tag{5}$$

where P is the vapor pressure in mm Hg and T the absolute (Kelvin) temperature.

Even in a sealed tube at room temperature, tetraphosphorus decasulfide is not entirely stable [1]: when heated above its melting point, it decomposes, probably to P_4S_7 and S initially [3]. When warmed in water, it forms mainly hydrogen sulfide and orthophosphoric acid (see § 8.3.3 for information on phosphoric acid), while at high pH it is hydrolyzed to a mixture that contains thiosulfates and sulfides [1].

Tetraphosphorus decasulfide is used in the manufacture of insecticides, lubricating oil additives, plasticizers, flame retardants, flotation agents for mineral processing, and some types of asphalts [5].

10.8.4 Hypophosphites

The hypophosphite ion, $H_2PO_2^{-1}$, forms salts with alkali, alkaline earth and heavier metals. The alkali metal hypophosphites include $LiH_2PO_2 \cdot H_2O$, $NaH_2PO_2 \cdot H_2O$, and KH_2PO_2; the alkali earth salts are $Mg(H_2PO_2)_2 \cdot 6H_2O$, $Ca(H_2PO_2)_2$, $Sr(H_2PO_2)_2$, $Ba(H_2PO_2)_2$, and $Ba(H_2PO_2)_2 \cdot H_2O$. Ammonium salts are also known to exist [1]. All of these salts are very soluble in water.

The metal hypophosphite salts are generally produced by mixing phosphinic acid, H_3PO_2 (also referred to as hypophosphorous acid) and the metal oxide or hydroxide [4]. As phosphinic acid is a strong monoprotic acid ($pK_a = 1.1$ [3]), the hypophosphite ion has little acid/base chemistry at the pH conditions typically encountered in the environment.

In alkaline solutions, hypophosphite salts are strong reducing agents. The electrode potential for the following reaction is -1.65 V at 25°C [6]:

$$HPO_3^{-2} + 2H_2O + 2e^- = H_2PO_2^- + 3OH^- \tag{6}$$

Their reducing ability gives them important industrial applications. The sodium salt is the most important hypophosphite salt in industry, where it is used mainly as a reducing agent in the electroless plating of nickel on metallic parts of steel and aluminum [3].

10.8.5 Literature Cited

1. Boeing, I.A., M.M. Crutchfield and C.W. Heitsch, "Phosphorus Compounds," in *Kirk-Othmer Encyclopedia of Chemical Technology*, 3rd ed., Vol. 17, John Wiley & Sons, New York (1982).
2. Dean, J.A. (ed.), *Lange's Handbook of Chemistry*, 12th ed., McGraw-Hill Book Co., New York (1979).
3. Toy, A.D.F., "Phosphorus," in *Comprehensive Inorganic Chemistry*, A.F. Trotman-Dickenson (exec. ed.), Pergamon Press, Oxford, England (1973).
4. Van Wazer, J.R., *Phosphorus and Its Compounds. Volume I: Chemistry*, Interscience Publishers, New York (1958).
5. Van Wazer, J.R., "Phosphorus and the Phosphides" in *Kirk-Othmer Encyclopedia of Chemical Technology*, 3rd ed., Vol.17, John Wiley & Sons, New York (1982).
6. Weast, R.C. (ed.), *Handbook of Chemistry and Physics*, 66th ed., CRC Press, Cleveland (1985).
7. White, W.E. and A.H. Bushey, "Aluminum Phosphide — Preparation and Composition," *J. Am. Chem. Soc.*, **66**, 1666-72 (1944).

10.9 BORON-CONTAINING COMPOUNDS

10.9.1 Introduction

Three major groups of boron-containing compounds are of commercial and environmental importance:

- Boric acid (H_3BO_3) and borates (e.g., $Na_2O \cdot 2B_2O_3 \cdot 10H_2O$, $K_2O \cdot 2B_2O_3 \cdot 4H_2O$)
- Boron halides (e.g., BCl_3, BBr_3)
- Boron hydrides (e.g., B_2H_6, B_5H_9) and metal tetrahydroborates (e.g., $NaBH_4$)

Each group has quite different properties and is discussed separately below.

Another group of boron-containing compounds, the peroxoborates, is discussed in section 10.16.

10.9.2 Boric Acid and Borates

USES

Boric oxide and borates have a wide variety of commercial applications. They are used in the manufacture of glass, including fiberglass insulation and textile fiberglass, as fluxing agents for porcelain enamels and ceramic glazes, as components of soap and cleaning agents, as trace nutrients in fertilizers, and as herbicides. In addition, boric acid is used as a catalyst for the air oxidation of hydrocarbons, as a preservative for wood, leather and rubber, as an antiseptic and as a fire retardant [3].

PROPERTIES

Boric acid, or orthoboric acid, H_3BO_3, is a solid at room temperature, normally melting at 170.9°C. When heated slowly, however, it loses water to form metaboric acid, HBO_2. Its solubility in water increases rapidly with temperature, from 2.52% (by weight of solution) at 0°C, to 4.72% at 20°C and 27.53% at 100°C. These solubilities may be either increased or decreased by the presence of inorganic salts [3].

Boric acid acts as a base acceptor rather than a proton donor, as in the following reaction with water:

$$B(OH)_3^0 + H_2O \rightleftharpoons B(OH)_4^- + H^+ \qquad (1)$$

The equilibrium constant (K) for equation 1 is 5.80×10^{-10} [3], but actual pH values differ considerably from those calculated on the basis of this constant, due to the formation of polyborate ions, as discussed below.

Boric oxide, B_2O_3, is a colorless, glassy solid that is manufactured by dehydration of boric acid (H_3BO_3) at high temperatures. It absorbs moisture from the air at room temperature, reacting with water to form the acid in the following highly exothermic reaction [3]:

$$B_2O_3\ \text{(glass)} + 3H_2O \longrightarrow 2B(OH)_3(s) \qquad \Delta H^0 = -75.94\ \text{kJ/mol} \tag{2}$$

Borates, the salts of boric acid, include both naturally occurring and synthetic compounds. Their nomenclature can be confusing, as the same compound may be called by different names depending on how its molecular formula is written. Thus, for clarity, it is important to use both the name and the formula in referring to a particular borate. Some examples of borates and their formulas are given in Table 10.9-1. One way of characterizing the borates and their aqueous solutions is the metal oxide/B_2O_3 mole ratio. For example, the Na_2O/B_2O_3 mole ratio for borax ($Na_2O \cdot 2B_2O_3 \cdot 10H_2O$) is 0.5, while for $Na_2O \cdot 5B_2O_3 \cdot 10H_2O$ it is 0.2.

Most borates are stable solids at room temperature. One of the most commercially important borates is borax (sodium diborate decahydrate), a naturally occurring mineral. It has a vapor pressure of 1.6 mm Hg at 10°C. If heated above 50°C, however, it decomposes to the pentahydrate ($Na_2O \cdot 2B_2O_3 \cdot 5H_2O$), which has a vapor pressure of 10 mm Hg at 19.8°C. Borax pentahydrate also occurs naturally, as a fine-grained deposit formed by dehydration of borax in the atmosphere [9].

Most borates are water-soluble. Table 10.9-2 lists the solubilities of some borates at 10°C and 25°C. Their solubilities generally increase with increasing temperature; for example, potassium pentaborate tetrahydrate ($K_2O \cdot 5B_2O_3 \cdot 8H_2O$ or $KB_5O_8 \cdot 4H_2O$) has a solubility of 1.56% at 0°C and 6.88% at 50°C (percent by weight, as the anhydrous salt) [9].

SPECIATION IN AQUEOUS SOLUTION

When boric acid or borates are dissolved in water, the monoborate ion $B(OH)_4^-$ (which may also be written $BO_2^- \cdot 2H_2O$ or simply BO_2^-) is formed. Although this ion is the predominant species in dilute aqueous solution, significant formation of polyborate ions occurs at concentrations greater than 0.1 *M* [9]. These polymeric ions are formed by the reaction of monoborate ions $B(OH)_4^-$ with free boric acid molecules $B(OH)_3$, as follows:

$$2B(OH)_3^0 + B(OH)_4^- \rightleftharpoons B_3O_3(OH)_4^- + 3H_2O \tag{3}$$

The equilibrium between monoborate and polyborates in solution is established rapidly [2].

As shown in Figure 10.9-1, the polyions account for an appreciable fraction of the total boron in the pH range between about 5 and 11. $B_3O_3(OH)_4^-$ is the most abundant of these, but the species $B_3O_3(OH)_5^{-2}$, $B_4O_5(OH)_4^{-2}$ and $B_5O_6(OH)_4^-$ are also significant [9]. In Figure 10.9-1, the fraction of the total boron present as a given ion at a

TABLE 10.9-1

Examples of Borates

Name	Formula	Equivalent Formula(s)
Disodium tetraborate decahydrate (Borax, Sodium diborate decahydrate, or Sodium tetraborate decahydrate)	$Na_2B_4O_7 \cdot 10H_2O$	$Na_2O \cdot 2B_2O_3 \cdot 10H_2O$
Sodium pentaborate pentahydrate	$NaB_5O_8 \cdot 5H_2O$	$Na_2O \cdot 2B_2O_3 \cdot 10H_2O$
Orthoboric acid (Boric acid)	H_3BO_3	$B(OH)_3$ or $HBO_2 \cdot H_2O$
Calcium metaborate	CaB_2O_4	$Ca(BO_2)_2$
Boric oxide	B_2O_3	—
Metaboric acid	HBO_2	—

Source: Doonan and Lower [3]

TABLE 10.9-2

Solubilities of Borates
(Expressed as % anhydrous salt by weight)

Compound	Solubility 10°C	Solubility 25°C
$Li_2O \cdot 2B_2O_3 \cdot 4H_2O$	2.55	2.90
$Li_2O \cdot B_2O_3 \cdot 16H_2O$	1.42	3.34
$Na_2O \cdot 5B_2O_3 \cdot 10H_2O$	8.10	12.20
$Na_2O \cdot 2B_2O_3 \cdot 10H_2O$	1.76	3.13
$Na_2O \cdot B_2O_3 \cdot 8H_2O$	17.0	21.7
$K_2O \cdot 5B_2O_3 \cdot 8H_2O$	2.11	3.28
$K_2O \cdot 2B_2O_3 \cdot 4H_2O$	9.02	13.6
$K_2O \cdot B_2O_3 \cdot 2\frac{1}{2}H_2O$	42.3	43.3
$Rb_2O \cdot 5B_2O_3 \cdot 8H_2O$	2.0	3.10
$Cs_2O \cdot 5B_2O_3 \cdot 8H_2O$	1.85	2.97
$Cs_2O \cdot B_2O_3 \cdot 7H_2O$	36.8[a]	
$(NH_4)_2O \cdot 2B_2O_3 \cdot 4H_2O$	5.26	9.00
$(NH_4)_2O \cdot 5B_2O_3 \cdot 8H_2O$	5.38	8.03

a. At 18°C

Source: Nies and Campbell [9], p. 100. *(Copyright 1964, John Wiley & Sons. Reprinted with permission.)*

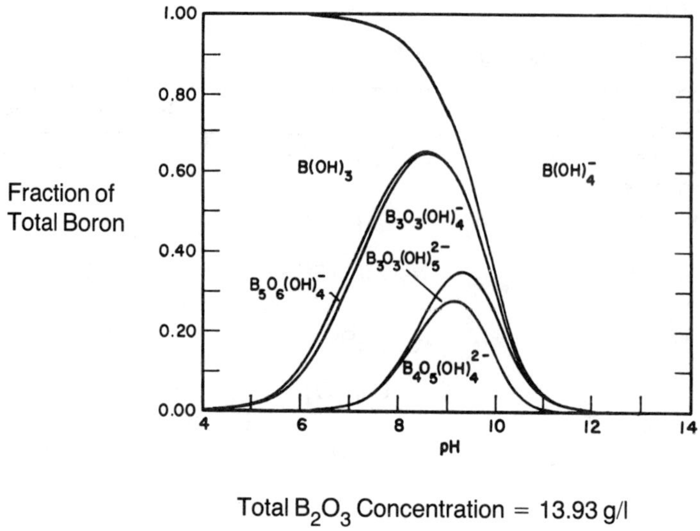

Source: Nies and Campbell [9]. *(Copyright 1964, John Wiley & Sons. Reprinted with permission.)*

FIGURE 10.9-1 Distribution of Boron Species in a Borate Solution at Various pH Values (see text for explanation)

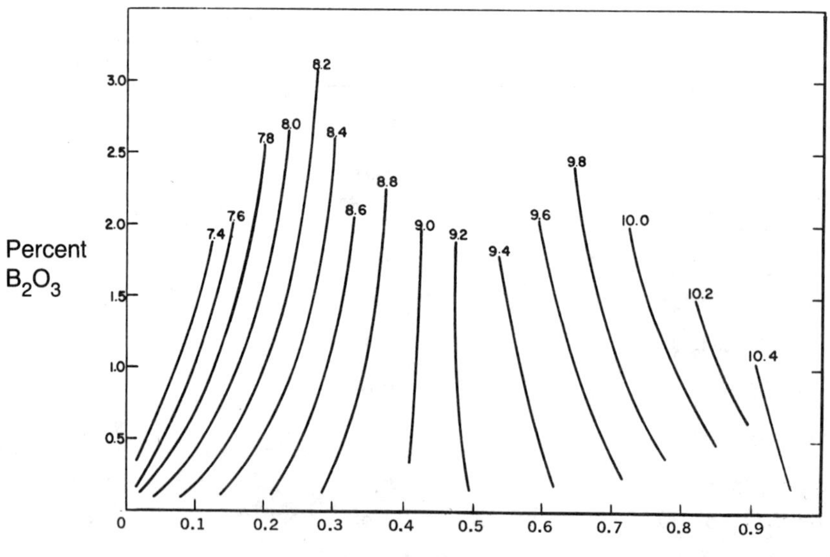

Source: Nies and Campbell [9]. *(Copyright 1964, John Wiley & Sons. Reprinted with permission.)*

FIGURE 10.9-2 pH Values in the System Na_2O – B_2O_3 – H_2O at 25°C

particular pH is represented by the portion of a vertical line falling within the corresponding range. For example, at pH = 9, approximately 28% of the boron is present as $B_4O_5(OH)_4^{-2}$, 5% as $B_3O_3(OH)_5^{-2}$, 30% as $B_3O_3(OH)_4^-$, 15% as $B(OH)_3$, and 22% as $B(OH)_4^-$.

As a result of polyion formation, the pH of a boric acid or borate solution cannot be predicted solely on the basis of the acid/base equilibrium constant of H_3BO_3. For pure boric acid solutions, the observed pH at high concentrations is lower than that calculated from the equilibrium constant; for example, while a 0.0904 *M* boric acid solution has a pH of 5.14, the same as the calculated value, a 0.753 *M* solution has a pH of 3.69, which is considerably below the calculated value of 4.54 [3].

Figure 10.9-2 shows the variation of pH with B_2O_3 concentration for borate solutions with various Na_2O/B_2O_3 ratios at 25°C. Borate solutions with a Na_2O/B_2O_3 ratio of about 0.4 have a pH value of 9.0 which is independent of concentration. However, solutions with a higher Na_2O/B_2O_3 ratio become less alkaline with dilution, while solutions with a lower Na_2O/B_2O_3 ratio become more alkaline with dilution. The H^+ concentration in a borate solution is equal to the ratio of boric acid concentration, $[B(OH)_3]$, to borate ion concentration, $[B(OH)_4^-]$, multiplied by the equilibrium constant for boric acid hydrolysis (eq. 1).

Thus, the variation in pH with concentration can be explained by the variation in the $[B(OH)_3]/[B(OH)_4^-]$ ratio. As a borate solution is diluted, the polyborate ions dissociate into boric acid molecules and monoborate ions. If the new equilibrium concentrations of these species result in an increase in the $[B(OH)_3]/[B(OH)_4^-]$ ratio, the pH of the solution will decrease, while if the $[B(OH)_3]/[B(OH)_4^-]$ ratio decreases, the pH will increase [9].

Another effect of polyion formation is to enhance the solubilities of boric acid and alkali borates in a solution containing both the acid and salt. Polymerization removes some of the boric acid molecules and borate ions from solution, allowing more solid borate or acid to dissolve [9].

Borate ions in solution may form complexes with a variety of metals. For example, they may combine with iron cations as follows:

$$Fe^{+3} + B(OH)_4^- \rightleftharpoons Fe[B(OH)_4]^{+2} \tag{4}$$

Table 10.9-3 gives stability constants for several metal-borate complexes. For further discussion of this process, see section 2.9.

MOBILITY IN SOILS

Because boron is an essential plant nutrient at low concentrations, but is toxic to plants at higher concentrations, the sorption and mobility of boron in soils has been widely researched. Rai *et al.* [10] have summarized available data on boron sorption. Since most known boron minerals are fairly soluble, sorption and desorption, rather

TABLE 10.9-3

Stability Constants of Metal-Borate Complexes[a]
(Log β for the equilibrium $M^{+x} + y\,[B(OH)_4^-] \rightleftharpoons (M[B(OH)_4]_y)^{+x-y}$)

Metal Ion	$M[B(OH)_4]$	$M[B(OH)_4]_2$	$M[B(OH)_4]_3$	$M[B(OH)_4]_4$
Na^+	1.87[b,c]	—	—	—
Fe^{+3}	8.5[b,d]	15.6[b,d]	20.6[b,d]	20.3[b,d]
Co^{+2}	—	—	—	10.03[e]
Ni^{+2}	—	—	8.44[e]	—
Cu^{+2}	7.13[b,e]	12.45[b,e]	15.17[b,e]	—
Ag^+	0.45[f]	—	—	—
Zn^{+2}	—	—	—	(11.8)[b,e]
Cd^{+2}	—	—	—	(10.6)[b,e]
Pb^{+2}	5.21[e]	—	11.17[e]	—

a. Values are for conditions of zero ionic strength and 25°C, unless otherwise noted.
b. Variable ionic strength.
c. 20°C
d. Temperature not given.
e. 22°C; influence of polyborates neglected.
f. I = 3.0

Source: Kotrlý and Šůcha [7], p. 111. *(Copyright 1985, John Wiley & Sons. Reprinted with permission.)*

than precipitation and dissolution, are expected to control the behavior of boron in soil. Boron sorption is a complex phenomenon, affected by a number of soil characteristics, including pH, aluminum and iron oxides, clay content, surface area, and salinity. Sorption is greatest at alkaline pH, perhaps because the borate ion $B(OH)_4^-$ is more strongly sorbed than $B(OH)_3$. Amorphous Al_2O_3 may be the most important soil component causing increased boron sorption. Other soil components that appear to enhance boron sorption include iron oxides, calcium, magnesium, and organic matter. Fixation by clay minerals or the formation and precipitation of aluminum and iron borates or hydroxy borates may be involved in irreversible boron sorption [10].

10.9.3 Boron Halides

Boron halides are a class of boron compounds in which boron and a halogen are combined. Some of the boron trihalides (BX_3, where X = F, Cl, Br) have important industrial applications: boron trichloride is used primarily in the manufacture of boron fibers, while the tribromide and trifluoride are used primarily as catalysts [8]. The boron halides with halogen-to-boron ratios of less than 3, such as B_2X_4(X = F, Cl, Br, I), are thermally unstable and therefore have almost no commercial application [8].

The boron trihalides are all very strong Lewis acids (i.e., electron acceptors), their Lewis acidity decreasing in the order $BI_3 \sim BBr_3 > BCl_3 > BF_3$. Much of their chemistry is dictated by their Lewis acidity. They form complexes with Lewis bases containing O, S, N, P or As donor atoms, and they react with compounds containing active hydrogen, such as water, alcohols, amines, and phosphines. At high temperatures they decompose, yielding toxic halogen-containing fumes [8].

Boron trifluoride, a gas at room temperature (b.p. −101°C), can react with water to form two hydrates, $BF_3 \cdot H_2O$ and $BF_3 \cdot 2H_2O$. These decompose above 20°C, however, releasing BF_3. When a small amount of BF_3 is combined with an excess of water, a solution of fluoroboric acid is produced:

$$4BF_3(g) + 6H_2O \longrightarrow 3H_3O^+ + 3BF_4^- + B(OH)_3^0 \quad (5)$$

The fluoroborate anion may then be hydrolyzed, yielding HF [2]:

$$BF_4^- + H_2O \rightleftharpoons [BF_3OH]^- + HF^0 \qquad K = 2.3 \times 10^{-3} \quad (6)$$

Boron trichloride is a liquid at room temperature under slight pressure (b.p. 12.5°C), and the tribromide is a liquid that boils at 90°C. In moist air both BCl_3 and BBr_3 fume. Hydrolysis of BCl_3 and BBr_3 goes to completion [2]:

$$BCl_3(g) + 3H_2O \longrightarrow B(OH)_3(s) + 3HCl(g) \quad (7)$$

$$BBr_3(l) + 3H_2O \longrightarrow B(OH)_3(s) + 3HBr(g) \quad (8)$$

Boron triiodide is a white solid (m.p. 43°C) that is explosively hydrolyzed by water [2]. It has no major commercial application [8].

10.9.4 Boron Hydrides (Boranes) and Sodium Tetrahydroborate

Boron hydrides, or boranes, are generally volatile, binary compounds of boron and hydrogen, containing boron atoms linked to each other. They may be grouped into a B_nH_{n+4} series and a B_nH_{n+6} series; the former are generally more stable than the latter [6], as exemplified by B_5H_9 and B_5H_{11} in Table 10.9-4. Alternatively, boranes may be classified as those that contain BH_2 groups and those that do not, the latter being both thermodynamically and chemically more stable [1]. Boron hydrides are used mainly as reducing agents in laboratories; they do not occur in nature and have little commercial production [11].

Reactions and transformations of boranes that are of environmental interest are volatilization, reaction with air, thermal decomposition, and reaction with water (hydrolysis). The tendencies of several boranes to undergo these reactions are summarized in Table 10.9-4. Thermal decomposition (pyrolysis) of boranes yields larger boranes, which decompose at progressively higher temperatures, ultimately yielding elemental boron and hydrogen at about 800°C. Reaction with oxygen (combustion)

TABLE 10.9-4

Environmental Properties of Some Boranes

Formula	Name	Melting Point (°C)	Boiling Point (°C)	Vapor Pressure at 0°C	Reaction with Air at 25°C	Thermal Stability	Reaction with H_2O
B_2H_6	Diborane(6)	−164.85	−92.6	(gas at 0°C)	Spontaneously flammable	Fairly stable at 25°C	Instant hydrolysis
B_4H_{10}	Tetraborane(10)	−120	18	—	Not spontaneously flammable if pure	Decomposes fairly rapidly at 25°C	Hydrolysis in 24 hrs.
B_5H_9	Pentaborane(9)	−46.6	48	65 mm Hg[a]	Spontaneously flammable	Stable at 25°C; slow decomp. at 150°C	Hydrolyzed only on heating
B_5H_{11}	Pentaborane(11)	−123	63	—	Spontaneously flammable	Decomposes very rapidly at 25°C	Rapid hydrolysis
B_6H_{10}	Hexaborane(10)	−62.3	108	7.2 mm Hg[b]	Stable	Slow decomposition at 25°C	Hydrolyzed only on heating
$B_{10}H_{14}$	Decaborane(14)	99.7	213	—	Very stable	Stable at 150°	Slow hydrolysis

a. Adams [1]
b. Kleinberg, Argersinger, and Griswold [6]

Source: Cotton and Wilkinson [2], p. 238, except as noted. *(Copyright 1972, John Wiley & Sons. Reprinted with permission.)*

generally yields boric oxides and water, as in the following reactions, both of which occur spontaneously when borane is exposed to air:

$$B_2H_6(g) + 3O_2(g) \longrightarrow B_2O_3(s) + 3H_2O \quad (9)$$

$$2B_5H_9(l) + 12O_2(g) \longrightarrow 5B_2O_3(s) + 9H_2O \quad (10)$$

The products of hydrolysis of boron hydrides are boric acid and hydrogen, according to the following reactions [1]:

$$B_2H_6(g) + 6H_2O \longrightarrow 2B(OH)_3(s) + 6H_2(g) \quad (11)$$

$$B_4H_{10}(g) + 12H_2O \longrightarrow 4B(OH)_3(s) + 11H_2(g) \quad (12)$$

$$B_5H_{11}(l) + 3H_2O \longrightarrow B_4H_{10}(g) + B(OH)_3(s) + 2H_2(g) \quad (13)$$

Rates of hydrolysis of boranes range from very slow to instantaneous, as summarized in Table 10.9-4.

A related compound of commercial importance is sodium tetrahydroborate, $NaBH_4$, commonly referred to as sodium borohydride. It is used largely as a specific reducing agent in the synthesis of pharmaceuticals, food additives, and other chemicals, as well as for textile dye reductions, the bleaching of wood pulp and clay [4], and the preparation of diborane [5]. Sodium tetrahydroborate is a crystalline solid, stable in dry air at environmental temperatures but slowly hydrolyzed in moist air. It is soluble in water (55 g/100 g H_2O at 25°C), forming a dihydrate, $NaBH_4 \cdot 2H_2O$, in aqueous solution. Aqueous $NaBH_4$ reacts slowly to form sodium metaborate ($NaBO_2$) and hydrogen. This hydrolysis occurs very slowly at pH > 13, proceeding more rapidly at lower pH or higher temperatures. Thus, the reaction only continues until the reaction products render the solution too alkaline. The rate of tetrahydroborate hydrolysis is also enhanced by certain transition metal ions in solution [5].

10.9.5 Literature Cited

1. Adams, R.M., "The Boranes or Boron Hydrides," in *Boron, Metallo-Boron Compounds and Boranes,* Chap. 7, R.M. Adams (ed.), Interscience Publishers, div. of John Wiley & Sons, New York (1964).
2. Cotton, F.A. and G. Wilkinson, *Advanced Inorganic Chemistry, A Comprehensive Text,* 3rd ed., Interscience Publishers, div. of John Wiley & Sons, New York (1972).
3. Doonan, D.J. and L.D. Lower, "Boron Oxides, Boric Acid, and Borates," in *Kirk-Othmer Encyclopedia of Chemical Technology,* 3rd ed., Vol. 4, John Wiley & Sons, New York (1982).
4. Dunks, G.B., "Boron Hydrides, Heteroboranes and their Metalloderivatives, Commercial Aspects," in *Kirk-Othmer Encyclopedia of Chemical Technology,* 3rd ed., Vol. 4, John Wiley & Sons, New York (1982).

5. James, B.D. and M.G.H. Wallbridge, "Metal Tetrahydroborates," *Prog. Inorg. Chem.*, **11**, 99-231 (1970).

6. Kleinberg, J., W.J. Argersinger, Jr. and E. Griswold, *Inorganic Chemistry,* D.C. Heath, Boston (1960).

7. Kotrlý, S. and S. Šůcha, *Handbook of Chemical Equilibria in Analytical Chemistry,* Halsted Press, div. of John Wiley & Sons, New York (1985).

8. Lower, L.D., "Boron Halides," in *Kirk-Othmer Encyclopedia of Chemical Technology,* 3rd ed., Vol. 4, John Wiley & Sons, New York (1982).

9. Nies, N.P. and G.W. Campbell, "Inorganic Boron-Oxygen Chemistry," in *Boron, Metallo-Boron Compounds and Boranes,* Chap. 3, R.M. Adams (ed.), Interscience Publishers, div. of John Wiley & Sons, New York (1964).

10. Rai, D., J.M. Zachara, A.P. Schwab, R.L. Schmidt, D.C. Girvin and J.E. Rogers, *Chemical Attenuation Rates, Coefficients, and Constants in Leachate Migration. Vol. 1: A Critical Review,* Report EA3356 to EPRI by Pacific Northwest Laboratories (Battelle Institute), Richland, Washington (1984).

11. Sprague, R.W., *The Ecological Significance of Boron,* U.S. Borax and Chemical Corporation, Anaheim, Calif. (1972).

10.10 SILICON-CONTAINING COMPOUNDS

10.10.1 Introduction

Silicon forms compounds with a wide range of commercial applications and widely different properties. This section addresses four classes of silicon-containing compounds: the soluble silicates (e.g., Na_4SiO_4, Na_2SiO_3), naturally occurring silicate minerals (e.g., talc, asbestos), silicones (synthetic organosilicon polymers), and silanes (e.g., SiH_2Cl_2, $SiHCl_3$). Silicon compounds of environmental interest discussed elsewhere in this report include silicon tetrahydride (silane) (§10.14, Hydrides), silicon carbide (§ 10.12, Carbides) and silicon dioxide (§ 6.6, Silicon).

10.10.2 Silica and Silicates

The term silica refers to the compound silicon dioxide, SiO_2. Pure silicon dioxide is a solid (m.p. ≈ 1723°C) that can exist in several crystalline forms, of which the mineral quartz is the stable form at environmental temperatures. Commercial silica products, however, include many amorphous, hydrated or hydroxylated forms of SiO_2, such as colloidal silica and silica gel [5]. The solubility and speciation of silica in water are discussed in § 6.6.

Silicates, which include both naturally occurring minerals and synthetic substances, are compounds containing silicate anions in combination with metal cations and water of hydration. The basic structural unit is the orthosilicate ion, a tetrahedron of a silicon atom surrounded by four oxygen atoms. These units may exist as discrete monomers (SiO_4^{-4}), dimers ($Si_2O_7^{-6}$), cyclic structures, infinite chains, sheets, or networks [5].

10.10.3 Soluble Silicates

Soluble silicates are synthetic inorganic compounds that are produced commercially for a variety of uses such as in soaps and detergents, for water treatment, as adhesives and binders, and for the manufacture of other silica-based products. They have the general formula $M_2O \cdot mSiO_2 \cdot nH_2O$, where M is an alkali metal. Sodium silicates (M = Na) are the most common, but potassium and lithium silicates are also produced. The value of m, the number of moles SiO_2 per mole M_2O, generally ranges from 0.5 to 4.0 in commercial materials [6]. The value of n, the number of moles of hydrate water per mole M_2O, may vary from zero to nine [15].

The soluble silicates are usually manufactured as glassy solids. They dissolve readily in water to form viscous, alkaline solutions. The rate of dissolution depends on many factors, including the ratio of SiO_2 to M_2O, the aqueous silicate concentration, and particle size. Solubility of the alkali silicates decreases with decreasing size of the alkali metal cation, i.e., in the order $K^+ > Na^+ > Li^+$ [6].

Solutions of soluble silicates are strongly alkaline, and they are excellent buffers. The pH value of a silicate solution is a function of both concentration and composition. With increasing ratio of SiO_2 to M_2O, solution pH decreases and buffering action increases. As shown in Figure 10.10-1, sodium silicate solutions with a SiO_2:Na_2O ratio of about 3 or greater reach a maximum pH at about 5% Na_2O, exhibiting lower pH values at both lower and higher concentrations, but solutions with lower SiO_2:Na_2O ratios show less buffering action [15]. For ratios between 2.0 and 4.0, the maximum solution pH value varies approximately linearly with the SiO_2:Na_2O ratio m, according to the equation pH = 13.4 − 0.69m [6]. The alkalinity and buffering ability of silicates leads to their use in soaps and detergents [6].

In solution, silicates form a complex mixture of more and less polymerized forms of silicate anions in a dynamic equilibrium. Equilibrium is attained rapidly in solutions with a pH greater than about 10, but slowly when the pH is less than about 2. The complex polymerization/depolymerization process may be represented empirically as follows:

$$\equiv SiOH + \equiv SiO^- \rightleftharpoons \equiv SiOSi \equiv + OH^-$$

The forward (condensation or polymerization) reaction is influenced by pH, proceeding most readily at a pH value equal to the pK_a of the participating silanol ($\equiv$SiOH) group. These pK_a values range from about 9.9 for monosilicic acid (H_4SiO_4) to 6.5 for high-molecular-weight polymers. The above reaction scheme is less valid at very low pH, when H^+ ions catalyze the forward reaction, and at pH values greater than about 10, when the rate of the reverse (depolymerization) reaction becomes

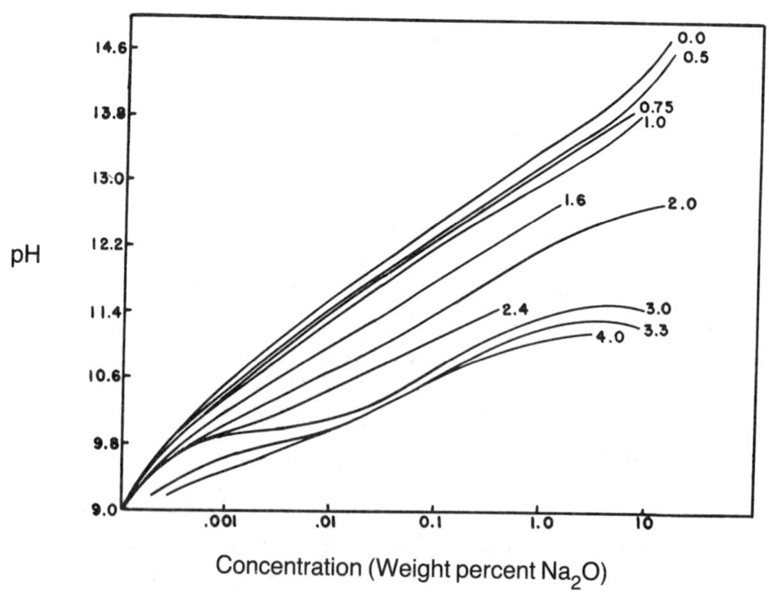

Numbers next to individual lines are SiO_2:Na_2O ratios by weight.

Source: Vail [15]. *(Copyright 1952, Reinhold Publishing Co. All rights reserved.)*

FIGURE 10.10-1 pH of Sodium Silicate Solutions as a Function of Concentration, for Various SiO_2:Na_2O Ratios

significant. Smaller particles tend to aggregate to form larger particles, and at high SiO_2 concentration (about 1 wt %), they ultimately form a continuous network, or gel, throughout the medium [6].

Highly polymerized soluble silicates interact with multivalent metal ions in solution via an ion-exchange (surface complexation) process. The interaction of highly polymerized silicates with metal ions appears to be similar to that of silica gels, for which a correlation has been observed between the stability of surface complexes and the pK_a of participating metal ions [6]. Highly polymerized silicates interact with metal ions to a greater extent than less polymerized forms, possibly due to the greater acidity of the oxygen atoms in larger polymers [6]. In addition, soluble silicates are strongly adsorbed on metal oxides. The ability of silicates to adsorb on oxide surfaces and to complex metal ions leads to their use in water treatment for the prevention of corrosion of metal surfaces in contact with water [6].

10.10.4 Naturally Occurring Silicate Minerals

Two naturally occurring silicate minerals, talc and asbestos, are discussed here. Both of these minerals have important commercial uses and are of environmental concern as airborne particulates.

TALC

The mineral talc is a hydrous magnesium silicate with the chemical formula $Mg_3SiO_{10}(OH)_2$. However, the term talc as used commercially can refer to a wide range of mineral mixtures containing mostly magnesium silicates, of which the mineral talc is usually a main constituent [9].

The major uses for talcs are in papermaking, ceramics, and protective coatings. Lower quality talcs (which contain a variety of mineral impurities) are used as insecticide carriers, rubber-dusting and textile-filling materials, and additives in asphalt roofing compounds; high-purity talcs are used in smaller quantities for cosmetics and pharmaceuticals [9].

Pure talc mineral is characterized by softness, hydrophobic surface properties, and slippery feel. It is thermally stable and inert in most chemical reagents [9].

ASBESTOS

Asbestos refers to a variety of naturally occurring hydrated silicates that separate into mineral fibers upon mechanical processing. The three asbestos minerals of major commercial importance are chrysotile ($3MgO \cdot 2SiO_2 \cdot 2H_2O$), amosite ($11FeO \cdot 3MgO \cdot 16SiO_2 \cdot 2H_2O$) and crocidolite ($Na_2O \cdot Fe_2O_3 \cdot 3FeO \cdot 8SiO_2 \cdot H_2O$). Since their chemical compositions vary, these formulas are approximate only. Chrysotile represents the variety of asbestos known as serpentine, while amosite and crocidolite are examples of a group of asbestos minerals called the amphiboles [13].

The largest use of asbestos is in asbestos-cement, which is formed into items such as pipes, ducts, and flat and corrugated sheets. Asbestos is also used for fire-resistant textiles, friction materials (such as brake linings), underlayment and roofing papers, and floor tiles [13].

The crystal structure of chrysotile consists of a silica sheet joined to a magnesium hydroxide-oxide layer. Strains and mismatches between the layers cause the sheets to curve, forming hollow cylindrical fibers. The amphiboles, on the other hand, have structures based on silicate chains, with bands of cations sandwiched between pairs of silicate chains. Fibers are formed by preferential cleavage in the direction parallel to these chains, between the silica-cation "sandwiches". Due to these structural differences, chrysotile fibers tend to be finer, softer, and more flexible than the amphibole fibers. In commercial grades of both types of asbestos, what looks like a single fiber is actually a bundle of many individual fibrils and can be subdivided further by mechanical processing. Chrysotile asbestos can be divided into finer fibrils than the amphiboles, attaining a greater specific surface area [13].

Asbestos particles and fibers vary widely in size. To provide comparable material for use by the many different researchers in asbestos toxicology, the International Union Against Cancer (UICC) prepared standard reference samples of the common types of asbestos. The physical and chemical characteristics of these samples have been determined; some characteristics relating to typical particle size are given in Table 10.10-1. Although these values vary with the method used to determine them, they represent the approximate characteristics of the asbestos minerals typically used in industry. Asbestos fibers in the atmosphere are likely to have a different size distribution, being generally smaller than those found in the workplace [16].

Asbestos minerals form flexible fibers of high tensile strength. They do not burn or rot, and they are resistant to heat and to alkaline solutions. Chrysotile asbestos is readily attacked by acid, which will dissolve its magnesium component, leaving a fragile silica framework. The amphiboles (e.g., amosite and crocidolite), however, are relatively acid-resistant [13]. The relatively rapid leaching of magnesium from chrysotile may explain why chrysotile fibers appear to be cleared from the lungs more quickly than amphibole fibers in animal inhalation experiments [3]. Asbestos is only slightly soluble in water; chrysotile specimens have solubility products ranging from 1.0×10^{-11} to 3×10^{-12}. Dispersions of chrysotile fibers in carbon dioxide-free distilled water are alkaline, reaching a pH of up to 10.33 [13].

The differences between chrysotile and the amphiboles have been emphasized in recent research on the toxic effects of asbestos. Some evidence indicates that chrysotile is less active than other fiber types in causing peritoneal mesothelioma (a type of cancer). However, the length and diameter of asbestos fibers may be at least as important as their chemical composition in determining their biological activity. The relative health risk associated with the different asbestos varieties remains a subject of debate [16].

Because both talc and asbestos are chemically and thermally stable and nearly insoluble in water, their major route of environmental transport is the dispersion of

TABLE 10.10-1

Physical Properties of UICC Standard Reference Samples of Asbestos

	Fiber Length Distribution[a] [% by Number in Given Size Range (μm)]										
Asbestos Type	**.2-.5**	**.5-1**	**1-2**	**2-5**	**5-10**	**10-25**	**25-50**	**50-100**	**100-200**	**Specific Surface Area[b] (m^2/g)**	**Respirable Fraction[b] (% by wt.)**
Amosite	23.0	31.1	25.5	14.7	4.40	1.08	0.16	0.03	0.02	2.5-6.0	67-83
Crocidolite	28.4	35.8	22.5	10.3	2.33	0.60	0.07	0.00	0.01	2.4-8.8	70-87
Rhodesian Chrysotile	20.7	34.9	23.1	15.2	2.83	2.49	0.62	0.15	0.00	5.4-22.8	67-86
Canadian Chrysotile	30.6	33.4	19.8	13.2	1.76	0.93	0.24	0.07	0.00	4.4-27.5	67-83

a. **Source**: Rendall [12]. *(Copyright 1970, Oxford Press. Reprinted with permission.)*

b. Ranges of values determined by various methods.
Sources: Rendall [12], Timbrell [14].

dust in the air. Levels of airborne talc and asbestos dusts in the workplace are regulated by OSHA (see Appendix D), and dust suppression or dust collection measures are required for talc mining and processing operations [9]. Environmental exposure to asbestos may also result from asbestos fibers suspended in water and food [4]; water quality criteria for asbestos have been established (see Appendix D).

10.10.5 Silicones

Silicones, or polyorganosiloxanes, are a group of synthetic organosilicon polymers that consist of a backbone of alternating silicon and oxygen atoms with organic groups attached to the silicon atoms. In commercially produced silicones, these organic substituents are mostly methyl groups. Other groups, such as longer carbon chain alkyl or fluoroalkyl, phenyl, chlorophenyl, vinyl, or hydrogen, can be substituted for the methyl groups to produce silicones with specific properties. Silicones may have linear (straight-chain), branched-chain, cyclic, or cross-linked structures. Some examples are shown below (R = OH or CH_3).

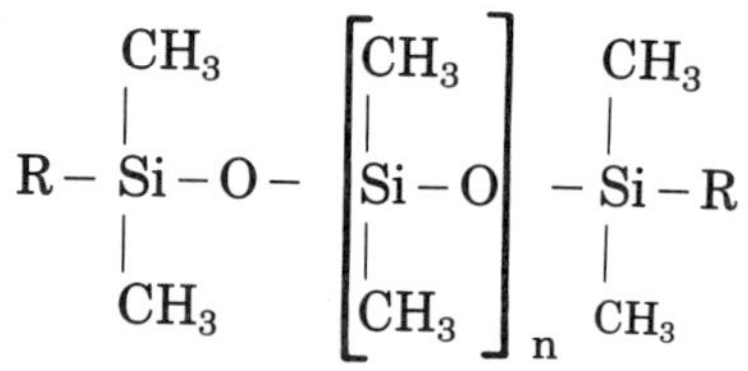

Polydimethylsiloxane

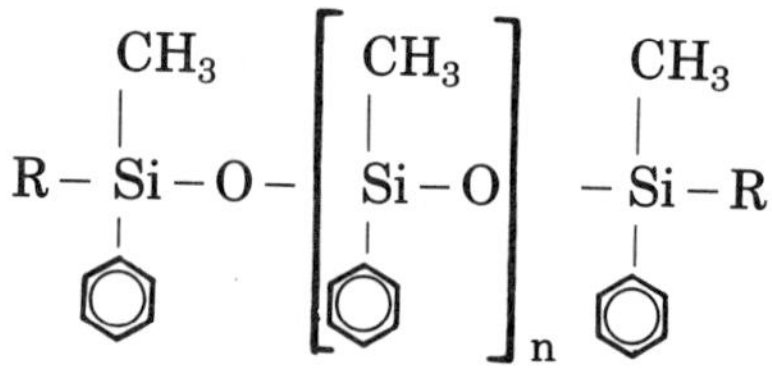

Polymethylphenylsiloxane

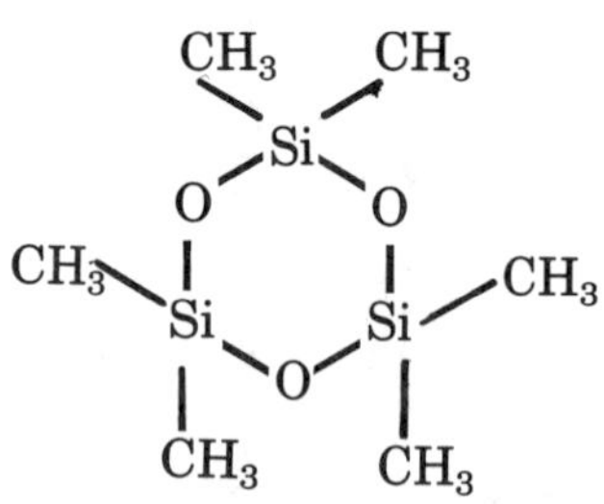

Hexamethylcyclotrisiloxane

Commercial silicone products include fluids (oils), resins, and rubbers (elastomers), as well as derivatives such as emulsions, greases, and coatings. These products have diverse applications, as cosmetic additives, medical implants, adhesives, sealants, molding materials, water repellents for textiles and paper, electrical insulating materials, polishes, lubricants, and antifoaming agents for petroleum oils [7].

Silicones have several distinctive characteristics responsible for their wide range of applications. They exhibit high thermal stability, chemical inertness, weather resistance, water repellency, and only slight variation of physical properties (such as viscosity or flexibility) with temperature. In addition, silicone fluids have low surface tension, enabling them to form very thin films; they are also surface-active and thus serve as foaming and antifoaming agents. Silicone resins and rubbers show outstanding release properties, applicable to molding operations [7,10].

Because silicones are stable in air to temperatures of at least 150°C and are water-repellent and weather-resistant, they undergo very few changes under environmental conditions. Silicone fluids can undergo some volatilization at environmental temperatures. Low-molecular-weight fluids have vapor pressures between 10^{-4} and 10^{-9} mm Hg at 20°C, while high-molecular-weight fluids have even lower vapor pressures [10].

Silicone resins may be attacked by boiling water, and most types of silicones can be destroyed by steam at elevated temperatures [10]. The degradation of silicone fluids also can occur as a result of catalysis under typical environmental conditions. Experiments with polydimethylsiloxane (PDMS) fluids have shown that soils catalyze their hydrolysis to low-molecular-weight linear and cyclic oligomers. The rate of hydrolysis depends on the clay content and moisture content of the soil, and on temperature. Since the catalytically active component of soils is clay, the higher the clay content, the more rapid the degradation. Degradation is considerably faster in dry soil than in moist soil and increases at higher temperatures. Half-lives of PDMS on soil can range from several minutes to weeks or longer, depending on these parameters [2].

Different types of clays catalyze rearrangement to different products. Some of the hydrolysis products are water-soluble and thus may enter into groundwater; other products are volatile and may therefore be released to air. Although the greatest degradation is expected to occur in the drier surface layer of soil, PDMS has been shown to migrate both upward and downward in soil, so silicone fluids several inches below the soil surface also may eventually be degraded [2].

Most silicones are attacked by strong acids and alkalis, especially at high temperatures, although fully hardened silicone resins are resistant to aqueous acids [10].

Silicones are also resistant to biological degradation. Although microorganisms may grow on the surface of silicone rubbers and resins, they have not been observed to degrade them. Similarly, silicone fluids are not biodegradable to any measurable extent [8].

Due to their general physical and chemical stability, silicones may be expected to accumulate in the environment. They have been detected in sediments and in the surface microlayer of water bodies at several locations. Because of their seemingly ubiquitous occurrence in these areas, they have been suggested for use as tracers for anthropogenic additions to the environment [11].

10.10.6 Silanes

Silanes, or silicon hydrides, are compounds containing a silicon-hydrogen bond. The simplest silicon hydride is silicon tetrahydride, or silane, SiH_4; its properties are described in § 10.14. the inorganic silanes also include derivatives of silane in which one or more of the hydrogen atoms is replaced with a halogen. Of these, the commercially important compounds are dichlorosilane (SiH_2Cl_2), a specialty chemical used in electronics applications, and trichlorosilane ($SiHCl_3$), which is used for the production of electronic-grade silicon and of other silane derivatives [1].

The inorganic silanes are all liquids or gases at room temperature. (For example, dichlorosilane is a gas with a boiling point of 8.2°C, and trichlorosilane is a liquid with a boiling point of 31.9°C.) They are all readily oxidized, and they are hydrolyzed in basic solution. The hydrolysis of chlorosilanes, yielding hydrogen chloride, occurs spontaneously in moist air. Additional hazards are posed by the flammability of trichlorosilane and the hydrolysis of dichlorosilane to a polymeric material that may ignite spontaneously in air [1].

10.10.7 Literature Cited

1. Arkles, B. and W.R. Peterson, Jr., "Silanes," in *Kirk-Othmer Encyclopedia of Chemical Technology*, 3rd ed., Vol. 22, John Wiley & Sons, New York (1982).
2. Buch, R.R. and D.N. Ingebrigtson, "Rearrangement of Poly(dimethylsiloxane) Fluids on Soil," *Environ. Sci. Technol.*, **13**, 676-79 (1979).
3. Chronic Hazard Advisory Panel, "Report to the U.S. Consumer Product Safety Commission by the Chronic Hazard Advisory Panel on Asbestos," U.S. Consumer Product Safety Commission, Directorate for Health Sciences, Washington, D.C. (July, 1983).
4. Cloag, D., "Asbestos Fibres and the Environment," *British Medical Journal*, **282**, 623-26 (1981).
5. Coyle, T.D., "Silica," in *Kirk-Othmer Encyclopedia of Chemical Technology*, 3rd ed., Vol. 20, John Wiley & Sons, New York (1982).
6. Falcone, J.S. "Synthetic Inorganic Silicates," in *Kirk-Othmer Encyclopedia of Chemical Technology*, 3rd ed., Vol. 20, John Wiley & Sons, New York (1982).
7. Hardman, B.B. and A. Torkelson, "Silicones," in *Kirk-Othmer Encyclopedia of Chemical Technology*, 3rd ed., Vol. 20, John Wiley & Sons, New York (1982).
8. Howard P.H., P.R. Durkin and A. Hanchett, "Assessment of Liquid Siloxanes (Silicones)," Report EPA-560/2-75-004 (September, 1975).

9. Mulryan, H.J., "Talc," in *Kirk-Othmer Encyclopedia of Chemical Technology*, 3rd ed., Vol. 22, John Wiley & Sons, New York (1982).

10. Noll, W., *Chemistry and Technology of Silicones*, Academic Press, New York (1968).

11. Pellenbarg, R., "Environmental Poly(organosiloxanes) (Silicones)," *Environ. Sci. Technol.*, **13**, 565-69 (1979).

12. Rendall, R.E.G., "The Data Sheets on the Chemical and Physical Properties of the U.I.C.C. Standard Reference Samples," *Proc. Int. Pneumoconiosis Conf.*, pp. 23-27, Oxford Press, Johannesburg, South Africa (1970).

13. Streib, W.C., "Asbestos," in *Kirk-Othmer Encyclopedia of Chemical Technology*, 3rd ed., Vol. 3, John Wiley & Sons, New York (1982).

14. Timbrell, V., "Characteristics of the International Union Against Cancer Standard Reference Samples of Asbestos," in *Proc. Int. Pneumoconiosis Conf.*, pp. 28-36, Oxford Press, Johannesburg, South Africa (1970).

15. Vail, J.G., *Chemistry*, Vol. 1 of *Soluble Silicates: Their Properties and Uses*, Reinhold Publishing Co., New York (1952).

16. Zurer, P.S., "Asbestos — The Fiber That's Panicking America," *Chem. Eng. News*, **63**, 28-41 (March 4, 1985).

10.11 CARBONYLS

10.11.1 Introduction

Carbonyls are compounds containing the C=O group. This section addresses two distinct groups of carbonyl compounds: the carbonyl halides and the metal carbonyls, in which the C=O group is bonded to a halogen or to a metal, respectively. The major reaction of environmental importance for carbonyl halides is hydrolysis in water, and the major environmental route for metal carbonyls is volatilization.

10.11.2 Carbonyl Halides

The carbonyl halides COX_2 (X = F, Cl, Br, I), also known as carbon oxohalides, are liquids or gases at environmental temperatures; their melting and boiling points are given in Table 10.11-1. Both carbonyl fluoride and carbonyl chloride are used as laboratory agents for organic synthesis. Carbonyl fluoride, COF_2, is a fluorinating agent, used in the preparation of organic fluoride compounds [5]. Carbonyl chloride, or phosgene, $COCl_2$, is used primarily in the preparation of isocyanates, which are employed in the manufacture of polyurethanes. It is also used in the synthesis of polycarbonates and of intermediates for the manufacture of pharmaceuticals and pesticides. Nearly all phosgene manufactured is used within the plant where it is produced, not sold or shipped elsewhere for use [4].

TABLE 10.11-1

Properties of Carbonyl Halides

Compound	Melting Point (°C)	Boiling Point (°C)
COF_2	−114	−83.1
COFCl	−138	−42
COFBr	−120	−20.6
COFI	−90	23.4
$COCl_2$	−127.8	7.56
$COBr_2$	—	64.5

Source: Holliday, *et al.* [5]. *(Copyright 1973, Pergamon Press. Reprinted with permission.)*

Phosgene is a highly toxic, colorless gas at ordinary temperature and pressure. It reacts with metal oxides and sulfides at elevated temperatures, yielding chlorides [4]. It acts as a Lewis base, forming adducts with $SnCl_4$, $AlCl_3$, $SbCl_5$ and $SOCl_2$ [5]. Phosgene reacts vigorously with ammonia in solution, yielding urea and other organic nitrogen compounds. It also reacts with a variety of organic compounds [4].

The carbonyl halides all are hydrolyzed by water, according to the reaction [5]:

$$COX_2 + H_2O \rightarrow CO_2 + 2HX \qquad (1)$$

10.11.3 Metal Carbonyls

Carbon monoxide, acting as a ligand, forms many complexes with transition metals. At least one carbonyl derivative is known for every transition metal. Nickel carbonyl is formed by reaction of finely divided nickel with carbon monoxide at room temperature. Iron, however, reacts with carbon monoxide only at elevated temperatures and pressures, and all the other metal carbonyls are prepared from metal compounds under reducing conditions [3]. The metal carbonyls are used for the preparation of high-purity metals, for catalysis, and for organic synthesis [6].

Both mononuclear metal carbonyls $M(CO)_x$ and polynuclear metal carbonyls $M_x(CO)_y$ are known. They are mostly volatile solids and sublime easily, although a few are liquids at room temperature [6]. Melting and boiling points of some metal carbonyls are listed in Table 10.11-2, and vapor pressures are given in Table 10.11-3.

TABLE 10.11-2

Melting and Boiling Points of Metal Carbonyls

Carbonyl Formula	Melting Point (°C)[a]
$V(CO)_6$	50 dec
$Cr(CO)_6$	149-155
$Mo(CO)_6$	150-151 dec
$W(CO)_6$	169-170
$Mn_2(CO)_{10}$	151-155
$Tc_2(CO)_{10}$	159-160
$Re_2(CO)_{10}$	177
$Fe(CO)_5$	−20 (bp 103)
$Fe_2(CO)_9$	100 dec
$Fe_3(CO)_{12}$	140 dec
$Ru_3(CO)_{12}$	150 dec
$Os_3(CO)_{12}$	224
$Co_2(CO)_8$	50-51
$Co_4(CO)_{12}$	60 dec
$Rh_4(CO)_{12}$	150 dec[b]
$Rh_6(CO)_{16}$	220 dec
$Ir_4(CO)_{12}$	210 dec
$Ni(CO)_4$	−25 (bp 43)

a. dec = decomposes; bp = boiling point.

b. **Source:** Calderazzo, Erocli, and Natta [2], p. 39

Source: Wagner [6], except as noted. *(Copyright 1982, John Wiley & Sons. Reprinted with permission.)*

TABLE 10.11-3

Vapor Pressures of Some Metal Carbonyls

	Vapor Pressure (mm Hg)			
Compound	**15°C**	**20°C**	**30°C**	**40°C**
$Ni(CO)_4$	262.7	322.9	478.9	692.0
$Fe(CO)_5$	15.7	20.8	35.4	58.2
$Ru(CO)_5$		~50		
$Os(CO)_5$		~4.5		
$Co_2(CO)_8$	0.07			
$Cr(CO)_6$			0.28[b]	0.69[b]
$Mo(CO)_6$			0.27[b]	0.63[b]
$W(CO)_6$			0.06[b]	0.14[b]
$Mn_2(CO)_{10}$ } a				
$Re_2(CO)_{10}$ }				

a. $Mn_2(CO)_{10}$ has a vapor pressure of 5.1 mm Hg at 110°C
$Re_2(CO)_{10}$ has a vapor pressure of 0.36 mm Hg at 100°C

b. Extrapolated

Source: Calderazzo, Ercoli, and Natta [2]. *(Copyright 1968, John Wiley & Sons. Reprinted with permission.)*

All the metal carbonyls are hydrophobic [3] and do not react with water at neutral pH. They dissolve in aqueous alkali, however, yielding solutions containing the carbonylate anions. For example, $Fe(CO)_5$ dissolves to form $HFe(CO)_4^-$, which upon acidification produces $H_2Fe(CO)_4$(g). In addition to being produced by reaction with aqueous or alcoholic solutions of alkali hydroxides, carbonylate anions may be obtained by reaction with amines, sulfoxides or other Lewis bases [3].

Most other reactions of the metal carbonyls are substitution reactions in which CO is replaced by other ligands. These reactions generally occur at high temperatures or are photochemically induced [3]; they are not likely to occur under environmental conditions.

Nickel carbonyl, one of the more common metal carbonyls, is extremely toxic in the gaseous state. At 20°C it is a colorless liquid with a vapor pressure of about 330 mm Hg. It reacts slowly with HCl and H_2SO_4 and vigorously with HNO_3 and the halogens, yielding CO and nickel salts [1]. An additional hazard posed by nickel carbonyl is that it explodes when heated to about 60°C [6].

10.11.4 Literature Cited

1. Antonsen, D.H., "Nickel Compounds," in *Kirk-Othmer Encyclopedia of Chemical Technology*, 3rd ed., Vol. 15, John Wiley & Sons, New York (1982).

2. Calderazzo, F., R. Erocli and G. Natta, "Metal Carbonyls: Preparation, Structure, and Properties," in I. Wender and P. Pino (eds.), *Organic Synthesis via Metal Carbonyls*, Vol. 1, Interscience Publishers, div. of John Wiley & Sons, New York (1968).

3. Cotton, F.A. and G. Wilkinson, *Advanced Inorganic Chemistry: A Comprehensive Text*, 3rd ed., Interscience Publishers, div. of John Wiley & Sons, New York (1972).

4. Hardy, E.E., "Phosgene," in *Kirk-Othmer Encyclopedia of Chemical Technology*, 3rd ed., Vol. 17, John Wiley & Sons, New York (1982).

5. Holliday, A.K., G. Hughes and S.M. Walker, "Carbon," Vol. 1 of *Comprehensive Inorganic Chemistry*, A.F. Trotman-Dickenson (exec. ed.), Pergamon Press, Oxford, England (1973).

6. Wagner, F.S., "Carbonyls," in *Kirk-Othmer Encyclopedia of Chemical Technology*, 3rd ed., Vol. 4, John Wiley & Sons, New York (1982).

10.11.5 Other Sources of Information

Chatt, J., P.J. Pauson and L.M. Venanzi, "Metal Carbonyls and Related Compounds," in H. Zeiss (ed.), *Organometallic Chemistry*, Reinhold Publishing Corp., New York (1960).

10.12 CARBIDES

10.12.1 Introduction

Carbides are defined as compounds in which carbon is combined with elements of lower or about equal electronegativity [1]. Carbides may be divided into three types: *salt-like* (e.g., CaC_2, Al_4C_3, Cu_2C_2, ZnC_2), *interstitial* (e.g., TiC, Cr_3C_2, Fe_3C) and *covalent* (SiC and B_4C). This section briefly describes the properties and behavior of each type, focusing on the chemistry of calcium carbide (CaC_2) and silicon carbide (SiC), two commercially important compounds. The major route for carbides in the environment is the reaction of ionic carbides with water to produce hydrocarbons. Most of the non-ionic carbides are relatively inert under environmental conditions.

10.12.2 Salt-like (Ionic) Carbides

The salt-like carbides are formed mainly by the metals of groups I, II, and III. They are colorless, transparent crystals that are decomposed by water or dilute acids at ordinary temperatures, producing hydrocarbons. Most ionic carbides contain carbon in either the C(−IV) or C_2^{-2} state. Those containing C(−IV) evolve methane when hydrolyzed, as in the following reaction [1]:

$$Al_4C_3 + 12H_2O \longrightarrow 4\,Al(OH)_3 + 3CH_4 \qquad (1)$$

The many carbides containing C_2^{-2} ions are known as acetylides because they yield acetylene upon hydrolysis [1]. Their reaction with water or moisture can be violent, evolving heat and producing acetylene, a flammable gas. One of the acetylides is calcium carbide, CaC_2, whose major industrial use is in the production of acetylene for the metalworking industry, via the following reaction [3]:

$$CaC_2 + 2H_2O \longrightarrow C_2H_2 + Ca(OH)_2 \qquad \Delta H = -130 \text{ kJ/mol} \qquad (2)$$

This highly exothermic reaction must be carried out in the presence of an excess of water to absorb the heat of reaction and avoid explosive conditions. When there is a deficiency of water, CaC_2 can react with the second reaction product, $Ca(OH)_2$ [3]:

$$CaC_2 + Ca(OH)_2 \longrightarrow C_2H_2 + 2\,CaO \qquad (3)$$

The second major industrial use of calcium carbide is for the production of calcium cyanamide by reaction with nitrogen [3]:

$$CaC_2 + N_2 \longrightarrow CaCN_2 + C \qquad \Delta H = -295 \text{ kJ/mol} \qquad (4)$$

Once the reaction is initiated by heating the carbide, it proceeds unaided because it is strongly exothermic. Calcium carbide is also used as a desulfurizing agent for iron and steel in blast furnaces. It reacts with the sulfur in molten metal as follows [3]:

$$CaC_2 + S \longrightarrow CaS + 2C \qquad (5)$$

10.12.3 Interstitial Carbides

Interstitial carbides are formed by most transition metals, especially those in groups IV, V, and VI. They are characterized by high melting points, hardness, and metallic electrical conductivity, and they are resistant to attack by water. These characteristics are attributed to the fact that the carbon atoms are closely packed into the spaces between metal atoms and thus do not alter some of the fundamental properties of the pure metal [1]. Many of the heavy metal carbides, e.g., WC, TiC, and TaC, are combined with cobalt or nickel to manufacture "cemented carbides," which are used for cutting tools and wear-resistant materials [2].

The carbides formed by the metals Cr, Mn, Fe, Co, and Ni are transitional between typical ionic and interstitial carbides, having structures in which carbon chains run through very distorted metal lattices [1]. They have lower hardness values and melting points than the typical interstitial carbides [2] and are easily hydrolyzed by water and dilute acids to yield hydrogen and a variety of hydrocarbons [1]. These carbides are used as hardening components of alloy steels and cast iron [2].

10.12.4 Covalent Carbides

The carbides of silicon, boron, and (to some extent) beryllium are covalent. Because of their extreme hardness, they are sometimes referred to as the diamond-like carbides. Silicon carbide, SiC, is also known as carborundum. It occurs in three crystalline forms and is an extremely hard substance, exhibiting high thermal conductivity and general chemical and physical stability. It is used mostly in abrasives, refractories, and heating elements for electric furnaces. Silicon carbide is generally stable, only undergoing violent reaction when it is heated with a mixture of potassium dichromate and lead chromate. It will also react with a variety of compounds at high temperatures [4].

10.12.5 Literature Cited

1. Cotton, F.A. and G. Wilkinson, *Advanced Inorganic Chemistry: A Comprehensive Text,* 3rd ed., Interscience Publishers, div. of John Wiley & Sons, New York (1972).
2. Kieffer, R. and F. Benesovsky, "Carbides," in *Kirk-Othmer Encyclopedia of Chemical Technology,* 3rd ed., Vol. 4, John Wiley & Sons, New York (1982).
3. Shine, N.B., "Calcium Carbide," in *Kirk-Othmer Encyclopedia of Chemical Technology,* 3rd ed., Vol. 4, John Wiley & Sons, New York (1982).
4. Smoak, R.H., T.M. Korzekwa, S.M. Kunz and E.D. Howell, "Silicon Carbide," in *Kirk-Othmer Encyclopedia of Chemical Technology,* 3rd. ed., Vol. 4, John Wiley & Sons, New York (1982).

10.13 CYANIDES

10.13.1 Introduction

Cyanides are compounds containing the C≡N group. They include cyanogen $(CN)_2$, the cyanogen halides XCN (X= F, Cl, Br, I), hydrogen cyanide HCN, and the metal cyanides. The metal cyanides include compounds that contain cyanide as an anion (e.g., KCN) and compounds in which cyanide is complexed to the metal as a ligand (e.g., $Na_3Fe(CN)_6$).

Reactions and transformations of cyanides that may occur under environmental conditions include the following:

- Polymerization of HCN,
- Dissolution of HCN or metal cyanides to form free CN^- and other species,
- Reaction of CN^- with water to form HCN,
- Volatilization of HCN from solution,
- Complexation of CN^- with metals,
- Photodecomposition of cyanide complexes (especially $Fe(CN)_6^{-4}$),
- Hydrolysis,
- Oxidation/chlorination
- Sorption to soils, and
- Biodegradation

10.13.2 Cyanogen and the Cyanogen Halides

One of the simplest of the compounds containing the cyanide group is cyanogen, $(CN)_2$. Cyanogen is used as a laboratory reagent for cyanation, ring formation and other organic processes [1]. It is a colorless, flammable gas and is very toxic [20]. It is highly water-soluble but is hydrolyzed by water, slowly at room temperature and faster with heating; the reaction yields a variety of products, including hydrogen cyanide, ammonium cyanate, oxamide, and urea [11,20]. Although it is not affected by moderate heating, it polymerizes at 300-500°C [4] to form paracyanogen, $(CN)_n$, a dark brown, water-insoluble solid [20].

Cyanogen resembles the halogens in some ways and undergoes many similar chemical reactions. For example, it reacts with hydroxide in a basic solution, forming cyanide and cyanate [4]:

$$(CN)_2(g) + 2OH^- \rightleftharpoons CN^- + OCN^- + H_2O \qquad (1)$$

The cyanogen halides, or halogen cyanides, FCN, ClCN, BrCN, and ICN, are volatile and behave similarly to the halogens. Cyanogen chloride is formed as an intermediate in the oxidation of cyanide by hypochlorous acid in wastewater treatment. It is used commercially in a process for the production of malononitrile [10]. Cyanogen chloride is a liquid at room temperature, while the bromide and iodide are solids. (The fluoride is unstable at room temperature.) All are very toxic, with irritant vapors. Although stable in pure form, ClCN, BrCN and ICN tend to polymerize in the presence of certain trace impurities, forming the cyanuric halides $(XCN)_3$ [20]. Properties of cyanogen and the cyanogen halides are summarized in Table 10.13-1.

TABLE 10.13-1

Properties of Cyanogen and the Cyanogen Halides

Compound	Melting Point (°C)[a]	Boiling Point (°C)[a]	Vapor Pressure (mm Hg)	Water Solubility	Comments
$(CN)_2$	−27.9	−21.17	—	450 ml/100 ml H_2O (20°C)[d]	Slowly hydrolyzed by water[b]
FCN	−82	−46	—	—	Polymerizes rapidly at room temp.[a]
ClCN	−6.9	13.0	451.5(0°C) 628.2 (7.75°C)[b]	2500 ml/100 ml H_2O (20°C)[c]	—
BrCN	51.3	61.3	21.2 (0°C) 109.5 (25°C)[b]	5-6% by wt.[b]	Instantly hydrolyzed by alkali hydroxide solutions[b]
ICN	146 (subl)	146	1 (25°C)[d]	Sparingly sol.[b]	Decomposes to $(CN)_2 + I_2$ at moderate heat[b]

Sources:

a. Holliday, Hughes and Walker [8]
b. Williams [20]
c. Dean [5]
d. Sax [16]

10.13.3 Hydrogen Cyanide

Hydrogen cyanide, or hydrocyanic acid, HCN, is a liquid or gas at ordinary temperatures (m.p. = −13.2°C, b.p. = 25.70°C) and is extremely poisonous [9]. The liquid has a vapor pressure of about 741 mm Hg at 25°C and 264 mm Hg at 0°C [5]. Hydrogen cyanide is used commercially for the production of a variety of chemicals, including methacrylate, cyanuric chloride, sodium cyanide, nitrilotriacetic acid, EDTA and other chelates [9].

Although HCN is stable as a gas, liquid HCN is unstable and can polymerize violently or even explosively, releasing heat. The reaction is promoted by the presence of base, water, and/or heat, and is inhibited by dilute acid [9]. In aqueous solution, the polymerization of HCN can be induced by ultraviolet light [4]. The many products of this reaction include aminomalononitrile, $HC(NH_2)(CN)_2$ (from three HCN units), diaminomaleonitrile, $H_2NC(CN)C(CN)NH_2$ (from four HCN units), and high-molecular-weight polymers [4].

In aqueous solution, HCN behaves as a weak acid, with a pK_a of 9.21 at 25°C [13]. Thus, at neutral or acid pH, over 99% of the total free (uncomplexed) cyanide in solution is present as undissociated HCN, with the remainder present as CN^-.

Undissociated HCN, whether released directly or formed by hydrolysis of CN^- released into natural waters, can volatilize readily from solution. The Henry's law constant (H) is given by:

$$\log H_{HCN} = \frac{-1272.9}{T} + 6.238$$

where H_{HCN} is the Henry's Law constant in mm Hg/*M* and T is the temperature in °K (determined for HCN concentrations of 0.01 to 0.5 *M*, at temperatures from 20° to 95°C) [12]. At 25°C, the value of H_{HCN} calculated from this equation is 93.1 mm Hg/*M*. Thus, for example, the equilibrium vapor pressure of HCN over a 0.05 *M* aqueous solution at 25°C is 4.65 mm Hg. (See § 10.6, Gases, for a discussion of gas volatilization and Henry's Law constants.)

Laboratory experiments have shown that volatilization of HCN from solution can occur fairly rapidly, indicating that volatilization is important as a fate of free cyanide in the environment [2]. The rate of volatilization from natural waters is influenced by a number of factors, including temperature, pH, mixing characteristics of the water, wind speed, and ice-cover [3].

10.13.4 Metal Cyanides

USES

Several of the metal cyanides, e.g., AgCN, CuCN, and $Zn(CN)_2$, are used commercially for electroplating with the respective metals. Both NaCN and KCN also have major uses as components of electroplating baths, as their presence in solution enhances the solubility of the transition-metal cyanides. Nickel cyanide, $Ni(CN)_2$, has been used as a catalyst for the conversion of acetylene to butadiene and other products. Calcium cyanide is used in the extraction or cyanidation of precious metal ores, in the froth flotation of minerals, and as a fumigant and rodenticide [10].

SOLUBILITY AND REACTION WITH WATER

Solubility data for some metal cyanides are given in Table 10.13-2. The alkali metal cyanides, e.g., NaCN and KCN, are solids that are soluble in water, dissociating to release free CN^- ions:

$$KCN(s) \longrightarrow K^+ + CN^- \quad (2)$$

The cyanide ion then reacts with water, forming HCN and producing a basic solution [11]:

$$CN^- + H_2O \rightleftharpoons HCN^0 + OH^- \quad (3)$$

TABLE 10.13-2

Solubility of Cyanides in Water

Compound	pK_{sp}	Solubility
NaCN		48 g/100 ml H_2O (10°C)
KCN		50 g/100 ml H_2O (cold water)
$Ba(CN)_2$		80 g/100 ml H_2O (14°C)
$Ni(CN)_2$		9.1×10^{-4} g/100 ml H_2O (25°C)[a]
CuCN	19.49[b]	2.6×10^{-4} g/100 ml H_2O (18°C)
AgCN	15.7[c]	2.3×10^{-5} g/100 ml H_2O (20°C)
$Zn(CN)_2$	15.9[c]	5×10^{-4} g/100 ml H_2O (20°C)
$Hg_2(CN)_2$	39.3[d]	8.1×10^{-15} g/100 ml saturated solution (25°C)[a]
$Hg(CN)_2$		9.3 g/100 ml H_2O (14°C)

Source: Weast [19], except as noted below.
a. Linke [14]
b. Dean [5]
c. Morel [15]
d. Kotrlý and Sucha [13]

The cyanides of the alkaline earth metals, e.g., $Ca(CN)_2$ and $Ba(CN)_2$, are also water-soluble, ionic solids that produce basic solutions in water. Solid calcium cyanide reacts readily with water, yielding slightly soluble $Ca(OH)_2$ and a black polymer of HCN. Magnesium cyanide has properties similar to those of $Ca(CN)_2$ [9].

Cyanide salts of the transition metals are typically very slightly soluble in water. (An exception is the soluble $Hg(CN)_2$.) However, they are usually soluble in aqueous solutions of the alkali cyanides, due to the formation of soluble complexes such as $Ag(CN)_2^-$, $Ni(CN)_4^{-2}$, and $Fe(CN)_6^{-3}$ [8].

10.13.5 Environmental Fate of Aqueous Cyanides

Cyanide in aqueous solution may be present as CN^-, as HCN, or in complexed form. The relative abundance of these species depends both on pH (as discussed above) and on the strength of metal complexation. The concentration of undissociated HCN is of particular importance, because undissociated HCN is believed to be the most toxic component of cyanide solutions [18]. When cyanide-containing waste is treated with chlorine gas (Cl_2) or hypochlorous acid (HOCl), cyanide is oxidized, via ClCN, to cyanate (OCN^-). Cyanate may then be further oxidized to nitrogen and carbon dioxide or carbonate [10]. Under environmental conditions, one possible fate of undissociated HCN is volatilization from solution, described above. Other possible environmental fates of free and complexed cyanide ions are discussed below.

COMPLEXATION

Dissolved cyanide ions can form complexes with a number of metal cations, especially those of the transition metals plus Zn, Cd and Hg. Formation of such complexes removes free CN^- ion from solution; this causes a shift in the HCN/CN^- equilibrium, ultimately reducing the concentration of undissociated HCN [17]. Most of these aqueous complexes are anionic and have the general formula $[M^{+n}(CN)_x]^{-(x-n)}$ [4]. Metal cyano complexes vary in stability, those of Zn and Cd being less stable and those of Fe and Co more so [18]. Stability constants for some metal cyano complexes are given in Table 10.13-3.

TABLE 10.13-3

Stability Constants of Cyano Complexes[a]
(log β for the equilibrium $M^+ + yCN^- \rightleftharpoons [M(CN)_y]^{+x-y)}$)

Metal Ion	M(CN)	$M(CN)_2$	$M(CN)_3$	$M(CN)_4$	$M(CN)_6$
Fe^{+2}	—	—	—	—	35.4
Fe^{+3}	—	—	—	—	43.6
Ni^{+2}	7.03[b]	—	—	30.22	—
Cu^+	—	16.26	21.6	23.1	—
Ag^+	—	20.48	21.4	20.8[c]	—
Au^+	—	38.3	—	—	—
Zn^{+2}	5.3[b]	11.07	16.05	19.62	—
Cd^{+2}	6.01	11.12	15.65	17.92	—
Hg^{+2}	17.00	32.75	36.31	38.97	—

a. Values are for conditions of 25°C and zero ionic strength unless otherwise noted.
b. I = 3.0
c. I = 1.0, 18°C

Source: Kotrlý and Šůcha [13]
(Copyright 1985, John Wiley & Sons. Reprinted with permission.)

Complexation with cyanide can stabilize metal ions in certain oxidation states. For example, copper is stabilized in the +1 state, as illustrated by the difference between the oxidation potentials for the formation of univalent copper in the absence and presence of cyanide ion [11]:

$$Cu = Cu^+ + e^- \qquad E^0 = -0.521V \qquad (4)$$

$$Cu + 2CN^- = Cu(CN)_2^- + e^- \qquad E^0 = \text{approx. } 0.43V \qquad (5)$$

PHOTODECOMPOSITION OF IRON CYANIDE COMPLEXES

Both Fe^{+2} and Fe^{+3} form stable complexes with cyanide. Ferrocyanide $Fe(CN)_6^{-4}$ and ferricyanide $Fe(CN)_6^{-3}$ — also known as hexacyanoferrate (II) and (III) — may be present in a variety of industrial effluents. These complexes decompose in the light, releasing free CN^-, and may re-form at night, resulting in a daily fluctuation of the HCN concentration in a body of water [18]. This process has been investigated by Broderius and Smith [2], whose findings are presented below.

When aqueous ferrocyanide solutions are irradiated with light of an appropriate wavelength, the complex decomposes, releasing free cyanide:

$$Fe(CN)_6^{-4} \overset{h\nu}{\rightleftharpoons} Fe(CN)_5^{-3} + CN^- \tag{6}$$

The released cyanide ions are subsequently hydrolyzed to HCN, and a molecule of water is substituted for the cyanide in the complex:

$$CN^- + H_2O \rightleftharpoons HCN^0 + OH^- \tag{7}$$

$$Fe(CN)_5^{-3} + H_2O \rightleftharpoons Fe(CN)_5H_2O^{-3} \tag{8}$$

Thus, the overall reaction results in the production of HCN and a raising of the pH of the solution:

$$Fe(CN)_6^{-4} + 2H_2O \overset{h\nu}{\rightleftharpoons} Fe(CN)_5H_2O^{-3} + HCN^0 + OH^- \tag{9}$$

This reaction may be reversed when the light is removed, if the solution has not been exposed to light for too long. After prolonged exposure to light, Fe^{+2} ions may be released, and the reaction becomes irreversible.

The iron (III) complex undergoes a similar photodecomposition, and it is reduced as well, with an overall reaction as follows:

$$2Fe(CN)_6^{-3} + H_2O + 2OH^- \xrightarrow{h\nu} 2Fe(CN)_5H_2O^{-3} + CN^- + CNO^- \tag{10}$$

The removal of light apparently does not cause this reaction to proceed in the opposite direction.

The maximum amount of total cyanide that could be photochemically released as HCN from hexacyanoferrate (II) and (III) solutions was experimentally determined to be about 85% and 49%, or 5 moles and 3 moles of CN respectively, per mole of iron (II) or (III) complex. These results are not explained by the photodecomposition mechanisms presented above; the reaction mechanisms are not yet fully understood.

The photodecomposition reaction follows approximately first-order kinetics for hexacyanoferrate concentrations up to 100 μg/l as total cyanide — i.e., the decomposition rate is proportional to the hexacyanoferrate concentration. However, at higher concentrations the rate constant tends to drop to some limiting value. The reaction rate also decreases somewhat with increasing pH and with increasing temperature.

The rate of photodecomposition is a direct function of natural light intensity. Since the amount of light penetrating natural waters is lessened by absorption and scattering, the rate of photolysis decreases with depth. (Specifically, the logarithm of the photolysis rate decreases linearly with depth.) Below a depth of about 50 to 100 cm in most well-mixed natural waters, very little HCN is likely to be produced through this process. Also as a result of decreasing light intensity, the rate of photodecomposition decreases from midday to sunset, from summer to winter, and from the tropics to higher latitudes.

HYDROLYSIS

Hydrogen cyanide or cyanide ion may be hydrolyzed to the formate ion ($HCOO^-$) and ammonia or ammonium ion (NH_3 or NH_4^+) [3]. Cited rate constants for the hydrolysis reactions vary. One study [6] calculated hydrolysis half-lives based on experimentally determined hydrolysis rates for cyanide in sterilized river water at a pH of 7-8; the calculated half-lives at 0° to 20°C ranged from 10 to 20 days. Other studies (cited in [3]) found hydrolysis half-lives for HCN of 0.4 to 40 days in strongly acidic solutions (pH < 1), and about 400 days at 33°C in alkaline solution (0.181 *M* KOH). The experimental data indicate that the fate of cyanide in the environment is more dependent on biodegradation and volatilization process than on hydrolysis [3,6].

SORPTION/MOBILITY IN SOIL

Cyanides are sorbed by a variety of materials, including clays, biological solids, and sediments. However, available data indicate that sorption is fairly weak [3].

Fuller [7] conducted experiments to determine the mobility of cyanide in soils. Solutions of KCN and of $K_3Fe(CN)_6$ in distilled water were found to be very mobile in soils, while a solution of KCN in landfill leachate was somewhat less mobile. This result suggests that although free cyanide and hexacyanoferrate are potentially quite mobile and easily leached from soils into ground water, complexation or reaction with species present in groundwater may result in significantly lower mobility. Cyanide mobility was found to vary considerably depending on soil type, with $K_3Fe(CN)_6$ behaving quite differently from KCN. While some explanations for the variation with soil type were offered, the experimental results were too limited to allow generalizations about the relationship of cyanide mobility to specific soil characteristics.

BIODEGRADATION

Almost all organisms can biodegrade hydrogen cyanide or its salts at low concentrations, and a bacterium has been isolated that can grow using HCN as its sole energy source [3]. The rate of biodegradation varies with cyanide concentration, pH, temperature, concentration of microbes, availability of nutrients, and whether the microbes are acclimated to cyanide. In general, biodegradation is considered an important fate process for aqueous cyanides [3]. However, in anaerobic environments, such as in water-saturated soil or sediments, biodegradation of cyanide may not occur. Although some anaerobic biodegradation is possible at low cyanide concentrations, the microorganisms responsible are very sensitive to high cyanide concentrations [7].

10.13.6 Literature Cited

1. "ACS Meeting Spotlights Syntheses and Separations," *Chem. Week,* April 11, 1979, p. 45.
2. Broderius, S.J. and L.L. Smith, Jr., *Direct Photolysis of Hexacyanoferrate Complexes: Proposed Applications to the Aquatic Environment*, EPA-600/3-80-003, U.S. Environmental Protection Agency, Washington, D.C. (January 1980).
3. Callahan, M., M. Slimak, N. Gabel, I. May, C. Fowler, J. Freed, P. Jennings, R. Durfee, F. Whitmore, B. Maestri, W. Mabey, B. Holt and C. Gould, *Water-Related Environmental Fate of 129 Priority Pollutants. Volume I: Introduction and Technical Background, Metals and Inorganics, Pesticides and PCBs,* EPA-440/4-79-029a, EPA Contracts 68-01-3852 and 68-01-3867, Office of Water Planning and Standards, U.S. Environmental Protection Agency, Washington, D.C. (1979).
4. Cotton, F.A. and G. Wilkinson, *Advanced Inorganic Chemistry: A Comprehensive Text,* Interscience Publishers, New York (1972).
5. Dean, J.A. (ed.), *Lange's Handbook of Chemistry,* 12th ed., McGraw-Hill Book Co., New York (1979).
6. Fiksel, J., C. Cooper, A. Eschenroeder, M. Goyer, J. Perwak, K. Scow, R. Thomas, W. Tucker and M. Wood, *An Exposure and Risk Assessment for Cyanide,* Final Draft Report, EPA Contracts 68-01-3857 and 68-01-5949, U.S. Environmental Protection Agency, Washington, D.C. (1981).
7. Fuller, W.H., *Investigation of Landfill Leachate Pollutant Attenuation by Soils,* EPA-600/2-78-156, Municipal Environmental Research Laboratory, Office of Research and Development, U.S. Environmental Protection Agency, Cincinnati (August 1978).
8. Holliday, A.K., G. Hughes and S.M. Walker, "Carbon," in *Comprehensive Inorganic Chemistry,* A.F. Trotman-Dickenson (exec. ed.), Pergamon Press, Oxford, England (1973).
9. Jenks, W.R., "Cyanides," in *Kirk-Othmer Encyclopedia of Chemical Technology,* 3rd ed., Vol. 7, John Wiley & Sons, New York (1982).
10. *Kirk-Othmer Encyclopedia of Chemical Technology*, 3rd ed., John Wiley & Sons, New York (1982).
11. Kleinberg, J., W.J. Argersinger, Jr. and E. Griswold, *Inorganic Chemistry,* D.C. Heath, Boston (1960).
12. Kotlik, S.B. and G.N. Lebedeva, "Equilibrium Pressures of HCN and NH_3 Over Aqueous Solutions," *Zh. Prikladnoi Khimii,* **47**, 444-46 (1974) [English translation by Consultants Bureau, a div. of Plenum Publishing Corp., New York (1974)].

13. Kotrlý, S. and L. Šůcha, *Handbook of Chemical Equilibria in Analytical Chemistry,* Halsted Press, a div. of John Wiley & Sons, New York (1985).
14. Linke, W.F., *Solubilities: Inorganic and Metal Organic Compounds,* 4th ed., D. Van Nostrand Co., Princeton, N.J. (1958).
15. Morel, F.M.M., *Principles of Aquatic Chemistry,* John Wiley & Sons, New York (1983).
16. Sax, N.I., *Dangerous Properties of Industrial Materials,* 6th ed., Van Nostrand Reinhold, New York (1984).
17. Snoeyink, V.L. and D. Jenkins, *Water Chemistry,* John Wiley & Sons, New York (1980).
18. U.S. Environmental Protection Agency, *Quality Criteria for Water,* U.S. EPA, Washington, D.C. (July 1976).
19. Weast, R.C. (ed.), *CRC Handbook of Chemistry and Physics,* 54th ed., CRC Press, Cleveland (1973).
20. Williams, H.E., *Cyanogen Compounds: Their Chemistry, Detection, and Estimation,* 2nd ed., Edward Arnold and Co., London (1948).

10.14 HYDRIDES

10.14.1 Introduction

Hydrides, as defined most broadly, refer to binary compounds containing hydrogen and one other element, as well as to complex derivatives of certain of these binary compounds. Some authorities feel that the term hydride should be limited to the hydrogen compounds of the metallic elements, thus excluding such compounds as NH_3, H_2S, and the hydrogen halides. This discussion will touch upon the binary hydrogen compounds of all elements except the halogens. The complex hydrides, such as $NaBH_4$ and $LiAlH_4$, will not be addressed here. For those elements that form two or more simple hydrides (e.g., SiH_4, Si_2H_6, and Si_3H_8), the most common or most stable hydride of each element is described. Properties of the hydrogen halides are discussed in sections 10.2 (Acids) and 10.6 (Gases), and the boron hydrides and sodium tetrahydroborate ($NaBH_4$) are discussed in section 10.9. The compounds NH_3 and H_2S are discussed both here and in section 10.6.

The binary hydrides include (a) salt-like or ionic hydrides, which are formed by the alkali metals and some of the alkaline earth metals (e.g., LiH and CaH_2), (b) polymeric covalent hydrides formed by Be, B, and Al, (c) volatile covalent hydrides, formed by many non-metals (e.g., N and P) and metalloids (e.g., Si, Ge, As, Sb and Se), (d) transition metal hydrides (e.g., TiH_2 and ZrH_2), and (e) some intermediate types.

A number of hydrides have important commercial uses; some of these are listed in Table 10.14-1. Several other hydrides are produced in smaller volumes for specialty uses.

Environmentally significant reactions of hydrides include:

- Volatilization,
- Reaction with air (combustion or oxidation),
- Reaction with water (hydrolysis or acid/base equilibrium), and
- Oxidation/reduction reactions in aqueous solution.

Environmental properties and reactions of some commercially important hydrides are summarized in Table 10.14-2.

10.14.2 Ionic Hydrides

The ionic hydrides, which are formed by the alkali metals (Li, Na, K, Rb, Cs) and some of the alkaline earth metals (Ca, Sr, Ba), are crystalline solids in which hydrogen is present as the hydride ion, H^- [1]. Magnesium hydride, MgH_2, is similar to the ionic hydrides but shows some properties intermediate between those of the ionic hydrides and the polymeric covalent hydrides [6].

The ionic hydrides react with water or moist air, yielding gaseous hydrogen and the metal hydroxides. Their chemical reactivity generally increases with increasing

TABLE 10.14-1

Commercial Uses of Some Hydrides

Compound	Name	Uses
LiH	Lithium hydride	Generation of silane for semiconductor uses; lightweight nuclear shielding material; condensation agent in organic chemistry.
NaH	Sodium hydride	Component of molten NaOH or fused eutectic salt baths as reducing systems for descaling metals; a powerful base in organic synthesis.
CaH_2	Calcium hydride	Reduction of refractory oxides of metal and alloy powders; portable source of hydrogen; drying agent for liquids and gases.
SiH_4	Silane	Preparation of semiconducting silicon by thermal decomposition.
GeH_4	Germane	Preparation of high-purity germanium metal or epitaxial deposits of germanium metal on substrates for electronics by thermal decomposition.
NH_3	Ammonia	Source of nitrogen in fertilizers; raw material for manufacture of explosives, fibers and plastics.
N_2H_4	Hydrazine	Rocket fuel; agent for removal of dissolved oxygen for corrosion control in boiler and hot-water heating systems; starting material for many derivatives, including pesticides and foaming agents for plastics.
PH_3	Phosphine	Grain fumigant.
AsH_3	Arsine	Dopant for silicon in semiconductors.
SbH_3	Stibine	Dopant for silicon in semiconductors.

Source: *Kirk-Othmer* [4]

atomic weight of the metal within a group, i.e., increasing from lithium to cesium and from calcium to barium, with calcium hydride somewhat less reactive than lithium hydride. For example, the reaction with water is "mild" for calcium hydride, "vigorous" for strontium, barium, and lithium hydrides, "violent" for sodium hydride, and increasingly violent for potassium, rubidium, and cesium hydrides. Similarly, although the hydrides of the lighter elements react slowly if at all with dry air, rubidium and cesium hydrides ignite in air [6].

TABLE 10.14-2

Environmental Properties and Reactions of Some Hydrides[a]

Compound	Physical State at 20°C (mp or bp in °C)	Reaction with Oxygen	Reaction with Water or Moisture
LiH	Solid (mp 688.7)	Stable in dry air at room temp.; ignites only at high temp.[b]	Vigorous reaction with water: $LiH + H_2O \rightarrow LiOH + H_2(g)$ Exothermic reaction with moist air[b] (normally slow; may ignite spontaneously at high humidity).
NaH	Solid (dec 425)	Oxidizes in dry air[b]; ignites only at high temp.	Violent reaction with water, generating enough heat to ignite the H_2 formed: $NaH + H_2O \rightarrow NaOH + H_2(g)$ Decomposes rapidly in moist air.
CaH_2	Solid (dec 600)[c]	No reaction with dry air at environmental temp.; will burn in air if ignited: $CaH_2 + O_2 \rightarrow CaO + H_2O$	Reacts readily with water: $CaH_2 + 2H_2O \rightarrow Ca(OH)_2 + 2H_2(g)$
SiH_4	Gas (bp −111.9)	Spontaneously flammable in air.	No reaction with pure water; rapid hydrolysis catalyzed by trace of alkali.
GeH_4	Gas (bp −88.5)	Not spontaneously flammable in air; will burn if ignited: $GeH_4 + 2O_2 \rightarrow GeO_2 + 2H_2O$	No reaction; resistant to hydrolysis by dilute acid or base.
NH_3	Gas (bp −33.42)	No reaction under normal conditions.	Highly soluble (52.9g/100g H_2O at 20°C, 1 atm); forms basic solution: $NH_3 + H_2O \rightleftharpoons NH_4^+ + OH^-$; $pK_b = 4.76$ (25°C)
N_2H_4	Liquid (mp 1.54, bp 113.8)[d]	Will burn in air if ignited.	Very soluble; somewhat weaker base than NH_3 $N_2H_4 + H_2O \rightleftharpoons N_2H_5^+ + OH^-$; $pK_b = 6.02$ (25°C) $N_2H_5^+ + H_2O \rightleftharpoons N_2H_6^{+2} + OH^-$; $pK_b = 14.9$ (20°C)
PH_3	Gas (bp −87.78)	Not spontaneously flammable when pure, but may ignite due to trace contaminants; burns in air when ignited.[e]	Soluble (26 ml/100 ml H_2O at 17°C, 1 atm); solutions neither acidic nor basic.[e]
AsH_3	Gas (bp −55)[c]	Will burn in air if ignited, yielding either free As or arsenic oxides.	Soluble (28 ml/100 ml H_2O at 20°C, 1 atm); forms slightly basic solutions.

(Continued)

TABLE 10.14-2 (Continued)

Compound	Physical State at 20°C (mp or bp in °C)[a]	Reaction with Oxygen	Reaction with Water or Moisture
SbH_3	Gas (bp −18.4)[f]	Slowly oxidized at lower temps., yielding Sb and water; burns in air if ignited, yielding Sb_2O_3 and water.[g]	Soluble (20 ml/100 ml H_2O at 0°C, 1 atm).
H_2S	Gas (bp −60.33)	Aqueous solution is slowly oxidized by atmospheric O_2 on exposure to light: $2H_2S + O_2 \rightarrow 2S + 2H_2O$	Soluble (0.3846 g/100 g H_2O at 20°C, 1 atm); forms weakly acidic solution: $H_2S + H_2O \rightleftharpoons HS^- + H_3O^+$; $pK_{a1} = 7.02$ (25°C) $HS^- + H_2O \rightleftharpoons S^{-2} + H_3O^+$; $pK_{a2} = 13.9$ (25°C)
H_2Se	Gas (bp −42)	Stable in dry air; will burn if ignited; rapid oxidation catalyzed by moisture; aqueous solutions decompose if exposed to air: $2H_2Se + O_2 \rightarrow 2Se + 2H_2O$	Soluble; solubility is greater at higher pH; forms acidic solution: $H_2Se + H_2O \rightleftharpoons HSe^- + H_3O^+$; $pK_{a1} = 3.89$ (25°C) $HSe^- + H_2O \rightleftharpoons Se^{-2} + H_3O^+$; $pK_{a2} = 15.0$ (22°C)

a. Sources: Melting points, boiling points and solubilities from Dean [2]; acid and base constants from Martell and Smith [7]; other information from Hurd [3], except as noted.
b. Source: Sullivan and Wade [8].
c. Source: Weast [9].
d. Vapor pressure of liquid N_2H_4 at 25°C = 14.4 mm Hg [4].
e. Source: Cotton and Wilkinson [1].
f. Decomposes slowly to Sb and H_2 at room temperature.
g. Source: *Kirk-Othmer* [4].

10.14.3 Covalent Hydrides

The hydrides of beryllium, boron, and aluminum are covalently bonded, nonvolatile, polymeric solids. Boron forms a series of hydrides B_xH_y with x ranging from 2 to 10 or more; these compounds are discussed in section 10.9. The copper and zinc group metals form "borderline" hydrides that are more similar to the covalent polymeric hydrides than to the transition metal hydrides, but they are mostly unstable or poorly characterized.

The covalent hydrides formed by the elements of Groups IV through VI of the periodic table are generally volatile compounds, and most are gases or liquids under normal conditions. They exhibit a wide range of chemical reactivities. (See Table 10.14-2 for a summary of their reactivity with air and water.) Some ignite spontaneously when exposed to air, others are stable in air but will burn if ignited, and a few react with oxygen only under extreme conditions. These combustion reactions usually yield water and the oxide of the parent element. Their reactions with water vary as well — from the hydrolysis of SiH_4 in water with a trace of base, to the formation of a basic solution by NH_3 or of an acidic solution by H_2Se. Most of these hydrides are good reducing agents and will precipitate heavy metals from solutions of their salts [3].

10.14.4 Transition Metal Hydrides

Of the transition metals, the elements of the scandium, titanium, and vanadium groups, the lanthanide and actinide elements, and chromium, nickel and palladium all form well-established binary hydrides. Their composition may be variable, not always following a simple numerical proportion of metal to hydrogen atoms. These compounds are dark or metallic in appearance, and most are brittle solids, although some are powders. They conduct electricity and have other metallic properties [5]. The lanthanide hydrides resemble the alkali and alkaline earth hydrides in their chemical reactivity [6]; they ignite spontaneously in air and react vigorously with water [1]. Uranium hydride (UH_3) also ignites spontaneously in air, but the titanium and vanadium group hydrides are fairly stable in air unless heated [1]. Some of the uses of these metal hydrides are in the production of metal powders, alloys, and coatings, in metal-ceramic and metal-metal bonding, and as moderators in nuclear reactors [8].

10.14.5 Literature Cited

1. Cotton, F.A. and G. Wilkinson, *Advanced Inorganic Chemistry; A Comprehensive Text,* 3rd ed., Interscience Publishers, div. of John Wiley & Sons, New York (1972).
2. Dean, J.A. (ed.), *Lange's Handbook of Chemistry,* 12th ed., McGraw-Hill Book Co., New York (1979).
3. Hurd, D.J., *An Introduction to the Chemistry of the Hydrides,* John Wiley & Sons, New York (1952).
4. *Kirk-Othmer Encyclopedia of Chemical Technology,* 3rd ed., John Wiley & Sons (1982).
5. Mackay, K.M., "Hydrides," in Trotman-Dickenson, A.F. (exec. ed.), *Comprehensive Inorganic Chemistry,* Vol. 1, Pergamon Press, Oxford, England (1973).
6. Mackay, K.M., *Hydrogen Compounds of the Metallic Elements,* E. & F.N. Spon Ltd., London (1966).
7. Martell, A.E. and R.M. Smith, *Inorganic Complexes*, Vol. 4 of *Critical Stability Constants*, Plenum Press, New York (1976).
8. Sullivan, E.A. and R.C. Wade, "Hydrides," in *Kirk-Othmer* [4], Vol. 12.
9. Weast, R.C. (ed.), *CRC Handbook of Chemistry and Physics,* 54th ed., CRC Press, Cleveland (1973).

10.15 HYPOCHLORITES AND CHLORATES

The hypochlorite ion (OCl^-) consists of a chlorine atom in a +1 formal oxidation state bonded to an oxygen atom. The hypochlorite ion is environmentally important because it readily oxidizes organic matter and thus is toxic to biota. Concern has also been raised over the formation of chlorinated organics (e.g., trihalomethanes) in waters to which hypochlorite or chlorine has been added.

In the chlorate ion (ClO_3^-), chlorine exists formally in the +5 oxidation state and is bonded to three oxygen atoms. Like hypochlorites, the environmental importance of chlorates stems from their reactions as oxidizing agents.

10.15.1 Uses

The major use of the most common hypochlorite salts, $Ca(OCl)_2$ and NaOCl, is in the disinfection of drinking water supplies, sewage, and swimming pool water. Sodium hypochlorite is also used as a household bleach and an intermediate in the manufacture of automatic dishwasher detergents and dry bleaching agents [11].

More than three quarters of the sodium chlorate ($NaClO_3$) produced is used in the pulp and paper industry, where it is converted to chlorine dioxide for bleaching wood pulp [1]. Textile manufacturers also use sodium chlorate for making bleach. Twelve percent of the sodium chlorate produced is used as an intermediate in the production of other chlorates and of perchlorates (e.g., $NaClO_4$) [1]. It is also used as an herbicide, a defoliant for cotton, a desiccant for soybean plants (to remove the leaves before mechanical harvesting), and an oxidizing agent in the processing of uranium ores. Potassium chlorate is used mostly for the production of matches.

10.15.2 Properties of Hypochlorites and Chlorates

The common chlorates and hypochlorites are solids at room temperature and are soluble in water to varying degrees. Table 10.15-1 lists the melting points and solubilities of serveral chlorate and hypochlorite salts.

All chlorates undergo thermal decomposition either to chloride and oxygen (at high temperatures) or perchlorate and chloride (moderate temperatures) [9]. In addition, sodium chlorate is slightly hygroscopic, while potassium chlorate is non-hygroscopic.

Hypochlorous acid and hypochlorites have varying degrees of stability. Calcium hypochlorite solutions are more stable than hypochlorous acid. Solid $Ca(OCl)_2$ is stable to a temperature of about 175°C (for the commercial grade product), at which point it decomposes exothermally, releasing O_2; solid NaOCl is relatively less stable [11].

10.15.3 Chemistry of Hypochlorites

The hypochlorite ion is formed by the chlorination of water with chlorine gas as well as by the dissolution of hypochlorite salts. (See section 10.6 for the values of the

TABLE 10.15-1

Melting Points and Solubilities of Hypochlorites and Chlorates

Compound	Melting Point (°C)	Solubility @ 20°C (g/100g H_2O)
Hypochlorites		
NaOCl	Decomposes	53.4
$Ca(OCl)_2$	Decomposes @ 100[a]	21.4% @ 25°C[b]
KOCl	Exists in solution only[a]	Very soluble[a]
Chlorates		
$NaClO_3$	248	95.9
$Ca(ClO_3)_2$	340	178
$KClO_3$	368	7.3

a. From Weast [9]

b. From Wojtowicz [11]

Source: Dean [3], except as indicated

Henry's Law constant and the solubility of $Cl_2(g)$ in water.) Aqueous chlorine first reacts to form hypochlorous acid, which then dissociates to form the hypochlorite ion:

$$Cl_2^0 + H_2O \rightleftharpoons HOCl^0 + H^+ + Cl^- \quad K = 4 \times 10^{-4} \text{ @ } 25°C \text{ [8]} \tag{1}$$

$$HOCl^0 \rightleftharpoons H^+ + OCl^- \quad pK_a = 7.537 \text{ @ } 25°C \text{ [5]} \tag{2}$$

The pK_a value for equation 2 was investigated for a variety of temperatures by Morris [5], who found the best fit line for temperatures, T (K), between 4 and 34°C to be $pK_a = 3000.00/T - 10.0686 + 0.0253\ T$. The dissolution of a hypochlorite salt results in the formation of OCl^-, which reacts with water to form hypochlorous acid:

$$OCl^- + H_2O \rightleftharpoons HOCl^0 + OH^- \quad pK_b = 6.4 \tag{3}$$

A comparison of equations 1 and 3 shows that the effect of chlorinating with $Cl_2(g)$ is the opposite of that of using a hypochlorite salt. The addition of $Cl_2(g)$ to an aqueous solution produces H^+ and reduces the pH, while the addition of a hypochlorite salt raises the pH by forming OH^-. At a pH of 7.5, approximately equal concentrations of HOCl and OCl^- exist in solution; at pH=8.5, only 10% of the OCl^- would exist as HOCl. The relative concentrations of HOCl and OCl^- are important, because HOCl is more effective than OCl^- in disinfection and biofouling control [4].

A protonated form of hypochlorous acid, the hypochloronium acidium ion (H_2OCl^+), is also known to exist in solution [6]. Its concentration in natural waters (pH between 5 and 9) would be extremely low, however, and thus it is of little environmental concern.

Hypochlorous acid is removed from solution in several ways. It may volatilize directly or after reacting to form chlorine monoxide as shown in equation 4.

$$2HOCl^0 \rightleftharpoons Cl_2O^0 + H_2O \quad K = 3.55 \times 10^{-3} \ @ \ 0°C \ [11] \tag{4}$$

Although the equilibrium of equation 4 favors the formation of HOCl over Cl_2O, the concentration of Cl_2O exceeds that of HOCl in the vapor phase above solutions of greater than approximately 1% HOCl [11]. The Henry's Law constant for Cl_2O (in terms of molarity) at 3.46°C has been given as 106.7 mm Hg/*M* [11]; thus, a solution of a hypochlorite salt (or hypochlorous acid) can be expected to have OCl^-, HOCl and Cl_2O in solution as well as HOCl and Cl_2O gases in equilibrium.

Hypochlorous acid is also consumed in solutions exposed to light. Ultraviolet light provides the energy for the decomposition of aqueous HOCl [8]:

$$2HOCl^0 \overset{h\nu}{\rightleftharpoons} 2H^+ + 2Cl^- + O_2 \tag{5}$$

The acid is also consumed through redox reactions. Hypochlorous acid rapidly oxidizes various reduced inorganic species such as S(-II), Mn(II), Fe(II), and NO_2^-. The oxidation of Fe^{+2}, for example, occurs as follows:

$$2Fe^{+2} + HOCl^0 + 5H_2O \rightleftharpoons 2Fe(OH)_3{}^0 + Cl^- + 5H^+ \tag{6}$$

The reaction with Mn(II) occurs only at pH >8.5. Oxidation of S(-II) at high pH tends to form polysulfides [8].

Hypochlorous acid reacts with aqueous ammonia to form mono-, di-, and trichloramines (nitrogen trichloride):

$$NH_3^0 + HOCl^0 \rightleftharpoons NH_2Cl^0 + H_2O \quad K = 1.7 \times 10^{11} \ [7] \tag{7}$$

$$NH_2Cl^0 + HOCl^0 \rightleftharpoons NHCl_2{}^0 + H_2O \quad K = 2.3 \times 10^{8} \ [11] \tag{8}$$

$$NHCl_2^0 + HOCl^0 \rightleftharpoons NCl_3{}^0 + H_2O \quad K = 1.1 \times 10^{5} \ [11] \tag{9}$$

At $HOCl/NH_3$ ratios ≤ 1, monochloramine is usually the only chloramine observed when the pH is greater than 8 [4]. Trichloramine and dichloramine may form at a lower pH or greater $HOCl/NH_3$ ratio. Since chloramines can act as disinfectants either directly (though with much less potency than HOCl) or through their reconversion to HOCl, their chemistry is important in determining chlorine dosages for disinfection.

A theoretical chlorination scheme is illustrated in Figure 10.15-1, which is a plot of "residual chlorine" concentration as a function of chlorine dosage. Residual chlorine refers to the total concentration of chloramines and "free chlorine" species (HOCl, OCl^-). The chlorine added to solution that is not reduced to Cl^- (through such reactions as equations 5 and 6 or by oxidizing organic matter) initially forms chloramines — primarily NH_2Cl with some $NHCl_2$. These species are known as the "combined chlorine" residual. Upon further chlorine addition, the chloramine concentration decreases as the chloramines are converted to $N_2(g)$ and NO_3^-. Once the chloramine concentration reaches the "breakpoint" (a minimum concentration near zero), added chlorine exists as free chlorine, and only traces of chloramines are present.

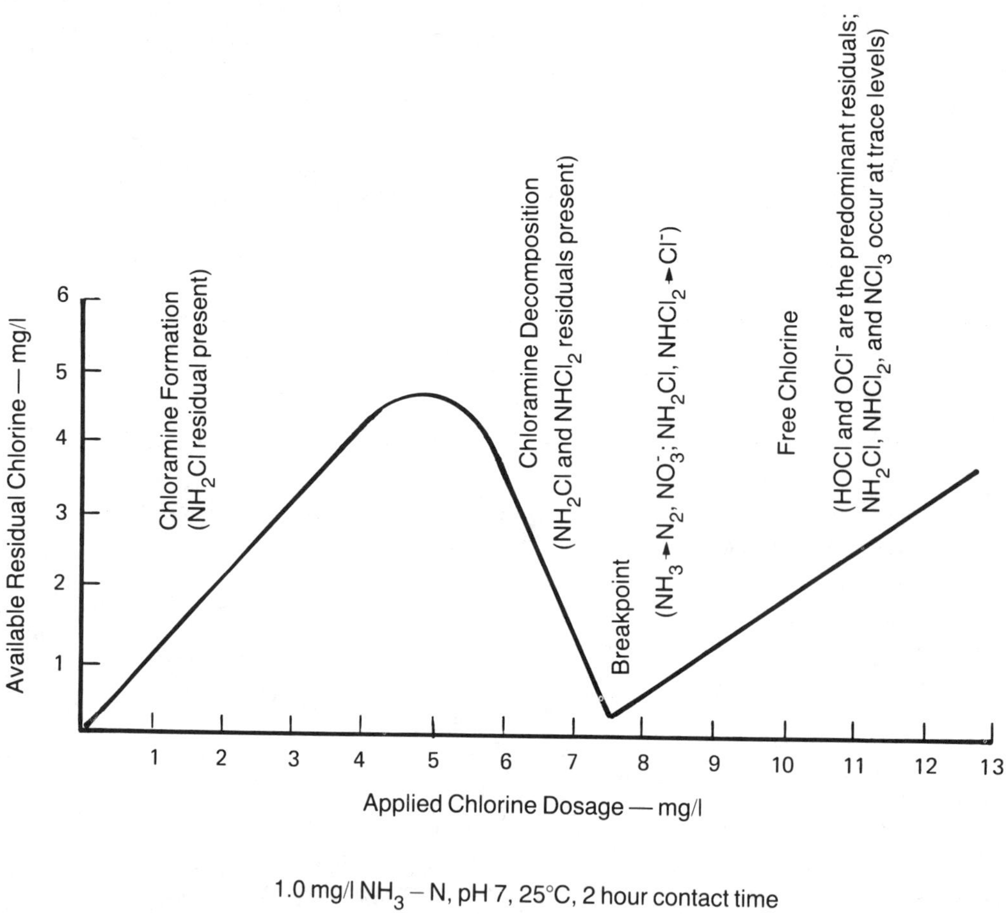

1.0 mg/l NH_3-N, pH 7, 25°C, 2 hour contact time

Source: Wolfe *et al.* [12]. *(Copyright 1984, American Water Works Assoc. Reprinted with permission.)*

FIGURE 10.15-1 Theoretical Breakpoint Chlorination Scheme

10.15.4 Chemistry of Chlorates

The chlorate ion can be formed in solution by the dissolution of chlorate salts or by the reaction of hypochlorous acid with the hypochlorite ion, as follows [10]:

$$2HOCl^0 + OCl^- \rightleftharpoons ClO_3^- + 2H^+ + 2Cl^- \quad (10)$$

Chloric acid ($HClO_3$) is a relatively strong acid, having a pK_a of -2.7. Thus, the chlorate ion has little acid/base interaction in the pH regions of waters of environmental concern. It is, however, a moderately strong oxidizing agent in acid solutions, according to the half-reaction (11), which has an E^0 of 1.45 V at 25°C [9].

$$ClO_3^- + 6H^+ + 6e^- \rightleftharpoons Cl^- + 3H_2O \quad (11)$$

10.15.5 Chlorates in Soil

As noted above, hypochlorites readily oxidize organic matter. For this reason they are expected to be short-lived in soils, and little data on hypochlorite retention by soils are available. However, the fate of chlorates in soil has been studied, because they are used as herbicides. Chlorate can be moderately persistent in soils where leaching does not occur. The application of 800 lb/acre of $NaClO_3$ was found to be effective as an herbicide for three to five years in one Texas study, while other findings have shown persistence for two or three years [7]. One investigation found that 40 cm of water was sufficent to leach chlorate from a clay loam and a fine sandy soil to below phytotoxic levels, while another found that 30, 40, and 30-70 cm of water were required to remove 95% of the chlorate from sandy, humus, and clay soils, respectively [7].

Chlorate can be removed biologically from soils, either through its uptake and reduction by plants or by microorganisms that use it as an oxygen supply in soils of moderate temperature and moisture content [4]. Plants may accumulate chlorate until tissue death occurs [2].

10.15.6 Literature Cited

1. Clapper, T.W., "Chloric Acid and Chlorates," in *Kirk-Othmer Encyclopedia of Chemical Technology*, 3rd ed., Vol. 5, John Wiley & Sons, New York (1982).
2. Crafts, A.S., "Physical Problems Connected with the Use of Sodium Chlorate in Weed Control," *Plant Physiol.*, **10**, 699-711 (1935), as cited in Rosenblatt *et al.* [7].
3. Dean, J.A. (ed.), *Lange's Handbook of Chemistry*, 12th ed., McGraw-Hill Book Co., New York (1979).
4. Jolley, R.L. and J.H. Carpenter, "Aqueous Chemistry of Chlorine: Chemistry, Analysis, and Environmental Fate of Reactive Oxidant Species," ORNL/TM 7788, Oak Ridge National Laboratory, Chemical Technology Division (January 1982).
5. Morris, J.C., "The Acid Ionization Constant of HOCl from 5 to 35°," *J. Phys. Chem.*, **70**, 3798-3805 (1966).

6. Rosenblatt, D.H., "Chlorine and Oxychlorine Species Reactivity with Organic Substances," in *Disinfection, Water and Wastewater*, J.D. Johnson (ed.), Ann Arbor Science Publishers, Ann Arbor, Mich. (1975), as cited in "The Chemistry of Disinfectants in Water: Reactions and Products," National Research Council, PB-292 776 (March 1979).

7. Rosenblatt, D.H., T.A. Miller, J.C. Dacre, I.M. Maj and D.R. Cogley (eds.), "Problem Definition Studies on Potential Pollutants. II: Physical, Chemical, Toxicological, and Biological Properties of 16 Substances," Technical Report 7509, U.S. Army Medical Bioengineering Research and Development Laboratory, Fort Detrick, Frederick, Md. (December 1975).

8. Snoeyink, V.L. and D. Jenkins, *Water Chemistry*, John Wiley & Sons, New York (1980).

9. Weast, R.C. (ed.), *Handbook of Chemistry and Physics*, 54th ed., CRC Press, Cleveland (1973).

10. Williams, F.A., J.D. Millar, F.E. Massoth, W.E. Hensel and I.A. Raben, "Chlorates and Perchlorates: Their Manufacture, Properties and Uses," Southwest Research Inst., San Antonio, Tex., NTIS PB-169 047 (May 1960).

11. Wojtowicz, J.A., "Chlorine Monoxide, Hypochlorous Acid and Hypochlorites," in *Kirk-Othmer Encyclopedia of Chemical Technology*, 3rd ed., Vol. 5, John Wiley & Sons, New York (1982).

12. Wolfe, R.L., N.R. Ward and B.H. Olson, "Inorganic Chloramines as Drinking Water Disinfectants: A Review," *J. Am. Water Works Assoc.*, **76**, 74-88 (1984).

10.16 PEROXIDES

10.16.1 Introduction

Peroxides and peroxo compounds contain the O_2^{-2} group, in which each oxygen atom has a formal oxidation state of -1.

The parent compound of this group is hydrogen peroxide, H_2O_2. Other commercially important inorganic peroxides and peroxo compounds are the metal peroxides (e.g., Na_2O_2 and CaO_2) and the peroxo acids and their salts (e.g., H_2SO_5 and $Na_2B_2H_4O_8$). The prefix peroxo, when used in conjunction with the name of an acid, indicates substitution of -O-O- for -O- groups in the compound.[1] In addition to the true peroxides, a number of compounds form peroxyhydrates, crystalline substances in which the hydrogen peroxide molecule as a whole is present in the crystal, in the same way as water is included in crystals to form hydrates.[2] Commercial uses of a number of peroxides, peroxo compounds, and peroxyhydrates are listed in Table 10.16-1.

Many of the above compounds form hydrogen peroxide upon dissolution in water, and all of them yield hydrogen peroxide by reaction with acid. The following description of their environmental chemistry covers their properties and reactions with water and the environmental reactions of hydrogen peroxide, the primary decomposition product of these compounds in water.

10.16.2 Metal Peroxides

All of the alkali metals and alkaline earth metals (except beryllium and radium) form peroxides. All are solid compounds that are generally stable at environmental temperatures but decompose with the liberation of O_2 at elevated temperatures, usually above 300°C. Zinc forms a peroxide, similar to those of the alkaline earth metals, that is stable up to 150°C. Peroxides of cadmium and mercury have also been prepared, but both are too unstable for practical use. Similarly, several of the transition metals form peroxides or peroxy complexes, but many are unstable and none have commercial applications [1].

The alkali metal peroxides M_2O_2 (where M is Li, Na, K, Rb, or Cs) are readily soluble in water and are therefore very reactive. They react with water, generating heat and forming hydrogen peroxide and the respective alkali metal hydroxide; thus, they form strongly alkaline solutions, are strong oxidants, and are corrosive to metals [1]. With dilute acid solutions, they react to form the alkali metal salt and hydrogen peroxide. In addition, they all react with moisture and carbon dioxide [4]. Sodium peroxide (Na_2O_2) readily absorbs carbon dioxide and water vapor from the air, forming an alkaline solution that decomposes to yield oxygen, sodium carbonate, and sodium hydroxide.

1. Peroxo compounds are also commonly referred to as peroxy compounds; e.g., H_2SO_5 may be called peroxomonosulfuric acid or peroxymonosulfuric acid.
2. The term "hydroperoxidate" has been suggested as an alternative to the term "peroxyhydrate" [3].

TABLE 10.16-1

Commercial Uses of Peroxides and Peroxo Compounds

Compound	Name[a]	Major Uses
Li_2O_2	Lithium peroxide	Production of high-purity, anhydrous lithium oxide
Na_2O_2	Sodium peroxide	Bleaching of wood pulp, cotton, cloth, and sea sponges
MgO_2	Magnesium peroxide	Antifermentative ingredient in some gastric antacid compositions
CaO_2	Calcium peroxide	Dough conditioner in baking industry; ingredient in dentifrices, chewing gums, and antacids; nontoxic disinfectant for seed and grain
SrO_2	Strontium peroxide	Pyrotechnics and tracer-bullet formulations
BaO_2	Barium peroxide	Pyrotechnics and tracer-bullet formulations
ZnO_2	Zinc peroxide	Ingredient of dusting powders; deodorants; pharmaceutical for treatment of infections and skin lesions; in rubber formulations as a cross-linking agent, filler, and pigment
$Na_2B_2H_4O_8 \cdot 3H_2O$	Sodium peroxoborate	Dry carrier of hydrogen peroxide in household laundry dry-bleach and detergent formulations; hard-surface cleansers; denture cleansers; cosmetics; vat and sulfur-dye oxidation; textile bleaching
$Na_2B_2H_4O_8$	Sodium peroxoborate	Denture cleansers; hard-surface cleansers; cosmetics
H_2SO_5	Peroxomonosulfuric acid	Treatment of wastewater (in Europe) (as oxidant for cyanide, thiocyanate, nitrate, phenol and formaldehyde)
$2KH_2SO_5 \cdot KHSO_4 \cdot K_2SO_4$	Potassium peroxomonosulfate (triple salt)	Bleaching agent in denture cleansers, toilet-bowl cleaners, and laundry dry bleaches; disinfectant
$(NH_4)_2S_2O_8$	Diammonium peroxodisulfate	Metal cleaning and etching; maturing agent for wheat flour; initiator of emulsion polymerizations; laboratory reagent; oxidation of dyestuffs
$K_2S_2O_8$	Dipotassium peroxodisulfate	Initiator of emulsion polymerization; hair bleaching; laboratory reagent
$Na_2S_2O_8$	Disodium peroxodisulfate	Metal cleaning and etching; initiator of emulsion polymerizations; starch modification; cosmetics
$2Na_2CO_3 \cdot 3H_2O_2$	Sodium carbonate peroxyhydrate	Dry carrier of hydrogen peroxide in household dry bleach and laundry detergent formulations

a. See text for structures of peroxoborate and peroxosulfate ions.

Source: Hall [1]

The alkaline earth metal peroxides MO_2 (where M is Mg, Ca, Sr, or Ba) and zinc peroxide (ZnO_2) are generally more stable toward moisture than the alkali peroxides, because they are insoluble (or only slightly soluble) in water [1]. However, they dissolve in dilute aqueous acids, reacting to form hydrogen peroxide and the corresponding salts in solution [4].

10.16.3 Peroxo Acids and Salts

Peroxoborate salts of several of the alkali and alkaline earth metals are known. Sodium peroxoborates are the most common and commercially important members of this group. Although the structure of the peroxoborate anion was uncertain long after its existence was discovered, the ion has now been shown to be a "true peroxoborate," with the following structure:

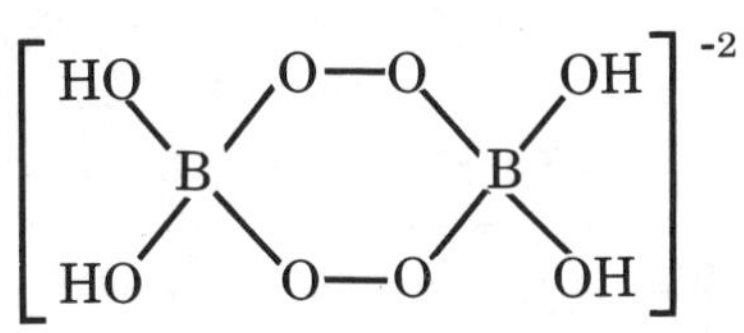

Thus, the compound known as sodium peroxoborate monohydrate, $NaBO_3 \cdot H_2O$, is actually an anhydrous salt, while the "tetrahydrate," $NaBO_3 \cdot 4H_2O$, actually contains only three moles of hydrate water per mole of boron. Both of these compounds are water-soluble, crystalline solids that form alkaline solutions in water. (Solubilities at 25°C are 3g/100g H_2O for the "tetrahydrate" and 2g/100g H_2O for the "monohydrate"; the pHs of the solutions are approximately 10.1 and 10.2, respectively.) Sodium peroxoborate is used in bleaching agents and cleansers, because it reacts to form hydrogen peroxide when dissolved in water [1].

Peroxomonosulfuric acid, H_2SO_5, and peroxodisulfuric acid, $H_2S_2O_8$, have the following structures:

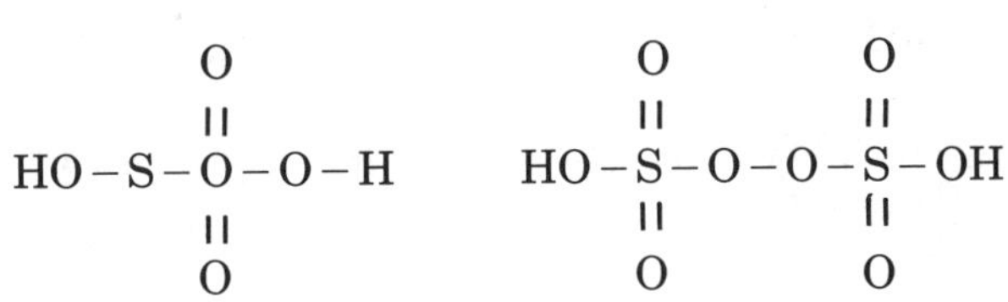

Both are strong acids and strong oxidants. In strongly acidic solutions, they are rapidly hydrolyzed, yielding hydrogen peroxide and sulfuric acid. In addition, peroxodisulfuric acid may decompose to yield sulfuric acid and oxygen, slowly at 20°C but measurably at 40° to 80°C. The peroxodisulfate ion in aqueous solution is one of the strongest oxidizing agents known, but it generally reacts slowly at 20°C. Salts of peroxodisulfate, such as $(NH_4)_2S_2O_8$, $K_2S_2O_8$, and $Na_2S_2O_8$, are water-soluble, crystalline solids. In aqueous solution they undergo the same hydrolysis, decomposition, and oxidation reactions as peroxodisulfuric acid [1].

10.16.4 Hydrogen Peroxide

PROPERTIES AND USES

Hydrogen peroxide (H_2O_2) is a clear, colorless liquid that is completely miscible with water. Pure H_2O_2 freezes at −0.41°C and boils at 150.2°C, while a 30% aqueous solution freezes at −25.7°C and boils at 106.2°C [2]. It is weakly acidic; that is, in aqueous solution, the following proton-transfer reaction occurs:

$$H_2O_2 + H_2O \rightleftharpoons H_3O^+ + O_2H^- \tag{1}$$

This reaction has an equilibrium constant of 1.78×10^{-12} at 20°C [3].

Hydrogen peroxide is available commercially in aqueous solutions with a wide range of concentrations. Its largest use is in the bleaching of textiles, but it has several other bleaching applications as well. Hydrogen peroxide is also used as an agent for the treatment of domestic and industrial waste (an ecologically desirable practice, because it yields only water and/or oxygen upon decomposition), in the manufacture of many organic and inorganic chemicals, in various mining applications, as a propellant, and in numerous specialty uses [2].

VOLATILIZATION

Volatilization of H_2O_2 from an aqueous solution can be significant. However, water is considerably more volatile and therefore accounts for the greater proportion of the vapor pressure above all but the most concentrated aqueous hydrogen peroxide solutions. Table 10.16-2 gives vapor pressures and vapor composition for hydrogen peroxide-water solutions at 20°C. The Henry's Law constant for H_2O_2, based on the data for 0.1 mole fraction of H_2O_2 in water, is 1.1×10^{-8} atm-m^3/mol (20°C).

CHEMICAL REACTIONS

The chemical reactions of hydrogen peroxides may be grouped into the following categories [2]:

- substitutions
- molecular additions
- oxidations and reductions
- decomposition

TABLE 10.16-2

Vapor Pressure and Vapor Composition Over Hydrogen Peroxide-Water Solutions at 20°C

Mole Fraction H_2O_2 in Liquid	Mole Fraction H_2O_2 in Vapor	Total Vapor Pressure (mm Hg)
0	0	17.5
0.1	0.003	15.6
0.2	0.009	13.3
0.3	0.020	10.9
0.4	0.041	8.69
0.5	0.077	6.68
0.6	0.138	5.00
0.7	0.238	3.66
0.8	0.397	2.64
0.9	0.640	1.89
1.0	1	1.36

Source: Schumb, Satterfield, and Wentworth [3], pp. 226-227

The last two are of greater environmental importance; the first two are described briefly below.

Substitution reactions of hydrogen peroxide with organic compounds are reactions in which the peroxy group is transferred to an organic molecule. These reactions are used commercially for the preparation of many organic peroxy compounds [2].

Molecular additions refer to the formation of peroxyhydrates, in which the hydrogen peroxide molecule as a whole is included in the crystalline form of a compound. Compounds that form peroxyhydrates include oxyacid salts, metal peroxides, and nitrogen compounds (e.g., NH_3, $CO(NH_2)_2$, and other amines). An aqueous solution of the peroxyhydrate of a given compound acts as a solution of the pure compound mixed with H_2O_2 [2].

OXIDATION AND REDUCTION

Hydrogen peroxide is a strong oxidizing agent, and most of the uses of hydrogen peroxide and its derivatives depend on this property. Hydrogen peroxide oxidizes a wide range of organic and inorganic substances, including both non-metal species such as sulfite, nitrite and cyanide, and metal ions such as Fe^{+2}. Many metals are dissolved by acidic H_2O_2 solutions as a result of this oxidative activity [3].

Although hydrogen peroxide acts as the oxidant in the majority of its redox reactions, it will reduce species that are stronger oxidizing agents than itself, such as chlorine, hypochlorite, and permanganate, releasing oxygen gas (O_2). The reduction of residual chlorine and hypochlorite in industrial waste streams is used for pollution control [2].

The half-reactions for hydrogen peroxide acting as an oxidizing agent or reducing agent are shown by equations 2 and 3, respectively [5].

$$H_2O_2 + 2H^+ + 2e^- = 2H_2O \qquad E^0 = 1.776V \tag{2}$$

$$O_2 + 2H^+ + 2e^- = H_2O_2 \qquad E^0 = 0.695V \tag{3}$$

The oxidizing potential of hydrogen peroxide is increased in acidic solution and decreased in basic solution. Conversely, its reducing power is decreased in acidic solution and increased in basic solution.

DECOMPOSITION

Although pure hydrogen peroxide of any concentration is very stable, a variety of trace impurities may catalyze its decomposition, according to the net reaction:

$$2H_2O_2 \rightarrow 2H_2O + O_2 \tag{4}$$

This decomposition generates heat (about 100.4 kJ/mol or 24 kcal/mol) and may occur violently; thus, controlling its occurrence is an important safety consideration [2]. The rate of decomposition of hydrogen peroxide is affected by temperature, pH, light, and the presence of catalytic or stabilizing substances.

The rate of decomposition of pure H_2O_2 solution increases markedly with temperature. In the absence of light or catalyzing species, a 90% solution decomposes at the rate of only 1% per year at 30°C, but at 100°C the rate increases to 2% per day, and at 140°C, decomposition is rapid and may generate enough heat to cause the solution to boil. If catalytic contaminants are present in solution, the increase in decomposition rate with temperature may be somewhat less [3].

Both strong acid and strong base promote the decomposition of hydrogen peroxide; the decomposition rate reaches a minimum at a pH of approximately 4.5 to 5.0. For solutions of 50% concentration by weight or greater, the effect of varying the pH in the range 2.5 to 5.5 is slight. The effect of pH is greater for increasingly dilute solutions. The dependence of decomposition rate on pH is also altered by the presence of traces of certain metal ions, such as Fe^{+3}, Cu^{+2}, Cr^{+3}, or Ag^+. When such ions are present, the decomposition rate may reach a sharp maximum at a particular pH, due to the formation of a catalytically active form of the metal [3].

Decomposition of hydrogen peroxide is also promoted by the ultraviolet light present in ordinary sunlight. Unstabilized samples of 90% hydrogen peroxide in glass containers may decompose as much as 1% per day at room temperature when exposed to direct sunlight [3].

Numerous inorganic species, when present in even trace amounts, can catalyze the decomposition of hydrogen peroxide. Even the container in which a hydrogen peroxide solution is placed may catalyze its decomposition. Decomposition catalysts include the halogens (except fluorine), selenium, arsenates, carbon, lead, mercury and several of the transition metals (e.g., Cr, Mn, Fe, Cu, Ag). Zinc also catalyzes decomposition in solutions of less than 40% concentration, but it acts as a stabilizer in concentrated solutions. Many of these catalysts promote decomposition by a cyclical oxidation-reduction process, in which the catalyst is continually interconverted between a more oxidized form and a more reduced form by reaction with hydrogen peroxide. For example, the catalytic decomposition of H_2O_2 by the halogens proceeds according to the following reactions [3]:

$$H_2O_2 + 2X^- + 2H^+ \longrightarrow X_2 + 2H_2O \tag{5}$$

$$H_2O_2 + X_2 \longrightarrow 2X^- + 2H^+ + O_2 \tag{6}$$

The catalytic behavior of many of the metals is strongly pH-dependent, generally being characterized by a pH range within which maximum decomposition occurs. In some cases the metal ion itself is the catalytically active species, while in other cases an oxide or hydroxide appears to cause the decomposition. The rate of H_2O_2 decomposition generally increases with increasing concentration of the catalyst. The effect of a combination of two catalysts is sometimes much greater than the sum of the effects of each catalyst when present separately [3].

Various stabilizing agents are used to prevent the decomposition of commercial hydrogen peroxide solutions. They work by inactivating catalytic species that may be present (as opposed to blocking some self-decomposition process of hydrogen peroxide); thus, the optimum type and amount of stabilizer for a given solution depends on the impurities present [3]. The most common stabilizers are combinations of stannates (tin salts) and phosphates; various inorganic and organic stabilizers have been developed for H_2O_2 solutions destined for specific uses [2].

Each stabilizing agent has an optimum pH at which it is most effective. The combined effect of two stabilizers may be greater than the sum of their separate effects, provided the pH of the solution is in the appropriate range for both stabilizers [3].

10.16.5 Literature Cited

1. Hall, R.E., "Peroxides and Peroxy Compounds, Inorganic," in *Kirk-Othmer Encyclopedia of Chemical Technology*, 3rd ed., Vol. 17, John Wiley & Sons, New York (1982).
2. Kirchner, J.R., "Hydrogen Peroxide," in *Kirk-Othmer Encylopedia of Chemical Technology*, 3rd ed., Vol. 13, John Wiley & Sons, New York (1982).
3. Schumb, W.C., C.N. Satterfield and R.L. Wentworth, *Hydrogen Peroxide*, Reinhold Publishing Corp., New York (1955).
4. Vol'nov, I.I., *Peroxides, Superoxides, and Ozonides of Alkali and Alkaline Earth Metals*, Plenum Press, New York (1966).
5. Weast, R.C. (ed.), *CRC Handbook of Chemistry and Physics*, 67th ed., CRC Press, Boca Raton, Florida (1986).

APPENDIX A
NOTATIONS AND SYMBOLS

The notation and symbols listed below are used in this report. The more infrequently used symbols are defined in the text and are not listed here.

Item	Notation or Symbol
Chemical Species and Phases	
Charged dissolved ions (note subscript "aq" has been omitted throughout)	Mg^{+2}, SO_4^{-2}
Dissolved neutral species	Cl_2^0, O_2^0
Ion pair, dissolved	$MgSO_4^0$
Complexed ion, dissolved	$Fe(CN)_5H_2O^{-3}$, $Mg(OH)_2^+$
Hydrates, mixed solids	$CaSO_4 \cdot 2H_2O(s)$
To distinguish between ion pair and complex when needed	(M,L) = ion pair (M-L) = complexed
General oxidation state (not referring to specific complex)	Cr(III), Tc(IV)
Specific referral to the aquo complexed ion (e.g., aquo-iron (III))	M^{+3} (e.g., Fe^{+3})
In text referral to oxidation state	II, III, IV, etc.
Pure liquid phase [equations only]	$H_2SO_4(l)$, except H_2O
Solid phase [equations only]	$CaSO_4(s)$ or $SiO_2(am)$
Gas phase [equations only]	$Cl_2(g)$
Acidity, hydrated proton	H^+
Electron in half reactions	e^-
Radionuclides	^{99}Tc or 99-technetium
Free Radicals	Free electron not shown (e.g., OH)

(Continued)

Item	Notation or Symbol
Concentration Symbols and Units	
Concentration brackets	$[Mg^{+2}]$, where appropriate
Activity brackets	(Mg^{+2}), where appropriate
Concentration Units:	
mg per liter	mg/l or $mg \cdot l^{-1}$
gram per liter	g/l or $g \cdot l^{-1}$
moles per liter	mol/l or $mol \cdot l^{-1}$ or M
molarity	M
parts per million	ppm
parts per billion	ppb
Other Symbols and Units	
Equilibrium constants	K_1, K_2, etc. } Some exceptions
Rate constants	k_1, k_2, etc. } Some exceptions
Overall stability constants	β_1, β_2 } Some exceptions
Indication of stoichiometric reaction or redox half-reaction. (Species listed do not necessarily exist in form as shown.)	$=$, $\longrightarrow$
Indication of equilibrium reactions. (Species shown are presumed to exist in form shown.)	$\rightleftharpoons$ (or $\rightleftarrows$)
Ionic strength	I (Units are M unless otherwise specified.)
Time:	
seconds	s
minutes	min
hours	hr
days	d
Half life	$t_{1/2}$ ($T_{1/2}$ for radioactivity)

(Continued)

Item	Notation or Symbol
Other Symbols and Units (cont'd.)	
Degrees Kelvin	K
Degrees Fahrenheit	°F
Degrees Celsius	°C
Electron Activity	pe [–log(e⁻)]
Redox potential (referenced to Pt electrode)	Eh
Electrode potential	E
Volts	V

a. Note that the convention for preparing Eh *vs.* pH diagrams (e.g., Pourbaix) is to use a total dissolved concentration of 10^{-6} *M* for a given valence state of an ion. Some authors, however, may deviate from this practice and the diagrams, for example in chapters 6, 7 and 8, may not conform to this practice. For more information on the assumptions used in preparation of these diagrams, see section 2.10.

APPENDIX B CHEMICAL COMPOSITION OF THE ENVIRONMENT

Christopher P. Loreti

The tables in this appendix indicate the chemical composition of various parts of the environment — water, land and air. As most of the data are mean or median values, large variations can be encountered locally. Furthermore, the values should not be interpreted as describing nature in its pristine state, since they are either generalized concentrations in particular environmental compartments or the results of analyses of specific samples.

The tables cover the following subjects:

Water

- Seawater (B-1)
- Fresh water (B-2)

Soil and Sediments

- Average soil composition (B-3)
- Composition of several soils (B-4)
- Composition of sediments (B-5)

Air

- Gaseous composition of the atmosphere (B-6)
- Atmospheric elements in aerosols (B-7)

Other Environmentally Important Substances

- Sewage treatment sludge (B-8)
- Various natural waters and wastewaters (B-9)
- Naturally occurring radionuclides (B-10)

Concentrations of inorganic substances in biota are discussed in Chapter 4 of this report.

The sources of the tabulated data are listed at the end of this appendix, following Table B-10.

TABLE B-1

Elemental Composition of Seawater

Element	Concentration ($\mu g/l$) Mean	Range	Theoretical Species	No. of Sources
Ag	0.04	0.03-2.7	$AgCl_2^-$	4
Al	2	1-8.4	Colloid, $Al(OH)_4^-$	1
Ar	450		Ar	1
As	3.7	0.5-3.7	AsO_4H^{-2}, Organic	3
Au	0.004	0.0005-0.027	$AuCl_2^-$	2
B	4440		$B(OH)_3$	1
Ba	13	2-63	Ba^{+2}	1
Be	0.0056	0.0006	$BeOH^+$	1
Bi	0.02	0.015-0.02	BiO^+?	1
Br	67,300		Br^-	1
C	28,000		HCO_3^-, CO_3^{-2}	1
Ca	412,000		Ca^{+2}	1
Cd	0.11	<0.01-9.4	Colloid, $CdCl_2$	4
Ce	0.0012		Ce^{+3}	1
Cl	19.35×10^6		Cl^-	1
Co	0.02	0.01-4.1	Co^{+2}, $CoCO_3$	5
Cr	0.3	0.2-50	$Cr(OH)_3$, CrO_4^{-2}	2
Cs	0.3	0.15-0.42	Cs^+	3
Cu	0.25	0.05-12	$CuOH^+$, $CuCO_3$	7
Dy	0.00091		$DyOH^{+2}$	1
Er	0.00087		$ErOH^{+2}$	1
Eu	0.00013		$EuOH^{+2}$	1
F	1300		F^-, MgF^+	1
Fe	2	0.03-70	Colloid, $Fe(OH)_2^+$	3
Ga	0.03		$Ga(OH)_4^-$	1
Gd	0.0007		$GdOH^{+2}$	1
Ge	0.05		$GeO_4H_3^-$	1
He	0.007		He	1
Hf	0.007		$Hf(OH)_4$	1
Hg	0.03	0.01-0.22	$HgCl_4^{-2}$, Organic	4

(Continued)

TABLE B-1 (Continued)

Element	Concentration (μg/l) Mean	Range	Theoretical Species	No. of Sources
Ho	0.00022		$HoOH^{+2}$	1
I	60	50-70	I^-, IO_3^-, CH_3I	1
In	0.00011			1
K	399,000		K^+	1
Kr	0.21		Kr	1
La	0.0034		La^{+3}	1
Li	180	170-194	Li^+	1
Lu	0.00015		$LuOH^{+2}$	1
Mg	1.29×10^6		Mg^{+2}	1
Mn	0.2	0.03-21	Mn^{+2}, $MnCl^+$, Colloid	3
Mo	10	4-10	MoO_4^{-2}	1
N	640/15,500		$NH_3 + NO_3^-/N_2$	1
Na	10.77×10^6		Na^+	1
Nb	0.01	0.01-0.015		1
Nd	0.0028		Nd^{+3}	1
Ne	0.12		Ne	1
Ni	0.56	0.13-43	Ni^{+2}, $NiCO_3$	5
P	60	60-88	HPO_4^{-2}, $MgPO_4^-$	1
Pa	$\leqslant 5 \times 10^{-8}$?	$5 \times 10^{-8} - 2 \times 10^{-10}$		1
Pb	0.03	0.03-13	$PbCO_3$, Colloid	4
Po	1.5×10^{-11}	$(0.7\text{-}2) \times 10^{-11}$		2
Pr	0.00064		Pr^{+3}	1
Ra	8.9×10^{-8}	$(3.2\text{-}9) \times 10^{-8}$	Ra^{+2}	1
Rb	120	67-195	Rb^+	3
Re	0.004	0.004-0.0084	ReO_4^-	2
Rn	6×10^{-13}		Rn	1
Ru	0.0007			1
S	905,000		SO_4^{-2}, $NaSO_4^-$	1
Sb	0.24	0.18-5.6	$Sb(OH)_6^-$	3

(Continued)

TABLE B-1 (Continued)

Element	Concentration (μg/l) Mean	Range	Theoretical Species	No. of Sources
Sc	0.0006	0.0006-0.12	$Sc(OH)_3$	2
Se	0.2	0.052-0.2	$SeO_3^{-2} + SeO_4^{-2}$	2
Si	2200	2200-2900	$Si(OH)_4$, Colloid	1
Sm	0.00045		$SmOH^{+2}$	1
Sn	0.004	0.002-0.81	$SnO_4H_3^-$, Organic	1
Sr	7900	7000-8500	Sr^{+2}	1
Ta	0.002			1
Tb	0.00014		$TbOH^{+2}$	1
Te			TeO_3H^-	1
Th	0.001	0.0001-0.22	$Th(OH)_4$	1
Ti	1			1
Tl	0.019			1
Tm	0.00017		$Tm OH^{+2}$	1
U	3.2	0.04-6	$UO_2(CO_3)_3^{-4}$	5
V	2.5	0.9-2.5	$H_2VO_4^-$, HVO_4^{-2}	1
W	0.1	0.001-0.7	WO_4^{-2}	1
Xe	0.05		Xe	1
Y	0.013		$Y(OH)_3$	1
Yb	0.00082		$YbOH^{+2}$	1
Zn	4.9	0.2-48	Zn^{+2}, $ZnOH^+$	7
Zr	0.03		$Zr(OH)_4$	1

Source: Bowen [6]. *(Copyright 1979, Academic Press. Reprinted with permission.)*

TABLE B-2

Elemental Composition of Fresh Water

Element	Concentration (μg/l) Median[a]	Range	Species[b]	No. of Sources
Ag	0.3	0.01-3.5		9
Al	300	8-3500	Colloid	9
Ar	600		Ar	2
As	0.5	0.2-230	Anion ± Methyl	11
Au	0.002	0.001-0.02	Anion ?	6
B	15	7-500	$B(OH)_3$	6
Ba	10	<3-150	Ba^{+2}	8
Be	0.3	0.01-1		5
Bi	0.02?			2
Br	14	0.05-55	Br^-	9
C	11 k	6 k-19 k	HCO_3^-	4
Ca	15 k	2 k-120 k	Ca^{+2}	4
Cd	0.1	0.01-3	e	3
Ce	0.2	0.1-0.2	Colloid	3
Cl	7000	1 k-35 k[c]	Cl^-	4
Co	0.2	0.04-8		10
Cr	1	0.1-6		10
Cs	0.02	0.005-1	Cs^+	8
Cu	3	0.2-30	e	11
Dy		0.005		1
Eu	0.006	0.002-0.009		4
F	100	50-2700[d]	F^-	7
Fe	500	10-1400[c]	Colloid	8
Ga	0.09			1
Hf	0.01	0.005-0.13		2
Hg	0.1	0.0001-2.8	e	13
I	2	0.5-7	I^-	7
K	2200	500-10 k	K^+	5
La	0.1	<0.05-0.8		7
Li	2	0.07-40	Li^+	6
Lu	0.003	0.002-0.005		1
Mg	4000	400-6 k[c]	Mg^{+2}	4
Mn	8	0.02-130		12

(Continued)

TABLE B-2 (Continued)

Element	Concentration (μg/l) Median[a]	Range	Species[b]	No. of Sources
Mo	0.5	0.03-10	MoO_4^{-2}	7
N	50	2-1800	NO_3^-	3
Na	6000	700-25 k[c]	Na^+	5
Nd	0.15	<0.06-0.25		1
Ni	0.5	0.02-27		8
P	20	1-300	$PO_4H_2^-$	4
Pb	3[f]	0.06-120		6
Ra	4×10^{-7}			1
Rb	1	0.6-9	Rb^+	7
Rn	1.7×10^{-12}		Rn	1
S	3700	200-40 k	SO_4^{-2}	4
Sb	0.2	0.01-5	Sb(V)	5
Sc	0.01	0.004-0.04		7
Se	0.2	0.02-1[c]	SeO_3^{-2}	4
Si	7000	500-12 k[c]		4
Sm	0.06?	0.01-0.12		2
Sn	0.009	0.004-0.09	Sn(IV) ± Methyl	4
Sr	70	3-1000	Sr^{+2}	7
Ta	<0.002			1
Tb	0.003	0.001-0.005		1
Th	0.03	0.007-0.1		5
Ti	5	3-18		3
U	0.4	0.002-5		10
V	0.5	0.01-20		9
W	0.03	<0.02-0.1		2
Yb	0.01	0.005-0.2		2
Zn	15	0.2-100	Zn^{+2} e	16
Zr	0.8	0.05-20		3

k = × 1000

a. Most medians are subjective estimates.
b. Evidence is generally weak.
c. Higher values have been reported for some African rivers.
d. Concentrations of 14,000 μg/l have been found in India.
e. Unknown organic complex; may not involve metal-carbon bond.
f. Prehistoric median was probably 0.5 μg/l.

Source: Bowen [6]. *(Copyright 1979, Academic Press. Reprinted with permission.)*

TABLE B-3

Typical Elemental Composition of Soils

Element	Concentration (mg/kg soil) Median	Range	No. of Sources	Element	Concentration (mg/kg soil) Median	Range	No. of Sources
Ag	0.05	0.01-8	7	Mg	5000	400-9000	6
Al	71,000	10 k-300 k	4	Mn	1000	20-10 k	13
As	6	0.1-40	11	Mo	1.2	0.1-40	8
Au		0.001-0.02?	2	N	2000	200-5000	1
B	20	2-270	6	Na	5000	150-25 k	7
Ba	500	100-3000	12	Nb	10	6-300	2
Be	0.3	0.01-40	6	Nd	35	4-63	5
Bi	0.2	0.1-13	3	Ni	50	2-750	12
Br	10	1-110	6	O	490,000		1
C (org.)	20,000	7 k-500 k	2	P	800	35-5300	4
Ca	15,000	700-500 k	6	Pb	35[a]	2-300	12
Cd	0.35	0.01-2	8	Pr	7	3-12	2
Ce	50	3-170	7	Ra	0.8 μ	0.2 μ-5 μ	3
Cl	100	8-1800	5	Rb	150	20-1000	8
Co	8	0.05-65	15	S	700	30-1600	2
Cr	70	5-1500	13	Sb	1	0.2-10	9
Cs	4	0.3-20	7	Sc	7	0.5-55	10
Cu	30	2-250	10	Se	0.4	0.1-2.0[b]	8
Dy	5	2-12	2	Si	330,000	250 k-410 k	3
Er	2	0.6-6	2	Sm	4.5	0.6-23	7
Eu	1	0.1-3.2	7	Sn	4	1-200	6
F	200	20-700	6	Sr	250	4-2000	8
Fe	40,000	2 k-550 k	11	Ta	2	0.4-6	3
Ga	20	2-100	8	Tb	0.7	0.1-1.6	6
Gd	4	2-6	2	Th	9	1-35	9
Ge	1	0.1-50	4	Ti	5000	150-25 k	7
Hf	6	0.5-34	6	Tl	0.2	0.1-0.8	2
Hg	0.06	0.01-0.5	6	Tm	0.6	0.3-1.2	2
Ho	0.6	0.4-2	2	U	2	0.7-9	7
I	5	0.1-25	6	V	90	3-500	9
In	1	0.7-3?	2	W	1.5	0.5-83	2
K	14,000	80-37 k	6	Y	40	10-250	4
La	40	2-180	9	Yb	3	0.04-12	8
Li	25	3-350	8	Zn	90	1-900	12
Lu	0.4	0.1-0.7	4	Zr	400	60-2000	7

k = × 1000; μ = $\times 10^{-6}$

a. 12 mg/kg prior to global contamination; up to 1000 mg/kg in urban soils.

b. From Bohn, McNeal and O'Connor [4]

Source: Bowen [6], unless otherwise noted. *(Copyright 1979, Academic Press. Reprinted with permission.)*

TABLE B-4

Composition of Representative United States Surface Soils
(in percent)

Constituent	Norfolk Fine Sand, Florida	Sassafras Sandy Loam, Virginia	Ontario Loam, New York	Loam from Ely, Nevada	Hagerstown Silt Loam, Tennessee	Cascade Silt Loam, Oregon	Marshall Silt Loam, Iowa	Summit Clay from Kansas
SiO_2	91.49	85.96	76.54	61.69	73.11	70.40	72.63	71.60
TiO_2	0.50	0.59	0.64	0.47	1.05	1.08	0.63	0.81
Fe_2O_3	1.75	1.74	3.43	3.87	6.12	3.90	3.14	3.56
Al_2O_3	4.51	6.26	9.38	13.77	8.30	13.14	12.03	11.45
MnO	0.007	0.04	0.08	0.12	0.44	0.07	0.10	0.06
CaO	0.01	0.40	0.80	5.48	0.37	1.78	0.79	0.97
MgO	0.02	0.36	0.75	2.60	0.45	0.97	0.82	0.86
K_2O	0.16	1.54	1.95	2.90	0.91	2.11	2.23	2.42
Na_2O	Trace	0.58	1.04	1.47	0.20	1.98	1.36	1.04
P_2O_5	0.05	0.02	0.10	0.18	0.16	0.16	0.12	0.09
SO_3	0.05	0.07	0.08	0.12	0.07	0.21	0.12	0.11
Ignition[a]	1.83	1.91	5.30	7.62	8.82	4.25	6.01	6.60
Nitrogen	0.02	0.02	0.16	0.10	0.27	0.08	0.17	0.09

a. Percent lost by ignition includes organic matter.

Source: Buckman and Brady [7]. *(Copyright 1969, Macmillan Co. Reprinted with permission.)*

TABLE B-5

Composition of Surface Sediments
(mg/kg dry weight)

	White Oak Creek, Ky. [1]	Ohio River [1]	Turkey Creek, Ohio & Pa. [2]	Lake Erie [2]	Elliot Bay (Puget Sound) [14]	So. Calif. Bight [8] Inner Shelves	So. Calif. Bight [8] Outer Shelves	National Median[a] [5]	National 95th Percentile[a] [5]
Total Kjeldahl Nitrogen	2160	1920	420	6600					
NO_2-N	0.12	0.44	5 (60 NH_3-N)	1 (45 NH_3-N)					
NO_3-N	≤0.64	2.80	20	3					
P	1370	1080	70 (PO_4-P)	90 (PO_4-P)					
Organic Matter	14.1%	3.66%	1700 (TOC)[b]	3300 (TOC)[b]					
Oil & Grease	27,200	1400	90 (oil)	140					
Cd	9.3	1.60	1	<1	0.81	0.57	0.52	1.0	12
Cr	560	34.6	7	19	47	56	53		
Cu			11	18	48	15	12	4.0[c]	32[c]
Fe	126,000	33,710	20,700	28,900					
Hg	0.113	0.143	<1	2	0.22			<1.0	1.0
Pb	246	25.6	16	31	94	17	9.1	16	199
Ni	27.5	65.5	21	32	37	19	16	13	99
Zn	6600	216	64	105	551	54	31	41	379

a. Based on analyses of stream, river, lake and reservoir sediments across the United States at thousands of sites (hundreds for Cu).
b. TOC = total organic carbon; the organic matter fraction is typically taken to be 1.724 times the organic carbon fraction.
c. Wet weight basis.

TABLE B-6

Mean Composition of the Atmosphere[a]

Gas	Percentage by Volume	Partial Pressure (atm)
N_2	78.1	0.781
O_2	20.9	0.209
Ar	0.93	0.0093
H_2O	0.1-2.8	0.028
CO_2	0.03	0.0003
Ne	1.8×10^{-3}	1.8×10^{-5}
He	5.2×10^{-4}	5.2×10^{-6}
CH_4	1.5×10^{-4}	1.5×10^{-6}
Kr	1.1×10^{-4}	1.1×10^{-6}
CO	$(0.06\text{-}1) \times 10^{-4}$	$(0.6\text{-}1) \times 10^{-6}$
SO_2	1×10^{-4}	1×10^{-6}
N_2O	5×10^{-5}	5×10^{-7}
H_2	5×10^{-5}	5×10^{-7}
O_3	$(0.1\text{-}1.0) \times 10^{-5}$	$(0.1\text{-}1.0) \times 10^{-7}$
Xe	8.7×10^{-6}	8.7×10^{-8}
NO_2	$(0.05\text{-}2) \times 10^{-5}$	$(0.05\text{-}2) \times 10^{-8}$
Rn	6×10^{-18}	6×10^{-20}

a. Values for minor components, especially those influenced by man's activities (such as SO_2 and NO_2) may vary greatly from place to place.

Source: Mirtov [16]. *(Copyright 1980, John Wiley & Sons. Reprinted with permission.)*

TABLE B-7

Elemental Content of Air from Various Sources
(ng/m^3 air at STP)

	Clean Air				Contaminated Air			
	Shetland 1976	N. Norway 1974	N.W. Canada 1971	South Pole 1977	Europe Median (Range)	North America Median (Range)	Japan (2 sites)	Hawaii or Etna Volcano
Ag	<0.07			<0.0004	1 (0.2-7)	1 (<0.04-2.4)		30
Al	60	32	66	0.82	600 (160-2900)	2000 (150-3500)	1600-6200	20 k
As	0.6	1.9	0.3	0.007	16 (1.5-53)	15 (1.7-40)	1.5-190	5.5-850
Au	0.005			0.00004	(0.001-0.006)	(<0.003-0.3)	0.05	8
B					3.5	4		680
Ba		0.73		0.016	(6-90)	12 (0.2-28)		
Be					(0.9-4)	0.2 (0.1-0.3)		
Bi					0.3	(<1-3)		
Br	15	4.4	0.5	1.4	460 (13-2500)	240 (8-1200)	90-600	31-21 k
Ca	820	42	40	0.5	1400 (430-4300)	2800 (100-7000)	2 k-15 k	460-69 k
Cd	<0.8			<0.015	(0.5-620)	(<1-41)	25	8-92
Ce	0.1	0.06	0.24	0.0023	0.9 (0.2-4)	6 (0.02-20)	2-18	
Cl	4400	440	9	2.6	3000 (1400-5400)	1800 (9-6000)	500-11 k	1 k-990 k
Co	0.06	0.06	0.04	0.0005	(0.2-37)	3 (0.13-23)	1.5-4	4.5-27
Cr	0.7	0.7	0.6	0.005	25 (1-140)	60 (1-300)	20-70	45-67
Cs	0.04	0.02		0.0001	0.5 (0.06-1.5)	0.2 (0.07-0.3)	1.4	14
Cu	20	2.5	0.9	0.036	340 (8-4900)	280 (5-1100)		200-3 k
Dy					0.04			
Er					0.026			
Eu	0.004			0.000017	0.02 (0.005-0.08)	0.35 (0.01-1.7)		
F					1.5	(10-400)		high
Fe	90	48	71	0.84	1400 (130-5900)	3600 (260-14 k)	1 k-14 k	1 k-10 k
Ga				<0.0001	0.23	1		
Gd					0.045			
Ge					2.8			
Hf				0.00006	0:3	(0.0005-0.04)		
Hg	<0.04	0.01	0.06		(<0.009-2.8)	(0.07-38)	1.6	18-250
Ho					0.013			
I	4	0.6	0.2	0.08	6 (3-15)	20 (0.4-60)	6-10	
In	<0.02	0.004	0.001	0.00005	0.2 (0.03-0.1)(*sic*)	0.09 (0.02-0.14)	1.2	
Ir					0.00027			
K	170	48	54	0.7	560 (140-1700)	8000 (200-40 k)	1400	94 k
La	<0.2	0.03	0.09	0.0005	1.5 (<0.1-10)	2.5 (0.5-6)	5	
Li					2.3			
Mg	480	77	16	1	430 (<160-1000)	3000 (60-11 k)	480-3 k	325
Mn	3	3	1.5	0.01	43 (9-210)	150 (6-900)	80-600	55-1300

(Continued)

TABLE B-7 (Continued)

	Clean Air				Contaminated Air			
	Shetland 1976	N. Norway 1974	N.W. Canada 1971	South Pole 1977	Europe Median (Range)	North America Median (Range)	Japan (2 sites)	Hawaii or Etna Volcano
Mo	<0.2				(<0.2-3.2)	(<1-10)		
Na	3300	550	18	3	1000 (350-1500)	1500 (120-5500)	1 k-7 k	110-87 k
Nb					0.12			
Nd					0.28	1		
Ni	<4	1.2	<2		25 (4-120)	90 (<1-120)	5-47	330
P					180	1100		
Pb	21		9-14	0.63	120 (55-340)	2700 (45-13 k)	200	28-1200
Pd					0.0007			
Pr					0.065			
Pt					<0.01			
Rb	<1			0.002	3 (<1-6.6)	1.1	0.6-20	
Rh					0.27			
Ru					0.27			
S				50	3200	8000 (2 k-13 k)		300 k
Sb	0.4	0.33	0.13	0.0008	8 (0.6-32)	12 (0.08-55)	2-63	45
Sc	0.015	0.005	0.044	0.00012	0.15 (0.03-0.7)	1.3 (0.1-3)	0.8	1-2
Se	0.5	0.26	0.043	0.0056	3 (0.15-11)	5 (0.06-30)	7.5	9-2100
Si		100			(1 k-6.3 k)	(5 k-63 k)		2700
Sm	<0.01	0.003	0.013	0.00009	0.08 (0.02-0.26)	0.4 (0.2-1)		
Sn					(1.5-800)	(<10-70)		
Sr				<0.5	2.2	(<7-50)	5-50	
Ta						0.011		
Tb					0.01			
Te					0.18			
Th	0.02	0.011	0.052	0.00014	0.16 (0.03-1)	(0.05-1.3)	1.3	
Ti	10	2.6	5	0.1	85 (22-210)	(<10-230)	140-1 k	
Tl					0.06	0.22		
Tm					0.004			
U					0.02	<0.5		
V	3	1.9	0.2	0.0015	30 (5-92)	400 (1-2000)	180	79
W	<0.5		0.016	0.0015	0.7 (0.35-1.5)	4 (0.03-6)	5-18	
Y					0.22		0.5-6	
Yb				<0.00005	(0.26-1.4)	(0.004-0.02)		
Zn	15	10	4	0.03	1200 (13-16 k)	500 (<10-1700)	100-6 k	10 k
Zr					0.7	5 (4-26)	3-25	

k = × 1000. Median values without ranges are often based on a single determination. Most of the data from Europe, North America and Japan came from large cities.

Source: Bowen [6]. *(Copyright 1979, Academic Press. Reprinted with permission.)*

TABLE B-8

Metal Concentrations in Sewage Sludge[a]

Component	Sample		Concentration		
	Type[b]	Number	Range	Median	Mean
			%		
Al	Anaerobic	73	0.1-13.5	0.5	1.7
	Aerobic	37	0.1-2.3	0.4	0.7
Ca	Anaerobic	87	1.9-20.0	4.9	5.8
	Aerobic	37	0.6-13.5	3.0	3.3
Fe	Anaerobic	96	0.1-15.3	1.2	1.6
	Aerobic	38	0.1-4.0	1.0	1.1
K	Anaerobic	86	0.02-2.64	0.30	0.52
	Aerobic	37	0.08-1.10	0.38	0.46
Mg	Anaerobic	87	0.03-1.92	0.48	0.58
	Aerobic	37	0.03-1.10	0.41	0.52
Na	Anaerobic	73	0.01-2.19	0.73	0.70
	Aerobic	36	0.03-3.07	0.77	1.11
			mg/kg		
As	Anaerobic	3	10-230	116	119
	Aerobic	—	—	—	—
Ba	Anaerobic	27	100-9000	500	800
	Aerobic	10	100-300	200	200
B	Anaerobic	62	12-760	36	97
	Aerobic	29	17-74	33	40
Cd	Anaerobic	98	3-3,410	16	106
	Aerobic	57	5-2,170	16	135
Co	Anaerobic	4	3-18	7.0	8.8
	Aerobic	—	—	—	—
Cr	Anaerobic	94	24-28,850	1,350	2,070
	Aerobic	53	10-13,600	260	1,270
Cu	Anaerobic	108	85-10,100	1,000	1,420
	Aerobic	58	85-2,900	970	940
Hg	Anaerobic	35	0.5-10,600	5	1,100
	Aerobic	20	1.0-22	5	7
Mn	Anaerobic	81	58-7,100	280	400
	Aerobic	38	55-1,120	340	420
Mo	Anaerobic	9	24-30	30	29
	Aerobic	3	30-30	30	30
Ni	Anaerobic	85	2-3,520	85	400
	Aerobic	46	2-1,700	31	150
Pb	Anaerobic	98	58-19,730	540	1,640
	Aerobic	57	13-15,000	300	720
Zn	Anaerobic	108	108-27,800	1,890	3,380
	Aerobic	58	108-14,900	1,800	2,170

a. From more than 150 treatment plants in 8 states.
b. Method of sludge digestion.
Souce: Sommers [18]. *(Copyright 1977, American Society of Agronomy, Inc. Reprinted with permission.)*

TABLE B-9

Composition of Several Natural Waters and Wastewaters
(mg/l, unless otherwise noted)

	Mississippi River [10]	Cowles Creek, Ohio [22]	Lake Erie [19]	Lake Erie [22]	Raw Well Water [11]	Raw Well Water Dayton, Ohio [17]
pH (pH units)	7.4-8.5 (min-max)	7	7.7	7.8	7.6	
D.O.[a]	10.0 (89% sat.)	7.5		11.0		
Conductivity (μmho/cm)	333	370			314[b]	
Hardness (as $CaCO_3$)		146			356	369
Alkalinity (as $CaCO_3$)	134	96			326	
TOC[c]	15					
Dissolved Organic C						
Suspended Solids	62	21		10.0		
TDS[d]	210	269		200	439	434
BOD_5[e]	3.7	2.0		1.7		
COD[f]	29			10.0		
SO_4		41	24.1	0.002 (sulfide)	83	84
NO_3	1.06 (NO_3-N)	0.32 (NO_3-N + NO_2-N)			5.0[b]	13
HCO_3			122		326	339
NH_4	0.21 (NH_4-N)	0.12 (NH_3-N)		0.04 (NH_3)		
Cl		26	8.9		19	9.6
P	0.22 (total)	0.056 (total)				
Na			9.2		11 (Na + K)	8.2
Ca	37		40.1		89	92
Mg	16		9.7		33	34
Fe	1.98	1.24		0.05 (dis)	0.23	0.09 [Fe(III)]
K			2.0		11 (Na + K)	1.4

(Continued)

TABLE B-9 (Continued)

	Rain & Snow		Sanitary Landfill Leachate		Secondary Sewage Treatment Effluent		
	[15][g]	[9][h]	[12]	[20]	[21][i]	[23][j]	[3][k]
pH (pH units)	4.1	4.1	4-8	3.7-8.5			7.0
D.O.[a]							
Conductivity (μmho/cm)				2810-16,800			
Hardness (as $CaCO_3$)				0-22,800			
Alkalinity (as $CaCO_3$)			500-10,000	0-20,850			
TOC[c]				256-28,000	65		26
Dissolved Organic C			200-30,000 (total)				
Suspended Solids				10-700	20	25	18
TDS[d]			5000-40,000	584-44,900		400	
BOD_5[e]				81-33,360	13	25	18
COD[f]			1000-90,000	40-89,520	68	70	68
SO_4	2.88	3.74	10-1000	1-1558			
NO_3	0.74	1.96	0.1-10	0.2-10.3 (NO_3-N + NO_2-N)		20[l]	
HCO_3							
NH_4	0.22	0.48	10-1000	0-1106 (NH_4-N)	5 (NH_3-N)	20[l]	16 (NH_3)
Cl	0.25	0.38	300-3000	4.7-2467		45	
P			1-100 (as PO_4)	0-130 (total)		10 (total)	
Na	0.11	0.26	200-1200	0-7700		50	
Ca	0.16	0.42	100-3000	60-7200	79	24	
Mg	0.049	0.09	100-1500	17-15,600	27	17	
Fe			1-1000 (total)	0-2820	0.392		
K	0.078	0.13	200-1000	28-3370		14	

a. Dissolved oxygen
b. After treatment
c. Total organic carbon
d. Total dissolved solids
e. 5-day biochemical oxygen demand
f. Chemical oxygen demand
g. Hubbard Brook, N.H.
h. 20 km north of Baltimore, Md.
i. Indianapolis, Ind.
j. Typical effluent
k. St. Louis, Mo.
l. Total N

TABLE B-10

Concentrations of Natural Radionuclides in the Environment

Radionuclide	Half-life	Atmosphere[a]		Marine Environment: Seawater		Marine Environment: Marine Sediment		Continental Earth's Crust	
		mg/m^3	pCi/m^3	ppm	pCi/l	ppm	pCi/kg	ppm	pCi/kg
Terrigenous radionuclides									
^{40}K	1.25×10^9y	—	—	4.8×10^{-2}	340	~3	$\sim 2 \times 10^4$	2.4	1.7×10^4
^{87}Rb	4.7×10^{10}y	—	—	3.4×10^{-2}	3	—	—	25	2.2×10^3
Thorium decay series									
^{228}Ra	6.7 y	—	—	1×10^{-14}	~0.002	2×10^{-9}	450	5×10^{-9}	1.1×10^3
^{228}Th	1.9 y	—	—	2×10^{-15}	2×10^{-3}	7×10^{-10}	560	1.4×10^{-9}	1.1×10^3
^{232}Th	1.42×10^{10}y	—	—	1×10^{-6}	1×10^{-4}	5	540	10	1.1×10^3
Actinium decay series									
^{227}Ac	21.6 y	—	—	$<1.5 \times 10^{-12}$	<0.1	6×10^{-9}	430	5.5×10^{-10}	40
^{231}Pa	3.43×10^4y	—	—	2×10^{-12}	1×10^{-4}	1×10^{-5}	450	1×10^{-6}	40
^{235}U	7.13×10^8y	—	—	2.1×10^{-5}	0.05	7×10^{-3}	15	0.02	40
Uranium decay series									
^{210}Pb	21 y	—	—	6×10^{-13}	0.05	5×10^{-8}	4×10^3	1×10^{-8}	900
^{210}Po	138 d	—	—	7×10^{-15}	0.03	9×10^{-10}	4×10^3	2×10^{-10}	900
^{222}Rn	3.8 d	?	?	1.3×10^{-15}	0.2	3×10^{-11}	5×10^3	$<6 \times 10^{-12}$	<900
^{226}Ra	1662 y	—	—	1×10^{-10}	0.1	4×10^{-6}	4×10^3	1×10^{-6}	900
^{230}Th	7.52×10^4y	—	—	2×10^{-11}	4×10^{-4}	2×10^{-4}	4×10^3	4×10^{-5}	900
^{234}Th	24.1 d	—	—	4×10^{-14}	1	1.4×10^{-11}	330	4×10^{-11}	900
^{234}U	2.48×10^5y	—	—	2×10^{-7}	1.2	8×10^{-5}	500	1.5×10^{-4}	900
^{238}U	4.5×10^9y	—	—	3×10^{-3}	1	1	300	3	900
Cosmogenic radionuclides									
^{3}H	12.3 y	6×10^{-14}	6×10^{-1}	1.7×10^{-15}	1.6×10^{-2}	—	—		
^{7}Be	53 d	5×10^{-16}	2×10^{-1}	$<5 \times 10^{-14}$	<20	—	—		
^{10}Be	2.5×10^6y	4×10^{-12}	6×10^{-5}	3×10^{-14}	4.5×10^{-7}	$(1\text{-}3) \times 10^{-7}$	1-4		
^{14}C	5730 y	4×10^{-10}	2	2×10^{-11}	0.1	$(0.1\text{-}1) \times 10^{-7}$	40-400		
^{26}Al	7.4×10^5y	4×10^{-16}	7×10^{-9}	3×10^{-16}	5×10^{-9}	—	—		
^{32}Si	~700 y	6×10^{-16}	1×10^{-5}	6×10^{-16}	1×10^{-5}	$<2 \times 10^{-10}$	<3		

a. Based on air at N.T.P., 0°C and 1 atm.

Source: Fukai and Yokoyama [13]. *(Copyright 1982, Springer-Verlag. Reprinted with permission.)*

Sources of Data for Appendix B

The bracketed numbers in Tables B-1 through B-10 indicate the applicable sources from among those listed below. An asterisk (*) before the entry indicates that the source version was modified for use in this report.

1. Arthur D. Little, Inc., *Environmental Impact Assessment — Proposed Coke Oven Facility*, Report No. C-79086 to ARMCO Steel Corp., Cambridge, Mass. (October 1977).

*2. Arthur D. Little, Inc., *Report on the Environmental Impacts of U.S. Steel Corporation's Proposed Lakefront Plant*, Vol. 3, Report to U.S. Steel Corp., Cambridge, Mass. (1978).

3. Arthur D. Little, Inc., *Sources of Toxic Pollutants Found in Influents to Sewage Treatment Plants: III. Coldwater Creek Drainage Basin, St. Louis, Missouri*, Report No. C-81099-16 to U.S. EPA, Cambridge, Mass. (1979).

4. Bohn, H.L., B.L. McNeal and G.A. O'Connor, *Soil Chemistry,* John Wiley & Sons, New York (1979).

5. Bolton, H.S., R.J. Breteler, B.W. Vigon, J.A. Scanlon and S.L. Clark, *National Perspective on Sediment Quality,* Battelle Washington Environmental Program Office, Report to Office of Water Regulations and Standards, U.S. EPA, under Contract 68-01-6986, Washington, D.C. (May 1985).

6. Bowen, H.J.M., *Environmental Chemistry of the Elements*, Academic Press, London (1979).

7. Buckman, H.O. and N.C. Brady, *The Nature and Properties of Soils*, 7th ed., Macmillan Co., New York (1969).

8. Chow, T.J. and J. Earl, *Trace Metals in Sediments: Southern California Baseline Study*, Report 4.1.1, Vol. III, Final Report to U.S. Dept. of Interior, Bureau of Land Management, Washington, D.C. (1978) as given in Arthur D. Little, Inc., *Point Arguello Field and Gaviota Processing Facility Area Study and Chevron/Texaco Development Plans EIR/EIS*, Appendix H, Public Draft Report to County of Santa Barbara, Calif. (July 1984).

9. Cleaves, E.T., D.W. Fisher and O.P. Bricker, "Chemical Weathering of Serpentinite in the Eastern Piedmont of Maryland," *Geol. Soc. Amer. Bull.,* **85** (1974), as given in Freeze and Cherry [12].

10. Delfino, J.J., "Effects of River Discharge and Suspended Sediment on Water Quality in the Mississippi River," *J. Environ. Sci. Health,* **AI2**, 79-94 (1977).

11. Durfor, C.N. and E. Becker, *Public Water Supplies of the 100 Largest Cities in the United States, 1962*, Water Supply Paper 1812, U.S. Geological Survey, Washington, D.C. (1964).

12. Freeze, R.A. and J.A. Cherry, *Groundwater*, Prentice-Hall, Englewood Cliffs, N.J. (1979).

13. Fukai, R. and Y. Yokoyama, "Natural Radionuclides in the Environment," in Vol. 1, Part B of *The Handbook of Environmental Chemistry*, O. Hutzinger (ed.), Springer-Verlag, Berlin (1982).

14. JRB Associates, *Initial Evaluation of Alternatives for Development of Sediment-related Criteria for Toxic Contaminants in Marine Waters (Puget Sound) —Phase II: Development and Testing of Sediment-Water Equilibrium Partitioning Approach*, Final Report to U.S. EPA, Bellevue, Wash. (1984).

*15. Likens, G.E., "Acid Precipitation," *Chem. Eng. News*, 29-44 (Nov. 22, 1976) as given in Lerman, A., *Geochemical Processes: Water and Sediment Environments*, John Wiley & Sons, New York (1979).

16. Mirtov, B.A., "Gaseous Composition of the Atmosphere and Its Analysis," Akad Nauk, SSSR, Inst. Prikl. Geofiz Moskva, as given in Snoeyink and Jenkins [17].

17. Snoeyink, V.L. and D. Jenkins, *Water Chemistry*, John Wiley & Sons, New York (1980).

*18. Sommers, L.E., "Chemical Composition of Sewage Sludges and Analysis of Their Potential Use as Fertilizers," *J. Environ. Qual.*, **6**, 225-32 (1977).

*19. Stumm, W. and J.J. Morgan, *Aquatic Chemistry*, 2nd ed., John Wiley & Sons, New York (1981).

20. U.S. Environmental Protection Agency, *Evaluation of Leachate Treatment: Vol. I — Characterization of Leachate*," Report No. EPA-600/2-77-186a, Washington, D.C. (September 1977).

21. U.S. Environmental Protection Agency, *Fate of Priority Pollutants in Publicly Owned Treatment Works*, Vol. I, Report No. EPA-440/1-82-303, Washington, D.C. (September 1982).

*22. U.S. Environmental Protection Agency, *Northeast Ohio Tributaries to Lake Erie — Waste Load Allocation Report* (March 1974), as given in Ref. [2].

23. U.S. Environmental Protection Agency, *Waste Treatment and Reuse by Land Application*, Vol. 1, Report No. EPA-660/2-73-006a, Washington, D.C. (August 1973), as given in Clark, J.W., W. Viessman, Jr. and M.J. Hammer, *Water Supply and Pollution Control*, 3rd ed., Harper and Row, New York (1977).

APPENDIX C. CLASSIFICATION AND PROPERTIES OF SOILS

Janice W. Green

LIST OF TABLES

LIST OF FIGURES

APPENDIX C — CLASSIFICATION AND PROPERTIES OF SOILS

Characterization of the site and soils of an area impacted by inorganic pollutants is helpful in understanding the migration potential and treatment possibilities. Site location, topography and soil properties are discussed in this appendix to aid the reader in generally predicting the fate of inorganic pollutants that enter the soil.

C.1 SITE LOCATION AND TOPOGRAPHY

The nature of the landscape on which an inorganic pollutant is located can dictate the type of action needed to remove or immobilize the pollutant. Also, information on the landform can assist in estimating depth to groundwater, tendency to erode, and flooding potential (Table C-1).

Warner [11] describes four types of sites that have especially high pollution potential:

(1) Sloping sites with relatively impermeable bedrock (e.g., shale, dense limestone, crystalline igneous rock) within 0.6 meter from the surface have a high potential for erosion, seepage and overland flow of contaminated effluent.

(2) Sites located in karst[1] topography (characterized by clayey residual soils overlying limestone or dolomite with moderate to high fracture and solution porosity and permeability) have a high potential for contamination of groundwater. (Although infiltration in the soil itself is slow, liquids can rapidly enter bedrock where soil is absent, creating sinkholes and paths for direct flow into groundwater systems.)

(3) Sites with little topographic relief, where the groundwater table is at or very near the surface (e.g., old lake beds, floodplains), have a high groundwater pollution potential.

(4) Sites with fractured bedrock and a shallow soil depth (e.g., in granitic areas) also have a high groundwater pollution potential.

C.2 SOIL CHARACTERISTICS

The hydraulic, physical and chemical properties of soil can affect the quantity of inorganic contaminants in runoff waters and suspended sediments. They can also affect the migration of these contaminants through the unsaturated zone into the groundwater. The unsaturated zone is the region extending from the ground surface to the upper surface of the principal water-bearing formation. In this zone, water in pore spaces primarily coexists with air, although there may be pockets of saturated

1. This and a number of other soil-science terms are defined in the Glossary (section C-5).

TABLE C-1

Influence of Landforms and Topography of Sites on Migration of Pollutant Constituents with Water

Landscape Descriptor: Topography	Migration Potential	
	Groundwater	Erosion
Eskers: long, low, narrow ridges	High permeability	High
Kames: low, steepsided ridges	High permeability	High
Outwash plains: near-level, broad tracts gently sloping from origin	Potential high water table, especially on fringe	Low
Terraces: flat areas separated by steps, commonly between river and upland :		
Glacial	High permeability	High at steps, low elsewhere
Marine	High to moderate permeability	
Lake	Low permeability	
Lake beds: broad, exceptionally flat	Low permeability, high water table	Very low
Glacial till plains: young — broad, gently rolling;	High water table	Low-moderate
old — broad, level areas	Moderately deep water table	Very low
Alluvial fans: smooth, moderate slopes, transitional area between highlands and lowlands	High permeability, potential high water table at bottom of fan	Low-moderate
Playas: broad, exceptionally flat surfaces	Low permeability	Low
Loess: undulating topography with smoothly rounded hills	Moderate permeability	High

(Continued)

TABLE C-1 (Continued)

Landscape Descriptor: Topography	Migration Potential	
	Groundwater	Erosion
Moraine	a	a
Floodplain		Low-moderate
Delta		Low-moderate
Beach ridges	High groundwater	High
Coastal plains	a	a
Tidal flat		Low
Sand dunes		High (wind)

a. Varies. Topography must be specified in greater detail.

Source: Adapted from Ryan and Loehr [9].

soil. Perched water tables may develop at interfaces of layers with greatly different textures. Prolonged infiltration may also result in saturated conditions. The top of the unsaturated zone usually consists of topsoils, which are weathered geological materials about 3 to 6 feet deep, arranged in more or less well-developed profiles. Water movement in the unsaturated topsoil is slower than when the soil is saturated, e.g., from a long rain. The pores are partially full of air, and the water is tightly held by the particles.

Weathered topsoil materials gradually merge with underlying sands. The topsoil differs from the material below it, in that it is more weathered, contains organic matter and the biological life associated with organic matter, and is the zone of plant-root growth. The entire unsaturated zone may be hundreds of feet thick, and the travel time of pollutants may be hundreds of thousands of years; other regions may be underlain by shallow, potable aquifers, which are especially susceptible to contamination due to short travel times and reduced potential for pollutant attenuation.

Table C-2 lists various site-soil properties that affect the migration of inorganic pollutants. The rates of infiltration and permeability are the most important factors with regard to water movement through the soil and, thus, contaminant mobility. Clearly, if information on all of the listed characteristics could be obtained, potential migration could be fairly accurately predicted. It should be kept in mind, however, that infiltration and permeability (as well as other properties) can be modified by some inorganic wastes.

TABLE C-2

Site and Soil Characteristics Related to Pollutant Migration
Site landform position and topography
Soil type and extent
Soil profile properties
Boundary characteristics
Depth
Texture and coarse-fragment content
Structure
Color and mottling
Bulk density
Colloidal fraction
Inorganic (clay) content
Organic content
pH
Cation exchange capacity
Porosity
Moisture content
Aeration status
Eh
Microbial population
Temperature
Hydraulic properties and conditions
Soil water characteristic curve
Field capacity/permanent wilting point
Water holding capacity
Permeability (hydraulic conductivity)
Infiltration rates
Depth to impermeable layer or bedrock
Depth to groundwater, including seasonal variations
Flooding frequency
Runoff potential
Geological and hydrogeological factors
Subsurface geological features
Groundwater flow patterns and characteristics

C.3 SOIL CLASSIFICATION

In the United States, two principal systems are presently used for classifying soils: (1) the Unified Soil Classification System (USCS) used by engineers and (2) the U.S. Department of Agriculture (USDA) System, used when a more complete classification synopsis is needed.

The USCS was developed to describe engineering properties of soils; the classification is based on particle (grain) sizes and response to physical manipulation at various water contents. An abbreviated description of the system, which does not include information on manipulation (i.e., liquid limit and plasticity index), is given in Table C-3.

The USDA system was developed for agricultural and other land management uses and considers both chemical and physical properties of the soil. This system, which is constantly being reviewed and refined, is in general use by U.S. soil scientists. The present system is based on quantitatively measurable properties of soils as they exist in the field. The ten major soil categories used in this system are described in Table C-4.

The two systems differ considerably. In attempting to relate the soil classifications used by these systems, Fuller [8] found that the elements most amenable to comparison are soil texture and associated modifiers (e.g., gravelly, mucky). However, the texture designation in the USDA system is based only on the relative amounts of sand, silt and clay (Figure C-1), while the "soil type" designation in the USCS also considers the response of the soil to physical manipulation at various water contents. Furthermore, each system uses a different size range when defining "sand," "silt" and "clay" (Table C-5). Fuller's correlations of USDA soil textures and USCS soil types (Tables C-6 and C-7) illustrate the difficulty of direct comparisons.

Correlations between USCS soil types and other parts of the USDA system are not possible. Texture, for example, is a major criterion in the USCS but only a minor criterion in the USDA system. Thus, in the former system, a soil of a given texture can be classified into only a limited number of 15 soil types; in contrast, soils of a given texture may be found in many of the USDA system's 10 orders and 43 suborders because of differences in their chemical properties or the climatic areas in which they are located.

The complete USDA soil classification system is as follows [10]:

(1) orders
(2) suborders
(3) great groups
(4) subgroups
(5) family
(6) series
(7) types

TABLE C-3

Major Divisions, Soil Type Symbols, and Type Descriptions for the Unified Soil Classification System (USCS)

Major Divisions			Symbol	Description
Coarse-grained Soils More than half of material is larger than #200 sieve size[a]	**Gravels** More than half of coarse fraction is larger than #4 sieve size[b]	**Clean Gravel** (little or no fines)	GW	Well-graded gravels, gravel-sand mixtures.
			GP	Poorly graded gravels or gravel-sand mixtures.
		Gravel + Fines (appreciable fines)	GM	Silty gravels, gravel-sand-silt mixture.
			GC	Clayey gravels, gravel-sand-clay mixtures.
	Sands More than half of coarse fraction is smaller than #4 sieve size[b]	**Clean Sands** (little or no fines)	SW	Well-graded sands, gravelly sands.
			SP	Poorly graded sands or gravelly sands.
		Sands + Fines (appreciable fines)	SM	Silty sands, sand-silt mixtures.
			SC	Clayey sands, sand-clay mixtures.
Fine-grained Soils More than half of material is smaller than #200 sieve size[a]	**Silts and Clays** (< 50% water)		ML	Inorganic silts & very fine sands, silty or clayey fine sands or clayey silts with slight plasticity. Includes rock flour.
			CL	Inorganic clays of low to medium plasticity, gravelly clays, sandy clays, silty clays, lean clays.
			OL	Organic silts and organic silty clays of low plasticity.
	Silts and Clays (> 50% water)		MH	Inorganic silts, micaceous or diatomaceous fine sandy or silty soils, elastic silts.
			CH	Inorganic clays of high plasticity.
			OH	Organic clays of medium to high plasticity, organic silts.
Highly Organic Soils			PT	Peat & other highly organic soils.

a. Sieve opening = 0.074 mm (0.0029 in.) b. Sieve opening = 4.76 mm (0.187 in.)

Source: Fuller [8]

TABLE C-4

Descriptions of Soils in the Highest (Most General) Level of the Present USDA Classification System

Order	Description
Alfisols	Alfisols have a clayey subsoil horizon and moderate to high base (cation) saturation. Water is held above the wilting point during at least three of the warm months of the year. Alfisols are higher in hydroxy-oxides (sesquioxides) than most soils, as the name implies, and therefore may have fragipan, duripan, sodium horizon, petrocalcic (lime), plinthite (iron oxides or sesquioxides) or other features that distinguish them from other soils. Where the temperature is moderate to cool, alfisols form a belt between the Mollisols of the grasslands and the Spodosols and Inceptisols of very humid climates. In warmer climates, Alfisols form a belt between Aridisols of arid regions and Inceptisols, Ultisols, and Oxisols of warmer climates. Leaching of bases from the soil may occur almost every year or may be infrequent.
Aridisols	Aridisols occur in arid climates. They do not have water available to mesophytic plants for long periods, as do the Alfisols. Water is held at a pressure of less than 15 bars, or it is salty. A few Aridisols occur in semiarid climates because they take up water slowly and most of the rainfall runs off. Aridisols have one or more pedogenic horizons that may have formed in the present dry environment or that may be relics of a former fluvial period. The pedogenic horizon may be the result of translocation and accumulation of salts, lime, or silicate clays or of cementation by carbonates or silica. The pH is usually neutral to alkaline but sometimes highly alkaline.
Entisols	Entisols do not have horizon or profile development, or at least no evidence of such. In many of the soils, time has been too short for distinct horizons to differentiate. Other Entisols are on soil slopes too steep for water to penetrate well, or where erosion rate exceeds development rate; still others are on flood or glacial outwash plains which continuously accumulate new alluvium. Some are wind-moved sand. Not all Entisols are young; some are very old, consisting mostly of quartz sand, which weathers very slowly. Such materials as organic matter, lime, gypsum, iron oxides, and clays accumulate only to a very small extent, if at all.

(Continued)

TABLE C-4 (Continued)

Order	Description
Histosols	Most Histosols are completely or nearly saturated with water most of the year. They are high in organic matter and represent what is often described as mucks, peat bogs, high moors, or raised peats.
Inceptisols	By USDA definition (1968), in part at least, "Inceptisols are soils of humid regions that have altered horizons which have lost bases or iron and aluminum but retain some weatherable minerals. They do not have an illuvial horizon enriched either with silicate clay that contains aluminum or with an amorphous mixture of aluminum and organic matter." This is a difficult soil order to visualize from the description. It represents more what other soils are *not* than what Inceptisols are. Inceptisols develop mainly in the more clayey parent materials, in contrast to Spodosols, which develop in materials which have little clay.
Mollisols	Very dark colored soils, rich in bases and naturally covered by grass (steppe land), are called Mollisols. Many soils of this order accumulate lime and/or sodium, and clay. Mollisols occupy extensive subhumid to semiarid areas of the grass plains in the United States. They are located generally between the Aridisols of arid climates and the Spodosols or Alfisols of humid climates. These soils are highly productive, constituting rich crop land in the breadbasket areas of the central and west central United States. Areas of Mollisols appear in nearly every state. Luxuriant, perennial grass seems to be essential to their formation. Where waste waters and leachates are applied to the surface, the high soil organic matter plays an important part in mobility of trace and heavy elements, in general absorbing most ions, especially cations. Where leachate and aqueous wastes pass through subsoils only, lime and bases influence mobility in addition to the effect of clay minerals.
Oxisols	Those soils which once were called red and yellow laterites are now named Oxisols. Reddish, yellowish, or grayish soils of tropical and subtropical climates that form on mostly gentle slopes on surfaces of great age are Oxisols. They are mixtures of quartz, kaolin, free oxides and organic matter. The boundaries of horizons blend into each other so gradually they are generally arbitrary. Weathering has proceeded to great depths. Water moves through these soils rapidly. Because of the high oxide (primarily iron) content that coats particles and forms granular particulates, these soils attenuate the trace and heavy metals very well. Oxisols occur in Hawaii.

(Continued)

TABLE C-4 (Continued)

Order	Description
Spodosols	Two well-defined horizons distinguish Spodosols: just below the surface layer of forest litter and partly decomposed dark organic matter is a bleached layer of uncoated quartz sand; the second layer, just below, usually is coffee color, caused by the accumulation of organic matter and iron complexes. However, this spodic is one in which amorphous mixtures of organic matter and aluminum may also occur with or without iron. Spodosols generally are coarse-textured, contain little or no clay, and usually permit rapid water movement. These soils occur under high-rainfall conditions and coniferous forest, though sometimes hardwoods are present. Attenuation is poor despite the spodic horizon where aluminum, and often iron, complexes with organic matter. However, the iron and aluminum content is usually low, even when a fragipan (soft when wet, brittle when dry) is present. The textures are mostly sandy, sandy-skeletal, coarse loam, loamy-skeletal, and coarse-silty. New England, New York, northern lake states, and Alaska are most noted for spodic (Podzolic) soils.
Ultisols	The concept of Ultisols is that of soils of middle to low latitudes that have a horizon containing an appreciable amount of translocated silicate clay but few bases. Highly humid conditions sometime during the year cause water to move through them. Ultisols are most commonly found in warm-humid climates that have a seasonal deficit of rain and on older surfaces. They develop from a wide variety of parent materials. Kaolin, gibbsite, and aluminum-interlaid clays are common in the soil clay fraction. They usually form under coniferous and hardwood-forest vegetation in the United States.
Vertisols	Vertisols are clayey soils that crack severely when dry and have high bulk densities between the cracks. The clay minerals are dominated by montmorillonite. Most are found in warm climates, i.e., thermic or warmer. In arid regions they form in closed depressions or playas. Vertisols often are referred to as "churning" soils because pressure is exerted during swelling, causing them to heave and turn over the soil.

Source: Adapted from Fuller [8]

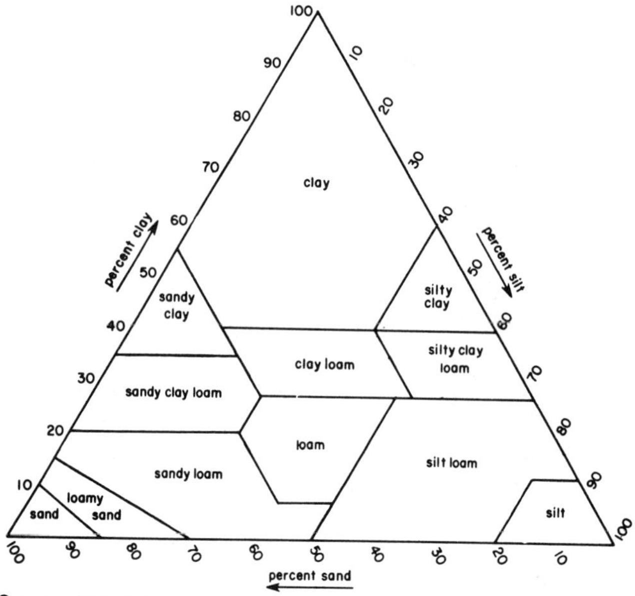

Source: Fuller [8]

FIGURE C-1 USDA Soil Textural Classification

TABLE C-5

Comparison of Particle Sizes: USDA System vs USCS

Particle	Size Range (mm)	
	USDA	USCS
Cobbles	76.2 -254	> 76.2
Gravel	2.0 - 76.2	4.76 -76.2
Coarse	12.7 - 76.2	19.1 -76.2
Fine	2.0 - 12.7	4.76 -19.1
Sand	0.05 - 2.0	0.074- 4.76
Very Coarse	1.0 - 2.0	n.a.
Coarse	0.5 - 1.0	2.0 - 4.76
Medium	0.25 - 0.5	0.42 - 2.0
Fine	0.1 - 0.25	0.074- 0.42
Very Fine	0.05 - 0.1	n.a.
Fines	n.a.	< 0.074
Silt	0.002- 0.05	a
Clay	< 0.002	a

a. USCS silt and clay designations are determined by response of the soil to manipulation at various water contents rather than by particle size.

Source: Adapted from Fuller [8]

TABLE C-6

Corresponding USCS and USDA Soil Classifications

Unified Soil Classification System (USCS) Soil Type	Corresponding United States Department of Agriculture (USDA) Soil Texture
GW	Same as GP – gradation of gravel sizes not a criterion.
GP	Gravel, very gravelly[a] sand less than 5% by weight silt and clay.
GM	Very gravelly[a] sandy loam, very gravelly[a] loamy sand, very gravelly[a] silt loam, and very gravelly[a] loam.[b]
GC	Very gravelly clay loam, very gravelly sandy clay loam, very gravelly silty clay loam, very gravelly silty clay, very gravelly clay.[b]
SW	Same – gradation of sand size not a criterion.
SP	Coarse to fine sand; gravelly sand[c] (less than 20% very fine sand).
SM	Loamy sands and sandy loams (with coarse to fine sand), very fine sand; gravelly loamy sand[c] and gravelly sandy loam.[c]
SC	Sandy clay loams and sandy clays (with coarse to fine sands); gravelly sandy clay loams and gravelly sandy clays.[c]
ML	Silt, silt loam, loam, very fine sandy loam.[d]
CL	Silty clay loam, clay loam, sandy clays with $<$ 50% sand.[d]
OL	Mucky silt loam, mucky loam, mucky silty clay loam, mucky clay loam.
MH	Highly micaceous or diatomaceous silts, silt loams – highly elastic.
CH	Silty clay and clay.[d]
OH	Mucky silty clay.
PT	Muck and peats.

a. Also includes cobbly, channery, and shaly.

b. Also includes all of textures with "gravelly" modifiers where more than half of total held on No. 200 sieve is of gravel size.

c. Gravelly textures included if more than half of total held on No. 200 sieve is of gravel size.

d. Also includes all of these textures with "gravelly" modifiers where more than half of the total soil passes the No. 200 sieve.

Source: Fuller [8]

TABLE C-7

Corresponding USDA and USCS Soil Classifications

United States Department of Agriculture (USDA) Soil Textures	Corresponding Unified Soil Classification System (USCS) Soil Types
Gravel, very gravelly loamy sand	GP, GW, GM
Sand, coarse sand, fine sand	SP, SW
Loamy gravel, very gravelly sandy loam, very gravelly loam	GM
Loamy sand, gravelly loamy sand, very fine sand	SM
Gravelly loam, gravelly sandy clay loam	GM, GC
Sandy loam, fine sandy loam, loamy very fine sand, gravelly sandy loam	SM
Silt loam, very fine sandy clay loam	ML
Loam, sandy clay loam	ML, SC
Silty clay loam, clay loam	CL
Sandy clay, gravelly clay loam, gravelly clay	SC, GC
Very gravelly clay loam, very gravelly sandy clay loam, very gravelly silty clay loam, very gravelly silty clay and clay	GC
Silty clay, clay	CH
Muck and peat	PT

Source: Fuller [8]

Each level is a subdivision of the preceding level. The orders are based on soil morphology and development similarity. Suborders emphasize common origin. Great groups are based on similarities in surface and subsurface horizons. A subgroup is based on general similarities of profiles of soils located within a large area. Families are based on properties important to the growth of plants, such as texture, mineral composition, and soil temperature. Members of soil series have similar but not identical profiles and different surface layer textures. A soil type consists of soils with similar surface texture but may be divided into phases according to some prominent deviation (e.g., slope, stoniness, erosion, or soluble salt content).

Identification of the kinds of soils at a site in terms of the higher categories of the classification system can help in characterizing pollutant attenuation and migration. The names of soils from the order to subgroup level are composed of a series of formative elements, which can be used to predict many soil properties relevant to pollutant migration potential. Table C-8 illustrates the kinds of information that can be inferred from soil names using the order Mollisol as an example. The reader is referred to *Soil Taxonomy* [10] for a comprehensive description of the USDA soil classification system.

C.4 SOIL PROFILE PROPERTIES

In climates where there is sufficient rainfall, soils become highly organized into "horizons" (layers). A soil profile is a vertical cross-section of the soil, made up of several horizons, each having its own distinctive characteristics. Descriptions of soil profiles made in the field provide valuable information for predicting inorganic migration potential. The following subsections give further details about each of the soil profile properties listed in Table C-2.

C.4.1 Boundary Characteristics

Boundary characteristics, especially abrupt changes in texture and structure (e.g., clay or sand layers, hard pans), can impede the vertical percolation of water. Horizontal flow may then predominate, resulting in contamination of adjacent areas.

C.4.2 Depth

The attenuation that occurs in soil typically increases with depth. Thus, in general, the deeper the soil overlying an aquifier, the greater the protection against migration of inorganic constituents to groundwater.

C.4.3 Texture and Coarse-fragment Content

Soil texture refers to the size distribution of various soil separates (particles less than 2 mm). The USDA system was diagrammed in Figure C-1 and compared with the USCS system in Table C-7. In general, soils high in silt and clay are called "fine-

TABLE C-8

Information That Can be Inferred from the USDA Soil Classification System, Using the Order Mollisol as an Example
(Formative elements for classification levels are italicized)

Soil Classification Level	Example	Information Relevant to Inorganic Pollutant Migration
Order	*Moll*isoll	Fertile, high in organic matter
Suborder	*Bor*oll	Long, cold winters; soil frozen for extended periods; only short time periods that soil is warm enough for biological activity
	*Aqu*oll	Naturally wet; often develops in low places where water collects and stands
	*Xer*oll	Dry for extended periods in summer; in most, moisture moves through soil in underlying layers in winter
Great Group	*Calc*ixeroll	Strong concentration of calcium carbonate or gypsum at some depth above 60 inches; calcareous in all parts above that depth unless texture is sandy
	*Arg*ixeroll	Horizon of clay accumulation in subsoil; most with horizon of carbonate accumulation below that
Subgroup	*Aquic* Argixeroll	Moderately shallow groundwater at some times of the year unless artificially drained
	Lithic Argixeroll	Hard rock at shallow depths
	Vertic Argixeroll	Fine-textured soils with deep, wide cracks at some period in most years
	Typic Argixeroll	Lacking features or combination of features specified for other subgroups

Source: Adapted from Agricultural Research Service [1]

textured" and are very plastic, cohesive, high in water-holding capacity, and relatively impermeable to water and air. Soils dominated by coarse separates, sand and gravel, are generally loose and friable, have low water-holding capacity, and provide good drainage, high permeability and aeration.

Coarse-textured soils usually remain oxidized, but fine-textured soils often become reduced when wetness excludes oxygen and in the presence of such reducing agents as sulfides and organic matter. The oxidation state affects the form in which various inorganic constituents mobilize and the speed with which they do so. However, the high adsorptive capacity of fine-textured soils enables them to immobilize various ions, such as heavy metals. Silty soils are often subject to crusting and thus have a high potential for runoff.

The soil textures that tend to present the least difficulties for management of inorganic pollutants are loam, silt loam, clay loam, sandy clay loam, silty clay loam, silty clay and sandy clay. Table C-9 summarizes the advantages and disadvantages of each type of soil texture in the USDA system.

C.4.4 Structure

Soil structure refers to the aggregation of primary soil particles (sand, silt, and clay) into compound particles or clusters of primary particles, which are called peds. The aggregates are commonly separated by surfaces of weakness or open planar voids (cracks) in the soil. The structures of the various horizons of a soil profile are essential characteristics of that profile, just as are texture and chemical composition. There are seven basic structure types, defined by the shape and arrangement of the peds (Figure C-2). A coarse-textured soil may have no structure, in which case it is called *single-grain*; fine-textured soils with no structure are called *massive*, because they agglomerate in large, uniform masses.

Aggregate formation is thought to result from the cementing action of soil colloidal matter (clay minerals, colloidal oxides of iron and manganese, and colloidal organic matter). The stability of the aggregate may vary in response to moisture content, chemical composition of the soil solution, biological activity, and management practices. Soils high in shrink/swell clays undergo particularly dramatic structural changes in response to changes in water content.

The type of structure determines the dominant direction of the pores and thus the direction of water movement. Platy structures restrict vertical percolation, while prismatic and columnar structures enhance it. Blocky and granular structures enhance percolation both vertically and horizontally.

C.4.5 Color and Mottling

Soil color and color pattern are good indicators of drainage characteristics. Soil colors can be described in general terms (brown, gray, yellow, etc.) or by the Munsell system, which characterizes a color in terms of hue, value, and chroma [10]. A uniform red,

TABLE C-9

Suitability of Various Textured Soils for Land Treatment of Hazardous Industrial Wastes

Texture	Advantages	Disadvantages
Sand	Very rapid infiltration Usually oxidized & dry Low runoff potential	Very low CEC[a] Very high hydraulic conductivity Low available water Little soil structure
Loamy sand	High infiltration Low to medium runoff	Low CEC Moderate to high hydraulic conductivity rate Low to medium available water
Loam	Moderate infiltration Fair oxidation Moderate runoff potential Workable in most seasons Good CEC	
Silt loam	Moderate infiltration Fair oxidation Moderate runoff potential Workable in most seasons Good CEC	Some crusting Fair to poor structure
Silt	Low infiltration Fair to poor oxidation Good CEC Good available water	High crusting potential Poor structure High runoff
Silty clay loam	Medium-low percolation Fair structure High CEC	Med-low infiltration Some crusting potential
Silty clay	Good to high available water	Often wet
Clay loam	Medium-low percolation Good structure Med-poor aeration High CEC High available water	Med-low infiltration Mod-high runoff Often wet

(Continued)

TABLE C-9 (Continued)

Texture	Advantages	Disadvantages
Clay	Low percolation High CEC High available water	Low infiltration Often massive structure High runoff Sometimes low aeration
Sandy clay	Med-low percolation Med-high CEC	Moderate-high runoff
Sandy clay loam	Med-high available water Good aeration	Medium infiltration

a. Cation exchange capacity

Source: Adapted from Brown & Assoc. [5]

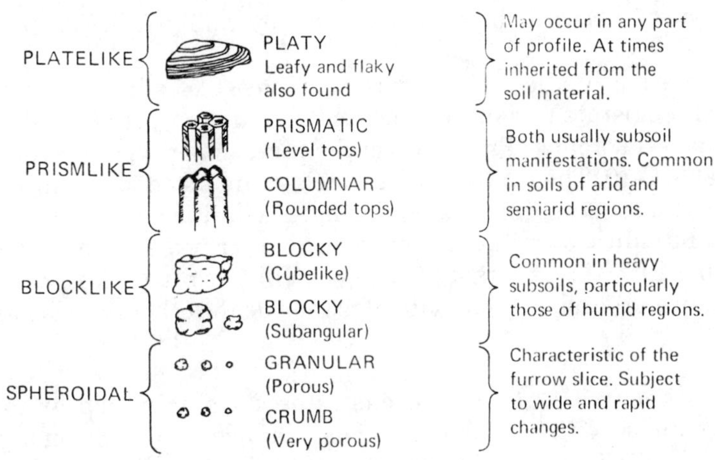

Source: Brady [4]. *(Copyright 1974, Macmillan Publishing Co. Reprinted with permission.)*

FIGURE C-2 Types of Mineral Soil Structure

yellow or brown color indicates that a soil is well drained and seldom or never saturated with water. A gray or blue color generally indicates that the soil is saturated continuously or for extended periods. Most soils with spots or streaks of red, yellow, or black in a gray matrix ("mottled") are periodically saturated.

C.4.6 Bulk Density

Bulk density is the mass of dry soil per unit bulk volume. As the density of a soil increases, the volume of pore space decreases. The relationship between density and pore space largely determines the amount of air and water that can be stored in the pore space and the ease with which they can move through the soil. The bulk density of sandy and loamy soils ranges from 1.2 to 1.8 g/cm^3. Soils of finer texture, which are usually more structured and contain more organic matter, have greater pore space and lower bulk densities (typically 1.0 to 1.6 g/cm^3). Very compact subsoils, regardless of their texture, may have bulk densities exceeding 2.0 g/cm^3 [4].

C.4.7 Colloidal Fraction

The colloidal fraction of a soil is of primary importance in the sorption and immobilization of inorganic compounds. It consists of organic and inorganic particles with a maximum size of 0.002 mm and is the most chemically active portion of a soil. Because of their small size, these particles have a large surface area, which gives them a capacity to adsorb and hold solids, gases, ions, and polar compounds and a tendency to hasten or retard chemical reactions by catalytic action [3].

The inorganic portion of the soil colloids can consist of silicate clays, which are characteristic of temperate regions, or iron and aluminum hydrous oxide clays, found in the tropics and semitropics. Properties of the three major types of silicate clays are compared in Table C-10. The montmorillonitic clays are noted for their high plasticity and cohesion and marked shrinkage on drying. Kaolinitic clays have low elasticity and do not expand. Illitic or micaceous clays have intermediate expansive properties between montmorillonitic and kaolinitic clays. Illitic soils typically contain a low percentage of clay and occur in cooler climates than do the other two clay soil types.

Although less is known about the hydrous oxide clays, they appear to have some properties in common with the silicates. For example, at least some of them are thought to have a definite crystalline structure. The small particles may carry negative charges and thus serve as central micelles that attract cations. However, they have a much smaller number of negative charges per micelle than do the silicate clays, so their cation adsorption is even lower than that of kaolinite. Also, most hydrous oxides are not as sticky, plastic, and cohesive as the silicates; thus, soils whose clay-size components are mostly hydrous oxides respond favorably to ploughing and other forms of mechanical manipulation.

The organic portion of the soil colloids is derived from (1) plant material, (2) animal matter, (3) microorganisms, both living and dead, (4) synthesized and secreted prod-

TABLE C-10

Comparative Properties of Three Major Types of Silicate Clay

Property	Montmorillonite[a]	Illite[b]	Kaolinite[c]
Size (μm)	0.01-1.0	0.1-2.0	0.1-5.0
Shape	Irregular flakes	Irregular flakes	Hexagonal crystals
Specific surface (m^2/g)	700-800	100-120	5-20
External surface	High	Medium	Low
Internal surface	Very high	Medium	None
Cohesion, plasticity	High	Medium	Low
Swelling capacity	High	Medium	Low
Cation exchange capacity (meq/100 g)	80-100	15-40	3-15

a. $Si_4Al_{1.5}Mg_{0.5}O_{10}(OH)_2$ with other cations
b. $Si_{3.34}Al_{0.66}O_{10}(OH)_2$ with interlayer K and octahedral layer cations
c. $Si_2Al_2O_5(OH)_4$

Source: Brady [4]

ucts of living plants and microorganisms, and (5) decomposition products of organic debris [3]. Most of these materials can interact with inorganic compounds to form water-soluble and water-insoluble complexes. The organic matter content of surface mineral soils in the United States typically ranges from less than 1% to 6% (Table C-11). The percentages are lower in the Southeast and higher in the northern Midwest. Figure C-3 shows the influence of the average annual temperature and the effective moisture on the organic matter content of grassland soils in the Midwest. In this example, the soils are more or less comparable in all respects except for climatic differences. Note that higher temperatures yield soils with lower organic content, while increasing moisture has the opposite effect. These climatic influences affect forest soils in much the same way.

Organic soils have a specific definition [10] and usually contain about 20% organic matter (except for peats, which have a much higher percentage). The surface soil horizon normally contains the highest percentage of organic matter. Below the top 30-40 cm, the percentage drops off dramatically. Organic horizons other than the surface horizon are occasionally found, but usually not below 40 cm.

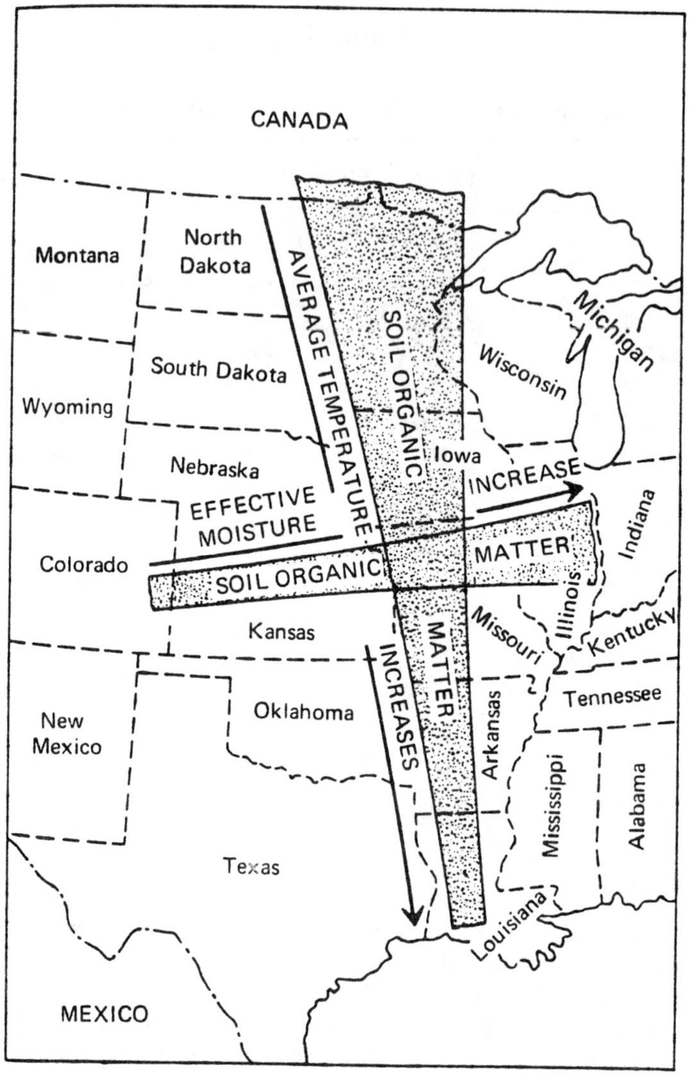

Width of shaded bands indicates relative content of organic matter.

Source: Brady [4]. *(Copyright 1974, Macmillan Publishing Co. Reprinted with permission.)*

FIGURE C-3 Effect of Moisture and Average Annual Temperature on Soil Organic Matter in the Midwestern United States

TABLE C-11

Organic Matter in Mineral Surface Soils in Several Areas of the United States

Location	No. of Samples	Organic Matter (%) Range	Avg.
West Virginia	240	0.74-15.1	2.88
Pennsylvania	15	1.70- 9.9	3.60
Kansas	117	0.11-3.62	3.38
Nebraska	30	2.43-5.29	3.83
Minnesota prairie	9	3.45-7.41	5.15
Southern Great Plains	21	1.16-2.16	1.55
Utah	21	1.54-4.93	2.69

Source: Adapted from Brady [4]

C.4.8 Soil pH

Soil pH, to some extent, affects the degree of surface charge on colloidal-sized soil particles: high pH is usually associated with negatively charged surfaces. Some soils may have a lot of negative charge at low pH; only with pure hydrous oxides would there be solely "+" charges at low pH. The tendency for adsorption of anions or cations is thus dependent on the pH of the soil water in the unsaturated zone. Figure C-4 shows the extreme range of pH for most mineral soils and the ranges commonly found in humid and arid regions. The pH of a calcareous soil or a calcareous soil horizon, which ranges from 7 to a maximum of 8.3, can greatly affect the solubility or ionization states of inorganic compounds. The optimum range for most soil micro-organisms is from 6.5 to 8.5. If the soil is sodic (high in sodium), the pH will be even higher than that of a calcareous soil, due to the hydrolysis of sodium carbonate and formation of a strong base, sodium hydroxide. When the cation exchange capacity is at least 15% saturated with sodium, or when there is a significant amount of sodium carbonate in the soil, the pH values range from 8.5 to 10.

C.4.9 Cation Exchange Capacity

The cation exchange capacity (CEC) of soils is defined as the total number of exchangeable cations that a soil can adsorb under defined conditions. It is sometimes referred to as "total exchange capacity," "base exchange capacity," or "cation adsorption capacity." The CEC of soil is expressed in milliequivalents per 100 grams

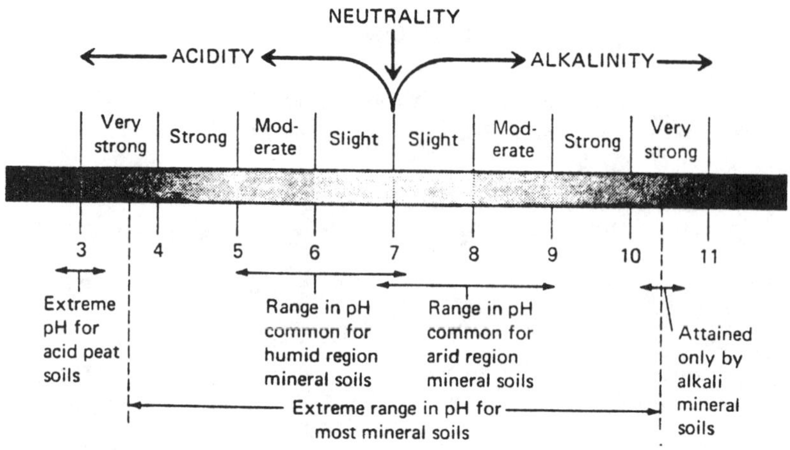

Source: Brady [4]. *(Copyright 1974, Macmillan Publishing Co. Reprinted with permission.)*

FIGURE C-4 Range in pH of Mineral Soils

of soil. The CEC of a total soil is always much less than that of its colloidal fraction. The following are some approximate CEC values for various soil components:

	meq/100 g
Humus (soil organic matter)	200
Vermiculite (mineral with Ca, Fe, K, Mg, Si, Al, O, H)	150
Montmorillonite	100
Hydrous mica (illite)	30
Chlorite ($Mg_5\,Al_2\,Si_3\,O_{10}\,(OH)_8$)	8
kaolinite and iron oxides	4

Table C-12 gives the CEC of a wide variety of soils in the United States. In any particular soil, the CEC will vary with pH, but should not vary with ionic strength, dielectric constant, and other characteristics of the soil water.

C.4.10 Moisture Content

Moisture content is another important soil parameter. As important as the current moisture content are the typical (recent) moisture status of the soil, and the aeration status and Eh of the moisture. Certain physical and chemical reactions are characteristic of saturated soil:

- The limited oxygen supply promotes anaerobic or reducing reactions and may retard root growth.
- Percolation (downward movement) of water is restricted, and lateral movement tends to predominate within the saturated zone.

TABLE C-12

Cation Exchange Capacities of Various Surface Soils
(Milliequivalents per 100 grams of dry soil)

Soil Type	Exchange Capacity	Soil Type	Exchange Capacity
Sand		Silt Loam	
Sassafras (N.J.)	2.0	Delta (Miss.)	9.4
Plainfield (Wis.)	3.5	Fayette (Minn.)	12.6
		Spencer (Wis.)	14.0
Sandy Loam		Dawes (Nebr.)	18.4
Greenville (Ala.)	2.3	Carrington (Minn.)	18.4
Sassafras (N.J.)	2.7	Penn (N.J.)	19.8
Norfolk (Ala.)	3.0	Miami (Wis.)	23.2
Cecil (S.C.)	5.5	Grundy (Ill.)	26.3
Coltneck (N.J.)	9.9		
Colma (Calif.)	17.1	Clay and Clay Loam	
		Cecil clay loam (Ala.)	4.0
Loam		Cecil clay (Ala.)	4.8
Sassafras (N.J.)	7.5	Coyuco sandy clay (Calif.)	20.2
Hoosic (N.J.)	11.4	Gleason clay loam (Calif.)	31.5
Dover (N.J.)	14.0	Susquehanna clay (Ala.)	34.2
Collington (N.J.)	15.9	Sweeney clay (Calif.)	57.5

Source: Brady [4]

- The specific heat of the soil is increased, so that, under conditions of fluctuating air temperatures, a saturated soil will have smaller temperature changes than an unsaturated soil.

The reducing conditions in a saturated soil can lead to Eh values of −400mV. Well-aerated, relatively dry soils, on the other hand, are highly oxidized and may have Eh values of 600 mV. Thus, many heavy metals that are highly mobile at low redox potentials are relatively immobile in well-drained soils. (See sections 2.3.5 and 2.4.1 for additional information.)

The moisture content of a soil is often reported in terms of a water potential (ψ) which is related to the relative humidity in the soil (rh) by a variation of the ideal gas law,

$$\psi = \frac{RT}{V} \log (rh)$$

where R is the gas constant (0.082 atm L/mol K), T the temperature (K), and V is the molar volume of water (23.2 L at 10°C). The potential is dimensionally equivalent to pressure and is commonly expressed in bars (1 bar = 0.9869 atm) [8a].

C.4.11 Microbial Population

The microbial population in the soil environment is highly variable, ranging from millions per gram of unpolluted soil to only a few hundred per gram in areas with toxic contaminants. The topsoil of fertile land (that is, the upper few inches or feet) has been aptly described as an entire universe [6]. True bacteria and actinomycetes, algae, fungi, viruses, and protozoa all live in the soil in numbers that vary with the pH, moisture content, salt content, inorganic nutrient and pollutant status, and organic food.

The microbial population reflects daily and seasonal variations in temperature and water content, in acidity, in availability of simple organic food substances and minerals, and in the amount of air trapped in the soil. It also decreases with depth, as shown in Table C-13. Microbes produce humus, the organic fraction of the soil. They also recycle plant nutrients.

C.4.12 Temperature

The temperature of soil rises or falls in keeping with the relationship between the energy absorbed and that being lost. The surface layers vary more or less according to air temperature and therefore exhibit a greater fluctuation than the subsoil. On the average, the top 15 cm of the soil are warmer than the ambient air just above the surface. Compared with air temperatures, the subsoil is warmer in autumn and winter but cooler in spring and summer because of its protected position and the lag in conduction. The range of variation over any given period decreases with depth and is only a few degrees below three meters in most parts of the United States. Figure C-5 shows the relationship between depth and average monthly soil temperature at two-month intervals in College Station, Texas. Note the lag in changes in soil temperature at the lower depths.

TABLE C-13

Distribution of Microorganisms in Various Soil Horizons
(spodosol soil)

Depth (cm)	Millions of Organisms/g Soil	Percentage of Total Organisms Counted
3-8	11.97	79
20-25	2.48	16
35-40	0.63	4
65-75	0.022	<1
135-145	0.004	<1

Source: Adapted from Alexander [2]

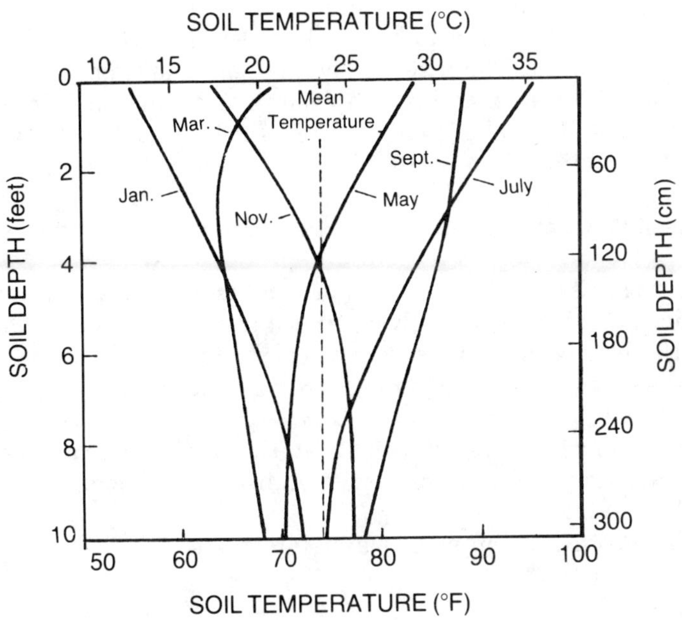

Source: Adapted from Fluker [7]. *(Copyright 1958, Williams & Wilkins Co. Reprinted with permission.)*

FIGURE C-5 Average Soil Temperatures at Different Soil Depths (College Station, Texas)

C.5 GLOSSARY

channery — Thin, flat fragments of limestone, sandstone, or schist up to 6 inches in major diameter.

dolomite — A calcium magnesium carbonate found in crystals and in extensive beds as a compact limestone.

duripan — A mineral soil horizon that is cemented by silica, usually opal or microcrystalline forms of silica, to the point that air-dry fragments will not slake in water or HCl. A duripan may also have accessory cement such as iron oxide or calcium carbonate.

esker — A winding, narrow ridge of sand or gravel, probably deposited by a stream flowing under glacial ice (see Table C-1).

fluvial — Deposited by waters such as streams and flash waters.

fracture porosity — The volume percentage of the total bulk occupied by fractures rather than solid particles.

fragipan — A natural subsurface horizon with high bulk density relative to the overlying soil profile, seemingly cemented when dry, but when moist showing a moderate to weak brittleness. The layer is low in organic matter, mottled, slowly or very slowly permeable to water, and usually shows occasional or frequent bleached cracks forming polygons. It may be found in profiles of either cultivated or virgin soils but not in calcareous material.

friable — A consistency term pertaining to the ease of crumbling of soils.

horizon — A layer of soil or soil material approximately parallel to the land surface and differing from adjacent genetically related layers in physical, chemical, and biological properties or characteristics such as color, structure, texture, consistency, kinds and numbers of organisms present, degree of acidity or alkalinity, etc.

illuvial — Pertaining to a soil layer in which material carried from an overlying layer has been precipitated from solution or deposited from suspension.

infiltration — The downward entry of water into the soil.

kame — see Table C-1

karst — A type of topography formed over limestone, dolomite, or gypsum by dissolving or solution and that is characterized by closed depressions or sinkholes, caves and underground drainage.

leachate — A suspension or solution of material removed from a layer or layers of a soil by percolating waters.

loess — Wind-deposited soil material of silt to silt loam texture that is highly fertile (see Table C-1).

mesophytic — Pertaining to plants that grow under medium conditions of moisture.

moraine — An accumulation of earth and stones carried and finally deposited by a glacier.

ped — A unit of soil structure such as an aggregate, crumb, prism, block, or granule, formed by natural processes (in contrast to a clod, which is formed artificially).

pedogenic — Pertaining to the formation of soil.

perched water table — Ground water separated from another, underlying body of ground water by a discontinuous zone of low permeability (e.g., clay lens).

permeability — Property of a bulk mass of soil or layer of soil that relates to the ease with which gases, liquids, or other substances can pass through it.

plasticity index — The numerical difference between the upper and lower plastic limits (q.v.).

plastic limit, lower — The minimum moisture percentage (by weight) at which a small sample of soil material can be deformed without rupture.

plastic limit, upper — The minimum moisture percentage (by weight) at which a small sample of soil will barely flow under a standard treatment.

playa — A desert basin that temporarily becomes a shallow lake after a heavy rain (see Table C-1).

separates — Mineral particles of less than 2 mm equivalent diameter, which are subdivided into sands of various coarseness, silt, and clay.

skeletal grains — Particles of mineral soil material that are larger than colloidal size.

solution porosity — Pores formed primarily by solution action. Chiefly found in limestone-derived soils.

spodic horizon — A mineral soil horizon characterized by the illuvial accumulation of amorphous materials composed of aluminum and organic carbon with or without iron. This horizon has a certain minimum thickness and a minimum quantity of extractable carbon plus iron plus aluminum in relation to its clay content.

till plain — see Table C-1

topographic relief — The elevations or inequalities of a land surface.

weathering — The action of climate and living organisms on geological materials over time. The effect of weathering varies with the nature of the parent material and the topography of the area.

wilting point — The water content of a soil when indicator plants growing in that soil wilt and fail to recover when placed in a humid chamber.

C.6 LITERATURE CITED

1. Agricultural Research Service, *Factors Involved in Land Application of Agricultural and Municipal Wastes,* ARS-USDA, Soil, Water and Air Sciences, Beltsville, Md. (1979).
2. Alexander, M., *Introduction to Soil Microbiology*, 2nd ed., John Wiley & Sons, New York (1977).
3. Anderson, W.P., *Weed Science: Principles*, West Publishing Co., St. Paul, Minn. (1977).

4. Brady, N.C., *The Nature and Properties of Soils*, 8th ed., Macmillan Publishing Co., New York (1974).
5. Brown, K.W., and Associates, Inc., "Hazardous Waste Land Treatment," (draft edition), SW-879, U.S. Environmental Protection Agency, Cincinnati (1980).
6. Burdon, K.L., *Microbiology*, 5th ed., Macmillan Publishing Co., New York (1964).
7. Fluker, B.J., "Soil Temperature," *Soil Sci.*, **86**, 35-46 (1958).
8. Fuller, W.H., "Investigations of Landfill Leachate Pollutant Attentuation by Soils," EPA-600/1-78-158, U.S. Environmental Protection Agency, Cincinnati (1978).

8a. Hornick, S.B., "The Interaction of Soils with Waste Constituents," in *Land Treatment of Hazardous Wastes*, J.F. Parr, P.B. Marsh and J.M. Kla (eds.), Noyes Data Corp., Park Ridge, N.J. (1983).

9. Ryan, J.R. and R.C. Loehr, "Site Selection Methodology for the Land Treatment of Wastewater," Special Report 81-28, U.S. Army Corps of Engineers, Cold Regions Research and Engineering Laboratory, Hanover, N.H. (1981).
10. Soil Survey Staff, Soil Conservation Service, *Soil Taxonomy*, Agricultural Handbook No. 436, U.S. Department of Agriculture, Washington, D.C. (1975).
11. Warner, D.L., "Hydrogeologic Aspects of Land Application of Wastewater" in *Land Treatment and Disposal of Wastewaters*, R.L. Sanks and T. Asano (eds.), Ann Arbor Science Publishers, Ann Arbor, Mich., pp. 133-46 (1976).

C.7 ADDITIONAL SOURCES OF INFORMATION

Bohn, H.L., B. McNeal and G. O'Connor, *Soil Chemistry,* 2nd ed., John Wiley & Sons, New York (1985).

Glossary of Soil Science Terms, Soil Science Society of America, Madison, WI (1984).

Hillel, D., *Fundamentals of Soil Physics,* Academic Press, New York (1980).

Iskandar, I.K., *Modeling Wastewater Renovation — Land Treatment,* John Wiley & Sons, New York (1981).

Jenny, H., *The Soil Resource — Origin and Behavior,* Springer-Verlag, New York (1980).

Overcash, M.R. and D. Pal, *Design of Land Treatment Systems for Industrial Waste — Theory and Practice,* Ann Arbor Science, Ann Arbor, MI (1979).

Reed, S. and R.W. Crites, *Handbook of Land Treatment Systems for Industrial and Municipal Waste,* Noyes Publications, Park Ridge, New Jersey (1984).

APPENDIX D. POLLUTANT CRITERIA AND STANDARDS

Anita E. Glazer

LIST OF TABLES

APPENDIX D — POLLUTANT CRITERIA AND STANDARDS

There are numerous regulations concerning inorganics and organometallics in the ambient air and water, in the workplace, and in food. Pollutant criteria or standards of concern are administered by several agencies, primarily the Environmental Protection Agency (EPA), the Occupational Safety and Health Administration (OSHA), and the Food and Drug Administration (FDA). Table D-1 shows the major regulations and criteria applicable to inorganic elements and compounds. This appendix briefly describes these regulations and criteria and tabulates values of the more important criteria and standards for particular inorganics/organometallics.

D.1 CLEAN AIR ACT (CAA)

D.1.1 National Ambient Air Quality Standards (NAAQS)

In accordance with Section 109 of the CAA, EPA establishes primary and secondary ambient air quality standards for air pollutants that may endanger public health or welfare and that have numerous or diverse sources. These standards reflect air quality criteria based on the latest scientific knowledge of the effects of the pollutants. *Primary* standards define levels "necessary, with an adequate margin of safety, to protect the public health." *Secondary* standards define levels "necessary to protect the public welfare from any known or anticipated adverse effects of a pollutant" (40 CFR 50).[1] (See Table D-2.) Implementation plans for the NAAQS (strategies for their implementation, maintenance, and enforcement) are developed on a state-by-state basis.

D.1.2 National Emissions Standards for Hazardous Air Pollutants (NESHAPs)

As defined under Section 112 of CAA, hazardous air pollutants are those to which no ambient air quality standards apply, and which may cause an increase in mortality or an increase in serious irreversible, or incapacitating reversible, illness. NESHAPs regulate emissions of these pollutants from particular sources; the standard for each pollutant includes different values for different types of facilities (40 CFR 61). In developing NESHAPs, EPA uses meteorological dispersion estimates to relate emissions to an ambient level determined sufficient to protect public health.

D.1.3 Standards of Performance for New, Stationary Sources

Standards of performance regulating emissions from stationary sources are established by EPA in accordance with Section 111 of CAA. A separate standard is issued for each established category of source (i.e., each type of industry); the list of regulated pollutants varies from industry to industry (40 CFR 60).

1. In this appendix, references to the U.S. Code of Federal Regulations (CFR) are given in the following format: (Volume No.)CFR(Part No.). References to the *Federal Register* (FR) are given in the format (Volume No.)FR(Page No.).

TABLE D-1

Standards/Criteria Applicable to Inorganics and Organometallics

						EPA Water Quality Criteria								WHO Guidelines for Drinking Water Quality			
	Ambient Air Quality Standards (40 CFR 50)	Hazardous Air Pollutants (40 CFR 61)	Stationary Source Performance Standards (40 CFR 60)	CWA List of Toxic Pollutants	CWA Hazardous Substance with RQ (40 CFR 116,117)	Freshwater Aquatic Life	Saltwater Aquatic Life	Human Health	Primary Drinking Water Std. (40 CFR 141)	Secondary Drinking Water Std. (40 CFR 143)	RCRA Hazardous Waste (40 CFR 261.33)	RCRA — EP Toxicity Criteria (40 CFR 261.24)	OSHA Std. for Air Contaminants in Workplace (29 CFR 1910.1000)	Health-related	Aesthetic Quality	IJC Objectives for Great Lakes Water Quality	Other
Ammonia					X,C	X	X	X			C		X,C			X	
Antimony				X	C	(X)	X	X					X,C				
Arsenic				X	C	X	X	X	X		C	X	X,C	X		X	
Asbestos		X		X				X	[X]				X			(X)	CPSA
Barium					C		X	X	X		C	X	C				
Beryllium		X		X	C	(X)	X	X			X		X,C				
Bromine					C		X				C		X,C				
Cadmium				X	C	X	X	X	X			X	X	X		X	FDA action level
Chlorine and Chlorides					X,C	X	X	X		X	C		X,C		X		
Chromium				X	C	X	X	X	X			X	X,C	X		X	
Copper				X	C	X	X	X	[X]	X	C		X		X	X	
Cyanides				X	C	X	X	X			X,C		X	X			
Fluorine and Fluorides			X		C		X	X	X		X,C		X,C	X		X	
Iron					C	X	X	X		X			C		X	X	
Lead	X		X	X	C	X	X	X	X		C	X	X,C	X		X	CPSA, FDA action level
Manganese							X	X		X			X		X		
Mercury		X		X	C	X	X	X	X		C	X	X,C	X		X	FDA action level
Nickel				X	C	X	X	X			C		X,C			X	
Nitrates					C			X	X		C		C	X			
Radium									X								
Selenium				X	C	X	X	X	X		C	X	C	X		X	
Silver				X	C	X	X	X	X		C	X	X,C				
Sulfides					C	X	X				C		C			C	
Sulfates					C					X	C		C		X		
Thallium				X	C	(X)	(X)	X			C		C				Health Advisory
Uranium					C		X						C				
Zinc				X	C	X	X	X		X	C		C		X	X	
Carbon Monoxide	X		X										X				
Nitrogen Oxides	X		X										X				
Ozone	X																
Sulfur Oxides	X		X										X				

X — Indicates that the indicated regulation, criterion, or standard applies to the listed substance and/or to all compounds containing it.
C — Indicates that the indicated regulation, criterion, or standard applies to a specific, named compound or compounds of the listed substance.
( ) Indicates that some data on toxicity, but no numerical criteria, are given.
[ ] Standard has been proposed, but is not yet implemented (50 FR 46936).

TABLE D-2

National Ambient Air Quality Standards (NAAQS) Applicable to Inorganics
(40 CFR 50)

Pollutant	Primary Standard $\mu g/m^3$ (ppm)	Secondary Standard $\mu g/m^3$ (ppm)
Sulfur oxides (as SO_2)	80 (0.03) annual avg. 365 (0.14) max. 24-hr. avg.[a]	1300 (0.5) max. 3-hr. avg.[a]
Carbon monoxide	10,000 (9) max. 8-hr. avg.[a] 40,000 (35) max. 1-hr. avg.[a]	(same as primary std.)
Nitrogen dioxide	100 (0.053) annual avg.	(same as primary std.)
Ozone	235 (0.12)[b]	(same as primary std.)
Lead	1.5 — max. arithmetic mean over a calender quarter	(same as primary std.)

a. Not to be exceeded more than once a year.

b. Standard is attained when the expected number of days per calendar year with maximum hourly average concentrations above the listed value is one or fewer.

D.2 FEDERAL WATER POLLUTION CONTROL ACT/CLEAN WATER ACT (CWA)

D.2.1 National Pollutant Discharge Elimination System (NPDES)

The NPDES (40 CFR 122 through 125), established under Section 402 of CWA, requires permits for discharging pollutants from any point source into waters of the United States. Applicants for NPDES permits must test for specified pollutants in their wastewater discharge. Quantitative data must be reported for some of these pollutants; for others, applicants need only report whether they are believed to be present.

D.2.2 Effluent Guidelines and Standards

The effluent guidelines and standards in 40 CFR 401 through 699 prescribe effluent limitations, standards of performance, and pretreatment standards for point-source discharges of pollutants, on an industry-by-industry basis. These guidelines, called for under Title III of the CWA, form the basis for limitations in permits issued under the NPDES.

D.2.3 Reportable Quantities for Discharges of Hazardous Substances (RQs)

Pursuant to Section 311 of CWA, EPA has designated 297 chemical compounds as "hazardous substances" (40 CFR 116) and assigned to each of them a reportable quantity (RQ) of 1, 10, 100, 1000, or 5000 pounds (40 CFR 117). These quantities were essentially determined on the basis of aquatic toxicity. If an amount equal to or greater than the RQ for any of these chemicals is discharged into the navigable waters of the United States in any 24-hour period, it must be reported to the EPA. This rule does not apply to discharges in compliance with NPDES permits or other such permitting and regulating systems, or to other "continuous or anticipated intermittent" discharges from a point source. Thus, the RQ ruling applies to non-routine discharges, such as accidental spills.

D.2.4 Water Quality Criteria

In accordance with Section 304(a)(1) of the CWA, EPA publishes and periodically updates ambient water quality criteria (WQC). These criteria were first published in 1972 in the "Blue Book" [4] and were updated in 1976 by the "Red Book" [5], and in 1980 (45 FR 79318) and 1985 (50 FR 30784) by individual water quality criteria documents. Criteria documents describe the effects of pollutants and give numerical criteria for maximum pollutant concentrations. Separate criteria are given for the protection of freshwater aquatic life, saltwater aquatic life, and human health.

While the earlier versions of the water quality criteria list acceptable concentrations for the total amount of a given substance in water, the criteria for metals as updated in 1985 (50 FR 30784) were developed based on amounts of "acid-soluble metal." However, since approved methods for a measurement such as "acid-soluble" are not yet available, EPA recommends applying the criteria using the total recoverable method. This has two impacts: (1) certain species of some metals cannot be analyzed directly because the total recoverable method does not distinguish between individual oxidation states, and (2) these criteria may be overly protective when based on the total recoverable method.

For each included substance for which sufficient data are available, both a long-term concentration and short-term or maximum concentration are given. These criteria are developed from scientific data according to the guidelines given in 45 FR 79341 and revised as summarized in 50 FR 30784. WQC have no regulatory force; they are intended to provide guidance in deriving regulatory requirements. (See Tables D-3 and D-4).

D.3 SAFE DRINKING WATER ACT (SDWA)

D.3.1 Drinking Water Regulations

Under Section 1412 of the SDWA, EPA has established standards for levels of contaminants in public drinking water supplies. *Primary* drinking water regulations (PDWRs) apply to contaminants that may have adverse effects on health. As required by the Act, EPA first issued interim PDWRs (40 CFR 141).

After further scientific study, these are to be replaced by revised PDWRs. The PDWRs specify a maximum contaminant level (MCL) allowable for each contaminant. (If it is not feasible to ascertain the level of a given contaminant in drinking water, the PDWR must specify instead treatment techniques for its removal.) The revised PDWRs, in

TABLE D-3
Water Quality Criteria and Toxicity Levels for Inorganics Relating to the Protection of Aquatic Life
(concentrations in µg/l)

	Freshwater Aquatic Life				Saltwater Aquatic Life			
	EPA Criteria[a]		Toxic Levels[b]		EPA Criteria[a]		Toxic Level[b]	IJC Objectives for Great Lakes Water Quality[i]
Substance	Long-term	Short-term	Chronic	Acute	Long-term	Short-term	(Acute)	
Aluminum	150[c,j] ← (for pH 6.5 to 9.0) →	950[c,j]	—	—	200[k]	1500[k]	—	—
Ammonia	d	d	—	—	10[k] (un-ionized NH_3)	400[k]	—	20[e] (un-ionized NH_3) 500[f] (total NH_3)
Antimony	—	—	1600[m] algae 610[m]	9000[m]	—	200[k]	—	—
Arsenic (III)	190[n]	360[n]	—	—	36[n]	69[n]		50[f] (total As)
(V)	—	—	(acute:chronic ratio – 28)[n]	850[n] Plants 48[n]	—	—	2319[n] Plants 5-50[n]	
Barium	—	—	—	—	500[k]	1000[k]	—	—
Beryllium	11(soft water), 1100(hard water)[p]		5.3[m]	130[m]	100[k]	1500[k]	—	—
Cadmium	exp[.7852 ln(hardness) − 3.490][n] (0.66,1.1,2.0)[g]	exp[1.128 ln(hardness) − 3.828][n] (1.8,3.9,8.6)[g]	—	—	9.3[n]	43[n]	—	0.2[e]
Chlorine	11[n] ← (Total residual Cl) →	19[n]	—	—	7.4[n] (Cl-produced oxidants)	13[n]	—	—
Chromium (VI)	11[n]	16[n]	—	—	54[n]	1100[n]	—	50[f]
(III)	exp[.819 ln(hardness) + 1.561][n] (120,210,370)[g]	exp[.819 ln(hardness) + 3.688][n] (980,1700,3100)[g]					(see n)	(total Cr)
Copper	exp[.8545 ln(hardness) − 1.465][n] (6.5,12,21)[g]	exp[.9422 ln(hardness) − 1.464][n] (9.2,18,34)[g]	—	—	—	2.9[n]	—	5[e]
Cyanide	5.2[n]	22[n]	—	—	—	1.0[n]	—	—
Fluoride	—	—	—	—	500[k]	1500[k]	—	1200[f]
Iron	1000[p]	—	—	—	50[k]	300[k]	—	300[e]
Lead	exp[1.266 ln(hardness) − 4.661][n] (1.3,3.2,7.7)[g]	exp[1.266 ln(hardness) − 1.416][n] (34,83,200)[g]	—	—	5.6[n]	140[n]	—	{ 10[e] L. Superior; 20[e] L. Huron; 25[e] All other Great Lakes
Manganese	—	—	—	—	20[k]	100[k]	—	—
Mercury	0.12[n]	2.4[n]	—	—	0.025[n]	2.1[n]	—	0.2[h]
Nickel	exp[.846 ln(hardness) + 1.1645][c,j] (88,160,280)[g]	exp[.846 ln(hardness) + 3.3612][c,j] (790,1400,2500)[g]	—	—	7.9[c,j]	71[c,j]	—	25[e]
Phosphorus (elemental)	—	—	—	—	0.1[p]		—	(see i)
Selenium (IV)	10[c,q]	190[c,q]	—	—	93[c,q]	300[c,q]	—	10[f] (total Se)
Silver	—	exp[1.72 ln(hardness) − 6.52][m] (1.2,4.1,13)[g]	0.12[m]	—		2.3[m]	—	—
Hydrogen Sulfide (undissociated)	2[p]	—	—	—	2[p]		—	2.0[e]
Thallium	—	—	40[m]	1400[m]			2130[m]	—
Uranium	—	—	—	—	100[k]	500[k]	—	—
Zinc	exp[.8195 ln(hardness) + .6881][c,r] (49,86,150)[g]	exp[.8195 ln(hardness) + .7871][c,r] (54,96,170)[g]	—	—	79[c,r]	87[c,r]	—	30[e]

NOTES TO TABLE D-3

a. "Long-term" and "short-term" criteria were redefined slightly each time the water quality criteria were updated. The 1972 "Blue Book" (footnote k) lists one level below which a given substance "presents minimal risk of deleterious effects" and a higher level that "constitutes a hazard in the marine environment" (listed here as long-term and short-term criteria, respectively). The 1976 "Red Book" (footnote p) lists only one value for each substance; these criteria are designed to provide long-term protection. The 1980 criteria documents (footnote m) specify 24-hour average concentrations (long-term) and maximum concentrations, not to be exceeded at any time (short-term). The 1985 and subsequent documents (footnotes n, j, q, r) list four-day average concentrations (long-term) and one-hour average concentrations (short-term), neither of which should be exceeded more than once every three years on the average.

   Note that the latest document does not always update all criterion values (freshwater, saltwater, and human health) for a given substance. This table includes all old values that have not been specifically superseded.

b. All listed values are prefixed with phrase "as low as". Even lower toxic concentrations may be associated with species more sensitive than those tested.

c. Proposed criteria, not officially established.

d. Criteria values for ammonia vary with pH and temperature, and different values apply depending on whether or not salmonids or other sensitive coldwater species are present (see n).

e. Not to be exceeded for the protection of aquatic life.

f. Not to be exceeded for the protection of raw water for public water supplies.

g. Criteria values for hardnesses of 50, 100, and 200 mg/l as $CaCO_3$.

h. Not to be exceeded for the protection of aquatic life and fish-consuming birds. Concentration in whole fish should not exceed 0.5 μg/g (wet weight basis).

Sources:

i. *Great Lakes Water Quality Agreement of 1978* [1]
j. 51 FR 8361 (March 11, 1986)
k. *Water Quality Criteria 1972* [4]
m. 45 FR 79318 (Nov. 28, 1980)
n. 50 FR 30784 (July 29, 1985)
p. *Quality Criteria for Water* (1976) [5]
q. 51 FR 16205 (May 1, 1986)
r. 51 FR 19269 (May 28, 1986)

TABLE D-4

Criteria and Standards Applicable to Inorganics in Water for Human Consumption
(concentrations in $\mu g/l$)

Substance	WHO Guidelines[j]		EPA Water Quality Criteria for Human Health		Drinking Water Regulations			EPA Office of Drinking Water Health Advisories Available
	Health-Related	Aesthetic Quality	Ingesting Aquatic Organisms and Water	Ingesting Aquatic Organisms Only	Primary (40 CFR 141) MCL	Primary (40 CFR 141) RMCL[a]	Secondary (40 CFR 143) SMCL	
Aluminum		200						
Ammonia (as N)			500[k]					
Antimony			146[m]	45,000[m]				
Arsenic	50		0.0022[b,m]	0.0175[b,m]	50	50		X
Asbestos			30,000 fibers/l[b,m]			7.1×10^6 fibers/l[c]		
Barium			1000[n]		1000	1500		X
Beryllium			0.0037[b,m]	0.0641[b,m]				
Cadmium	5		10[m]		10	5		X
Chloride		250,000	250,000[k] (for taste)				250,000	
Chromium	50		50[m] (Cr VI) 170[m] (Cr III)	3433[m] (Cr III)	50	120		X
Copper		1000	1000 (for taste and odor)[m]			1300	1000	
Cyanide	100		200[m]					X
Fluoride	1500[d]		1400-2400[e,k]		4000	4000	2000	
Iron		300	300[n] (for taste)				300	
Lead	50		50[m]		50	20		X
Manganese		100	50[n] (for taste)	100[n]			50	
Mercury	1		0.144[m]	0.146[m]	2	3		X
Nickel			13.4[m]	100[m]				X
Nitrate (as N)	10,000		10,000[n]		10,000	10,000		X
Nitrite						1000		X
Radium					5 pCi/l[f]			
Selenium	10		10[m]		10	45		
Silver			50[m]		50			
Sulfate		400,000					250,000	
Thallium			13[m]	48[m]				
Zinc		5,000	5,000[m] (for taste and odor)				5,000	
Gross Alpha-Particle Activity	0.1 Bq/l		(discussed in k)		15 pCi/l[g]			
Beta Activity	1 Bq/l[h]		(discussed in k)		i			

a. Recommended maximum contaminant levels. Of the values listed, only the RMCL for fluoride has been officially promulgated. The other RMCLs were proposed on November 13, 1985 (50 FR 46936).
b. For maximum protection from the potential carcinogenicity of this substance, the ambient water concentration should be zero. The listed value is estimated for a 10^{-6} incremental increase in cancer risk.
c. Applies to fibers longer than 10 μm. Based on a 10^{-6} cancer risk level.
d. Local or climatic conditions may necessitate variation.
e. Recommended fluoride values vary depending on climate.
f. Combined ^{226}Ra and ^{228}Ra.
g. Including ^{226}Ra, excluding Rn and U.
h. Gross beta activity.
i. Beta particle and photon radioactivity: dose equivalent to the total body or any internal organ of 4 millirem per year.

Sources:
j. *Guidelines for Drinking-Water Quality* (1984) [3]
k. *Water Quality Criteria 1972* [4]
m. 45 FR 79318 (Nov. 28, 1980)
n. *Quality Criteria for Water (1976)* [5]

addition to revising MCLs, must set "recommended maximum contaminant levels" (RMCLs) at levels that would result in no known or anticipated adverse health effects, with an adequate margin for safety. RCMLs are non-enforceable health goals; MCLs, which are enforceable standards, are to be set as close to the RMCLs as is technologically and economically feasible. The revised PDWRs are being developed in phases. A final RMCL and MCL for fluoride have been established; for other inorganic chemicals, RMCLs were proposed in November, 1985, and revised MCLs will be proposed when the RMCLs are officially promulgated. (See Table D-4.)

The Office of Drinking Water, in its proposal to set RMCLs for several inorganic chemicals, also suggested guidance levels for a number of chemicals for which RMCLs were not proposed: aluminum, cyanide, molybdenum, nickel, silver, sodium, and sulfate. (See 51 FR 46936.)

Secondary drinking water regulations (40 CFR 143) apply to contaminants that primarily affect the odor or appearance of drinking water. Secondary maximum contaminant levels (SMCLs) are not Federally enforceable; they are intended to be guidelines for the states. (See Table D-4.)

D.3.2 Health Advisories

In addition to establishing RMCLs and MCLs, the Office of Drinking Water has developed health advisories for various chemical contaminants in drinking water (See Table D-4). Health advisories suggest contaminant levels that are not expected to cause adverse health effects over a short time period. Levels are usually calculated for 1-day, 10-day, and longer-term (several months to several years) exposures. They are non-enforceable levels, intended as guidance values for short-term exposure situations such as spills and accidents. Levels are derived both for a 70-kg adult (assumed to consume 2 liters of water per day) and for a 10-kg child (assumed to consume 1 liter of water per day). Health advisory levels for inorganics for which RMCLs have not been proposed are as follows (51 FR 46936):

- Cyanide — 0.22 mg/l for a child and 0.75 mg/l for an adult, for a 10-day exposure.
- Nickel — 1.0 mg/l for a child and 3.5 mg/l for an adult, for a 10-day exposure.
- Uranium — 10 pCi/l for an adult, for chronic exposure.

D.3.3 Underground Injection Control (UIC) Program

The UIC Program (40 CFR 144), established under RCRA and part C of the SDWA, regulates the disposal of fluids by underground injection in order to prevent the movement of contaminants into underground sources of drinking water.

D.4 RESOURCE CONSERVATION AND RECOVERY ACT (RCRA)

Regulations governing hazardous waste management are established under RCRA. Lists of substances designated as hazardous wastes, along with the criteria for so listing them, are published in 40 CFR 261. Both specific chemicals that become hazardous wastes when discarded and wastes from specified sources or processes may be designated as hazardous wastes.

D.4.1 EP Toxicity

One of the possible criteria for listing a substance as a hazardous waste is EP toxicity. As initially proposed, a substance was considered "EP toxic" if an extract produced by a specified extraction procedure (EP) performed on the substance contained any of 14 listed components at or above their listed maximum concentrations (40 CFR 261.24). These maximum concentration levels were set at 100 times the MCLs for drinking water established under the SDWA, since the intent of the EP is to identify wastes likely to leach toxic chemicals into groundwater at hazardous levels. In June 1986, the EPA proposed to amend the extraction procedure by: (1) including an additional 38 compounds; (2) applying compound-specific dilution/attenuation factors generated from a groundwater transport model; and (3) introducing a second generation leaching procedure, the Toxicity Characteristic Leaching Procedure (TCLP), that has been developed to address the mobility of both organic and inorganic compounds (51 FR 21648).

D.4.2 Criteria for Listing Hazardous Wastes

In general, a substance may be listed as a hazardous waste if it exhibits the characteristics of ignitability, corrosivity, reactivity, or EP toxicity, or if it contains any of a specified list of toxic constituents. In addition, substances meeting specified criteria for high toxicity are designated as "acutely hazardous wastes" and are subject to more stringent regulations. (Specific hazardous wastes are listed in 40 CFR 261.33). Residue or contaminated material resulting from a spill of a listed hazardous waste is itself considered a hazardous waste, as are containers or container liners that have held acutely hazardous waste. (See Tables D-5 and D-6.)

D.4.3 Exclusion Limits

Generators of hazardous waste who produce less than an established "exclusion limit" of hazardous waste in a given month and do not accumulate that amount on-site at any time are not subject to the bulk of the RCRA regulations. This exclusion limit is set at 1000 kg for listed hazardous wastes, 1 kg for acutely hazardous wastes, and 100 kg for residue or contaminated materials from the cleanup of a spill of an acutely hazardous waste (40 CFR 261.5).

D.5 COMPREHENSIVE ENVIRONMENTAL RESPONSE, COMPENSATION, AND LIABILITY ACT (CERCLA, OR SUPERFUND)

Under CERCLA, reportable quantities (RQs), like those designated under the Clean Water Act (see § D.2.3), are assigned for the release of hazardous substances into any medium (not only water). Hazardous substances as defined under CERCLA include

TABLE D-5

Inorganics and Organometallics Listed as Acute Hazardous Wastes Under the Resource Conservation and Recovery Act (RCRA)
(40 CFR 261.33(e))

Hazardous Waste No.	Substance	Hazardous Waste No.	Substance
P006	aluminum phosphide	P074	nickel (II) cyanide
P009	ammonium picrate (R)	P076	nitric oxide (nitrogen (II) oxide)
P119	ammonium vanadate	P078	nitrogen dioxide (nitrogen (IV) oxide)
P010	arsenic acid	P087	osmium tetroxide
P012	arsenic (III) oxide (arsenic trioxide)	P092	phenylmercuric acetate
P011	arsenic (V) oxide (arsenic pentoxide)	P096	phosphine (hydrogen phosphide)
P013	barium cyanide	P098	potassium cyanide
P015	beryllium dust	P099	potassium silver cyanide
P021	calcium cyanide	P103	selenourea (carbamimidoselenoic acid)
P022	carbon disulfide	P104	silver cyanide
P095	carbonyl chloride (phosgene)	P105	sodium azide
P033	chlorine cyanide (cyanogen chloride)	P106	sodium cyanide
P029	copper cyanides	P058	sodium fluoroacetate
P030	cyanides (soluble cyanide salts), not elsewhere specified	P107	strontium sulfide
P031	cyanogen	P110	tetraethyl lead
P036	dichlorophenylarsine	P111	tetraethyl pyrophosphate
P038	diethyl arsine	P113	thallium (III) oxide (thallic oxide)
P101	ethyl cyanide	P114	thallium (I) selenite
P056	fluorine	P115	thallium (I) sulfate
P063	hydrogen cyanide	P120	vanadium (V) oxide (vanadium pentoxide)
P065	mercury fulminate (R,T)	P121	zinc cyanide
P073	nickel carbonyl (nickel tetracarbonyl)	P122	zinc phosphide, when present at concentrations greater than 10%

Note: The primary hazardous properties of these materials are indicated by the letters T (toxicity) and R (reactivity). Absence of a letter indicates that the compound is listed for acute toxicity only.

TABLE D-6

Inorganics and Organometallics Listed As Hazardous Wastes Under the Resource Conservation and Recovery Act (RCRA)
(40 CFR 261.33(f))

Hazardous Waste No.	Substance
U246	bromine cyanide (cyanogen bromide)
U032	calcium chromate
U033	carbon oxyfluoride (carbonyl fluoride) (R,T)
U211	carbon tetrachloride
U103	dimethyl sulfate
U134	hydrogen fluoride (hydrofluoric acid) (C,T)
U136	hydroxydimethylarsine oxide
U139	iron dextran (ferric dextran)
U144	lead acetate
U145	lead phosphate
U146	lead subacetate
U151	mercury
U189	phosphorus sulfide (sulfur phosphide) (R)
U204	selenium dioxide (selenious acid)
U205	selenium disulfide (sulfur selenide) (R,T)
U135	sulfur hydride (hydrogen sulfide)
U214	thallium (I) acetate
U215	thallium (I) carbonate
U216	thallium (I) chloride
U217	thallium (I) nitrate
U249	zinc phosphide, when present at concentrations of 10% or less

Note: The primary hazardous properties of these materials are indicated by the letters T (toxicity), R (reactivity), and C (corrosivity). Absence of a letter indicates that the compound is listed for toxicity only.

all of those designated and assigned RQs under Section 311 of CWA, plus the hazardous wastes listed under RCRA, the toxic pollutants identified under Section 307(a) of CWA, the hazardous air pollutants listed under Section 112 of CAA, and any imminently hazardous substance with respect to which action has been taken under Section 7 of TSCA (see below), as well as any additional designations that may be made in Section 102 of CERCLA. Superfund RQs were initially set at the same levels as those assigned under Section 311 of CWA for all substances there included, and at 1 pound for all other hazardous substances. Many of these RQs have since been adjusted (see 40 CFR 302); the CWA Section 311 RQs were simultaneously revised to be identical to them.

The National Oil and Hazardous Substances Pollution Contingency Plan (NCP) (40 CFR 300) is called for by Section 105 of CERCLA and Section 311(c) of CWA. The Plan provides for efficient and coordinated response to discharges of oil and releases of hazardous substances, pollutants and contaminants. It includes the division of responsibilities among different levels of government, requirements for regional and local Plans, and policies and procedures for removal and response operations.

D.6 OCCUPATIONAL HEALTH STANDARDS

In accordance with Section 6 of the Occupational Safety and Health Act, the Occupational Safety and Health Administration (OSHA) establishes exposure standards for air contaminants in the workplace (29 CFR 1910.1000 to 1910.1047). These standards are based on a combination of health effects and technical/economic feasibility. They are designed to protect the health and functional capacity of workers for up to 10-hour workdays and 40-hour workweeks over a working lifetime. They are usually issued as 8-hour time-weighted averages, but ceiling (maximum) values are sometimes listed in addition or instead. Table D-7 lists the exposure standards for inorganic and organometallic chemicals. OSHA standards are also established for many organic chemicals, and for a number of mineral dusts, including silica, mica, soapstone, talc, Portland cement, graphite, coal dust, and inert dust.

The National Institute for Occupational Safety and Health (NIOSH) provides recommended standards to OSHA. These recommendations, based on health effects, have no regulatory force; OSHA uses them along with considerations of feasibility to develop regulatory standards.

The American Conference of Governmental and Industrial Hygienists (ACGIH) issues threshold limit values (TLVs) for airborne concentrations of chemical substances in the work environment. It is believed that nearly all workers can be repeatedly exposed to such conditions without adverse effect. The values may be based on protection against health impairment, irritation, narcosis, nuisance or other forms of stress. TLVs are listed primarily as time-weighted averages (TWAs) for an 8-hour workday and 40-hour workweek. In addition, for substances reported to cause toxic effects after short-term exposure, ACGIH establishes 15-minute time-weighted average exposure limits that should not be exceeded at any time during a workday. Ceiling concentration values, which should not be exceeded even momentarily, are also listed for some substances.

TABLE D-7

OSHA Standards for Airborne Contaminants in the Workplace (Inorganics and Organometallics)
(29 CFR 1910.1000)

	8-Hour Time Weighted Average		Acceptable Ceiling Concentration	
Substance	**ppm[a]**	**mg/m[3 b]**	**ppm[a]**	**mg/m[3 b]**
Ammonia	50	35		
Ammonium sulfamate (Ammate)		15		
Antimony and compounds (as Sb)		0.5		
Arsenic, inorganic[c]		0.010		
Arsenic organic compounds (as As)		0.5		
Arsine	0.05	0.2		
Asbestos	d			
Barium (soluble compounds)		0.5		
Beryllium and beryllium compounds[e]		0.002		0.005
Boron oxide		15		
Boron trifluoride			1	3
Bromine	0.1	0.7		
tert-Butyl chromate (as CrO_3)— Skin		0.1		
Cadmium fume		0.1		0.3
Cadmium dust		0.2		0.6
Calcium oxide		5		
Carbon black		3.5		
Carbon dioxide	5,000	9,000		
Carbon disulfide	20		30	
Carbon monoxide	50	55		
Carbon tetrachloride[e]	10		25	
Chlorine			1	3
Chlorine dioxide	0.1	0.3		
Chlorine trifluoride			0.1	0.4
Chromic acid and chromates				1 mg/10m^3
Chromium				
Sol. chromic & chromous salts (as Cr)		0.5		
Metal and insol. salts		1		
Cobalt, metal fume and dust		0.1		
Copper fume		0.1		
Dusts and Mists		1		

(Continued)

TABLE D-7 (Continued)

Substance	8-Hour Time Weighted Average		Acceptable Ceiling Concentration	
	ppm[a]	mg/m^3 [b]	ppm[a]	mg/m^3 [b]
Cyanide (as CN) — Skin		5		
Dibutyl phosphate	1	5		
Dimethylsulfate — Skin	1	5		
Ferrovanadium dust		1		
Fluoride (as F)		2.5		
Fluoride as dust		2.5		
Fluorine	0.1	0.2		
Hafnium		0.5		
Hydrogen bromide	3	10		
Hydrogen chloride			5	7
Hydrogen cyanide — Skin	10	11		
Hydrogen fluoride	3			
Hydrogen peroxide (90%)	1	1.4		
Hydrogen selenide	0.05	0.2		
Hydrogen sulfide[e]			20	
Iodine			0.1	1
Iron oxide fume		10		
Lead[f]		0.050		
Lithium hydride		0.025		
Magnesium oxide fume		15		
Manganese				5
Mercury				1 mg/10m^3
Organo (alkyl) mercury		0.01		0.04
Molybdenum:				
Soluble compounds		5		
Insoluble compounds		15		
Nickel carbonyl	0.001	0.007		
Nickel, metal and soluble cmpds, as Ni		1		
Nitric acid	2	5		
Nitric oxide	25	30		
Nitrogen dioxide			5	9

(Continued)

TABLE D-7 (Continued)

Substance	8-Hour Time Weighted Average		Acceptable Ceiling Concentration	
	ppm[a]	mg/m^3 [b]	ppm[a]	mg/m^3 [b]
Nitrogen trifluoride	10	29		
Osmium tetroxide		0.002		
Oxygen difluoride	0.05	0.1		
Ozone	0.1	0.2		
Perchloryl fluoride	3	13.5		
Phosgene (carbonyl chloride)	0.1	0.4		
Phosphine	0.3	0.4		
Phosphoric acid		1		
Phosphorus (yellow)		0.1		
Phosphorus pentrachloride		1		
Phosphorus pentasulfide		1		
Phosphorus trichloride	0.5	3		
Platinum (soluble salts) as Pt		0.002		
Rhodium				
Metal fume and dusts (as Rh)		0.1		
Soluble salts		0.001		
Selenium compounds (as Se)		0.2		
Selenium hexafluoride	0.05	0.4		
Silver, metal and soluble cmpds		0.01		
Sodium fluoroacetate — Skin		0.05		
Sodium hydroxide		2		
Stibine	0.1	0.5		
Sulfur dioxide	5	13		
Sulfur hexafluoride	1,000	6,000		
Sulfuric acid		1		
Sulfur monochloride	1	6		
Sulfur pentafluoride	0.025	0.25		
Sulfuryl fluoride	5	20		
Tantalum		5		
Tellurium		0.1		
Tellurium hexafluoride	0.02	0.2		
Tetraethyl lead (as Pb) — Skin		0.075		
Tetramethyl lead (as Pb) — Skin		0.075		

(Continued)

TABLE D-7 (Continued)

Substance	8-Hour Time Weighted Average		Acceptable Ceiling Concentration	
	ppm[a]	mg/m³ [b]	ppm[a]	mg/m³ [b]
Thallium (soluble compounds) — Skin, as Tl		0.1		
Tin (inorganic cmpds, except oxides)		2		
Organic compounds		0.1		
Titanium dioxide		15		
Tributyl phosphate		5		
Triphenyl phosphate		3		
Uranium (soluble compounds)		0.05		
Insoluble compounds		0.25		
Vanadium pentoxide (dust)				0.5
Fume				0.1
Yttrium		1		
Zinc chloride fume		1		
Zinc oxide fume		5		
Zirconium compounds (as Zr)		5		

a. Parts of vapor or gas per million parts of contaminated air by volume at 25°C and 1 atm
b. Approximate milligrams of particulate per cubic meter of air
c. Additional regulations are in 29 CFR 1910.1018
d. Five fibers (longer than 5 μm) per cm^3; additional regulations are in 29 CFR 1910.1001
e. Acceptable maximum peak above the acceptable ceiling concentration for an 8-hour shift:

Material	Concentration	Maximum Duration
Beryllium and beryllium compounds	25 μg/m³	30 minutes
Carbon disulfide	100 ppm	30 minutes
Carbon tetrachloride	200 ppm	5 minutes in any 4-hour period
Hydrogen sulfide	50 ppm	10 minutes once, only if no other measurable exposure occurs

f. Additional regulations are in 29 CFR 1910.1025

D.7 FOOD, DRUG, AND COSMETIC ACT (FDCA)

All food additives except those "generally recognized as safe" (GRAS) must be certified by the Food and Drug Administration (FDA), and conditions or concentrations for safe use must be issued. These regulations are published in 21 CFR 170 to 189.

In addition, FDA may regulate an added "poisonous or deleterious substance" in food by issuing a tolerance or an action level (21 CFR 109) specifying a maximum permissible level for the substance. Tolerances or action levels may be established only for substances that "cannot be avoided by good manufacturing practice"; thus, they do not represent levels that are permissible if contamination is avoidable. A *tolerance* may be established only when no foreseeable "technological or other changes" might affect its appropriateness in the near future. If foreseeable changes might affect the appropriateness of the level, an *action level*, which may be changed at any time, may be established, to be revoked when a tolerance for the same substance goes into effect. Tolerances are established as regulations in 21 CFR 109 Subpart B.

Regulatory action may also be taken against foods that contain a substance in excess of its action level, but the action level itself must be defended in court during such proceedings.

Action levels have been established for mercury in aquatic animals (1.0 ppm, in the edible portion only) and in wheat (1.0 ppm, in the pink kernels only). Action levels have also been established for lead and cadmium in ceramics used for food and for lead in silver-plated holloware used to hold food. These levels, based on the amount of the contaminant leached out of these items by a specified method, are as follows:

- For silver-plated holloware intended for use by adults, 7 μg lead/ml leaching solution (average of 6 units examined);
- For silver-plated holloware intended exclusively for use by children, 0.5 μg lead/ml leaching solution (any one of 6 units examined);
- For ceramic flatware, 7.0 μg lead or 0.5 μg cadmium per ml leaching solution (average of 6 units examined);
- For small ceramic holloware, 5.0 μg lead or 0.5 μg cadmium per ml leaching solution (any one of 6 units examined); and
- For large ceramic holloware, 2.5 μg lead or 0.25 μg cadmium per ml leaching solution (any one of 6 units examined).

D.8 OTHER FEDERAL REGULATIONS

Under the Consumer Product Safety Act (CPSA), the Consumer Product Safety Commission regulates consumer products that present an unreasonable risk of injury (16 CFR 1201 to 1404). These regulations ban certain products and specify design criteria or require special labeling for others. Included among these are regulations applying to asbestos in consumer patching compounds and artificial emberizing materials (16 CFR 1304-1305) and lead in paint (16 CFR 1303).

The Department of Transportation regulates the transportation of hazardous materials, stipulating labeling, packaging and other requirements. Lists of the hazardous materials so regulated are published in 49 CFR 172.101 and 172.102.

Under the Federal Insecticide, Fungicide, and Rodenticide Act (FIFRA), any pesticide that is marketed or shipped in the United States must be registered with the EPA. Regulations for the enforcement of FIFRA are published in 40 CFR 160. Under the Food, Drug, and Cosmetic Act (FDCA), EPA establishes tolerances (maximum permissible levels) for pesticide chemicals on raw agricultural commodities (40 CFR 180) and other foods (21 CFR 193).

Under the Marine Protection, Research, and Sanctuaries Act (MPRSA), an EPA-issued permit is required for the dumping of any material (including hazardous chemicals) into ocean waters. Regulations pertaining to ocean dumping are published in 40 CFR 220-234; criteria for the issuance of ocean dumping permits are in 40 CFR 227.

The Toxic Substances Control Act (TSCA) governs the manufacture, processing, and distribution of chemical substances and mixtures that "present an unreasonable risk of injury to health or the environment." Testing is required for such substances or for substances to which there will be significant environmental or human exposure. Manufacturers and processors must give advance notice to EPA before producing a new chemical substance. EPA can regulate substances under TSCA by prohibiting or limiting their manufacture or by requiring special labeling, recordkeeping, or disposal procedures. In addition, special action may be taken with respect to substances that are "imminent hazards." Regulations for the enforcement of TSCA are published in 40 CFR 702-792.

In response to recently heightened awareness of the hazards posed by exposure to indoor radon gas, the EPA has issued *A Citizen's Guide to Radon* (August, 1986), a set of guidelines discussing the risks associated with various levels of radon exposure. The level given by the EPA as a goal for radon mitigation efforts is 4 pCi of radon per liter of air.

D.9 WORLD HEALTH ORGANIZATION GUIDELINES

The World Health Organization (WHO) publishes guidelines for drinking water quality for use by countries as a basis for the development of standards [3]. A guideline value represents the level of a constituent that "ensures an aesthetically pleasing water and does not result in any significant risk to the health of the consumer." Guideline values have been derived to protect health based on lifelong consumption, assuming a daily per capita consumption of two liters of water. Guidelines for chemical contaminants are based on toxicity, levels and frequency of occurrence, and available control technology. Health-related guidelines are recommended for nine inorganic constituents, and brief explanations of why guidelines could not be set are provided for seven others. In addition, guidelines are set for several chemical constituents that may affect the aesthetic quality of water. For radioactivity in water, values are listed "below which water can be considered potable without any further radiological examination." (See Table D-4.)

D.10 U.S./CANADA INTERNATIONAL JOINT COMMISSION GREAT LAKES WATER QUALITY AGREEMENT

In the Great Lakes Water Quality Agreement of 1978, a U.S./Canada International Joint Commission (IJC) agreed to make a maximum effort to restore and maintain the water quality of the Great Lakes system [1]. This goal is to be met by eliminating or reducing the discharge of pollutants and by promoting programs for a better understanding of the Great Lakes Basin ecosystem. Water quality standards and other regulatory requirements established by the two governments must be consistent with the water quality objectives set forth in the Agreement. The IJC is responsible for collecting data concerning Great Lakes water quality and for providing advice and assistance on research and other programs established pursuant to the Agreement. Specific objectives for pollutant levels that should not be exceeded in water are based on cause/effect relationships between pollutants and receptors, to protect the recognized most sensitive use in all waters. Thus, criteria for some substances are based on the protection of raw water for public water supplies, while others are based on the protection of aquatic life. (See Table D-3.)

Among the programs called for by the Agreement are "measures for the abatement and control of pollution from all dredging activities, including the development of criteria for the identification of polluted sediments and compatible programs for disposal of polluted dredged material." Although rigid criteria for determining whether sediments are polluted have not been adopted, in 1982 the Dredging Subcommittee of the Great Lakes Water Quality Board published guidelines for evaluating the environmental acceptability of proposed dredging projects and dredged material disposal in the Great Lakes [2]. Included in this document are numerical guidelines developed by the U.S. EPA, Region V, and by the Ontario Ministry of Environment (MOE) for classifying Great Lakes sediments as polluted or nonpolluted (Table D-8). These criteria are best used only for a rough screening of sediments for pollution, however, since they are based on total pollutant concentrations, which do not correlate well with bioavailability of contaminants. A number of efforts are currently underway to develop more reliable and meaningful sediment quality criteria (see below).

D.11 DEVELOPMENT OF SEDIMENT QUALITY CRITERIA

In addition to the guidelines mentioned above (§ D.10), sediment quality criteria have been or are being developed by several state and Federal agencies. They are often designed for use in determining the acceptability of dredged material disposal options. Most criteria values are based on total contaminant concentrations, but one recently developed set of guidelines is based instead on the ratio of heavy metal concentration to aluminum concentration. To date, criteria values have most commonly been determined by reference to reported ambient or "background" levels of contaminants. More recently, values have been determined by using bioassay studies to relate contaminant concentrations to biological effects, or by using sediment-water partitioning coefficients to estimate contaminant concentrations in sediments that will not result in concentrations in interstitial water exceeding the EPA water quality criteria.

TABLE D-8

Guidelines for Classification of Great Lakes Sediments[a]
(concentrations in mg/kg dry weight)

	U.S. EPA, Region V			Ontario MOE[b]
	Nonpolluted	Moderately Polluted	Heavily Polluted	
Chemical Oxygen Demand	<40,000	40,000-80,000	>80,000	50,000
Total Kjeldahl Nitrogen	<1,000	1,000- 2,000	>2,000	2,000
Lead	<40	40-60	>60	50
Zinc	<90	90-200	>200	100
Mercury	<1	—	≥1	0.3
Ammonia	<75	75-200	>200	100
Cyanide	< 0.10	0.10-0.25	>0.25	0.1
Phosphorus	<420	420-650	>650	1,000
Iron	<17,000	17,000-25,000	>25,000	10,000
Nickel	<20	20-50	>50	25
Manganese	<300	300-500	>500	—
Arsenic	<3	3-8	>8	8
Cadmium	—	—	>6	1
Chromium	<25	25-75	>75	25
Barium	<20	20-60	>60	—
Copper	<25	25-50	>50	25

a. The applicability and limitations of these guidelines are discussed in Ref. 2.

b. Concentrations indicative of contaminated sediments.

Source: *Guidelines and Register for Evaluation of Great Lakes Dredging Projects* [2].

D.12 LITERATURE CITED

1. *Great Lakes Water Quality Agreement of 1978*, International Joint Commission, Canada and the United States, Ottawa, Ontario (1978).
2. *Guidelines and Register for Evaluation of Great Lakes Dredging Projects*, Report of the Dredging Subcommittee to the Water Quality Programs Committee of the Great Lakes Water Quality Board, International Joint Commission, Windsor, Ontario (January, 1982).
3. *Guidelines for Drinking Water Quality: Vol. 1 — Recommendations*, World Health Organization, Geneva (1984).
4. National Academy of Sciences/National Academy of Engineering Environmental Studies Board, *Water Quality Criteria 1972*, U.S. Environmental Protection Agency, Washington, D.C. (1972).
5. *Quality Criteria for Water*, U.S. Environmental Protection Agency, Washington, D.C. (1976).

SUBJECT INDEX